These safety symbols are used in laboratory and field investigations in this book ing of each symbol and refer to this page often. *Remember to wash your hands*

PROTECTIVE EQUIPMENT

Do not begin any lab without the proper

GOGGLES 	Proper eye protection must be worn when performing or observing science activities that involve items or conditions as listed below.	**APRON** 	Wear an approved apron when using substances that could stain, wet, or destroy cloth.	**SOAP** 	Wash soap and water before removing goggles and after all lab activities.		working with biological materials, chemicals, animals, or materials that can stain or irritate hands.

LABORATORY HAZARDS

Symbols	Potential Hazards	Precaution	Response
DISPOSAL	contamination of classroom or environment due to improper disposal of materials such as chemicals and live specimens	• DO NOT dispose of hazardous materials in the sink or trash can. • Dispose of wastes as directed by your teacher.	• If hazardous materials are disposed of improperly, notify your teacher immediately.
EXTREME TEMPERATURE	skin burns due to extremely hot or cold materials such as hot glass, liquids, or metals; liquid nitrogen; dry ice	• Use proper protective equipment, such as hot mitts and/or tongs, when handling objects with extreme temperatures.	• If injury occurs, notify your teacher immediately.
SHARP OBJECTS	punctures or cuts from sharp objects such as razor blades, pins, scalpels, and broken glass	• Handle glassware carefully to avoid breakage. • Walk with sharp objects pointed downward, away from you and others.	• If broken glass or injury occurs, notify your teacher immediately.
ELECTRICAL	electric shock or skin burn due to improper grounding, short circuits, liquid spills, or exposed wires	• Check condition of wires and apparatus for fraying or uninsulated wires, and broken or cracked equipment. • Use only GFCI-protected outlets	• DO NOT attempt to fix electrical problems. Notify your teacher immediately.
CHEMICAL	skin irritation or burns, breathing difficulty, and/or poisoning due to touching, swallowing, or inhalation of chemicals such as acids, bases, bleach, metal compounds, iodine, poinsettias, pollen, ammonia, acetone, nail polish remover, heated chemicals, mothballs, and any other chemicals labeled or known to be dangerous	• Wear proper protective equipment such as goggles, apron, and gloves when using chemicals. • Ensure proper room ventilation or use a fume hood when using materials that produce fumes. • NEVER smell fumes directly. • NEVER taste or eat any material in the laboratory.	• If contact occurs, immediately flush affected area with water and notify your teacher. • If a spill occurs, leave the area immediately and notify your teacher.
FLAMMABLE	unexpected fire due to liquids or gases that ignite easily such as rubbing alcohol	• Avoid open flames, sparks, or heat when flammable liquids are present.	• If a fire occurs, leave the area immediately and notify your teacher.
OPEN FLAME	burns or fire due to open flame from matches, Bunsen burners, or burning materials	• Tie back loose hair and clothing. • Keep flame away from all materials. • Follow teacher instructions when lighting and extinguishing flames. • Use proper protection, such as hot mitts or tongs, when handling hot objects.	• If a fire occurs, leave the area immediately and notify your teacher.
ANIMAL SAFETY	injury to or from laboratory animals	• Wear proper protective equipment such as gloves, apron, and goggles when working with animals. • Wash hands after handling animals.	• If injury occurs, notify your teacher immediately.
BIOLOGICAL	infection or adverse reaction due to contact with organisms such as bacteria, fungi, and biological materials such as blood, animal or plant materials	• Wear proper protective equipment such as gloves, goggles, and apron when working with biological materials. • Avoid skin contact with an organism or any part of the organism. • Wash hands after handling organisms.	• If contact occurs, wash the affected area and notify your teacher immediately.
FUME	breathing difficulties from inhalation of fumes from substances such as ammonia, acetone, nail polish remover, heated chemicals, and mothballs	• Wear goggles, apron, and gloves. • Ensure proper room ventilation or use a fume hood when using substances that produce fumes. • NEVER smell fumes directly.	• If a spill occurs, leave area and notify your teacher immediately.
IRRITANT	irritation of skin, mucous membranes, or respiratory tract due to materials such as acids, bases, bleach, pollen, mothballs, steel wool, and potassium permanganate	• Wear goggles, apron, and gloves. • Wear a dust mask to protect against fine particles.	• If skin contact occurs, immediately flush the affected area with water and notify your teacher.
RADIOACTIVE	excessive exposure from alpha, beta, and gamma particles	• Remove gloves and wash hands with soap and water before removing remainder of protective equipment.	• If cracks or holes are found in the container, notify your teacher immediately.

EARTH & SPACE iSCIENCE

GLENCOE

McGraw Hill Education

COVER: Gary Ombler/Dorling Kindersley/Getty Images

mheducation.com/prek-12

Send all inquiries to:
McGraw-Hill Education
8787 Orion Place
Columbus, OH 43240

ISBN: 978-0-07-677385-5
MHID: 0-07-677385-X

Printed in the United States of America.

4 5 6 7 8 9 QVS 22 21 20 19 18 17

Contents in Brief

Authors and Contributors

Authors

American Museum of Natural History
New York, NY

Michelle Anderson, MS
Lecturer
The Ohio State University
Columbus, OH

Juli Berwald, PhD
Science Writer
Austin, TX

John F. Bolzan, PhD
Science Writer
Columbus, OH

Rachel Clark, MS
Science Writer
Moscow, ID

Patricia Craig, MS
Science Writer
Bozeman, MT

Randall Frost, PhD
Science Writer
Pleasanton, CA

Lisa S. Gardiner, PhD
Science Writer
Denver, CO

Jennifer Gonya, PhD
The Ohio State University
Columbus, OH

Mary Ann Grobbel, MD
Science Writer
Grand Rapids, MI

Whitney Crispen Hagins, MA, MAT
Biology Teacher
Lexington High School
Lexington, MA

Carole Holmberg, BS
Planetarium Director
Calusa Nature Center and Planetarium, Inc.
Fort Myers, FL

Tina C. Hopper
Science Writer
Rockwall, TX

Jonathan D. W. Kahl, PhD
Professor of Atmospheric Science
University of Wisconsin-Milwaukee
Milwaukee, WI

Nanette Kalis
Science Writer
Athens, OH

S. Page Keeley, MEd
Maine Mathematics and Science Alliance
Augusta, ME

Cindy Klevickis, PhD
Professor of Integrated Science and Technology
James Madison University
Harrisonburg, VA

Kimberly Fekany Lee, PhD
Science Writer
La Grange, IL

Michael Manga, PhD
Professor
University of California, Berkeley
Berkeley, CA

Devi Ried Mathieu
Science Writer
Sebastopol, CA

Elizabeth A. Nagy-Shadman, PhD
Geology Professor
Pasadena City College
Pasadena, CA

William D. Rogers, DA
Professor of Biology
Ball State University
Muncie, IN

Donna L. Ross, PhD
Associate Professor
San Diego State University
San Diego, CA

Marion B. Sewer, PhD
Assistant Professor
School of Biology
Georgia Institute of Technology
Atlanta, GA

Julia Meyer Sheets, PhD
Lecturer
School of Earth Sciences
The Ohio State University
Columbus, OH

Michael J. Singer, PhD
Professor of Soil Science
Department of Land, Air and Water Resources
University of California
Davis, CA

Karen S. Sottosanti, MA
Science Writer
Pickerington, Ohio

Paul K. Strode, PhD
I.B. Biology Teacher
Fairview High School
Boulder, CO

Jan M. Vermilye, PhD
Research Geologist
Seismo-Tectonic Reservoir Monitoring (STRM)
Boulder, CO

Judith A. Yero, MA
Director
Teacher's Mind Resources
Hamilton, MT

Dinah Zike, MEd
Author, Consultant, Inventor of Foldables
Dinah Zike Academy; Dinah-Might Adventures, LP
San Antonio, TX

Margaret Zorn, MS
Science Writer
Yorktown, VA

Consulting Authors

Alton L. Biggs
Biggs Educational Consulting
Commerce, TX

Ralph M. Feather, Jr., PhD
Assistant Professor
Department of Educational Studies and Secondary Education
Bloomsburg University
Bloomsburg, PA

Douglas Fisher, PhD
Professor of Teacher Education
San Diego State University
San Diego, CA

Edward P. Ortleb
Science/Safety Consultant
St. Louis, MO

Series Consultants

Science

Solomon Bililign, PhD
Professor
Department of Physics
North Carolina Agricultural and Technical State University
Greensboro, NC

John Choinski
Professor
Department of Biology
University of Central Arkansas
Conway, AR

Anastasia Chopelas, PhD
Research Professor
Department of Earth and Space Sciences
UCLA
Los Angeles, CA

David T. Crowther, PhD
Professor of Science Education
University of Nevada, Reno
Reno, NV

A. John Gatz
Professor of Zoology
Ohio Wesleyan University
Delaware, OH

Sarah Gille, PhD
Professor
University of California San Diego
La Jolla, CA

David G. Haase, PhD
Professor of Physics
North Carolina State University
Raleigh, NC

Janet S. Herman, PhD
Professor
Department of Environmental Sciences
University of Virginia
Charlottesville, VA

David T. Ho, PhD
Associate Professor
Department of Oceanography
University of Hawaii
Honolulu, HI

Ruth Howes, PhD
Professor of Physics
Marquette University
Milwaukee, WI

Jose Miguel Hurtado, Jr., PhD
Associate Professor
Department of Geological Sciences
University of Texas at El Paso
El Paso, TX

Monika Kress, PhD
Assistant Professor
San Jose State University
San Jose, CA

Mark E. Lee, PhD
Associate Chair & Assistant Professor
Department of Biology
Spelman College
Atlanta, GA

Linda Lundgren
Science writer
Lakewood, CO

Keith O. Mann, PhD
Ohio Wesleyan University
Delaware, OH

Charles W. McLaughlin, PhD
Adjunct Professor of Chemistry
Montana State University
Bozeman, MT

Katharina Pahnke, PhD
Research Professor
Department of Geology and Geophysics
University of Hawaii
Honolulu, HI

Jesús Pando, PhD
Associate Professor
DePaul University
Chicago, IL

Hay-Oak Park, PhD
Associate Professor
Department of Molecular Genetics
Ohio State University
Columbus, OH

David A. Rubin, PhD
Associate Professor of Physiology
School of Biological Sciences
Illinois State University
Normal, IL

Toni D. Sauncy
Assistant Professor of Physics
Department of Physics
Angelo State University
San Angelo, TX

Series Consultants, continued

Malathi Srivatsan, PhD
Associate Professor of Neurobiology
College of Sciences and Mathematics
Arkansas State University
Jonesboro, AR

Cheryl Wistrom, PhD
Associate Professor of Chemistry
Saint Joseph's College
Rensselaer, IN

Reading

ReLeah Cossett Lent
Author/Educational Consultant
Blue Ridge, GA

Math

Vik Hovsepian
Professor of Mathematics
Rio Hondo College
Whittier, CA

Series Reviewers

Thad Boggs
Mandarin High School
Jacksonville, FL

Catherine Butcher
Webster Junior High School
Minden, LA

Erin Darichuk
West Frederick Middle School
Frederick, MD

Joanne Hedrick Davis
Murphy High School
Murphy, NC

Anthony J. DiSipio, Jr.
Octorara Middle School
Atglen, PA

Adrienne Elder
Tulsa Public Schools
Tulsa, OK

Carolyn Elliott
Iredell-Statesville Schools
Statesville, NC

Christine M. Jacobs
Ranger Middle School
Murphy, NC

Jason O. L. Johnson
Thurmont Middle School
Thurmont, MD

Felecia Joiner
Stony Point Ninth Grade Center
Round Rock, TX

Joseph L. Kowalski, MS
Lamar Academy
McAllen, TX

Brian McClain
Amos P. Godby High School
Tallahassee, FL

Von W. Mosser
Thurmont Middle School
Thurmont, MD

Ashlea Peterson
Heritage Intermediate Grade Center
Coweta, OK

Nicole Lenihan Rhoades
Walkersville Middle School
Walkersvillle, MD

Maria A. Rozenberg
Indian Ridge Middle School
Davie, FL

Barb Seymour
Westridge Middle School
Overland Park, KS

Ginger Shirley
Our Lady of Providence Junior-Senior High School
Clarksville, IN

Curtis Smith
Elmwood Middle School
Rogers, AR

Sheila Smith
Jackson Public School
Jackson, MS

Sabra Soileau
Moss Bluff Middle School
Lake Charles, LA

Tony Spoores
Switzerland County Middle School
Vevay, IN

Nancy A. Stearns
Switzerland County Middle School
Vevay, IN

Kari Vogel
Princeton Middle School
Princeton, MN

Alison Welch
Wm. D. Slider Middle School
El Paso, TX

Linda Workman
Parkway Northeast Middle School
Creve Coeur, MO

Teacher Advisory Board

The Teacher Advisory Board gave the authors, editorial staff, and design team feedback on the content and design of the Student Edition. They provided valuable input in the development of *Glencoe Earth iScience*.

Frances J. Baldridge
Department Chair
Ferguson Middle School
Beavercreek, OH

Jane E. M. Buckingham
Teacher
Crispus Attucks Medical Magnet High School
Indianapolis, IN

Elizabeth Falls
Teacher
Blalack Middle School
Carrollton, TX

Nelson Farrier
Teacher
Hamlin Middle School
Springfield, OR

Michelle R. Foster
Department Chair
Wayland Union Middle School
Wayland, MI

Rebecca Goodell
Teacher
Reedy Creek Middle School
Cary, NC

Mary Gromko
Science Supervisor K–12
Colorado Springs District 11
Colorado Springs, CO

Randy Mousley
Department Chair
Dean Ray Stucky Middle School
Wichita, KS

David Rodriguez
Teacher
Swift Creek Middle School
Tallahassee, FL

Derek Shook
Teacher
Floyd Middle Magnet School
Montgomery, AL

Karen Stratton
Science Coordinator
Lexington School District One
Lexington, SC

Stephanie Wood
Science Curriculum Specialist, K–12
Granite School District
Salt Lake City, UT

Welcome to iSCIENCE

We are your partner in learning by meeting your diverse 21st century needs. Designed for today's tech-savvy middle school students, the Glencoe *iScience* program offers hands-on investigations, rigorous science content, and engaging, real-world applications to make science fun, exciting, and stimulating.

Quick Start Guide
Glencoe iScience | Student Center

Login information

1. Go to **connected.mcgraw-hill.com.**
2. Enter your registered Username and Password.
3. For **new users** click here to create a new account.
4. Get **ConnectED Help** for creating accounts, verifying master codes, and more.

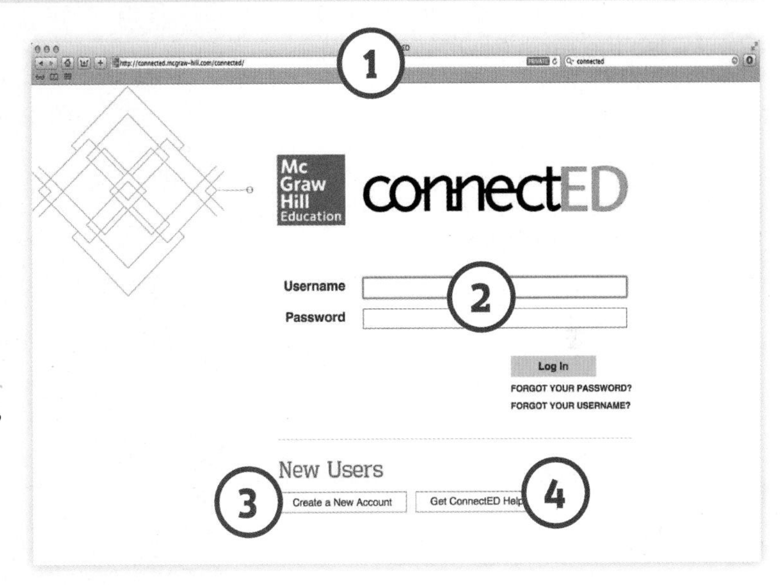

Your ConnectED Center

5. Scroll down to find the program from which you would like to work.

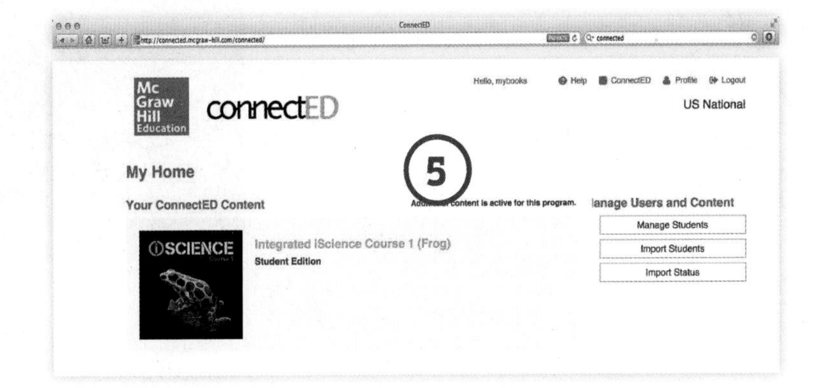

Quick Start Guide
iScience | Student Center

1. The Menu allows you to easily jump to anywhere you need to be.
2. Click the **program icon** at the top left to **return to the main page** from any screen.
3. **Select a Chapter and Lesson** Use the drop down boxes to quickly jump to any lesson in any chapter.
4. Return to your **My Home** page for all your **ConnectED** content.
5. The **Help** icon will guide you to online help. It will also allow for a quick logout.
6. The **Search Bar** allows you to search content by topic or standard.
7. **Access the eBook** **Use** the **Student Edition** to see content.

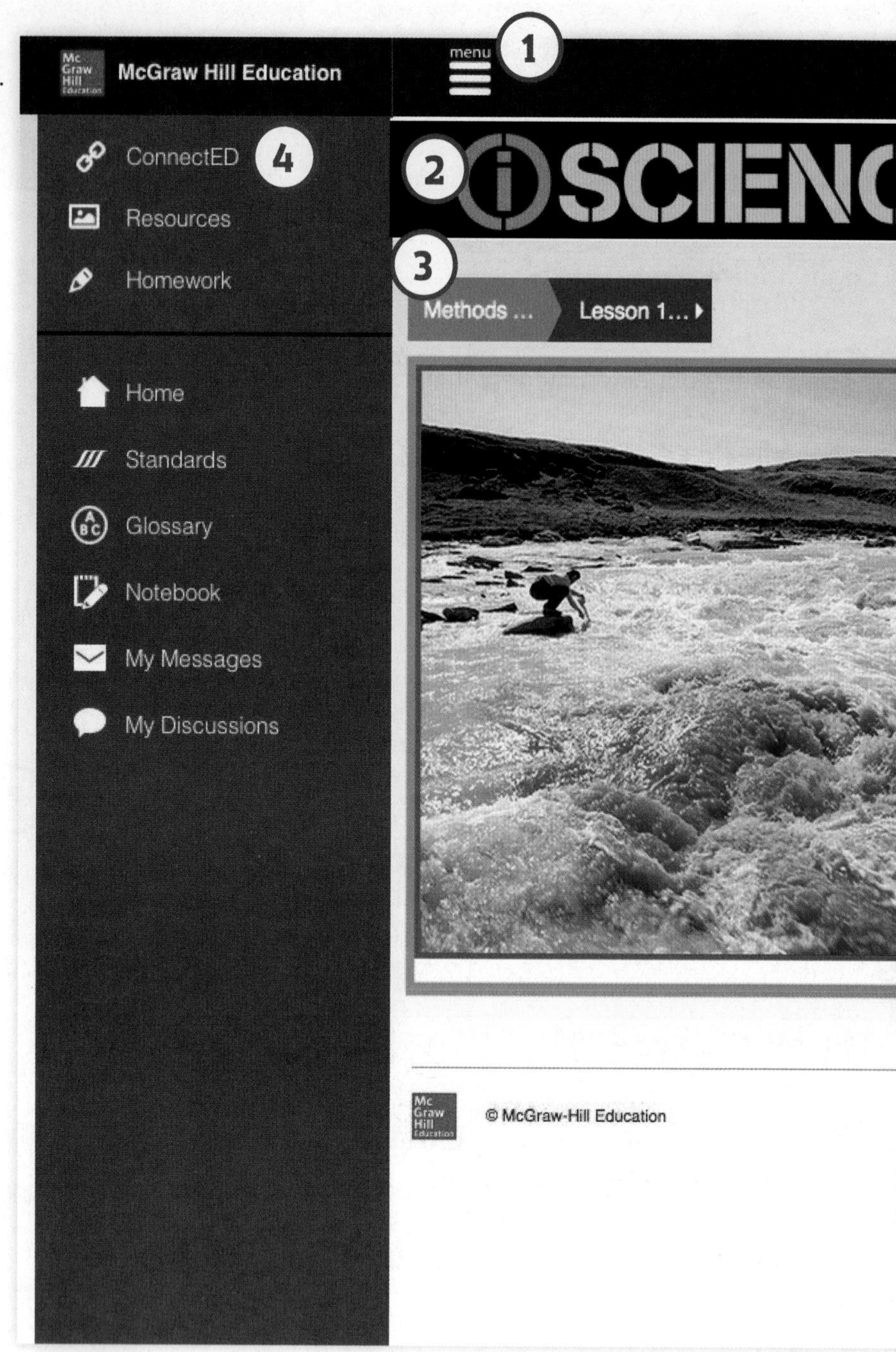

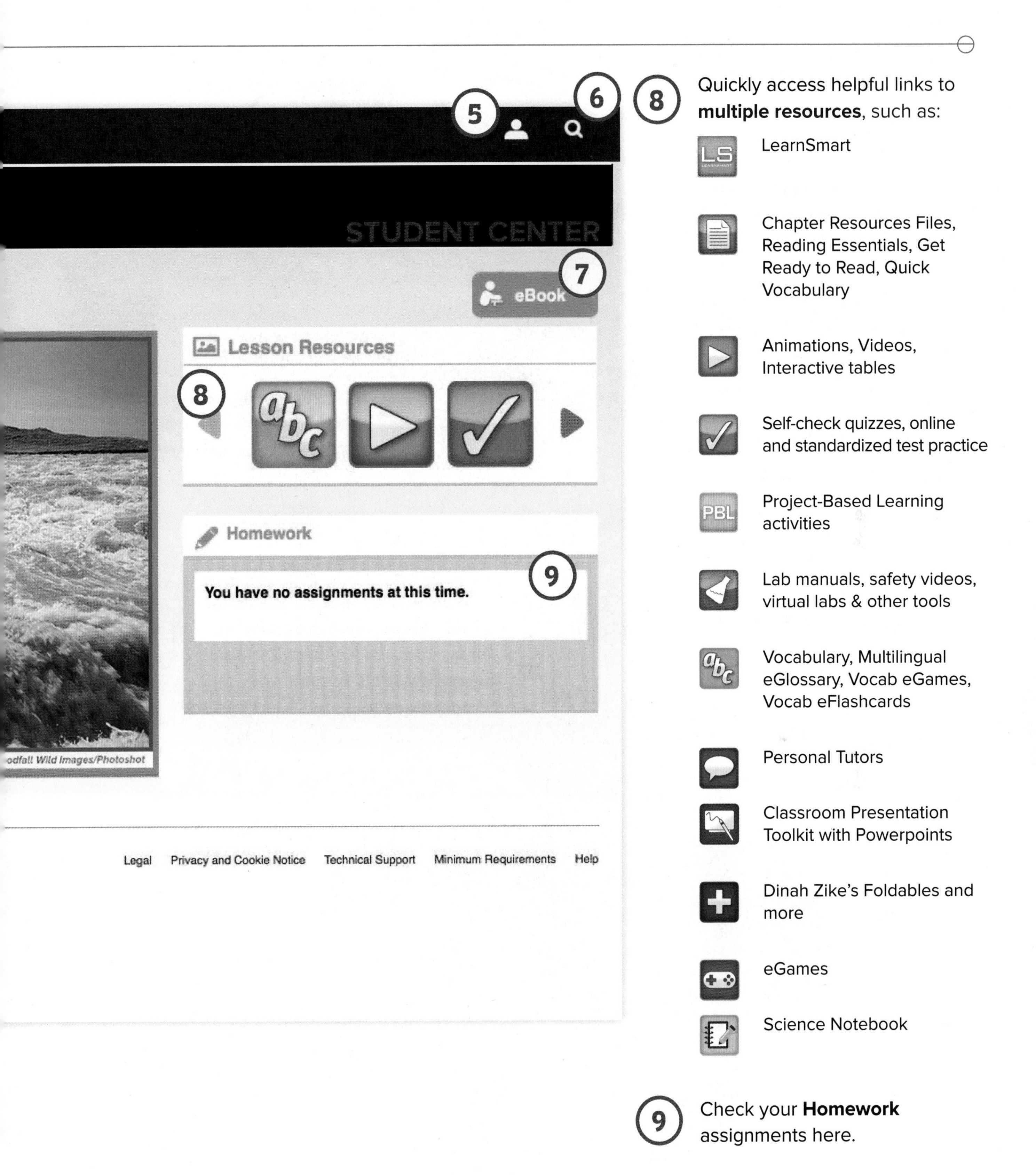

8 Quickly access helpful links to **multiple resources**, such as:

- LearnSmart
- Chapter Resources Files, Reading Essentials, Get Ready to Read, Quick Vocabulary
- Animations, Videos, Interactive tables
- Self-check quizzes, online and standardized test practice
- Project-Based Learning activities
- Lab manuals, safety videos, virtual labs & other tools
- Vocabulary, Multilingual eGlossary, Vocab eGames, Vocab eFlashcards
- Personal Tutors
- Classroom Presentation Toolkit with Powerpoints
- Dinah Zike's Foldables and more
- eGames
- Science Notebook

9 Check your **Homework** assignments here.

Treasure Hunt

Your science book has many features that will aid you in your learning. Some of these features are listed below. You can use the activity at the right to help you find these and other special features in the book.

- THE BIG IDEA can be found at the start of each chapter.
- The Reading Guide at the start of each lesson lists **Key Concepts,** vocabulary terms, and online supplements to the content.
- connectED icons direct you to online resources such as animations, personal tutors, math practices, and quizzes.
- Inquiry Labs and Skill Practices are in each chapter.
- Your FOLDABLES help organize your notes.

START

1 What four margin items can help you build your vocabulary?

2 On what page does the glossary begin? What glossary is online?

3 In which Student Resource at the back of your book can you find a listing of Laboratory Safety Symbols?

4 Suppose you want to find a list of all the Launch Labs, MiniLabs, Skill Practices, and Labs, where do you look?

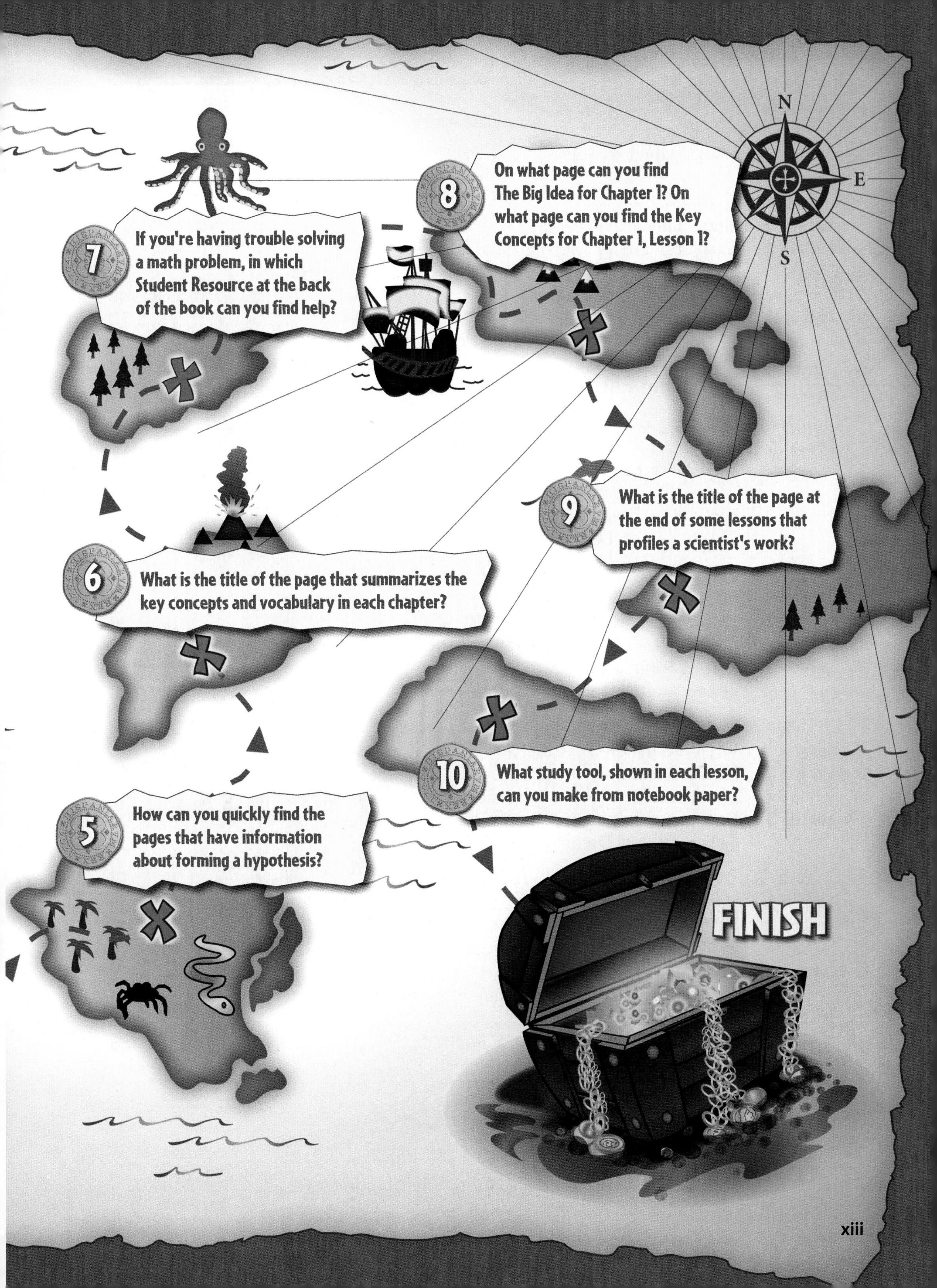
N
E
S
8
On what page can you find The Big Idea for Chapter 1? On what page can you find the Key Concepts for Chapter 1, Lesson 1?
7
If you're having trouble solving a math problem, in which Student Resource at the back of the book can you find help?
9
What is the title of the page at the end of some lessons that profiles a scientist's work?
6
What is the title of the page that summarizes the key concepts and vocabulary in each chapter?
10
What study tool, shown in each lesson, can you make from notebook paper?
5
How can you quickly find the pages that have information about forming a hypothesis?
FINISH

Table of Contents

TABLE OF CONTENTS

Student Resources

 Launch Labs

MiniLabs

Skill Practice

TABLE OF CONTENTS

Labs

Features

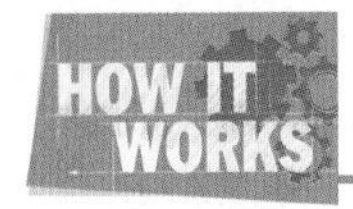

HOW NATURE WORKS

GREEN SCIENCE

SCIENCE & SOCIETY

CAREERS in SCIENCE

Nature of Science

Methods of Science

THE BIG IDEA What processes do scientists use when they perform scientific investigations?

Inquiry Pink Water?

This scientist is using pink dye to measure the speed of glacier water in the country of Greenland. Scientists are testing the hypothesis that the speed of the glacier water is increasing because amounts of meltwater, caused by climate change, are increasing.

- What is a hypothesis?
- What other ways do scientists test hypotheses?
- What processes do scientists use when they perform scientific investigations?

Nature of SCIENCE

This chapter begins your study of the nature of science, but there is even more information about the nature of science in this book. Each unit begins by exploring an important topic that is fundamental to scientific study. As you read these topics, you will learn even more about the nature of science.

Mc Graw Hill Education **connectED**

Your one-stop online resource
connectED.mcgraw-hill.com

LearnSmart®

Chapter Resources Files, Reading Essentials, Get Ready to Read, Quick Vocabulary

Animations, Videos, Interactive Tables

Self-checks, Quizzes, Tests

Project-Based Learning Activities

Lab Manuals, Safety Videos, Virtual Labs & Other Tools

Vocabulary, Multilingual eGlossary, Vocab eGames, Vocab eFlashcards

Personal Tutors

Lesson 1 Understanding Science

Reading Guide

Key Concepts

ESSENTIAL QUESTIONS

- What is scientific inquiry?
- How do scientific laws and scientific theories differ?
- What is the difference between a fact and an opinion?

Vocabulary

science p. NOS 4
observation p. NOS 6
inference p. NOS 6
hypothesis p. NOS 6
prediction p. NOS 6
technology p. NOS 8
scientific theory p. NOS 9
scientific law p. NOS 9
critical thinking p. NOS 10

 Multilingual eGlossary

 BrainPOP®
What's Science Got to do With It?

What is science?

Did you ever hear a bird sing and then look in nearby trees to find the singing bird? Have you ever noticed how the Moon changes from a thin crescent to a full moon each month? When you do these things, you are doing science. **Science** *is the investigation and exploration of natural events and of the new information that results from those investigations.*

For thousands of years, men and women of all countries and cultures have studied the natural world and recorded their observations. They have shared their knowledge and findings and have created a vast amount of scientific information. Scientific knowledge has been the result of a great deal of debate and confirmation within the science community.

People use science in their everyday lives and careers. For example, firefighters, as shown in **Figure 1**, wear clothing that has been developed and tested to withstand extreme temperatures and not catch fire. Parents use science when they set up an aquarium for their children's pet fish. Athletes use science when they use high-performance gear or wear high-performance clothing. Without thinking about it, you use science or the results of science in almost everything you do. Your clothing, food, hair products, electronic devices, athletic equipment, and almost everything else you use are results of science.

Figure 1 Firefighters' clothing, oxygen tanks, and equipment are all results of science.

Thomas Del Brase/Getty Images

Branches of Science

There are many different parts of the natural world. Because there is so much to study, scientists often focus their work in one branch of science or on one topic within that branch of science. There are three main branches of science–Earth science, life science, and **physical** science.

WORD ORIGIN

physical
from Latin *physica,* means "study of nature"

Earth Science

The study of Earth, including rocks, soils, oceans, and the atmosphere is Earth science. The Earth scientist to the right is collecting lava samples for research. Earth scientists might ask other questions such as

- How do different shorelines react to tsunamis?
- Why do planets orbit the Sun?
- What is the rate of climate change?

Life Science

The study of living things is life science, or biology. These biologists are attaching a radio collar to a tiger to help track its movements and learn more about its behavior. They are also weighing and measuring the tiger to gain information about this species. Biologists also ask questions such as

- Why do some trees lose their leaves in winter?
- How do birds know which direction they are going?
- How do mammals control their body temperature?

Physical Science

The study of matter and energy is physical science. It includes both physics and chemistry. This research chemist is preparing chemical solutions for analysis. Physicists and chemists ask other questions such as

- What chemical reactions must take place to launch a spaceship into space?
- Is it possible to travel faster than the speed of light?
- What makes up matter?

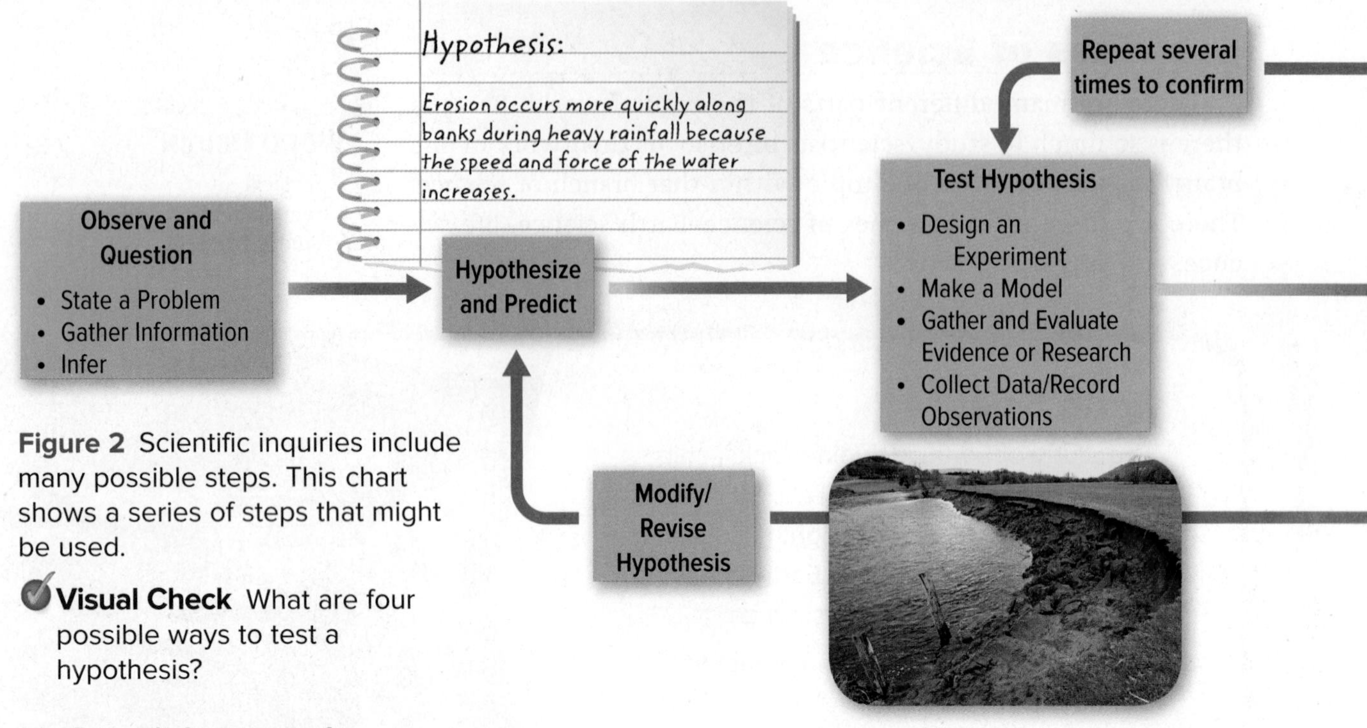

Figure 2 Scientific inquiries include many possible steps. This chart shows a series of steps that might be used.

Visual Check What are four possible ways to test a hypothesis?

Scientific Inquiry

When scientists conduct scientific investigations, they use scientific inquiry. Scientific inquiry is a process that uses a set of skills to answer questions or to test ideas about the natural world. There are many kinds of scientific investigations, and there are many ways to conduct them. The series of steps used in each investigation often varies. The flow chart in **Figure 2** shows an example of the skills used in scientific inquiry.

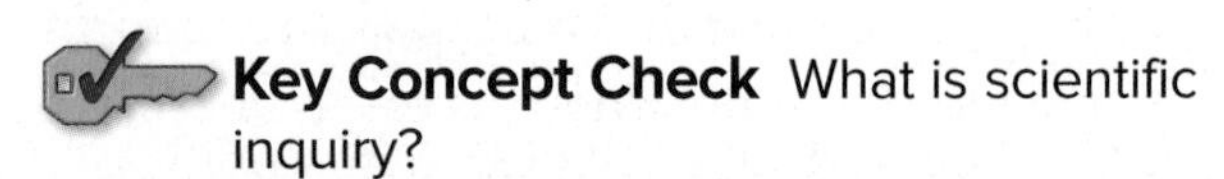
Key Concept Check What is scientific inquiry?

Ask Questions

One way to begin a scientific inquiry is to observe the natural world and ask questions. **Observation** *is the act of using one or more of your senses to gather information and taking note of what occurs.* Suppose you observe that the banks of a river have eroded more this year than in the previous year, and you want to know why. You also note that there was an increase in rainfall this year. After these observations, you make an inference based on these observations. *An* **inference** *is a logical explanation of an observation that is drawn from prior knowledge or experience.*

You infer that the increase in rainfall caused the increase in erosion. You decide to investigate further. You develop a hypothesis and a method to test it.

Hypothesize and Predict

A **hypothesis** *is a possible explanation for an observation that can be tested by scientific investigations.* A hypothesis states an observation and provides an explanation. For example, you might make the following hypothesis: More of the riverbank eroded this year because the amount, the speed, and the force of the river water increased.

When scientists state a hypothesis, they often use it to make predictions to help test their hypothesis. *A* **prediction** *is a statement of what will happen next in a sequence of events.* Scientists make predictions based on what information they think they will find when testing their hypothesis. For example, predictions for the hypothesis above could be: If rainfall increases, then the amount, the speed, and the force of river water will increase. If the amount, the speed, and the force of river water increase, then there will be more erosion.

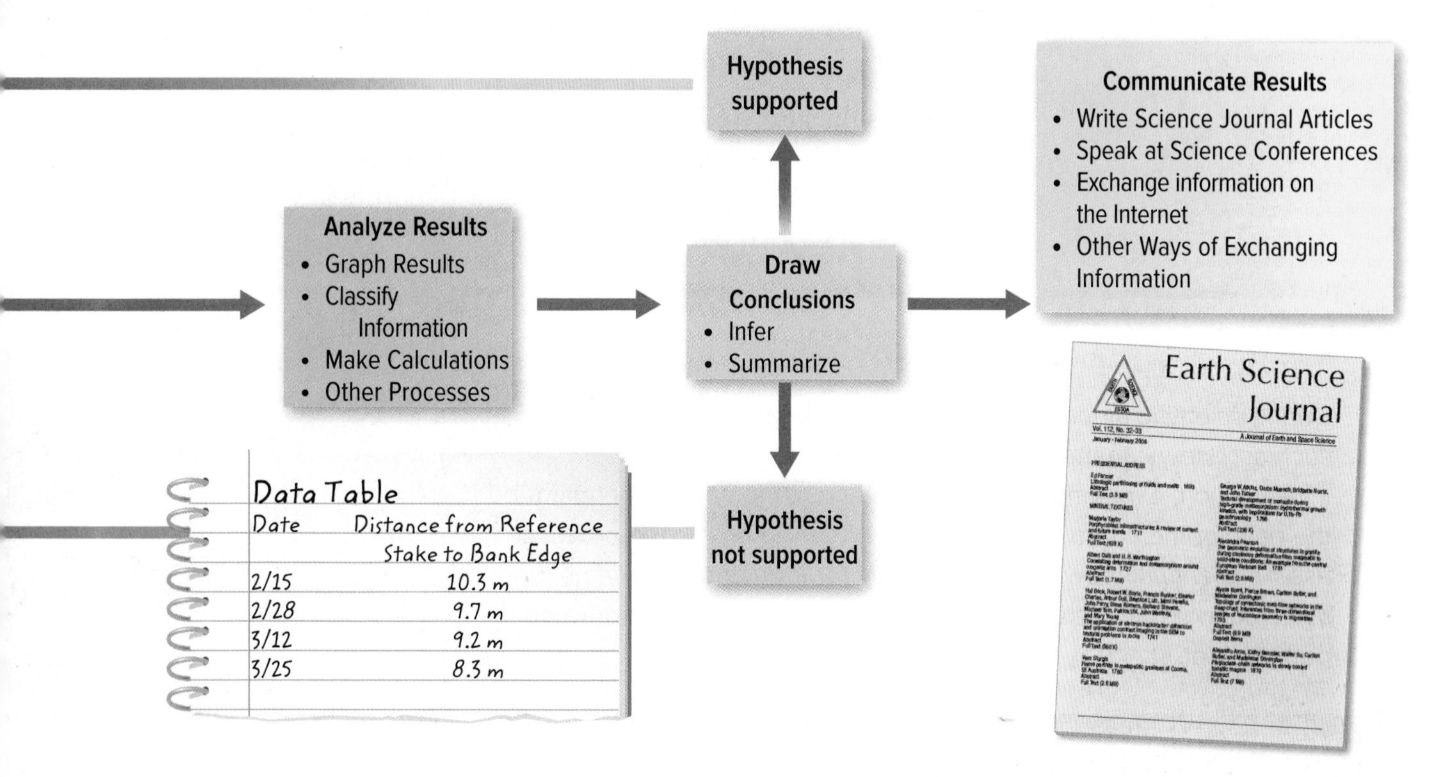

Test Hypothesis

When you test a hypothesis, you often test whether your predictions are true. If a prediction is confirmed, then it supports your hypothesis. If your prediction is not confirmed, you might need to modify your hypothesis and retest it.

There are several ways to test a hypothesis when performing a scientific investigation. Four possible ways are shown in **Figure 2.** For example, you might make a model of a riverbank in which you change the speed and the amount of water and record results and observations.

Analyze Results

After testing your hypothesis, you analyze your results using various methods, as shown in **Figure 2.** Often, it is hard to see trends or relationships in data while collecting it. Data should be sorted, graphed, or classified in some way. After analyzing the data, additional inferences can be made.

Draw Conclusions

Once you find the relationships among data and make several inferences, you can draw conclusions.

A conclusion is a summary of the information gained from testing a hypothesis. Scientists study the available information and draw conclusions based on that information.

Communicate Results

An important part of the scientific inquiry process is communicating results. Several ways to communicate results are listed in **Figure 2.** Scientists might share their information in other ways, too. Scientists communicate results of investigations to inform other scientists about their research and their conclusions. When a scientist uses that information to repeat another scientist's experiment, he or she is replicating the experiment to confirm results.

Further Scientific Inquiry

After finishing an experiment, a scientist must verify his or her results. If the hypothesis is supported, the scientist will repeat the experiment several times to make sure the conclusions are the same–this is called experimental repetition. If the hypothesis is not supported, any new information gained can be used to revise the hypothesis. Hypotheses can be revised and tested many times.

Results of Science

The results and conclusions from an investigation can lead to many outcomes, such as the answers to a question, more information on a specific topic, or support for a hypothesis. Other outcomes are described below.

Technology

During scientific inquiry, scientists often look for answers to questions such as, "How can the hearing impaired hear better?" After investigation, experimentation, and research, the conclusion might be the development of a new technology. **Technology** *is the practical use of scientific knowledge, especially for industrial or commercial use.* Technology, such as the cochlear implant, can help some deaf people hear.

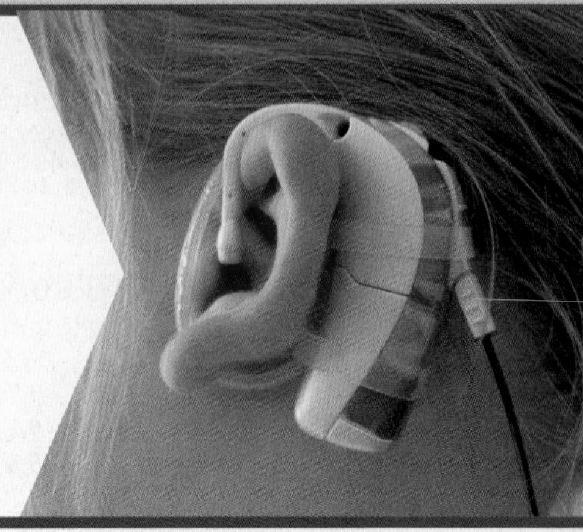

New Materials

Space travel has unique challenges. Astronauts must carry oxygen to breathe. They also must be protected against temperature and pressure extremes, as well as small, high-speed flying objects. Today's spacesuit, a result of research, testing, and design changes, consists of layers of material. The outer layer is made of a blend of materials. One material is waterproof and another material is heat and fire-resistant.

Possible Explanations

Scientists often perform investigations to find explanations as to why or how something happens. NASA's *Spitzer Space Telescope,* which has aided in our understanding of star formation, shows a cloud of gas and dust with newly formed stars.

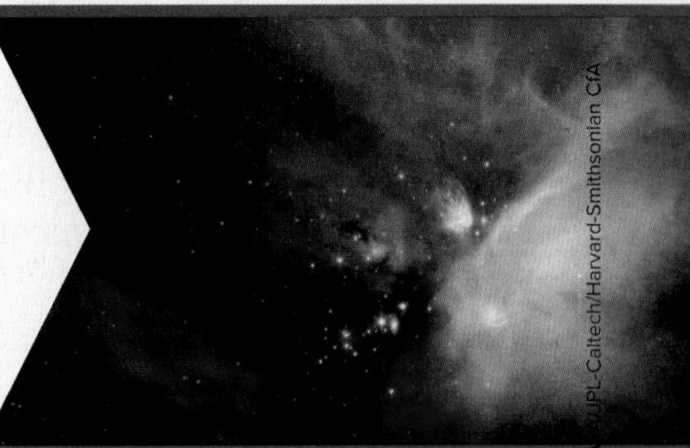

 **Reading Check** What are some results of science?

Scientific Theory and Scientific Law

Another outcome of science is the development of scientific theories and laws. Recall that a hypothesis is a possible explanation about an observation that can be tested by scientific investigations. What happens when a hypothesis or a group of hypotheses has been tested many times and has been supported by the repeated scientific investigations? The hypothesis can become a scientific theory.

Scientific Theory

Often, the word *theory* is used in casual conversations to mean an untested idea or an opinion. However, scientists use *theory* differently. *A* **scientific theory** *is an explanation of observations or events that is based on knowledge gained from many observations and investigations.*

Scientists regularly question scientific theories and test them for validity. A scientific theory generally is accepted as true until it is disproved. An example of a scientific theory is the theory of plate tectonics. The theory of plate tectonics explains how Earth's crust moves and why earthquakes and volcanoes occur. Another example of a scientific theory is discussed in **Figure 3.**

▲ **Figure 3** Scientists once believed Earth was the center of the solar system. In the 16th century, Nicolaus Copernicus hypothesized that Earth and the other planets actually revolve around the Sun.

Scientific Law

A scientific law is different from a social law, which is an agreement among people concerning a behavior. *A* **scientific law** *is a rule that describes a pattern in nature.* Unlike a scientific theory that explains why an event occurs, a scientific law only states that an event will occur under certain circumstances. For example, Newton's law of gravitational force implies that if you drop an object, it will fall toward Earth. Newton's law does not explain why the object moves toward Earth when dropped, only that it will.

Key Concept Check How do scientific laws and theories differ?

New Information

Scientific information constantly changes as new information is discovered or as previous hypotheses are retested. New information can lead to changes in scientific theories, as explained in **Figure 4.** When new facts are revealed, a current scientific theory might be revised to include the new facts, or it might be disproved and rejected.

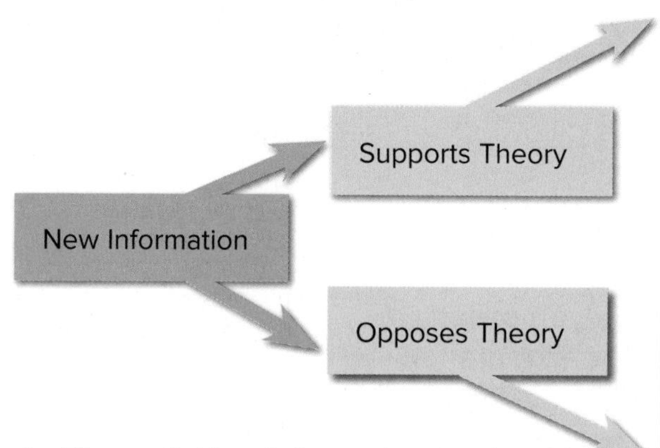

If new information supports a current scientific theory, then the theory is not changed. The information might be published in a scientific journal to show further support of the theory. The new information might also lead to advancements in technology or spark new questions that lead to new scientific investigations.

If new information opposes, or does not support a current scientific theory, the theory might be modified or rejected altogether. Often, new information will lead scientists to look at the original observations in a new way. This can lead to new investigations with new hypotheses. These investigations can lead to new theories.

▲ **Figure 4** New information can lead to changes in scientific theories.

FOLDABLES®

Make a six-tab book, and label it as shown. Use it to organize your notes about a scientific inquiry.

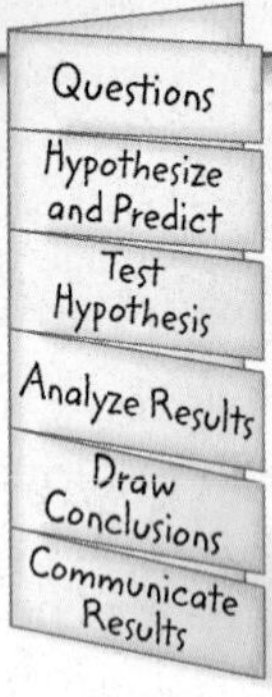

Evaluating Scientific Evidence

Did you ever read an advertisement, such as the one below, that made extraordinary claims? If so, you have practiced **critical thinking**–*comparing what you already know with the information you are given in order to decide whether you agree with it.* To determine whether information is true and scientific or pseudoscience (information incorrectly represented as scientific), you should be skeptical and identify facts and opinions. This helps you evaluate the strengths and weaknesses of information and make informed decisions. Critical thinking is important in all decision making–from everyday decisions to community, national, and international decisions.

 **Key Concept Check** How do a fact and an opinion differ?

Learn Algebra *While You Sleep!*

Have you struggled to learn algebra? Struggle no more.

Math-er-ific's new algebra pillow is scientifically proven to transfer math skills from the pillow to your brain while you sleep. This revolutionary scientific design improved the algebra test scores of laboratory mice by 150 percent.

Dr. Tom Equation says, "I have never seen students or mice learn algebra so easily. This pillow is truly amazing."

For only $19.95, those boring hours spent studying are a thing of the past. So act fast! If you order today, you can get the algebra pillow and the equally amazing geometry pillow for only $29.95. That is a $10 savings!

Skepticism

To be skeptical is to doubt the truthfulness or accuracy of something. Because of skepticism, science can be self-correcting. If someone publishes results or if an investigation gives results that don't seem accurate, a skeptical scientist usually will challenge the information and test the results for accuracy.

Identifying Opinions

An opinion is a personal view, feeling, or claim about a topic. Opinions are neither true nor false.

Identifying Facts

The prices of the pillows and the savings are facts. A fact is a measurement, observation, or statement that can be strictly defined. Many scientific facts can be evaluated for their validity through investigations.

Mixing Facts and Opinions

Sometimes people mix facts and opinions. You must read carefully to determine which information is fact and which is opinion.

Science cannot answer all questions.

Scientists recognize that some questions cannot be studied using scientific inquiry. Questions that deal with opinions, beliefs, values, and feelings cannot be answered through scientific investigation. For example, questions that cannot be answered through scientific investigation might include

- Are comedies the best kinds of movies?
- Is it ever okay to lie?
- Which food tastes best?

The answers to all of these questions are based on opinions, not facts.

Figure 5 Always use safe lab practices when doing scientific investigations.

Safety in Science

It is very important for anyone performing scientific investigations to use safe practices, such as the student shown in **Figure 5.** You should always follow your teacher's instructions. If you have questions about potential hazards, use of equipment, or the meaning of safety symbols, ask your teacher. Always wear protective clothing and equipment while performing scientific investigations. If you are using live animals in your investigations, provide appropriate care and ethical treatment to them. For more information on practicing safe and ethical science, consult the Science Safety Skill Handbook in the back of this book.

ACADEMIC VOCABULARY

potential
(adjective) possible, likely, or probable

Online Quiz
Virtual Lab

Lesson 1 Review

Use Vocabulary

1. The practical use of science, especially for industrial or commercial use, is ________.
2. **Distinguish** between a hypothesis and a prediction.
3. **Define** *observation* in your own words.

Understand Key Concepts

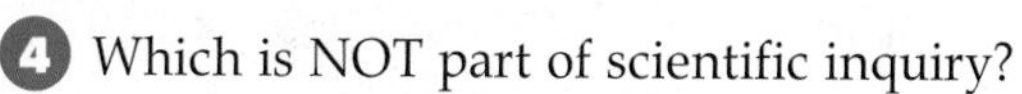

4. Which is NOT part of scientific inquiry?
   - **A.** analyze results
   - **B.** falsify results
   - **C.** make a hypothesis
   - **D.** make observations
5. **Explain** the difference between a scientific theory and a scientific law. Give an example of each.
6. **Write** an example of a fact and an example of an opinion.

Interpret Graphics

7. **Organize** Draw a graphic organizer similar to the one below. List four ways a scientist can communicate results.

Critical Thinking

8. **Identify** a real-world problem related to your home, your community, or your school that could be investigated scientifically.
9. **Design** a scientific investigation to test one possible solution to the problem you identified in the previous question.

Lesson 2 Measurement and Scientific Tools

Reading Guide

Key Concepts
ESSENTIAL QUESTIONS

- Why is it important for scientists to use the International System of Units?
- What causes measurement uncertainty?
- What are mean, median, mode, and range?

Vocabulary

description p. NOS 12
explanation p. NOS 12
International System of Units (SI) p. NOS 12
significant digits p. NOS 14

 Multilingual eGlossary

Description and Explanation

The scientist in **Figure 6** is observing a volcano. He describes in his journal that the flowing lava is bright red with a black crust, and it has a temperature of about 630°C. *A* **description** *is a spoken or written summary of observations.* There are two types of descriptions. When making a qualitative description, such as *bright red,* you use your senses (sight, sound, smell, touch, taste) to describe an observation. When making a quantitative description, such as *630°C,* you use numbers and measurements to describe an observation. Later, the scientist might explain his observations. *An* **explanation** *is an interpretation of observations.* Because the lava was bright red and about 630°C, the scientist might explain that these conditions indicate the lava is cooling and the volcano did not recently erupt.

The International System of Units

At one time, scientists in different parts of the world used different units of measurement. Imagine the confusion when a British scientist measured weight in pounds-force, a French scientist measured weight in Newtons, and a Japanese scientist measured weight in momme (MOM ee). Sharing scientific information was difficult, if not impossible.

In 1960, scientists adopted a new system of measurement to eliminate this confusion. *The* **International System of Units (SI)** *is the internationally accepted system for measurement.* SI uses standards of measurement, called base units, which are shown in **Table 1** on the next page. A base unit is the most common unit used in the SI system for a given measurement.

Figure 6 Scientists use descriptions and explanations when observing natural events.

Interactive Table

Table 1 SI Base Units		
Quantity Measured	**Unit**	**Symbol**
Length	meter	m
Mass	kilogram	kg
Time	second	s
Electric current	ampere	A
Temperature	Kelvin	K
Amount of substance	mole	mol
Intensity of light	candela	cd

◀ **Table 1** You can use SI units to measure the physical properties of objects.

SI Unit Prefixes

In addition to base units, SI uses prefixes to identify the size of the unit, as shown in **Table 2.** Prefixes are used to indicate a fraction of ten or a multiple of ten. In other words, each unit is either ten times smaller than the next larger unit or ten times larger than the next smaller unit. For example, the prefix *deci-* means 10^{-1}, or 1/10. A decimeter is 1/10 of a meter. The prefix *kilo-* means 10^3, or 1,000. A kilometer is 1,000 m.

Converting Between SI Units

Because SI is based on ten, it is easy to convert from one SI unit to another. To convert SI units, you must multiply or divide by a factor of ten. You also can use proportions as shown below in the Math Skills activity.

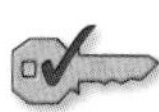

Key Concept Check Why is it important for scientists to use the International System of Units (SI)?

Table 2 Prefixes are used in SI to indicate the size of the unit. ▼

Table 2 Prefixes	
Prefix	**Meaning**
Mega- (M)	1,000,000 (10^6)
Kilo- (k)	1,000 (10^3)
Hecto- (h)	100 (10^2)
Deka- (da)	10 (10^1)
Deci- (d)	0.1 (10^{-1})
Centi- (c)	0.01 (10^{-2})
Milli- (m)	0.001 (10^{-3})
Micro- (μ)	0.000 001 (10^{-6})

Math Skills — Use Proportions

Math Practice

Personal Tutor

A book has a mass of **1.1 kg**. Using a proportion, find the mass of the book in grams.

1. Use the table to determine the correct relationship between the units. One kg is 1,000 times greater than 1 g. So, there are 1,000 g in 1 kg.
2. Then set up a proportion.

$$\left(\frac{x}{\mathbf{1.1\ kg}}\right) = \left(\frac{1{,}000\ \text{g}}{1\ \text{kg}}\right)$$

$$x = \left(\frac{(1{,}000\ \text{g})(\mathbf{1.1\ \cancel{kg}})}{\cancel{1\ \text{kg}}}\right) = 1{,}100\ \text{g}$$

3. Check your units. The answer is 1,100 g.

Practice

1. Two towns are separated by 15,328 m. What is the distance in kilometers?
2. A dosage of medicine is 325 mg. What is the dosage in grams?

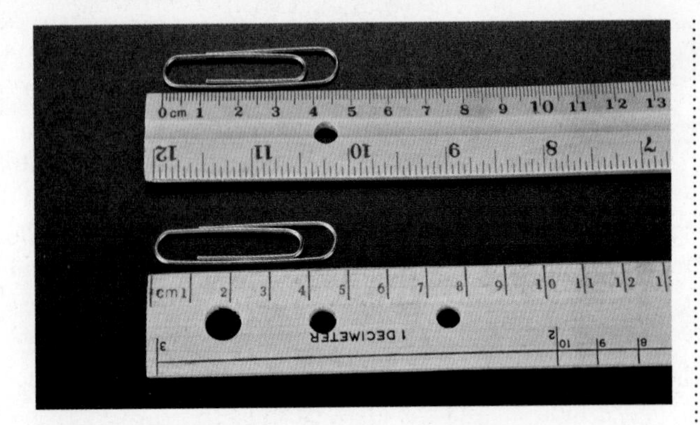

Figure 7 All measurements have some uncertainty.

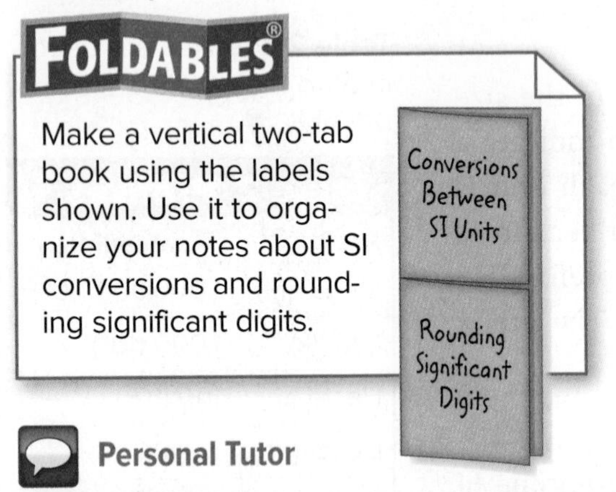

Make a vertical two-tab book using the labels shown. Use it to organize your notes about SI conversions and rounding significant digits.

Personal Tutor

Table 3 Significant Digits Rules

1. All nonzero numbers are significant.
2. Zeros between significant digits are significant.
3. All final zeros to the right of the decimal point are significant.
4. Zeros used solely for spacing the decimal point are NOT significant. The zeros only indicate the position of the decimal point.

* The blue numbers in the examples are the significant digits.

Number	Significant Digits	Applied Rules
1.234	4	1
1.02	3	1, 2
0.023	2	1, 4
0.200	3	1, 3
1,002	4	1, 2
3.07	3	1, 2
0.001	1	1, 4
0.012	2	1, 4
50,600	3	1, 2, 4

Measurement and Uncertainty

Have you ever measured an object, such as a paper clip? The tools used to take measurements can limit the accuracy of the measurements. Look at the bottom ruler in **Figure 7.** Its measurements are divided into centimeters. The paper clip is between 4 cm and 5 cm. You might guess that it is 4.7 cm long. Now, look at the top ruler. Its measurements are divided into millimeters. You can say with more precision that the paper clip is about 4.75 cm long. This measurement is more precise than the first measurement.

Key Concept Check What causes measurement uncertainty?

Significant Digits and Rounding

Because scientists duplicate each other's work, they must record numbers with the same degree of precision as the original data. Significant digits allow scientists to do this. **Significant digits** *are the number of digits in a measurement that you know with a certain degree of reliability.* **Table 3** lists the rules for expressing and determining significant digits.

In order to achieve the same degree of precision as a previous measurement, it often is necessary to round a measurement to a certain number of significant digits. Suppose you have the number below, and you need to round it to four significant digits.

1,348.527 g

To round to four significant digits, you need to round the 8. If the digit to the right of the 8 is 0, 1, 2, 3, or 4, the digit being rounded (8) remains the same. If the digit to the right of the 8 is 5, 6, 7, 8, or 9, the digit being rounded (8) increases by one. The rounded number is 1,349 g.

What if you need to round 1,348.527 g to two significant digits? You would look at the number to the right of the 3 to determine how to round. 1,348.527 rounded to two significant digits would be 1,300 g. The 4 and 8 become zeros.

Mean, Median, Mode, and Range

A rain gauge measures the amount of rain that falls on a location over a period of time. A rain gauge can be used to collect data in scientific investigations, such as the data shown in **Table 4a.** Scientists often need to analyze their data to obtain information. Four values often used when analyzing numbers are median, mean, mode, and range.

 **Key Concept Check** What are mean, median, and mode?

Median

The median is the middle number in a data set when the data are arranged in numerical order. The rainfall data are listed in numerical order in Table 4b. If you have an even number of data items, add the two middle numbers together and divide by two to find the median.

$$\text{median} = \frac{8.18 \text{ cm} + 8.84 \text{ cm}}{2}$$

$$= 8.51 \text{ cm}$$

Table 4a Rainfall Data

Month	Rainfall
January	7.11 cm
February	11.89 cm
March	9.58 cm
April	8.18 cm
May	7.11 cm
June	1.47 cm
July	18.21 cm
August	8.84 cm

Mean

The mean, or average, of a data set is the sum of the numbers in a data set divided by the number of entries in the set. To find the mean, add the numbers in your data set and then divide the total by the number of items in your data set.

$$\text{mean} = \frac{(\text{sum of numbers})}{(\text{number of items})}$$

$$= \frac{72.39 \text{ cm}}{8 \text{ months}}$$

$$= \frac{9.05 \text{ cm}}{\text{month}}$$

Mode

The mode of a data set is the number or item that appears most often. The number in blue in Table 4b appears twice. All other numbers appear only once.

mode = 7.11 cm

Table 4b Rainfall Data (numerical order)

Rainfall
1.47 cm
7.11 cm
7.11 cm
8.18 cm
8.84 cm
9.58 cm
11.89 cm
18.21 cm

Range

The range is the difference between the greatest number and the least number in the data set.

$$\text{range} = 18.21 \text{ cm} - 1.47 \text{ cm}$$

$$= 16.74 \text{ cm}$$

Scientific Tools

As you engage in scientific inquiry, you will need tools to help you take quantitative measurements. Always follow appropriate safety procedures when using scientific tools. For more information about the proper use of these tools, see the Science Skill Handbook at the back of this book.

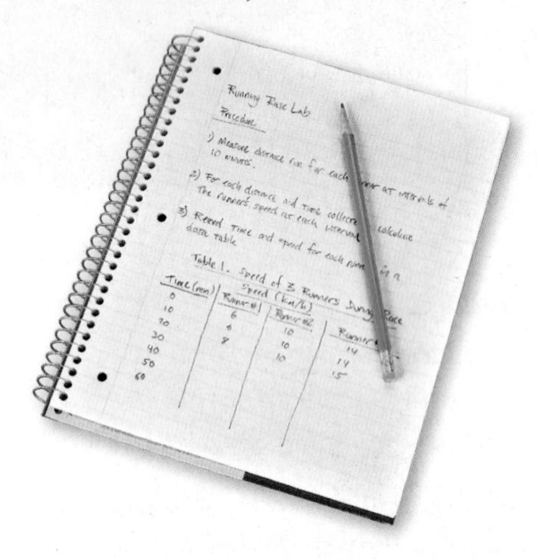

◂ Science Journal

Use a science journal to record observations, questions, hypotheses, data, and conclusions from your scientific investigations. A science journal is any notebook that you use to take notes or record information and data while you conduct a scientific investigation. Keep it organized so you can find information easily. Write down the date whenever you record new information in the journal. Make sure you are recording your data honestly and accurately.

Rulers and Metersticks ▸

Use rulers and metersticks to measure lengths and distances. The SI unit of measurement for length is the meter (m). For small objects, such as pebbles or seeds, use a metric ruler with centimeter and millimeter markings. To measure larger objects, such as the length of your bedroom, use a meterstick. To measure long distances, such as the distance between cities, use an instrument that measur es in kilometers. Be careful when carrying rulers and metersticks, and never point them at anyone.

◂ Glassware

Use beakers to hold and pour liquids. The lines on a beaker do not provide accurate measurements. Use a graduated cylinder to measure the volume of a liquid. Volume is typically measured in liters (L) or milliliters (mL).

(t)Matt Meadows; (c)By Ian Miles-Flashpoint Pictures/Alamy; (b)Joseph Clark/Photodisc/Getty Images

Triple-Beam Balance ▸

Use a triple-beam balance to measure the mass of an object. The mass of a small object is measured in grams. The mass of large object is usually measured in kilograms. Triple-beam balances are instruments that require some care when using. Follow your teacher's instructions so that you do not damage the instrument. Digital balances also might be used.

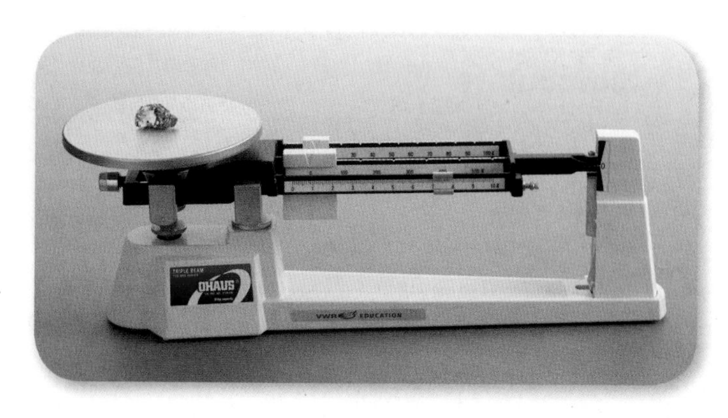

◂ Thermometer

Use a thermometer to measure the temperature of a substance. Kelvin is the SI unit for temperature, but you will use a thermometer to measure temperature in degrees Celsius (°C). To use a thermometer, place a room-temperature thermometer into the substance for which you want to measure temperature. Do not let the thermometer touch the bottom of the container that holds the substance or you will get an inaccurate reading. When you finish, remember to place your thermometer in a secure place. Do not lay it on a table, because it can roll off the table. Never use a thermometer as a stirring rod.

Computers and the Internet ▸

Use a computer to collect, organize, and store information about a research topic or scientific investigation. Computers are useful tools to scientists for several reasons. Scientists use computers to record and analyze data, to research new information, and to quickly share their results with others worldwide over the Internet.

(t)Hutchings Photography/Digital Light Source; (cl)McGraw-Hill Education; (br)Steve Cole/Getty Images

Tools Used by Earth Scientists

Binoculars

Binoculars are instruments that enable people to view faraway objects more clearly. Earth scientists use them to view distant landforms, animals, or even incoming weather.

Compass

A compass is an instrument that shows magnetic north. Earth scientists use compasses to navigate when they are in the field and to determine the direction of distant landforms or other natural objects.

Wind Vane and Anemometer

A wind vane is a device, often attached to the roofs of buildings, that rotates to show the direction of the wind. An anemometer, or wind-speed gauge, is used to measure the speed and the force of wind.

Streak Plate

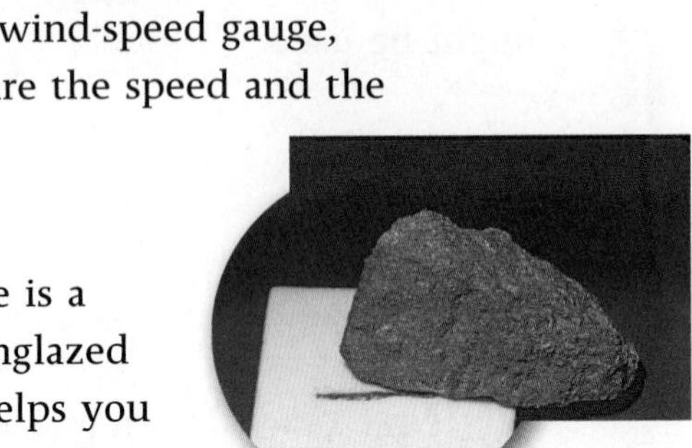

A streak plate is a piece of hard, unglazed porcelain that helps you identify minerals. When you scrape a mineral along a streak plate, the mineral leaves behind powdery marks. The color of the mark is the mineral's streak.

Lesson 2 Review

Online Quiz

Use Vocabulary

1. **Distinguish** between description and explanation.
2. **Define** *significant digits* in your own words.

Understand Key Concepts

3. Which base unit is NOT part of the International System of Units?
   A. ampere
   B. meter
   C. pound
   D. second
4. **Give an example** of how scientific tools cause measurement uncertainty.
5. Differentiate among mean, median, mode, and range.

Interpret Graphics

6. **Change** Copy the graphic organizer below, and change the number shown to have the correct number of significant digits indicated.

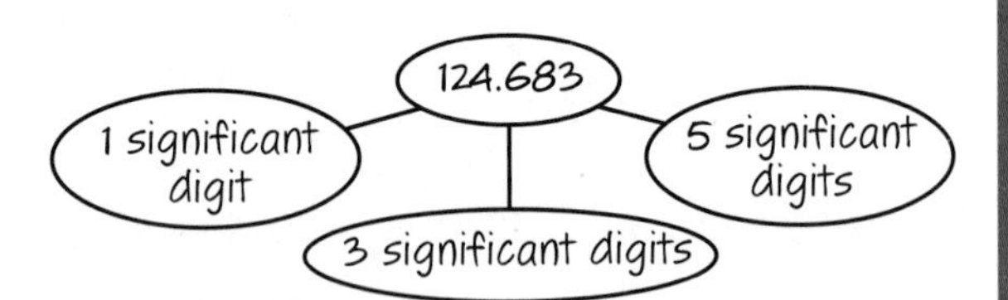

Critical Thinking

7. **Write** a short essay explaining why the United States should consider adopting SI as the measurement system used by supermarkets and other businesses.

Math Skills

Math Practice

8. **Convert** 52 m to kilometers. Explain how you got your answer.

Skill Practice — Collect and Analyze Data — 45 minutes

What can you learn by collecting and analyzing data?

People who study ancient cultures often collect and analyze data from soil samples. Soil samples contain bits of pottery, bones, seeds, and other clues to how ancient people lived and what they ate. In this activity, you will separate and analyze a simulated soil sample from an ancient civilization.

Materials

250-mL beaker

large piece of newsprint

1-L containers

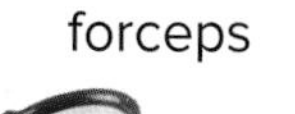

forceps

strainer

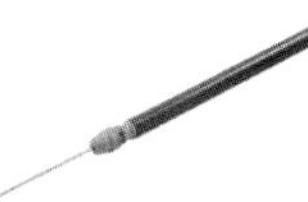

probe

Also needed: soil mixture, balance, plastic containers

Safety

Learn It

Data includes observations you can make with your senses and observations based on measurements of some kind. **Collecting and analyzing data** includes collecting, classifying, comparing and contrasting, and interpreting (looking for meaning in the data).

Try It

1. Read and complete a lab safety form.
2. Obtain a 200-mL sample of "soil."
3. Spread the newsprint over your workspace. Slowly pour the soil through a strainer over a plastic container. Shake the strainer gently so that all of the soil enters the container.
4. Pour the remaining portion of the soil sample onto the newsprint. Use a probe and forceps to separate objects. Classify different types of objects, and place them into the other plastic containers.
5. Copy the data tables from the board into your Science Journal.
6. Use the balance to measure and record the masses of each group of objects found in your soil sample. Write your group's data in the data table on the board.
7. When all teams have finished, use the class data from the board to find the mean, the median, the mode, and the range for each type of object.

Apply It

8. **Make Inferences** Assuming that the plastic objects represented animal bones, how many different types of animals were indicated by your analysis? Explain.
9. **Evaluate** Archaeologists often include information about the depth at which soil samples are taken. If you received a soil sample that kept the soil and other objects in their original layers, what more might you discover?
10. **Key Concept** Why didn't everyone in the class get the same data? What were some possible sources of uncertainty in your measurements?

Lesson 3 Case Study

Reading Guide

Key Concepts
ESSENTIAL QUESTIONS

- How are independent variables and dependent variables related?
- How is scientific inquiry used in a real-life scientific investigation?

Vocabulary

variable p. NOS 21

independent variable p. NOS 21

dependent variable p. NOS 21

 **Multilingual eGlossary**

 Go to the resource tab in ConnectED to find the PBL *Solutions for Pollution.*

The Iceman's Last Journey

The Tyrolean Alps border western Austria, northern Italy, and eastern Switzerland, as shown in **Figure 8.** They are popular with tourists, hikers, mountain climbers, and skiers. In 1991, two hikers discovered the remains of a man, also shown in **Figure 8,** in a melting glacier on the border between Austria and Italy. They thought the man had died in a hiking accident. They reported their discovery to the authorities.

Initially authorities thought the man was a music professor who disappeared in 1938. However, they soon learned that the music professor was buried in a nearby town. Artifacts near the frozen corpse indicated that the man died long before 1938. The artifacts, as shown in **Figure 9,** were unusual. The man, nicknamed the Iceman, was dressed in leggings, a loincloth, and a goatskin jacket. A bearskin cap lay nearby. He wore shoes made of red deerskin with thick bearskin soles. The shoes were stuffed with grass for insulation. In addition, investigators found a copper ax, a partially constructed longbow, a quiver containing 14 arrows, a wooden backpack frame, and a dagger at the site.

Figure 8 Excavators used jackhammers to free the man's body from the ice, which caused serious damage to his hip. Part of a longbow also was found nearby.

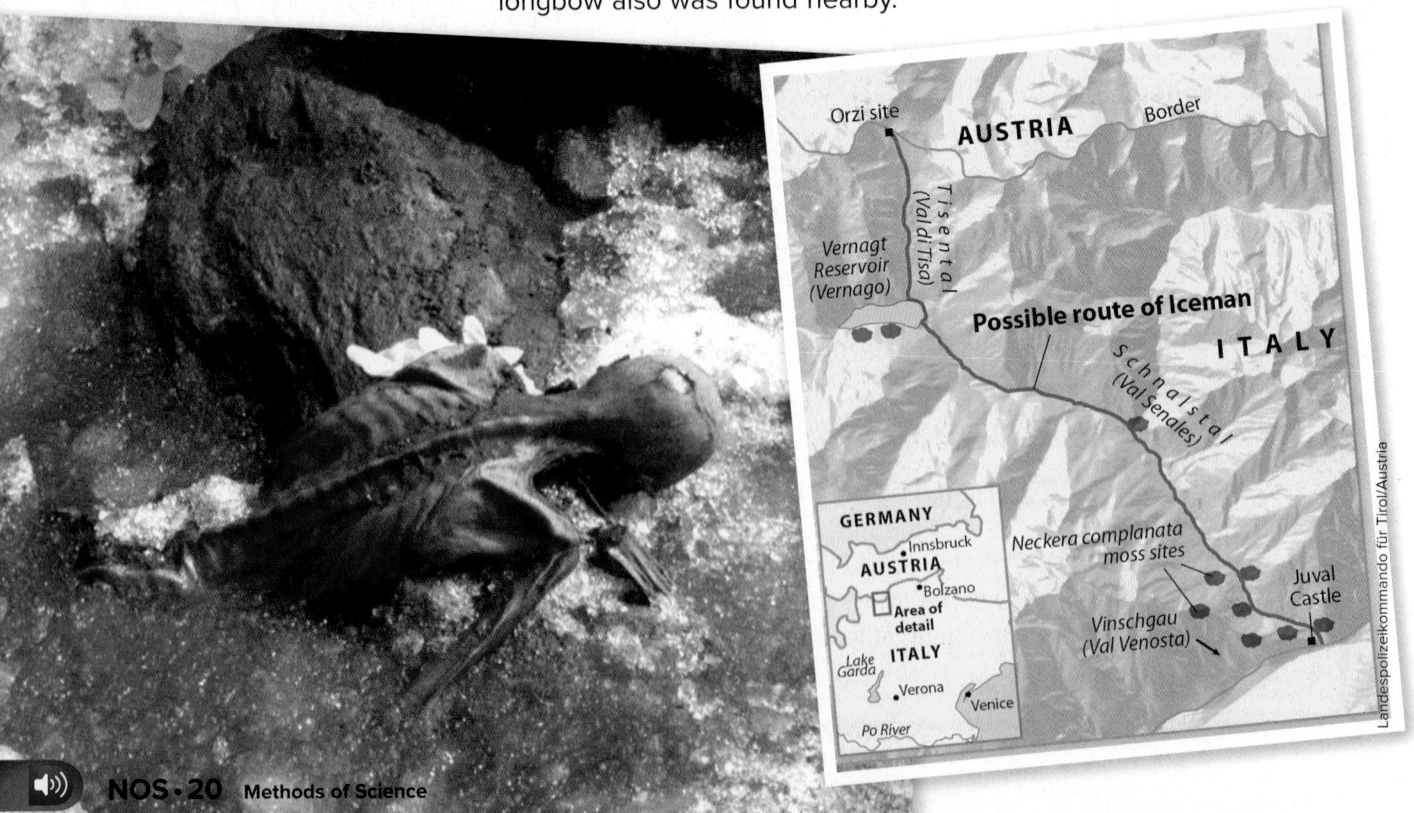

A Controlled Experiment

The identity of the corpse was a mystery. Several people hypothesized about his identity, but controlled experiments were needed to unravel the mystery of who the Iceman was. Scientists and the public wanted to know the identity of the man, why he had died, and when he had died.

Identifying Variables and Constants

When scientists design a controlled experiment, they have to identify factors that might affect the outcome of an experiment. *A* **variable** *is any factor that can have more than one value.* In controlled experiments, there are two kinds of variables. *The* **independent variable** *is the factor that you want to test. It is changed by the investigator to observe how it affects a dependent variable. The* **dependent variable** *is the factor you observe or measure during an experiment.* When the independent variable is changed, it causes the dependent variable to change.

A controlled experiment has two groups–an experimental group and a control group. The experimental group is used to study how a change in the independent variable changes the dependent variable. The control group contains the same factors as the experimental group, but the independent variable is not changed. Without a control, it is difficult to know if your experimental observations result from the variable you are testing or from another factor.

Scientists used inquiry to investigate the mystery of the Iceman. As you read the rest of the story, notice how scientific inquiry was used throughout the investigation. The blue boxes in the margins point out examples of the scientific inquiry process. The notebooks in the margin identify what a scientist might have written in a journal.

Figure 9 These models show what the Iceman and the artifacts found with him might have looked like.

Scientific investigations often begin when someone asks a question about something observed in nature.

Observation: A corpse was found buried in ice in the Tyrolean Alps.
Hypothesis: The corpse found in the Tyrolean Alps is the body of a missing music professor because he disappeared in 1938, and had not been found.
Observation: Artifacts near the body suggested that the body was much older than the music professor would have been.
Revised Hypothesis: The corpse found was dead long before 1938 because the artifacts found near him appear to date before the 1930s.
Prediction: If the artifacts belong to the corpse, and date back before 1930, then the corpse is not the music professor.

An inference is a logical explanation of observations based on past experiences.

Inference: Based on its construction, the ax is at least 4,000 years old.
Prediction: If the ax is at least 4,000 years old, then the body found near it is also at least 4,000 years old.
Test Results: Radiocarbon dating showed the man to be 5,300 years old.

After many observations, revised hypotheses, and tests, conclusions often can be made.

Conclusion: The Iceman is about 5,300 years old. He was a seasonal visitor to the high mountains. He died in autumn. When winter came the Iceman's body became buried and frozen in the snow, which preserved his body.

An Early Conclusion

Konrad Spindler was a professor of archeology at the University of Innsbruck in Austria when the Iceman was discovered. Spindler estimated that the ax, shown in **Figure 10,** was at least 4,000 years old based on its construction. If the ax was that old, then the Iceman was also at least 4,000 years old. Later, radiocarbon dating showed that the Iceman actually lived about 5,300 years ago.

The Iceman's body was in a mountain glacier 3,210 m above sea level. What was this man doing so high in the snow- and ice-covered mountains? Was he hunting for food, shepherding his animals, or looking for metal ore?

Spindler noted that some of the wood used in the artifacts was from trees that grew at lower elevations. He concluded that the Iceman was probably a seasonal visitor to the high mountains.

Spindler also hypothesized that shortly before the Iceman's death, the Iceman had driven his herds from their summer high mountain pastures to the lowland valleys. However, the Iceman soon returned to the mountains where he died of exposure to the cold weather.

The Iceman's body was extremely well preserved. Spindler inferred that ice and snow covered the Iceman's body shortly after he died. Spindler concluded that the Iceman died in autumn and was quickly buried and frozen, which preserved his body and all his possessions.

Figure 10 This ax, bow and quiver, and dagger and sheath were found with the Iceman's body.

More Observations and Revised Hypotheses

When the Iceman's body was discovered, Klaus Oeggl was an assistant professor of botany at the University of Innsbruck. His area of study was plant life during prehistoric times in the Alps. He was invited to join the research team studying the Iceman.

Upon close examination of the Iceman and his belongings, Professor Oeggl found three plant materials–grass from the Iceman's shoe, as shown in **Figure 11,** a splinter of wood from his longbow, and a tiny fruit called a sloe berry.

Over the next year, Professor Oeggl examined bits of charcoal wrapped in maple leaves that had been found at the discovery site. Examination of the samples revealed the charcoal was from the wood of eight different types of trees. All but one of the trees grew only at lower elevations than where the Iceman's body was found. Like Spindler, Professor Oeggl suspected that the Iceman had been at a lower elevation shortly before he died. From Oeggl's observations, he formed a hypothesis and made some predictions.

Oeggl realized that he would need more data to support his hypothesis. He requested that he be allowed to examine the contents of the Iceman's digestive tract. If all went well, the study would show what the Iceman had swallowed just hours before his death.

Figure 11 Professor Oeggl examined the Iceman's belongings along with the leaves and grass that were stuck to his shoe.

South Tyrol Museum of Archaeology, Italy (www.iceman.it)

Scientific investigations often lead to new questions.

Observations: Plant matter near body to study—grass on shoe, splinter from longbow, sloe berry fruit, charcoal wrapped in maple leaves, wood in charcoal from 8 different trees—7 of 8 types of wood in charcoal grow at lower elevations

Hypothesis: The Iceman had recently been at lower elevations before he died because the plants identified near him grow only at lower elevations.

Prediction: If the identified plants are found in the digestive tract of the corpse, then the man actually was at lower elevations just before he died.

Question: What did the Iceman eat the day before he died?

Experiment to Test Hypothesis

There is more than one way to test a hypothesis. Scientists might gather and evaluate evidence, collect data and record their observations, create a model, or design and perform an experiment. They also might perform a combination of these skills.

Test Plan:

- Divide a sample of the Iceman's digestive tract into four sections.
- Examine the pieces under microscopes.
- Gather data from observations of the pieces and record observations.

The research teams provided Professor Oeggl with a tiny sample from the Iceman's digestive tract. He was determined to study it carefully to obtain as much information as possible. Oeggl carefully planned his scientific inquiry. He knew that he had to work quickly to avoid the decomposition of the sample and to reduce the chances of contaminating the samples.

His plan was to divide the material from the digestive tract into four samples. Each sample would undergo several chemical tests. Then, the samples would be examined under an electron microscope to see as many details as possible.

Professor Oeggl began by adding a saline solution to the first sample. This caused it to swell slightly, making it easier to identify particles using the microscope at a relatively low magnification. He saw particles of a wheat grain known as einkorn, which was a common type of wheat grown in the region during prehistoric times. He also found other edible plant material in the sample.

Oeggl noticed that the sample also contained pollen grains in the digestive tract of the Iceman, who is shown in **Figure 12**. To see the pollen grains more clearly, he used a chemical that separated unwanted substances from the pollen grains. He washed the sample a few times with alcohol. After each wash, he examined the sample under a microscope at a high magnification. The pollen grains became more visible. Many more microscopic pollen grains could now be seen. Professor Oeggl identified these pollen grains as those from a hop-hornbeam tree.

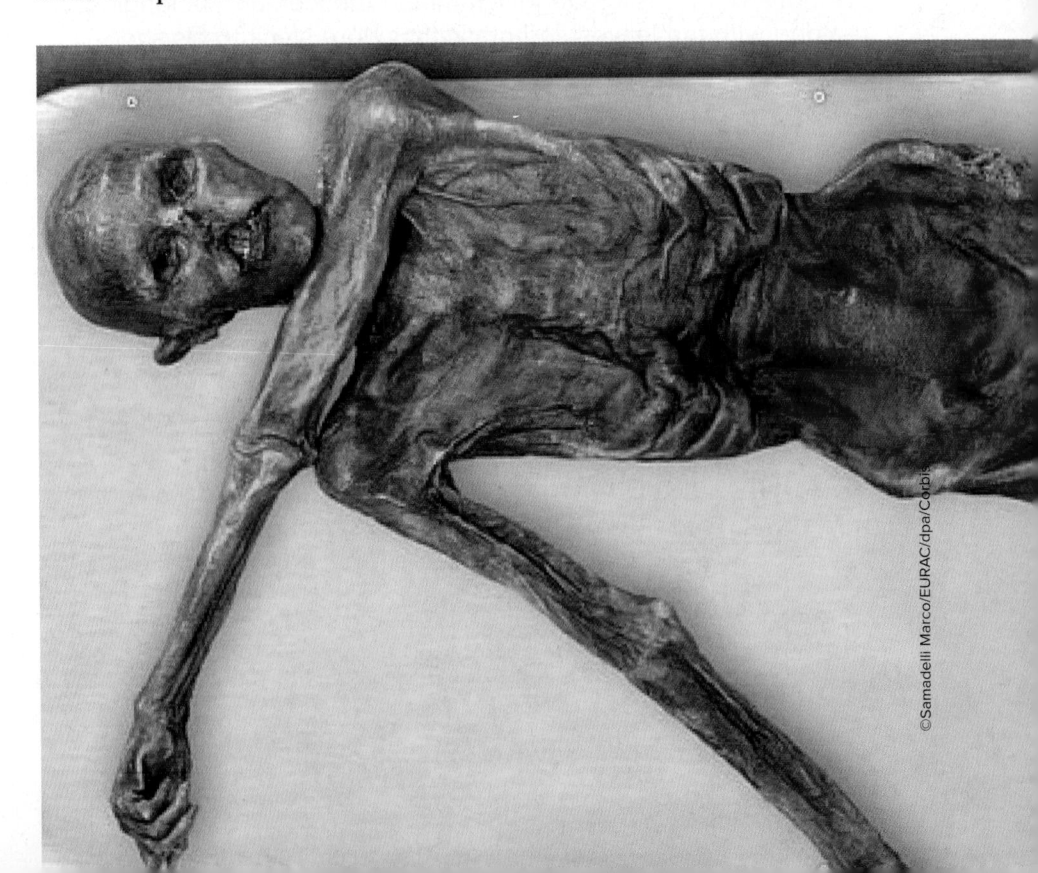

Figure 12 The Iceman, shown here, had pollen grains from hop hornbeam trees in his digestive tract.

Analyzing Results

Professor Oeggl observed that the hop-hornbeam pollen grains had not been digested. Therefore, the Iceman must have swallowed them within hours before his death. But, hop-hornbeam trees only grow in lower valleys. Oeggl was confused. How could pollen grains from trees at low elevations be ingested within a few hours of this man dying in high, snow-covered mountains? Perhaps the samples from the Iceman's digestive tract had been contaminated. Oeggl knew he needed to investigate further.

Further Experimentation

Oeggl realized that the most likely source of contamination would be Oeggl's own laboratory. He decided to test whether his lab equipment or saline solution contained hop-hornbeam pollen grains. To do this, he prepared two identical, sterile slides with saline solution. Then, on one slide, he placed a sample from the Iceman's digestive tract. The slide with the sample was the experimental group. The slide without the sample was the control group.

The independent variable, or the variable that Oeggl changed, was the presence of the sample on the slide. The dependent variable, or the variable Oeggl measured, was whether hop-hornbeam pollen grains showed up on the slides. Oeggl examined the slides carefully.

Analyzing Additional Results

The experiment showed that the control group (the slide without the digestive tract sample) contained no hop-hornbeam pollen grains. Therefore, the pollen grains had not come from his lab equipment or solutions. Each sample from the Iceman's digestive tract was closely re-examined. All of the samples contained the same hop-hornbeam pollen grains. The Iceman had indeed swallowed the hop-hornbeam pollen grains.

Error is unavoidable in scientific research. Scientists are careful to document procedures and any unanticipated factors or accidents. They also are careful to document possible sources of error in their measurements.

Procedure:
- Sterilize laboratory equipment.
- Prepare saline slides.
- View saline slides under electron microscope. Results: no hop-hornbeam pollen grains
- Add digestive tract sample to one slide.
- View this slide under electron microscope. Result: hop-hornbeam pollen grains present

Controlled experiments contain two types of variables.

Dependent Variables: amount of hop-hornbeam pollen grains found on slide
Independent Variable: digestive tract sample on slide

Without a control group, it is difficult to determine the origin of some observations.

Control Group: sterilized slide
Experimental Group: sterilized slide with digestive tract sample

An inference is a logical explanation of an observation that is drawn from prior knowledge or experience. Inferences can lead to predictions, hypotheses, or conclusions.

Observation: The Iceman's digestive tract contains pollen grains from the hop-hornbeam tree and other plants that bloom in spring.
Inference: Knowing the rate at which food and pollen decompose after swallowed, it can be inferred that the Iceman ate three times on the day that he died.
Prediction: The Iceman died in the spring within hours of digesting the hop-hornbeam pollen grains.

Mapping the Iceman's Journey

The hop-hornbeam pollen grains were helpful in determining the season the Iceman died. Because the pollen grains were whole, Professor Oeggl inferred that the Iceman swallowed the pollen grains during their blooming season. Therefore, the Iceman must have died between March and June.

After additional investigation, Professor Oeggl was ready to map the Iceman's final trek up the mountain. Because Oeggl knew the rate at which food travels through the digestive system, he inferred that the Iceman had eaten three times in the final day and a half of his life. From the digestive tract samples, Oeggl estimated where the Iceman was located when he ate.

First, the Iceman ingested pollen grains native to higher mountain regions. Then he swallowed hop-hornbeam pollen grains from the lower mountain regions several hours later. Last, the Iceman swallowed other pollen grains from trees of higher mountain areas again. Oeggl proposed the Iceman traveled from the southern region of the Italian Alps to the higher, northern region as shown in **Figure 13**, where he died suddenly. He did this all in a period of about 33 hours.

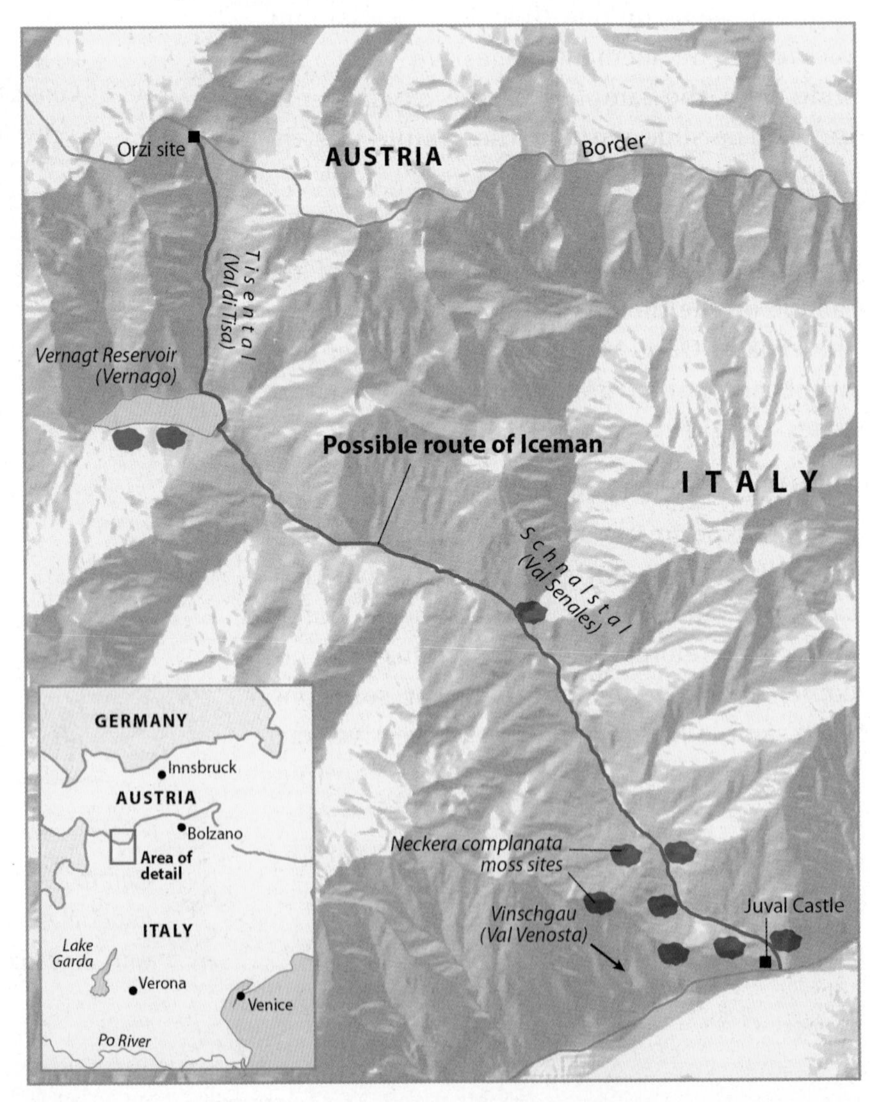

Figure 13 By examining the contents of the Iceman's digestive tract, Professor Oeggl was able to reconstruct the Iceman's last journey.

Conclusion

Researchers from around the world worked on different parts of the Iceman mystery and shared their results. Analysis of the Iceman's hair revealed his diet usually contained vegetables and meat. Examining the Iceman's one remaining fingernail, scientists determined that he had been sick three times within the last six months of his life. X-rays revealed an arrowhead under the Iceman's left shoulder. This suggested that he died from that serious injury rather than from exposure.

Finally, scientists concluded that the Iceman traveled from the high alpine region in spring to his native village in the lowland valleys. There, during a conflict, the Iceman sustained a fatal injury. He retreated back to the higher elevations, where he died. Scientists recognize their hypotheses can never be proved, only supported or not supported. However, with advances in technology, scientists are able to more thoroughly investigate mysteries of nature.

Scientific investigations may disprove early hypotheses or conclusions. However, new information can cause a hypothesis or conclusion to be revised many times.

Revised Conclusion:
In spring, the Iceman traveled from the high country to the valleys. After he was involved in a violent confrontation, he climbed the mountain into a region of permanent ice where he died of his wounds.

Lesson 3 Review

Use Vocabulary

1. A factor that can have more than one value is a(n) ________.
2. **Differentiate** between independent and dependent variables.

Understand Key Concepts 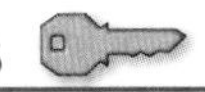

3. Which part of scientific inquiry was NOT used in this case study?
   A. Draw conclusions.
   B. Make observations.
   C. Hypothesize and predict.
   D. Make a computer model.
4. **Determine** which is the control group and which is the experimental group in the following scenario: Scientists are testing a new kind of aspirin to see whether it will relieve headaches. They give one group of volunteers the aspirin. They give another group of volunteers pills that look like aspirin but are actually sugar pills.

Interpret Graphics

5. **Summarize** Copy and fill in the flow chart below summarizing the sequence of scientific inquiry steps that was used in one part of the case study. Draw the number of boxes needed for your sequence.

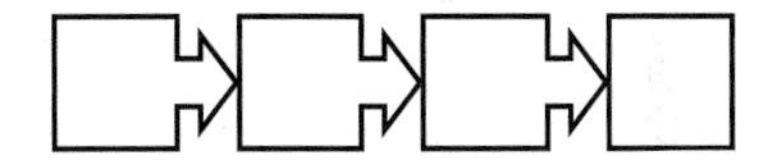

6. **Explain** What is the significance of the hop-hornbeam pollen found in the Iceman's digestive tract?

Critical Thinking

7. **Formulate** more questions about the Iceman. What would you want to know next?
8. **Evaluate** the hypotheses and conclusions made during the study of the Iceman. Do you see anything that might be an assumption? Are there holes in the research?

Inferring from Indirect Evidence

Materials

owl pellet

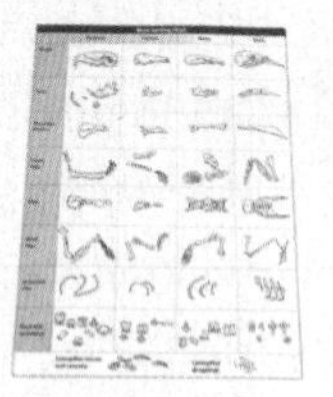

bone identification chart

probe

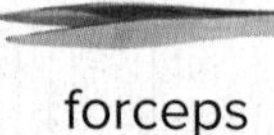

forceps

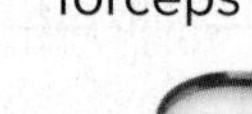

magnifying lens

Also needed: toothpicks, small brush, paper plate, ruler

Safety

In the case study about the Iceman, you learned how scientists used evidence found in or near the body to learn how the Iceman might have lived and what he ate. In this investigation, you will use similar indirect evidence to learn more about an owl.

An owl pellet is a ball of fur and feathers that contains bones, teeth, and other undigested parts of animals eaten by the owl. Owls and other birds, such as hawks and eagles, swallow their prey whole. Stomach acids digest the soft parts of the food. Skeletons and body coverings are not digested and form a ball. When the owl coughs up the ball, it might fall to the ground. Feathers, straw, or leaves often stick to the moist ball when it strikes the ground.

Ask a Question

What kinds of information can I learn about an owl by analyzing an owl pellet?

Make Observations

1. Read and complete a lab safety form.
2. Carefully measure the length, the width, and the mass of your pellet. Write the data in your Science Journal.
3. Gently examine the outside of the pellet using a magnifying lens. Do you see any sign of fur or feathers? What other substances can you identify? Record your observations.
4. Use a probe, toothpicks, and forceps to gently pull apart the pellet. Try to avoid breaking any of the tiny bones. Spread out the parts on a paper plate.
5. Copy the table into your Science Journal. Use the bone identification chart to identify each of the bones and other materials found in your pellet. Make a mark in the table for each part you identify.

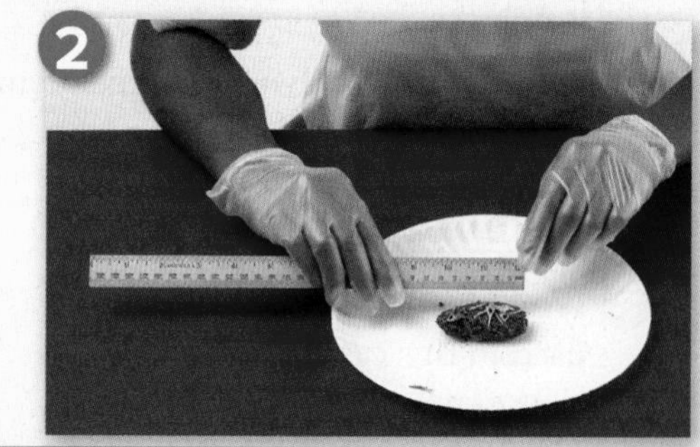

Bone Identification Chart

Bone	Animal	Number
Skull		
Jaw		
Shoulder blade		
Forelimb		
Hind limb		
Hip/pelvis		
Rib		
Vertebrae		
Insect parts		

Analyze and Conclude

6. **Assemble** the bones you find into a skeleton. You may need to locate pictures of rodents, shrews, moles, and birds.
7. **Discuss** with your teammates why parts of an animal skeleton might be missing.
8. **Write** a report that includes your data and conclusions about the owl's diet.
9. **Identify Cause and Effect** Is every bone you found in the pellet necessarily from the owl's prey? Why or why not?
10. **Analyze** What conclusions can you reach about the diet of the particular owl from which your pellet came? Can you extend this conclusion to the diets of all owls? Why or why not?
11. **The Big Idea** How did the scientific inquiry you used in the investigation compare to those used by the scientists studying the Iceman? In what ways were they the same? In what ways were they different?

Communicate Your Results

Compare your results with those of several other teams. Discuss any evidence to support that the owl pellets did or did not come from the same area.

Inquiry Extension

Put your data on the board. Use the class data to determine a mean, median, mode, and range for each type of bone.

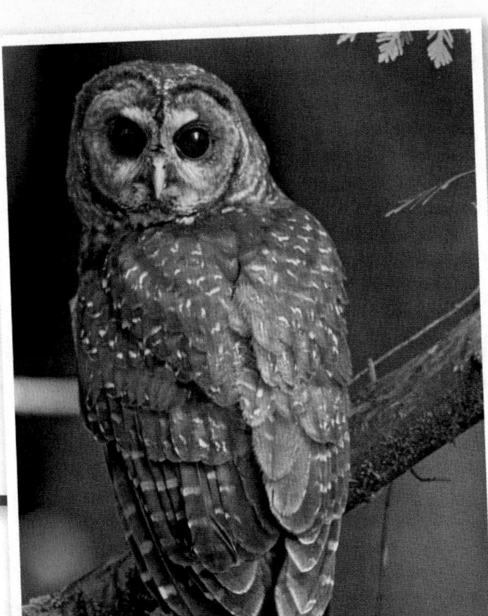

Lab Tips

- ☑ When using your forceps, squeeze the sides very lightly so that you don't crush fragile bones.
- ☑ Use the brush to clean each bone. Try rotating the bones as you match them to the chart.
- ☑ Lay the bones on the matching box on the chart as you separate them. Then count them when you are finished.

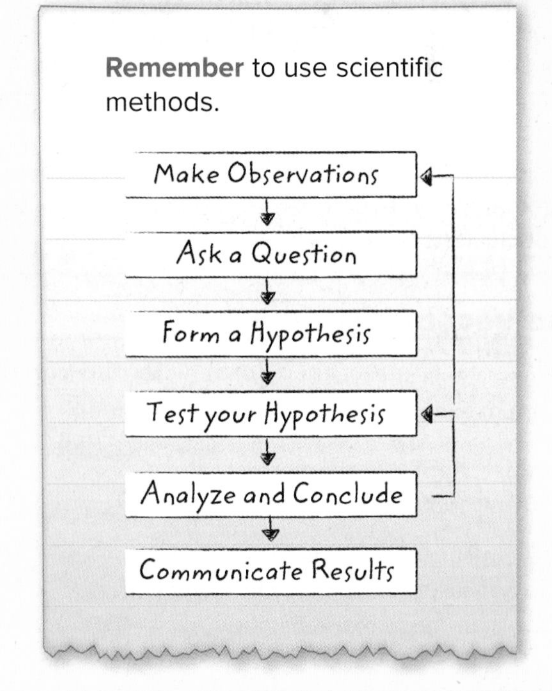

Chapter Study Guide and Review

Scientists use the process of scientific inquiry to perform scientific investigations.

Key Concepts Summary

Lesson 1: Understanding Science

- Scientific inquiry is a process that uses a set of skills to answer questions or to test ideas about the natural world.
- A **scientific law** is a rule that describes a pattern in nature. A **scientific theory** is an explanation of things or events that is based on knowledge gained from many **observations** and investigations.
- Facts are measurements, observations, and theories that can be evaluated for their validity through objective investigation. Opinions are personal views, feelings, or claims about a topic that cannot be proven true or false.

Vocabulary

science p. NOS 4
observation p. NOS 6
inference p. NOS 6
hypothesis p. NOS 6
prediction p. NOS 6
technology p. NOS 8
scientific theory p. NOS 9
scientific law p. NOS 9
critical thinking p. NOS 10

Lesson 2: Measurement and Scientific Tools

- Scientists worldwide use the **International System of Units (SI)** because their work is easier to confirm and repeat by their peers.
- Measurement uncertainty occurs because no scientific tool can provide a perfect measurement.
- Mean, median, mode, and range are statistical calculations that are used to evaluate sets of data.

Vocabulary

description p. NOS 12
explanation p. NOS 12
International System of Units (SI) p. NOS 12
significant digits p. NOS 14

Lesson 3: Case Study: The Iceman's Last Journey

- The **independent variable** is the factor a scientist changes to observe how it affects a **dependent variable.** A dependent variable is the factor a scientist measures or observes during an experiment.
- Scientific inquiry was used throughout the investigation of the Iceman when hypotheses, predictions, tests, analysis, and conclusions were developed.

Vocabulary

variable p. NOS 21
independent variable p. NOS 21
dependent variable p. NOS 21

Use Vocabulary

Replace each underlined term with the correct vocabulary word.

1. A description is an interpretation of observations.
2. The means are the numbers of digits in a measurement that you know with a certain degree of reliability.
3. The act of watching something and taking note of what occurs is a(n) inference.
4. A scientific theory is a rule that describes a pattern in nature.

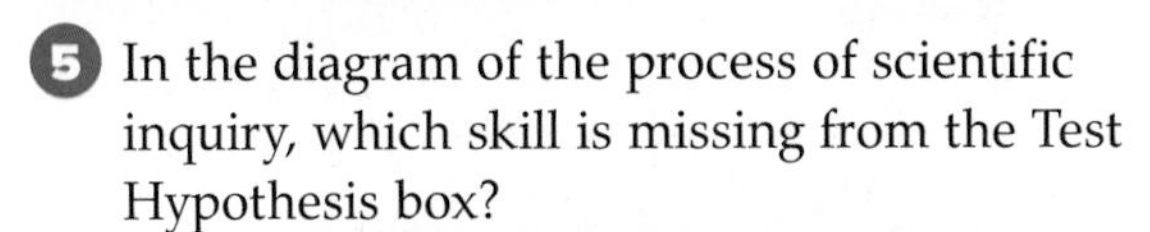

Understand Key Concepts

5. In the diagram of the process of scientific inquiry, which skill is missing from the Test Hypothesis box?

Test Hypothesis
- Design an Experiment
- Gather and Evaluate Evidence
- Collect Data/Record Observations

A. Analyze results.

B. Communicate results.

C. Make a model.

D. Make observations.

6. You have the following data set: 2, 3, 4, 4, 5, 7, and 8. Is 6 the mean, the median, the mode, or the range of the data set?

A. mean

B. median

C. mode

D. range

7. Which best describes an independent variable?

A. It is a factor that is not in every test.

B. It is a factor the investigator changes.

C. It is a factor you measure during a test.

D. It is a factor that stays the same in every test.

Critical Thinking

8. **Predict** what would happen if every scientist tried to use all the skills of scientific inquiry in the same order in every investigation.

9. **Assess** the role of measurement uncertainty in scientific investigations.

10. **Evaluate** the importance of having a control group in a scientific investigation.

Writing in Science

11. **Write** a five-sentence paragraph explaining why the International System of Units (SI) is an easier system to use than the English system of measurement. Be sure to include a topic sentence and a concluding sentence in your paragraph.

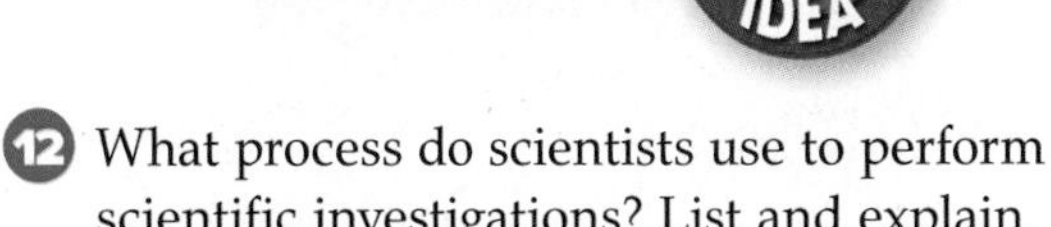

12. What process do scientists use to perform scientific investigations? List and explain three of the skills involved.

13. Infer the purpose of the pink dye in the scientific investigation shown in the photo.

Math Skills

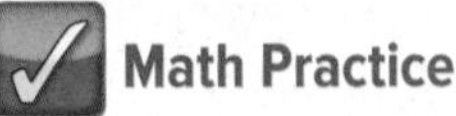

Use Numbers

14. Convert 162.5 hg to grams.

15. Convert 89.7 cm to millimeters.

Unit 1

Exploring Earth

SIR, we're lost and our GPS unit is dead.

LOST?! Nonsense, I have just what we need.

Or NOT. On second thought, I guess we already knew we were in Mexico, so this globe won't help.

And we certainly don't need a road map. We're nowhere near a road.

What to do? What to do?

Ahem

10000 B.C. 1500 1900

12000 B.C.
A map scratched into a mammoth jawbone, the oldest surviving map, depicts a group of settlements and the surrounding countryside in what is now Mehirich, Ukraine.

2300 B.C.
The oldest surviving city map, a map of the Mesopotamian city of Lagash that includes the layout of the city, is created.

150 A.D.
Ptolemy illustrates a world map with Earth as a sphere from 60°N to 30°S latitudes.

1506
Francesco Rosselli produces the first map to show the "New World."

1930s
Maps become increasingly accurate and factual due to the widespread use of aerial photography after World War I.

1950

2000

1960s
Geographic Information Systems (GIS) are developed. GIS displays large amounts of information and includes computer software and hardware as well as digital data and storage.

1993
The space-based Global Positioning System achieves initial operational capability.

2005
An internet mapping tool displays satellite images of Earth's surface.

Patterns

You might sometimes see the Moon as a large, glowing disk in the night sky. At other times, the Moon appears as different shapes. These shapes are the Moon's phases, or the changing portions of the Moon that are seen from Earth. The Moon's phases occur as a repeating pattern every 29.5 days. A **pattern** is a consistent plan or model used as a guide for understanding and predicting things. You can predict the next phase of the Moon or you can determine the previous phase of the Moon if you know the pattern.

Patterns in Earth Science

Patterns help scientists understand observations. This allows them to predict future events or understand past events. For instance, geologists are Earth scientists who measure the chemical composition, age, and location of rocks. They look for patterns in these measurements. The patterns allow geologists to propose what processes formed the rocks millions or billions of years ago. Geologists also use patterns to draw conclusions about how Earth has changed over time and to estimate how it will change in the future.

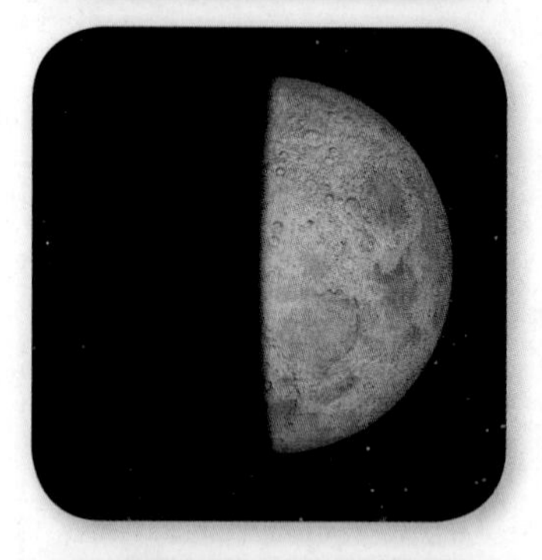

Meteorologists are scientists who study weather and climate. They study patterns of fronts, winds, cloud formation, precipitation, and ocean temperatures to make weather forecasts. For example, meteorologists track patterns in hurricanes, such as wind speed, movements, and rotation velocity. These patterns help meteorologists understand the conditions under which a hurricane can form. Predicting the strength and the path of a storm can help save lives, buildings, and property. When meteorologists see weather patterns similar to those of past hurricanes, they can predict the severity of the storm and when and where it will hit. Then meteorologists can send advance warnings for people to safely prepare.

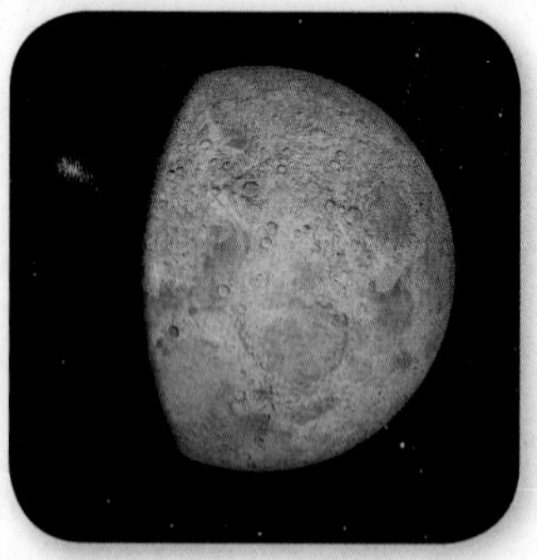

Jason Reed/Photodisc/Getty Images

Types of Patterns

Physical Patterns

A pattern you observe using your eyes or other instruments is a physical pattern. Earthquake uplift and erosion reveal physical patterns in layers of rock, as shown in the photo. Patterns in exposed rock layers tell geologists many things, including the order in which the rocks formed, the different minerals and fossils the rocks contain, and the age and movement of landforms.

Patterns in Graphs

Scientists plot data on graphs and then analyze the graphs for patterns. Patterns on graphs can appear as straight lines, curved lines, or waves. The graph to the right shows a pattern in sea level as it increased between 1994 and 2008. Scientists analyze graphic patterns to predict events in the future. For example, a scientist might predict that in 2019, the sea level will be 20 mm higher than it was in 2013.

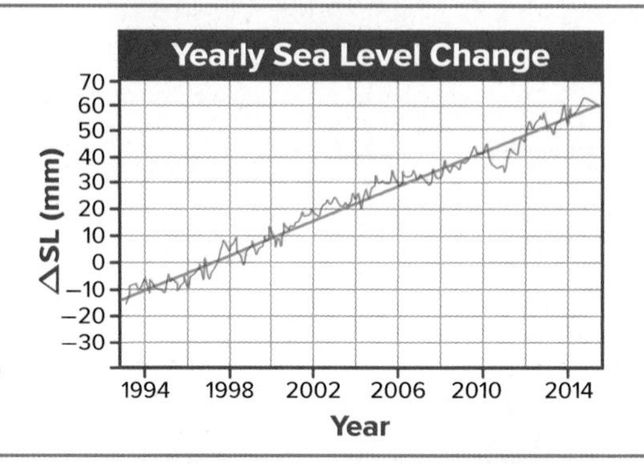

Cyclic Patterns

An event that repeats many times in a predictable order, such as the phases of the Moon, has a cyclic pattern. As shown in the graph, water temperatures in both the North and South Atlantic Ocean rise and fall equally each year. The annual changes in the water temperature follow a cyclic pattern. How do the temperature patterns in the two oceans differ?

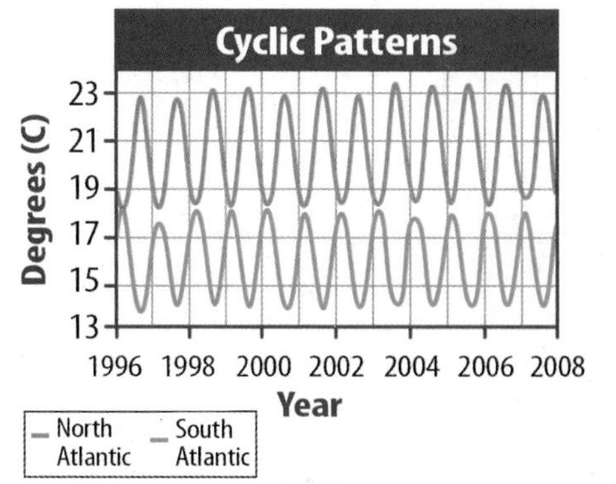

MiniLab

15 minutes

What patterns are in your year?

What are some of the cyclic patterns in your life?

1. On a sheet of **notebook paper,** draw four concentric circles with diameters of 20 cm, 18 cm, 10 cm, and 4 cm. Write your name in the innermost circle.
2. Divide the two outermost circles into 12 equal sections. Write each month of the year (one month per section) in each section of the outermost circle. In the next circle, write personal events or activities that take place in each corresponding month.
3. Divide the 10-cm circle into four sections. Write the weather conditions and the plant patterns that correspond to the months in the outermost circle.

Analyze and Conclude

Observe If you start at one month and move inward through the rings, what patterns do you observe? How do these observations fit into the yearly cycle?

Chapter 1

Mapping Earth

THE BIG IDEA **How are Earth's surface features measured and modeled?**

Inquiry Do these maps show the same area?

Maps show the features of Earth's surface, such as mountains, roads, or different rock types. Notice that these maps show different features of the same area.

- What features are shown in each map?
- How were these maps made?
- How would you measure and model features on Earth's surface?

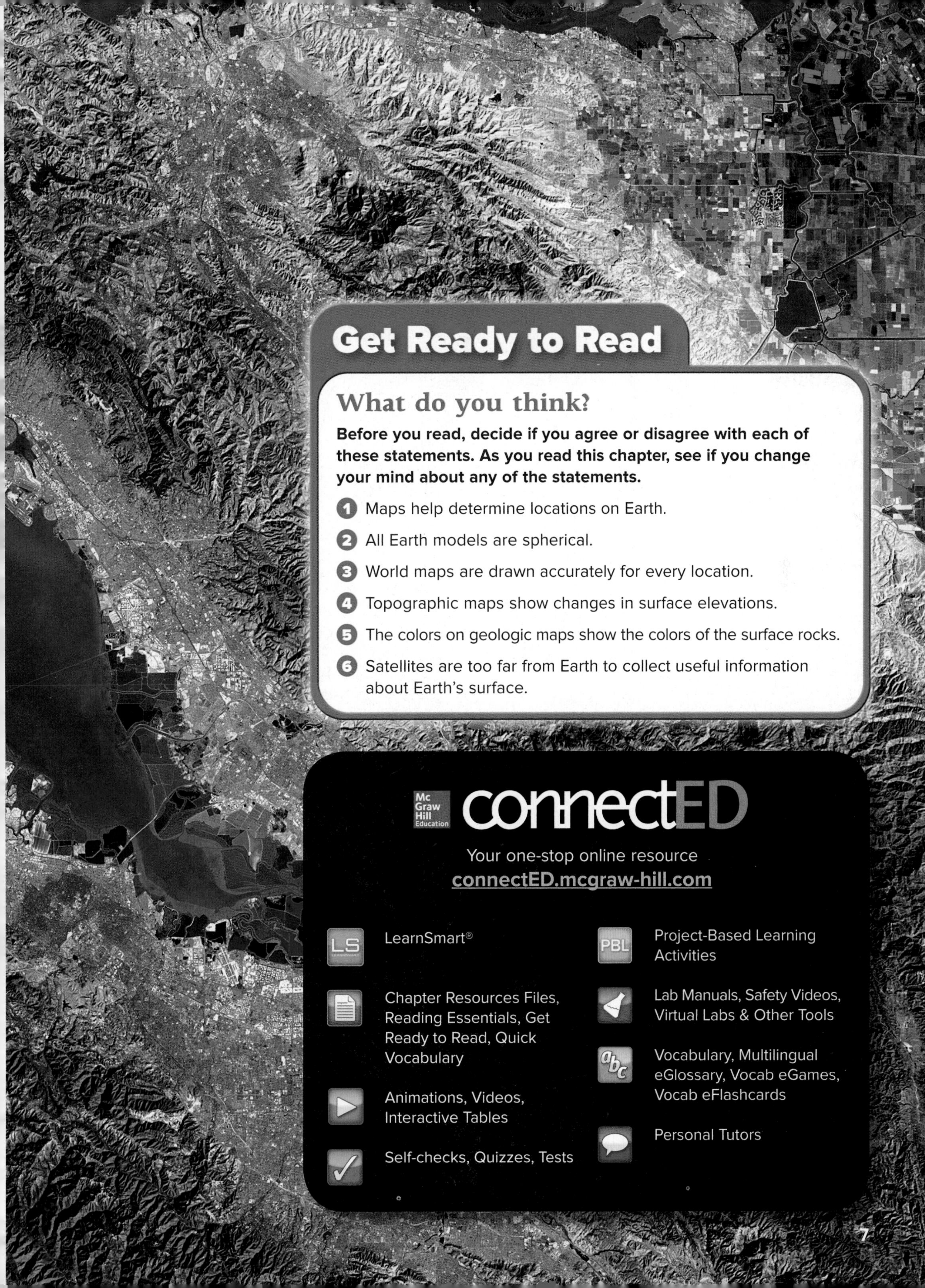

Get Ready to Read

What do you think?

Before you read, decide if you agree or disagree with each of these statements. As you read this chapter, see if you change your mind about any of the statements.

1. Maps help determine locations on Earth.
2. All Earth models are spherical.
3. World maps are drawn accurately for every location.
4. Topographic maps show changes in surface elevations.
5. The colors on geologic maps show the colors of the surface rocks.
6. Satellites are too far from Earth to collect useful information about Earth's surface.

Lesson 1

Reading Guide

Key Concepts 

ESSENTIAL QUESTIONS

- How can a map help determine a location?
- Why are there different map projections for representing Earth's surface?

Vocabulary

map view p. 9

profile view p. 9

map legend p. 10

map scale p. 11

longitude p. 12

latitude p. 12

time zone p. 14

International Date Line p. 14

Multilingual eGlossary

Maps

Inquiry Where are they?

Look at the horizon—the place where the blue sky and the blue water come together. What do you see? Have you ever had to figure out where you were without using any landmarks? Suppose you are sailing in the South Pacific. How would you navigate without landmarks to use as reference points?

Launch Lab

15 minutes

How will you get from here to there?

When you need to get to a place you have never visited, you might use a map to help you find your way. Maps help people get where they are going without getting lost.

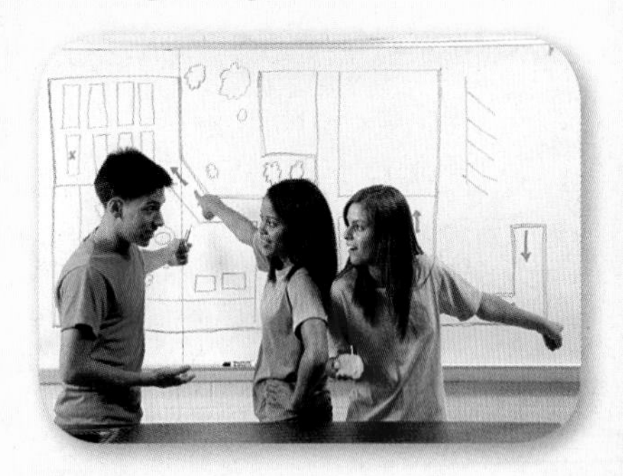

1. Suppose it is a new student's first day at your school. Write directions for the student to get from the science classroom to the cafeteria.
2. Now draw a map for the student to get from the science classroom to the cafeteria.

Think About This

1. How were the written instructions different from the map?
2. **Key Concept** How are maps useful?

Understanding Maps

When was the last time you used a map to find information? Maybe you looked at a map of your school to find all your classrooms. Or, maybe you reviewed the map of the school to practice for a fire drill or a disaster drill. A map might show all the exits or the safest room to go to if there were a tornado. There are many kinds of maps, such as road maps, trail maps, and weather maps. Each type of map contains different information and serves a different purpose.

A map can be used to model Earth. In order to model Earth's surface, you can make a flat representation of an area of Earth on a piece of paper. In order to model the entire planet and its shape, you can make a globe.

Map Views

Most maps are drawn in **map view**–*drawn as if you were looking down on an area from above Earth's surface.* Map view also can be referred to as plan view, and is shown in **Figure 1.**

A **profile view** *is a drawing that shows an object as though you were looking at it from the side.* A profile view is like a side view of a house. To help you visualize this concept, a map view and a profile view of a house are shown in **Figure 1.** Map views and profile views will be used to describe topographic maps and geologic maps at the end of this chapter. Also, you will use profile views when you study cross sections, or models of the inner structures of Earth.

Map Views

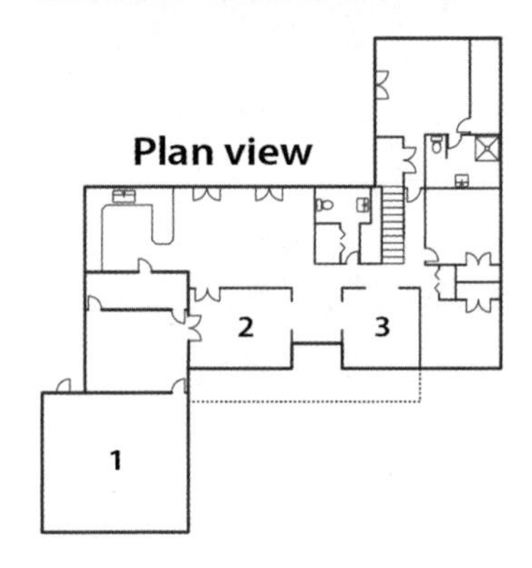

Figure 1 A map view, or plan view, looks down on an object, while a profile view looks from the side.

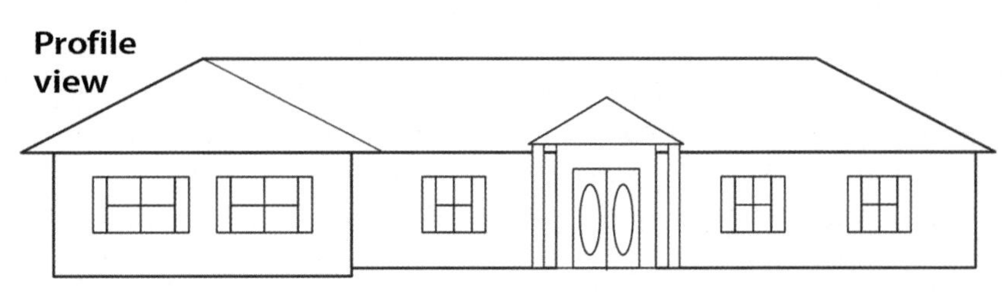

Hutchings Photography/Digital Light Source

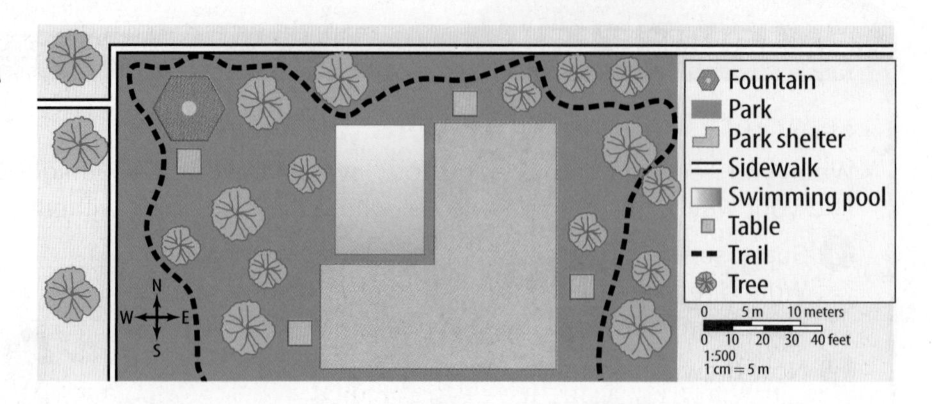

Figure 2 The legend on this map explains what the symbols mean.

SCIENCE USE V. COMMON USE

legend

Science Use part of a map that explains the map symbols

Common Use a story coming down from the past

Map Legends and Scales

Maps have two features to help you read and understand the map. One feature is a series of symbols called a map **legend**. The other is a ratio, which establishes the map scale.

Map Legends Maps use specific symbols to represent certain features on Earth's surface, such as roads in a city or restrooms in a park. These symbols allow mapmakers to fit many details on a map without making it too cluttered. All maps include a **map legend**–*a key that lists all the symbols used on the map*–so you can interpret the symbols. It also explains what each symbol means. For example, in the map legend shown in **Figure 2**, a dashed line represents a trail.

Reading Check What is the purpose of a map legend?

Road Map with Scale

Personal Tutor

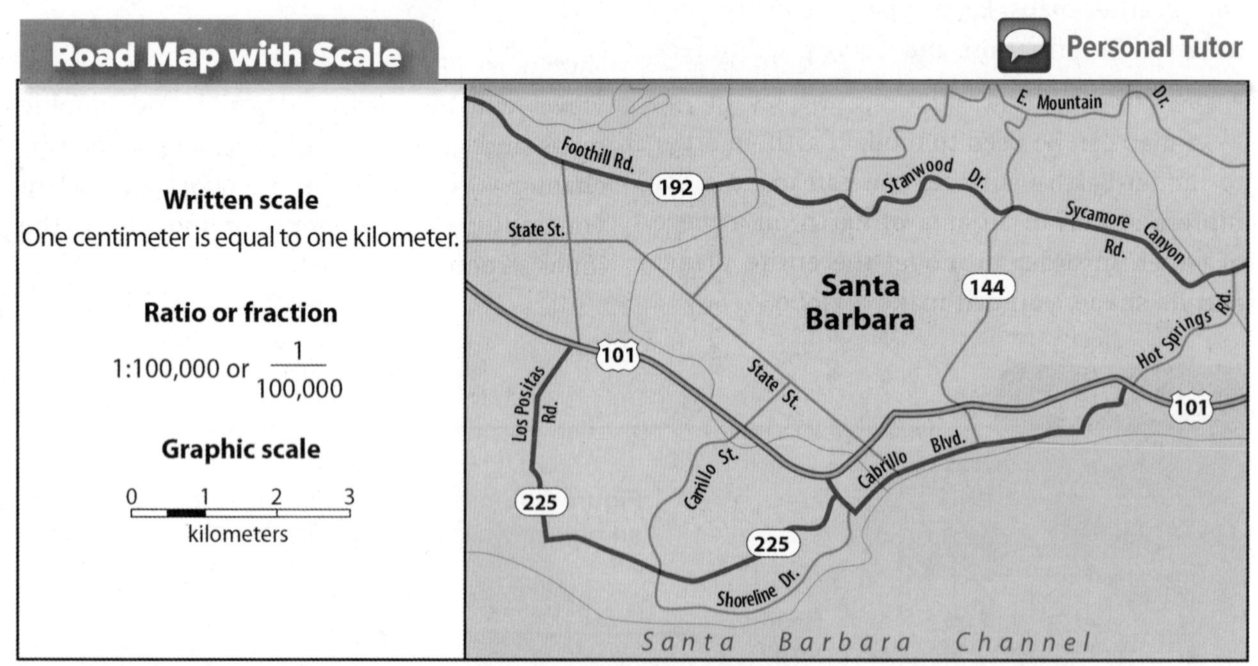

Figure 3 Different types of scales can be used with maps. For example, the graphic scale compares map distance to actual distance.

Visual Check Which scale would you use to measure the distance between the two rivers that intersect Route 192?

Map Scales When mapmakers draw a map, they need to decide how big or small to make the map. They need to decide on the map's scale. **Map scale** *is the relationship between a distance on the map and the actual distance on the ground.* The scale can be verbally written such as "one centimeter is equal to 1 kilometer." The scale also can be written as a ratio, such as 1:100. Because this is a ratio, there are no units. Verbally, you would say, "every unit on the map is equal to 100 units on the ground." If your unit were 1 cm on the map, it would be equal to 100 cm on the ground. If you drew a map of your school at a scale of 1:1, your map would be as large as your school! **Figure 3** gives you a written scale, a ratio scale, and a graphic scale in the map legend. Each one can be useful in different ways. For example, the graphic scale, or scale bar, would be useful in measuring distances on the map. You would have to measure it, however, to find that 1 cm is equal to 1 km.

Figure 4 shows another way in which scales are useful. Models are built with scaled measurements that can be increased or decreased relative to the measurements of real objects. Models have the same relative proportions as the objects they represent, similar to a map scale.

Math Skills 

Ratio Scale

A ratio is a comparison of two quantities by division. For example, a map scale is the ratio of the distance on the map to the actual distance. A map might use a scale in which 1 cm on the map represents **5 km** of actual distance. This may be written as a ratio:

1 cm to **5 km** or

1 cm:**5 km** or

$\frac{1 \text{ cm}}{5 \text{ km}}$

This ratio is the map scale.

Practice

A map uses a scale of 1 cm: **1 km**. If the distance between two points on the map is 3 cm, what is the actual distance between the points?

 **Math Practice**

 **Personal Tutor**

Figure 4 These images have different scales. In the large photo, the scale is 1:25. In the smaller photo to the right, the scale is 12:1.

Reading Maps

To find your way to a specific place, you need a way to determine where you are on Earth. Imagine telling someone your exact position on the snow-covered continent of Antarctica. It would be difficult to describe. Ship captains and airplane pilots experience the same problems as they plot their courses across the oceans or above a cloud-covered Earth.

A Grid System for Plotting Locations

Have you ever played a game of chess? If you have, you know that the board is set up with grid lines to help you choose your moves and the position of the pieces on the board. Long ago mapmakers created a system for identifying locations on Earth that uses a similar grid system. This system uses two sets of imaginary lines that encircle Earth. The two sets of lines are called latitude and longitude. The intersection of a line of latitude and a line of longitude can pinpoint a location on a map or a globe.

WORD ORIGIN

longitude
from Latin *longitudo,* means "length"

Longitude Mapmakers started the grid system with a line that circled Earth and passed through the North Pole and the South Pole. The half of the circle from the North Pole to the South Pole passes through Greenwich, England, and is known as the prime meridian. The prime meridian is shown in **Figure 5.** The other half of the circle is the 180° meridian. Similar circles are drawn at every degree east and west of the prime meridian. **Longitude** *is the distance in degrees east or west of the prime meridian.* The prime meridian and the 180° meridian combine to divide Earth into east and west halves, or hemispheres–the eastern hemisphere and the western hemisphere. East of the prime meridian, longitude is measured in degrees east, and west of the prime meridian, longitude is measured in degrees west. They both meet at the 180° meridian. All the meridians pass through the North Pole and the South Pole.

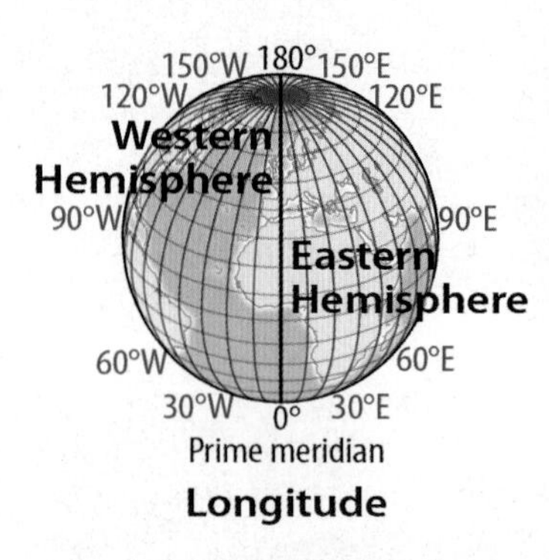

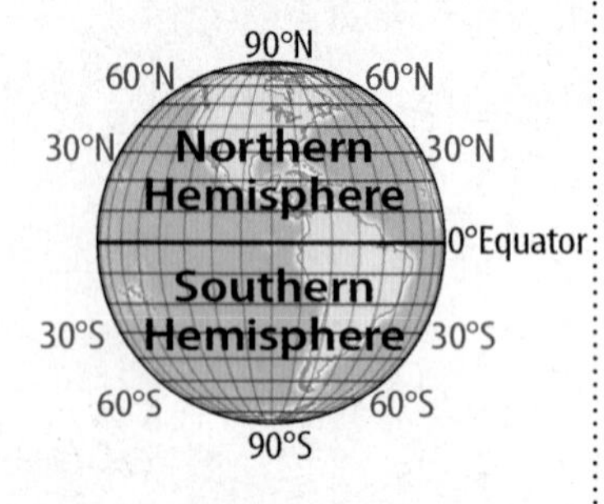

Figure 5 Longitude and latitude are imaginary lines used to pinpoint places on Earth.

Latitude Mapmakers also drew lines east to west around Earth. These lines, called lines of latitude, are somewhat perpendicular to lines of longitude. The center line, called the equator, divides Earth into the northern hemisphere and the southern hemisphere. **Latitude** *is the distance in degrees north or south of the equator.* Unlike lines of longitude, lines of latitude are parallel, as shown in **Figure 5.** The equator is the largest circle. All the other circles become smaller and smaller the closer they are to Earth's poles.

Key Concept Check What relationship do lines of longitude and lines of latitude have?

Plotting Locations

How can you use Earth's grid system to plot locations? First, think about why longitude and latitude are measured in degrees. Earth is a sphere–a ball-shaped object. If you look straight down on a sphere, it looks like a circle. Like a circle, a sphere can be divided into 360 degrees. Look back at **Figure 5.** The latitude at the equator is 0°. All other lines of latitude are measured in degrees north and south of the equator. The North Pole is located at 90 degrees north latitude (90°N), and the South Pole is located at 90 degrees south latitude (90°S).

Lines of longitude are measured in degrees east or west of the prime meridian. There are 180 degrees of east longitude and 180 degrees of west longitude. Any location on Earth can be described by the intersection of the closest line of latitude and the closest line of longitude. Latitude is always stated before longitude.

Minutes and Seconds Longitude and latitude lines are far apart. To better help determine locations, each degree is divided into 60 parts. These parts are called minutes ('). Each minute is divided into 60 parts. These parts are called seconds ("). Degrees, minutes, and seconds allow you to accurately locate places on Earth.

Key Concept Check How do latitude and longitude describe a location on Earth?

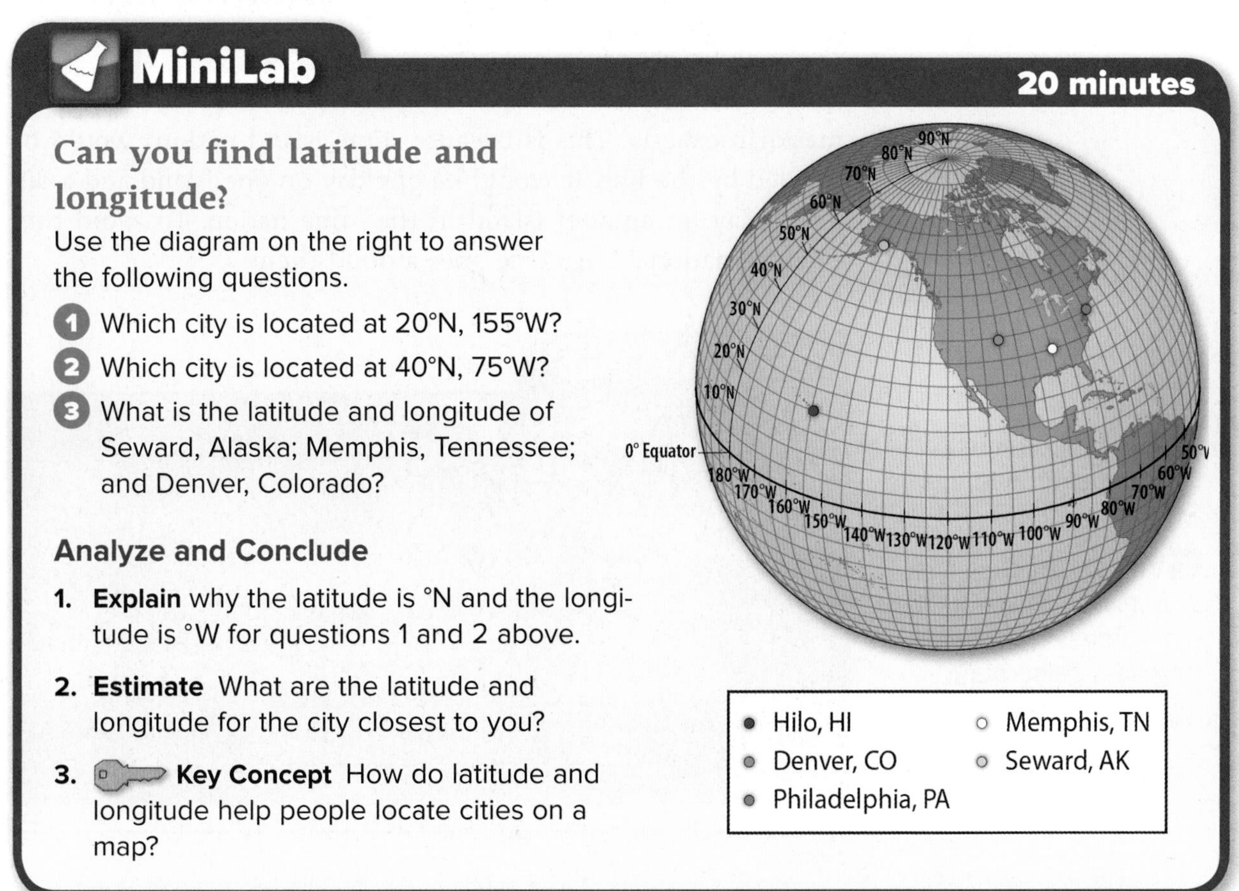

MiniLab

20 minutes

Can you find latitude and longitude?

Use the diagram on the right to answer the following questions.

1. Which city is located at 20°N, 155°W?
2. Which city is located at 40°N, 75°W?
3. What is the latitude and longitude of Seward, Alaska; Memphis, Tennessee; and Denver, Colorado?

- Hilo, HI
- Denver, CO
- Philadelphia, PA
- Memphis, TN
- Seward, AK

Analyze and Conclude

1. **Explain** why the latitude is °N and the longitude is °W for questions 1 and 2 above.
2. **Estimate** What are the latitude and longitude for the city closest to you?
3. **Key Concept** How do latitude and longitude help people locate cities on a map?

Time Zones When it is high noon at your location, the Sun is directly overhead. But as Earth rotates, the Sun is directly above different locations at different times. Businesses in cities many miles apart would have a hard time doing business with each other if every city had its own time. Time zones were created to make travel, communicating, and doing business easier for everyone. *A* **time zone** *is the area on Earth's surface between two meridians where people use the same time.* The reference or starting point for time zones is the **prime** meridian. Earth is divided into 24 time zones, and the width of a time zone is 15° longitude. But, as shown in **Figure 6,** the time zones do not follow the meridians exactly. Their locations are sometimes altered at political boundaries. Notice how the time changes by one hour at the boundary of each time zone. What happens then when you go halfway around the globe?

ACADEMIC VOCABULARY

prime
(adjective) first in rank

Reading Check Why do we need a starting point for time zones?

International Date Line *The line of longitude 180° east or west of the prime meridian is called the* **International Date Line.** Recall that there are 24 time zones, and 24 hours in a day. Because one location can't have two different times on the same day, the day changes as you cross the date line. If you cross from east to west, it is the next day in the west. And if you cross from west to east, it is the day before in the east.

Notice that the International Date Line does not follow the 180° meridian exactly. This is because some island nations would be divided by the line. It would be one day on one island and a different day for another island in the same nation. To avoid this, the International Date Line goes around them.

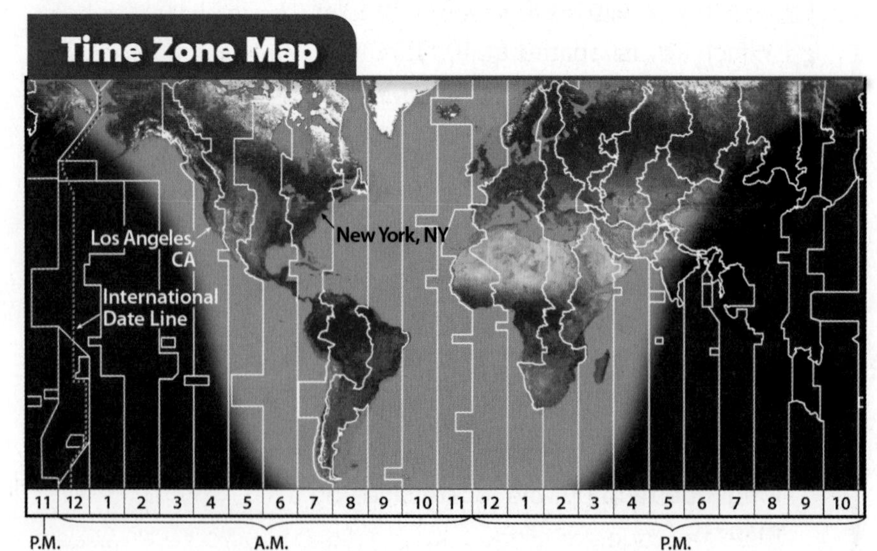

Figure 6 There are 24 time zones around the world.

Visual Check If it is 2:00 P.M. in New York City, what time is it in Los Angeles, California?

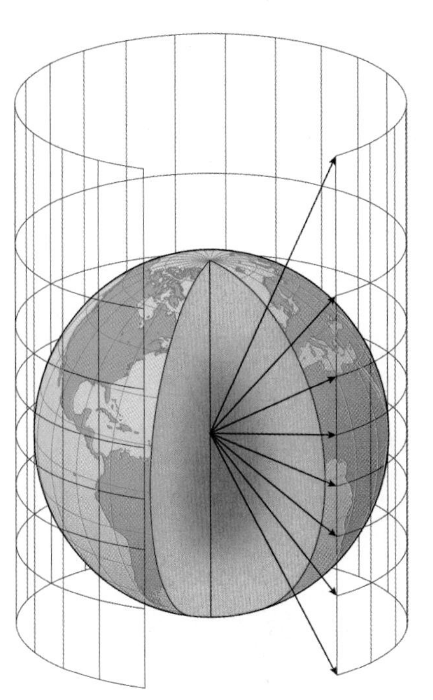

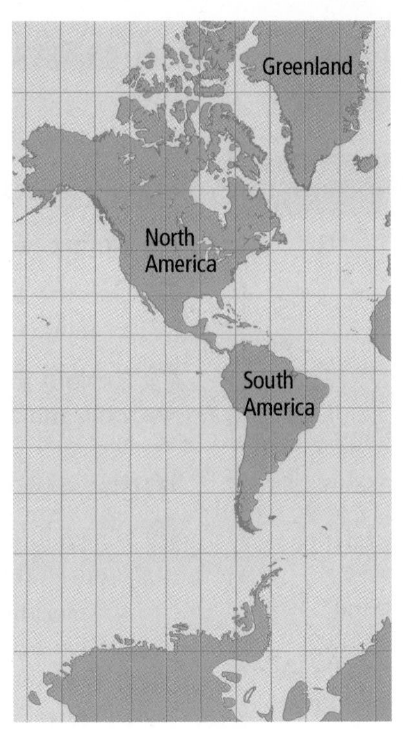

Cylindrical Projection

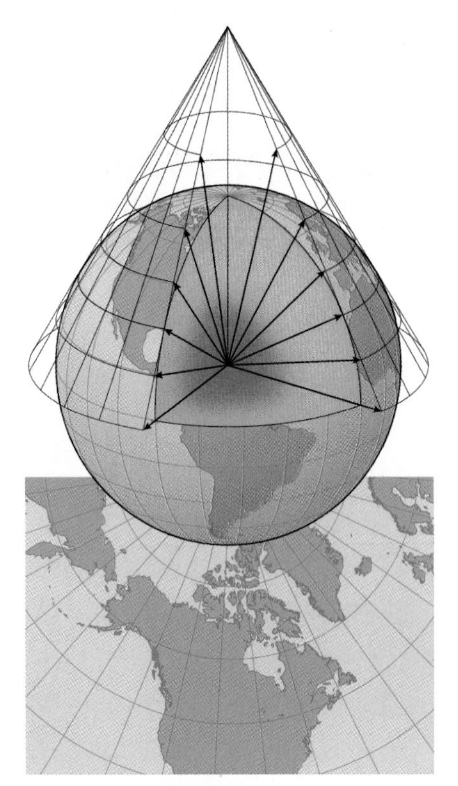

Conical Projection

Map Projections

Since a globe is spherical like Earth, Earth's features are not distorted on a globe. Maps, however, are flat. How can a flat map be made from a sphere? One way to transfer features from a globe to a flat map is to make a projection.

Cylindrical Projections Imagine a light at the center of a globe. It would throw shadows of the continents and the latitude and longitude lines onto a sheet of paper if it were wrapped around the globe. Because the paper is shaped like a cylinder, as shown in **Figure 7**, this is called a cylindrical projection. The resulting map represents shapes near the equator very well. However, shapes near the poles are enlarged. Notice that in the cylindrical projection, Greenland appears to be larger than South America. However, Greenland is about one-eighth the size of South America.

Conical Projections Wrapping a cone around the globe makes a conical projection. It has little distortion near the line of latitude where the cone touches the globe, but it is distorted elsewhere. All types of projections distort the shapes observed on a sphere. In other projections, the continents are represented accurately only because the other areas, such as the oceans, are distorted or cut away.

Key Concept Check What are the advantages and disadvantages of cylindrical projections and conical projections?

Figure 7 The cylindrical and conical projections transfer features located on a sphere to a flat map. This process always results in some distortion.

FOLDABLES®

Make a folded book from a sheet of paper. Label it as shown. Use it to record information about map projections. Label the outside of the book Map Projections.

Conical Projections | Cylindrical Projections

Lesson 1 Review

Visual Summary

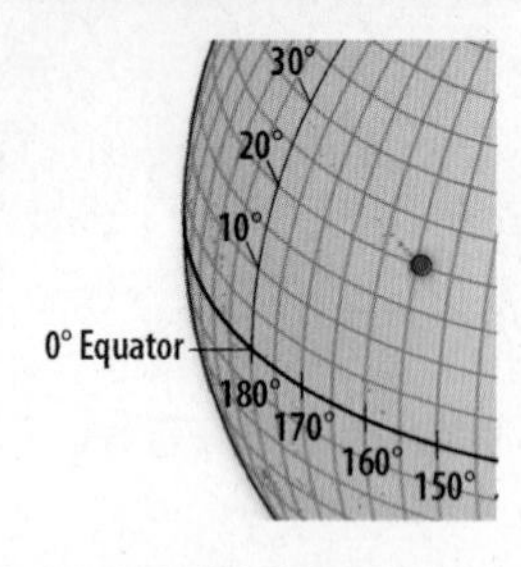

Finding locations on a map or a globe can be done accurately by using grid lines called longitude and latitude.

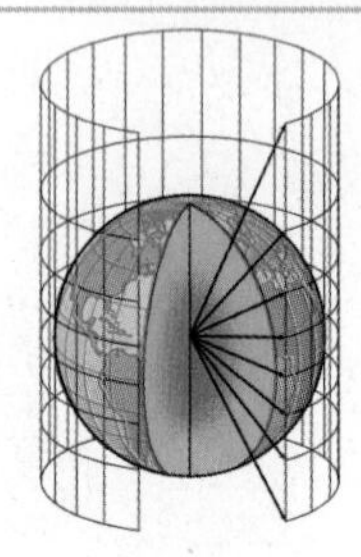

Different projections offer different solutions to the distortion problem of transferring three dimensions into two dimensions.

FOLDABLES®

Use your lesson Foldable to review the lesson. Save your Foldable for the project at the end of the chapter.

What do you think NOW?

You first read the statements below at the beginning of the chapter.

1. Maps help determine locations on Earth.
2. All Earth models are spherical.
3. World maps are drawn accurately for every location.

Did you change your mind about whether you agree or disagree with the statements? Rewrite any false statements to make them true.

Use Vocabulary

1. **Define** *profile view* in your own words.
2. **Use the terms** *latitude* and *longitude* in a sentence.
3. **Explain** the difference between a map scale and a map legend.

Understand Key Concepts

4. Which lines are used to measure the distance south of the equator?
   - **A.** meridians
   - **B.** lines of latitude
   - **C.** the International Date Line
   - **D.** lines of longitude
5. **Compare** a globe to a map. Explain why distortions occur.
6. **Explain** why the International Date Line does not match the 180° meridian exactly.

Interpret Graphics

7. **Identify** Copy and fill in the graphic organizer below to identify the three units used to measure latitude and longitude.

Critical Thinking

8. **Suggest** a reason that the time zones do not exactly follow meridians.
9. **Evaluate** Which type of projection—conical or cylindrical—would show less distortion of central Africa? Explain your choice.

Math Skills

10. The distance between two towns on a map is 7 cm. The map scale is 1 cm:100 km. What is the actual distance between the two towns?

Skill Practice — Compare and Contrast **25 minutes**

How can you fit your entire classroom on a single sheet of paper?

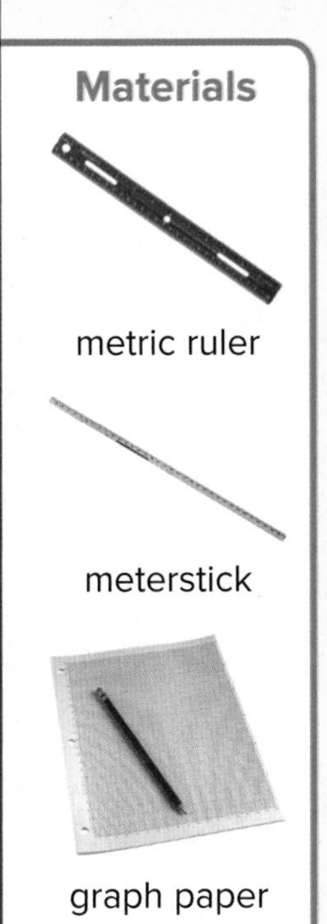

Materials

metric ruler

meterstick

graph paper

Mapmakers must measure objects and distances very carefully to produce accurate maps. Without detailed and accurate measurements, maps would not be useful. Most maps are scaled down. This means that the map and details in it are smaller than what they represent. Sizes and ditances on a scaled map are proportions of the actual values. For example, if a map has a 1 cm to 1 m scale, 5 cm on the map represents 5 m.

Learn It

When you look for similarities between two things, you **compare** them. When you find differences between two things, you **contrast** them. Creating a ratio in order to scale down the dimensions of a room to make a map compares the room's actual dimensions to the map's scale dimensions. The difference between the map and the room are the units of measurement (cm:m).

Try It

1. Read and complete a lab safety form.
2. On a blank sheet of graph paper, sketch your classroom as if you were looking down on it. Do not worry about accuracy right now.
3. Select several objects or structures lining the classroom, such as windows or doors. Measure how far each is from the corners of the walls. Record your data in your Science Journal.

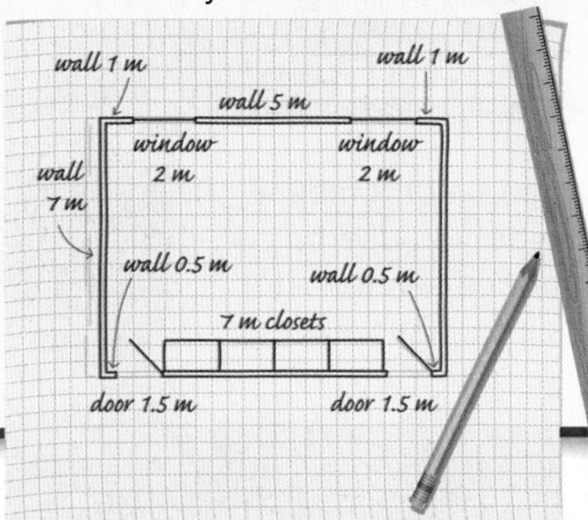

4. Your teacher will tell you the dimensions of the classroom. Choose a scale for a map of the room. Use the dimensions and your scale to draw a map of the classroom on a single piece of graph paper.
5. Make sure to include all the features from your sketch. Also, include a scale, legend, and the total area.

1 2 5 2 1
7
1.5 7
Total area is 77 m²
Key
Wall
Window
Door
Scale 1 cm = 1 meter

Apply It

6. What scale did you use in your map? Explain why you chose that scale.
7. How is your map similar to a map of Earth? How is it different?
8. **Key Concept** Would the sketch or the map you made be more useful to help someone locate an object in the room? Support your reasoning.

Lesson 2

Reading Guide

Key Concepts
ESSENTIAL QUESTIONS

- What can a topographic map tell you about the shape of Earth's surface?
- What can you learn from geologic maps about the rocks near Earth's surface?
- How can modern technology be used in mapmaking?

Vocabulary

topographic map p. 20
elevation p. 20
relief p. 20
contour line p. 20
contour interval p. 21
slope p. 21
geologic map p. 23
cross section p. 23
remote sensing p. 27

Multilingual eGlossary

What's Science Got to do With It?

Technology and Mapmaking

Inquiry Mountains or Molehills?

Have you ever hiked to the top of a mountain or a mesa? How did you know how high it was? Maybe you used a map with information about the height above sea level for locations along the trail. Why would this information be helpful? How is this information shown on a map?

Launch Lab

20 minutes

Will this be an easy hike or a challenging hike?

If you were going for a hike, you would probably want to know if it would be easy or hard. Would you have to climb a steep hill or is the area flat? How could you find this information?

1. Obtain a map with elevation information on it.
2. Plan two hikes that cover the same distance on the map. Plan one easy hike over flat terrain and one challenging hike in which a hill will be climbed.
3. Share with a partner how both hikes would be different. How are the elevations of locations on your map shown?

Think About This

1. What are the benefits of knowing where there are steep and gentle slopes on a map?
2. **Key Concept** How would you describe elevation information on a map?

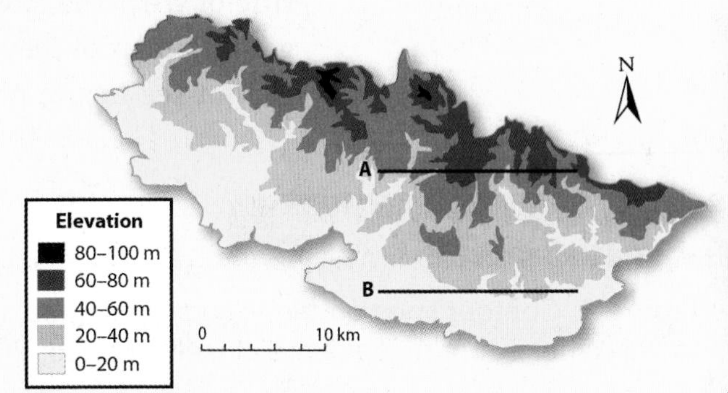

Types of Maps

If you were going to join two pieces of wood together, you might use a hammer and nails. To scramble eggs you could use a whisk and a skillet. Just as there are tools for doing different jobs, there are maps for different purposes.

General-Use Maps

The first maps were hand-drawn by explorers and sailors to record their trading routes. Today we use maps in a variety of situations. You might use a map to help a friend find your house or the quickest route to the mall. If you go to a park there might be a trail map outlining the route you will hike. Some everyday maps you might use include:

- **Physical maps** use lines, shading, and color to indicate features such as mountains, lakes, and streams.
- **Relief maps** use shading and shadows to identify mountains and flat areas.
- **Political maps** show the boundaries between countries, states, counties, or townships. The boundaries can be shown as a variety of solid or dashed lines. Different colors might be used to indicate areas within the boundaries.
- **Road maps,** as shown in **Figure 8,** can show interstates or a range of roads from four-lane expressways to gravel roads. Maps all are useful in helping you find your way.

Figure 8 Road maps of counties or cities can be very detailed, but an atlas of maps of the 50 states might show only the important or main roads.

Topographic Maps

If you were hiking across the United States, you might want to follow level terrain. If you were piloting an airplane across the United States, you would definitely want to fly higher than a mountain. Showing you how high or low land features are is a feature of one kind of specialty map.

WORD ORIGIN

topography
from Greek *topos,* means "place"; and *graphein,* means "to write"

The shape of the land surface is called topography. *A* **topographic map** *shows the detailed shapes of Earth's surface, along with its natural and human-made features.* It helps give you a picture of what the landscape looks like without seeing it. The topographic map of Devil's Tower in **Figure 9** shows the details you cannot see in the photo.

Topographic Map

Figure 9 Contour lines on the topographic map show differences in elevation on this volcanic tower. Where contour lines are closely spaced, the topography is much steeper.

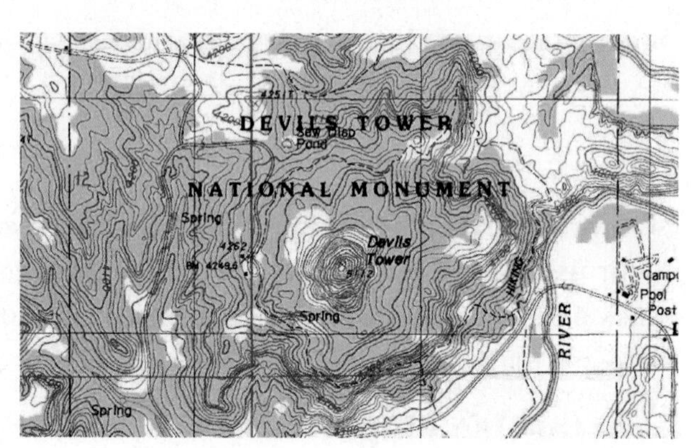

Elevation and Relief *The height above sea level of any point on Earth's surface is its* **elevation.** For example, Mt. Rainier in Washington is 4,392 m above sea level. The city of Olympia, Washington, is about 43 m above sea level. *The difference in elevation between the highest and lowest point in an area is called* **relief.** For example, the relief between Mt. Rainier and Olympia is calculated 4,392 m − 43 m = 4,349 m.

Contour lines *are lines on a topographic map that connect points of equal elevation.* Similar to lines of latitude and longitude, contour lines do not really exist on Earth's surface. Using contour lines, you can measure both elevation and relief on a topographic map. If the top of Devil's Tower is 5,112 ft and the base is 4,400 ft, what is the relief?

Reading Check How are contour lines similar to lines of latitude and longitude?

(l)Robert Glusic/Getty Images; (r)USGS

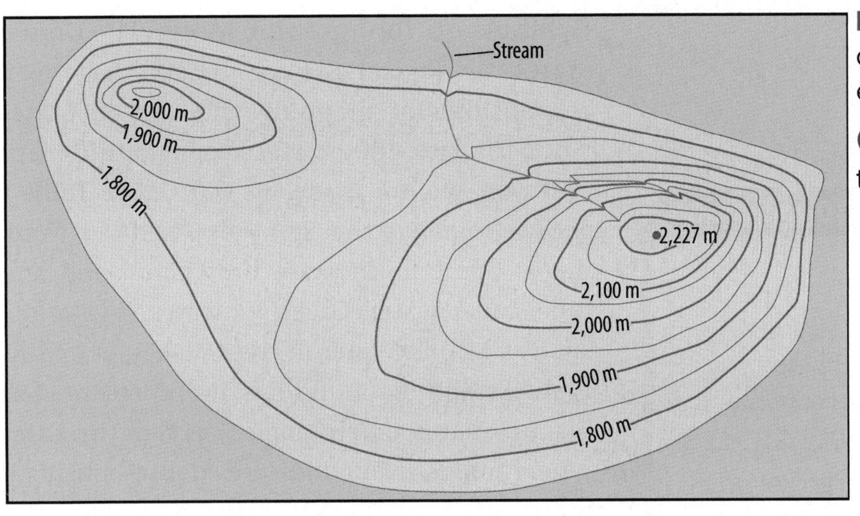

Figure 10 Contour lines connect points of equal elevation.

Visual Check Where is the slope gentle or steep?

Interpreting Contours Contour lines that represent a mountain are shown in **Figure 10.** Notice that the elevation is not written on every contour line. The darker contour lines, called index contours, are labeled with the elevation. How do you find the elevation of contours other than by using index contours? You need to know the elevation difference between the lines.

The elevation difference between contours that are next to each other is called the **contour interval.** The map in **Figure 10** has a contour interval of 50 m. You can find the elevation of an unlabeled contour by using the numbered index contours. First, find the closest index contour below the contour you are identifying. Then, count up to it by 50s from the index contour.

Study **Figure 10** again. Notice that a contour line at the top of the mountain forms a small enclosed loop with a dot in the middle of it. This dot represents the highest point on the mountain–2,227 m. The V-shaped contours pointing downhill indicate ridges. A small V pointing uphill indicates a stream valley or drainage.

The spacing of contours indicates slope. **Slope** *is a measure of the steepness of the land.* If the contours are spaced far apart, the slope is gradual or flat. If the contours are close together, the slope is steep.

Key Concept Check What can you learn about the features at Earth's surface from studying contour lines?

Topographic Profiles You read about map views and profile views in Lesson 1. The information contour lines provided on a topographic map can be used to draw an accurate profile of the topography. Making profiles like this can help you determine the easiest path to take when crossing the land.

FOLDABLES®

Make a vertical trifold book. Label the front *Topographic Maps,* and label the inside of the book as shown. Use it to collect information about what a topographic map can show you.

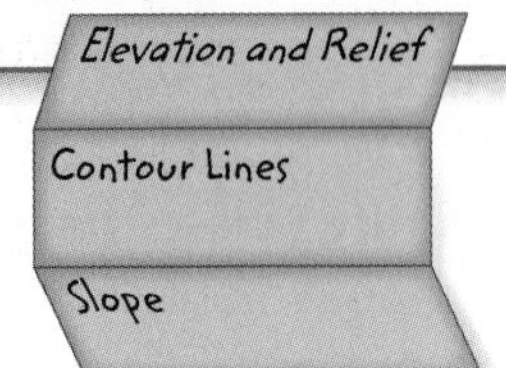

MiniLab

20 minutes

Can you construct a topographic profile?

A topographic profile of line AB helps you identify geological features of a contour map.

1. Use a piece of **graph paper** to set up your topographic profile graph. Label the x-axis *Distance Between A and B.* Label the y-axis *Elevation (m).*
2. Measure the length of line AB on the contour map below. Use a **ruler** to measure the distance from point A to the intersection of the first contour line. Plot the point on your graph.
3. Plot distance and elevation pairs for each contour line where it intersects line AB.
4. Connect the points on your graph and observe the topographic profile.

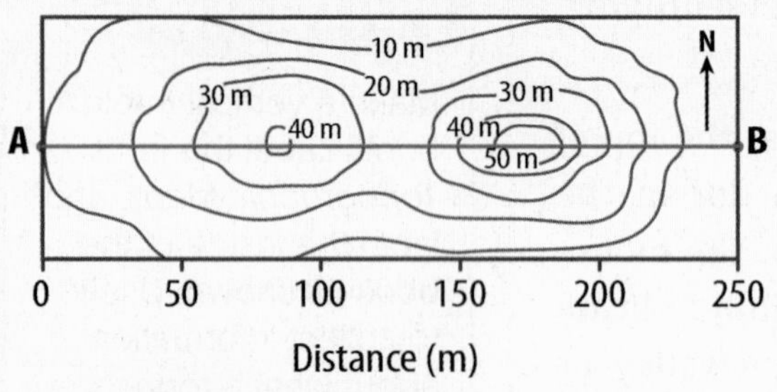

Analyze and Conclude

1. **Analyze** At what distance from point A is the highest point on line AB? The lowest?
2. **Identify** where the topography is the steepest along line AB. Explain how you know this.
3. **Predict** how a contour map and topographic profile would be useful as you design a skateboard park.
4. **Key Concept** Describe three topographic features depicted in your topographic profile.

Symbols on Topographic Maps The United States Geological Survey (USGS) has been responsible for mapping the United States since the late 1800s. Most topographic maps that you see are made by the USGS. **Table 1** shows some of the symbols used on these maps. Contour lines are brown on land and blue under water. Green indicates vegetation, such as woods. Water in rivers, lakes, or oceans is shown in blue. Buildings are represented as black squares or rectangles, except in cities where pink shading indicates dense housing.If information has been updated since the original map was made, it is added in purple.

Reading Check Why is it important for a topographic map to have a legend?

Interactive Table

Table 1 USGS Topographic Map Symbols

Description	Symbol
Primary highway	
Secondary highway	
Unimproved road	
Railroad	
Buildings	
Urban area	
Index contour	100
Intermediate contour	
Perennial streams	
Intermittent streams	
Wooded marsh	
Woods or brushwood	

Visual Check How are primary and secondary highways differentiated on a USGS topographic map?

Geologic Maps

Another kind of specialty map is a geologic map. **Geologic maps** *show the surface geology of the mapped area.* This can include the rock types, their ages, and locations of faults. The geologic map in **Figure 11** shows the geology of the Grand Canyon.

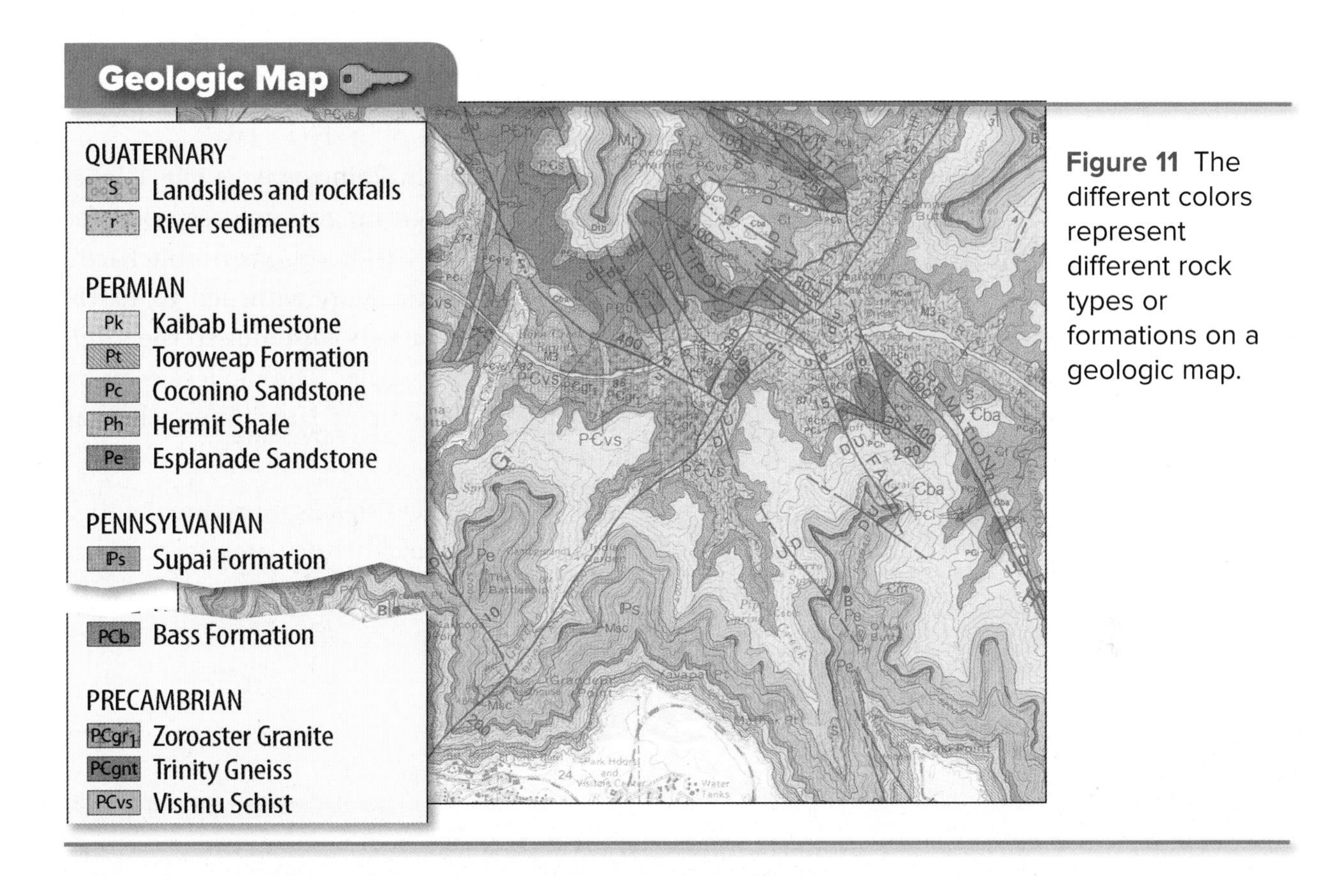

Figure 11 The different colors represent different rock types or formations on a geologic map.

Geologic Formations On a geologic map, different colors and symbols represent different geologic formations. A geologic formation is a rock unit with similar origins, rock type, and age. The map legend lists the colors and symbols along with the age of the rock formation. The colors do not indicate the rock's true colors; they show the many formations on the map. Find the Kaibab formation in the map legend of **Figure 11.** It tells you this limestone rock was made during the Permian period.

Geologic Cross Sections Sometimes geologists need to know what the rocks are like underground as well as on the surface. Information can be gathered by drilling for samples, studying earthquake waves, or looking at cliffs. A cliff face is like a profile view of the ground. Geologists use this information to produce a profile view of the rocks below the ground. The resulting diagram is called a **cross section**–*a profile view that shows a vertical slice through rocks below the surface.* **Figure 12** shows a cross section of a geologic map.

Figure 12 A cross section of a geologic map shows a vertical slice through the rocks below the surface. ▼

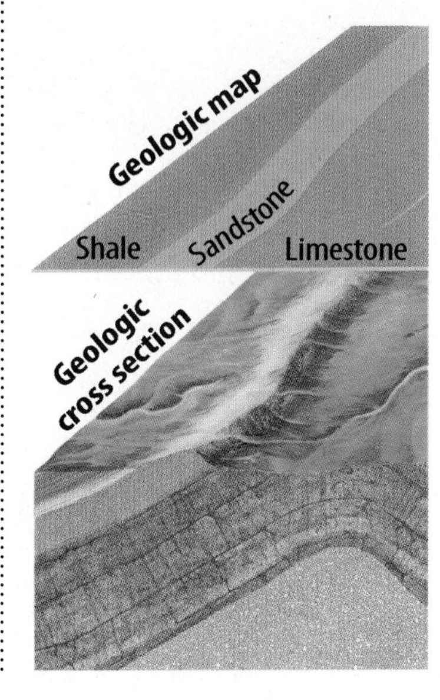

 **Key Concept Check** How is color used in a geologic map?

Making Maps Today

For centuries mapmakers made observations of Earth and gathered information from explorers. First, mapmakers and explorers used instruments such as a compass, a telescope, and a sextant, which is used to find latitude, to make and record measurements. Then, mapmakers carefully drew new maps by hand. Today, mapmakers use computers and data from satellites to make maps.

Global Positioning System

One important resource for mapmakers today is the Global Positioning System (GPS). GPS is a group of satellites used for navigation. As shown in **Figure 13,** 24 GPS satellites orbit Earth. Signals sent from devices on the surface are returned to Earth. The relayed signals are used to calculate the distance to the satellite based on the average time of the signal. The devices may be hand-held units the size of a cell phone, or larger units such as the one shown in **Figure 14.**

At any given time, a GPS unit receives signals from three or four different satellites. Then the receiver quickly calculates its location– its latitude, longitude, and altitude. GPS is used by mapmakers to accurately locate reference points.

Global Positioning System

Figure 13 GPS receivers detect signals from the 24 GPS satellites orbiting Earth. Using signals from at least three satellites, the receiver can calculate its location within 10 m.

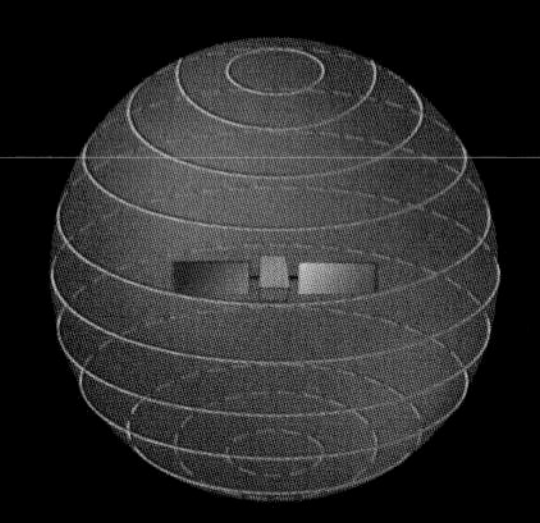

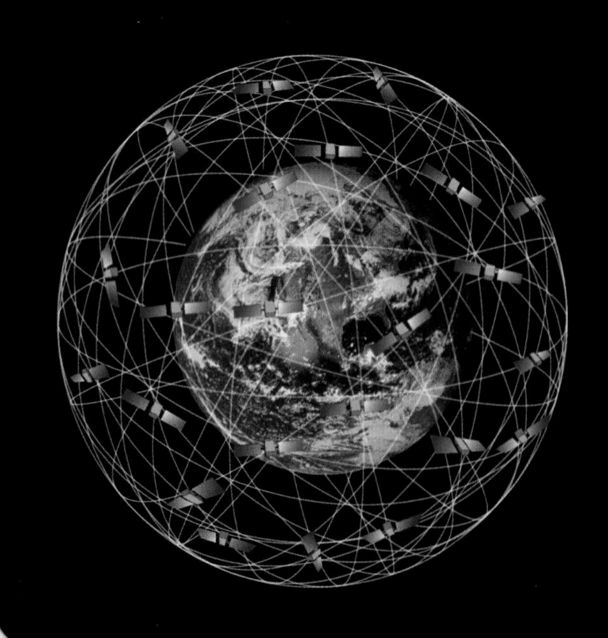

1 The information from one satellite tells the GPS receiver that it is somewhere on a sphere surrounding that satellite. Suppose the distance to the satellite is 20,000 km. This limits the possible location of the receiver to a spherical radius of 20,000 km from the satellite. If the receiver is on Earth, that limits the location to a large circle somewhere on Earth.

Originally designed for military purposes, it is now a continuously available service for everyone worldwide. Airplanes, ships, and cars have navigational systems that use GPS technology. People can find their way to restaurants, hotels, or sporting events and home again. Other uses include tracking wildlife for scientific data collection, detecting earthquakes, hiking, biking, and land surveying. **Figure 14** shows a portable GPS receiver you might use while traveling in a car.

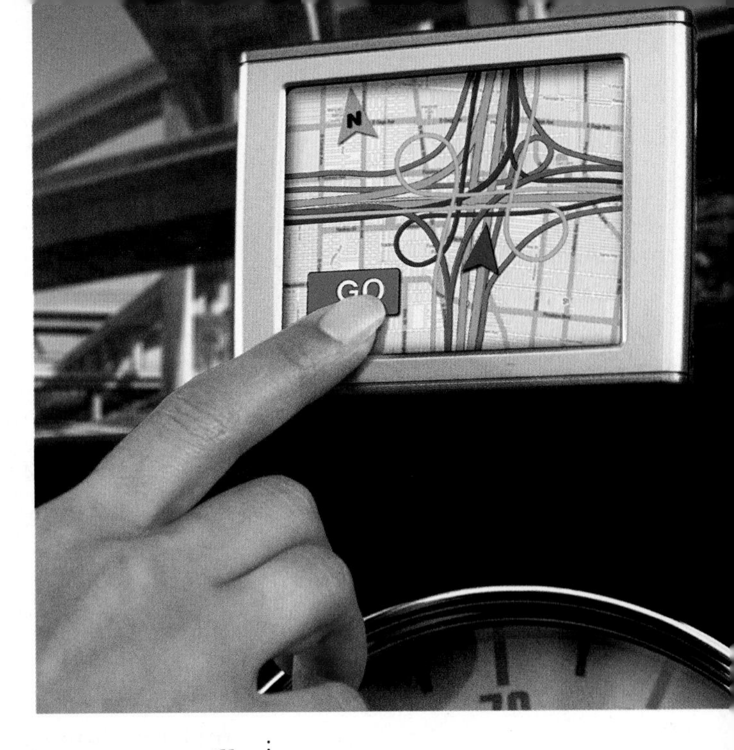

Reading Check What are some common uses of GPS?

GPS technology continues to improve. Land-based units used in combination with satellites are already being used to pinpoint people and places to the centimeter. Future improvement will include guiding self-driven automobiles and additional civil channels projected to improve safety and rescue operations.

Key Concept Check How can GPS technology be used in mapmaking?

▲ **Figure 14** Portable GPS receivers like this one can be used to help find your way.

Nick Koudis/Photodisc/Getty Images

Animation

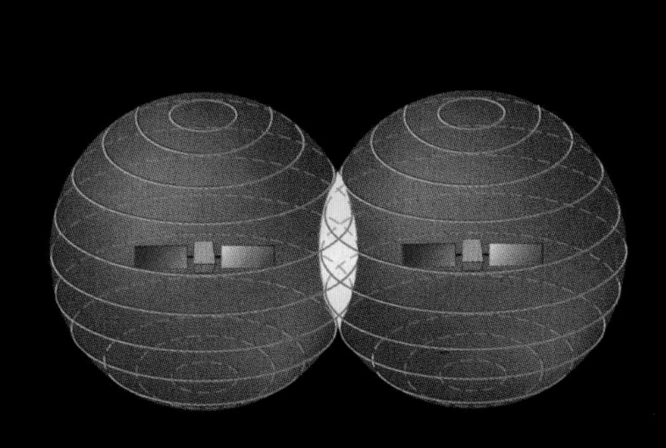

2 Next, the receiver measures the distance to a second satellite. Suppose this distance is calculated to be 21,000 km away. The location of the receiver has to be somewhere within the area where the two spheres intersect, shown here in yellow.

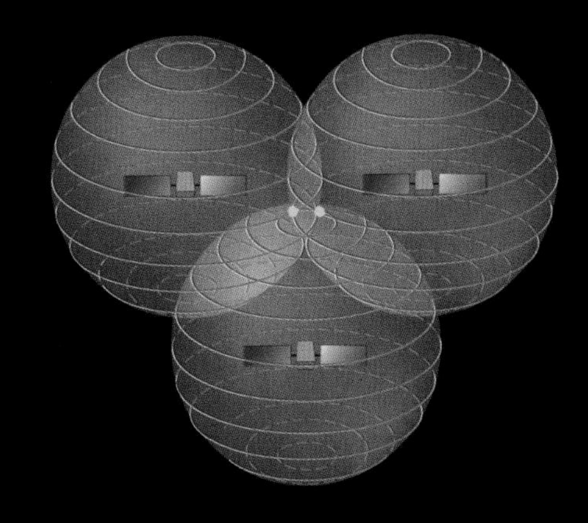

3 Finally, the distance to a third satellite is calculated. Using this information, the location of the receiver can be narrowed even further. By adding a third sphere, the location can be calculated to be one of two points as shown. Often one of these points can be rejected as an improbable or impossible location. Information from a fourth satellite can be used to tell elevation above Earth's surface. A pilot or a climber might find this useful.

Geographic Information Systems

Geographic Information Systems (GIS) are computerized information systems used to store and analyze map data. GIS combine data collected from many different sources, including satellites, scanners, and **aerial** photographs. Aerial photographs are taken from above the ground. Data collection that at one time took months now takes hours or minutes.

ACADEMIC VOCABULARY

aerial
(adjective) operating or occurring overhead

Mapmakers use GIS to analyze and organize those data and then create digital maps. One of the features of GIS is that it creates different map layers of the same location. As shown in **Figure 15**, the map layers are like the layers of a cake. However, when the map layers are placed on top of each other, you can see through to the lower layers. Different layers can show land use, elevation, roads, streams and lakes, or the type of soil on the ground.

Three Views Imagine setting up a model for an airplane landing under certain weather conditions using GIS.

- Database view begins the process by assembling information from existing databases on winds, airplane flight, landing procedures, and airport layouts.
- Map view would draw from a set of interactive, digital maps that show features and their relationship to Earth's surface.
- Model view then would pull all the information together so you could run simulations under changing weather conditions.

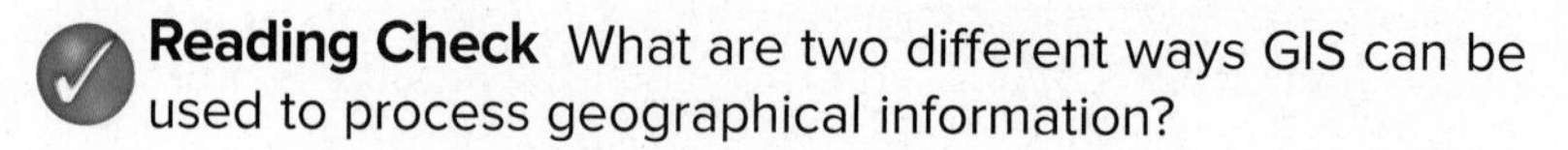

Reading Check What are two different ways GIS can be used to process geographical information?

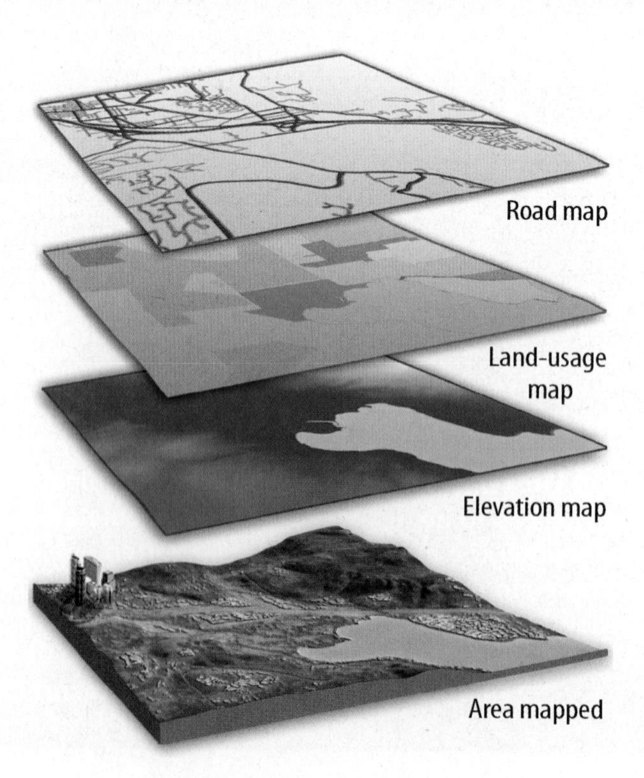

Figure 15 GIS combines the data from many maps to give detailed information about a mapped area.

Remote Sensing

A cup of hot chocolate looks very hot. You place your hand over it and feel the heat from the liquid. Without even touching it, you know it's still too hot to drink. You have just avoided burning your mouth by using remote sensing.

Remote sensing *is the process of collecting information about an area without coming into physical contact with it.* There are many applications for remote sensing. This process produces maps that show detailed information about agriculture, forestry, geology, land use, and many other subjects. Often these maps cover huge areas.

Mapmaking was transformed when it became possible to take aerial photographs from airplanes. Now an even more powerful type of remote sensing is being used. Since the 1970s, satellites orbiting thousands of kilometers above Earth's surface have been used to collect data.

Monitoring Change with Remote Sensing Satellites orbit Earth repeatedly. This means images of a location made at different times can be used to study change. For example, a 3-year drought in northern California dropped water levels in many local reservoirs. Before and after images of the Shasta Lake reservoir are shown in **Figure 16.** Images like these help mapmakers to quickly make maps of areas affected by natural disasters. The maps are then used to monitor damage and help organize rescue efforts.

Key Concept Check How can remote sensing be an advantage to mapmakers?

Figure 16 These satellite images show the changing shoreline of the Shasta Lake reservoir after a 3-year drought. The after image shows the increase in shoreline that occurred due to periods of extreme drought.

Before

After

NASA images courtesy the MODIS Rapid Response Team at NASA GSFC.

Landsat One series of satellites used to collect data about Earth's surface is the Landsat group. *Landsat 7*, launched in 1999, and *Landsat 8*, launched in 2013, scan Earth's entire surface. Comparing today's data to similar data collected years ago, scientists can recognize global changes in agriculture, forestry, geology, and changes caused by natural disasters such as hurricanes, flooding, and forest fires. Images obtained from *Landsat 7* and *Landsat 8* can be used for emergency response and disaster relief. Landsat has been used to contribute to the GIS database as well.

OSTM/*Jason-2* and *Jason-3* A series of satellite missions, including the OSTM/*Jason-2* and *Jason-3*, have been used to determine ocean topography and circulation, sea level, tides, and climate changes. **Figure 17** shows ocean surface changes due to El Niño and La Niña as captured by the *Jason-2*.

SeaBeam A device that uses sonar to map the bottom of the ocean is SeaBeam. SeaBeam is mounted onboard a ship. Computers calculate the time a sound wave takes to bounce off the ocean floor and return to the ship. This gives the operators an accurate image of the seafloor and the depth of the ocean at that point. SeaBeam is used by fishing fleets, drilling operations, and various scientists.

Reading Check What are some methods used to collect remote sensing data?

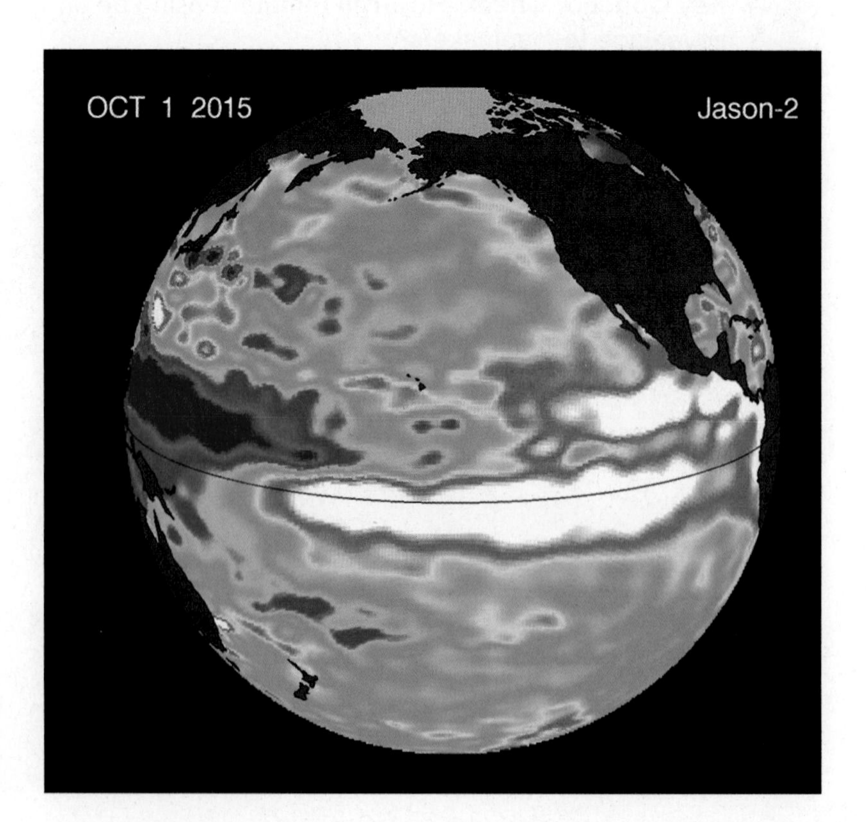

Figure 17 This image of the Pacific was captured by *Jason-2* during an El Niño event. The white area shows an increase in the ocean surface temperature compared to normal.

NASA/JPL Ocean Surface Topography Team

Lesson 2 Review

Visual Summary

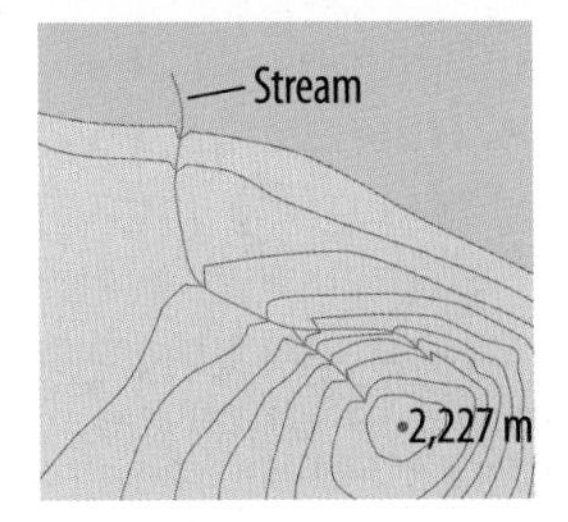

Topographic maps have contour lines that help describe the elevation and relief of the surface of Earth at a particular location.

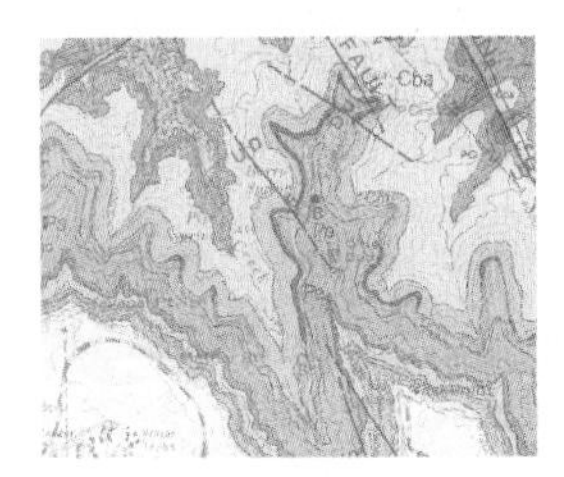

Geologic maps are useful in determining rock type, rock age, and the rock formations in an area.

FOLDABLES®

Use your lesson Foldable to review the lesson. Save your Foldable for the project at the end of the chapter.

What do you think NOW?

You first read the statements below at the beginning of the chapter.

4. Topographic maps show changes in surface elevations.

5. The colors on geologic maps show the colors of the surface rocks.

6. Satellites are too far from Earth to collect useful information about Earth's surface.

Did you change your mind about whether you agree or disagree with the statements? Rewrite any false statements to make them true.

Use Vocabulary

1. **Define** *cross section* in your own words.
2. Change in elevation is called ________.
3. **Use the terms** *contour lines* and *contour interval* in a complete sentence.

Understand Key Concepts 

4. Which type of map would be more useful for determining the quickest route by car?
   - **A.** geologic map
   - **B.** political map
   - **C.** road map
   - **D.** topographic map
5. **Illustrate** a mountaintop that is 850 m high using a contour interval of 50 m.
6. **Separate** the symbols shown on a topographic map legend into groups of natural and cultural features.

Interpret Graphics

7. **Summarize** Copy and fill in the graphic organizer below to identify three things you can learn about the shape of Earth's surface from contour lines.

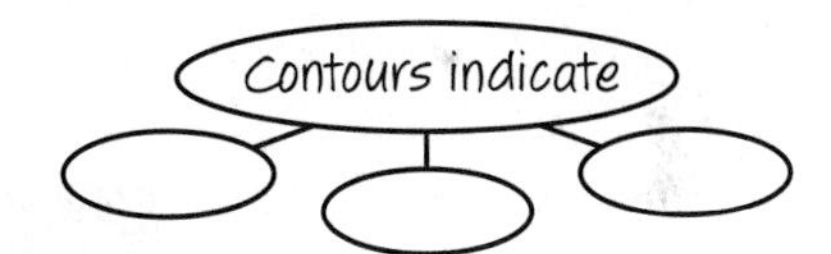

8. **Determine** the contour interval for the contour map below.

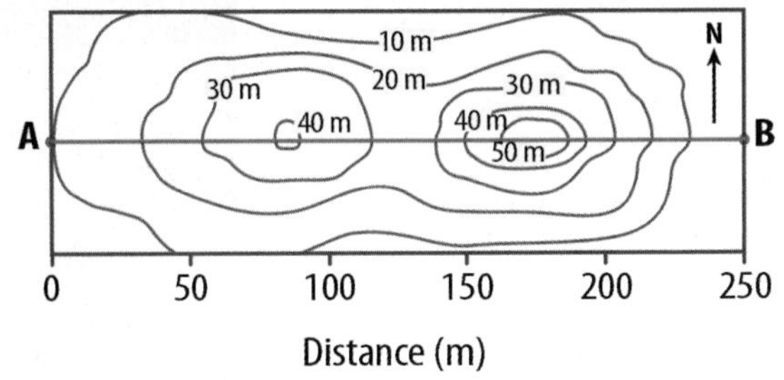

Critical Thinking

9. **Suggest** how a photograph of the Grand Canyon could be used to make a geologic map of Arizona.

 Lab

40 minutes

Materials

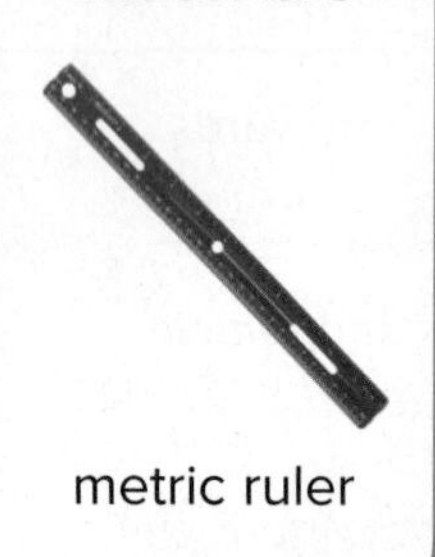

metric ruler

Seeing Double?

The satellite images shown here are called stereo photographs. At first glance, the photos in each set might appear the same. However, if you look very closely, you should see slight differences between the pictures in each set. These differences will allow your eyes to change the 2-D images into 3-D views of the land. Let's see if your eyes can create these 3-D optical illusions so that you can study these features of Earth's surface.

Question

How can stereo photographs be used to study Earth's surface features?

Procedure

1. Look at the images on this page. Find the thin, gray lines in the image. They are rivers that were filled with ash from a volcanic eruption.
2. Now locate the black shapes near the center of each image. These are lakes.
3. Much of the white and gray area in the photo is one landform made of a volcano and its ash and debris deposits.
4. Now study the images on the next page. The bright blue areas on the images are lakes.
5. The brown and reddish-brown colored areas on this set of images are rocks and soil.
6. Now locate the different green-colored areas on this set of images. These are trees and other vegetation.

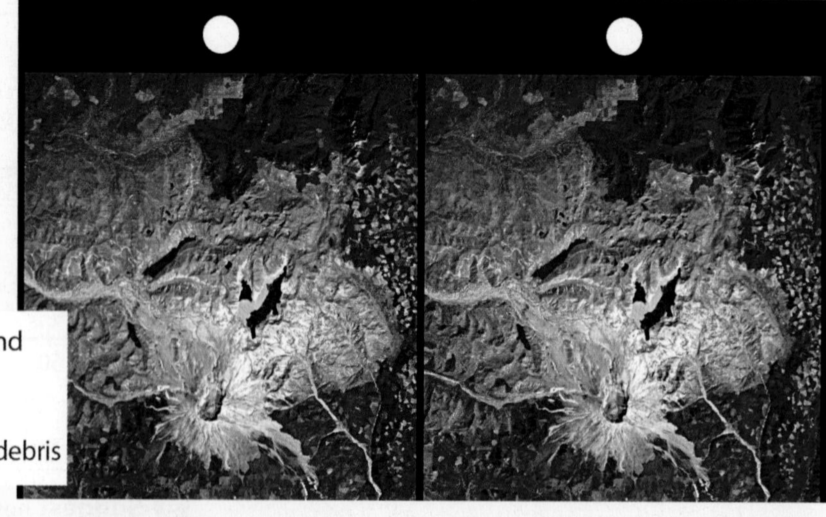

Forest-covered land
Lake water
Volcanic ash and debris

(t)Michael Scott/McGraw-Hill Education; (b)NASA/JPL

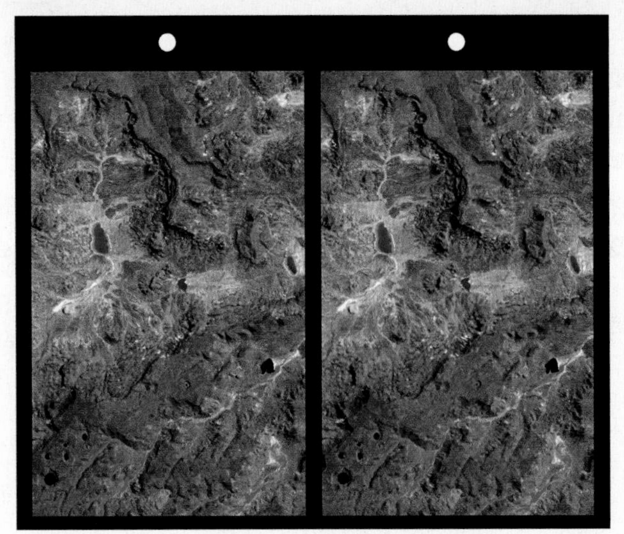

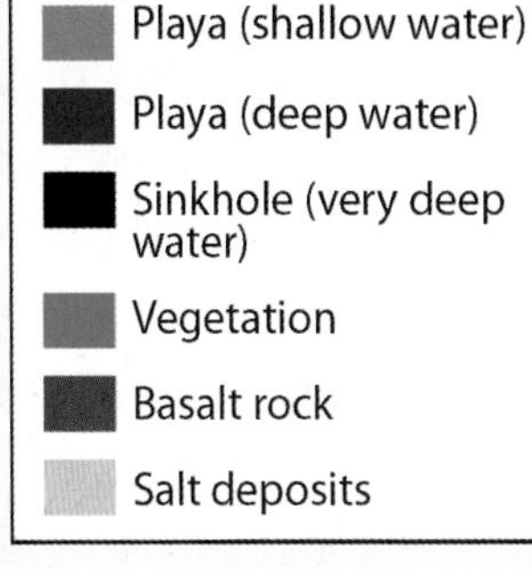

7. View each landscape in 3-D. To do this, slightly cross your eyes while looking at the two white dots above the images. A third white dot will appear between the two dots. At this point, you should see a 3-D view of the landscape. Study the landscape. Then repeat this procedure for the other set of images.

Lab Tips

- ☑ Cross your eyes only slightly to help form the 3-D view.
- ☑ Relax! Sometimes, if you try too hard to see in 3-D, your eyes will not be able to form the optical illusion.

Analyze and Conclude

8. **Explain** What features changed when you viewed the photos in 3-D?
9. **Measure** The scale for the image above is 1 cm = 2.39 km. In the left-hand image, what is the actual distance from the western edge of the largest lake to the eastern tip of the large mesa at the eastern edge?
10. **The Big Idea** What types of models are satellite images, and why are they sometimes used to study Earth's features?

Communicate Your Results

Write several sentences to describe the general topography of each landscape.

Which landscape is best for your favorite outdoor activity? Justify your answer.

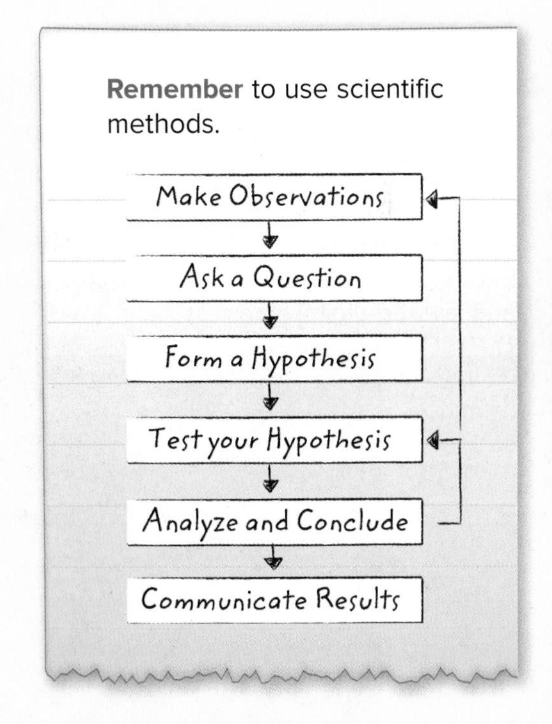

Chapter 1 Study Guide

Geologists model Earth's features using map projections, topographic maps, and geologic maps. They measure Earth's features using remote sensing, primarily from satellites.

Key Concepts Summary

Lesson 1: Maps

- Maps represent the features of Earth's surface and have symbols, scales, and a grid of **latitude** and **longitude** lines that identify locations.
- Distortion occurs because maps are flat representations of Earth. Mapmakers use different map projections to reduce distortion in certain areas.

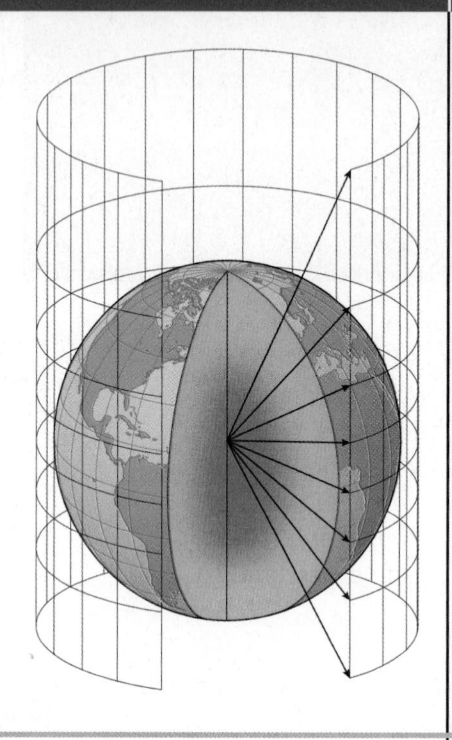

Lesson 2: Technology and Map Making

- Topographic maps show **elevation** through **contour lines.**
- **Geologic maps** contain information about rocks, such as rock types, rock age, and faults.
- **Remote sensing** techniques use satellites and create maps of Earth's surface features. **GIS** integrates data and creates detailed and layered digital maps.

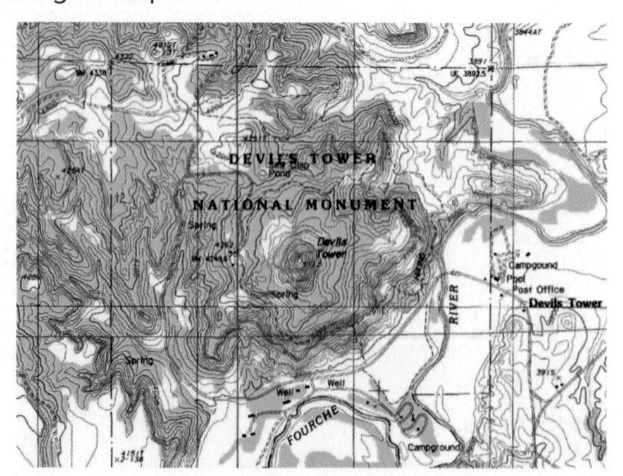

USGS

Vocabulary

map view p. 9
profile view p. 9
map legend p. 10
map scale p. 11
longitude p. 12
latitude p. 12
time zone p. 14
International Date Line p. 14

topographic map p. 20
elevation p. 20
relief p. 20
contour line p. 20
contour interval p. 21
slope p. 21
geologic map p. 23
cross section p. 23
remote sensing p. 27

Personal Tutor

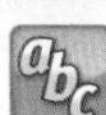
Vocabulary eFlashcards
Vocabulary eGames

FOLDABLES® Chapter Project

Assemble your lesson Foldables as shown to make a Chapter Project. Use the project to review what you have learned in this chapter.

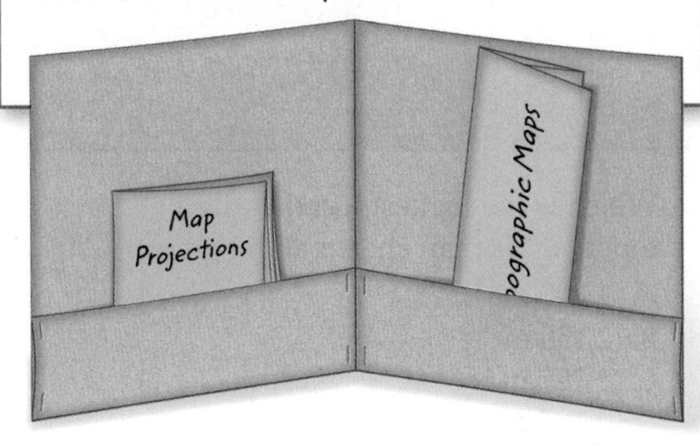

Use Vocabulary

1. Both a map scale and a ________ are added to a map to help interpret the features shown on the map.
2. In order to simplify timekeeping in locations that are close to each other, Earth is divided into 24 ________.
3. In order to learn about geologic formations underground, you need to look at a ________.
4. The ________ indicates the difference in elevation between two adjacent contour lines.
5. Aerial photographs are part of the map-making technology known as ________.
6. A map that shows the shape and features of Earth's surface is a(n) ________.

Link Vocabulary and Key Concepts

Interactive Concept Map

Copy this concept map, and then use vocabulary terms from the previous page to complete the concept map.

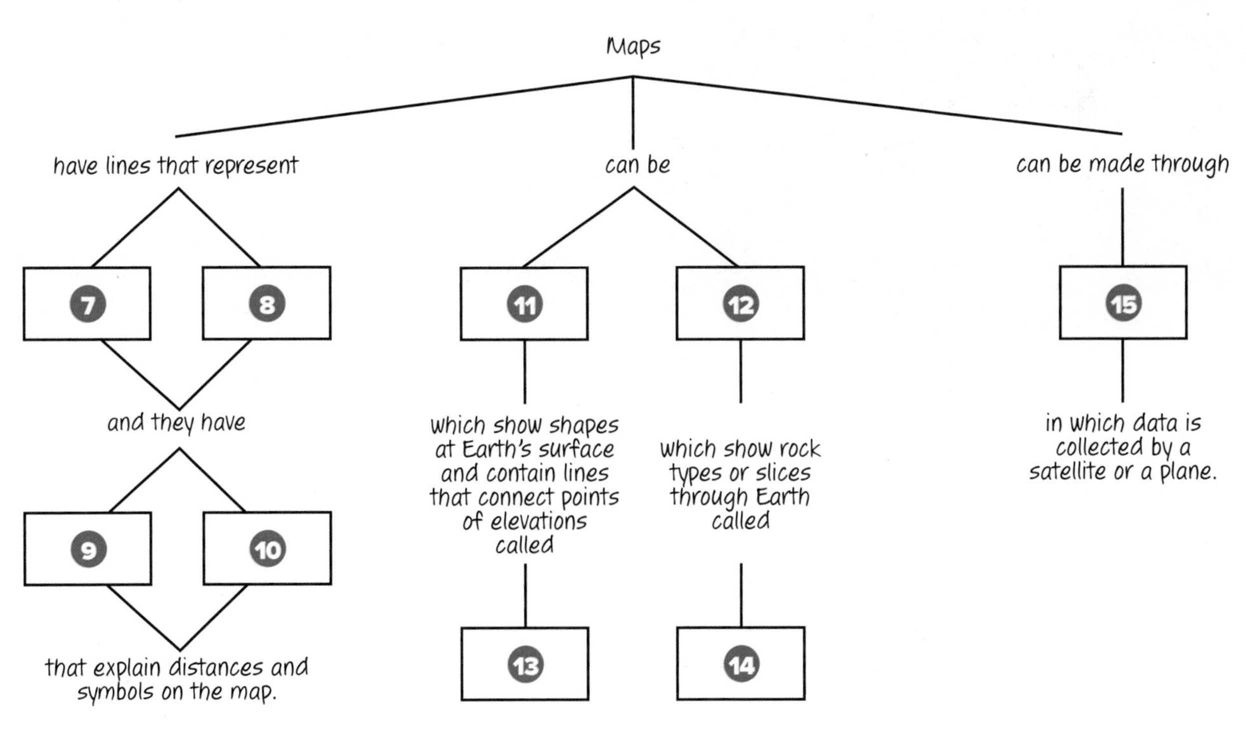

Chapter 1 Review

Understand Key Concepts 

1. Which group of terms describes lines on a map?
   A. latitude, meridians, International Date Line
   B. parallels, profiles, time zones
   C. legends, meridians, International Date Line
   D. longitude, latitude, legends

2. If you traveled west from one time zone to another, what would the time difference be?
   A. one minute C. one hour
   B. two minutes D. two hours

3. Which model of Earth does not distort surface features?
   A. a conical projection
   B. a cylindrical projection
   C. a globe
   D. a map

4. What do the white lines represent in the figure below?

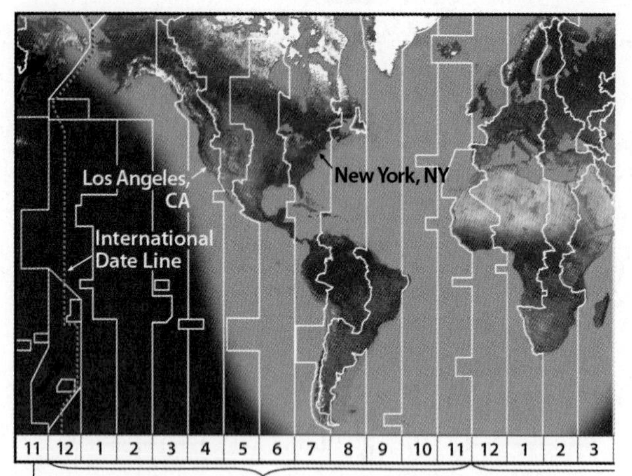

   A. time zones C. lines of latitude
   B. meridians D. lines of longitude

5. A diagram that represents a slice of Earth is called
   A. a cross section. C. relief.
   B. topography. D. an elevation.

6. Study the topographic map below.

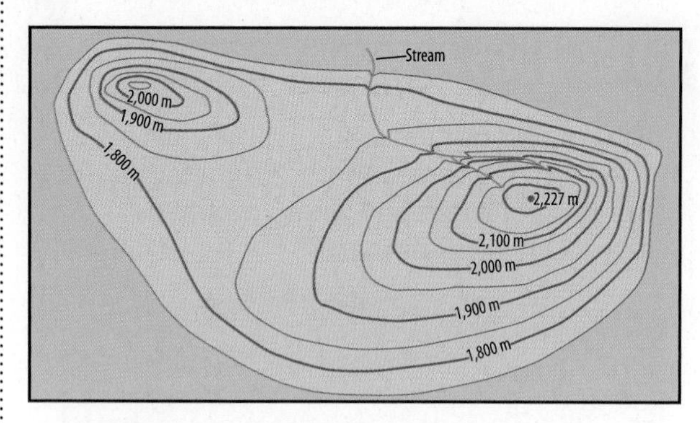

   Which is the highest location on the topographic map shown above?
   A. 1,800 m
   B. 2,150 m
   C. 2,227 m
   D. 2,300 m

7. What information on a topographic map would be helpful if you were making a geologic cross section?
   A. formations
   B. streams
   C. mountaintops
   D. cliffs

8. What is the minimum number of satellites needed to find your exact location using GPS?
   A. 1
   B. 3
   C. 12
   D. 24

9. Which of these legend items would you use to measure the distance between two cities on a map?
   A. contour interval
   B. graphic scale
   C. map projection
   D. road symbol

Critical Thinking

10 **Compare** GIS to GPS.

11 **Construct** a map of an imaginary city. Include a scale and a legend.

12 **Distinguish** between political maps and geologic maps.

13 **Suggest** a reason that the contour interval used on a map of a mountain might be different from the contour interval used on a map of a plain.

14 **Evaluate** the benefit of creating a topographic profile.

15 **Analyze** Which projection in the image below has less distortion? What explanation can you give for this difference?

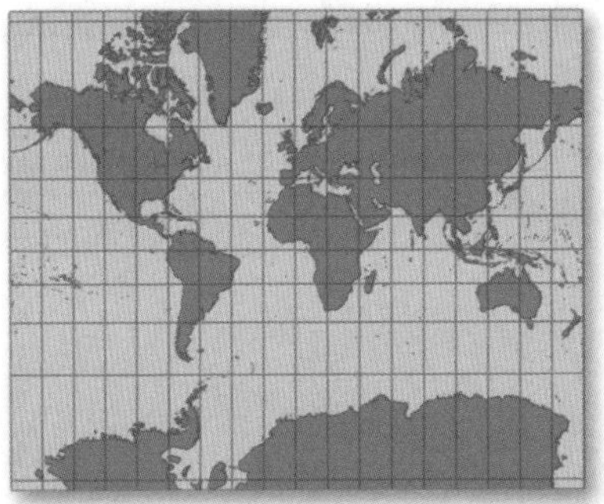

Mercator

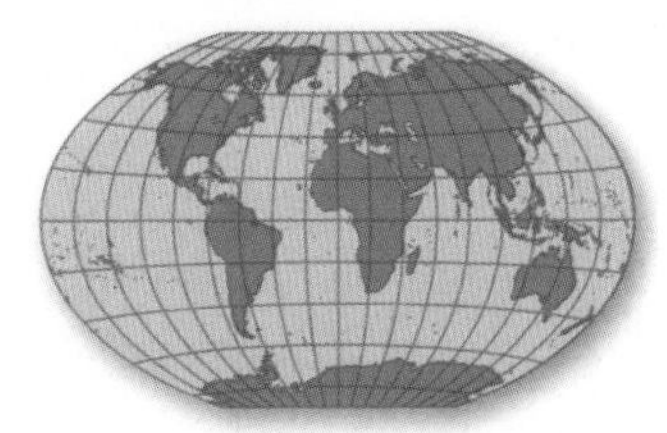

Winkel

16 **Justify** using remote sensing technology during a natural disaster.

Writing in Science

17 **Write** a paragraph that justifies the expense of making and placing satellites in orbit for observing Earth.

REVIEW THE BIG IDEA

18 How do different types of maps model the features of Earth? Explain how the method of data collection or the way in which a map is constructed affects the information on the map.

19 What is the value of having different types of maps of the same area?

20 How can you determine the scale of the chair in the photo below?

Math Skills

Math Practice

Ratio Scale

21 When making a map of the school, you decide to let 1 cm on your map represent 10 m of actual distance. Write this ratio in three different ways.

22 A large wall map of a museum uses a scale of 1 cm:2 m. If the length of one room on the map measures 25 cm, what is the actual length of the room?

23 You are making a map of your city with a scale of 1 cm:4 km. If two buildings in the city are 6 km apart, how far apart will they be on your map?

Standardized Test Practice

Record your answers on the answer sheet provided by your teacher or on a sheet of paper.

Multiple Choice

1 What is the relationship between map distance and actual distance?

A location latitude

B location longitude

C map legend

D map scale

Use the diagram below to answer questions 2 and 3.

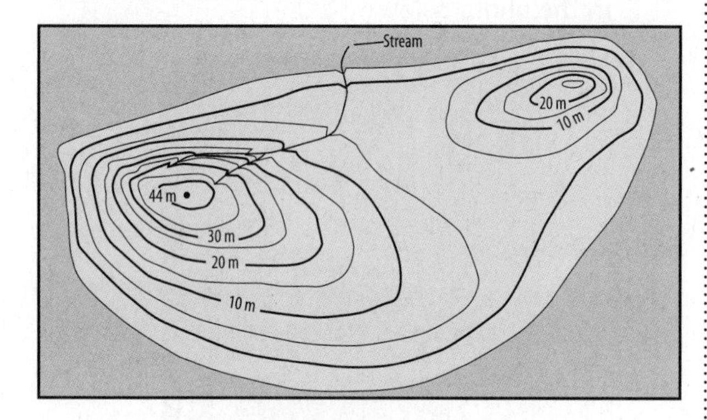

2 What is the approximate height of the lower peak in the diagram above?

A 20 m

B 27 m

C 30 m

D 32 m

3 What is the relief on the map?

A 27 m

B 30 m

C 32 m

D 44 m

4 How many minutes comprise each degree of longitude and latitude?

A 60

B 90

C 180

D 360

Use the diagram below to answer question 5.

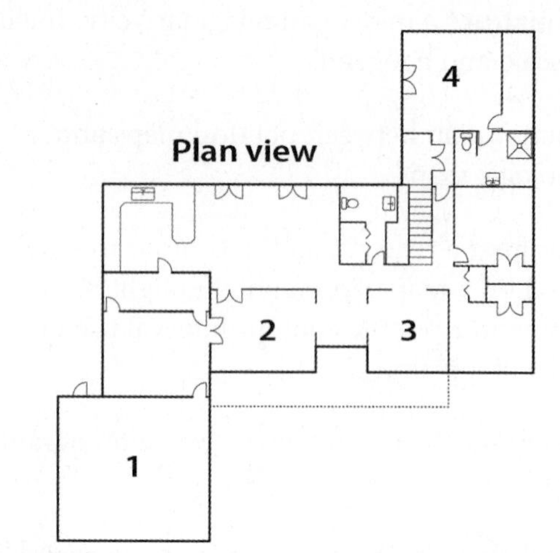

5 In the diagram above, which number on the plan view corresponds to the shaded area on the profile view?

A 1

B 2

C 3

D 4

6 Which is a feature of GIS?

A different map layers of the same location

B navigational satellites

C rock types and locations

D surface geology of a particular area

7 Which map mainly shows boundaries between states and counties?

A physical

B political

C relief

D road

Use the diagram below to answer question 8.

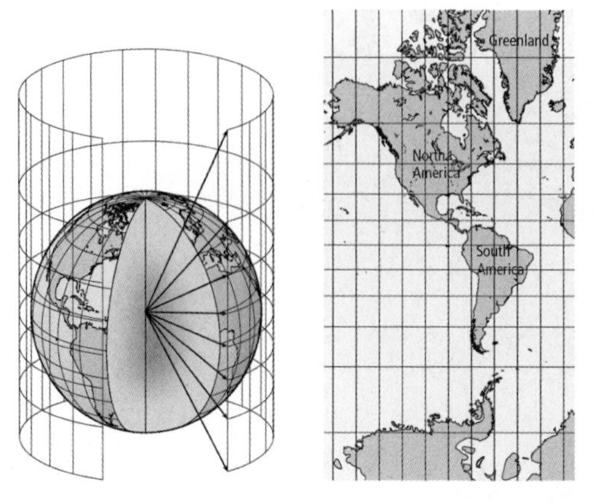

8 Where is map distortion greatest in the type of projection shown above?

A along the mid-latitudes

B along the prime meridian

C at the equator

D at the poles

9 What do contour lines on a topographic map connect?

A areas with similar climates

B highest and lowest points

C points of equal elevation

D regions under the same rule

10 If Earth science students want to find a sandstone rock layer beneath the surface layer on which they stand, which map should they study?

A geologic

B physical

C relief

D road

Constructed Response

Use the diagram below to answer questions 11 and 12.

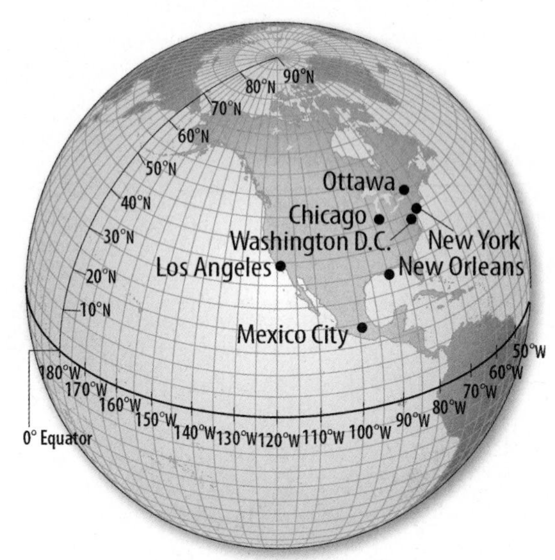

11 In the diagram above, what are the coordinates of Mexico City and New Orleans to the nearest 10 degrees? What is the difference in latitude between the two cities?

12 What are the coordinates of New York and Los Angeles to the nearest 5 degrees? What is the difference in longitude between the two cities?

13 How do modern technologies, such as remote sensing and global positioning systems, help mapmakers?

14 Why do hikers often use topographic maps? Describe map symbols, colors, and features that hikers might find helpful.

NEED EXTRA HELP?														
If You Missed Question...	1	2	3	4	5	6	7	8	9	10	11	12	13	14
Go to Lesson...	1	2	2	1	1	2	2	1	2	2	1	1	2	2

Chapter 2

Earth's Structure

THE BIG IDEA

How is Earth structured?

Inquiry What is in the sky?

These dancing lights in the night sky are called an aurora. Interactions between Earth's atmosphere and charged particles from the Sun cause an aurora. Conditions deep in Earth's interior structure create a magnetic field that attracts the charged particles to Earth's North Pole and South Pole.

- How is Earth structured?
- How does Earth's core create Earth's magnetic field?

Get Ready to Read

What do you think?

Before you read, decide if you agree or disagree with each of these statements. As you read this chapter, see if you change your mind about any of the statements.

1. People have always known that Earth is round.
2. Earth's hydrosphere is made of hydrogen gas.
3. Earth's interior is made of distinct layers.
4. Scientists discovered that Earth's outer core is liquid by drilling deep wells.
5. All ocean floors are flat.
6. Most of Earth's surface is covered by water.

Your one-stop online resource
connectED.mcgraw-hill.com

LearnSmart®

Chapter Resources Files, Reading Essentials, Get Ready to Read, Quick Vocabulary

Animations, Videos, Interactive Tables

Self-checks, Quizzes, Tests

Project-Based Learning Activities

Lab Manuals, Safety Videos, Virtual Labs & Other Tools

Vocabulary, Multilingual eGlossary, Vocab eGames, Vocab eFlashcards

Personal Tutors

Lesson 1

Reading Guide

Key Concepts 
ESSENTIAL QUESTIONS

- What are Earth's major systems and how do they interact?
- Why does Earth have a spherical shape?

Vocabulary

sphere p. 41
geosphere p. 42
gravity p. 43
density p. 45

 Multilingual eGlossary

Science Video

Spherical Earth

PBL Go to the resource tab in ConnectED to find the PBL *Gravity Glue.*

Inquiry Why is Earth spherical?

This image of Earth was taken from space. Notice Earth's shape and the wispy clouds that surround part of the planet. What else do you notice about Earth?

Launch Lab

10 minutes

How can you model Earth's systems?

Earth has different systems made of water, solid materials, air, and life. Each system has unique characteristics.

1. Read and complete a lab safety form.
2. Set a **clear plastic container** on your table, and add **gravel** to a depth of about 2 cm.
3. Pour equal volumes of **corn syrup** and **colored water** into the container.
4. Observe the container for 2 minutes. Record your observations in your Science Journal.

Think About This

1. What happened to the materials?
2. **Key Concept** Which Earth system did each material represent?

Describing Earth

Imagine standing on a mountaintop. You can probably see that the land stretches out beneath you for miles. But you cannot see all of Earth–it is far too large. People have tried to determine the shape and size of Earth for centuries. They have done so by examining the parts they can see.

Many years ago, people believed that Earth was a flat disk with land in the center and water at the edges. Later they used clues to determine Earth's true shape, such as studying Earth's shadow on the Moon during an eclipse.

The Size and Shape of Earth

Now there are better ways to get a view of Earth–the largest of the four rocky planets closest to the Sun. Using satellites and other technology, scientists know that Earth is a sphere. *A* **sphere** *is shaped like a ball, with all points on the surface at an equal distance from the center.* But Earth is not a perfect sphere. As illustrated in **Figure 1,** Earth is somewhat flattened at the poles with a slight bulge around the equator. This means the diameter of Earth is larger around the equator than at the poles. Earth has an average diameter of almost 13,000 km.

Figure 1 The red dashes around the image of Earth illustrate a perfect sphere. The blue dashes represent Earth's axis. Earth is shaped like a sphere that is somewhat flattened.

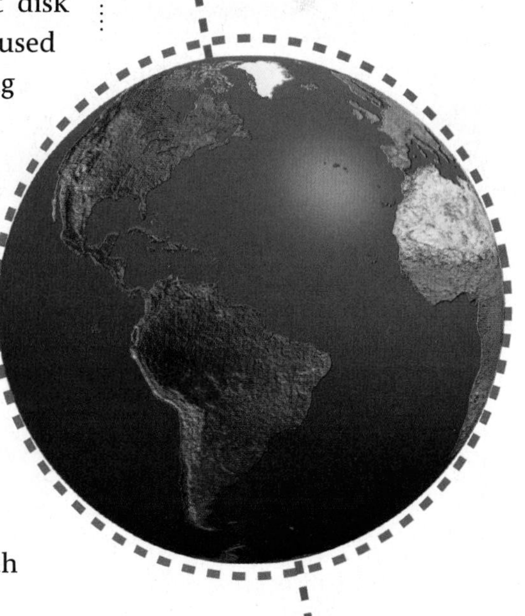

Hutchings Photography/Digital Light Source

Earth Systems

Earth is large and complex. To simplify the task of studying Earth, scientists describe Earth systems, as shown in **Figure 2.** All of these systems interact by exchanging matter and energy. For example, water from the ocean evaporates and enters the atmosphere. Later, the water precipitates onto land and washes salts into the ocean.

The Atmosphere, the Hydrosphere, and the Cryosphere The atmosphere is the layer of gases surrounding Earth. It is Earth's outermost system. This layer is hundreds of kilometers thick. It is a mixture of nitrogen, oxygen, carbon dioxide, and traces of other gases. The hydrosphere is water on Earth's surface, water underground, and liquid water in the atmosphere.

Most of the water in the hydrosphere is in salty oceans. Freshwater is in most rivers and lakes and underground. Frozen water, such as glaciers, is part of the cryosphere, also. Water continually moves between the atmosphere and the hydrosphere. This is one example of how Earth systems interact.

The Geosphere and the Biosphere *The* **geosphere** *is Earth's entire solid body.* It contains the thin layer of soil and sediments on Earth's surface, down to it's rocky center. It is the largest Earth system. The biosphere includes all living things on Earth. Organisms in the biosphere live within and interact with the atmosphere, hydrosphere, and even the geosphere.

Key Concept Check Identify Earth's major systems.

Earth's Systems

Figure 2 Earth's systems interact. A change in one Earth system affects all other Earth systems. They exchange energy and matter, making Earth suitable for life.

How did Earth form?

Earth formed about 4.6 billion years ago (bya), along with the Sun and the rest of our solar system. Materials from a large cloud of gas and dust came together, forming the Sun and all the planets. In order to understand how this happened, you first need to know how gravity works.

The Influence of Gravity

Gravity *is the force that every object exerts on all other objects because of their masses.* The force of gravity between two objects depends on the objects' masses and the distance between them. The more mass either object has, or the closer together they are, the stronger the gravitational force. You can see an example of this in **Figure 3.**

FOLDABLES®

Make a half book from a sheet of paper. Label it as shown. Use it to organize your notes about Earth's formation.

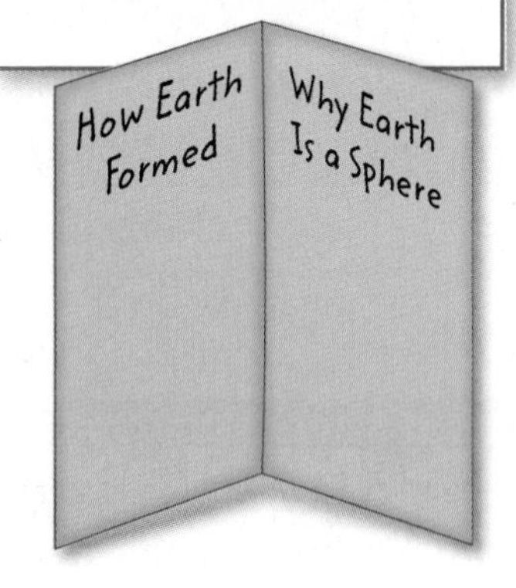

Force of Gravity

The two objects in row A are the same distance apart as the two objects in row B. One of the objects in row B has more mass, creating a stronger gravitational force between the two objects in row B.

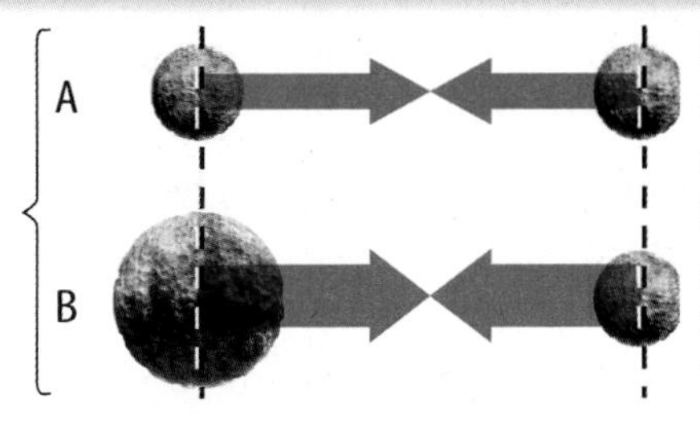

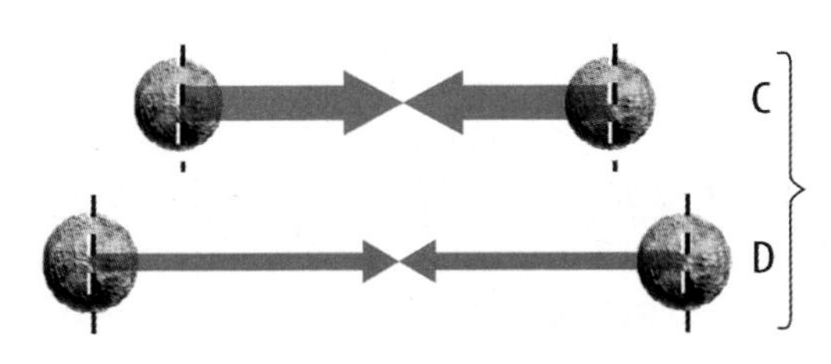

All four objects have the same mass. The two objects in row C are closer to each other than the two objects in row D and, therefore, have a stronger gravitational force between them.

Figure 3 Mass and distance affect the strength of the gravitational force between objects. This strength is represented by the thickness of the arrows. The thicker the arrows, the greater the force between the two objects.

Visual Check Why does Earth exert a greater gravitational force on you than other objects do?

The force of gravity is strongest between the objects in row B. Even though the objects in row A are the same distance apart as those in row B, the force of gravity between them is weaker because they have less mass. The force of gravity is weakest between the objects in row D.

Reading Check What factors affect the strength of the gravitational force between objects?

As illustrated in **Figure 4,** all objects on or near Earth are pulled toward Earth's center by gravity. Earth's gravity holds us on Earth's surface. Since Earth has more mass than any object near you, it exerts a greater gravitational force on you than other objects do. You don't notice the gravitational force between less massive objects.

Figure 4 Earth's gravity pulls objects toward the center of Earth.

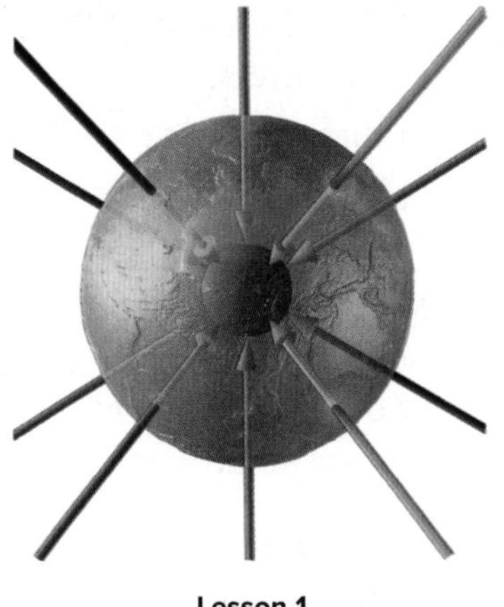

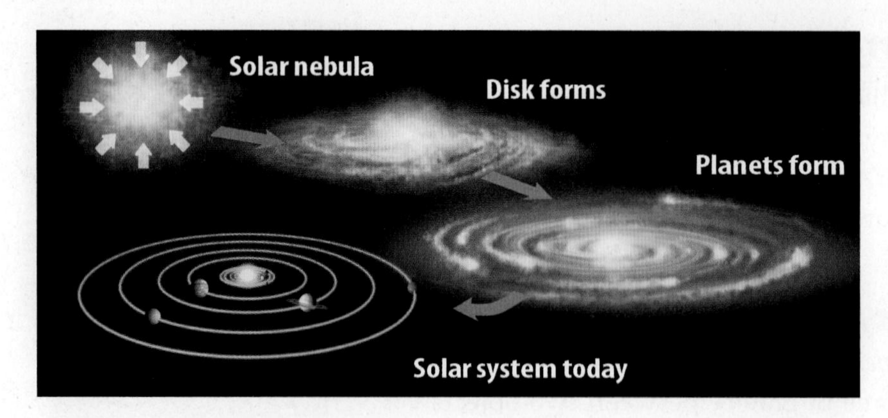

Figure 5 Gravity helped change a cloud of dust, gas, and ice, called a nebula, into our solar system. The Sun formed first, and the planets formed from the swirling disk of particles that remained.

Visual Check Our solar system formed from what type of cloud?

MiniLab — 15 minutes

Which materials will sink?

You can investigate the density of a material by comparing it to the density of water.

1. Read and complete a lab safety form.
2. Add water to a **clear, glass bowl** until it is about three-quarters full.
3. Hold a piece of **balsa wood** just under the surface of the water, then release it. Record your observations in your Science Journal. Remove the wood from the bowl.
4. Repeat step 2, using a piece of **granite, pumice,** and then **ironwood.**

Analyze and Conclude

1. **Summarize** Which materials sank? Which materials floated? Hypothesize why this happened.
2. **Key Concept** Use the concept of density to infer why the hydrosphere is above the geosphere but below the atmosphere.

The Solar Nebula

The force of gravity played a major role in the formation of our solar system. As shown in **Figure 5,** the solar system formed from a cloud of gas, ice, and dust called a nebula. Gravity pulled the materials closer together. The nebula shrank and flattened into a disk. The disk began to rotate. The materials in the center of the disk became denser, forming a star–the Sun.

Next, the planets began to take shape from the remaining bits of material. Earth formed as gravity pulled these small particles together. As they collided, they stuck to each other and formed larger, unevenly shaped objects. These larger objects had more mass and attracted more particles. Eventually enough matter collected and formed Earth. But how did the unevenly shaped, young planet become spherical?

Early Earth

Eventually the newly formed Earth grew massive and generated thermal energy, commonly called heat, in its interior. The rocks of the planet softened and began to flow.

Gravity pulled in the irregular bumps on the surface of the newly formed planet. As a result, Earth developed a relatively even spherical surface.

Key Concept Check How did Earth develop its spherical shape?

The Formation of Earth's Layers

Thermal energy from Earth's interior affected Earth in other ways, as well. Before heating up, Earth was a mixture of solid particles. The thermal energy melted some of this material and it began to flow. As it flowed, Earth developed distinct layers of different materials.

The different materials formed layers according to their densities. **Density** *is the amount of mass in a material per unit volume.* Density can be described mathematically as

$$D = m/V$$

where D is the density of the material, m is the material's mass, and V is its volume. If two materials have the same volume, the denser material will have more mass.

Reading Check Can a small object have more mass than a larger object? Explain your answer.

There is a stronger gravitational force between Earth and a denser object than there is between Earth and a less dense object. You can see this if you put an iron block and a pinewood block with the same volumes in a pan of water. The wooden block, which is less dense than water, will float on the water's surface. The iron block, which is denser than water, will be pulled through the water to the bottom of the pan.

When ancient Earth started melting, something much like this happened. The densest materials sank and formed the innermost layer of Earth. The least dense materials stayed at the surface, and formed a separate layer. The materials with intermediate densities formed layers in between the top layer and the bottom layer. Earth's three major layers are shown in **Figure 6**.

Math Skills

Solve One-Step Equations

Comparing the masses of substances is useful only if the same volume of each substance is used. To calculate density, divide the mass by the volume. The unit for density is a unit of mass, such as g, divided by a unit of volume, such as cm^3. For example, an aluminum cube has a mass of 27 g and a volume of 10 cm^3. The density of aluminum is 27 g / 10 cm^3 = 2.7 g/cm^3.

Practice

A chunk of gold with a volume of 5.00 cm^3 has a mass of 96.5 g. What is the density of gold?

Math Practice

Personal Tutor

WORD ORIGIN

density
from Latin *densus*, means "thick, crowded"

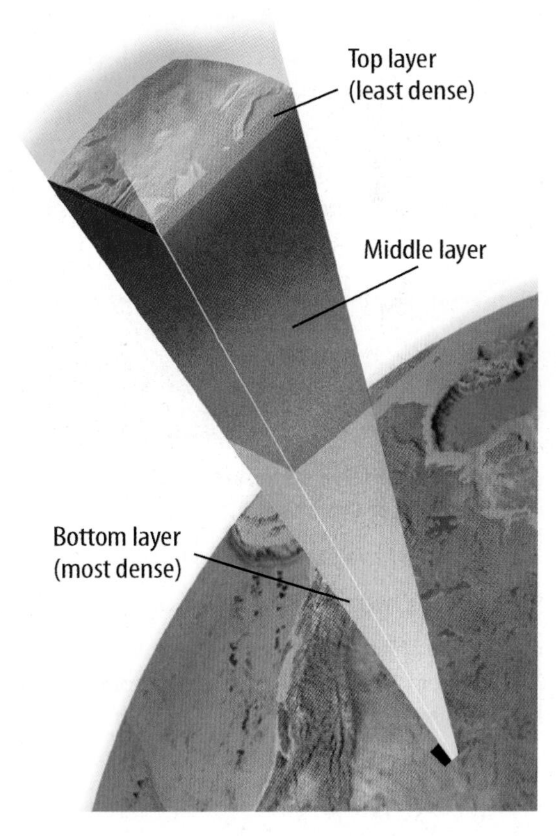

Figure 6 Earth's geosphere is divided into three major layers.

Lesson 1 Review

Visual Summary

Earth's systems, including the atmosphere, hydrosphere, cryosphere, biosphere, and geosphere, interact with one another.

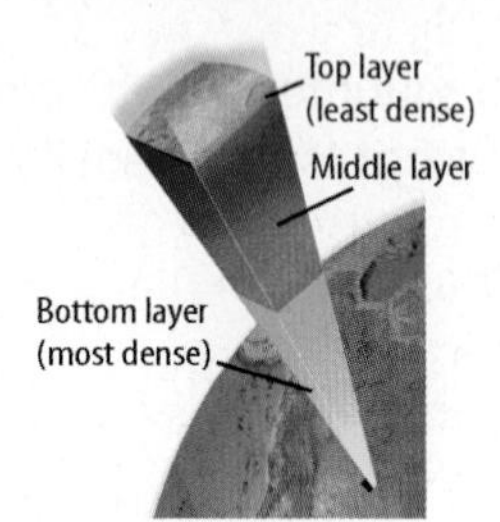

The geosphere is the solid body of Earth.

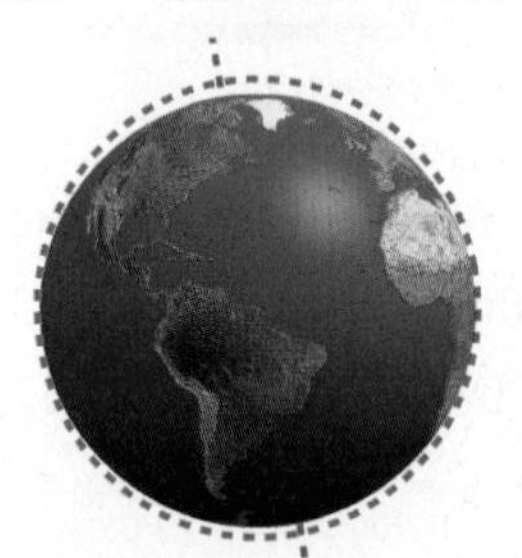

The solar system, including Earth, formed about 4.6 bya. Gravity caused particles to come together and formed a spherical Earth.

FOLDABLES®

Use your lesson Foldable to review the lesson. Save your Foldable for the project at the end of the chapter.

What do you think NOW?

You first read the statements below at the beginning of the chapter.

1. People have always known that Earth is round.
2. Earth's hydrosphere is made of hydrogen gas.

Did you change your mind about whether you agree or disagree with the statements? Rewrite any false statements to make them true.

Use Vocabulary

1. The Earth system made mainly of surface water is called the __________.
2. **Use the term** *density* in a sentence.

Understand Key Concepts 

3. Which is part of the atmosphere?
   - **A.** a rock
   - **B.** a tree
   - **C.** oxygen gas
   - **D.** the ocean
4. **Describe** how gravity affected Earth's shape during Earth's formation.

Interpret Graphics

5. **Organize** Copy and complete the graphic organizer below to show each of Earth's systems.

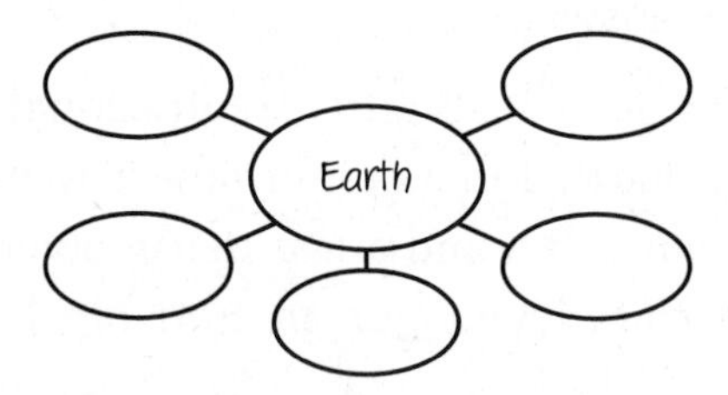

Critical Thinking

6. **Combine** your understanding of how Earth became spherical and observations of the Moon. Then form a hypothesis about the formation of the Moon.
7. **Explain** As the newly formed Earth became spherical, did it grow larger or become smaller? Explain your answer.

Math Skills

Math Practice

8. At a given temperature, 3.00 m^3 of carbon dioxide has a mass of 5.94 kg. What is the density of carbon dioxide at this temperature?

Gary Vestal/Getty Images

CAREERS in SCIENCE

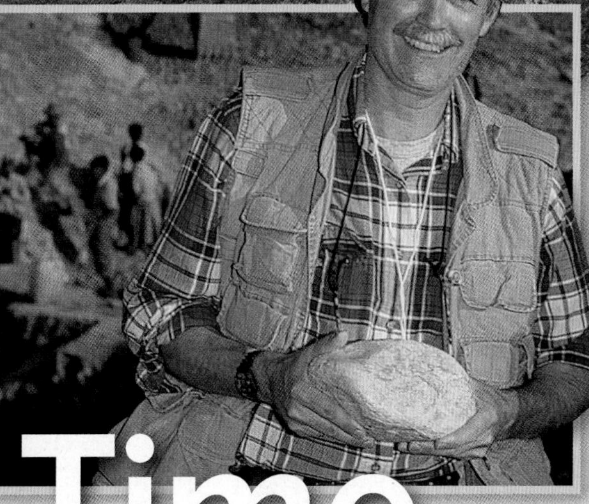

◀ George Harlow studies diamonds to learn more about Earth's interior.

Time Capsules

Formed billions of years ago in Earth's mantle, diamonds hold important clues about our planet's mysterious interior.

George Harlow is fascinated by diamonds. Not because of their dazzling shine or their value, but because of what they can reveal about Earth. He considers diamonds to be tiny time capsules that capture a picture of the ancient mantle, where they became crystals.

Most diamonds we find today formed billions of years ago deep within Earth's mantle, over 161 km below Earth's surface. Tiny bits of mantle, called inclusions, were trapped inside these extremely hard crystals as they formed. Millions of years later, the inclusions' diamond cases still protect them.

Harlow collects these diamonds from places such as Australia, Africa, and Thailand. Back in the lab, Harlow and his colleagues remove inclusions from diamonds. First, they break open a diamond with a tool similar to a nutcracker. Then they use a microscope and a pinlike tool to sift through the diamond rubble. They look for an inclusion, which is about the size of a grain of sand. When they find one, they use an electron microprobe and a laser to analyze the inclusion's composition, or chemical makeup. The sample might be tiny, but it's enough for scientists to learn the temperature, pressure, and composition of the mantle in which the diamond formed.

Next time you see a diamond, you might wonder if it too has a tiny bit of ancient mantle from deep inside Earth.

Going Up?

Diamond crystals form deep within the mantle under intense pressures and temperatures. They come to Earth's surface in molten rock, or magma. The magma pulls diamonds from rock deep underground and rapidly carries them to the surface. The magma erupts onto Earth's surface in small, explosive volcanoes. Diamonds and other crystals and rocks from the mantle are in deep, carrot-shaped cones called kimberlite pipes that are part of these rare volcanoes.

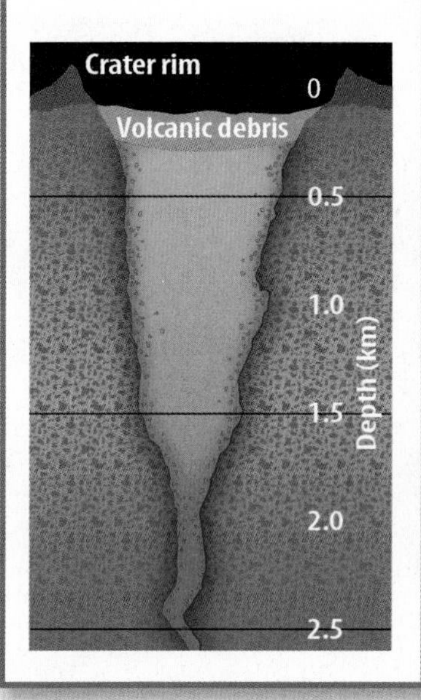

It's Your Turn

RESEARCH Diamonds are the world's most popular gemstone. What other uses do diamonds have? Research the properties of diamonds and how they are used in industry. Report your findings to your class.

Lesson 2

Reading Guide

Key Concepts
ESSENTIAL QUESTIONS

- What are the interior layers of Earth?
- What evidence indicates that Earth has a solid inner core and a liquid outer core?

Vocabulary
crust p. 51
mantle p. 52
lithosphere p. 52
asthenosphere p. 52
core p. 54
magnetosphere p. 55

Multilingual eGlossary

Science Video

Earth's Interior

Inquiry What is inside Earth?

Earth is thousands of kilometers thick. The deepest caves, mines, and wells in the world barely scratch Earth's surface. How do you think scientists learn about Earth's interior?

Stephen Alvarez/National Geographic/Getty Images

Launch Lab

10 minutes

How can you model Earth's layers?

Earth is made of three main layers: the thin outer crust, the thick mantle, and the central core. You can use different objects to model these layers.

1. Read and complete a lab safety form.
2. Place a **hard-cooked egg** on a **paper towel.** Use a **magnifying lens** to closely examine the surface of the egg. Is its shell smooth or rough? Record your observations in your Science Journal.
3. Carefully peel away the shell from the egg.
4. Use the **plastic knife** to cut the egg in half. Observe the characteristics of the shell, the egg white, and the yolk.
5. Make a drawing of the egg's layers in your Science Journal. Which layers could represent layers of Earth? Label the layers as *crust, mantle,* or *core*.

Think About This

1. What other objects could be used to model Earth's layers?
2. **Key Concept** Explain why a hard-cooked egg is a good model for Earth's layers.

Clues to Earth's Interior

Were you ever given a gift and had to wait to open it? Maybe you tried to figure out what was inside by tapping on it or shaking it. Using methods such as these, you might have been able to determine the gift's contents. Scientists can't see what is inside Earth, either. But they can use indirect methods to discover what Earth's interior is like.

What's below Earth's surface?

Deep mines and wells give scientists hints about Earth's interior. The deepest mine ever constructed is a gold mine in South Africa. It is nearly 4 km deep. People can go down the mine to explore the geosphere.

Drilled wells are even deeper. The deepest well is on the Kola Peninsula in Russia. It is more than 12 km deep. Drilling to such great depths is extremely difficult–it took more than 20 years to drill the Kola well. Even though people cannot go down in the well, they can send instruments down to make **observations** and bring samples to the surface. What have scientists learned about Earth's interior by studying mines and wells like the two mentioned above?

REVIEW VOCABULARY

observation
an act of recognizing and noting a fact or an occurrence

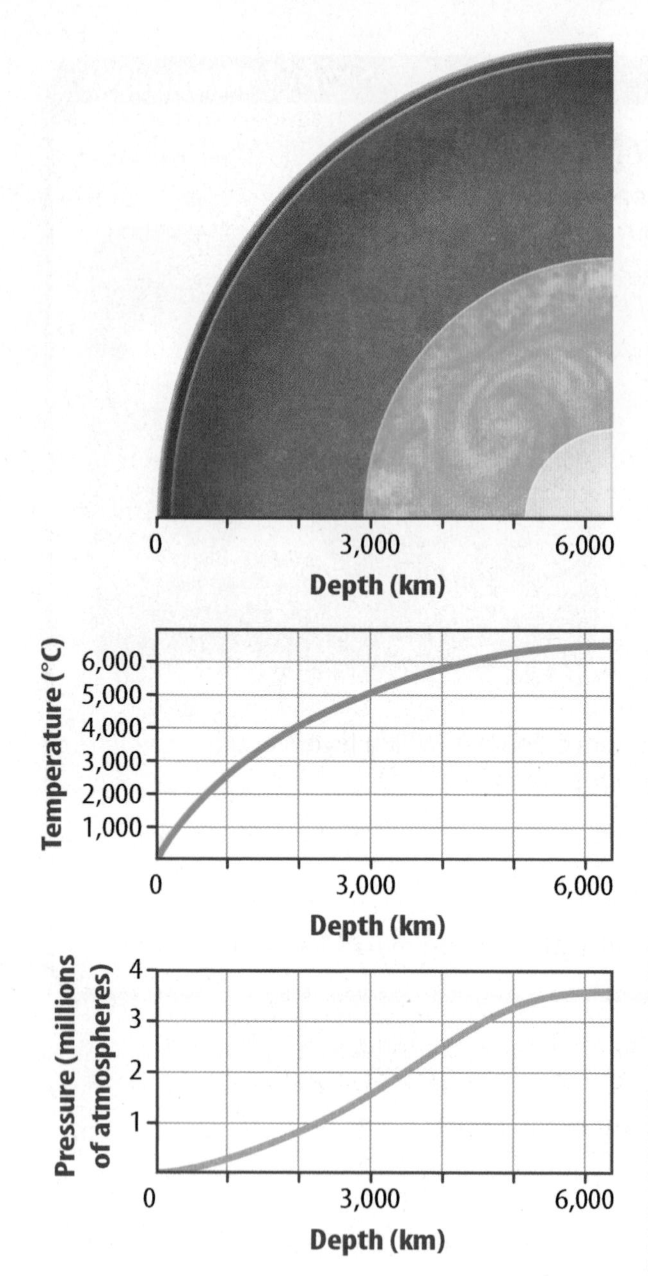

Figure 7 Temperature and pressure increase with depth in the geosphere.

Temperature and Pressure Increase with Depth

One thing that workers notice in deep mines or wells is that it is hot inside Earth. In the South African gold mines, 3.9 km below Earth's surface, the temperature is about 53°C (127°F). The temperature at the bottom of the Kola well is 190°C (374°F). That's hot enough to bake cookies! No one has ever recorded the temperature of Earth's center, but it is estimated to be about 6,000°C (10,832°F). As shown in **Figure 7**, temperature within Earth increases with increasing depth.

Not only does temperature increase, but pressure also increases as depth increases inside Earth. This is due to the weight of the overlying rocks. The high pressure squeezes the rocks and makes them much denser than surface rocks.

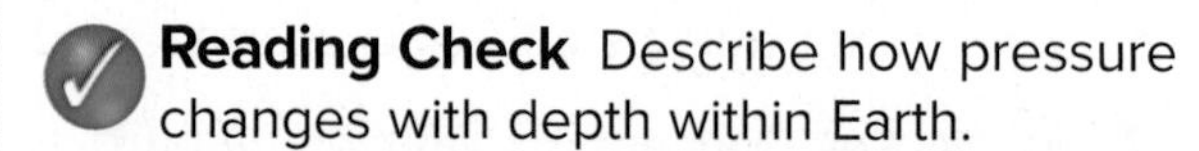

Reading Check Describe how pressure changes with depth within Earth.

High temperatures and pressures make it difficult to drill deep wells. The depth of the Kola well is less than 1 percent of the distance to Earth's center. Therefore, only a small part of the geosphere has been sampled. How can scientists learn about what is below the deepest wells?

Using Earthquake Waves

As you read earlier, scientists use indirect methods to study Earth's interior. They get most of their evidence by analyzing earthquake waves. Earthquakes release energy in the form of three types of waves. As these waves move through Earth, they are affected by the different materials they travel through. Some waves cannot travel through certain materials. Other waves change direction when they reach certain materials. By studying how the waves move, scientists are able to infer the density and composition of materials within Earth.

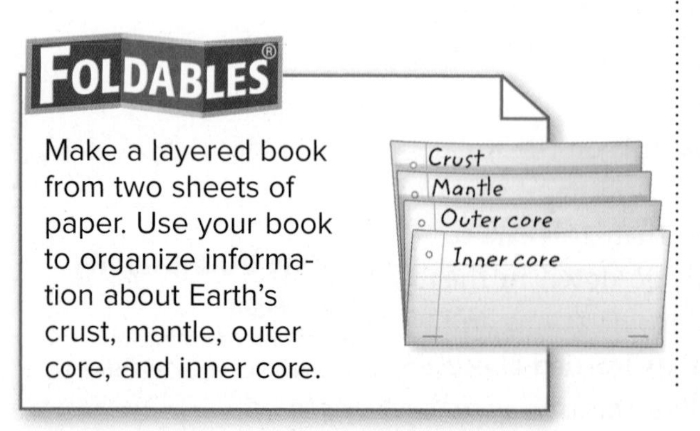

Make a layered book from two sheets of paper. Use your book to organize information about Earth's crust, mantle, outer core, and inner core.

Earth's Layers

Recall that differences in density resulted in materials within Earth forming layers. Each layer has a different composition, with the densest materials in the center of Earth and the least dense materials at the surface.

Crust

The brittle, rocky outer layer of Earth is called the **crust.** It is much thinner than the other layers, like the shell on a hard-cooked egg. It is the least dense layer of the geosphere. It is made mostly of elements of low mass, such as silicon and oxygen.

Crustal rocks are under oceans and on land. The crust under oceans is called oceanic crust. It is made of dense rocks that contain iron and magnesium. The crust on land is called continental crust. It is less dense and about four times thicker than oceanic crust. Continental crust is thickest under tall mountains. **Figure 8** shows a comparison of the two types of crust.

There is a distinct boundary between the crust and the layer beneath it. When earthquake waves cross this boundary, they speed up. This indicates that the layer beneath the crust, called the mantle, is denser than the crust.

Reading Check How does oceanic crust differ from continental crust?

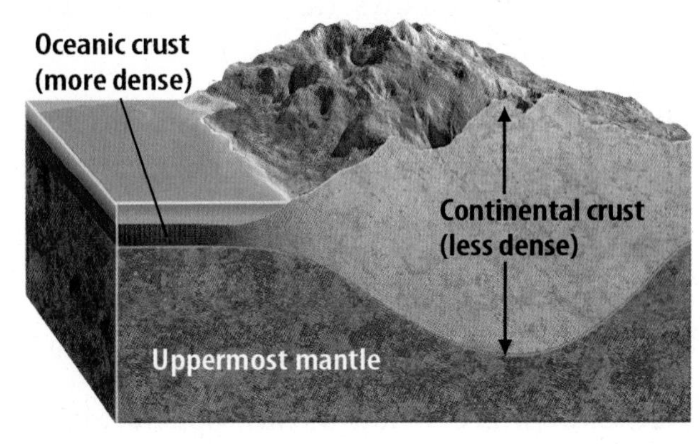

Personal Tutor

Figure 8 Oceanic crust is thin and dense compared to continental crust.

MiniLab

Which liquid is densest?

Earth's layers were determined by density. The iron in the inner core makes up Earth's densest layer. The silicon and oxygen in Earth's crust are much less dense.

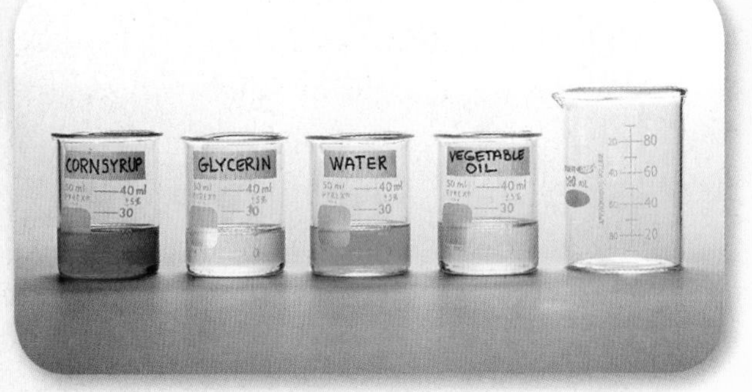

1. Read and complete a lab safety form.
2. Pour 50 mL of **corn syrup** into a **100-mL beaker.** Label the beaker.
3. Fill the remaining three beakers with 50 mL of **glycerin, water,** and **vegetable oil,** respectively. Label them.
4. Stir a few drops of **blue food coloring** into the water using a **spoon.**
5. Rinse the spoon. Then stir a few drops of **red food coloring** into the corn syrup.
6. Pour the corn syrup into a **250-mL beaker.**
7. Use a **funnel** to gently pour the glycerin on top of the corn syrup. Hold the funnel along the side of the beaker.
8. Repeat step 7 using the vegetable oil, then the water.

Analyze and Conclude

1. **Describe** what happened to the liquids. Why did this occur?
2. **Key Concept** How are the layers of liquid in the beaker similar to Earth's layers?

Mantle

Earth's mantle is immediately below the crust. *The* **mantle** *is the thick middle layer in the solid part of Earth.* It contains more iron and magnesium than oceanic crust does. This makes it denser than either type of crust. Like the crust, the mantle is made of rock. Scientists group the mantle into layers according to the way rocks react when forces push or pull on them. **Figure 9** shows the mantle and the other layers of Earth.

Uppermost Mantle The rocks in the uppermost layer of the mantle are brittle and rigid, like the rocks in the crust. Because of this, *scientists group the crust and the uppermost mantle into a rigid layer called the* **lithosphere** (LIH thuh sfihr).

Asthenosphere Below the lithosphere, rocks in the upper mantle are so hot that, while they are solid, the rocks are no longer brittle. Instead, they begin to flow. Scientists use the term *plastic* to describe rocks that flow in this way. *This plastic layer within the mantle is called the* **asthenosphere** (as THEN uh sfihr).

Reading Check Compare the lithosphere and the asthenosphere.

The asthenosphere does not resemble the plastics used to make everyday products. The word *plastic* refers to materials that are soft enough to flow. The asthenosphere flows very slowly. Even if it were possible to visit the mantle, you could never see this flow. Rocks in the asthenosphere move about as slowly as your fingernails grow.

Word Origin

asthenosphere
from Greek *asthenes,* means "weak"; and *spharia,* means "sphere"

Upper Mantle and Lower Mantle The rock in the upper and lower mantle below the asthenosphere is hotter than the rock in the asthenosphere, yet it does not flow. How can this be? The pressure at this depth is so great that no melting occurs. While increased temperature tends to melt rock, high pressure tends to prevent melting. High pressure squeezes the rock into a solid. The solid rock of the upper mantle and the lower mantle forms the largest layer of Earth.

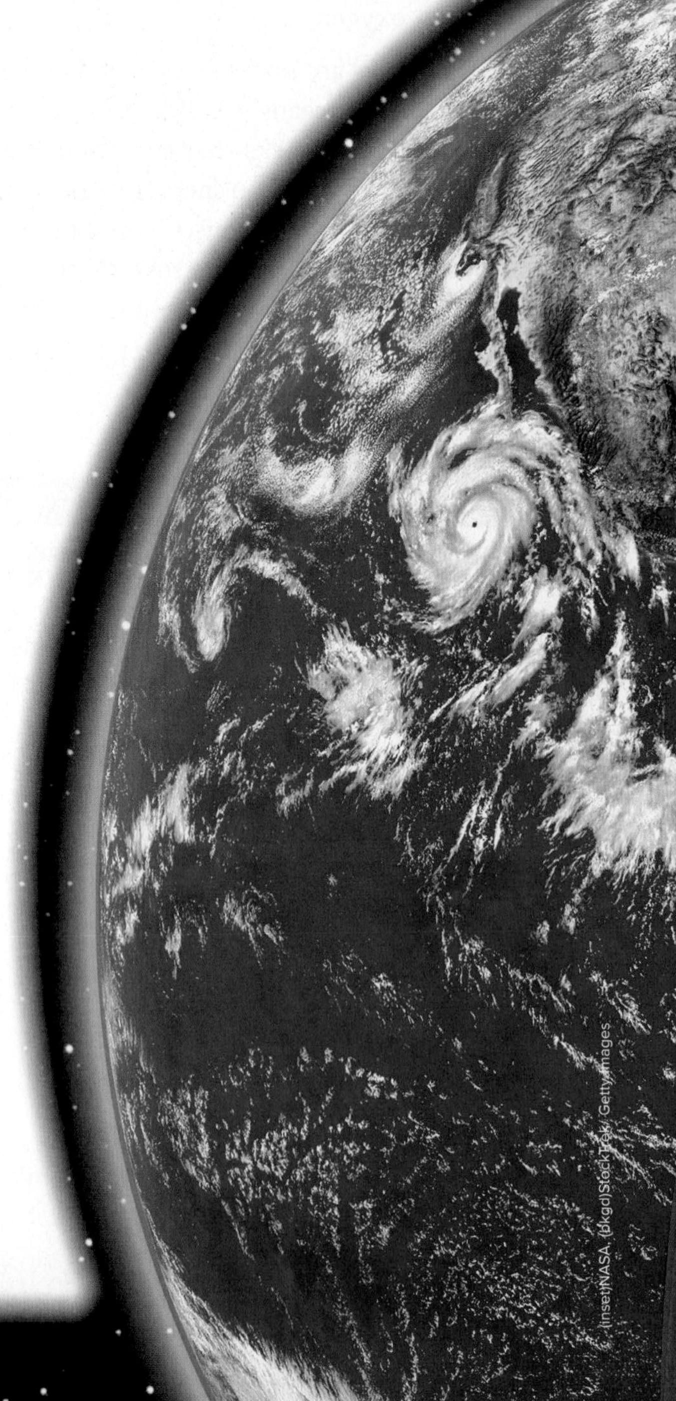

Animation

Figure 9 Earth's main layers include the crust, mantle, and core. The layers are subdivided according to chemical and physical characteristics.

Visual Check Which of Earth's layers is the most dense?

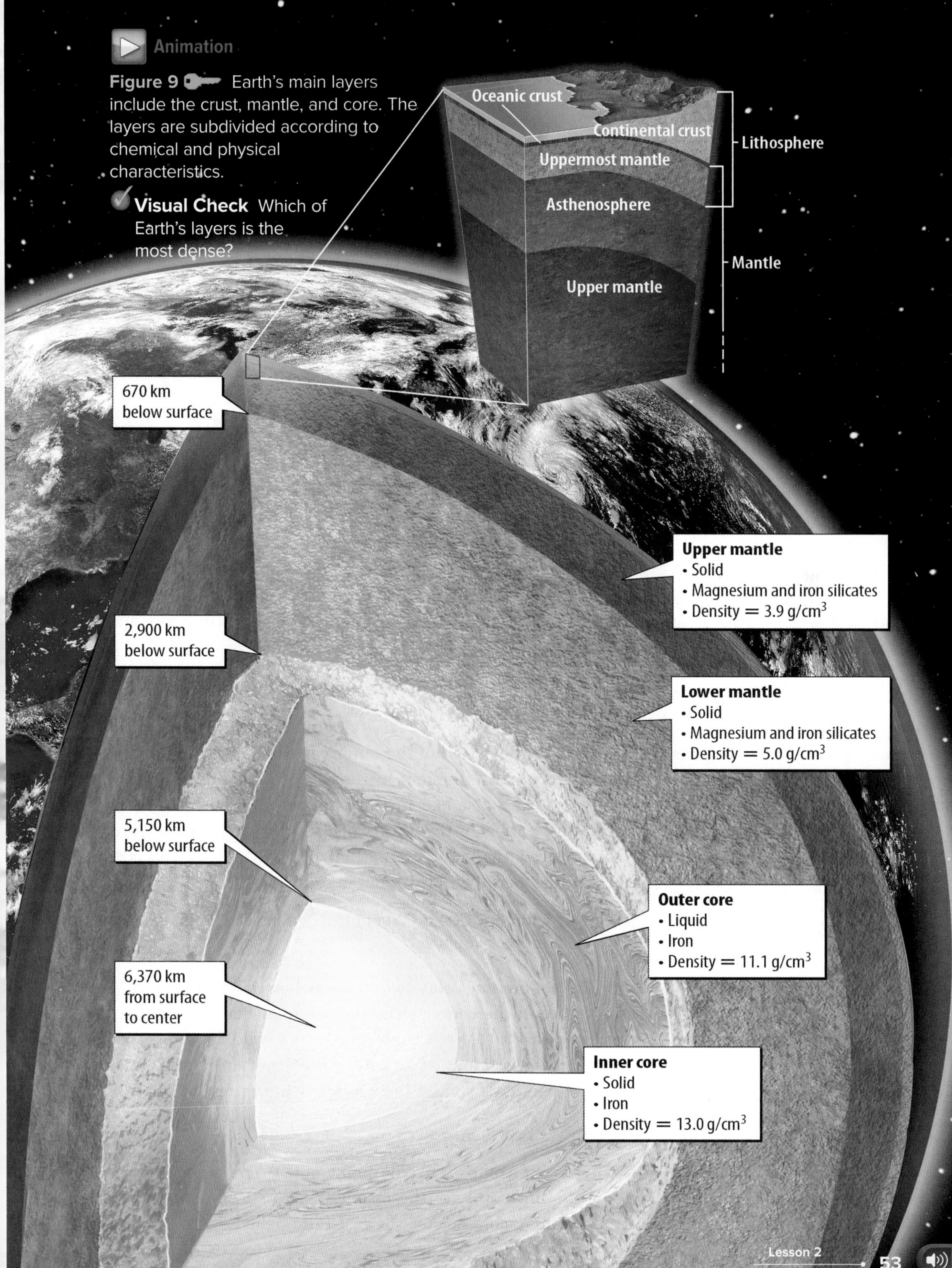

Earth's Core

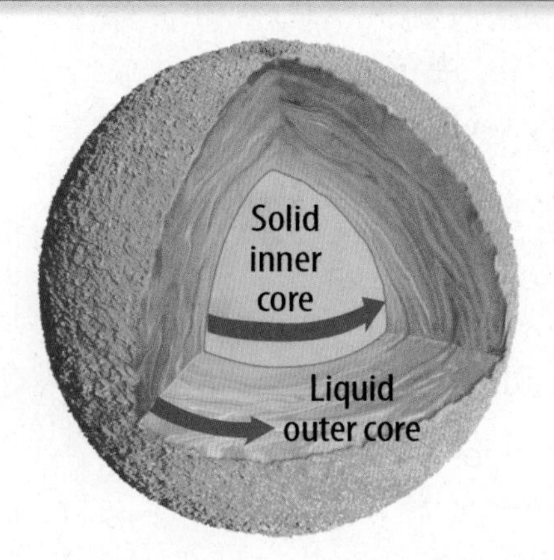

Figure 10 Earth's core has a liquid outer layer surrounding a solid inner layer of iron. The inner core spins a little faster than the outer core.

Visual Check How do the arrows in this figure indicate that the inner core spins faster than the outer core?

Core

The dense metallic center of Earth is the **core,** as shown in **Figure 10.** If you imagine Earth as a hard-cooked egg, the core is the yolk. Earth's crust and mantle are made of rock. Why is the core made of metal? Recall that in Earth's early history, the planet was much hotter than it is now. Earth materials flowed, like they do in the asthenosphere today. Scientists don't know how much of Earth melted. But they do know that it was soft enough for gravity to pull the densest material down to the center. This dense material is metal. It is mostly iron with small amounts of **nickel** and other elements. The core has a liquid outer core and a solid inner core.

SCIENCE USE V. COMMON USE
nickel
Science Use a specific type of metal
Common Use a coin worth five cents

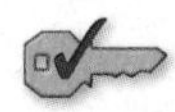 **Key Concept Check** What are the interior layers of Earth?

Outer Core If pressure is great enough to keep the lower mantle in a solid state, how can the outer core be liquid? The mantle and core are made of different materials, and have different melting temperatures. Just like in the asthenosphere, the effects of temperature outweigh the effects of pressure in the outer core. Scientists used the indirect method of analyzing earthquake waves and learned that Earth's outer core is liquid.

 **Key Concept Check** What evidence indicates that the outer core is liquid?

Inner Core The inner core is a dense ball of solid iron crystals. The pressure in the center of Earth is so high that even at temperatures of about 6,000°C, the iron is in a solid state. Because the outer core is liquid, it is not rigidly attached to the inner core. The inner core spins a little faster than the rest of Earth.

Earth's Core and Geomagnetism

Why does a compass needle point north? The metallic compass needle lines up with the magnetic field surrounding Earth. Earth's spinning core creates the magnetic field.

Earth's Magnetic Field

Recall that Earth's inner core spins faster than the outer core. This produces streams of flowing, molten iron in the outer core. Earth's magnetic field is a region of magnetism produced in part by the flow of molten materials in the outer core. The magnetic field acts much like a giant bar magnet. It has opposite poles, as shown in **Figure 11.**

For centuries, people have used compasses and Earth's magnetic field to navigate. But, the magnetic field is not completely stable. Over geologic time, its strength and direction vary. At several times in Earth's history, the direction has even reversed.

Magnetosphere

Earth's magnetic field protects Earth from cosmic rays and charged particles coming from the Sun. It pushes away some charged particles and traps others. *The outer part of the magnetic field that interacts with these particles is called the* **magnetosphere.** Examine **Figure 12** to see how the shape of the magnetosphere is produced by the flow of these charged particles.

Figure 11 Earth's magnetic field is produced by the movement of molten materials in the outer core.

WORD ORIGIN

magnetosphere
from Latin *magnes*, means "lodestone"; and *spharia*, means "sphere"

The Magnetosphere

Figure 12 Trapped particles and Earth's magnetic field form a shield around Earth.

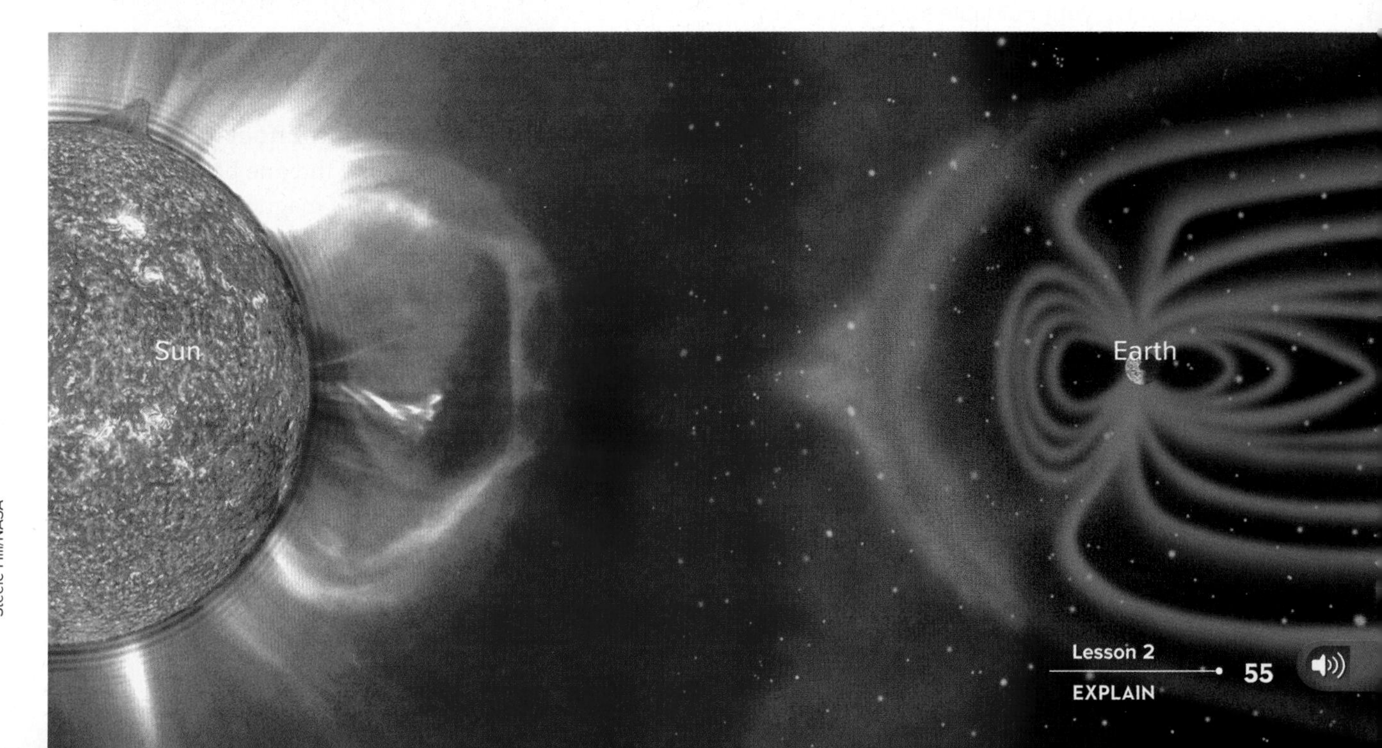

Steele Hill/NASA

Lesson 2 Review

Visual Summary

Earth's layers include the crust, mantle, and core. Oceanic crust is under oceans. The continents are made of continental crust.

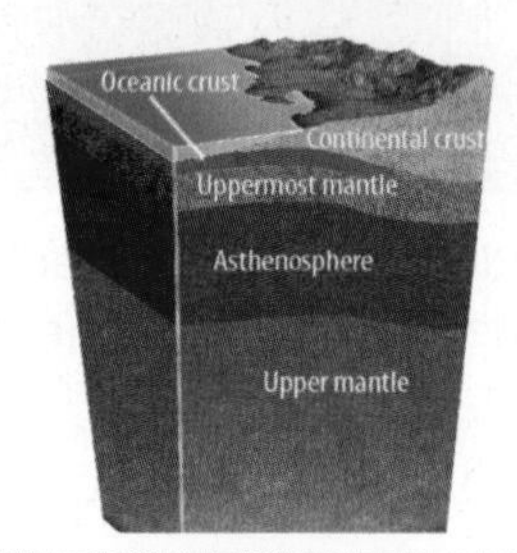

The mantle is Earth's thickest layer. It includes part of the lithosphere and the asthenosphere.

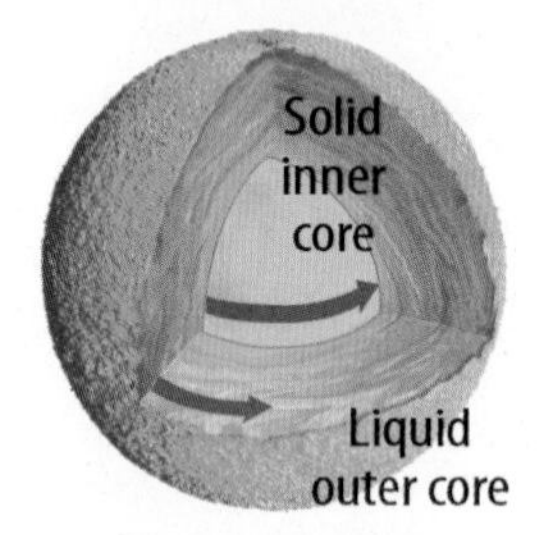

Earth's core has a liquid outer core and a solid inner core.

FOLDABLES®

Use your lesson Foldable to review the lesson. Save your Foldable for the project at the end of the chapter.

What do you think NOW?

You first read the statements below at the beginning of the chapter.

3. Earth's interior is made of distinct layers.

4. Scientists discovered that Earth's outer core is liquid by drilling deep wells.

Did you change your mind about whether you agree or disagree with the statements? Rewrite any false statements to make them true.

Use Vocabulary

1. The layer of Earth made of metal is the ________.
2. **Distinguish** between the crust and the lithosphere.
3. **Use the terms** *core* and *mantle* in a complete sentence.

Understand Key Concepts 

4. Which of Earth's layers is made of melted materials?
   - **A.** the crust
   - **B.** the inner core
   - **C.** the lithosphere
   - **D.** the outer core
5. **Design** a model of Earth's layers. List the materials needed to make your model.
6. **Classify** Earth's layers based on their physical properties.

Interpret Graphics

7. **Identify** and compare the two types of crust shown below.

8. **Determine Cause and Effect** Draw a graphic organizer like the one below and list two facts about Earth's magnetic field.

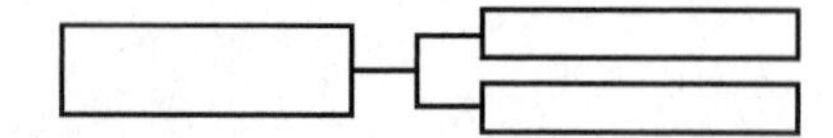

Critical Thinking

9. **Evaluate** Earthquakes produce waves that help scientists study Earth's interior, but earthquake waves can also cause damage. Discuss whether earthquake waves are good or bad.

How can you find the density of a liquid?

Earth's interior is made of solids and liquids that have different densities. You can measure volume and mass and then calculate density using the equation:

$$\text{Density} = \frac{\text{Mass}}{\text{Volume}}$$

Materials

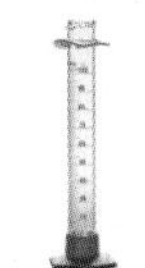
beaker

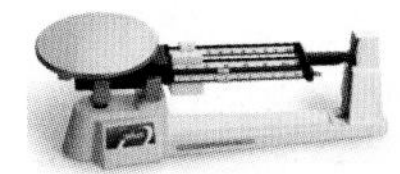
balance

graduated cylinder

vegetable oil

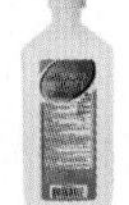
corn syrup

isopropyl alcohol

Safety

Learn It

Scientists **measure** to learn how much they have of a particular type of matter. Recall that matter is anything that has mass and volume. You can measure mass using a triple-beam balance. The unit of mass you will use most often is the gram (g). You can measure liquid volume using a graduated cylinder. Milliliter (mL) is the unit of volume you will use most often.

Try It

1. Read and complete a lab safety form.
2. Measure the mass of a 50-mL graduated cylinder. Record the mass in your Science Journal.
3. Pour about 15 mL of alcohol into a clean beaker.
4. Slowly pour the alcohol into the graduated cylinder until the alcohol measures 10 mL.
5. Measure and record the mass of the alcohol and the graduated cylinder.
6. In your Science Journal, subtract the mass recorded in step 2 from the mass recorded in step 5.
7. Empty and clean the graduated cylinder as instructed by your teacher.
8. Repeat steps 3–7 using the corn syrup and then the vegetable oil.

Apply It

9. Calculate and record the density of each liquid using your mass and volume measurements and the equation shown above.
10. Which fluid has the greatest density? Which has the least? Explain your answer.
11. 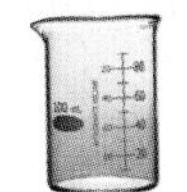 **Key Concept** Based on what you have learned, describe the relative density of Earth's layers.

4

(t to b, 2-3, 5-6, br)Hutchings Photography/Digital Light Source; (4)Jacques Cornell/McGraw-Hill Education

Lesson 3

Earth's Surface

Reading Guide

Key Concepts
ESSENTIAL QUESTIONS

- What are Earth's major landforms and how do they compare?
- What are the major landform regions of the United States?

Vocabulary
landform p. 60
plain p. 62
plateau p. 63
mountain p. 63

Multilingual eGlossary

BrainPOP®

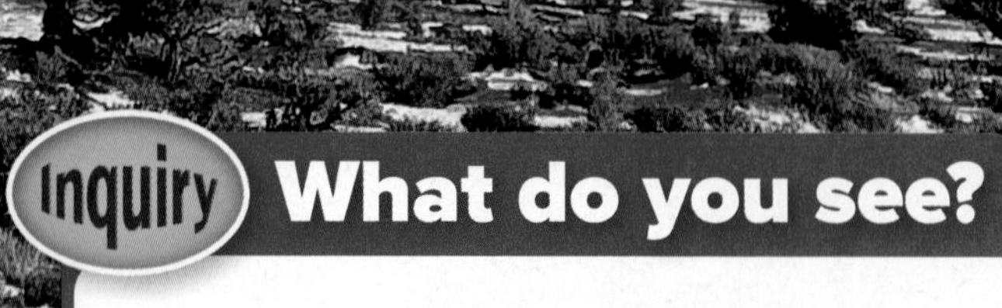

Inquiry What do you see?

Some features on Earth's surface are flat and low. Other features are steep and high. What else is different about these features?

Launch Lab

15 minutes

How can you measure topographic relief?

Relief describes differences in elevation for a given area. The area might have tall mountains or deep valleys. In this lab, you will use simple materials to measure relief on a model landscape.

1. Read and complete a lab safety form.
2. Form some **salt dough** into a thick disk slightly larger than your hand.
3. With your fingers spread apart, press your hand firmly into the dough so that some of the dough squeezes up between your fingers.
4. Stretch **dental floss** across the finger impressions in the dough. Slice off a section of the dough model by pressing the dental floss down through the dough.
5. Also make a slice through the palm section of your dough model.
6. Observe the profiles of your two cross sections. Use a **ruler** to measure the difference between the highest and lowest points within the palm section.
7. Measure the difference between the highest and lowest points within the fingers section.

Think About This

1. What is the difference in elevation between the highest and lowest points of your hand print?
2. **Key Concept** Compare and contrast your model features. How are they similar to features on Earth?

Oceans and Continents

Earth's surface is made of oceans and continents. Oceans cover more than 70 percent of Earth's surface. The surface of the ocean is relatively smooth. But what is below the water's surface? Imagine that you can explore the ocean floor as easily as you travel on dry land. What do you think you would see there?

Many of the features that appear on dry land, such as mountains, valleys, and canyons, also appear on the ocean floor. For example, the longest mountain ranges on Earth are near the centers of the oceans. Monterey Canyon, illustrated in **Figure 13**, is a submarine canyon which is comparable in size to the Grand Canyon on land.

Figure 13 From its rim to the canyon floor, Monterey Canyon reaches a maximum depth of about 1,920 m, making it slightly deeper than the Grand Canyon.

Santa Cruz Mountains
Monterey Bay
Monterey Canyon

Common Landforms

Figure 14 Earth's common landforms are characterized by size, shape, slope, elevation, and relief.

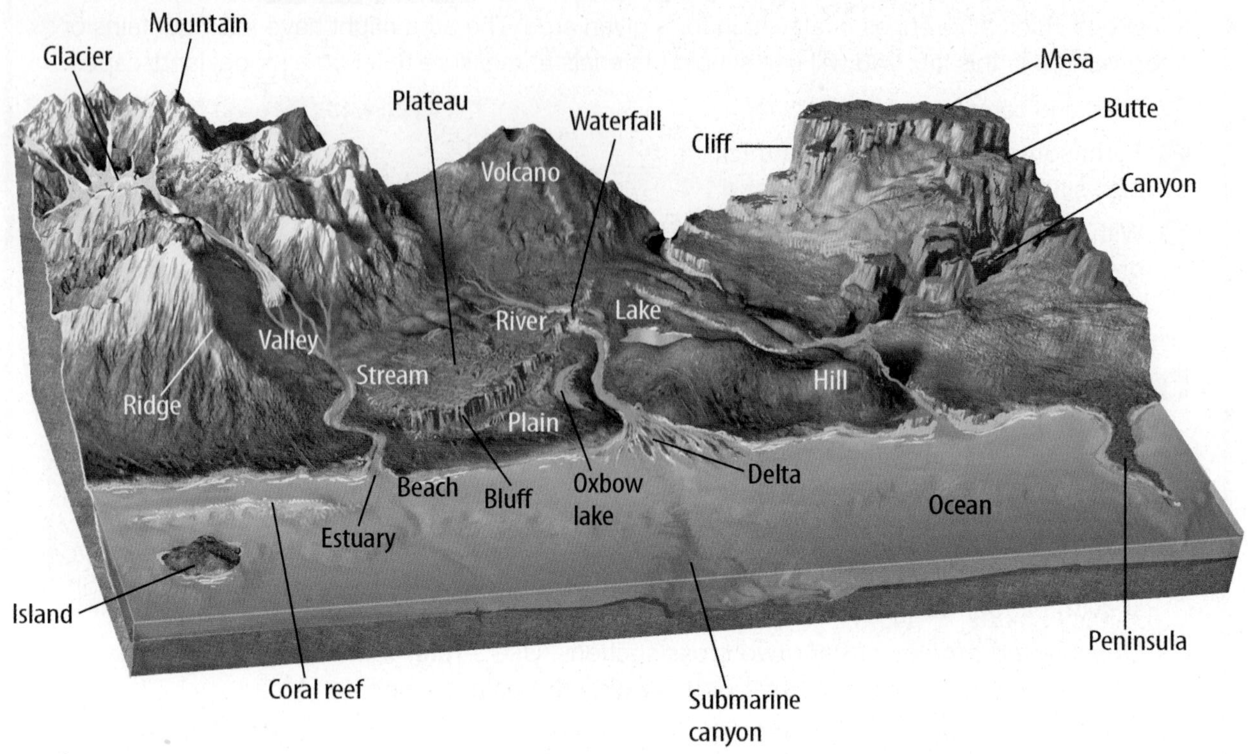

Visual Check Which landforms are most familiar to you?

Landforms

ACADEMIC VOCABULARY
feature
(noun) the structure, form, or appearance of something

Mountains, plains, plateaus, canyons, and other **features** are called landforms. Some examples of Earth's landforms are shown in **Figure 14. Landforms** *are topographic features formed by processes that shape Earth's surface.* They can be as big as mountains or as small as ant hills. Characteristics such as size, shape, slope, elevation, relief, and orientation to the surrounding landscape are often used to describe landforms. A landform is usually identified by its form and location.

 **Key Concept Check** What are landforms?

Landforms are not permanent. Their characteristics change over time. Many factors such as erosion or uplift of Earth's surface can create and affect landforms.

Elevation

Scientists use the term *elevation* to describe the height above sea level of a particular feature. Some landforms have high elevation. Other landforms have low elevation. For example, elevation is one of the major characteristics that is used to distinguish a plain from a plateau.

Relief

Do you recall how you measured relief in the Launch Lab at the beginning of this lesson? *Relief* is a term that scientists use to describe differences in elevation. Some landforms or geographic areas are described as having low relief. This means that there is a relatively small difference between the lowest elevation and the highest elevation in an area. Landforms or areas with high relief have a relatively large difference between the lowest elevation and the highest elevation. For example, if you were to climb out of the Grand Canyon, you would say it has high relief.

Topography

Scientists use the term *topography* to describe the shape of a geographic area. You can describe the topography of a small location or you can think about the general topography of a large region. Relief and topography can be used to describe features on continents and on the ocean floor. Next, you will read how relief and elevation are used to describe the most common landforms on Earth–plains, plateaus, and mountains.

MiniLab

20 minutes

How do landforms compare?

The terms *gully, ravine,* and *canyon* all describe an elongated depression formed by erosion from water. But how do these landforms differ?

1. Read and complete a lab safety form.
2. Working with a partner, use a **dictionary** to find the definition of the landforms in one of the lists below.

List 1	List 2	List 3	List 4
butte	hill	bay	channel
mesa	knoll	cove	strait
plateau	mountain	gulf	sound

3. Use **modeling clay** to represent the landforms in the list you chose.
4. Use **scissors** to cut different colors of **construction paper** and make scenes with your landforms.
5. Label each part of the scene.

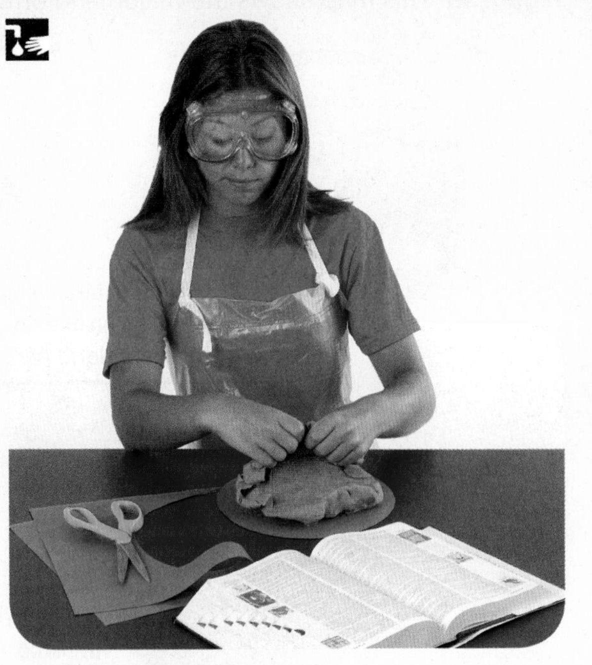

Analyze and Conclude

Key Concept Compare and contrast the model landforms.

Figure 15 Plains, plateaus, and mountains differ in terms of elevation and relief. ▶

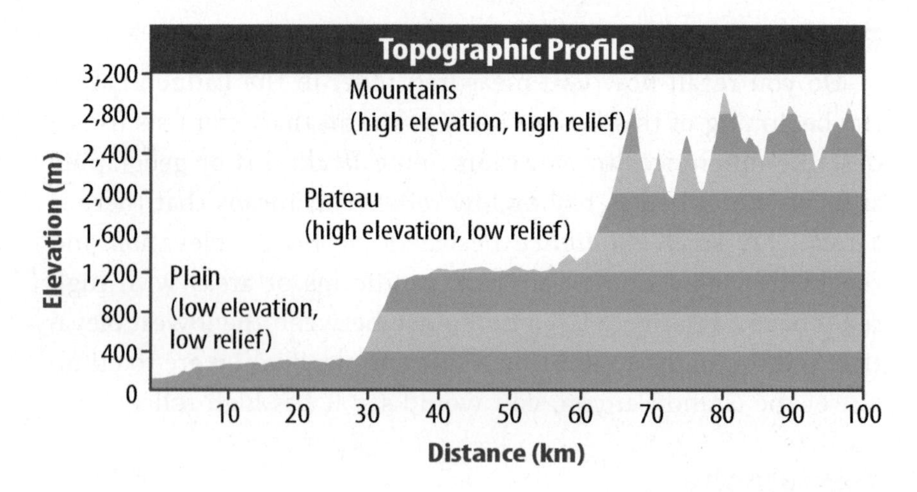

WORD ORIGIN

plain
from Latin *planus*, means "flat, level"

Plains

The features that cover most of Earth's surface are plains. **Plains** *are landforms with low relief and low elevation,* as illustrated in **Figure 15.** The broad, flat area in the center of North America is called the interior plains, as shown in **Figure 16.**

Plains can form when sediments are deposited by water or wind. Their soil is often rich. For this reason, many plains are used for growing crops or grazing animals.

Major Landform Regions

Figure 16 This map shows the major landform regions on Earth—plains, plateaus, and mountains.

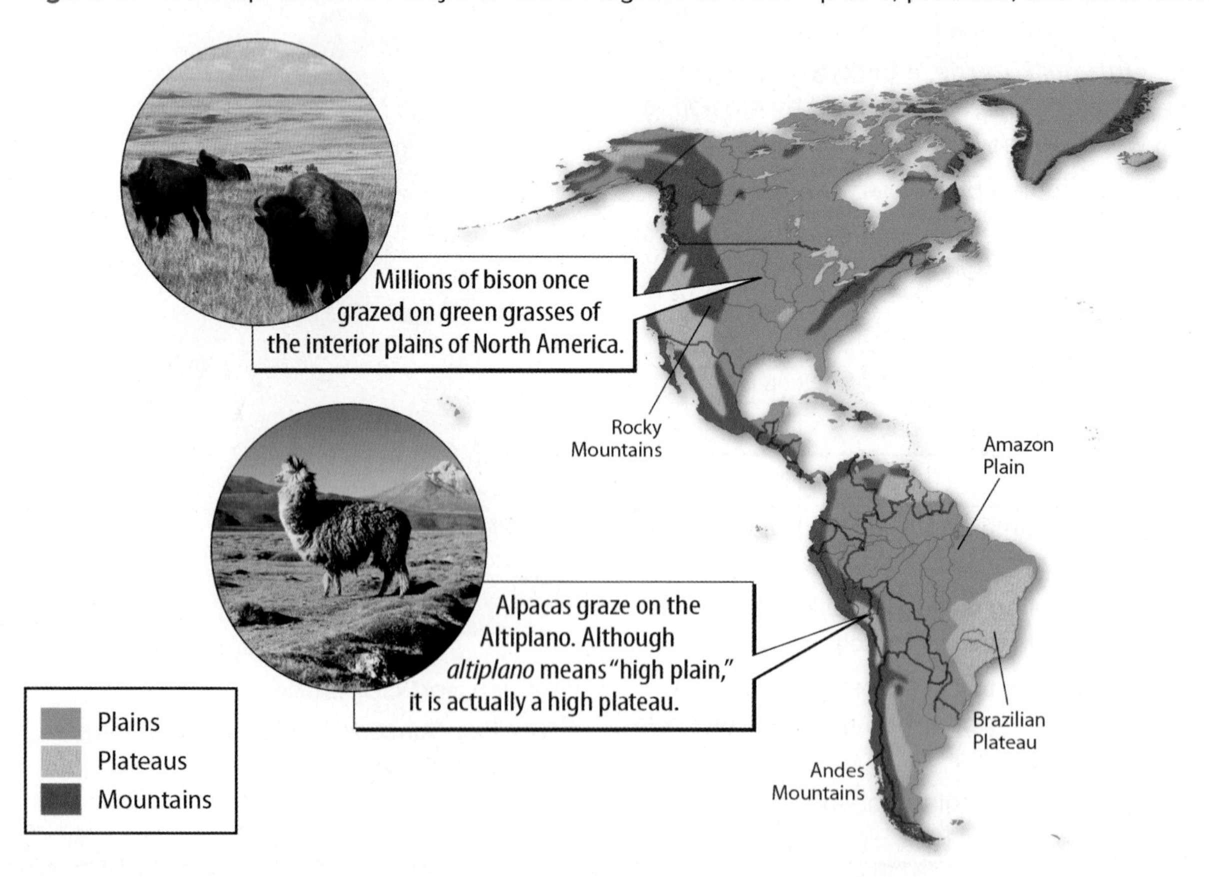

Plateaus

As you just read, plains are relatively flat and low. In contrast, plateaus are flat and high. **Plateaus** *are areas with low relief and high elevation.* Look again at **Figure 15** to see how a plateau differs from a plain.

Plateaus are much higher than the surrounding land and often have steep, rugged sides. They are less common than plains, but they are on every continent. Find some plateaus in different parts of the world in **Figure 16.**

Reading Check Describe a plateau.

Plateaus can form when forces within Earth uplift rock layers or cause collisions between sections of Earth's crust. For example, the highest plateau in the world is the Tibetan Plateau, called the "roof of the world." It is still being formed by collisions between India and Asia.

Plateaus also can be formed by volcanic activity. For example, the Columbia Plateau in the western United States is the result of the buildup of many successive lava flows.

Mountains

The tallest landforms of all are mountains. **Mountains** *are landforms with high relief and high elevation.* Look again at the world map in **Figure 16.** How many of Earth's well-known mountains can you find?

Mountains can form in several different ways. Some mountains form from the buildup of lava on the ocean floor. Eventually, the mountain grows tall enough to rise above the ocean's surface. The Hawaiian Islands are mountains that formed this way. Other mountains form when forces inside Earth fold, push, or uplift huge blocks of rocks. The Himalayas, the Rocky Mountains, and the Appalachian Mountains all formed from tremendous forces within Earth.

Visual Check Which of Earth's three major types of landforms—plains, plateaus, or mountains—covers most of Earth's land surface?

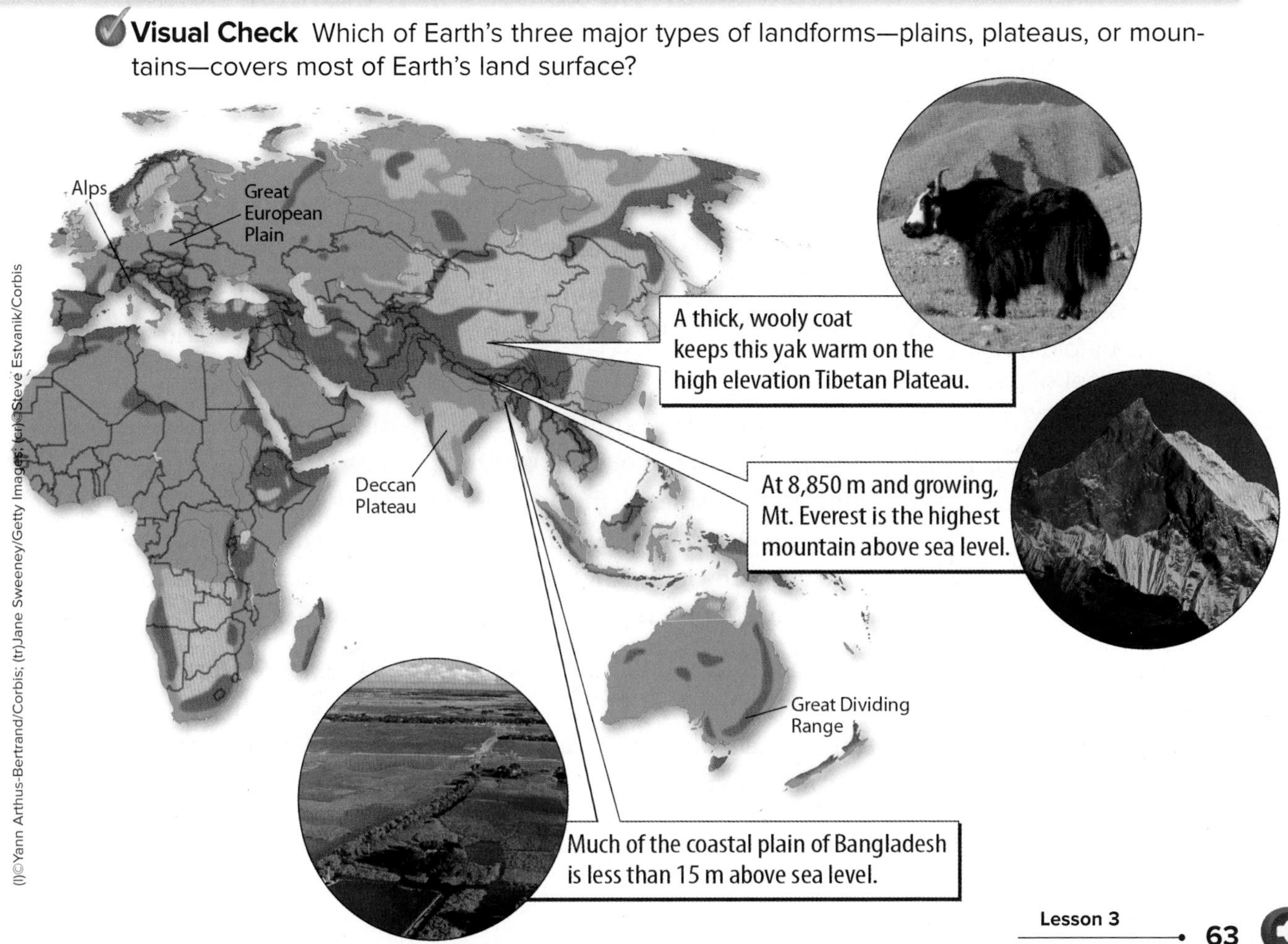

Major U.S. Landform Regions

Animation

Figure 17 The United States consists of several major landform regions.

Visual Check Which landform region do you live in?

Major Landform Regions in the United States

From flat plains to towering mountains, the United States has a variety of landforms. The major landform regions in the United States are shown in **Figure 17**.

Coastal plains are along much of the East Coast and the Gulf Coast. These plains formed millions of years ago when sediments were deposited on the ocean floor.

The interior plains make up much of the central part of the United States. This flat, grassy area has thick soils and is well suited for growing crops and grazing animals.

The Appalachian Mountains, in the eastern United States, began forming about 480 million years ago (mya). They were once much taller than they are today. Erosion has reduced their average elevation to about 2,000 m. The Rocky Mountains are in the western United States and western Canada. They are younger, taller, and more rugged than the Appalachians.

The Colorado Plateau is also a rugged region. It formed when forces within Earth lifted up huge sections of Earth's crust. Over time, the Colorado River cut through the plateau, forming the Grand Canyon.

FOLDABLES®

Make a tri-fold book from a sheet of paper. Label it as shown. Use it to organize your notes about Earth's major landforms.

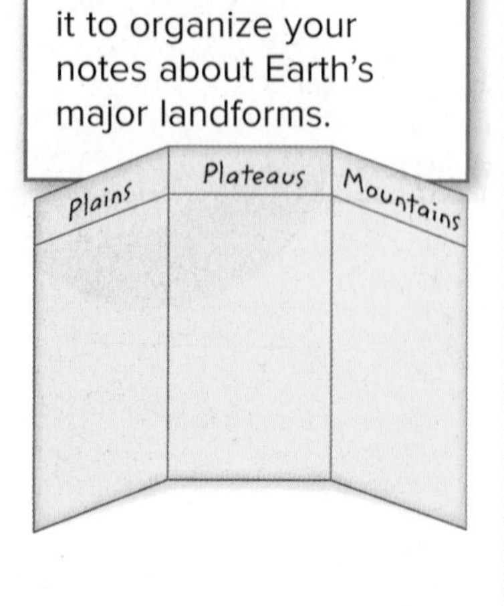

Key Concept Check Describe at least three major landform regions in the United States.

Lesson 3 Review

Visual Summary

Landforms are topographic features formed by processes that shape Earth's surface.

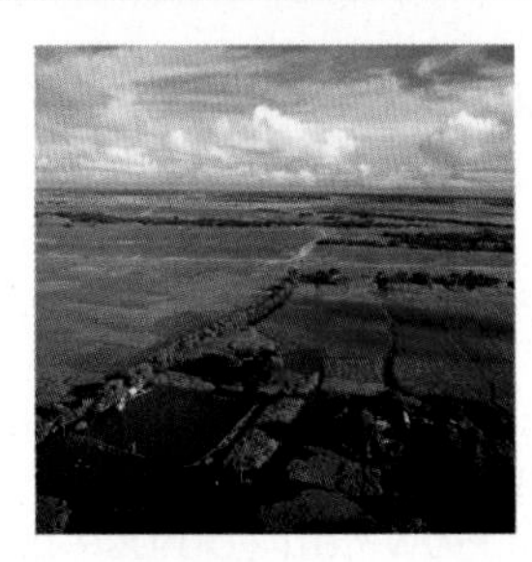

Major landforms include flat plains, high plateaus, and rugged mountains.

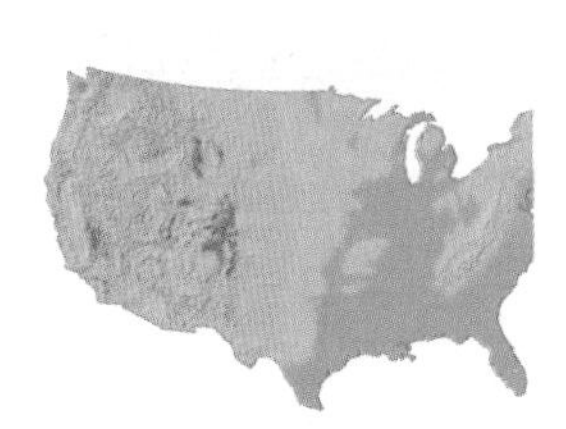

Major landform regions in the United States include the Appalachian Mountains, the Great Plains, the Colorado Plateau, and the Rocky Mountains.

FOLDABLES®

Use your lesson Foldable to review the lesson. Save your Foldable for the project at the end of the chapter.

What do you think 

You first read the statements below at the beginning of the chapter.

5. All ocean floors are flat.

6. Most of Earth's surface is covered by water.

Did you change your mind about whether you agree or disagree with the statements? Rewrite any false statements to make them true.

Use Vocabulary

1 Plains and mountains are examples of _________ formed by processes that shape Earth's surface.

2 A(n) _________ is a landform with high relief and high elevation.

3 **Distinguish** between a plain and a plateau.

Understand Key Concepts 

4 A landform with low relief and high elevation is a

A. mountain. **C.** plateau.
B. plain. **D.** topography.

5 **Describe** any landforms that are near your school.

Interpret Graphics

6 **Compare** Study the illustration below. How do plains compare to plateaus in terms of relief?

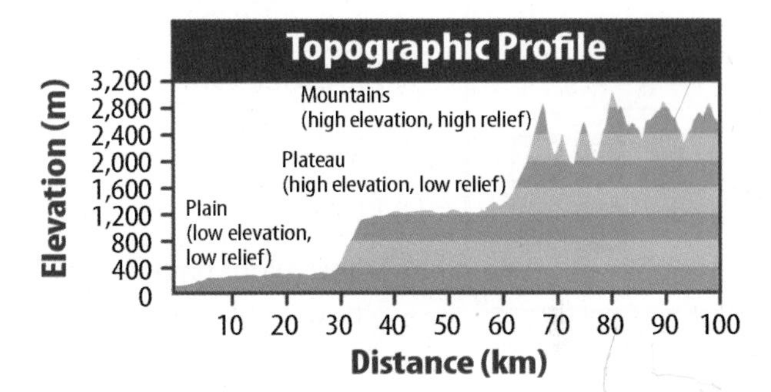

7 **Summarize** Copy and fill in the graphic organizer below to identify the major types of landforms.

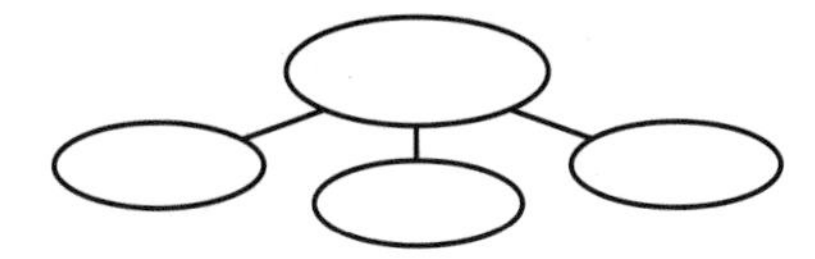

Critical Thinking

8 **Suggest** a way to model plains, plateaus, and mountains by using sheets of cardboard.

9 **Evaluate** the drawbacks and benefits of living in the mountains.

Lab

40 minutes

Modeling Earth and Its Layers

Materials

salt dough

food coloring

waxed paper

centimeter ruler

plastic knife

rolling pin or can

Safety

Earth has distinct layers. Each layer has a specific relative volume. You can use those volumes to build a model of Earth with each of the layers in proportion.

Question

Knowing the relative volume of Earth's inner core, outer core, mantle, and crust, how can you build an accurate scale model of these layers?

Procedure

1. Read and complete a lab safety form.
2. Obtain a piece of salt dough from your teacher. Study the chart below showing the relative volume of each layer of Earth. How can you use that data to turn your lump of dough into a model of Earth's layers?

The Relative Volumes of Each Layer of Earth

Layer	Relative Volume (%)
Inner core	0.7
Outer core	15.7
Mantle	82.0
Crust	1.6

3. You might have lots of ideas about how to divide your dough into the correct proportions to build your model. Here is one way you could try:
   - Work on a sheet of waxed paper so the dough won't stick. Roll your dough into a cylinder that measures 10 cm long. The cylinder represents 100 percent of the volume.
   - Now use your centimeter ruler to measure and mark off each of the percentages listed in the chart.
   - Cut off each piece and roll it into a sphere.

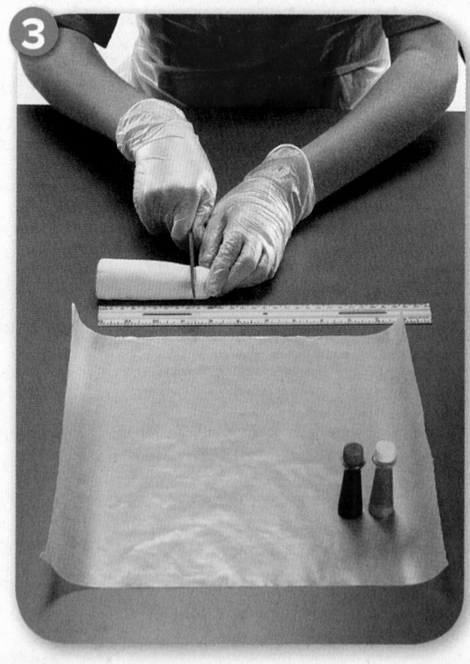

4. Use the data from the chart to figure out how you can build an accurate model.
5. Make a model of Earth's layers using the spheres that represent the relative volume of each layer. Add some food coloring to make each of the four spheres a different color. Work the salt dough so that the color is evenly distributed. Form each lump of dough into a sphere again. Your spheres should look similar to the ones shown in the photo below.

6. Cut in half the sphere representing the outer core.
7. Gently make a small depression in the flat side of each half of the outer core. Then place the inner core inside the outer core and seal the sphere.
8. Cut in half the sphere representing the mantle.
9. Gently make a small depression in the flat side of each half of the mantle. Fit the sphere representing the inner and outer cores into the mantle.
10. The last sphere represents Earth's crust. Put it on a piece of waxed paper and use a rolling pin (or a can) to spread out the sphere enough to make it fit onto the outside of the mantle.
11. Cut your model in half.

Hutchings Photography/Digital Light Source

Analyze and Conclude

12. **The Big Idea** Describe how each layer of Earth is represented on your model.
13. **Think Critically** Do you think your model accurately shows the volumes of the different layers? Why or why not?
14. **Draw a Conclusion** What can you conclude about the relative volumes of the different layers? Why couldn't you stretch the crust out far enough to cover the mantle? Why couldn't you just add more dough to the crust?

Communicate Your Results

Draw and label Earth's layers. Display the drawing next to your model and use both to explain what you have discovered about Earth's layers.

Inquiry Extension

How could you make an edible model of Earth's layers? Hint: Think about using ice cream or gelatin molds.

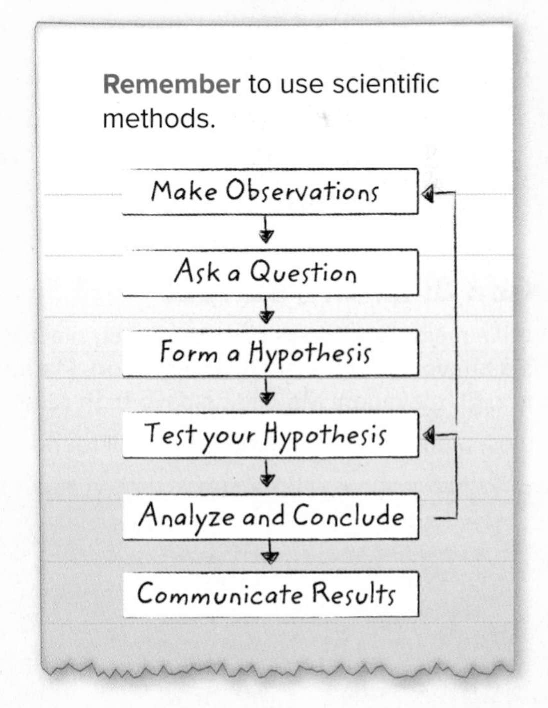

Chapter 2 Study Guide

Earth's three major layers are the crust, the mantle, and the core.

Key Concepts Summary	Vocabulary
Lesson 1: Spherical Earth • Earth's major systems include the atmosphere, hydrosphere, cryosphere, biosphere, and **geosphere.** • All major Earth systems interact by exchanging matter and energy. A change in one Earth system affects all other Earth systems. • **Gravity** caused particles to come together to form a spherical Earth. 	**sphere** p. 41 **geosphere** p. 42 **gravity** p. 43 **density** p. 45
Lesson 2: Earth's Interior  • Earth's interior layers include the **crust, mantle,** and **core.** • By analyzing earthquake waves, scientists have determined that the outer core is liquid and the inner core is solid.	**crust** p. 51 **mantle** p. 52 **lithosphere** p. 52 **asthenosphere** p. 52 **core** p. 54 **magnetosphere** p. 55
Lesson 3: Earth's Surface • Earth's major **landforms** include **plains, plateaus,** and **mountains.** Plains have low relief and low elevation. Plateaus have low relief and high elevation. Mountains have high relief and high elevation. • Plains, plateaus, and mountains are all found in the United States. 	**landform** p. 60 **plain** p. 62 **plateau** p. 63 **mountain** p. 63

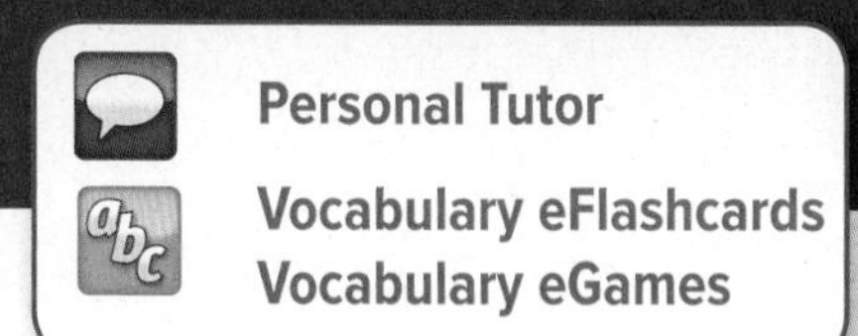

FOLDABLES® Chapter Project

Assemble your lesson Foldables as shown to make a Chapter Project. Use the project to review what you have learned in this chapter.

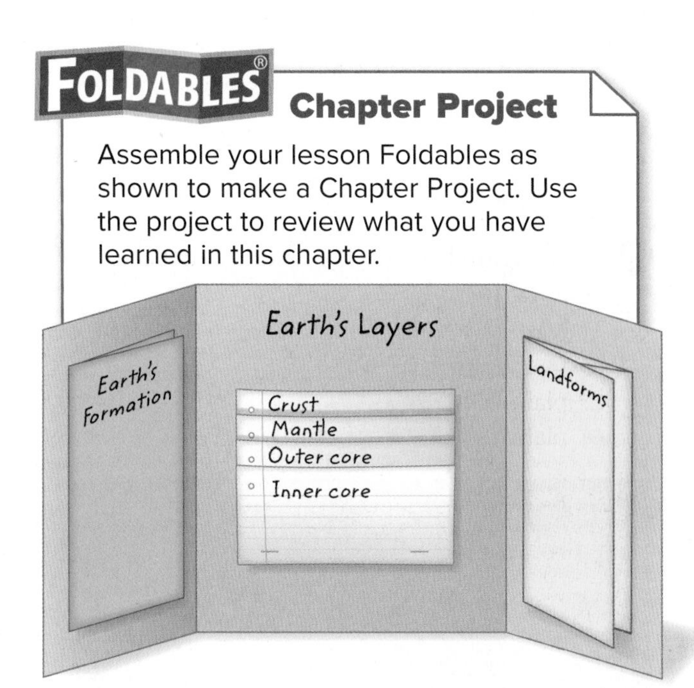

Use Vocabulary

1. Earth formed when _________ pulled together gas and dust that was spinning around the Sun.
2. The gravitational force is greater between similar-sized objects that have a higher _________.
3. The _________ is the largest of Earth's systems.
4. Small amounts of melted material in the _________ produce flow in the mantle.
5. The least dense rocks on Earth are in the _________.
6. Liquid in the _________ produces Earth's magnetic field.
7. A topographic feature formed by processes that shape Earth's surface is a _________.
8. A(n) _________ has low relief and low elevation.
9. A landform that is high and flat is a(n) _________.

Link Vocabulary and Key Concepts

Copy this concept map, and then use vocabulary terms from the previous page to complete the concept map.

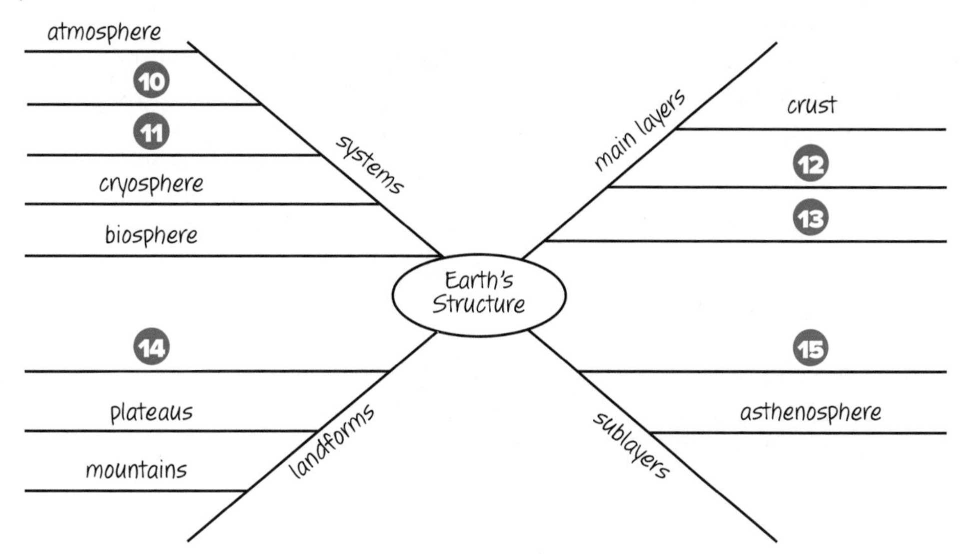

Chapter 2 Review

Understand Key Concepts 

1. What does the biosphere contain?
   - A. air
   - B. living things
   - C. rocks
   - D. water

2. What affects the strength of gravity between two objects?
   - A. the density of the objects
   - B. the mass of the objects
   - C. the distance between the objects
   - D. both the mass and the distance between the objects

3. The figure below shows Earth's layers. What does the red layer represent?
   - A. asthenosphere
   - B. crust
   - C. lithosphere
   - D. mantle

4. What is the shape of Earth?
   - A. disklike
   - B. slightly flattened sphere
   - C. sphere
   - D. sphere that bulges at the poles

5. Which do scientists use to learn about Earth's core?
   - A. earthquake waves
   - B. mines
   - C. temperature measurements
   - D. wells

6. What does the magnetosphere protect people from?
   - A. asteroids
   - B. cosmic rays
   - C. global warming
   - D. sun spots

7. In the figure below, what feature is the arrow pointing to?
   - A. core
   - B. mountain
   - C. plain
   - D. plateau

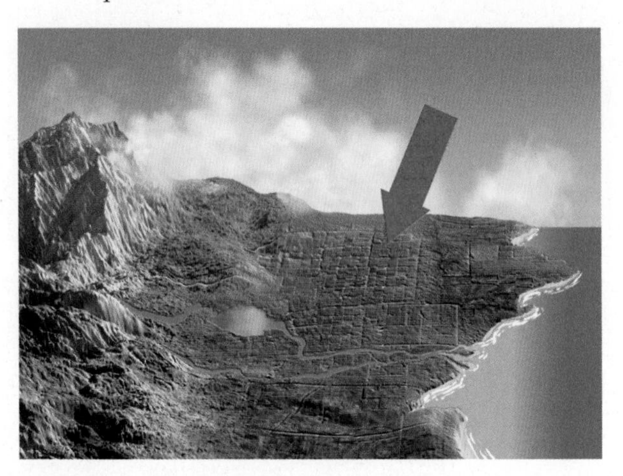

8. What does topography describe?
   - A. depth of an ocean feature
   - B. height of a landform
   - C. shape of a given area
   - D. width of an area

9. What is true of landforms?
   - A. They are all flat.
   - B. They are permanent.
   - C. They change over time.
   - D. They are only on continents.

10. A box sitting on the floor models what type of landform?
    - A. mountain
    - B. plain
    - C. plateau
    - D. relief

Critical Thinking

11. **Explain** how gravity would affect you differently on a planet with less mass than Earth, such as Mercury.
12. **Compare** materials in the geosphere to materials in the atmosphere.
13. **Consider** How would Earth's layers be affected if all the materials that make up Earth had the same density?
14. **Relate** How do Earth's systems interact?
15. **Explain** why everything on or near Earth is pulled toward Earth's center.
16. **State** how the crust and the uppermost mantle are similar.
17. **Summarize** Earth's crust, mantle, and core on the basis of relative position, density, and composition.
18. **Create** a model of Earth's magnetosphere.
19. **Explain** how a plateau differs from a plain.
20. **Summarize** the characteristics of the landform regions labeled in the map below.

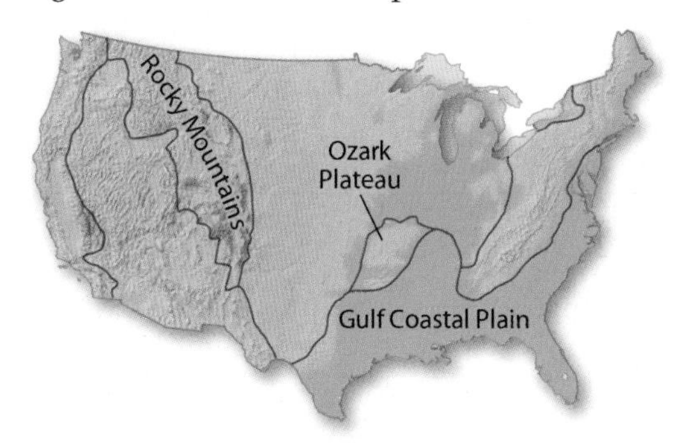

21. **Evaluate** which type of landform is best suited for agriculture.

Writing in Science

22. A song includes lyrics that usually rhyme. Write the lyrics to a song about the elevation of landforms.

REVIEW THE BIG IDEA

23. Identify and describe the different layers of Earth.
24. How does Earth's core create Earth's magnetic field?
25. Hypothesize what might happen to life on Earth if Earth did not have a magnetosphere.

Math Skills

Math Practice

Solve One-Step Equations

26. A large weather balloon holds 3.00 m^3 of air. The air in the balloon has a mass of 3.75 kg. What is the density of the air in the balloon?
27. A pine board has a volume of 18 cm^3. The mass of the board is 9.0 g. What is the density of the pine board?
28. 100 cm^3 of water has a mass of 100 g. Will the pine board in the previous question float or sink in the water?

Robert Postma/age fotostock

Standardized Test Practice

Record your answers on the answer sheet provided by your teacher or on a sheet of paper.

Multiple Choice

1 Density equals

A mass divided by volume.

B mass times volume.

C volume divided by mass.

D volume times mass.

2 Which force gave Earth its spherical shape?

A electricity

B friction

C gravity

D magnetism

Use the diagram below to answer question 3.

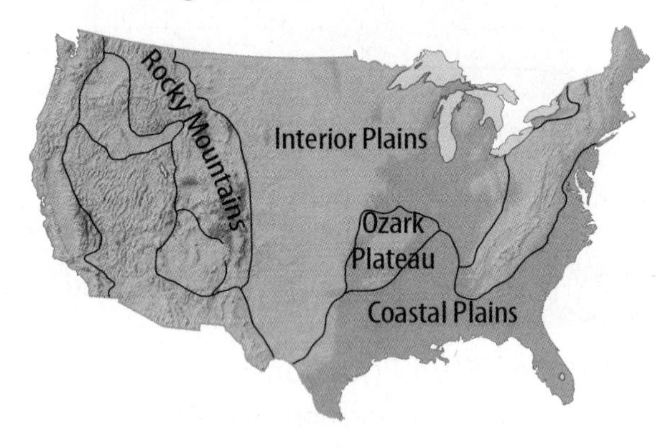

3 Which landform covers the largest area of the central United States?

A Coastal Plains

B Interior Plains

C Ozark Plateau

D Rocky Mountains

4 Which describes Earth's asthenosphere?

A brittle

B fast-moving

C freeze-dried

D plastic

Use the diagram and the graphs below to answer question 5.

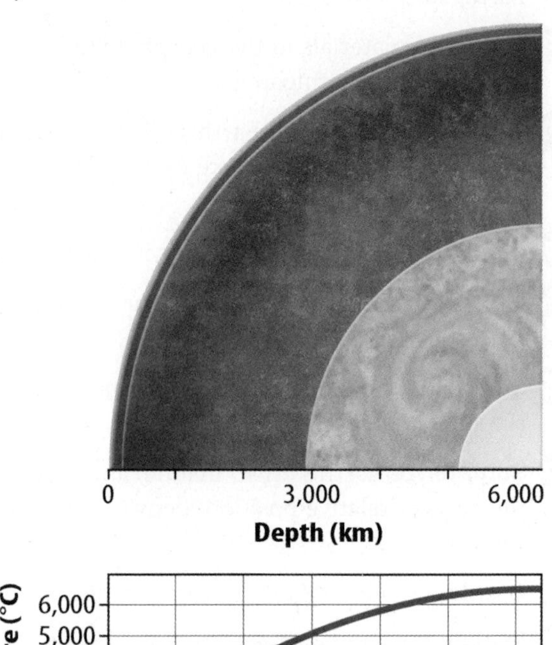

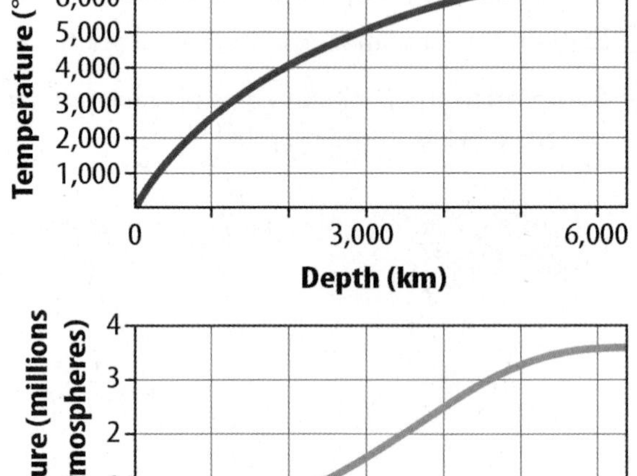

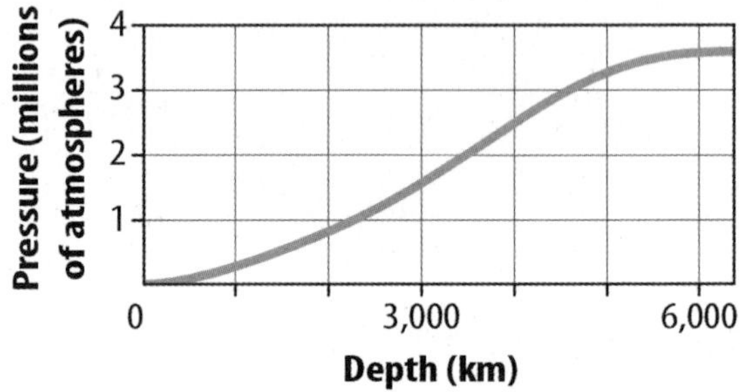

5 Which describes temperature and pressure at Earth's center?

A high pressure and high temperature

B high pressure and low temperature

C low pressure and high temperature

D low pressure and low temperature

6 Which explains the term *topography*?

A the geological ages of features

B the heights and locations of features

C the seasonal variations of features

D the travel routes between features

7 Which is the correct order of Earth's layers from the surface to the center?

A crust, core, mantle

B crust, mantle, core

C mantle, core, crust

D mantle, crust, core

Use the diagram below to answer question 8.

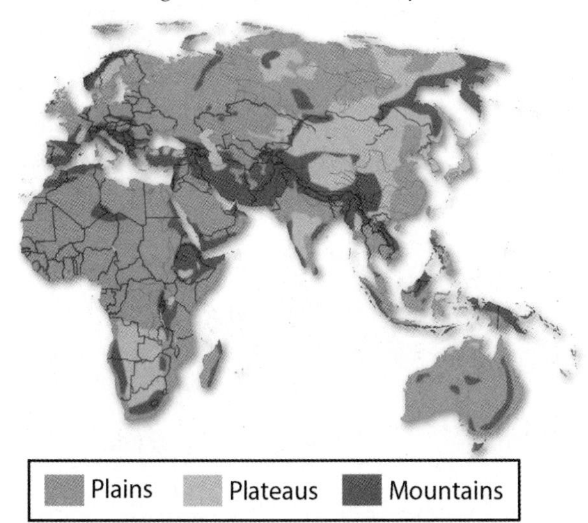

Plains	Plateaus	Mountains

8 Which continent has the greatest area of plateaus?

A Africa

B Asia

C Australia

D Europe

Constructed Response

9 Compare and contrast plateaus and plains.

Use the diagrams below to answer questions 10–12.

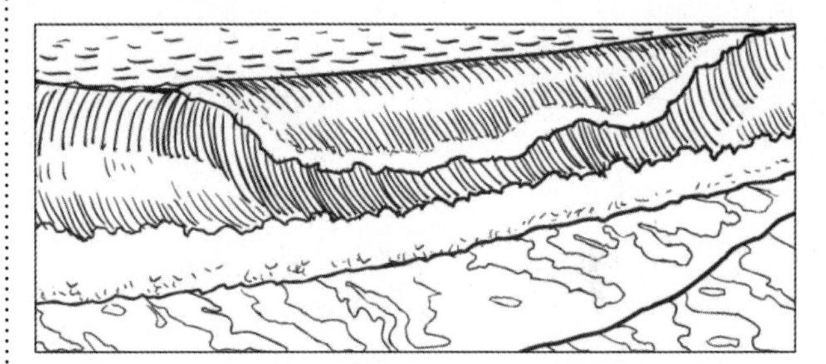

10 Three Earth systems are highlighted above. Name each system and describe its features.

11 Describe how these three systems interact.

12 Draw a diagram of the fourth major Earth system. Describe its features.

NEED EXTRA HELP?												
If You Missed Question...	1	2	3	4	5	6	7	8	9	10	11	12
Go to Lesson...	1	1	3	2	2	3	2	3	3	1	1	1

Chapter 3

Minerals

THE BIG IDEA **What are minerals and why are they useful?**

Inquiry How did these giant crystals form?

These gypsum crystals, in Mexico's Cave of Crystals, are the largest in the world. How did they form? They crystallized from solids dissolved in superheated water. Over time, the solids combined and formed gigantic crystals of the mineral gypsum. These crystals are beautiful, but can they be used for something?

- What is a mineral?
- How are minerals classified?
- How do you use minerals in your everyday life?

Get Ready to Read

What do you think?

Before you read, decide if you agree or disagree with each of these statements. As you read this chapter, see if you change your mind about any of the statements.

1. A mineral is anything solid on Earth.
2. Some minerals form when water evaporates from Earth's surface.
3. The best way to identify a mineral is by color.
4. Hardness, streak, and luster are among the properties used to identify minerals.
5. An ore is a concentration of minerals that contains only iron.
6. Gemstone and ore deposits are evenly distributed around the world.

Lesson 1

Reading Guide

Key Concepts
ESSENTIAL QUESTIONS

- What is a mineral?
- What are the common rock-forming minerals?
- How do minerals form?

Vocabulary

mineral p. 77
silicate p. 81
crystallization p. 81
magma p. 83
lava p. 83

Multilingual eGlossary

What is a mineral?

Castles on Mono Lake?

These rocky towers form from the salty water of Mono Lake in California. The lake is saturated with dissolved salts. The salts crystallize from the lake, and over time they grow into tall towers and columns. Are mineral deposits like these common?

Justin Bailie/Getty Images

Launch Lab

20 minutes

Are rocks and minerals the same?

Rocks and minerals are often mistaken for one another, but there are differences between them. Rocks commonly contain two or more minerals. However, a mineral is made of one uniform substance. In this lab, you will investigate a variety of objects and try to determine whether each object is a rock or a mineral based on its physical properties.

1. Read and complete a lab safety form.
2. Observe the **group of objects** or **pictures** on your table.
3. Try to organize some of the objects or pictures into one group according to a common characteristic. Do not reveal this characteristic to your classmates. Place a **loop of string** around your group. Ask classmates to try to guess the common characteristic of all the objects in the group.
4. Have a classmate think of another way to group the objects. Ask him or her to arrange the objects into a new group and place a loop of string around them. Now it's your turn to guess the common characteristic these objects share.
5. Group all objects together that seem to be made of one uniform substance. Make another group of objects that contain two or more substances.

Think About This

1. Decide with your class which group contains rocks and which group contains only minerals. Explain your answer.
2. **Key Concept Check** As a class, write a definition for *rocks* and one for *minerals*.

What is a mineral?

When you woke up this morning, minerals probably weren't the first thing that came to mind. As you climbed out of bed to get ready for school, without knowing it, you used materials made from minerals. For example, deodorant, shampoo, and makeup are made from minerals. Anything made of metal--your belt buckle, jewelry, or zippers on your clothes--came from a mineral. Even the salt you put on your eggs at breakfast is a mineral. A **mineral** *is a naturally occurring, inorganic solid with a definite chemical composition and an orderly arrangement of atoms or ions.* **Figure 1** shows some common household objects that are made from minerals.

Key Concept Check What is a mineral?

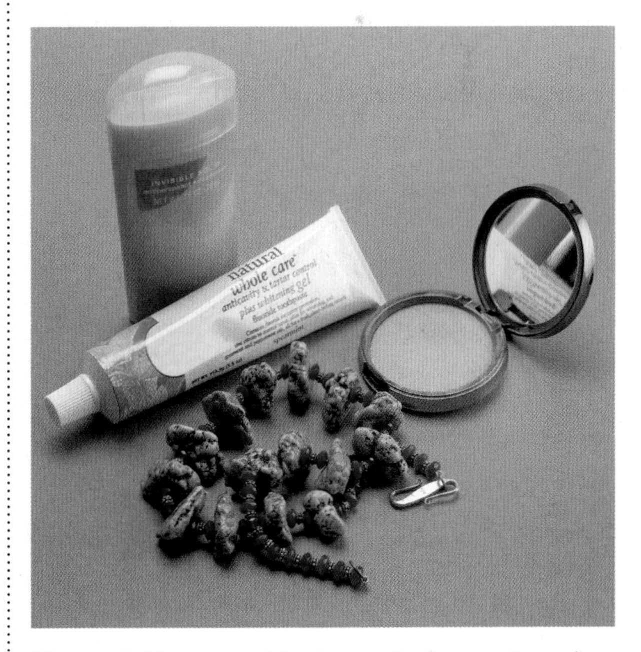

Figure 1 You use objects made from minerals every day.

Naturally Occurring

From the tops of the tallest mountains to sediments on the seafloor, minerals are everywhere. There are approximately 4,000 minerals on Earth. These minerals occur naturally and are not made in a laboratory. Of the 4,000 naturally occurring minerals, only about 30 are common. Ten of these are called the rock-forming minerals. Quartz, feldspar, and olivine are a few of the rock-forming minerals.

Definite Chemical Composition

Minerals have a definite chemical composition. For example, the mineral hematite has the chemical formula Fe_2O_3. If you look at a periodic table of the elements, you can see that these symbols represent the elements iron (Fe) and oxygen (O). Any material made of two parts iron and three parts oxygen is called hematite. Some minerals, such as silver (Ag) and sulfur (S), are composed of just one element. These are called native elements. Other minerals, such as potassium feldspar ($KAlSi_3O_8$), are made up of a combination of several elements.

SCIENCE USE V. COMMON USE

crystal

Science Use a solid of a chemical substance with a regular, repeating arrangement of its atoms

Common Use a clear, colorless glass of superior quality

Crystalline Form

Minerals form predictable **crystal** patterns. The internal arrangement of atoms or **ions** determines the shape of a crystal. Have you ever looked at salt crystals up close or with a magnifying lens? You might notice that these small salt crystals form cubes. The crystal shape of salt is not random.

REVIEW VOCABULARY

ion

an atom or group of atoms with an electric charge

The salt that you shake on your food often contains the mineral halite. Halite's internal atomic arrangement is shown in **Figure 2.** Notice that sodium and chlorine ions repeat over and over in all three dimensions of space. A repeating arrangement of atoms or ions in three directions makes it a crystal. Compare halite's cubic crystal pattern to the arrangement of atoms in glass, also shown in **Figure 2.**

Personal Tutor

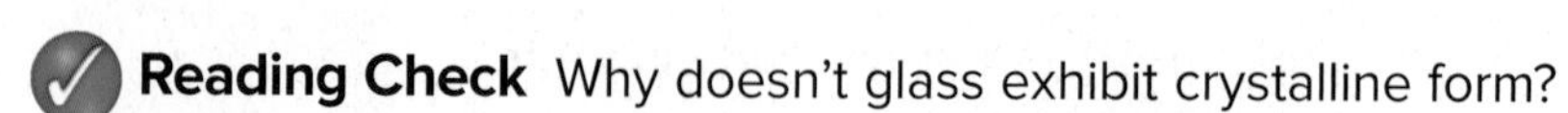
Reading Check Why doesn't glass exhibit crystalline form?

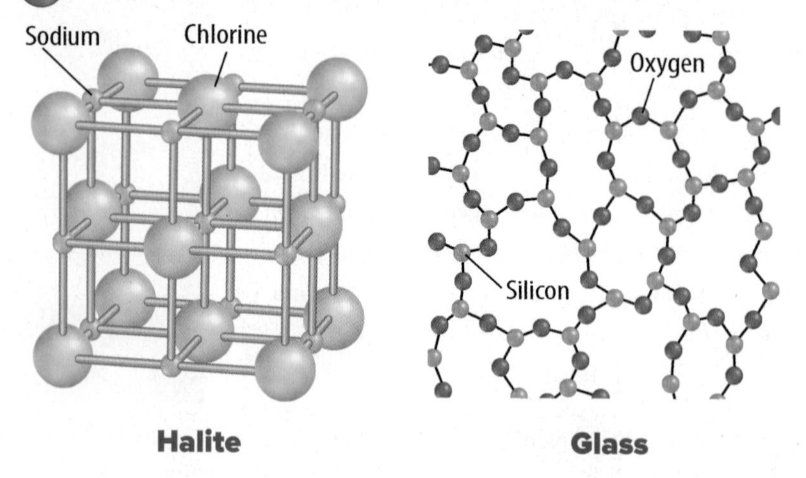

Figure 2 The sodium and chlorine atoms in the mineral halite are arranged in a three-dimensional, orderly, repeating pattern. The silicon and oxygen atoms in glass are not.

Solid

All minerals are solids, but not all solids are minerals. A solid is matter with tightly packed atoms or ions. It has a definite shape and volume. To be a mineral, a solid must have a crystal form. Solids without crystal form, liquids, and gases, are not minerals.

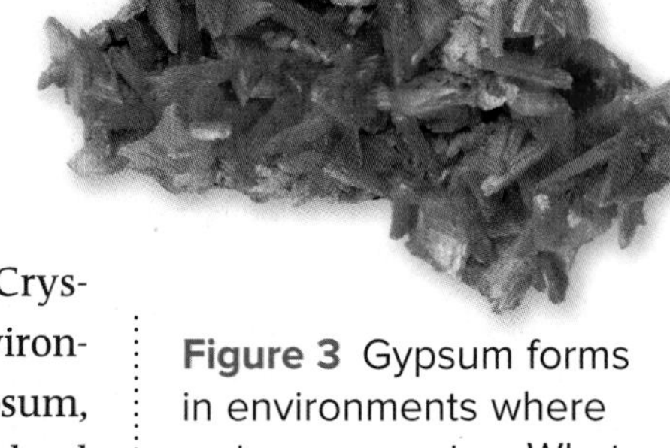

Inorganic

Minerals are inorganic, or not from biologic origins. Crystallization is the formation of crystals. It occurs in different environments and can be caused by evaporation. The mineral gypsum, shown in **Figure 3,** formed as water evaporated, leaving dissolved solids behind.

Figure 3 Gypsum forms in environments where water evaporates. What evidence suggests that gypsum is a mineral?

Despite being inorganic, some minerals can form as a result of organic processes. For example, marine organisms can extract dissolved solids from seawater and make their shells.

Key Concept Check Identify the five main characteristics of a mineral.

The Structure of Minerals

Recall that minerals have definite crystalline form and that the shape reflects the internal arrangement of atoms or ions. Compare the shape of the quartz (SiO_2) crystals to the shape of the calcite ($CaCO_3$) crystals in **Figure 4.** Well-formed quartz crystals are long and typically have distinctly pointed ends. Well-formed calcite crystals are somewhat diamond shaped and have sets of parallel sides. Do you think it would be easy to identify a mineral based on crystal structure alone?

Reading Check What can geologists infer from the shape of a mineral?

Figure 4 Quartz (SiO_2) and calcite ($CaCO_3$) form very different crystal shapes.

Figure 5 This scanning electron microscope image shows the cubic crystal faces of galena.

Visual Check Can you identify the 90° angles that form between Pb and S ions in the galena crystal?

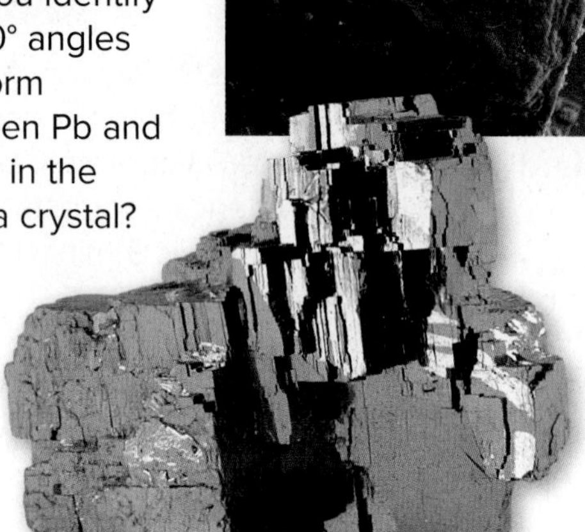

Galena

Crystal Shape

As you will observe, minerals occur in many different shapes. Minerals are often small and difficult to identify. Occasionally when a crystal forms under the right conditions and has time to grow, it will develop a characteristic crystal shape. **Figure 5** illustrates a galena (PbS) crystal in which the arrangement of lead and sulfur ions forms cubes. The angles between the lead and sulphur ions are 90°. Thus the galena crystal has a characteristic cubic shape.

Minerals do not always exist in large, well-developed geometric shapes. Most of the time, minerals grow in tiny clusters. Scientists can examine the shape of tiny crystals using scanning electron microscope images. The high magnifications reveal an orderly arrangement of atoms or ions inside the crystal despite the crystal's tiny size, as shown in **Figure 5.**

Reading Check When do crystals grow into large and well-developed shapes?

MiniLab

20 minutes

How can you tell crystals apart?

Geologists use crystal shapes to help identify unknown minerals. Halite, or rock salt, is a common mineral in table salt. Epsom salt (magnesium sulfate) is found in mineral water.

1. Read and complete a lab safety form.
2. Pour some crystals of **rock salt** onto a small piece of dark-colored **construction paper.**
3. Use a **magnifying lens** to observe crystal shape.
4. Draw a sketch of a single salt crystal in your Science Journal.
5. Repeat steps 2–4 using **Epsom salt.**

Analyze and Conclude

1. **Describe** the shapes of the rock salt and the Epsom salt.
2. **Key Concept** How can you distinguish between a crystal of table salt and Epsom salt?

(t)Image courtesy of Ellery Frahm, Electron Microprobe Laboratory, Department of Geology and Geophysics, University of Minnesota-Twin Cities; (c)Jacques Cornell/McGraw-Hill Education; (b)Hutchings Photography/Digital Light Source

Common Minerals

You have read that every type of mineral is unique due to its chemical composition and the arrangement of its atoms or ions. Of the 30 most abundant minerals on Earth, only a few are referred to as the common rock-forming minerals. A few of these minerals are shown in **Table 1.** The common-rock forming minerals are composed of combinations of elements that are abundant in Earth's crust. The two most abundant elements in the crust are oxygen and silicon. Pure quartz is composed of only oxygen and silicon. Quartz is a member of one of the two main families of common rock-forming minerals.

The two main families of rock-forming minerals are the silicates and the nonsilicates. *A* **silicate** *is a member of the mineral group that has silicon and oxygen in its crystal structure.* Quartz, with the chemical composition of SiO_2, is a silicate mineral. Feldspar is the most common silicate mineral in Earth's crust. Other silicates include olivine, pyroxene, amphibole, and mica, as shown in **Table 1.** Common nonsilicate minerals, or minerals that do not contain silicon, include calcite and halite.

Key Concept Check What are the two main families of rock-forming minerals?

How do minerals form?

Minerals form in a variety of environments. No matter the environment, all minerals form through a process called crystallization. *The process of* **crystallization** *occurs when particles dissolved in a liquid or a melt solidify and form crystals.* Minerals can crystallize from either hot or cool solutions. For example, the mineral halite forms from cool solutions, where water with dissolved solids evaporates leaving halite crystals behind. Minerals can also crystallize as molten rock cools. The chemical and physical properties of minerals can help geologists infer the type of environment where these minerals formed.

Table 1 Common Minerals

Silicates	
Quartz	
Potassium feldspar	
Plagioclase feldspar	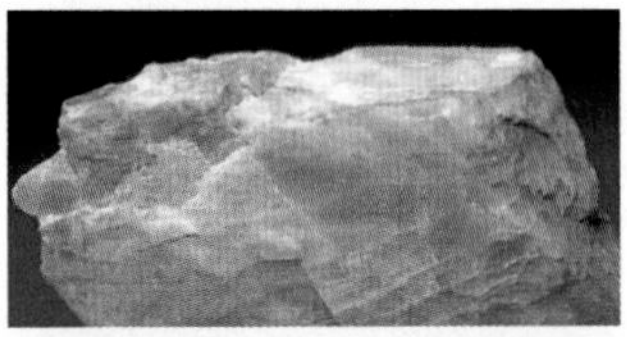
Olivine	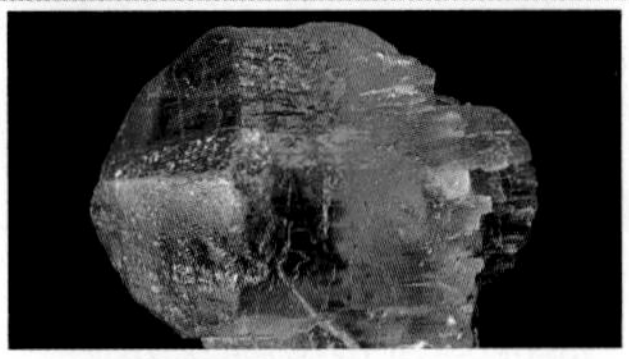
Pyroxene	
Amphibole	
Mica	 
Nonsilicates	
Calcite	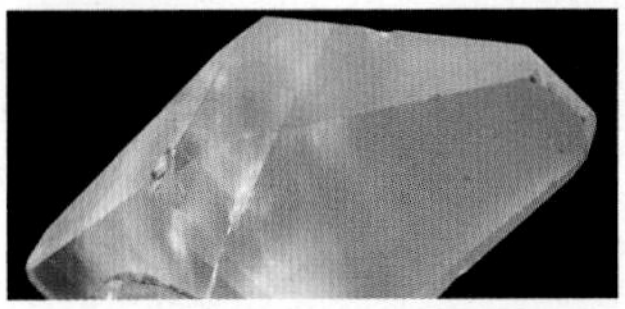
Halite	

(l to r, t to b, 9)©José Manuel Sanchis Calvete/Corbis; (2)Phil Robinson/age fotostock; (3)©Doug Sherman/Geofile; (4)Mark A. Schneider/Photo Researchers, Inc.; (5–8)Andrew Silver/U.S. Geological Survey; (10)Bob Coyle/McGraw-Hill Education

Mineral Formation

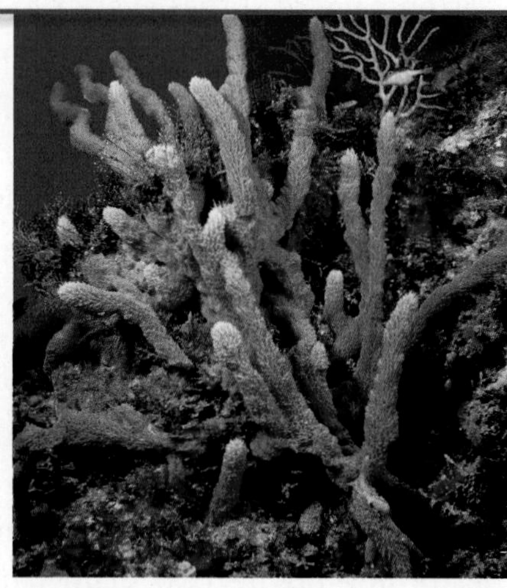

Figure 6 The deposit of halite on the left formed when water on Earth's surface evaporated. The coral reef on the right formed as organisms extracted dissolved minerals from the ocean.

Minerals from Cool Solutions

When it rains or as snow melts, the resulting water either seeps into the ground or flows over Earth's surface. As the water moves, it interacts with minerals in rocks and the soil. The water dissolves some of these minerals and picks up elements such as potassium, calcium, iron, and silicon from the rock and soil. These elements become dissolved solids.

REVIEW VOCABULARY

dissolve
to disperse or disappear in solution

Water can only hold a certain amount of dissolved solids. During dry conditions, as water evaporates, solids crystallize out of the water and form minerals. A deposit of the mineral halite–common rock salt–formed when water from a shallow lake evaporated, as shown in **Figure 6.**

Sometimes minerals can crystallize from water in environments that aren't dry. For example, the dissolved salts in seawater make the ocean salty. Seawater can become saturated with dissolved salts. In other words, the water cannot hold any more salt. Certain marine organisms can remove these salts from seawater and produce protective shells or build a reef, such as the one shown in **Figure 6.**

Minerals from Hot Solutions

Water on Earth's surface can flow through cracks in the crust into deep and hot environments. These hot solutions sometimes carry with them large concentrations of dissolved solids. Some of these dissolved solids eventually form valuable mineral deposits. For example, the vein of gold, shown in **Figure 7,** formed this way. When conditions were just right, gold crystallized from the hot-water solution and filled the cracks in the rock. The giant crystals shown on the first page of this chapter are extreme examples of this process.

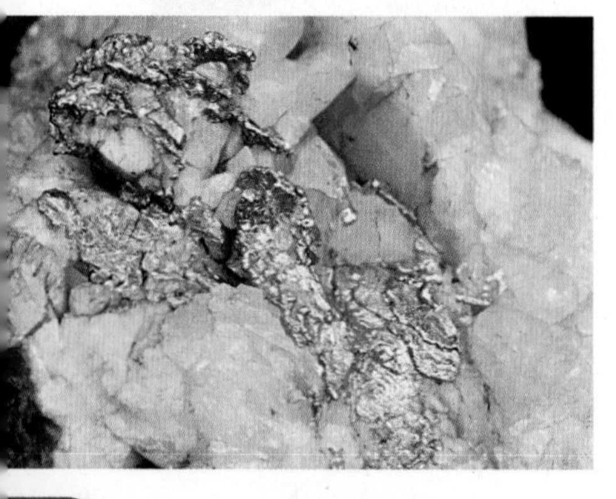

Figure 7 Hot fluids can flow through cracks in Earth's crust and crystallize to form new minerals.

Reading Check How does a vein of gold form?

Figure 8 The volcanic rock on the left has very small crystals. The volcanic rock on the right has larger crystals.

Visual Check Which rock in this figure cooled more quickly?

Minerals from Molten Rock

If you've ever watched a volcano erupt, you've seen molten rock in action. **Magma** *is molten rock stored beneath Earth's surface. When molten rock erupts on or near Earth's surface, it is called* **lava** or ash. As lava or ash cools above ground or magma cools underground, atoms and ions arrange themselves and form mineral crystals. Crystals differ in size depending on the cooling rate of magma, lava, or ash.

Small crystals—some barely visible with a light microscope—form as lava cools quickly on or near Earth's surface. Large crystals sometimes form as magma cools and crystallizes slowly below Earth's surface. The basalt and the andesite shown in **Figure 8** cooled at different rates. Compare the crystal sizes to determine which one cooled more quickly, the basalt or the andesite.

WORD ORIGIN

lava
from Latin *lavare,* means "to wash"

Changes in Minerals

Some minerals form deep within Earth's crust and mantle. These minerals are stable under high pressure and high temperature conditions. Tectonic activity can uplift minerals from great depths onto Earth's surface. Because pressure and temperature conditions on Earth's surface are less extreme, the minerals become unstable. Changes in pressure and temperature combined with agents of erosion, such as water, wind, and ice, cause the minerals to break down. Eventually new minerals form.

Minerals can be used to interpret the conditions of their formation. It is highly unlikely that you will find a rock that contains both the minerals olivine and quartz. Olivine forms under high temperatures and high pressures. The mineral quartz, however, forms in less extreme conditions.

Key Concept Check Identify the ways minerals can form.

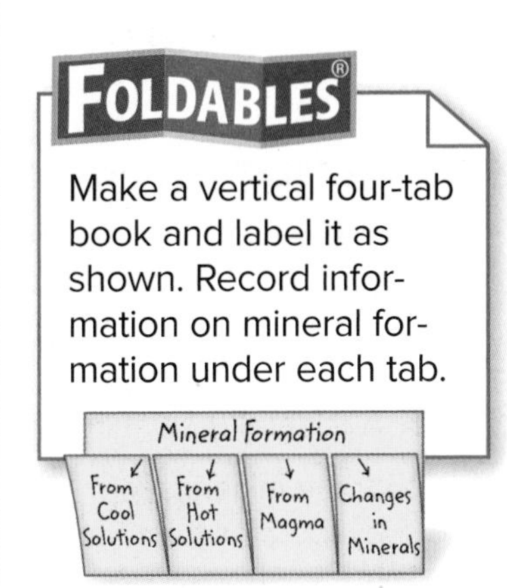

Lesson 1 Review

Visual Summary

A mineral is a naturally occurring, inorganic solid with a definite chemical composition and crystalline form.

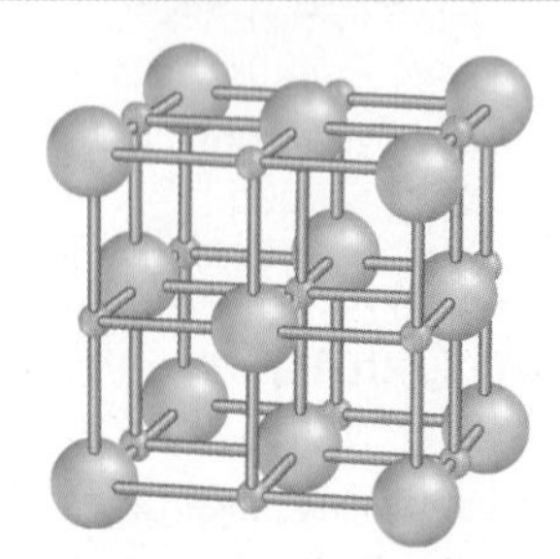

Crystal shape reflects the internal arrangement of atoms or ions.

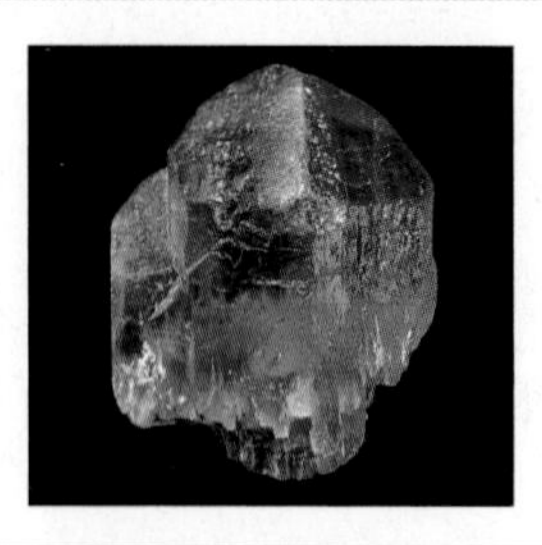

The most common rock-forming minerals are the silicates.

FOLDABLES®

Use your lesson Foldable to review the lesson. Save your Foldable for the project at the end of the chapter.

What do you think NOW?

You first read the statements below at the beginning of the chapter.

1. A mineral is anything solid on Earth.
2. Some minerals form when water evaporates from Earth's surface.

Did you change your mind about whether you agree or disagree with the statements? Rewrite any false statements to make them true.

Use Vocabulary

1. **Use the term** *crystallization* in a sentence to describe how minerals form.
2. Molten rock material that exists below Earth's surface is called ________.
3. A ________ mineral contains the elements silicon and oxygen.

Understand Key Concepts 

4. What process causes minerals to form in a shallow lake during dry conditions?
   A. burial
   B. evaporation
   C. high pressure
   D. melting
5. **Compare** the formation of minerals by evaporation with those that crystallize directly from seawater and those that crystallize from magma.
6. What process causes crystallization of minerals from magma?
   A. cooling
   B. eruption
   C. evaporation
   D. melting

Interpret Graphics

7. **Create** Draw a graphic organizer like the one below. Identify three environments where minerals form.

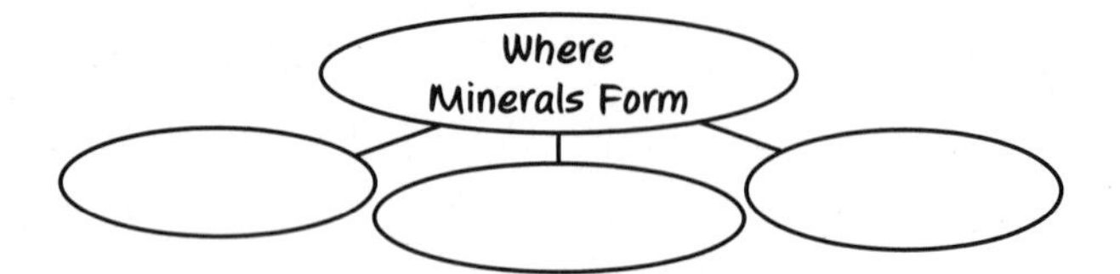

Critical Thinking

8. **Design and Evaluate** Design an experiment to grow sugar crystals from a sugary solution. Keep a log in your Science Journal to describe how the sugar crystals change size and shape over time as they grow. Evaluate your design. What variable(s) would you change, if any, to create large crystals more successfully?
9. **Assess** Are sugar crystals minerals? Why or why not?

Careers in Science

Ed Mathez works in mines such as this one near the Bushveld Igneous Complex in South Africa. He examines layered igneous rocks in search of rare metals. ▶

Billions of Years in the Making

Ed Mathez investigates ore deposits to determine the source of rare metals.

Platinum—it is a precious metal, rarer than gold and silver. In fact, most of the world's platinum is mined from just three locations: one in Montana, one in Russia, and one in South Africa. To Ed Mathez, these sites contain more than just precious metals—they contain clues to Earth's past. Mathez, a geologist for the American Museum of Natural History in New York City, studies platinum ore deposits and the geologic processes that led to their formation long ago.

Mathez makes maps of layered igneous rocks in the Bushveld Igneous Complex, South Africa. Layers formed as magma, or molten rock, cooled and crystallized within Earth's crust. The layers at the bottom of the Bushveld Igneous Complex contain high concentrations of dense sulfide deposits: iron, sulfur, and chromium-rich minerals. Sulfide deposits also contain tiny concentrations of platinum. Mathez has determined that the densest layers near the bottom of the formation hold an average of 6 to 8 parts per million (ppm) of platinum. That means for every 1 million atoms, only about 6 are platinum. It's a huge job to extract platinum from the rocks in order to produce profitable amounts of platinum. But it's worth the effort. Today, many important products and new technologies rely on this precious metal.

From Mine to Market

1. **Extraction** Miners extract the metal-rich rock from the ore deposit, crush it, and mix it with a liquid called froth. Metal sticks to the froth, and is skimmed off the top.
2. **Concentration** The froth is dried. The remains are melted in a furnace to separate the metals from the nonmetals.
3. **Refining** Platinum is separated from other metals, such as nickel and gold. Refinement requires chemical reactions to separate metals.
4. **Manufacturing** Platinum is typically mixed with other metals to produce jewelry, electronic equipment, and catalytic converters in automobiles.

▲ Platinum's durability and shiny surface make it a popular choice for jewelry. Platinum helps to convert toxic gas emissions into less harmful gases in catalytic converters.

It's Your Turn

MAKE A LIST With a partner, identify items in your home and school that are made of platinum or other metals. Provide a description of the item, and try to determine what property makes the metal useful. Compare your list with your classmates' lists.

(t)American Museum of Natural History/Jacob Mey; (cr)Tetra Images/Getty Images; (br)Clive Streeter/Getty Images; (bkgd)Jill Vantongeren/AMNH

Lesson 2

Reading Guide

Key Concepts
ESSENTIAL QUESTIONS

- Why is it necessary to use more than one property for mineral identification?
- What properties can you use to identify minerals?

Vocabulary

Multilingual eGlossary

BrainPOP®

How are minerals identified?

Inquiry Two Minerals or One?

This is a variety of the mineral tourmaline called watermelon tourmaline—can you guess why? Tourmaline occurs in many colors—yellow, pink, green, and blue. Why isn't color a reliable property for mineral identification? What other physical and chemical properties can you use to identify this mineral?

Launch Lab

20 minutes

Can you grow crystals from solution?

As solutions evaporate, the dissolved substances in them can crystallize. Try growing crystals of different substances from saturated solutions.

1. Read and complete a lab safety form.
2. Label a **small beaker** *Salt.* Add 20 mL of **hot** (not boiling) **water.** Add one **teaspoon of salt** and stir until dissolved.
3. Continue adding salt slowly until no additional salt dissolves.
4. Repeat steps 2–3 using **alum, Epsom salt,** and **washing soda.** Label each beaker appropriately.
5. Remove 5 mL of each solution with a **dropper.** Use a clean dropper for each solution.
6. Place 5–10 drops of each solution onto a **jar lid** making four separate puddles.
7. Place 2 drops of each solution into a fifth "mixed" puddle.
8. Place the lid in a warm place. Check your solution at regular time intervals for two days.

Think About This

1. Did the substances form crystals? Did all the crystals look the same?
2. **Key Concept** Identify a property that helped you differentiate between crystals in the mixed solution.

Physical Properties

You use minerals every day. How do you suppose scientists discovered the valuable uses for these mineral resources? **Mineralogists,** *scientists who study the distribution of minerals, mineral properties, and their uses,* have identified simple tests to help classify unknown minerals. Using the physical and chemical properties of minerals, you too can discover a mineral's identity.

Color

As you learn how to identify minerals, you will discover that color alone cannot be used for mineral identification. Many different minerals can be the same color. For example, olivine and pyroxene both are green. In contrast, the same mineral can be different colors. For example, quartz can be clear, white, smoky gray, purple, orange, or pink. The watermelon tourmaline on the previous page is pink and green but it can also be yellow and blue. Variations in color reflect the presence of different types of chemical impurities, such as iron, chromium, or manganese. What problems would occur if mineralogists used color alone to identify an unknown mineral?

Key Concept Check Why can't the mineral quartz be classified based on color alone?

FOLDABLES®

Make a half book. Fold it in half again. Label the front of the book as shown. Describe the physical properties used for mineral identification inside the book.

Hutchings Photography/Digital Light Source

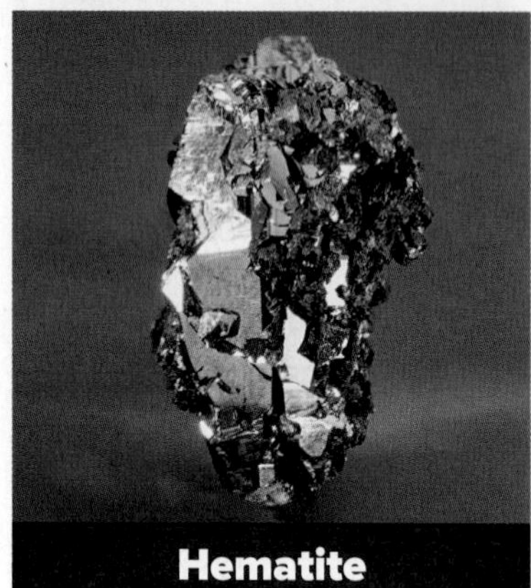

Figure 9 The luster of a mineral is caused by the way light interacts with its surface.

Visual Check
Describe the luster of the minerals talc and hematite.

Luster

What do you notice first when you see a bright metal object, such as the chrome wheel on a bicycle, or a brand new car? Is it the shine? *The way a mineral reflects or absorbs light at its surface is called* **luster.**

Minerals that are also metals, such as copper, silver, and gold, reflect light. This produces the shiniest luster, called metallic luster. Nonmetallic minerals have luster types that might be shiny, but are not reflective like a metal. For example, a cut and polished diamond has brilliant luster. Other descriptions of a mineral's luster include waxy, silky, pearly, and vitreous (VIH tree us), which means glassy. Minerals that lack shiny luster are often called earthy or dull. Luster is directly related to the chemical composition of minerals. **Figure 9** shows the luster of two different minerals, one with metallic luster and one with nonmetallic luster.

Streak

Some minerals, such as graphite, produce a powdery residue when scratched. Rubbing a mineral across an unglazed porcelain plate, called a scratch plate, will sometimes leave a colored streak on its surface. **Streak** *is the color of a mineral in powdered form.* Streak is only useful for identifying minerals that are softer than porcelain.

Nonmetallic minerals generally produce a white streak. Many metallic minerals, however, produce characteristic streak colors. In fact, different samples of the same metallic mineral can vary in color but have the same streak color. As shown in **Figure 10,** all hematite samples (Fe_2O_3) produce a reddish/brown streak, even though some hematite samples are silver or gray while others are red.

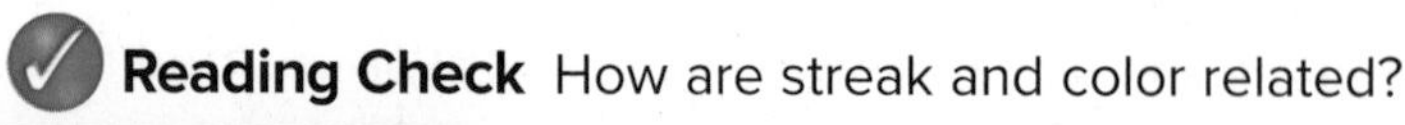

Reading Check How are streak and color related?

Figure 10 These two samples of hematite are different colors, but they both produce the same reddish/brown streak color when scratched on a porcelain plate.

Hardness

Streak relates to the mineral's composition and hardness. **Hardness** *is the resistance of a mineral to being scratched.* Friedrich Mohs, a German mineralogist, developed a scale to compare the hardness of different minerals. The Mohs hardness scale ranges from 1 to 10, as shown in **Table 2.** A hardness value of 1 is assigned to the softest mineral on the scale, talc. Diamond is the hardest mineral on the scale with a hardness value equal to 10.

The mineral quartz has a hardness of 7. If a piece of quartz is rubbed across the surface of a mineral that is softer than quartz (with a hardness less than 7), the quartz will scratch the mineral. Quartz will scratch feldspar, calcite, and talc because each has a hardness less than 7. Quartz will not scratch topaz, corundum, or diamond because they have a hardness greater than 7.

Mineralogists often use ordinary objects to compare the hardness of unknown minerals, as shown in **Table 2.** The hardness of a steel file, a piece of glass, a penny, and your fingernail are known. Mineralogists have added these values to the Mohs hardness scale. For example, a mineral that scratches a penny but not quartz has a hardness between 3 and 7. A mineral that can be scratched by your fingernail has a hardness less than 2.5. As displayed in **Figure 11,** objects of known hardness can help you estimate the hardness of an unknown mineral sample.

Figure 11 A piece of glass can be used in the field or in the lab to test a mineral for hardness.

Table 2 Mohs Hardness Scale

Hardness	Mineral or Ordinary Object
10	Diamond
9	Corundum
8	Topaz
7	Quartz
6.5	Steel file
6	Feldspar
5.5	Glass
5	Apatite
4.5	Iron nail
4	Fluorite
3.5	Penny
3	Calcite
2.5	Fingernail
2	Gypsum
1	Talc

Cleavage and Fracture

Figure 12 Notice that the mineral with cleavage forms flat surfaces where it breaks. Minerals fracture unevenly because their bonds are equal strength in all directions.

Mineral with Cleavage

Mineral with Fracture

Cleavage and Fracture

Sometimes the way a mineral breaks provides clues to a mineral's identity, as shown in **Figure 12.** The arrangement of atoms or ions and the strengths of their chemical bonds determines how a mineral breaks. Minerals break where bonds between atoms or ions are weak. *If a mineral breaks with smooth, flat surfaces, it has* **cleavage.** Minerals such as calcite, the image on the left in **Figure 12,** break along three directions of cleavage, identified as three sets of parallel sides.

WORD ORIGIN

cleavage
from Old English *cleofan,* means "to split, separate"

Other minerals such as quartz, on the right in **Figure 12,** break unevenly because bonds are equally strong in all directions. *If a mineral breaks and forms uneven surfaces, it has* **fracture.** Fracture patterns can be unpredictable. Some minerals, such as asbestos, break into splinters or fibers. Others, such as quartz, break like thick glass with smooth and curved surfaces.

MiniLab

20 minutes

How are cleavage and fracture different?

When a mineral breaks into tiny pieces, it will break where bonds are weakest. Sometimes when bonds between atoms or ions are weak, the mineral breaks and produces a smooth and flat surface. Sometimes minerals break in random and rough patterns.

1. Read and complete a lab safety form.
2. Obtain a **set of minerals** from your teacher.
3. Separate them into two groups—those with cleavage and those with fracture.
4. Determine how many sets of parallel sides each mineral has. Each set of parallel sides is equal to one cleavage direction. Record your observations in your Science Journal.
5. Look at the minerals without cleavage. Describe their surfaces in your Science Journal.

Analyze and Conclude

1. **Evaluate** the number of cleavage directions for each mineral in step 4.
2. **Identify** Did you identify any minerals that exhibit fracture?
3. **Key Concept** How can cleavage or fracture be useful in mineral identification?

Density

Before you pick up a bowling ball, you probably expect it to feel heavy. But when you pick up a volleyball, you know it will feel light. If the bowling ball and the volleyball are about the same size, why is the volleyball lighter? The volleyball is lighter because it is less dense than the bowling ball. *The* **density** *of an object is equal to its mass divided by its volume* (g/cm^3). Since the volumes of the bowling ball and volleyball are approximately equal and the mass of the bowling ball is greater, the bowling ball is more dense than the volleyball. By measuring the mass and the volume of any object, you can calculate an object's density.

Just as you can compare the densities of a bowling ball to a volleyball, you can compare the densities of different minerals without having to measure their mass and volume. If you pick up two different minerals that are about the same volume and hold one in each of your hands, the one that feels heavier is the one with a greater mass and therefore a greater density. With practice, you will be able to identify certain minerals simply based on how heavy they feel.

Special Properties

Some minerals have special properties that help you to identify them. A mineral's texture, or how it feels, might be greasy or smooth to the touch. Graphite feels greasy. Talc is smooth. Some minerals react. Calcite fizzes and produces a gas when it comes in contact with hydrochloric acid. Some minerals have distinctive odors. Sulfur smells like a match, and kaolinite smells like clay. Fluorescence, shown in **Figure 13,** is a mineral's ability to glow when exposed to ultraviolet light. Calcite and fluorite are two common minerals that fluoresce. Some minerals, such as the magnetite shown in **Figure 13,** are magnetic.

Key Concept Check Identify all the properties used to classify an unknown mineral.

Special Properties

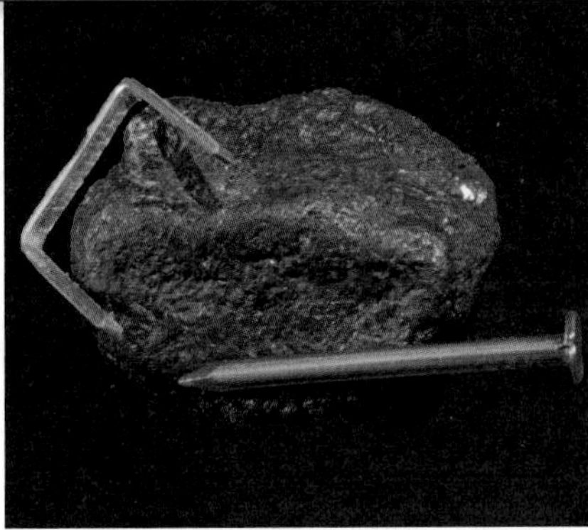

Figure 13 Calcite is fluorescent when exposed to ultraviolet light. Magnetite is magnetic due to the presence of iron in its chemical formula.

Lesson 2 Review

Visual Summary

Streak is the color of a mineral in powdered form.

Minerals vary in hardness. Hardness is the resistance of a mineral to being scratched.

Minerals with special properties such as fluorescence can be easier to identify.

FOLDABLES®

Use your lesson Foldable to review the lesson. Save your Foldable for the project at the end of the chapter.

What do you think NOW?

You first read the statements below at the beginning of the chapter.

3. The best way to identify a mineral is by color.

4. Hardness, streak, and luster are among the properties used to identify minerals.

Did you change your mind about whether you agree or disagree with the statements? Rewrite any false statements to make them true.

Use Vocabulary

1 **Define** *hardness* in your own words.

2 **Distinguish** between cleavage and fracture used in mineral identification.

3 **Use the term** *luster* in a sentence.

Understand Key Concepts 

4 What do symmetrical crystal shapes indicate about a mineral?

A. a hardness of 8 on Mohs hardness scale
B. a metallic luster
C. an orderly arrangement of atoms
D. a tendency to vary in color

5 **Observe** Look at the minerals in the photos below. Are they the same mineral? Or are they different minerals? Explain.

Interpret Graphics

6 **Complete** a graphic organizer that describes the five main properties used to determine an unknown mineral's identity.

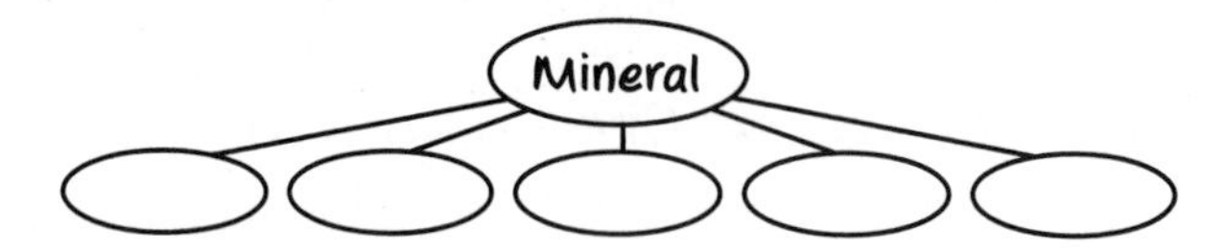

Critical Thinking

7 **Describe** how to distinguish between a mineral with a hardness of 6 and one with a hardness of 4 using only a glass plate and a copper penny.

8 **Critique** the following passage: Most samples of the mineral kaolinite do not have well-formed crystals. These samples of kaolinite do not have an orderly internal arrangement of atoms.

Materials

mineral samples

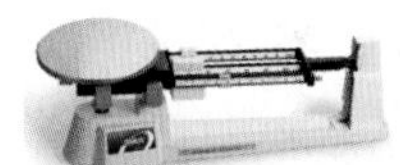

balance

100-mL graduated cylinder

Safety

How do you determine the density of a mineral?

Density is a physical property often used in mineral identification. To compare the densities of different minerals, you must measure the mass of each mineral. Mass refers to how much matter a substance contains. Density is equal to the mass of an object divided by its volume (g/cm^3). The volume of an irregular solid, such as a mineral, is equal to the amount of water it displaces. 1 mL of water = 1 cm^3.

Learn It

To determine the identity of an unknown mineral, mineralogists examine and **measure** the physical and chemical properties of the object. Measurement involves noting characteristics such as the mass and volume of an object. Equations such as the one for density, $D = m \div V$, can be used to help identify an unknown mineral.

Try It

1. Read and complete a lab safety form.
2. Use the balance to measure the mass of each mineral sample. Record the mass for each mineral in a table like the one shown below in your Science Journal.
3. Place about 50 mL of water in a 100-mL graduated cylinder. Record the exact volume of water in your data table.
4. Tie a string around an unknown mineral and carefully lower it into the graduated cylinder until it is below the water's surface but not touching the bottom. Read the final volume of water and record this volume in your data table.
5. Remove the mineral from the water. Calculate the volume of your mineral sample and record in your data table.
6. Repeat steps 3-5 for each of the mineral samples.

Apply It

7. Calculate the density of each of your mineral samples. How could you use the densities to identify the minerals?
8. **Key Concept** Would density or color be more useful in identifying a particular mineral? Explain your answer.

Sample	Mass (g)	Volume of Water (mL)	Volume of Water + Mineral (mL)	Volume of Sample mL=cm^3	Density (mass/volume)
1					
2					

Lesson 3

Sources and Uses of Minerals

Reading Guide

Key Concepts
ESSENTIAL QUESTIONS

- How are minerals used in your daily life?
- Why are minerals a valuable resource?

Vocabulary

ore p. 95

gemstone p. 98

Multilingual eGlossary

Science Video

Is this lava?

The material in this photo is clearly molten. This material is not lava. It is molten iron ore. Iron is mixed with carbon in large furnaces to produce this hot, molten mixture. The result is a cleaner, stronger iron metal. Iron ore is a metal resource that you use daily. Do you use any other minerals? Where do these minerals come from? What makes mineral resources valuable?

Launch Lab

20 minutes

What are common uses of minerals?

We use minerals for many things every day. Create a card game to learn the uses of minerals.

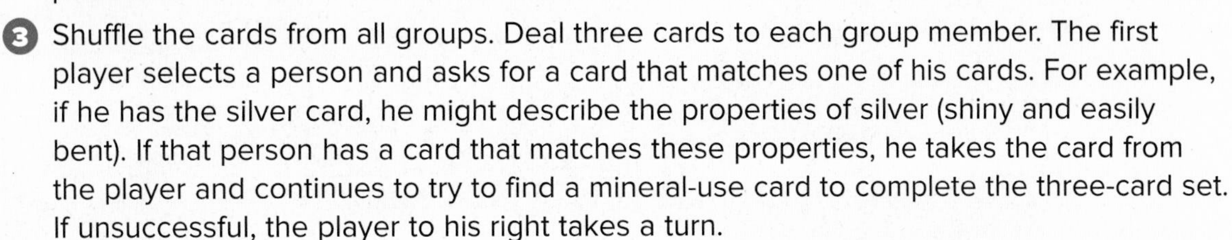

1. Observe the materials on your lab table. Are they strong, easily bent, or brittle? Recall the properties of minerals and use these properties to describe materials.
2. Work in groups of three. Make a three-card set for each material: Card 1—write the name of a mineral; Card 2 —describe the mineral's properties; Card 3—describe how the mineral is used in the material provided.
3. Shuffle the cards from all groups. Deal three cards to each group member. The first player selects a person and asks for a card that matches one of his cards. For example, if he has the silver card, he might describe the properties of silver (shiny and easily bent). If that person has a card that matches these properties, he takes the card from the player and continues to try to find a mineral-use card to complete the three-card set. If unsuccessful, the player to his right takes a turn.
4. The game ends when all cards are in sets on the table. The person with the most sets wins.

Think About This

1. Identify a few minerals, their properties, and their common uses in everyday life.
2. **Key Concept** How do you think minerals are used in your daily life?

Mineral Resources

Think about all the rock, brick, mineral, and metal resources that were used to build your home and your school. Where did these resources come from?

The average person uses 22,000 kg of mineral resources each year. For example, copper is used in electric wiring and plumbing fixtures, and quartz is used to make glass and ceramics. The automotive industry; agriculture and food production; and road, home, and building construction use mineral resources.

Recall that minerals form in a variety of different environments. Sometimes these environments are deep within Earth and difficult to find. Other times, the resource might be a surface feature, such as a salt lake deposit. People mine these mineral resources because they are useful in everyday life. These mined materials, however, must contain enough of the mineral or rock resource to produce a profit. *Rock that contains high enough concentrations of a desired substance, such as a metal, so that it can be mined for a profit is called an* **ore.** For example, the ore bauxite shown in **Figure 14** is a profitable source of aluminum that is used in electronics, transportation, and the food industry.

Figure 14 Aluminum is mined from the mineral bauxite and is a profitable ore in a variety of industries.

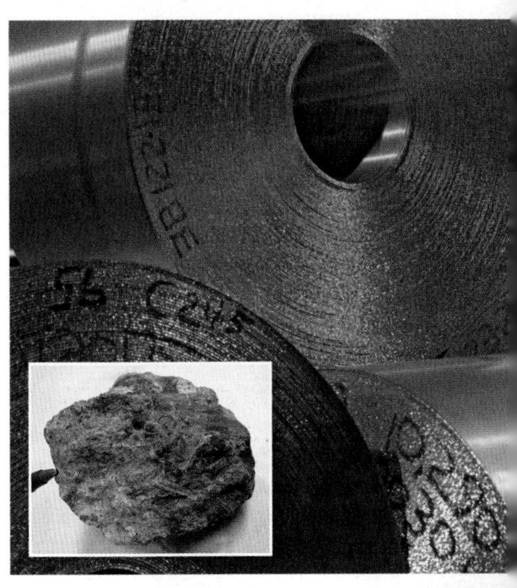

Mineral Uses

Figure 15 Minerals are valuable resources used to construct many parts of a home.

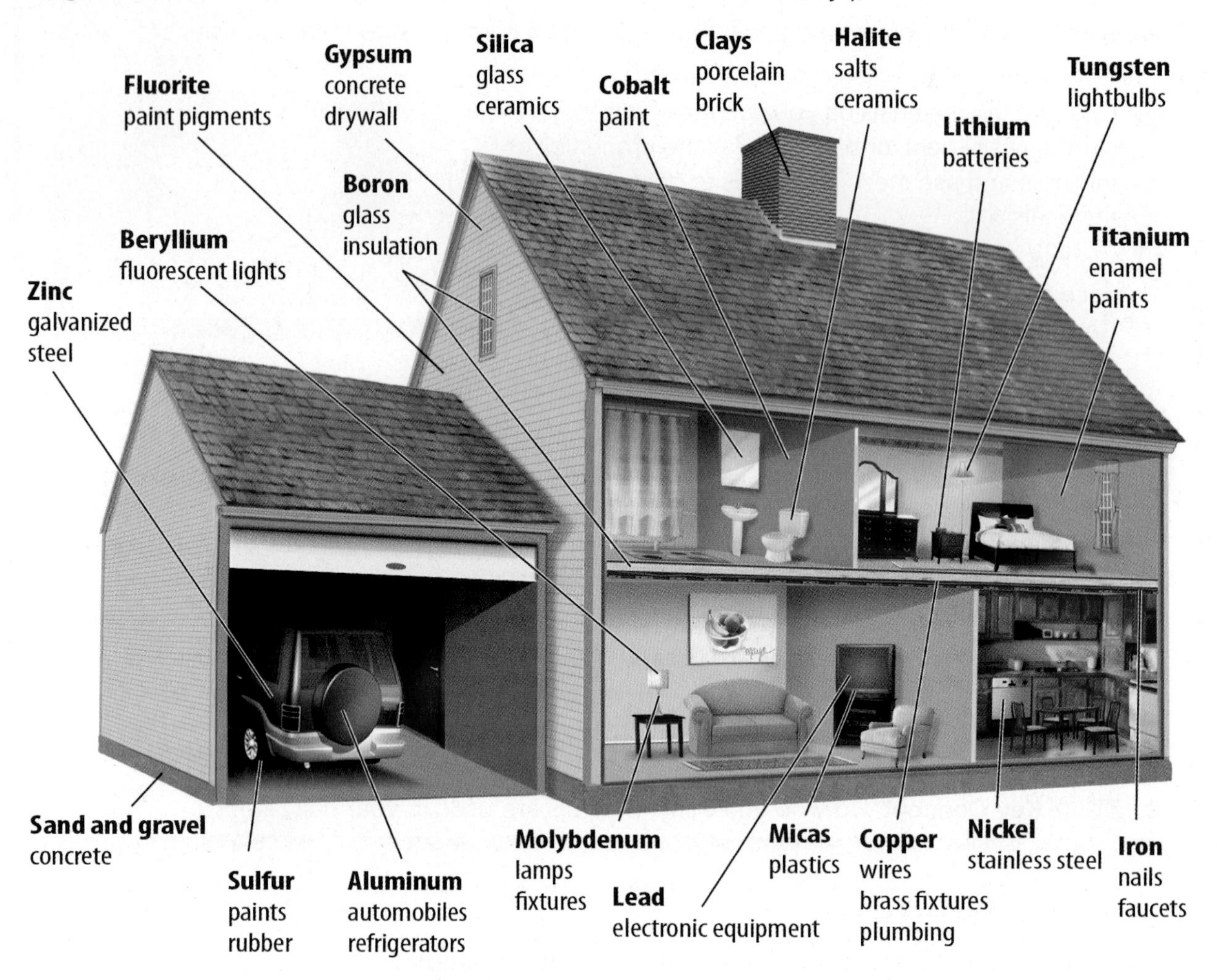

Metallic Mineral Resources

Ores of the elements iron (Fe) and aluminum (Al) are among the most abundant of the metallic mineral resources used every day. For example, the minerals hematite (Fe_2O_3) and magnetite (Fe_3O_4) are important sources of iron. Iron is the main ingredient in the steel used to construct buildings and bridges and to manufacture automobiles and trains. Iron is also a common ingredient in the nails, screws, and fixtures in your home, as shown in **Figure 15.**

Metals are also used in the food industry, such as in the manufacture of aluminum cans for food and beverages. Aluminum is abundant in Earth's crust, but rarely occurs as a native element. The mineral bauxite is a mixture of aluminum and other elements. Bauxite is mined and aluminum is extracted to manufacture a variety of products. The mining industry processes more than 2.7×10^{10} kg of aluminum each year.

Reading Check What element is a common ingredient of steel?

Make a two-tab book and label as shown. Record information about the common uses of minerals in your everyday life.

Rare Metals

Gold occurs in a ratio of 1 part gold to 4 billion parts rock in Earth's crust. Gold occurs in concentrations that are, however, large enough to be mined for a profit. Its lustrous yellow color and metallic properties make gold a desirable metal for making jewelry. Gold also conducts electricity and does not corrode. It has many scientific and industrial uses.

The technology industry is dependent upon other metallic mineral resources. Platinum is used in catalytic converters to help regulate harmful gas emissions from automobiles. The converters change these gases into CO_2 and water. Scientists are currently researching the use of platinum in fuel cells for electric cars. These new cars will not produce harmful gas emissions and will be better for the environment as a result.

Nonmetallic Mineral Resources

Every day, humans use minerals that are not ores. Raw materials used for road construction, ceramic products, building stone, and fertilizers are all examples of nonmetallic mineral resources. As shown in **Figure 16,** the construction of a building and a parking lot requires many mineral resources.

The sand you might have played in when you were young is also a nonmetallic mineral resource. Sand is commonly composed of particles of the mineral quartz (SiO_2).

Key Concept Check List at least five examples of minerals and their common use.

Other Mineral Resources

Figure 16 The construction of this building and parking lot requires several different rock and mineral resources. Can you identify any of these resources?

Math Skills

Use Percentages

The amount of metal that can be obtained from an ore is called the percent yield. For example, when 500 kg of iron oxide (Fe_3O_4) is processed, 308 kg of pure iron (Fe) is produced. What is the percent yield?

1. Express the numbers as a fraction.

$$\frac{308 \text{ kg Fe}}{500 \text{ kg } Fe_3O_4}.$$

2. Convert the fraction to a decimal.

$$\frac{308}{500} = 0.616$$

3. Multiply by 100 and add %.

$$0.616 \times 100 = 61.6\%$$

Practice

If the 500 kg of ore in the previous example is ground up before processing, 410 kg of iron (Fe) is produced. How much does grinding improve the percent yield?

Math Practice

Personal Tutor

Interactive Table

Gemstones

A **gemstone** *is a rare and attractive mineral that can be worn as jewelry.* Minerals such as diamonds and rubies take on special characteristics when they are cut and polished. The brilliant luster of a cut and polished diamond is what makes diamonds valuable gemstones. But the physical properties of gemstones also make them useful in industry. For example, on Mohs' hardness scale a diamond has a hardness of 10 and corundum has a hardness of 9. Because of their hardness, they are commonly used in abrasives and in cutting tools. Of course large gem-quality diamonds would not be used for these purposes. In fact, many industrial gemstones are actually synthetic stones manufactured by humans. Sometimes human-made gems are less expensive than the same natural gems. **Table 3** shows examples of common gemstones and their mineral names.

Reading Check What are diamonds used for?

Table 3 Some Natural Gemstones

Gemstone	Mineral Name and Chemical Formula
Emerald	beryl $Be_3Al_2Si_6O_{18}$
Sapphire	corundum Al_2O_3
Ruby	corundum Al_2O_3
Diamond	diamond C
Peridot	olivine $(Mg,Fe)_2SiO_4$
Amethyst	quartz SiO_2

MiniLab

20 minutes

How are minerals used in our daily lives?

Minerals are natural resources that you use daily. Metal pots, ceramic dishes, and toothpaste are a few examples of things that are made from minerals.

1. Read and complete a lab safety form.
2. Place a small scoop of **talc** on a **black piece of paper.** Observe the talc and rub it between your fingers. Record your observations in your Science Journal.
3. Dip a **damp paper towel** into the talc. Rub a **nail** with the talc using 20 strokes. Record your observations.
4. Repeat steps 2 and 3 with **sand.**

Analyze and Conclude

1. **Compare** how the talc and the sand felt between your fingers.
2. Describe the effect of the talc and the sand on the nail.
3. **Key Concept** Explain why talc is used as an ingredient in body powder and why sand is better than talc as an abrasive. Explain why minerals are a valuable resource in your everyday life.

Lesson 3 Review

Visual Summary

An ore contains high enough concentrations of a desired substance that it can be mined at a profit.

Metallic mineral resources are used in the construction of buildings, cars, and planes.

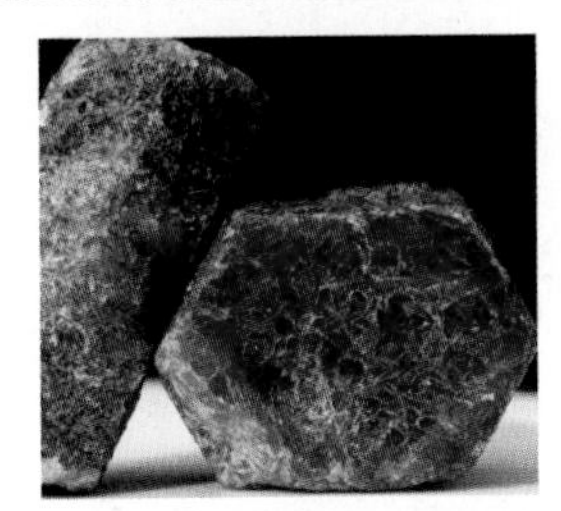

A gemstone is a valuable mineral known for its beauty, rarity, or durability.

FOLDABLES®

Use your lesson Foldable to review the lesson. Save your Foldable for the project at the end of the chapter.

What do you think NOW?

You first read the statements below at the beginning of the chapter.

5. An ore is a concentration of minerals that contains only iron.

6. Gemstone and ore deposits are evenly distributed around the world.

Did you change your mind about whether you agree or disagree with the statements? Rewrite any false statements to make them true.

Use Vocabulary

1. **Use the term** *ore* in a sentence.
2. **Describe** gemstones in your own words.
3. **Compare and contrast** metallic and nonmetallic mineral resources.

Understand Key Concepts

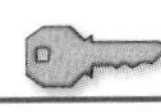

4. Aluminum is extracted from ______ ore.
   A. bauxite
   B. hematite
   C. magnetite
   D. quartz
5. **Identify** five products derived from mineral resources.
6. Which mineral is an important source of iron ore?
   A. feldspar
   B. hematite
   C. mica
   D. quartz

Interpret Graphics

7. **Organize** Copy and fill in the graphic organizer below. In each oval, list one product made from nonmetallic mineral resources.

Critical Thinking

8. **Interpret** the following statement: A large gem-quality diamond grown in a laboratory is useful as a gemstone, but it is not a mineral.

Math Skills

9. During the California Gold Rush in the 1870s, miners obtained an average of 2,700 kg of gold from every 5,000 kg of ore. What was the percent yield of gold?
10. A 3,000-kg sample of bauxite (aluminum ore) discovered in Australia was mined and processed to produce 750 kg of aluminum. What is the percent yield?

Mineral Detective

Detectives gather physical evidence to determine what took place during a crime. Much like detectives, geologists gather physical and chemical evidence to help classify unknown minerals. Once minerals are identified, geologists can often interpret how they formed.

Question

What physical and chemical properties can you use to identify a mineral?

Procedure

1. Read and complete a lab safety form.
2. Examine the mineral samples. With your group, discuss which physical and chemical properties might be most useful to identify each mineral.

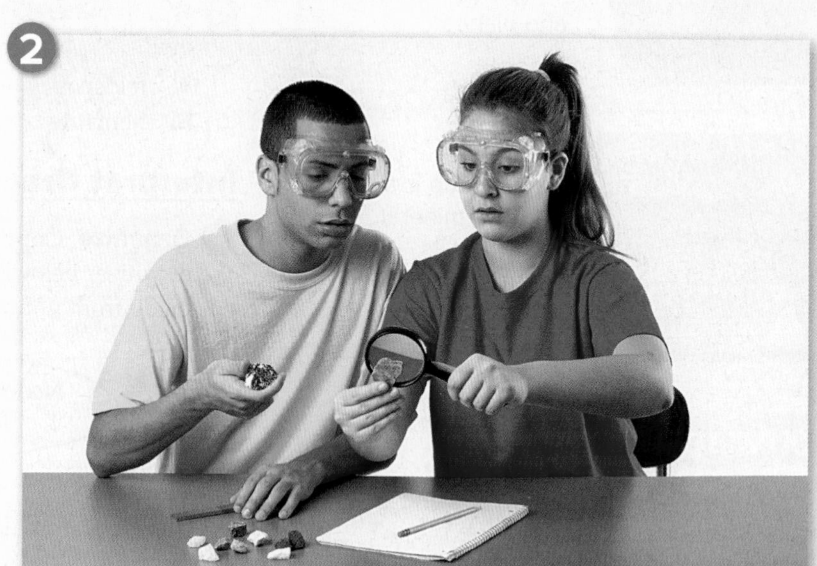

3. Copy the table below in your Science Journal.

Sample ID	Luster	Hardness	Streak	Cleavage or Fracture	Color	Other Properties	Mineral Name
A	metallic	5.5	red brown	no cleavage	rust red		hematite

Materials

6–8 mineral samples

magnifying lens

magnet

balance

steel nail

penny

Also needed: 5% HCl, glass plate, mineral identification flow chart, porcelain tile, 100-mL graduated cylinder

Safety

WARNING: *If an HCl spill occurs, notify your teacher and rinse with cool water.*

4. Complete the following steps for each mineral sample and record your observations in the table in your Science Journal.

- Is the luster of the sample metallic or nonmetallic?
- Use your fingernail, a penny, an iron nail, and the glass plate to determine and record the hardness of each sample. Refer to Mohs hardness scale in Lesson 2 of this chapter.
- Use the porcelain tile to determine the sample's streak color.
- Does the sample show cleavage or fracture?
- What color is the sample?
- Record any other properties such as smell, magnetism, fluorescence, and so on.

5. Read through the mineral identification flow chart. Then use the table and the flow chart to identify each sample. Record the mineral's name in the table in your Science Journal.

Analyze and Conclude

6. **Compare and Contrast** Which properties were the most useful for mineral identification? Which were the least useful? Explain.
7. **The Big Idea** Describe at least three ways in which the physical properties of minerals contribute to their everyday uses.
8. **Describe** Which samples were the most difficult to identify? Explain.

Communicate Your Results

In small groups, create a graphic organizer that outlines the steps you used to identify the unknown minerals in this lab activity.

How might you more accurately measure the hardness of a mineral, particularly one with a hardness greater than 6? Are there any additional tests you could have used to identify minerals in lab?

Lab Tips

☑ Metallic luster is often described as shiny. Nonmetallic luster may be described as vitreous (like glass), pearly, greasy, silky, dull, or brilliant, like a diamond.

☑ Cleavage planes can be identified by looking for sets of smooth and flat, parallel sides.

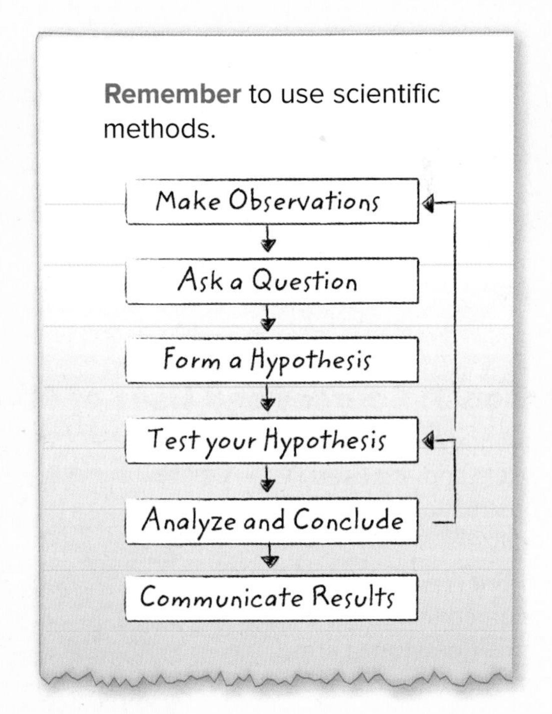

Chapter 3 Study Guide

Minerals are naturally occurring, inorganic solids with a definite chemical composition and an orderly arrangement of atoms or ions. Minerals are used in everyday materials and as gemstones.

Key Concepts Summary

Lesson 1: What is a mineral?

- A **mineral** is a naturally occurring, inorganic solid with a definite chemical composition and crystalline form.
- The common rock-forming minerals come from the **silicate** family and the non-silicate family.
- Minerals form from **crystallization** of hot and cool solutions above and below Earth's surface. They also form from cooling **magma** and **lava.**

Vocabulary

mineral p. 77
silicate p. 81
crystallization p. 81
magma p. 83
lava p. 83

Lesson 2: How are minerals identified?

- The same mineral can exist in many different colors due to chemical impurities. More than one mineral can be the same color.
- Minerals are identified by their physical properties: color, **luster, streak, hardness, cleavage, fracture, density,** and other special properties.

Vocabulary

mineralogist p. 87
luster p. 88
streak p. 88
hardness p. 89
cleavage p. 90
fracture p. 90
density p. 91

Lesson 3: Sources and Uses of Minerals

- Minerals are sources of metals and are used in construction materials and fertilizers.
- An **ore** is a metallic mineral resource mined for a profit. Some **gemstones** are valuable because they are rare, beautiful, and durable.

Vocabulary

ore p. 95
gemstone p. 98

(t)Phil Robinson/age fotostock; (cl)Dr Parvinder Sethi; (c)Jesus Ayala/Getty Images; (cr)RF Company/Alamy; (b)Matteo Chinellato-ChinellatoPhoto/Getty Images

 Personal Tutor

 Vocabulary eFlashcards
Vocabulary eGames

FOLDABLES® Chapter Project

Assemble your lesson Foldables as shown to make a Chapter Project. Use the project to review what you have learned in this chapter.

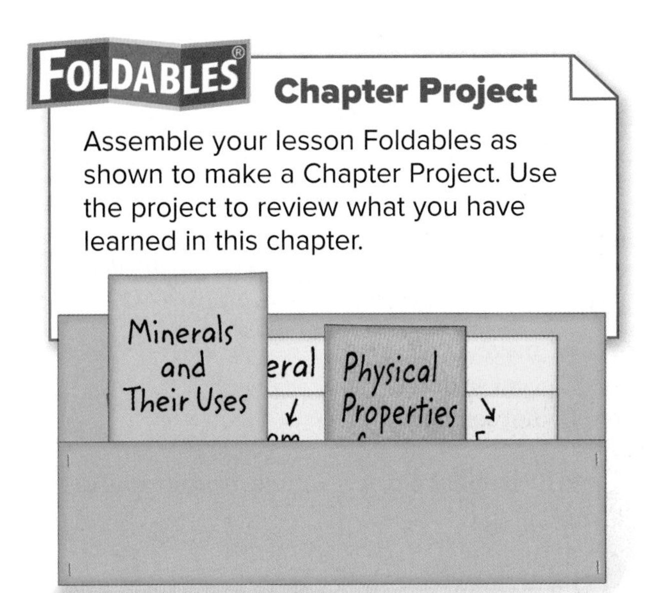

Use Vocabulary

1. The ________ of a mineral can be calculated by dividing the mass of the mineral by its volume.
2. Mohs scale is used to describe the relative ________ of a mineral.
3. How does streak differ from color?
4. When particles of dissolved solids in water arrange themselves in an orderly repeating pattern, the process of ________ forms a mineral.
5. Use the word *lava* in a sentence.
6. Define the term *ore* in your own words.
7. A(n) ________ is a solid that is beautiful or rare, yet durable enough to be used for adornment or as an art object.
8. Minerals that break along planes of weakness exhibit ________.
9. Another word for the molten material beneath Earth's surface is ________.

 Interactive Concept Map

Link Vocabulary and Key Concepts

Copy this concept map, and then use vocabulary terms from the previous page to complete the concept map.

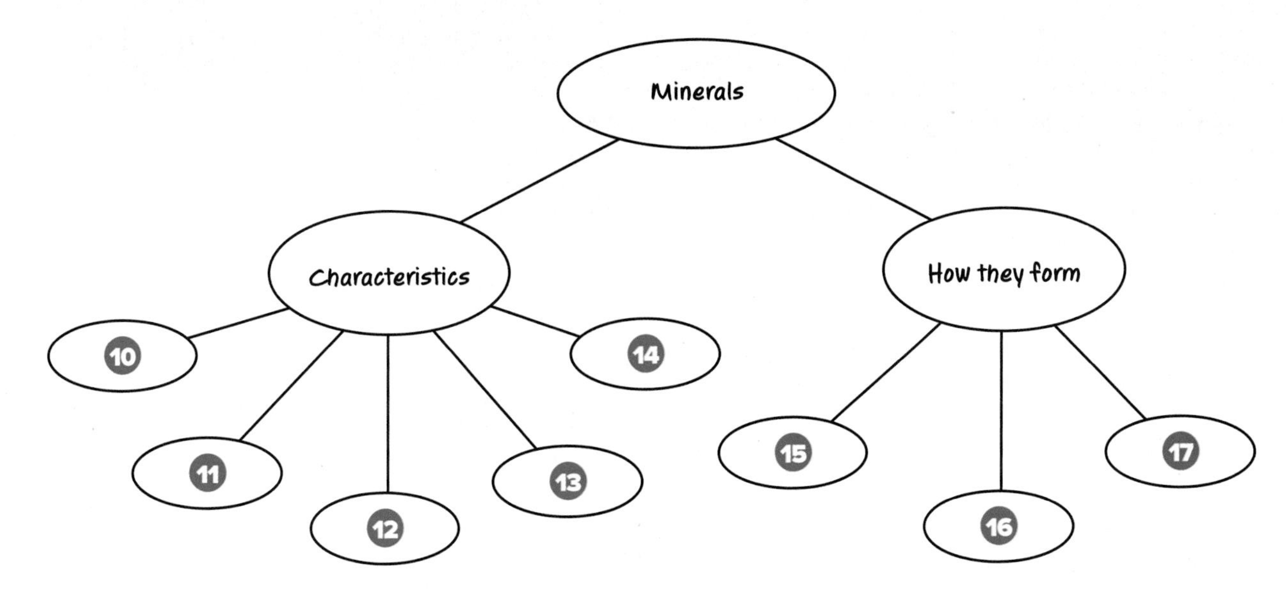

Chapter 3 Review

Understand Key Concepts 

1 A student observes that an unknown mineral scratches a glass plate (hardness 5.5) and a sample of quartz (hardness 7). What else does the student know about the unknown mineral?

A. Its hardness is between 5.5 and 7.
B. Its hardness is greater than 7.
C. Its hardness is less than 7.
D. Its hardness is less than 5.5.

2 Examine the mineral below. What feature should you look for to verify that a mineral exhibits cleavage?

A. hexagonal-shaped crystals
B. uneven, dull surfaces
C. smooth, flat surfaces
D. wavy lines running through the sample

3 Which tool is used to determine a mineral's streak?

A. copper penny
B. glass plate
C. unglazed porcelain tile
D. your fingernail

4 What physical properties result from the way light interacts with a mineral?

A. color and density
B. color and luster
C. density and luster
D. hardness and luster

5 Minerals cannot be

A. crystalline.
B. naturally occurring.
C. organic.
D. solids.

6 Which property causes halite to break into cubes?

A. cleavage
B. density
C. hardness
D. luster

7 A gold vein forms in which of the following crystallization environments?

A. lava
B. magma
C. cool solution
D. hot solution

8 Which mineral reacts with hydrochloric acid?

A. calcite
B. fluorite
C. gypsum
D. quartz

9 Examine the minerals below. Dull, vitreous, pearly, earthy, and metallic are terms that best describe a mineral's

A. cleavage
B. density
C. luster
D. streak

10 Which group of minerals is composed of mainly silicon and oxygen?

A. carbonates
B. halides
C. oxides
D. silicates

Critical Thinking

11. **Identify** a crystal that is not a mineral.

12. **Design** a flowchart to help you identify 10 common minerals using at least 3 of the physical properties you learned about in Lesson 2.

13. **Observe** the arrangement of ions in the mineral galena below. Note the angles the ions make with one another. Predict the crystal shape of galena.

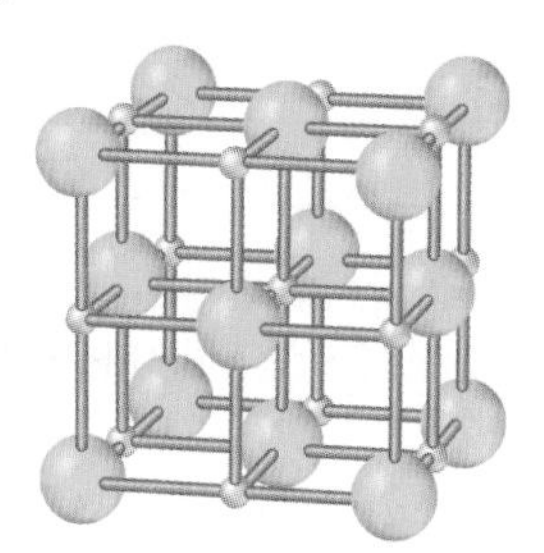

14. **Compare and contrast** how a mineral forms from cool solutions to how a mineral forms from hot solutions.

15. **Infer** What can you infer about the formation of a cluster of well-formed quartz crystals?

16. **Infer** Why are minerals that form deep below Earth's surface unstable at Earth's surface?

17. **Create** a booklet of illustrations to show environments where minerals may form.

Writing in Science

18. **Defend** the following statement: Many minerals form in environments that people cannot see. Scientists design experiments to make minerals at the high pressures and temperatures that cannot be observed directly.

19. **Hypothesize** Minerals are considered to be nonrenewable. With this in mind, why do you think recycling of products like glass and aluminum is important?

REVIEW THE BIG IDEA

20. What are minerals? Distinguish among minerals and at least one other type of solid that forms naturally.

21. How are minerals used in daily life? List two examples of how you use minerals each day.

Math Skills

Math Practice

Use Percentages

22. The longer a metal is mined in a certain location, the lower the yield from the remaining ore. In 1955, copper ore from a mine in Butte, Montana, produced 6,000 kg of copper (Cu) from 20,000 kg of copper ore. What was the percent yield? Today 100 kg of Cu can be produced from 20,000 kg of ore. What is the current percent yield?

23. The average percent yield of silver from ore discovered in California during the 1870s, was 48 percent. Use the percent yield to calculate the mass of silver that could be obtained from 70,000 kg of silver ore.

Standardized Test Practice

Record your answers on the answer sheet provided by your teacher or on a sheet of paper.

Multiple Choice

1 Which is one of the two main families of rock-forming minerals?

A carbonates

B elements

C oxides

D silicates

Use the diagram below to answer question 2.

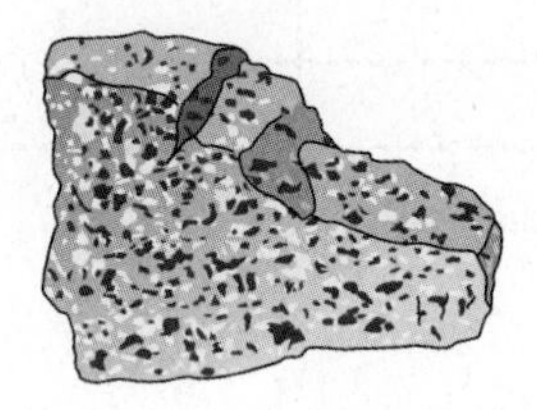

2 The igneous rock above most likely

A contains very few minerals.

B cooled quickly.

C cooled slowly.

D formed on Earth's surface.

3 The process of crystallization

A breaks off particles from solids.

B forms ALL of Earth's minerals.

C is limited to cool solutions.

D only occurs in dry environments.

4 Which is the softest mineral?

A calcite

B diamond

C glass

D talc

5 Which is a gemstone?

A coral

B gypsum

C quartz

D ruby

Use the table below to answer questions 6 and 7.

Substances	Uses
Aluminum	automobiles, soft drink containers
	automobiles, nails, faucets
Lead	batteries, electronic equipment
Sulfur	tires, matches
Nickel	stainless steel
Copper	

6 Refer to the table above. The unknown mineral in the *Substances* column is likely

A clay.

B cobalt.

C iron.

D silica.

7 Identify one of the common uses of copper to complete the table above.

A concrete

B insulation

C plastics

D plumbing

8 A penny has a hardness of 3. Glass has a hardness of 5.5. Based on this information, which statement is true?

A Each can scratch the other.

B Glass can scratch the penny.

C Neither can scratch the other.

D The penny can scratch glass.

9 Minerals can form crystals as ________ cools.

A conditioned air

B gasoline

C magma

D purified water

Use the diagram below to answer question 10.

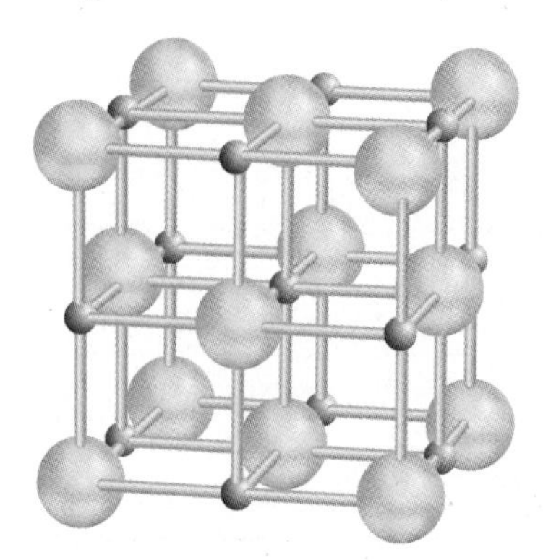

10 The diagram above shows the internal arrangement of ions in the mineral halite. What is halite's crystal shape?

A circular

B cubic

C cylindrical

D elliptical

11 Which type of rock contains enough of a desired substance that it can be mined for a profit?

A diamond

B lava

C ore

D stone

12 Minerals are considered inorganic because they

A are nonliving.

B exist in liquids.

C form crystals.

D originate underground.

Constructed Response

Use the table below to answer questions 13 and 14.

Properties of Minerals	Description
Color	
Special properties	

13 In the table above, identify all of the properties scientists use to classify minerals. Describe each property.

14 Why is knowledge of mineral properties important? Which properties might be most useful in construction? In jewelry-making? Explain.

Use the diagram below to answer question 15.

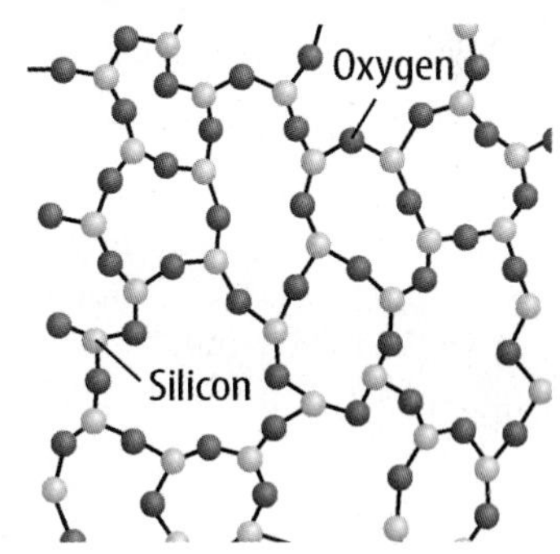

15 The arrangement of atoms in glass is shown above. Explain why glass cannot be a mineral.

NEED EXTRA HELP?															
If You Missed Question...	1	2	3	4	5	6	7	8	9	10	11	12	13	14	15
Go to Lesson...	1	1	1	2	3	3	3	2	1	1	3	1	2	2	1

Chapter 4

Rocks

THE BIG IDEA How do the three main types of rocks form?

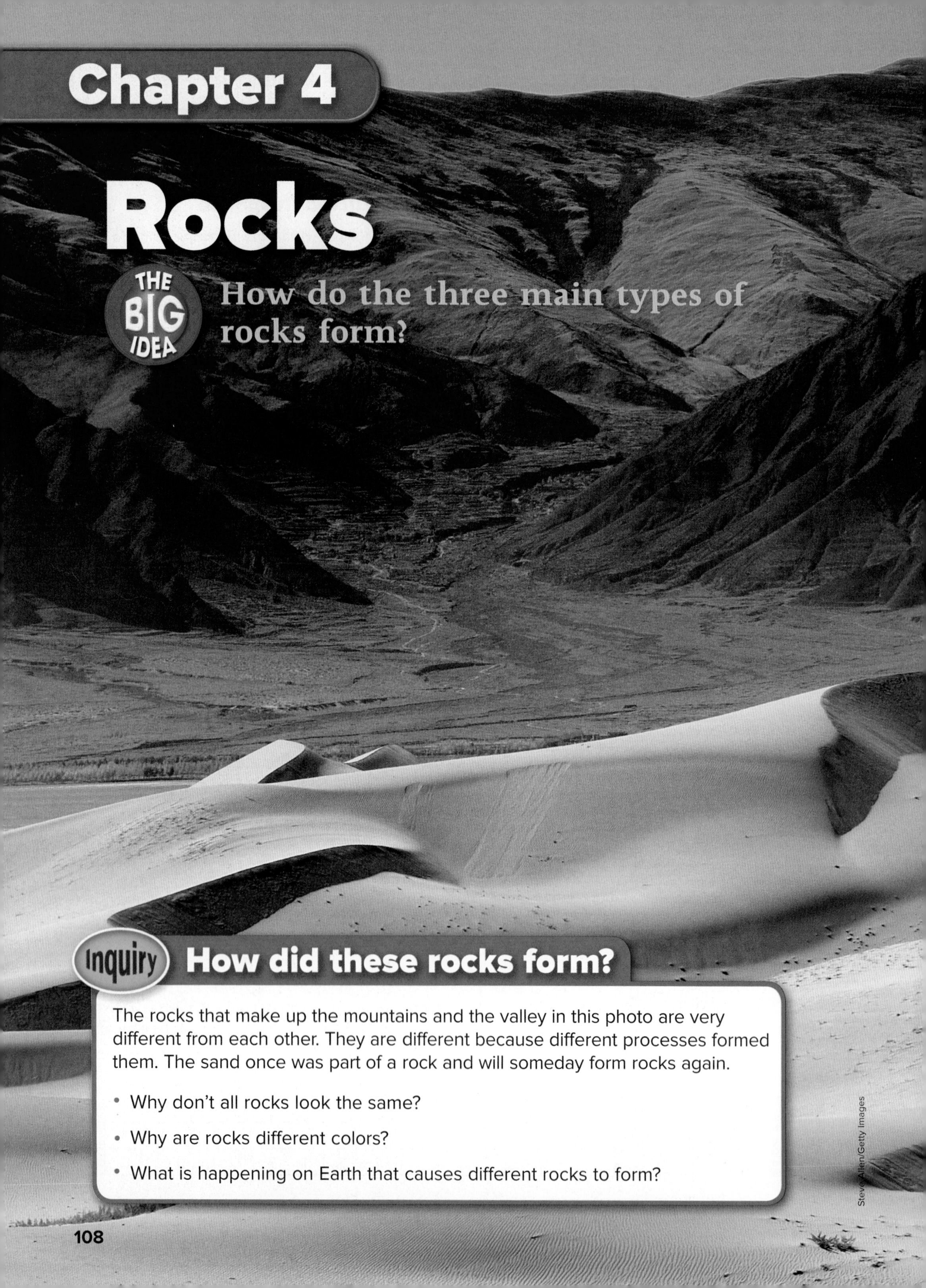

Inquiry How did these rocks form?

The rocks that make up the mountains and the valley in this photo are very different from each other. They are different because different processes formed them. The sand once was part of a rock and will someday form rocks again.

- Why don't all rocks look the same?
- Why are rocks different colors?
- What is happening on Earth that causes different rocks to form?

Steve Allen/Getty Images

Get Ready to Read

What do you think?

Before you read, decide if you agree or disagree with each of these statements. As you read this chapter, see if you change your mind about any of the statements.

1. Once a rock forms as part of a mountain, it does not change.
2. Some rocks, when exposed on Earth's surface, undergo weathering and erosion.
3. Large crystals form when lava cools quickly on Earth's surface.
4. Igneous rocks form when cooling magma crystallizes.
5. Water can dissolve rock.
6. All sedimentary rocks on Earth formed from the remains of organisms that lived in oceans.
7. With the right pressure and temperature conditions, minerals in a rock can change shape without breaking or melting.
8. Metamorphic rocks have layers that form as minerals melt and then recrystallize.

Your one-stop online resource
connectED.mcgraw-hill.com

LearnSmart®

Chapter Resources Files, Reading Essentials, Get Ready to Read, Quick Vocabulary

Animations, Videos, Interactive Tables

Self-checks, Quizzes, Tests

Project-Based Learning Activities

Lab Manuals, Safety Videos, Virtual Labs & Other Tools

Vocabulary, Multilingual eGlossary, Vocab eGames, Vocab eFlashcards

Personal Tutors

Lesson 1

Reading Guide

Key Concepts
ESSENTIAL QUESTIONS
- How are rocks classified?
- What is the rock cycle?

Vocabulary

rock p. 111
grain p. 111
texture p. 112
magma p. 113
lava p. 113
sediment p. 113
rock cycle p. 114

Multilingual eGlossary

BrainPOP®
Science Video

Rocks and the Rock Cycle

Inquiry What formed this feature?

Over time, this stream has slowly carved a channel into layers of rock and ash from a volcanic eruption. Notice the sediment in the foreground. Where did all of this sediment come from? What will happen to this sediment over time?

Launch Lab

20 minutes

What's in a rock?

You've probably seen different types of rock, either outside or in photographs. Rocks have different colors and textures, and they can contain a combination of minerals, shells, or grains. In this activity, you will observe differences among rock samples.

1. Read and complete a lab safety form.
2. Obtain a few **rock samples** from your teacher.
3. Examine each rock, both with and without a **magnifying lens.**
4. Describe each rock sample in detail. Record the color and texture, and describe the minerals or grains in the rock for each sample in your Science Journal.

Think About This

1. Write a brief description for each rock sample in your Science Journal. Identify the ways in which your samples are similar and different.
2. **Key Concept** Do you think all rocks form in the same way? Explain.

Rocks

Rocks are everywhere. Mountains, valleys, the seafloor, and beaches like the one shown in **Figure 1**, contain rocks. Rock and mineral resources even make up parts of your home. Today it is common for floors, countertops, and even tabletops to be made of some type of rock.

A **rock** *is a natural, solid mixture of minerals or grains.* Individual mineral crystals, broken bits of minerals, or rock fragments make up these grains. Sometimes a rock contains the remains of organisms or volcanic glass. Processes on Earth's surface can cause rocks to break apart into many different-sized fragments. *Geologists call the fragments that make up a rock* **grains.** They use a grain's size, shape, and chemical composition to classify rocks.

REVIEW VOCABULARY

mineral
a naturally occurring, inorganic solid, with a definite chemical composition, and an orderly arrangement of atoms

Figure 1 Rocks are everywhere on Earth. By studying rocks, geologists can gain a better understanding of the processes that create different rock types and the environments in which they form.

Granite

Conglomerate

Figure 2 Geologists use texture and composition to classify these rocks as granite and conglomerate.

Visual Check
Compare and contrast the grain shape and size in each of these rocks.

Texture

Geologists use two important observations to identify rocks: texture and composition. *The grain size and the way grains fit together in a rock are called* **texture.** When a geologist examines a rock's texture, he or she looks at the size of minerals or grains in the rock, the arrangement of these individual grains, and the overall feel of the rock.

Texture can be used to determine the environment in which a rock formed. The granite shown in **Figure 2** has large mineral crystals. This colorful, crystalline texture helps a geologist classify this rock as an igneous rock. The conglomerate (kun GLAHM uh rut) shown in **Figure 2** has rounded rock fragments. Well-rounded rock fragments imply that strong forces, such as water or ice, carved the individual clasts and produced smooth surfaces. This is a sedimentary rock. You will learn more about igneous, sedimentary, and a third rock type–metamorphic rocks–in the lessons that follow.

Composition

The minerals or grains present in a rock help geologists identify the rock's composition. This information can be used to determine where the rock formed, such as inside a volcano or alongside a river. Geologists conduct fieldwork using maps, a field journal, a compass, a rock hammer, and other tools to examine a rock's composition and texture, as shown in **Figure 3.** These tools also help geologists to interpret the specific conditions under which the rock formed. For example, the presence of certain minerals might suggest that the rock formed under extreme temperature and pressure. Other minerals indicate that the rock formed from molten material deep beneath Earth's surface.

Key Concept Check How are rocks classified?

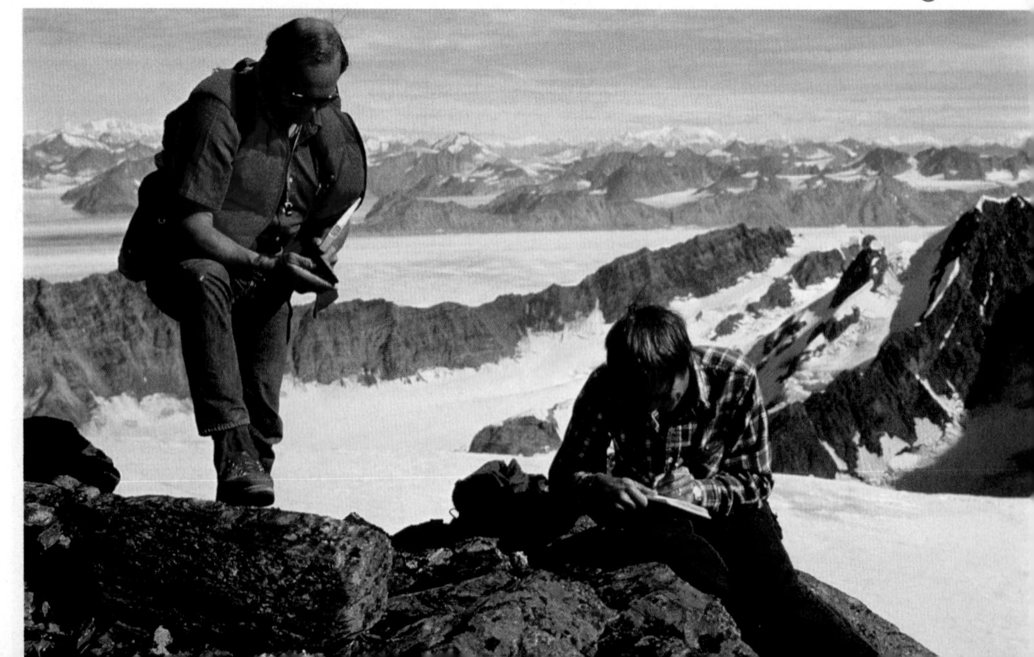

Figure 3 A geologist in the field uses tools such as a field journal, a rock hammer, and maps to interpret the conditions of rock formation.

Three Major Rock Types

Geologists classify rocks, or place them into groups, based on how they form. The three major groups of rocks are igneous, sedimentary, and metamorphic rocks. Geologists can interpret the environment where these rocks formed based on the physical and chemical characteristics of each rock type.

Igneous Rocks

You might remember that when **magma,** *molten or liquid rock underground,* cools, mineral crystals form. *Molten rock that erupts on Earth's surface is called* **lava.** When magma or lava cools and crystallizes, it creates igneous rock. As mineral crystals grow, they connect much like pieces of a jigsaw puzzle. These crystals become the grains in an igneous rock.

The texture and composition of these grains help geologists to classify the type of igneous rock and the environment where this rock may have formed. Igneous rocks form in a variety of environments including subduction zones, mid-ocean ridges, and hot spots where volcanoes are common.

Sedimentary Rocks

When rocks are exposed on Earth's surface, they can break down and be transported to new environments. Forces such as wind, running water, ice, and even gravity cause rocks on Earth's surface to break down. **Sediment** *is rock material that forms where rocks are broken down into smaller pieces or dissolved in water as rocks erode.* These materials, which include rock fragments, mineral crystals, or the remains of certain plants and animals, are the building blocks of sedimentary rocks.

Sedimentary rocks form where sediment is **deposited.** Sedimentary environments include rivers and streams, deserts, and valleys like the one shown in **Figure 4.** Even the loose sediment in the picture at the beginning of this lesson will someday turn into rock. Sedimentary rocks can be found in mountain valleys, along river banks, on the beach, or even in your backyard.

SCIENCE USE V. COMMON USE

deposit

Science Use sediment or rock added to a landform

Common Use to put money in a bank

Figure 4 Wind, water, ice, and the force of gravity can deposit sediments in environments like the mountain valley shown here.

Metamorphic Rocks

Figure 5 Metamorphic rocks form from preexisting rocks that react to changes in temperature and pressure or the addition of chemical fluids.

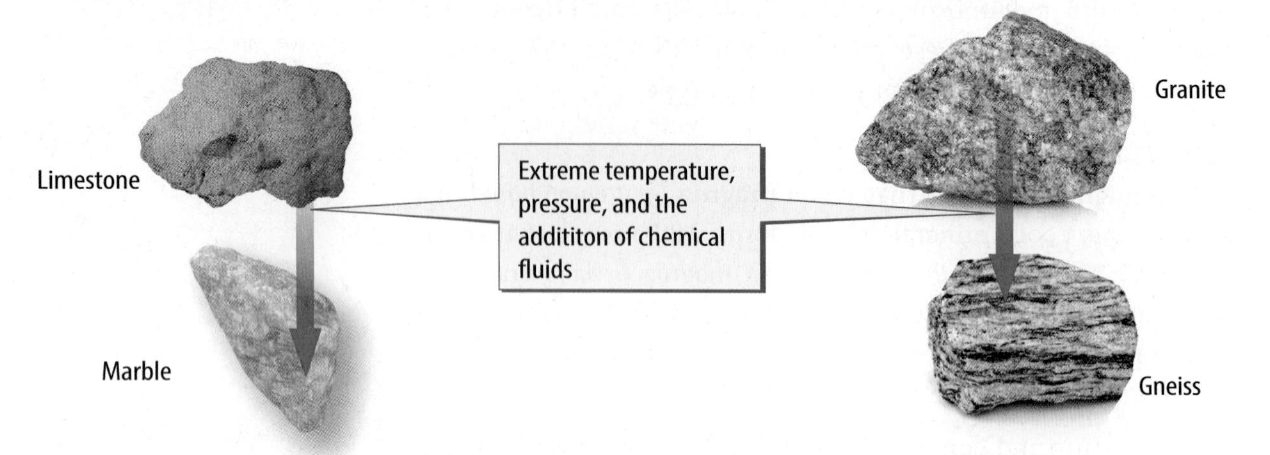

Visual Check What type of rock results when granite is subjected to extreme temperature and pressure?

Metamorphic Rocks

When rocks are exposed to extreme temperature and pressure, such as along plate boundaries, they can change to metamorphic rocks. The addition of chemical fluids can also cause rocks to become metamorphic rocks. The minerals that make up the rock's composition change as well as the texture, or arrangement of the individual mineral grains. In many cases, the change is so intense that the arrangement of the grains appears as bent or twisted layers, as shown in the gneiss in **Figure 5.** Metamorphic rocks can form from any igneous or sedimentary rock or even another metamorphic rock. For example, the igneous rock granite metamorphoses into gneiss, and the sedimentary rock limestone metamorphoses into marble, as shown in **Figure 5.**

FOLDABLES®

Use a sheet of paper to make a horizontal half-book to illustrate and explain the rock cycle.

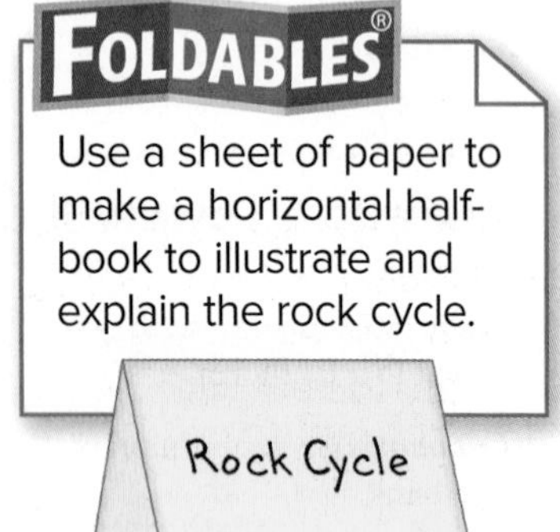

The Rock Cycle

When you look at a mountain of rock, it is hard to imagine it can ever change. But rocks are changing all the time. You usually don't see this change because it happens so slowly. *The series of processes that change one type of rock into another type of rock is called the* **rock cycle.** Forces on Earth's surface and deep within Earth drive this cycle. This cycle describes how one rock type can change into another rock type through natural processes. Imagine an igneous rock that begins as lava. The lava cools and crystallizes. Over time, the igneous rock is exposed on Earth's surface. Water can erode this rock and form sediments that eventually cement together and become sedimentary rock.

Key Concept Check What is the rock cycle?

(tl)Siim Sepp/Alamy, (tr)Givaga/Alamy, (bl)Ken Cavanagh/McGraw-Hill Education, (br)Tyler Boyes/iStock/Getty Images

Rocks in Action

Figure 6 shows how igneous, sedimentary, and metamorphic rocks originate and change throughout the rock cycle. The rectangles represent different Earth materials: magma, sediment, and the three rock types. The ovals represent natural processes that change one type of rock into another. The arrows indicate the many different pathways within the cycle both above and below the ground.

Some rock cycle processes typically occur beneath Earth's surface, such as those that involve extreme temperature, pressure, and melting. Uplift is a tectonic process that forces these rocks onto Earth's surface. On the surface, rocks can change due to natural processes, such as weathering, erosion, deposition, compaction, and cementation.

Can you trace a complete pathway through the rock cycle using rock types and processes? Start anywhere on the cycle and see how many different pathways you can make.

Animation

Figure 6 The rock cycle describes how Earth materials and processes continually form and change rocks.

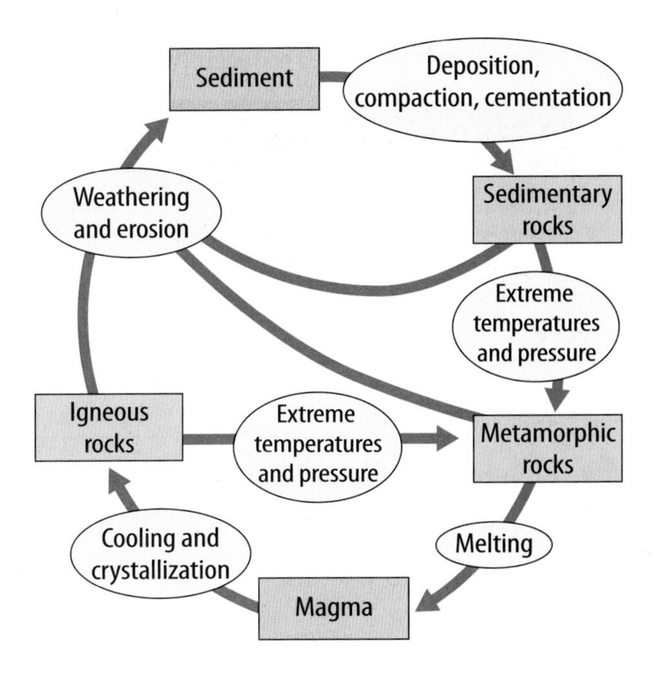

MiniLab 20 minutes

Can you model the rock cycle?

The rock cycle includes all the changes that can occur in rocks. You can use a crayon model of a rock to learn about some of these changes.

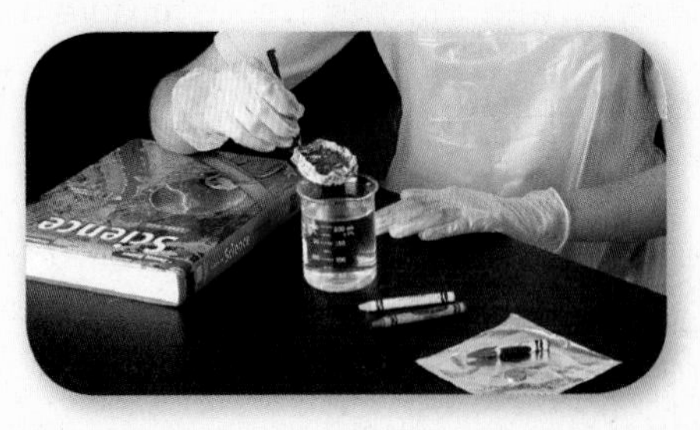

1. Read and complete a lab safety form.
2. Scrape a **coin** against the side of two or three different colors of **crayons.** Layer your scrapings on a piece of **aluminum foil.**
3. Fold the foil around the scrapings and press down hard on it with your hands. Open the package. Record your observations of the crayon rock in your Science Journal. Try to fold your crayon rock in half. It might break. Repackage your crayon rock.
4. Get a **beaker** of **hot water** from your teacher. Using **tongs,** put the foil package in the water for about 10 s. Remove it and dry it on a **paper towel.** Press your **textbook** down on top of the foil package. Open it and record your observations in your Science Journal.
5. Repackage your crayon rock. Give it to your teacher to **iron.** Allow your package to cool, then open it. Record your observations in your Science Journal.

Analyze and Conclude

1. **Recognize Cause and Effect** What part of the rock cycle did ironing your crayon rock represent?
2. Model What type of rock did you model in steps 3, 4, and 5?
3. **Key Concept** How could you continue the rock cycle using the crayon rock you created in step 4?

Lesson 1 Review

Online Quiz

Virtual Lab

Visual Summary

Rocks are a natural solid mixture of minerals or grains.

Texture describes the size and arrangement of minerals or grains in a rock.

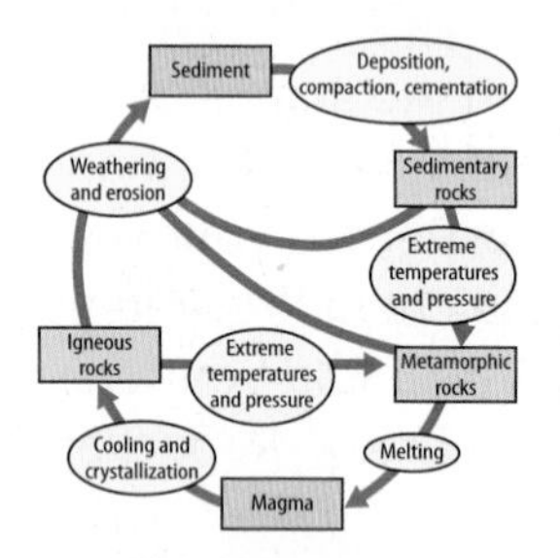

The rock cycle represents a series of processes that change one rock type into another.

FOLDABLES®

Use your lesson Foldable to review the lesson. Save your Foldable for the project at the end of the chapter.

What do you think NOW?

You first read the statements below at the beginning of the chapter.

1. Once a rock forms as part of a mountain, it does not change.

2. Some rocks, when exposed on Earth's surface, undergo weathering and erosion.

Did you change your mind about whether you agree or disagree with the statements? Rewrite any false statements to make them true.

Use Vocabulary

1 Use the terms *grain* and *sediment* in a sentence.

2 Distinguish A rock that forms as magma solidifies is a(n) ________ rock.

3 Use the term *metamorphic rock* in a complete sentence.

Understand Key Concepts 

4 What type of rock forms on Earth's surface from pieces of other rocks?

A. extrusive rock
B. intrusive rock
C. metamorphic rock
D. sedimentary rock

5 Explain why there is no beginning or end to the rock cycle.

Interpret Graphics

6 Compare the rocks shown below. How does the texture of each rock provide information about how they formed?

7 Relate Copy the table below. Fill in the Earth materials and processes for each type of rock.

Rock Type	Earth Material	Processes
Igneous		
Sedimentary		
Metamorphic		

Critical Thinking

8 Critique the following statement: When igneous, sedimentary, or metamorphic rock is exposed to high temperatures and pressures, metamorphic rock forms.

(t)Sodapix AG, Switzerland/Glow Images; (cl)George Bernard/Photo Researchers; (bl)Ken Cavanagh/McGraw-Hill Education; (br)sonsam/Getty Images

CAREERS in SCIENCE

A supervolcano quietly simmers.

Volcanic rocks tell a story about a supervolcano's explosive past and provide clues about future eruptions.

Yellowstone National Park in Wyoming is home to thousands of natural wonders such as erupting geysers, simmering steam vents, gurgling mud pots, and colorful hot springs. The source of the thermal energy that drives these geologic marvels—superheated magma—is stored in a magma chamber a few kilometers below the park. Yellowstone is home to the largest active volcanic area in North America. Some of its past eruptions were so explosive that ash spread across the North American continent earning Yellowstone the title of supervolcano.

How do supervolcanoes form? Sarah Fowler, a geologist with the American Museum of Natural History, is searching for clues. She studies magma chambers under supervolcanoes such as Yellowstone to determine the causes of explosive eruptions. Since she can't sample the magma chamber directly, she analyzes rocks that formed from lava that erupted from the supervolcano in the past.

Pumice

What do the rocks tell her? Fowler studies pumice, a lightweight volcanic rock filled with tiny holes. These holes were left behind as gas escaped from molten material during cooling and crystallization. Lava and ash that contain trapped gas erupt explosively. Therefore, Fowler knows the presence of pumice indicates an explosive past. Fowler also studies a volcanic rock called tuff. During a gas-rich eruption, a volcano ejects ash into the atmosphere. The ash settles and eventually accumulates in layers, fusing together and forming tuff. Fowler examines the size of the ash to determine where the blast originated. Larger fragments fall closer to the source. Smaller ones are carried by wind and fall farther away. When Fowler finds rocks such as pumice and tuff, she records their location and studies their texture and composition. With these data, she can produce computer models that simulate past eruptions.

Tuff

It is unlikely that Yellowstone will erupt any time soon; however, geologists monitor earthquake activity and other indicators for signs of a future eruption.

It's Your Turn

WRITE Imagine you are a geologist, and you discover an igneous rock, such as basalt or granite. Describe the rock in your Science Journal. Conduct some research to explain where this rock formed and the processes that led to its formation.

Lesson 2

Igneous Rocks

Reading Guide

Key Concepts

ESSENTIAL QUESTIONS

- How do igneous rocks form?
- What are the common types of igneous rocks?

Vocabulary

extrusive rock p. 120

volcanic glass p. 120

intrusive rock p. 121

 Multilingual eGlossary

 BrainPOP®

Inquiry Can rock be liquid?

The composition and temperature of lava influence whether it will be thick and pasty or thin and fluid. How did this lava form? When it cools and crystallizes, what type of rock will it become? Where does this type of rock commonly form?

Launch Lab

15 minutes

How does igneous rock form?

One way igneous rock forms is through the cooling and crystallization of lava. You might have seen video of molten lava flowing into the ocean. What do you think happens when the lava hits the cool ocean water?

1. Read and complete a lab safety form.
2. Observe as your teacher drips **hot, melted sugar** slowly into a **beaker** of cold water. Record what happens in your Science Journal.
3. Observe as your teacher quickly pours hot, melted sugar into a beaker of cold water. Record what happens in your Science Journal.
4. Examine each of the "candy rocks" that formed in the cold water.

Think About This

1. What is the difference between the candy rocks that formed in step 2 and the candy rocks that formed in step 3?
2. **Key Concept** How does this activity model the formation of igneous rocks?

Igneous Rock Formation

Do you remember what the difference is between magma and lava? Magma is molten rock below Earth's surface, and lava is molten rock that has erupted onto Earth's surface. When you hear the word *lava,* you might picture a hot, gooey liquid that flows easily. When lava cools and crystallizes, it becomes igneous rock. The lava shown in the picture on the previous page is already on its way to becoming solid igneous rock. It cools quickly after coming in contact with the cooler air around it. You can see where the lava has started to crystallize. It is the darker material on top of the red-hot, molten material below.

Not all molten rock makes it to Earth's surface. Large volumes of magma cool and crystallize beneath Earth's surface. Under these conditions, cooling and crystallization takes a long time. The rock that results from magma cooling below the surface is different from the rock that results from lava cooling on Earth's surface. Over time wind, rain, and other factors can wear away materials on Earth's surface. The rock that was once deep underground may now be exposed on Earth's surface. Stone Mountain, shown in **Figure 7**, is an example of igneous rock that formed from magma cooling slowly underground.

Figure 7 Stone Mountain in Georgia is made of igneous rocks that formed underground and are now exposed on Earth's surface.

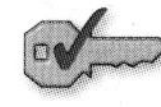

Key Concept Check How do igneous rocks form?

Figure 8 Geologists study the texture and composition of extrusive igneous rocks to determine how they formed.

Obsidian

Pumice

Extrusive Rocks

When volcanic material erupts and cools and crystallizes on Earth's surface, it forms a type of igneous rock called **extrusive rock.** Materials, such as lava and ash, solidify and form extrusive igneous rocks.

Lava can cool rapidly on Earth's surface. This means that there might not be enough time for any crystals to grow. Extrusive igneous rocks, therefore, have fine-grained texture. **Volcanic glass** *is rock that forms when lava cools too quickly to form crystals,* such as the obsidian shown in **Figure 8.**

Magma stored underground can contain dissolved gases. As magma moves toward the surface, pressure decreases, and the gases separate from the molten mixture. This is similar to the carbon dioxide that escapes when you open a carbonated beverage. When gas-rich lava erupts from a volcano, gases escape. Among the most noticeable features of some extrusive igneous rocks, such as pumice (PUH mus), are holes that are left after gas escapes, shown in **Figure 8.**

Reading Check Why are there holes in some igneous rocks?

MiniLab

20 minutes

How are cooling rate and crystal size related?

Crystal size is directly related to the crystallization rate. In this lab, you will model crystal formation under different temperature conditions.

1. Read and complete a lab safety form.
2. Mix 10 mL of warm water with 10 mg of **Epsom salt** ($MgSO_4$). Dissolve completely.
3. Completely fill three **beakers** with hot, warm, and cold water. Label the beakers.
4. Place a **watch glass** on top of each beaker so its bottom is touching the water.
5. Measure 3 mL of the Epsom salt solution in a **graduated cylinder.** Pour this amount into each watch glass.
6. Leave overnight. Record your observations in your Science Journal.

Analyze and Conclude

1. **Describe** the crystals in each watch glass.
2. **Infer** In which watch glass did crystals form first?
3. **Hypothesize** How does your answer to question 2 relate to the cooling rate and crystal size of igneous rocks?
4. **Key Concept** Which watch glass represented the type of crystals found in extrusive igneous rocks? Which one represented intrusive igneous rocks?

Intrusive and Extrusive Rocks

Figure 9 Magma that cools and crystallizes beneath Earth's surface forms intrusive igneous rock. Lava or ash that erupts and cools on Earth's surface forms extrusive igneous rock.

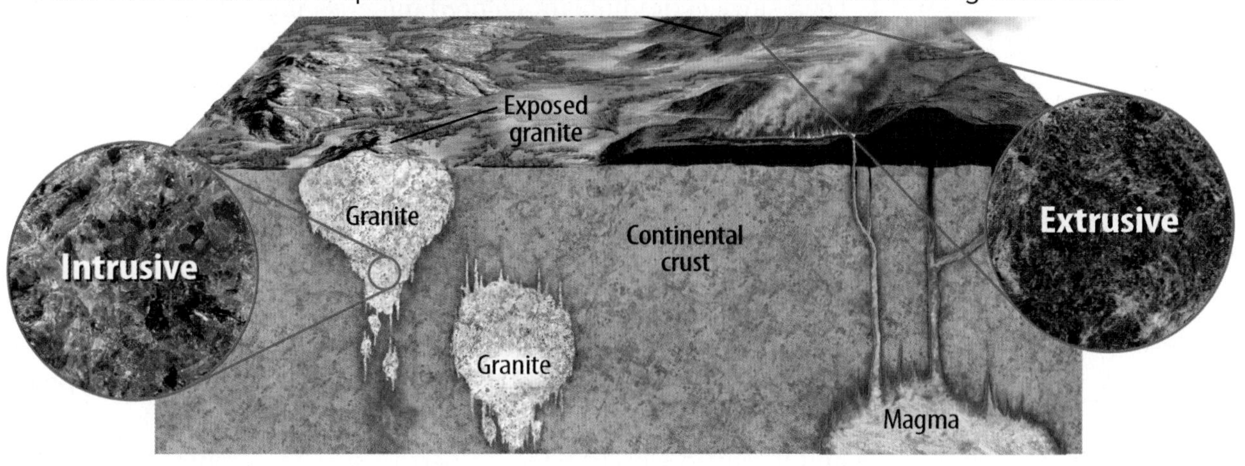

Visual Check Where is magma cooling slowly? Where is lava cooling quickly?

Intrusive Rocks

Igneous rocks that form as magma cools underground are called **intrusive rocks.** Because magma within Earth is insulated by solid rock, it cools more slowly than lava on Earth's surface. When magma cools slowly, large, well-defined crystals form.

Figure 9 shows a cross section of Earth's crust where a magma chamber has solidified and formed **intrusive** rock. The arrangement of crystals in intrusive rocks is random. Crystals interlock like jigsaw puzzle pieces. A random arrangement and large crystals are typical of intrusive igneous rocks.

Word Origin
intrusive
from Latin *intrudere,* means "to push in"

Reading Check Where do intrusive rocks form?

Igneous Rock Identification

As you read in Lesson 1, two characteristics can help to identify all rocks: texture and composition. Geologists identify an igneous rock using the arrangement and size of mineral crystals in the rock. Mineral composition can also be used for igneous rock identification.

FOLDABLES®

Use a sheet of paper to make a horizontal two-tab book. Collect information on extrusive and intrusive igneous rocks.

Extrusive Rocks | Intrusive Rocks

Texture

Geologists determine whether an igneous rock is extrusive or intrusive by studying the rock's texture. If the crystals are small or impossible to see without a magnifying lens, the rock is extrusive. If all the crystals are large enough to see and have an interlocking texture, the rock is intrusive.

(tl)Ken Cavanagh/McGraw-Hill Education; (tr)Jacques Cornell/McGraw-Hill Education

Composition

In addition to texture, geologists study the mineral composition of igneous rocks. Igneous rocks are classified, in part, based on their silica content. Light-colored minerals such as quartz and feldspar contain greater amounts of silica. Dark-colored minerals such as olivine and pyroxene contain less silica and greater amounts of elements like magnesium and iron. If minerals are difficult to identify, you can sometimes estimate the composition by observing how dark in color the rock is. Lighter-colored rocks are similar to granite in mineral composition. Darker-colored rocks are similar to basalt in composition.

Magma composition, the location where the lava or magma cools and crystallizes, and the cooling rate determine the type of igneous rock that forms. For example, granite is high in silica, and it cooled slowly beneath Earth's surface. Granite is an intrusive igneous rock. Basalt is an extrusive igneous rock that has low silica content. It formed as lava rapidly cooled on Earth's surface.

Table 1 organizes common igneous rocks according to their texture and mineral composition. Notice that an extrusive igneous rock can have the same mineral composition as an intrusive igneous rock, but their textures differ. Also notice that the minerals present in the rock affect the rock color.

Key Concept Check How are extrusive and intrusive rocks different?

Interactive Table

Table 1 Texture indicates whether an igneous rock is intrusive or extrusive. The color of the minerals gives clues to a rock's composition.

Table 1 Common Igneous Rocks

Important Rock-Forming Minerals Present	Intrusive Texture *(all crystals visible with unaided eye)*	Extrusive Texture *(some or no crystals visible with unaided eye)*
quartz, feldspar, mica, amphibole	granite	rhyolite
pyroxene, feldspar, mica, amphibole, some quartz	diorite	andesite
olivine, pyroxene, feldspar, mica, amphibole, little or no quartz	gabbro	basalt

Lesson 2 Review

Visual Summary

An extrusive igneous rock cools and crystallizes from volcanic material erupted on Earth's surface.

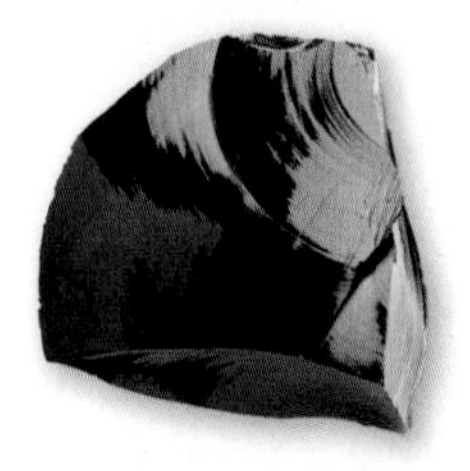

When lava cools fast, volcanic glass forms.

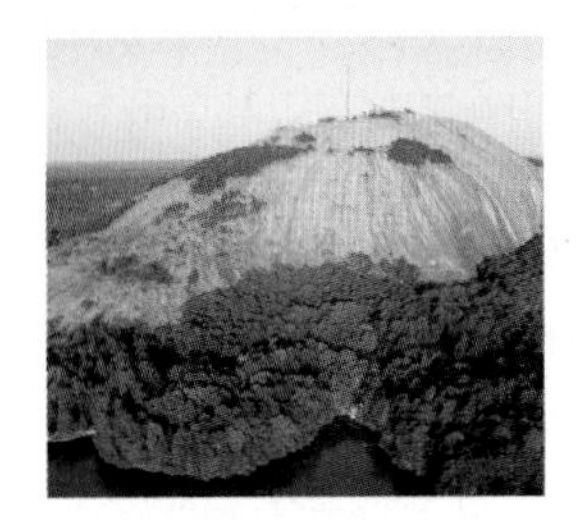

An intrusive igneous rock forms as magma cools and crystallizes deep inside Earth.

FOLDABLES®

Use your lesson Foldable to review the lesson. Save your Foldable for the project at the end of the chapter.

What do you think NOW?

You first read the statements below at the beginning of the chapter.

3. Large crystals form when lava cools quickly on Earth's surface.

4. Igneous rocks form when cooling magma crystallizes.

Did you change your mind about whether you agree or disagree with the statements? Rewrite any false statements to make them true.

Use Vocabulary

1. **Use the terms** *intrusive rock* and *extrusive rock* in a sentence.
2. **Recall** which type of igneous rock has the largest crystals.
3. **Describe** the formation of volcanic glass.

Understand Key Concepts

4. What causes holes to form in extrusive igneous rock?
   - **A.** crystals
   - **B.** gases
   - **C.** magma
   - **D.** water
5. Compare the texture of igneous rocks that crystallize deep inside Earth with those that crystallize on Earth's surface.
6. What process is required for minerals to crystallize from magma?
   - **A.** cooling
   - **B.** eruption
   - **C.** evaporation
   - **D.** melting
7. Which igneous rock contains the greatest amount of quartz?
   - **A.** basalt
   - **B.** gabbro
   - **C.** granite
   - **D.** scoria

Interpret Graphics

8. **Analyze** Which intrusive igneous rock has the same mineral composition as basalt?

9. **Organize** Draw a graphic organizer similar to the one below to identify different textures in igneous rocks.

Critical Thinking

10. **Predict** the texture of an igneous rock formed from an explosive volcanic eruption.

(tl)Joyce Photographics/Photo Researchers, Inc.; (cl)Harry Taylor/Dorling Kindersley/Getty Images; (bl)©Kevin Fleming/Corbis; (r)Slim Sepp/Alamy

Skill Practice

Compare and Contrast

30 minutes

How do you identify igneous rocks?

Igneous rocks can be classified based on texture and mineral composition. The texture is dependent upon cooling environment. When magma cools slowly beneath Earth's surface, large crystals form. When lava cools quickly on Earth's surface, tiny crystals form. Color can be used to determine whether a rock is rich in silica. Geologists compare and contrast the texture and mineral composition of igneous rocks to determine the processes that formed them.

Materials

igneous rocks (granite, pumice, basalt, gabbro, rhyolite, obsidian)

magnifying lens

Safety

Learn It

Comparisons help scientists to classify unknowns when given only a description of their properties. In this activity, you will **compare and contrast** a variety of igneous rocks and classify these rocks based on detailed descriptions of their texture and mineral composition.

Try It

1. Read and complete a lab safety form.
2. Copy the data table below in your Science Journal.
3. Obtain samples of granite and gabbro. These are both intrusive igneous rocks.
4. Describe the crystal size and color of granite and gabbro and record observations in your data table.
5. Now, obtain samples of pumice, basalt, and rhyolite. These are all extrusive igneous rocks.
6. Describe the crystal size and color of the extrusive rocks and record observations in your data table.
7. Finally, examine a sample of obsidian. Describe how this rock is different from other extrusive rocks.

Apply It

8. **Think Critically** Why do you think that obsidian (volcanic glass) differs from the other extrusive igneous rocks?
9. **Infer** Is pumice less dense than rhyolite? Explain your answer.
10. **Key Concept Check** How do the intrusive and extrusive igneous rocks differ?

Igneous Rock Characteristics		
Rock	Texture: Crystal Size	Color
Granite		
Gabbro		
Pumice		
Basalt		
Rhyolite		
Obsidian		

Lesson 3

Reading Guide

Key Concepts 

ESSENTIAL QUESTIONS

- How do sedimentary rocks form?
- What are the three types of sedimentary rocks?

Vocabulary

compaction p. 126

cementation p. 126

clastic rock p. 127

clast p. 127

chemical rock p. 128

biochemical rock p. 129

Multilingual eGlossary

Sedimentary Rocks

Inquiry How did these broken rock fragments form?

This river contributes to the formation of sedimentary rocks. The flowing water erodes rock and deposits broken fragments in the river bed. Some of these rock fragments could have originated in the mountains above. What will happen to all this material?

Launch Lab

20 minutes

How do sedimentary rocks differ?

Sedimentary rocks are made from mixtures of mineral grains, rock fragments, and sometimes organic material. Can you compare grain sizes and determine types of sedimentary rock?

1. Read and complete a lab safety form.
2. Obtain a set of **labeled samples** from your teacher.
3. Use a **hand lens** to closely examine the sediment that makes up rock sample A. Record your observations in your Science Journal.
4. Repeat step 3 with the other samples in your set.
5. Review your notes and determine how many different types of sedimentary rocks you have. Check with your teacher to see if you are correct.

Think About This

1. What characteristics did you use to distinguish between the rock samples?
2. **Key Concept** Why do you think sedimentary rocks are so common on Earth's surface?

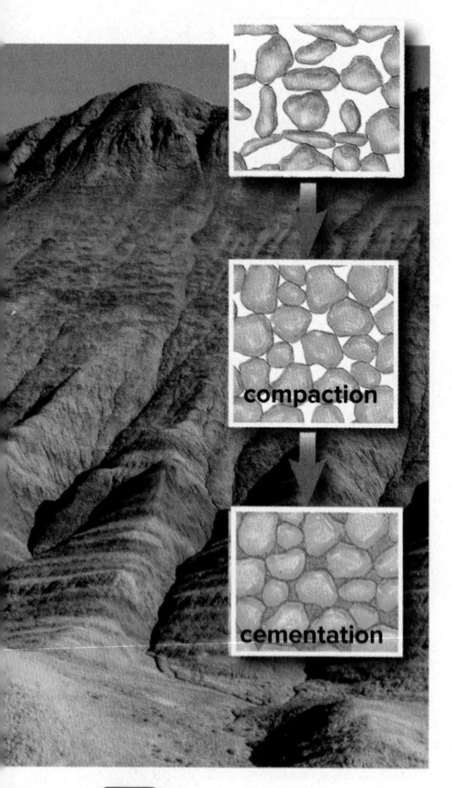

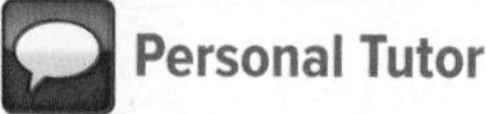

Figure 10 After sediments are deposited, the process of compaction and cementation begins.

Sedimentary Rock Formation

Like igneous rocks, sedimentary rocks can form in different environments through a series of natural steps. Water and air can change the physical or chemical properties of rock. This change can cause rock to break apart, to dissolve, or to form new minerals. When water travels through rock, some of the elements in the rock can dissolve and be transported to new locations. Mineral and rock fragments can also be transported by water, glacial ice, gravity, or wind. The sediments eventually are deposited, or laid down, where they can then accumulate in layers.

Imagine sediment deposits becoming thicker over time. Younger sediment layers bury older sediment layers. Eventually, the old and young layers of sediment can be buried by even younger sediment deposits. *The weight from the layers of sediment forces out fluids and decreases the space between grains during a process called* **compaction.** Compaction can lead to a process called cementation. *When minerals dissolved in water crystallize between sediment grains, the process is called* **cementation.** Mineral cement holds the grains together, as shown in **Figure 10.** Common minerals that cement sediment together include quartz, calcite, and clay.

Key Concept Check What is the difference between compaction and cementation?

(t)Jacques Cornell/McGraw-Hill Education; (b)sihasakprachum/iStock/Getty Images

Sedimentary Rock Identification

Like igneous rocks, sedimentary rocks are classified according to how they form. Sedimentary rocks form when sediments, rock fragments, minerals, or organic materials are deposited, compacted, and then cemented together. They also form when minerals crystallize from water or when organisms remove minerals from the water to make their shells or skeletons.

Clastic Sedimentary Rocks

Some rocks, such as sandstone, have a gritty texture that is similar to sugar. Sandstone is a common clastic sedimentary rock. *Sedimentary rocks that are made up of broken pieces of minerals and rock fragments are known as* **clastic (KLAH stik) rocks.** *The broken pieces and fragments are called* **clasts.**

Geologists identify clastic rocks according to clast size and shape. The conglomerate in **Figure 11** is an example of a rock that was deposited in a river channel. The large sediment pieces were polished and rounded as they bounced along the bottom of the channel. However, the angular fragments in the breccia in **Figure 11** probably weren't transported far, because their sharp edges were not worn away.

Sediment size alone cannot be used to determine the environment where a clastic rock formed. For example, sediment deposited by a glacier can be the size of a car or as small as grains of flour. That's because ice can move both large and small clasts. Geologists study the shape of clasts to help determine the environment where a rock formed. For example, a fast-flowing river and ocean waves tend to move large sediment. Small, gritty sediment is typically deposited in calm environments such as the seafloor or the bottom of a lake.

Reading Check Why can't sediment size alone be used to identify a sedimentary rock environment?

FOLDABLES®

Use a sheet of paper to make a vertical two-tab book. Collect information on clastic and chemical sedimentary rocks.

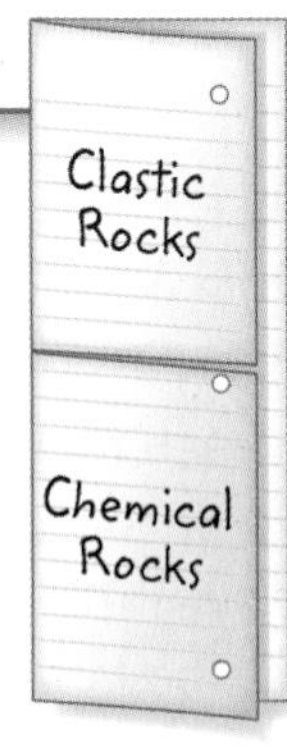

WORD ORIGIN

clastic
from Greek *klastos,* means "broken"

Conglomerate

Breccia

Figure 11 The clasts in the conglomerate on the left were rounded by a fast flowing river. The forces that created the angular fragments of the breccia on the right may not have been as strong or as long-lived.

MiniLab 15 minutes

Where did these rocks form?

Sedimentary rocks can be classified based on the size and shape of their grains. In this activity, you will identify sedimentary rocks and use their grains to infer the environment in which they likely formed.

1. Read and complete a lab safety form.
2. Obtain a set of **rock samples** from your teacher.
3. For each rock, think about where you have seen a similar rock or the grains that this rock is made of.
4. Copy the table below in your Science Journal. Estimate and record the relative grain size and shape for each sample. Then, propose a possible environment where each rock may have formed.

	Grain Size/Shape	Environment
A		
B		
C		

Analyze and Conclude

1. **Organize** Imagine yourself standing on the beach. Arrange your samples according to the place they might have formed, from shoreline outward.
2. **Describe** the grain size for each sample.
3. **Key Concept** What types of sedimentary rocks are in your sample set?

Chemical Sedimentary Rocks

Remember that as water flows through cracks or empty spaces in rock, it can dissolve minerals in the rock. Eventually rivers carry these dissolved minerals to the oceans. Dissolved minerals entering the ocean contribute to the saltiness of seawater.

Water can become saturated with dissolved minerals. When this occurs, particles can crystallize out of the water and form minerals. **Chemical rocks** *form when minerals crystallize directly from water.* Rock salt, shown in **Figure 12,** rock gypsum, and limestone are examples of common chemical sedimentary rocks.

Reading Check How do chemical rocks form?

Chemical sedimentary rocks often have an interlocking crystalline texture, similar to the textures of many igneous rocks. One difference between intrusive igneous rocks and chemical sedimentary rocks is that igneous rocks are composed of a variety of minerals and they appear multicolored. Chemical sedimentary rocks are generally composed of one dominant mineral and are uniform in color. For example, granite is made of quartz, feldspar, and mica, but rock salt is composed only of the mineral halite.

Figure 12 The water that once filled this lake bed was saturated, or filled, with dissolved halite. The water evaporated and crystalline rock salt formed.

Table 2 Common Chemical and Biochemical Rocks						
Rock Name	chemical limestone	rock gypsum	rock salt	fossiliferous limestone	chert	coal
Mineral Composition	calcite	gypsum	halite	aragonite or calcite	quartz	carbon**
Type	chemical	chemical	chemical	biochemical	biochemical*	biochemical
Example						

*Some chert is not biochemical. **Carbon in coal is not a mineral.

Table 2 Chemical and biochemical sedimentary rocks are common on Earth's surface.

Biochemical Sedimentary Rocks

Biochemical rock *is a sedimentary rock that was formed by organisms or contains the remains of organisms.* The most common biochemical sedimentary rock is limestone. Marine organisms make their shells from dissolved minerals in the ocean. When these organisms die, their shells settle onto the seafloor. This sediment is compacted and cemented and forms limestone. Sometimes the remains or traces of these organisms are preserved as fossils in sedimentary rock. Geologists call limestone that contains fossils fossiliferous (FAH suh LIH fuh rus) limestone, shown in **Table 2.** Limestone is classified as a type of carbonate rock because it contains the elements carbon and oxygen. Carbonate rocks will fizz when they come into contact with hydrochloric acid. Geologists use this chemical property to help identify different varieties of limestone.

Not all biochemical sedimentary rocks are carbonates. Some microscopic ocean organisms make their shells by removing silicon and oxygen from seawater. When these organisms die and settle on the ocean floor, compaction and cementation turns this sediment into the sedimentary rock chert.

Coal is another type of biochemical sedimentary rock. It is composed of the remains of plants and animals from prehistoric swamps. Over time, these organic remains were buried. Burial led to compression, which eventually changed the remains into a sedimentary rock.

Key Concept Check How do chemical and biochemical sedimentary rocks form?

(l to r)Ilan Rosen/Alamy; (2)Frank Blackburn/Alamy; (3)DEA/C.BEVILACQUA/Getty Images; (4)Corbin17/Alamy; (5)Nearby/Alamy; (6)Adam88xx/Getty Images

Lesson 3 Review

Visual Summary

A clastic sedimentary rock is made of clasts of minerals or rock fragments.

When minerals crystallize directly from water, a chemical sedimentary rock results.

A biochemical sedimentary rock contains the remains living organisms or was formed by organisms.

FOLDABLES®

Use your lesson Foldable to review the lesson. Save your Foldable for the project at the end of the chapter.

What do you think NOW?

You first read the statements below at the beginning of the chapter.

5. Water can dissolve rock.

6. All sedimentary rocks on Earth formed from the remains of organisms that lived in oceans.

Did you change your mind about whether you agree or disagree with the statements? Rewrite any false statements to make them true.

Use Vocabulary

1. **Use the term** *compaction* in a sentence.
2. Coal is an example of a type of sedimentary rock called a(n) ________.
3. **Distinguish** among clastic, chemical, and biochemical sedimentary rock.

Understand Key Concepts

4. Which is a clastic rock?
   - **A.** coal
   - **B.** limestone
   - **C.** rock gypsum
   - **D.** sandstone
5. **Classify** the following sedimentary rocks as clastic, chemical, or biochemical: conglomerate, rock gypsum, fossiliferous limestone, and rock salt.
6. Identify a factor that is NOT responsible for the formation of sedimentary rocks.
   - **A.** glacier
   - **B.** magma
   - **C.** river
   - **D.** wind

Interpret Graphics

7. **Organize** Arrange the terms below in the correct order to describe the formation of clastic sedimentary rocks: *transportation, cementation, deposition, erosion.*

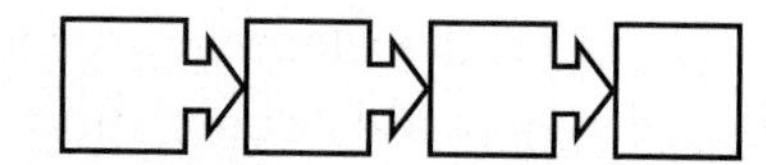

Critical Thinking

8. **Compare and contrast** the textures of conglomerate and breccia.
9. **Hypothesize** Over time, limestone dissolves in the presence of acid rain on Earth's surface. Relate this chemical property to the use of limestone for the construction of buildings and monuments.
10. **Analyze** this statement: As rock erodes, rivers carry dissolved minerals to the ocean, increasing the saltiness of seawater.

How are sedimentary rocks classified?

Materials

sedimentary rocks (limestone, sandstone, shale, conglomerate)

vinegar

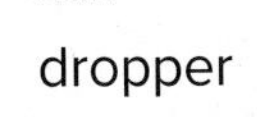

dropper

magnifying lens

Safety

For millions of years, rocks on Earth's surface have eroded with the help of water, wind, ice, and gravity. Sediments are transported and deposited, settling to the bottom of rivers, lakes, and oceans. Layers of sediment accumulate and undergo compaction and cementation and become a sedimentary rock. In this activity, you will use a flow chart to identify different sedimentary rocks.

Learn It

Scientists make observations to help develop hypotheses. In this activity, you will **observe** the chemical and physical properties of sedimentary rocks to identify different rock types.

Try It

1. Read and complete a lab safety form.
2. Obtain a sample set of sedimentary rocks.
3. Copy the flowchart in your Science Journal.
4. Start at the top of the flowchart. If there are particles present, determine the size of the particles, and identify the sample.
5. If the rock is smooth to the touch, test the sample with vinegar. If the sample fizzes with vinegar, it is limestone. If it does not fizz, the sample is shale.
6. Repeat steps 4-5 for four different sedimentary rock samples.

Apply It

7. **Infer** Why was vinegar used in the lab?
8. **Identify** Are the samples you identified clastic, chemical, or biochemical?
9. **Key Concept Check** What characteristics can be used to organize sedimentary rocks for identification?

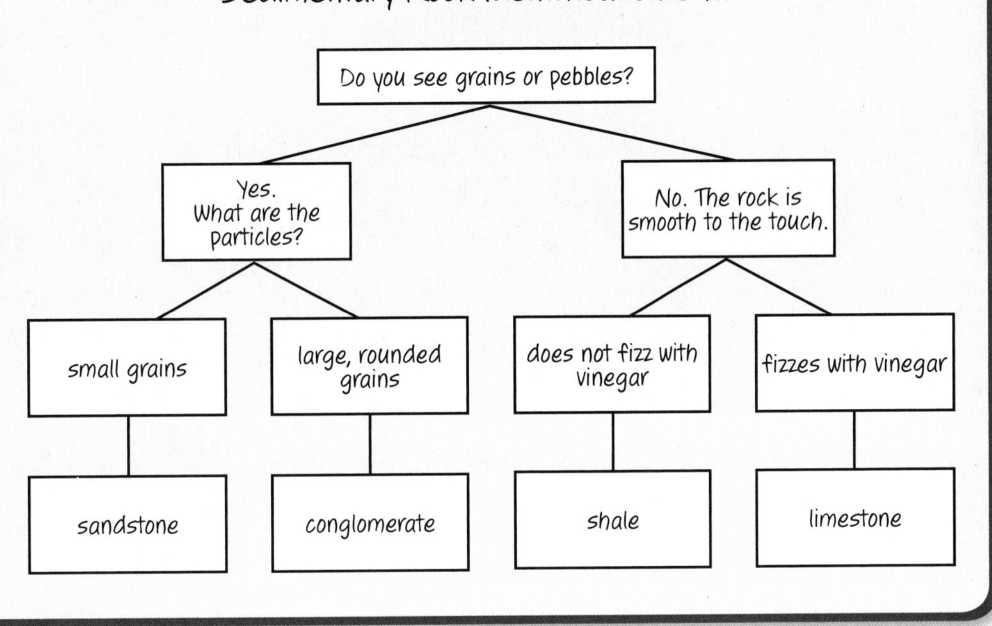

Lesson 4

Metamorphic Rocks

Reading Guide

Key Concepts

ESSENTIAL QUESTIONS

- How do metamorphic rocks form?
- How do types of metamorphic rock differ?

Vocabulary

metamorphism p. 133
plastic deformation p. 134
foliated rock p. 135
nonfoliated rock p. 135
contact metamorphism p. 136
regional metamorphism p. 136

Multilingual eGlossary

Go to the resource tab in ConnectED to find the PBL *Rockin' Around the Park.*

Inquiry How did this wrinkle?

Rocks can form from layers of sediment or hardened magma. Under the right conditions, those rocks can bend and twist. Imagine the incredible pressure required to cause solid rock, such as this, to bend!

Launch Lab — 20 minutes

How does pressure affect rock formation?

How does pressure affect the minerals in a rock? The arrangement of minerals in a metamorphic rock can be used to help classify the rock.

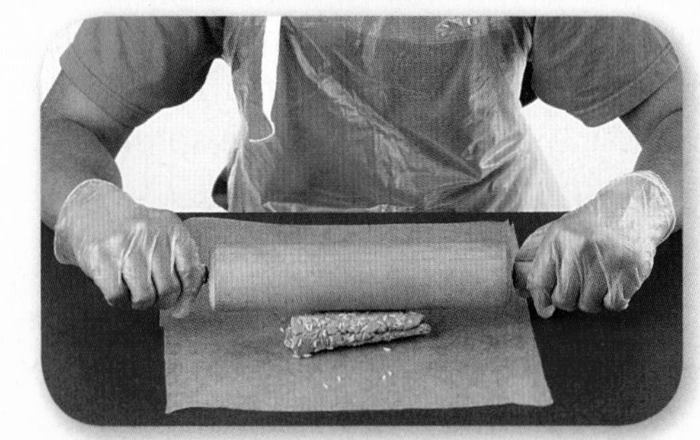

1. Read and complete a lab safety form.
2. Place some rice grains on the table.
3. Roll a **ball of clay** on top of the **rice**. Knead the ball until the rice is evenly mixed in the clay.
4. Use a **rolling pin** or a **round can** to roll the clay to a thickness of about 0.5 cm. Draw and label a picture of the sheet of clay and the rice grains in your Science Journal.
5. Fold the edge of the clay closest to you toward the edge away from you. Roll the clay in the direction you folded it. Repeat and flatten the clay to a thickness of 0.5 cm again. Draw and label a picture of the sheet of clay and the rice grains in your Science Journal.

Think About This

1. Describe the differences you observed in the orientation of rice grains between steps 4 and 5.
2. **Key Concept** What force caused the orientation of rice grains in the clay to change? How might this process be similar to the formation of metamorphic rocks?

Metamorphic Rock Formation

Imagine you left a cheese sandwich in your backpack on a hot day and then threw your backpack into your locker. Would the sandwich look the same after school? Changes in the temperature during the day would likely cause the cheese to soften. Pressure from the weight of your backpack would squish the sandwich. Like the sandwich, rocks are also affected by changes in temperatures and pressure. These rocks are called metamorphic rocks. **Metamorphism** *is any process that affects the structure or composition of a rock in a solid state as a result of changes in temperature, pressure, or the addition of chemical fluids.*

Most metamorphic rocks form deep within Earth's crust. Like igneous rocks, metamorphic rocks form under high temperature conditions. But unlike igneous rocks, metamorphic rocks do not crystallize from magma. Like many sedimentary rocks, metamorphic rocks can form layers. But unlike sedimentary rocks, the layers are not a result of deposition. The metamorphic rocks shown on the facing page have changed shape. Exposed to extreme temperatures and pressure, the rocks were bent and twisted into wrinkly layers and are classified as metamorphic rocks.

Reading Check What is metamorphism?

ACADEMIC VOCABULARY

expose
(verb) to uncover or subject

MiniLab 20 minutes

Can you model metamorphism?

Extreme temperatures and pressure can cause metamorphism. In this activity, you will model the formation of a metamorphic rock using bread and cheese spread.

1. Read and complete a lab safety form.
2. Get two pieces of **white bread**, two pieces of **wheat bread**, some **cheese spread**, and a **plastic knife** from your teacher.
3. Place a **paper towel** on the lab table.
4. Stack the bread on the paper towel in this order: white bread, wheat bread, cheese spread, white bread, wheat bread.
5. Place another paper towel on top of your stack and press down on your stack with a heavy **book.**
6. Remove the book and slowly fold your stack in half.
7. Place a paper towel on top and push down on your sandwich layers again.
8. Heat the layers in an **oven** or a **microwave oven** for about 2 minutes.

Analyze and Conclude

1. **Describe** What represents the parent rock in your model of metamorphism?
2. **Interpret** In which step did you model plastic deformation? Explain.
3. **Think Critically** In what ways is your model different from metamorphism?
4. **Key Concept** Explain how changes in temperature and pressure play a role in the formation of a metamorphic rock.

Temperature and Pressure

When rocks experience an increase in temperature and pressure, they behave like a bendable plastic. Without melting, the rocks bend or fold. This *permanent change in shape by bending and folding is called* **plastic deformation.** It's one way the texture of a rock changes during metamorphism. Plastic deformation occurs during uplift events when tectonic plates collide and form mountains, such as the Himalayas in Asia. Changes in composition and structure are clues that a rock has been metamorphosed.

The rock that changes during metamorphism is called the parent rock. Parent rocks can be exposed to high temperatures and pressures as tectonic processes bury them deeper in Earth. Recall that temperature increases with depth in Earth's interior. Pressure also increases with depth, as shown in **Figure 13.** Metamorphism of the parent rock occurs at temperatures higher than 200°C and pressures higher than 3 kb.

Key Concept Check Under what conditions do metamorphic rocks form?

Figure 13 Pressure increases with depth in Earth.

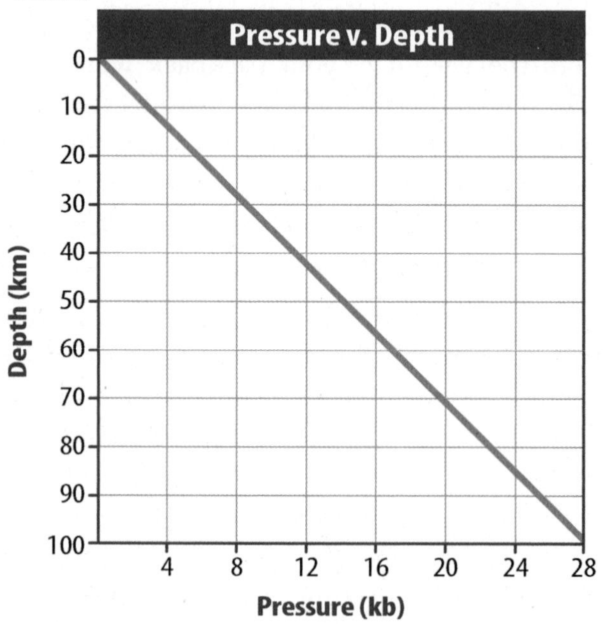

Hutchings Photography/Digital Light Source

Metamorphic Rock Identification

Changes in temperature, pressure, or the addition of chemical fluids can result in the rearrangement of minerals or the formation of new minerals in a metamorphic rock. Geologists study the texture and composition of minerals to identify metamorphic rocks.

Metamorphic rocks are classified into two groups based on texture. In many cases, changes in pressure cause minerals to align and form layers in metamorphic rocks. This layering can appear similar to the layers associated with clastic sedimentary rocks. However, the crystalline minerals present in a metamorphic rock distinguish it from a clastic sedimentary rock. In other cases, the rock can have blocky, interlocking crystals that appear uniform in color.

Math Skills 

Use Graphs

The line graph in **Figure 13** represents pressure below Earth's surface. What is the pressure at a depth of 50 km?

a. Read the title of the graph to determine what data are represented.

b. Read the labels on the *x*- and *y*-axis to determine the units.

c. Move horizontally from 50 km to the orange line. Move vertically from the orange line to the *x*-axis. The pressure is 14 kb.

Practice

At what depth is the pressure 20 kb?

 **Math Practice**

 **Personal Tutor**

Foliated Rocks

The metamorphic rock gneiss, shown in **Figure 14**, is an example of a foliated rock. **Foliated rocks** *contain parallel layers of flat and elongated minerals.* Look closely at the layers, or bands, of dark and light minerals in the gneiss. High pressure during metamorphism causes some minerals to align perpendicular to the pressure. Foliation is a common feature in metamorphic rocks.

Word Origin

foliate

from Latin *foliatus,* means "consisting of thin, leaf-like layers"

 **Reading Check** What type of metamorphic rock has layers?

Nonfoliated Rocks

Metamorphic rocks that have mineral grains with a random, interlocking texture are **nonfoliated rocks.** There is no obvious alignment of the mineral crystals in a nonfoliated metamorphic rock. Instead, the individual crystals are blocky and approximately equal in size. This crystalline texture differs from an igneous rock in that the minerals are generally uniform in color as opposed to multicolored, as in igneous rocks such as granite.

Figure 14 Elongated or flat minerals in foliated rocks line up in response to pressure.

Visual Check Can you determine the direction that pressure was applied?

Siim Sepp/Alamy

Figure 15 Nonfoliated rocks don't show obvious orientation of minerals.

Contact and Regional Metamorphism

One way a nonfoliated metamorphic rock can form is when magma intrudes rock. *During* **contact metamorphism,** *magma comes in contact with existing rock, and its thermal energy and gases interact with the surrounding rock, forming new metamorphic rock.* Contact metamorphism can increase crystal size or form new minerals and change rock. A common example of a nonfoliated rock, marble, is shown in **Figure 15.** Notice the uniform color and crystal size in this specimen. **Table 3** illustrates other examples of nonfoliated and foliated metamorphic rocks.

Regional metamorphism *is the formation of metamorphic rock bodies that are hundreds of square kilometers in size.* This process can create an entire mountain range of metamorphic rock. Changes in temperature and pressure and the presence of chemical fluids act on large volumes of rock and produce metamorphic textures. These textures can help unravel the mysteries of a mountain-building event. The Himalayas in Asia and the Appalachian Mountains of the eastern United States exhibit structures associated with regional metamorphism.

Key Concept Check Compare and contrast contact metamorphism and regional metamorphism.

Table 3 Metamorphic rocks are classified into two groups based on texture.

FOLDABLES®

Make a vertical two-tab book. Use it to organize your notes on contact and regional metamorphism.

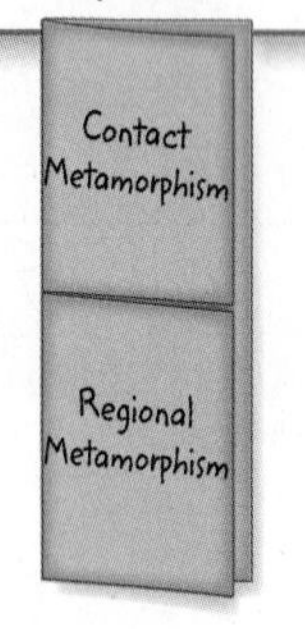

Table 3 Metamorphic Rocks

Texture		Composition	Rock Name	Example
Foliated	layered	quartz, mica, clay minerals	slate	
	layered	quartz, mica, clay minerals	phyllite	
	layered	quartz, feldspar, amphibole, mica	schist	
	banded	quartz, feldspar, amphibole, pyroxene	gneiss	
Nonfoliated	blocky crystals	quartz	quartzite	
	blocky crystals	calcite	marble	

Lesson 4 Review

Visual Summary

Foliated metamorphic rocks have distinct layers of flat and elongated minerals.

A nonfoliated metamorphic rock has minerals arranged in a random, interlocking texture.

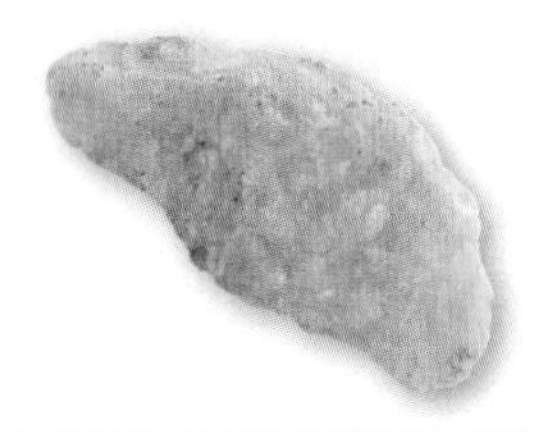

Contact metamorphism occurs when rocks come in contact with magma without melting.

Use your lesson Foldable to review the lesson. Save your Foldable for the project at the end of the chapter.

What do you think NOW?

You first read the statements below at the beginning of the chapter.

7. With the right pressure and temperature conditions, minerals in a rock can change shape without breaking or melting.

8. Metamorphic rocks have layers that form as minerals melt and then recrystallize.

Did you change your mind about whether you agree or disagree with the statements? Rewrite any false statements to make them true.

Use Vocabulary

1 **Use the term** *plastic deformation* in a sentence.

2 Stacks of paper resemble ________ texture in metamorphic rocks.

3 Crystals in a ________ metamorphic rock are blocky and equal in size.

Understand Key Concepts

4 Which force contributes to the formation of metamorphic rocks?

A. compaction
B. cementation
C. crystallization
D. pressure

5 **Classify** the following rocks as either foliated or nonfoliated: quartzite, schist.

Interpret Graphics

6 **Identify** Create a graphic organizer to identify the three possible causes of metamorphism.

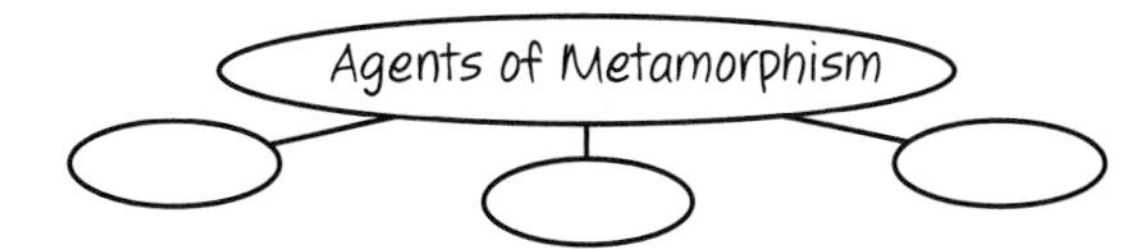

Critical Thinking

7 **Explain** how to differentiate between the igneous rock granite and the metamorphic rock gneiss.

Math Skills

8 **Based** on the graph below, what is the pressure at a depth of 40 km? At what depth would the pressure be 30 kb?

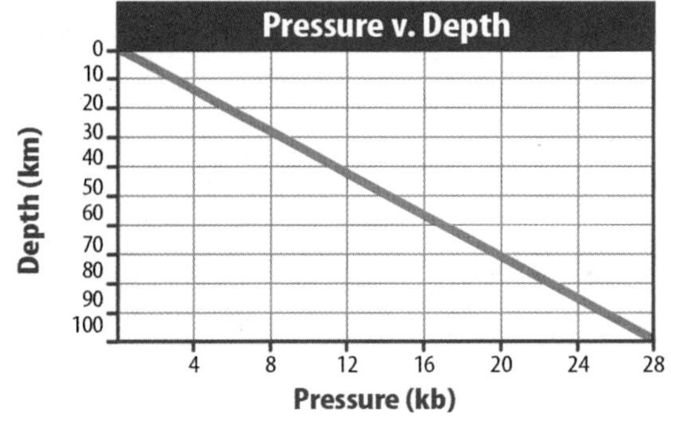

(t)Siim Sepp/Alamy; (c)Andrew J. Martinez/Photo Researchers, Inc.; (b)Jacques Cornell/McGraw-Hill Education

Lab

45 minutes

Identifying the Type of Rock

Materials

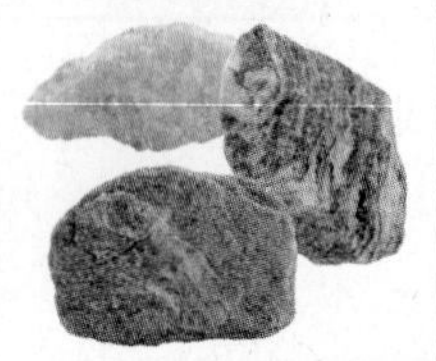

metamorphic rocks (marble, gneiss, schist)

vinegar

dropper

magnifying lens

rocks

Safety

Rocks can be classified into three major groups: igneous, sedimentary, and metamorphic. Geologists examine rock texture and mineral composition to classify rocks. Igneous rocks can be coarse or fine crystalline and multicolored or glassy. Sedimentary rocks are often layered and contain a mix of rock fragments, shells, minerals, and fossils. Metamorphic rocks can reflect a change in shape due to an increase in temperature and pressure or the addition of chemical fluids. Foliation is common in metamorphic rocks. In this activity, you will be given a variety of rock samples to try and classify based on physical and chemical properties.

Question

How can the texture and mineral composition of a rock be used to classify the rock as igneous, sedimentary, or metamorphic?

Procedure

1. Read and complete a lab safety form.
2. Obtain a metamorphic rock and examine its texture and grain size.
3. Record your observations in your Science Journal.
4. Does the rock have distinct, parallel layers? If so, it is a foliated metamorphic rock. If not, it is nonfoliated.
5. Use a dropper to place 1–2 drops of vinegar on the rock. If it fizzes, the rock contains the mineral calcite.
6. Repeat the classification steps with another metamorphic rock.

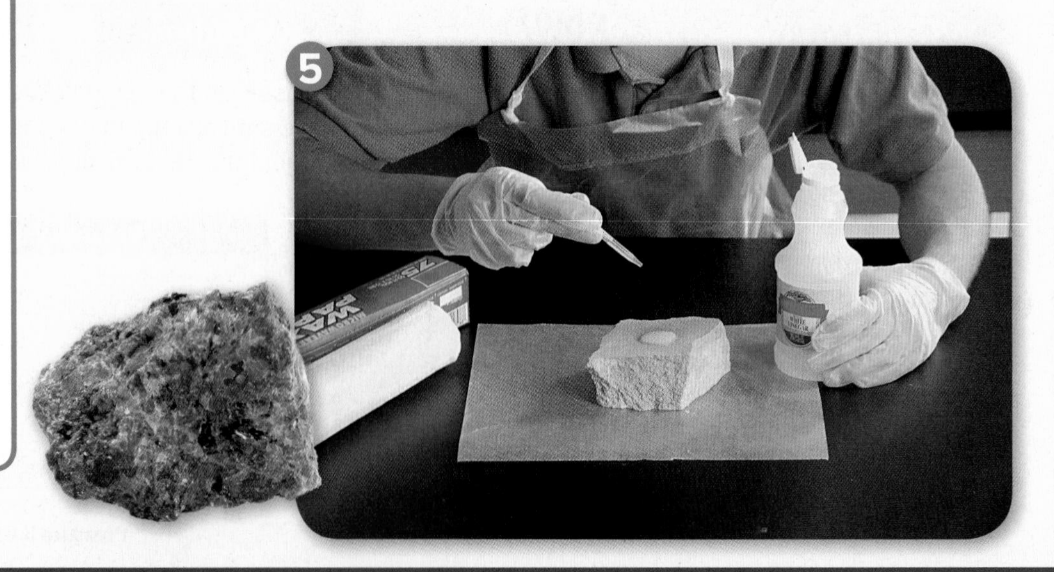

t to b, 2, 3, rocks) Jacques Cornell/McGraw-Hill Education; (6, br)Hutchings Photography/Digital Light Source

7. Design a flowchart for rock identification that incorporates the three rock types and all of the characteristics that you examined in each rock identification lab. Draw the flowchart in your Science Journal.
8. Experiment with your flowchart as you classify several unknown rock samples.
9. Refine the classification flowchart as needed so that it works for all samples that you are given.
10. Identify the unknown samples and incorporate their names into the flowchart that you created.

Analyze and Conclude

11. **Describe** why some samples were more difficult to classify than others.
12. **Explain** which characteristic was the least helpful in your classification scheme?
13. **The Big Idea** What characteristics made it easier to classify rock samples?

Communicate Your Results

Construct a poster of your final flowchart to share with your class. Be sure to label the characteristics and the choices at each step in the chart so that it is easy to follow.

Inquiry Extension

Research the rocks you identified in this activity. Explain how each rock formed, and describe the similarities and differences among the rock types.

Lab Tips

- Remember that the light or dark colors of a rock can lead to the identification of minerals in the rock.
- Rocks that fizz when acid, such as vinegar, is added to them contain the mineral calcite. Examples of such rocks include limestone and marble.

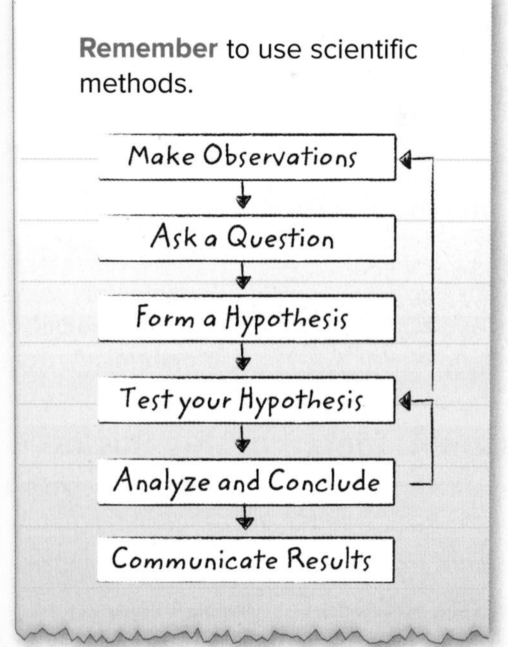

Chapter 4 Study Guide

Igneous rocks form from molten rock that cools and crystallizes. Sedimentary rocks form from compaction and cementation of sediments or evaporation and crystallization of minerals dissolved in water. Metamorphic rocks form from exposure of existing rocks to high pressures, temperatures, or the addition of chemical fluids.

Key Concepts Summary

Lesson 1: Rocks and the Rock Cycle

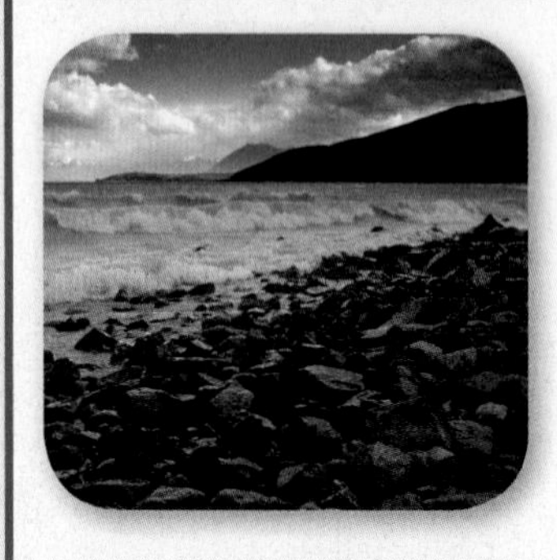

- The three major **rock** types are igneous, sedimentary, and metamorphic. Geologists classify rocks based on how the rocks formed.
- The **rock cycle** is a series of processes that continually change rocks and Earth materials into different types of rocks.

Vocabulary

rock p. 111
grain p. 111
texture p. 112
magma p. 113
lava p. 113
sediment p. 113
rock cycle p. 114

Lesson 2: Igneous Rocks

- Igneous rocks form when volcanic material cools and crystallizes.
- Most igneous rocks are either **intrusive rocks** or **extrusive rocks.** The size of crystals in igneous rocks depends on how quickly the magma or lava cools.

Vocabulary

extrusive rock p. 120
volcanic glass p. 120
intrusive rock p. 121

Lesson 3: Sedimentary Rocks

- Sedimentary rocks form through processes of weathering, erosion, transportation, deposition, **compaction, cementation** and crystallization.
- Sedimentary rocks are usually classified as **clastic rocks, chemical rocks,** or **biochemical rocks.**

Vocabulary

compaction p. 126
cementation p. 125
clastic rock p. 127
clast p. 127
chemical rock p. 128
biochemical rock p. 129

Lesson 4: Metamorphic Rocks

- Metamorphic rocks form from a parent rock that has been exposed to increases in temperature, pressure, or the addition of chemical fluids.
- **Foliated rocks** contain parallel layers of minerals. **Nonfoliated rocks** contain coarse, blocky crystals that are uniform in color.

Vocabulary

metamorphism p. 133
plastic deformation p. 134
foliated rock p. 135
nonfoliated rock p. 136
contact metamorphism p. 136
regional metamorphism p. 136

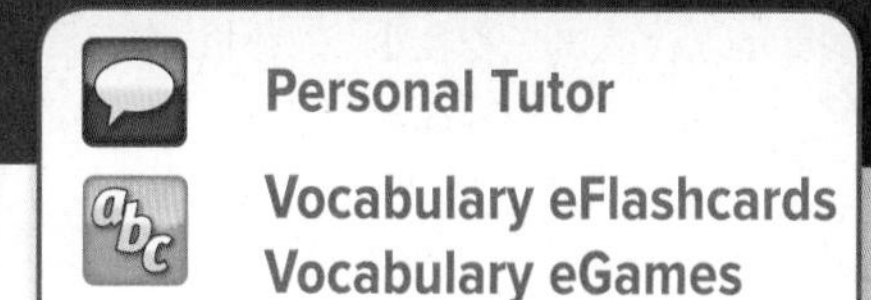

Chapter Project

Assemble your lesson Foldables as shown to make a Chapter Project. Use the project to review what you have learned in this chapter.

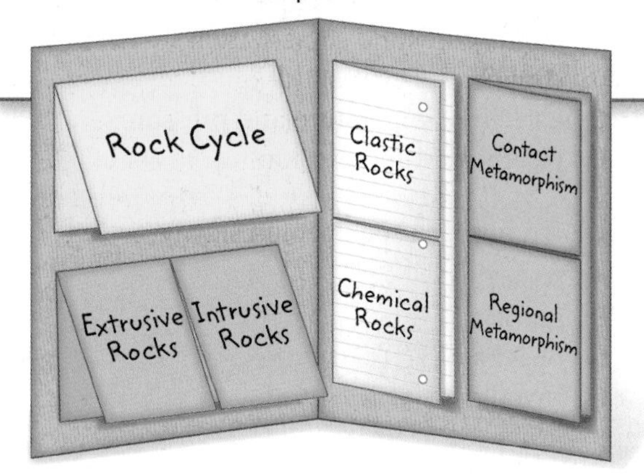

Use Vocabulary

1. Define *igneous rock.*
2. Use the word *sediment* in a sentence.
3. While remaining solid, a ________ forms at high pressure or temperatures.
4. Define *intrusive rock.*
5. Use the phrase *extrusive rock* in a sentence.
6. Identify two textures common in metamorphic rocks.
7. During ________, minerals such as calcite or quartz crystallize between grains of clastic rock.
8. A ________ sedimentary rock is made of mineral and rock fragments.
9. Use the term *chemical rock* in a sentence.
10. Describe a *nonfoliated rock.*
11. Folded layers are examples of ________ in a metamorphic rock.
12. Use the term *regional metamorphism* in a sentence.

Interactive Concept Map

Link Vocabulary and Key Concepts

Copy this concept map, and then use vocabulary terms from the previous page to complete the concept map.

Rock Cycle

rocks formed from

molten rock → 13 → solidifies → on surface → 16; below ground → 17

solids → erosion and deposition → 14 → is made by compaction and → 18

solids → heat and pressure → 15 → rocks have been → 19

Chapter 4 Review

Understand Key Concepts 

1. Which rock type forms from cooling lava on Earth's surface?
   A. extrusive igneous
   B. intrusive igneous
   C. granite
   D. limestone

2. Which process squeezes fluids from between individual grains?
   A. cementation
   B. compaction
   C. erosion
   D. transportation

3. Basalt is an example of a(n)
   A. extrusive igneous rock.
   B. intrusive igneous rock.
   C. metamorphic rock.
   D. sedimentary rock.

4. What can be determined by studying the shape of clastic grains?
   A. distance they have been transported
   B. how they were eroded
   C. mineral content of the parent rock
   D. what the parent rock was

5. Examine the rock above. What property indicates that it is a biochemical rock?
   A. foliated layers
   B. fossilized shells
   C. large rounded clasts
   D. minerals of different colors

6. What is true about volcanic glass?
   A. It contains no crystals.
   B. It cools slowly.
   C. It fractures easily.
   D. It is both intrusive and extrusive.

7. Which characteristics are used to classify a sedimentary rock such as sandstone?
   A. glass content and texture
   B. grain size
   C. luster and hardness
   D. texture and mineral composition

8. How do metamorphic rocks form?
   A. compaction and cementation
   B. cooling and crystallization
   C. extreme temperature and pressure
   D. weathering and erosion

9. Which igneous rock cooled slowly?
   A. basalt
   B. granite
   C. obsidian
   D. rhyolite

10. What is the general term for a rock fragment present in a sedimentary rock?
    A. clast
    B. glass
    C. mineral
    D. pore

11. What process is occurring in this photo?
    A. cementation
    B. condensation
    C. crystallization
    D. evaporation

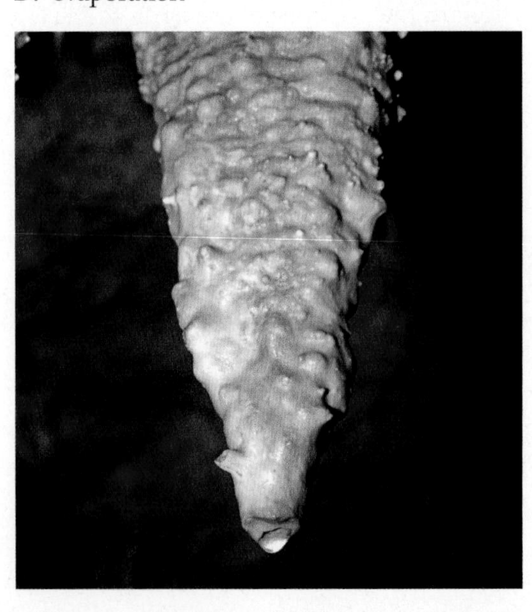

Critical Thinking

12 **Decide** which of the three types of sedimentary rocks forms when water in a shallow sea evaporates.

13 **Complete** the chart below with at least three common rock names for each major rock type.

Rock Type	Rock Name
Igneous	
Sedimentary	
Metamorphic	

14 **Relate** What rock cycle process is the opposite of crystallization of magma?

15 **Compare** a rock cycle process on Earth's surface with a process that typically occurs below the surface.

16 **Hypothesize** Imagine the temperature inside Earth was no longer hot. How might this affect the rock cycle?

17 **Deduce** how a rock formed from an explosive volcanic eruption could resemble a clastic sedimentary rock.

18 **Hypothesize** how the direction of stress applied affects the arrangement of minerals in a metamorphic rock.

19 **Relate** the presence of the holes to the ability of pumice to float in water.

20 **Compare and contrast** the formation of gneiss to the formation of its parent rock, granite.

Writing in Science

21 **Write** a paragraph that distinguishes among the rock cycle processes that form metamorphic rocks and igneous rocks deep beneath Earth's surface.

REVIEW THE BIG IDEA

22 Use the rock cycle to explain how each rock type forms.

23 What might happen to the sand in the valley if more sand is deposited on top of it?

Steve Allen/Getty Images

Math Skills

Math Practice

Use the graph to answer the following questions.

24 At about what depth does the pressure reach 20 kb?

25 Use the trend on the graph to predict the approximate pressure at a depth of 200 km.

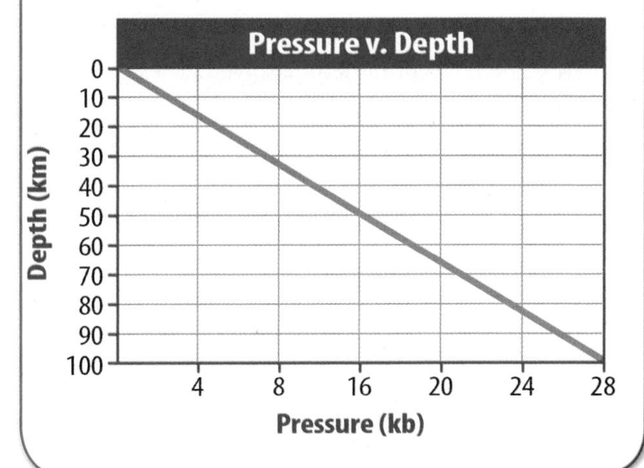

Standardized Test Practice

Record your answers on the answer sheet provided by your teacher or on a sheet of paper.

Multiple Choice

1 Light-colored minerals, such as quartz, contain greater amounts of

A iron.

B magnesium.

C manganese.

D silica.

Use the diagram below to answer question 2.

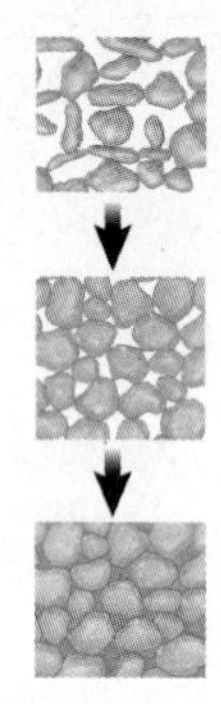

2 Which process is illustrated in the last part of the diagram above?

A cementation

B compaction

C metamorphism

D transport

3 Why don't crystals form in volcanic glass?

A The lava contained dissolved gases.

B The lava cooled deep within Earth.

C The lava cooled too quickly.

D The lava failed to erupt.

4 What rock contains the hard remains of marine organisms made from minerals in seawater?

A basalt

B granite

C limestone

D marble

Use the graph below to answer question 5.

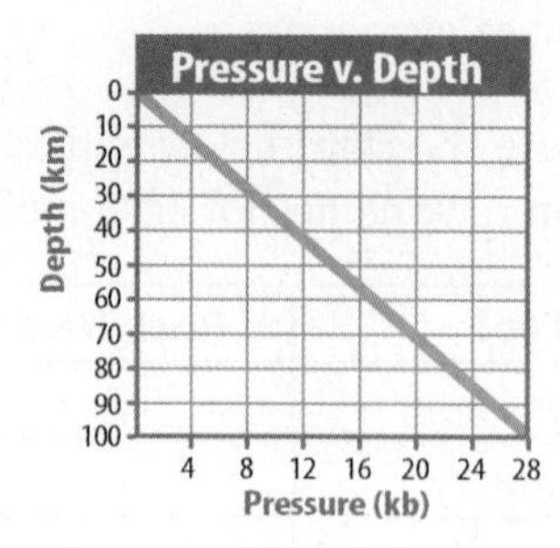

5 According to the graph above, how much greater is the pressure in Earth's interior at a depth of 100 km compared to a depth of 50 km?

A 12 kb

B 14 kb

C 16 kb

D 20 kb

6 Solid rocks exposed to changes in thermal energy, pressure, or the addition of chemical fluids form metamorphic rocks. Igneous, NOT metamorphic, rocks will form if

A the heat is withdrawn.

B the pressure decreases.

C the rock contains minerals.

D the rock melts.

7 A rock permanently changes shape by bending or folding during

A contact metamorphism.

B foliation.

C plastic deformation.

D sedimentation.

8 What is the term for broken pieces of rock?

A clasts

B crystals

C glass

D layers

Use the diagram below to answer question 9.

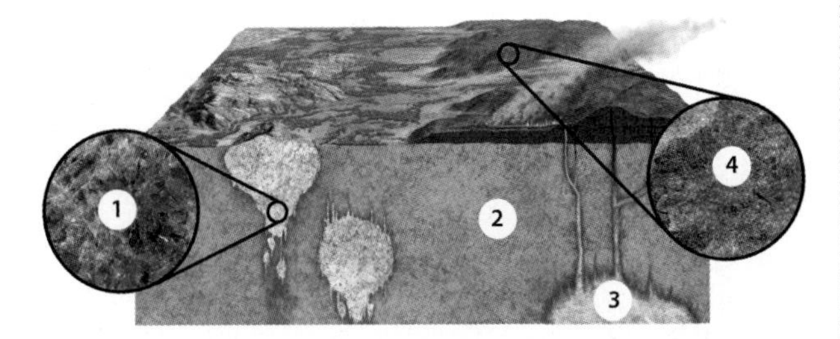

9 In the diagram above, which number represents extrusive igneous rock?

A 1

B 2

C 3

D 4

10 On what basis are rocks classified into three main groups?

A their texture

B the way they form

C their age

D their color

11 What condition produces the distinct layers of flat and elongated minerals in foliated metamorphic rocks?

A drastic changes in temperature

B random, interlocking texture

C extremely high pressure

D uniformity of mineral colors

Constructed Response

Use the diagram below to answer questions 12 and 13.

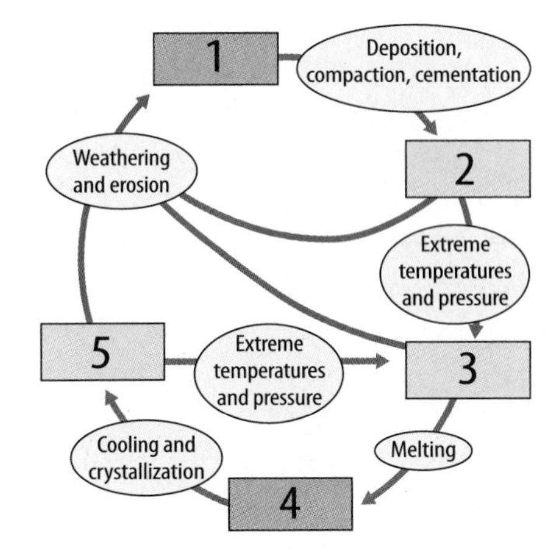

12 What do the numbers in the rock cycle above represent? Explain your reasoning.

13 Use the diagram above to identify and describe at least three processes that change rock as it travels through the rock cycle.

Use the table below to answer question 14.

Type of Rock	Process of Formation

14 What three main types of sedimentary rocks belong in column 1 of the table above? How does each type form?

NEED EXTRA HELP?														
If You Missed Question...	1	2	3	4	5	6	7	8	9	10	11	12	13	14
Go to Lesson...	2	3	2	3	4	1	4	3	2	1	4	1	1	3

Chapter 5

Weathering and Soil

What natural processes break down rocks and begin soil formation?

Inquiry What is dust?

Dust is weathered rock or rock broken into tiny pieces. These tiny pieces of rock make up a large part of soil. Sometimes they are so small that they are easily blown by the wind.

- How does rock break into tiny pieces of dust?
- What natural processes break down rocks and begin soil formation?

Get Ready to Read

What do you think?

Before you read, decide if you agree or disagree with each of these statements. As you read this chapter, see if you change your mind about any of the statements.

1. Any two rocks weather at the same rate.
2. Humans are the main cause of weathering.
3. Plants can break rocks into smaller pieces.
4. Air and water are present in soil.
5. Soil that is 1,000 years old is young soil.
6. Soil is the same in all locations.

McGraw Hill Education connectED

Your one-stop online resource

connectED.mcgraw-hill.com

LearnSmart®

Chapter Resources Files, Reading Essentials, Get Ready to Read, Quick Vocabulary

Animations, Videos, Interactive Tables

Self-checks, Quizzes, Tests

Project-Based Learning Activities

Lab Manuals, Safety Videos, Virtual Labs & Other Tools

Vocabulary, Multilingual eGlossary, Vocab eGames, Vocab eFlashcards

Personal Tutors

Lesson 1 Weathering

Reading Guide

Key Concepts
ESSENTIAL QUESTIONS

- How does weathering break down or change rock?
- How do mechanical processes break rocks into smaller pieces?
- How do chemical processes change rocks?

Vocabulary
weathering p. 149
mechanical weathering p. 150
chemical weathering p. 152
oxidation p. 153

Multilingual eGlossary

Science Video

Inquiry What carved this rock?

Rocks carved like this can be along ocean shores and rivers, in deserts, and even underground. What carved them? What do they have in common?

Launch Lab

10 minutes

How can rocks be broken down?

Have you ever looked at the rocks in a stream? What makes some rocks look different from other rocks?

1. Read and complete a lab safety form.
2. Obtain 12 pieces of **candy-coated chocolate candies.** Put four of them in a **plastic cup.** Place the rest into a **container with a lid.**
3. Fasten the lid tightly. Shake the container vigorously 300 times.
4. Remove about half of the pieces. Place them in another plastic cup.
5. Replace the lid, and shake the container 300 more times. Remove the remaining "rocks," and place them in another cup.

Think About This

1. Compare and contrast the "rocks" from each cup.
2. **Key Concept** What do you think caused your "rocks" to change?

Weathering and Its Effects

Everything around you changes over time. Brightly painted walls and signs slowly fade. Shiny cars become rusty. Things made of wood dry out and change color. These changes are some examples of weathering. *The mechanical and chemical processes that change objects on Earth's surface over time are called* **weathering.**

Weathering also changes Earth's surface. Earth's surface today is different from what it was in the past and what it will be in the future. Weathering processes break, wear, abrade, and chemically alter rocks and rock surfaces. Weathering can produce strangely shaped rocks like those on the previous page as well.

Over thousands of years, weathering can break rock into smaller and smaller pieces. These pieces, also known as sediment, are called sand, silt, and clay. The largest soil pieces are sand grains and the smallest ones are clay. Weathering also can change the chemical makeup of a rock. Often, chemical changes can make a rock easier to break down.

Key Concept Check How does weathering break down or change rock?

SCIENCE USE V. COMMON USE

weather

Science Use to change from the action of the environment

Common Use the state of the atmosphere

FOLDABLES®

Make a two-tab book and label it as shown. Use it to organize your notes about how mechanical and chemical weathering affect rocks.

Math Skills

Use Geometry

The area (A) of a rectangular surface is the product of its length and its width.

$$A = \ell \times w$$

Area has square units, such as square centimeters (cm^2).

The surface area (SA) of a regular solid is the sum of the areas of all of its sides.

Practice

A rock sample is a cube and measures 3 cm on each side.

1. What is the surface area of the rock?
2. If you break the sample into two equal parts, what is the total surface area now?

Math Practice

Personal Tutor

Mechanical Weathering

When physical processes naturally break rocks into smaller pieces, **mechanical weathering** occurs. The chemical makeup of a rock is not changed by mechanical weathering. For example, if a piece of granite undergoes mechanical weathering, the smaller pieces that result are still granite.

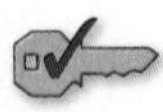

Examples of Mechanical Weathering

An example of mechanical weathering is when the intense temperature of a forest fire causes nearby rocks to expand and crack. Other causes of mechanical weathering are described in **Table 1** on the next page.

Key Concept Check What is the result of a rock undergoing mechanical weathering?

Surface Area

As shown in **Figure 1,** when something is broken into smaller pieces, the total surface area increases. Surface area is the amount of space on the outside of an object. The rate of weathering depends on a rock's surface area that is exposed to the environment.

Sand and clay are both the result of mechanical weathering. If you pour water on sand, some of the water sticks to the surface. Suppose you pour the same amount of water on an equal volume of clay. Clay particles are only about one-hundredth the size of sand. The greater total surface area of clay particles means more water sticks to its surfaces, along with any substances the water contains. The increased surface area means that weathering has a greater effect on soil with smaller particles. It also increases the rate of chemical weathering.

Reading Check Why is the surface area of a rock important?

Surface Area

Figure 1 The surface area of an object is all of the area on its exposed surfaces.

Surface area of cube = 6 equal squares
urface area = 6 squares × 64 cm^2/square
Surface area = 384 cm^2

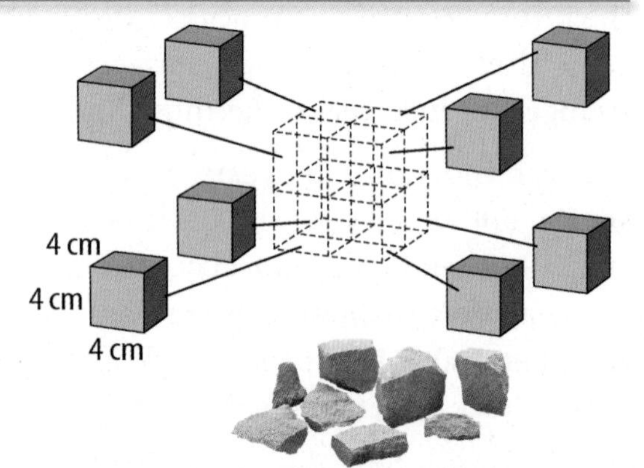

Surface area of 8 cubes = 48 equal squares
Surface area = 48 squares × 16 cm^2/square
Surface area = 768 cm^2

Hutchings Photography/Digital Light Source

Table 1 Causes of Mechanical Weathering

Ice Wedging
One of the most effective weathering processes is ice wedging—also called frost wedging. Water enters cracks in rocks. When the temperature reaches 0°C, the water freezes. Water expands as it freezes and the expansion widens the crack. As shown in the photo, repeated freezing and thawing can break rocks apart.

Abrasion
Another effective mechanical weathering process is abrasion—the grinding away of rock by friction or impact. For example, a strong current in a stream can carry loose fragments of rock downstream. The rock fragments tumble and grind against one another. Eventually, the fragments grind themselves into smaller and smaller pieces. Glaciers, wind, and waves along ocean or lake shores can also cause abrasion.

Plants
Plants can cause weathering by crumbling rocks. Imagine a plant growing into a crack in a rock. Roots absorb minerals from the rock, making it weaker. As the plant grows, its stem and roots not only get longer, they also get wider. The growing plant pushes on the sides of the crack. Over time, the rock breaks.

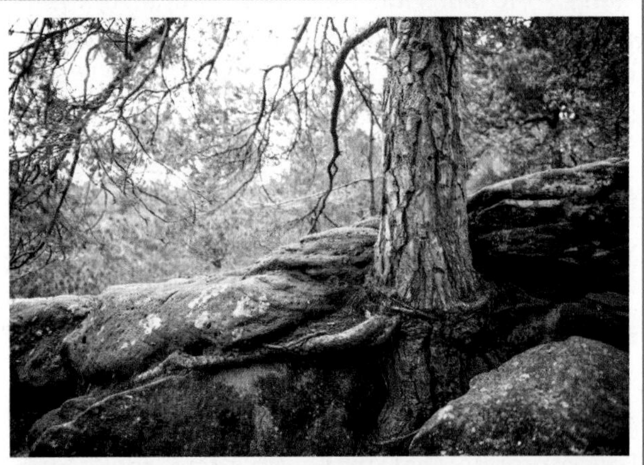

Animals
Animals that live in soil create holes in the soil where water enters and causes weathering. Animals burrowing through loose rock can also help to break down rocks as they dig.

Personal Tutor

Figure 2 These granite obelisks were carved in a dry climate, then one was moved to a different, wetter climate.

Visual Check What is the evidence that chemical weathering occurred?

Chemical Weathering

Figure 2 shows how chemical weathering can affect some rock. Both obelisks were carved in Egypt about 3,500 years ago. One was moved to New York City in the 1800s. There it has been exposed to more agents of chemical weathering. **Chemical weathering** *changes the materials that are part of a rock into new materials.* If a piece of granite weathers chemically, the composition and size of the granite changes.

Reading Check How does chemical weathering differ from mechanical weathering?

Water and Chemical Weathering

Water is important in chemical weathering because most substances dissolve in water. The minerals that make up most rocks dissolve very slowly in water. Sometimes the amount that dissolves over several years is so small that it seems as though the mineral does not dissolve at all.

For a rock, the process of dissolving happens when minerals in the rock break into smaller parts in solution. For example, table salt is the mineral sodium chloride. When table salt dissolves in water, it breaks into smaller sodium ions and chlorine ions.

Dissolving by Acids

Acids increase the rate of chemical weathering more than rain or water does. The action of acids attracts atoms away from rock minerals and dissolves them in the acid.

Scientists use pH, which is a property of solutions, to learn if a solution is acidic, basic, or neutral. They rate the pH of a solution on a scale from 0 to 14, where 7 is neutral. The pH of an acid is between 0 and 7. Vinegar has a pH of 2 to 3, so it is an acid.

Normal rain is slightly acidic, around 5.6, because carbon dioxide in the air forms a weak acid when it reacts with rain. This means rain can dissolve rocks, as it did to the obelisk in **Figure 2.**

Acid-forming chemicals enter the air from natural sources such as volcanoes. Pollutants in the air also react with rain and make it more acidic. For example, when coal burns, sulfur oxides form and enter the atmosphere. When these oxides dissolve in rain, they ionize the water to produce acid rain. Acid rain has a pH of 4.5 or less. It can cause more chemical weathering than normal rain causes.

Reading Check How can pollutants create acid rain?

Oxidation

Another process that causes chemical weathering is called oxidation. **Oxidation** *combines the element oxygen with other elements or molecules.* Most of the oxygen needed for oxidation comes from the air.

The addition of oxygen to a substance produces an oxide. Iron oxide is a common oxide of Earth materials. Useful ores, such as bauxite and hematite, are oxides of aluminum and iron, respectively.

Do all parts of an iron-containing rock oxidize at the same rate? The outside of the rock has the most contact with oxygen in the air. Therefore, this outer part oxidizes the most. When rocks that contain iron oxidize, a layer of red iron oxide forms on the outside surface, as shown in **Figure 3.**

Key Concept Check How does chemical weathering change rock?

Figure 3 The thin, red outer layer of this rock was created by oxidation. The oxidized minerals in this layer are different from the minerals in other layers of the rock.

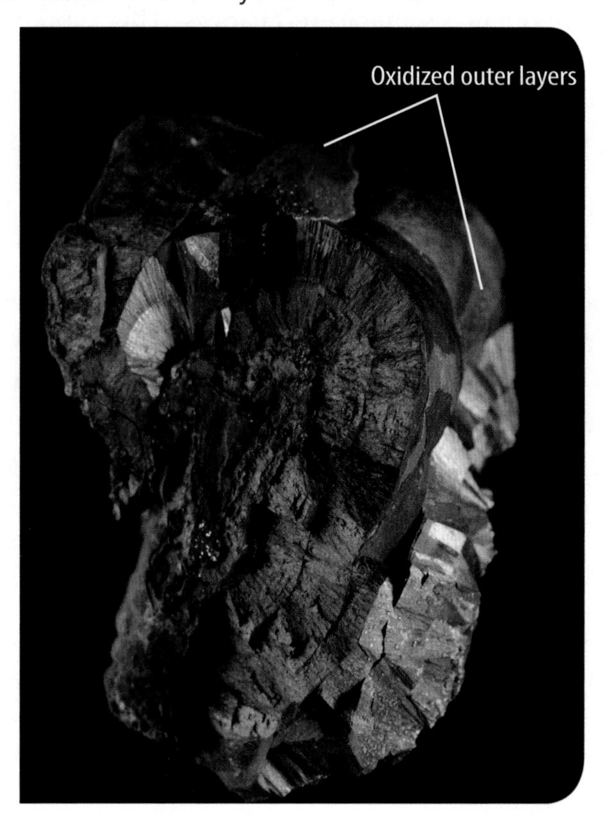

MiniLab 20 minutes

How are rocks weathered?

Chemical weathering can be caused by weak acids. These acids react with minerals in the rock and produce new substances.

1. Read and complete a lab safety form.
2. Use a **magnifying lens** to carefully examine the **rocks** provided by your teacher. Note details such as color, texture, and size of grains.
3. Use a **thin-stem pipette** to place several drops of **water** on each rock.
4. Observe what happens to each rock. Record your observations in your Science Journal.
5. Use the pipette to place several drops of dilute **hydrochloric acid** on each rock. Again, record your observations.

Analyze and Conclude

1. **Recognize Cause and Effect** Which substance reacted with the rock? How do you know a reaction occurred?
2. **Key Concept** What might happen to rocks exposed to such a substance in the environment?

Figure 4 The NIST wall is constructed of rock from almost every US state and several foreign countries. The wall has been exposed to continuous weathering since 1948.

Visual Check Point out which rocks have been weathered.

What affects weathering rates?

You saw in **Figure 2** that similar rocks can weather at different rates. What causes this difference?

The **environment** in which weathering occurs helps determine the rate of weathering. Both types of weathering depend on water and temperature. Mechanical weathering occurs fastest in locations that have frequent temperature changes. This type of weathering requires cycles of either wetting and drying or freezing and thawing. Chemical weathering is fastest in warm, wet places. As a result, weathering often occurs fastest in the regions near the equator.

ACADEMIC VOCABULARY
(noun) environment
the physical, chemical, and biotic factors acting in a community

Reading Check Why is weathering slow in cold, dry places?

The type of rock being weathered also affects the rate of weathering. The National Institute of Standards and Technology (NIST) constructed the wall shown in **Figure 4** to observe how different rocks weather under the same conditions.

Rocks can be made of one mineral or many minerals. The most easily weathered mineral determines the rate that the entire rock weathers. For example, rocks containing minerals with low hardness undergo mechanical weathering more easily. This increases the surface area of the rock. As a rocks surface area increases, more of its sides are exposed to the agents of weathering. This will cause the rock to weather more quickly. The size and number of holes in a rock also affect the rate at which a rock weathers.

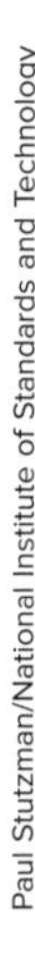
Paul Stutzman/National Institute of Standards and Technology

Lesson 1 Review

Visual Summary

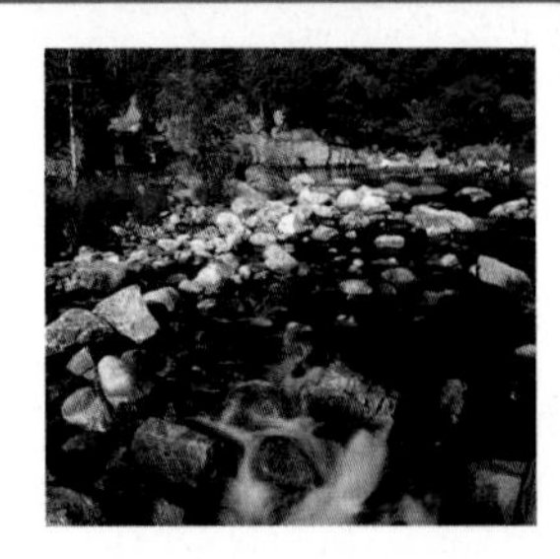

Weathering is the mechanical and chemical processes that change things over time.

Mechanical weathering does not change the identity of the materials that make up rocks. It breaks up rocks into smaller pieces.

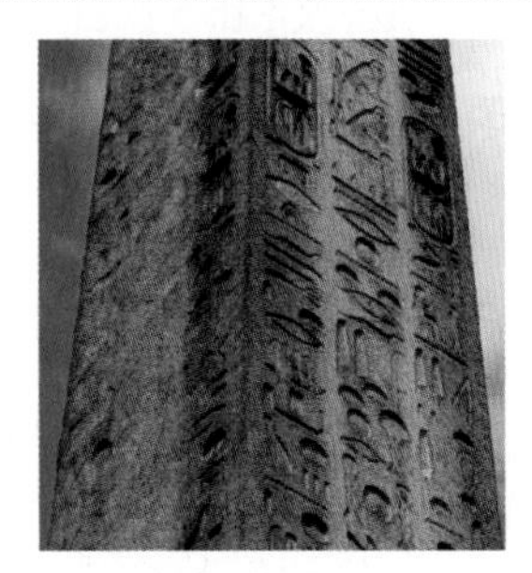

Chemical weathering is the process that changes the minerals in rock into different materials. Oxidation is a type of chemical weathering, as is reaction with an acid.

FOLDABLES®

Use your lesson Foldable to review the lesson. Save your Foldable for the project at the end of the chapter.

What do you think NOW?

You first read the statements below at the beginning of the chapter.

1. Any two rocks weather at the same rate.
2. Humans are the main cause of weathering.
3. Plants can break rocks into smaller pieces.

Did you change your mind about whether you agree or disagree with the statements? Rewrite any false statements to make them true.

Use Vocabulary

1. The chemical and physical processes that change things over time are called ______.
2. **Define** *mechanical weathering* in your own words.
3. **Use the term** *oxidation* in a sentence.

Understand Key Concepts 

4. **Identify** What kinds of rocks weather most rapidly?
5. What conditions produce the fastest weathering?
   - **A.** cold and dry
   - **B.** hot and dry
   - **C.** hot and wet
   - **D.** cold and wet
6. **Summarize** How does weathering change rocks and minerals?

Interpret Graphics

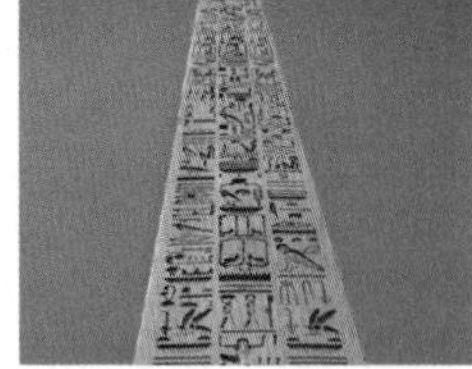

7. **Explain** How might chemical weathering change the appearance of this obelisk?
8. **Compare and contrast** types of weathering by copying and completing this table.

Weathering	Alike	Different
Chemical and Mechanical		

Critical Thinking

9. **Explain** how rates of chemical weathering change as temperature increases.

Math Skills

Math Practice

10. A block of stone measures 15 cm × 15 cm × 20 cm. What is the total surface area of the stone? Hint: A block has six sides.

What causes weathering?

Over time, rocks that are exposed at Earth's surface undergo mechanical and chemical weathering. You have already seen how mechanical processes break down a rock into small particles called sediment. Now you will model the mechanical weathering of rock and determine how much rock is weathered.

Materials

balance

rock chips

wide-mouthed plastic bottle with lid

water

timer

paper towels

Safety

Learn It

Scientists use **models** in a lab for many reasons. One use of a model is to study processes that happen too slowly to study them efficiently outside of the lab. Weathering is such a process.

Try It

1. Read and complete a lab safety form.
2. Copy the data table into your Science Journal.
3. Soak some rock chips in water. Then drain off the water and pat dry. Use a balance to measure 10.0 g of the soaked rock chips.
4. Place the rock chips in a bottle. Add enough water to cover the chips. Put the lid on the bottle. Shake the bottle vigorously for 3 minutes.
5. Drain the water and carefully remove the rock chips. Pat off the water with a paper towel. Measure the mass of the damp rock chips to the nearest tenth of a gram. Record the results in the data table.
6. Repeat steps 4 and 5 four additional times.
7. Calculate the percent of mass lost in each trial. Use the following steps. Record each answer in your data table.
   a. **Find the amount of mass lost.** Subtract the mass at the end of the trial from the mass at the start of the trial.
   b. **Find the percent of mass lost.** Divide the amount of mass lost (step a) by the mass at the start of the trial. Your answer should be to three decimal places. Then multiply by 100 to change the answer to a percent.

Apply It

8. How did this percentage change during the experiment?
9. **Key Concept** What type of weathering did you model in this experiment? How is this model similar to the natural process that it represents? How does it differ?

Data Table

Trial	Mass of Rocks at End of Trial (g)	Amount of Mass Lost (g)	Percent of Mass Lost
Start	10.0	None	None
1			
2			
3			
4			
5			

Lesson 2

Soil

Reading Guide

Key Concepts 

ESSENTIAL QUESTIONS

- How is soil created?
- What are soil horizons?
- Which soil properties can be observed and measured?
- How are soils and soil conditions related to life?

Vocabulary

soil p. 158
organic matter p. 158
pore p. 158
decomposition p. 159
parent material p. 160
climate p. 160
topography p. 161
biota p. 161
horizon p. 162

Multilingual eGlossary

William Yu Photography/Getty Images

Inquiry Why is the soil so red?

Soils have different colors because of what they contain. This soil contains iron, which makes it red. Why do iron-rich soils turn red? Is it red underground, too? What color is the soil where you live?

Launch Lab

10 minutes

What is in your soil?

Soils are different in different places. Suppose you look at the soil along a river bank. Is this soil like the soil in a field? Are either of these soils like the soil near your home? What is in the soil where you live?

1. Read and complete a lab safety form.
2. Place about a cup of **local soil** in a **jar** that has a **lid.** Add a few drops of **liquid detergent.**
3. Add **water** to the jar until it is almost full. Firmly attach the lid.
4. Shake for 1 minute and place it on your desk.
5. Observe the contents of the jar after 2 minutes and again after 5 minutes.

Think About This

1. How many different layers did your sample form?
2. **Key Concept** From your observations, what do you think makes up each layer?

What is soil?

A soil scientist might think of soil as the "active skin of Earth." Soil is full of life, and life on Earth depends on soil.

If you were to dig into soil, what would you find? About half the volume of soil is solid materials. The other half is liquids and gases. **Soil** *is a mixture of weathered rock, rock fragments, decayed organic matter, water, and air.*

As you read in Lesson 1, weathering gradually breaks rocks into smaller and smaller fragments. These fragments, however, do not become good soil until plants and animals live in them. Plants and animals add organic matter to the rock fragments. **Organic matter** *is the remains of something that was once alive.*

The amounts of water and air vary in the small holes and spaces in soil. *These small holes and spaces are called* **pores.** Pores are important because they enable water to flow into and through soil. Pores can vary greatly in size depending on the particles that make up the soil. **Figure 5** shows three particles common in soil–sand, silt and clay. As particle size increases, pore size also increases. The pores between clay particles are smaller than the pores between sand particles.

WORD ORIGIN
pore
from Greek *poros,* means "passage"

Reading Check What is in a pore?

Hutchings Photography/Digital Light Source

The Organic Part of Soil

Recall that the solid part of soil that was once part of an organism is called organic matter. Pieces of leaves, dead insects, and waste products of animals that are in the soil are examples of organic matter.

How does organic matter form? Soil is home to many organisms, from roots of plants to tiny bacteria. Over time, roots die, and leaves and twigs fall to the ground. Organisms living in the soil decompose these materials for food. **Decomposition** *is the process of changing once-living material into dark-colored organic matter.* In the end, something that was once recognizable as a pine needle becomes organic matter.

Reading Check How is decomposition related to organic matter?

Organic matter gives soil important properties. Dark soil absorbs sunlight while organic matter holds water and provides plant nutrients. Organic material holds minerals together in clusters. This helps keep soil pores open for the movement of water and air in soil.

The Inorganic Part of Soil

The term *inorganic* describes materials that have never been alive. Mechanical and chemical weathering of rocks into fragments forms inorganic matter in soil. Soil scientists classify the soil fragments according to their sizes. Rock fragments can be boulders, cobbles, gravel, sand, silt, or clay. **Figure 5** shows a magnified image of the three smallest sizes of soil particles. Between large particles are large pores, which affect soil properties such as drainage and water storage.

Sand feels rough.	**Silt** feels smooth.	**Clay** feels sticky.

Figure 5 Inorganic matter contributes different properties to soil. Large pores occur between large particles, which drain rapidly; small particle pores retain more water in the soil.

MiniLab

20 minutes

How can you determine soil composition?

Scientists can sometimes feel soil to help identify the soil's composition. Can you identify soil composition by how it feels?

1. Read and complete a lab safety form.
2. Carefully observe your **soil** sample with a **magnifying lens.** In your Science Journal, record the sizes of the particles you observe.
3. Fill a **spray bottle** or a **sprinkling can** with **water.** Use the water to moisten the soil.
4. Rub some moist soil between your fingers.
5. Use **Figure 5** and your observations to classify your soil as mostly sand, mostly silt, or mostly clay.

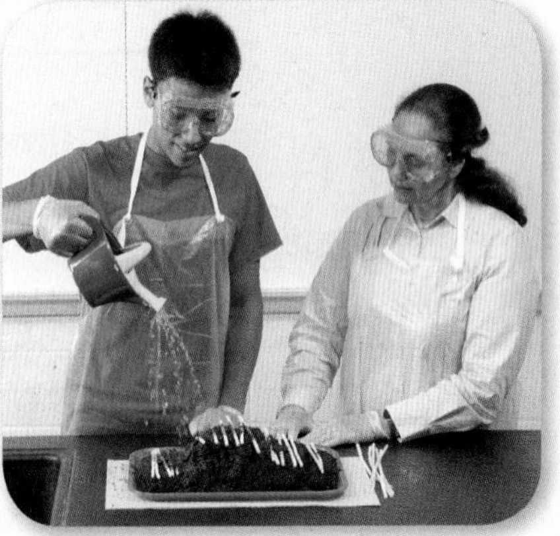

Hutchings Photography/Digital Light Source

Analyze and Conclude

1. **Classify** What texture does the soil have?
2. **Key Concept** What other properties of your soil sample did you observe?

FOLDABLES®

Divide a circle into five parts and label it as shown. Use the circle organizer to record information about the factors of soil formation.

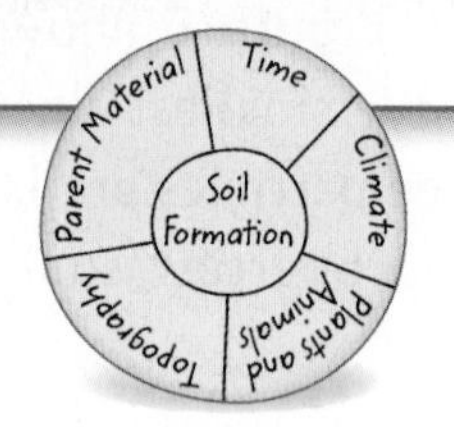

Formation of Soil

Why is the soil near your school different from the soil along a river bank or soil in a desert? The many kinds of soils that form depend on five factors, called the factors of soil formation. The five factors are parent material, climate, topography, biota, and time.

Parent Material

The starting material of soil is **parent material**. It is made of the rock or **sediment** that weathers and forms the soil, as shown in **Figure 6.** Soil can develop from rock that weathered in the same place where the rock first formed. This rock is known as bedrock. Soil also can develop from weathered pieces of rock that were carried by wind or water from another location. The particle size and the type of the parent material can determine the properties of the soil that develops.

Key Concept Check What is the role of parent material in creating soil?

Animation

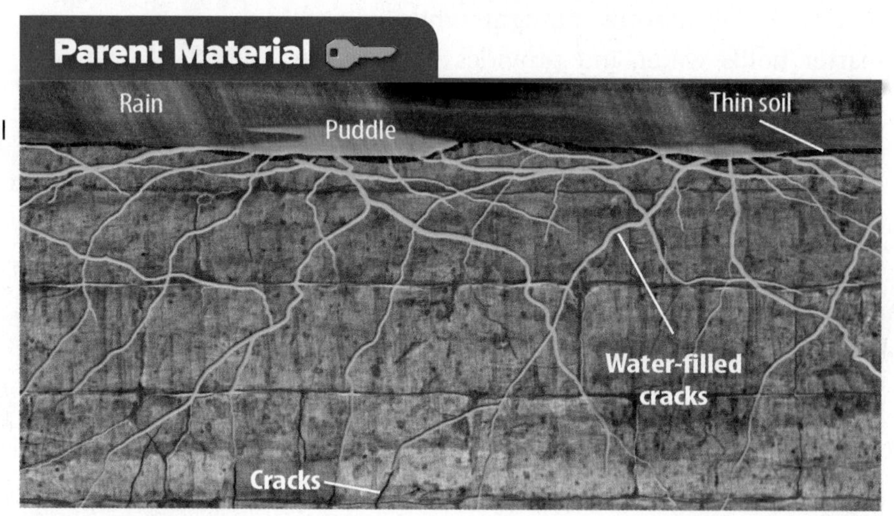

Figure 6 Parent material is broken down by mechanical and chemical weathering.

Climate

The average weather of an area is its **climate.** How can you describe the climate where you live? The amount of precipitation and the daily and average annual temperatures are some measures of climate. If the parent material is in a warm, wet climate, soil formation can be rapid. Large amounts of rain can speed up weathering as it contacts the surface of the rock. Warm temperatures also speed up weathering by increasing the rate of chemical changes. Weathering rates also increase in locations where freezing and thawing occur.

REVIEW VOCABULARY

sediment
rock material that has been broken down or dissolved in water

Reading Check Why do soils form rapidly in warm, moist climates?

Topography

Is the land where you live flat or hilly? If it is hilly, are the hills steep or gentle? **Topography** *is the shape and steepness of the landscape.* The topography of an area determines what happens to water that reaches the soil surface. For example, in flat landscapes, most of the water enters the soil. Water speeds weathering. In steep landscapes, much of the water runs downhill. Water running downhill can carry soil with it, leaving some slopes bare of soil. **Figure 7** shows that broken rock and sediments collect at the bottom of a steep slope. There, they undergo further weathering.

 **Reading Check** What is topography?

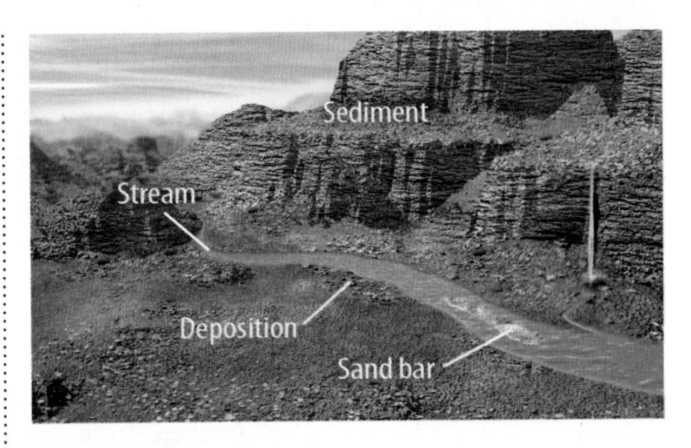

Figure 7 Broken rock and sediment collects at the bottom of steep slopes. Sediment is redistributed by streams and moving water as sandbars and shoreline deposition.

Biota

Soil is home to a large number and variety of organisms. They range from the smallest bacteria to small rodents. *All of the organisms that live in a region are called* **biota** (bi OH tuh). Biota in the soil help speed up the process of soil formation. Some soil biota form passages for water to move through. Most soil organisms are involved in the decomposition of materials that form organic matter. As **Figure 8** shows, rock and soil are affected by organism activity.

 **Key Concept Check** How do biota aid in soil formation?

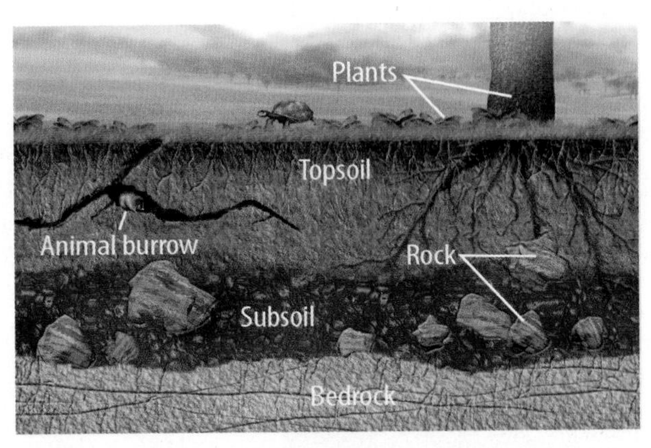

Figure 8 Mature soils form over thousands of years as plants, animals, and other processes break down the bedrock and subsoil.

Time

As time passes, weathering is constantly acting on rock and sediment. Therefore, soil formation is a constant, but slow, process. A 90 year-old person is considered old, but soil is still young after a thousand years. It is difficult to see all the soil-producing changes in one human lifetime.

As **Figure 8** shows, mature soils develop layers as new soil forms on top of older soil. Each layer has different characteristics as organic matter is added or as water carries elements and nutrients downward.

Horizons

You know that soil is more than what you see when you look at the ground. If you dig into the soil, you see that it is different as you dig deeper. You might see dark soil on or near the surface. The soil you see deeper down is lighter in color and probably contains larger pieces of rock. Soil might be loosely packed on the surface, but deeper soil is more tightly packed.

WORD ORIGIN
horizon
from Latin *horizontem,* means "bounding circle"

Soil has layers, called horizons. **Horizons** *are layers of soil formed from the movement of the products of weathering.* Each horizon has characteristics based on the type of materials it contains. The three horizons common to most soils are identified as A-horizon, B-horizon, and C-horizon, as shown in **Figure 9**. Each horizon can appear quite different depending on where the soil forms. The top, organic layer is called the O-horizon and the unweathered, bedrock layer is the R-horizon.

Key Concept Check What are soil horizons?

Common Soil Horizons

Figure 9 A-, B-, and C-horizons are commonly found in soil. Some soils contain other kinds of horizons. Not every kind of horizon is found in every soil.

Visual Check One horizon contains a lot of clay, and another horizon is dark. Of these two horizons, which is on top? Explain your answer.

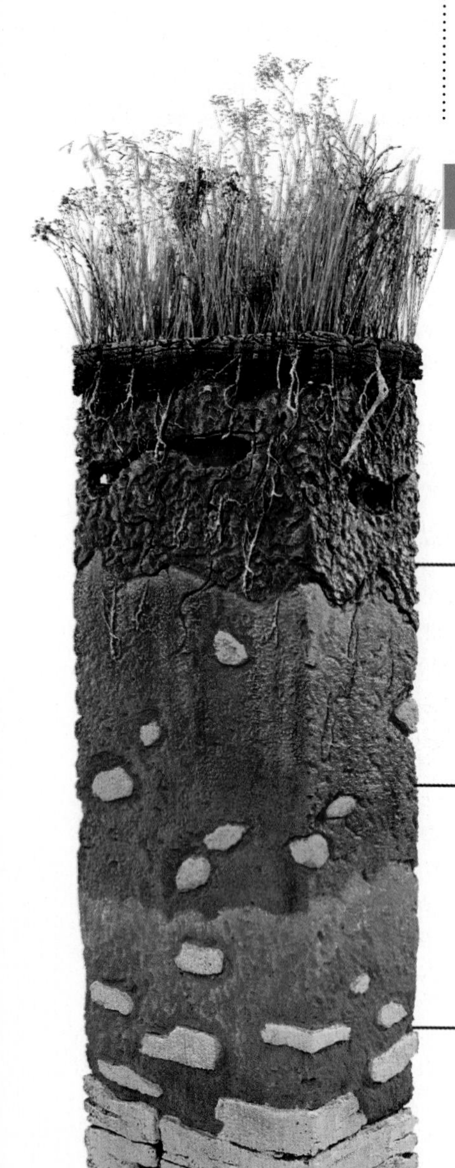

A-horizon
The A-horizon is the part of the soil that you are the most likely to see when you dig a shallow hole in the soil with your fingers. Organic matter from the decay of roots and the action of soil organisms often makes this horizon excellent for plant growth. Because the A-horizon contains most of the organic matter in the soil, it is usually darker than other horizons.

B-horizon
When water from rain or snow seeps through pores in the A-horizon, it carries clay particles. The clay is then deposited below the upper layer, forming a B-horizon. Other materials also accumulate in B-horizons.

C-horizon
The layer of weathered parent material is called the C-horizon. Parent material can be rock or sediments.

Soil Properties and Uses

Soil horizons in different locations have different properties. Recall that properties are characteristics used to describe something. Several soil properties are listed and described in **Table 2.** The properties of a soil determine the best use of that soil. For example, soil that is young, deep, and has few horizons is good for plant growth.

Observing and Measuring Soil Properties

Some properties of soil can be determined just by observation. The amount of sand, silt, and clay in a soil can be estimated by feeling the soil. The types of horizons also provide information about the soil. The color of a soil is easily observed and shows how much organic matter it contains.

Many soil properties can be measured more accurately in a laboratory. Laboratory measurements can determine exactly what is in each sample of soil. Measuring nutrient content and soil pH to determine the suitability for farming or gardening requires careful laboratory analysis.

Key Concept Check List soil properties that can be observed and measured.

Soil Properties That Support Life

Plants depend on the nutrients that come from organic matter and the weathering of rocks. Plant growers can observe how well plants grow in the soil to get information about soil nutrients. Crop plants depend less on weathering for nutrients because farmers usually use fertilizers that add nutrients to the soil.

It takes thousands of years to form soil from parent material. Soil that is damaged or misused is slow to replenish its nutrients. The restoration can take many human lifetimes.

Key Concept Check How are soil nutrients related to life?

Table 2 Soil Properties

Color	Soil can be described based on the color, such as how yellow, brown, or red it is; how light or dark it is; and how intense the color is.
Texture	The texture of soil ranges from boulder-sized pieces to very fine clay.
Structure	Soil structure describes the shape of soil clumps and how the particles are held together. Structure can look grainy, blocky, or prism shaped.
Consistency	The hardness or softness of a soil is the measure of its consistency. Consistency varies with moisture. For example, some soils have a soft, slippery consistency when they are moist.
Infiltration	Infiltration describes how fast water enters a soil.
Soil moisture	The amount of water in soil pores is its moisture content. Soil scientists determine weight loss by drying samples in an oven at 100°C. The weight difference is the amount of moisture in the soil.
pH	Most soils have a pH between 5.5 and 8.2. Soils can be more acidic in humid environments.
Fertility	Soil fertility is the measure of the ability of a soil to support plant growth. Soil fertility includes the amount of certain elements that are essential for good plant growth.
Temperature	On the ground surface, soil temperature changes with daily cycles and the weather. Soil temperature in lower layers changes less.

Table 2 Many soil properties are observed. Others are more likely to be measured. These properties can predict soil quality.

Soil Types of North America

Figure 10 The properties of soil are different in different climates. There are 12 major soil types on Earth. North America contains almost every type of soil.

Visual Check What soil property might be typical of a desert in the Southwest?

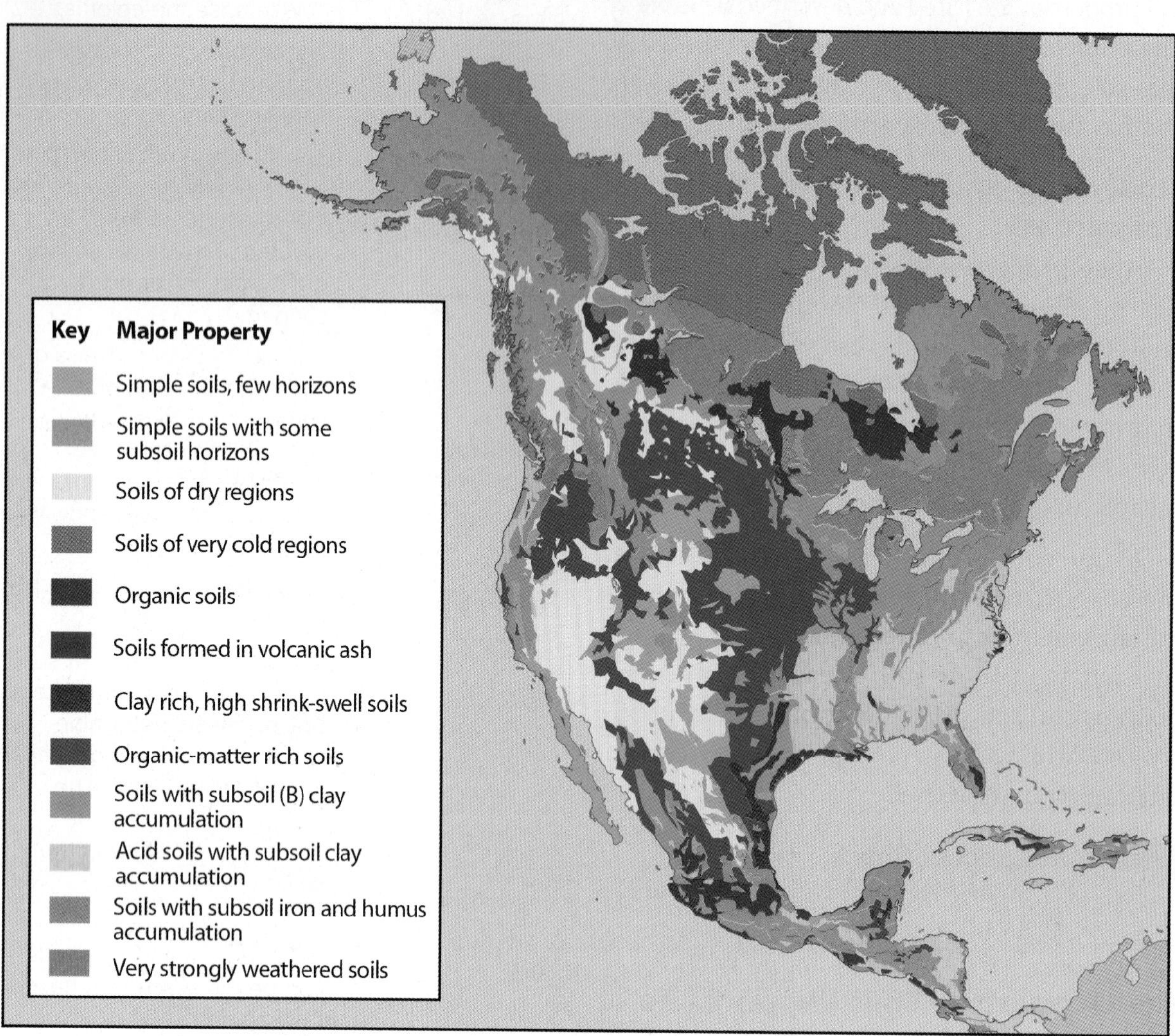

Soil Types and Locations

Recall that the type of soil formed depends partly on climate. Can you see how the soil types shown in **Figure 10** depend on the climate where they form? For example, in northern parts of Canada and Alaska, and along mountain ranges, some soils stay frozen throughout the year. These soils are very simple and have few horizons. In the mid-latitudes, you can see a wide variety of soil types and depths. Farther toward the warm and wet climate of the tropics, soils are deeply weathered. Soils formed near volcanoes, such as those in Alaska and California, are acidic and have fine ash particles from volcanic activity.

 **Key Concept Check** Are soils the same everywhere?

Lesson 2 Review

Visual Summary

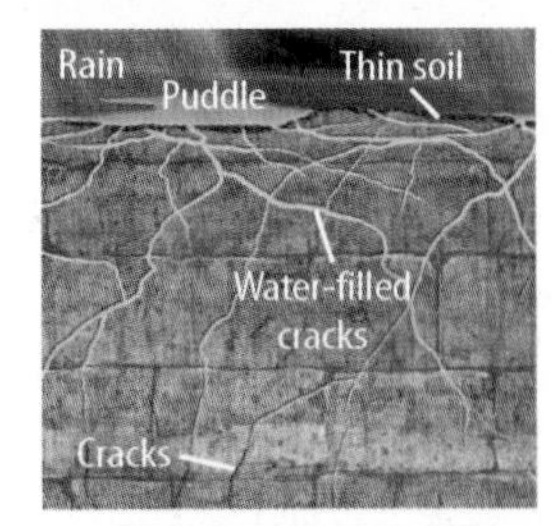

The inorganic matter in soil is made up of weathered parent material. The organic matter in soil is made by the decomposition of things that once lived.

The five factors that contribute to soil formation are parent material, topography, climate, biota, and time.

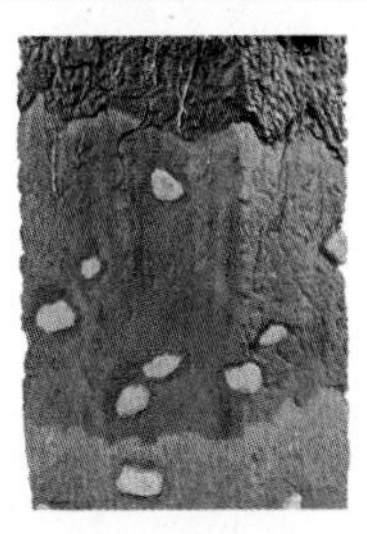

Soil contains horizons, which are layers formed from the movement of the products of weathering. Most soil contains A-, B-, and C-horizons.

FOLDABLES®

Use your lesson Foldable to review the lesson. Save your Foldable for the project at the end of the chapter.

What do you think NOW?

You first read the statements below at the beginning of the chapter.

4. Air and water are present in soil.

5. Soil that is 1,000 years old is young soil.

6. Soil is the same in all locations.

Did you change your mind about whether you agree or disagree with the statements? Rewrite any false statements to make them true.

Use Vocabulary

1. **Use the term** *decomposition* correctly in a sentence.
2. **Explain** how a leaf is organic matter.
3. **Define** *biota* in your own words.

Understand Key Concepts 

4. What is in the C-horizon?
   - **A.** bedrock
   - **B.** clay
   - **C.** weathered stone
   - **D.** organic material
5. **Contrast** rocks and soil. List three differences.
6. **Describe** what fills soil pores.

Interpret Graphics

7. **Identify** Use the diagram to identify the soil horizon that contains the most organic matter.

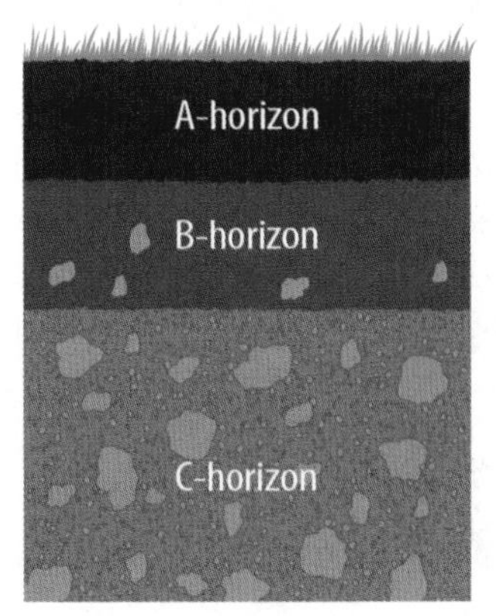

8. **Sequence** Copy the graphic organizer below. Starting with parent material, list steps that lead to the formation of an A-horizon.

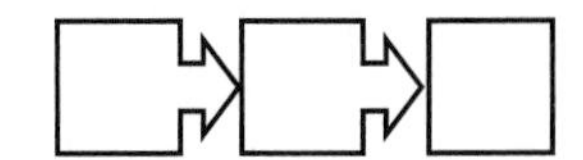

Critical Thinking

9. **Explain** What three things does soil provide for plants?
10. **Apply** Describe the soil-forming factors around your school.

Lab

40 minutes

Materials

index cards

glue

colored pencils

silt

clay

sand

topsoil

Safety

Soil Horizons and Soil Formation

Soil, the complex mixture of weathered rock and partially decayed organic matter, covers most of Earth's land surfaces. Soil is different in different locations because it forms from different rocks and in different climates and topography. As soil develops, it forms horizontal layers that have different properties. These layers vary in color and thickness. Together, they form a soil profile. How can you model a soil profile and relate it to how soil formed at that location?

Question

How is a soil profile in a certain location determined by the soil-forming factors there?

Procedure

1. Discuss the types of rocks, the climate, and the topography of Minnesota, Colorado, and Florida. You can use reference materials to obtain this information. Record some similarities and differences in your Science Journal.
2. Examine the soil profile from each of the samples shown on these pages. Record some similarities and differences.
3. Draw the sample profiles and mark the A-, B-, and C-horizons that are present on each drawing.
4. Use what you know about soil formation and the sample profiles to state how each soil horizon relates to factors of soil formation.

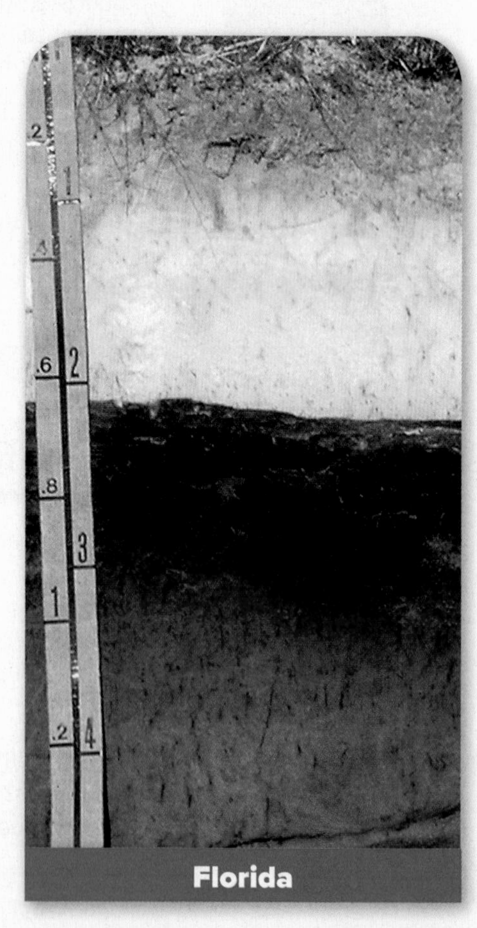
Florida

(t to b)Janette Beckman/McGraw-Hill Education; (2)Joe Polillio/McGraw-Hill Education; (3)Jacques Cornell/McGraw-Hill Education; (4-5, 7)Hutchings Photography/Digital Light Source; (6)Ken Cavanagh/McGraw-Hill Education; (r)USDA/NRCS and SWSD-UnFL

5 Choose one of the three soil profiles shown in this activity. Use the provided materials to model this profile. Label the model with the state name and the horizons you see.

6 Examine the information about parent material, climate, and topography for the state you chose. Make generalizations about how soil profiles are affected by soil-forming factors.

Analyze and Conclude

7 Were any of the profiles missing an A-, B-, or C-horizon? Explain why a horizon might not be present in a profile.

8 Was one of the horizons thicker in any of the profiles? What could explain this?

9 **The Big Idea** What did your conclusions show about how a soil profile relates to soil-forming factors?

Minnesota

(l)USDA; (r)Tom Reinsch/USDA-NRCS

Communicate Your Results

As a class, place a soil-profile model for each listed state on a map of the United States. For each profile, discuss what other states might have a similar soil profile.

Inquiry Extension

Choose a location on another continent that you think would have a similar soil profile to one of the profiles examined in this activity. Research the soil in that location, and report the similarities and differences between the soils in these two locations. In your report, explain why the soils might differ.

Lab Tips

☑ Review where silt, clay, sand, and topsoil appear in soil horizons before modeling a soil profile.

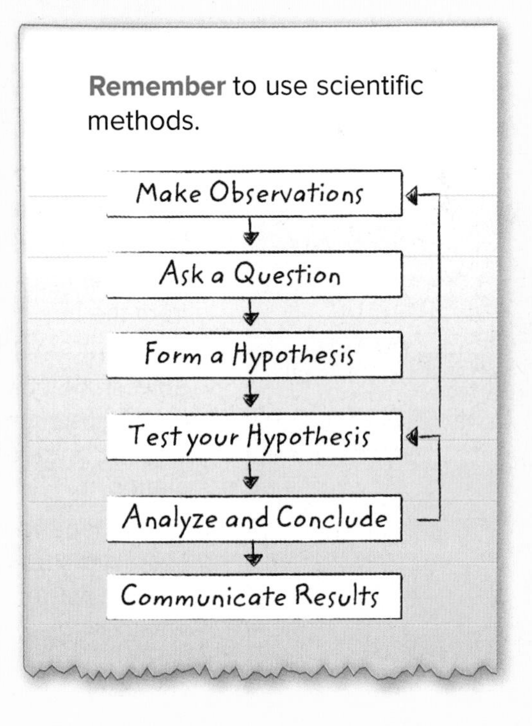

Chapter 5 Study Guide

Mechanical and chemical weathering break down rocks, which begins the formation of soil.

Key Concepts Summary

Lesson 1: Weathering

- **Weathering** acts mechanically and chemically and breaks down rocks.
- Through the action of Earth processes such as freezing and thawing, **mechanical weathering** breaks rocks into smaller pieces.
- **Chemical weathering** by water and acids changes the materials in rocks into new materials.

Vocabulary

weathering p. 149
mechanical weathering p. 150
chemical weathering p. 152
oxidation p. 153

Lesson 2: Soil

- Five factors—**parent material, climate, topography, biota,** and time—affect the formation of soil.
- **Horizons** are soil layers formed from the movement of the various products of weathering.
- Soil can be characterized by properties such as the amount of **organic matter** and inorganic matter.
- Plants depend on certain characteristics of soil, such as organic matter and amount of weathering.

Vocabulary

soil p. 158
organic matter p.158
pore p.158
decomposition p. 159
parent material p.160
climate p. 160
topography p. 161
biota p. 161
horizon p. 162

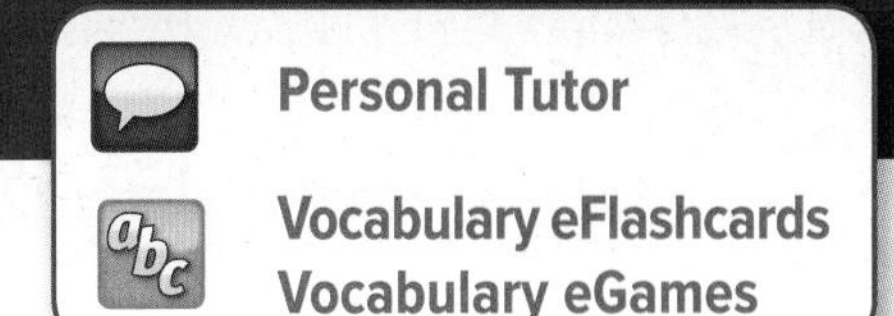

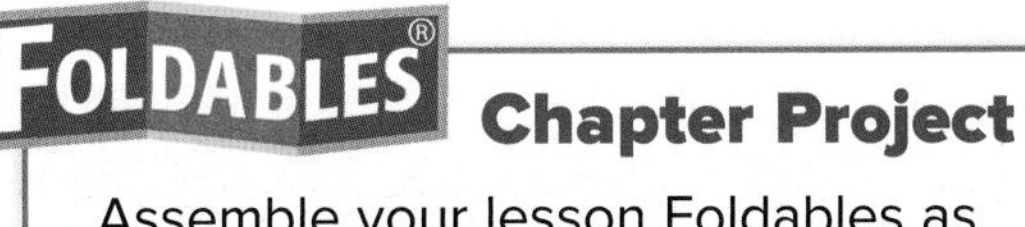

Assemble your lesson Foldables as shown to make a Chapter Project. Use the project to review what you have learned in this chapter.

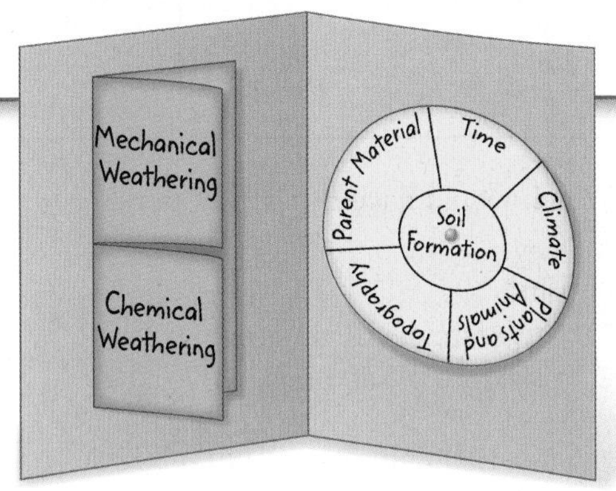

Use Vocabulary

1. When rock undergoes ________, the product is smaller pieces of the same kind of rock.
2. Rock fragments and other materials combine to form ________.
3. The part of soil that comes from plants and animals is ________.
4. An important soil-forming factor that includes trees and microorganisms is ________.
5. Oxygen combines with other elements or compounds during the process of ________.
6. The shape of the land is its ________.

Link Vocabulary and Key Concepts

Copy this concept map, and then use vocabulary terms from the previous page to complete the concept map.

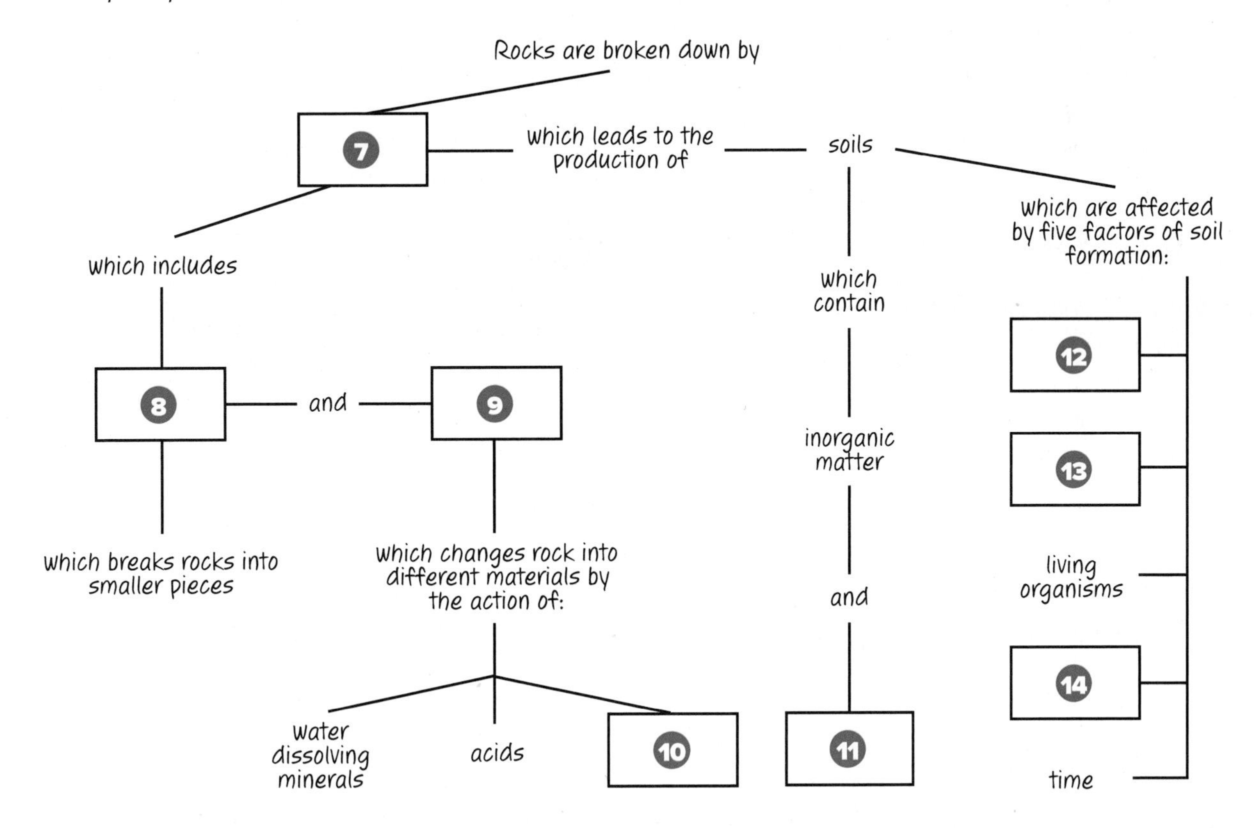

Chapter 5 Review

Understand Key Concepts 

1. Which is an example of chemical weathering?
   - A. abrasion
   - B. ice wedging
   - C. organisms
   - D. oxidation

2. A statue made of limestone is damaged by its environment. What most likely caused this damage?
   - A. acid
   - B. a root
   - C. topography
   - D. wind

3. The picture below shows how mechanical and chemical weathering changes a rock.

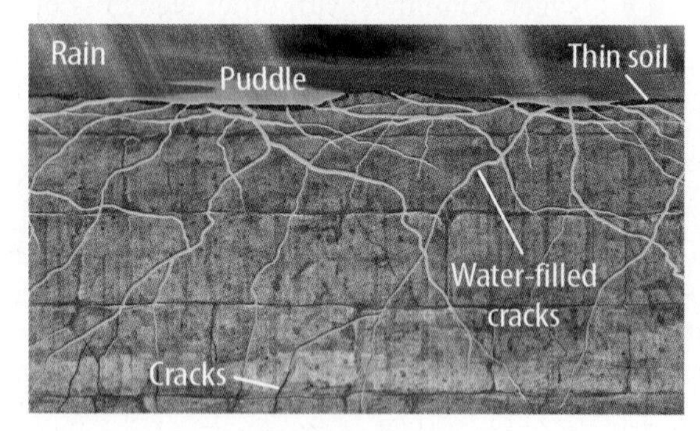

   What kind of chemical weathering is most likely illustrated above?
   - A. acid rain reactions
   - B. ice wedging
   - C. mineral absorption
   - D. root pressure

4. What kind of climate has the fastest weathering?
   - A. cold and dry
   - B. cold and wet
   - C. hot and dry
   - D. hot and wet

5. How does organic matter help soil?
   - A. It decomposes bacteria in the soil.
   - B. It holds water.
   - C. It weathers and forms clay.
   - D. It weathers nearby rocks.

6. The table below shows different sizes of soil particles.

Sand feels rough.	**Silt** feels smooth.	**Clay** feels sticky.

   Which would have the largest pores?
   - A. clay
   - B. sand
   - C. a mixture of clay and silt
   - D. a mixture of sand and silt

7. What is the main material in a B-horizon?
   - A. clay
   - B. iron
   - C. organic matter
   - D. parent material

8. Which statement is true about soils worldwide?
   - A. They are the same color.
   - B. They are the same age.
   - C. They are different in many ways.
   - D. They differ only in thickness.

9. Which process causes river gravel to have rounded edges?
   - A. abrasion
   - B. acid rain
   - C. ice wedging
   - D. oxidation

10. Which is NOT a soil property?
    - A. color
    - B. pH
    - C. texture
    - D. topography

Online Test Practice

Critical Thinking

11 **Infer** A student notices that when it rains, most of the water that falls on her yard runs off instead of soaking in. Is it more likely that the soil in her yard contains mostly clay or mostly sand? Explain.

12 **Explain** How do the biota shown in the image below help form soil?

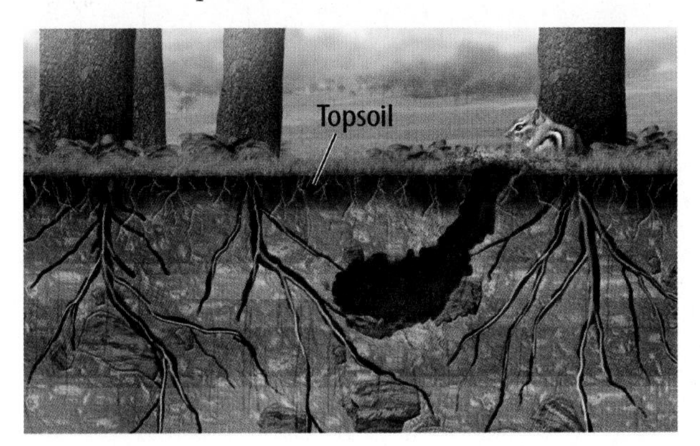

13 **Explain** how climate helps to create soil.

14 **Describe** how soil horizons are produced and identified.

15 **Compare** Stone buildings near cities usually undergo more chemical weathering than buildings away from cities. Explain why this is true.

16 **Summarize** how soil is important to life.

17 **Identify** how chemical weathering and mechanical weathering make soil.

18 **Describe** how ice wedging and plant roots are similar in breaking rocks down.

Writing in Science

19 **Write** a short story that explains how a large boulder becomes sand through weathering. In your story, include both mechanical and chemical weathering. Include main ideas and supporting details.

REVIEW THE BIG IDEA

20 What processes might have created the dust in the chapter opener photo?

21 How might dust become an agent of soil formation?

Math Skills

Math Practice

Use the following data to answer the questions.

Rock Sample	Length	Width	Height
X	8 cm	8 cm	8 cm
Y	2 cm	16 cm	16 cm

22 How do the surface areas of rock sample X and rock sample Y compare?

23 What is the surface area of each face of rock X? Rock Y?

24 Rock sample X breaks into 8 equal cubes.

a. What is the surface area of each cube?

b. What is the total surface area of the broken rock?

c. How does this area compare with the original surface area?

Standardized Test Practice

Record your answers on the answer sheet provided by your teacher or on a sheet of paper.

Multiple Choice

1 Which is true of oxidation?

A It is a mechanical process.

B No change occurs in the makeup of rock.

C Rock parts weather at different rates.

D Water enters cracks in rock.

2 What does the term *biota* describe?

A all of the organisms living in a region

B how burrowing animals change soil and rock

C the ability of a certain type of soil to support plant life

D the remains of once-living things in soil

Use the table below to answer question 3.

Rain sample	pH
1	5.3
2	4.7
3	5.5
4	4.3

3 Students collected and recorded the pH of four samples of rainwater in the table above. Which sample is the most acidic?

A 1

B 2

C 3

D 4

4 Which soil property is a measure of the consistency of soil?

A the moisture content

B its ability to support plant growth

C its hardness or softness

D the size of its particles

Use the diagram below to answer question 5.

5 At which spot in the landscape above would you most likely find a pile of broken, weathering rocks?

A 1

B 2

C 3

D 4

6 What is the pH range of most soils?

A 2.0–3.0

B 4.4–7.0

C 5.5–8.2

D 7.5–10.5

7 The grinding of rock by friction or impact is called

A abrasion.

B decomposition.

C erosion.

D infiltration.

8 Which is NOT organic matter?

A animal wastes

B dead insects

C decayed leaves

D mineral fragments

Use the diagram below to answer question 9.

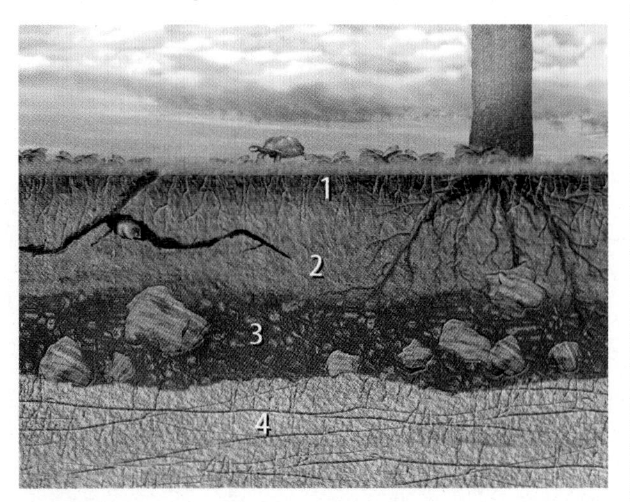

9 Which area pictured in the diagram above contains the most organic matter?

A 1

B 2

C 3

D 4

10 Which is LEAST likely to weather bedrock buried beneath layers of soil?

A abrasion

B acidic water

C ice

D plant roots

11 If volume were the same, which would have the greatest surface area?

A clay

B gravel

C sand

D silt

Constructed Response

Use the table below to answer questions 12 and 13.

Soil Horizon	Description
O	
A	
B	the clay-rich layer beneath the A-horizon
C	
R	unweathered bedrock that makes up the parent material for the soil

12 Describe the O-, A-, and C- horizons to complete the table above.

13 Why is the B-horizon rich in clay?

14 What are the five factors of soil formation? Describe each.

15 What are pores in soil? Why are they important?

NEED EXTRA HELP?															
If You Missed Question...	1	2	3	4	5	6	7	8	9	10	11	12	13	14	15
Go to Lesson...	1	2	1	2	2	2	1	2	2	1	1	2	2	2	2

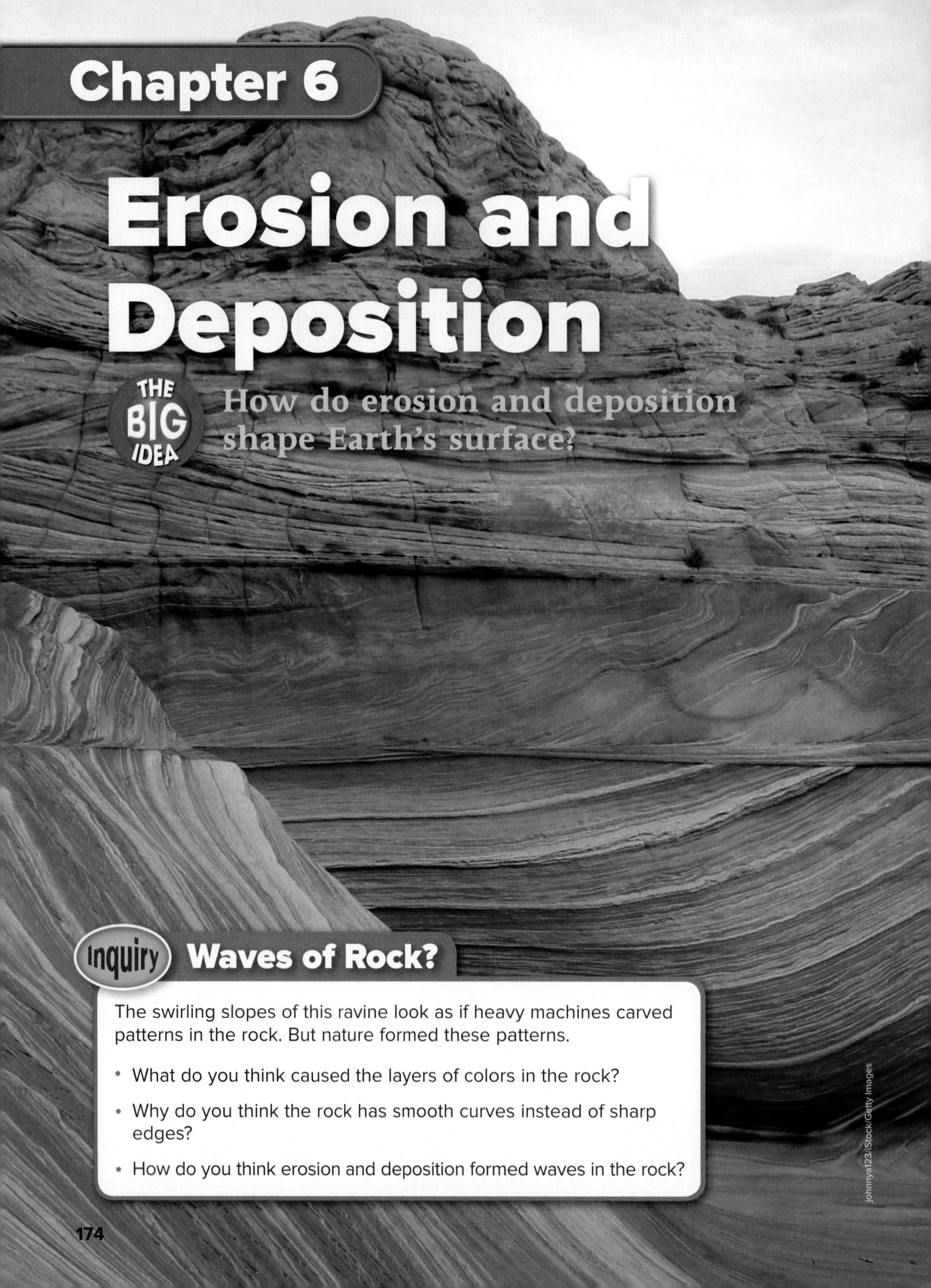

Chapter 6

Erosion and Deposition

THE BIG IDEA

How do erosion and deposition shape Earth's surface?

Inquiry Waves of Rock?

The swirling slopes of this ravine look as if heavy machines carved patterns in the rock. But nature formed these patterns.

- What do you think caused the layers of colors in the rock?
- Why do you think the rock has smooth curves instead of sharp edges?
- How do you think erosion and deposition formed waves in the rock?

johnnya123/iStock/Getty Images

Get Ready to Read

What do you think?

Before you read, decide if you agree or disagree with each of these statements. As you read this chapter, see if you change your mind about any of the statements.

1. Wind, water, ice, and gravity continually shape Earth's surface.
2. Different sizes of sediment tend to mix when being moved along by water.
3. A beach is a landform that does not change over time.
4. Windblown sediment can cut and polish exposed rock surfaces.
5. Landslides are a natural process that cannot be influenced by human activities.
6. A glacier leaves behind very smooth land as it moves through an area.

Your one-stop online resource
connectED.mcgraw-hill.com

LearnSmart®

Chapter Resources Files, Reading Essentials, Get Ready to Read, Quick Vocabulary

Animations, Videos, Interactive Tables

Self-checks, Quizzes, Tests

Project-Based Learning Activities

Lab Manuals, Safety Videos, Virtual Labs & Other Tools

Vocabulary, Multilingual eGlossary, Vocab eGames, Vocab eFlashcards

Personal Tutors

Lesson 1

Reading Guide

Key Concepts
ESSENTIAL QUESTIONS

- How can erosion shape and sort sediment?
- How are erosion and deposition related?
- What features suggest whether erosion or deposition created a landform?

Vocabulary
erosion p. 179

deposition p. 181

Multilingual eGlossary

The Erosion-Deposition Process

Inquiry Stripes and Cuts?

Long ago, this area was at the bottom of an ocean. Today, it is dry land in Badlands National Park, South Dakota. Why do you think these hills are striped? What do you think caused such deep cuts in the land? What natural processes created landforms such as these?

Launch Lab

10 minutes

How do the shape and size of sediment differ?

Sediment forms when rocks break apart. Wind, water, and other factors move the sediment from place to place. As the sediment moves, its shape and size can change. In this activity, you will observe the different shapes and sizes of sediment.

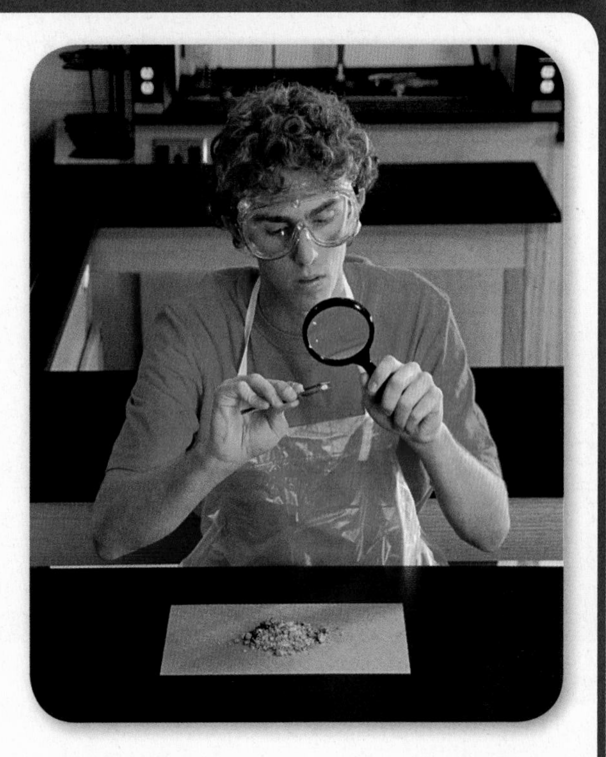

1. Read and complete a lab safety form.
2. Obtain a **bag of sediment** from your teacher. Pour the sediment onto a sheet of **paper.**
3. Use a **magnifying lens** to observe the differences in shape and size of the sediment.
4. Divide the sediment into groups according to its size and whether it has rounded or sharp edges.

Think About This

1. What were the different groups you used to sort the sediment?
2. **Key Concept** How do you think movement by wind and water might affect the shape and size of the sediment?

Reshaping Earth's Surface

Have you ever seen bulldozers, backhoes, and dump trucks at the construction site of a building project? You might have seen a bulldozer smoothing the land and making a flat surface or pushing soil around and forming hills. A backhoe might have been digging deep trenches for water or sewer lines. The dump trucks might have been dumping gravel or other building materials into small piles. The changes that people make to a landscape at a construction site are small examples of those that happen naturally to Earth's surface.

A combination of constructive processes and destructive processes produce landforms. Constructive processes build up features on Earth's surface. For example, lava erupting from a volcano hardens and forms new land on the area where the lava falls. Destructive processes tear down features on Earth's surface. A strong hurricane, for example, can wash part of a shoreline into the sea. Constructive and destructive processes continually shape and reshape Earth's surface.

Academic Vocabulary

process
(noun) an ongoing event or a series of related events

Personal Tutor

Figure 1 The continual weathering, erosion, and deposition of sediment occurs from the top of a mountain and across Earth's surface to the distant ocean.

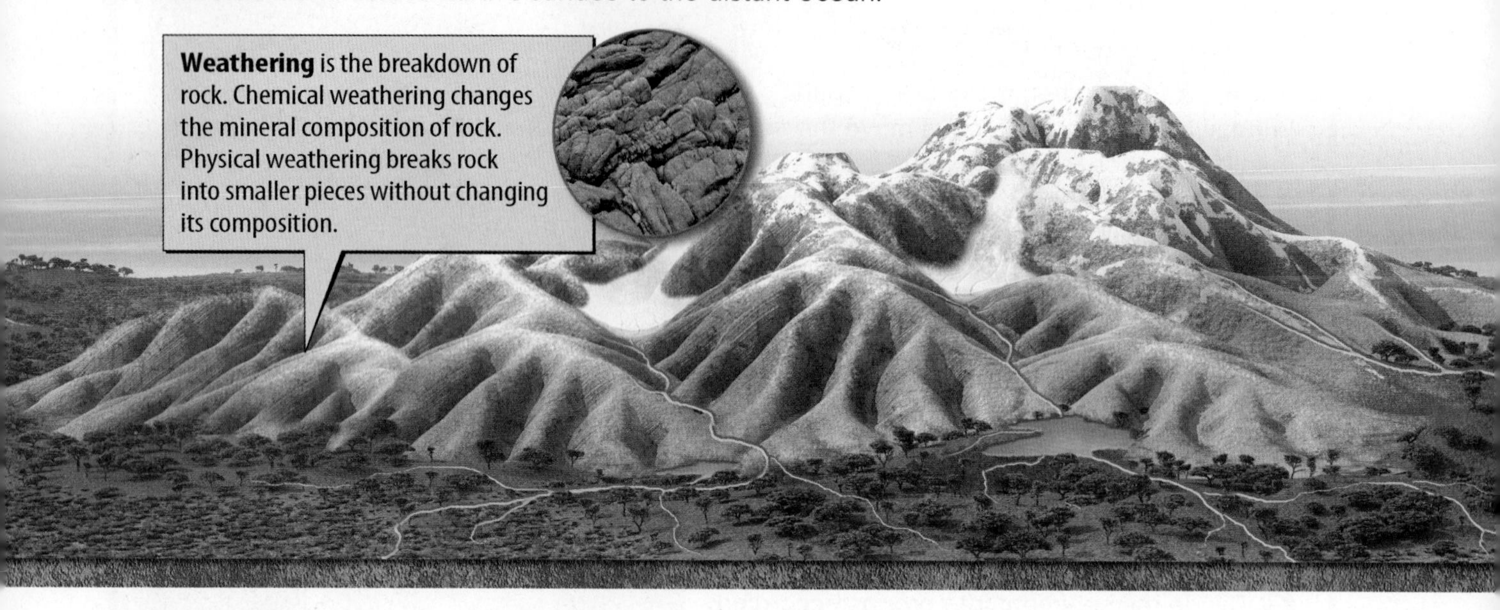

Visual Check How do you think weathering and erosion will affect the mountains over the next thousand years?

A Continual Process of Change

Imagine standing on a mountain, such as one shown in **Figure 1.** In the distance you might see a river or an ocean. What was this area like thousands of years ago? Will the mountains still be here thousands of years from now? Landforms on Earth are constantly changing, but the changes often happen so slowly that you do not notice them. What causes these changes?

Weathering

One destructive process that changes Earth's surface is weathering, the breakdown of rock. Chemical weathering changes the chemical composition of rock. Physical weathering breaks rock into pieces, called sediment, but it does not change the chemical composition of rock. Gravel, sand, silt, and clay are different sizes of sediment.

Weathering Agents Water, wind, and ice are called agents, or causes, of weathering. Water, for example, can dissolve minerals in rock. Wind can grind and polish rocks by blowing particles against them. Also, a rock can break apart as ice expands or as plant roots grow within cracks in the rock.

Different Rates of Weathering The mineral composition of some rocks makes them more resistant to weathering than other rocks. The differences in weathering rates can produce unusual landforms, as shown in **Figure 2.** Weathering can break away less resistant parts of the rock and leave behind the more resistant parts.

Figure 2 Different rates of weathering of rock can produce unusual rock formations.

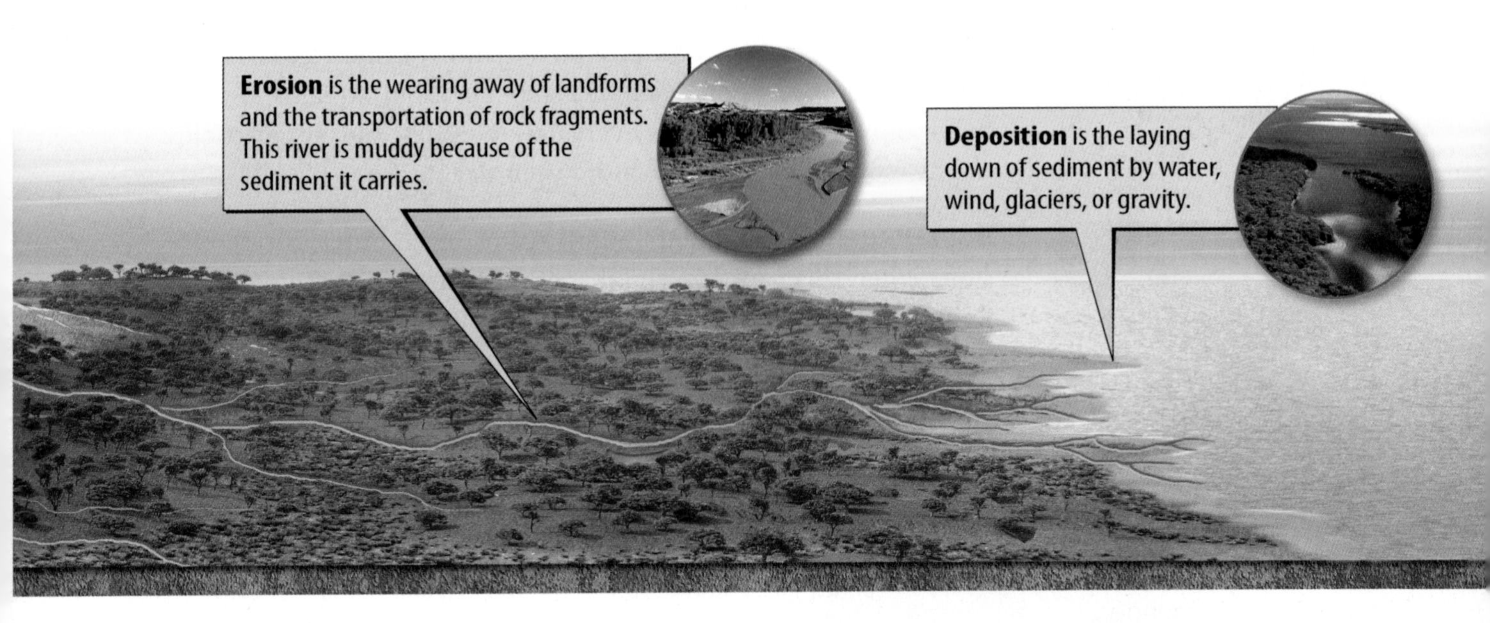

Erosion

What happens to weathered material? This material is often transported away from its source rock in another destructive process called erosion. **Erosion** *is the removal of weathered material from one location to another.* Agents of erosion include water, wind, glaciers, and gravity. The muddy water shown in **Figure 1,** for example, is evidence of erosion.

The Rate of Erosion Like weathering, erosion occurs at different rates. For example, a rushing stream can erode a large quantity of material quickly. However, a gentle stream might erode a small amount of material slowly. Factors that affect the rate of erosion include weather, climate, topo-graphy, and type of rock. For example, strong wind transports weathered rock more easily than a gentle breeze does. Weathered rock moves faster down a steep hill than across a flat area. The presence of plants and the way humans use the land also affect the rate of erosion. Erosion occurs faster on barren land than on land covered with vegetation.

Reading Check What are some factors that affect the rate of erosion?

MiniLab

15 minutes

Can weathering be measured?

You can measure the weathering of rocks.

1. Read and complete a lab safety form.
2. Obtain **pieces of broken rock.** Rinse the rocks and pat completely dry with **paper towels.**
3. Measure the rocks' mass using a **balance.** Record your data in your Science Journal.
4. Place the rocks in a **plastic bottle.** Cover the rocks with water, and seal the bottle. Shake the bottle vigorously for 5 minutes.
5. Rinse the rocks and pat completely dry with paper towels. Record the mass again.

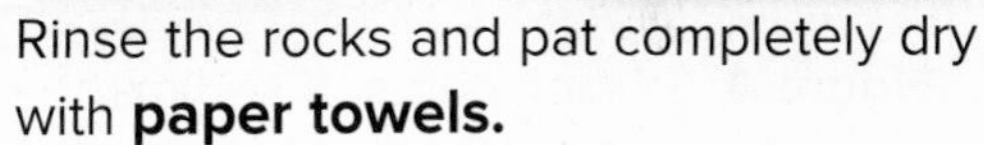

Analyze and Conclude

1. **Compare and contrast** the mass of the rocks before and after shaking.
2. **Key Concept** What evidence suggests that weathering has occurred?

Figure 3 Erosion can change poorly rounded rocks (top) to well-rounded rocks (bottom).

Rate of Erosion and Rock Type The rate of erosion sometimes depends on the type of rock. Weathering can break some types of rock, such as sandstone, into large pieces. Other rock types, such as shale or siltstone, can easily break into smaller pieces. These smaller pieces can be removed and transported faster by agents of erosion. For example, large rocks in streams usually move only short distances every few decades, but silt particles might move a kilometer or more each day.

Rounding Rock fragments bump against each other during erosion. When this happens, the shapes of the fragments can change. Rock fragments can range from poorly rounded to well-rounded. The more spherical and well-rounded a rock is, the more it has been polished during erosion. Rough edges break off as the rock fragments bump against each other. Differences in sediment rounding are shown in **Figure 3**.

Key Concept Check How can erosion affect the shape of sediment?

Sorting Erosion also affects the level of sorting of sediment. Sorting is the separating of items into groups according to one or more properties. As sediment is transported, it can become sorted by grain size, as shown in **Figure 4.** Sediment is often well-sorted when it has been moved a lot by wind or waves. Poorly sorted sediment often results from rapid transportation, perhaps by a storm, a flash flood, or a volcanic eruption. Sediment left at the edges of glaciers is also poorly sorted.

Key Concept Check How can erosion sort sediment?

Sediment Sorting by Size

Figure 4 Erosion can sort sediment according to its size.

Poorly sorted sediment has a wide range of sizes.

Moderately sorted sediment has a small range of sizes.

Well-sorted sediment is all about the same size.

Deposition

You have read about two destructive processes that shape Earth's surface–weathering and erosion. After material has been eroded, a constructive process takes place. **Deposition** *is the laying down or settling of eroded material.* As water or wind slows down, it has less energy and can hold less sediment. Some of the sediment can then be laid down, or deposited.

WORD ORIGIN
deposition
from French *deposer,* means "put down"

Key Concept Check How are erosion and deposition related?

Depositional Environments Sediment is deposited in locations called depositional environments. These locations are on land, along coasts, or in oceans. Examples include swamps, deltas, beaches, and the ocean floor.

REVIEW VOCABULARY
swamp
a wetland occasionally or partially covered with water

Reading Check What is a depositional environment?

Environments where sediment is transported and deposited quickly are high-energy environments. Examples include rushing rivers, ocean shores with large waves, and deserts with strong winds. Large grains of sediment tend to be deposited in high-energy environments.

Small grains of sediment are often deposited in low-energy environments. Deep lakes and areas of slow-moving air or water are low-energy environments. The swamp shown in **Figure 5** is an example of a low-energy environment. The material that makes up a fine-grained sedimentary rock, such as shale, was probably deposited in a low-energy environment.

Sediment Layers Sediment deposited in water typically forms layers called beds. Some examples of beds appear as "stripes" in the photo at the beginning of this lesson. Beds often form as layers of sediment at the bottom of rivers, lakes, and oceans. These layers can be preserved in sedimentary rocks.

A Low-Energy Depositional Environment

Figure 5 Silt and clay are deposited in low-energy environments such as swamps. Swamp deposits also include dark, organic material from decaying trees and other plants.

DEA/F. BARBAGALLO/Getty Images

Figure 6 The tall, steep, somewhat sharp features shown in these photographs are common in landforms carved by erosion.

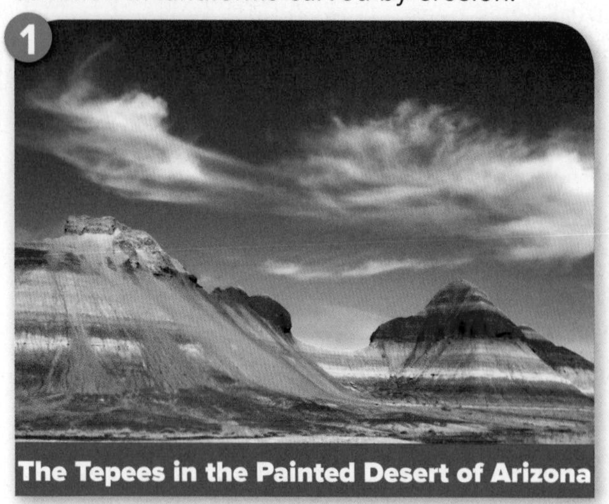

The Tepees in the Painted Desert of Arizona

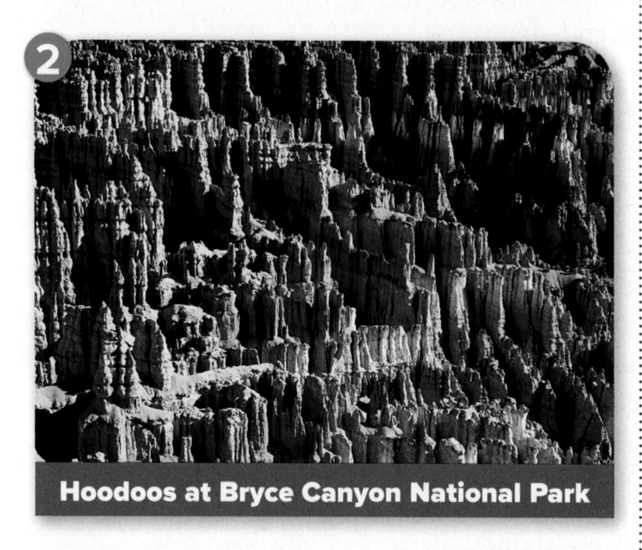

Hoodoos at Bryce Canyon National Park

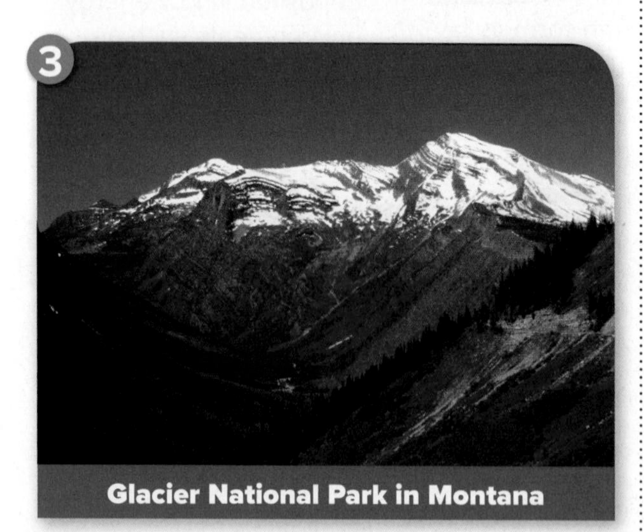

Glacier National Park in Montana

Visual Check How did the passage of glaciers through these mountains change the shape of the valleys?

Interpreting Landforms

What do landform characteristics, such as structure, elevation, and rock exposure, suggest about the development of landforms? Examples of landforms include mountains, valleys, plains, sea cliffs, and beaches. These landforms are always changing, although you might not observe these changes in your lifetime. Landform characteristics can be observed to determine whether destructive forces, such as erosion, or constructive forces, such as deposition, produced the landforms.

Landforms Created by Erosion

Landforms can have features that are clearly produced by erosion. These landforms are often tall, jagged structures with cuts in layers of rock, as shown in the photographs in **Figure 6.**

1 Landforms formed by erosion can expose several layers of rock. The Tepees in the Painted Desert of Arizona contain several layers of different materials. Over time, erosion wore away parts of the land, leaving behind multicolored mounds.

2 Recall that different rates of erosion can result in unusual landforms when some rocks erode and leave more erosion-resistant rocks behind. For example, tall, protruding landforms called hoodoos are shown in the middle photograph of **Figure 6.** Over time, water and ice eroded the less-resistant sedimentary rock. The remaining rocks are more resistant. If you would like to examine hoodoos more closely, look back at **Figure 2.**

3 Glacial erosion and coastal erosion also form unique landforms. Glacial erosion can produce ice-carved features in mountains. The U-shaped valleys of Glacier National Park in Montana, shown in the bottom photograph, formed by glacial erosion. Coastal erosion forms picturesque landforms, such as sea cliffs, caves, and sea arches.

(t)Marc Crumpler/Getty Images; (c)©Mark Karrass/Corbis; (b)Photography by R.G. McGimsey, U.S. Geological Survey

Landforms Created by Deposition

Landforms created by deposition are often flat and low-lying. Wind deposition, for example, can gradually form deserts of sand. Deposition also occurs where mountain streams reach the gentle slopes of wide, flat valleys. An apron of sediment, called an alluvial fan, often forms where a stream flows from a steep, narrow canyon onto a flat plain at the foot of a mountain, as shown in **Figure 7.**

 **Reading Check** How does an alluvial fan develop?

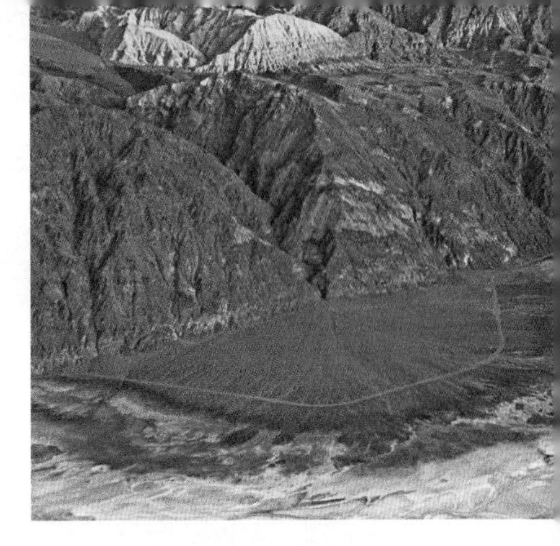

▲ **Figure 7** An alluvial fan is a gently sloping mass of sediment that forms where a stream empties onto flat land at the foot of a steep slope.

Water traveling in a river can slow due to friction with the edges and the bottom of the river channel. An increase in channel width or depth also can slow the current and promote deposition. Deposition along a riverbed occurs where the speed of the water slows. This deposition can form a sandbar, as shown in **Figure 8.** The endpoint for most rivers is where they reach a lake or an ocean and deposit sediment under water. Wave action along shorelines also moves and deposits sediment.

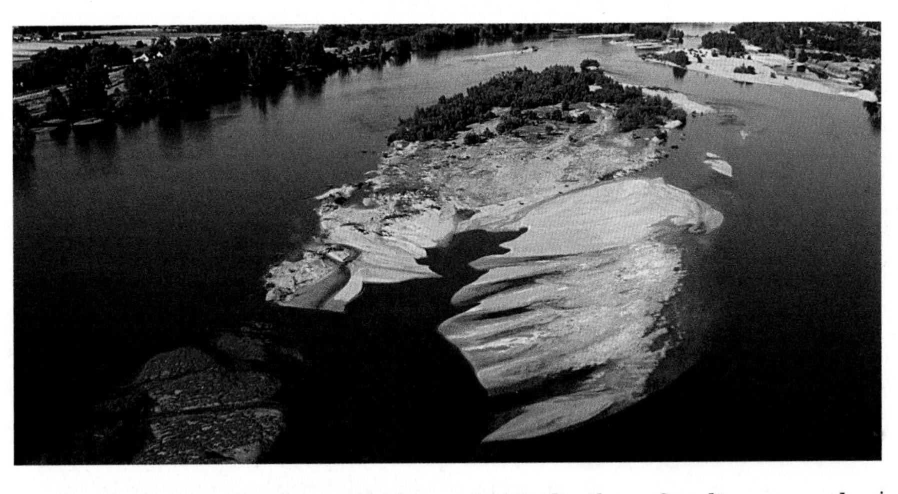

◀ **Figure 8** A sandbar is a depositional feature in rivers and near ocean shores.

As glaciers melt, they can leave behind piles of sediment and rock. For example, glaciers can create long, narrow deposits called eskers and moraines. In the United States, these features are best preserved in northern states such as Wisconsin and New York. You will read more about glacial deposition in Lesson 3.

Comparing Landforms

Look again at the landforms shown in **Figure 6, Figure 7,** and **Figure 8.** Notice how landforms produced by erosion and deposition are different. Erosion produces landforms that are often tall and jagged, but deposition usually produces landforms on flat, low land. By observing the features of a landform, you can infer whether erosion or deposition produced it.

 **Key Concept Check** What features suggest whether erosion or deposition produced a landform?

FOLDABLES®

Make a two-tab book and label it as shown. Use your book to describe and identify some landforms created by the processes of erosion and deposition.

Landforms created by

Erosion	*Deposition*

Lesson 1 Review

Visual Summary

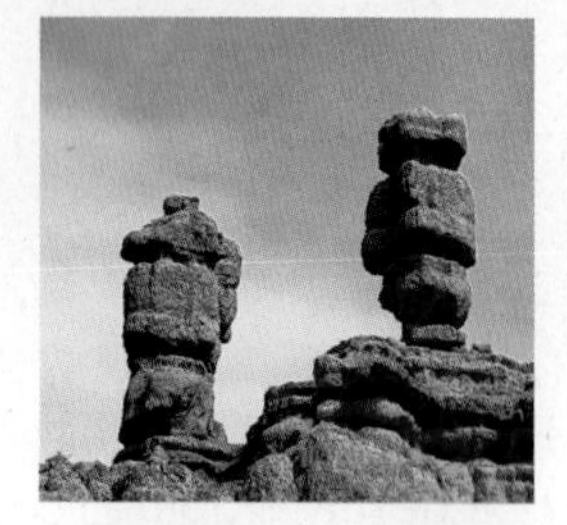

Weathering occurring at different rates can carve rock into interesting landforms.

Rock fragments with rough edges are rounded during transportation.

Landforms produced by deposition are often flat and low-lying.

FOLDABLES®

Use your lesson Foldable to review the lesson. Save your Foldable for the project at the end of the chapter.

What do you think NOW?

You first read the statements below at the beginning of the chapter.

1. Wind, water, ice, and gravity continually shape Earth's surface.

2. Different sizes of sediment tend to mix when being moved along by water.

Did you change your mind about whether you agree or disagree with the statements? Rewrite any false statements to make them true.

Use Vocabulary

1. **Define** *deposition* in your own words.

2. **Use the term** *erosion* in a complete sentence.

Understand Key Concepts 

3. Which would most likely leave behind well-sorted sediment?
   - **A.** flash flood
   - **B.** melting glacier
   - **C.** ocean waves
   - **D.** volcanic eruption

4. **Describe** some features of an alluvial fan that suggest that it was formed by deposition.

5. **Explain** how erosion and deposition by a stream are related.

Interpret Graphics

6. **Examine** the illustration of sediment particle sizes shown below.

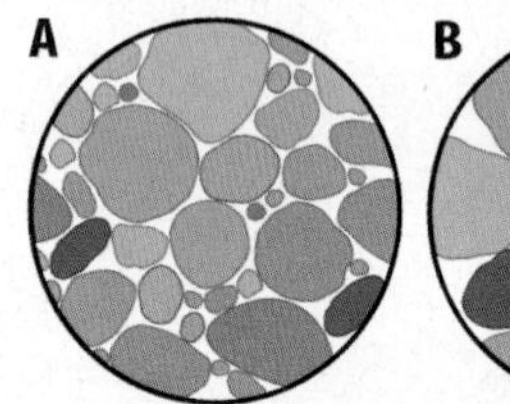

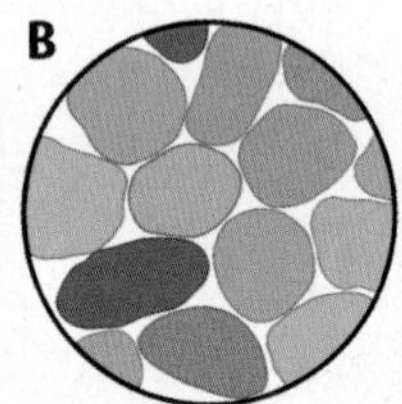

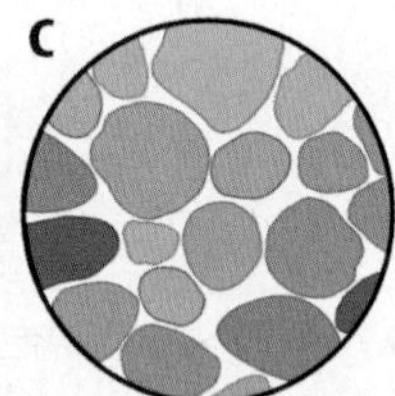

Classify each set of particles as well-sorted, moderately sorted, or poorly sorted. Explain.

7. **Sequence** Copy and fill in the graphic organizer below to describe a possible history of a grain of the mineral quartz that begins in a boulder at the top of a mountain and ends as a piece of sand on the coast.

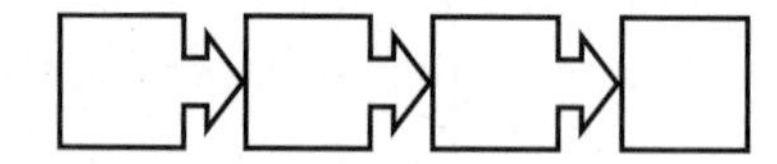

Critical Thinking

8. **Decide** Imagine a river that deposits only small particles where it flows into a sea. Is the river current most likely fast or slow? Why?

(t)Image Source; (c)©Stephen Reynolds; (b)Digital Vision/SuperStock

Clues from the Canyon

HOW NATURE WORKS

Rocks of the majestic Grand Canyon tell a story about Earth's past.

Visitors to the Grand Canyon in Arizona are awestruck by its magnificent size and depth. But to many scientists, the canyon's walls are even more impressive. The soaring walls hold about 40 layers of colorful rocks in shades of red, yellow, brown, and gray. Each layer is like a page in a history book about Earth's past—and the deeper the layer, the older it is. The different layers reflect the particular types of environments in which they formed.

Weathering The canyon walls continue to weather and erode today. Rockfalls and landslides are common. Harder rock such as sandstone weathers in big chunks that break off, forming steep cliffs. The softer rocks weather and erode more easily. This forms gentle slopes.

Deposition These rock layers formed between 280 million and 260 million years ago. During the early part of this period, the region was covered by sand dunes and wind-deposited layers of sand. Later, shallow seas covered this area and layers of shells settled on the seafloor. Gradually, the sediments were compacted and cemented together and these multicolored layers of sedimentary rock were formed.

Erosion Several million years ago, the movement of tectonic plates pushed up the layers of rock. This formed what is called the Colorado Plateau. As the rocks rose higher, the slope of the Colorado River became steeper and its waters flowed faster and with greater force. The Colorado River cut through the weathered rock and carried away sediment. Over millions of years, this erosion formed the canyon.

Jeff Foott/Getty Images

It's Your Turn

DIAGRAM With a partner, find a photo of a local natural land formation. Research and write short descriptions explaining how parts of the formation were created. Attach your descriptions to the appropriate places on the photo.

Lesson 2

Reading Guide

Key Concepts

ESSENTIAL QUESTIONS

- What are the stages of stream development?
- How do water erosion and deposition change Earth's surface?
- How do wind erosion and deposition change Earth's surface?

Vocabulary

meander p. 188

longshore current p. 189

delta p. 190

abrasion p. 192

dune p. 192

loess p. 192

Multilingual eGlossary

What's Science Got to do With It?

Landforms Shaped by Water and Wind

Inquiry Twisted River?

As a river flows down a mountain, it usually flows in the same general direction. What causes this river to flow side-to-side? Why doesn't it flow in a straight path?

Theo Allofs/Getty Images

Launch Lab

15 minutes

How do water and wind shape Earth?

Imagine a fast-moving river rushing over rocks or a strong wind blowing across a field. What changes on Earth do the water and the wind cause?

1. Form into groups and discuss the pictures below with others in your group.
2. Can you recognize evidence of ways water and wind have changed the land—through both erosion and deposition?

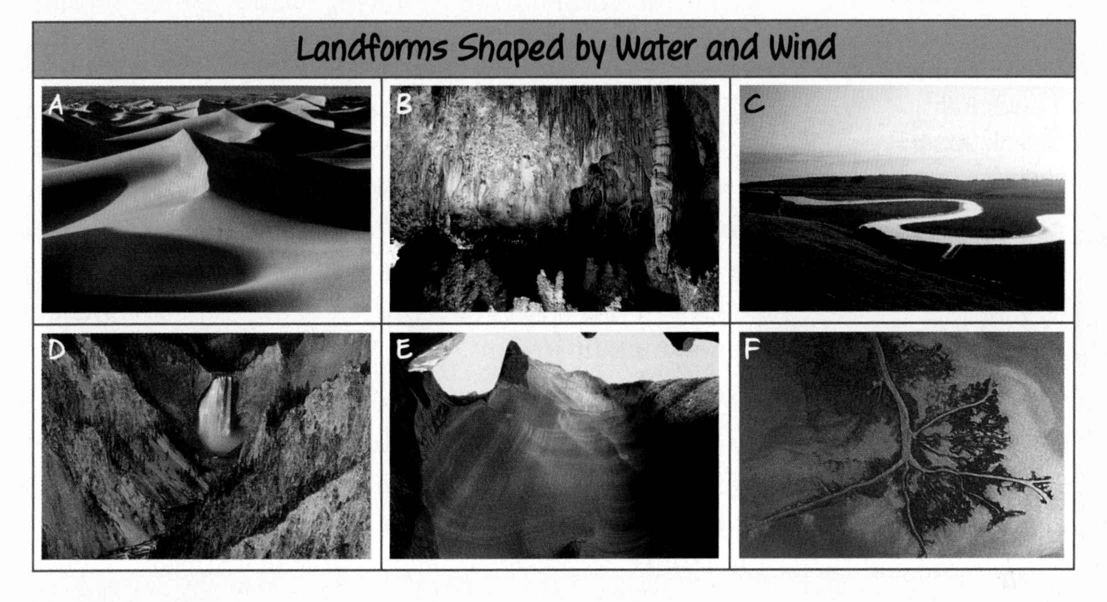

Think About This

1. What are some examples of erosion and deposition in the pictures?
2. **Key Concept** Describe ways you think water might have changed the land in the pictures. What are some ways wind might have changed the land?

Shaping the Land with Water and Wind

Recall that landforms on Earth's surface undergo continual change. Weathering and erosion are destructive processes that shape Earth's surface. These destructive processes often produce tall, jagged landforms. Deposition is a constructive process that also shapes Earth's surface. Constructive processes often produce flat, low-lying landforms.

What causes these processes that continually tear down and build up Earth's surface? In this lesson, you will read that water and wind are two important agents of weathering, erosion, and deposition. The cliffs shown in **Figure 9** are an example of how erosion by water and wind can change the shape of landforms. In the next lesson you will read about ways Earth's surface is changed by the downhill movement of rocks and soil and by the movement of glaciers.

Figure 9 Erosion by water and wind formed these cliffs along Lake Superior.

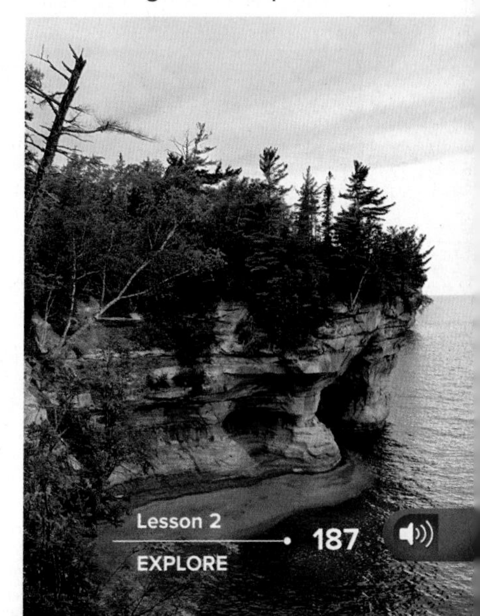

(l to r, t to b)Harald Sund/Photographer's Choice/Getty Images; (2)©Steve Hamblin/Corbis; (3)Image Source/Getty Images; (4)Michael Melford/Getty Images; (5)Robert Glusic/Photodisc/Getty Images; (6)©Royalty-Free/Corbis; (7)Dirk Anschutz/Getty Images

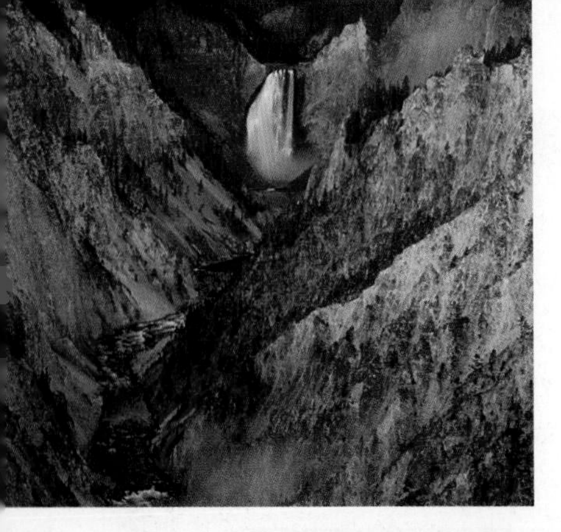

Figure 10 Water erosion carved this V-shaped valley at Lower Falls, Yellowstone National Park, in Wyoming.

Water Erosion and Deposition

Water can shape landforms on and below Earth's surface. The speed of water movement and the depositional environment often affect the shape of landforms.

Water Erosion

If you have ever had a chance to wade into an ocean and feel the waves rushing toward shore, you know that moving water can be incredibly strong. Moving water causes erosion along streams, at beaches, and underground.

Stream Erosion Streams are active systems that erode land and transport sediment. The erosion produced by a stream depends on the stream's energy. This energy is usually greatest in steep, mountainous areas where young streams flow rapidly downhill. The rushing water often carves V-shaped valleys, such as the one shown in **Figure 10.** Waterfalls and river rapids are common in steep mountain streams.

Water in a young stream slows as it reaches gentler slopes. The stream is then called a mature stream, such as the one shown in **Figure 11.** Slower moving water erodes the sides of a stream channel more than its bottom, and the stream develops curves. *A* **meander** *is a broad, C-shaped curve in a stream.*

When a stream reaches flat land, it moves even slower and is called an old stream. Over time, meanders change shape. More erosion occurs on the outside of bends where water flows faster. More deposition occurs on the inside of bends where water flows slower. Over time, the meander's size increases.

FOLDABLES®

Make a two-tab book and label it as shown. Use your book to organize information about landforms and features created by erosion and deposition by water and wind.

Erosion and Deposition

Water	Wind

Key Concept Check Describe the stream development stages.

Stages of Stream Development

Figure 11 Streams change as they flow from steep slopes to gentle slopes and finally to flat plains.

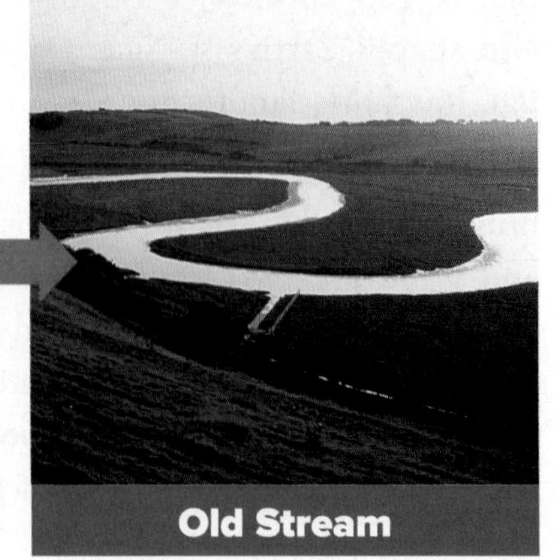

Erosion by Longshore Current

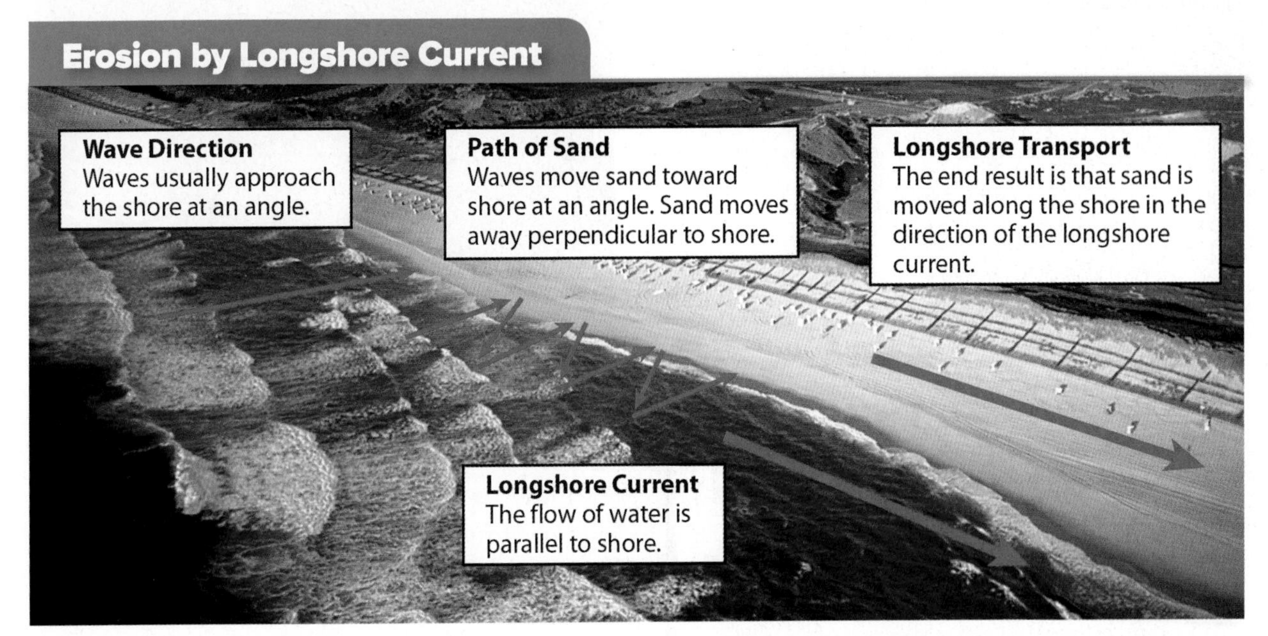

▲ **Figure 12** A longshore current erodes and deposits large amounts of sediment along a shoreline.

Visual Check What causes the sand to move away perpendicular to shore?

Coastal Erosion Like streams, coastlines continually change. Waves crashing onto shore erode loose sand, gravel, and rock along coastlines. One type of coastal erosion is shown in **Figure 12.** *A* **longshore current** *is a current that flows parallel to the shoreline.* This current moves sediment and continually changes the size and shape of beaches. Coastal erosion also occurs when the cutting action of waves along rocky shores forms sea cliffs. Erosional features such as sea caves, sea stacks (tall pillars just offshore), and sea arches (rock bridges extending into the sea) can form when waves erode less resistant rocks along the shore.

Key Concept Check How does water erosion change Earth's surface?

Groundwater Erosion Water that flows underground can also erode rock. Have you ever wondered how caves form? When carbon dioxide in the air mixes with rainwater, a weak acid forms. Some of this rainwater becomes groundwater. As acidic groundwater seeps through rock and soil, it can pass through layers of limestone. The acidic water dissolves and washes away the limestone, forming a cave, as shown in **Figure 13.**

Reading Check How does water erosion form a cave?

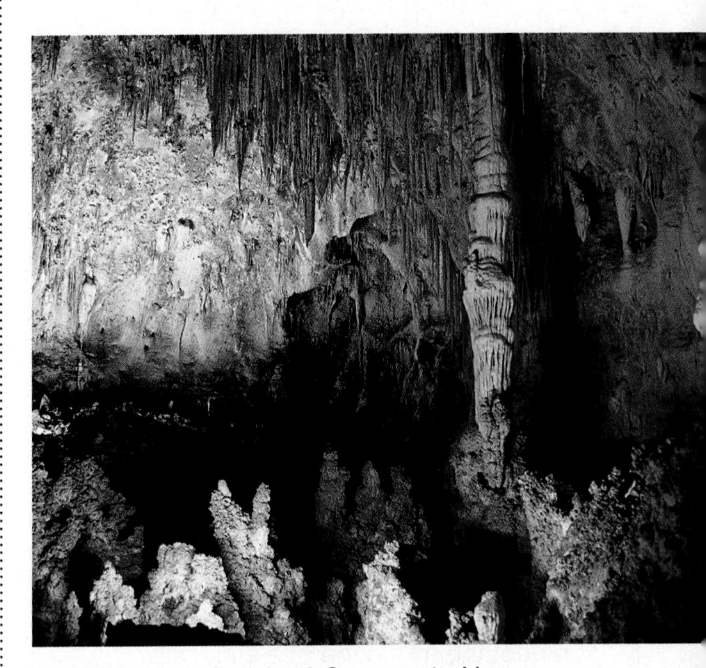

▲ **Figure 13** Carlsbad Caverns in New Mexico was formed by water erosion.

Figure 14 This delta formed by deposition of sediment when water flowed from a river into an ocean.

MiniLab

20 minutes

How do stalactites form?

A stalactite forms when minerals are deposited as crystals. In this lab, you will model the formation of a stalactite.

1. Read and complete a lab safety form.
2. Use **scissors** to poke a hole in the bottom of a **small paper cup.**

3. Tie a **washer** to one end of a 25-cm length of **yarn.** Thread the other end through the hole in the cup and a hole in the top of a **box.** Place the cup on the box with the holes aligned.
4. Half-fill **another cup** with **Epsom salts.** Add **warm water** until the cup is full. Stir with a **spoon.** Pour the salt water into the cup with yarn so that it drips down the yarn into a **bowl.**
5. Record in your Science Journal observations of your model each day for one week.

Analyze and Conclude

1. **Describe** daily changes in your model.
2. **Key Concept** How did this activity model the formation of a stalactite?

Water Deposition

Flowing water deposits sediment as the water slows. A loss of speed reduces the amount of energy that the water has to carry sediment.

Deposition Along Streams Deposition by a stream can occur anywhere along its path where the water's speed decreases. As you read earlier, slower-moving water deposits sediment on the inside curves of meanders. A stream also slows and deposits sediment when it reaches flat land or a large body of water, such as a lake or an ocean. An example is the delta shown in **Figure 14.** *A* **delta** *is a large deposit of sediment that forms where a stream enters a large body of water.*

Deposition Along Coastlines Much of the sand on most ocean beaches was originally deposited by rivers. Longshore currents transport the sand along ocean coasts. Eventually, sand is deposited where currents are slower and have less energy. Sandy beaches often develop at those locations.

Key Concept Check How does water deposition change Earth's surface?

Groundwater Deposition Weathering and erosion produce caves, but deposition forms many structures within caves. Look again at **Figure 13.** The cave contains landforms that dripping groundwater formed as it deposited minerals. Over time, the deposits developed into stalactites and stalagmites. Stalactites hang from the ceiling. Stalagmites build up on the cave's floor.

Land Use Practices

Damage caused by water erosion can be affected by the ways people use land. Two areas of concern are beaches along coasts and surface areas within continental interiors.

Beach Erosion Ocean waves can erode beaches by removing sediment. To reduce this erosion, people sometimes build structures such as retaining walls, or groins, like those shown in **Figure 15.** A row of groins is constructed at right angles to the shore. They are built to trap sediment and reduce the erosive effects of longshore currents.

▲ **Figure 15** These shoreline groins prevent beach erosion by trapping sediment.

Some ways people affect beaches are unintended. For example, people build dams on rivers for purposes of flood control and other reasons. However, dams on rivers prevent river sand from reaching beaches. Beach sand that is washed out to sea by waves is not replaced.

Surface Erosion Reducing the amount of vegetation or removing it from the land increases surface erosion. Agricultural production, construction activities, and cutting trees for lumber and paper production are some reasons that people remove vegetation.

Reading Check What are some ways human activities affect water erosion?

A floodplain is a wide, flat area next to a river. It is usually dry land but can be flooded when the river overflows. Heavy rain or rapid melting of snow can cause a river to flood. Building within a floodplain is risky, as shown in **Figure 16.** However, floods supply mineral-rich soil that is ideal for farming. One way to decrease flooding on a floodplain is to build a levee. A levee is a long, low ridge of soil along a river. However, decreasing flooding also decreases the renewed supply of mineral-rich soil for farming.

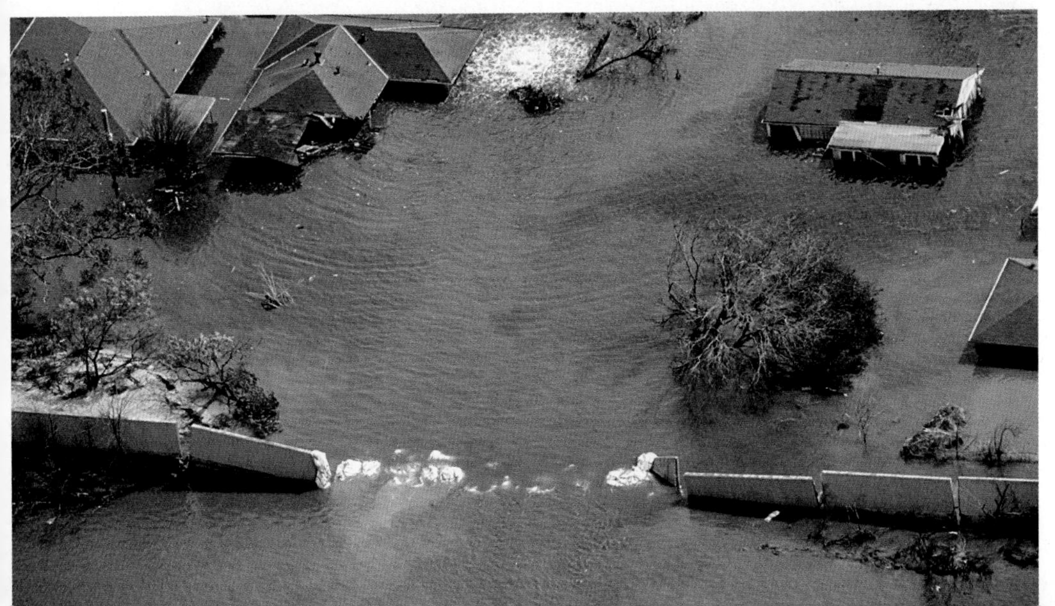

◄ **Figure 16** This 2005 levee break in New Orleans caused extensive flood damage.

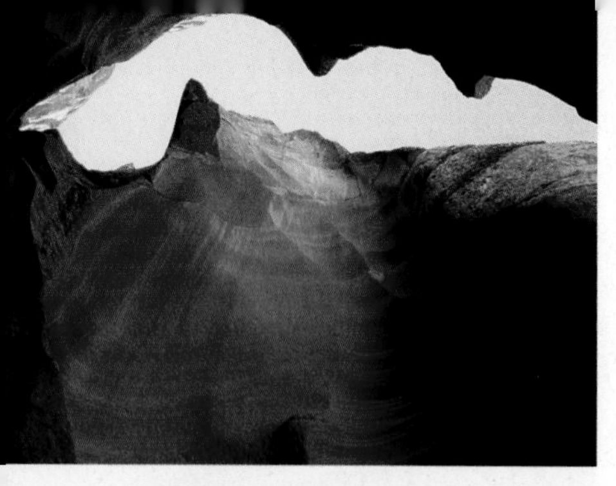

▲ **Figure 17** Wind abrasion carved this unusual landform in the red sandstone of Nevada's Valley of Fire region.

Wind Erosion and Deposition

If you think about a gentle wind that blows leaves in the autumn, it seems unlikely that the wind can cause land erosion and deposition. But strong or long-lasting winds can significantly change the land.

Wind Erosion

Wind can be a major erosional agent, especially in regions that have little vegetation to hold soil in place. Wind can easily pick up and move fine, dry particles. The abrasive action of windblown particles can also carve natural landforms. **Abrasion** *is the grinding away of rock or other surfaces as particles carried by wind, water, or ice scrape against them.* Examples of rock surfaces carved by wind abrasion are shown in **Figure 17** and at the beginning of this chapter.

Wind Deposition

WORD ORIGIN

loess
from Swiss German *Lösch,* means "loose"

Two common types of wind-blown deposits are dunes and loess (LUHS). *A* **dune** *is a pile of windblown sand.* Over time, entire fields of dunes can travel across the land as wind continues to blow the sand. Some dunes are shown in **Figure 18.** **Loess** *is a crumbly, windblown deposit of silt and clay.* One type of loess forms from rock that was ground up and deposited by glaciers. Wind picks up this fine-grain sediment and redeposits it as thick layers of dust called loess.

Key Concept Check How do wind erosion and deposition change Earth's surface?

Land Use Practices

People contribute to wind erosion. For example, plowed fields and dry, overgrazed pastures expose soil. Strong winds can remove topsoil that is not held in place by plants. One way to slow the effects of wind erosion is to leave fields unplowed after harvesting crops. Farmers can also plant rows of trees to slow wind and protect the farmland.

Wind Erosion and Deposition

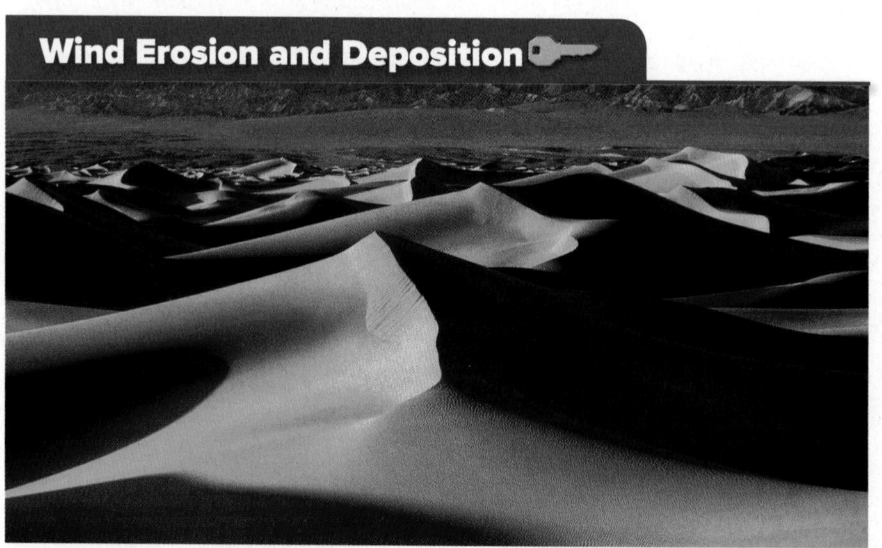

Figure 18 Dunes, such as these in Death Valley, California, formed by the deposition of wind-blown sand. ►

Visual Check What are two effects wind has had on this landscape?

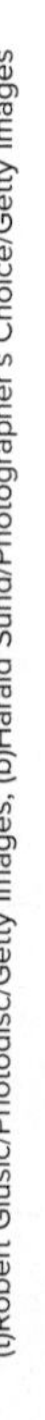

Lesson 2 Review

Online Quiz
Virtual Lab

Visual Summary

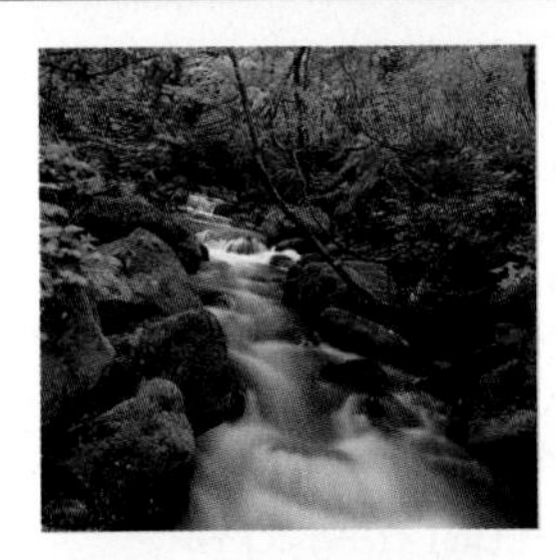

Water erosion changes Earth's surface. An example of this is the change in features of a stream over time.

Water transports sediment and deposits it in places where the speed of the water decreases.

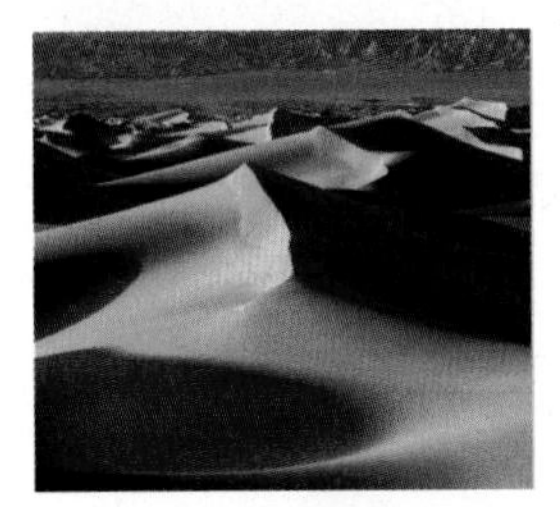

Wind erosion can change Earth's surface by moving sediment. A dune and loess are two types of wind deposition.

FOLDABLES®

Use your lesson Foldable to review the lesson. Save your Foldable for the project at the end of the chapter.

What do you think NOW?

You first read the statements below at the beginning of the chapter.

3. A beach is a landform that does not change over time.

4. Windblown sediment can cut and polish exposed rock surfaces.

Did you change your mind about whether you agree or disagree with the statements? Rewrite any false statements to make them true.

Use Vocabulary

1. **Distinguish** between loess and a dune.
2. **Use the term** *delta* in a complete sentence.
3. Sediment is transported parallel to the shoreline by a ________.

Understand Key Concepts

4. Which feature would a young river most likely have?
   A. meander
   B. slow movement
   C. waterfall
   D. wide channel
5. **Explain** how wind erosion might affect exposed rock.
6. **Compare and contrast** the advantages and disadvantages of farming on a floodplain.

Interpret Graphics

7. **Determine Cause and Effect** Copy and fill in the graphic organizer below to identify two ways waves cause seashore erosion.

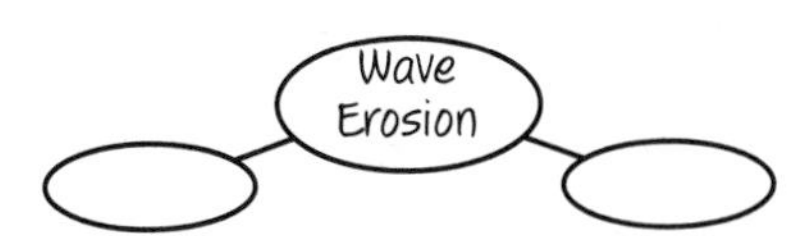

8. **Examine** the image below.

How have erosion and deposition shaped the stream?

Critical Thinking

9. **Suppose** the amount of sand in front of a large, beachfront hotel is slowly disappearing. Explain the process that is likely causing this problem. Suggest a way to avoid further loss of sand.
10. **Recommend** What are some steps a farmer could take to avoid wind erosion and water erosion of farmland?

40 minutes

How do water erosion and deposition occur along a stream?

Materials

sand

paper cup

craft sticks

tub

stream table

small rock

Also needed: drain tube

Safety

Water flowing in a stream erodes the land it flows over. As stream water slows down, it deposits sediments. You can learn about this type of erosion and deposition by analyzing how water shapes land.

Learn It

When you **analyze** an event, such as erosion or deposition, you observe the different things that happen. You also consider the effects of changes. In this activity, you will analyze how erosion and deposition occur along a stream.

Try It

1. Read and complete a lab safety form.
2. Half-fill a stream table with sand. Add water to dampen the sand. Tilt the table slightly, and put the drain tube in a tub.
3. Flatten the sand into a gentle slope. Slowly pour water from a paper cup onto the high end of the sand. Notice the movement of sand along the water's path. Record your observations in your Science Journal.
4. Flatten the sand again. Use a craft stick to make a straight channel for the water. Pour water into the channel slowly and then faster. Analyze the movement of sand along the channel.

Apply It

5. Test the effect of having an object, such as a rock, in the water's path. Analyze how this affects the path of the water and the movement of sand.
6. Think about how flowing water affects the shape of a meander. Test this with your damp sand and water. Describe your results.
7. **Key Concept** How did water erosion and deposition occur along the stream?

3

Lesson 3

Mass Wasting and Glaciers

Reading Guide

Key Concepts
ESSENTIAL QUESTIONS

- What are some ways gravity shapes Earth's surface?
- How do glaciers erode Earth's surface?

Vocabulary

mass wasting p. 196
landslide p. 197
talus p. 197
glacier p. 199
till p. 200
moraine p. 200
outwash p. 200

Multilingual eGlossary

BrainPOP®

Inquiry River of Mud?

Heavy rains loosened the sediment on this mountain. Eventually the land collapsed and caused a river of mud to flow downhill. Events such as this can seriously damage land as well as homes and businesses.

Launch Lab

15 minutes

How does a moving glacier shape Earth's surface?

A glacier is a huge mass of slow-moving ice. The weight of a glacier is so great that its movement causes significant erosion and deposition along its path. In this lab, you will use a model glacier to observe these effects.

1. Read and complete a lab safety form.
2. Half-fill an **aluminum pan** with **dirt** and **gravel.** Mix enough water so that the dirt holds together easily. Use **two books** to raise one end of the pan.
3. Sprinkle **colored sand** at the top of the dirt hill.
4. Place a **model glacier** at the top of the hill. Slowly move the glacier downhill, pressing down gently.

Think About This

1. What happened to the colored sand as the glacier moved downhill?
2. **Key Concept** What kinds of erosion and deposition did your model glacier cause?

Mass Wasting

Have you ever seen or heard a news report about a large pile of boulders that has fallen down a mountain onto a road? This is an example of a mass wasting event. **Mass wasting** *is the downhill movement of a large mass of rocks or soil because of the pull of gravity.* There are two important parts to this definition:

- material moves in bulk as a large mass
- gravity is the dominant cause of movement. For example, the mass moves all at once, rather than as separate pieces over a long period of time. Also, the mass is not moved by, in, on, or under a transporting agent such as water, ice, or air.

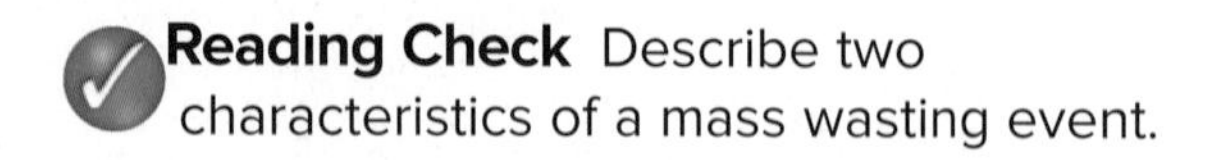

Reading Check Describe two characteristics of a mass wasting event.

Look again at the photo on the previous page. It is a photo of a mass wasting event called a mud flow. Even though water did not transport the mud, it did contribute to this mass wasting event. Mass wasting commonly occurs when soil on a hillside is soaked with rainwater. The water-soaked soil becomes so heavy that it breaks loose and slides down the hillside.

Recall that vegetation on a steep slope reduces the amount of water erosion during a heavy rainfall. The presence of thick vegetation on a slope also reduces the likelihood of a mass wasting event. Root systems of plants help hold sediment in place. Vegetation also reduces the force of falling rain. This minimizes erosion by allowing water to gently soak into the soil.

Hutchings Photography/Digital Light Source

Examples of Mass Wasting

Rockfall

Slump

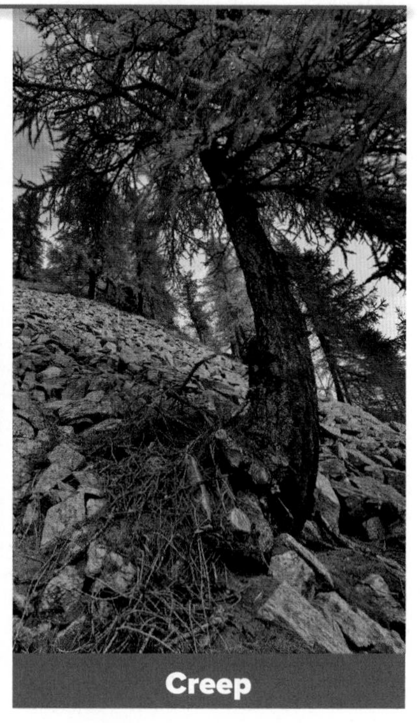

Creep

Figure 19 A rockfall, slump, and creep are examples of mass wasting.

Visual Check What evidence do you see in the figure that mass wasting has occurred?

Erosion by Mass Wasting

There are many types of mass wasting events. For example, *a* **landslide** *is the rapid downhill movement of soil, loose rocks, and boulders.* Two types of landslides are a rockfall, such as the one shown in **Figure 19**, and a mudslide, shown on the first page of this lesson. Slump is a type of mass wasting where blocks of material move down slope on a curved surface. If the material moves too slowly to be noticeable, causing trees and other objects to lean over, the event is called creep, also shown in **Figure 19**.

The amount of erosion that occurs during a mass wasting event depends on factors such as the type of rock, the amount of water in the soil, and how strongly the rock and soil are held together. Erosion also tends to be more destructive when the mass wasting occurs on steep slopes. For example, landslides on a steep hillside can cause extensive damage because they transport large amounts of material quickly.

Key Concept Check What are some ways gravity shapes Earth's surface?

Deposition by Mass Wasting

The erosion that occurs during mass wasting continues as long as gravity is greater than other forces holding the rock and soil in place. But when the material reaches a stable location, such as the base of a mountain, the material is deposited. **Talus** *is a pile of angular rocks and sediment from a rockfall,* like the pile of rock at the base of the hill in **Figure 19**.

FOLDABLES

Make a two-tab book and label it as shown. Use your book to organize information about landforms and features created by erosion and deposition by mass wasting and by glaciers.

Erosion and Deposition

Mass Wasting | Glaciers

Math Skills 

Use Ratios

Slope is the ratio of the change in vertical height over the change in horizontal distance. The slope of the hill in the drawing is

$$\frac{(108\text{ m} - 100\text{ m})}{40\text{ m}} = \frac{8\text{ m}}{40\text{ m}} = 0.2$$

Multiply the answer by 100 to calculate percent slope.

$$0.2 \times 100 = 20\%$$

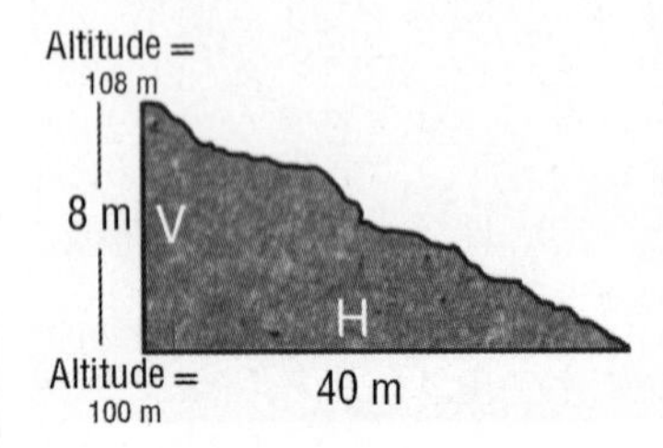

Practice

A mountain rises from 380 m to 590 m over a horizontal distance of 3,000 m. What is its percent slope?

 Math Practice

 Personal Tutor

Figure 20 Building on steep slopes can increase the risk of a landslide. Construction or removal of vegetation makes the hillside even less stable.

Land Use Practices

Human activities can affect both the severity of mass wasting and the tendency for it to occur. The homes in **Figure 20** were built on steep and unstable slopes and were damaged during a landslide. Removing vegetation increases soil erosion and can promote mass wasting. The use of heavy machines or blasting can shake the ground and trigger mass wasting. In addition, landscaping can make a slope steeper. A steep slope is more likely to undergo mass wasting.

 **Reading Check** What are some ways human activities can increase or decrease the risk of mass wasting?

MiniLab

20 minutes

How does the slope of a hill affect erosion?

1. Read and complete a lab safety form.
2. Use **scissors** to poke holes in one end of an **aluminum pan.** Prop the other end up with a **book.** Place a **second pan** under the low end. Pile **300 mL of soil** in the high end.
3. Quickly pour **400 mL of water** over the soil. Drain the water from the second pan. Use a **balance** to measure the mass of the soil that was washed into the second pan.
4. Clean the pans. Using fresh soil, repeat steps 2 and 3 with **three books** holding up the pan.

Analyze and Conclude

1. **Predict** what your results would have been if you had sprinkled the water on slowly.
2. **Key Concept** How did the slope of the hill affect the amount of erosion?

Glacial Erosion and Deposition

You have read about erosion and deposition caused by mass wasting events. Glaciers can also cause erosion and deposition. *A* **glacier** *is a large mass of ice that formed on land and moves slowly across Earth's surface.* Glaciers form on land in areas where the amount of snowfall is greater than the amount of snowmelt. Although glaciers appear to be motionless, they can move several centimeters or more each day.

▲ **Figure 21** The Mendenhall Glacier in Alaska is an alpine glacier.

There are two main types of glaciers–alpine glaciers and ice sheets. Alpine glaciers, like the one shown in **Figure 21**, form in mountains and flow downhill. More than 100,000 alpine glaciers exist on Earth today. Ice sheets cover large areas of land and move outward from central locations. Continental ice sheets were common in past ice ages but only exist today on Antarctica and Greenland.

Glacial Erosion

Glaciers erode Earth's surface as they slide over it. They act as bulldozers, carving the land as they move. Rocks and grit frozen within the ice create grooves and scratches on underlying rocks. This is similar to the way sandpaper scratches wood. Alpine glaciers produce distinctive erosional features like the ones shown in **Figure 22**. Notice the U-shaped valleys that glaciers carved through the mountains.

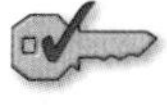

Key Concept Check How do glaciers erode Earth's surface?

Figure 22 Alpine glaciers produce distinctive erosion features.

Visual Check How would the mountains and the valley be different if a glacier had not passed through? ▼

Glacial Erosion

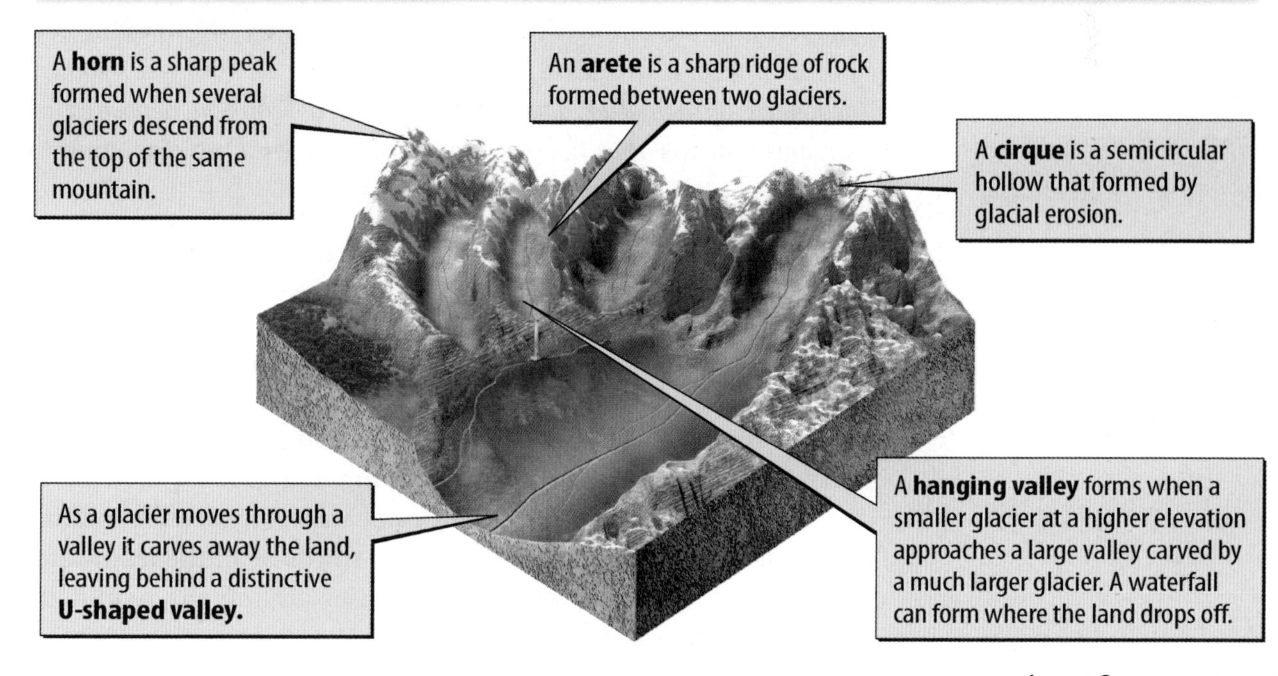

Glacial Deposition

Figure 23 Melting glaciers form various land features as they deposit rock and sediment.

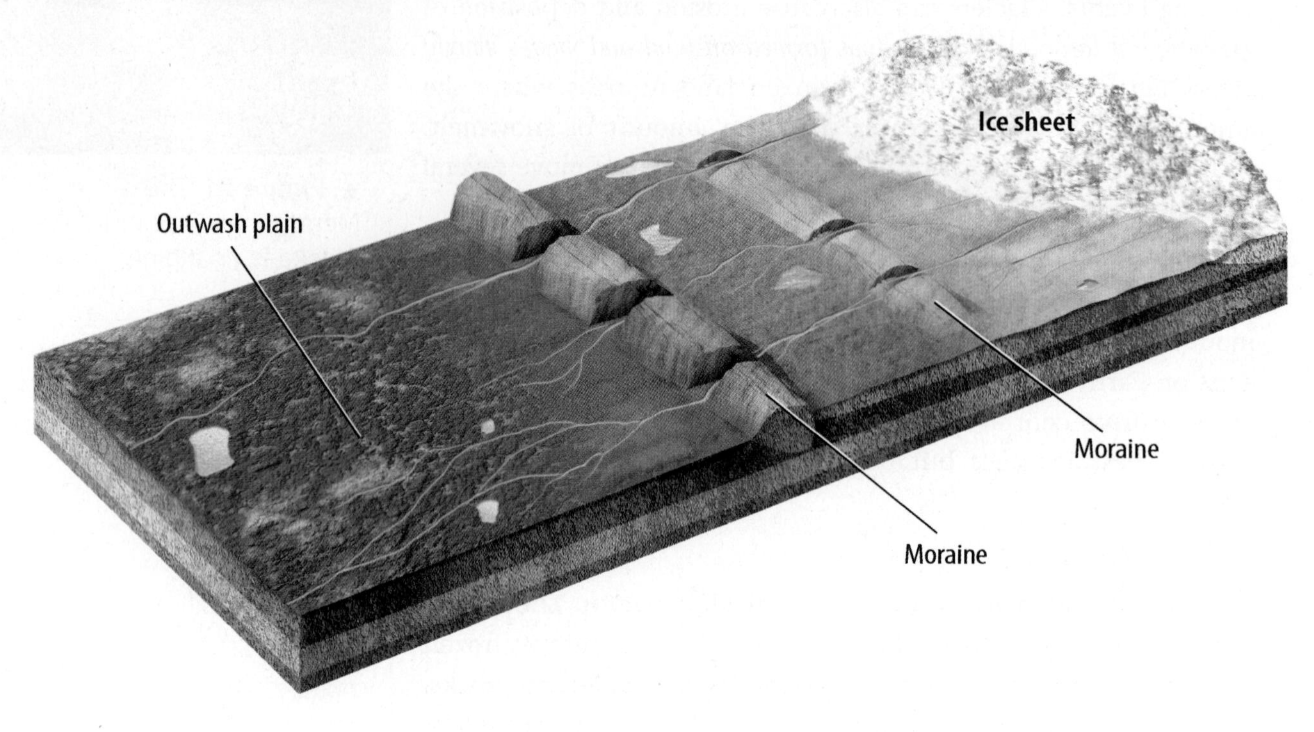

SCIENCE USE V. COMMON USE

till

Science Use rock and sediment deposited by a glacier

Common Use to work by plowing, sowing, and raising crops

WORD ORIGIN

moraine

from French *morena,* means "mound of earth"

Glacial Deposition

Glaciers slowly melt as they move down from high altitudes or when the climate in the area warms. Sediment that was once frozen in the ice eventually is deposited in various forms, as illustrated in **Figure 23.** **Till** *is a mixture of various sizes of sediment deposited by a glacier.* Deposits of till are poorly sorted. They commonly contain particles that range in size from boulders to silt. Till often piles up along the sides and fronts of glaciers. It can be shaped and streamlined into many features by the moving ice. For example, *a* **moraine** *is a mound or ridge of unsorted sediment deposited by a glacier.* **Outwash** *is layered sediment deposited by streams of water that flow from a melting glacier.* Outwash consists mostly of well sorted sand and gravel.

Reading Check How does outwash differ from a moraine?

Land Use Practices

At first, it might not seem that human activities affect glaciers. But in some ways, the effects are more significant than they are for other forms of erosion and deposition. For example, human activities contribute to global warming–the gradual increase in Earth's average temperature. This can cause considerable melting of glaciers. Glaciers contain about two-thirds of all the freshwater on Earth. As glaciers melt, sea level rises around the world and coastal flooding is possible.

Lesson 3 Review

Visual Summary

Mass wasting can occur very fast, such as when a landslide occurs, or slowly over many years.

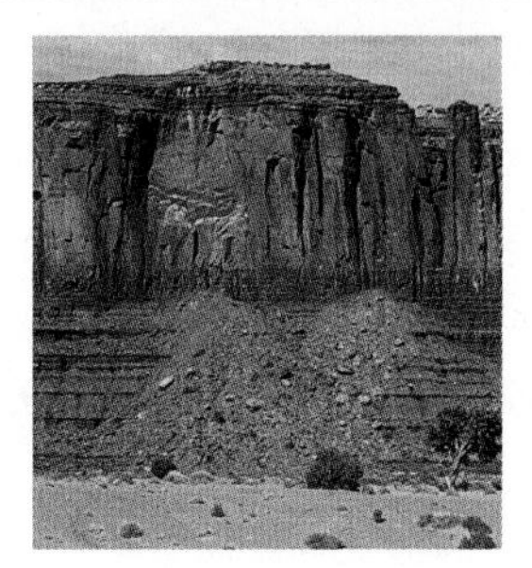

Material moved by a mass wasting event is deposited when it reaches a relatively stable location. An example is talus deposited at the base of this hill.

A glacier erodes Earth's surface as it moves and melts. Glaciers can form U-shaped valleys when they move past mountains.

FOLDABLES®

Use your lesson Foldable to review the lesson. Save your Foldable for the project at the end of the chapter.

What do you think NOW?

You first read the statements below at the beginning of the chapter.

5. Landslides are a natural process that cannot be influenced by human activities.

6. A glacier leaves behind very smooth land as it moves through an area.

Did you change your mind about whether you agree or disagree with the statements? Rewrite any false statements to make them true.

Use Vocabulary

1. **Define** *mass wasting* in your own words.
2. **Use the term** *talus* in a complete sentence.
3. Erosion by the movement of a __________ can produce a U-shaped valley.

Understand Key Concepts

4. Which is the slowest mass wasting event?
   A. creep
   B. landslide
   C. rockfall
   D. slump
5. **Classify** each of the following as features of either erosion or deposition: (a) arete, (b) outwash, (c) cirque, and (d) till.

Interpret Graphics

6. **Examine** the drawing. What feature formed by the glacier is indicated by the arrow?

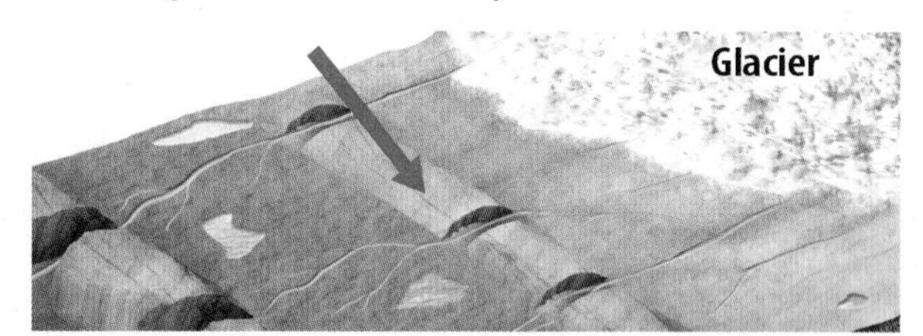

7. **Compare and Contrast** Copy and fill in the table below to compare and contrast moraine and outwash.

Similarities	Differences

Critical Thinking

8. **Compose** a list of evidence for erosion and deposition that you might find in a mountain park that would indicate that glaciers once existed in the area.

Math Skills

9. A mountain's base is 2,500 m high. The peak is 3,500 m high. The horizontal distance covers 4,000 m. What is the percent slope?

Lab **40 minutes**

Avoiding a Landslide

Materials

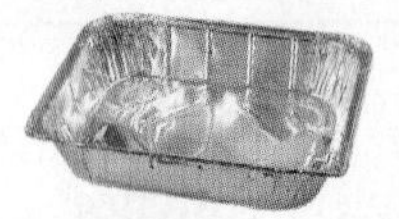

aluminum pan

sand

cup

model house

paper

collection of grass, small sticks, and pebbles

Safety

The damage caused by landslides can be costly to humans. Sometimes landslides are even deadly. Landslides occur most often after a period of heavy rain in regions prone to earthquakes. In this lab, you will analyze ways to protect a house from a landslide.

Ask a Question

What are some ways to reduce the risk of a landslide?

Make Observations

1. Read and complete a lab safety form.
2. In a pan, mix two parts sand to one part water. There should be 2–3 cm of damp sand in the pan.
3. Shape the damp sand into a hill. Place a model house on top of the hill.

4. Using a cup, pour water over the hill, as if it were raining. Record your observations in your Science Journal.
5. Rebuild the hill and the house. This time, gently shake the pan, as if there were an earthquake. Record your observations.

Landslide Test Observations		
Setup	Action	Observations
damp sand hill, no ground cover	pour on water with no shaking	
damp sand hill, no ground cover	pour water and shake the pan	

(t to b)Jacques Cornell/McGraw-Hill Education; (2)Ken Cavanagh/McGraw-Hill Education; (3-6, r)Hutchings Photography/Digital Light Source

Form a Hypothesis

6. Suppose someone built a house on the top of a hill. What are three ways to reduce the risk of a landslide? For each way, develop a hypothesis to save the house from a landslide.

Test Your Hypothesis

7. Develop a plan for testing each hypothesis. Present your plans to your teacher. When they are approved, obtain additional materials from your teacher to implement the plans.
8. Test your plans with both rain and an earthquake. Rebuild the hill and replace the house between tests, if necessary.

Lab TIPS

☑ Mix the sand and water completely, but allow water to drain out to make a strong hill.

☑ Before testing your hypotheses, predict which method will be most effective in reducing the risk of a landslide.

Analyze and Conclude

9. **Describe** the results of your tests. For each test, was your hypothesis correct? What might have worked better?
10. **Analyze** What is the relationship between the amount of water in the soil and the likelihood of a landslide? Use specific examples from the lab in your explanation.
11. **The Big Idea** What are some ways people can alter Earth's surface to reduce the risk of a landslide?

Communicate Your Results

People who live in areas prone to landslides need to take precautions to protect their homes. Write and perform a 30-second public service announcement that describes your results and how they can help people protect their homes.

Extension

Evaluate your home for risk of a landslide. Is it on a slope? Do you get a lot of rain? Do you live in an area prone to earthquakes?

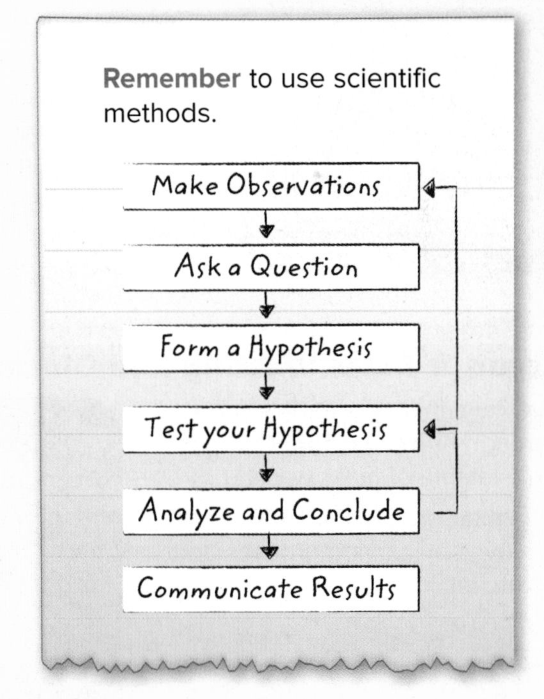

J.T. McGill/U.S. Geological Survey

Erosion and deposition shape Earth's surface by building up and tearing down landforms.

Key Concepts Summary

Lesson 1: The Erosion-Deposition Process

- **Erosion** is the wearing away and transportation of weathered material. **Deposition** is the laying down of the eroded material.
- Erosion tends to make rocks more rounded. Erosion can sort sediment according to its grain size.
- Landforms produced by deposition are usually on flat, low land. Landforms produced by erosion are often tall and/or jagged.

Vocabulary

erosion p. 179
deposition p. 181

Lesson 2: Landforms Shaped by Water and Wind

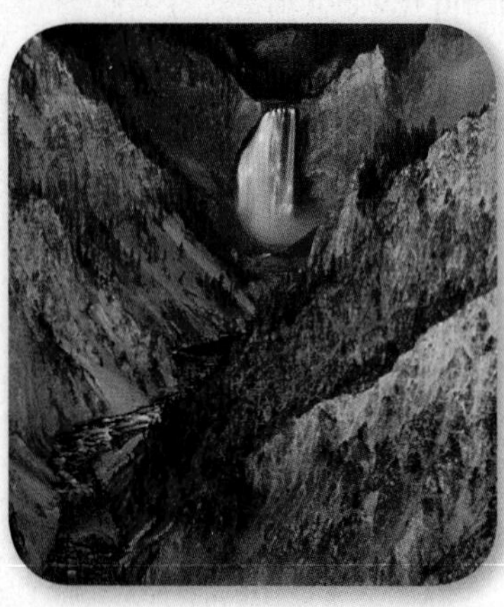

- A young stream moves quickly down steep slopes. A mature stream moves more slowly and develops **meanders.** An old stream is wider and moves slowly.
- Water erosion can form V-shaped valleys. **Longshore currents** reshape beaches. Deposition of sediment from water can form **deltas.**
- Wind **abrasion** can change the shape of rock. Wind deposition can form a **dune** or **loess.**

Vocabulary

meander p. 188
longshore current p. 189
delta p. 190
abrasion p. 192
dune p. 192
loess p. 192

Lesson 3: Mass Wasting and Glaciers

- Gravity can shape Earth's surface through **mass wasting.** Creep is an example of mass wasting.
- A **glacier** erodes Earth's surface as it moves by carving grooves and scratches into rock.

Vocabulary

mass wasting p. 196
landslide p. 197
talus p. 197
glacier p. 199
till p. 200
moraine p. 200
outwash p. 200

Personal Tutor

Vocabulary eFlashcards
Vocabulary eGames

FOLDABLES® Chapter Project

Assemble your lesson Foldables as shown to make a Chapter Project. Use the project to review what you have learned in this chapter.

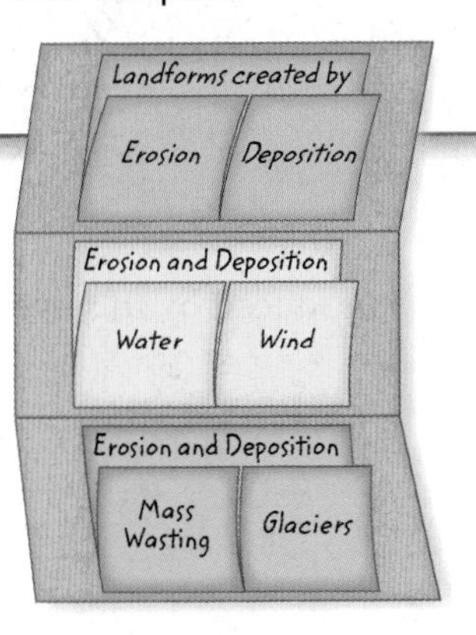

Use Vocabulary

1. Water moving sediment down slopes and a glacier forming a U-shaped valley as it moves past mountains are examples of ________.
2. Wind has less energy as it slows, and ________ of sediment occurs.
3. The grinding of rock as water, wind, or glaciers move sediment is ________.
4. An apron of sediment known as a(n) ________ forms where a stream enters a lake or an ocean.
5. A landslide and creep are types of ________.
6. A large pile of rocks formed from a rockfall is ________.

Interactive Concept Map

Link Vocabulary and Key Concepts

Copy this concept map, and then use vocabulary terms from the previous page to complete the concept map.

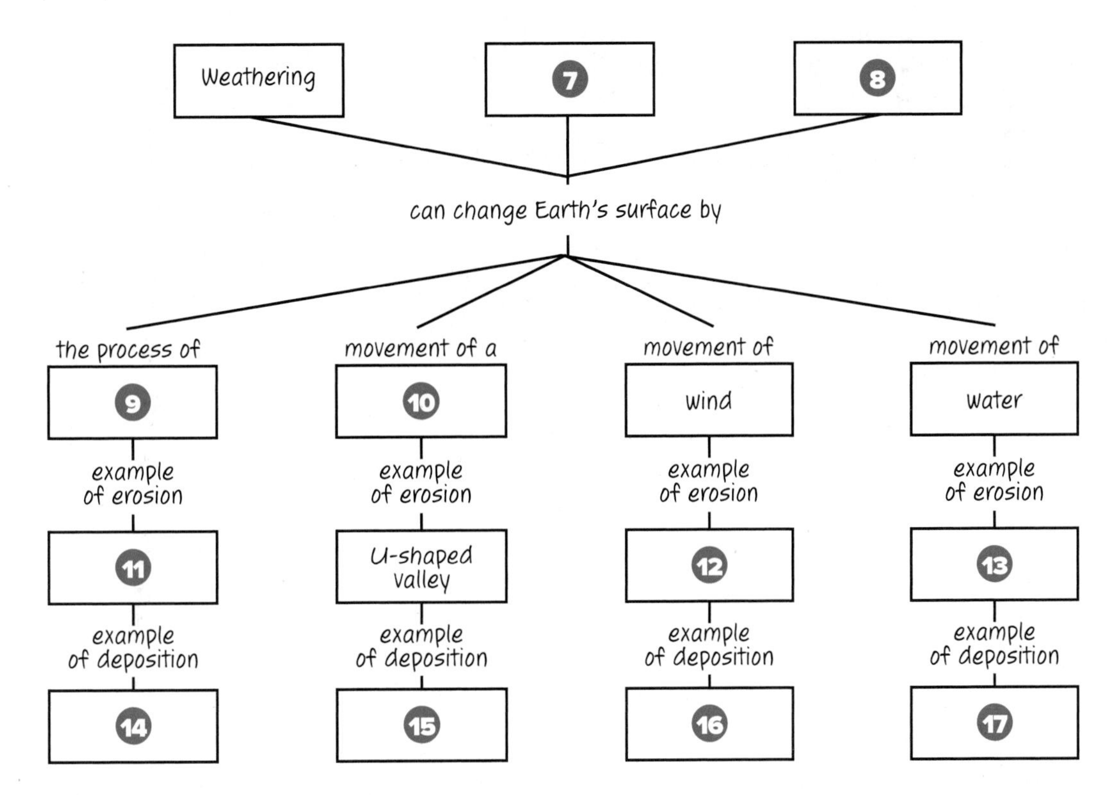

Chapter 6 Review

Understand Key Concepts 

1. Which is a structure created mostly by deposition?
   A. cirque
   B. hoodoo
   C. sandbar
   D. slump

2. Which shows an example of sediment that is both poorly rounded and well-sorted?

A.

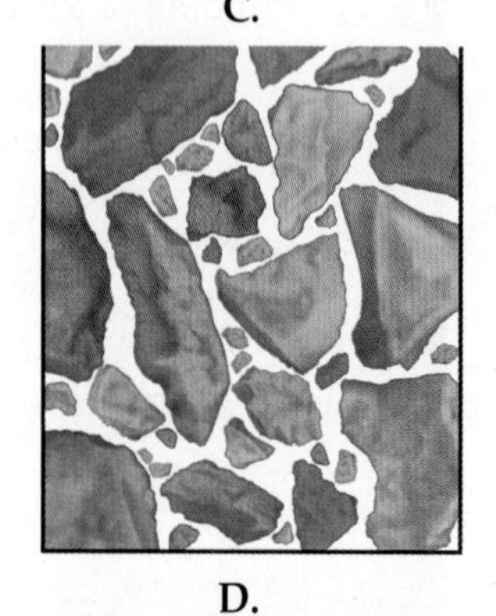

C.

B.

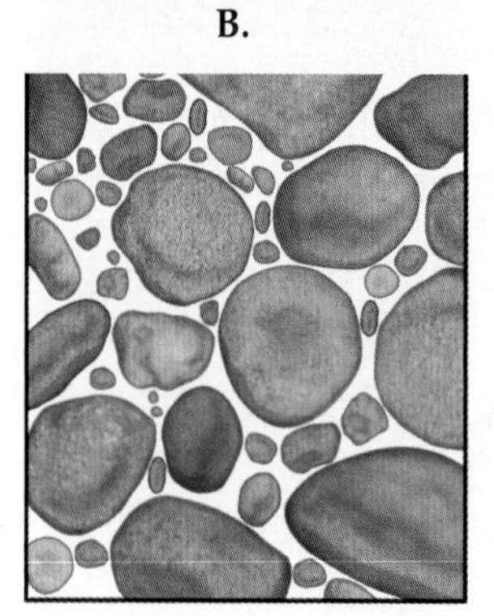

D.

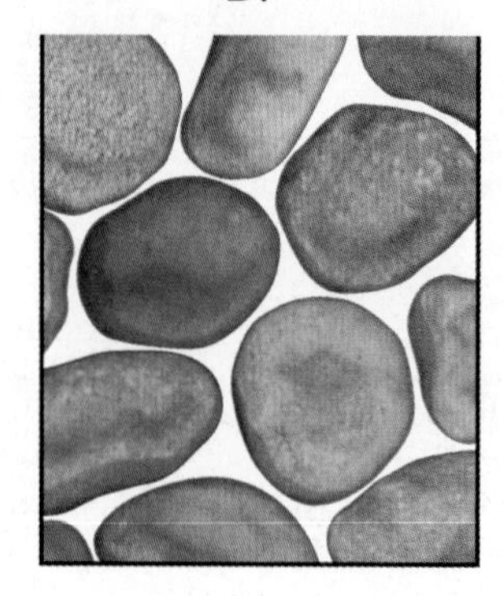

3. Which is typically a low-energy depositional environment?
   A. a fast-moving river
   B. an ocean shore with waves
   C. a stream with meanders
   D. a swamp with decaying trees

4. Which would most likely produce a moraine?
   A. a glacier
   B. an ocean
   C. a river
   D. the wind

5. The illustration below shows a type of mass wasting.

Which was produced by this event?
   A. cirque
   B. moraine
   C. talus
   D. till

6. What is the main difference between slump and creep?
   A. the type of land that is affected
   B. the place where they occur
   C. the speed at which they occur
   D. the amount of rain that causes them

7. Which best describes the difference between a dune and loess?
   A. They are produced in different places.
   B. One is erosion, and the other is deposition.
   C. They are deposits of different-sized particles.
   D. One is caused by wind, and the other is caused by water.

8. Where would you most likely find a meander?
   A. in a cave
   B. in a mature stream
   C. under a glacier
   D. beside a waterfall

9. Which is built to prevent beach erosion?
   A. delta
   B. groin
   C. levee
   D. sandbar

Critical Thinking

10 **Describe** one erosion feature and one deposition feature you might expect to find (a) in a valley, (b) in a desert, and (c) high in the mountains.

11 **Classify** these landforms as formed mostly by erosion or deposition: (a) cirque, (b) sand dune, (c) alluvial fan, (d) hoodoo.

12 **Construct** a chart that lists three careless land uses that result in mass wasting that could be dangerous to humans. Include in your chart details about how each land use could be changed to be safer.

13 **Produce** a list of at least three hazardous erosion or deposition conditions that would be worse during a particularly stormy, rainy season.

14 **Predict** several ways the mountains and the valleys shown below might change as the glaciers slide down slopes.

15 **Contrast** the rounding and sorting of sediment caused by a young stream to that caused by an old stream.

Writing in Science

16 **Write** Imagine you are planning to build a home on a high cliff overlooking the sea. Write a paragraph that assesses the potential for mass wasting along the cliff. Describe at least four features that would concern you.

REVIEW THE BIG IDEA

17 How do erosion and deposition shape Earth's surface?

18 The photo below shows a landform known as The Wave in the southwestern United States. Explain how erosion and deposition might have produced this landform.

Math Skills

Math Practice

Use Ratios

19 Calculate the average percent slope of the mountains in parts a and b.

a. Mountain A rises from 3,200 m to 6,700 m over a horizontal distance of 10,000 m.

b. Mountain B rises from 1,400 m to 9,400 m over a horizontal distance of 2.5 km.

c. If mountains A and B are composed of the same materials, which mountain is more likely to experience mass wasting?

20 If the slope of a hill is 10 percent, how many meters does the hill rise for every 10 m of horizontal distance?

Standardized Test Practice

Record your answers on the answer sheet provided by your teacher or on a sheet of paper.

Multiple Choice

1 Which landform is created by deposition?

A alluvial fan

B glacial valley

C mountain range

D river channel

Use the diagram below to answer question 2.

2 Which process formed the features shown in the diagram above?

A A stream eroded and deposited sediment.

B Groundwater deposited minerals in a cave.

C Groundwater dissolved several layers of rock.

D Wind and ice wore away soft sedimentary rock.

3 Which causes movement in mass wasting?

A gravity

B ice

C magnetism

D wind

4 Which typically is NOT a depositional environment?

A delta

B mountain peak

C ocean floor

D swamp

Use the diagram below to answer questions 5 and 6.

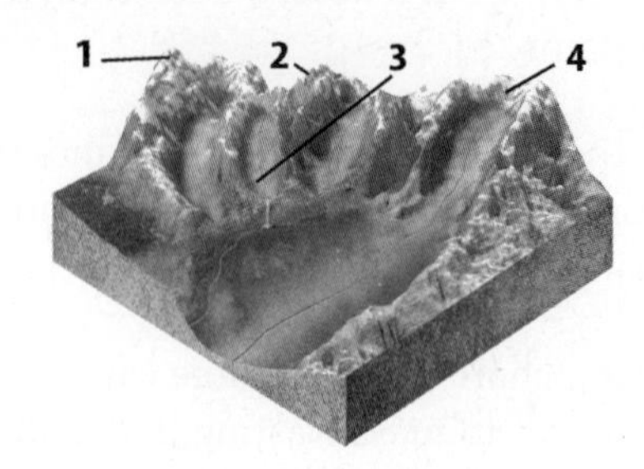

5 Which landform on the diagram above is a cirque?

A 1

B 2

C 3

D 4

6 How did structure *1* form in the diagram above?

A A glacier deposited a large amount of land as it moved.

B A small glacier approached a valley carved by a large glacier.

C Several glaciers descended from the top of the same mountain.

D Two glaciers formed on either side of a ridge.

7 Which agent of erosion can create a limestone cave?

A acidic water

B freezing and melting ice

C growing plant roots

D gusty wind

8 Which deposit does mass wasting create?

A loess

B outwash

C talus

D till

Use the diagram below to answer question 9.

9 Which river feature does the arrow point to in the diagram above?

A a current

B a meander

C a valley

D an alluvial fan

10 Which is true of a longshore current?

A It ALWAYS flows perpendicular to the shoreline.

B It can form large underground caves.

C It continually changes the size and shape of beaches.

D It creates stretches of sand dunes along the beach.

11 Which geological process is often caused by the growth of plant roots?

A deposition

B erosion

C sorting

D weathering

Constructed Response

Use the diagram below to answer questions 12 and 13.

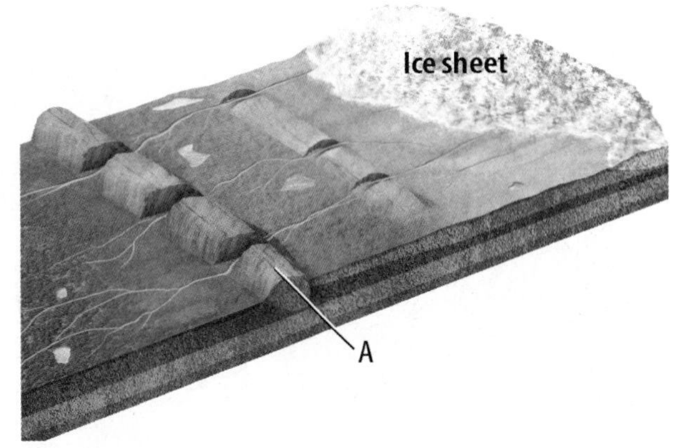

12 Describe the characteristics of deposits found in the feature labeled *A*.

13 How did feature *A* form?

14 A sedimentary rock formation contains alternating layers of fine-grained rock and conglomerate rock, which contains smooth pebble-sized sediments. What is the process that most likely deposited the sediments that make up this rock formation?

15 What factors determine the amount of erosion that occurs during a mass wasting event? How does slope affect the destructive power of this event?

16 What is the typical appearance of a landform formed by erosion?

NEED EXTRA HELP?																
If You Missed Question...	1	2	3	4	5	6	7	8	9	10	11	12	13	14	15	16
Go to Lesson...	1	2	3	1	3	3	2	3	2	2	1	3	3	1, 2	3	1

Unit 2

Geologic Changes

5 Billion B.C. — 1700 — 1800

4.57 billion years ago
The Sun forms.

4.56 billion years ago
Earth forms.

1778
French naturalist Comte du Buffon creates a small globe resembling Earth and measures its cooling rate to estimate Earth's age. He concludes that Earth is approximately 75,000 years old.

1830
Geologist Charles Lyell begins publishing *The Principles of Geology*; his work popularizes the concept that the features of Earth are perpetually changing, eroding, and reforming.

1862
Physicist William Thomson publishes calculations that Earth is approximately 20 million years old. He claims that Earth had formed as a completely molten object, and he calculates the amount of time it would take for the surface to cool to its present temperature.

1900

2000

1899–1900
John Joly releases his findings from calculating the age of Earth using the rate of oceanic salt accumulation. He determines that the oceans are about 80–100 million years old.

1905
Ernest Rutherford and Bertrand Boltwood use radiometric dating to determine the age of rock samples. This technique would later be used to determine the age of Earth.

1956
Today's accepted age of Earth is determined by C.C. Patterson using uranium-lead isotope dating on several meteorites.

Nature of SCIENCE

Science and History

About 500,000 years ago, early humans used stone to make tools, weapons, and small decorative items. Then, about 8,000 years ago, someone might have spied a shiny object among the rocks. It was gold–thought to be the first metal discovered by humans. Gold was very different from stone. It did not break when it was struck. It could easily be shaped into useful and beautiful objects. Over time, other metals were discovered. Each metal helped advance human civilization. Metals from Earth's crust have helped humans progress from the Stone Age to the Moon, to Mars, and beyond.

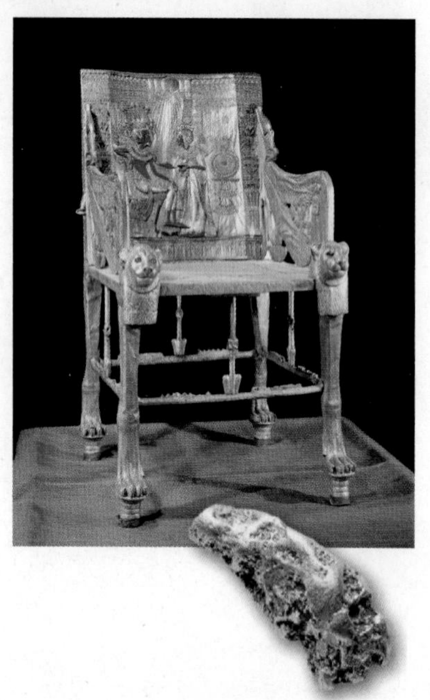

Gold

Since the time of its discovery, gold has been a symbol of wealth and power. It is used mainly in jewelry, coins, and other valuable objects. King Tut's coffin was made of pure gold. Tut's body was surrounded by the largest collection of gold objects ever discovered–chariots, statues, jewelry, and a golden throne. Because gold is so valuable, much of it is recycled. If you own a piece of gold jewelry, it might contain gold that was mined thousands of years ago!

Lead

Ancient Egyptians used the mineral lead sulfide, also called galena, as eye paint. About 5,500 years ago, metalworkers found that galena melts at a low temperature, forming puddles of the lead. Lead bends easily, and the Romans shaped it into pipes for carrying water. Over the years, the Romans realized that lead was entering the water and was toxic to humans. Despite possible danger, lead water pipes were common in modern homes for decades. Finally, however, in 2004 the use of lead pipes in home construction was banned.

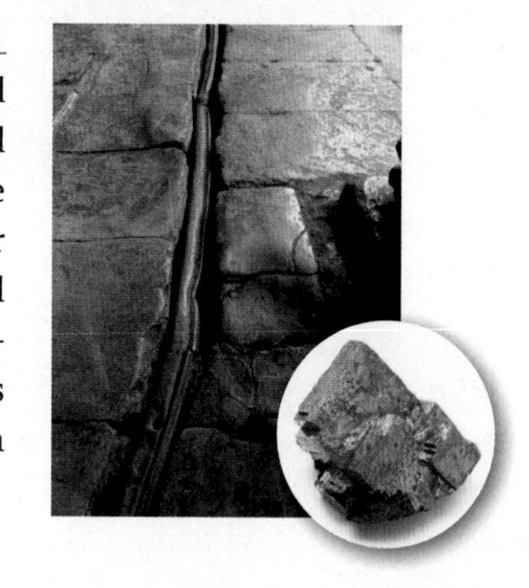

Copper

The first metal commonly traded was copper. About 5,000 years ago, Native Americans mined more than half a million tons of copper from the area that is now Michigan. Copper is stronger than gold. Back then, it was shaped into saws, axes, and other tools. Stronger saws made it easier to cut down trees. The wood from trees then could be used to build boats, which allowed trade routes to expand. Many cultures today still use methods to shape copper that are similar to those used by ancient peoples.

(t to b)Robert Harding/Robert Harding World Imagery/Getty Images; (2)©Don Mason/Corbis; (3)Christine Strover/Alamy; (4)Photolibrary/age fotostock; (5)maurizio grimaldi/age fotostock; (6)Neal & Molly Jansen/age fotostock

Tin and Bronze

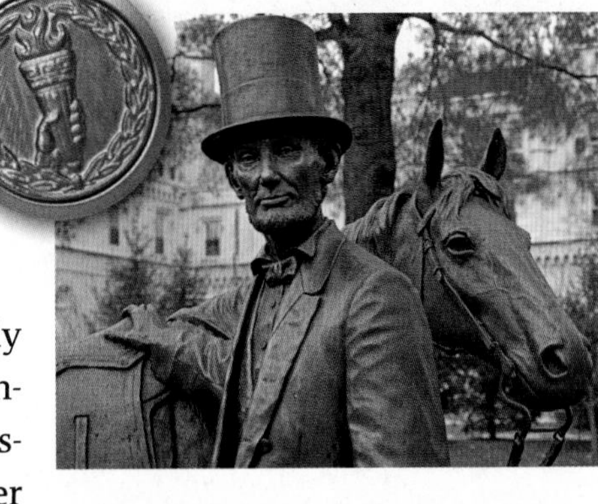

Around 4,500 years ago, the Sumerians noticed differences in the copper they used. Some flowed more easily when it melted and was stronger after it hardened. They discovered that this harder copper contained another metal–tin. Metalworkers began combining tin and copper to produce a metal called bronze. Bronze eventually replaced copper as the most important metal to society. Bronze was strong and cheap enough to make everyday tools. It could easily be shaped into arrowheads, armor, axes, and sword blades. People admired the appearance of bronze. It continues to be used in sculptures. Bronze, along with gold and silver, is used in Olympic medals as a symbol of excellence.

Iron and Steel

Although iron-containing rock was known centuries ago, people couldn't build fires hot enough to melt the rock and separate out the iron. As fire-building methods improved, iron use became more common. It replaced bronze for all uses except art. Iron farm tools revolutionized agriculture. Iron weapons became the choice for war. Like metals used by earlier civilizations, iron increased trade and wealth, and improved people's lives.

In the 17th century, metalworkers developed a way to mix iron with carbon. This process formed steel. Steel quickly became valued for its strength, resistance to rusting, and ease of use in welding. Besides being used in the construction of skyscrapers, bridges, and highways, steel is used to make tools, ships, vehicles, machines, and appliances.

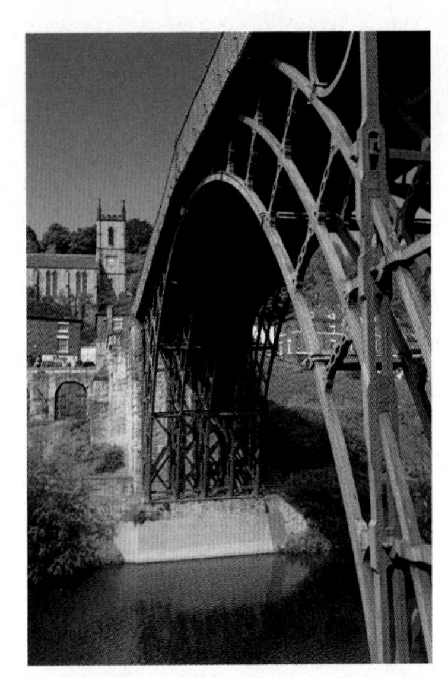

Try to imagine your world without metals. Throughout history, metals changed society as people learned to use them.

(tl)Photodisc/Getty Images; (tr)Carol M. Highsmith's America, Library of Congress, Prints and Photographs Division; (c)Tom Grill/age fotostock; (b)Neale Clark/Robert Harding World Imagery/Getty Images

MiniLab
20 minutes

How do a metal's properties affect its uses?

Why are different common objects made of a variety of different metals?

1. Read and complete a lab safety form.
2. Examine a **lead fishing weight,** a piece of **copper tubing,** and an **iron bolt.**
3. Create a table comparing characteristics of the objects in your Science Journal.
4. Use a **hammer** to tap on each item. Record your observations in your table.

Analyze and Conclude

1. **Infer** Why was lead, not copper or iron, used to make the fishing weight?
2. **Compare** What similarities do all three objects share?
3. **Infer** Why do you think ancient peoples used lead for pipes and iron for weapons?

Chapter 7

Plate Tectonics

THE BIG IDEA What is the theory of plate tectonics?

Inquiry Is this a volcano?

Iceland is home to many active volcanoes like this one. This eruption is called a fissure eruption. This occurs when lava erupts from a long crack, or fissure, in Earth's crust.

- Why is the crust breaking apart here?
- What factors determine where a volcano will form?
- How are volcanoes associated with plate tectonics?

Get Ready to Read

What do you think?

Before you read, decide if you agree or disagree with each of these statements. As you read this chapter, see if you change your mind about any of the statements.

1. India has always been north of the equator.
2. All the continents once formed one supercontinent.
3. The seafloor is flat.
4. Volcanic activity occurs only on the seafloor.
5. Continents drift across a molten mantle.
6. Mountain ranges can form when continents collide.

Your one-stop online resource
connectED.mcgraw-hill.com

 LearnSmart®

 Chapter Resources Files, Reading Essentials, Get Ready to Read, Quick Vocabulary

 Animations, Videos, Interactive Tables

 Self-checks, Quizzes, Tests

 Project-Based Learning Activities

 Lab Manuals, Safety Videos, Virtual Labs & Other Tools

 Vocabulary, Multilingual eGlossary, Vocab eGames, Vocab eFlashcards

 Personal Tutors

Lesson 1

Reading Guide

Key Concepts
ESSENTIAL QUESTIONS

- What evidence supports continental drift?
- Why did scientists question the continental drift hypothesis?

Vocabulary

Pangaea p. 217

continental drift p. 217

Multilingual eGlossary

BrainPOP®

The Continental Drift Hypothesis

Inquiry How did this happen?

In Iceland, elongated cracks called rift zones are easy to find. Why do rift zones occur here? Iceland is above an area of the seafloor where Earth's crust is breaking apart. Earth's crust is constantly on the move. Scientists realized this long ago, but they could not prove how or why this happened.

Launch Lab

20 minutes

Can you put together a peel puzzle?

Early map makers observed that the coastlines of Africa and South America appeared as if they could fit together like pieces of a puzzle. Scientists eventually discovered that these continents were once part of a large landmass. Can you use an orange peel to illustrate how continents may have fit together?

1. Read and complete a lab safety form.
2. Carefully peel an **orange,** keeping the orange-peel pieces as large as possible.
3. Set the orange aside.
4. Refit the orange-peel pieces back together in the shape of a sphere.
5. After successfully reconstructing the orange peel, disassemble your pieces.
6. Trade the entire orange peel with a classmate and try to reconstruct his or her orange peel.

Think About This

1. Which orange peel was easier for you to reconstruct? Why?
2. Look at a world map. Do the coastlines of any other continents appear to fit together?
3. **Key Concept** What additional evidence would you need to prove that all the continents might have once fit together?

Pangaea

Did you know that Earth's surface is on the move? Can you feel it? Each year, North America moves a few centimeters farther away from Europe and closer to Asia. That is several centimeters, or about the thickness of a textbook. Even though you don't necessarily feel this motion, Earth's surface moves slowly every day.

Nearly 100 years ago Alfred Wegener (VAY guh nuhr), a German scientist, began an important investigation that continues today. Wegener wanted to know whether Earth's continents were fixed in their positions. He proposed that *all the continents were once part of a supercontinent called* **Pangaea** (pan JEE uh). Over time Pangaea began breaking apart, and the continents slowly moved to their present positions. Wegener proposed the hypothesis of **continental drift,** *which suggested that continents are in constant motion on the surface of Earth.*

Alfred Wegener observed the similarities of continental coastlines now separated by oceans. Look at the outlines of Africa and South America in **Figure 1.** Notice how they could fit together like pieces of a puzzle. Hundreds of years ago mapmakers noticed this jigsaw-puzzle pattern as they made the first maps of the continents.

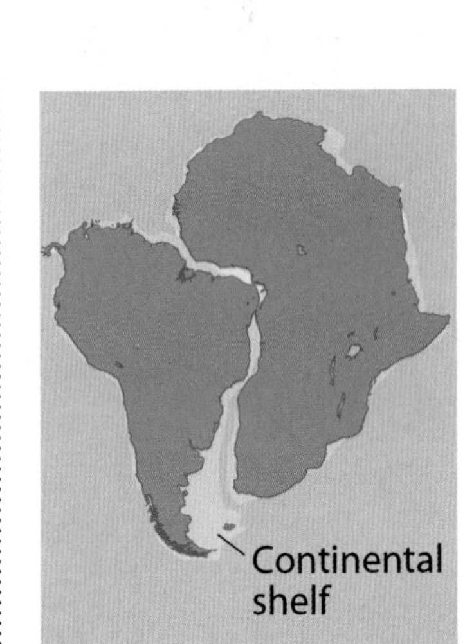

Figure 1 The eastern coast of South America mirrors the shape of the west coast of Africa.

Evidence That Continents Move

If you had discovered continental drift, how would you have tested your hypothesis? The most obvious evidence for continental drift is that the continents appear to fit together like pieces of a puzzle. But scientists were skeptical, and Wegener needed additional evidence to help support his hypothesis.

Climate Clues

Wegener used climate clues to support his continental drift hypothesis. He studied the sediments deposited by glaciers in South America and Africa, as well as in India and Australia. Beneath these sediments, Wegener discovered glacial grooves, or deep scratches in rocks made as the glaciers moved across land. **Figure 2** shows where these glacial features are found on neighboring continents today. Because these regions are too warm for glaciers to develop today, Wegener proposed that they were once located near the South Pole.

When Wegener pieced Pangaea together, he proposed that South America, Africa, India, and Australia were located closer to Antarctica 280 million years ago. He suggested that the climate of the southern hemisphere was much cooler at the time. Glaciers covered large areas that are now parts of these continents. These glaciers would have been similar to the ice sheet that covers much of Antarctica today.

Climate Clues

Animation

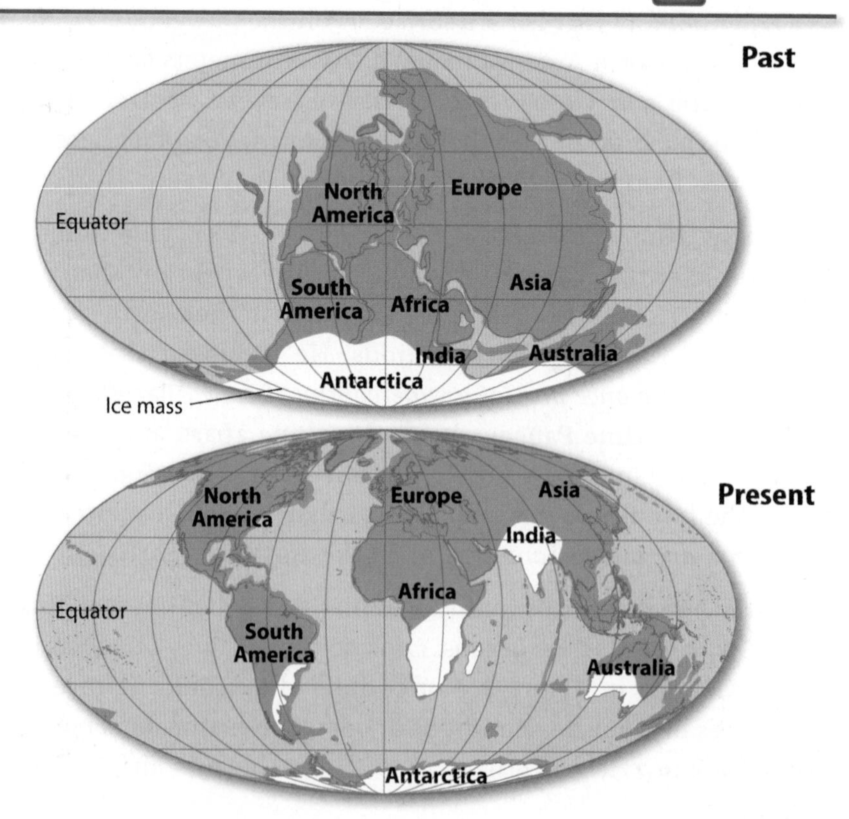

Figure 2 If the southern hemisphere continents could be reassembled into Pangaea, the presence of an ice sheet would explain the glacial features on these continents today.

Fossil Clues

Animals and plants that live on different continents can be unique to that continent alone. Lions live in Africa but not in South America. Kangaroos live in Australia but not on any other continent. Because oceans separate continents, these animals cannot travel from one continent to another by natural means. However, **fossils** of similar organisms have been found on several continents separated by oceans. How did this happen? Wegener argued that these continents must have been connected some time in the past.

REVIEW VOCABULARY

fossil
the naturally preserved remains, imprints, or traces of organisms that lived long ago

Fossils of a plant called *Glossopteris* (glahs AHP tur us) have been discovered in rocks from South America, Africa, India, Australia, and Antarctica. These continents are far apart today. The plant's seeds could not have traveled across the vast oceans that separate them. **Figure 3** shows that when these continents were part of Pangaea 225 million years ago, *Glossopteris* lived in one region. Evidence suggests these plants grew in a swampy environment. Therefore, the climate of this region, including Antarctica, was different than it is today. Antarctica had a warm and wet climate. The climate had changed drastically from what it was 55 million years earlier when glaciers existed.

Reading Check How did climate in Antarctica change between 280 and 225 million years ago?

Fossil Clues

Figure 3 Fossils of *Glossopteris* have been found on many continents that are now separated by oceans. The orange area in the image on the right represents where *Glossopteris* fossils have been found.

Visual Check Which of the continents would not support *Glossopteris* growth today?

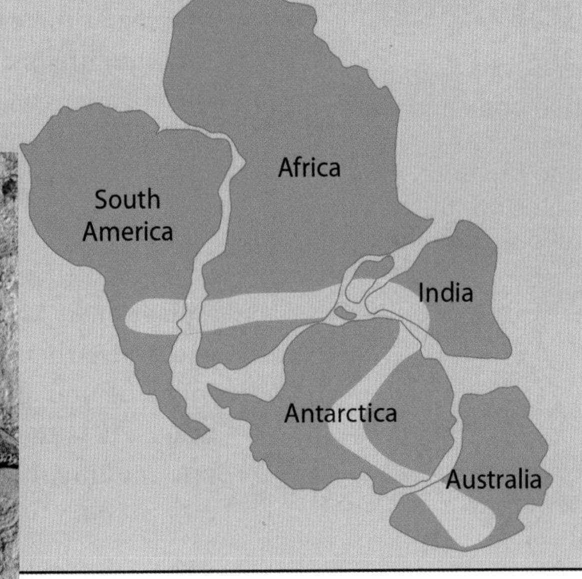

Rock Clues

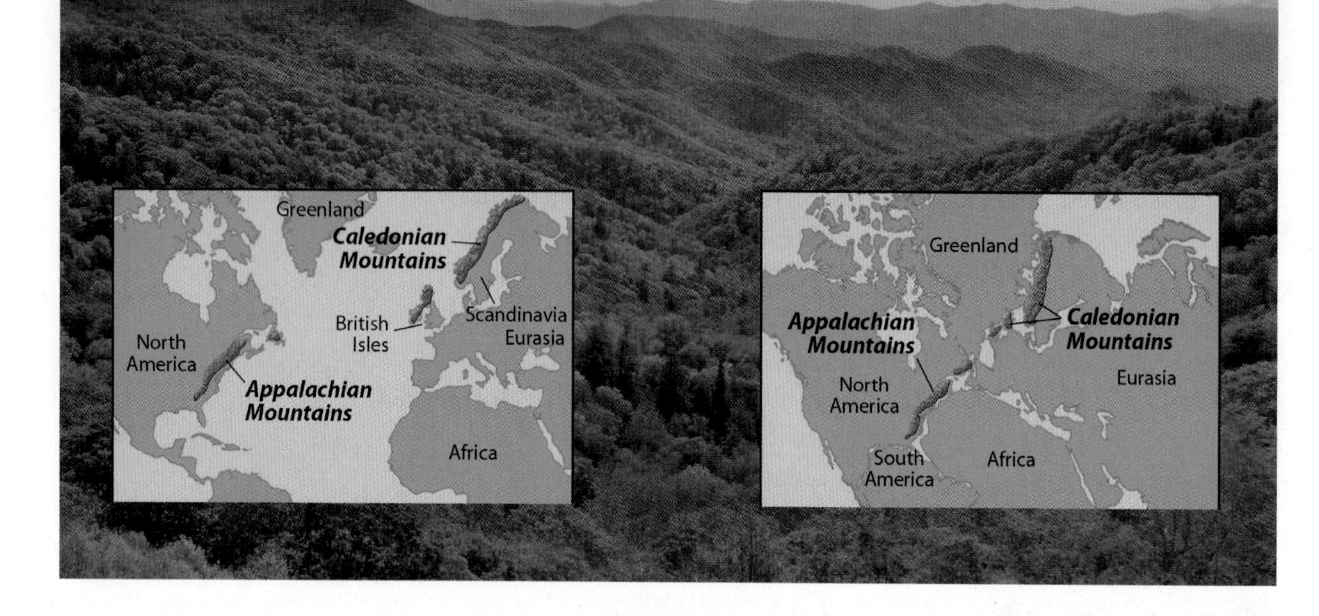

Figure 4 If you could move North America and Europe next to each other, the Appalachian Mountains and the Caledonian mountains would appear to form one continuous mountain range with similar formations.

FOLDABLES®

Make a horizontal half-book and write the title as shown. Use it to organize your notes on the continental drift hypothesis.

Evidence for the Continental Drift Hypothesis

Rock Clues

Wegener realized he needed more evidence to support the continental drift hypothesis. He observed that mountain ranges like the ones shown in **Figure 4** and rock formations on different continents had common origins. Today, geologists can determine when these rocks formed. For example, geologists suggest that large-scale volcanic eruptions occurred on the western coast of Africa and the eastern coast of South America at about the same time hundreds of millions of years ago. The volcanic rocks from the eruptions are identical in both chemistry and age. Refer back to **Figure 1.** If you could superimpose similar rock types onto the maps, these rocks would be in the area where Africa and South America fit together.

The Caledonian mountain range in northern Europe and the Appalachian Mountains in eastern North America are similar in age and structure. They are also composed of the same rock types. If you placed North America and Europe next to each other, these mountains would meet and form one long, continuous mountain belt. **Figure 4** illustrates where this mountain range would be.

Key Concept Check How were similar rock types used to support the continental drift hypothesis?

What was missing?

Wegener continued to support the continental drift hypothesis until his death in 1930. Wegener's ideas were not widely accepted until nearly four decades later. Why were scientists skeptical of Wegener's hypothesis? Although Wegener had evidence to suggest that continents were on the move, he could not explain how they moved.

One reason scientists questioned continental drift was because it is a slow process. It was not possible for Wegener to measure how fast the continents moved. The main objection to the continental drift hypothesis, however, was that Wegener could not explain what forces caused the continents to move. The mantle beneath the continents and the seafloor is made of solid rock. How could continents push their way through solid rock? Wegener needed more scientific evidence to prove his hypothesis. However, this evidence was hidden on the seafloor between the drifting continents. The evidence necessary to prove continental drift was not discovered until long after Wegener's death.

SCIENCE USE V. COMMON USE

mantle

Science Use the middle layer of Earth, situated between the crust above and the core below

Common Use a loose, sleeveless garment worn over other clothes

Key Concept Check Why did scientists argue against Wegener's continental drift hypothesis?

MiniLab

20 minutes

How do you use clues to put puzzle pieces together?

When you put a puzzle together, you use clues to figure out which pieces fit next to each other. How did Wegener use a similar technique to piece together Pangaea?

1. Read and complete a lab safety form.
2. Using **scissors,** cut a piece of **newspaper** or a page from a **magazine** into an irregular shape with a diameter of about 25 cm.
3. Cut the piece of paper into at least 12 but not more than 20 pieces.
4. Exchange your puzzle with a partner and try to fit the new puzzle pieces together.
5. Reclaim your puzzle and remove any three pieces. Exchange your incomplete puzzle with a different partner. Try to put the incomplete puzzles back together.

Analyze and Conclude

1. **Summarize** Make a list of the clues you used to put together your partner's puzzle.
2. **Describe** How was putting together a complete puzzle different from putting together an incomplete puzzle?
3. **Key Concept** What clues did Wegener use to hypothesize the existence of Pangaea? What clues were missing from Wegener's puzzle?

Hutchings Photography/Digital Light Source

Lesson 1 Review

Visual Summary

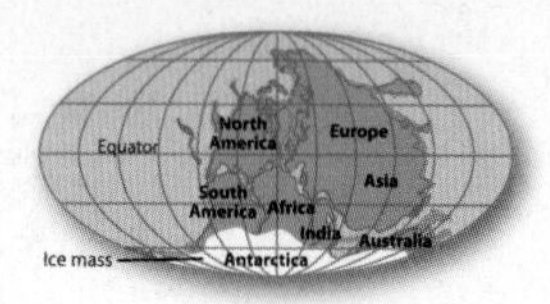

Past

All continents were once part of a supercontinent called Pangaea.

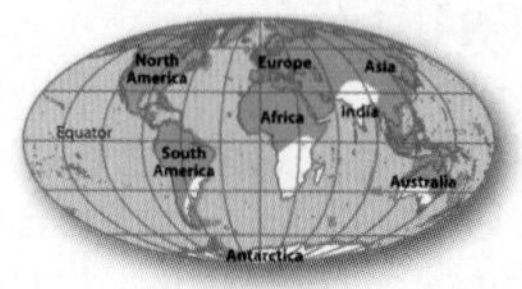

Present

Evidence found on present-day continents suggests that the continents have moved across Earth's surface.

Use your lesson Foldable to review the lesson. Save your Foldable for the project at the end of the chapter.

What do you think NOW?

You first read the statements below at the beginning of the chapter.

1. India has always been north of the equator.
2. All the continents once formed one supercontinent.

Did you change your mind about whether you agree or disagree with the statements? Rewrite any false statements to make them true.

Use Vocabulary

1. **Define** *Pangaea.*
2. **Explain** the continental drift hypothesis and the evidence used to support it.

Understand Key Concepts 

3. **Identify** the scientist who first proposed that the continents move away from or toward each other.
4. Which can be used as an indicator of past climate?
   - **A.** fossils
   - **B.** lava flows
   - **C.** mountain ranges
   - **D.** tides

Interpret Graphics

5. **Interpret** Look at the map of the continents below. What direction has South America moved relative to Africa?

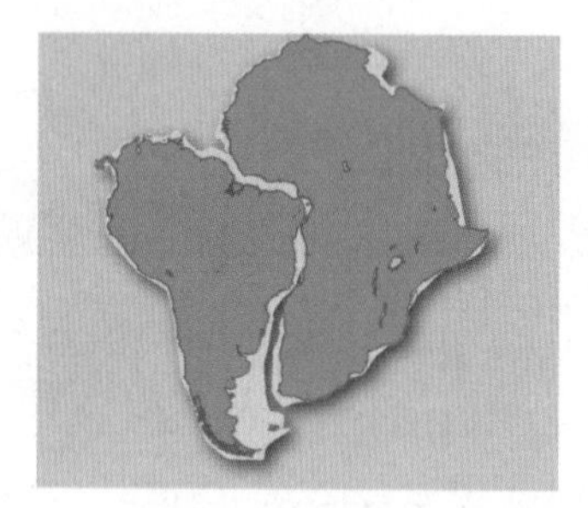

6. **Summarize** Copy and fill in the graphic organizer below to show the evidence Alfred Wegener used to support his continental drift hypothesis.

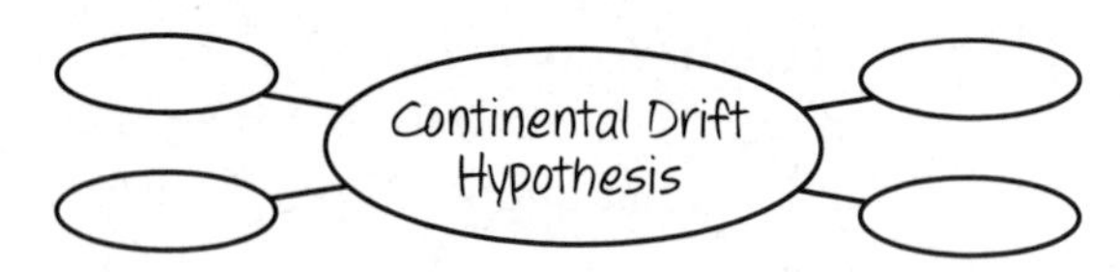

Critical Thinking

7. **Recognize** The shape and age of the Appalachian Mountains are similar to the Caledonian mountains in northern Europe. What else could be similar?
8. **Explain** If continents continue to drift, is it possible that a new supercontinent will form? Which continents might be next to each other 200 million years from now?

AMERICAN MUSEUM OF NATURAL HISTORY

▼ This small mammal is a close living relative of an animal that once roamed Antarctica.

CAREERS in SCIENCE

▲ Ross MacPhee is a paleontologist working for the American Museum of Natural History in New York City. Here, he is searching for fossils in Antarctica.

Gondwana

A Fossil Clue from the Giant Landmass that Once Dominated the Southern Hemisphere

If you could travel back in time 120 million years, you would probably discover that Earth looked very different than it does today. Scientists believe that instead of seven continents, there were two giant landmasses, or supercontinents, on Earth at that time. Scientists named the landmass in the northern hemisphere *Laurasia*. The landmass in the southern hemisphere is known as *Gondwana*. It included the present-day continents of Antarctica, South America, Australia, and Africa.

How do scientists know that Gondwana existed? Ross MacPhee is a paleontologist—a scientist who studies fossils. MacPhee recently traveled to Antarctica where he discovered the fossilized tooth of a small land mammal. After carefully examining the tooth, he realized that it resembled fossils from ancient land mammals found in Africa and North America. MacPhee believes that these mammals are the ancient relatives of a mammal living today on the African island-nation of Madagascar.

How did the fossil remains and their present-day relatives become separated by kilometers of ocean? MacPhee hypothesizes that the mammal migrated across land bridges that once connected parts of Gondwana. Over millions of years, the movement of Earth's tectonic plates broke up this supercontinent. New ocean basins formed between the continents, resulting in the arrangement of landmasses that we see today.

▲ *Gondwana* and *Laurasia* formed as the supercontinent Pangaea broke apart.

It's Your Turn

RESEARCH Millions of years ago, the island of Madagascar separated from the continent of Gondwana. In this environment, the animals of Madagascar changed and adapted. Research and report on one animal. Describe some of its unique adaptations.

Lesson 2

Reading Guide

Key Concepts

ESSENTIAL QUESTIONS

- What is seafloor spreading?
- What evidence is used to support seafloor spreading?

Vocabulary

mid-ocean ridge p. 225

seafloor spreading p. 226

normal polarity p. 228

magnetic reversal p. 228

reversed polarity p. 228

Multilingual eGlossary

Development of a Theory

Inquiry What do the colors represent?

The colors in this satellite image show topography. The warm colors, red, pink, and yellow, represent landforms above sea level. The greens and blues indicate changes in topography below sea level. Deep in the Atlantic Ocean there is a mountain range, shown here as a linear feature in green. Is there a connection between this landform and the continental drift hypothesis?

Launch Lab

15 minutes

Can you guess the age of the glue?

The age of the seafloor can be determined by measuring magnetic patterns in rocks from the bottom of the ocean. How can similar patterns in drying glue be used to show age relationships between rocks exposed on the seafloor?

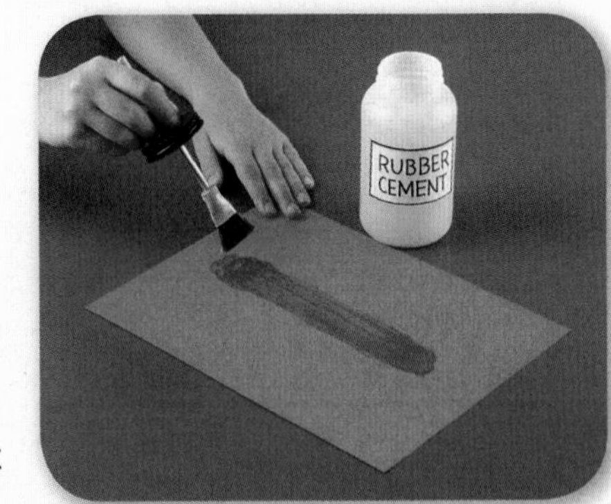

1. Read and complete a lab safety form.
2. Carefully spread a thin layer of **rubber cement** on a sheet of **paper.**
3. Observe for 3 minutes. Record the pattern of how the glue dries in your Science Journal.
4. Repeat step 2. After 1 minute, exchange papers with a classmate.
5. Ask the classmate to observe and tell you which part of the glue dried first.

Think About This

1. What evidence helped you to determine the oldest and youngest glue layers?
2. How is this similar to a geologist trying to estimate the age of rocks on the seafloor?
3. **Key Concept** How could magnetic patterns in rock help predict a rock's age?

Mapping the Ocean Floor

During the late 1940s after World War II, scientists began exploring the seafloor in greater detail. They were able to determine the depth of the ocean using a device called an echo sounder, as shown in **Figure 5.** Once ocean depths were determined, scientists used these data to create a topographic map of the seafloor. These new topographic maps of the seafloor revealed that vast mountain ranges stretched for many miles deep below the ocean's surface. *The mountain ranges in the middle of the oceans are called* **mid-ocean ridges.** Mid-ocean ridges, shown in **Figure 5,** are much longer than any mountain range on land.

Hutchings Photography/Digital Light Source

Figure 5 An echo sounder produces sound waves that travel from a ship to the seafloor and back. The deeper the ocean, the longer the time this takes. Depth can be used to determine seafloor topography.

Seafloor Topography

Mid-ocean ridge
Sediment
Magma

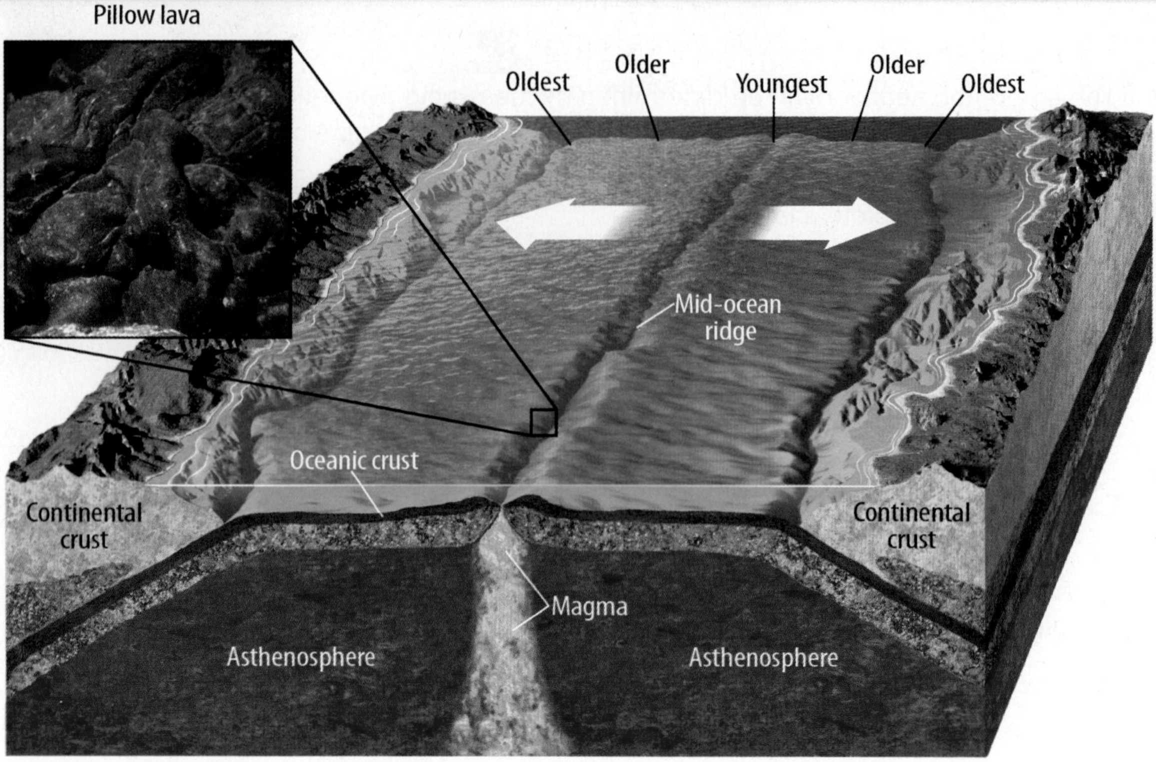

Figure 6 When lava erupts along a mid-ocean ridge, it cools and crystallizes, forming a type of rock called basalt. Basalt is the dominant rock on the seafloor. The youngest basalt is closest to the ridge. The oldest basalt is farther away from the ridge.

Visual Check Looking at the image above, can you propose a pattern that exists in rocks on either side of the mid-ocean ridge?

Seafloor Spreading

By the 1960s scientists discovered a new process that helped explain continental drift. This process, shown in **Figure 6**, is called seafloor spreading. **Seafloor spreading** *is the process by which new oceanic crust forms along a mid-ocean ridge and older oceanic crust moves away from the ridge.*

When the seafloor spreads, the mantle below melts and forms magma. Because magma is less dense than solid mantle material, it rises through cracks in the crust along the mid-ocean ridge. When magma erupts on Earth's surface, it is called lava. As this lava cools and crystallizes on the seafloor, it forms a type of rock called basalt. Because the lava erupts into water, it cools rapidly and forms rounded structures called pillow lavas. Notice the shape of the pillow lava shown in **Figure 6.**

As the seafloor continues to spread apart, the older oceanic crust moves away from the mid-ocean ridge. The closer the crust is to a mid-ocean ridge, the younger the oceanic crust is. Scientists argued that if the seafloor spreads, the continents must also be moving. A mechanism to explain continental drift was finally discovered long after Wegener proposed his hypothesis.

Key Concept Check What is seafloor spreading?

Topography of the Seafloor

The rugged mountains that make up the mid-ocean ridge system can form in two different ways. For example, large amounts of lava can erupt from the center of the ridge, cool and build up around the ridge. Or, as the lava cools and forms new crust, it cracks. The rocks move up or down along these cracks in the seafloor, forming jagged mountain ranges.

Reading Check How do mountains form along the mid-ocean ridge?

Over time, sediment accumulates on top of the oceanic crust. Close to the mid-ocean ridge there is almost no sediment. Far from the mid-ocean ridge, the layer of sediment becomes thick enough to make the seafloor smooth. This part of the seafloor, shown in **Figure 7**, is called the abyssal (uh BIH sul) plain.

Moving Continents Around

The theory of seafloor spreading provides a way to explain how continents move. Continents do not move through the solid mantle or the seafloor. Instead, continents move as the seafloor spreads along a mid-ocean ridge.

MiniLab **20 minutes**

How old is the Atlantic Ocean?

If you measure the width of the Atlantic Ocean and you know the rate of seafloor spreading, you can calculate the age of the Atlantic.

1. Use a **ruler** to measure the horizontal distance between a point on the eastern coast of South America and a point on the western coast of Africa on a **world map.** Repeat three times and calculate the average distance in your Science Journal.
2. Use the map's legend to convert the average distance from centimeters to kilometers.
3. If Africa and South America have been moving away from each other at a rate of 2.5 cm per year, calculate the age of the Atlantic Ocean.

Analyze and Conclude

1. **Measure** Did your measurements vary?
2. **Key Concept** How does the age you calculated compare to the breakup of Pangaea 200 million years ago?

Abyssal Plain

Figure 7 The abyssal plain is flat due to an accumulation of sediments far from the ridge.

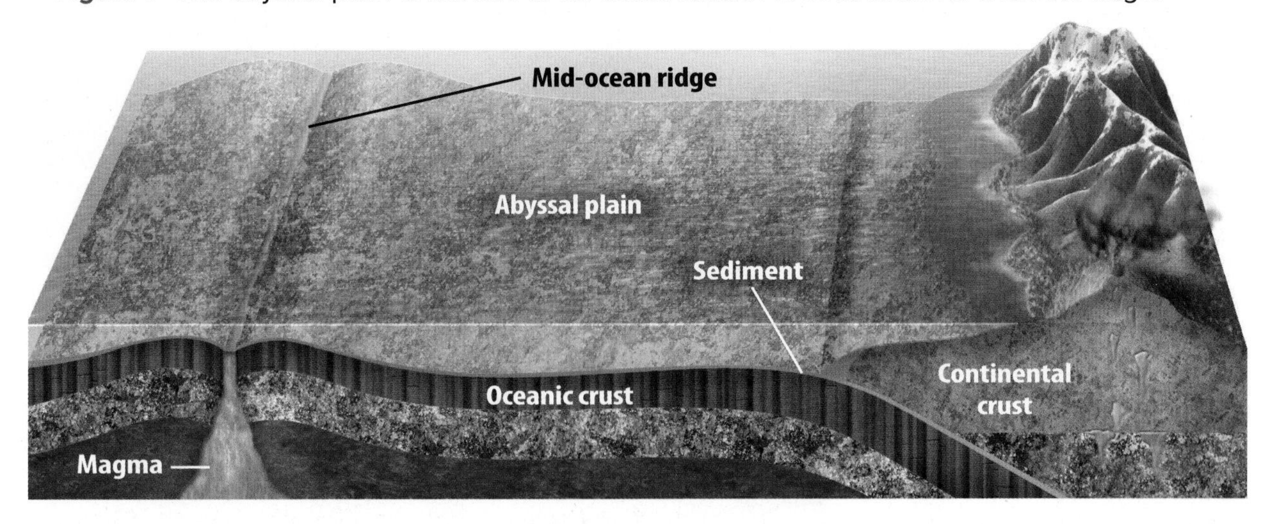

Visual Check Compare and contrast the topography of a mid-ocean ridge to an abyssal plain.

Development of a Theory

The first evidence used to support seafloor spreading was discovered in rocks on the seafloor. Scientists studied the magnetic signature of minerals in these rocks. To understand this, you need to understand the direction and orientation of Earth's magnetic field and how rocks record magnetic information.

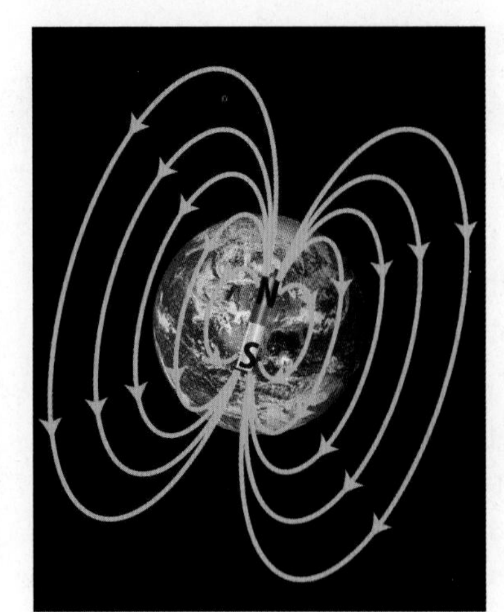

Reversed magnetic field

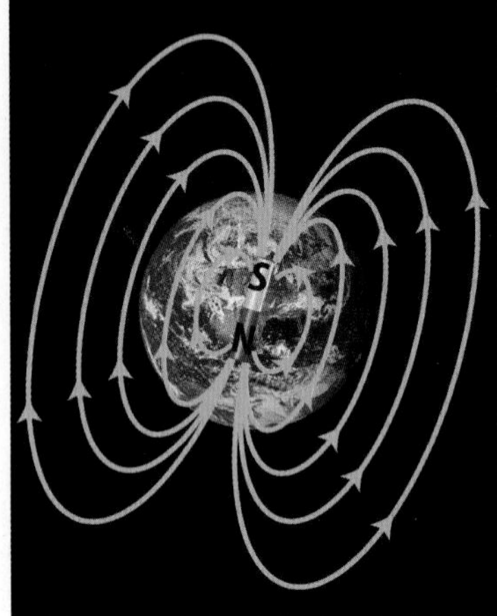

Normal magnetic field

▲ **Figure 8** Earth's magnetic field is like a large bar magnet. Therefore, the north end of a compass magnet is drawn to the south pole of Earth's magnetic field.

Magnetic Reversals

Recall that the iron-rich, liquid outer core is like a giant magnet that creates Earth's magnetic field. The direction of the magnetic field is not constant. Today's magnetic field, shown in **Figure 8**, is described as having **normal polarity**–*a state in which magnetized objects, such as compass needles, will orient themselves to point north.* Sometimes a **magnetic reversal** *occurs and the magnetic field reverses direction.* The opposite of normal polarity is **reversed polarity**–*a state in which magnetized objects would reverse direction and orient themselves to point south,* as shown in **Figure 8.** Magnetic reversals occur every few hundred thousand to every few million years.

Reading Check Is Earth's magnetic field currently normal or reversed polarity?

Rocks Reveal Magnetic Signature

Basalt on the seafloor contains iron-rich minerals that are magnetic. Each mineral acts like a small magnet. **Figure 9** shows how magnetic minerals align themselves with Earth's magnetic field. When lava erupts along a mid-ocean ridge, it cools and crystallizes. This permanently records the direction and orientation of Earth's magnetic field at the time of the eruption. Scientists have discovered parallel patterns in the magnetic signature of rocks on either side of a mid-ocean ridge.

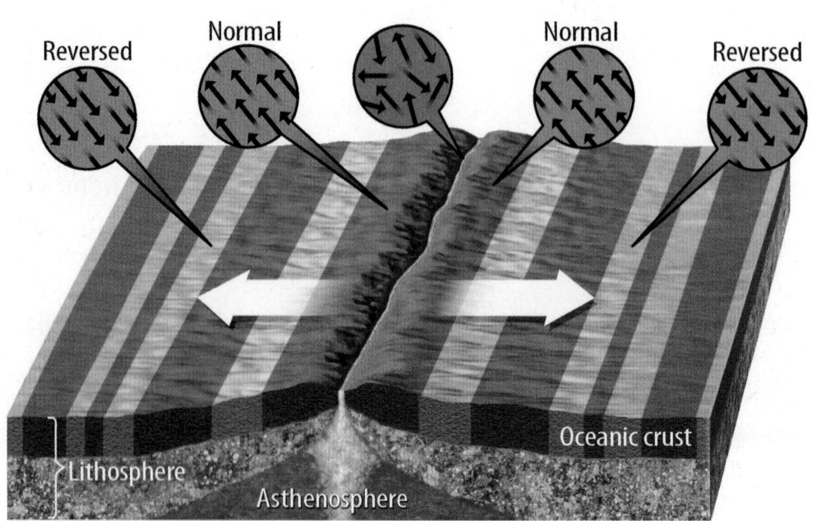

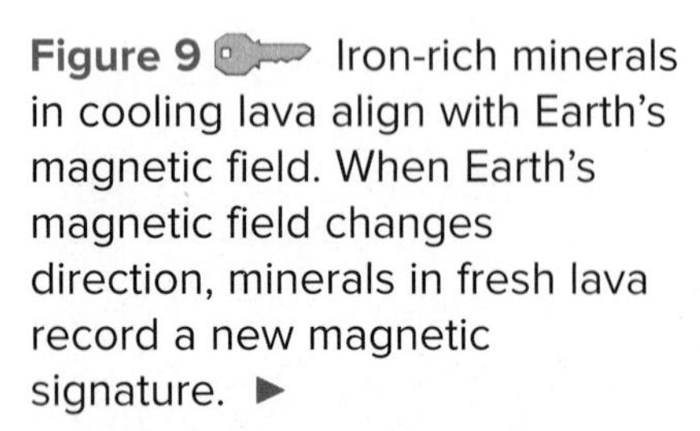

Figure 9 Iron-rich minerals in cooling lava align with Earth's magnetic field. When Earth's magnetic field changes direction, minerals in fresh lava record a new magnetic signature. ▶

Visual Check Describe the pattern in the magnetic stripes shown in the image to the right.

Seafloor Spreading Theory

Figure 10 A mirror image in the magnetic stripes on either side of the mid-ocean ridge shows that the crust formed at the ridge is carried away in opposite directions.

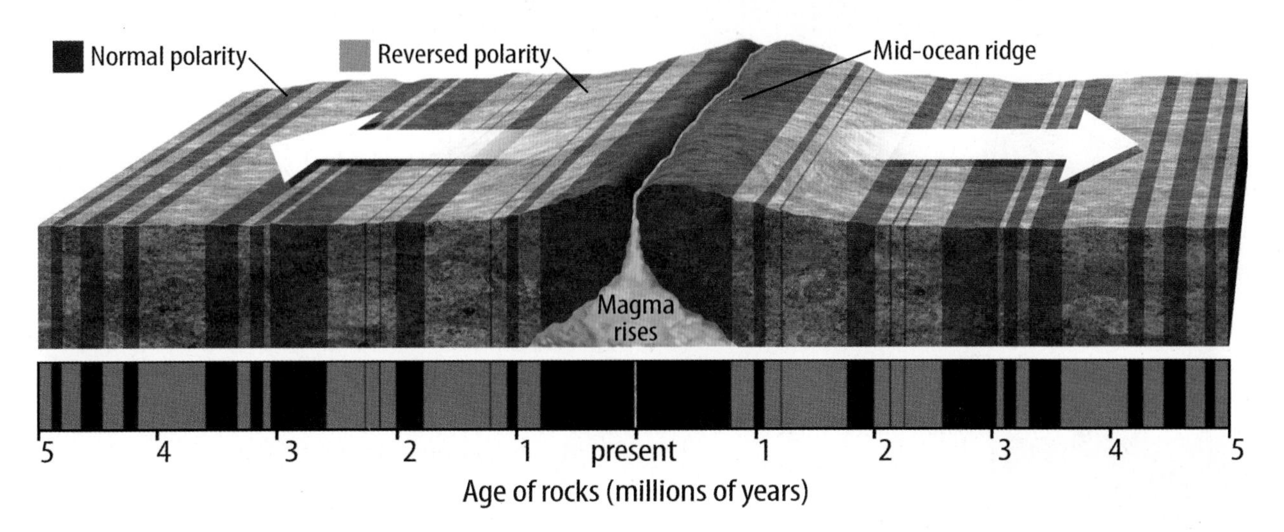

Evidence to Support the Theory

How did scientists prove the theory of seafloor spreading? Scientists studied magnetic minerals in rocks from the seafloor. They used a magnetometer (mag nuh TAH muh tur) to measure and record the magnetic signature of these rocks. These measurements revealed a surprising pattern. Scientists have discovered parallel magnetic stripes on either side of the mid-ocean ridge. Each pair of stripes has a similar composition, age, and magnetic character. Each magnetic stripe in **Figure 10** represents crust that formed and magnetized at a mid-ocean ridge during a period of either normal or reversed polarity. The pairs of magnetic stripes confirm that the ocean crust formed at mid-ocean ridges is carried away from the center of the ridges in opposite directions.

Reading Check How do magnetic minerals help support the theory of seafloor spreading?

Other measurements made on the seafloor confirm seafloor spreading. By drilling a hole into the seafloor and measuring the temperature beneath the surface, scientists can measure the amount of thermal energy leaving Earth. The measurements show that more thermal energy leaves Earth near mid-ocean ridges than is released from beneath the abyssal plains.

Additionally, sediment collected from the seafloor can be dated. Results show that the sediment closest to the mid-ocean ridge is younger than the sediment farther away from the ridge. Sediment ages as it is carried away. Sediment thickness also increases with distance away from the mid-ocean ridge.

ACADEMIC VOCABULARY

normal
(adjective) conforming to a type, standard, or regular pattern

FOLDABLES®

Make a layered book using two sheets of notebook paper. Use the two pages to record your notes and the inside to illustrate seafloor spreading.

Lesson 2 Review

Visual Summary

Lava erupts along mid-ocean ridges.

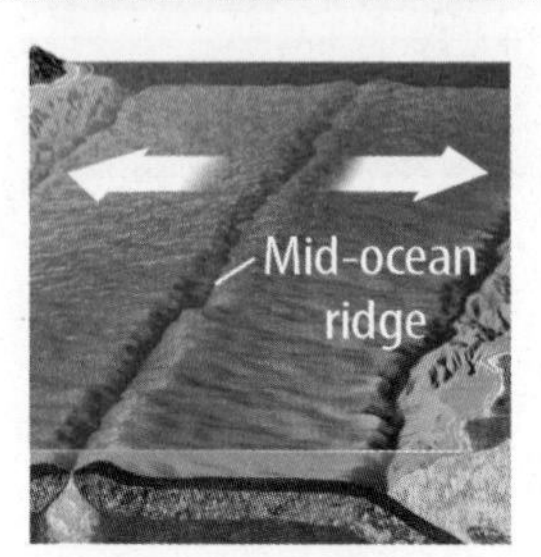

Mid-ocean ridges are large mountain ranges that extend throughout Earth's oceans.

A magnetic reversal occurs when Earth's magnetic field changes direction.

FOLDABLES®

Use your lesson Foldable to review the lesson. Save your Foldable for the project at the end of the chapter.

What do you think NOW?

You first read the statements below at the beginning of the chapter.

3. The seafloor is flat.

4. Volcanic activity occurs only on the seafloor.

Did you change your mind about whether you agree or disagree with the statements? Rewrite any false statements to make them true.

Use Vocabulary

1. **Explain** how rocks on the seafloor record magnetic reversals over time.
2. **Diagram** the process of seafloor spreading.
3. **Use the term** *seafloor spreading* to explain how a mid-ocean ridge forms.

Understand Key Concepts

4. Oceanic crust forms
   - **A.** at mid-ocean ridges.
   - **B.** everywhere on the seafloor.
   - **C.** on the abyssal plains.
   - **D.** by magnetic reversals.
5. **Explain** why magnetic stripes on the seafloor are parallel to the mid-ocean ridge.
6. **Describe** how scientists can measure the depth to the seafloor.

Interpret Graphics

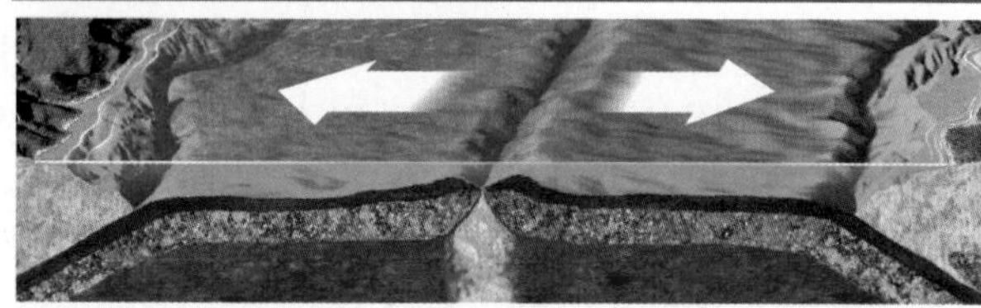

7. **Determine** Refer to the image above. Where is the youngest crust? Where is the oldest crust?
8. **Describe** how seafloor spreading helps to explain the continental drift hypothesis.
9. **Sequence Information** Copy and fill in the graphic organizer below to explain the steps in the formation of a mid-ocean ridge.

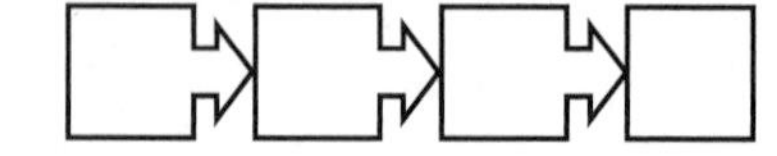

Critical Thinking

10. **Infer** why magnetic stripes in the Pacific Ocean are wider than in the Atlantic Ocean.
11. **Explain** why the thickness of seafloor sediments increases with increasing distance from the ocean ridge.

Skill Lab Model

30 minutes

How do rocks on the seafloor vary with age away from a mid-ocean ridge?

Materials

vanilla yogurt
berry yogurt

foam board
(10 cm × 4 cm)

waxed paper

plastic spoon

Safety

Do not eat anything used in this lab.

Scientists discovered that new ocean crust forms at a mid-ocean ridge and spreads away from the ridge slowly over time. This process is called seafloor spreading. The age of the seafloor is one component that supports this theory.

Learn It

Scientists use **models** to represent real-world science. By creating a three-dimensional model of volcanic activity along the Mid-Atlantic Ridge, scientists can model the seafloor spreading process. They can then compare this process to the actual age of the seafloor. In this skill lab, you will investigate how the age of rocks on the seafloor changes with distance away from the ridge.

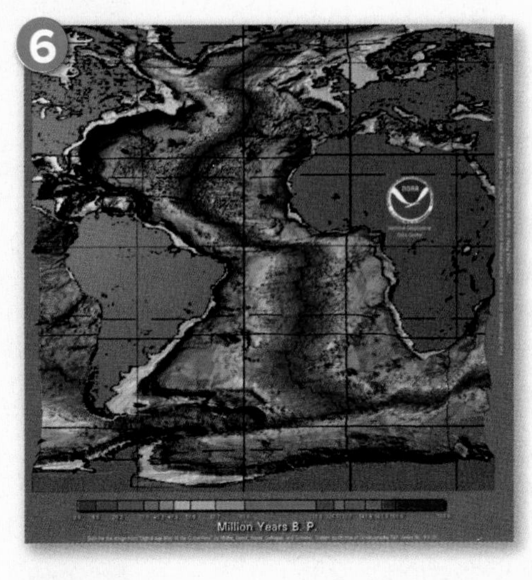

Try It

1. Read and complete a lab safety form.
2. Lay the sheet of waxed paper flat on the lab table. Place two spoonfuls of vanilla yogurt in a straight line near the center of the waxed paper, leaving it lumpy and full.
3. Lay the two pieces of foam board over the yogurt, leaving a small opening in the middle. Push the foam boards together and down, so the yogurt oozes up and over each of the foam boards.
4. Pull the foam boards apart and add a new row of two spoonfuls of berry yogurt down the middle. Lift the boards and place them partly over the new row. Push them together gently. Observe the outer edges of the new yogurt while you are moving the foam boards together.
5. Repeat step 4 with one more spoonful of vanilla yogurt. Then repeat again with one more spoonful of berry yogurt.

Apply It

6. Compare the map and the model. Where is the Mid-Atlantic Ridge on the map? Where is it represented in your model?
7. Which of your yogurt strips matches today on this map? And millions of years ago?
8. How do scientists determine the ages of different parts of the ocean floor?
9. **Conclude** What happened to the yogurt when you added more?
10. **Key Concept** What happens to the material already on the ocean floor when magma erupts along a mid-ocean ridge?

(t to b, 2, 4)Hutchings Photography/ Digital Light Source; (3)Jacques Cornell/McGraw-Hill Education; (r)Dr. Peter Sloss, formerly of NGDC/NOAA/NGDC

Lesson 3

The Theory of Plate Tectonics

Reading Guide

Key Concepts
ESSENTIAL QUESTIONS

- What is the theory of plate tectonics?
- What are the three types of plate boundaries?
- Why do tectonic plates move?

Vocabulary

plate tectonics p. 233
lithosphere p. 234
divergent plate boundary p. 235
transform plate boundary p. 235
convergent plate boundary p. 235
subduction p. 235
convection p. 238
ridge push p. 239
slab pull p. 239

Multilingual eGlossary

Go to the resource tab in ConnectED to find the PBL *Movin' Mountains.*

Inquiry How did these islands form?

The photograph shows a chain of active volcanoes. These volcanoes make up the Aleutian Islands of Alaska. Just south of these volcanic islands is a 6-km deep ocean trench. Why did these volcanic mountains form in a line? Can you predict where volcanoes are? Are they related to plate tectonics?

Launch Lab

15 minutes

Can you determine density by observing buoyancy?

Density is the measure of an object's mass relative to its volume. Buoyancy is the upward force a liquid places on objects that are immersed in it. If you immerse objects with equal densities into liquids that have different densities, the buoyant forces will be different. An object will sink or float depending on the density of the liquid compared to the object. Earth's layers differ in density. These layers float or sink depending on density and buoyant force.

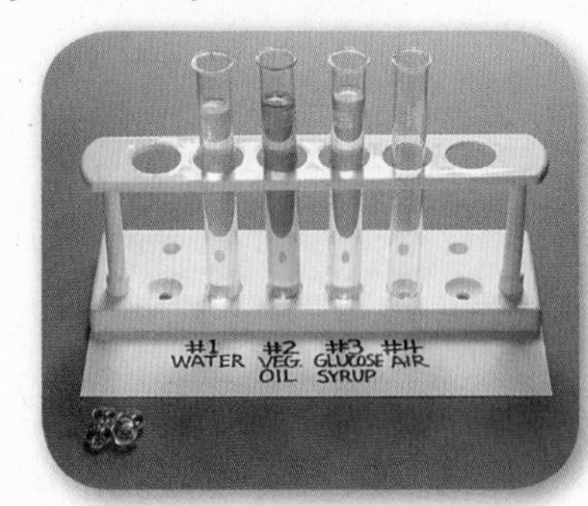

1. Read and complete a lab safety form.
2. Obtain four **test tubes.** Place them in a **test-tube rack.** Add **water** to one test tube until it is ¾ full.
3. Repeat with the other test tubes using **vegetable oil** and **glucose syrup.** One test tube should remain empty.
4. Drop **beads** of equal density into each test tube. Observe what the object does when immersed in each liquid. Record your observations in your Science Journal.

Think About This

1. How did you determine which liquid has the highest density?
2. **Key Concept** What happens when layers of rock with different densities collide?

The Plate Tectonics Theory

When you blow into a balloon, the balloon expands and its surface area also increases. Similarly, if oceanic crust continues to form at mid-ocean ridges and is never destroyed, Earth's surface area should increase. However, this is not the case. The older crust must be destroyed somewhere–but where?

By the late 1960s a more complete theory, called plate tectonics, was proposed. The theory of **plate tectonics** states that *Earth's surface is made of rigid slabs of rock, or plates, that move with respect to each other.* This new theory suggested that Earth's surface is divided into large plates of rigid rock. Each plate moves over Earth's hot and semi-plastic mantle.

Key Concept Check What is plate tectonics?

Geologists use the word *tectonic* to describe the forces that shape Earth's surface and the rock structures that form as a result. Plate tectonics provides an explanation for the occurrence of earthquakes and volcanic eruptions. When plates separate on the seafloor, earthquakes result and a mid-ocean ridge forms. When plates come together, one plate can dive under the other, causing earthquakes and creating a chain of volcanoes. When plates slide past each other, earthquakes can result.

Earth's Tectonic Plates

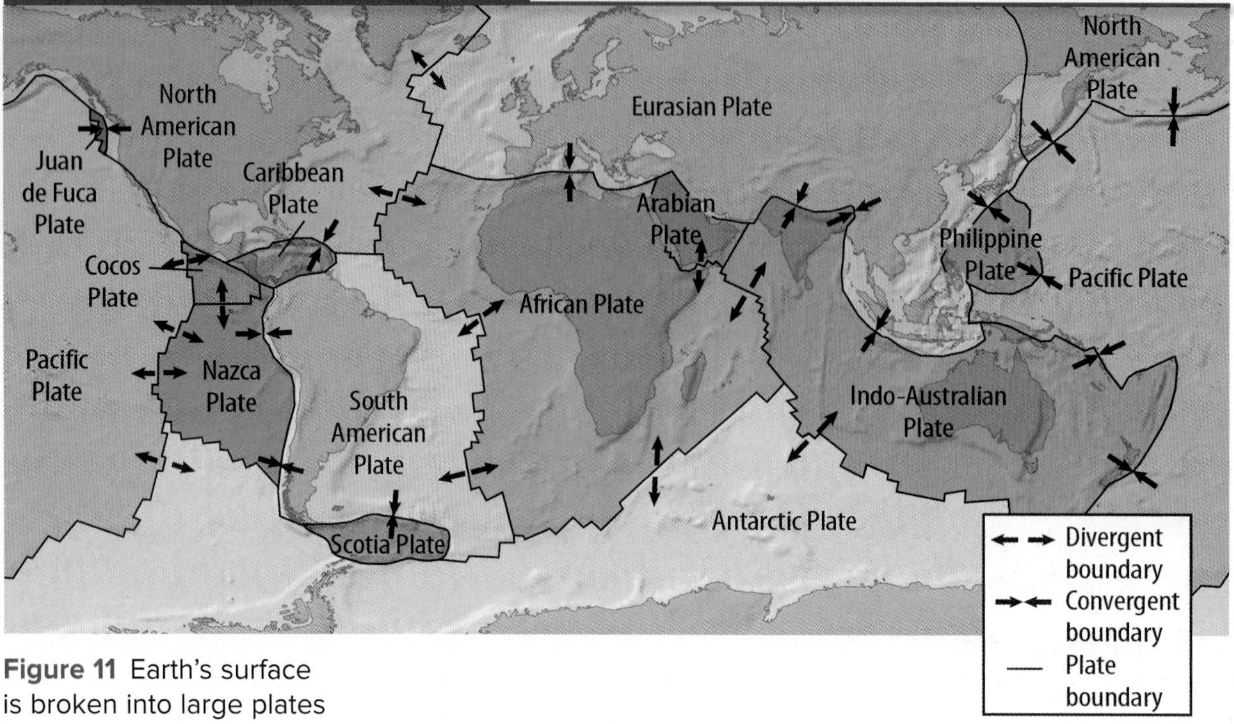

Figure 11 Earth's surface is broken into large plates that fit together like pieces of a giant jigsaw puzzle. The arrows show the general direction of movement of each plate.

Tectonic Plates

You read on the previous page that the theory of plate tectonics states that Earth's surface is divided into rigid plates that move relative to one another. These plates are "floating" on top of a hot and semi-plastic mantle. The map in **Figure 11** illustrates Earth's major plates and the boundaries that define them. The Pacific Plate is the largest plate. The Juan de Fuca Plate is one of the smallest plates. It is between the North American and Pacific Plates. Notice the boundaries that run through the oceans. Many of these boundaries mark the positions of the mid-ocean ridges.

Earth's outermost layers are cold and rigid compared to the layers within Earth's interior. *The cold and rigid outermost rock layer is called the* **lithosphere.** It is made up of the crust and the uppermost mantle. The lithosphere is thin below mid-ocean ridges and thick below continents. Earth's tectonic plates are large pieces of lithosphere. These lithospheric plates fit together like the pieces of a giant jigsaw puzzle.

The layer of Earth below the lithosphere is called the asthenosphere (as THEN uh sfihr). This layer is so hot that although it is solid, it behaves like a **plastic** material. This enables Earth's plates to move because the hotter, plastic mantle material beneath them can flow. The interactions between lithosphere and asthenosphere help to explain plate tectonics.

Science Use v. Common Use

plastic

Science Use capable of being molded or changing shape without breaking

Common Use any of numerous organic, synthetic, or processed materials made into objects

 **Reading Check** What are Earth's outermost layers called?

Plate Boundaries

Place two books side by side and imagine each book represents a tectonic plate. A plate boundary exists where the books meet. How many different ways can you move the books with respect to each other? You can pull the books apart, you can push the books together, and you can slide the books past one another. Earth's tectonic plates move in much the same way.

Divergent Plate Boundaries

Mid-ocean ridges are located along divergent plate boundaries. A **divergent plate boundary** *forms where two plates separate.* When the seafloor spreads at a mid-ocean ridge, lava erupts, cools, and forms new oceanic crust. Divergent plate boundaries can also exist in the middle of a continent. They pull continents apart and form rift valleys. The East African Rift is an example of a continental rift.

Transform Plate Boundaries

The famous San Andreas Fault in California is an example of a transform plate boundary. A **transform plate boundary** *forms where two plates slide past each other.* As they move past each other, the plates can get stuck and stop moving. Stress builds up where the plates are "stuck." Eventually, the stress is too great and the rocks break, suddenly moving apart. This results in a rapid release of energy as earthquakes.

Convergent Plate Boundaries

Convergent plate boundaries *form where two plates collide. The denser plate sinks below the more buoyant plate in a process called* **subduction.** The area where a denser plate descends into Earth along a convergent plate boundary is called a **subduction** zone.

When two oceanic plates collide, the older and denser oceanic plate will subduct beneath the younger oceanic plate. This creates a deep ocean trench and a line of volcanoes called an island arc. This process can also occur when an oceanic plate and a continental plate collide. The denser oceanic plate subducts under the edge of the continent. This creates a deep ocean trench. A line of volcanoes forms above the subducting plate on the edge of the continent.

Over time, an oceanic plate can be completely subducted, dragging an attached continent behind it. When two continents collide, neither plate is subducted, and mountains such as the Himalayas in southern Asia form from uplifted rock. **Table 1** on the next page summarizes the interactions of Earth's tectonic plates.

Key Concept Check What are the three types of plate boundaries?

FOLDABLES®

Make a layered book using two sheets of notebook paper. Use it to organize information about the different types of plate boundaries and the features that form there.

Plate Boundaries
Divergent
Convergent
Transform

WORD ORIGIN

subduction
from Latin *subductus,* means "to lead under, removal"

Table 1 The direction of motion of Earth's plates creates a variety of features at the boundaries between the plates.

Animation

Table 1 Interactions of Earth's Tectonic Plates

Plate Boundary	Relative Motion	Example
Divergent plate boundary When two plates separate a divergent plate boundary forms. This process occurs where the seafloor spreads along a mid-ocean ridge, forming new crust and hydrothermal vents, as shown to the right. This process can also occur in the middle of continents and is referred to as continental rifting.	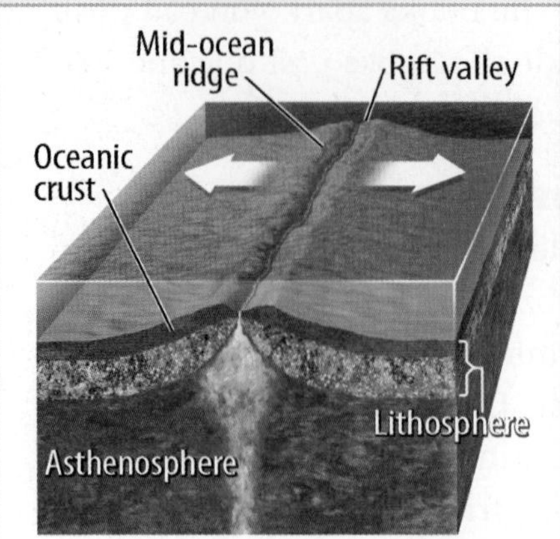	
Transform plate boundary Two plates slide horizontally past one another along a transform plate boundary. Earthquakes are common along this type of plate boundary. The San Andreas Fault, shown to the right, is part of the transform plate boundary that extends along the coast of California.	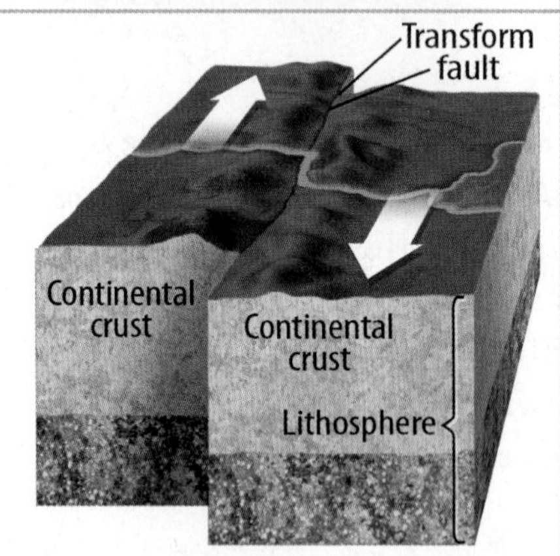	
Convergent plate boundary (ocean-to-continent) When an oceanic and a continental plate collide, they form a convergent plate boundary. The denser plate will subduct. A volcanic mountain, such as Mount Rainier in the Cascade Mountains, forms along the edge of the continent. This process can also occur where two oceanic plates collide, and the denser plate is subducted.	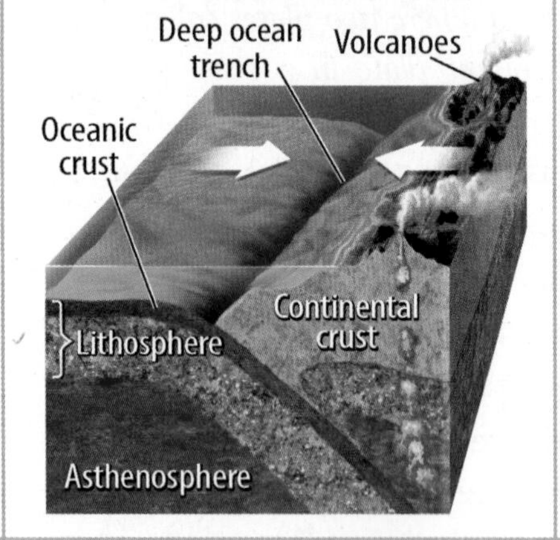	
Convergent plate boundary (continent-to-continent) Convergent plate boundaries can also occur where two continental plates collide. Because both plates are equally dense, neither plate will subduct. Both plates uplift and deform. This creates huge mountains like the Himalayas, shown to the right.	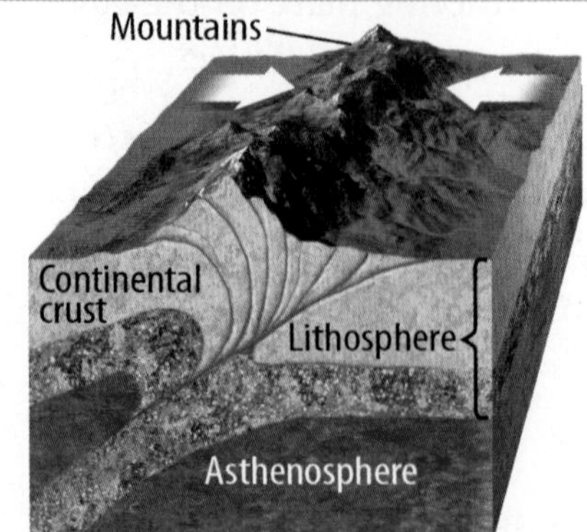	

(t to b)Dr. Ken MacDonald/Photo Researchers, Inc.; (2)©Lloyd Cluff/Corbis; (3)©Jim Richardson/Corbis; (4)Tony Waltham/Getty Images

Evidence for Plate Tectonics

When Wegener proposed the continental drift hypothesis, the technology used to measure how fast the continents move today wasn't yet available. Recall that continents move apart or come together at speeds of a few centimeters per year. This is about the length of a small paperclip.

Today, scientists can measure how fast continents move. A network of satellites orbiting Earth monitors plate motion. By keeping track of the distance between these satellites and Earth, it is possible to locate and determine how fast a tectonic plate moves. This network of satellites is called the Global Positioning System (GPS).

The theory of plate tectonics also provides an explanation for why earthquakes and volcanoes occur in certain places. Because plates are rigid, tectonic activity occurs where plates meet. When plates separate, collide, or slide past each other along a plate boundary, stress builds. A rapid release of energy can result in earthquakes. Volcanoes form where plates separate along a mid-ocean ridge or a continental rift or collide along a subduction zone. Mountains can form where two continents collide. **Figure 12** illustrates the relationship between plate boundaries and the occurrence of earthquakes and volcanoes. Refer back to the lesson opener photo. Find these islands on the map. Are they located near a plate boundary?

Key Concept Check How are earthquakes and volcanoes related to the theory of plate tectonics?

Figure 12 Notice that most earthquakes and volcanoes occur near plate boundaries.

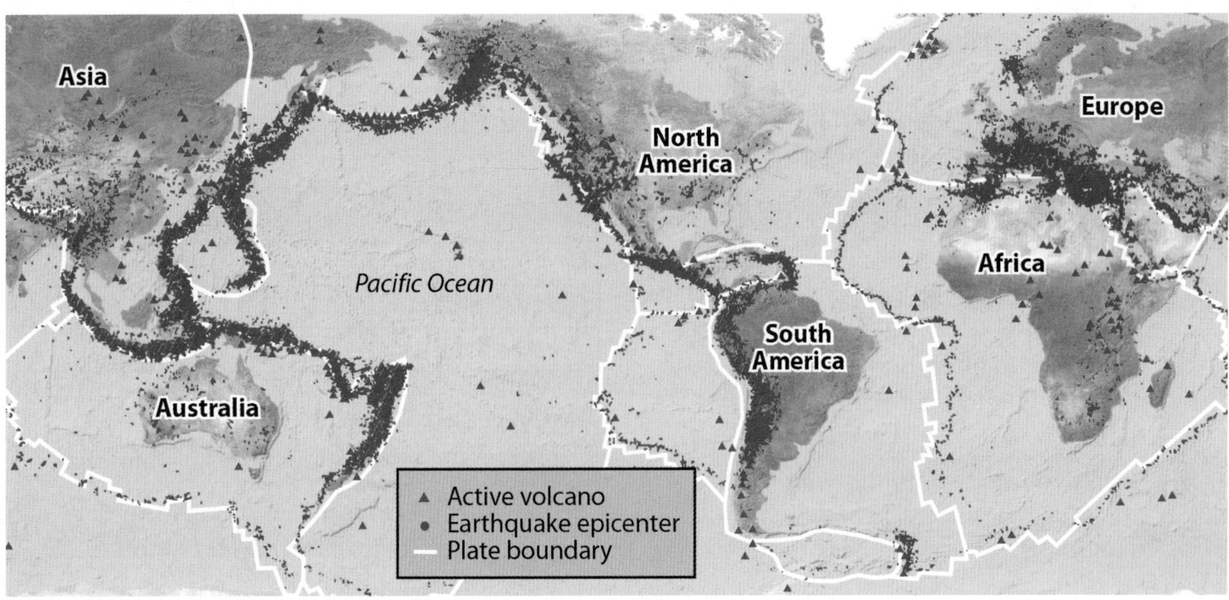

Visual Check Do earthquakes and volcanoes occur anywhere away from plate boundaries?

 Personal Tutor

Figure 13 When water is heated, it expands. Less dense heated water rises because the colder water sinks, forming convection currents.

Plate Motion

The main objection to Wegener's continental drift hypothesis was that he could not explain why or how continents move. Scientists now understand that continents move because the asthenosphere moves underneath the lithosphere.

Convection Currents

You are probably already familiar with the process of **convection,** *the circulation of material caused by differences in temperature and density.* For example, the upstairs floors of homes and buildings are often warmer. This is because hot air rises while dense, cold air sinks. Look at **Figure 13** to see convection in action.

 **Reading Check** What causes convection?

Plate tectonic activity is related to convection in the mantle, as shown in **Figure 14.** Radioactive elements, such as uranium, thorium, and potassium, heat Earth's interior. When materials such as solid rock are heated, they expand and become less dense. Hot mantle material rises upward and comes in contact with Earth's crust. Thermal energy is transferred from hot mantle material to the colder surface above. As the mantle cools, it becomes denser and then sinks, forming a convection current. These currents in the asthenosphere act like a conveyor belt moving the lithosphere above.

 **Key Concept Check** Why do tectonic plates move?

MiniLab

20 minutes

How do changes in density cause motion?

Convection currents drive plate motion. Material near the base of the mantle is heated, which decreases its density. This material then rises to the base of the crust, where it cools, increasing in density and sinking.

1. Read and complete a lab safety form.
2. Copy the table to the right into your Science Journal and add a row for each minute. Record your observations.
3. Pour 100 mL of **carbonated water** or **clear soda** into a **beaker** or a **clear glass.**
4. Drop five **raisins** into the water. Observe the path that the raisins follow for 5 minutes.

Time Interval	Observations
First minute	
Second minute	
Third minute	

Analyze and Conclude

1. **Observe** Describe each raisin's motion.
2. **Key Concept** How does the behavior of the raisin model compare to the motion in Earth's mantle?

Convection Currents 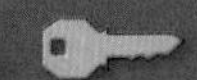

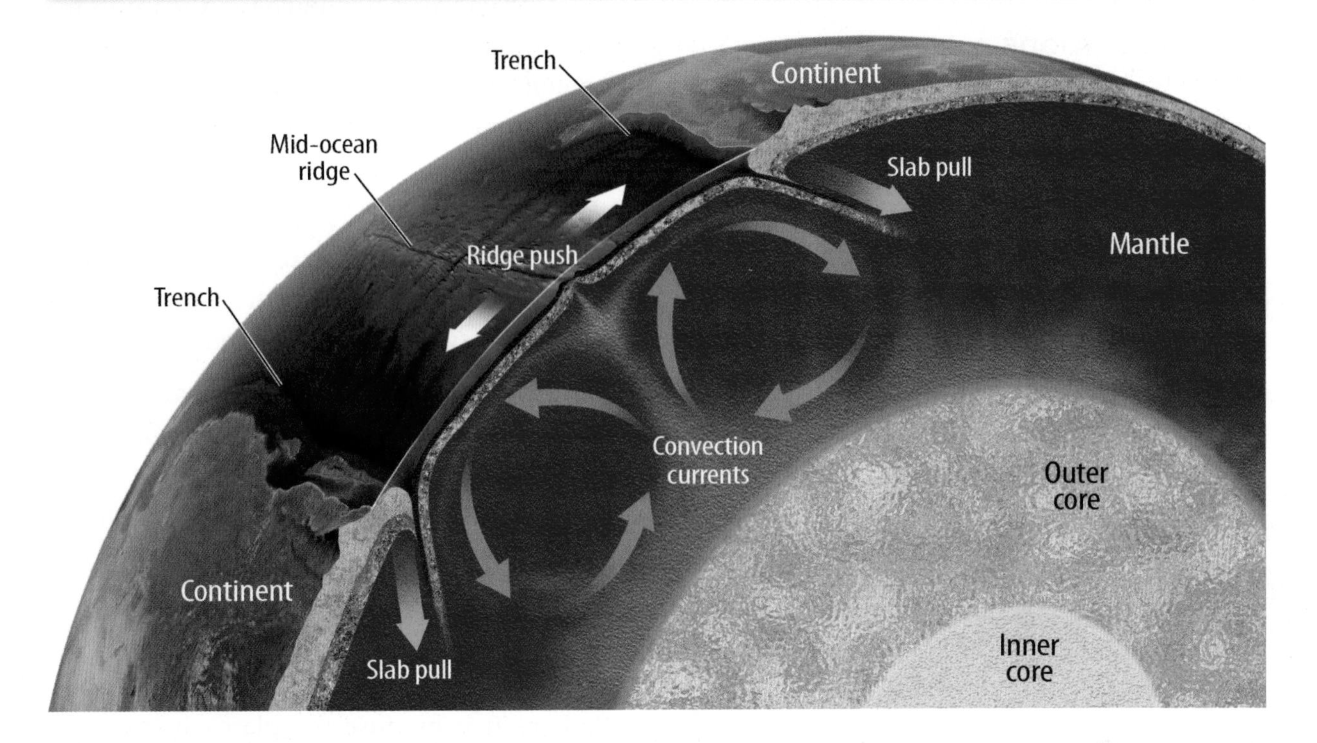

Forces Causing Plate Motion

How can something as massive as the Pacific Plate move? **Figure 14** shows the processes that determine how convection affects the movement of tectonic plates.

Basal Drag Convection currents in the mantle produce a force that causes motion called basal drag. Notice in **Figure 14** how convection currents in the asthenosphere circulate and drag the lithosphere similar to the way a conveyor belt moves items along at a supermarket checkout.

Ridge Push Recall that mid-ocean ridges have greater elevation than the surrounding seafloor. Because mid-ocean ridges are higher, gravity pulls the surrounding rocks down and away from the ridge. *Rising mantle material at mid-ocean ridges creates the potential for plates to move away from the ridge with a force called* **ridge push.** Ridge push moves lithosphere in opposite directions away from the mid-ocean ridge.

Slab Pull As you read earlier in this lesson, when tectonic plates collide, the denser plate will sink into the mantle along a subduction zone. This plate is called a slab. Because the slab is old and cold, it is denser than the surrounding mantle and will sink. *As a slab sinks, it pulls on the rest of the plate with a force called* **slab pull.** Slab pull is thought to be a more significant force than ridge push in moving tectonic plates.

Figure 14 Convection occurs in the mantle underneath Earth's tectonic plates. Three forces act on plates to make them move: basal drag from convection currents, ridge push at mid-ocean ridges, and slab pull from subducting plates.

Visual Check What is happening to a plate that is undergoing slab pull?

Math Skills

Use Proportions

The plates along the Mid-Atlantic Ridge spread at an average rate of 2.5 cm/y. How long will it take the plates to spread 1 m? Use proportions to find the answer.

1. **Convert the distance to the same unit.**

   1 m = 100 cm

2. **Set up a proportion:**

$$\frac{2.5 \text{ cm}}{1 \text{ y}} = \frac{100 \text{ cm}}{x \text{ y}}$$

3. **Cross multiply and solve for *x* as follows:**

$$2.5 \text{ cm} \times x\text{y} = 100 \text{ cm} \times 1 \text{ y}$$

4. **Divide both sides by 2.5 cm.**

$$x = \frac{100 \text{ cm y}}{2.5 \text{ cm}}$$

$$x = 40 \text{ y}$$

Practice

The Eurasian plate travels the slowest, at about 0.7 cm/y. How long would it take the plate to travel 3 m?

(1 m = 100 cm)

Math Practice

Personal Tutor

A Theory in Progress

Plate tectonics has become the unifying theory of geology. It explains the connection between continental drift and the formation and destruction of crust along plate boundaries. It also helps to explain the occurrence of earthquakes, volcanoes, and mountains.

The investigation that Wegener began nearly a century ago is still being revised. Several unanswered questions remain.

- Why is Earth the only known planet in the solar system that currently has plate tectonic activity? Different hypotheses have been proposed to explain this. Scientists have found evidence of plate tectonics in Mars's past.
- Why do some earthquakes and volcanoes occur far away from plate boundaries? Perhaps it is because the plates are not perfectly rigid. Different thicknesses and weaknesses exist within the plates. Also, the mantle is much more active than scientists originally understood.
- What forces dominate plate motion? Currently accepted models suggest that convection currents occur in the mantle. However, there is no way to measure or observe them.
- What will scientists investigate next? **Figure 15** shows an image produced by a new technique called anisotropy that creates a 3-D image of seismic wave velocities in a subduction zone. This developing technology might help scientists better understand the processes that occur within the mantle and along plate boundaries.

Reading Check Why does the theory of plate tectonics continue to change?

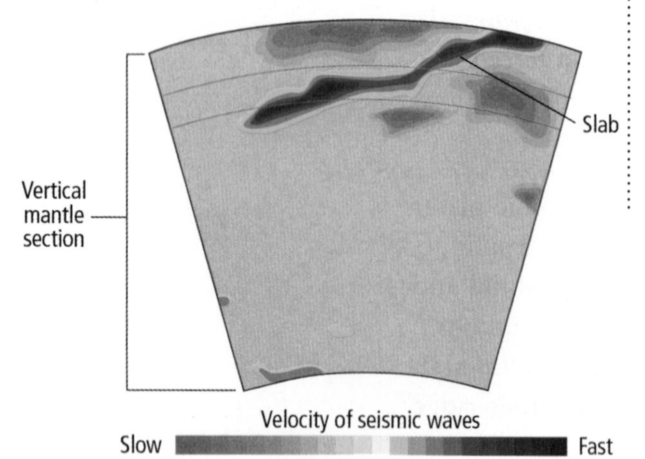

Figure 15 Seismic waves were used to produce this tomography scan. These colors show a subducting plate. The blue colors represent rigid materials with faster seismic wave velocities.

Lesson 3 Review

Visual Summary

Tectonic plates are made of cold and rigid slabs of rock.

Mantle convection—the circulation of mantle material due to density differences—drives plate motion.

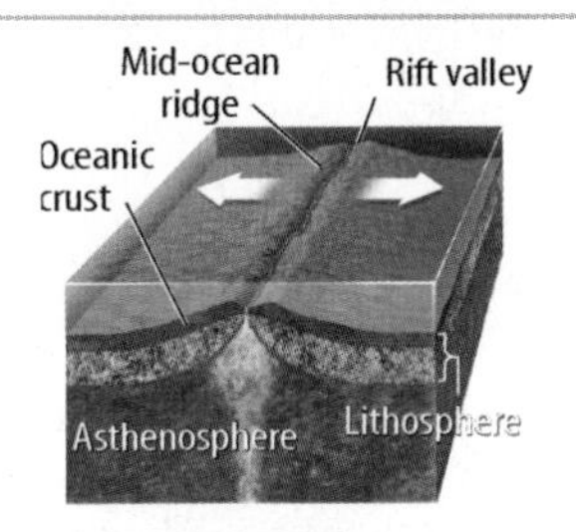

The three types of plate boundaries are divergent, convergent, and transform boundaries.

FOLDABLES®

Use your lesson Foldable to review the lesson. Save your Foldable for the project at the end of the chapter.

What do you think NOW?

You first read the statements below at the beginning of the chapter.

5. Continents drift across a molten mantle.

6. Mountain ranges can form when continents collide.

Did you change your mind about whether you agree or disagree with the statements? Rewrite any false statements to make them true.

Use Vocabulary

1 The theory that proposes that Earth's surface is broken into moving, rigid plates is called ________.

Understand Key Concepts 

2 **Compare and contrast** the geological activity that occurs along the three types of plate boundaries.

3 **Explain** why mantle convection occurs.

4 Tectonic plates move because of

- **A.** convection currents.
- **B.** Earth's increasing size.
- **C.** magnetic reversals.
- **D.** volcanic activity.

Interpret Graphics

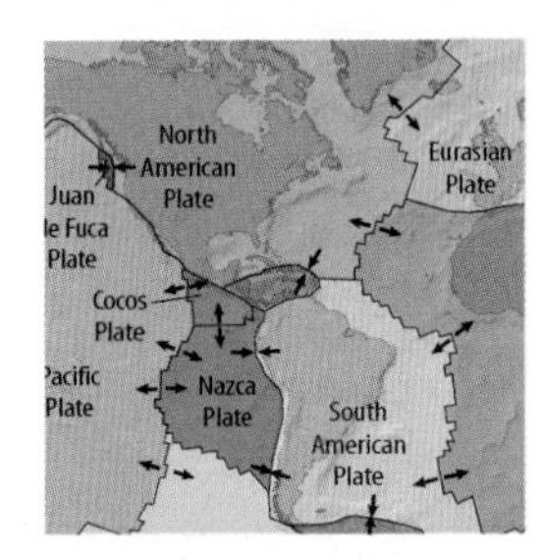

5 **Identify** Name the type of boundary between the Eurasian Plate and the North American Plate and between the Nazca Plate and South American Plate.

6 **Determine Cause and Effect** Copy and fill in the graphic organizer below to list the cause and effects of convection currents.

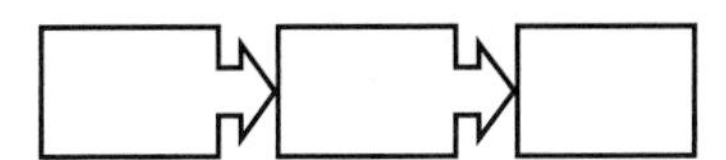

Critical Thinking

7 **Explain** why earthquakes occur at greater depths along convergent plate boundaries.

Math Skills  Math Practice

8 Two plates in the South Pacific separate at an average rate of 15 cm/y. How far will they have separated after 5,000 years?

Hutchings Photography/Digital Light Source

Movement of Plate Boundaries

Materials

graham crackers

waxed paper (four 10×10-cm squares)

dropper

frosting

plastic spoon

Safety

Earth's surface is broken into 12 major tectonic plates. Plates may collide and crumple or fold to make mountains. One plate may subduct under another, forming volcanoes. They may move apart and form a mid-ocean ridge, or they may slide past each other causing earthquakes. This investigation models plate movements.

Question

What happens where two plates come together?

Procedure

Part I

1. Read and complete a lab safety form.
2. Obtain the materials from your teacher.
3. Break a graham cracker along the perforation line into two pieces.
4. Lay the pieces side by side on a piece of waxed paper.
5. Slide crackers in opposite directions so that the edges of the crackers rub together.

Part II

6. Place two new graham crackers side by side but not touching.
7. In the space between the crackers, add several drops of water.
8. Slide the crackers toward each other and observe what happens.

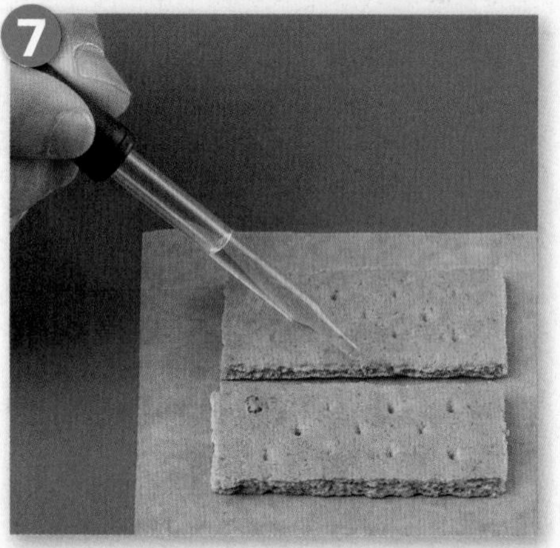

Part III

9. Place a spoonful of frosting on the waxed-paper square.
10. Place two graham crackers on top of the frosting so that they touch.
11. Push the crackers down and spread them apart in one motion.

Analyze and Conclude

12. Analyze the movement of the crackers in each of your models.

Part I

13. What type of plate boundary do the graham crackers in this model represent?
14. What do the crumbs in the model represent?
15. Did you feel or hear anything when the crackers moved past each other? Explain.
16. How does this model simulate an earthquake?

Part II

17. What does the water in this model represent?
18. What type of plate boundary do the graham crackers in this model represent?
19. Why didn't one graham cracker slide beneath the other in this model?

Part III

20. What type of plate boundary do the graham crackers in this model represent?
21. What does the frosting represent?
22. What shape does the frosting create when the crackers move?
23. What is the formation formed from the crackers and frosting?

Communicate Your Results

Create a flip book of one of the boundaries to show a classmate who was absent. Show how each boundary plate moves and the results of those movements.

Extension

Place a graham cracker and a piece of cardboard side by side. Slide the two pieces toward each other. What type of plate boundary does this model represent? How is this model different from the three that you observed in the lab?

Lab Tips

- ☑ Use fresh graham crackers.
- ☑ Slightly heat frosting to make it more fluid for experiments.

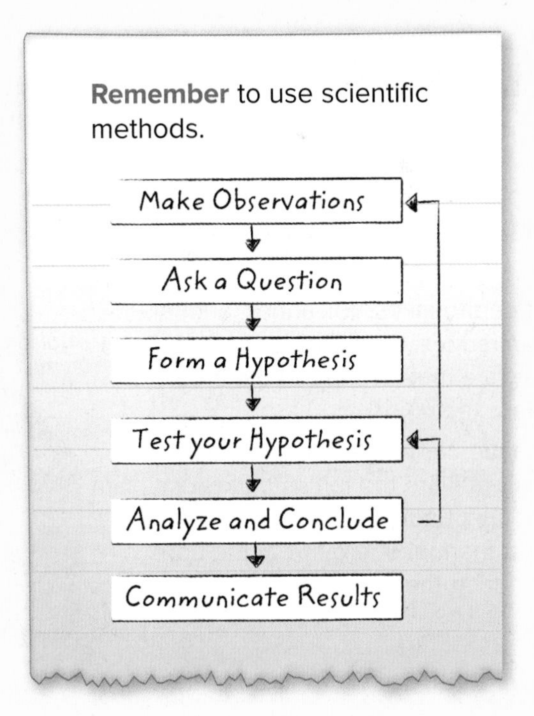

Chapter 7 Study Guide

The theory of plate tectonics states that Earth's lithosphere is broken up into rigid plates that move over Earth's surface.

Key Concepts Summary

Lesson 1: The Continental Drift Hypothesis

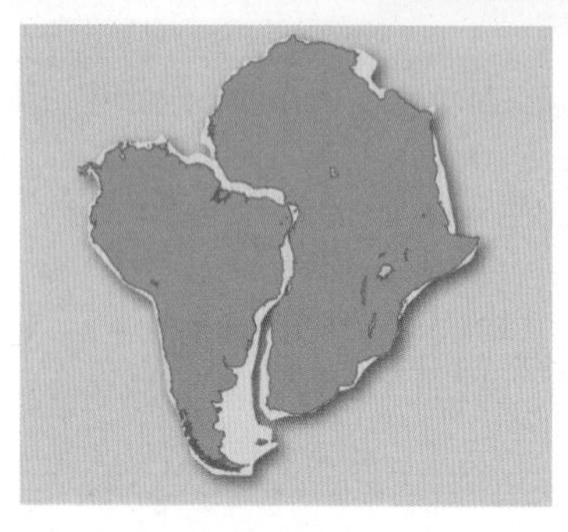

- The puzzle piece fit of continents, fossil evidence, climate, rocks, and mountain ranges supports the hypothesis of **continental drift.**
- Scientists were skeptical of continental drift because Wegener could not explain the mechanism for movement.

Vocabulary

Pangaea p. 217
continental drift p. 217

Lesson 2: Development of a Theory

- **Seafloor spreading** provides a mechanism for continental drift.
- Seafloor spreading occurs at **mid-ocean ridges.**
- Evidence of **magnetic reversal** in rock, thermal energy trends, and the discovery of seafloor spreading all contributed to the development of the theory of plate tectonics.

Vocabulary

mid-ocean ridge p. 225
seafloor spreading p. 226
normal polarity p. 228
magnetic reversal p. 228
reversed polarity p. 228

Lesson 3: The Theory of Plate Tectonics

- Types of plate boundaries, the location of earthquakes, volcanoes, and mountain ranges, and satellite measurement of plate motion support the theory of **plate tectonics.**
- Mantle **convection, ridge push,** and **slab pull** are the forces that cause plate motion. Radioactivity in the mantle and thermal energy from the core produce the energy for convection.

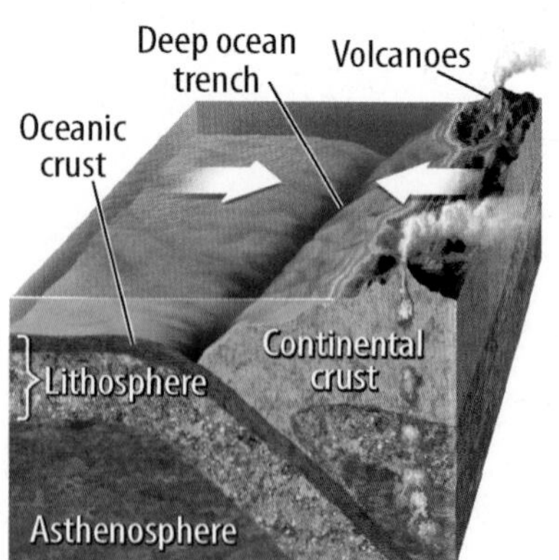

Vocabulary

plate tectonics p. 233
lithosphere p. 234
divergent plate boundary p. 235
transform plate boundary p. 235
convergent plate boundary p. 235
subduction p. 235
convection p. 238
ridge push p. 239
slab pull p. 239

Personal Tutor

Vocabulary eFlashcards
Vocabulary eGames

FOLDABLES® Chapter Project

Assemble your lesson Foldables as shown to make a Chapter Project. Use the project to review what you have learned in this chapter.

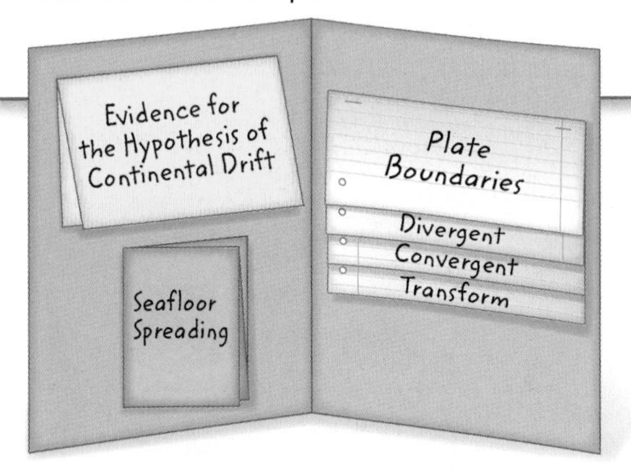

Use Vocabulary

1. The process in which hot mantle rises and cold mantle sinks is called ________.
2. What is the plate tectonics theory?
3. What was Pangaea?
4. Identify the three types of plate boundaries and the relative motion associated with each type.
5. Magnetic reversals occur when ________.
6. Explain seafloor spreading in your own words.

Interactive Concept Map

Link Vocabulary and Key Concepts

Copy this concept map, and then use vocabulary terms from the previous page to complete the concept map.

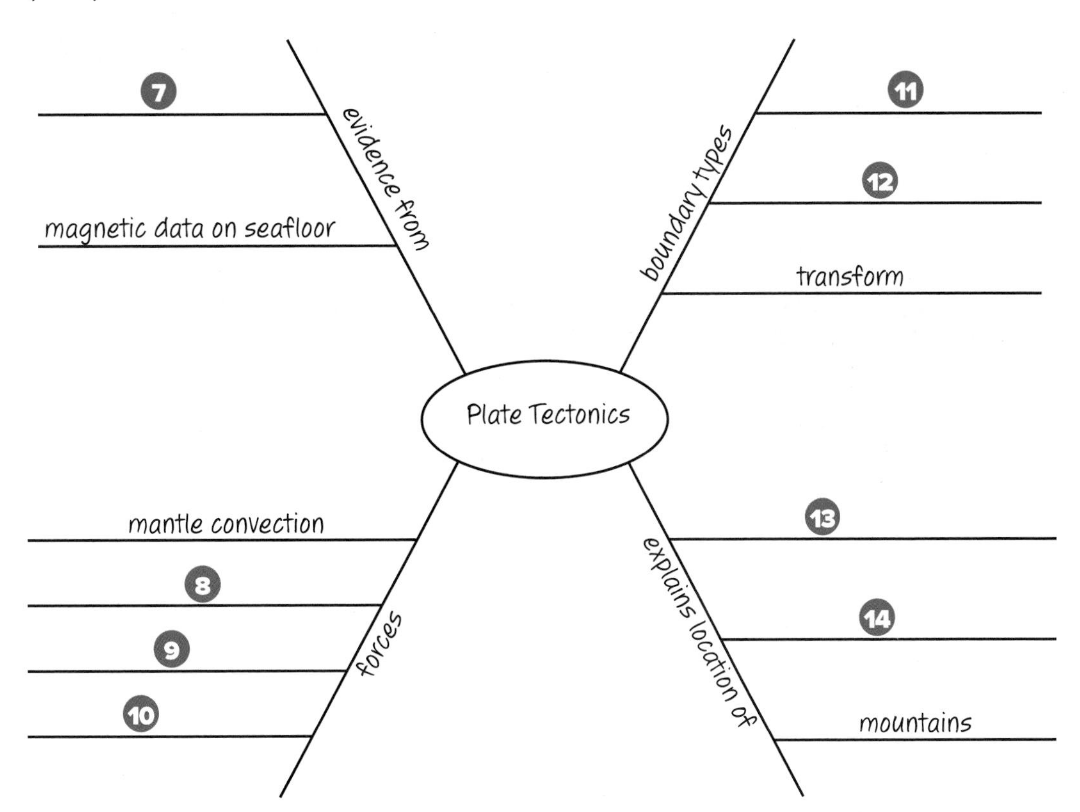

Chapter 7 Review

Understand Key Concepts 

1. Alfred Wegener proposed the _________ hypothesis.
   A. continental drift
   B. plate tectonics
   C. ridge push
   D. seafloor spreading

2. Ocean crust is
   A. made from submerged continents.
   B. magnetically produced crust.
   C. produced at the mid-ocean ridge.
   D. produced at all plate boundaries.

3. What technologies did scientists NOT use to develop the theory of seafloor spreading?
   A. echo-sounding measurements
   B. GPS (global positioning system)
   C. magnetometer measurements
   D. seafloor thickness measurements

4. The picture below shows Pangaea's position on Earth approximately 280 million years ago. Where did geologists discover glacial features associated with a cooler climate?
   A. Antarctica
   B. Asia
   C. North America
   D. South America

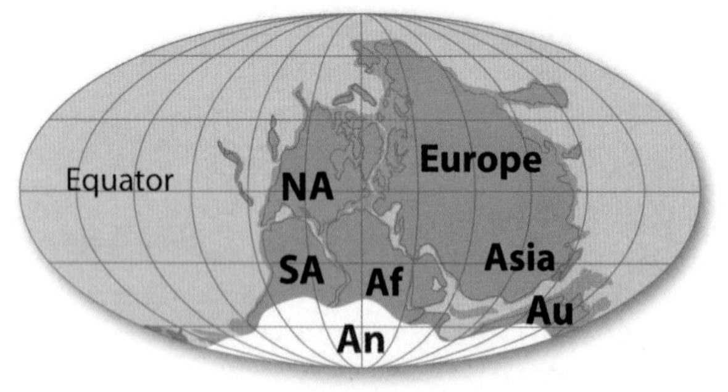

Pangaea

5. Mid-ocean ridges are associated with
   A. convergent plate boundaries.
   B. divergent plate boundaries.
   C. hot spots.
   D. transform plate boundaries.

6. Two plates of equal density form mountain ranges along
   A. continent-to-continent convergent boundaries.
   B. ocean-to-continent convergent boundaries.
   C. divergent boundaries.
   D. transform boundaries.

7. Which type of plate boundary is shown in the figure below?
   A. convergent boundary
   B. divergent boundary
   C. subduction zone
   D. transform boundary

8. What happens to Earth's magnetic field over time?
   A. It changes polarity.
   B. It continually strengthens.
   C. It stays the same.
   D. It weakens and eventually disappears.

9. Which of Earth's outermost layers includes the crust and the upper mantle?
   A. asthenosphere
   B. lithosphere
   C. mantle
   D. outer core

Critical Thinking

10. **Evaluate** The oldest seafloor in the Atlantic Ocean is located closest to the edge of continents, as shown in the image below. Explain how this age can be used to figure out when North America first began to separate from Europe.

11. **Examine** the evidence used to develop the theory of plate tectonics. How has new technology strengthened the theory?

12. **Explain** Sediments deposited by glaciers in Africa are surprising because Africa is now warm. How does the hypothesis of continental drift explain these deposits?

13. **Draw** a diagram to show subduction of an oceanic plate beneath a continental plate along a convergent plate boundary. Explain why volcanoes form along this type of plate boundary.

14. **Infer** Warm peanut butter is easier to spread than cold peanut butter. How does knowing this help you understand why the mantle is able to deform in a plastic manner?

Writing in Science

15. **Predict** If continents continue to move in the same direction over the next 200 million years, how might the appearance of landmasses change? Write a paragraph to explain the possible positions of landmasses in the future. Based on your understanding of the plate tectonic theory, is it possible that new supercontinents will form in the future?

REVIEW THE BIG IDEA

16. What is the theory of plate tectonics? Distinguish between continental drift, seafloor spreading, and plate tectonics. What evidence was used to support the theory of plate tectonics?

17. Use the image below to interpret how the theory of plate tectonics helps to explain the formation of huge mountains like the Himalayas.

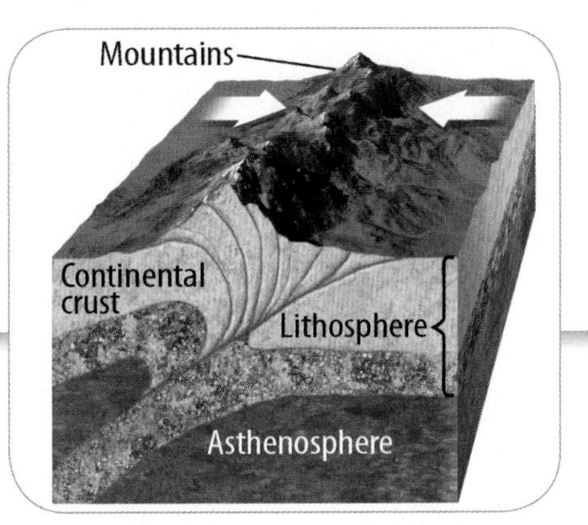

Math Skills | Math Practice

Use Proportions

18. Mountains on a convergent plate boundary may grow at a rate of 3 mm/y. How long would it take a mountain to grow to a height of 3,000 m? (1 m = 1,000 mm)

19. The North American Plate and the Pacific Plate have been sliding horizontally past each other along the San Andreas fault zone for about 10 million years. The plates move at an average rate of about 5 cm/y.

    a. How far have the plates traveled, assuming a constant rate, during this time?

    b. How far has the plate traveled in kilometers? (1 km = 100,000 cm)

Standardized Test Practice

Record your answers on the answer sheet provided by your teacher or on a sheet of paper.

Multiple Choice

Use the diagram below to answer questions 1 and 2.

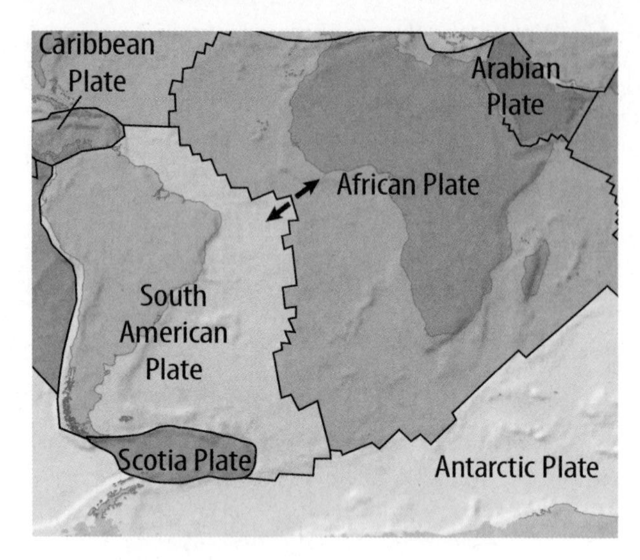

1 In the diagram above, what does the irregular line between tectonic plates represent?

A abyssal plain

B island chain

C mid-ocean ridge

D polar axis

2 What do the arrows indicate?

A magnetic polarity

B ocean flow

C plate movement

D volcanic eruption

3 What evidence helped to support the theory of seafloor spreading?

A magnetic equality

B magnetic interference

C magnetic north

D magnetic polarity

4 Which plate tectonic process creates a deep ocean trench?

A conduction

B deduction

C induction

D subduction

5 What causes plate motion?

A convection in Earth's mantle

B currents in Earth's oceans

C reversal of Earth's polarity

D rotation on Earth's axis

6 New oceanic crust forms and old oceanic crust moves away from a mid-ocean ridge during

A continental drift.

B magnetic reversal.

C normal polarity.

D seafloor spreading.

Use the diagram below to answer question 7.

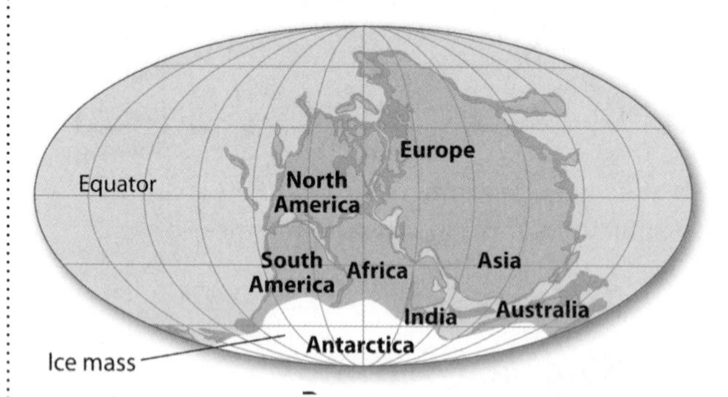

7 What is the name of Alfred Wegener's ancient supercontinent pictured in the diagram above?

A Caledonia

B continental drift

C *Glossopteris*

D Pangaea

Use the diagram below to answer question 8.

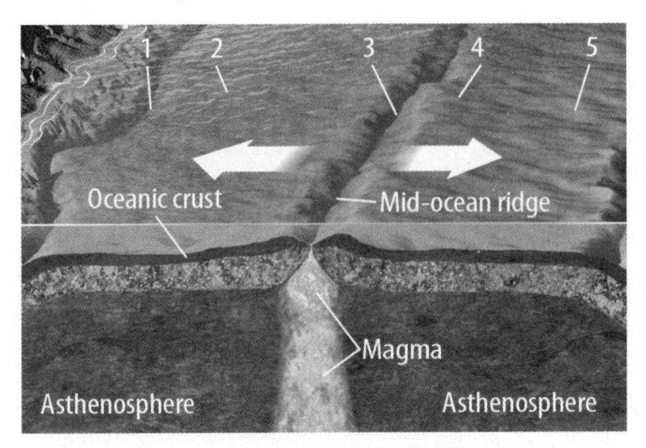

8 The numbers in the diagram represent seafloor rock. Which represent the oldest rock?

A 1 and 5

B 2 and 4

C 3 and 4

D 4 and 5

9 Which part of the seafloor contains the thickest sediment layer?

A abyssal plain

B deposition band

C mid-ocean ridge

D tectonic zone

10 What type of rock forms when lava cools and crystallizes on the seafloor?

A a fossil

B a glacier

C basalt

D magma

Constructed Response

Use the table below to answer questions 11 and 12.

Plate Boundary	Location

11 In the table above, identify the three types of plate boundaries. Then describe a real-world location for each type.

12 Create a diagram to show plate motion along one type of plate boundary. Label the diagram and draw arrows to indicate the direction of plate motion.

13 Identify and explain all the evidence that Wegener used to help support his continental drift hypothesis.

14 Why was continental drift so controversial during Alfred Wegener's time? What explanation was necessary to support his hypothesis?

15 How did scientists prove the theory of seafloor spreading?

16 If new oceanic crust constantly forms along mid-ocean ridges, why isn't Earth's total surface area increasing?

NEED EXTRA HELP?																
If You Missed Question...	1	2	3	4	5	6	7	8	9	10	11	12	13	14	15	16
Go to Lesson...	3	3	2	3	3	2	1	2	2	2	3	3	1	1	2	3

Chapter 8

Earth Dynamics

THE BIG IDEA **How is Earth's surface shaped by plate motion?**

Inquiry Why is Mount Everest different?

You might think that seashells are only found near oceans. But some of the rocks in Mount Everest contain seashells from the ocean floor!

- How do you think seashells got to the top of Mount Everest?
- What are the different ways tectonic plates move to make mountains such as these—or deep-sea trenches, valleys, and plateaus?
- How is Earth's surface shaped by plate motion?

©Imageshop/Alamy

Get Ready to Read

What do you think?

Before you read, decide if you agree or disagree with each of these statements. As you read this chapter, see if you change your mind about any of the statements.

1. Forces created by plate motion are small and do not deform or break rocks.
2. Plate motion causes only horizontal motion of continents.
3. New landforms are created only at plate boundaries.
4. The tallest and deepest landforms are created at plate boundaries.
5. Metamorphic rocks formed deep below Earth's surface sometimes can be located near the tops of mountains.
6. Mountain ranges can form over long periods of time through repeated collisions between plates.
7. The centers of continents are flat and old.
8. Continents are continually shrinking because of erosion.

Mc Graw Hill Education connectED

Your one-stop online resource

connectED.mcgraw-hill.com

LearnSmart®

Project-Based Learning Activities

Chapter Resources Files, Reading Essentials, Get Ready to Read, Quick Vocabulary

Lab Manuals, Safety Videos, Virtual Labs & Other Tools

Vocabulary, Multilingual eGlossary, Vocab eGames, Vocab eFlashcards

Animations, Videos, Interactive Tables

Personal Tutors

Self-checks, Quizzes, Tests

Lesson 1

Reading Guide

Key Concepts
ESSENTIAL QUESTIONS

- How do continents move?
- What forces can change rocks?
- How does plate motion affect the rock cycle?

Vocabulary

 Multilingual eGlossary

 Science Video

Forces That Shape Earth

Inquiry Can rocks talk?

This campsite in Thingvellir, Iceland, can tell a story about Earth if you ask the right questions. Why is this cliff next to a flat, grassy valley? How did it get like this? Has it always been this way? You can find some answers by looking at the forces that shape Earth.

Rex A. Stucky/Getty Images

Launch Lab

10 minutes

Do rocks bend?

As Earth's continents move, rocks get smashed between them and bend or break. Land can take on different shapes, depending on the temperature and composition of the rocks and the size and direction of the force.

1. Read and complete a lab safety form.
2. Spread out a **paper towel** on your work area, and place an unwrapped **candy bar** on the paper towel.
3. Gently pull on the edges of your candy bar. Observe any changes to the candy bar. Draw your observations in your Science Journal.
4. Reassemble your candy bar and gently squeeze the two ends of your candy bar together. Draw your observations.

Think About This

1. How are the results of pulling and pushing different?
2. What would be different if the candy bar were warm? What if it were cold?
3. **Key Concept** What kinds of forces do you think can change rocks?

Plate Motion

How far is your school from the nearest large mountain? If you live in the west or along the east coast of the United States, you are probably close to mountains. In contrast, the central region of the United States is flat. Why are these regions so different?

The Rocky Mountains in the west are high and have sharp peaks, but the Appalachian Mountains in the east are lower and gently rounded, as shown in **Figure 1**.

Reading Check How are the Rocky Mountains different from the Appalachian Mountains?

Mountains do not last forever. Weathering and erosion gradually wear them down. The Appalachian Mountains are shorter and smoother than the Rocky Mountains because they are older. They formed hundreds of millions of years ago. The Rockies formed just 50 to 100 million years ago.

Mountain ranges are produced by plate tectonics. The theory of plate tectonics states that Earth's surface is broken into rigid plates that move horizontally on Earth's more fluid upper mantle. Mountains, valleys, and other features form where plates collide, move away from each other, or slide past each other.

Rocky Mountains

Appalachian Mountains

Figure 1 The younger Rocky Mountains are high and have sharp peaks. The older Appalachian Mountains are low and gently rounded.

Figure 2 The massive lower portion of an iceberg is under water. Similarly, the root of a mountain extends deep into the mantle.

WORD ORIGIN

isostasy
from Greek *iso,* means "equal"; and Greek *stasy,* means "standing"

Vertical Motion

To understand how massive pieces of Earth can rise vertically and form mountainous regions, you need to understand the forces that produce vertical motion.

Balance in the Mantle

Think of an iceberg floating in water. The iceberg floats with its top above the water, but most of it is under the surface of the water, as shown in **Figure 2.** It floats this way because ice is less dense than water and because the mass of the ice equals the mass of the water it displaces, or pushes out of the way.

Similarly, continents rise above the seafloor because continental crust is made of rocks that are less dense than Earth's mantle. Continental crust displaces some of the mantle below it until an equilibrium, or balance, is reached. **Isostasy** (i SAHS tuh see) *is the equilibrium between continental crust and the denser mantle below it.* A continent floats on top of the mantle because the mass of the continent is equal to the mass of the mantle it displaces. Mountains act the same way on a smaller scale.

Reading Check What is isostasy?

Continental crust changes over time due to plate tectonics and erosion. If a part of the continental crust becomes thicker, it sinks deeper into the mantle, as shown in **Figure 3.** But it also rises higher until a balance is reached. This is why mountains are taller than the continental crust around them. Although the mountain is massive, it is still less dense than the mantle, so it "floats." Below Earth's surface, the mountain extends deep into the mantle. Above Earth's surface, the mountain rises above the surrounding continental crust. As illustrated in **Figure 3,** as a mountain erodes, the continental crust rises.

Maintaining Balance

Personal Tutor

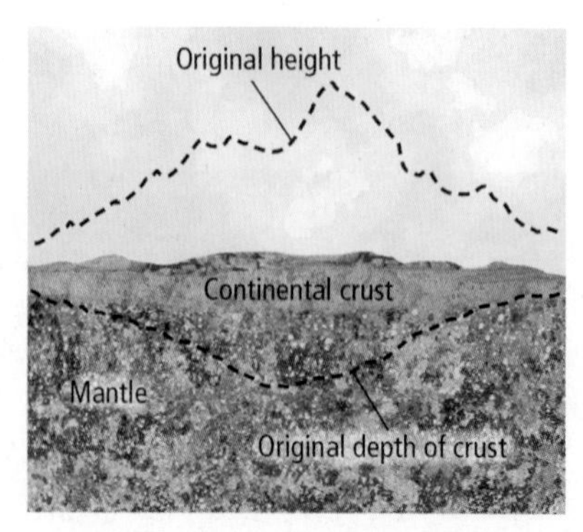

Figure 3 Over time, erosion and weathering remove the top of a mountain. To maintain isostasy, continents move up or down until the mass of the continent equals the mass of mantle it displaces.

Subsidence and Uplift

Much of North America was covered by glaciers more than 1 km thick 20,000 years ago. The weight of the ice pushed the crust downward into the mantle, as shown in **Figure 4.** *The downward vertical motion of Earth's surface is called* **subsidence.** When the ice melted and the water ran off, the isostatic balance was upset again. In response, the crust moved upward. *The upward vertical motion of Earth's surface is called* **uplift.** In the center of Hudson Bay in Canada, the land surface is still rising 1 cm each year as it moves toward isostatic balance.

Key Concept Check What can cause Earth's surface to move up or down?

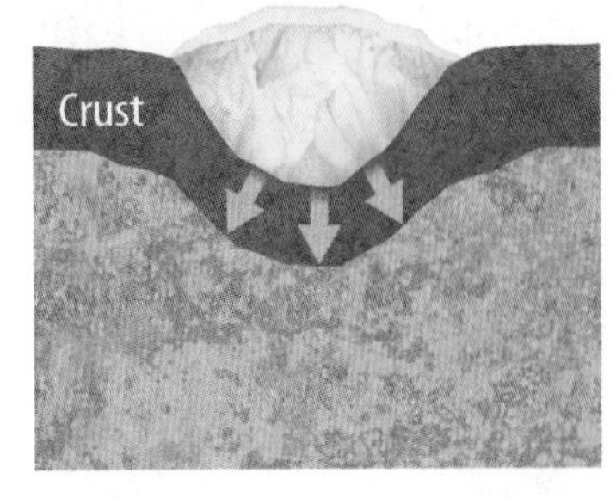

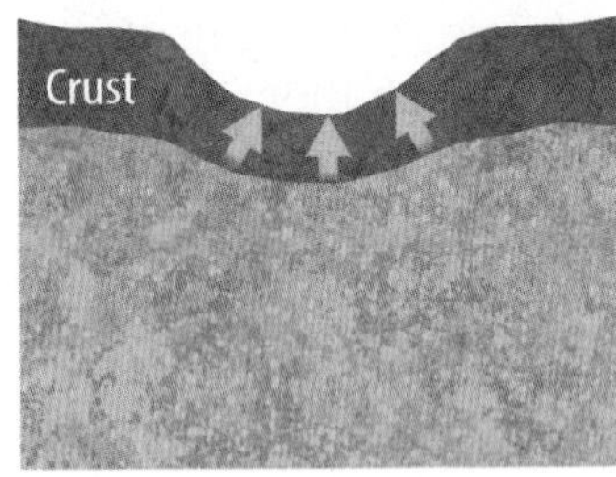

Figure 4 The weight of a glacier pushes down on the land. When the glacier melts, the land rises until isostasy is restored.

Horizontal Motion

Find a small rock and squeeze it. You've just applied force to the rock. Did its shape change? Did it break? Horizontal motion at plate boundaries applies much greater forces to rocks. Forces at plate boundaries are strong enough to break rocks or change the shape of rocks. The same forces also can form mountains.

Types of Stress

Stress is the force acting on a surface. There are three types of stress, as illustrated in **Figure 5.** *Squeezing stress is* **compression.** *Stress that pulls something apart is* **tension.** *Parallel forces acting in opposite directions are* **shear.** These are all stresses that can change rock as plates move horizontally.

FOLDABLES®

Use a sheet of paper to make a three-tab book. Label the tabs as illustrated, and describe the forces that shape Earth.

Stresses

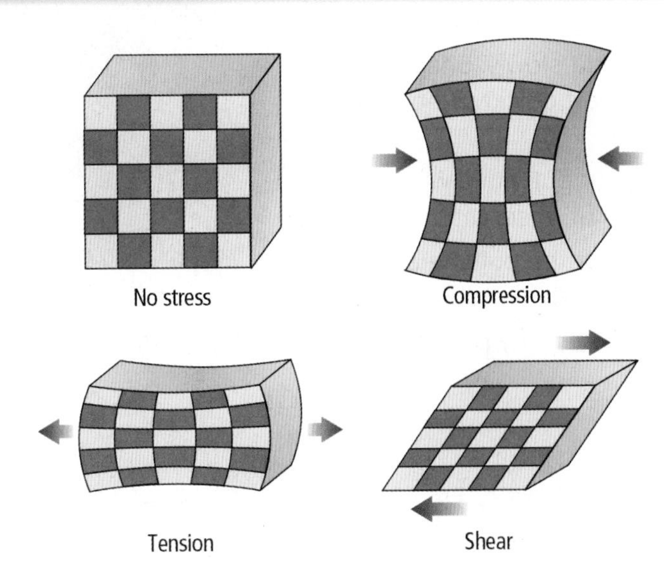

Figure 5 Compression, tension, and shear stresses cause rocks to change shape.

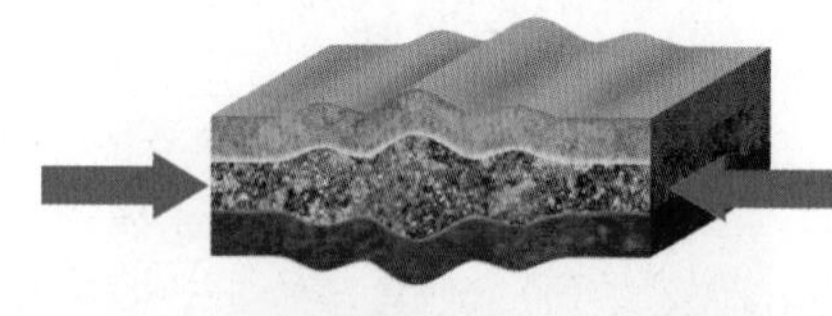

Figure 6 Compression can fold rocks. Tension can stretch them. Whether rocks go back to their original shape depends on the type of strain.

Visual Check Which of these two illustrations shows tension?

Types of Strain

Rocks can change when stress acts on them. *A change in the shape of rock caused by stress is called* **strain.** There are two main types of strain.

Elastic strain does not permanently change, or deform, rocks. When stress is removed, rocks return to their original shapes. Elastic strain occurs when stresses are small or rocks are very strong. **Plastic** strain creates a permanent change in shape. Even if the stress is removed, the rocks do not go back to their original shapes. Plastic strain occurs when rocks are weak or hot.

Reading Check Which type of strain permanently changes rocks?

SCIENCE USE V. COMMON USE

plastic
Science Use capable of being molded

Common Use a commonly used synthetic material

Deformation in the Crust

In the hotter lower crust and upper mantle, rocks tend to deform plastically like putty. As illustrated in **Figure 6,** compression thickens and folds layers of rock. Tension stretches and thins layers of rock. In the colder, upper part of the crust, rocks can break before they deform plastically. When strain breaks rocks rather than just changing their shape, it is called failure. When rocks fail, fractures–or faults–form.

Key Concept Check What can cause rocks to thicken or fold?

MiniLab

10 minutes

What will happen?

If enough force is put on a rock, it will begin to strain, or change shape. Depending on the nature of the force and the rock, sometimes the rock will bend and sometimes it will break.

1. Read and complete a lab safety form.
2. Knead a piece of **putty,** and pull it apart slowly. Shape the putty into an oval ball. Try to pull it apart quickly. Record your observations in your Science Journal.
3. Shape your putty into an oval. Put your putty in a warm water bath for 2 min. Pull it apart. Record your observations.
4. Shape the putty into an oval shape. Put the putty in an **ice water** bath for 2 min. Pull it apart. Record your observations.
5. Try to break your putty by pulling on it and by pushing on it. Record your observations in your Science Journal.

Analyze and Conclude

1. **Summarize** the effects of rate of strain, temperature, and type of stress on the putty.
2.  **Key Concept** Relate your experience with your putty model to the forces that can change rocks and to the conditions in Earth that will cause them to change.

Plate Tectonics and the Rock Cycle

Although it might seem as if rocks are always the same, rocks are moving around–usually very slowly. Rocks never stop moving through the rock cycle, as illustrated in **Figure 7.** The theory of plate tectonics combined with uplift and subsidence explain why there is a rock cycle on Earth.

The forces that cause plate tectonics produce horizontal motion. Isostasy results in vertical motion within continents. Together, plate motion, uplift, and subsidence keep rocks moving through the rock cycle.

Uplift brings metamorphic and igneous rocks from deep in the crust up to the surface. At the surface, erosion breaks down rocks into sediment. Sediment gets buried by still more sediment. Buried sediment becomes sedimentary rocks. Pressure and temperature increase as rocks are buried, and eventually sedimentary rocks become metamorphic rocks. Subduction takes all types of rocks deep into Earth, where they can create new igneous or metamorphic rocks.

Key Concept Check How does plate motion affect the rock cycle?

The Rock Cycle

Figure 7 Horizontal tectonic motion and vertical motion by uplift and subsidence help move rocks through the rock cycle.

Visual Check What happens to eroded sediment?

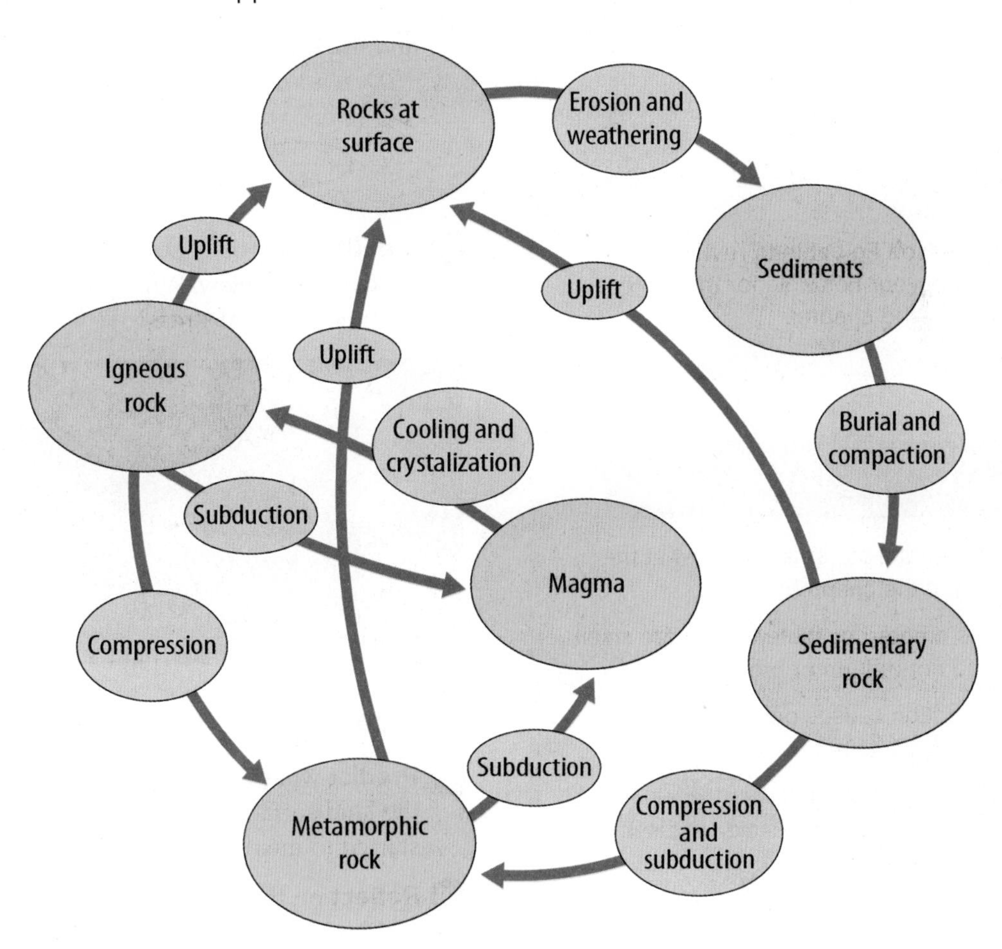

Lesson 1 Review

Online Quiz

Virtual Lab

Visual Summary

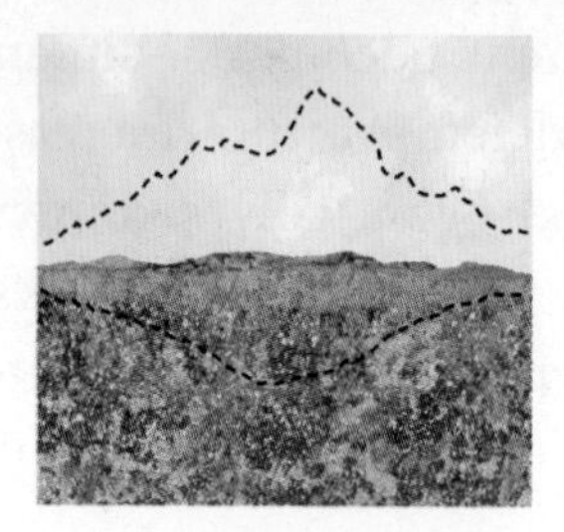

As a mountain is eroded away, the continent will rise until isostatic balance is restored.

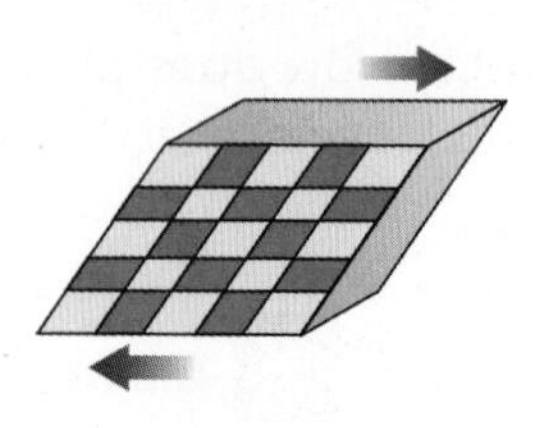

Different types of stress change rocks in different ways.

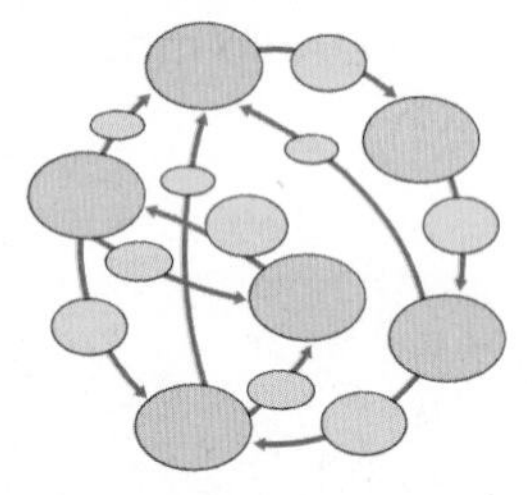

Horizontal and vertical motions are part of what keep rocks moving through the rock cycle.

FOLDABLES®

Use your lesson Foldable to review the lesson. Save your Foldable for the project at the end of the chapter.

What do you think NOW?

You first read the statements below at the beginning of the chapter.

1. Forces created by plate motion are small and do not deform or break rocks.
2. Plate motion causes only horizontal motion of continents.

Did you change your mind about whether you agree or disagree with the statements? Rewrite any false statements to make them true.

Use Vocabulary

1. The opposite of uplift is ________.
2. The balance between the crust and the mantle below it is ________.
3. **Explain** the difference between the two types of strain.

Understand Key Concepts 

4. **Describe** what happens to the elevation of the land surface when crust thickens.
5. **Name** one result of rock failure.
6. Which type of deformation is produced by compression of plastic crust?
   A. failure
   B. faults
   C. folds
   D. shear

Interpret Graphics

7. **List** Copy the graphic organizer below, and use it to show three types of stress.

8. **Identify** the type of stress that deformed the rocks shown below. What kind of strain resulted from the stress?

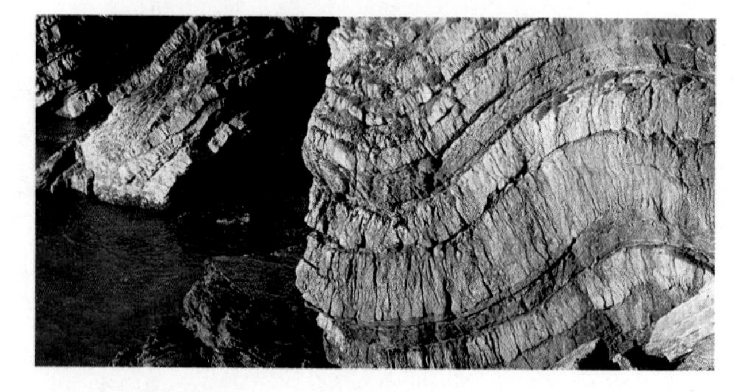

Critical Thinking

9. **Predict** what would happen to the height of the land surface of Antarctica if the ice sheet started to melt.
10. **Reflect** on the relationship between vertical and horizontal motion. How are they related? When are they not related?

Skill Practice Design an Experiment

40 minutes

Materials

assorted weights

cardboard

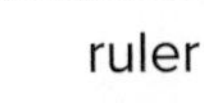

ruler

stopwatch

putty

Safety

Can you measure how stress deforms putty?

Scientists who study rocks study how the stress, or force applied to a rock, causes the rock to change shape, or deform. Because rocks are hard and they only deform with enormous forces at slow speeds, special equipment is needed to study rocks. In this lab, you will apply forces to putty and take measurements similar to ones scientists take on rocks.

Learn It

A scientist makes many decisions before beginning an investigation. Some decisions involve figuring out how to make the needed measurements. In this lab, you will **design** instruments and an **experiment** to measure stress and deformation of putty.

Try It

1. Read and complete a lab safety form.
2. Determine how you will use the materials provided to measure stress, deformation, and the time it takes to deform the putty.
3. Write a procedure in your Science Journal, and ask your teacher to approve your plan.
4. Test your procedures. Modify them if necessary.
5. Collect your data, and record them in a data table such as the one shown below.

Apply It

6. **Summarize** the relationship between stress and rate of deformation.
7. **Key Concept** Relate the deformation of putty to how forces change rocks.

	Stress Applied	Measured Deformation	Measured Time	Rate of Deformation (deformation / time)
Trial 1				
Trial 2				

(t to b, 2, 4, r)Hutchings Photography/Digital Light Source, (3)Michael Scott/McGraw-Hill Education; (5)McGraw-Hill Education

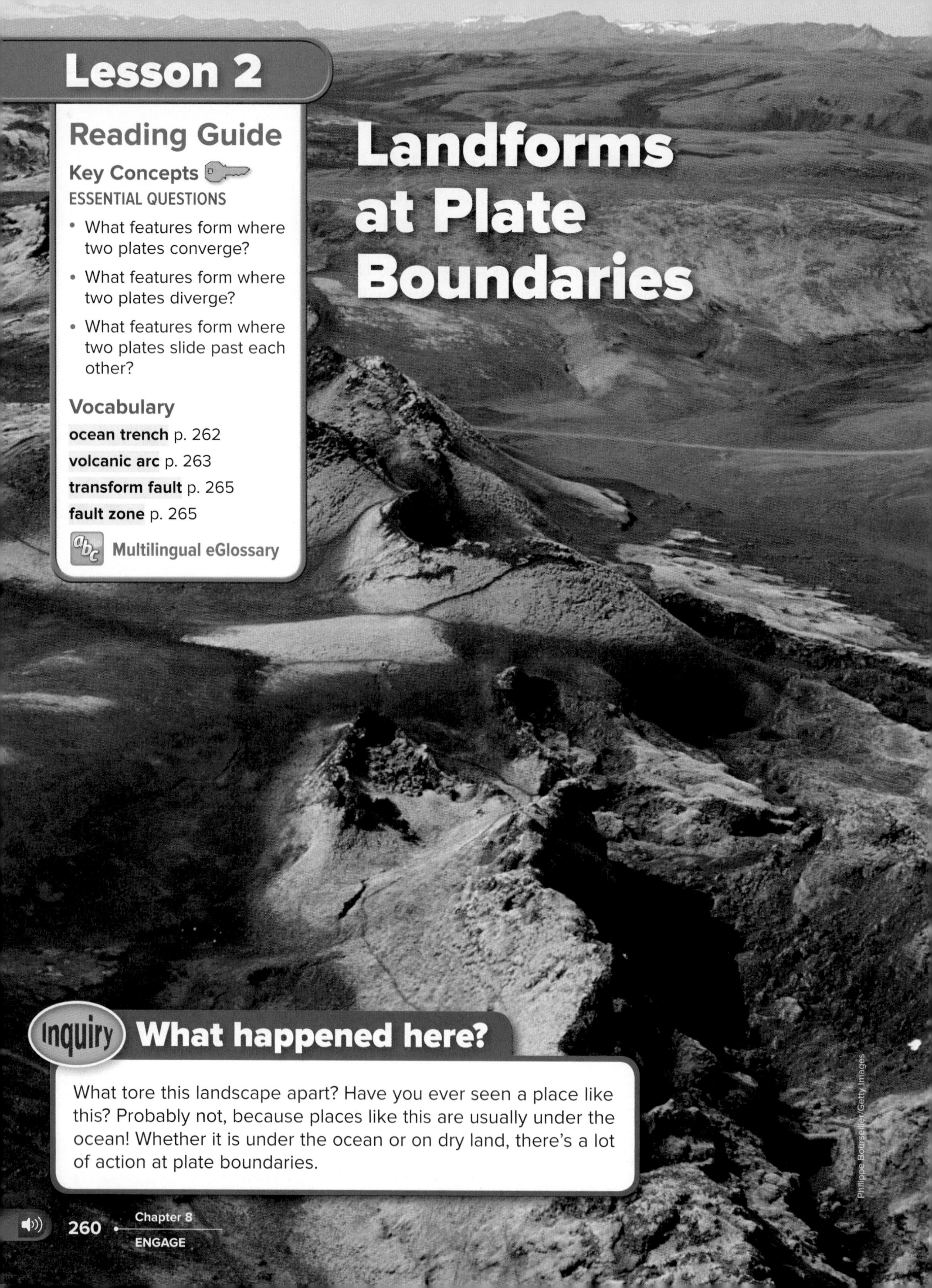

Phillippe Bourseiller/Getty Images

Lesson 2

Reading Guide

Key Concepts

ESSENTIAL QUESTIONS

- What features form where two plates converge?
- What features form where two plates diverge?
- What features form where two plates slide past each other?

Vocabulary

ocean trench p. 262
volcanic arc p. 263
transform fault p. 265
fault zone p. 265

Multilingual eGlossary

Landforms at Plate Boundaries

Inquiry What happened here?

What tore this landscape apart? Have you ever seen a place like this? Probably not, because places like this are usually under the ocean! Whether it is under the ocean or on dry land, there's a lot of action at plate boundaries.

Launch Lab

10 minutes

What happens when tectonic plates collide?

As Earth's continents move, tectonic plates can come together, pull apart, or slide past each other. Each of these interactions produces different landforms.

1. Using **construction paper** and **index cards,** set up a model plate boundary. Draw your model in your Science Journal.
2. Label the two tectonic plates, the fault between them, and the mantle. Title it *Before Stress.*
3. Model tension by pulling the plates apart. Draw and label your model. Title it *Tension.*
4. Model shear by sliding one plate forward and the other backward. Draw and label your model. Title it *Shear.*
5. Model compression by pushing the plates together. Experiment until you get two different results. Draw and label both results. Title it *Compression.*

Think About This

1. What might happen if compression and shear occurred together?
2. **Key Concept** How are the features that form under the different types of stresses different? How are they similar?

Landforms Created by Plate Motion

Tectonic plates move slowly, only 1–9 cm per year. But these massive, slow-moving plates have so much force they can build tall mountains, form deep valleys, and rip Earth's surface apart.

Compression, tension, and shear stresses are at work at plate boundaries. Each type of stress produces different types of landforms. For example, the San Andreas Fault on the west coast of the United States is the result of shear stresses where plates move past each other. Tall mountains, such as the Dolomites shown in **Figure 8,** are created by compression stresses where plates collide.

Reading Check How fast do tectonic plates move?

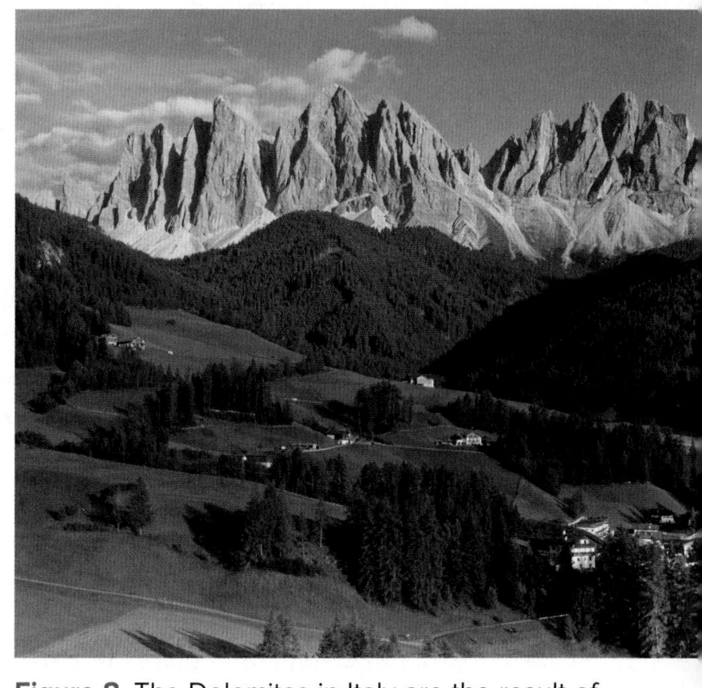

Figure 8 The Dolomites in Italy are the result of a collision long ago between landmasses that are now the continents of Europe and Africa.

Formation of the Himalayas

Figure 9 Three stages in the growth of the Himalayas are illustrated. The plates beneath India and Asia started colliding almost 50 million years ago and continue colliding today. Because the plates are still colliding, the Himalayas grow a few millimeters each year due to compression.

Visual Check Which two landforms collided?

FOLDABLES®

Use a sheet of paper to make a vertical three-tab book. As you read, describe how different features form at plate boundaries. Include specific examples of each landform.

Landforms Created by Compression

The largest landforms on Earth are produced by compression at convergent plate boundaries. The types of landforms that form depend on whether the plates are oceanic or continental.

Mountain Ranges

A collision between two continental plates can produce tall mountains. But the mountains form slowly and in stages over millions of years. The history of the Himalayas is illustrated in **Figure 9.** The Himalayas continue to grow even now as continental collision pushes them higher. Note that although the plates move horizontally, the collision causes the crust to move vertically also.

Ocean Trenches

When two plates collide, one can go under the other and be forced into the mantle in a process called subduction. As shown in **Figure 10,** a deep trench forms where the two plates meet. **Ocean trenches** *are deep, underwater troughs created by an oceanic plate subducting under another plate at a convergent plate boundary.* Ocean trenches, also called deep-sea trenches, are the deepest places in Earth's oceans.

Key Concept Check What are two landforms that can form where two plates converge?

Volcanic Arcs

Volcanic mountains can form in the ocean where oceanic plates converge and one plate subducts under another one. These volcanoes emerge as islands. *A curved line of volcanoes that forms parallel to a plate boundary is called a* **volcanic arc.** Most of the active volcanoes in the United States are part of the Aleutian volcanic arc in Alaska. There are about 40 active volcanoes there. They formed as a result of the Pacific Plate subducting under the North American Plate.

Volcanic arcs in the ocean are also called island arcs. But a volcanic arc can also form where an oceanic plate subducts under a continental plate. Because the continent is above sea level, the volcanoes sit on top of the continent, as does Mount Shasta in California, shown in **Figure 10.**

Reading Check Where do volcanic arcs form?

Figure 10 Volcanic arcs also can form on continents. Mount Shasta in California is part of the Cascade volcanic arc.

Landforms Created by Tension

Where plates move apart, tension stresses stretch Earth's crust. Distinct landforms are produced by tension.

Mid-Ocean Ridges

It might surprise you that tension stresses under the ocean can produce long mountain ranges more than 2 km tall. They form under water at divergent boundaries as oceanic plates move away from each other.

As tension stresses cause oceanic crust to spread apart, hot rock from the mantle rises. Because hot rock is less dense than cold rock, the hot mantle pushes the seafloor upward. In this way, long, high ridges are created in Earth's oceans. You might have already learned that a long, tall mountain range that forms where oceanic plates diverge is called a mid-ocean ridge. **Figure 11** shows a mid-ocean ridge near the North American west coast.

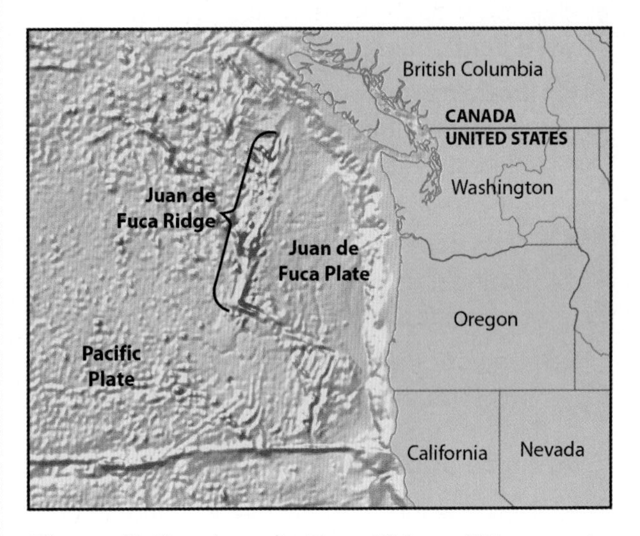

Figure 11 The Juan de Fuca Ridge off the coast of Washington and Oregon is a mid-ocean ridge.

Visual Check What direction is the Juan de Fuca Plate moving relative to the Pacific Plate?

Figure 12 A divergent boundary has created a continental rift in Africa. This rift eventually will separate Africa into two parts.

Continental Rifts

When divergent boundaries occur within a continent, they can form continental rifts, or enormous splits in Earth's crust. Tension stresses in the cold upper part of the crust create faults. At these faults, large blocks of crust move downward, creating valleys between two ridges.

The East African Rift, pictured in **Figure 12,** is an example of an active continental rift that is beginning to split the African continent into two parts. Each year, the two parts move 3-6 mm farther from each other. One day, millions of years from now, the divergent boundary will have created two separate landmasses. Water will fill the space between them.

Reading Check Where on Earth is a continental rift forming now?

The valley at this rift also is subsiding. The warm lower part of the crust acts like putty. As the crust stretches, it becomes thinner and subsides, as shown in **Figure 12.**

Key Concept Check What features form at divergent boundaries?

MiniLab

20 minutes

What is the relationship between plate motion and landforms?

As tectonic plates move, they create landforms in predictable patterns. Can you analyze the motion of the plates and predict what landforms will form?

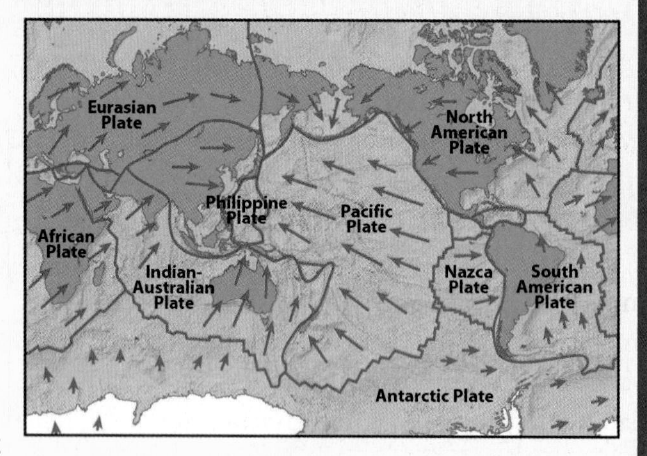

1. Examine the world map shown here indicating the movement of tectonic plates. Determine the meaning of the arrows and the lines.
2. On a **copy of the map,** label plate boundaries as divergent, convergent, or transform.
3. On your map, predict the landforms that will form at each plate boundary.

Analyze and Conclude

1. **Describe** the landform that is found around the edge of the Pacific Ocean. Hypothesize why this is called the Ring of Fire.
2. **Compare and contrast** the landforms on the east and west coasts of South America. Support your answer with data.
3. **Key Concept** Create a table that relates the type of stress (compression, tension, shear) and the location of the plate boundary (middle of a continent, edge of a continent, middle of an ocean). Fill in the table with the landforms found in each situation. List one place on Earth where each is occurring.

Transform Faults

Figure 13 On the left, a transform fault forms as crust moves past each other. On the right, the yellow line shows the mid-ocean ridge. The red lines are transform faults.

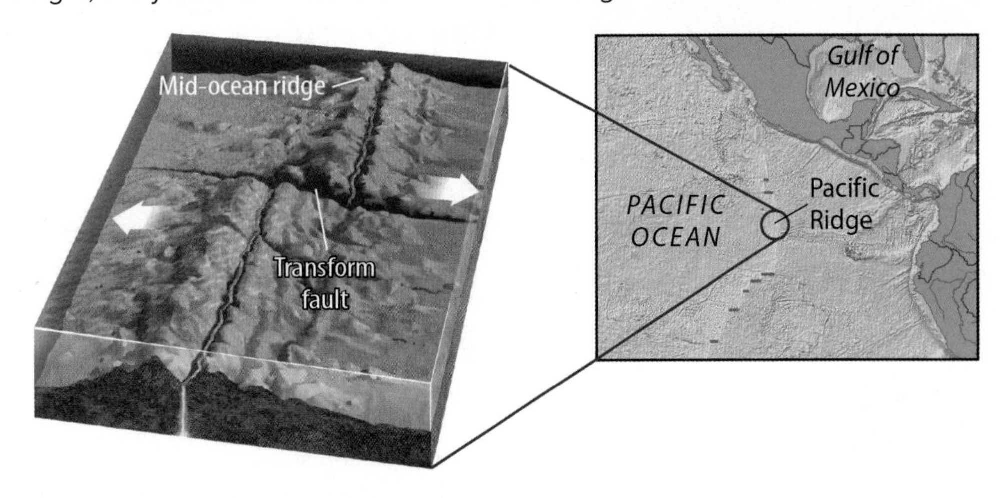

Visual Check In the figure on the right, where are the fracture zones?

Landforms Created by Shear Stresses

Recall that as plates slide horizontally past each other, shear stresses produce transform boundaries. Landforms created by shear stresses are not as obvious as landforms created by tension or compression. Transform boundaries are characterized by long faults and fault zones.

WORD ORIGIN

transform
from Latin *trans,* means "across"; and *formare,* means "to form"

Transform Faults

Where blocks of crust slide horizontally past each other, they form **transform faults.** Some transform faults form perpendicular to mid-ocean ridges, as shown on the left in **Figure 13.** Recall that tension produces mid-ocean ridges at divergent boundaries. As the plates slide away from each other, transform faults also form and can separate sections of mid-ocean ridges. The map in **Figure 13** shows the transform faults along the Pacific Ridge.

Key Concept Check What features form where plates slide past each other?

Fault Zones

Some transform faults can be seen at Earth's surface. For example, the San Andreas Fault in California is visible in many places. Although the San Andreas Fault is visible at Earth's surface, much of this fault system is underground. As shown in **Figure 14,** the San Andreas Fault is not a single fault. Many smaller faults exist in the area around the San Andreas Fault. *An area of many fractured pieces of crust along a large fault is called a* **fault zone.**

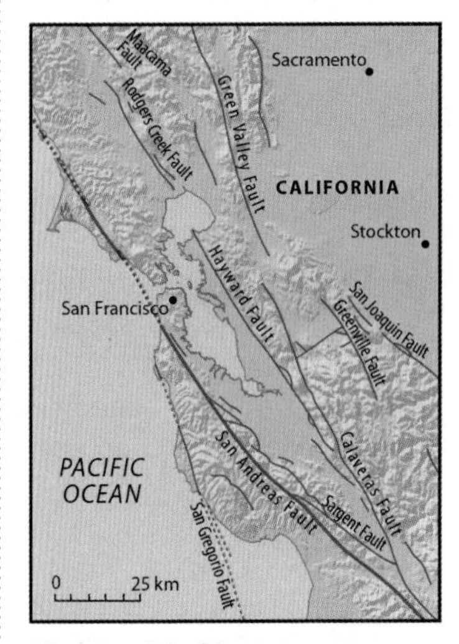

Figure 14 Shear stresses create faulting at Earth's surface. Below the surface there might be many other faults that are part of the same fault zone.

Lesson 2 Review

Visual Summary

The deepest and tallest landforms on Earth are created at plate boundaries.

Tension stresses within continents can produce enormous splits in Earth's surface.

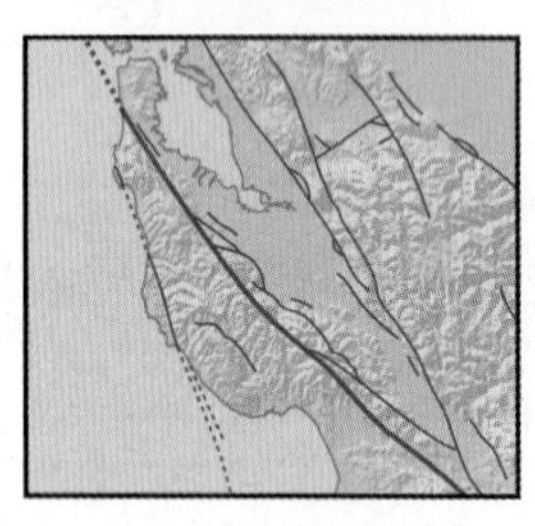

Faults at Earth's surface can be part of much larger fault zones that have many underground faults.

FOLDABLES®

Use your lesson Foldable to review the lesson. Save your Foldable for the project at the end of the chapter.

What do you think NOW?

You first read the statements below at the beginning of the chapter.

3. New landforms are created only at plate boundaries.

4. The tallest and deepest landforms are created at plate boundaries.

Did you change your mind about whether you agree or disagree with the statements? Rewrite any false statements to make them true.

Use Vocabulary

1. **Use the terms** *volcanic arc* and *trench* in a sentence.
2. **Relate** a transform fault to a fault zone.
3. **Define** *fault zone* in your own words.

Understand Key Concepts 

4. Which type of stress is currently producing the East African Rift?
   - **A.** shear stress
   - **B.** tension stress
   - **C.** compression and tension stresses
   - **D.** shear and compression stresses
5. **Compare** the development of tall mountains to how ocean trenches form.
6. **Summarize** the processes involved in the formation of a volcanic arc.

Interpret Graphics

7. **Connect** Copy the graphic organizer below. Use it to relate stresses at plate boundaries to the landforms associated with each type of boundary.

Stresses	Landforms
Compression	
Tension	
Shear	

Critical Thinking

8. **Apply** The rocks at the top of Mount Everest contain marine fossils. Describe the tectonic processes that brought sediments and fossils once buried at the bottom of the sea to the top of the tallest mountain.
9. **Relate** a continental rift to a mid-ocean ridge.
10. **Create** a diagram that illustrates how mountains form where two continents collide.

Hot Spots!

HOW NATURE WORKS

Volcanoes on a Plate

Not all volcanoes form at plate boundaries. Some, called hot spot volcanoes, pop up in the middle of a tectonic plate. A hot spot volcano forms over a rising column of magma called a mantle plume. The origin of mantle plumes is still uncertain, but evidence shows they probably rise up from the boundary between Earth's mantle and core.

As a tectonic plate passes over a mantle plume, a volcano forms above the plume. The tectonic plate continues to move, and a chain of volcanoes forms. If the volcanoes are in the ocean and if they get large enough, they become islands, such as the Hawaiian Islands. Here is how this happens:

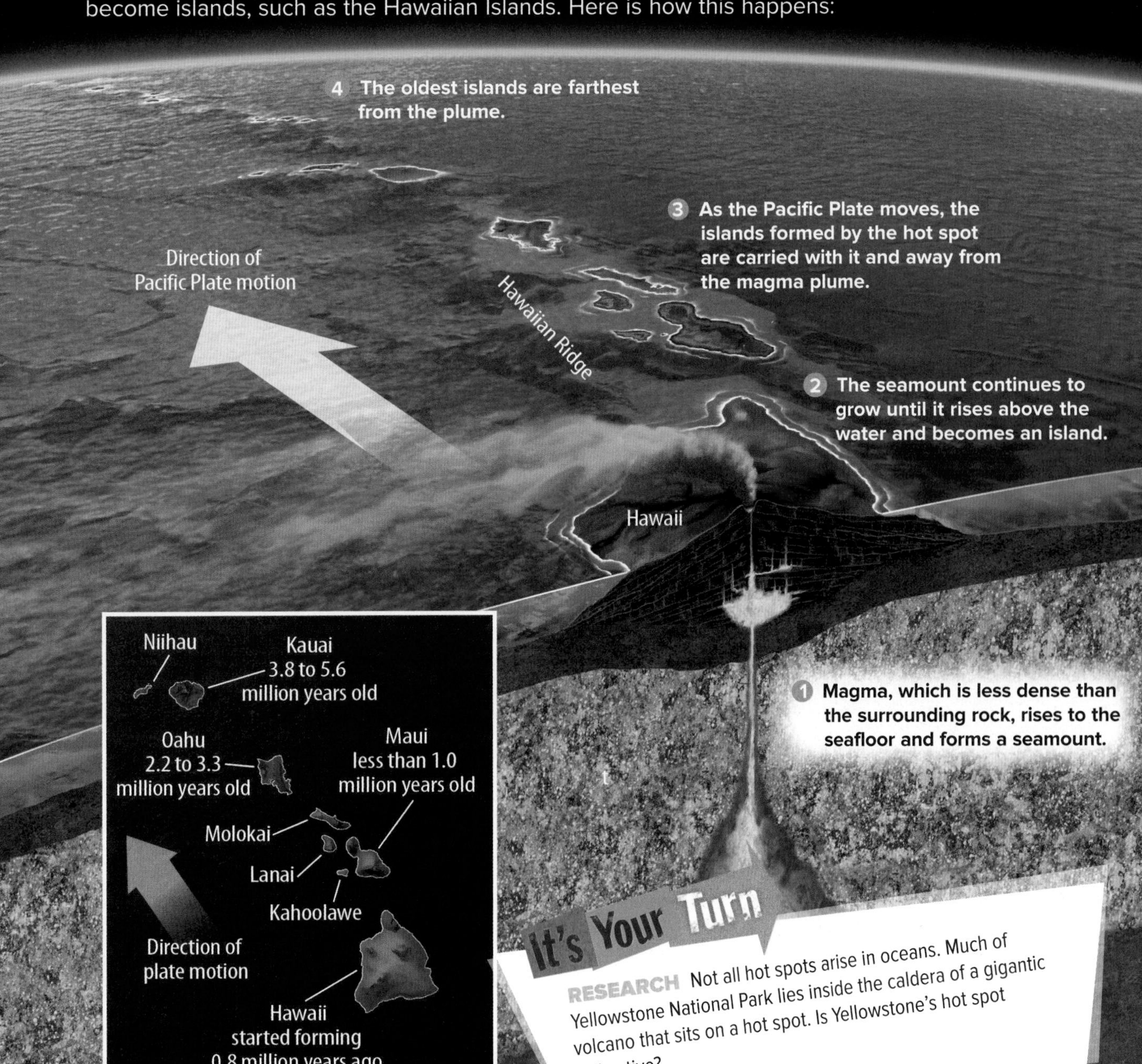

It's Your Turn

RESEARCH Not all hot spots arise in oceans. Much of Yellowstone National Park lies inside the caldera of a gigantic volcano that sits on a hot spot. Is Yellowstone's hot spot still active?

Lesson 3

Reading Guide

Key Concepts
ESSENTIAL QUESTIONS

- How do mountains change over time?
- How do different types of mountains form?

Vocabulary

folded mountain p. 271
fault-block mountain p. 272

Multilingual eGlossary

BrainPOP®

Mountain Building

Inquiry Is this a safe place to live?

Is it safe to live next to this mountain? Will lava erupt from it? Will there be earthquakes nearby? Not all mountains are the same. Once you know how a mountain formed, you can predict what is likely to happen in the future.

Launch Lab

10 minutes

What happens when Earth's tectonic plates diverge?

When tectonic plates diverge, the crust gets thinner. Sometimes large blocks of crust subside and form valleys. Blocks next to them move up and become fault-block mountains. This is how the Basin and Range Province in the western United States formed.

1. Stand 5–6 **hardbound books** on a desk with the bindings vertical.
2. Using a **ruler,** measure the width and the height of the books, as shown. Record the results in a table in your Science Journal.
3. Holding the books together, tilt them sideways at the same time to about a 30° angle. Measure the width and the height of the books. Record the results in your table.
4. Tilt the books to about a 60° angle. Measure the width and the height of the books. Record the results in your table.
5. Draw a diagram of the tilted books in your Science Journal, and label mountains, valleys, and faults.

Think About This

1. How does the thickness of the crust relate to the height of a mountain?
2. **Key Concept** How do you think fault-block mountains form?

The Mountain-Building Cycle

Mountain ranges are built slowly, and they change slowly. Because they are the result of many different plate collisions over many millions of years, they are made of many different types of rocks. The processes of weathering and erosion can remove part or all of a mountain.

Reading Check What processes can remove part of a mountain?

Converging Plates

Recall that when plates collide at a plate boundary, a combination of folds, faults, and uplift creates mountains. Eventually, after millions of years, the forces that originally caused the plates to move together can become inactive. As shown in **Figure 15,** a single new continent is created from two old ones, and the plate boundary becomes inactive. With no compression at a convergent plate boundary, the mountains stop increasing in size.

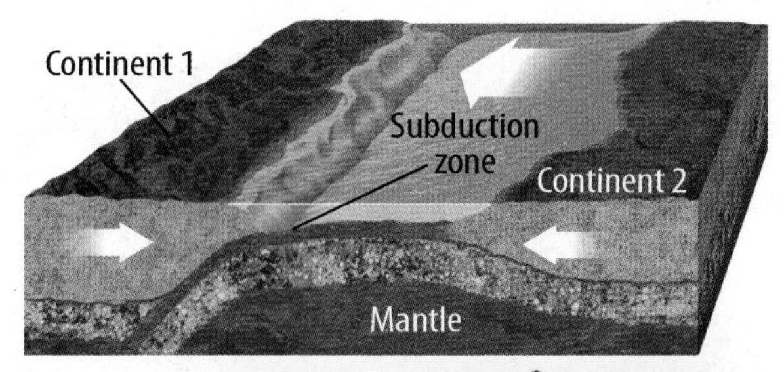

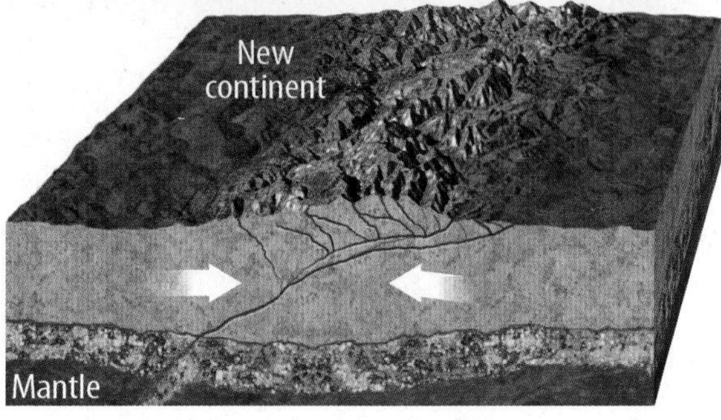

Figure 15 The forces that originally caused plates to move together eventually become inactive. A single continent is created from the two old ones, and the plate boundary becomes inactive.

FOLDABLES®

Fold a sheet of paper to make a vertical three-tab book. Label the tabs as illustrated. Describe how different types of mountains form. Identify a specific example of each type.

WORD ORIGIN

Appalachian
from the Apalachee *abalahci,* means "other side of the river"

Collisions and Rifting

Continents are continuously changing because Earth's tectonic plates are always moving. When continents split at a divergent plate boundary, they often break close to the place where they first collided. First a large split, or rift, forms. The rift grows, and seawater flows into it, forming an ocean.

Eventually plate motion changes again, and the continents collide. New mountain ranges form on top of or next to older mountain ranges. The cycle of repeated collisions and rifting can create old and complicated mountain ranges, such as the **Appalachian** Mountains.

Reading Check Where do plates tend to break apart?

Figure 16 illustrates the history of the plate collisions and rifting that produced the mountain range as it is today. Rocks that make up mountain ranges such as the Appalachian Mountains record the history of plate motion and collisions that formed the mountains.

Weathering

The Appalachian Mountains are an old mountain range that stretches along most of the eastern United States. They are not as high and rugged as the Rocky Mountains in the west because they are much older. They are no longer growing. Weathering has rounded the peaks and lowered the elevations.

Formation of the Appalachians

Figure 16 The Appalachian Mountains formed over several hundred million years.

Visual Check Which mountain range is between Valley and Ridge and Piedmont?

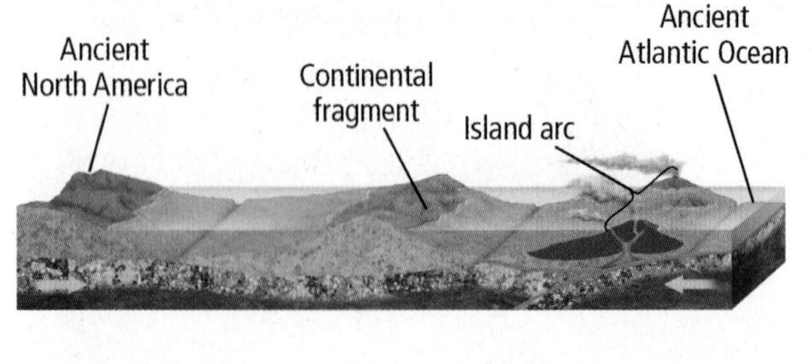

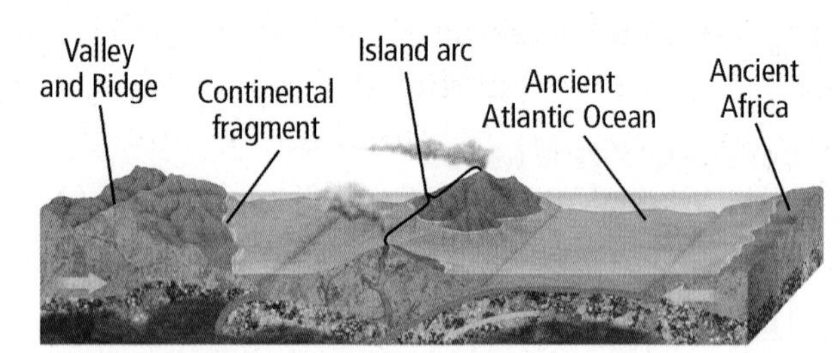

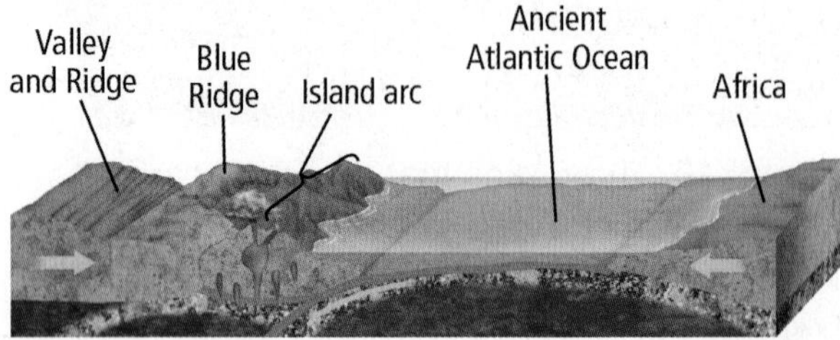

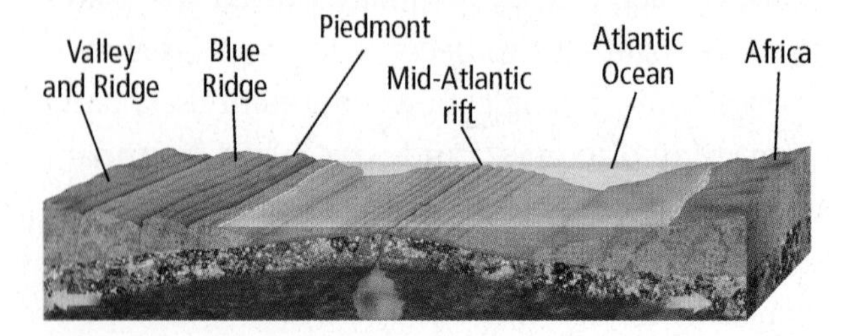

Erosion and Uplift

Over time, natural processes wear down mountains, smooth their peaks, and reduce their height. But some mountain ranges are hundreds of millions of years old. How do they last so long? Recall how isostasy works. As a mountain erodes, the crust under it must rise to restore the balance between what is left of the mountain and how it floats on the mantle. Therefore, rocks deep under continents rise slowly toward Earth's surface. In old mountain ranges, metamorphic rocks that formed deep below the surface are exposed on the top of mountains, such as the rocks in **Figure 17.**

Key Concept Check How can mountains change over time?

Figure 17 Metamorphic rocks, such as these, formed deep below Earth's surface. After the material above them eroded, the rock rose due to isostasy. Now they are on Earth's surface.

Types of Mountains

You learned in the first lesson that stresses caused by plate movement can pull or compress crust. This is one way plate motion is involved in creating many types of mountains. But the effect of plate movement is also responsible for changing the positions of rocks and the rocks themselves within a mountain range.

Folded Mountains

Rocks that are deeper in the crust are warmer than rocks closer to Earth's surface. Deeper rocks are also under much more pressure. When rocks are hot enough or under enough pressure, folds form instead of faults, as shown in **Figure 18.** **Folded mountains** *are made of layers of rocks that are folded.* Folded mountains form as continental plates collide, folding and uplifting layers of rock. When erosion removes the upper part of the crust, folds are exposed on the surface.

ACADEMIC VOCABULARY

perpendicular
(adj.) being at right angles to a line or plane

The arrangement of the folds is not accidental. You can demonstrate this by taking a piece of paper and gently pushing the ends toward one another to form a fold. The fold is a long ridge that is **perpendicular** to the direction in which you pushed. Folded mountains are similar. The folds are perpendicular to the direction of the compression that created them.

Figure 18 Compression stresses folded these rocks. Because the folds run up and down, the compression must have come from the sides.

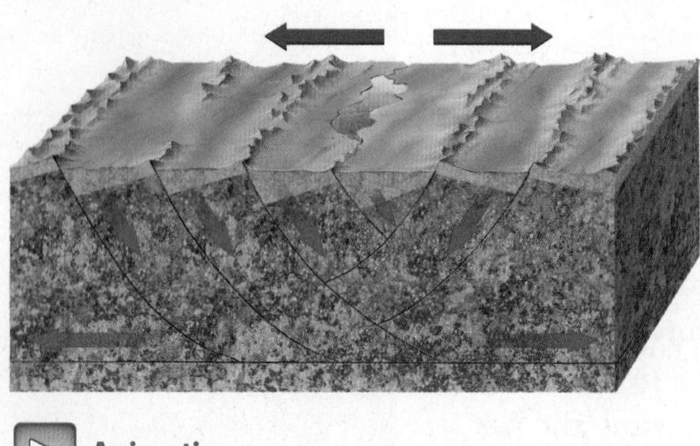

Animation

Figure 19 In the middle of a continent, tension can pull crust apart. Where the crust breaks, fault-block mountains and valleys can form as huge blocks of Earth rise or fall.

Visual Check Which way is the tension pulling?

Fault-Block Mountains

Sometimes tension stresses within a continent create mountains. As tension pulls crust apart, faults form, as shown in **Figure 19.** At the faults, some blocks of crust fall and others rise. **Fault-block mountains** *are parallel ridges that form where blocks of crust move up or down along faults.*

The Basin and Range Province in the western United States consists of dozens of parallel fault-block mountains that are oriented north to south. The tension that created the mountains pulled in the east-west directions. One of these mountains is shown at the beginning of this lesson. Notice how a high, craggy ridge is right next to a valley. Somewhere between the two, there is a fault where huge movement once occurred.

Key Concept Check How do folded and fault-block mountains form?

MiniLab

15 minutes

How do folded mountains form?

When two continental plates converge, rocks crumple and fold, forming folded mountains. If the rocks formed in layers, such as sedimentary rocks, the folds can be visible.

1. Read and complete a lab safety form.
2. On a piece of **waxed paper,** shape four balls of different **colored dough** into rectangles about 1 cm thick.
3. Stack the rectangles on top of each other. Using a **plastic knife,** trim the edges so that all layers are clearly visible. Draw a side view of the unfolded layers in your Science Journal.
4. Compress the dough by pushing the short ends together into an S-shape. Try to get at least one upward and one downward fold. Draw a side view of the folded layers in your Science Journal.
5. Using the knife, simulate erosion by slicing off the top of your folded mountains. Draw a top view of the eroded mountains in your Science Journal.

Analyze and Conclude

1. **Relate** the direction of compression to the direction of the peaks of the mountains.
2. **Key Concept** Describe how folded mountains form and change over time.

Uplifted Mountains

To the west of the Basin and Range Province is the Sierra Nevada range. The rocks in the Sierra Nevada are made of granite, which is an igneous rock originally formed several kilometers below Earth's surface. The granite on top of the Sierra Nevada's Mount Whitney was once 10 km below Earth's surface. Now it is on top of a 4,420-m-tall mountain. When large regions rise vertically with very little deformation, uplifted mountains form.

Reading Check What type of rocks are found in the Sierra Nevada?

The entire Sierra Nevada can be thought of as a gigantic tilted fault block. Uplift caused by faulting, and erosion, have exposed it on Earth's surface. But why is it continuing to rise? Scientists do not fully understand how uplifted mountains like the Sierra Nevada range form. One hypothesis proposes that cold mantle under the crust detaches from the crust and sinks deeper into the mantle, as shown in **Figure 20.** The sinking mantle pulls the crust and creates compression closer to the surface. As the crust thickens, the upper part of the crust rises to maintain isostasy.

Volcanic Mountains

You might not think of volcanoes as mountains, but scientists consider volcanoes to be special types of mountains. In fact, some of the largest mountains on Earth are made by volcanic eruptions. As molten rock and ash erupt onto Earth's surface, they harden. Over time, many eruptions can build huge volcanic mountains such as the ones that make up the Hawaiian Islands.

Not all volcanic mountains erupt all the time. Some volcanic mountains are dormant, which means they might erupt again someday. Some volcanic mountains will never erupt again.

Key Concept Check How do uplifted and volcanic mountains form?

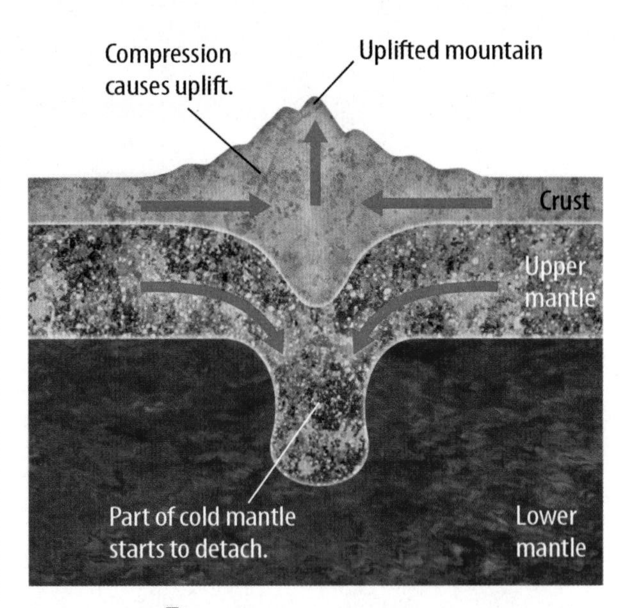

Figure 20 One possible explanation for how uplifted mountains form is that sinking mantle creates compression of the crust. The crust rises to regain isostasy, forming mountains.

Math Skills

Use Proportions

An equation showing two equal ratios is a proportion. Some mountains in the Himalayas are rising 0.001 m/y. How long would it take the mountains to reach a height of 7,000 m?

1. Set up a proportion.

$$\frac{0.001 \text{ m}}{1\text{y}} = \frac{7{,}000 \text{ m}}{x\text{y}}$$

2. Cross multiply.

$$0.001x = 7{,}000$$

3. Divide both sides by 0.001.

$$\frac{0.001x}{0.001} = \frac{7{,}000}{0.001}$$

4. Solve for x.

$$x = 7{,}000{,}000 \text{ y}$$

Practice

If the uplift rate of Mount Everest is 0.0006 m/y, how long did it take Mount Everest to reach a height of 8,848 m?

Math Practice

Personal Tutor

Lesson 3 Review

Visual Summary

Mountain ranges can be the result of repeated continental collision and rifting.

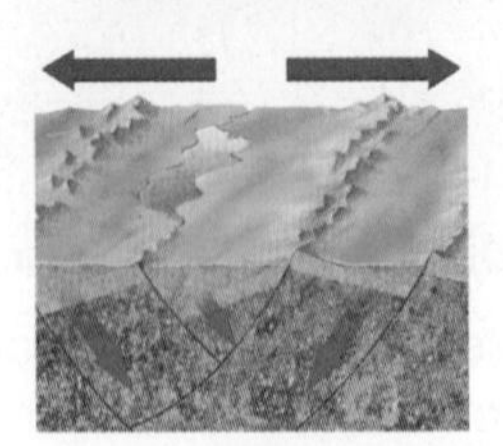

Tension stresses create mountain ranges that are a series of faults, ridges, and valleys.

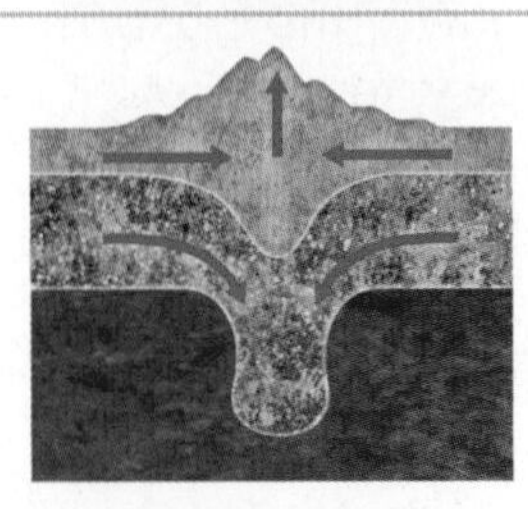

Uplifted mountains form as a result of compression near Earth's surface.

FOLDABLES®

Use your lesson Foldable to review the lesson. Save your Foldable for the project at the end of the chapter.

What do you think NOW?

You first read the statements below at the beginning of the chapter.

5. Metamorphic rocks formed deep below Earth's surface sometimes can be located near the tops of mountains.

6. Mountain ranges can form over long periods of time through repeated collisions between plates.

Did you change your mind about whether you agree or disagree with the statements? Rewrite any false statements to make them true.

Use Vocabulary

1. Compression stress can create ________.
2. **Name** two types of mountains that can form far from plate boundaries.
3. Rocks formed deep inside Earth can be found at the surface as ________.

Understand Key Concepts 

4. **Contrast** folded and fault-block mountains.
5. Which type of mountains form with little deformation?
   - **A.** fault-block mountains
   - **B.** folded mountains
   - **C.** uplifted mountains
   - **D.** volcanic mountains
6. **Identify** the type of plate boundary where the Appalachian Mountains formed.

Interpret Graphics

7. **Summarize** the plate tectonic events that built the Appalachian Mountains, using a graphic organizer like the one below.

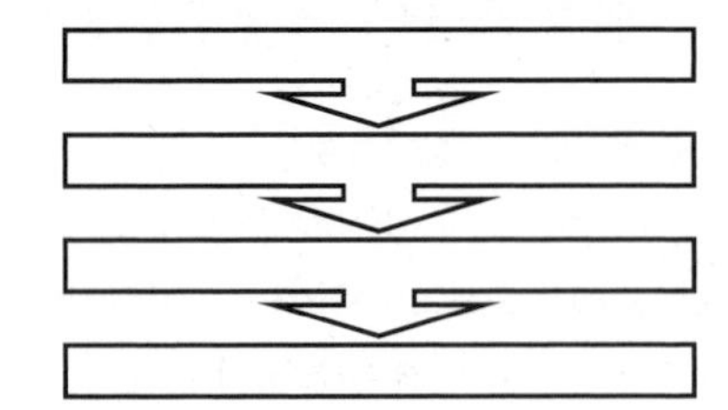

Critical Thinking

8. **Critique** the generalization that mountains only form at convergent boundaries. Explain how other processes can produce mountains.

Math Skills

Math Practice

9. Volcanoes in Hawaii began forming on the seafloor, about 5,000 m below the surface. If a volcano reaches the surface in 300,000 years, what was its rate of vertical growth per year?

Skill Practice **Research Information** 2 class periods

What tectonic processes are most responsible for shaping North America?

Materials

metric ruler

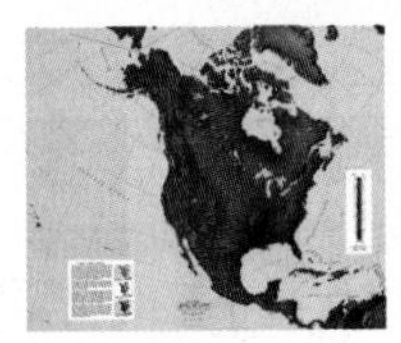

North America map

Mountains are important structures of the North American landscape. By studying the types of mountains, scientists can figure out what processes have shaped the continent over the last several hundred million years.

Learn It

Before scientists can make conclusions about the formation of a continent, they have to know what happened in all of its parts. Scientists **research information** so they can answer questions and draw conclusions about the continent as a whole.

Try It

1. Study the map shown. Choose a mountain range to research. With your teacher's approval, you may research a range that is not shown on this map.
2. Using sources approved by your teacher, research your mountain range. Be sure to make a list of the sources you used for your research. Answer the following questions.

- What is the name and location of your mountain range?
- What type of plate boundary is near your mountain range?
- What tectonic plates form the boundary near your mountain range?
- What type(s) of mountains make up your mountain range?
- How did the mountains form?
- What type(s) of rocks make up the mountains?
- How tall are the mountains in your range?
- How old is your mountain range?
- What other factors have affected the height or the shape of the mountains?

3. Record your results in your Science Journal.

Apply It

4. **Create** a visual presentation about the mountain range you researched. Include photographs or sketches to support your research.
5. **Compare and Contrast** Combine your research with the other groups' research. How are your mountain ranges similar? How are they different?
6. **Key Concept** What forces have shaped North America?

(t)Michael Scott/McGraw-Hill Education; (b)U.S. Geological Survey

Lesson 4

Continent Building

Key Concepts

ESSENTIAL QUESTIONS

- What are two ways continents grow?
- What are the differences between interior plains, basins, and plateaus?

Vocabulary

plains p. 279

basin p. 279

plateau p. 280

Multilingual eGlossary

Go to the resource tab in ConnectED to find the PBL *When on Earth...?*

Inquiry What is it really?

You might have heard that the Grand Canyon is just a big hole in the ground. In fact, it's not a hole at all. What do you think it might be? How did it form? You may be surprised at the answer.

Michael Busselle/Digital Vision/SuperStock

Launch Lab

20 minutes

How do continents grow?

Over the history of Earth, continents have been slowly increasing in size. Continents can grow when fragments of crust that formed in other parts of the world stick to the edges of the continent at convergent plate boundaries.

1. Read and complete a lab safety form.
2. Place **waxed paper** on the lab bench, and place a **block of wood** on one end of the paper.
3. Using **shaving cream,** create a volcanic arc on the waxed paper.
4. Pull the waxed paper under the wood and observe what happens. Record your observations in your Science Journal.

Think About This

1. Using the vocabulary from the chapter (words such as *compression*, *convergence*, *folded mountains*, *volcanic mountains*), describe what occurred as you completed the lab.
2. Create a labeled diagram showing the motion of the ocean plate and the continental plate. Include the volcanic arc and describe what happened to it when it ran into the continent.
3. **Key Concept** How do you think continents grow?

The Structure of Continents

If you look at the map shown on the left in **Figure 21**, you will notice that most of the highest elevations are located near the edges of continents. Why do you think that is?

In contrast, the interiors of most continents are flat. Usually, the middle of a continent is only a few hundred meters above sea level. Continental interiors have very few mountains. In these regions, the rocks are old igneous and metamorphic rocks. A map showing the old, stable interiors of the world's continents is on the right in **Figure 21.** Notice that they usually lie near the middle of the continent. These areas are usually smooth and flat because millions or even billions of years of erosion have smoothed them out.

WORD ORIGIN

continent
from Latin *terra continens*, means "continuous land"

Reading Check Where are high elevations usually located? Where are the low elevations?

Figure 21 The map on the left shows areas of high elevation in white. They are usually near the edges of continents. The map on the right shows old, stable continental interiors.

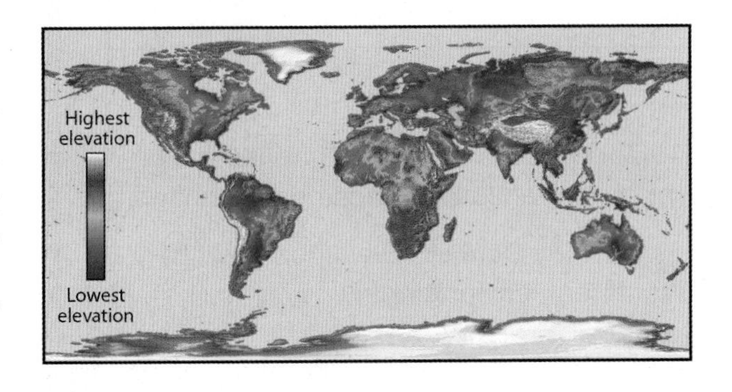

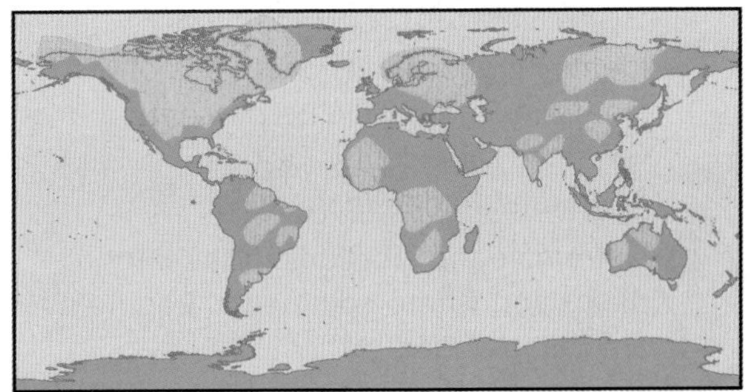

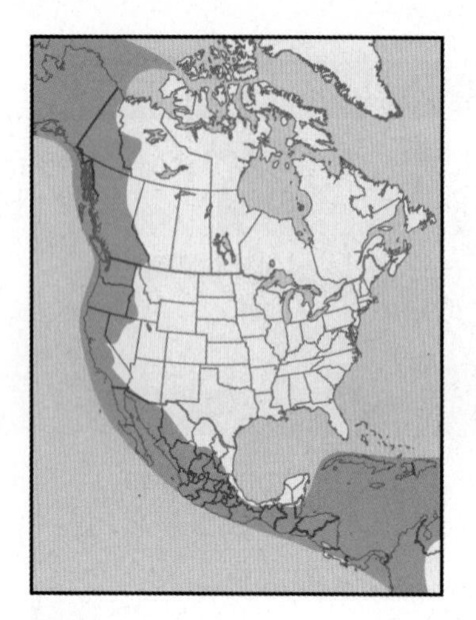

 Animation

Figure 22 The green areas show parts of present-day North America that were once attached to other continents in other parts of the world. 

How Continents Grow

The shapes and the sizes of the continents have changed many times over Earth's history. Continents can break up and get smaller, or they can get bigger. One way continents get bigger is through the addition of igneous rocks by erupting volcanoes. A second way is when tectonic plates carry island arcs, whole continents, or fragments of continents with them.

When a plate carrying fragments reaches a continent at a convergent boundary, the least dense fragments get pushed onto the edge of the continent. **Figure 22** is a map showing fragments that have been added to the west coast of North America within the last 600 million years. Fragments in the western United States include volcanic arcs, ancient seafloor, and small pieces from other continents.

Key Concept Check What are two ways continents grow?

If rock were not being added to continents, what do you think would happen to the size of continents? Rifting would change their sizes and shapes. Weathering and erosion gradually would wear them down.

Inquiry MiniLab

20 minutes

Can you analyze a continent?

As tectonic plates move across Earth's surface, they interact in predictable patterns. Suppose you could get a glimpse of a continent on Earth at some other time. You can use what you know to figure out what that continent is like.

1. Read and complete a lab safety form.
2. Copy the imaginary continent shown at right into your Science Journal. Arrow length is proportional to the speed of the plates.
3. Use **colored pencils** to differentiate regions of compression, tension, and shear.
4. Identify the locations and types of landforms that would be present on Gigantia. Label fault-block mountains, faults, and folded mountains.
5. Determine the locations of the interior plains and where continental fragments are being added.

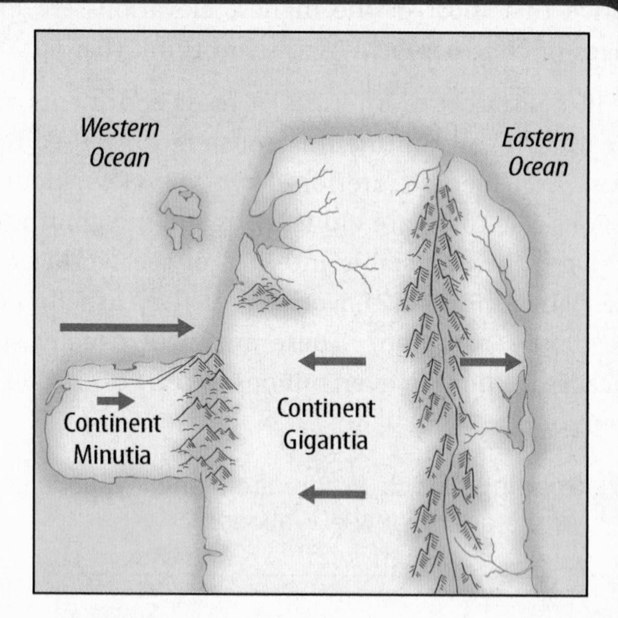

Analyze and Conclude

1. **Select** a region of Earth that has similar plate interactions to those on Gigantia.
2. **Key Concept** Describe how the continent is changing.

Continental Interiors

Rocks in continental interiors tend to be stable, flat, very old, and very strong. They are usually more than 500 million years old. In some continental interiors, the rocks are much older than that! **Figure 23** shows rocks that might be the oldest on Earth's surface. Although they might not look very exciting, it is incredible to think they are more than 4.2 billion years old!

Figure 23 The Canadian rocks pictured here have existed throughout much of Earth's history. They are more than 4.2 billion years old.

Formation of Interior Plains

A **plain** *is an extensive area of level or rolling land.* Most of the central region of North America is referred to as the Interior Plains. The rocks in these plains came from collisions of several smaller plates about 1 billion years ago. At different times in Earth's history, the plains were covered by shallow seas. The plains have been flattened by millions of years of weathering and erosion.

Key Concept Check What is a plain?

Formation of Basins

Just as plate motion and isostasy create mountains, they also can cause subsidence. *Areas of subsidence and regions with low elevation are called* **basins.** Sediments eroded from mountains accumulate in basins. **Figure 24** is a map showing the largest basins in North America. Can you find a relationship between the locations of basins and large mountain ranges?

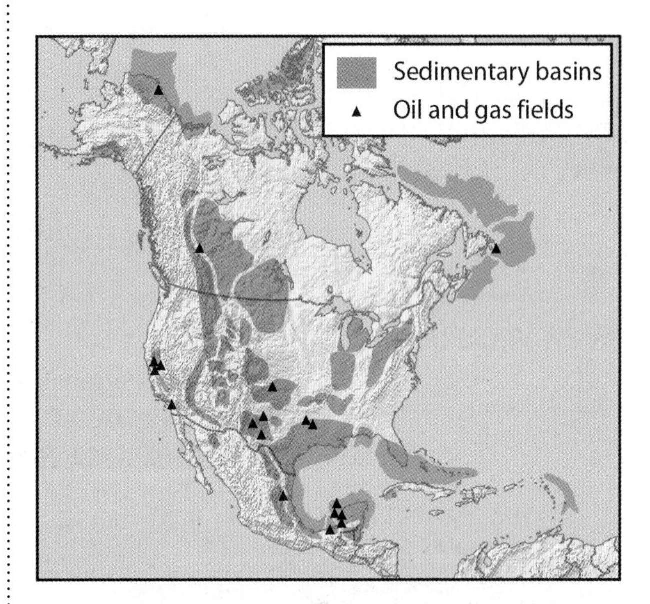

Figure 24 Ancient sedimentary basins are important because oil, natural gas, and coal usually are found in basins.

Visual Check Where are oil and gas fields in relation to sedimentary basins?

Reading Check What is the name of the feature where sediment accumulates?

Basins can have great economic importance. Under the right conditions, the remains of plants and animals are buried in the sediments that accumulate in basins. Over millions of years, heat and pressure convert the plant and animal remains into oil, natural gas, and coal. Most of our energy resources are extracted from sedimentary basins. The world's largest oil and natural gas fields also lie in sedimentary basins.

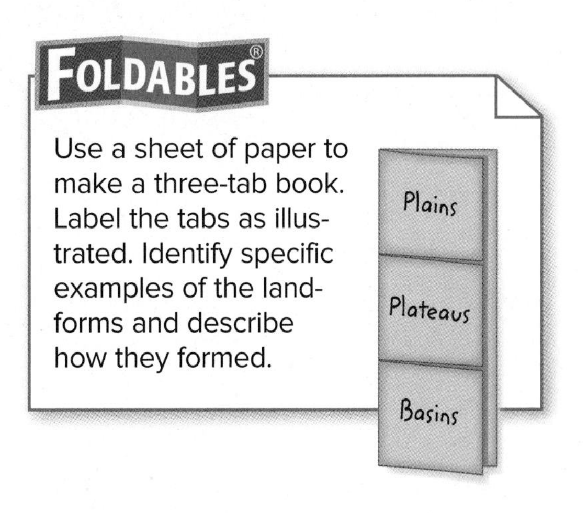

Use a sheet of paper to make a three-tab book. Label the tabs as illustrated. Identify specific examples of the landforms and describe how they formed.

Formation of Plateaus

Some regions are high above sea level but are flat. *Flat regions with high elevations are called* **plateaus.** Some plateaus form through uplift. An example of an uplifted plateau is the Colorado Plateau, shown in **Figure 25.** In the last 5 million years, this region has been uplifted by more than 1 km.

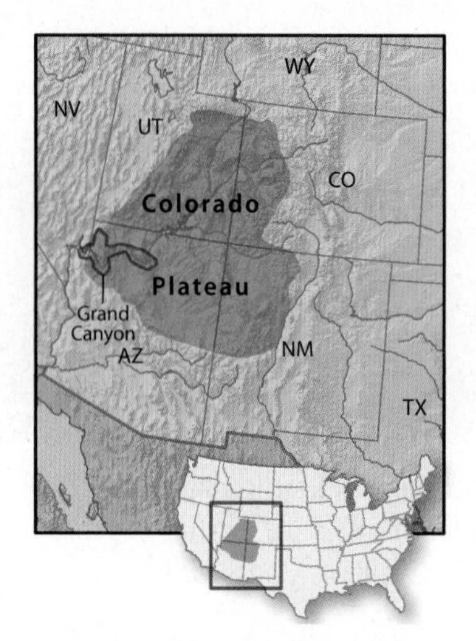

Figure 25 The Colorado Plateau is an example of an uplifted plateau.

Visual Check Which states are partly covered by the Colorado Plateau?

Notice in **Figure 25** that the Grand Canyon is only a small part of the Colorado Plateau. It was created as the Colorado River cut through and eroded the uplifting plateau. So, the Grand Canyon was created by water!

The eruption of lava also can create large plateaus. For example, more than 200,000 km^3 of lava flooded the Columbia Plateau shown in **Figure 26.** Over 2 million years, multiple eruptions built up layers of rock. In some places, the plateau is more than 3 km thick!

Key Concept Check What are the differences between plains, basins, and plateaus?

REVIEW VOCABULARY

lava
molten rock that erupts on Earth's surface.

Dynamic Landforms

When you started reading this chapter, you might have thought Earth had always looked the same. Now you know that Earth's surface is constantly changing. Mountains form only to be eroded away. Continents grow, shift, and shrink. Nothing stays the same for long on dynamic Earth.

The Columbia Plateau

Figure 26 The map shows the area that was covered by multiple eruptions of lava over millions of years. The lava cooled and formed the Columbia River basalt. The layers of basalt are visible in the photograph. Some parts of the Columbia Plateau are more than 3 km thick.

Lesson 4 Review

Visual Summary

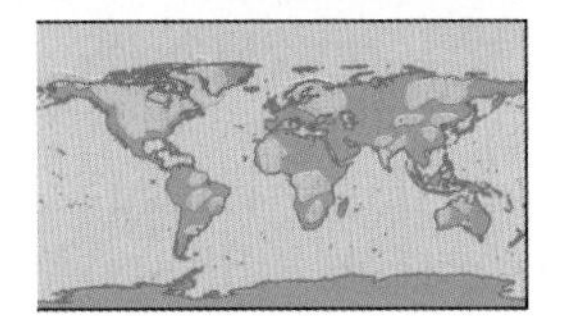

Rocks at the center of most continents are very old, very strong, and flat.

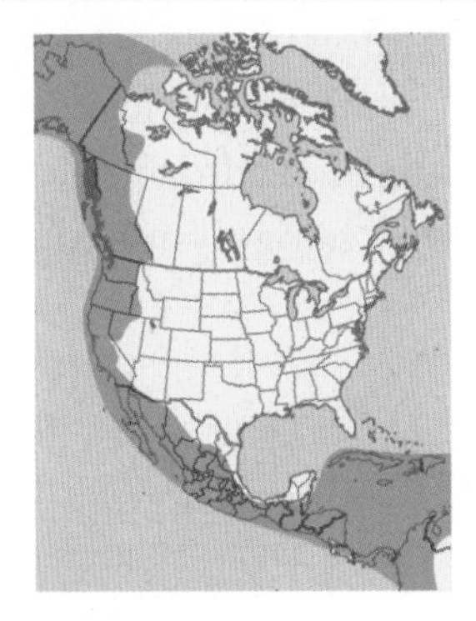

Fragments of crust are added to continents at convergent boundaries.

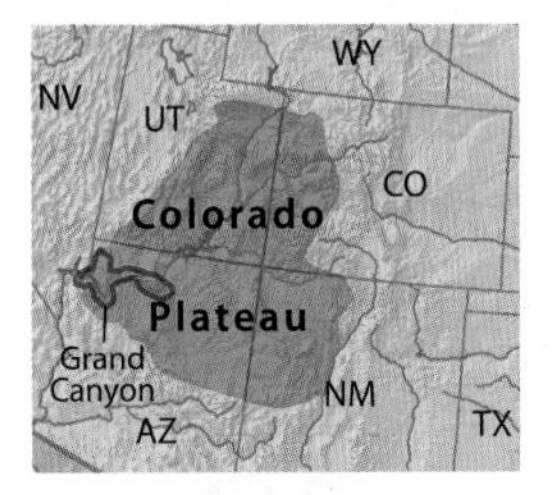

Large, elevated plateaus are created through uplift and lava flows.

FOLDABLES®

Use your lesson Foldable to review the lesson. Save your Foldable for the project at the end of the chapter.

What do you think NOW?

You first read the statements below at the beginning of the chapter.

7. The centers of continents are flat and old.

8. Continents are continually shrinking because of erosion.

Did you change your mind about whether you agree or disagree with the statements? Rewrite any false statements to make them true.

Use Vocabulary

1. As a mountain erodes, sediment can accumulate in a nearby ________.
2. The central, flat region of North America is known as the ________.
3. The Grand Canyon was eroded out of a large ________.

Understand Key Concepts 

4. Which term best describes the center of North America?
   A. basin
   B. lava
   C. plateau
   D. plain
5. **Describe** how continents change over time.
6. **Contrast** basins and plateaus.

Interpret Graphics

7. **Summarize** Use a graphic organizer like the one below to show the different stages involved in continent growth. Begin with an old continental interior and end with a new continent.

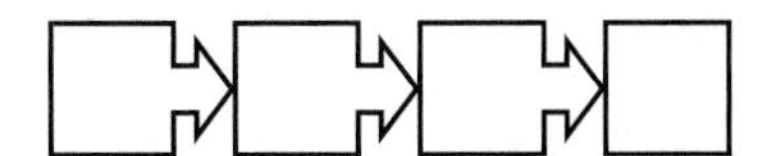

Critical Thinking

8. **Infer** In this lesson you learned that fragments of other continents were added to the west coast of North America. Where in the United States have other continental fragments been added?
9. **Generalize** How are landforms near the edges of continents different from landforms in continental interiors? How are these landforms related to plate tectonics or processes in the rock cycle?
10. **Infer** What would happen to the Grand Canyon if there were further uplift of the Colorado Plateau?

Lab

2 class periods

Design Landforms

Suppose you are a museum designer and you want to show people that different landforms form under different circumstances. Sometimes rocks fold, sometimes they break, sometimes they form mountains, and sometimes they sink into Earth and create trenches. What they do depends on the properties of the rock and the type of stress. Unfortunately, rocks do all of these things so slowly that it is hard to see rocks in motion. What materials would you use to model the formation of landforms? What factors affect how rocks behave? How could you change your materials to model how rocks behave?

Materials

tub

measuring cup

cornstarch

hooked weights

thermometer

flour

hot plate

Also needed: stopwatch, spoon, metric ruler, ice

Safety

Ask a Question

What materials could represent rocks? How are the materials different from rocks?

Make Observations

1. Read and complete the lab safety form.
2. Mix some ingredients in your plastic bin or mixing bowl. Try different combinations until you make a material you can use to model landforms.
3. Experiment with the materials, and try to create different landforms.
4. Record your observations in your Science Journal. How do the materials behave like rocks, and how are they different?

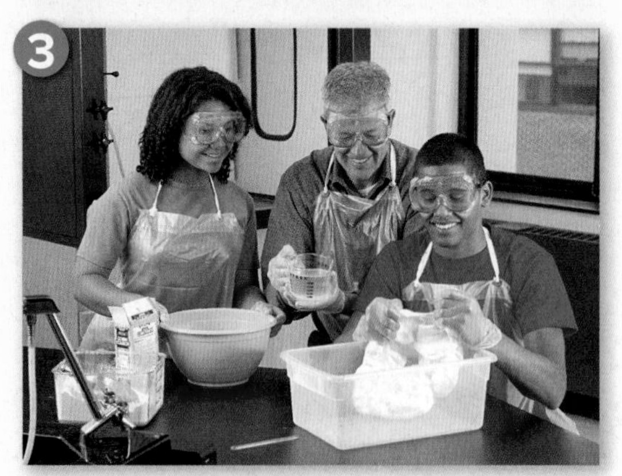

(t to b, 4-5, 7, r)Hutchings Photography/Digital Light Source; (2, 6)Michael Scott/McGraw-Hill Education; (3)Jacques Cornell/McGraw-Hill Education

Form a Hypothesis

5. After observing the behavior of your material, think of factors that cause rocks to behave differently. How might you recreate these different situations? Pick one factor and develop a hypothesis about how you can use the materials to model the behavior of rocks.

Test Your Hypothesis

6. Develop a procedure to test your hypothesis. What is your dependent variable and your independent variable? How will you make quantitative measurements of both variables?
7. Create a table to record your results.
8. Have your teacher approve your procedure and your table.
9. Conduct your experiment, and record your results.

Analyze and Conclude

10. **Create** a graph displaying your results.
11. **Interpret** your graph and explain the relationship between the variables.
12. **Critique** your procedure and your results.
13. THE BIG IDEA **The Big Idea** Relate your results to how Earth's surface is shaped by plate motion.

Communicate Your Results

Design a museum exhibit that models the formation of one or more landforms.

Extension

What materials could represent Earth's crust? Now that you have modeled Earth's landforms, put the landforms on tectonic plates. Model plate motion, and describe how your landforms change at different types of plate boundaries.

Lab Tips

- This lab might be messy! Clean up dry cornstarch and flour with a broom and dustpan.
- Be quantitative! Figure out concrete ways to measure both the variable you are changing (the independent variable) and the variable you are measuring (the dependent variable).
- If your first try is not successful, try something different! Science rarely works on the first try.

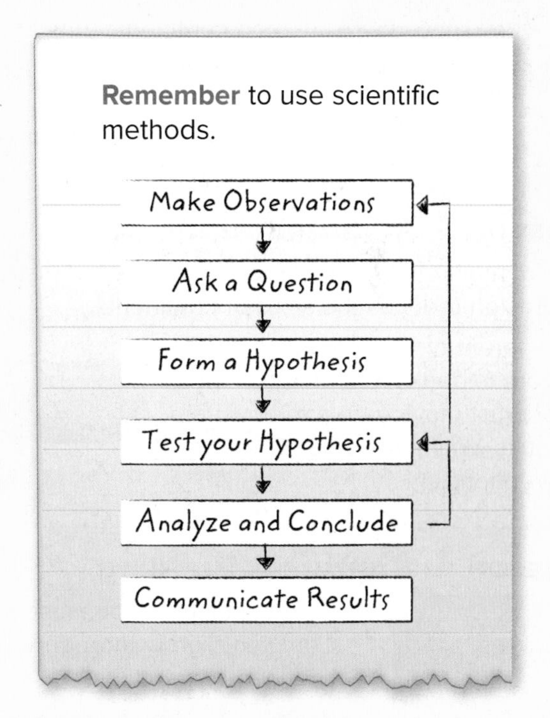

Chapter 8 Study Guide

The forces created by the movement of tectonic plates are responsible for the variety of Earth's constantly changing landforms.

Key Concepts Summary

Lesson 1: Forces That Shape Earth

- As continents float in the mantle, they rise and fall to maintain the balance of **isostasy.**
- **Compression, tension,** and **shear** stresses can deform or break rocks.
- **Uplift** and plate motion move rocks through the rock cycle.

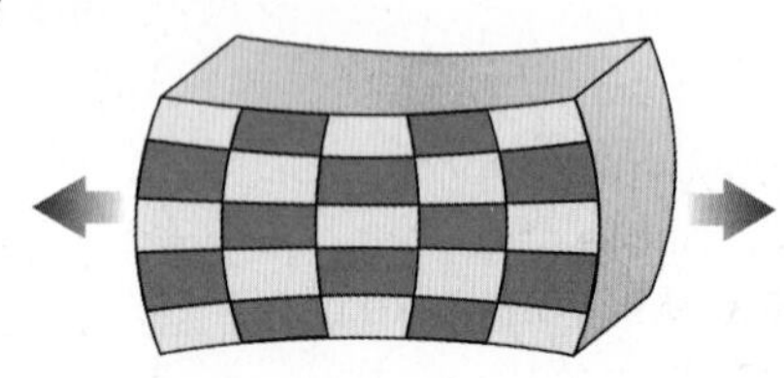

Vocabulary

isostasy p. 254
subsidence p. 255
uplift p. 255
compression p. 255
tension p. 255
shear p. 255
strain p. 256

Lesson 2: Landforms at Plate Boundaries

- When two continental plates collide, tall mountain ranges form. When an oceanic plate subducts below another one, an **ocean trench** and a **volcanic arc** form.
- At divergent boundaries, mid-ocean ridges and continental rifts form.
- **Transform faults** can create large areas of faulting and fracturing, not all of which can be seen at Earth's surface.

Vocabulary

ocean trench p. 262
volcanic arc p. 263
transform fault p. 265
fault zone p. 265

Lesson 3: Mountain Building

- Mountain ranges can grow from repeated plate collisions. Erosion reduces the sizes of continents.
- Different types of mountains form from folded layers of rock, blocks of crust moving up and down at faults, uplift, and volcanic eruptions.

Vocabulary

folded mountain p. 271
fault-block mountain p. 272

Lesson 4: Continent Building

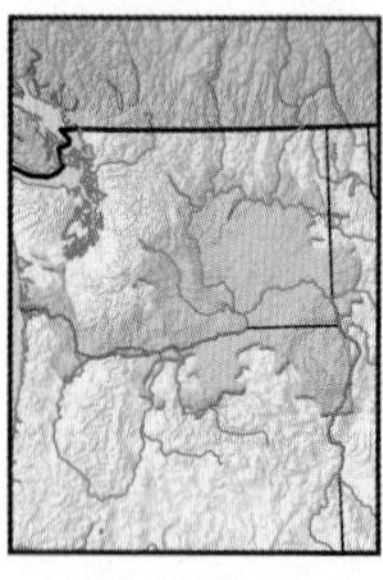

- Continents shrink because of erosion and rifting. Continents grow through volcanic activity and continental collisions.
- **Plains** are generally flat areas of land, usually in the center of continents. **Basins** are regions at low elevation where sediment accumulates or once accumulated. **Plateaus** are large, flat regions at high elevation.

Vocabulary

plains p. 279
basin p. 279
plateau p. 280

Personal Tutor

Vocabulary eFlashcards

Vocabulary eGames

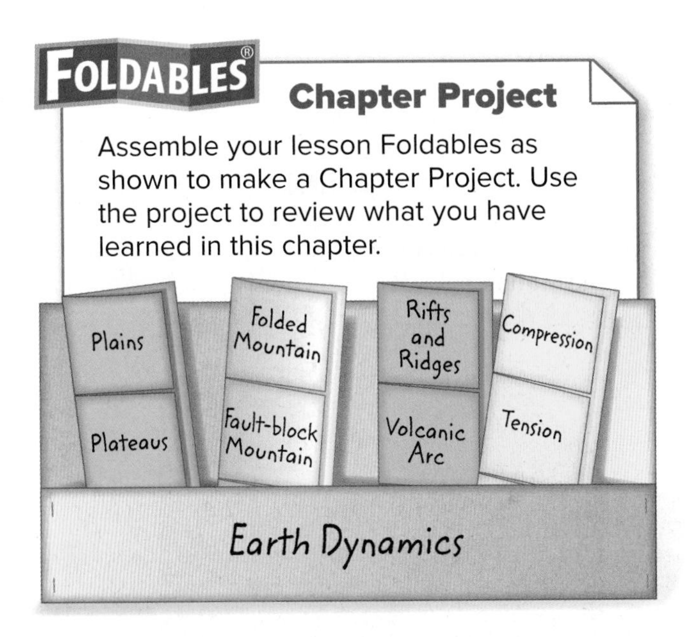

Use Vocabulary

1. Plastic and elastic deformation are types of ________.
2. Areas of fractured crust along a fault are called ________.
3. Mountains that rise with little deformation of rock are ________.
4. Repeated volcanic eruptions on land can create large ________.
5. The downward vertical motion of Earth's surface is ________.
6. Parallel ridges separated by faults and valleys are ________.

Interactive Concept Map

Link Vocabulary and Key Concepts

Copy this concept map, and then use vocabulary terms from the previous page and other terms from the chapter to complete the concept map.

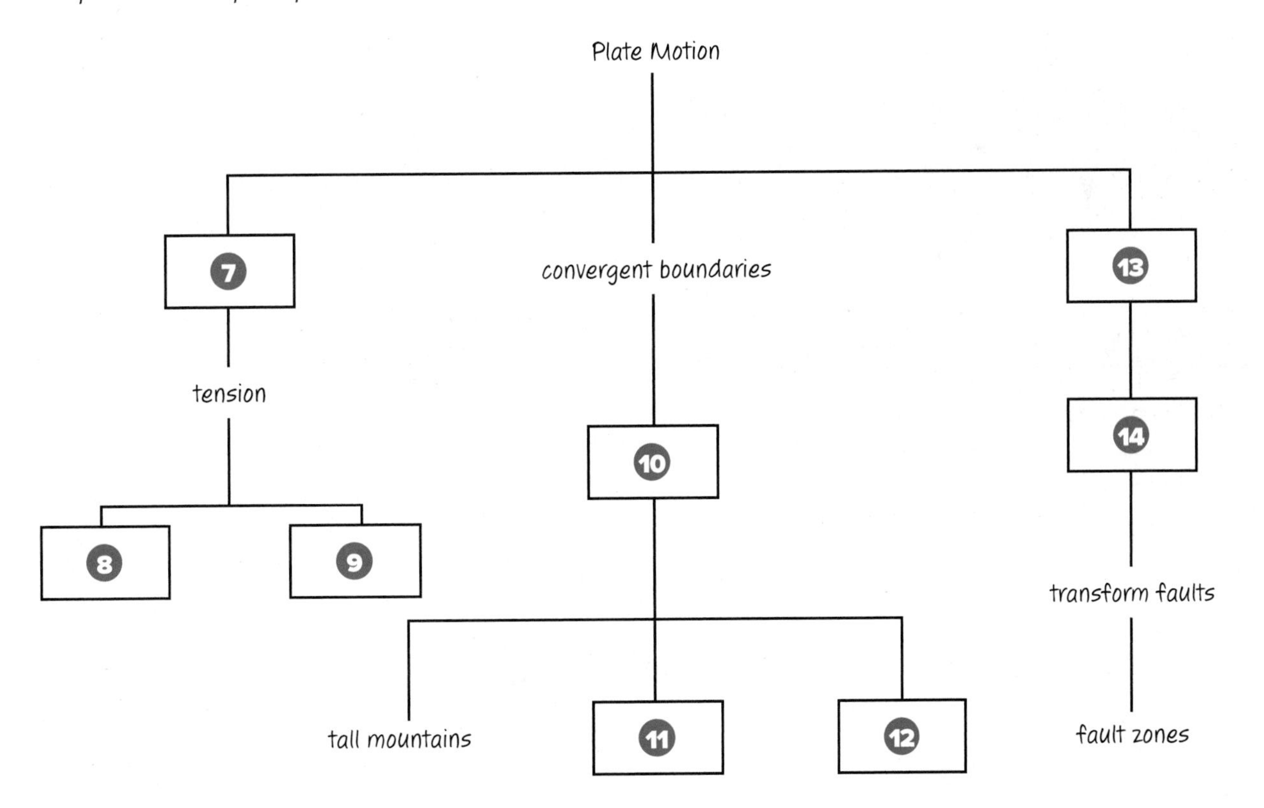

Chapter 8 Review

Understand Key Concepts 

1 The fact that land surface is high where crust is thick is due to what?

A. isostasy
B. subduction
C. shear stresses
D. tension stresses

2 The highest mountains form at which type of plate boundary?

A. convergent
B. divergent
C. oceanic
D. transform

3 Why does plastic deformation occur in the lower crust?

A. Rocks are hot.
B. Rocks are strong.
C. Tension occurs in the lower crust.
D. The mantle is plastic.

4 Lake Baikal in Siberia, pictured below, fills a continental rift valley. What type of stress is creating the rift?

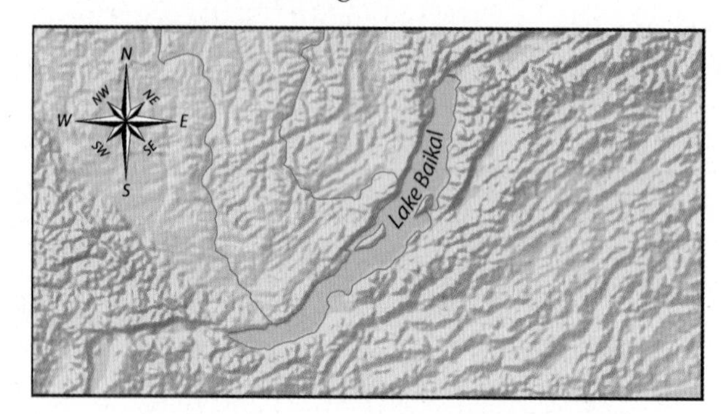

A. compression in the north-south direction
B. shear in the northeast-southwest direction.
C. tension in the north-south direction
D. tension in the northwest-southeast direction

5 When an oceanic plate converges with a continental plate, an island arc

A. does not form.
B. forms on both plates.
C. forms on the continental plate.
D. forms on the larger plate.

6 Which feature is indicated by the arrow in the illustration below?

A. a fault zone
B. an ocean trench
C. an uplifted mountain
D. a volcanic arc

7 Which of the following locations in the United States has grown due to the addition of colliding fragments?

A. in the center
B. along the west coast
C. the Hawaiian Islands
D. along the Gulf of Mexico

8 Which are the cause of rocks being exposed at Earth's surface?

A. erosion and subsidence
B. erosion and uplift
C. faulting and folding
D. folding and subsidence

Critical Thinking

9 **Compare** a floating iceberg with a continent floating on the mantle.

10 **Explain** how seashells got on top of Mount Everest.

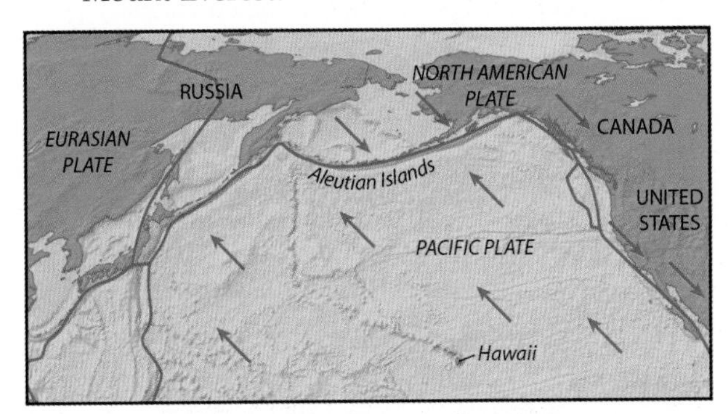

11 **Infer** Look at the illustration above. Where will the Hawaiian Islands be added to another continent?

12 **Assess** the statement, "Isostasy never stops causing uplift and subsidence."

13 **Suggest** the source for the sediment that filled the basins in North Dakota and Colorado.

14 **Defend** the statement that the large mountains of the Hawaiian Islands did not form at a plate boundary.

15 **Predict** where on Earth the crust is thickest.

16 **Infer** Why is the area in southern Africa flat but elevated?

17 **Illustrate** a possible future for the Appalachian Mountains. How do you think the Appalachian Mountains will look in 200 million years? What processes will change these mountains?

Writing in Science

18 **Write** a paragraph describing the history of an imaginary mountain range. Use the terms *fold, volcanic arc, convergent,* and *divergent* in your paragraph.

REVIEW

19 If plate tectonics were suddenly to stop, how would Earth's surface change?

20 The photo below shows Mount Everest in the Himalayas. How did it get to be so tall?

Math Skills

Math Practice

Use Proportions

21 The Himalayas formed when the Indian subcontinent collided with the Eurasian Plate. The Indian subcontinent moved about 10 cm/y.

a. How far would it have moved in 24,000,000 years?

b. How many kilometers did the plate move? (1 km = 100,000 cm)

22 A continent travels 0.006 m/y. How long would it take the continent to travel 100 m?

23 Mount Whitney is 4,421 m high. It started as a hill with an elevation of only 457 m about 40 mya. What was the rate of uplift for Mount Whitney in m/y? (Hint: Figure out the total elevation gain first.)

Standardized Test Practice

Record your answers on the answer sheet provided by your teacher or on a sheet of paper.

Multiple Choice

1 Which is the result of isostasy?

A a basin filling with sediment

B an iceberg floating in the ocean

C magma rising beneath a mountain

D one plate subducting under another one

2 The San Andreas Fault is classified as a transform fault. Which type of stress can create transform faults?

A compression

B shear

C fracture

D tension

Use the figure below to answer question 3.

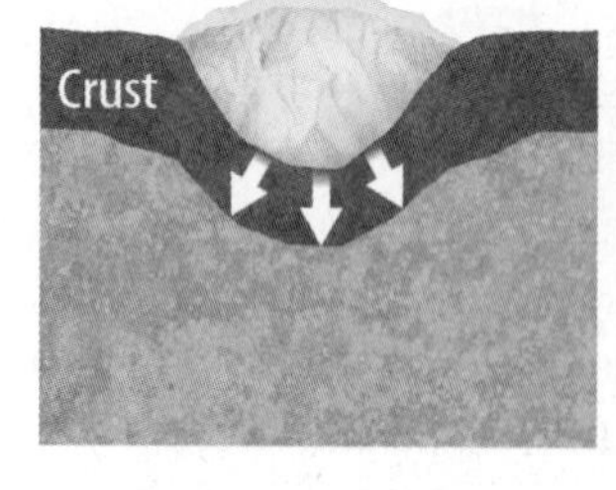

3 Which processes are shown in the figure?

A compression and divergence

B shearing and tension

C subsidence and uplift

D uplifting and subduction

4 What happens when rock fails?

A It breaks.

B It deforms elastically.

C It deforms plastically.

D It folds.

5 What part does subduction play in the rock cycle?

A It breaks rocks into sediment.

B It decreases pressure on buried rocks.

C It pulls rocks deep into Earth.

D It pushes rocks up to Earth's surface.

Use the figure below to answer questions 6 and 7.

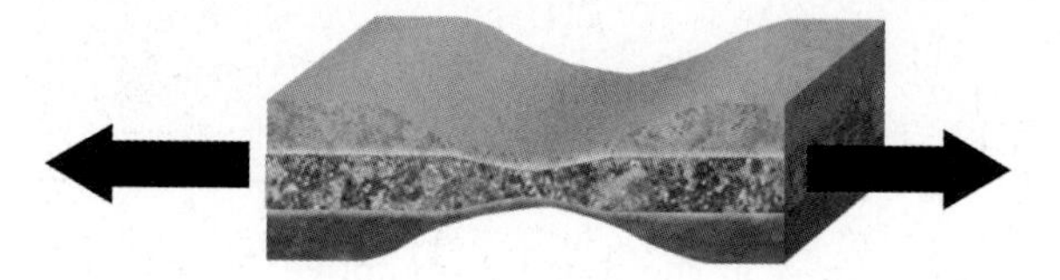

6 Which force is shown in the figure?

A compression

B shearing

C tension

D uplift

7 What type of mountain results from the force shown in the figure?

A folded

B fault-block

C uplifted

D volcanic

8 Which continental landform can result when two plates diverge?

A basin

B continental rift

C mid-ocean ridge

D transform fault

Use the figure below to answer question 9.

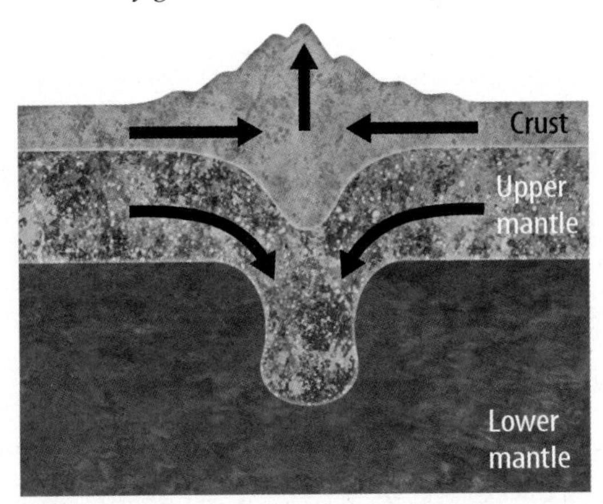

9 Which kind of mountain is shown in the figure?

A fault-block

B folded

C uplifted

D volcanic

10 Which feature is a flat region at a high elevation?

A a basin

B a mountain

C a plain

D a plateau

11 Which landforms are most likely to have coal, oil, and gas deposits?

A basins

B mountains

C plains

D plateaus

Constructed Response

Use the figure below to answer questions 12 and 13.

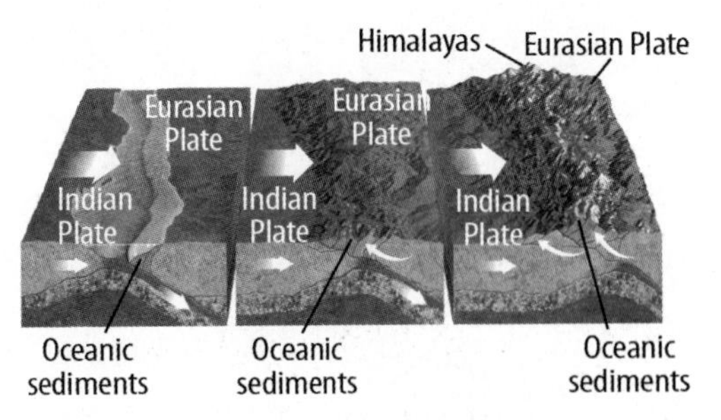

12 Use the figure to explain how the Himalayas formed. Identify the forces involved.

13 What would happen if the plate motion were reversed? Describe possible scenarios for the stages shown on the left and the right of the figure.

Use the figure below to answer question 14.

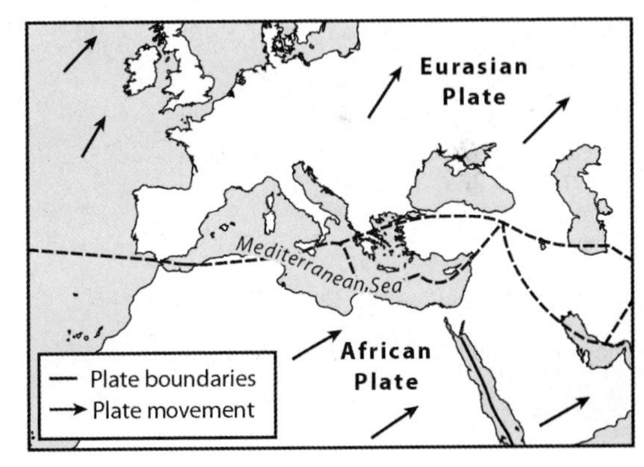

14 Each year, the African Plate moves closer to the Eurasian Plate. Predict how the Mediterranean Sea and the continents, shown in the figure, will change in 100 million years.

NEED EXTRA HELP?														
If You Missed Question...	1	2	3	4	5	6	7	8	9	10	11	12	13	14
Go to Lesson...	1	2	1	2	1	1	2	3	4	4	2	3	4	4

Chapter 9

Earthquakes and Volcanoes

THE BIG IDEA What causes earthquakes and volcanic eruptions?

Why do volcanoes erupt?

Mount Pinatubo, a volcano in the Philippines, ejected superheated particles of ash and dust in June 1991. This truck is trying to outrun a pyroclastic flow produced during this eruption. *Pyroclastic* means "fire fragments." Why do you suppose this eruption was so dangerous?

- Why did Mount Pinatubo erupt explosively?
- Can scientists predict earthquakes and volcanic eruptions?
- What causes earthquakes and volcanic activity?

Get Ready to Read

What do you think?

Before you read, decide if you agree or disagree with each of these statements. As you read this chapter, see if you change your mind about any of the statements.

1. Earth's crust is broken into rigid slabs of rock that move, causing earthquakes and volcanic eruptions.
2. Earthquakes create energy waves that travel through Earth.
3. All earthquakes occur on plate boundaries.
4. Volcanoes can erupt anywhere on Earth.
5. Volcanic eruptions are rare.
6. Volcanic eruptions only affect people and places close to the volcano.

Your one-stop online resource
connectED.mcgraw-hill.com

LearnSmart®

Chapter Resources Files, Reading Essentials, Get Ready to Read, Quick Vocabulary

Animations, Videos, Interactive Tables

Self-checks, Quizzes, Tests

Project-Based Learning Activities

Lab Manuals, Safety Videos, Virtual Labs & Other Tools

Vocabulary, Multilingual eGlossary, Vocab eGames, Vocab eFlashcards

Personal Tutors

Lesson 1

Earthquakes

Reading Guide

Key Concepts
ESSENTIAL QUESTIONS

- What is an earthquake?
- Where do earthquakes occur?
- How do scientists monitor earthquake activity?

Vocabulary

earthquake p. 293
fault p. 295
seismic wave p. 296
focus p. 296
epicenter p. 296
primary wave p. 297
secondary wave p. 297
surface wave p. 297
seismologist p. 298
seismometer p. 299
seismogram p. 299

Multilingual eGlossary

Science Video

Go to the resource tab in ConnectED to find the PBL *Shake, Rattle, and Roll!*

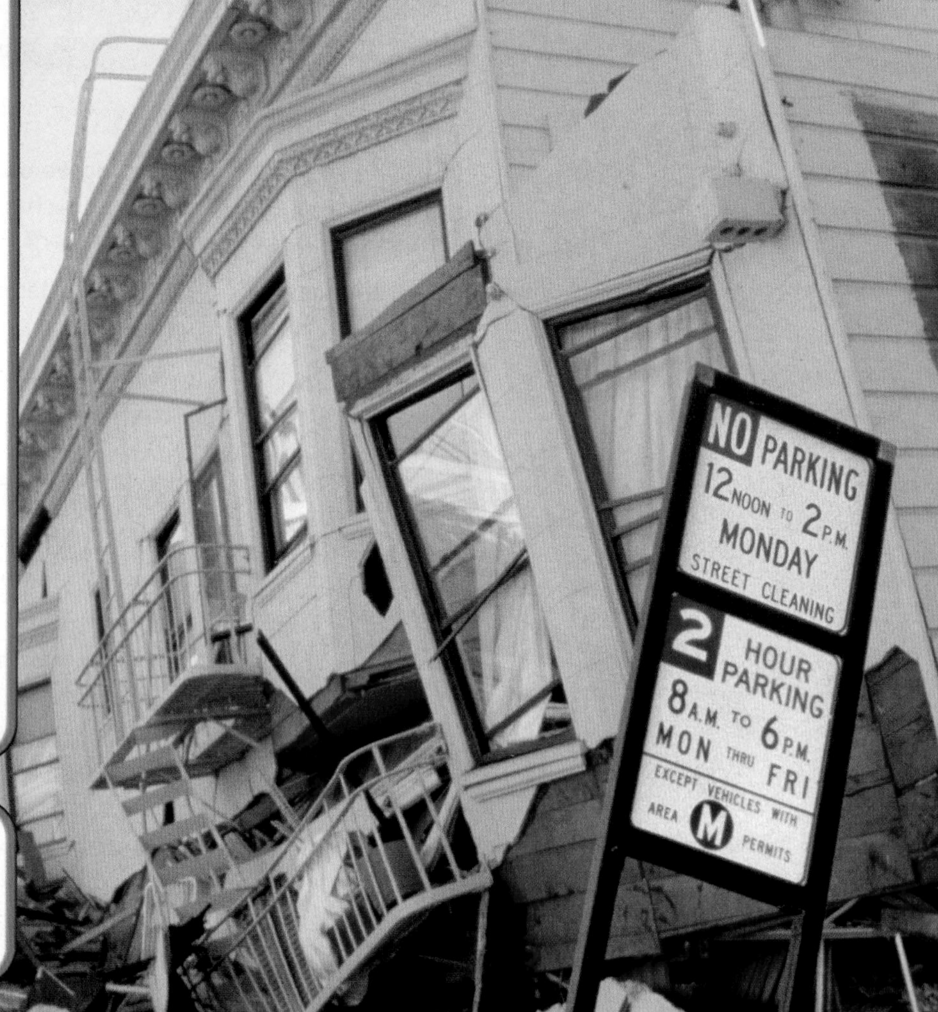

Inquiry Why did this building collapse?

This building collapsed during the Loma Prieta earthquake that shook the San Francisco Bay area of California in 1989. The magnitude 7.1 earthquake produced severe shaking and damage. Freeways and buildings collapsed and a number of injuries and fatalities occurred. Why are earthquakes common in California?

Launch Lab

15 minutes

What causes earthquakes?

Earthquakes occur every day. On average, approximately 35 earthquakes happen on Earth every day. These earthquakes vary in severity. What causes the intense shaking of an earthquake? In this activity, you will simulate the energy released during an earthquake and observe the shaking that results.

1. Read and complete a lab safety form.
2. Tie two **large, thick rubber bands** together.
3. Loop one rubber band lengthwise around a **textbook.**
4. Use **tape** to secure a sheet of **medium-grained sandpaper** to the tabletop.
5. Tape a second sheet of sandpaper to the cover of the textbook.
6. Place the book on the table so that the sheets of sandpaper touch.
7. Slowly pull on the end of the rubber band until the book moves.
8. Observe and record what happens in your Science Journal.

Think About This

1. How does this experiment model the buildup of stress along a fault?
2. **Key Concept** Why does the rapid movement of rocks along a fault result in an earthquake?

What are earthquakes?

Have you ever tried to bend a stick until it breaks? When the stick snaps, it vibrates, releasing energy. Earthquakes happen in a similar way. **Earthquakes** *are the vibrations in the ground that result from movement along breaks in Earth's lithosphere.* These breaks are called faults.

Key Concept Check What is an earthquake?

Why do rocks move along a fault? The forces that move tectonic plates also push and pull on rocks along the fault. If these forces become large enough, the blocks of rock on either side of the fault can move horizontally or vertically past each other. The greater the force applied to a fault, the greater the chance of a large and destructive earthquake. **Figure 1** shows earthquake damage from the Haiti earthquake in 2010.

Figure 1 In 2010, the Haiti earthquake along the Enriquillo–Plantain Garden fault caused $14 billion in damage.

(t)Hutchings Photography/Digital Light Source, (b)William S Helsel/Getty Images

Earthquake Distribution

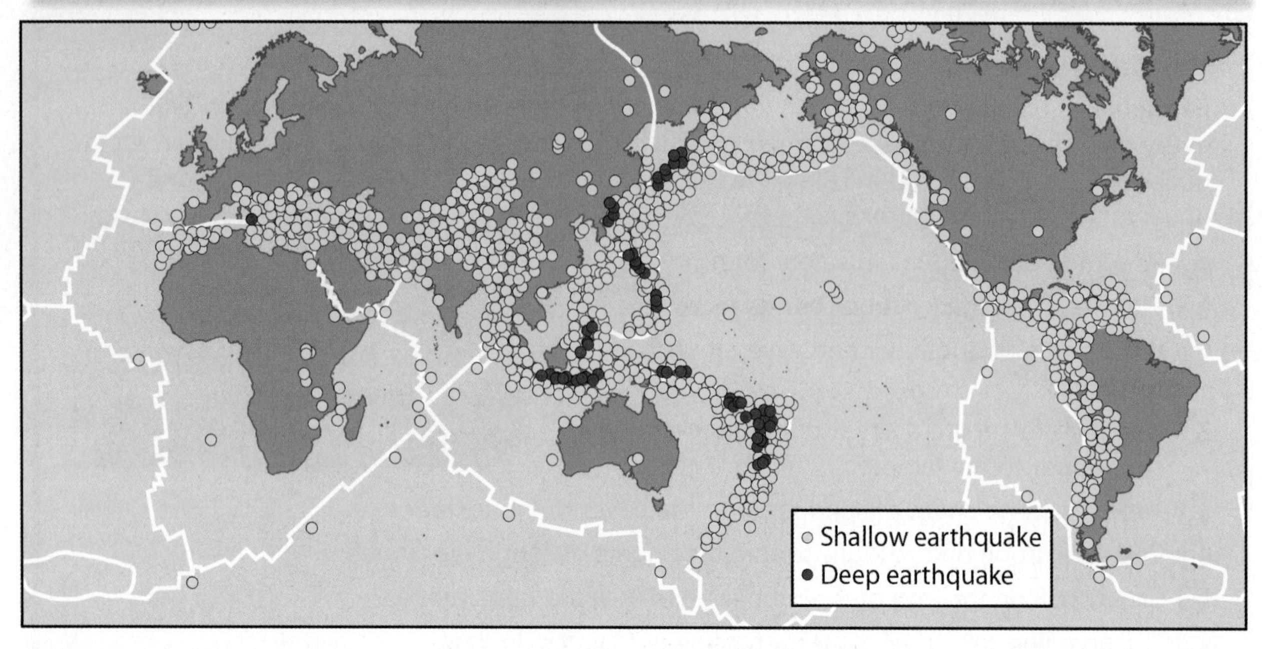

Figure 2 Notice that most earthquakes occur along plate boundaries.

Where do earthquakes occur?

The locations of major earthquakes that occurred between 2000 and 2008 are shown in **Figure 2.** Notice that only a few earthquakes occurred in the middle of a continent. Records show that most earthquakes occur in the oceans and along the edges of continents. Are there any exceptions?

Earthquakes and Plate Boundaries

Compare the location of earthquakes in **Figure 2** with tectonic plate boundaries. What is the relationship between earthquakes and plate boundaries? Earthquakes result from the buildup and release of stress along active plate boundaries.

REVIEW VOCABULARY

plate boundary
area where Earth's lithospheric plates move and interact with each other, producing earthquakes, volcanoes, and mountain ranges

Some earthquakes occur more than 100 km below Earth's surface, as shown in **Figure 2.** Which plate boundaries are associated with deep earthquakes? The deepest earthquakes occur where plates collide along a convergent plate boundary. Here, the denser oceanic plate subducts into the mantle. Earthquakes that occur along convergent plate boundaries typically release tremendous amounts of energy. They can also be disastrous.

Shallow earthquakes are common where plates separate along a divergent plate boundary, like the mid-ocean ridge system. Shallow earthquakes can also occur along transform plate boundaries like the San Andreas Fault in California. Earthquakes of varying depths occur where continents collide. Continental collisions result in the formation of large and deformed mountain ranges such as the Himalayas in Asia.

 **Key Concept Check** Where do most earthquakes occur?

Rock Deformation

At the beginning of this lesson, you read that earthquake energy is similar to bending and breaking a stick. Rocks below Earth's surface behave the same way. When a force is applied to a body of rock, depending on the properties of the rock and the force applied, the rock might bend or break.

When a force such as pressure is applied to rock along plate boundaries, the rock can change shape. This is called rock deformation. Eventually the rocks can be deformed so much that they break and move. **Figure 3** illustrates how rock deformation can result in ground displacement. Notice that rock deformation has resulted in ground displacement and has caused the stream to change direction.

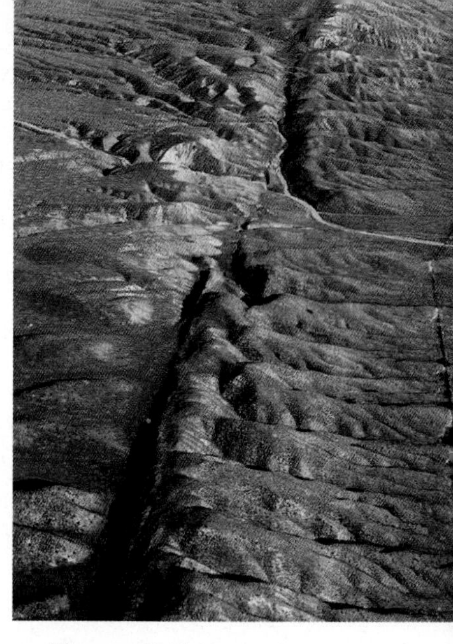

▲ **Figure 3** Forces at work along the San Andreas Fault in California changed drainage patterns and the course of this stream along the fault.

Faults

When stress builds in places like a plate boundary, rocks can form faults. *A* **fault** *is a break in Earth's lithosphere where one block of rock moves toward, away from, or past another.* When rocks move in any direction along a fault, an earthquake occurs. The direction that rocks move on either side of the fault depends on the forces applied to the fault. **Table 1** lists three types of faults that result from motion along plate boundaries. These faults are called strike-slip, normal, and reverse faults.

 **Reading Check** What is a fault?

Table 1 The three types of faults are defined based on relative motion along the fault. ▼

Table 1 Types of Faults

Strike-slip	• Two blocks of rock slide horizontally past each other in opposite directions. • Location: transform plate boundaries	
Normal	• Forces pull two blocks of rock apart. The block of rock above the fault moves down relative to the block of rock below the fault. • Location: divergent plate boundaries	
Reverse	• Forces push two blocks of rock together. The block of rock above the fault moves up relative to the block of rock below the fault. • Location: convergent plate boundaries	

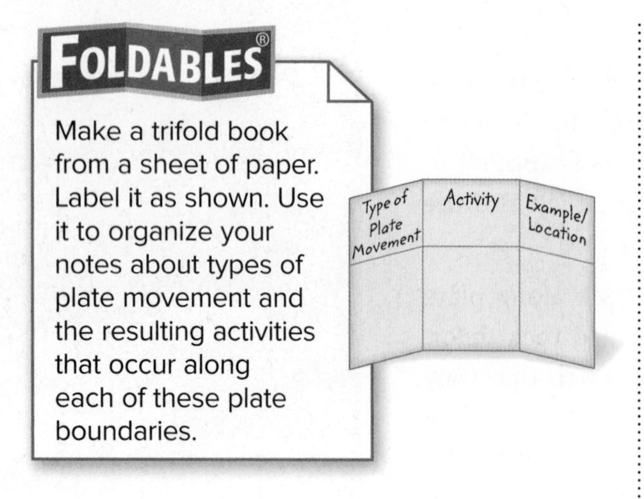

Make a trifold book from a sheet of paper. Label it as shown. Use it to organize your notes about types of plate movement and the resulting activities that occur along each of these plate boundaries.

Types of Faults Strike-slip faults can form along transform plate boundaries. There, forces cause rocks to slide horizontally past each other in opposite directions. In contrast, normal faults can form when forces pull rocks apart along a divergent plate boundary. At a normal fault, one block of rock moves down relative to the other. Forces push rocks toward each other at a convergent plate boundary and a reverse fault can form. There, one block of rock moves up relative to another block of rock.

Reading Check What are the three types of faults?

Earthquake Focus and Epicenter

When rocks move along a fault, they release *energy that travels as vibrations on and in Earth called* **seismic waves.** *These waves originate where rocks first move along the fault, at a location inside Earth called the* **focus.** Earthquakes can occur anywhere between Earth's surface and depths of greater than 600 km. When you watch a news report, the reporter often will identify the earthquake's epicenter. *The* **epicenter** *is the location on Earth's surface directly above the earthquake's focus.* **Figure 4** shows the relationship between an earthquake's focus and its epicenter.

SCIENCE USE V. COMMON USE

focus

Science Use the place of origin of an earthquake

Common Use to concentrate

Figure 4 An earthquake epicenter is above a focus, where the motion along the fault first occurs.

Visual Check What is the relationship between an earthquake focus and an epicenter?

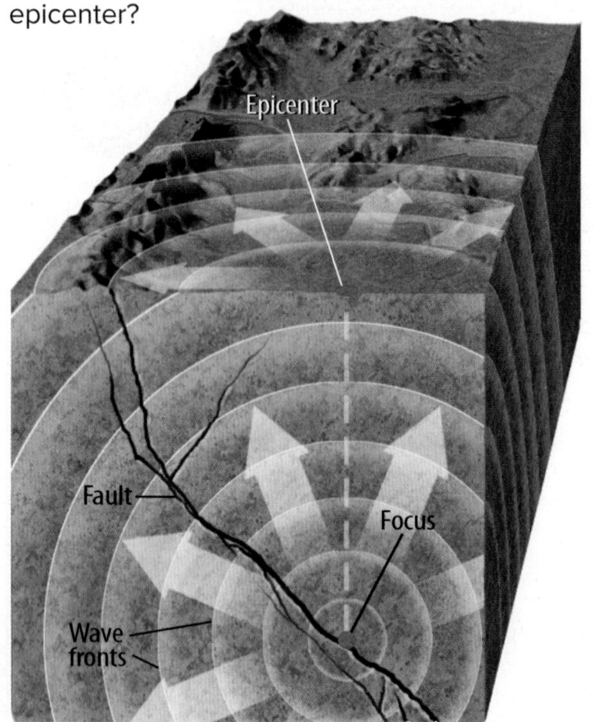

Seismic Waves

During an earthquake, a rapid release of energy along a fault produces seismic waves. Seismic waves travel outward in all directions through rock. It is similar to what happens when you drop a stone into water. When the stone strikes the water's surface, ripples move outward in circles. Seismic waves transfer energy through the ground and produce the motion that you feel during an earthquake. The energy released is strongest near the epicenter. As seismic waves move away from the epicenter, they decrease in energy and intensity. The farther you are from an earthquake's epicenter, the less the ground moves.

Types of Seismic Waves

When an earthquake occurs, particles in the ground can move back and forth, up and down, or in an elliptical motion parallel to the direction the seismic wave travels. Scientists use wave motion, wave speed, and the type of material that the waves travel through to classify seismic waves. The three types of seismic waves are primary waves, secondary waves, and surface waves.

As shown in **Table 2**, **primary waves,** *also called P-waves, cause particles in the ground to move in a push-pull motion similar to a coiled spring.* P-waves are the fastest-moving seismic waves. They are the first waves that you feel following an earthquake. **Secondary waves,** *also called S-waves,* are slower than P-waves. *They cause particles to move up and down at right angles relative to the direction the wave travels.* This movement can be demonstrated by shaking a coiled spring side to side and up and down at the same time. **Surface waves** *cause particles in the ground to move up and down in a rolling motion,* similar to ocean waves. Surface waves travel only on Earth's surface closest to the epicenter. P-waves and S-waves can travel through Earth's interior. However, scientists have discovered that S-waves cannot travel through liquid.

Reading Check Describe the three types of seismic waves.

WORD ORIGIN

primary
from Latin *primus,* means "first"

Animation

Table 2 The three types of seismic waves are classified by wave motion, wave speed, and the types of materials they can travel through.

Table 2 Properties of Seismic Waves

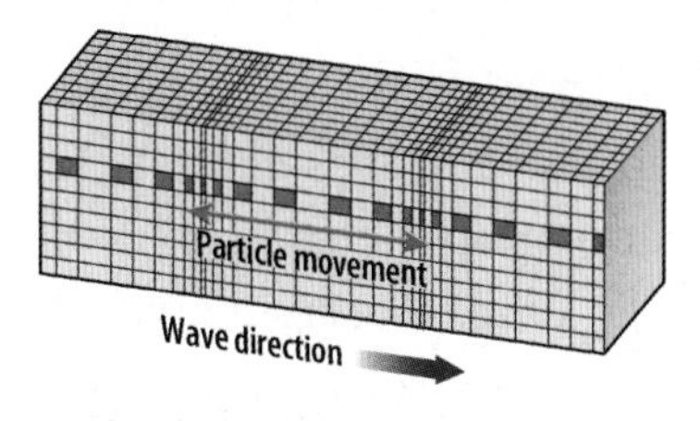	**Primary wave** • Cause rock particles to vibrate in the same direction that waves travel • Fastest seismic waves • First to be detected and recorded • Travel through solids and liquids
Secondary wave • Cause rock particles to vibrate perpendicular to the direction that waves travel • Slower than P-waves, faster than surface waves • Detected and recorded after P-waves • Only travel through solids	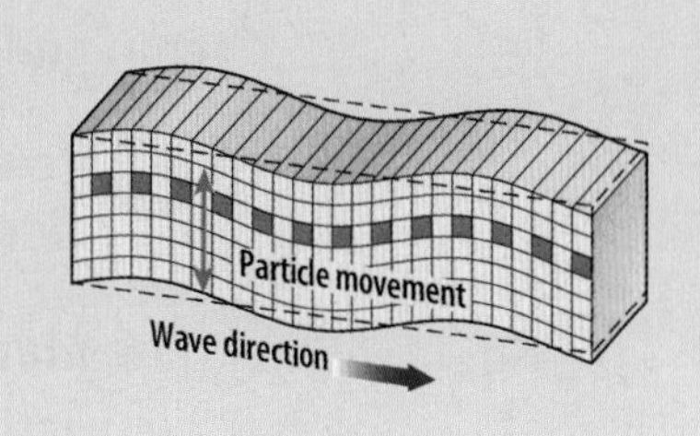
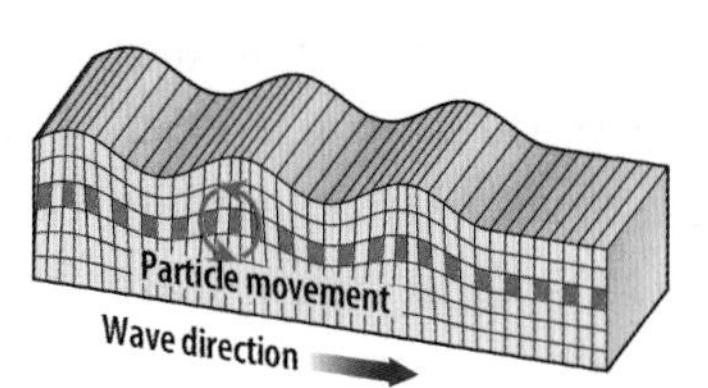	**Surface wave** • Cause rock particles to move in a rolling or elliptical motion in the same direction that waves travel • Slowest seismic wave • Generally cause the most damage at Earth's surface

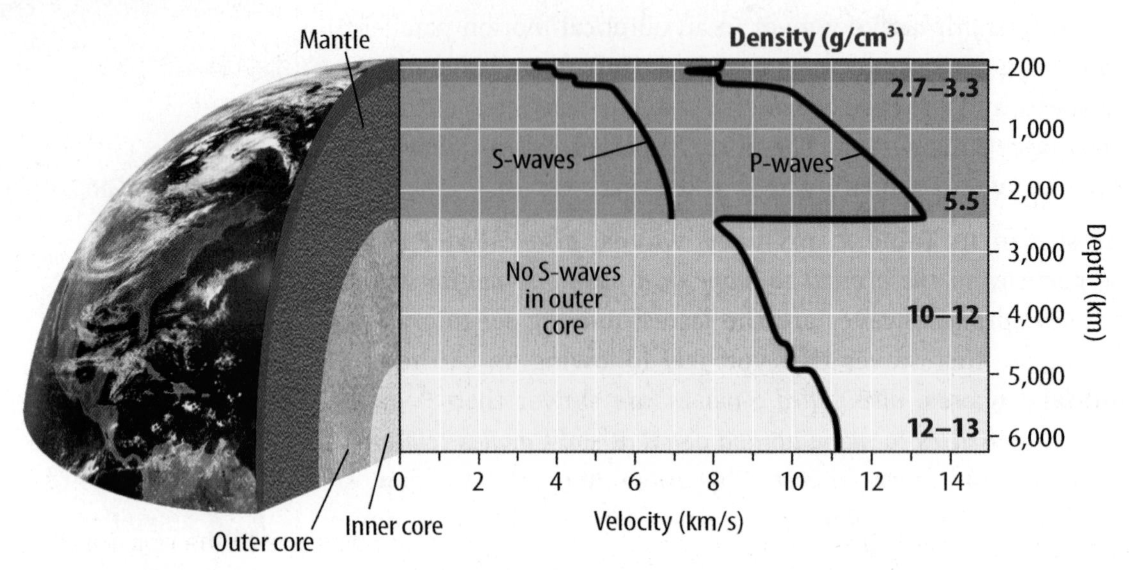

Figure 5 Seismic waves change speed and direction as they travel through Earth's interior. S-waves do not travel through Earth's outer core because it is liquid.

Visual Check What happens to P-waves and S-waves at a depth of 2500 km?

Mapping Earth's Interior

Scientists that study earthquakes are called **seismologists** (size MAH luh just). They use the properties of seismic waves to map Earth's interior. P-waves and S-waves change speed and direction depending on the material they travel through. **Figure 5** shows the speed of P-waves and S-waves at different depths within Earth's interior. By comparing these measurements to the densities of different Earth materials, scientists have determined the composition of Earth's layers.

Inner and Outer Core Through extensive earthquake studies, seismologists have discovered that S-waves cannot travel through the outer core. This discovery proved that Earth's outer core is liquid unlike the solid inner core. By analyzing the speed of P-waves traveling through the core, seismologists also have discovered that the inner and outer cores are composed of mostly iron and nickel.

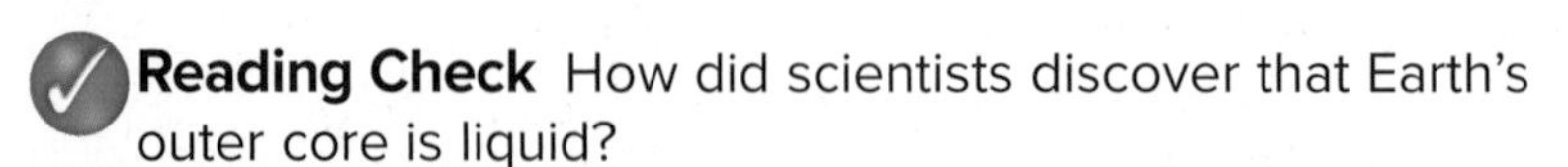

Reading Check How did scientists discover that Earth's outer core is liquid?

The Mantle Seismologists also have used seismic waves to model convection currents in the mantle. The speeds of seismic waves depend on the temperature, pressure, and chemistry of the rocks that the seismic waves travel through. Seismic waves tend to slow down as they travel through hot material. For example, seismic waves are slower in areas of the mantle beneath mid-ocean ridges or near hotspots. Seismic waves are faster in cool areas of the mantle near subduction zones.

Locating an Earthquake's Epicenter

An instrument called a **seismometer** (size MAH muh ter) *measures and records ground motion and can be used to determine the distance seismic waves travel.* Ground motion is recorded as a **seismogram,** *a graphical illustration of seismic waves,* shown in **Figure 6.**

Seismologists use a method called triangulation to locate an earthquake's epicenter. This method uses the speeds and travel times of seismic waves to determine the distance to the earthquake epicenter from at least three different seismometers.

1 Find the arrival time difference.

First, determine the number of seconds between the arrival of the first P-wave and the first S-wave on the seismogram. This time difference is called lag time. Using the time scale on the bottom of the seismogram, subtract the arrival time of the first P-wave from the arrival time of the first S-wave.

2 Find the distance to the epicenter.

Next, use a graph showing the P-wave and S-wave lag time plotted against distance. Look at the *y*-axis and locate the place on the solid blue line that intersects with the lag time that you calculated from the seismogram. Then, read the corresponding distance from the epicenter on the *x*-axis.

3 Plot the distance on a map.

Next, use a ruler and a map scale to measure the distance between the seismometer and the earthquake epicenter. Draw a circle with a radius equal to this distance by placing the compass point on the seismometer location. Set the pencil at the distance measured on the scale. Draw a complete circle around the seismometer location. The epicenter is somewhere on the circle. When circles are plotted for data from at least three seismic stations, the epicenter's location can be found. This location is the point where the three circles intersect.

Triangulation

1 Find the arrival time difference.

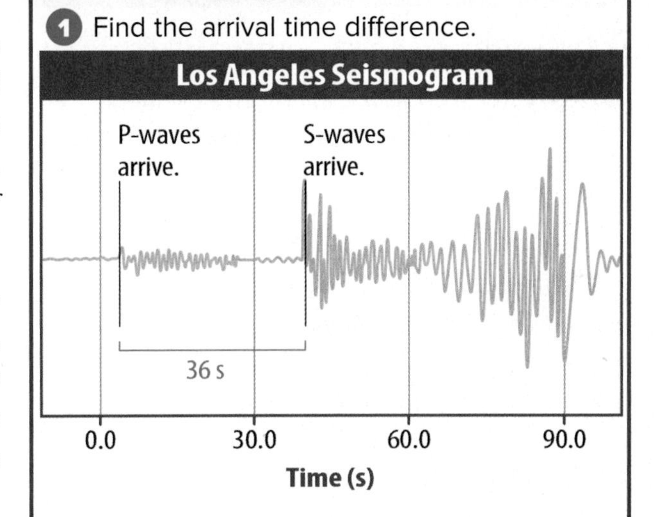

2 Find the distance to the epicenter.

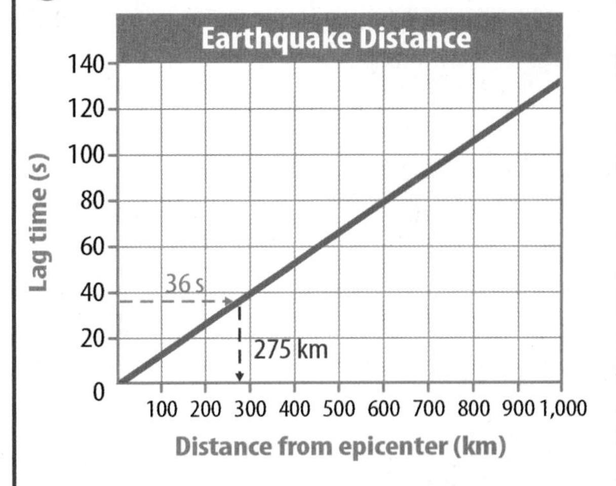

3 Plot the distance on the map.

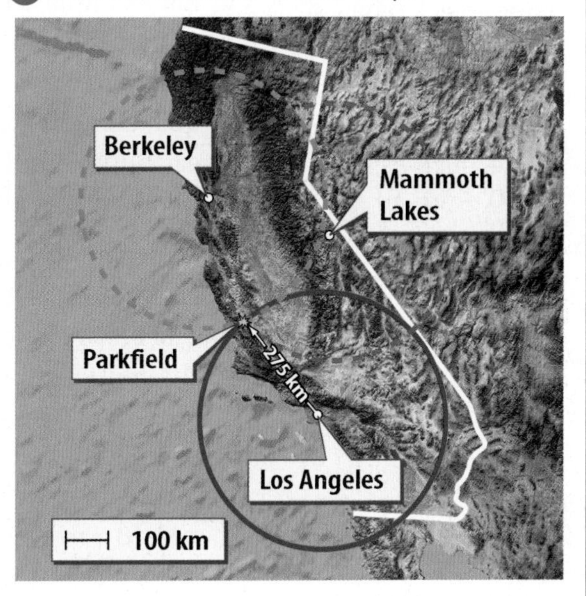

Figure 6 Seismograms provide the information necessary to locate an earthquake epicenter.

Personal Tutor

MiniLab

15 minutes

Can you use the Mercalli scale to locate an epicenter?

Isoseismic (I soh SIZE mihk) lines connect areas that experience equal intensity during an earthquake. In this activity, you will observe trends in intensity and use the Mercalli scale to locate an earthquake epicenter.

1. Obtain a **map** of Mercalli ratings for the San Francisco Bay area.
2. Draw a line that connects all the points of equal intensity, making a closed loop. This is your first isoseismic line.
3. Continue drawing isoseismic lines for each Mercalli rating on the map. Just like contour lines, these lines should never cross.

Analyze and Conclude

1. **Interpret Data** Identify two cities that experienced similar effects during the earthquake.
2. **Infer** What were some of the experiences people in San Francisco might have had during the earthquake?
3.  **Key Concept** Can you identify the earthquake's epicenter on your map? Why did you choose this location?

Math Skills

Use Roman Numerals

Use the following rules to evaluate Roman numerals.

1. Values: X = 10; V = 5; I = 1
2. Add similar values that are next to one another, such as III (1 + 1 + 1 = 3)
3. Add a smaller value that comes after a larger value, such as XV (10 + 5 = 15)
4. Subtract a smaller value that precedes a larger value, such as IX (10 − 1 = 9)
5. Use the fewest possible numerals to express the value (X rather than VV)

Practice

What is the value of the Roman numeral XVI? XIV?

 Math Practice

 Personal Tutor

Determining Earthquake Magnitude

Scientists can use three different scales to measure and describe earthquakes. The Richter magnitude scale uses the amount of ground motion at a given distance from an earthquake to determine magnitude. The Richter magnitude scale is used when reporting earthquake activity to the general public.

The Richter scale begins at zero, but there is no upper limit to the scale. Each increase of 1 unit on the scale represents ten times the amount of ground motion recorded on a seismogram. For example, a magnitude 8 earthquake produces 10 times greater shaking than a magnitude 7 earthquake and 100 times greater shaking than a magnitude 6 earthquake does. The largest earthquake ever recorded was a magnitude 9.5 in Chile in 1960. The earthquake and the tsunamis that followed left nearly 2,000 people dead and 2 million people homeless.

Seismologists use the moment magnitude scale to measure the total amount of energy released by the earthquake. The energy released depends on the size of the fault that breaks, the motion that occurs along the fault, and the strength of the rocks that break during an earthquake. The units on this scale are exponential. For each increase of one unit on the scale, the earthquake releases 31.5 times more energy. That means that a magnitude 8 earthquake releases more than 992 times the amount of energy than that of a magnitude 6 earthquake.

 **Reading Check** Compare the Richter scale to the moment magnitude scale.

Describing Earthquake Intensity

Another way to measure and describe an earthquake is to evaluate the damage that results from shaking. Shaking is directly related to earthquake intensity. The Modified Mercalli scale measures earthquake intensity based on descriptions of the earthquake's effects on people and structures. The Modified Mercalli scale, shown in **Table 3**, ranges from I, when shaking is not noticeable, to XII, when everything is destroyed.

Local geology also contributes to earthquake damage. In an area covered by loose sediment, ground motion is exaggerated. The intensity of the earthquake will be greater there than in places built on solid bedrock even if they are the same distance from the epicenter. Recall the lesson opener. The 1989 Loma Prieta earthquake produced severe shaking in an area called the Marina District in the San Francisco Bay area. This area had been built on loose sediment susceptible to shaking.

Table 3 The Modified Mercalli scale is used to evaluate earthquake intensity based on the damage that results.

Table 3 Modified Mercalli Scale

I	Not felt except under unusual conditions.
II	Felt by few people; suspended objects might swing.
III	Most noticeable indoors; vibrations feel like the effects of a truck passing by.
IV	Felt by many people indoors but by few people outdoors; dishes and windows rattle; standing cars rock noticeably.
V	Felt by nearly everyone; some dishes and windows break and some walls crack.
VI	Felt by all; furniture moves; some plaster falls from walls and some chimneys are damaged.
VII	Everybody runs outdoors; some chimneys break; damage is light in well-built structures but considerable in weak structures.
VIII	Chimneys, smokestacks, and walls fall; heavy furniture is overturned; partial collapse of ordinary buildings occurs.
IX	Great general damage occurs; buildings shift off foundations; ground cracks; underground pipes break.
X	Most ordinary structures are destroyed; rails are bent; landslides are common.
XI	Few structures remain standing; bridges are destroyed; railroad rails are greatly bent; broad fissures form in the ground.
XII	Total destruction; objects are thrown upward into the air.

Earthquake Hazard Map

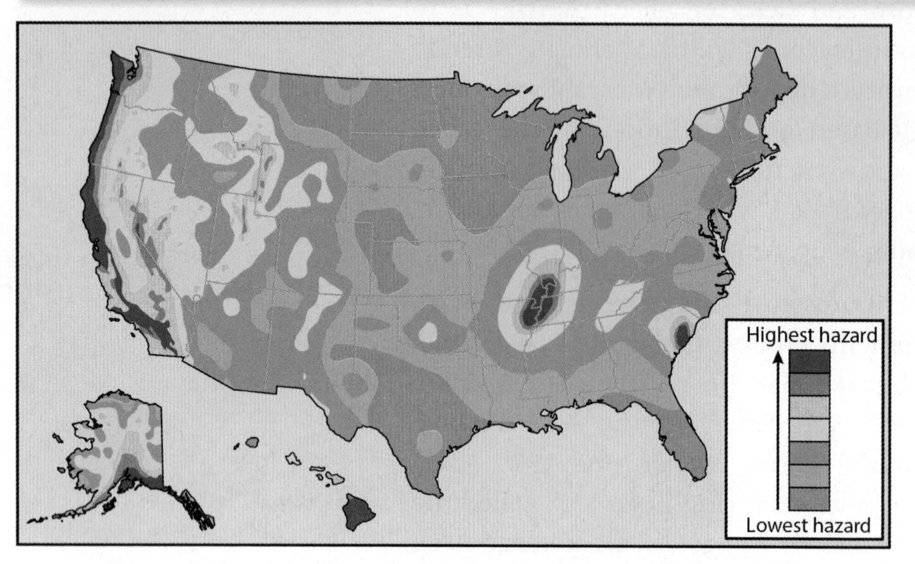

Figure 7 Areas that experienced earthquakes in the past will likely experience earthquakes again. Notice that even some parts of the central and eastern United States have high earthquake risk because of past activity.

REVIEW VOCABULARY

convergent
tending to move toward one point or approaching each other

Earthquake Risk

Recall that most earthquakes occur near tectonic plate boundaries. The transform plate boundary in California and the **convergent** plate boundaries in Oregon, Washington, and Alaska have the highest earthquake risks in the United States. However, not all earthquakes occur near plate boundaries. Some of the largest earthquakes in the United States have occurred far from plate boundaries.

From 1811-1812, three earthquakes with magnitudes between 7.8 and 8.1 occurred on the New Madrid Fault in Missouri. In contrast, the 1989 Loma Prieta earthquake had a magnitude of 7.1. **Figure 7** illustrates earthquake risk in the United States. Fortunately, high energy, destructive earthquakes are not very common. On average, only about 10 earthquakes with a magnitude greater than 7.0 occur worldwide each year. Earthquakes with magnitudes greater than 9.0, such as the Indian Ocean earthquake that caused the Asian tsunami in 2004, are rare.

Because earthquakes threaten people's lives and property, seismologists study the probability that an earthquake will occur in a given area. Probability is one of several factors that contribute to earthquake risk assessment. Seismologists also study past earthquake activity, the geology around a fault, the population density, and the building design in an area to evaluate risk. Engineers use these risk assessments to design earthquake-safe structures that are able to withstand the shaking during an earthquake. City and state governments use risk assessments to help plan and prepare for future earthquakes.

 **Key Concept Check** How do seismologists evaluate risk?

Lesson 1 Review

Visual Summary

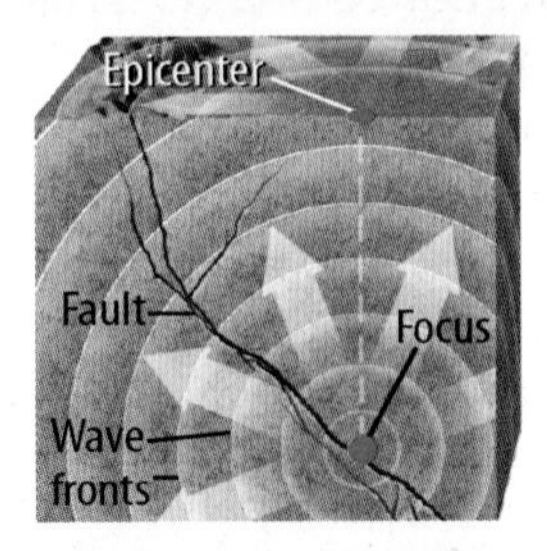

The focus is the area on a fault where an earthquake begins.

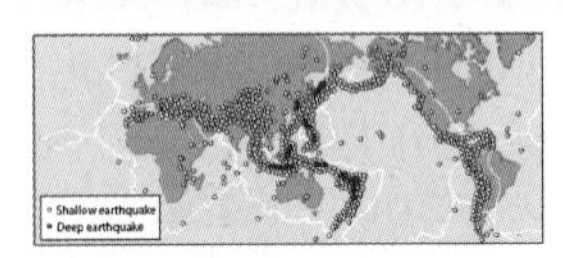

Earthquakes can occur along plate boundaries.

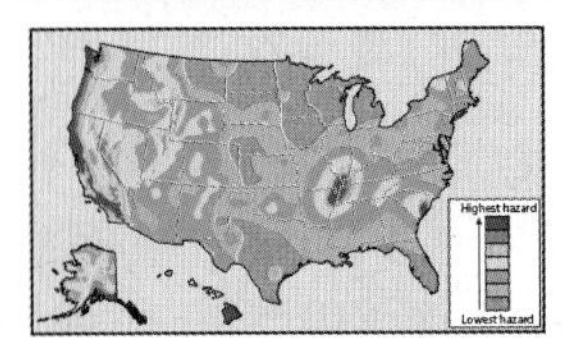

Seismologists assess earthquake risk by studying past earthquake activity and local geology.

FOLDABLES®

Use your lesson Foldable to review the lesson. Save your Foldable for the project at the end of the chapter.

What do you think NOW?

You first read the statements below at the beginning of the chapter.

1. Earth's crust is broken into rigid slabs of rock that move, causing earthquakes and volcanic eruptions.

2. Earthquakes create energy waves that travel through Earth.

3. All earthquakes occur on plate boundaries.

Did you change your mind about whether you agree or disagree with the statements? Rewrite any false statements to make them true.

Use Vocabulary

1. **Compare and contrast** the three types of faults.
2. **Distinguish** between an earthquake focus and an earthquake epicenter.
3. **Use the terms** *seismogram* and *seismometer* in a sentence.

Understand Key Concepts 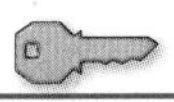

4. **Identify** areas in the United States that have the highest earthquake risk.
5. Approximately how much more energy is released in a magnitude 7 earthquake compared to a magnitude 5 earthquake?

   A. 30 **C.** 90
   B. 60 **D.** 1000

Interpret Graphics

6. **Compare and contrast** Create a table with the column headings for wave type, wave motion, and wave properties. Use the table to compare and contrast the three types of seismic waves.
7. Describe Use the image below to describe Earth's interior.

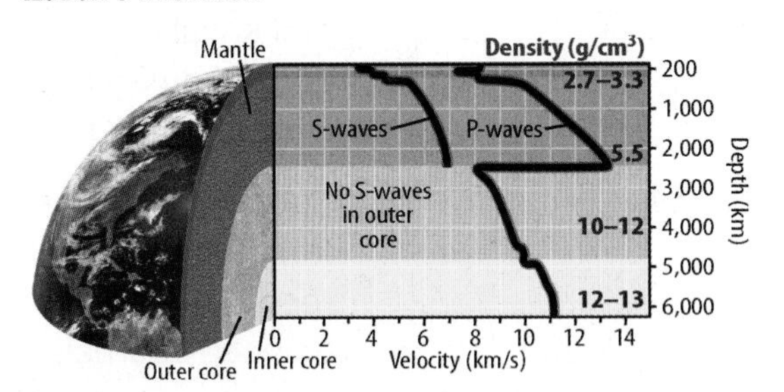

Critical Thinking

8. **Determine** what measurements you would make to evaluate earthquake risk in your hometown.

Math Skills

9. What is the value of Roman numeral XXVI?

Can you locate an earthquake's epicenter?

Materials

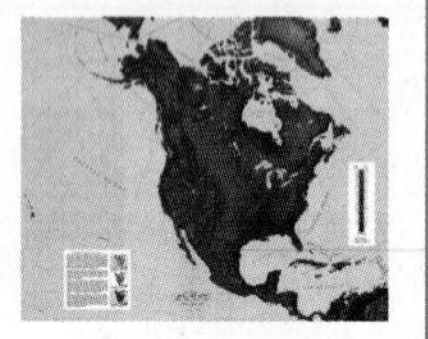

map of North America

drawing compass

Safety

Imagine the room where you are sitting suddenly begins to shake. This movement lasts for about 10 seconds. Based on the shaking that you have felt, it might seem like an earthquake happened nearby. But, to locate the epicenter, you need to analyze P-wave and S-wave data recorded for the same earthquake in at least three different locations.

Learn It

When scientists conduct experiments, they make measurements and collect and **analyze** data. For example, seismologists measure the difference in arrival times between P-waves and S-waves following an earthquake. They collect seismic wave data from at least three different locations. Using the difference in arrival times, or lag time, seismologists can determine the distance to an earthquake epicenter.

Try It

1. Read and complete a lab safety form.
2. Obtain a map of the United States from your teacher.
3. Study the three seismograms. Determine the arrival times, to the nearest second, of the P- and S-waves for each seismometer station: Berkeley, CA; Parkfield, CA; and Kanab, UT. Record the location and the arrival times for P- and S-waves in your Science Journal.
4. Subtract the P-wave arrival time from the S-wave arrival time and record the lag time in your Science Journal.
5. Use the lag time and the Earthquake Distance graph to determine the distance to the epicenter for each seismometer station.

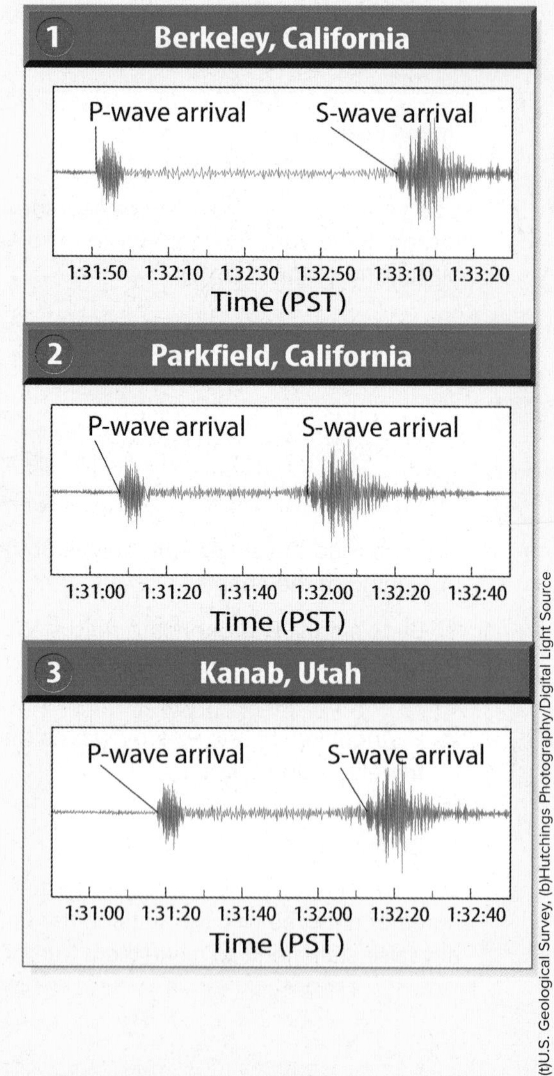

6 Use the map scale to set the spacing between the pencil and the point on the compass equal to the distance to the first seismometer. Draw a circle with a radius equal to the distance around the seismic station on the map.

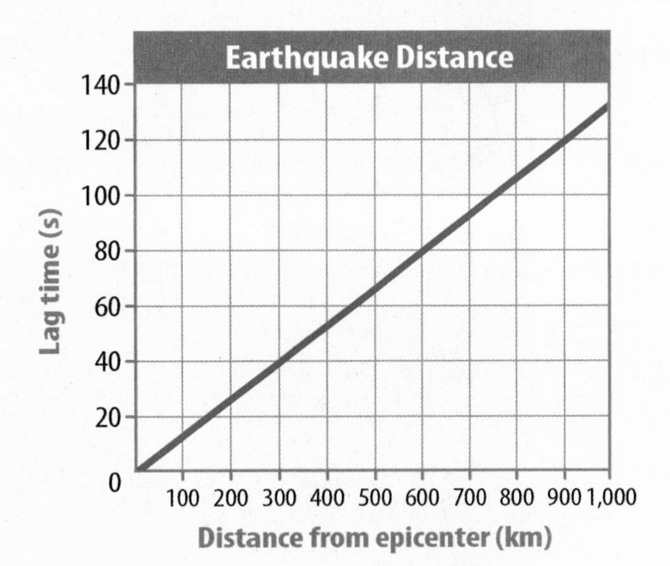

7 Repeat for the two other seismometer locations. The point where the three circles intersect marks the earthquake epicenter.

Apply It

8 Consider the difference between the arrival times of P-waves for all three seismometer locations. Why does this difference occur?

9 **Examine** the calculated lag times for all three seismograms. Why do you think the arrival-time differences are greater for the stations that are furthest from the epicenter?

10 Where did the earthquake occur?

11 **Key Concept** Why does it take three seismograms to locate an earthquake epicenter? What is this process called?

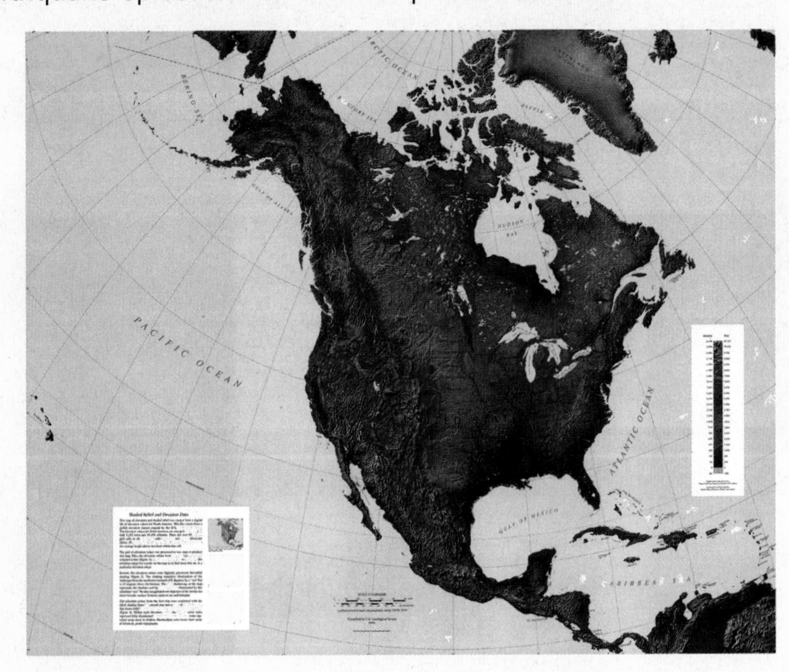

U.S. Geological Survey

Lesson 2

Reading Guide

Key Concepts
ESSENTIAL QUESTIONS

- How do volcanoes form?
- What factors contribute to the eruption style of a volcano?
- How are volcanoes classified?

Vocabulary

volcano p. 307
magma p. 307
lava p. 308
hot spot p. 308
shield volcano p. 310
composite volcano p. 310
cinder cone p. 310
volcanic ash p. 311
viscosity p. 311

Multilingual eGlossary

Science Video
What's Science Got to do With It?

Volcanoes

Inquiry What makes an eruption explosive?

Notice the red, hot "fire fountain" erupting from Kilauea volcano in Hawaii. Kilauea is the most active volcano in the world. Now recall the ash eruption pictured in the chapter opener. What makes volcanoes erupt so differently? The answer can be found in magma chemistry.

Launch Lab

15 minutes

What determines the shape of a volcano?

Not all volcanoes look the same. The location of a volcano and the magma chemistry play an important part in determining the shape of a volcano.

1. Read and complete a lab safety form.
2. Obtain a **tray, a beaker of sand, a beaker with a mixture of flour and water, waxed paper,** and a **plastic spoon.**
3. Lay the waxed paper inside the tray.
4. Hold the beaker of sand about 30 cm above the tray. Slowly pour the sand onto the waxed paper and observe how it piles up.
5. Fold the paper in half and use it to carefully pour the sand back into the beaker.
6. Stir the flour and water mixture. It should be about the consistency of oatmeal. Add water if necessary.
7. Repeat steps 4 and 5 with the flour and water mixture. Record your observations for each trial in your Science Journal.

Think About This

1. What do the sand and the flour and water mixture represent?
2. **Key Concept** How do you think volcanoes get their shape?

What is a volcano?

Perhaps you have heard of some famous volcanoes such as Mount St. Helens, Kilauea, or Mount Pinatubo. All of these volcanoes have erupted within the last 30 years. *A* **volcano** *is a vent in Earth's crust through which melted–or molten–rock flows. Molten rock below Earth's surface is called* **magma.** Volcanoes are in many places worldwide. Some places have more volcanoes than others. In this lesson, you will learn about how volcanoes form, where they form, and about their structure and eruption style.

 **Reading Check** What is magma?

How do volcanoes form?

Volcanic eruptions constantly shape Earth's surface. They can form large mountains, create new crust, and leave a path of destruction behind. Scientists have learned that the movement of Earth's tectonic plates causes the formation of volcanoes and the eruptions that result.

Hutchings Photography/Digital Light Source

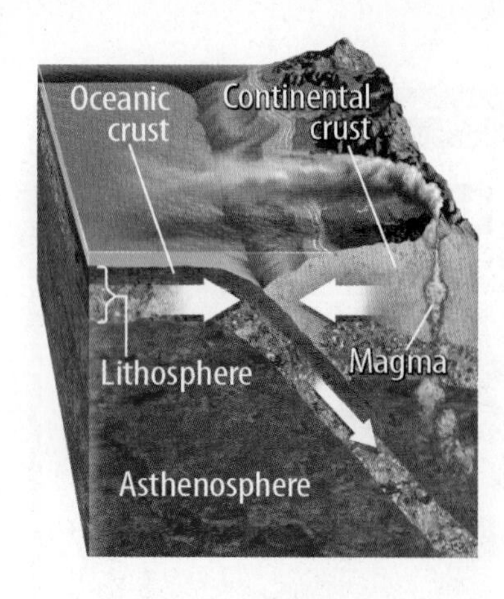

◄ **Figure 8** During subduction, magma forms when one plate sinks beneath another plate.

Figure 9 When plates spread apart, it forces magma to the surface and creates new crust. The pillow lava shown in the photograph formed at the mid-ocean ridge. ▼

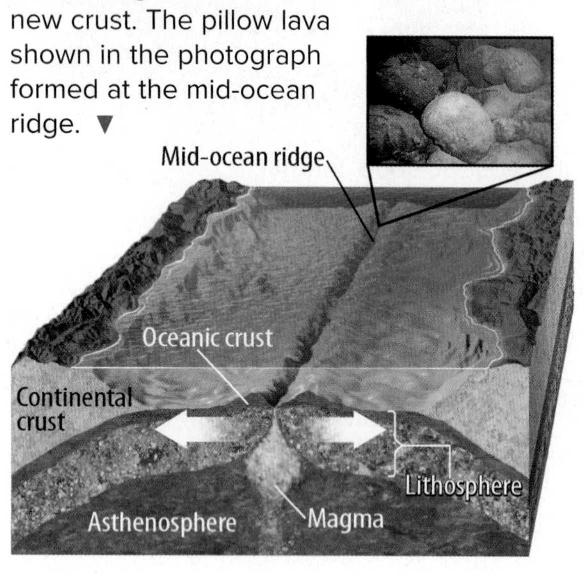

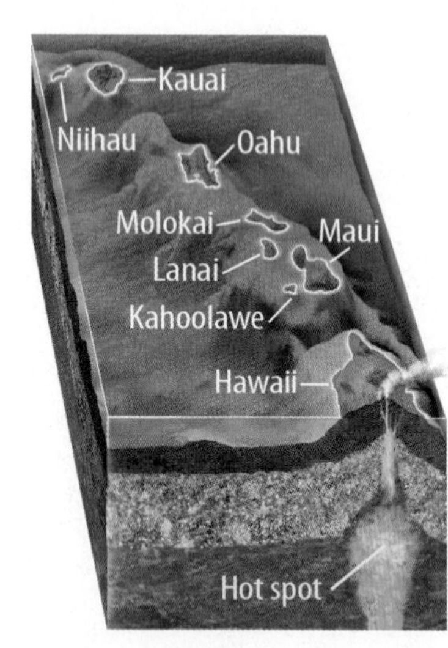

◄ **Figure 10** The farther each of the Hawaiian Islands is from the hot spot, the older the island is.

Convergent Boundaries

Volcanoes can form along convergent plate boundaries. Recall that when two plates collide, the denser plate sinks, or subducts, into the mantle, as shown in **Figure 8.** The thermal energy below the surface and fluids driven off the subducting plate melt the mantle and form magma. Magma is less dense than the surrounding mantle and rises through cracks in the crust. This forms a volcano. *Molten rock that erupts onto Earth's surface is called* **lava.**

Divergent Boundaries

Lava erupts along divergent plate boundaries too. Recall that two plates spread apart along a divergent plate boundary. As the plates separate, magma rises through the vent or opening in Earth's crust that forms between them. This process commonly occurs at a mid-ocean ridge and forms new oceanic crust, as shown in **Figure 9.** More than 60 percent of all volcanic activity on Earth occurs along mid-ocean ridges.

Hot spots

Not all volcanoes form on or near plate boundaries. Volcanoes in the Hawaiian Island-Emperor Seamount chain are far from plate boundaries. *Volcanoes that are not associated with plate boundaries are called* **hot spots.** Geologists hypothesize that hot spots originate above a rising convection current from deep within Earth's mantle. They use the word *plume* to describe these rising currents of hot mantle material.

Figure 10 illustrates how a new volcano forms as a tectonic plate moves over a plume. When the plate moves away from the plume, the volcano becomes dormant, or inactive. Over time, a chain of volcanoes forms as the plate moves. The oldest volcano will be farthest away from the hot spot. The youngest volcano will be directly above the hot spot.

Key Concept Check How do volcanoes form?

Volcano Distribution

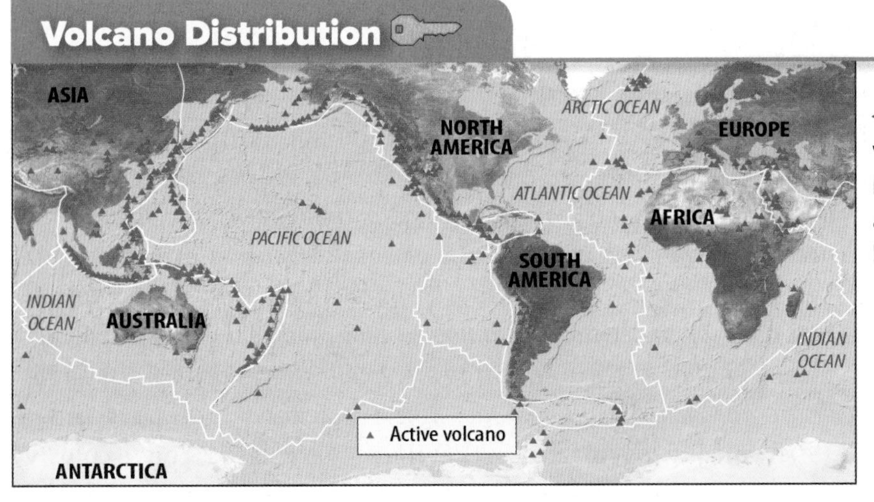

◄ **Figure 11** Most of the world's active volcanoes are located along convergent and divergent plate boundaries and hot spots.

Where do volcanoes form?

The world's active volcanoes are shown in **Figure 11.** The volcanoes all erupted within the last 100,000 years. Notice that most volcanoes are close to plate boundaries.

Ring of Fire

The Ring of Fire represents an area of earthquake and volcanic activity that surrounds the Pacific Ocean. When you compare the locations of active volcanoes and plate boundaries in **Figure 11,** you can see that volcanoes are mostly along convergent plate boundaries where plates collide. They also are located along divergent plate boundaries where plates separate. Volcanoes also can occur over hot spots, like Hawaii, the Galapagos Islands, and Yellowstone National Park in Wyoming.

Reading Check Where is the Ring of Fire?

Volcanoes in the United States

There are 60 potentially active volcanoes in the United States. Most of these volcanoes are part of the Ring of Fire. Alaska, Hawaii, Washington, Oregon, and northern California all have active volcanoes, such as Mount Redoubt in Alaska. A few of these volcanoes have produced violent eruptions, like the explosive eruption of Mount St. Helens in 1980.

The United States Geological Survey (USGS) has established three volcano observatories to monitor the potential for future volcanic eruptions in the United States. Because large populations of people live near volcanoes such as Mount Rainier in Washington, shown in **Figure 12,** the USGS has developed a hazard assessment program. Scientists monitor earthquake activity, changes in the shape of the volcano, gas emissions, and the past eruptive history of a volcano to evaluate the possibility of future eruptions.

Figure 12 Mount Rainier is an active volcano in the Cascade Mountains of the Pacific Northwest. Many people live in close proximity to the volcano. ▼

FOLDABLES®

Fold a sheet of paper to make a pyramid book. Use it to illustrate the three main types of volcanoes. Organize your notes inside the pyramid.

Shield
Composite
Cinder Cone

Types of Volcanoes

Volcanoes are classified based on their shapes and sizes, as shown in **Table 4.** Magma composition and eruptive style of the volcano contribute to the shape. **Shield volcanoes** *are common along divergent plate boundaries and oceanic hot spots. Shield volcanoes are large with gentle slopes of basaltic lavas.* **Composite volcanoes** *are large, steep-sided volcanoes that result from explosive eruptions of andesitic and rhyolitic lava and ash along convergent plate boundaries.* **Cinder cones** *are small, steep-sided volcanoes that erupt gas-rich, basaltic lavas.* Some volcanoes are classified as supervolcanoes–volcanoes that have very large and explosive eruptions. Approximately 630,000 years ago, the Yellowstone Caldera in Wyoming ejected more than 1,000 km^3 of rhyolitic ash and rock in one eruption. This eruption produced nearly 2,500 times the volume of material erupted from Mount St. Helens in 1980.

Key Concept Check What determines the shape of a volcano?

Table 4 Geologists classify volcanoes based on their size, shape, and eruptive style.

Table 4 Volcanic Features — Interactive Table

Shield volcano	Composite volcano
 Large, shield-shaped volcano with gentle slopes made from basaltic lavas.	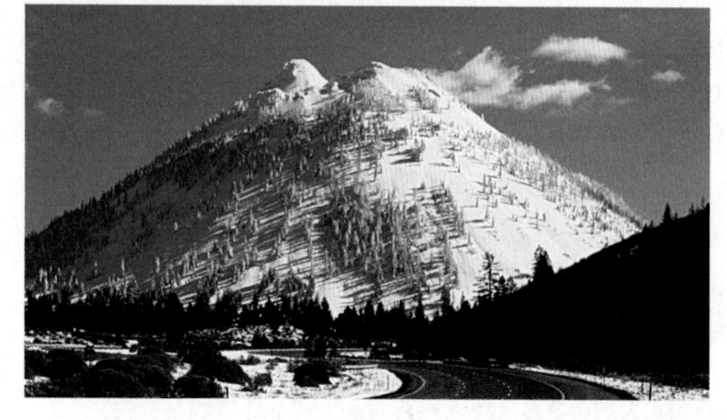 Large, steep-sided volcano made from a mixture of andesitic and rhyolitic lava and ash.
Cinder cone volcano	**Caldera**
 Small, steep-sided volcano; made from moderately explosive eruptions of basaltic lavas.	 Large volcanic depression formed when a volcano's summit collapses or is blown away by explosive activity.

Volcanic Eruptions

When magma surfaces, it might erupt as a lava flow, such as the lava shown in **Figure 13** erupting from Kilauea volcano in Hawaii. Other times, magma might erupt explosively, sending **volcanic ash**–*tiny particles of pulverized volcanic rock and glass*–high into the atmosphere. **Figure 13** also shows Mount St. Helens in Washington, erupting violently in 1980. Why do some volcanoes erupt violently while others erupt quietly?

Eruption Style

Magma chemistry determines a volcano's eruptive style. The explosive behavior of a volcano is affected by the amount of dissolved gases, specifically the amount of water vapor, a magma contains. It is also affected by the silica, SiO_2, content of magma.

Magma Chemistry Magmas that form in different volcanic environments have unique chemical compositions. Silica is the main chemical compound in all magmas. Differences in the amount of silica affect magma thickness and its **viscosity**–*a liquid's resistance to flow.*

Magma that has a low silica content also has a low viscosity and flows easily like warm maple syrup. When the magma erupts, it flows as fluid lava that cools, crystallizes, and forms the volcanic rock basalt. This type of lava commonly erupts along mid-ocean ridges and at oceanic hot spots, such as Hawaii.

Magma that has a high silica content has a high viscosity and flows like sticky toothpaste. This type of magma forms when rocks rich in silica melt or when magma from the mantle mixes with continental crust. The volcanic rocks andesite and rhyolite form when intermediate and high silica magmas erupt from subduction zone volcanoes and continental hot spots.

Key Concept Check What factors affect eruption style?

Quiet Eruption

Violent Eruption

Figure 13 Lavas that are low in silica and the amount of dissolved gases erupt quietly. Explosive eruptions result from lava and ash that are high in silica and dissolved gases.

Academic Vocabulary

dissolve
(verb) to cause to disperse or disappear

Dissolved Gases The presence of **dissolved** gases in magma contributes to how explosive a volcano can be. This is similar to what happens when you shake a can of soda and then open it. The bubbles come from the carbon dioxide that is dissolved in the soda. The pressure inside the can decreases rapidly when you open it. Trapped bubbles increase in size rapidly and escape as the soda erupts from the can.

All magmas contain dissolved gases. These gases include water vapor and small amounts of carbon dioxide and sulfur dioxide. As magma moves toward the surface, the pressure from the weight of the rock above decreases. As pressure decreases, the ability of gases to stay dissolved in the magma also decreases. Eventually, gases can no longer remain dissolved in the magma and bubbles begin to form. As the magma continues to rise to the surface, the bubbles increase in size and the gas begins to escape. Because gases cannot easily escape from high-viscosity lavas, this combination often results in explosive eruptions. When gases escape above ground, the lava, ash, or volcanic glass that cools and crystallizes has holes. These holes, shown in **Figure 14**, are a common feature in the volcanic rock pumice.

Figure 14 The holes in this pumice were caused by gas bubbles that escaped during a volcanic eruption.

MiniLab

20 minutes

Can you model the movement of magma?

Magma erupts because it is less dense than Earth's crust. Similarly, oil is less dense than water and can be used to model magma.

1. Read and complete a lab safety form.
2. Half-fill a **clear plastic cup** with **pebbles.**
3. Fill the cup with **water** to a level just above the top of the pebbles.
4. Fill a **syringe** with 5 mL of **olive oil.**
5. Insert the syringe between the pebbles and the side of the cup until it touches the bottom.
6. Inject the oil slowly, 1 mL at a time.
7. Observe and record your results in your Science Journal.
8. Repeat the procedures using **motor oil.**

Analyze and Conclude

1. **Observe** What happens to the oil when you inject it into the water?
2. Compare How did the movement of the two oils differ?
3. **Key Concept** Which oil behaves like magma that will become basalt? Which behaves like magma that will become rhyolite? Explain.

(t)Tony Lilley/Alamy, (b)Hutchings Photography/Digital Light Source

Effects of Volcanic Eruptions

On average, about 60 different volcanoes erupt each year. The effects of lava flows, ash fall, pyroclastic flows, and mudflows can affect all life on Earth. Volcanoes enrich rock and soil with valuable nutrients and help to regulate climate. Unfortunately, they also can be destructive and sometimes even deadly.

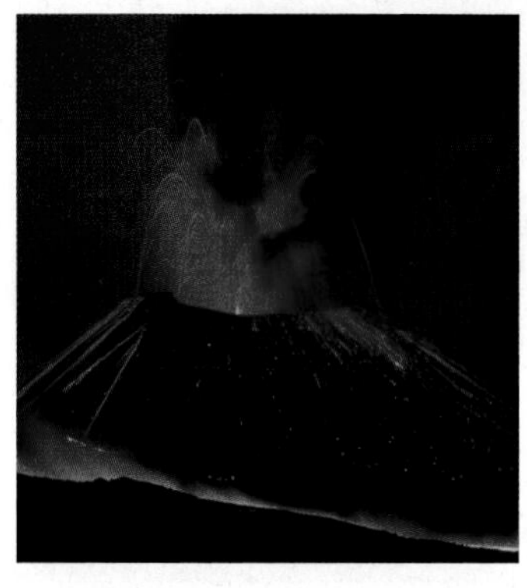

▲ **Figure 15** Mount Etna is one of the world's most active volcanoes. People that live near the volcano are accustomed to frequent eruptions of both lava and ash.

Lava Flows Because lava flows are relatively slow moving, they are rarely deadly. But lava flows can be damaging. Mount Etna in Sicily, Italy, is Europe's most active volcano. **Figure 15** shows a fountain of fluid, hot lava erupting from one of the volcano's many vents. In May 2008, the volcano began spewing lava and ash in an eruption lasting over six months. Although lavas tend to be slow moving, they threaten communities nearby. People who live on Mount Etna's slopes are used to evacuations due to frequent eruptions.

Ash Fall During an explosive eruption, volcanoes can erupt large volumes of volcanic ash. Ash columns can reach heights of more than 40 km. Recall that ash is a mixture of particles of pulverized rock and glass. Ash can disrupt air traffic and cause engines to stop mid-flight as shards of rock and ash fuse onto hot engine blades. Ash can also affect air quality and can cause serious breathing problems. Large quantities of ash erupted into the atmosphere can also affect climate by blocking out sunlight and cooling Earth's atmosphere.

Mudflows The thermal energy a volcano produces during an eruption can melt snow and ice on the summit. This meltwater can then mix with mud and ash on the mountain to form mudflows. Mudflows are also called lahars. Mount Redoubt in Alaska erupted on March 23, 2009. Snow and meltwater mixed to form the mudflows shown in **Figure 16**.

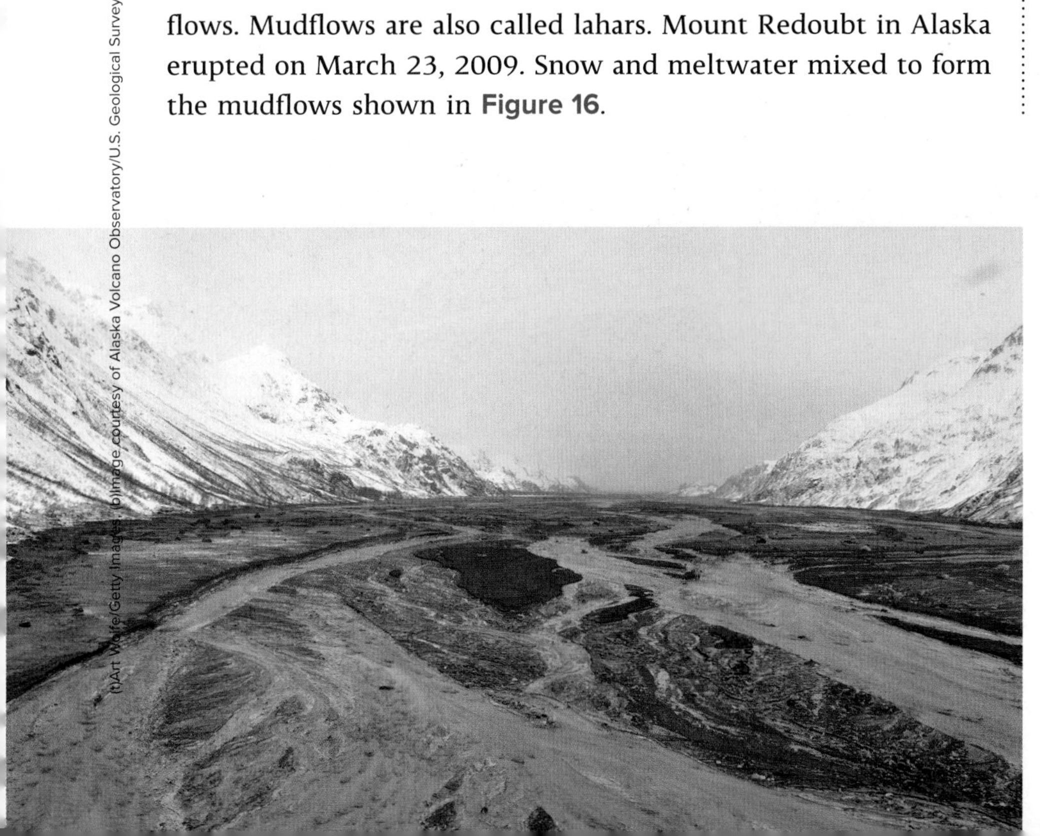

◄ **Figure 16** Many of the steep-sided composite volcanoes are covered with seasonal snow. When a volcano becomes active, the snow can melt and mix with mud and ash to form a mudflow like the one shown here in the Cook Inlet, Alaska.

▲ **Figure 17** A pyroclastic flow travels down the side of Mount Mayon in the Philippines. Pyroclastic flows are made of hot (*pyro*) volcanic particles (*clast*).

Pyroclastic Flow Explosive volcanoes can produce fast-moving avalanches of hot gas, ash, and rock called pyroclastic (pi roh KLAS tihk) flows. Pyroclastic flows travel at speeds of more than 100 km/hr and with temperatures greater than 1000°C. In 1980, Mount St. Helens produced a pyroclastic flow that killed 58 people and destroyed 1 billion km^3 of forest. Mount Mayon in the Phillippines erupts frequently producing pyroclastic flows like the one shown in **Figure 17.**

Predicting Volcanic Eruptions

Unlike earthquakes, volcanic eruptions can be predicted. Moving magma can cause ground deformation, a change in shape of the volcano, and a series of earthquakes called an earthquake swarm. Volcanic gas emissions can increase. Ground and surface water near the volcano can become more acidic. Geologists study these events, in addition to satellite and aerial photographs, to assess volcanic hazards.

Volcanic Eruptions and Climate Change

Volcanic eruptions affect climate when volcanic ash in the atmosphere blocks sunlight. High-altitude wind can move ash around the world. In addition, sulfur dioxide gases released from a volcano form sulfuric acid droplets in the upper atmosphere. These droplets reflect sunlight into space, resulting in lower temperatures as less sunlight reaches Earth's surface. **Figure 18** shows the result of sulfur dioxide gas in the atmosphere from the 1991 eruption of Mt. Pinatubo.

 **Key Concept Check** How do volcanoes affect climate?

Figure 18 In 1991, Mount Pinatubo erupted more than 20 million tons of gas and volcanic ash into the atmosphere. The greatest concentration of sulfur dioxide gas from the eruption is shown below in blue. The eruption caused temperatures to decrease by almost one degree Celsius in one year. ▼

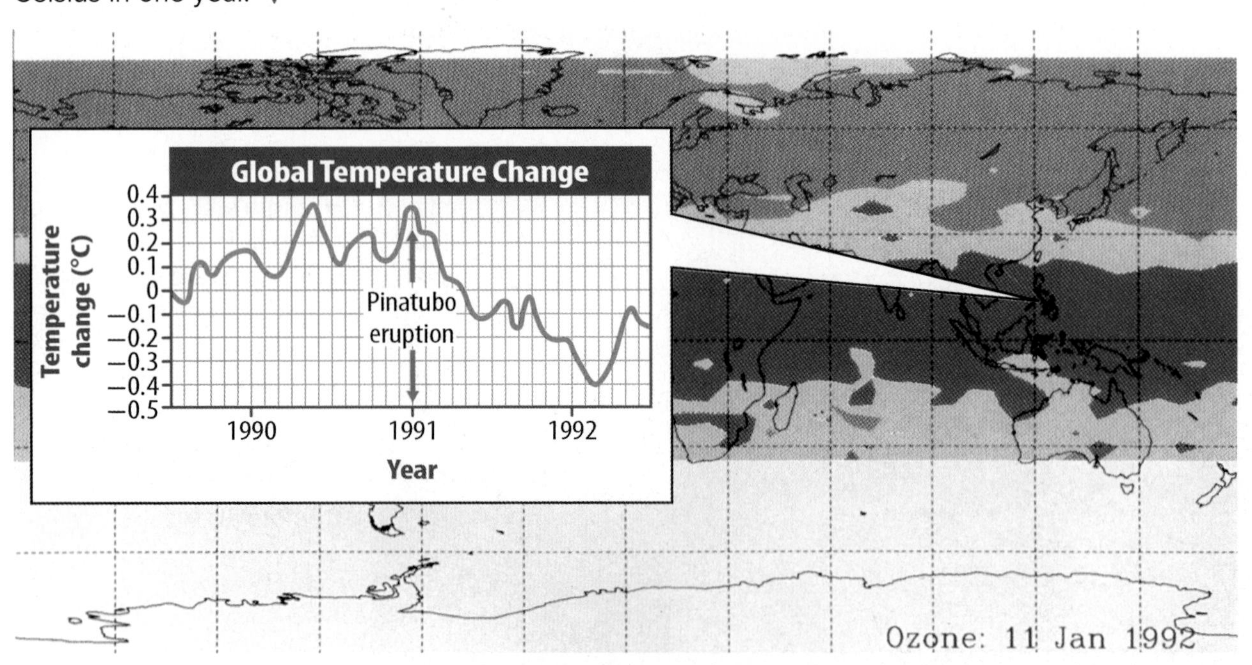

(t)Chris G. Newhall/U.S. Geological Survey; (b)Science Source

Lesson 2 Review

Visual Summary

Volcanoes form when magma rises through cracks in the crust and erupts from vents on Earth's surface.

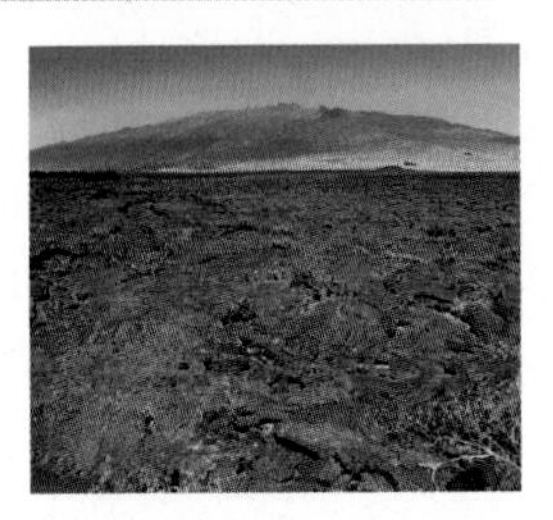

Magma with low amounts of silica and low viscosity erupts to form shield volcanoes.

Magma with high amounts of silica and high viscosity erupts explosively to form composite cones.

FOLDABLES®

Use your lesson Foldable to review the lesson. Save your Foldable for the project at the end of the chapter.

What do you think NOW?

You first read the statements below at the beginning of the chapter.

4. Volcanoes can erupt anywhere on Earth.

5. Volcanic eruptions are rare.

6. Volcanic eruptions only affect people and places close to the volcano.

Did you change your mind about whether you agree or disagree with the statements? Rewrite any false statements to make them true.

Use Vocabulary

1. **Compare and contrast** lava and magma.
2. **Explain** the term *viscosity.*
3. Pulverized rock and ash that erupts from explosive volcanoes is called ________.

Understand Key Concepts

4. **Identify** places where volcanoes form.
5. **Compare** the three main types of volcanoes.
6. What type of lava erupts from shield volcanoes?

   A. andesitic
   B. basaltic
   C. granitic
   D. rhyolitic

Interpret Graphics

7. **Analyze** the image below and explain what factors contribute to explosive eruptions.

8. **Create** a graphic organizer to illustrate the four types of eruptive products that can result from a volcanic eruption.

Critical Thinking

9. **Compare** the shapes of composite volcanoes and shield volcanoes. Why are their shapes and eruptive styles so different?
10. **Explain** how explosive volcanic eruptions can cause climate change. What might happen if Yellowstone Caldera erupted today?

Lab

45 minutes

Materials

colored pencils

metric ruler

drawing compass

topographic map of Mount Rainier

Safety

The Dangers of Mount Rainier

If you have ever visited the area near Seattle or Tacoma, Washington, it is difficult to miss majestic snow-capped Mount Rainier on the horizon. Mount Rainier, at nearly 4.4 km, is the highest active volcano in the Cascade Mountains of western Washington. More than 3.6 million people live within 100 km to the north and west of Mount Rainier.

Mount Rainier last erupted in 1895, but historical records show that it erupts with a frequency between 100 to 500 years. Mount Rainier's explosive past is evident from the pyroclastic flows, mudflows, and ash deposits that surround the volcano. Geologists predict that Mount Rainier will erupt in the future, but when? In this lab, you will assess the volcanic dangers of Mount Rainier.

Ask a Question

Imagine that you decide to open a mountain bike shop in either Sunrise, Longmire, or Ashford, Washington. Before you make your final decision about location, you must examine volcanic dangers of Mount Rainier. Which town is the safest choice?

Make Observations

1. Obtain a topographic map of Mount Rainier from your teacher.
2. Use the topographic map and the table below to indicate where mudflows might occur. Locate this area on your map and color it yellow.
3. Use the information in the table provided, the map scale, and a compass to identify the area that might be affected by
   - lava and pyroclastic flows; color this area orange on your map;
   - ash fall; color this area blue on your map.

   Be sure to include a legend on your map.

Volcanic Hazards

Type of Hazard	Range	Notes
Mudflow	Up to 64 km	Contained within river valleys
Lava and pyroclastic flow	Within 16 km of the summit	Will likely remain within boundary of Mount Rainier National Park
Falling ash	96 km downwind	Wind generally blows to the east of Mount Rainier

Form a Hypothesis

4. Use your observations to form a hypothesis about which town—Sunrise, Longmire, or Ashford—would be the safest choice for opening a bike shop. Base your hypothesis on your assessment of the volcanic hazards associated with Mount Rainier.

Test your Hypothesis

5. Compare your map to a classmate's map. If your assessments differ, explain how you developed your hypothesis.

Analyze and Conclude

6. **Calculate** If a mudflow from Mount Rainier traveled down the Nisqually River Valley, how much time would the towns of Longmire and Ashford have to prepare? *(Hint: Mudflows can move at speeds of 80 km/hr.)*
7. **Predict** Based on the extent of the volcanic hazards you mapped, would it be possible for a mudflow to reach Tacoma, Washington? Support your answer.
8. **The Big Idea** Mount Rainier is in the Cascade subduction zone. Why are hazards such as earthquakes and volcanic eruptions common along a subduction zone?

Communicate Your Results

As the owner of a bike shop, you want your clients to have a great visit to Mount Rainier. However, you want them to understand the risks associated with recreation on a volcano. Create a pamphlet that describes Mount Rainier's volcanic hazards. Include a map of the hazards. You might want to include names and contact information for local emergency response agencies.

Extension

Imagine you are riding your mountain bike on trails high above the Nisqually River Valley when a mudflow floods the valley below. Write a story describing your experience. Describe what you saw, what you heard, and what you felt. Explain how the mudflow might change the way people think about the volcanic dangers of Mount Rainier.

Lab TIPS

- ☑ Use the distance scale on the map to determine the extent of volcanic hazards.
- ☑ Mudflows that originate on Mount Rainier follow topography and flow down river valleys.

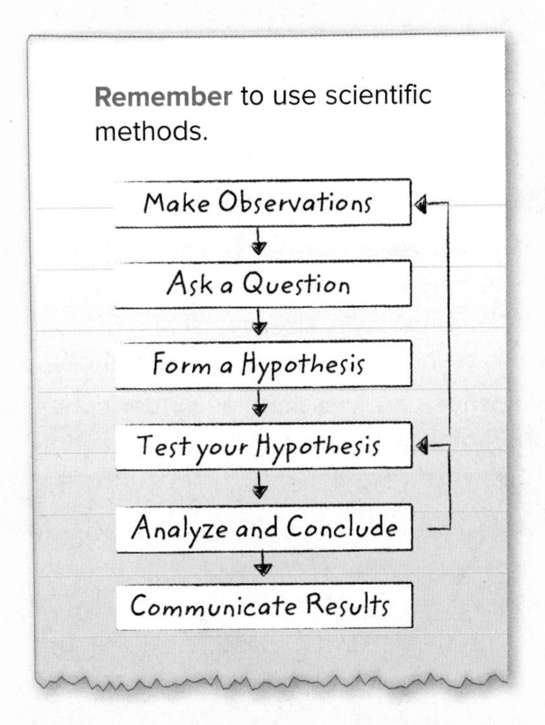

Chapter 9 Study Guide

Most earthquakes occur along plate boundaries where plates slide past each other, collide, or separate. Volcanoes form at subduction zones, mid-ocean ridges, and hot spots.

Key Concepts Summary

Lesson 1: Earthquakes

- Earthquakes commonly occur on or near tectonic plate boundaries.
- Earthquakes are used to study the composition and structure of Earth's interior and to identify the location of active faults.
- Earthquakes are monitored using **seismometers** and described using the Richter magnitude scale, the moment magnitude scale, and the Modified Mercalli scale.

○ Shallow earthquake
● Deep earthquake

Vocabulary

earthquake p. 293
fault p. 295
seismic wave p. 296
focus p. 296
epicenter p. 296
primary wave p. 297
secondary wave p. 297
surface wave p. 297
seismologist p. 298
seismometer p. 299
seismogram p. 299

Lesson 2: Volcanoes

- Molten **magma** is forced upward through cracks in the crust, erupting from volcanoes.
- The eruption style, size, and shape of a volcano depends on the composition of the magma, including the amount of dissolved gas.
- Volcanoes are classified as **cinder cones, shield volcanoes,** and **composite cones.**

Robert Glusic/Getty Images

Vocabulary

volcano p. 307
magma p. 307
lava p. 308
hot spot p. 308
shield volcano p. 310
composite volcano p. 310
cinder cone p. 310
volcanic ash p. 311
viscosity p. 311

Personal Tutor

Vocabulary eFlashcards
Vocabulary eGames

FOLDABLES® Chapter Project

Assemble your lesson Foldables as shown to make a Chapter Project. Use the project to review what you have learned in this chapter.

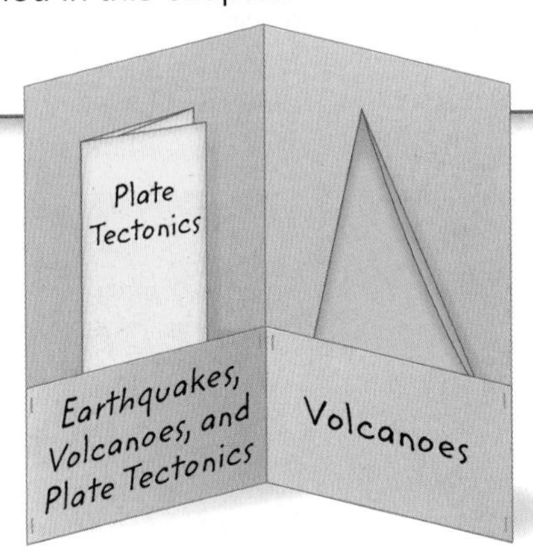

Use Vocabulary

1. A volcano with gently sloping sides is a(n) ________.
2. Write a sentence using the terms *seismic waves, P-waves,* and *S-waves.*
3. Magma that erupts quietly is ________. Magma most likely to erupt explosively is ________.
4. Volcanic activity that does not occur near a plate boundary happens at a(n) ________.
5. Molten rock inside Earth is called ________.
6. ________ are used to record ground motion during an earthquake.
7. The ________ marks the exact location where an earthquake occurs. The ________ is the place on Earth's surface directly above it.
8. A type of seismic wave that has movement similar to an ocean wave is a(n) ________.
9. A mixture of pulverized ash, rock, and gas ejected during explosive eruptions is called a(n) ________.

Link Vocabulary and Key Concepts

Interactive Concept Map

Copy this concept map, and then use vocabulary terms from the previous page to complete the concept map.

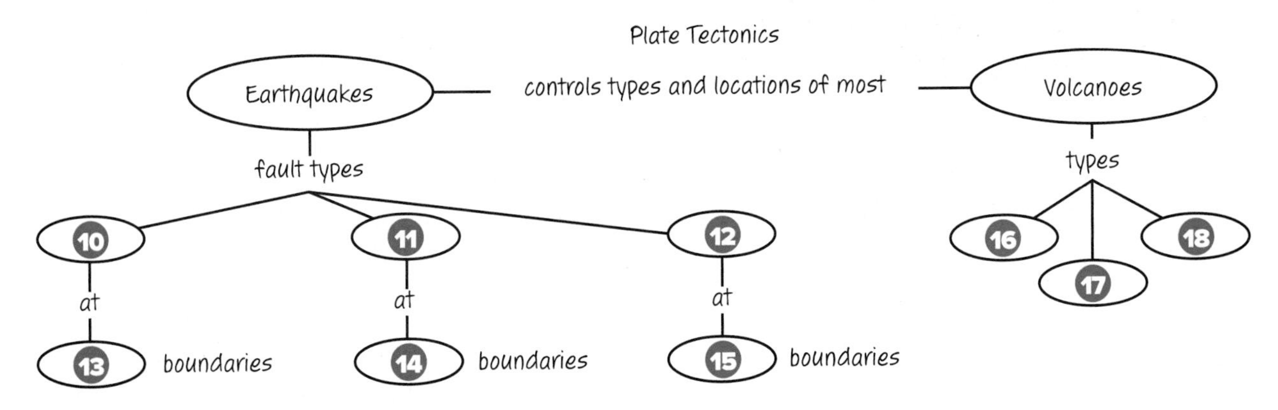

Chapter 9 Review

Understand Key Concepts

1. Most of the volcanic activity on Earth occurs
   A. along mid-ocean ridges.
   B. along transform plate boundaries.
   C. at hot spots.
   D. within the crust.

2. At a divergent plate boundary such as a mid-ocean ridge, you should expect to find
   A. low viscosity lava and normal faults.
   B. low viscosity lava and reverse faults.
   C. high viscosity lava and normal faults.
   D. high viscosity lava and reverse faults.

3. High energy earthquakes occur
   A. away from plate boundaries.
   B. away from divergent plate boundaries.
   C. on convergent plate boundaries.
   D. on transform plate boundaries.

4. Large and explosive volcanic eruptions, such as the one shown below, can change climate because
   A. ash and gas that erupt high into the atmosphere can reflect sunlight.
   B. the magma that erupts is hot.
   C. volcanic ash keeps Earth from losing its heat.
   D. volcanic mountains block solar radiation.

5. What is an earthquake?
   A. a fault at a convergent plate boundary
   B. a wave of water in the crust
   C. energy released as rocks break and move along a fault
   D. the elastic strain stored in rocks

6. Approximately how much more ground motion is recorded on a seismogram from a magnitude 6 earthquake compared to a magnitude 4 earthquake?
   A. 10 times more
   B. 50 times more
   C. 100 times more
   D. 1,000 times more

7. The figure below shows the Hawaiian Islands, formed by a hot spot. Which island is the oldest?
   A. Hawaii
   B. Kauai
   C. Maui
   D. Oahu

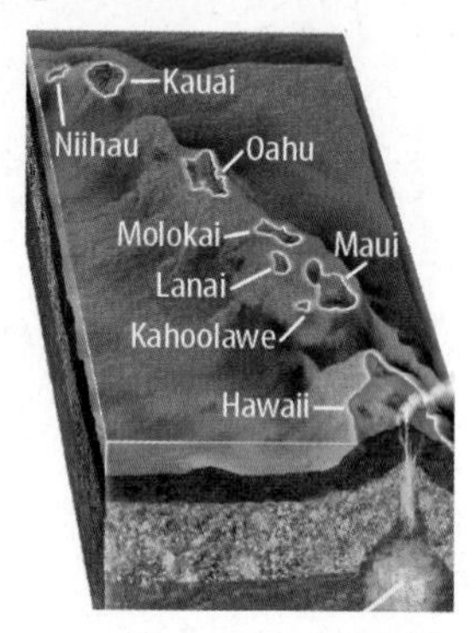

8. A lag-time graph illustrates the relationship between the time it takes a seismic wave to travel from the earthquake epicenter to a seismometer and the
   A. distance between the earthquake and the seismometer.
   B. earthquake intensity.
   C. earthquake magnitude.
   D. size of the fault.

9. Which can show the amount of energy released by an earthquake?
   A. a lag-time graph
   B. the Modified Mercalli scale
   C. the moment magnitude scale
   D. the Richter magnitude scale

10. The location of an earthquake can be determined from seismic data recorded by at least
    A. one seismometer.
    B. two seismometers.
    C. three seismometers.
    D. five seismometers.

Critical Thinking

11. **Explain** why Alaska has such a high risk associated with earthquakes.

12. **Analyze** the various types of volcanoes shown in **Table 4.** Which type of volcano is most likely to form at a hot spot in the ocean? Explain your answer.

13. **Evaluate** the following statement: "Yellowstone is a caldera that has erupted more than 1,000 km^3 of magma three times over the past 2.2 million years." Suggest how you might test the hypothesis that there is hot molten material beneath Yellowstone today.

14. **Hypothesize** Use the map below to identify evidence to suggest that Africa is splitting into two continents.

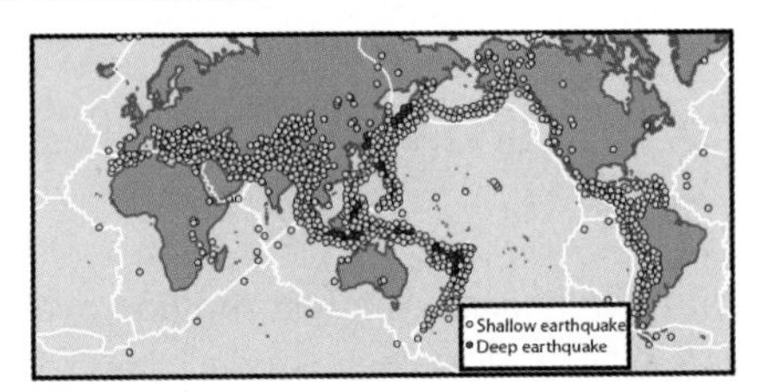

15. **Describe** how seismologists discovered that most of the mantle is solid.

16. **Identify** several reasons why a magnitude 6 earthquake in New Orleans might be more damaging than a magnitude 7 earthquake in San Francisco.

17. **Explain** why pyroclastic flows are responsible for more deaths than lava flows.

18. **Describe** Look at a map of the Hawaiian Island–Emperor Seamount chain formed by an active hot spot. Describe the relationship between these two chains. What do you think changed to form two chains instead of one?

Writing in Science

19. **Hypothesize** how scientists might be able to determine the composition of the Moon's interior given what you know about Earth's interior.

REVIEW THE BIG IDEA

20. How does the theory of plate tectonics explain the location of most earthquakes and volcanoes?

21. The photo below shows a pyroclastic flow from Mount Pinatubo in the Philippines. Why was this eruption so explosive?

Math Skills

Math Practice

22. **Identify** What is the value of Roman numeral XXXIX?

23. **Evaluate** How would you write number 38 in Roman numerals?

24. **Evaluate** In Roman numerals, L = 50. What is the value of the Roman numeral XL?

25. **Determine** How would you write the number 83 in Roman numerals?

Standardized Test Practice

Record your answers on the answer sheet provided by your teacher or on a sheet of paper.

Multiple Choice

1 Along which type of plate boundary do the deepest earthquakes occur?

A convergent

B divergent

C passive

D transform

2 The Richter scale registers the magnitude of an earthquake by determining the

A amount of energy released by the earthquake.

B amount of ground motion measured at a given distance from the earthquake.

C descriptions of damage caused by the earthquake.

D type of seismic waves produced by the earthquake.

3 Which state has no active volcanoes?

A California

B Hawaii

C New York

D Washington

Use the diagram below to answer question 4.

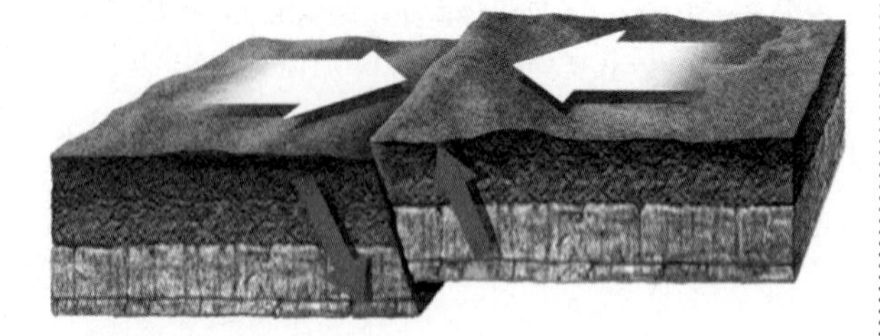

4 Which type of fault is shown in the diagram above?

A normal

B reverse

C shallow

D strike-slip

Use the diagram below to answer question 5.

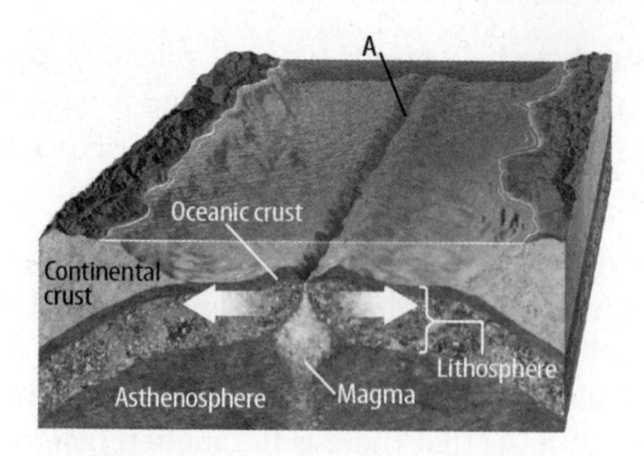

5 Which feature is labeled with the letter *A* in the diagram above?

A a caldera

B a chain of hot spot volcanoes

C a mid-ocean ridge

D a subducting tectonic plate

6 Which term describes a fast-moving avalanche of hot gas, ash, and rock that erupts from an explosive volcano?

A ash fall

B cinder cone

C lahar

D pyroclastic flow

7 Earthquakes occur along the San Andreas Fault. Which is an example of this type of plate boundary?

A convergent

B divergent

C passive

D transform

8 Hot spot volcanoes ALWAYS

A appear at plate boundaries.

B erupt in chains.

C form above mantle plumes.

D remain active.

Use the map below to answer questions 9 and 10.

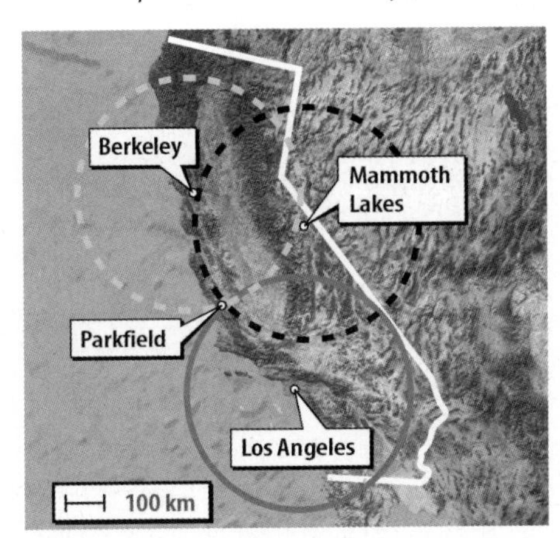

9 What do the circles represent in the map of seismic activity illustrated above?

A the distance between waves

B the distance to an earthquake epicenter

C the seismic wave speeds

D the wave travel times

10 According to the map, where is the earthquake epicenter?

A Berkeley

B Los Angeles

C Mammoth Lakes

D Parkfield

11 Where do seismic waves originate?

A above ground

B epicenter

C focus

D seismogram

Constructed Response

Use the diagram below to answer questions 12 and 13.

12 The diagram above shows one way volcanoes form. Explain the process shown in the diagram and why volcanoes form as a result of this process.

13 What type of volcano results from the process shown in the diagram? Describe it. What is the eruptive style of this type of volcano? Why?

Use the table below to answer question 14.

Wave Type	Characteristics

14 Re-create the table above and identify the three types of seismic waves. Then, describe wave characteristics such as movement, speed, and difference in arrival time for each type.

NEED EXTRA HELP?														
If You Missed Question...	1	2	3	4	5	6	7	8	9	10	11	12	13	14
Go to Lesson...	1	1	2	1	2	2	1	2	1	1	1	2	2	1

Chapter 10

Clues to Earth's Past

THE BIG IDEA What evidence do scientists use to determine the ages of rocks?

Inquiry Always a Canyon?

The Colorado River started cutting through the rock layers of the Grand Canyon only about 6 million years ago. Hundreds of millions of years earlier, these rock layers were deposited at the bottom of an ancient sea. Even before that, a huge mountain range existed here.

- What evidence do scientists use to learn about past environments?
- What evidence do scientists use to determine the ages of rocks?

Get Ready to Read

What do you think?

Before you read, decide if you agree or disagree with each of these statements. As you read this chapter, see if you change your mind about any of the statements.

1. Fossils are pieces of dead organisms.
2. Only bones can become fossils.
3. Older rocks are always located below younger rocks.
4. Relative age means that scientists are relatively sure of the age.
5. Absolute age means that scientists are sure of the age.
6. Scientists use radioactive decay to determine the ages of some rocks.

Lesson 1

Fossils

Reading Guide

Key Concepts
ESSENTIAL QUESTIONS

- What are fossils and how do they form?
- What can fossils reveal about Earth's past?

Vocabulary

fossil p. 327
catastrophism p. 327
uniformitarianism p. 328
carbon film p. 330
mold p. 331
cast p. 331
trace fossil p. 331
paleontologist p. 332

Multilingual eGlossary

BrainPOP®
Science Video

Inquiry Fossils?

These insects are fossils. Millions of years ago, they became stuck in sticky tree sap. The sap fell to the ground, where it was buried by mud or sand. Over time, the sap became amber, and the insects were preserved as fossils.

Howard Grey/Getty Images

Launch Lab

20 minutes

What can trace fossils show?

Did you know that a fossil can be a footprint or the imprint of an ancient nest? These are examples of trace fossils. Although trace fossils do not contain any part of an organism, they do hold clues about how organisms lived, moved, or behaved.

1. Read and complete a lab safety form.
2. Flatten some **clay** into a pancake shape.
3. Think about a behavior or movement you would like your fossil to model. Use available tools, such as a **plastic knife,** a **chenille stem,** or a **toothpick,** to make a fossil showing that behavior or movement.
4. Exchange your fossil with another student. Try to figure out what behavior or movement he or she modeled.

Think About This

1. Were you able to determine what behavior or movement your classmate's fossil modeled? Was he or she able to determine yours? Why or why not?
2. **Key Concept** What do you think scientists can learn by studying trace fossils?

Evidence of the Distant Past

Have you ever looked through an old family photo album? Each photo shows a little of your family's history. You might guess the age of the photographs based on the clothes people are wearing, the vehicles they are driving, or even the paper the photographs are printed on.

Just as old photos can provide clues to your family's past, rocks can provide clues to Earth's past. Some of the most obvious clues found in rocks are the remains or traces of ancient living things. **Fossils** *are the preserved remains or evidence of ancient living things.*

WORD ORIGIN
fossil
from Latin *fossilis,* means "dug up"

Catastrophism

Many fossils represent plants and animals that no longer live on Earth. Ideas about how these fossils formed have changed over time. Some early scientists thought that great, sudden, catastrophic disasters killed the organisms that became fossils. These scientists explained Earth's history as a series of disastrous events occurring over short periods of time. **Catastrophism** (kuh TAS truh fih zum) *is the idea that conditions and organisms on Earth change in quick, violent events.* The events described in catastrophism include volcanic eruptions and widespread flooding. Scientists eventually disagreed with catastrophism because Earth's history is full of violent events.

Figure 1 Hutton realized that erosion happens on small or large scales.

Uniformitarianism

Most people who supported catastrophism thought that Earth was only a few thousand years old. In the 1700s, James Hutton rejected this idea. Hutton was a naturalist and a farmer in Scotland. He observed how the landscape on his farm gradually changed over the years. Hutton thought that the processes responsible for changing the landscape on his farm could also shape Earth's surface. For example, he thought that erosion caused by water, such as that shown in **Figure 1,** could also wear down mountains. Because he realized that this would take a long time, Hutton proposed that Earth is much older than a few thousand years.

Hutton's ideas eventually were included in a principle called uniformitarianism (yew nuh for muh TER ee uh nih zum). *The principle of* **uniformitarianism** *states that geologic processes that occur today are similar to those that have occurred in the past.* According to this view, Earth's surface is constantly being reshaped in a steady, **uniform** manner.

ACADEMIC VOCABULARY

uniform
(adjective) having always the same form, manner, or degree; not varying or variable

 **Reading Check** What is uniformitarianism?

Today, uniformitarianism is the basis for understanding Earth's past. But scientists also know that catastrophic events do sometimes occur. Huge volcanic eruptions and giant meteorite impacts can change Earth's surface very quickly. These catastrophic events punctuate Earth's slow, gradual story.

MiniLab

15 minutes

How is a fossil a clue?

Fossils provide clues about once-living organisms. Sometimes those clues are hard to interpret.

1. Read and complete a lab safety form.
2. Select an **object** from a bag provided by your teacher. Do not let anyone see your object.
3. Make a fossil impression of your object by pressing only part of it into a piece of **clay.**
4. Place your clay fossil and object in separate locations indicated by your teacher.
5. Make a chart in your Science Journal that matches your classmates' objects and fossils.

Analyze and Conclude

1. Did you correctly match the objects with their fossils?
2. Why might scientists need more than one fossil of an organism to understand what it looked like?
3. **Key Concept** What do you think you could learn from fossils?

Fossil Formation

Figure 2 A fossil can form if an organism with hard parts, such as a fish, is buried quickly after it dies.

1 A dead fish falls to a river bottom during a flood. Its body is rapidly buried by mud, sand, or other sediment.

2 Over time, the body decomposes but the hard bones become a fossil.

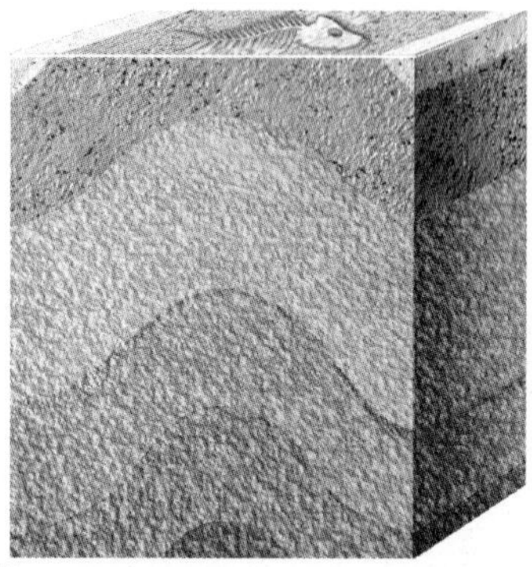

3 The sediments, hardened into rock, are uplifted and eroded, which exposes the fossil fish on the surface.

Visual Check How did the fish fossil reach the surface?

Formation of Fossils

Recall that fossils are the remains or traces of ancient living organisms. Not all dead organisms become fossils. Fossils form only under certain conditions.

Conditions for Fossil Formation

Most plants and animals are eaten or decay when they die, leaving no evidence that they ever lived. Think about the chances of an apple becoming a fossil. If it is on the ground for many months, it will decay into a soft, rotting lump. Eventually, insects and bacteria consume it.

However, some conditions increase the chances of fossil formation. An organism is more likely to become a fossil if it has hard parts, such as shells, teeth, or bones, like the fish in **Figure 2.** Unlike a soft apple, hard parts do not decay easily. Also, an organism is more likely to form a fossil if it is buried quickly after it dies. If layers of sand or mud bury an organism quickly, decay is slowed or stopped.

Key Concept Check What conditions increase the chances of fossil formation?

Fossils Come in All Sizes

You might have seen pictures of dinosaur fossils. Many dinosaurs were large animals, and large bones were left behind when they died. Not all fossils are large enough for you to see. Sometimes it is necessary to use a microscope to see fossils. Tiny fossils are called microfossils. The microfossils in **Figure 3** are each about the size of a speck of dust.

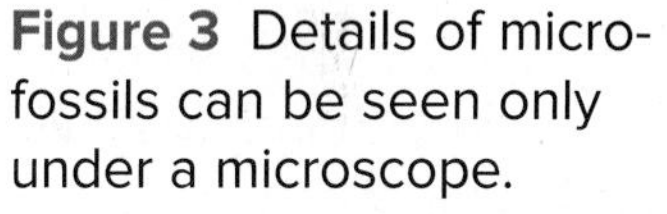

Figure 3 Details of microfossils can be seen only under a microscope.

Dark Field LM Magnification: 100×

Michler Hanns-Frieder/age fotostock

Types of Preservation

Fossils are preserved in different ways. As shown in **Figure 4,** there are many ways fossils can form.

Preserved Remains

Sometimes the actual remains of organisms are preserved as fossils. For this to happen, an organism must be completely enclosed in some material over a long period of time. This would prevent it from being exposed to air or bacteria. Generally, preserved remains are 10,000 or fewer years in age. However, insects preserved in amber–shown in the photo at the beginning of this lesson–can be millions of years old.

Carbon Films

Sometimes when an organism is buried, exposure to heat and pressure forces gases and liquids out of the organism's tissues. This leaves only the carbon behind. *A* **carbon film** *is the fossilized carbon outline of an organism or part of an organism.*

Mineral Replacement

Replicas, or copies, of organisms can form from minerals in groundwater. They fill in the pore spaces or replace the tissues of dead organisms. Petrified wood is an example.

SCIENCE USE V. COMMON USE

petrified

Science Use turned into stone by the replacement of tissue with minerals

Common Use made rigid with fear

Figure 4 Fossils can form in many different ways.

Types of Preservation

Preserved Remains Organisms trapped in amber, tar pits, or ice can be preserved over thousands of years. This baby mammoth was preserved in ice for more than 10,000 years before it was discovered. ▶

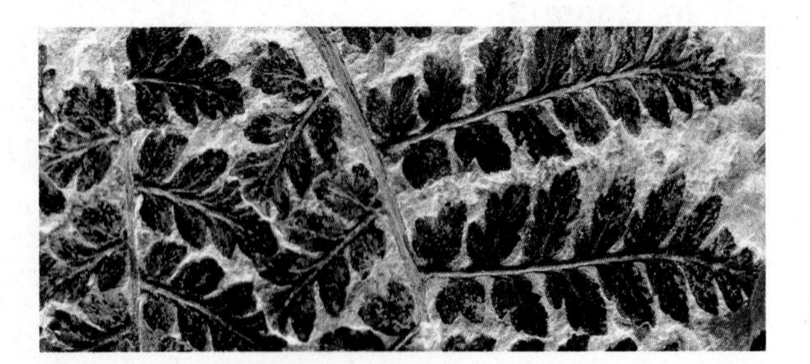

◀ **Carbon Film** Only a carbon film remains of this ancient fern. Carbon films are usually shiny black or brown. Fish, insects, and plant leaves are often preserved as carbon films.

Mineral Replacement Rock-forming minerals dissolved in groundwater can fill in pore spaces or replace the tissues of dead organisms. This petrified wood formed when silica (SiO_2) filled in the spaces between the cell walls in a dead tree. The wood petrified when the SiO_2 crystallized. ▶

Molds

Sometimes all that remains of an organism is its fossilized imprint or impression. *A* **mold** *is the impression in a rock left by an ancient organism.* A mold can form when sediment hardens around a buried organism. As the organism decays over time, an impression of its shape remains in the sediment. The sediment eventually turns to rock.

Casts

Sometimes, after a mold forms, it is filled with more sediment. *A* **cast** *is a fossil copy of an organism made when a mold of the organism is filled with sediment or mineral deposits.* The process is similar to making a gelatin dessert using a molded pan.

Trace Fossils

Some animals leave fossilized traces of their movement or activity. *A* **trace fossil** *is the preserved evidence of the activity of an organism.* Trace fossils include tracks, footprints, and nests. These fossils help scientists learn about characteristics and behaviors of animals. The dinosaur tracks in **Figure 4** reveal clues about the dinosaur's size, its speed, and whether it was traveling alone or in a group.

Reading Check What are some examples of trace fossils?

Make a tri-fold book from a sheet of paper. Label it as shown. Use it to organize your notes about the different types of fossils.

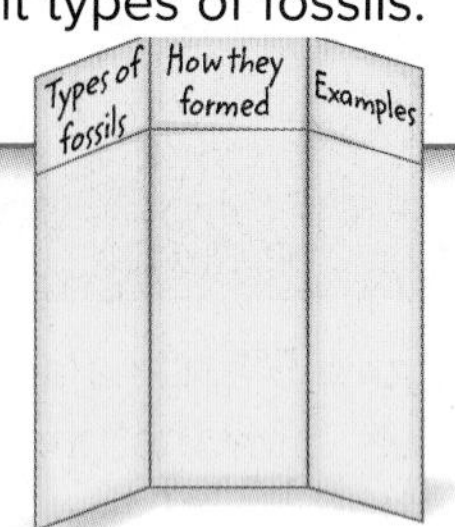

Mold This mold of an ancient trilobite formed after it was buried by sediment and then decayed. The sediment hardened, leaving an impression of its shape in the rock. ▼

▲ **Cast** This cast was formed when the mold was later filled with sediment that then hardened. Molds and casts show only the exterior, or outside, features of organisms.

Trace Fossil These trace fossils formed when dinosaur tracks in soft sediments were later filled in by other sediments, which then hardened. Trace fossils reveal information about the behavior of organisms. ▶

(t, c)José Enrique Molina/age fotostock; (b)age fotostock/SuperStock

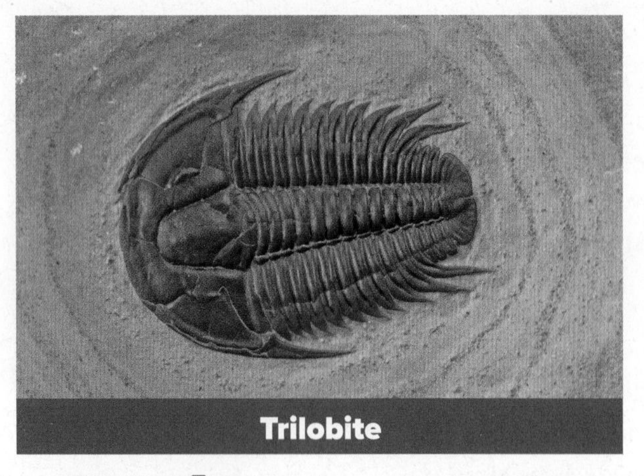

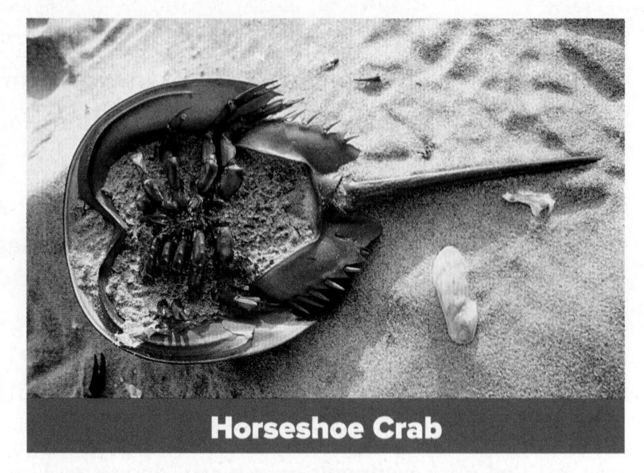

▲ **Figure 5** Partly because a trilobite fossil looks like a present-day horseshoe crab, scientists infer that the trilobite lived in an environment similar to the environment where a horseshoe crab lives.

Ancient Environments

Scientists who study fossils are called **paleontologists** (pay lee ahn TAH luh jihstz). Paleontologists use the principle of uniformitarianism to learn about ancient organisms and the environments where ancient organisms lived. For example, they can compare fossils of ancient organisms with organisms living today. The trilobite fossil and the horseshoe crab in **Figure 5** look alike. Horseshoe crabs today live in shallow water on the ocean floor. Partly because trilobite fossils look like horseshoe crabs, paleontologists infer that trilobites also lived in shallow ocean water.

Shallow Seas

Today, Earth's continents are mostly above sea level. But sea level has risen, flooding Earth's continents, many times in the past. For example, a shallow ocean covered much of North America 450 million years ago, as illustrated in the map in **Figure 6.** Fossils of organisms that lived in that shallow ocean, like those shown in **Figure 6,** help scientists reconstruct what the seafloor looked like at that time.

Key Concept Check What can fossils tell us about ancient environments?

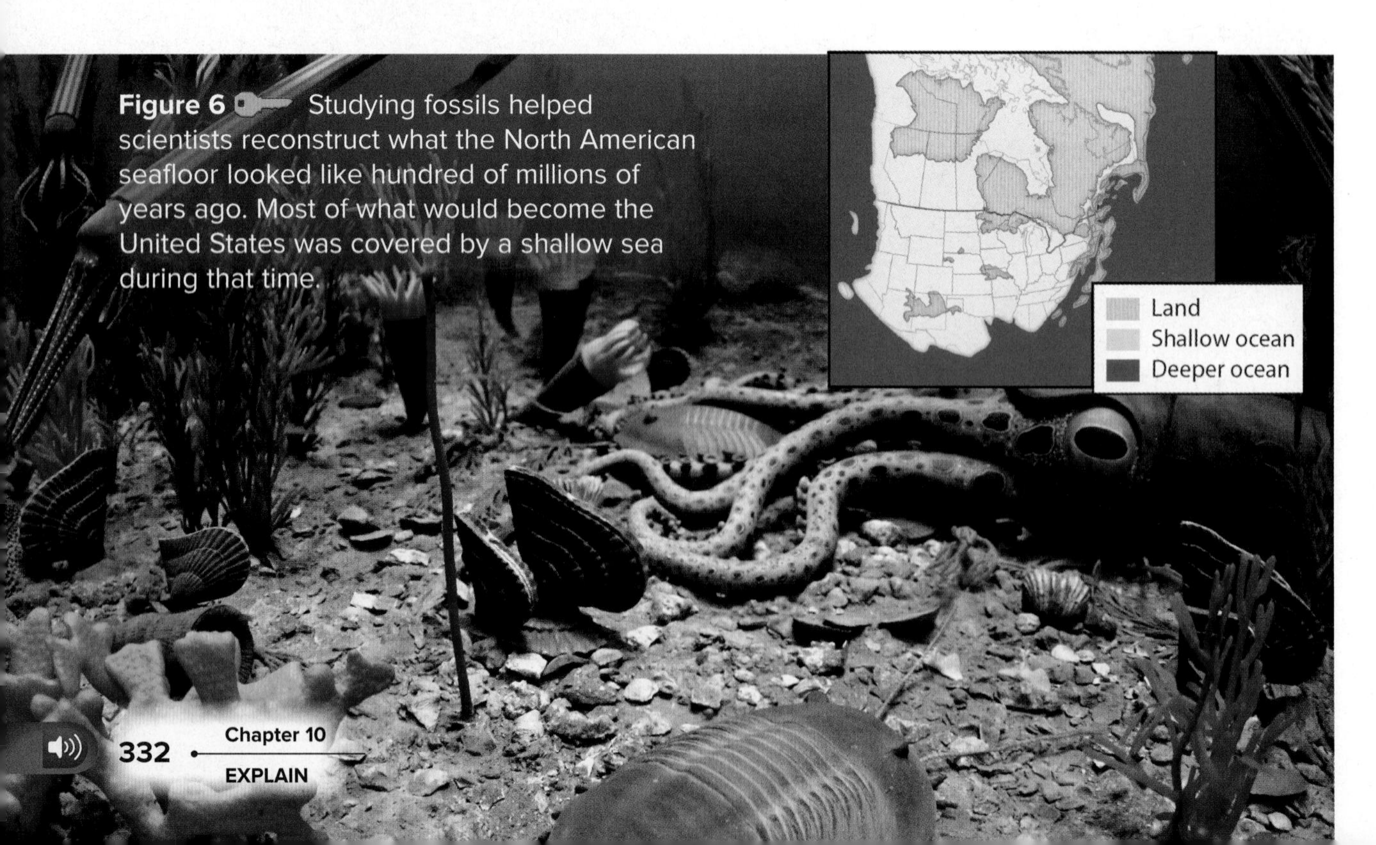

Figure 6 Studying fossils helped scientists reconstruct what the North American seafloor looked like hundred of millions of years ago. Most of what would become the United States was covered by a shallow sea during that time.

Figure 7 About 100 million years ago, tropical forests and swamps covered much of North America. Dinosaurs also lived on Earth at that time.

Past Climates

You might have heard people talking about global climate change, or maybe you've read about climate change. Evidence indicates that Earth's present-day climate is warming. Fossils show that Earth's climate has warmed and cooled many times in the past.

Plant fossils are especially good indicators of climate change. For example, fossils of ferns and other tropical plants dating to the time of the dinosaurs reveal that Earth was very warm 100 million years ago. Tropical forests and swamps covered much of the land, as illustrated in **Figure 7.**

Key Concept Check What was Earth's climate like when dinosaurs lived?

Millions of years later, the swamps and forests were gone, but coarse grasses grew in their place. Huge sheets of ice called glaciers spread over parts of North America, Europe, and Asia. Fossils suggest that some species that lived during this time, such as the woolly mammoth shown in **Figure 8,** were able to survive in the colder climate.

Fossils of organisms such as ferns and mammoths help scientists learn about ancient organisms and past environments. In the following lessons, you will read how scientists use fossils and other clues, such as the order of rock layers and radioactivity, to learn about the ages of Earth's rocks.

Figure 8 The woolly mammoth was well adapted to a cold climate.

Lesson 1 Review

Visual Summary

The principle of uniformitarianism is the basis for understanding Earth's past.

Fossils can form in many different ways.

Fossils help scientists learn about Earth's ancient organisms and past environments.

FOLDABLES®

Use your lesson Foldable to review the lesson. Save your Foldable for the project at the end of the chapter.

What do you think NOW?

You first read the statements below at the beginning of the chapter.

1. Fossils are pieces of dead organisms.

2. Only bones can become fossils.

Did you change your mind about whether you agree or disagree with the statements? Rewrite any false statements to make them true.

Use Vocabulary

1. **Distinguish** between catastrophism and uniformitarianism.
2. Plant leaves are often preserved as ________.
3. **Use the terms** *cast* and *mold* in a complete sentence.

Understand Key Concepts

4. Which conditions aid in the formation of fossils?
   - **A.** hard parts and slow burial
   - **B.** hard parts and rapid burial
   - **C.** soft parts and rapid burial
   - **D.** soft parts and slow burial
5. What human body system could be fossilized? Explain.
6. **Determine** what type of an environment a fossil palm tree would indicate.

Interpret Graphics

7. **Compare** the two sets of dinosaur footprints below. Which dinosaur was running? How can you tell?

8. **Organize Information** Copy and fill in the graphic organizer below to list types of fossil preservation.

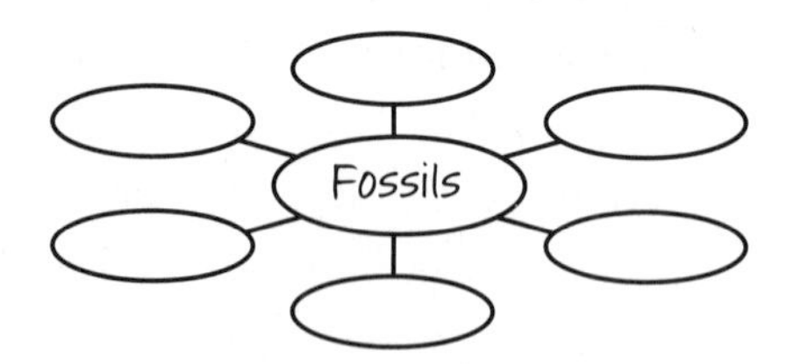

Critical Thinking

9. **Invent** a process for the formation of ocean basins consistent with catastrophism.
10. **Evaluate** how the following statement relates to what you have read in this lesson: "The present is the key to the past."

AMERICAN MUSEUM OF NATURAL HISTORY

▼ This Byronosaurus skull was discovered at Ukhaa Tolgod. It gave scientists important clues about how birds and dinosaurs are related.

Perfect Fossils— A Rare Find

Spectacular fossils formed when ancient organisms were buried quickly.

The key to a well-preserved fossil is what happens to an organism right after it dies. When organisms are buried swiftly, their remains are protected from scavengers and natural events. Over time, their bones and teeth form amazingly well-preserved fossils.

As a boy growing up in Los Angeles, Michael Novacek loved visiting the La Brea tar pits. The fossils from these tar pits are remarkably intact because animals became stuck in gooey tar puddles and quickly became submerged. These fossils inspired Novacek to become a paleontologist.

Years later, on an expedition for the American Museum of Natural History, Novacek discovered another extraordinary site. In the Gobi Desert, in Mongolia, Novacek and his team uncovered a rich collection of fossils at a site called Ukhaa Tolgod.

The fossils were astonishingly complete. One fossil was of a 2-inch mammal skeleton, still with its microscopic ear bones. Like the animals at the La Brea tar pits, those that died here were buried quickly. But after examining the evidence, scientists have determined that these animals were likely killed and covered by a devastating avalanche or landslide. Spectacular fossils formed when ancient organisms were buried quickly.

Discovering What Happened at Ukhaa Tolgod

Scientists have looked for clues in the rocks where the fossils were found. Most of the rocks are made of sandstone. One hypothesis was that the animals were buried alive by drifting dunes during a sandstorm. But then scientists noticed the rocks near the fossils held large pebbles that were too big to be carried by wind.

To find an explanation, they turned to Nebraska's Sand Hills—a region with giant, stable dunes similar to those that existed at Ukhaa Tolgod. In the Sand Hills, heavy rains can set off avalanches of wet sand. An enormous slab of heavy, wet sand can bury everything in its path. The current hypothesis about Ukhaa Tolgod is that heavy rains triggered an avalanche of sand that slid down the dunes and buried the animals below.

◀ Novacek and his team have found amazing fossils in Mongolia's Gobi Desert, including this velociraptor.

Today, the dunes at Ukhaa Tolgod are sandy and barren. But long ago, many plants and animals lived in the region. ▶

It's Your Turn

DRAW With a partner, draw a comic strip showing how animals in Ukhaa Tolgod might have died and been buried. Use the comic strip to explain how the almost-perfect fossils were preserved.

Lesson 2

Relative-Age Dating

Reading Guide

Key Concepts
ESSENTIAL QUESTIONS

- What does relative age mean?
- How can the positions of rock layers be used to determine the relative ages of rocks?

Vocabulary

relative age p. 337
superposition p. 338
inclusion p. 339
unconformity p. 340
correlation p. 340
index fossil p. 341

Multilingual eGlossary

Go to the resource tab in ConnectED to find the PBL *Puzzles Rock!*

Inquiry How did this happen?

Hundreds of millions of years ago, hot magma from deep in Earth was forced into these red, horizontal rock layers in the Grand Canyon. As the magma cooled, it formed this dark gash. How do you think features such as this help scientists determine the relative ages of rock layers?

Launch Lab

15 minutes

Which rock layer is oldest?

Scientists study rock layers to learn about the geologic history of an area. How do scientists determine the order in which layers of rock were deposited?

1. Read and complete a lab safety form.
2. Break a **disposable polystyrene meat tray** in half. Place the two pieces on a flat surface so that the broken edges touch one another.
3. Break **another meat tray** in half. Place the two pieces directly on top of the first broken meat tray.
4. Place a **third, unbroken meat tray** on top of the two broken meat trays.

Think About This

1. If you observed rock layers that looked like your model, what would you think might have caused the break only in the two bottom layers?
2. **Key Concept** How do you think your model resembles a rock formation? Which layer in your model is youngest? Which is oldest?

Relative Ages of Rocks

You just remembered where you left the money you have been looking for. It is in the pocket of the pants you wore to the movies last Saturday. Look at your pile of dirty clothes. How can you tell where your money is? There really is some order in that pile of dirty clothes. Every time you add clothes to the pile, you place them on top, like the clothes you wore last night. And the clothes from last Saturday are on the bottom. That's where your money is.

Just as there is order in a pile of clothes, there is order in a rock formation. In the rock formation shown in **Figure 9,** the oldest rocks are in the bottom layer and the youngest rocks are in the top layer.

Maybe you have brothers and sisters. If you do, you might describe your age by saying, "I'm older than my sister and younger than my brother." In this way, you compare your age to others in your family. Geologists–the scientists who study Earth and rocks–have developed a set of principles to compare the ages of rock layers. They use these principles to organize the layers according to their relative ages. **Relative age** *is the age of rocks and geologic features compared with other rocks and features nearby.*

Key Concept Check How might you define your relative age?

Figure 9 Just as there is order in a pile of clothes, there is order in this rock formation.

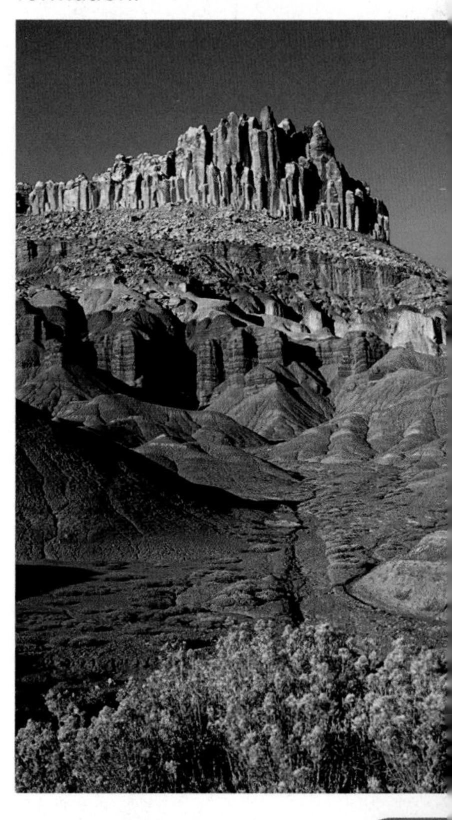

Principles of Relative-Age Dating

Figure 10 Geologic principles help scientists determine the relative order of rock layers.

Visual Check Which rock layer is the oldest?

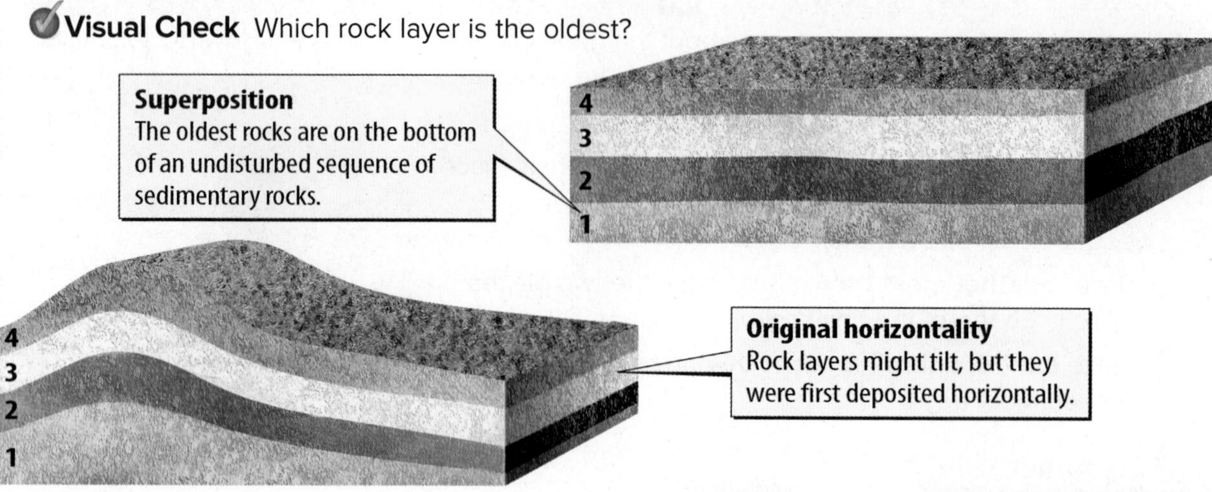

Lateral continuity
Layers are deposited in continuous sheets in all directions until they thin out or hit a barrier. A river might cut through the layers, but the order of layers does not change.

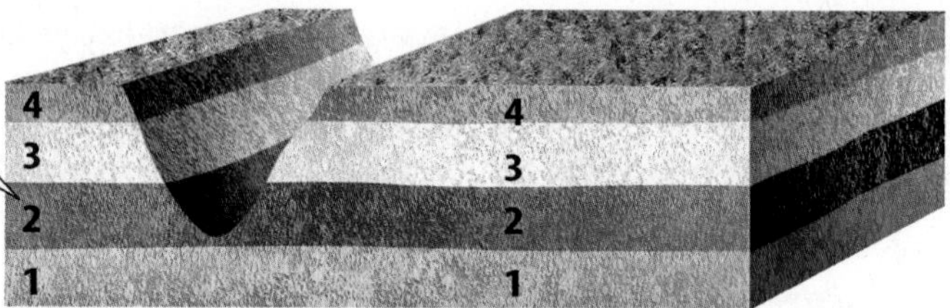

Superposition

Your pile of dirty clothes demonstrates the first principle of relative-age dating–superposition. **Superposition** *is the principle that in undisturbed rock layers, the oldest rocks are on the bottom.* Unless some force disturbs the layers after they were deposited, each layer of rocks is younger than the layer below it, as shown in **Figure 10.**

FOLDABLES®

Make a five-tab book and label it as shown. Use it to organize information about the principles of relative-age dating.

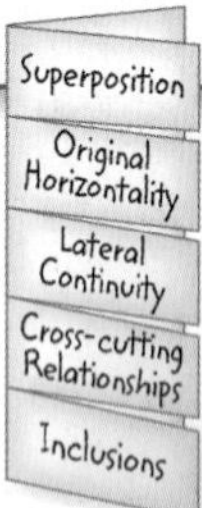

Original Horizontality

An example of the second principle of relative-age dating–original horizontality–is also shown in **Figure 10.** According to the principle of original horizontality, most rock-forming materials are deposited in horizontal layers. Sometimes rock layers are deformed or disturbed after they form. For example, the layers might be tilted or folded. Even though they might be tilted, all the layers were originally deposited horizontally.

Reading Check How might rock layers be disturbed?

Lateral Continuity

Another principle of relative-age dating is that sediments are deposited in large, continuous sheets in all lateral directions. The sheets, or layers, continue until they thin out or meet a barrier. This principle, called the principle of lateral continuity, is illustrated in the bottom image of **Figure 10.** A river might erode the layers, but their placements do not change.

WORD ORIGIN

lateral
from Latin *lateralis,* means "belonging to the side"

Figure 11 Dikes and faults help scientists determine the order in which rock layers were deposited.

Inclusions

Occasionally when rocks form, they contain pieces of other rocks. This can happen when part of an existing rock breaks off and falls into soft sediment or flowing magma. When the sediment or magma becomes rock, the broken piece becomes a part of it. *A piece of an older rock that becomes part of a new rock is called an* **inclusion.** According to the principle of inclusions, if one rock contains pieces of another rock, the rock containing the pieces is younger than the pieces. The vertical intrusion in **Figure 11**, called a dike, is younger than the pieces of rock inside it.

Cross-Cutting Relationships

Sometimes, forces within Earth cause rock formations to break, or fracture. When rocks move along a fracture line, the fracture is called a fault. Faults and dikes cut across existing rock. According to the principle of cross-cutting relationships, if one geologic feature cuts across another feature, the feature that it cuts across is older, as shown in **Figure 11.** This principle is illustrated in the photo at the beginning of this lesson. The black rock layer formed as magma cut across pre-existing red rock layers and crystallized.

Key Concept Check What geologic principles are used in relative-age dating?

MiniLab — 20 minutes

Can you model rock layers?

Can a classmate determine the order of your three-dimensional model of rock layers?

1. Read and complete a lab safety form.
2. Cut out a **cube template** as instructed by your teacher.
3. On the sides and top, use **colored pencils** to draw a rock formation that contains 4–5 layers. Include faults, dikes, inclusions, and other disturbances.
4. **Glue** your cube to make a three-dimensional model.
5. Exchange models with another student and determine the order of the layers.

Analyze and Conclude

Key Concept Summarize how positions of rock layers can be used to determine the relative ages of rocks.

Hutchings Photography/Digital Light Source

Unconformities

After rocks form, they are sometimes uplifted and exposed at Earth's surface. When rocks are exposed, wind and rain start to weather and erode them. These eroded areas represent a gap in the rock record.

Often, new rock layers are deposited on top of old, eroded rock layers. When this happens, an unconformity (un kun FOR muh tee) occurs. *An* **unconformity** *is a surface where rock has eroded away, producing a break, or gap, in the rock record.*

An unconformity is not a hollow gap in the rock. It is a surface on a layer of eroded rocks where younger rocks have been deposited. However, an unconformity does represent a gap in time. It could represent a few hundred years, a million years, or even billions of years. Three major types of unconformities are shown in **Table 1**.

Key Concept Check How does an unconformity represent a gap in time?

Animation

Correlation

You have read that rock layers contain clues about Earth. Geologists use these clues to build a record of Earth's geologic history. Many times the rock record is incomplete, such as happens in an unconformity. Geologists fill in gaps in the rock record by matching rock layers or fossils from separate locations. *Matching rocks and fossils from separate locations is called* **correlation** (kor uh LAY shun).

Matching Rock Layers

Another word for correlation is *connection.* Sometimes it is possible to connect rock layers simply by walking along rock formations and looking for similarities. At other times, soil might cover the rocks, or rocks might be eroded away. In these cases, geologists correlate rocks by matching exposed rock layers in different locations. Through correlation, geologists have established a historical record for part of the southwestern United States, as shown in **Figure 12.**

Table 1 Types of Unconformities

Disconformity Younger sedimentary layers are deposited on top of older, horizontal sedimentary layers that have been eroded.		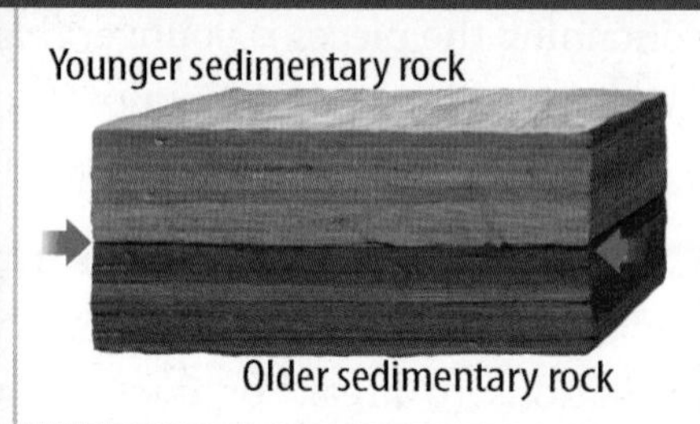
Angular Unconformity Sedimentary layers are deposited on top of tilted or folded sedimentary layers that have been eroded.		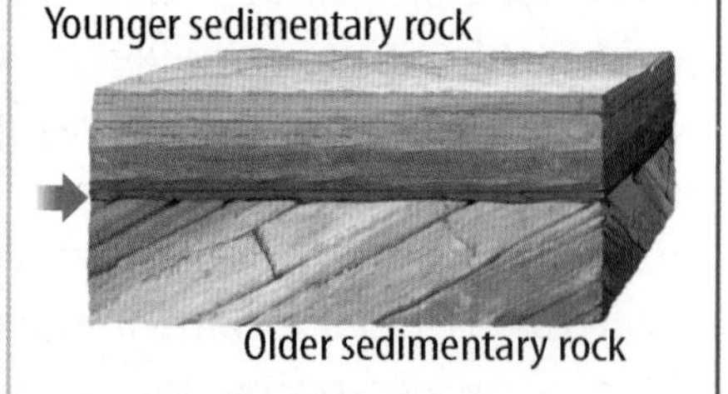
Nonconformity Younger sedimentary layers are deposited on older igneous or metamorphic rock layers that have been eroded.		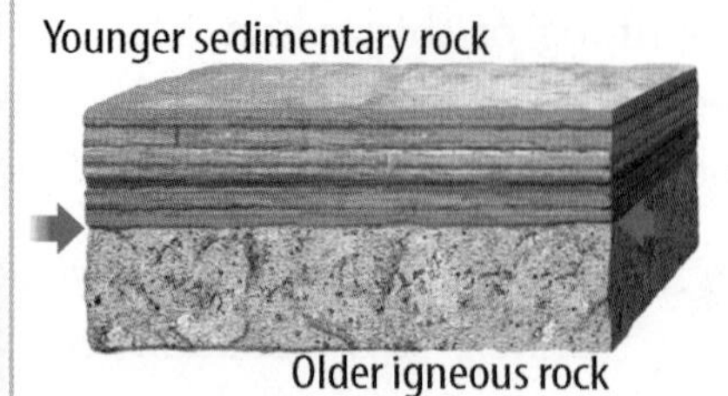

Correlation

Personal Tutor

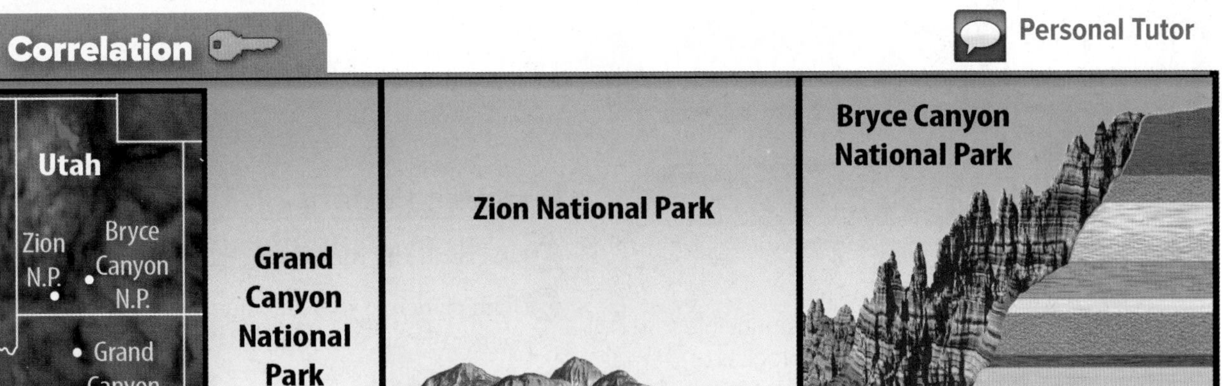

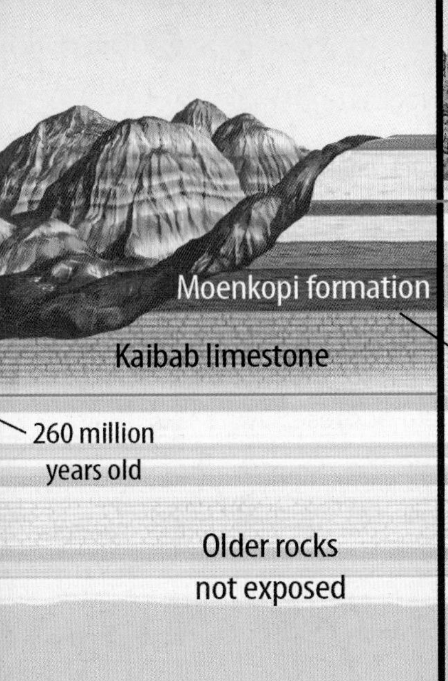
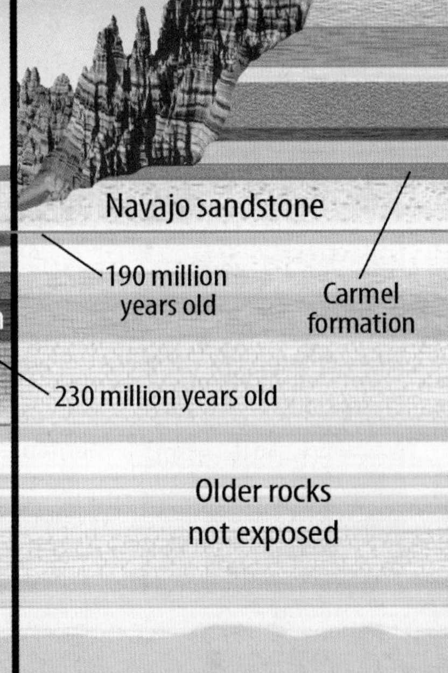

Figure 12 Exposed rock layers from three national parks have been correlated to make a historical record.

Visual Check Which geologic principles must be assumed in order to correlate these layers?

Index Fossils

The rock formations in **Figure 12** are correlated based on similarities in rock type, structure, and fossil evidence. They exist within a few hundred kilometers of one another. If scientists want to learn the relative ages of rock formations that are very far apart or on different continents, they often use fossils. If two or more rock formations contain fossils of about the same age, scientists can infer that the formations are also about the same age.

Not all fossils are useful in determining the relative ages of rock layers. Fossils of species that lived on Earth for hundreds of millions of years are not helpful. They represent time spans that are too long. The most useful fossils represent species, like certain trilobites, that existed for only a short time in many different areas on Earth. These fossils are called index fossils. **Index fossils** *represent species that existed on Earth for a short length of time, were abundant, and inhabited many locations.* When an index fossil is found in rock layers at different locations, geologists can infer that the layers are of similar age.

Key Concept Check How are index fossils useful in relative-age dating?

Lesson 2 Review

Visual Summary

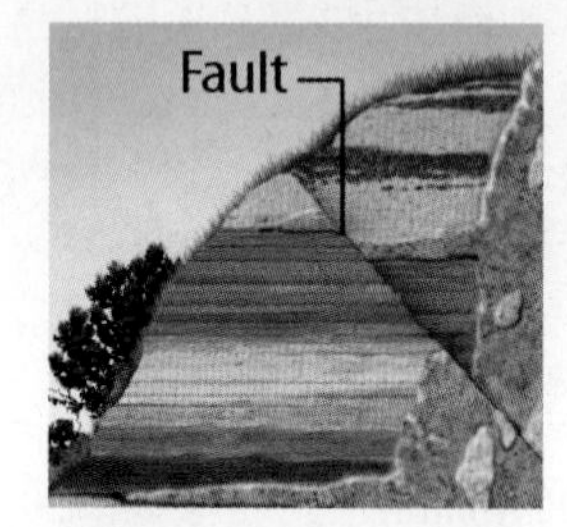

Geologic principles help geologists learn the relative ages of rock layers.

The rock record is incomplete because some of it has eroded away.

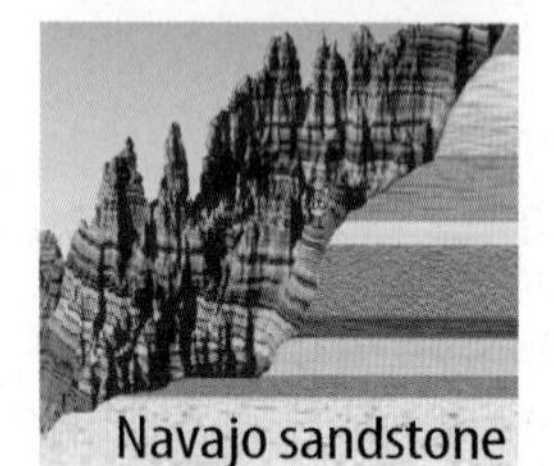

Geologists fill in gaps in the rock record by correlating rock layers.

Use your lesson Foldable to review the lesson. Save your Foldable for the project at the end of the chapter.

What do you think NOW?

You first read the statements below at the beginning of the chapter.

3. Older rocks are always located below younger rocks.

4. Relative age means that scientists are relatively sure of the age.

Did you change your mind about whether you agree or disagree with the statements? Rewrite any false statements to make them true.

Use Vocabulary

1. A gap in the rock record is a(n) ________.
2. The principle that the oldest rocks are generally on the bottom is ________.
3. **Use the terms** *correlation* and *index fossil* in a complete sentence.

Understand Key Concepts

4. Which might be useful in correlation?
   - **A.** amber
   - **B.** inclusion
   - **C.** trilobite
   - **D.** unconformity
5. **Draw** and label a sequence of rock layers showing how an unconformity might form.
6. **Relate** uniformitarianism to principles of relative-age dating.

Interpret Graphics

Use the diagram below to answer question 7.

7. **Decide** Which is older—the rock layers or the dike? Explain which geologic principle you used to arrive at your answer.
8. **Summarize** Copy and fill in the graphic organizer below to identify five geologic principles useful in relative-age dating.

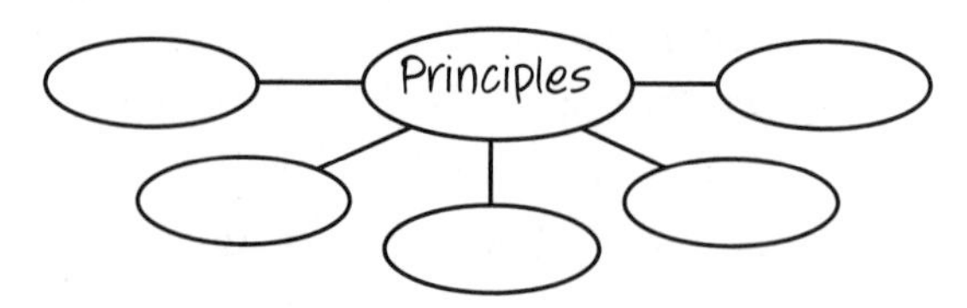

Critical Thinking

9. **Evaluate** why fossils might be more useful than rock types in correlating rock layers on two different continents.
10. **Debate** whether you think humans might be useful as index fossils in the future.

Skill Practice **Interpret Scientific Illustrations** **30 minutes**

Can you correlate rock formations?

Most rocks have been buried in Earth for thousands, millions, or even billions of years. Occasionally, rock layers become exposed on Earth's surface. To correlate rock layers exposed at different locations, it is sometimes necessary to **interpret scientific illustrations** of the layers.

Materials

pencil

colored pencils

large, soft eraser

ruler

Learn It

Drawings and photos can make complex scientific data easier to understand. Use the drawings below to represent rock formations. As you correlate the layers, use the key to **interpret each illustration.**

Try It

1. As well as you can, copy the drawings of the four rock columns shown below into your Science Journal. *Do not write in this book.*
2. Color your drawings so that each rock layer is one color in each of the four rock columns. Use the key to determine what type of rocks each layer contains.
3. Carefully study your drawings. Try to determine which rock columns correlate the best.

Apply It

4. Which rock columns correlated the best? Which principle of relative-age dating did you use when correlating the rock layers?
5. Which rock layer is the oldest in rock column C? The youngest? Which geologic principle did you use to determine this?
6. Identify the type of unconformity that exists in rock column B.
7. **Key Concept** How can you use types of rocks to correlate rock layers? What other type of evidence could you use to determine the relative ages of rock layers?

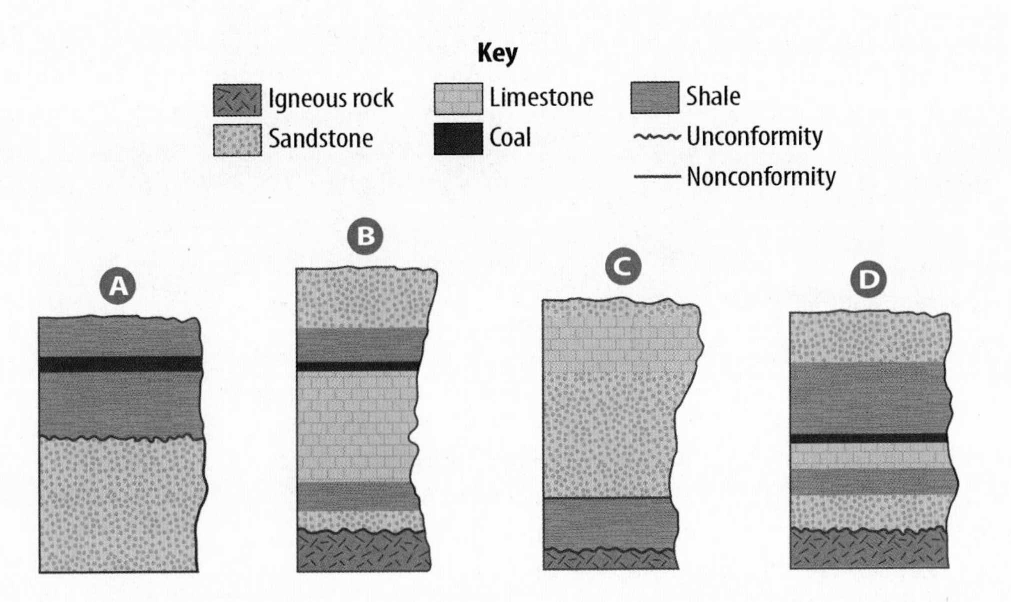

(t to b)Hutchings Photography/Digital Light Source; (2, 4)Jacques Cornell/McGraw-Hill Education; (3)Ken Cavanagh/McGraw-Hill Education

Lesson 3

Reading Guide

Key Concepts
ESSENTIAL QUESTIONS

- What does absolute age mean?
- How can radioactive decay be used to date rocks?

Vocabulary

absolute age p. 345
isotope p. 346
radioactive decay p. 346
half-life p. 347

Multilingual eGlossary

Absolute-Age Dating

Inquiry How old are they?

These mammoth bones are dry and fragile. They have not yet turned to rock. Scientists analyze samples of the bones to discover their ages. Absolute-age dating requires precise measurements in very clean laboratories. What techniques can be used to learn the age of an ancient organism simply by analyzing its bones?

Launch Lab

10 minutes

How can you describe your age?

If you described your relative age compared to your classmates', how would you do it? How do you think your actual, or absolute, age differs from your relative age?

1. One student will write down his or her birth date on an **index card.** The student will hold the card while everyone else files by and looks at it.
2. Form two groups depending on whether your birth date falls before or after the date on the card.
3. Remaining in your group, write down your own birth date on an index card. Quietly form a line in order of your birth dates.

Think About This

1. When you were in two groups, what did you know about everyone's age? When you lined up, what did you know about everyone's age? Which is your relative age? Your absolute age?
2. Can you think of a situation where it would be important to know your absolute age?
3. **Key Concept** Why do you think scientists would want to know the absolute age of a rock?

Absolute Ages of Rocks

Recall from Lesson 2 that you have a relative age. You might be older than your sister and younger than your brother–or you might be the youngest in your family. You also can describe your age by saying your age in years, such as "I am 13 years old." This is not a relative age. It is your age in numbers–your numerical age.

Similarly, scientists can describe the ages of some kinds of rocks numerically. Scientists use the term **absolute age** *to mean the numerical age, in years, of a rock or object.* By measuring the absolute ages of rocks, geologists have developed accurate historical records for many geologic formations.

Key Concept Check How is absolute age different from relative age?

Scientists have been able to determine the absolute ages of rocks and other objects only since the beginning of the twentieth century. That is when radioactivity was discovered. Radioactivity is the release of energy from unstable atoms. The image in **Figure 13** was made using X-rays. How can radioactivity be used to date rocks? In order to answer this question, you need to know about the internal structure of the atoms that make up elements.

Figure 13 The release of radioactive energy can be used to make an X-ray.

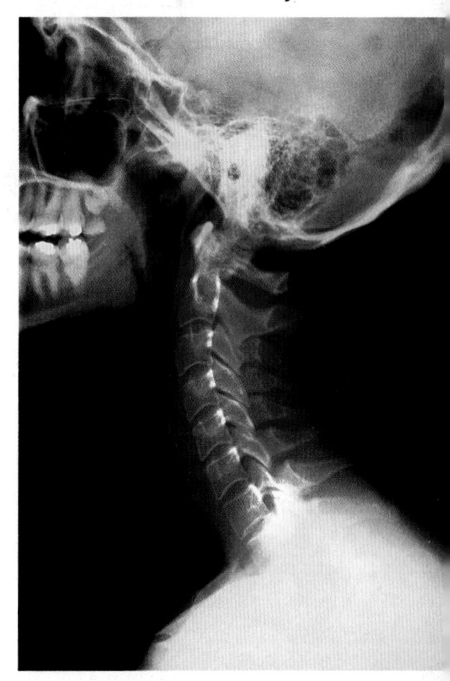

(t)Hutchings Photography/Digital Light Source, (b)Stockbyte/PunchStock

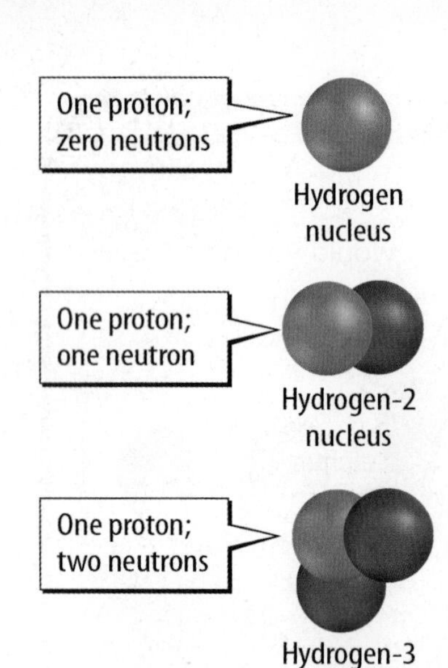

Figure 14 All forms of hydrogen contain only one proton regardless of the number of neutrons.

Atoms

You are probably familiar with the periodic table of the elements, which is shown inside the back cover of this book. Each element is made up of atoms. An atom is the smallest part of an element that has all the properties of the element. Each atom contains smaller particles called protons, neutrons, and electrons. Protons and neutrons are in an atom's nucleus. Electrons surround the nucleus.

Isotopes

All atoms of a given element have the same number of protons. For example, all hydrogen atoms have one proton. But an element's atoms can have different numbers of neutrons. The three atoms shown in **Figure 14** are all hydrogen atoms. Each has the same number of protons–one. However, one of the hydrogen atoms has no neutrons, one has one neutron, and the other has two neutrons. The three different forms of hydrogen atoms are called hydrogen isotopes (I suh tohps). **Isotopes** *are atoms of the same element that have different numbers of neutrons.*

Reading Check How do an element's isotopes differ?

Radioactive Decay

Most isotopes are stable. Stable isotopes do not change under normal conditions. But some isotopes are unstable. These isotopes are known as radioactive isotopes. Radioactive isotopes decay, or change, over time. As they decay, they release energy and form new, stable atoms. **Radioactive decay** *is the process by which an unstable element naturally changes into another element that is stable.* The unstable isotope that decays is called the parent isotope. The new element that forms is called the daughter isotope. **Figure 15** illustrates an example of radioactive decay. The atoms of an unstable isotope of hydrogen (parent) decay into atoms of a stable isotope of helium (daughter).

WORD ORIGIN

isotope
from Greek *isos*, means "equal"; and *topos*, means "place"

Radioactive Decay

Figure 15 An unstable parent hydrogen isotope produces the stable daughter helium isotope.

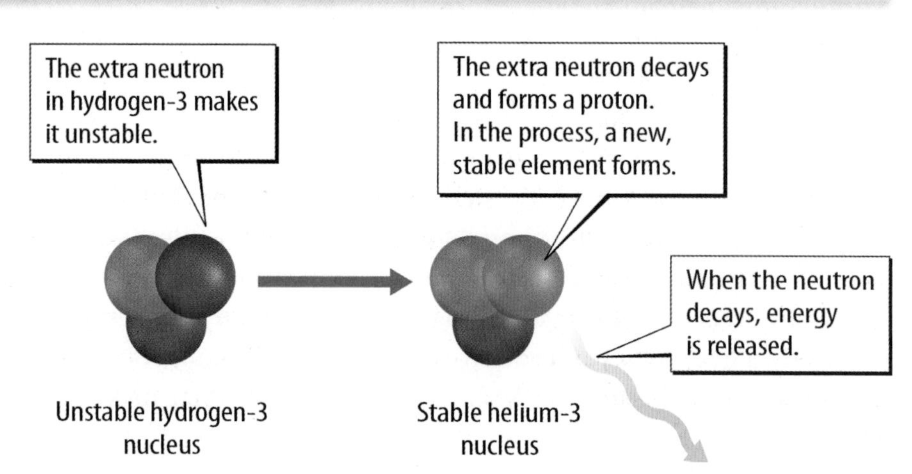

Half-Life of Radioactive Decay

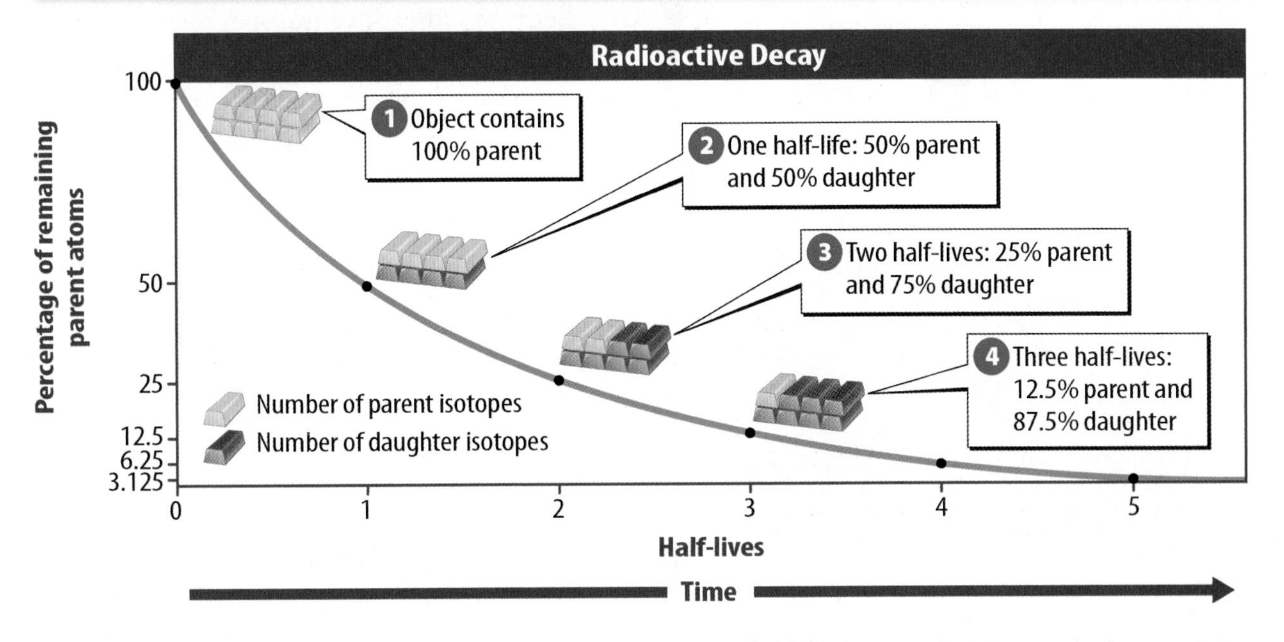

Figure 16 The half-life is the time it takes for one-half of the parent isotopes to change into daughter isotopes.

Visual Check What percentages of parent isotopes and daughter isotopes will there be after four half-lives?

Half-Life

The rate of decay from parent isotopes into daughter isotopes is different for different radioactive elements. But the rate of decay is constant for a given isotope. This rate is measured in time units called half-lives. *An isotope's* **half-life** *is the time required for half of the parent isotopes to decay into daughter isotopes.* Half-lives of radioactive isotopes range from a few microseconds to billions of years.

Reading Check What is half-life?

The graph in **Figure 16** shows how half-life is measured. As time passes, more and more unstable parent isotopes decay and form stable daughter isotopes. That means the ratio between the numbers of parent and daughter isotopes is always changing. When half the parent isotopes have decayed into daughter isotopes, the isotope has reached one half-life. At this point, 50 percent of the isotopes are parents and 50 percent of the isotopes are daughters. After two half-lives, one-half of the remaining parent isotopes have decayed so that only one-quarter as much parent remains as at the start. At this point, 25 percent of the isotopes are parent and 75 percent of the isotopes are daughter. After three half-lives, half again of the remaining parent isotopes have decayed into daughter isotopes. This process continues until nearly all parent isotopes have decayed into daughter isotopes.

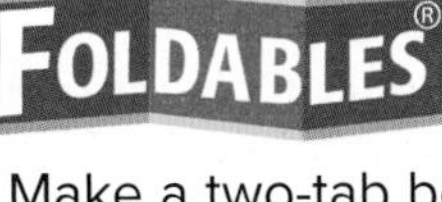

Make a two-tab book from a sheet of paper. Use it to compare how the absolute ages of organic materials and rocks are determined.

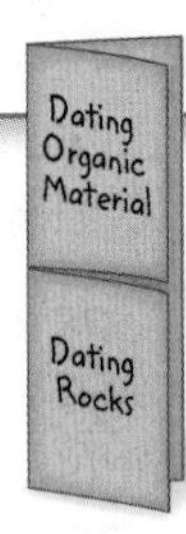

MiniLab 10 minutes

What is the half-life of a drinking straw? 

You can model half-life with a drinking straw.

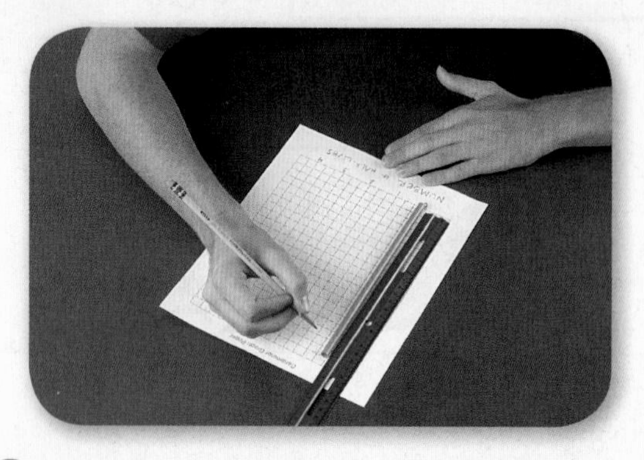

1. Read and complete a lab safety form.
2. On a piece of **graph paper,** draw an x- axis and a y-axis. Label the x-axis *Number of half-lives,* from 0 to 4 in equal intervals. Leave the y-axis blank.
3. Use a **metric ruler** to measure a **drinking straw.** Mark its height on the y-axis, as shown in the photo. Use **scissors** to cut the straw in half and discard half of it. Mark the height of the remaining half as the first half-life.
4. Repeat four times, each time cutting the straw in half and each time adding a measurement to your graph's y-axis.

Analyze and Conclude

1. Compare your graph to the graph in **Figure 16.** How is it similar? How is it different?
2. **Key Concept** Explain how your disappearing straw represents the decay of a radioactive element.

Radiometric Ages

Because radioactive isotopes decay at a constant rate, they can be used like clocks to measure the age of the material that contains them. In this process, called radiometric dating, scientists measure the amount of parent isotope and daughter isotope in a sample of the material they want to date. From this ratio, they can determine the material's age. Scientists make these very precise measurements in laboratories.

Reading Check What is measured in radiometric dating?

Radiocarbon Dating

One important radioactive isotope used for dating is an isotope of carbon called radiocarbon. Radiocarbon is also known as carbon-14, or C-14, because there are 14 particles in its nucleus–six protons and eight neutrons. Radiocarbon forms in Earth's upper atmosphere. There, it mixes in with a stable isotope of carbon called carbon-12, or C-12. The ratio of C-14 to C-12 in the atmosphere is constant.

All living things use carbon as they build and repair tissues. As long as an organism is alive, the ratio of C-14 to C-12 in its tissues is identical to the ratio in the atmosphere. However, when an organism dies, it stops taking in C-14. The C-14 already present in the organism starts to decay to nitrogen-14 (N-14). As the dead organism's C-14 decays, the ratio of C-14 to C-12 changes. Scientists measure the ratio of C-14 to C-12 in the remains of the dead organism to determine how much time has passed since the organism died.

The half-life of carbon-14 is 5,730 years. That means radiocarbon dating is useful for measuring the age of the remains of organisms that died up to about 60,000 years ago. In older remains, there is not enough C-14 left to measure accurately.Too much of it has decayed to N-14.

Animation

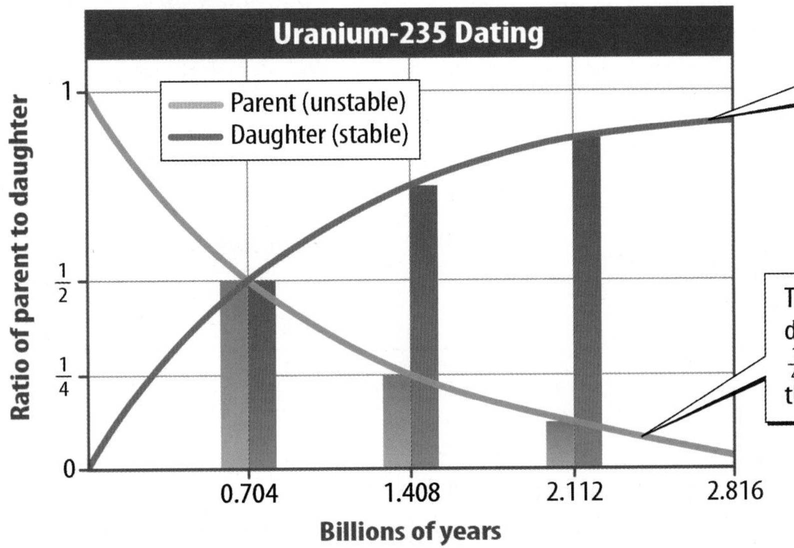

An unstable parent isotope (U-235) will decay at a constant rate and form a daughter product (Pb-207). After one half-life, the concentrations of parent and daughter isotopes are equal.

The parent isotope will continue to decay over time. After two half-lives, $\frac{1}{4}$ of the original parent remains. After three half-lives, $\frac{1}{8}$ remains, and so on.

Figure 17 Scientists determine the absolute age of an igneous rock by measuring the ratio of uranium-235 isotopes (parent) to lead-207 isotopes (daughter) in the rock's minerals.

Visual Check How old is a mineral that contains 25 percent U-235?

Dating Rocks

Radiocarbon dating is useful only for dating organic material—material from once-living organisms. This material includes bones, wood, parchment, and charcoal. Most rocks do not contain organic material. Even most fossils are no longer organic. In most fossils, living tissue has been replaced by rock-forming **minerals.** For dating rocks, geologists use different kinds of radioactive isotopes.

Dating Igneous Rock One of the most common isotopes used in radiometric dating is uranium-235, or U-235. U-235 is often trapped in the minerals of igneous rocks that crystallize from hot, molten magma. As soon as U-235 is trapped in a mineral, it begins to decay to lead-207, or Pb-207, as shown in **Figure 17.** Scientists measure the ratio of U-235 to Pb-207 in a mineral to determine how much time has passed since the mineral formed. This provides the age of the rock that contains the mineral.

REVIEW VOCABULARY

mineral
a naturally occurring, inorganic solid with a definite chemical composition and an orderly arrangement of atoms

Dating Sedimentary Rock In order to be dated by radiometric means, a rock must have U-235 or other radioactive isotopes trapped inside it. The grains in many sedimentary rocks come from a variety of weathered rocks from different locations. The radioactive isotopes within these grains generally record the ages of the grains—not the time when the sediment was deposited. For this reason, sedimentary rock is not as easily dated as igneous rock in radiometric dating.

Key Concept Check Why are radioactive isotopes not useful for dating sedimentary rocks?

Table 2 Radioactive Isotopes Used for Dating Rocks		
Parent Isotope	**Half-Life**	**Daughter Product**
Uranium-235	704 million years	lead-207
Potassium-40	1.25 billion years	argon-40
Uranium-238	4.5 billion years	lead-206
Thorium-232	14.0 billion years	lead-208
Rubidium-87	48.8 billion years	strontium-87

Table 2 Radioactive isotopes useful for dating rocks have long half-lives.

Math Skills 

Use Significant Digits

The answer to a problem involving measurement cannot be more precise than the measurement with the fewest number of significant digits. For example, if you begin with 36 grams (2 significant digits) of U-235, how much U-235 will remain after 2 half-lives?

1. After the first half-life, $\frac{36\text{ g}}{2} = 18$ g of U-235 remain.
2. After the second half-life, $\frac{18\text{ g}}{2} = 9.0$ g of U-235 remain. Add the zero to retain two significant digits.

Practice

The half-life of rubidium-87 (Rb-87) is 48.8 billion years. What is the length of three half-lives of Rb-87?

 Math Practice

 Personal Tutor

Different Types of Isotopes The half-life of uranium-235 is 704 million years. This makes it useful for dating rocks that are very old. **Table 2** lists five of the most useful radioactive isotopes for dating old rocks. All of them have long half-lives. Radioactive isotopes with short half-lives cannot be used for dating old rocks. They do not contain enough parent isotope to measure. Geologists often use a combination of radioactive isotopes to measure the age of a rock. This helps make the measurements more accurate.

 **Key Concept Check** Why is a radioactive isotope with a long half-life useful in dating very old rocks?

The Age of Earth

The oldest known rock formation dated by geologists using radiometric means is in Canada. It is estimated to be between 4.03 billion and 4.28 billion years old. However, individual crystals of the mineral zircon in igneous rocks in Australia have been dated at 4.4 billion years.

With rocks and minerals more than 4 billion years old, scientists know that Earth must be at least that old. Radiometric dating of rocks from the Moon and meteorites indicate that Earth is 4.56 billion years old. Scientists accept this age because evidence suggests that Earth, the Moon, and meteorites all formed at about the same time.

Radiometric dating, the relative order of rock layers, and fossils all help scientists understand Earth's long history. Understanding Earth's history can help scientists understand changes occurring on Earth today–as well as changes that are likely to occur in the future.

Lesson 3 Review

Online Quiz

Visual Summary

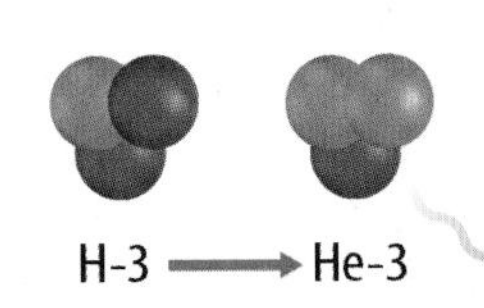

When the unstable atoms of radioactive isotopes decay, they form new, stable isotopes.

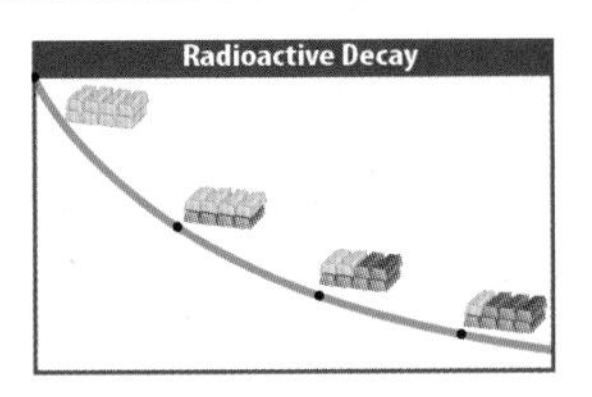

Because radioactive isotopes decay at constant rates, they can be used to determine absolute ages.

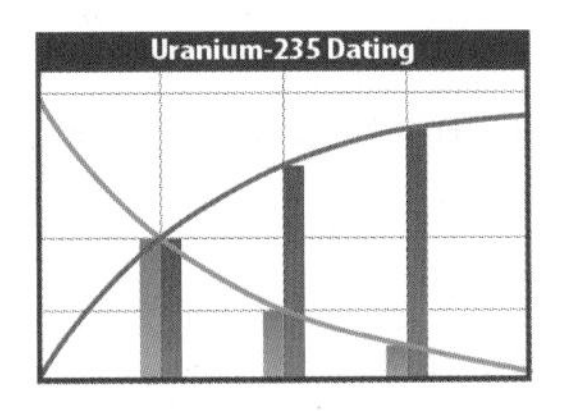

Isotopes with long half-lives are the most useful for dating old rocks.

FOLDABLES

Use your lesson Foldable to review the lesson. Save your Foldable for the project at the end of the chapter.

What do you think NOW?

You first read the statements below at the beginning of the chapter.

5. Absolute age means that scientists are sure of the age.

6. Scientists use radioactive decay to determine the ages of some rocks.

Did you change your mind about whether you agree or disagree with the statements? Rewrite any false statements to make them true.

Use Vocabulary

1. **Compare** absolute age and relative age.
2. The rate of radioactive decay is expressed as an isotope's ________.
3. **Use the terms** *atom* and *isotope* in a complete sentence.

Understand Key Concepts 

4. Which could you date with carbon-14?
   - **A.** a fossilized shark's tooth
   - **B.** an arrowhead carved out of rock
   - **C.** a petrified tree
   - **D.** charcoal from an ancient campfire
5. **Explain** why radioactive isotopes are more useful for dating igneous rocks than they are for dating sedimentary rocks.
6. **Differentiate** between parent isotopes and daughter isotopes.

Interpret Graphics

7. **Identify** Copy and fill in the graphic organizer below to identify the three parts of an atom.

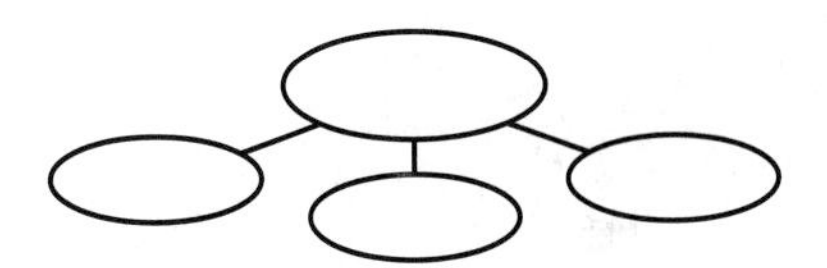

Critical Thinking

8. **Evaluate** the importance of radioactive isotopes in determining the age of Earth.

Math Skills

 Math Practice

9. The half life of potassium-40 (K-40) is 1.25 billion years. If you begin with 130 g of K-40, how much remains after 2.5 billion years? Use the correct number of significant digits in your answer.

Lab

40 minutes

Correlate Rocks Using Index Fossils

Imagine you are a geologist and you have been asked to correlate the rock columns below in order to determine the relative ages of the layers. Recall that geologists can correlate rock layers in different ways. In this lab, use index fossils to correlate and date the layers.

Question

How can index fossils be used to determine the relative ages of Earth's rocks?

Procedure

1. Carefully examine the three rock columns on this page. Each rock layer can be identified with a letter and a number. For example, the second layer down in column A is layer A-2.
2. In your Science Journal, correlate the layers using only the fossils—not the types of rock. Before you begin, look at the fossil key on the next page. It shows the time intervals during which each organism or group of organisms lived on Earth. Refer to the key as you correlate.

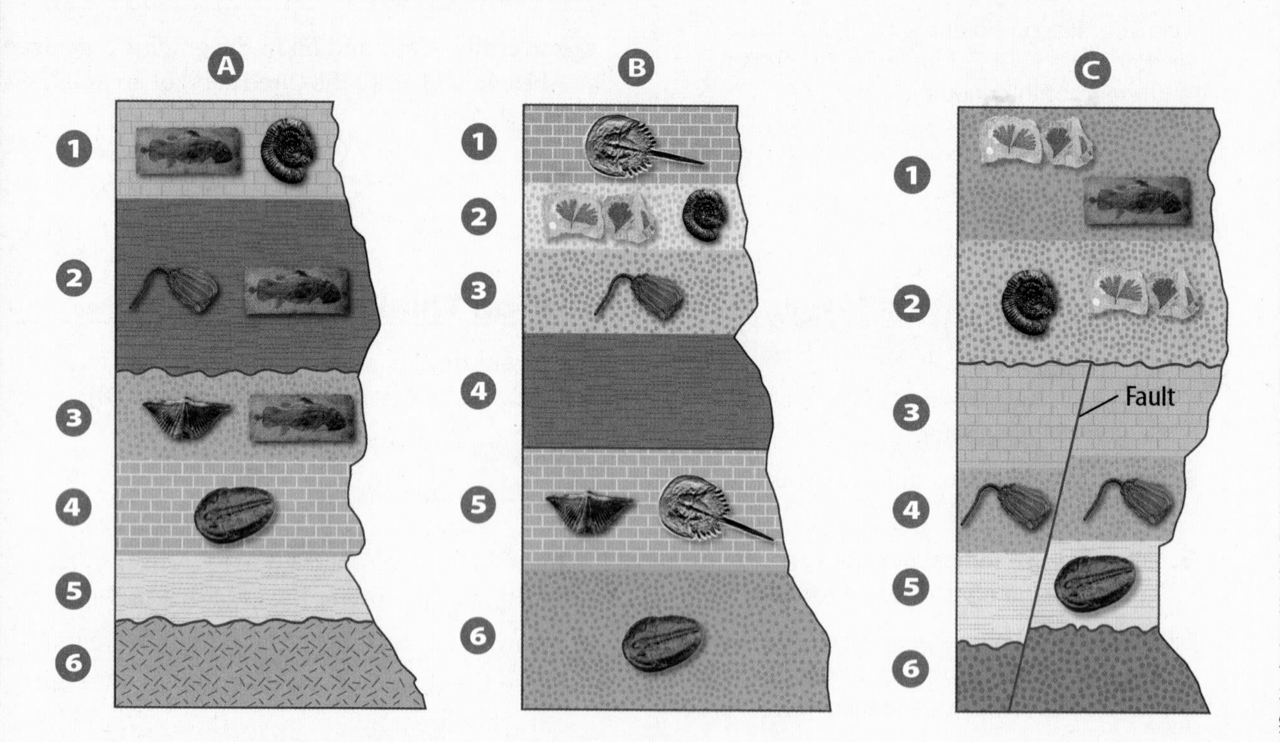

(l to r, t to b)DK Limited/Corbis, Andrew Ward/Life File/Getty Images, koi88/Alamy, Kjell B. Sandved/Science Source, Kjell B. Sandved/Science Source, Andrew Ward/Life File/Getty Images, DK Limited/Corbis, The Natural History Museum/Alamy, DK Limited/Corbis, The Natural History Museum/Alamy, Andrew Ward/Life File/Getty Images, Kjell B. Sandved/Science Source, DK Limited/Corbis, Jacob Hamblin/iStock/360/Getty Images, The Natural History Museum/Alamy, Jacob Hamblin/iStock/360/Getty Images, koi88/Alamy, Jacob Hamblin/iStock/360/Getty Images, DK Limited/Corbis,The Natural History Museum/Alamy

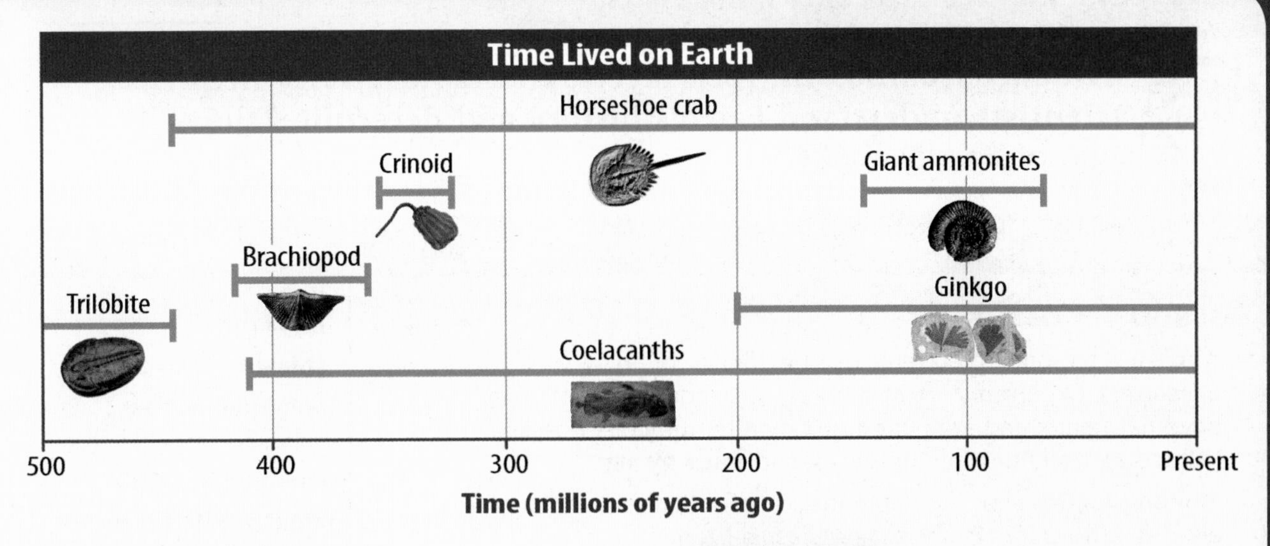

Analyze and Conclude

3. **Differentiate** Which fossils in the key appear to be index fossils? Explain your choices.
4. **Match** Correlate layer A-2 to one layer in each of the other two columns. Approximately how old are these layers? How do you know?
5. **Infer** What is the approximate age of layer B-4? *Hint: It lies between two index fossils.*
6. **Infer** How old is the fault in column C?
7. **Compare and Contrast** How does correlating rocks using fossils differ from correlating rocks using types of rock?
8. **The Big Idea** How can fossils be used to determine the relative ages of rocks?

Communicate Your Results

Choose a partner. One of you is a reporter and one is a geologist. Conduct an interview about what kinds of fossils are best used to date rocks.

Extension

Choose one of the three rock formations you correlated. Based on your results, provide a range of dates for each of the layers within it.

Lab Tips

☑ You might want to copy the rock layers in your Science Journal and correlate them by drawing lines connecting the layers.

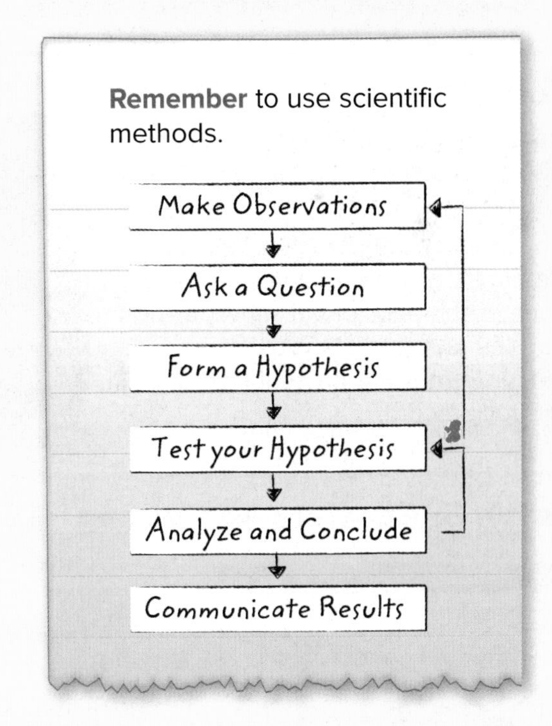

Chapter 10 Study Guide

Evidence from fossils, rock layers, and radioactivity help scientists understand Earth's history and determine the ages of Earth's rocks.

Key Concepts Summary 

Lesson 1: Fossils

- A **fossil** is the preserved remains or evidence of ancient organisms. Organisms are more likely to become fossils if they have hard parts and are buried quickly after they die. Fossils include **carbon films, molds, casts,** and **trace fossils.**
- **Paleontologists** use clues from fossils to learn about ancient life and the environments ancient organisms lived in.

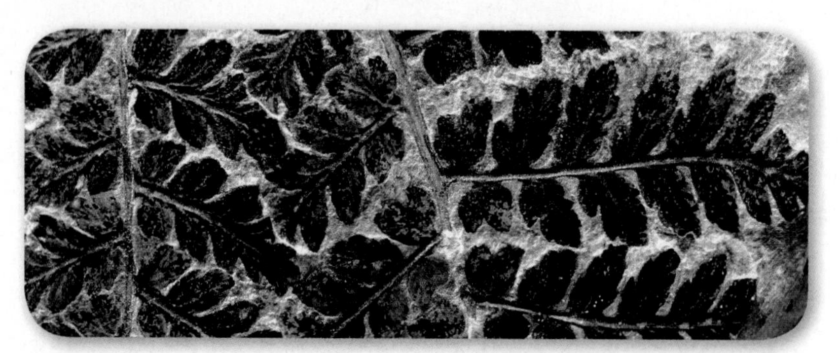

Vocabulary

fossil p. 327
catastrophism p. 327
uniformitarianism p. 328
carbon film p. 330
mold p. 331
cast p. 331
trace fossil p. 331
paleontologist p. 332

Lesson 2: Relative-Age Dating

- **Relative age** is the age of rocks and geologic features compared with rocks and features nearby.
- The relative age of rock layers can be determined using geologic principles, such as the principle of **superposition** and the principle of **inclusion. Unconformities** represent time gaps in the rock record.

relative age p. 337
superposition p. 338
inclusion p. 339
unconformity p. 340
correlation p. 340
index fossil p. 341

Lesson 3: Absolute-Age Dating

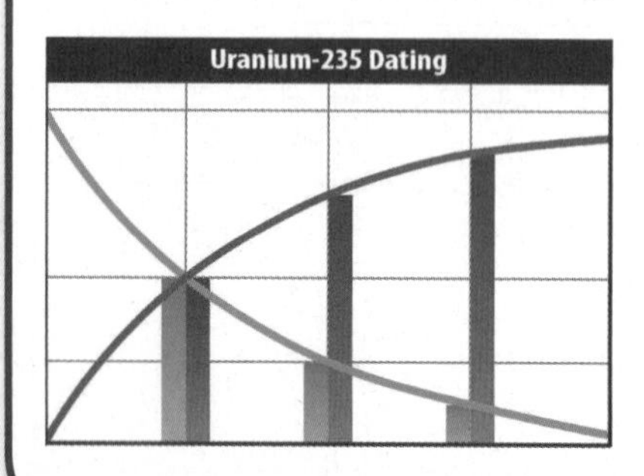

- **Absolute age** is the age in years of a rock or object.
- The **radioactive decay** of unstable **isotopes** occurs at a constant rate, measured as **half-life.** To date a rock or object, scientists measure the ratios of its parent and daughter isotopes.

absolute age p. 345
isotope p. 346
radioactive decay p. 346
half-life p. 347

(t)David Lyons/Alamy; (b)©Hemis/Corbis

Personal Tutor

Vocabulary eFlashcards
Vocabulary eGames

FOLDABLES® Chapter Project

Assemble your lesson Foldables® as shown to make a Chapter Project. Use the project to review what you have learned in this chapter.

Use Vocabulary

1. An ancient dinosaur track is a(n) ________ .
2. ________ use the principle of ________ to reconstruct ancient environments.
3. The principle of ________ states that the oldest layers are generally at the bottom.
4. In ________, geologists use ________ to match rock layers on separate continents.
5. A(n) ________ is an eroded surface.
6. The process of ________ can be used like a clock to determine a rock's ________ .
7. A Uranium-235 ________ decays with a constant ________ of 704 million years.

Interactive Concept Map

Link Vocabulary and Key Concepts

Copy this concept map, and then use vocabulary terms from the previous page to complete the concept map.

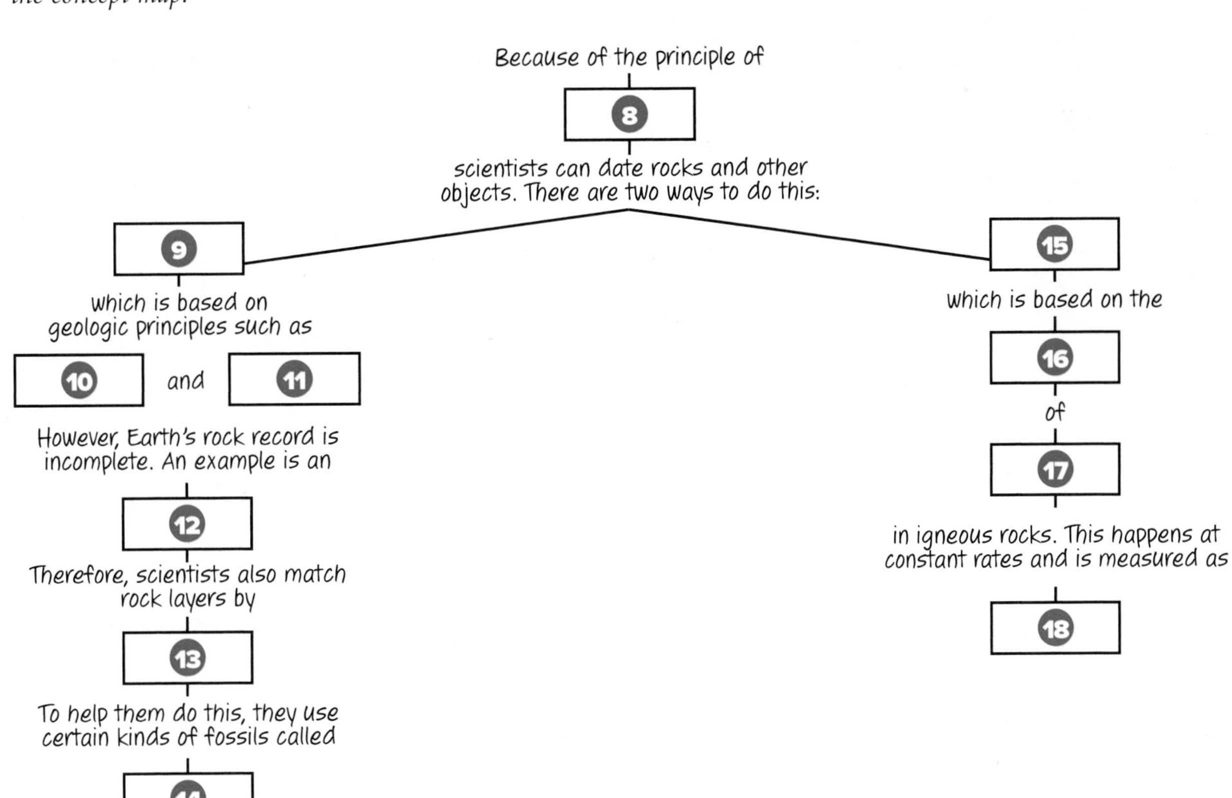

Chapter 10 Review

Understand Key Concepts 

1. Which idea explains Earth's history by examining present conditions on Earth?
   A. absolute-age dating
   B. catastrophism
   C. relative-age dating
   D. uniformitarianism

2. Which part of a dinosaur is least likely to be fossilized?
   A. bone
   B. brain
   C. horn
   D. tooth

3. Which makes a species a good index fossil?
   A. lived a long time and was abundant
   B. lived a long time and was scarce
   C. lived a short time and was scarce
   D. lived a short time and was abundant

4. In the drawing below, what is the order of rock layers from oldest to youngest?

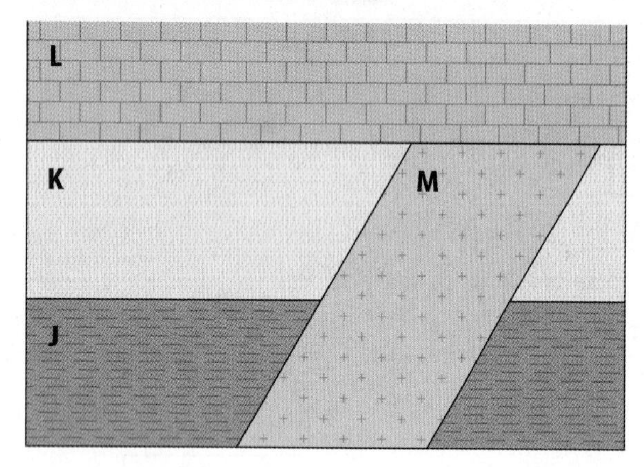

   A. J, K, L, M
   B. J, K, M, L
   C. L, K, J, M
   D. M, J, K, L

5. What do geologists look for in order to correlate rocks in different locations?
   A. different rock types and similar fossils
   B. many rock types and many fossils
   C. similar rock types and lack of fossils
   D. similar rock types and similar fossils

6. What is the half-life on the graph below?

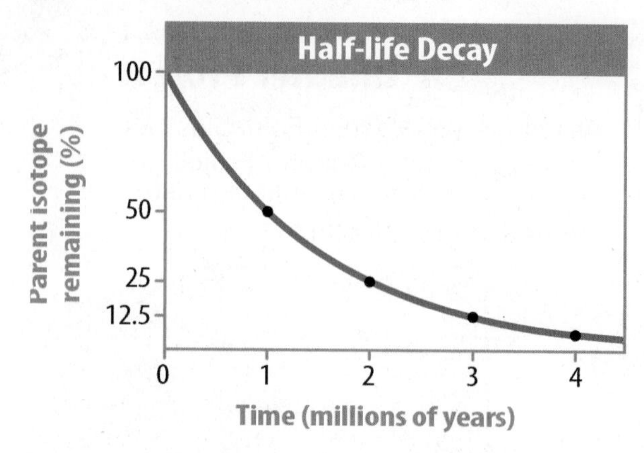

   A. 1 million years
   B. 2 million years
   C. 3 million years
   D. 4 million years

7. What are isotopes?
   A. atoms of the same element with different numbers of electrons but the same number of protons
   B. atoms of the same element with different numbers of electrons but the same number of neutrons
   C. atoms of the same element with different numbers of neutrons but the same number of protons
   D. atoms of the same element with equal numbers of neutrons and protons.

8. What do scientists measure when determining the absolute age of a rock?
   A. amount of radioactivity
   B. number of uranium atoms
   C. ratio of neutrons and electrons
   D. ratio of parent and daughter isotopes

9. Why is radiometric dating less useful to date sedimentary rocks than igneous rocks?
   A. Sedimentary rocks are more eroded.
   B. Sedimentary rocks contain fossils.
   C. Sedimentary rocks contain grains formed from other rocks.
   D. Sedimentary rocks contain grains less than 60,000 years old.

Critical Thinking

10. **Give** an example of superposition from your own life.

11. **Suggest** a way that an ancient human might have been preserved as a fossil.

12. Explain why scientists use a combination of uniformitarianism and catastrophism ideas to understand Earth.

13. **Reason** You are studying a rock formation that includes layers of folded sedimentary rocks cut by faults and dikes. Describe the geologic principles you would use to determine the relative order of the layers.

14. **Construct** a graph showing the radioactive decay of an unstable isotope with a half-life of 250 years. Label three half-lives.

15. **Assess** The ash layers in the drawing below have been dated as shown. What conclusions can you draw about the ages of each of the layers A, B, and C?

C
Ash deposited 540 mya
B
Ash deposited 730 mya
A

Writing in Science

16. **Write** a paragraph of at least five sentences explaining why absolute-age dating has been more useful than relative-age dating in determining the age of Earth. Include a main idea, supporting details, and concluding sentence.

REVIEW

17. What evidence do scientists use to determine the ages of rocks?

18. The photo below shows many rock layers of the Grand Canyon. Explain how the development of the principle of uniformitarianism might have changed earlier ideas about the age of the Grand Canyon and how it formed.

Math Skills

Math Practice

Use Significant Figures

19. If you begin with 68 g of an isotope, how many grams of the original isotope will remain after four half-lives?

20. The half-life of radon-222 (Rn-222) is 3.823 days.
    a. How long would it take for three half-lives?
    b. What percentage of the original sample would remain after three half-lives?

21. The half-life of Rn-222 is 3.823 days. What was the original mass of a sample of this isotope if 0.0500 g remains after 7.646 days?

Standardized Test Practice

Record your answers on the answer sheet provided by your teacher or on a sheet of paper.

Multiple Choice

1 Which is a copy of a dead organism formed when its impression fills with mineral deposits or sediments?

A carbon film

B cast

C mold

D trace fossil

Use the diagram below to answer question 2.

2 In the diagram above, which rock layer typically is youngest?

A 1

B 2

C 3

D 4

3 Which characteristic of rocks does radioactive decay measure?

A absolute age

B lateral continuity

C relative age

D unconformity

4 Which increases the likelihood that a dead organism will be fossilized?

A fast decay of bones

B presence of few hard body parts

C quick burial after death

D vast amounts of skin

Use the diagram below to answer question 5.

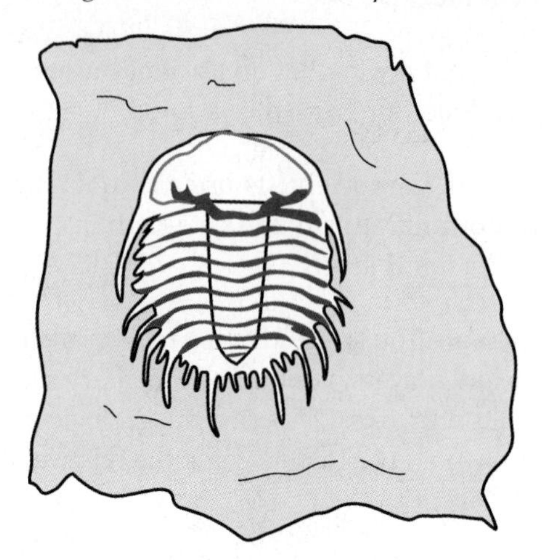

5 Which fossilized ancient organism is pictured in the diagram above?

A clam

B mammoth

C mastodon

D trilobite

6 Which explains most of Earth's geological features as a result of short periods of earthquakes, volcanoes, and meteorite impacts?

A catastrophism

B evolution

C supernaturalism

D uniformitarianism

7 Which fossil type helps geologists infer that rock layers in different geographic locations are similar in age?

A carbon film

B index fossil

C preserved remains

D trace fossil

8 Which pie chart shows the ratio of parent to daughter atoms after four half-lives?

A

Parent

Daughter

B

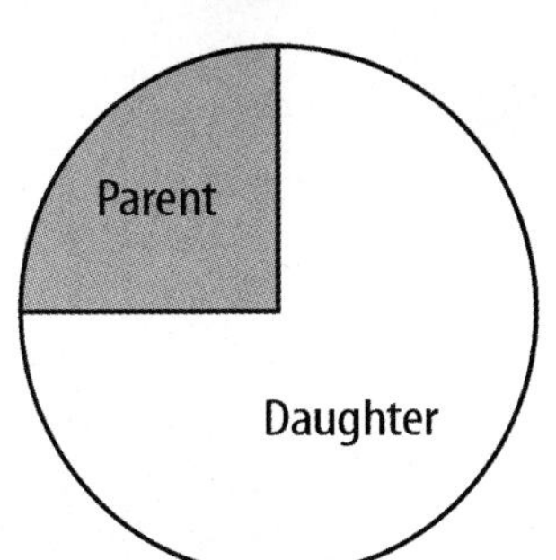

C

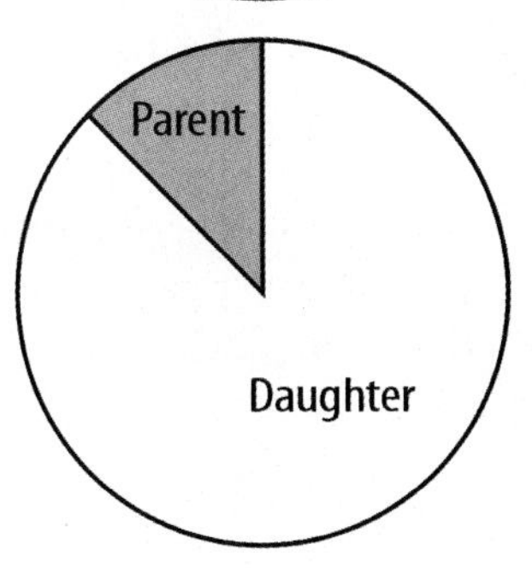

D

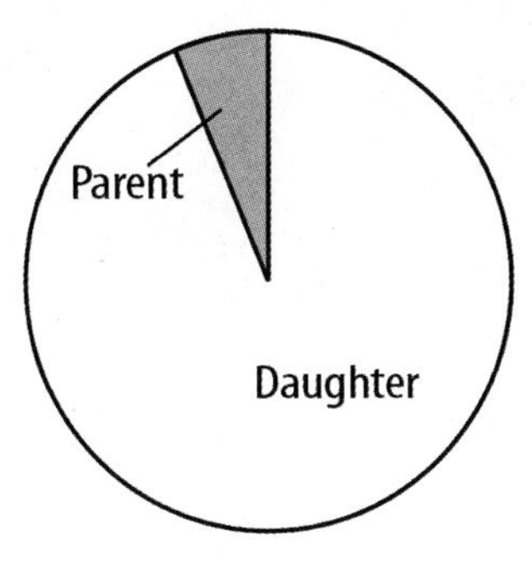

Constructed Response

Use the diagram below to answer questions 9 and 10.

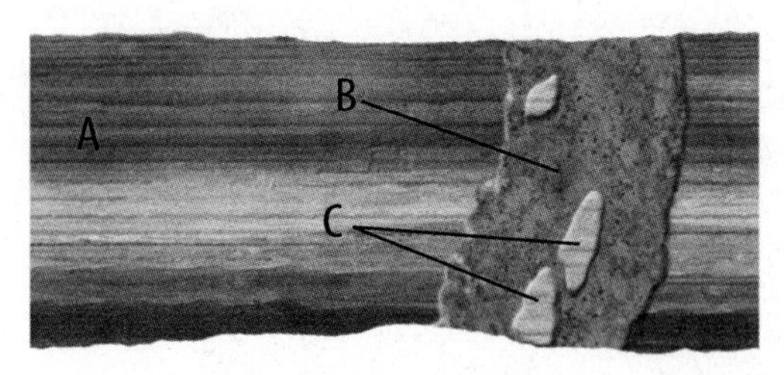

9 Are the sedimentary rock layers (A) older or younger than the dike (B)? How do you know?

10 Is the dike (B) older or younger than the inclusions (C)? How do you know?

Use the diagram below to answer question 11.

11 Identify the type of unconformity that exists in the diagram above. Hypothesize how this could have happened.

12 What is C-14? What role does it play in radiocarbon dating? Why does time limit the effectiveness of radiocarbon dating as a tool for measuring age?

NEED EXTRA HELP?												
If You Missed Question...	1	2	3	4	5	6	7	8	9	10	11	12
Go to Lesson...	1	2	3	1	1	1	2	3	2	2	2	3

Chapter 11

Geologic Time

THE BIG IDEA

What have scientists learned about Earth's past by studying rocks and fossils?

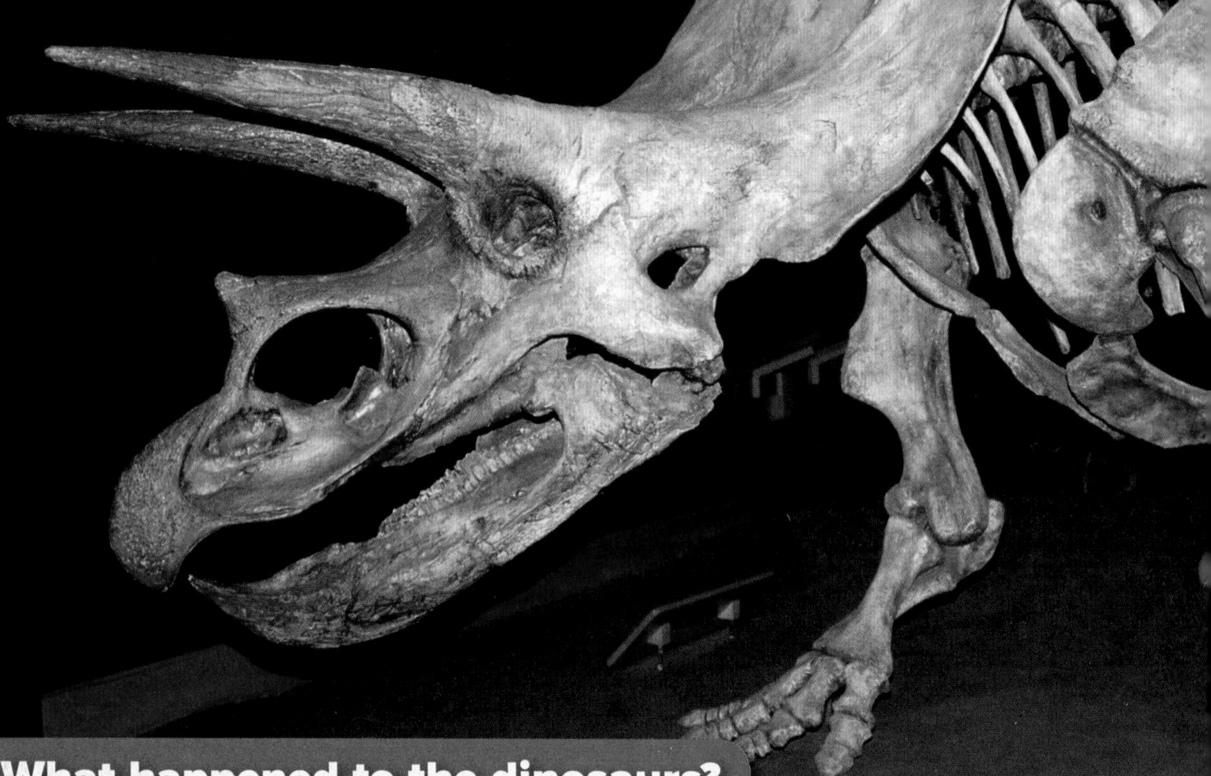

Inquiry What happened to the dinosaurs?

This Triceratops lived millions of years ago. Hundreds of other kinds of dinosaurs lived at the same time. Some were as big as houses; others were as small as chickens. Scientists learn about dinosaurs by studying their fossils. Like many organisms that have lived on Earth, dinosaurs disappeared suddenly. Why did the dinosaurs disappear?

- How has Earth changed over geologic time?
- How do geologic events affect life on Earth?
- What have scientists learned about Earth's past by studying rocks and fossils?

Kevin Schafer/Corbis

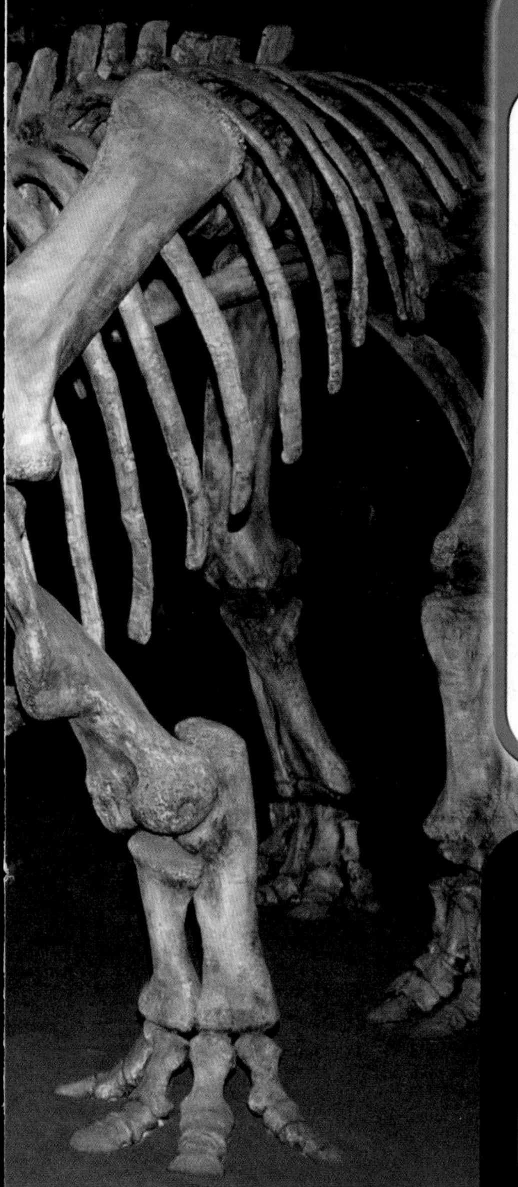

Get Ready to Read

What do you think?

Before you read, decide if you agree or disagree with each of these statements. As you read this chapter, see if you change your mind about any of the statements.

1. All geologic eras are the same length of time.
2. Meteorite impacts cause all extinction events.
3. North America was once on the equator.
4. All of Earth's continents were part of a huge supercontinent 250 million years ago.
5. All large Mesozoic vertebrates were dinosaurs.
6. Dinosaurs disappeared in a large mass extinction event.
7. Mammals evolved after dinosaurs became extinct.
8. Ice covered nearly one-third of Earth's land surface 11,700 years ago.

Your one-stop online resource
connectED.mcgraw-hill.com

LearnSmart®

Chapter Resources Files, Reading Essentials, Get Ready to Read, Quick Vocabulary

Animations, Videos, Interactive Tables

Self-checks, Quizzes, Tests

Project-Based Learning Activities

Lab Manuals, Safety Videos, Virtual Labs & Other Tools

Vocabulary, Multilingual eGlossary, Vocab eGames, Vocab eFlashcards

Personal Tutors

Lesson 1

Reading Guide

Key Concepts
ESSENTIAL QUESTIONS

- How was the geologic time scale developed?
- What are some causes of mass extinctions?
- How is evolution affected by environmental change?

Vocabulary

eon p. 363
era p. 363
period p. 363
epoch p. 363
mass extinction p. 365
land bridge p. 366
geographic isolation p. 366

Multilingual eGlossary

BrainPOP®
Science Video

Geologic History and the Evolution of Life

Inquiry What happened here?

A meteorite 50 m in diameter crashed into Earth 50,000 years ago. The force of the impact created this crater in Arizona and threw massive amounts of dust and debris into the atmosphere. Scientists hypothesize that a meteorite 200 times this size—the size of a small city—struck Earth 66 million years ago. How might it have affected life on Earth?

Launch Lab

20 minutes

Can you make a time line of your life?

How would you organize a time line of your life? You might include regular events, such as birthdays. But you might also include special events, such as a weekend camping trip or a summer vacation.

1. Read and complete a lab safety form.
2. Use **scissors** to cut two pieces of **graph paper** in half. **Tape** them together to make one long piece of paper. Write down the years of your life in horizontal sequence, marked off at regular intervals.
3. Choose up to 12 important events or periods of time in your life. Mark those events on your time line.

Think About This

1. Do the events on your time line appear at regular intervals?
2. **Key Concept** How do you think the geologic time scale is like a time line of your life?

Developing a Geologic Time Line

Think about what you did over the last year. Maybe you went on vacation during the summer or visited relatives in the fall. To organize events in your life, you use different units of time, such as weeks, months, and years. Geologists organize Earth's past in a similar way. They developed a time line of Earth's past called the geologic time scale. As shown in **Figure 1**, time units on the geologic time scale are thousands and millions of years long–much longer than the units you use to organize events in your life.

Units in the Geologic Time Scale

Eons *are the longest units of geologic time.* Earth's current eon, the Phanerozoic (fan er oh ZOH ihk) eon, began 541 million years ago (mya). *Eons are subdivided into smaller units of time called* **eras.** *Eras are subdivided into* **periods.** *Periods are subdivided into* **epochs** (EH pocks). Epochs are not shown on the time line in **Figure 1.** Notice that the time units are not equal. For example, the Paleozoic era is longer than the Mesozoic and Cenozoic eras combined.

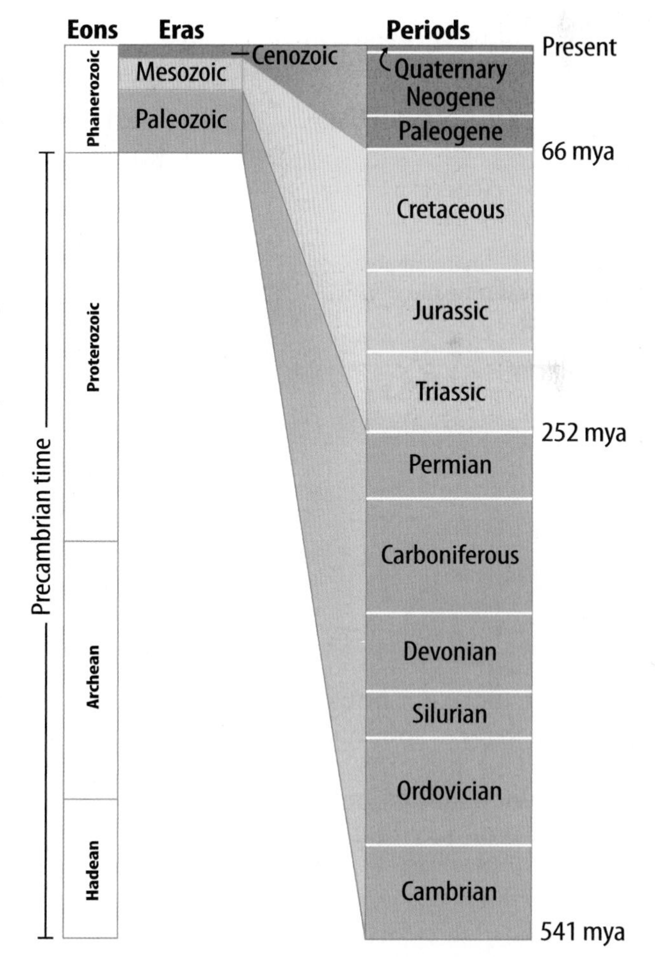

Figure 1 In the geologic time scale, the 4.6 billion years of Earth's history are divided into time units of unequal length.

Fossils in Rock Layers

Animation

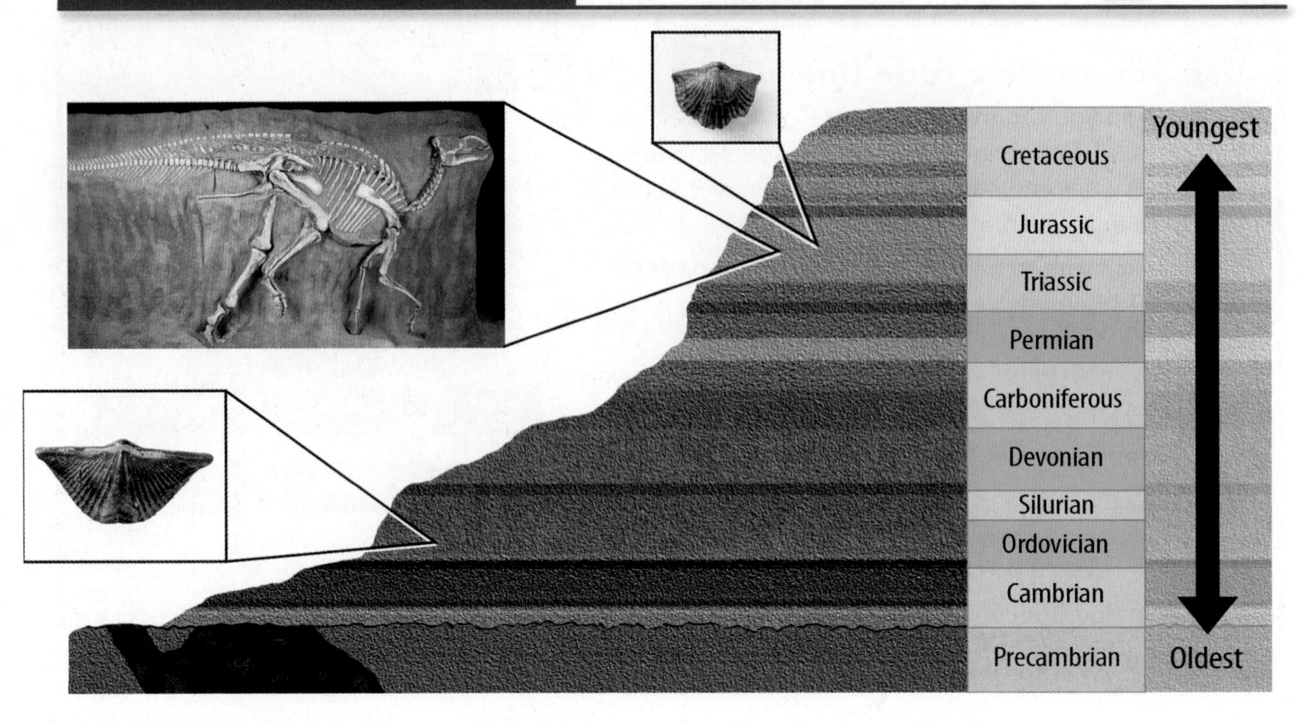

Figure 2 Both older and younger rocks contain fossils of small, relatively simple life-forms. Only younger rocks contain larger, more complex fossils.

SCIENCE USE V. COMMON USE

scale

Science Use a series of marks or points at known intervals

Common Use an instrument used for measuring the weight of an object

The Time Scale and Fossils

Hundreds of years ago, when geologists began developing the geologic time **scale**, they chose the time boundaries based on what they observed in Earth's rock layers. Different layers contained different fossils. For example, older rocks contained only fossils of small, relatively simple life-forms. Younger rocks contained these fossils as well as fossils of other, more complex organisms, such as dinosaurs, as illustrated in **Figure 2.**

Major Divisions in the Geologic Time Scale

While studying the fossils in rock layers, geologists often saw abrupt changes in the types of fossils within the layers. Sometimes, fossils in one rock layer did not appear in the rock layers right above it. It seemed as though the organisms that lived during that period of time had disappeared suddenly. Geologists used these sudden changes in the fossil record to mark divisions in geologic time. Because the changes did not occur at regular intervals, the boundaries between the units of time in the geologic time scale are irregular. This means the time units are of unequal length.

The time scale is a work in progress. Scientists debate the placement of the boundaries as they make new discoveries.

Key Concept Check Why are fossils important in the development of the geologic time scale?

FOLDABLES®

Make a four-door book from a vertical sheet of paper. Use it to organize information about the units of geologic time.

Responses to Change

Sudden changes in the fossil record represent times when large populations of organisms died or became extinct. A **mass extinction** *is the extinction of many species on Earth within a short period of time.* As shown in **Figure 3,** there have been several mass extinction events in Earth's history.

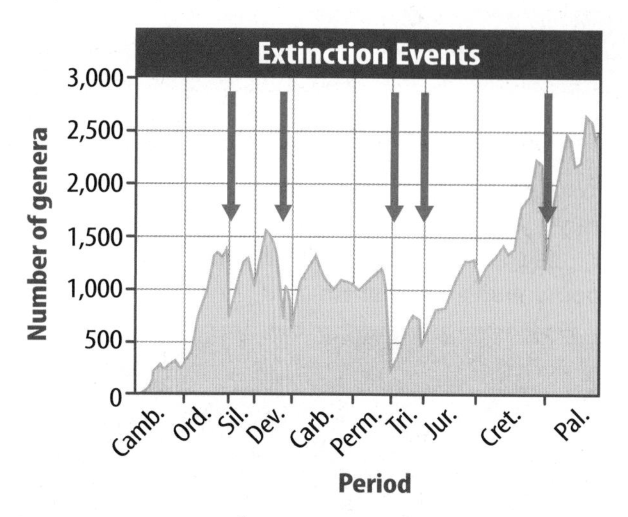

Figure 3 There have been five major mass extinctions in Earth's history. In each one, the number of genera—groups of species—decreased sharply.

Visual Check When was Earth's greatest mass-extinction event?

Changes in Climate

What could cause a mass extinction? All species of organisms depend on the environment for their survival. If the environment changes quickly and species do not adapt to the change, they die.

Many things can cause a climate change. For example, gas and dust from volcanoes can block sunlight and reduce temperatures. As you read on the first page of this lesson, the results of a meteorite crashing into Earth would block sunlight and change climate.

Scientists hypothesize that a meteorite impact might have caused the mass extinction that occurred when dinosaurs became extinct. Evidence for this impact is in a clay layer containing the element iridium in rocks around the world. Iridium is rare in Earth rocks but common in meteorites. No dinosaur fossils have been found in rocks above the iridium layer. A sample of rock containing this layer is shown in **Figure 4.**

WORD ORIGIN

extinct
from Latin *extinctus,* means "dying out"

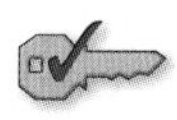

Key Concept Check Describe a possible event that could cause a mass extinction.

Meteorite Impact

Figure 4 An iridium-enriched clay layer in Earth's rocks is evidence that a large meteorite crashed into Earth 66 million years ago. A meteorite impact can contribute to a mass extinction event.

MiniLab — 10 minutes

How does geographic isolation affect evolution?

Have you ever played the phone game? How is this game similar to what happens when populations of organisms are separated?

1. Form two groups.
2. One person in each group should whisper a sentence—provided by your teacher—into the ear of his or her neighbor. Each person in turn will whisper the sentence to his or her neighbor until it returns to the first person.

Analyze and Conclude

1. **Observe** Did the sentence change? Did it change in the same way in each group?
2. **Key Concept** How is this activity similar to organisms that are geographically isolated?

Geography and Evolution

When environments change, some species of organisms are unable to adapt. They become extinct. However, other species do adapt to environmental changes. Evolution is the change in species over time as they adapt to their environments. Sudden, catastrophic changes in the environment can affect evolution. So can the slow movement of Earth's tectonic plates.

Land Bridges When continents collide or when sea level drops, landmasses can join together. *A* **land bridge** *connects two continents that were previously separated.* Over time, organisms move across land bridges and evolve as they adapt to new environments.

Geographic Isolation The movement of tectonic plates or other slow geologic events can cause geographic areas to move apart. When this happens, populations of organisms can become isolated. **Geographic isolation** *is the separation of a population of organisms from the rest of its species due to some physical barrier, such as a mountain range or an ocean.* Separated populations of species evolve in different ways as they adapt to different environments. Even slight differences in environments can affect evolution, as shown in **Figure 5**.

Key Concept Check How can geographic isolation affect evolution?

Geographic Isolation

Figure 5 A population of squirrels was gradually separated as the Grand Canyon developed. Each group adapted to a slightly different environment and evolved in a different way.

Kaibab squirrel

Abert's squirrel

Precambrian time

Hadean Eon | Archean Eon | Proterozoic Eon | Phanerozoic Eon

Earth's origin

4.5 4.0 3.5 3.0 2.5 2.0 1.5 1.0 0.5 0

Time (in billions of years)

Increasing diversity

600 mya 541 mya

Precambrian Cambrian

Personal Tutor

Figure 6 Precambrian time is nearly 90 percent of Earth's history. An explosion of life-forms appeared at the beginning of the Phanerozoic eon, during the Cambrian period.

Precambrian Time

Life has been evolving on Earth for billions of years. The oldest fossil evidence of life on Earth is in rocks that are about 3.5 billion years old. These ancient life-forms were simple, unicellular organisms, much like present-day bacteria. The oldest possible fossil evidence of multicellular life is 2.1 billion years old, however the oldest undisputed fossils of multicellular organisms are about 600 million years old. These fossils are rare, and early geologists did not know about them. They hypothesized that multicellular life first appeared in the Cambrian (KAM bree un) period, at the beginning of the Phanerozoic eon 541 mya. Time before the Cambrian was called Precambrian time. Scientists have determined that Precambrian time is nearly 90 percent of Earth's history, as shown in **Figure 6.**

Precambrian Life

The rare fossils of multicellular life-forms in Precambrian rocks are from soft-bodied organisms different from organisms on Earth today. Many of these species became extinct at the end of the Precambrian.

Cambrian Explosion

Precambrian life led to a sudden appearance of new types of multicellular life-forms in the Cambrian period. This sudden appearance of new, complex life-forms, indicated on the right in **Figure 6,** is often referred to as the Cambrian explosion. Some Cambrian life-forms, such as trilobites, were the first to have hard body parts. Because of their hard body parts, trilobites were more easily preserved. Evidence of trilobites, like the fossils shown in **Figure 7,** are in the fossil record. Scientists hypothesize that some of them are distant ancestors of organisms alive today.

Reading Check What is the Cambrian explosion?

Figure 7 The hard body parts of these trilobites were preserved as fossils.

Lesson 1 Review

Visual Summary

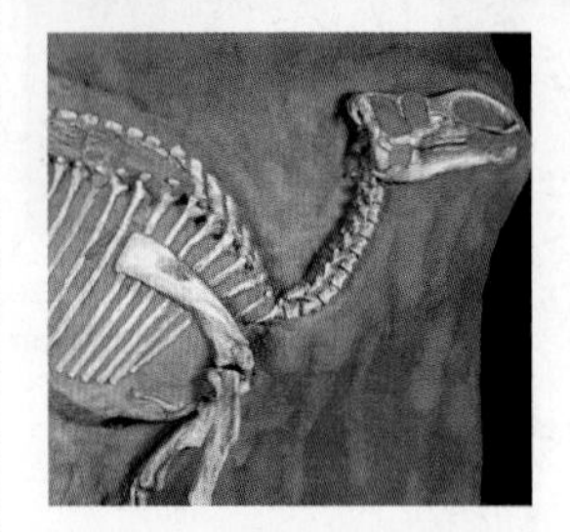

Earth's history is organized into eons, eras, periods, and epochs.

Climate change caused by the results of a meteorite impact could contribute to a mass extinction event.

Slow changes in geography affect evolution.

Use your lesson Foldable to review the lesson. Save your Foldable for the project at the end of the chapter.

What do you think NOW?

You first read the statements below at the beginning of the chapter.

1. All geologic eras are the same length of time.
2. Meteorite impacts cause all extinction events.

Did you change your mind about whether you agree or disagree with the statements? Rewrite any false statements to make them true.

Use Vocabulary

1. **Distinguish** between an eon and an era.
2. A(n) ________ might form when continents move close together.
3. A(n) ________ might occur if an environment changes suddenly.

Understand Key Concepts 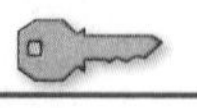

4. Which could contribute to a mass-extinction event?
   A. an earthquake
   B. a hot summer
   C. a hurricane
   D. a volcanic eruption
5. **Explain** how geographic isolation can affect evolution.
6. **Distinguish** between a calendar and the geologic time scale.

Interpret Graphics

7. **Explain** what the graph below represents. What happened at this time in Earth's past?

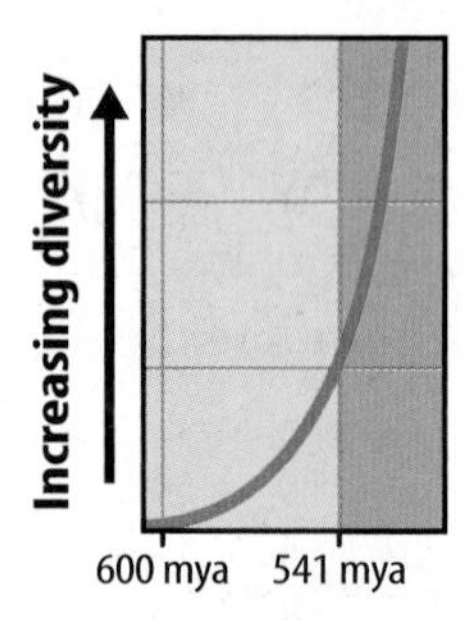

8. **Organize Information** Copy and fill in the graphic organizer below to show units of the geologic time scale from longest to shortest.

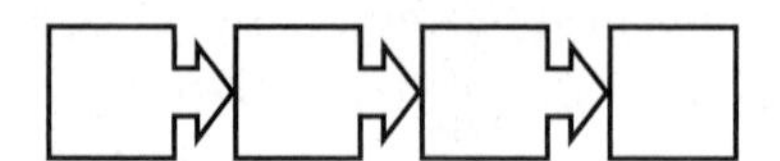

Critical Thinking

9. **Suggest** how humans might contribute to a mass extinction event.
10. **Propose** why Precambrian rocks contain few fossils.

(t)Andy Crawford/Getty Images, (b)kojihirano/Getty Images

Skill Practice Interpret Data

30 minutes

How has life changed over time?

Fossil evidence indicates that there have been wide fluctuations in the types, or diversity, of organisms that have lived on Earth over geologic time.

Learn It

Line graphs compare two variables and show how one variable changes in response to another variable. Line graphs are particularly useful in presenting data that change over time. The first line graph below shows how the diversity of genera has changed over time. The second graph shows how extinction rates, presented as percentages of genera, have changed over time. **Interpret data** in these graphs to learn how they relate to each another.

Try It

1. Carefully study each graph. Note that time, the independent variable, is plotted on the *x*-axis of each graph. The dependent variable of each graph—the diversity, or number of genera, in one graph and the extinction rate in the other graph—are plotted on the *y*-axes.
2. Use the graphs to answer questions 3–7.

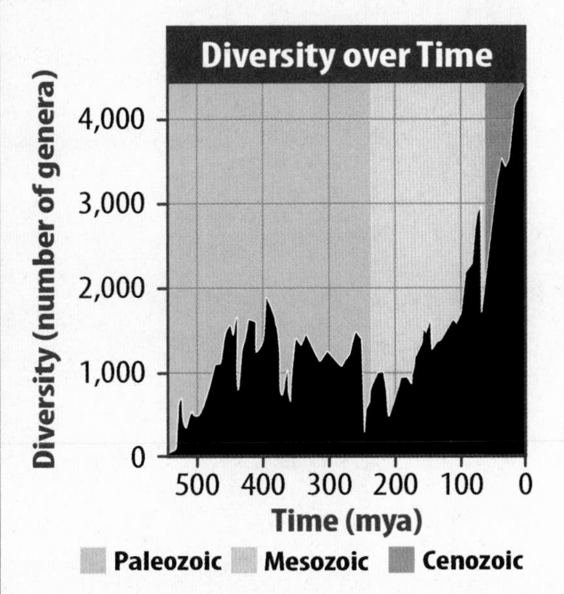

Apply It

3. According to the graph on the left, at what time in Earth's past was diversity the lowest? At what time was diversity the highest?
4. Approximately what percentage of genera became extinct 250 million years ago?
5. Approximately when did each of Earth's major mass extinctions take place?
6. What is the relationship between diversity and extinction rate?
7. **Key Concept** How have mass extinctions helped scientists develop the geologic time scale?

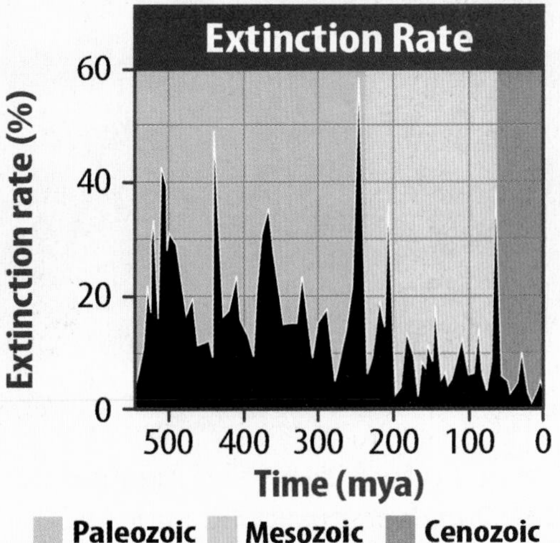

Lesson 2

Reading Guide

Key Concepts 
ESSENTIAL QUESTIONS

- What major geologic events occurred during the Paleozoic era?
- What does fossil evidence reveal about the Paleozoic era?

Vocabulary

Paleozoic era p. 371
Mesozoic era p. 371
Cenozoic era p. 371
inland sea p. 372
coal swamp p. 374
supercontinent p. 375

Multilingual eGlossary

The Paleozoic Era

Inquiry What animal was this?

Imagine going for a swim and meeting up with this Paleozoic monster. *Dunkleosteus* (duhn kuhl AHS tee us) was one of the largest and fiercest fish that ever lived. Its head was covered in bony armor 5 cm thick—even its eyes had bony armor. It had razor-sharp teethlike plates that bit with a force like that of present-day alligators.

Mark Steinmetz

Launch Lab

20 minutes

What can you learn about your ancestors?

Scientists use fossils and rocks to learn about Earth's history. What could you use to research your past?

1. Write as many facts as you can about one of your grandparents or other older adult family members or friends.
2. What items, such as photos, do you have that can help you?

Think About This

1. If you wanted to know about a great-great-great grandparent, what clues do you think you could find?
2. How does knowledge about past generations in your family benefit you today?
3. **Key Concept** How do you think learning about distant relatives is like studying Earth's past?

Early Paleozoic

In many families, three generations–grandparents, parents, and children–live closely together. You could call them the old generation, the middle generation, and the young generation. These generations are much like the three eras of the Phanerozoic eon. *The* **Paleozoic** (pay lee uh ZOH ihk) **era** *is the oldest era of the Phanerozoic eon. The* **Mesozoic** (mez uh ZOH ihk) **era** *is the middle era of the Phanerozoic eon. The* **Cenozoic** (sen uh ZOH ihk) **era** *is the youngest era of the Phanerozoic eon.*

As shown in **Figure 8**, the Paleozoic era lasted for more than half the Phanerozoic eon. Because it was so long, it is often divided into three parts: early, middle, and late. The Cambrian and Ordovician periods make up the Early Paleozoic.

The Age of Invertebrates

The organisms from the Cambrian explosion were invertebrates (ihn VUR tuh brayts) that lived only in the oceans. Invertebrates are animals without backbones. So many kinds of invertebrates lived in Early Paleozoic oceans that this time is often called the age of invertebrates.

WORD ORIGIN

Paleozoic
from Greek *palai*, means "ancient"; and Greek *zoe*, means "life"

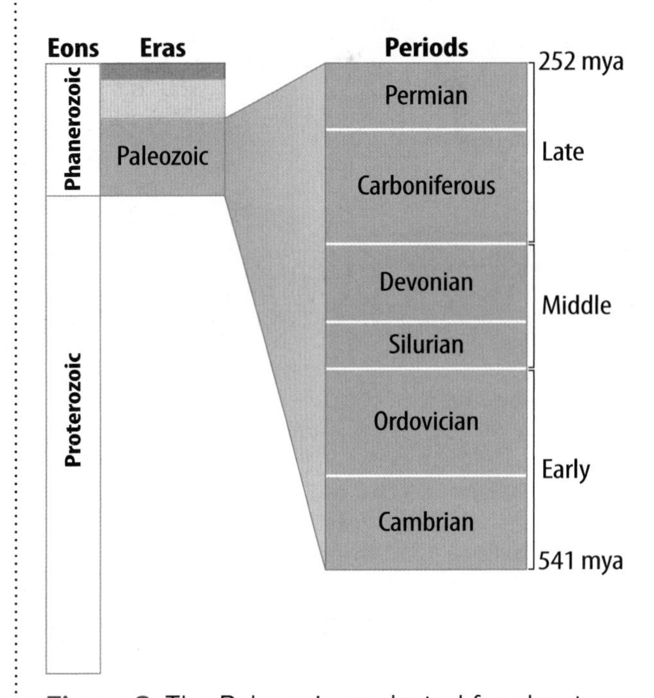

Figure 8 The Paleozoic era lasted for about 290 million years. It is divided into six periods.

Hutchings Photography/Digital Light Source

Figure 9 Earth's continents and life-forms changed dramatically during the Paleozoic era.

Visual Check In what period did life first appear on land?

Geology of the Early Paleozoic

If you could have visited Earth during the Early Paleozoic, it would have seemed unfamiliar to you. As shown in **Figure 9**, there was no life on land. All life was in the oceans. The shapes and locations of Earth's continents also would have been unfamiliar, as shown in **Figure 10**. Notice that the landmass that would become North America was on the equator.

Earth's climate was warm during the Early Paleozoic. Rising seas flooded the continents and formed many shallow inland seas. *An* **inland sea** *is a body of water formed when ocean water floods continents.* Most of North America was covered by an inland sea.

FOLDABLES®

Make a horizontal, three-tab book. Label it as shown. Use your book to record information about changes during the Paleozoic Era.

Early Paleozoic | Middle Paleozoic | Late Paleozoic

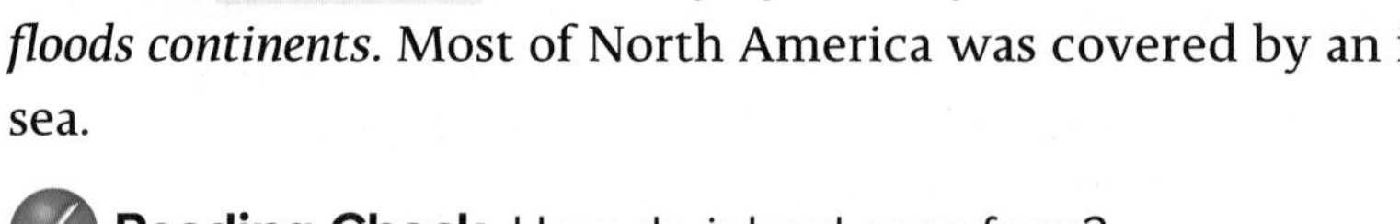

Reading Check How do inland seas form?

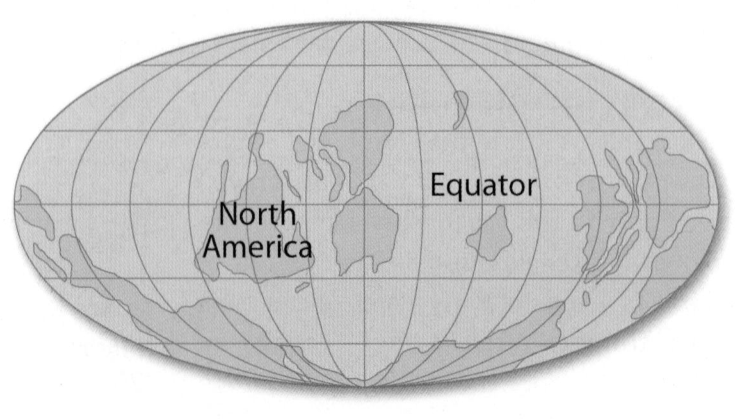

Figure 10 During the Early Paleozoic, North America straddled the equator.

Middle Paleozoic

The Early Paleozoic ended with a mass extinction event, but many invertebrates survived. New forms of life lived in huge coral reefs along the edges of the continents. Soon, animals with backbones, called vertebrates, evolved.

The Age of Fishes

Some of the earliest vertebrates were fishes. So many types of fishes lived during the Silurian (suh LOOR ee un) and Devonian (dih VOH nee un) periods that the Middle Paleozoic is often called the age of fishes. Some fishes, such as the *Dunkleosteus* pictured at the beginning of this lesson, were heavily armored. **Figure 11** also shows what a *Dunkleosteus* might have looked like. On land, cockroaches, dragonflies, and other insects evolved. Earth's first plants appeared. They were small and lived in water.

Geology of the Middle Paleozoic

Middle Paleozoic rocks contain evidence of major collisions between moving continents. These collisions created mountain ranges. When several landmasses collided with the eastern coast of North America, the Appalachian (ap uh LAY chun) Mountains began to form. By the end of the Paleozoic era, the Appalachians were probably as high as the Himalayas are today.

Key Concept Check How did the Appalachian Mountains form?

Figure 11 *Dunkleosteus* was a top Devonian predator.

Late Paleozoic

Like the Early Paleozoic, the Middle Paleozoic ended with a mass extinction event. Many marine invertebrates and some land animals disappeared.

The Age of Amphibians

In the Late Paleozoic, some fishlike organisms spent part of their lives on land. *Tiktaalik* (tihk TAH lihk) was an organism that had lungs and could breathe air. It was one of the earliest amphibians. Amphibians were so common in the Late Paleozoic that this time is known as the age of amphibians.

Ancient amphibian species adapted to land in several ways. As you read, they had lungs and could breathe air. Their skins were thick, which slowed moisture loss. Their strong limbs enabled them to move around on land. However, all amphibians, even those living today, must return to the water to mate and lay eggs.

Reptile species evolved toward the end of the Paleozoic era. Reptiles were the first animals that did not require water for reproduction. Reptile eggs have tough, leathery shells that protect them from drying out.

Key Concept Check How did different species adapt to land?

Coal Swamps

During the Late Paleozoic, dense, tropical forests grew in swamps along shallow inland seas. When trees and other plants died, they sank into the swamps, such as the one illustrated in **Figure 12.** *A* **coal swamp** *is an oxygen-poor environment where, over time, plant material changes into coal.* The coal swamps of the Carboniferous (car buhn IF er us) and Permian periods eventually became major sources of coal that we use today.

Figure 12 Plants buried in ancient coal swamps became coal.

Formation of Pangaea

Geologic evidence indicates that many continental collisions occurred during the Late Paleozoic. As continents moved closer together, new mountain ranges formed. By the end of the Paleozoic era, Earth's continents had formed a giant supercontinent– Pangaea. A **supercontinent** *is an ancient landmass which separated into present-day continents.* Pangaea formed close to Earth's equator, as shown in **Figure 13.** As Pangaea formed, coal swamps dried up and Earth's climate became cooler and drier.

The Permian Mass Extinction

The largest mass extinction in Earth's history occurred at the end of the Paleozoic era. Fossil evidence indicates that 95 percent of marine life-forms and 70 percent of all life on land became extinct. This extinction event is called the Permian mass extinction.

Key Concept Check What does fossil evidence reveal about the end of the Paleozoic era?

Scientists debate what caused this mass extinction. The formation of Pangaea likely decreased the amount of space where marine organisms could live. It would have contributed to changes in ocean currents, making the center of Pangaea drier. But Pangaea formed over many millions of years. The extinction event occurred more suddenly.

Some scientists hypothesize that a large meteorite impact caused drastic climate change. Others propose that massive volcanic eruptions changed the global climate. Both a meteorite impact and large-scale eruptions would have ejected ash and rock into the atmosphere, blocking out sunlight, reducing temperatures, and causing a collapse of food webs.

Whatever caused it, Earth had fewer species after the Permian mass extinction. Only species that could adapt to the changes survived.

Pangaea

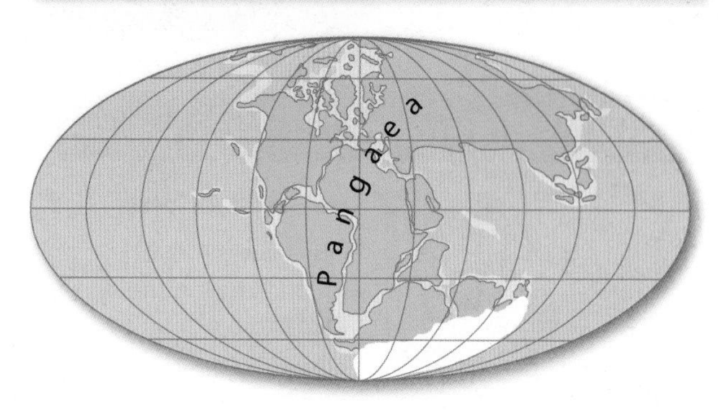

Figure 13 The supercontinent Pangaea formed at the end of the Paleozoic era.

MiniLab — 20 minutes

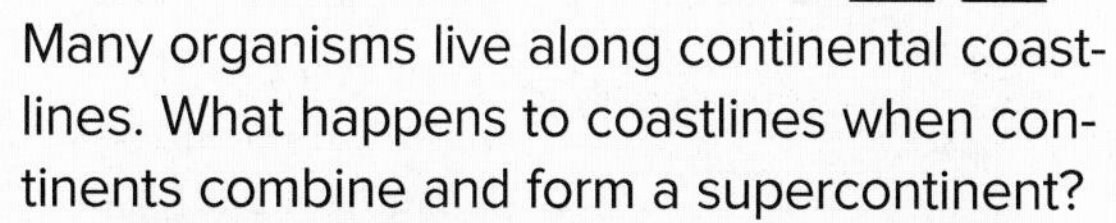

What would happen if a supercontinent formed?

Many organisms live along continental coastlines. What happens to coastlines when continents combine and form a supercontinent?

1. Read and complete a lab safety form.
2. Form a stick of **modeling clay** into a flat pancake shape. Form three pancake shapes from an identical stick of clay. Make all four shapes equal thicknesses.
3. With a **flexible tape measure,** measure the perimeter of each shape.

Analyze and Conclude

1. **Compare** Is the perimeter of the larger shape more or less than the combined perimeters of the three smaller shapes?
2. **Key Concept** How might the formation of Pangaea have affected life on Earth?

Lesson 2 Review

Visual Summary

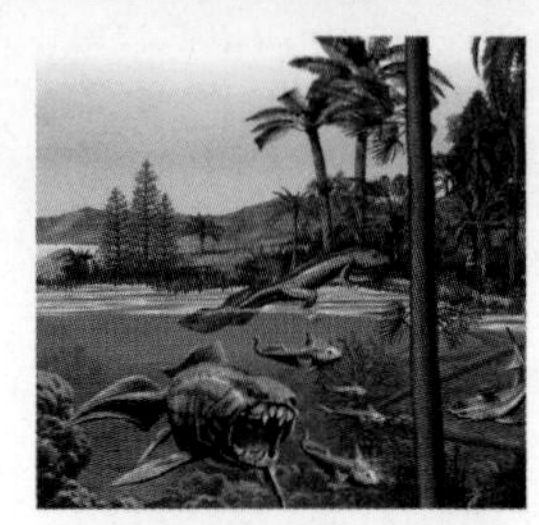

Life slowly moved to land during the Paleozoic era as amphibians and reptiles evolved.

In the Late Paleozoic, massive coal swamps formed along inland seas.

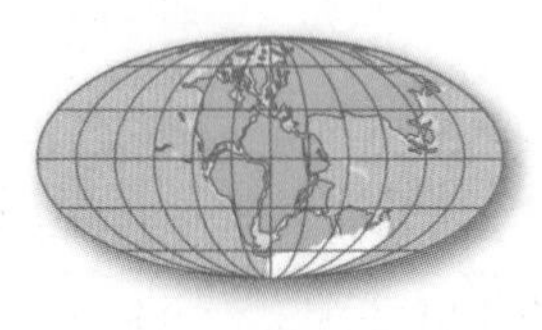

At the end of the Paleozoic era, a mass extinction event coincided with the final stages of the formation of Pangaea.

FOLDABLES®

Use your lesson Foldable to review the lesson. Save your Foldable for the project at the end of the chapter.

What do you think NOW?

You first read the statements below at the beginning of the chapter.

3. North America was once on the equator.

4. All of Earth's continents were part of a huge supercontinent 250 million years ago.

Did you change your mind about whether you agree or disagree with the statements? Rewrite any false statements to make them true.

Use Vocabulary

1. **Distinguish** between the Paleozoic era and the Mesozoic era.
2. When ocean water covers part of a continent, a(n) ________ forms.
3. **Use the term** *supercontinent* in a complete sentence.

Understand Key Concepts

4. Which was true of North America during the Early Paleozoic?
   - **A.** It had many glaciers.
   - **B.** It was at the equator.
   - **C.** It was part of a supercontinent.
   - **D.** It was populated by reptiles.
5. **Compare** ancient amphibians and reptiles and explain how each group adapted to live on land.
6. **Draw** a cartoon that shows how the Appalachian Mountains formed.

Interpret Graphics

7. **Organize** A time line of the Paleozoic era is pictured below. Copy the time line and fill in the missing periods.

Paleozoic					
	Ordovician	Silurian	Devonian	Carboniferous	

8. **Sequence** Copy and fill in the graphic organizer below. Start with Precambrian time, then list the eras in order.

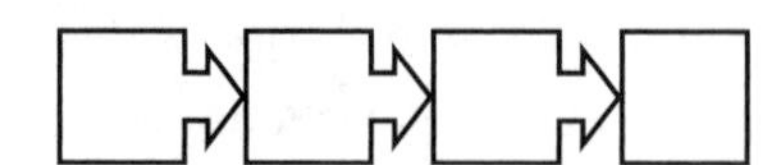

Critical Thinking

9. **Consider** What if 100 percent of organisms had become extinct at the end of the Paleozoic era?
10. **Evaluate** the possible effects of climate change on present-day organisms.

Arthur Dorety/Stocktrek Images/ Getty Images

Skill Practice Interpret Data

30 minutes

When did coal form?

Coal is fossilized plant material. When swamp plants die, they become covered by oxygen-poor water and change to peat. Over time, high temperatures and pressure from sediments transform the peat into coal. When did the plants live that formed the coal we use today?

Learn It

A bar graph can display the same type of information as a line graph. However, instead of data points and a line that connects them, a bar graph uses rectangular bars to show how values compare. **Interpret the data** below to learn when most coal formed.

Try It

1. Carefully study the bar graph. Notice that time is plotted on the *x*-axis (as geologic periods), and that coal deposits (as tons accumulated per year) are plotted on the *y*-axis.
2. Use the graph and what you know about coal formation to answer the following questions.

Apply It

3. Which coal deposits are oldest? Which are youngest?
4. During which geologic period did most of the coal form?
5. Approximately how much coal accumulated during the Paleozoic era? The Mesozoic era?
6. Why are there no data on the graph for the Cambrian, Ordovician, and Silurian periods of geologic time?
7. **Key Concept** What does fossil evidence reveal about the Paleozoic era?

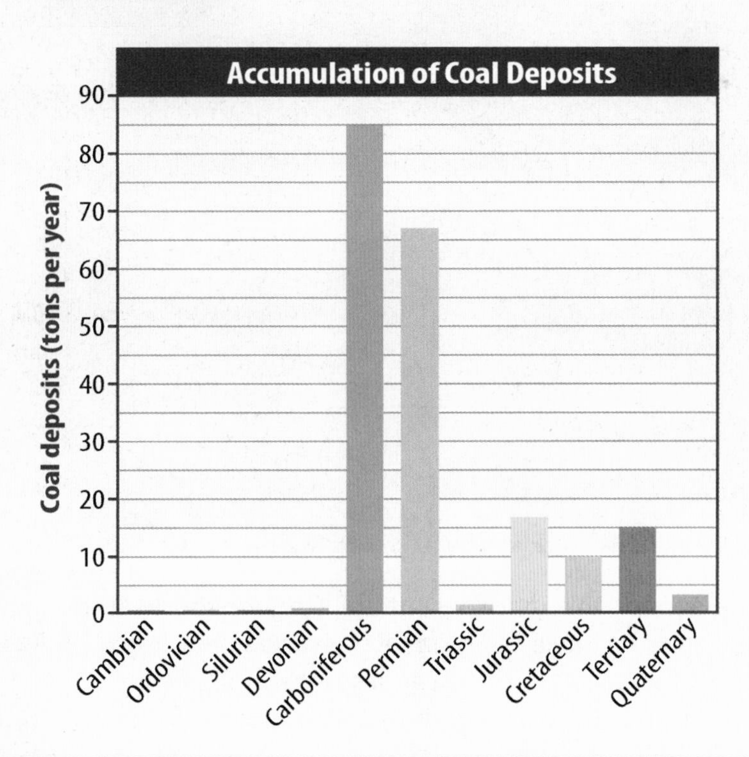

Lesson 3

Reading Guide

Key Concepts 
ESSENTIAL QUESTIONS

- What major geologic events occurred during the Mesozoic era?
- What does fossil evidence reveal about the Mesozoic era?

Vocabulary

dinosaur p. 382
plesiosaur p. 383
pterosaur p. 383

Multilingual eGlossary

The Mesozoic Era

Inquiry Mesozoic Thunder?

Can you imagine the sounds this dinosaur made? *Corythosaurus* had a tall, bony crest on top of its skull. Long nasal passages extended into the crest. Scientists suspect these nasal passages amplified sounds that could be used for communicating over long distances.

Launch Lab

20 minutes

How diverse were dinosaurs?

How many different dinosaurs were there?

1. Read and complete a lab safety form.
2. Your teacher will give you an **index card** listing a species name of a dinosaur, the dinosaur's dimensions, and the time when it lived.
3. Draw a picture of what you imagine your dinosaur looked like. Before you begin, decide with your classmates what common scale you should use.
4. **Tape** your dinosaur drawing to the Mesozoic time line your teacher provides.

Think About This

1. What was the biggest dinosaur? The smallest? Can you see any trends in size on the time line?
2. Did all the dinosaurs live at the same time?
3. **Key Concept** Dinosaurs were numerous and diverse. Do you think any dinosaurs could swim or fly?

Geology of the Mesozoic Era

When people imagine what Earth looked like millions of years ago, they often picture a scene with dinosaurs, such as the *Corythosaurus* shown on the opposite page. Dinosaurs lived during the Mesozoic era. The Mesozoic era lasted from 252 mya to 66 mya. As shown in **Figure 14,** it is divided into three periods: the Triassic (tri A sihk), the Jurassic (joo RA sihk), and the Cretaceous (krih TAY shus).

Breakup of Pangaea

Recall that the supercontinent Pangaea formed at the end of the Paleozoic era. The breakup of Pangaea was the dominant geologic event of the Mesozoic era. Pangaea began to break apart in the Late Triassic. Eventually, Pangaea split into two separate landmasses–Gondwanaland (gahn DWAH nuh land) and Laurasia (la RAY shzah). Gondwanaland was the southern continent. It included the future continents of Africa, Antarctica, Australia, and South America. Laurasia, the northern continent, included the future continents of North America, Europe, and Asia.

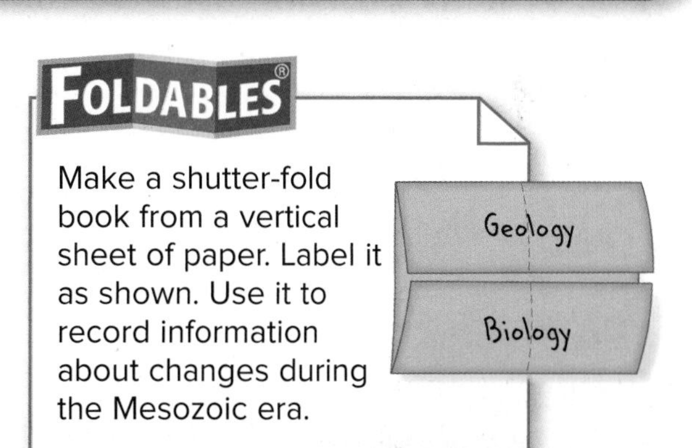

Make a shutter-fold book from a vertical sheet of paper. Label it as shown. Use it to record information about changes during the Mesozoic era.

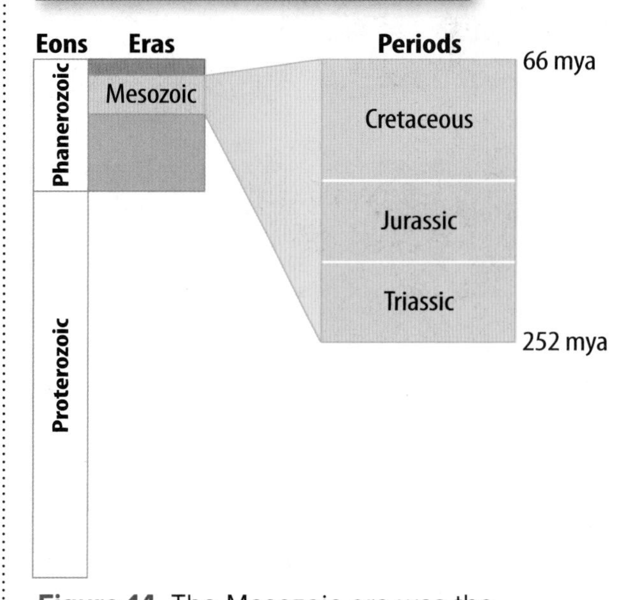

Figure 14 The Mesozoic era was the middle era of the Phanerozoic eon. It lasted for about 185 million years.

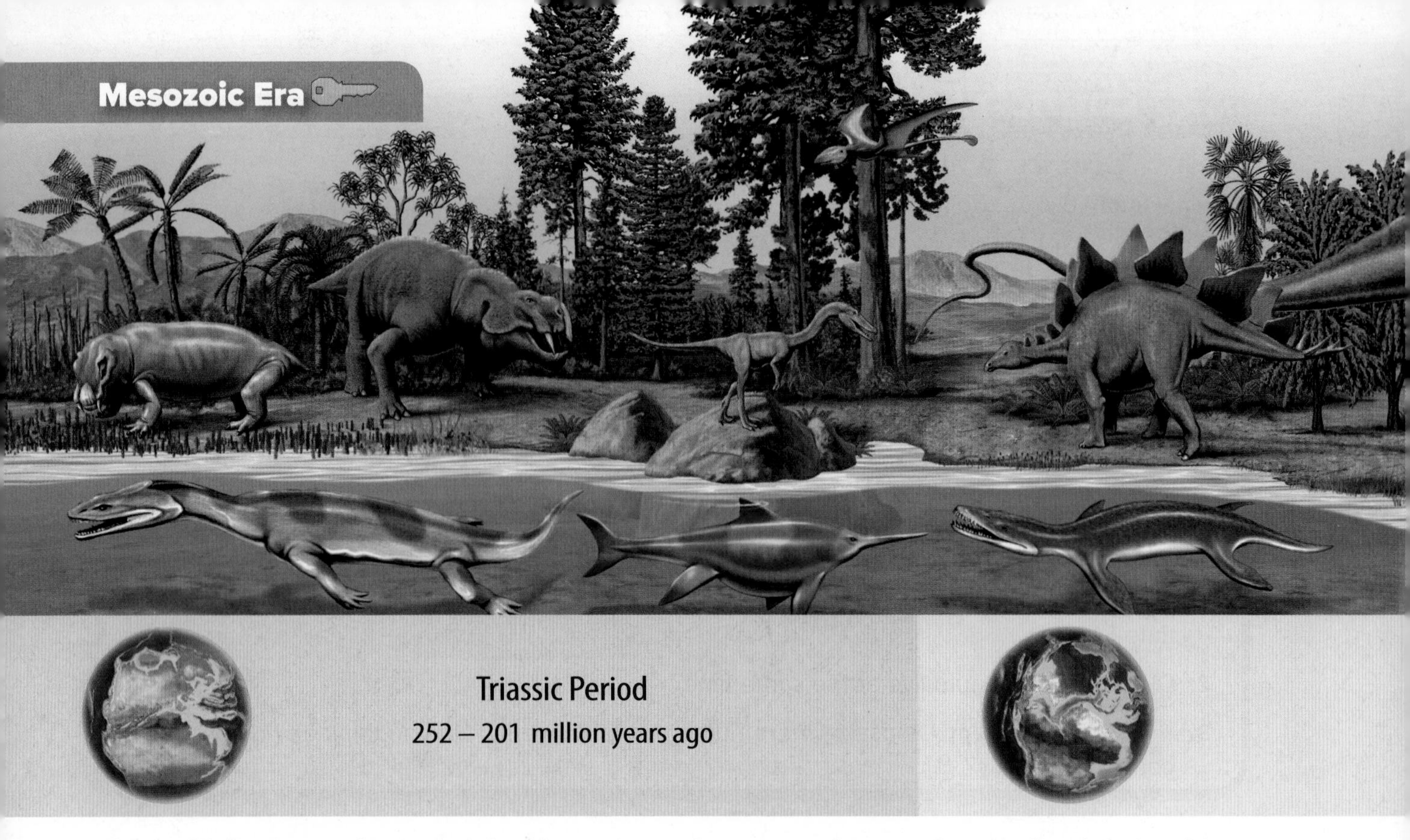

Figure 15 Dinosaurs dominated the Mesozoic era, but many other species also lived during this time in Earth's history.

Return of Shallow Seas

The type of species represented in Figure 15 adapted to an environment of lush tropical forests and warm ocean waters. That is because the climate of the Mesozoic era was warmer than the climate of the Paleozoic era. It was so warm that, for most of the era, there were no ice caps, even at the poles. With no glaciers, the oceans had more water. Some of this water flowed onto the continents as Pangaea split apart. This created narrow channels that grew larger as the continents moved apart. Eventually, the channels became oceans. The Atlantic Ocean began to form at this time.

Key Concept Check When did the Atlantic Ocean begin to form?

Sea level rose during most of the Mesozoic era, as shown in Figure 16. Toward the end of the era, sea level was so high that inland seas covered much of Earth's continents. This provided environments for the evolution of new organisms.

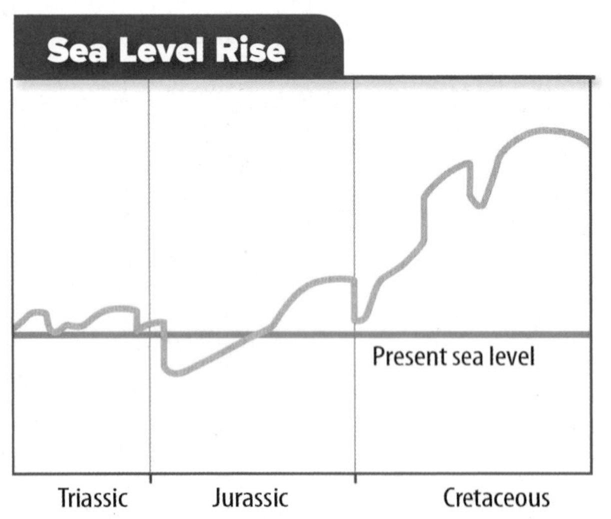

Figure 16 Earth's sea level rose during the Mesozoic era.

Visual Check In which period was sea level at its highest?

Mesozoic North America

Along North America's eastern coast and the Gulf of Mexico, sea level rose and receded over millions of years. As this happened, seawater evaporated, leaving massive salt deposits behind. Some of these salt deposits are sources of salt today. Other salt deposits later became traps for oil. Today, salt traps in the Gulf of Mexico are an important source of oil.

Throughout the Mesozoic era, the North American continent moved slowly and steadily westward. Its western edge collided with several small landmasses carried on an ancient oceanic plate. As this plate subducted beneath the North American continent, the crust buckled inland, slowly pushing up the Rocky Mountains, shown on the map in **Figure 17.** In the dry southwest, windblown sand formed huge dunes. In the middle of the continent, a warm inland sea formed.

Key Concept Check How did the Rocky Mountains form?

REVIEW VOCABULARY
evaporated
changed from liquid to gas

Figure 17 The Rocky Mountains began forming during the Mesozoic era. By the end of the era, an inland sea covered much of the central part of North America.

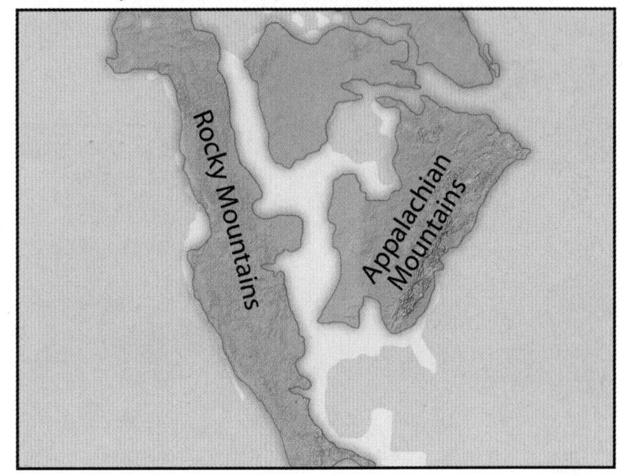

MiniLab 20 minutes

Can you run like a reptile?

Unlike the limbs of crocodiles and other modern reptiles, dinosaur limbs were positioned directly under their bodies. What did this mean?

1. Pick a partner. One of you—the dinosaur—will run on all fours with arms held straight directly below the shoulders. The other—the crocodile—will run with arms bent and positioned out from the body.
2. Race each other, then reverse positions.

Analyze and Conclude

1. **Compare** Which could move faster—the dinosaur or the crocodile?
2. **Infer** Which posture do you think could support more weight?
3. **Key Concept** How might their posture have enabled dinosaurs to become so successful? How might it have helped them become so large?

Mesozoic Life

The species of organisms that survived the Permian mass extinction event lived in a world with few species. Vast amounts of unoccupied space were open for organisms to inhabit. New types of cone-bearing trees, such as pines and cycads, began to appear. Toward the end of the era, the first flowering plants evolved. Dominant among vertebrates living on land were the dinosaurs. Hundreds of species of many sizes existed.

Dinosaurs

Though dinosaurs have long been considered reptiles, scientists today actively debate dinosaur classification. Dinosaurs share a common ancestor with present-day reptiles, such as crocodiles. However, dinosaurs differ from present-day reptiles in their unique hip structure, as shown in **Figure 18.** **Dinosaurs** *were dominant Mesozoic land vertebrates that walked with legs positioned directly below their hips.* This meant that many walked upright. In contrast, the legs of a crocodile stick out sideways from its body. It appears to drag itself along the ground.

Scientists hypothesize that some dinosaurs are more closely related to present-day birds than they are to present-day reptiles. Dinosaur fossils with evidence of feathery exteriors have been found. For example, *Archaeopteryx* (ar kee AHP tuh rihks), a small bird the size of a pigeon, had wings and feathers but also claws and teeth. Many scientists suggest it was an ancestor to birds.

Figure 18 Fossils provide evidence that the hip structure of a dinosaur enabled it to walk upright.

Dinosaur Posture

Hutchings Photography/Digital Light Source

Figure 19 Not all large Mesozoic vertebrates were dinosaurs.

Visual Check How did the limbs of these reptiles compare to the limbs of dinosaurs?

Other Mesozoic Vertebrates

Dinosaurs dominated land. But, fossils indicate that other large vertebrates swam in the seas and flew in the air, as shown in Figure 19. **Plesiosaurs** (PLY zee oh sorz) *were Mesozoic marine reptiles with small heads, long necks, and flippers.* Through much of the Mesozoic, these reptiles dominated the oceans. Some were as long as 14 m.

Other Mesozoic reptiles could fly. **Pterosaurs** (TER oh sorz) *were Mesozoic flying reptiles with large, batlike wings.* One of the largest pterosaurs, the *Quetzalcoatlus* (kwetz oh koh AHT lus), had a wingspread of nearly 12 m. Though they could fly, pterosaurs were not birds. As you have read, birds are more closely related to dinosaurs.

WORD ORIGIN
pterosaur
from Greek *pteron,* means "wing"; and *sauros,* means "lizard"

Key Concept Check How could you distinguish fossils of plesiosaurs and pterosaurs from fossils of dinosaurs?

Appearance of Mammals

Dinosaurs and reptiles dominated the Mesozoic era, but another kind of animal also lived during this time–mammals. Mammals evolved early in the Mesozoic and remained small in size throughout the era. Few were larger than present-day cats.

Cretaceous Extinction Event

The Mesozoic era ended 66 mya with a mass extinction called the Cretaceous extinction event. You read in Lesson 1 that scientists propose a large meteorite impact contributed to this extinction. This crash would have produced enough dust to block sunlight for a long time. There is evidence that volcanic eruptions also occurred at the same time. These eruptions would have added more dust to the atmosphere. Without light, plants died. Without plants, animals died. Dinosaur species and other large Mesozoic vertebrate species could not adapt to the changes. They became extinct.

Lesson 3 Review

Visual Summary

As Pangaea broke up, the continents began to move into their present-day positions.

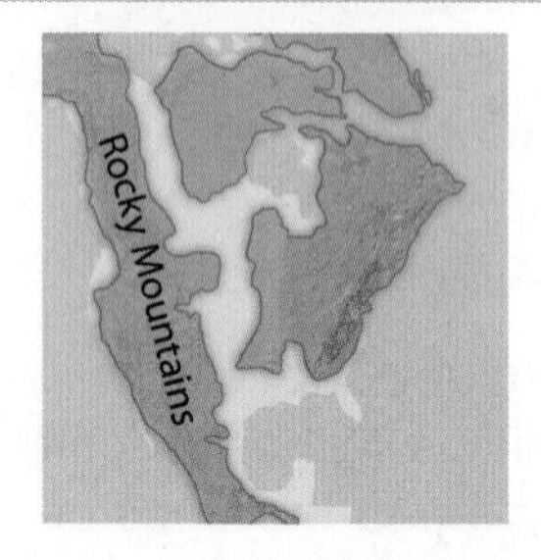

The Mesozoic climate was warm and sea level was high.

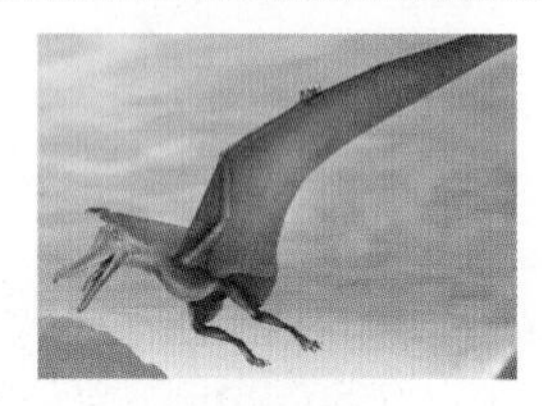

Dinosaurs were not the only large vertebrates that lived during the Mesozoic era.

FOLDABLES®

Use your lesson Foldable to review the lesson. Save your Foldable for the project at the end of the chapter.

What do you think NOW?

You first read the statements below at the beginning of the chapter.

5. All large Mesozoic vertebrates were dinosaurs.

6. Dinosaurs disappeared in a large mass extinction event.

Did you change your mind about whether you agree or disagree with the statements? Rewrite any false statements to make them true.

Use Vocabulary

1. A(n) ________ was a marine Mesozoic reptile.
2. A(n) ________ was a Mesozoic reptile that could fly.

Understand Key Concepts 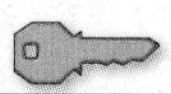

3. Which major event happened during the Mesozoic era?
   - **A.** Humans evolved.
   - **B.** Life moved onto land.
   - **C.** The Appalachian Mountains formed.
   - **D.** The Atlantic Ocean formed.
4. **Compare** the sizes of reptiles and mammals during the Mesozoic era.
5. **Explain** how the Rocky Mountains formed.

Interpret Graphics

6. **Identify** Which type of vertebrate does each skeletal figure below represent?

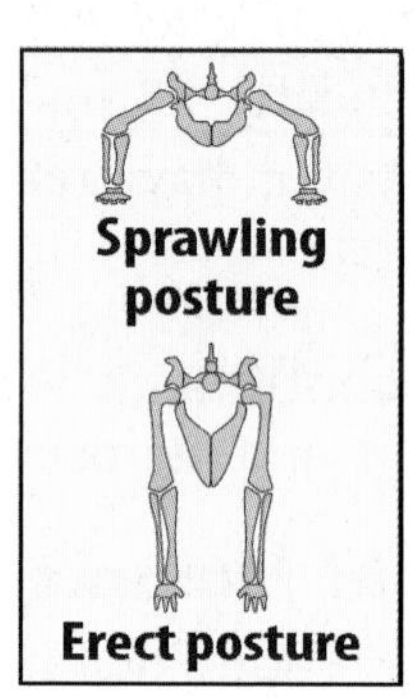

7. **Sequence** Copy and fill in the graphic organizer below to list the periods of the Mesozoic era in order.

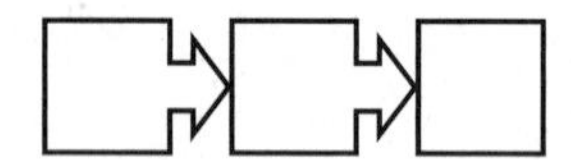

Critical Thinking

8. **Infer** how Earth might be different if there had been no extinction event at the end of the Mesozoic era.
9. **Propose** how the breakup of Pangaea might have affected evolution.

AMERICAN MUSEUM of NATURAL HISTORY

Digging Up a Surprise

A fossil discovery in China reveals some unexpected clues about early mammals.

The Mesozoic era, 252 to 66 million years ago, was the age of the dinosaurs. Many species of dinosaurs roamed Earth, from the ferocious tyrannosaurs to the giant, long-necked brachiosaurs. What other animals lived among the dinosaurs? For years, paleontologists assumed that the only mammals that lived at that time were no bigger than mice. They were no match for the dinosaurs.

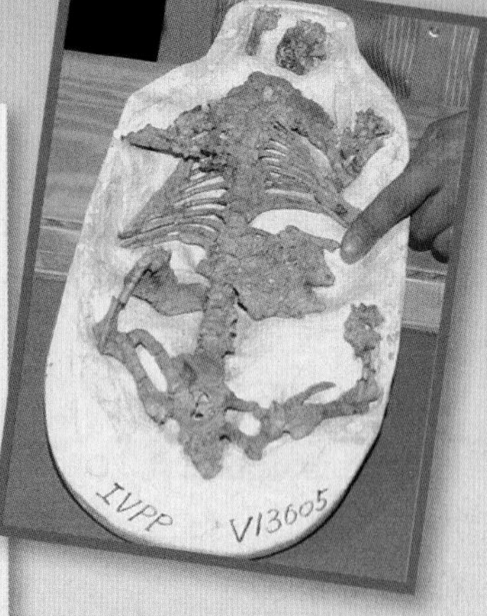

▲ **Paleontologists studying a fossil of the mammal *Repenomamus robustus* found tiny *Psittacosaurus* bones in its stomach.**

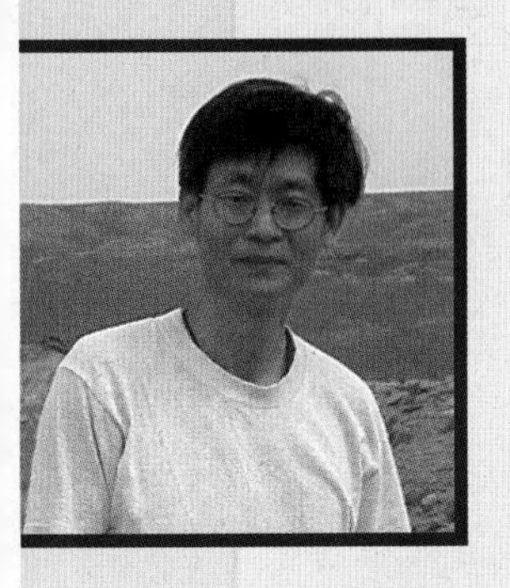

Recent fossil discoveries revealed new information about these early mammals. Jin Meng is a paleontologist at the American Museum of Natural History in New York City. In northern China, Meng and other paleontologists discovered fossils of animals that probably died in volcanic eruptions 130 million years ago. Among these fossils were the remains of a mammal over 1 foot long—about the size of a small dog. A representation of the mammal, *Repenomamus robustus* (reh peh noh MA muhs • roh BUS tus), is shown to the right.

This fossil would reveal an even bigger surprise. When examined under microscopes in the lab, scientists discovered small bones in the fossil's rib cage where its stomach had been. The bones were the tiny limbs, fingers, and teeth of a young plant-eating dinosaur. The mammal's last meal had been a young dinosaur!

This was an exciting discovery. Meng and his team learned that early mammals were larger than they thought and were meat eaters, too. Those tiny bones proved to be a huge find. Paleontologists now have a new picture of how animals interacted during the age of dinosaurs.

This is a representation of a young *Psittacosaurus*—only 12 cm long. ▶

It's Your Turn

DIAGRAM With a group, research the plants and the animals that lived in the same environment as *Repenomamus*. Create a drawing showing the relationships among the organisms. Compare your drawing to those of other groups.

Lesson 4

Reading Guide

Key Concepts

ESSENTIAL QUESTIONS

- What major geologic events occurred during the Cenozoic era?
- What does fossil evidence reveal about the Cenozoic era?

Vocabulary

Holocene epoch p. 387

Pleistocene epoch p. 389

ice age p. 389

glacial groove p. 389

mega-mammal p. 390

Multilingual eGlossary

The Cenozoic Era

Inquiry Is this animal alive?

No, this is a statue in a Los Angeles, California, pond that has been oozing tar for thousands of years. It shows how a mammoth might have become stuck in a tar pit. Mammoths lived at the same time as early humans. What do you think it was like to live alongside these animals?

Launch Lab

10 minutes

What evidence do you have that you went to kindergarten?

Rocks and fossils provide evidence about Earth's past. The more recent the era, the more evidence exists. Is this true for you, too?

1. Make a list of items you have, such as a diploma, that could provide evidence about what you did and what you learned in kindergarten.
2. Make another list of items that could provide evidence about your school experience during the past year.

Think About This

1. Which list is longer? Why?
2. **Key Concept** How do you think the items on your lists are like evidence from the first and last eras of the Phanerozoic eon?

Geology of the Cenozoic Era

Have you ever experienced a severe storm? What did your neighborhood look like afterward? Piles of snow, rushing water, or broken trees might have made your neighborhood seem like a different place. In a similar way, the landscapes and organisms of the Paleozoic and Mesozoic eras might have been strange and unfamiliar to you. Though some unusual animals lived during the Cenozoic era, this era is more familiar. People know more about the Cenozoic era than they know about any other era because we live in the Cenozoic era. Its fossils and its rock record are better preserved.

As shown in **Figure 20**, the Cenozoic era spans the time from the end of the Cretaceous period, 66 mya, to present day. Geologists divide it into three periods–the Paleogene period, the Neogene period, and the Quaternary (KWAH tur nayr ee) period. These periods are further subdivided into epochs. *The most recent epoch, the* **Holocene** (HOH luh seen) **epoch,** *began 11,700 years ago.* You live in the Holocene epoch.

FOLDABLES®

Make a shutter-fold book from a vertical sheet of paper. Label it as shown. Use it to record information about changes during the Cenozoic era.

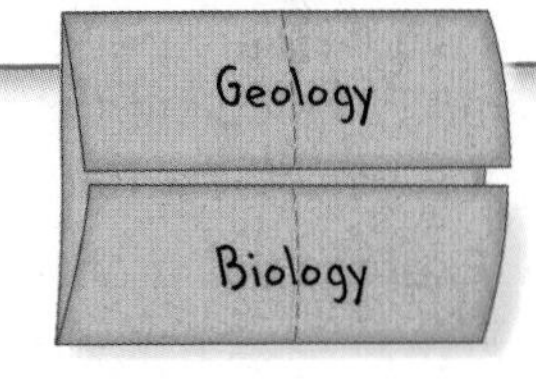

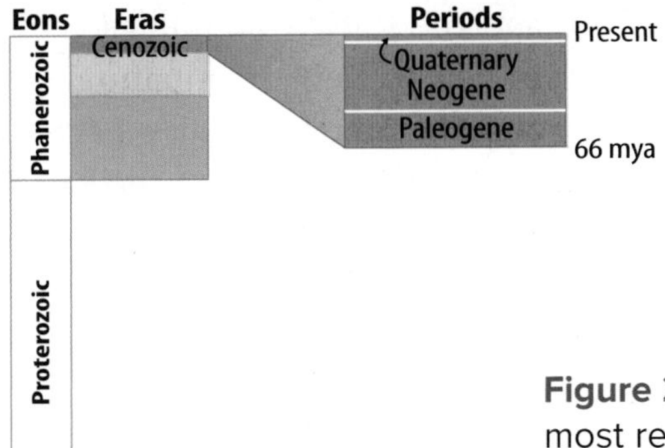

Figure 20 The Cenozoic era is Earth's most recent era. It began 66 mya.

Figure 21 Mammals dominated the landscapes of the Cenozoic era.

WORD ORIGIN

Cenozoic
From Greek *kainos*, means "new"; and *zoic*, means "life"

 Math Practice

 Personal Tutor

Math Skills

Use Percentages

The Cenozoic era began 66 mya. What percentage of the Cenozoic era is taken up by the Quaternary period, which began 2.6 mya? To calculate the percentage of a part to the whole, perform the following steps:

a. Express the problem as a fraction.

$$\frac{2.6 \text{ mya}}{66 \text{ mya}}$$

b. Convert the fraction to a decimal. 2.6 mya divided by 66 mya = 0.039

c. Multiply by 100 and add %.
$0.039 \times 100 = 3.9\%$

Practice

What percent of the Cenozoic era is represented by the Neogene period, which lasted from 23 mya to 2.6 mya? [Hint: Subtract to find the length of the Neogene period.]

Cenozoic Mountain Building

As shown in the globes in **Figure 21**, Earth's continents continued to move apart during the Cenozoic era, and the Atlantic Ocean continued to widen. As the continents moved, some landmasses collided. In the Paleogene period, India crashed into Asia. This collision began to push up the Himalayas–the highest mountains on Earth today. At about the same time, Africa began to push into Europe, forming the Alps. These mountains continue to get higher today.

In North America, the western coast continued to push against the seafloor next to it, and the Rocky Mountains continued to grow in height. New mountain ranges–the Cascades and the Sierra Nevadas–began to form along the western coast. On the eastern coast, there was little tectonic activity. The Appalachian Mountains, which formed during the Paleozoic era, continue to erode today.

 **Reading Check** Why are the Appalachian Mountains relatively small today?

Pleistocene Ice Age

Like the Mesozoic era, the early part of the Cenozoic era was warm. Toward the end of the Paleogene period, the climate began to cool. By the Pliocene (PLY oh seen) epoch, ice covered the poles as well as many mountaintops. It was even colder during the next epoch–the Pleistocene (PLY stoh seen).

The **Pleistocene epoch** *was the first epoch of the Quaternary period.* During this time, glaciers advanced and retreated many times. They covered as much as 30 percent of Earth's land surface. *An* **ice age** *is a time when a large proportion of Earth's surface is covered by glaciers.* Sometimes, rocks carried by glaciers created deep gouges or grooves, as shown in **Figure 22**. **Glacial grooves** *are grooves made by rocks carried in glaciers.*

The glaciers contained huge amounts of water. This water originated in the oceans. With so much water in glaciers, sea level dropped. As sea level dropped, inland seas drained away, exposing dry land. When sea level was at its lowest, the Florida peninsula was about twice as wide as it is today.

Pleistocene Ice Age

Figure 22 Glacial grooves in Ohio are evidence that glaciers extended far into North America during the Pleistocene ice age.

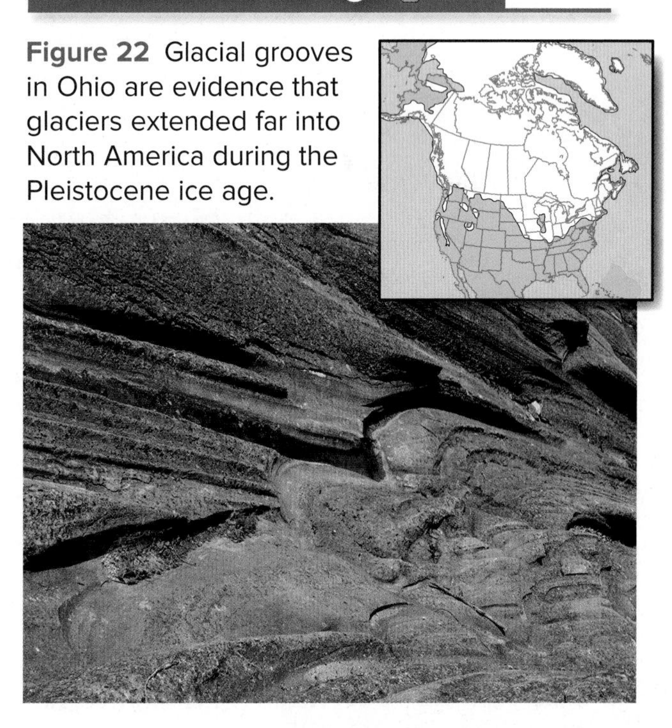

joel zatz/Alamy

Visual Check Approximately what percentage of the United States was covered with ice?

Figure 23 These mega-mammals lived at different times during the Cenozoic era. They are all extinct today. The human is included for reference.

Cenozoic Life—The Age of Mammals

The mass extinction event at the end of the Mesozoic era meant that there was more space for each surviving species. Flowering plants, including grasses, evolved and began to dominate the land. These plants provided new food sources. This enabled the evolution of many types of animal species, including mammals. Mammals were so successful that the Cenozoic era is sometimes called the age of mammals.

Mega-Mammals

Recall that mammals were small during the Mesozoic era. Many new types of mammals appeared during the Cenozoic era. Some were very large, such as those shown in **Figure 23.** *The large mammals of the Cenozoic era are called* **mega-mammals.** Some of the largest lived during the Oligocene and Miocene periods, from 34 mya to 5 mya. Others, such as woolly mammoths, giant sloths, and saber-toothed cats, lived during the cool climate of the Pliocene and Pleistocene periods, from 5 mya to 11,700 years ago. Many fossils of these animals have been discovered. The saber-toothed cat skull in **Figure 24** was discovered in the Los Angeles tar pits pictured at the beginning of this lesson. A few mummified mammoth bodies also have been discovered preserved for thousands of years in glacial ice.

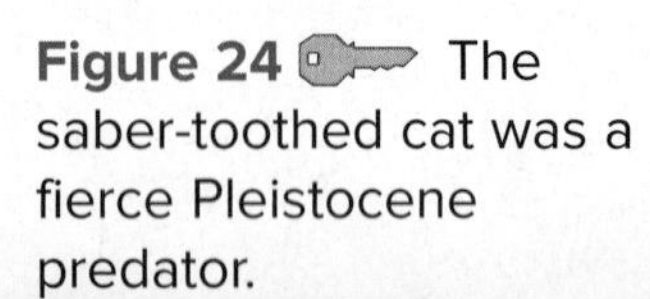

Figure 24 The saber-toothed cat was a fierce Pleistocene predator.

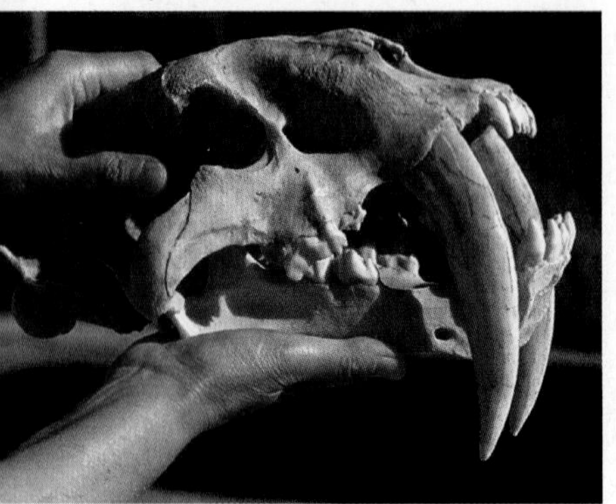

Key Concept Check How do scientists know that mega-mammals lived during the Cenozoic era?

Isolated Continents and Land Bridges

The mammals depicted in **Figure 23** lived in North America, South America, Europe, and Asia. Different mammal species evolved in Australia. This is mostly because of the movement of Earth's tectonic plates. You read earlier that land bridges can connect continents that were once separated. You also read that when continents are separated, species that once lived together can become geographically isolated.

Most of the mammals that live in Australia today are marsupials (mar SOO pee ulz). These mammals, like kangaroos, carry their young in pouches. Some scientists suggest that marsupials did not evolve in Australia. Instead, they **hypothesize** that marsupial ancestors migrated to Australia from South America when South America and Australia were connected to Antarctica by land bridges, as shown in **Figure 25.** After ancestral marsupials arrived in Australia, Australia moved away from Antarctica, and water covered the land bridges between South America, Antarctica, and Australia. Over time, the ancestral marsupials evolved into the types of marsupials that live in Australia today.

ACADEMIC VOCABULARY

hypothesize
(verb) To make an assumption about something that is not positively known

Reading Check What major geologic events affected the evolution of marsupials in Australia?

Land Bridges

Figure 25 At the beginning of the Cenozoic era, Australia was linked to South America via Antarctica, which was then warm. This provided a route for animal migration.

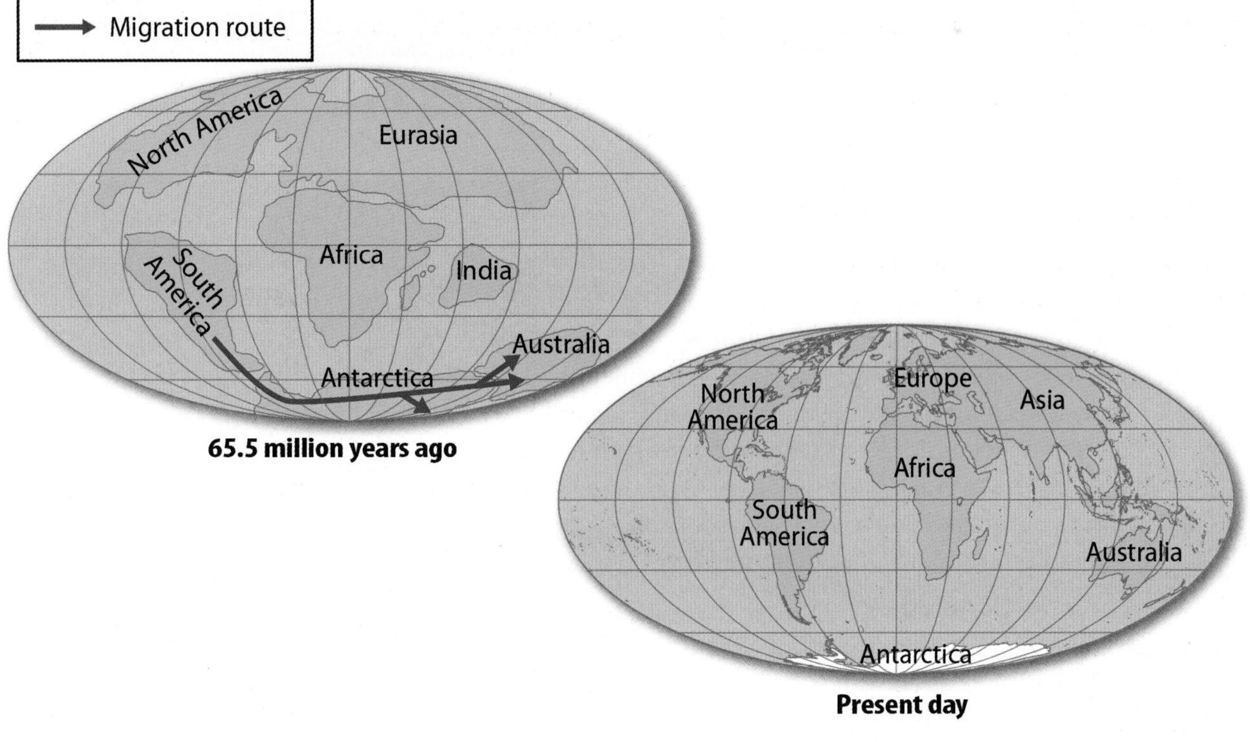

Rise of Humans

The oldest fossil remains of human ancestors have been found in Africa, where scientists think humans first evolved. These fossils are nearly 6 million years old. A replica skeleton of a 3.2-million-year-old human ancestor is shown in **Figure 26.**

Modern humans–called *Homo sapiens*–didn't evolve until the Pleistocene epoch. Early *Homo sapiens* migrated to Europe, Asia, and eventually North America. Early humans likely migrated to North America from Asia using a land bridge that connected the continents during the Pleistocene ice age. This land bridge is now covered with water.

Pleistocene Extinctions

Climate changed at the close of the Pleistocene epoch 11,700 years ago. The Holocene epoch was warmer and drier. Forests replaced grasses. The mega-mammals that lived during the Pleistocene became extinct. Some scientists suggest that mega-mammal species could not adapt fast enough to survive the environmental changes.

Key Concept Check How did climate change at the end of the Pleistocene epoch?

Future Changes

There is evidence that present-day Earth is undergoing a global-warming climate change. Many scientists suggest that humans have contributed to this change because of their use of coal, oil, and other fossil fuels over the past few centuries.

Figure 26 *Lucy* is the name scientists have given this 3.2-million-year-old human ancestor.

MiniLab

20 minutes

What happened to the Bering land bridge?

Pleistocene animals and humans likely crossed into North America from Asia using the Bering land bridge. Why did this bridge disappear?

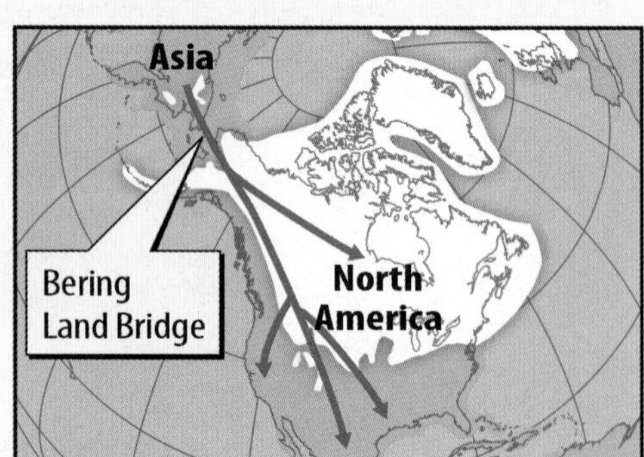

1. Read and complete a lab safety form
2. Form two pieces of **modeling clay** into continents, each with a continental shelf.
3. Place the clay models into a **watertight container** with the continental shelves touching. Add water, leaving the continental shelves exposed. Place a dozen or more **ice cubes** on the continents.
4. During your next science class, observe the container and record your observations.

Analyze and Conclude

Key Concept How does your model represent what happened at the end of the Pleistocene epoch?

Lesson 4 Review

Visual Summary

The mega-mammals that lived during most of the Cenozoic era are extinct.

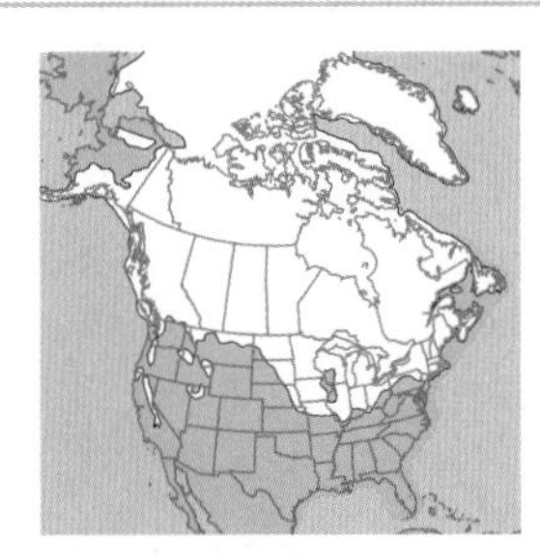

Glaciers extended well into North America during the Pleistocene ice age.

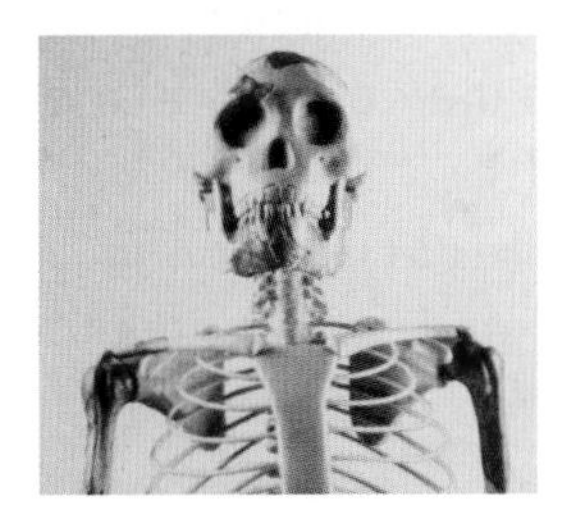

Lucy is a 3.2-million-year-old human ancestor.

FOLDABLES®

Use your lesson Foldable to review the lesson. Save your Foldable for the project at the end of the chapter.

What do you think NOW?

You first read the statements below at the beginning of the chapter.

7. Mammals evolved after dinosaurs became extinct.

8. Ice covered nearly one-third of Earth's land surface 11,700 years ago.

Did you change your mind about whether you agree or disagree with the statements? Rewrite any false statements to make them true.

Use Vocabulary

1. Gouges made by ice sheets are ________.
2. You live in the ________ epoch.

Understand Key Concepts

3. Which organism lived during the Cenozoic era?
   A. Brachiosaurus
   B. Dunkleosteus
   C. saber-toothed cats
   D. trilobites
4. **Classify** Which terms are associated with the Cenozoic era: *Homo sapiens,* mammoth, dinosaur, grass?

Interpret Graphics

5. **Determine** The map below shows coastlines of the southeastern U.S. at three times during the Cenozoic era. Which choice represents the coastline at the height of the Pleistocene ice age?

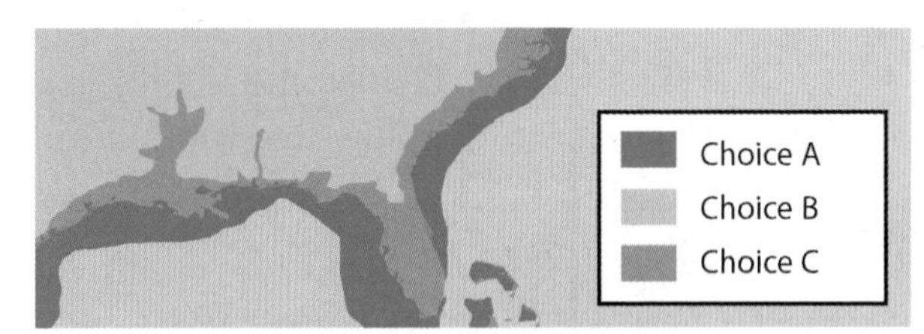

6. **Summarize** Copy and fill in the graphic organizer below to list living mammals that might be considered mega-mammals today.

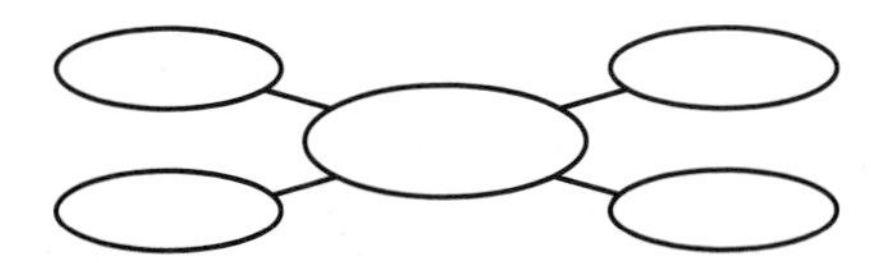

Critical Thinking

7. **Suggest** what might happen if the Australian continent crashed into Asia.

Math Skills

Math Practice

8. The Cenozoic era began 66 mya. The Oligocene and Miocene epochs extended from 34 mya to 5 mya. What percentage of the Cenozoic era is represented by the Oligocene and Miocene epochs?

Lab

90 minutes

Modeling Geologic Time

Materials

meterstick

tape measure

poster board

colored markers

colored paper

string

maps

Evidence suggests that Earth formed approximately 4.6 billion years ago. But how long is 4,600,000,000 years? It is difficult to comprehend time that extends so far into the past unless you can relate it to your own experience. In this activity, you will develop a metaphor for geologic time using a scale that is familiar to you. Then, you will create a model to share with your class.

Question

How can you model geologic time using a familiar scale?

Procedure

1. Think of something you are familiar with that can model a long period of time. For example, you might choose the length of a football field or the distance between two U.S. cities on a map—one on the east coast and one on the west coast.
2. Make a model of your metaphor using a metric scale. On your model, display the events listed in the table on the next page. Use the equation below to generate true-to-scale dates in your model.

$$\frac{\text{Known age of past event (years before present)}}{\text{Known age of Earth (years before present)}} = \frac{X \text{ time scale unit location}}{\text{Maximum distance or extent of metaphor}}$$

Example: To find where "first fish" would be placed on your model if you used a meterstick (100 cm), set up your equation as follows:

$$\frac{500{,}000{,}000 \text{ years}}{4{,}600{,}000{,}000 \text{ years}} = \frac{X \text{ (location on meterstick)}}{100 \text{ cm}}$$

3. In your Science Journal, keep a record of all the math equations you used. You can use a calculator, but show all equations.

(t to b, 3, 5-7)Hutchings Photography/Digital Light Source; (2)Ken Cavanagh/McGraw-Hill Education; (4)Ken Karp/McGraw-Hill Education

Analyze and Conclude

4. **Calculate** What percentage of geologic time have modern humans occupied? Set up your equation as follows:

$$\frac{100{,}000}{4{,}600{,}000{,}000} \times 100 = \text{\% of time occupied by } \textit{H. sapiens}$$

5. **Estimate** Where does the Precambrian end on your model? Estimate how much of geologic time falls within the Precambrian.
6. **Evaluate** What other milestone events in Earth's history, other than those listed in the table, could you include on your model?
7. **Appraise** the following sentence as it relates to your life: "Time is relative."
8. **The Big Idea** The Earth events on your model are based mostly on fossil evidence. How are fossils useful in understanding Earth's history? How are they useful in the development of the geologic time scale?

Communicate Your Results

Share your model with the class. Explain why you chose the model you did, and demonstrate how you calculated the scale on your model.

Inquiry Extension

Imagine that you were asked to teach a class of kindergartners about Earth's time. How would you do it? What metaphor would you use? Why?

Some Important Approximate Dates in the History of Earth:

MYA	Event
4,600	Origin of Earth
3,500	Oldest evidence of life
500	First fish
375	*Tiktaalik* appears
320	First reptiles
252	Permian extinction event
220	Mammals and dinosaurs appear
155	*Archaeopteryx* appears
145	Atlantic Ocean forms
66	Cretaceous extinction event
6	Human ancestors appear
2.6	Pleistocene Ice Age begins
0.1	*Homo sapiens* appear
0.00052	Columbus lands in New World
??	Your birth date

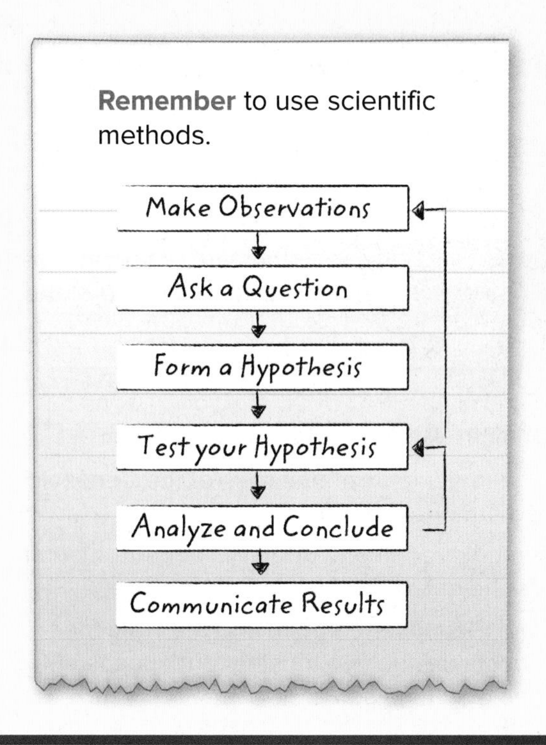

Chapter 11 Study Guide

The geologic changes that have occurred during the billions of years of Earth's history have strongly affected the evolution of life.

Key Concepts Summary 

Lesson 1: Geologic History and the Evolution of Life

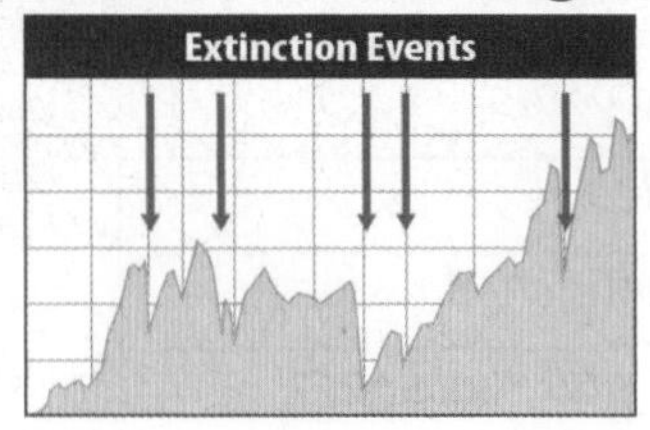

- Geologists organize Earth's history into **eons, eras, periods,** and **epochs.**
- Life evolves over time as Earth's continents move, forming **land bridges** and causing **geographic isolation.**
- **Mass extinctions** occur if many species of organisms cannot adapt to sudden environmental change.

Vocabulary

eon p. 363
era p. 363
period p. 363
epoch p. 363
mass extinction p. 365
land bridge p. 366
geographic isolation p. 366

Lesson 2: The Paleozoic Era

- Life diversified during the **Paleozoic era** as organisms moved from water to land.
- **Coal swamps** formed along **inland seas.** Later, land became drier as the **supercontinent** Pangaea formed.
- The largest mass extinction in Earth's history occurred at the end of the Permian period.

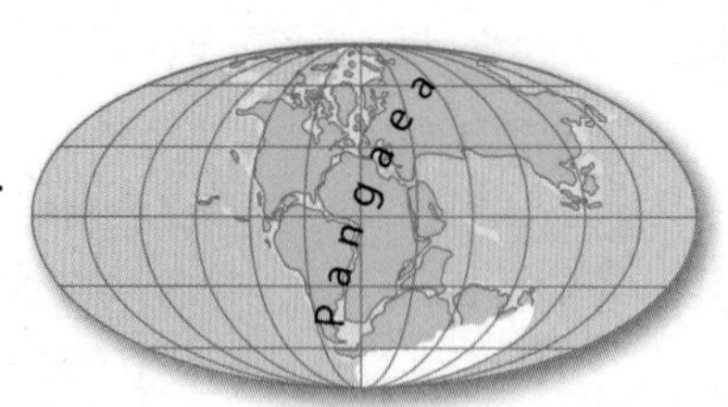

Vocabulary

Paleozoic era p. 371
Mesozoic era p. 371
Cenozoic era p. 371
inland sea p. 372
coal swamp p. 374
supercontinent p. 375

Lesson 3: The Mesozoic Era

- Sea level rose as the climate warmed.
- The Atlantic Ocean and the Rocky Mountains began to form as Pangaea broke apart.
- **Dinosaurs, plesiosaurs, pterosaurs,** and other large Mesozoic vertebrates became extinct at the end of the era.

Vocabulary

dinosaur p. 382
plesiosaur p. 383
pterosaur p. 383

Lesson 4: The Cenozoic Era

- The large, extinct mammals of the Cenozoic were **mega-mammals.**
- Ice covered nearly one-third of Earth's land at the height of the Pleistocene **ice age.**
- The **Pleistocene epoch** and the **Holocene epoch** are the two most recent epochs of the geologic time scale.

Vocabulary

Holocene epoch p. 387
Pleistocene epoch p. 389
ice age p. 389
glacial groove p. 389
mega-mammal p. 390

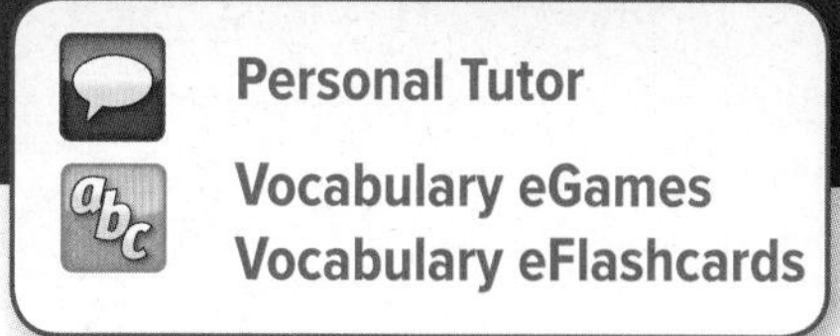

FOLDABLES® Chapter Project

Assemble your lesson Foldables as shown to make a Chapter Project. Use the project to review what you have learned in this chapter.

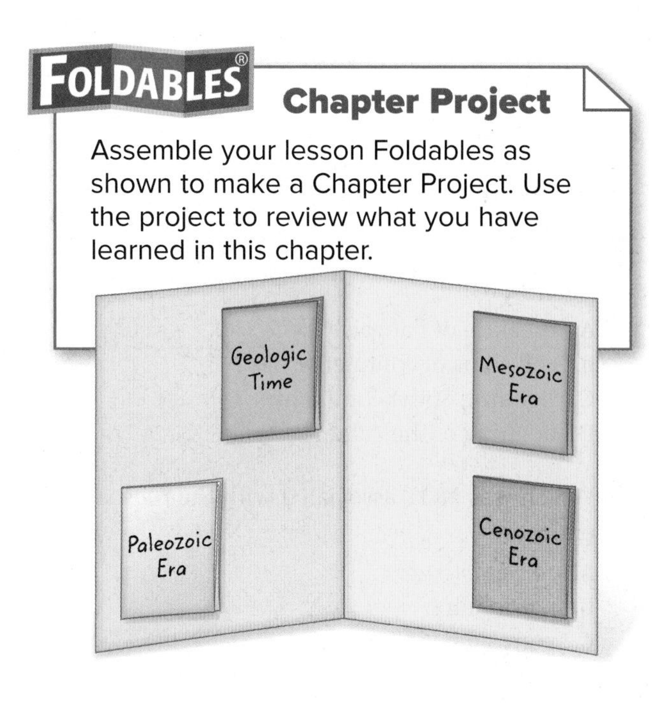

Use Vocabulary

1. The longest time unit in the geologic time scale is the ________.
2. Eras are subdivided into ________.
3. Many boundaries in the geologic time scale are marked by the occurrence of ________.
4. When glaciers melt, shallow ________ form in the interiors of continents.
5. The ________ was the first era of the Phanerozoic eon.
6. A(n) ________ can form when plants are buried in an oxygen-poor environment.
7. Marine Mesozoic reptiles included ________.
8. Modern humans evolved during the ________.

Link Vocabulary and Key Concepts

Copy this concept map, and then use vocabulary terms from the previous page and other terms from the chapter to complete the concept map.

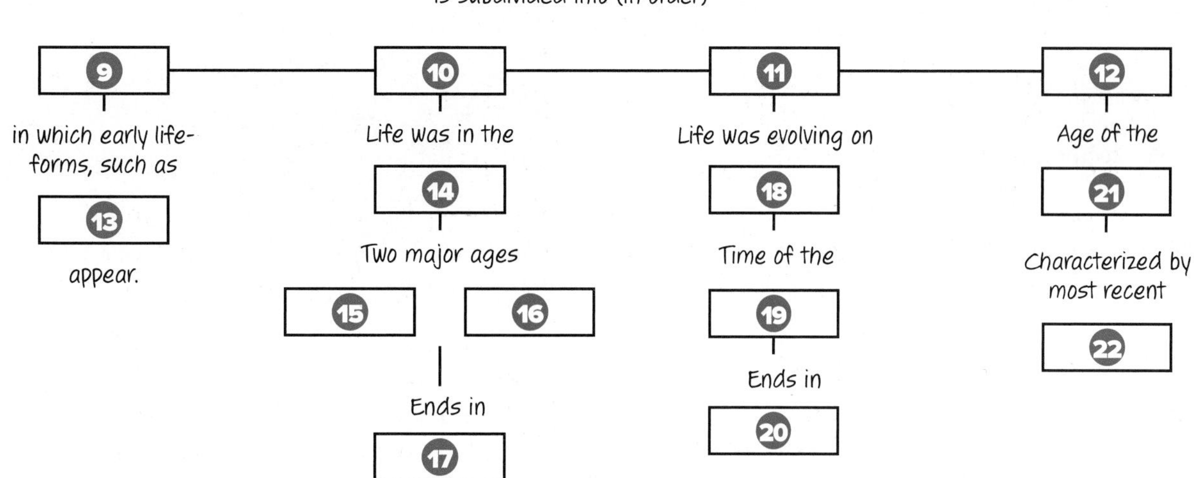

Chapter 11 Review

Understand Key Concepts 

1. The trilobite fossil below represents an organism that lived during the Cambrian period.

What distinguished this organism from organisms that lived earlier in time?
   - A. It had hard parts.
   - B. It lived on land.
   - C. It was a reptile.
   - D. It was multicellular.

2. What are the many divisions in the geologic time scale based on?
   - A. changes in the fossil record every billion years
   - B. changes in the fossil record every million years
   - C. gradual changes in the fossil record
   - D. sudden changes in the fossil record

3. Which is NOT a cause of a mass extinction event?
   - A. meteorite collision
   - B. severe hurricane
   - C. tectonic activity
   - D. volcanic activity

4. Which is the correct order of eras, from oldest to youngest?
   - A. Cenozoic, Mesozoic, Paleozoic
   - B. Mesozoic, Cenozoic, Paleozoic
   - C. Paleozoic, Cenozoic, Mesozoic
   - D. Paleozoic, Mesozoic, Cenozoic

5. Which were the first organisms to inhabit land environments?
   - A. amphibians
   - B. plants
   - C. reptiles
   - D. trilobites

6. Which event(s) produced the Appalachian Mountains?
   - A. breakup of Pangaea
   - B. collisions of continents
   - C. flooding of the continent
   - D. opening of the Atlantic Ocean

7. Which was NOT associated with the Mesozoic era?
   - A. *Archaeopteryx*
   - B. plesiosaurs
   - C. pterosaurs
   - D. *Tiktaalik*

8. Which is true for the beginning of the Cenozoic era?
   - A. Mammals and dinosaurs lived together.
   - B. Mammals first evolved.
   - C. Dinosaurs had killed all mammals.
   - D. Dinosaurs were extinct.

9. What is unrealistic about the picture on this stamp?

   - A. Dinosaurs were not this large.
   - B. Dinosaurs did not have long necks.
   - C. Humans did not live with dinosaurs.
   - D. Early humans did not use stone tools.

Jacob Hamblin/iStock/360/Getty Images

 Online Test Practice

Critical Thinking

10 **Hypothesize** how a major change in global climate could lead to a mass extinction.

11 **Evaluate** how the Permian-Triassic mass extinction affected the evolution of life.

12 **Predict** what Earth's climate might be like if sea level were very low.

13 **Differentiate** between amphibians and reptiles. What feature enabled reptiles–but not amphibians–to be successful on land?

14 **Hypothesize** how the bone structure of dinosaur limbs might have contributed to the success of dinosaurs during the Mesozoic era.

15 **Debate** Some scientists argue that humans have changed Earth so much that a new epoch–the Anthropocene epoch–should be added to the geologic time scale. Explain whether you think this is a good idea and, if so, when it should begin.

16 **Interpret Graphics** What is wrong with the geologic time line shown below?

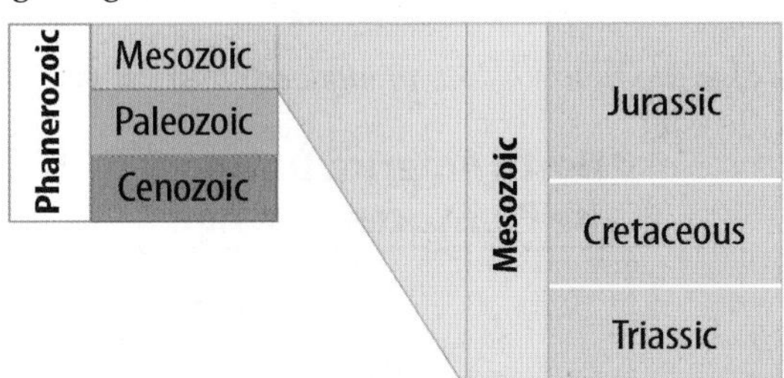

Writing in Science

17 **Decide** which period of Earth's history you would want to visit if you could travel back in time. Write a letter to a friend about your visit, describing the climate, the organisms, and the positions of Earth's continents at the time of your visit. Include a main idea, supporting details and examples, and a concluding sentence.

REVIEW THE BIG IDEA

18 What have scientists learned about Earth's past by studying rocks and fossils? How is the evolution of Earth's life-forms affected by geologic events? Provide examples.

19 The photo below shows an extinct dinosaur. What changes on Earth can cause organisms to become extinct?

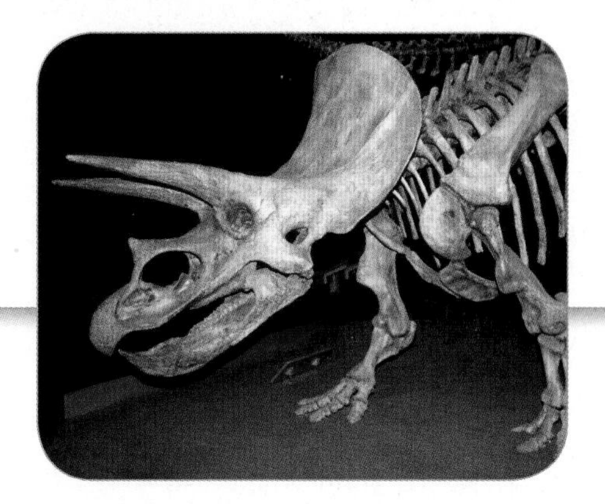

Math Skills

 Math Practice

Use Percentages

Use the table to answer the questions.

Era	Period	Epoch	Time Scale
Cenozoic	Quaternary	Holocene	11,700 years ago
		Pleistocene	2.6 mya
	Neogene	Pliocene	5.3 mya
		Miocene	23.0 mya
	Paleogene	Oligocene	33.9 mya
		Eocene	56.0 mya
		Paleocene	66 mya

20 What percentage of the Quaternary period is represented by the Holocene epoch?

21 What percentage of the Neogene period is represented by the Pliocene epoch?

Standardized Test Practice

Record your answers on the answer sheet provided by your teacher or on a sheet of paper.

Multiple Choice

Use the figure below to answer question 1.

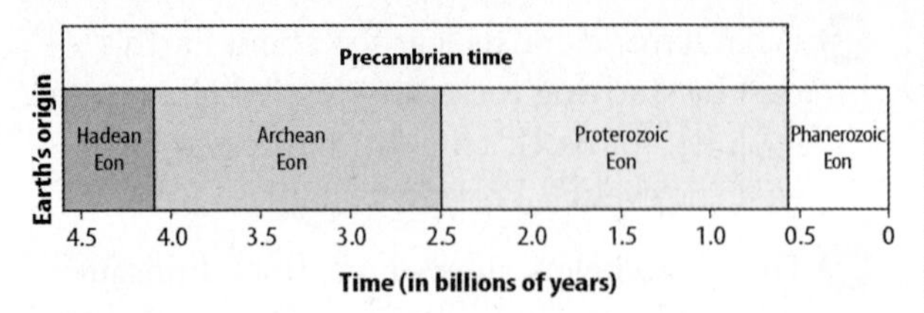

1 Approximately how long did Precambrian time last?

A 0.5 billion years

B 3.5 billion years

C 4.0 billion years

D 4.25 billion years

2 Which is the smallest unit of geologic time?

A eon

B epoch

C era

D period

3 Which is known as the age of invertebrates?

A Early Cenozoic

B Early Paleozoic

C Late Mesozoic

D Late Precambrian

4 Which made dinosaurs different from modern-day reptiles?

A head shape

B hip structure

C jaw alignment

D tail length

5 What is the approximate age of the oldest fossils of early human ancestors?

A 10,000 years

B 6 million years

C 65 million years

D 1.5 billion years

6 Which was NOT an adaptation that enabled amphibians to live on land?

A ability to breathe oxygen

B ability to lay eggs on land

C strong limbs

D thick skin

7 Which is considered a mega-mammal?

A *Archaeopteryx*

B plesiosaur

C *Tiktaalik*

D woolly mammoth

Use the figure below to answer question 8.

North America During the Pleistocene Ice Age

8 The figure above is a map of glacial coverage in North America. Which section of the United States would most likely have the greatest number of glacial grooves?

A the Northeast

B the Northwest

C the Southeast

D the Southwest

Use the graph below to answer question 9.

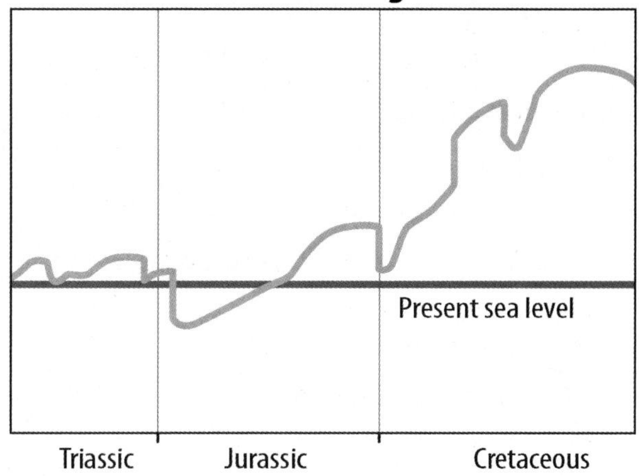

9 Based on the graph above, when might inland seas have covered much of Earth's continents?

A Early Cretaceous

B Early Jurassic

C Middle Triassic

D Late Cretaceous

10 Which did NOT occur in the Paleozoic era?

A appearance of mammals

B development of coal swamps

C evolution of invertebrates

D formation of Pangaea

11 What do geologists use to mark divisions in geologic time?

A abrupt changes in the fossil record

B frequent episodes of climate change

C movements of Earth's tectonic plates

D rates of radioactive mineral decay

Constructed Response

Use the graph below to answer questions 12 and 13.

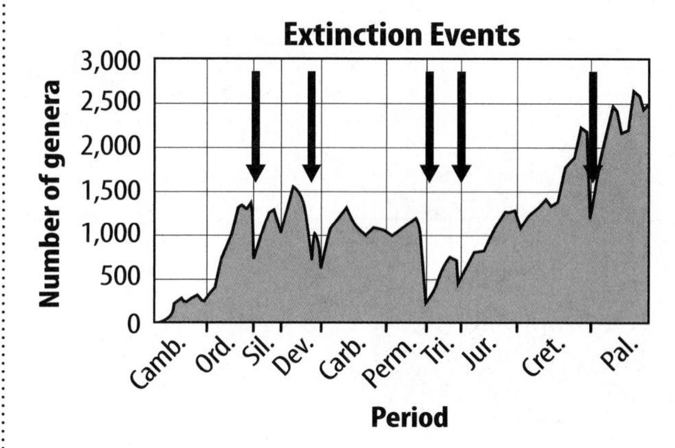

12 In the graph above, what events do the arrows mark? What happens during these events?

13 What event appears to have had the greatest impact? Explain your answer in terms of the graph.

14 What are two possible reasons why large populations of organisms die?

15 What is the relationship between the evolution of marsupials and the movement of Earth's tectonic plates?

16 Why did new and existing aquatic organisms flourish during the Mesozoic era? Use the terms glaciers, Pangaea, and sea level in your explanation.

17 What is the link between iridium and the mass extinction of dinosaurs?

NEED EXTRA HELP?																	
If You Missed Question...	1	2	3	4	5	6	7	8	9	10	11	12	13	14	15	16	17
Go to Lesson...	1	1	2	3	4	2	4	4	3	2	1	1	1	1-3	4	3	1

1441
Prince Munjong of Korea invents the first rain gauge to gather and measure the amount of liquid precipitation over a period of time.

1450
The first anemometer, a tool to measure wind speed, is developed by Leone Battista Alberti.

1643
Italian physicist Evangelista Torricelli invents the barometer to measure pressure in the air. This tool improves meteorology, which relied on simple sky observations.

1714
German physicist Daniel Fahrenheit develops the mercury thermometer, making it possible to measure temperature.

1752
Swedish astronomer Andres Celsius proposes a centigrade temperature scale where 0° is the freezing point of water and 100° is the boiling point of water.

1800 1900 2000

1806
Francis Beaufort creates a system for naming wind speeds and aptly names it the Beaufort Wind Force Scale. This scale is used mainly to classify sea conditions.

1960
TIROS 1, the world's first weather satellite, is sent into space equipped with a TV camera.

1964
The U.S. National Severe Storms Laboratory begins experimenting with the use of Doppler radar for weather-monitoring purposes.

2006
Meteorologists hold 8,800 jobs in the United States alone. These scientists work in government and private agencies, in research services, on radio and television stations, and in education.

Unit 3

Nature of SCIENCE

Models

In 2004 over 200,000 people died as a tsunami swept across the Indian Ocean, shown in **Figure 1.** How can scientists predict future tsunamis to help save lives? Researchers around the world have developed different models to study tsunami waves and their effects. A **model** is a representation of an object, a process, an event, or a system that is similar to the physical object or idea being explained. Scientists use models to study something that is too big or too small, happens too quickly or too slowly, or is too dangerous or too expensive to study directly.

Models of tsunamis help predict how future tsunamis might impact land. Information from these models can help save ecosystems, buildings, and lives.

▲ **Figure 1** A massive wave approaches the shore in the 2004 Indian Ocean tsunami.

Types of Models

Mathematical Models and Computer Simulations

A mathematical model represents an event, a process, or a system, using equations. A mathematical model can be one equation, for example: speed = distance/time. Or, they can be several hundred equations, such as those used to calculate tsunami effects.

A computer simulation is a model that combines many mathematical models. Computer simulations allow the user to easily change variables. Simulations often show a change over time or a sequence of events. Computer programs that include animations and graphics are used to visually display mathematical models.

Researchers from Texas A&M University constructed a tsunami simulation using many mathematical models of Seaside, Oregon, as shown in **Figure 2.** Simulations that use equations to model the force of waves hitting buildings are displayed on a computer screen. Researchers change variables, such as the size, the force, or the shape of tsunami waves, to determine how Seaside might be damaged by a tsunami.

Figure 2 This series of images is from an animated simulation model of a tsunami approaching Seaside, Oregon. ▼

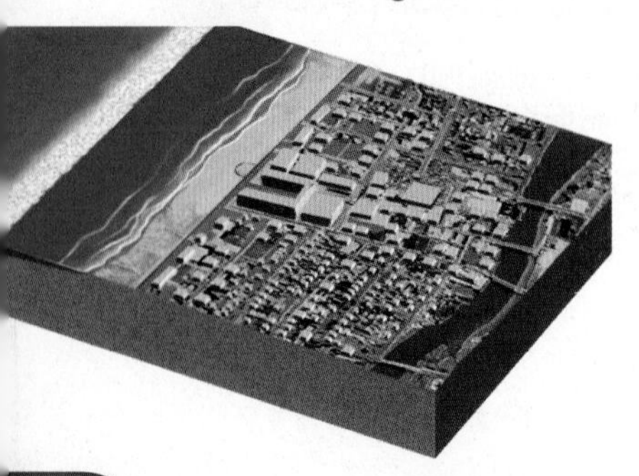

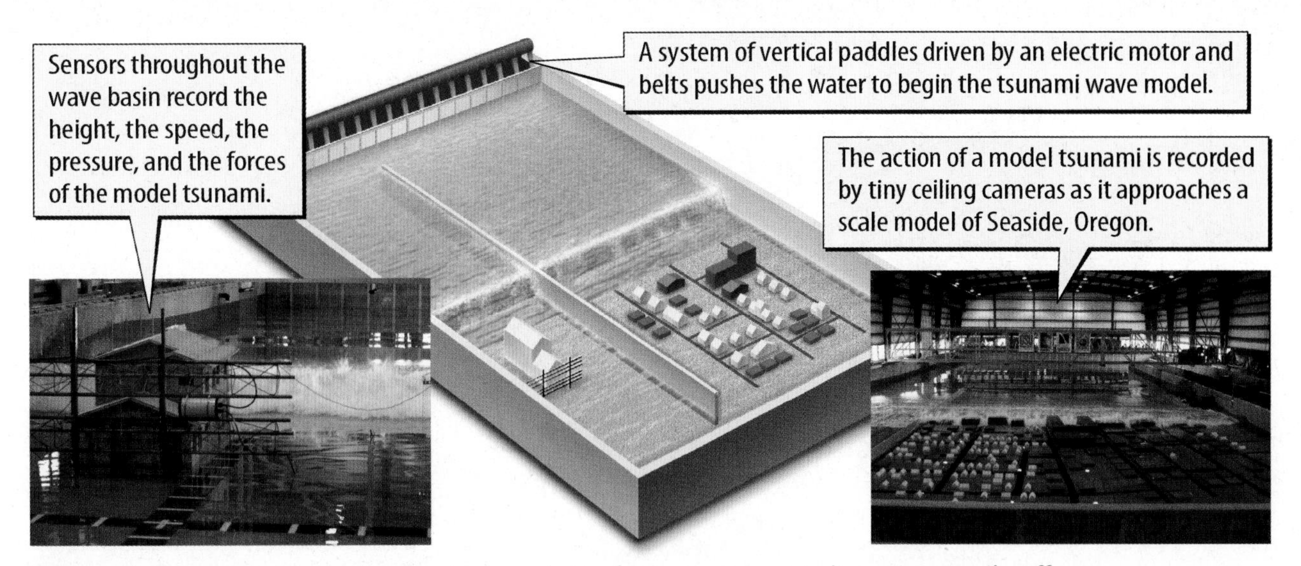

Figure 3 Researchers study physical models of tsunamis to predict a tsunami's effects.

(l)John W. van de Lindt/Colorado State University; (r)Daniel Cox/O.H. Hinsdale Research Laboratory/Oregon State University; (b) Pacific Marine Environmental Laboratory/NOAA

Physical Models

A physical model is a model that you can see and touch. It shows how parts relate to one another, how something is built, or how complex objects work. Scientists at Oregon State University built physical scale models of Seaside, Oregon, as shown in **Figure 3.** They placed the model at the end of a long wave tank. Sensors in the wave tank and on the model buildings measure and record velocities, forces, and turbulence created by a model tsunami wave. Scientists use these measurements to predict the effects of a tsunami on a coastal town, and to make recommendations for saving lives and preventing damage.

Conceptual Models

Images that represent a process or relationships among ideas are conceptual models. The conceptual model below shows that the United States has a three-part plan for minimizing the effects of tsunamis. Hazard assessment involves identification of areas that are in high risk of tsunamis. Response involves education and public safety. Warning includes a system of sensors that detect the approach of a tsunami.

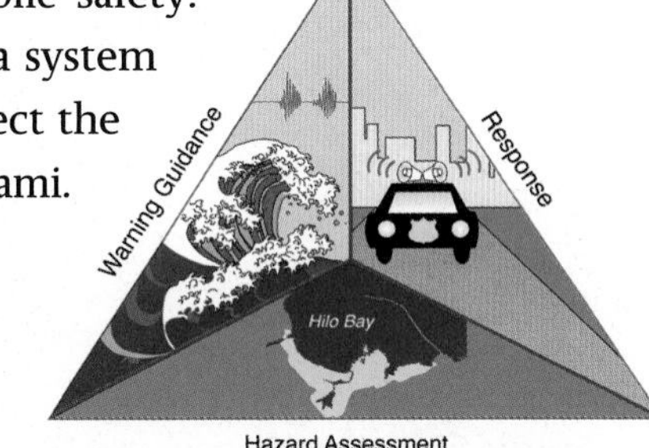

MiniLab

30 minutes

How can you model a tsunami?

What tsunami behaviors can you observe in your own model of a wave tank?

1. Read and complete a lab safety form.
2. Pour **sand** into a **glass pan,** creating a slope from one end of the pan to the other. Fill the pan with water. Place a **cork** in the center of the pan. Draw your setup in your Science Journal.
3. Use a **dowel** to create a wave at the deep end of the pan. Record your observations.
4. Place several **common objects** on end at the shallow end of the pan. Record your observations of the behaviors of the cork and the different objects when you create a wave.

Analyze and Conclude

1. **Describe** What do the different parts of your physical model represent?
2. **Explain** What are some limitations of your physical model?

Chapter 12

Earth's Atmosphere

THE BIG IDEA **How does Earth's atmosphere affect life on Earth?**

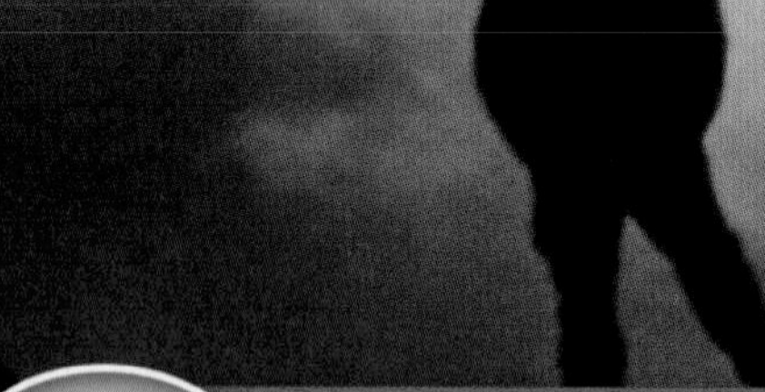

Inquiry What's in the atmosphere?

Earth's atmosphere is made up of gases and small amounts of liquid and solid particles. Earth's atmosphere surrounds and sustains life.

- What type of particles make up clouds in the atmosphere?
- How do conditions in the atmosphere change as height above sea level increases?
- How does Earth's atmosphere affect life on Earth?

Daniel H. Bailey/Alamy

Get Ready to Read

What do you think?

Before you read, decide if you agree or disagree with each of these statements. As you read this chapter, see if you change your mind about any of the statements.

1. Air is empty space.
2. Earth's atmosphere is important to living organisms.
3. All the energy from the Sun reaches Earth's surface.
4. Earth emits energy back into the atmosphere.
5. Uneven heating in different parts of the atmosphere creates air circulation patterns.
6. Warm air sinks and cold air rises.
7. If no humans lived on Earth, there would be no air pollution.
8. Pollution levels in the air are not measured or monitored.

Your one-stop online resource
connectED.mcgraw-hill.com

LearnSmart®

Chapter Resources Files, Reading Essentials, Get Ready to Read, Quick Vocabulary

Animations, Videos, Interactive Tables

Self-checks, Quizzes, Tests

Project-Based Learning Activities

Lab Manuals, Safety Videos, Virtual Labs & Other Tools

Vocabulary, Multilingual eGlossary, Vocab eGames, Vocab eFlashcards

Personal Tutors

Lesson 1

Describing Earth's Atmosphere

Reading Guide

Key Concepts

ESSENTIAL QUESTIONS

- How did Earth's atmosphere form?
- What is Earth's atmosphere made of?
- What are the layers of the atmosphere?
- How do air pressure and temperature change as altitude increases?

Vocabulary

atmosphere p. 409
water vapor p. 410
troposphere p. 412
stratosphere p. 412
ozone layer p. 412
ionosphere p. 413

Multilingual eGlossary

Inquiry Why is the atmosphere important?

What would Earth be like without its atmosphere? Earth's surface would be scarred with craters created from the impact of meteorites. Earth would experience extreme daytime-to-nighttime temperature changes. How would changes in the atmosphere affect life? What effect would atmospheric changes have on weather and climate?

Launch Lab

20 minutes

Where does air apply pressure?

With the exception of Mercury, most planets in the solar system have some type of atmosphere. However, Earth's atmosphere provides what the atmospheres of other planets cannot: oxygen and water. Oxygen, water vapor, and other gases make up the gaseous mixture in the atmosphere called air. In this activity, you will explore air's effect on objects on Earth's surface.

1. Read and complete a lab safety form.
2. Add **water** to a **cup** until it is two-thirds full.
3. Place a large **index card** over the opening of the cup so that it is completely covered.
4. Hold the cup over a tub or a large bowl.
5. Place one hand on the index card to hold it in place as you quickly turn the cup upside down. Remove your hand.

Think About This

1. What happened when you turned the cup over?
2. How did air play a part in your observation?
3. **Key Concept** How do you think these results might differ if you repeated the activity in a vacuum?

Importance of Earth's Atmosphere

The photo on the previous page shows Earth's atmosphere as seen from space. How would you describe the atmosphere? *The* **atmosphere** (AT muh sfihr) *is a thin layer of gases surrounding Earth.* Earth's atmosphere is divided into five layers. The first four layers extend to about 600 km, and the outermost layer extends to about 10,000 km above Earth.

WORD ORIGIN

atmosphere
from Greek *atmos,* means "vapor"; and Latin *sphaera,* means "sphere"

The atmosphere contains the oxygen, carbon dioxide, and water necessary for life on Earth. Earth's atmosphere also acts like insulation on a house. It helps keep temperatures on Earth within a range in which living organisms can survive. Without it, daytime temperatures would be extremely high and nighttime temperatures would be extremely low.

The atmosphere helps protect living organisms from some of the Sun's harmful rays. It also helps protect Earth's surface from being struck by asteroids. Most asteroids that fall toward Earth burn up before reaching Earth's surface. Friction with the atmosphere causes them to burn. Only the very largest asteroids strike Earth.

Reading Check Why is Earth's atmosphere important to life on Earth?

Origins of Earth's Atmosphere

Most scientists agree that when Earth formed, it was a ball of molten rock. As Earth slowly cooled, its outer surface hardened. Erupting volcanoes emitted hot gases from Earth's interior. These gases surrounded Earth, forming its atmosphere.

Ancient Earth's atmosphere was thought to be water vapor with a little carbon dioxide (CO_2) and nitrogen. **Water vapor** *is water in its gaseous form.* This ancient atmosphere did not have enough oxygen to support life as we know it. As Earth and its atmosphere cooled, the water vapor condensed into liquid. Rain fell and then evaporated from Earth's surface repeatedly for thousands of years. Eventually, water accumulated on Earth's surface, forming oceans. Most of the original CO_2 that dissolved in rain is in rocks on the ocean floor. Today the atmosphere has more nitrogen than CO_2.

REVIEW VOCABULARY

liquid
matter with a definite volume but no definite shape that can flow from one place to another

Earth's first organisms could undergo photosynthesis, which changed the atmosphere. Recall that photosynthesis uses light energy to produce sugar and oxygen from carbon dioxide and water. The organisms removed CO_2 from the atmosphere and released oxygen into it. Eventually the levels of CO_2 and oxygen supported the development of other organisms.

Key Concept Check How did Earth's present atmosphere form?

MiniLab

20 minutes

Why does the furniture get dusty?

Have you ever noticed that furniture gets dusty? The atmosphere is one source for dirt and dust particles. Where can you find dust in your classroom?

1. Read and complete a lab safety form.
2. Choose a place in your classroom to collect a sample of dust.
3. Using a **duster,** collect dust from about a 50-cm^2 area.
4. Examine the duster with a **magnifying lens.** Observe any dust particles. Some might be so small that they only make the duster look gray.
5. Record your observations in your Science Journal.
6. Compare your findings with those of other members of your class.

Analyze and Conclude

1. **Analyze** how the area surrounding your collection site might have influenced how much dust you observed on the duster.

2. **Infer** the source of the dust.
3. **Key Concept** Other than gases and water droplets, predict what Earth's atmosphere might contain.

Hutchings Photography/Digital Light Source

Figure 1 Oxygen and nitrogen make up most of the atmosphere, with the other gases making up only 1 percent. ▼

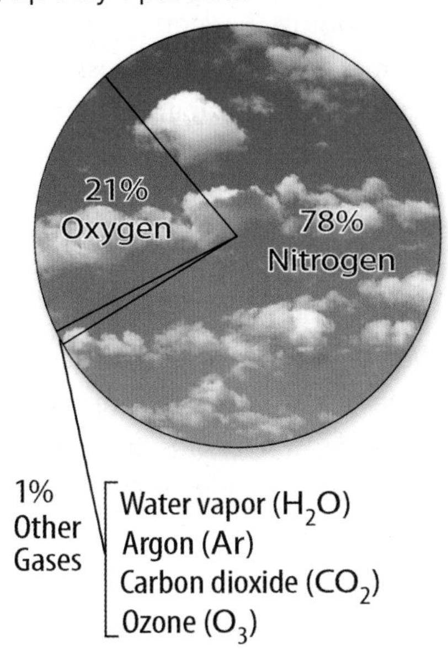

Visual Check What percent of the atmosphere is made up of oxygen and nitrogen?

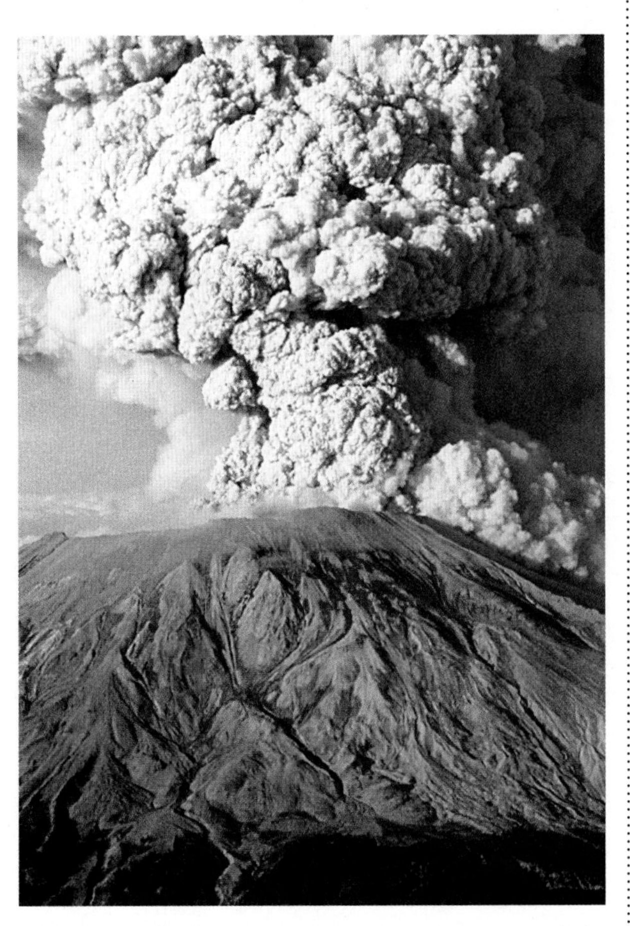

▲ **Figure 2** One way solid particles enter the atmosphere is from volcanic eruptions.

(t)PhotoLink/Getty Images; (b)C. Sherburne/PhotoLink/Getty Images

Composition of the Atmosphere

Today's atmosphere is mostly made up of invisible gases, including nitrogen, oxygen, and carbon dioxide. Some solid and liquid particles, such as ash from volcanic eruptions and water droplets, are also present.

Gases in the Atmosphere

Study **Figure 1**. Which gas is the most abundant in Earth's atmosphere? Nitrogen makes up about 78 percent of Earth's atmosphere. About 21 percent of Earth's atmosphere is oxygen. Other gases, including argon, carbon dioxide, and water vapor, make up the remaining 1 percent of the atmosphere.

The amounts of water vapor, carbon dioxide, and ozone vary. The concentration of water vapor in the atmosphere ranges from 0 to 4 percent. Carbon dioxide is 0.038 percent of the atmosphere. A small amount of ozone is at high altitudes. Ozone also occurs near Earth's surface in urban areas.

Solids and Liquids in the Atmosphere

Tiny solid particles are also in Earth's atmosphere. Many of these, such as pollen, dust, and salt, can enter the atmosphere through natural processes. **Figure 2** shows another natural source of particles in the atmosphere–ash from volcanic eruptions. Some solid particles enter the atmosphere because of human activities, such as driving vehicles that release soot.

The most common liquid particles in the atmosphere are water droplets. Although microscopic in size, water droplets are visible when they form clouds. Other atmospheric liquids include acids that result when volcanoes erupt and fossil fuels are burned. Sulfur dioxide and nitrous oxide combine with water vapor in the air and form the acids.

Key Concept Check What is Earth's atmosphere made of?

Animation

Layers of Atmosphere

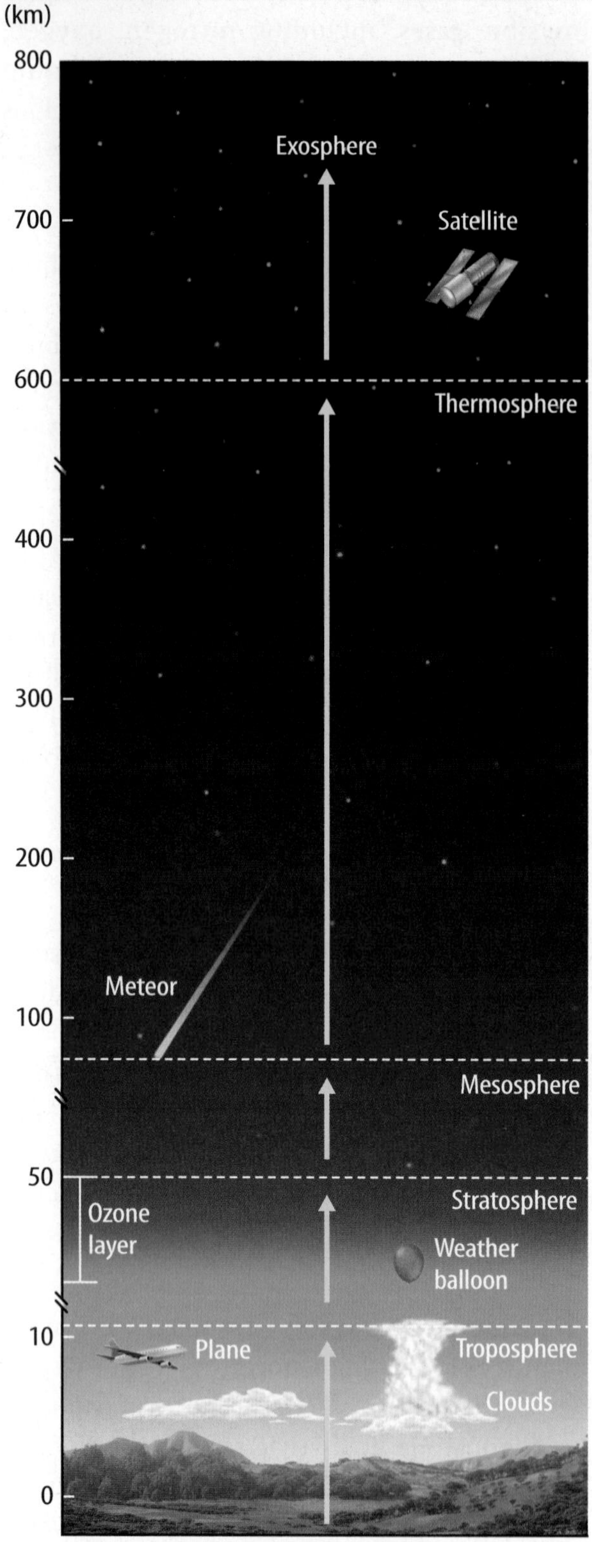

(km)
800
700
600
400
300
200
100
50
10
0
Exosphere
Satellite
Thermosphere
Meteor
Mesosphere
Stratosphere
Ozone layer
Weather balloon
Troposphere
Plane
Clouds

Figure 3 Scientists divide Earth's atmosphere into different layers.

Visual Check In which layer of the atmosphere do planes fly?

Layers of the Atmosphere

The atmosphere has several different layers, as shown in **Figure 3**. Each layer has unique properties, including the composition of gases and how temperature changes with altitude. Notice that the scale varies in **Figure 3**. This is so all the layers can be shown in one image.

Troposphere

The atmospheric layer closest to Earth's surface is called the **troposphere** (TRO puh sfihr). Most people spend their entire lives within the troposphere. It extends from Earth's surface to altitudes between 8–15 km. Its name comes from the Greek word *tropos,* which means "change." The temperature in the troposphere decreases as you move away from Earth. The warmest part of the troposphere is near Earth's surface. This is because most sunlight passes through the atmosphere and warms Earth's surface. The warmth is radiated to the troposphere, causing weather.

Reading Check Describe the troposphere.

Stratosphere

The atmospheric layer directly above the troposphere is the **stratosphere** (STRA tuh sfihr). The stratosphere extends from about 15 km to about 50 km above Earth's surface. The upper half of the stratosphere contains the greatest amount of ozone gas. *The area of the stratosphere with a high concentration of ozone is referred to as the* **ozone layer.** The presence of the ozone layer causes increasing stratospheric temperatures with increasing altitude.

An ozone (O_3) molecule differs from an oxygen (O_2) molecule. Ozone has three oxygen atoms instead of two. This difference is important because ozone absorbs the Sun's ultraviolet rays more effectively than oxygen does. Ozone protects Earth from ultraviolet rays that can kill plants, animals, and other organisms and cause skin cancer in humans.

Mesosphere and Thermosphere

As shown in **Figure 3,** the mesosphere extends from the stratosphere to about 85 km above Earth. The thermosphere can extend from the mesopshere to more than 600 km above Earth. Combined, these layers are much broader than the troposphere and the stratosphere, yet only 1 percent of the atmosphere's gas molecules are found in the mesosphere and the thermosphere. Most meteors burn up in these layers instead of striking Earth.

Ionosphere *The* **ionosphere** *is a region within the mesosphere and thermosphere that contains ions.* Between 60 km and 300 km above Earth's surface, the ionosphere's ions reflect AM radio waves transmitted at ground level. After sunset when ions recombine, this reflection increases. **Figure 4** shows how AM radio waves can travel long distances, especially at night, by bouncing off Earth and the ionosphere.

FOLDABLES®

Make a vertical four-tab book using the titles shown. Use it to record similarities and differences among these four layers of the atmosphere. Fold the top half over the bottom and label the outside *Layers of the Atmosphere.*

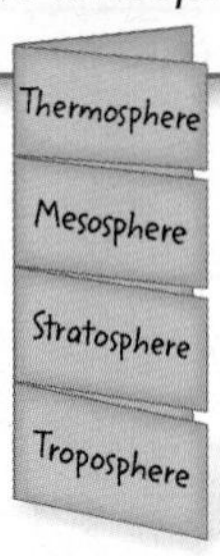

Radio Waves and the Ionosphere

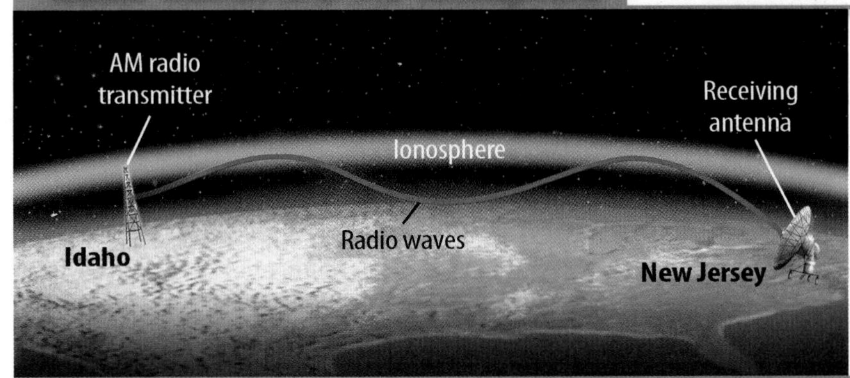

Figure 4 Radio waves can travel long distances in the atmosphere.

Auroras The ionosphere is where stunning displays of colored lights called auroras occur, as shown in **Figure 5.** Auroras are most frequent in the spring and fall, but are best seen when the winter skies are dark. Auroras occur when ions from the Sun strike air molecules, causing them to emit vivid colors of light. People who live in the higher latitudes, nearer to the North Pole and the South Pole, are most likely to see auroras.

Exosphere

The exosphere is the atmospheric layer farthest from Earth's surface. Here, pressure and density are so low that individual gas molecules rarely strike one another. The molecules move at incredibly fast speeds after absorbing the Sun's radiation. The atmosphere does not have a definite edge, and molecules that are part of it can escape the pull of gravity and travel into space.

 **Key Concept Check** What are the layers of the atmosphere?

▲ **Figure 5** Auroras occur in the ionosphere.

Figure 6 Molecules in the air are closer together near Earth's surface than they are at higher altitudes. ▼

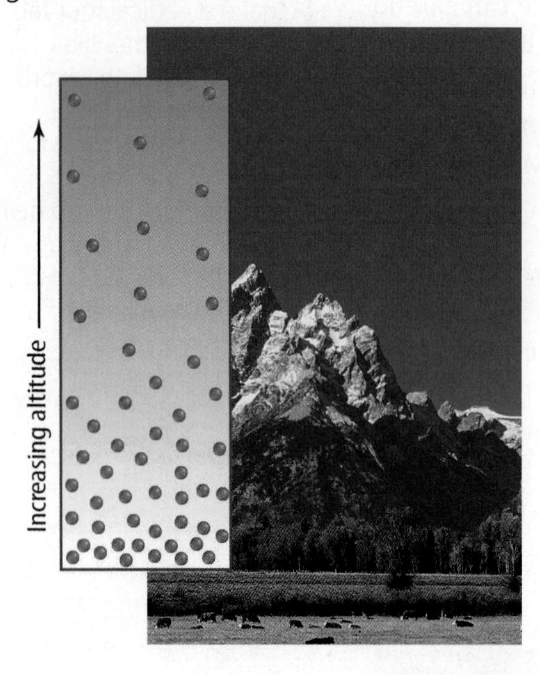

Air Pressure and Altitude

Gravity is the force that pulls all objects toward Earth. When you stand on a scale, you can read your weight. This is because gravity is pulling you toward Earth. Gravity also pulls the atmosphere toward Earth. The pressure that a column of air exerts on anything below it is called air pressure. Gravity's pull on air increases its density. At higher altitudes, the air is less dense. **Figure 6** shows that air pressure is greatest near Earth's surface because the air molecules are closer together. This dense air exerts more force than the less dense air near the top of the atmosphere. Mountain climbers sometimes carry oxygen tanks at high altitudes because fewer oxygen molecules are in the air at high altitudes.

Reading Check How does air pressure change as altitude increases?

Temperature and Altitude

Figure 7 shows how temperature changes with altitude in the different layers of the atmosphere. If you have ever been hiking in the mountains, you have experienced the temperature cooling as you hike to higher elevations. In the troposphere, temperature decreases as altitude increases. Notice that the opposite effect occurs in the stratosphere. As altitude increases, temperature increases. This is because of the high concentration of ozone in the stratosphere. Ozone absorbs energy from sunlight, which increases the temperature in the stratosphere.

In the mesosphere, as altitude increases, temperature again decreases. In the thermosphere and exosphere, temperatures increase as altitude increases. These layers receive large amounts of energy from the Sun. This energy is spread across a small number of particles, creating high temperatures.

Figure 7 Temperature differences occur within the layers of the atmosphere. ▼

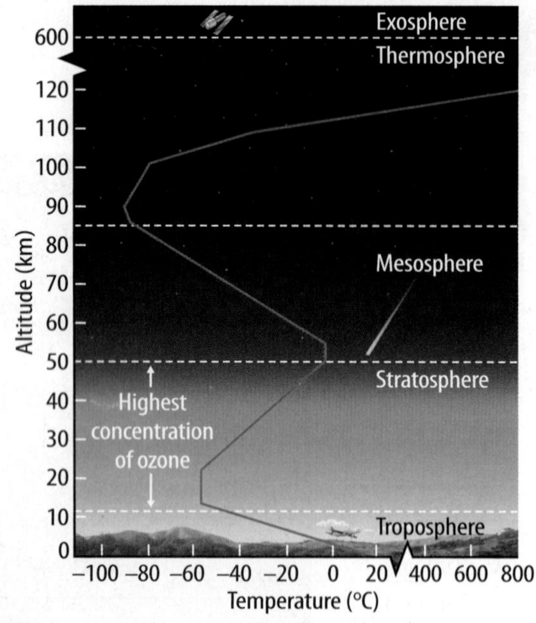

Visual Check Which temperature pattern is most like the troposphere's?

Key Concept Check How does temperature change as altitude increases?

Lesson 1 Review

Online Quiz
Virtual Lab

Visual Summary

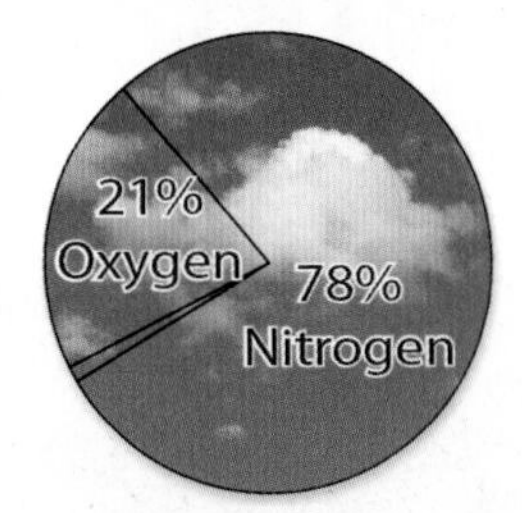

Earth's atmosphere consists of gases that make life possible.

Layers of the atmosphere include the troposphere, the stratosphere, the mesosphere, the thermosphere, and the exosphere.

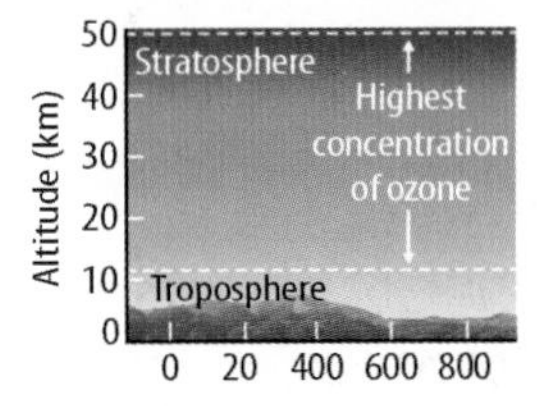

The ozone layer is the area in the stratosphere with a high concentration of ozone.

FOLDABLES®

Use your lesson Foldable to review the lesson. Save your Foldable for the project at the end of the chapter.

What do you think NOW?

You first read the statements below at the beginning of the chapter.

1. Air is empty space.

2. Earth's atmosphere is important to living organisms.

Did you change your mind about whether you agree or disagree with the statements? Rewrite any false statements to make them true.

Use Vocabulary

1 The __________ is a thin layer of gases surrounding Earth.

2 The area of the stratosphere that helps protect Earth's surface from harmful ultraviolet rays is the __________.

3 **Define** Using your own words, define *water vapor.*

Understand Key Concepts 

4 Which atmospheric layer is closest to Earth's surface?

A. mesosphere
B. stratosphere
C. thermosphere
D. troposphere

5 **Identify** the two atmospheric layers in which temperature decreases as altitude increases.

Interpret Graphics

6 **Contrast** Copy and fill in the graphic organizer below to contrast the composition of gases in Earth's early atmosphere and its present-day atmosphere.

Atmosphere	Gases
Early	
Present-day	

7 **Determine** the relationship between air pressure and the water in the glass in the photo below.

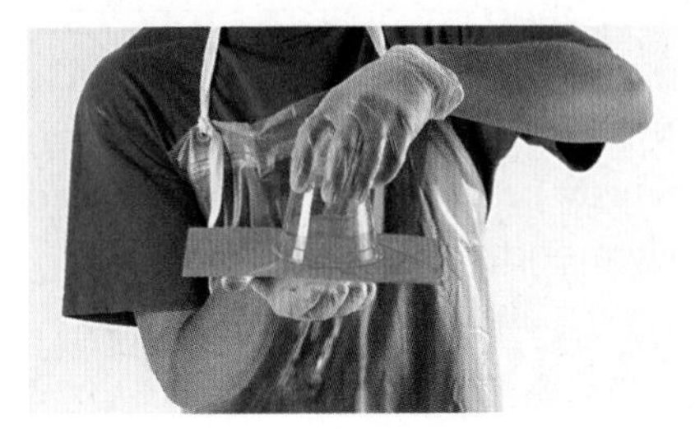

Critical Thinking

8 **Explain** three ways the atmosphere is important to living things.

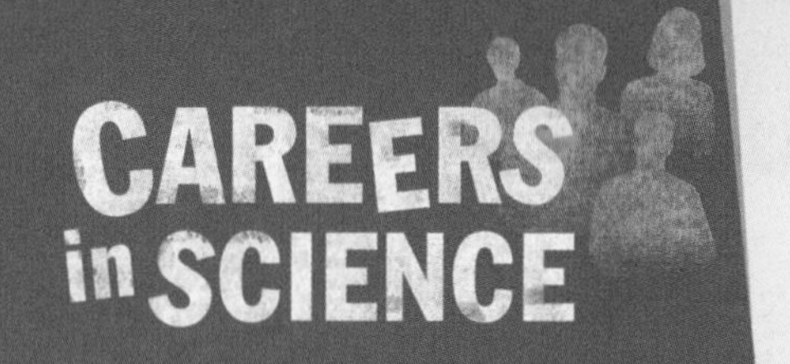

AMERICAN MUSEUM of NATURAL HISTORY

A Crack in Earth's Shield

Scientists discover an enormous hole in the ozone layer that protects Earth.

The ozone layer is like sunscreen, protecting Earth from the Sun's ultraviolet rays. But not all of Earth is covered. Every spring since 1985, scientists have been monitoring a growing hole in the ozone layer above Antarctica.

This surprising discovery was the outcome of years of research from Earth and space. The first measurements of polar ozone levels began in the 1950s, when a team of British scientists began launching weather balloons in Antarctica. In the 1970s, NASA started using satellites to measure the ozone layer from space. Then, in 1985 a close examination of the British team's records indicated a large drop in ozone levels during the Antarctic spring. The levels were so low that the scientists checked and rechecked their instruments before they reported their findings. NASA scientists quickly confirmed the discovery—an enormous hole in the ozone layer over the entire continent of Antarctica. They reported that the hole might have originated as far back as 1976.

Human-made compounds found mostly in chemicals called chlorofluorocarbons, or CFCs, are destroying the ozone layer. During cold winters, molecules released from these compounds are transformed into new compounds by chemical reactions on ice crystals that form in the ozone layer over Antarctica. In the spring, warming by the Sun breaks down the new compounds and releases chlorine and bromine. These chemicals break apart ozone molecules, slowly destroying the ozone layer.

In 1987, CFCs were banned in many countries around the world. Since then, the loss of ozone has slowed and possibly reversed, but a full recovery will take a long time. One reason is that CFCs stay in the atmosphere for more than 40 years. Still, scientists predict the hole in the ozone layer will eventually mend.

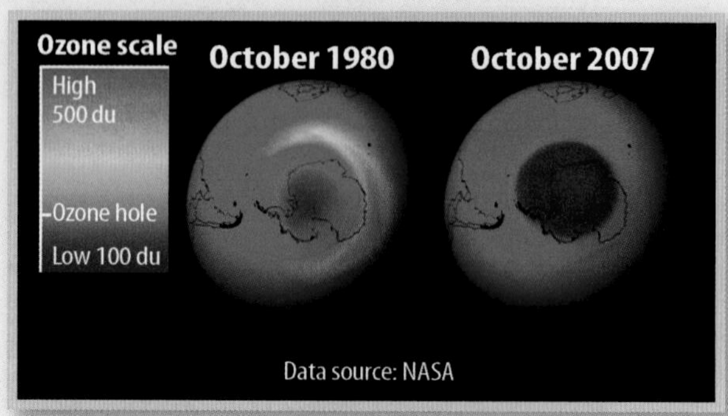

▲ **A hole in the ozone layer has developed over Antarctica. Even though it has gotten worse over the years, the hole has not grown as fast as scientists initially thought it would.**

Global Warming and the Ozone

Drew Shindell is a NASA scientist investigating the connection between the ozone layer in the stratosphere and the buildup of greenhouse gases throughout the atmosphere. Surprisingly, while these gases warm the troposphere, they are actually causing temperatures in the stratosphere to become cooler. As the stratosphere cools above Antarctica, more clouds with ice crystals form—a key step in the process of ozone destruction. While the buildup of greenhouse gases in the atmosphere may slow the recovery, Shindell still thinks that eventually the ozone layer will heal itself.

It's Your Turn

NEWSCAST Work with a partner to develop three questions about the ozone layer. Research to find the answers. Take the roles of reporter and scientist. Present your findings to the class in a newscast format.

Lesson 2

Reading Guide

Key Concepts 

ESSENTIAL QUESTIONS

- How does energy transfer from the Sun to Earth and the atmosphere?
- How are air circulation patterns within the atmosphere created?

Vocabulary

Multilingual eGlossary

Energy Transfer in the Atmosphere

Inquiry What's really there?

Mirages are created as light passes through layers of air that have different temperatures. How does energy create the reflections? What other effects does energy have on the atmosphere?

John King/Alamy

Launch Lab

15 minutes

What happens to air as it warms?

Light energy from the Sun is converted to thermal energy on Earth. Thermal energy powers the weather systems that impact your everyday life.

1. Read and complete a lab safety form.
2. Turn on a **lamp** with an incandescent lightbulb.
3. Place your hands under the light near the lightbulb. What do you feel?
4. Dust your hands with **powder.**
5. Place your hands below the lightbulb and clap them together once.
6. Observe what happens to the particles.

Think About This

1. How might the energy in step 3 move from the lightbulb to your hand?
2. How did the particles move when you clapped your hands?
3. **Key Concept** How did particle motion show you how the air was moving?

ACADEMIC VOCABULARY

process
(noun) an ordered series of actions

Energy from the Sun

The Sun's energy travels 148 million km to Earth in only 8 minutes. How does the Sun's energy get to Earth? It reaches Earth through the process of radiation. **Radiation** *is the transfer of energy by electromagnetic waves.* Ninety-nine percent of the radiant energy from the Sun consists of visible light, ultraviolet light, and infrared radiation.

Visible Light

The majority of sunlight is visible light. Recall that visible light is light that you can see. The atmosphere is like a window to visible light, allowing it to pass through. At Earth's surface it is converted to thermal energy, commonly called heat.

Near-Visible Wavelengths

The wavelengths of ultraviolet (UV) light and infrared radiation (IR) are just beyond the range of visibility to human eyes. UV light has short wavelengths and can break chemical bonds. Excess exposure to UV light will burn human skin and can cause skin cancer. Infrared radiation (IR) has longer wavelengths than visible light. You can sense IR as thermal energy or warmth. Earth absorbs energy from the Sun and then radiates it into the atmosphere as IR.

Reading Check Contrast visible light and ultraviolet light.

Hutchings Photography/Digital Light Source

Energy on Earth

As the Sun's energy passes through the atmosphere, some of it is absorbed by gases and particles, and some of it is reflected back into space. As a result, not all the energy coming from the Sun reaches Earth's surface.

Absorption

Study **Figure 8.** Gases and particles in the atmosphere absorb about 20 percent of incoming solar radiation. Oxygen, ozone, and water vapor all absorb incoming ultraviolet light. Water and carbon dioxide in the troposphere absorb some infrared radiation from the Sun. Earth's atmosphere does not absorb visible light. Visible light must be converted to infrared radiation before it can be absorbed.

Reflection

Bright surfaces, especially clouds, **reflect** incoming radiation. Study **Figure 8** again. Clouds and other small particles in the air reflect about 25 percent of the Sun's radiation. Some radiation travels to Earth's surface and is then reflected by land and sea surfaces. Snow-covered, icy, or rocky surfaces are especially reflective. As shown in **Figure 8,** this accounts for about 5 percent of incoming radiation. In all, about 30 percent of incoming radiation is reflected into space. This means that, along with the 20 percent of incoming radiation that is absorbed in the atmosphere, Earth's surface only receives and absorbs about 50 percent of incoming solar radiation.

SCIENCE USE V. COMMON USE

reflect

Science Use to return light, heat, sound, and so on, after it strikes a surface

Common Use to think quietly and calmly

Figure 8 Some of the energy from the Sun is reflected or absorbed as it passes through the atmosphere.

Incoming Radiation

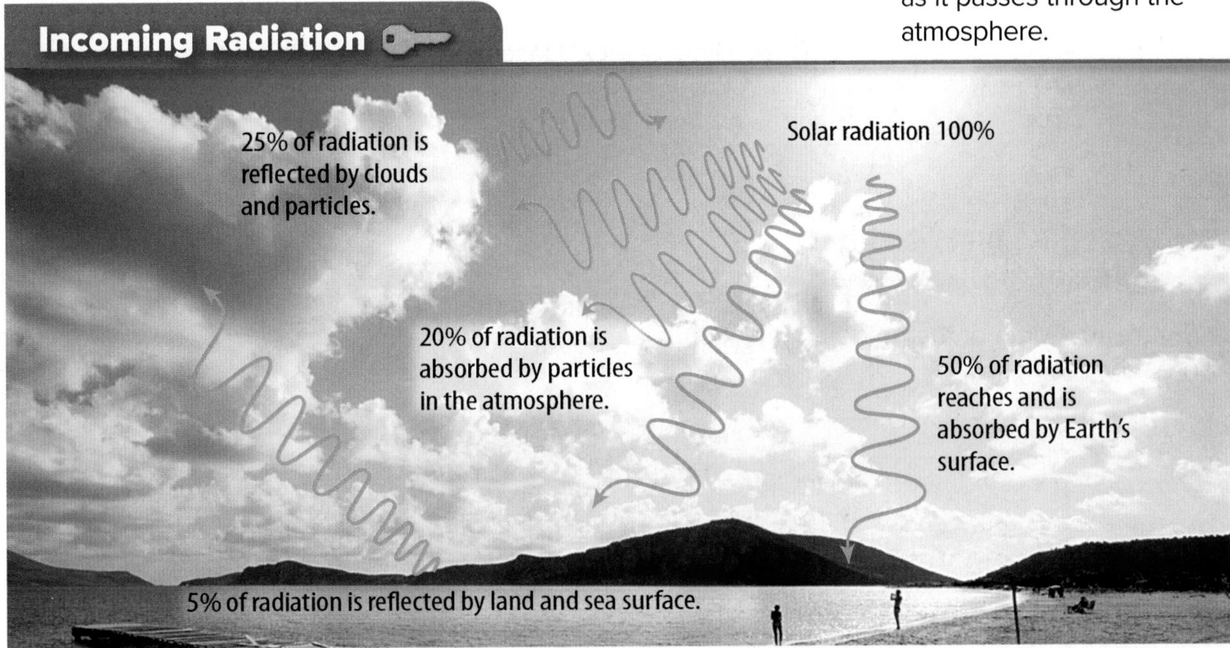

Visual Check What percent of incoming radiation is absorbed by gases and particles in the atmosphere?

Eric James/Alamy

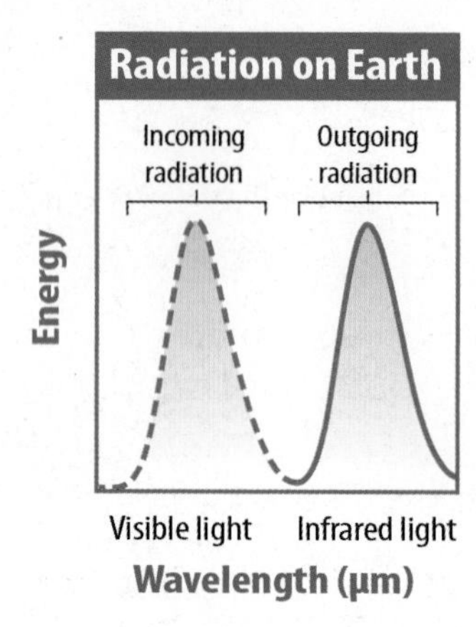

▲ **Figure 9** The amount of solar energy absorbed by Earth and its atmosphere is equal to the amount of energy Earth radiates back into space.

Radiation Balance

The Sun's radiation heats Earth. So, why doesn't Earth get hotter and hotter as it continues to receive radiation from the Sun? There is a balance between the amount of incoming radiation from the Sun and the amount of outgoing radiation from Earth.

The land, water, plants, and other organisms absorb solar radiation that reaches Earth's surface. The radiation absorbed by Earth is then re-radiated, or bounced back, into the atmosphere. Most of the energy radiated from Earth is infrared radiation, which heats the atmosphere. **Figure 9** shows that the amount of radiation Earth receives from the Sun is the same as the amount Earth radiates into the outer atmosphere. Earth absorbs the Sun's energy and then radiates that energy away until a balance is achieved.

The Greenhouse Effect

As shown in **Figure 10**, the glass of a greenhouse allows light to pass through, where it is converted to infrared energy. The glass prevents the IR from escaping and it warms the greenhouse. Some of the gases in the atmosphere, called greenhouse gases, act like the glass of a greenhouse. They allow sunlight to pass through, but they prevent some of Earth's IR energy from escaping. Greenhouse gases in Earth's atmosphere trap IR and direct it back to Earth's surface. This causes an additional buildup of thermal energy at Earth's surface. The gases that trap IR best are water vapor (H_2O), carbon dioxide (CO_2), and methane (CH_4).

Reading Check Describe the greenhouse effect.

The Greenhouse Effect

Animation

Figure 10 Some of the outgoing radiation is directed back toward Earth's surface by greenhouse gases.

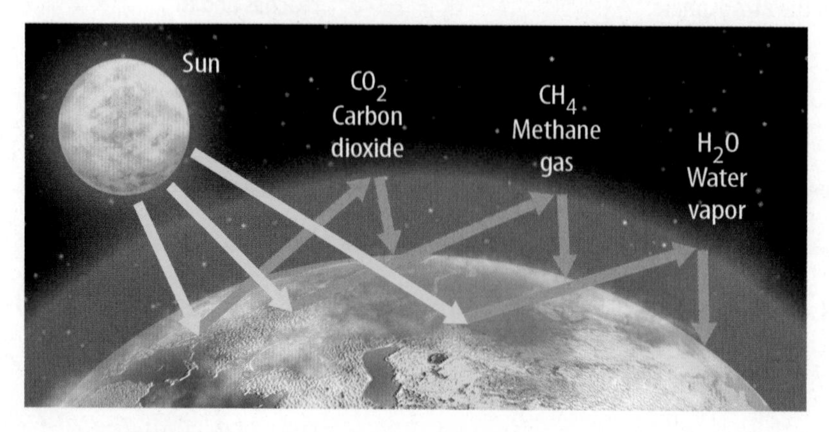

Thermal Energy Transfer

Recall that there are three types of thermal energy transfer–radiation, conduction, and convection. All three occur in the atmosphere. Recall that radiation is the process that transfers energy from the Sun to Earth.

Conduction

Thermal energy always moves from an object with a higher temperature to an object with a lower temperature. **Conduction** *is the transfer of thermal energy by collisions between particles of matter.* Particles must be close enough to touch to transfer energy by conduction. Touching the pot of water, shown in **Figure 11,** would transfer energy from the pot to your hand. Conduction occurs where the atmosphere touches Earth.

Convection

As molecules of air close to Earth's surface are heated by conduction, they spread apart, and air becomes less dense. Less dense air rises, transferring thermal energy to higher altitudes. *The transfer of thermal energy by the movement of particles within matter is called* **convection.** Convection can be seen in **Figure 11** as the boiling water circulates and steam rises.

Latent Heat

More than 70 percent of Earth's surface is covered by a highly unique substance–water! Water is the only substance that can exist as a solid, a liquid, and a gas within Earth's temperature ranges. Recall that latent heat is exchanged when water changes from one phase to another, as shown in **Figure 12.** Latent heat energy is transferred from Earth's surface to the atmosphere.

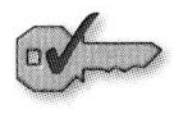

Key Concept Check How does energy transfer from the Sun to Earth and the atmosphere?

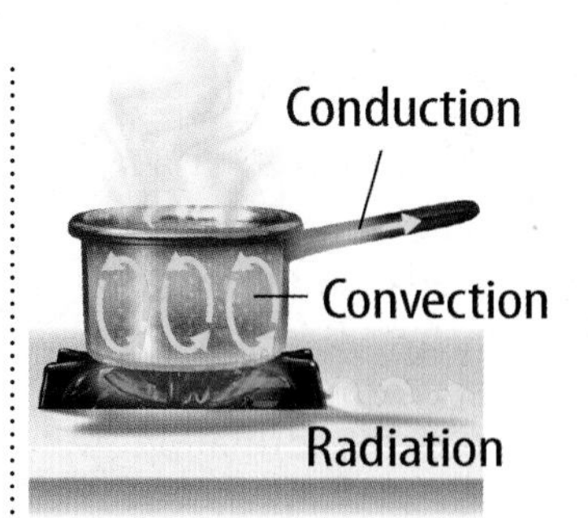

Personal Tutor

▲ **Figure 11** Energy is transferred through conduction, convection, and radiation.

WORD ORIGIN

conduction
from Latin *conducere,* means "to bring together"

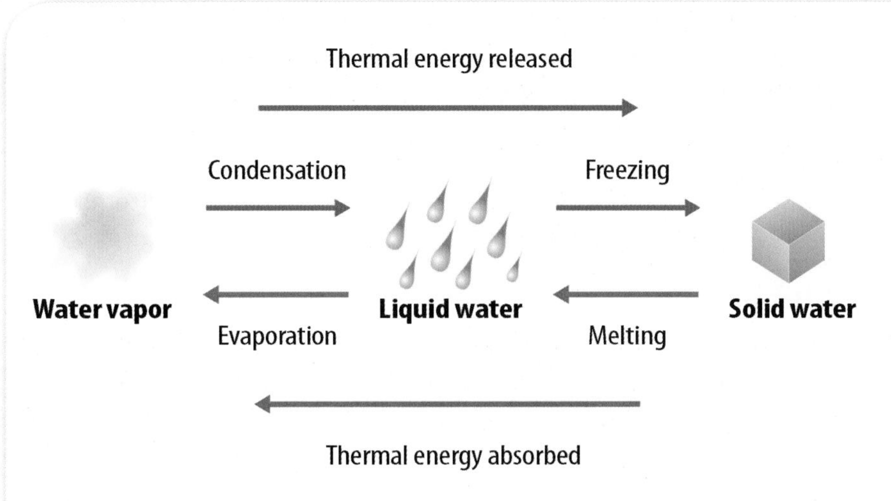

Figure 12 Water releases or absorbs thermal energy during phase changes.

FOLDABLES®

Fold a sheet of paper to make a four-column, four-row table and label as shown. Use it to record information about thermal energy transfer.

Energy Transfer by	Description	Everyday Example	Effect on the Atmosphere
Radiation			
Convection			
Conduction			

Circulating Air

Figure 13 Rising warm air is replaced by cooler, denser air that sinks beside it.

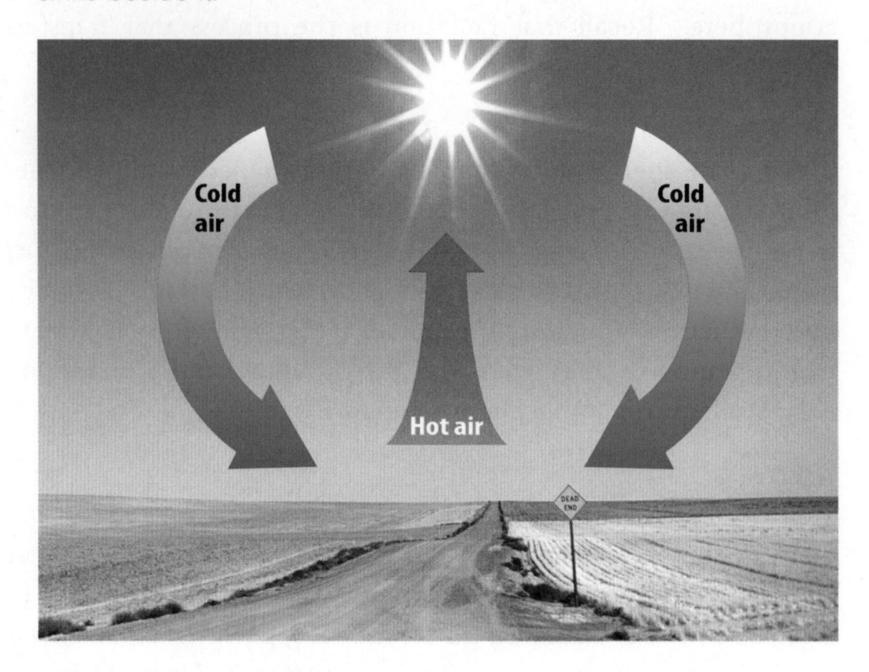

Figure 14 Lens-shaped lenticular clouds form when air rises with a mountain wave. ▼

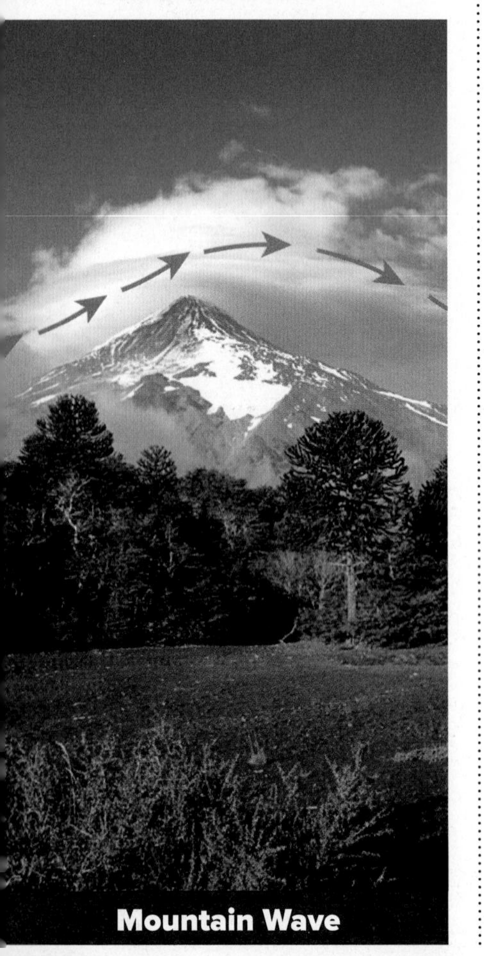

Circulating Air

You've read that energy is transferred through the atmosphere by convection. On a hot day, air that is heated becomes less dense. This creates a pressure difference. Cool, denser air pushes the warm air out of the way. The warm air is replaced by the more dense air, as shown in **Figure 13.** The warm air is often pushed upward. Warmer, rising air is always accompanied by cooler, sinking air.

Air is constantly moving. For example, wind flowing into a mountain range rises and flows over it. After reaching the top, the air sinks. This up-and-down motion sets up an atmospheric phenomenon called a mountain wave. The upward moving air within mountain waves creates lenticular (len TIH kyuh lur) clouds, shown in **Figure 14.** Circulating air affects weather and climate around the world.

 **Key Concept Check** How are air circulation patterns within the atmosphere created?

Stability

When you stand in the wind, your body forces some of the air to move above you. The same is true for hills and buildings. Conduction and convection also cause air to move upward. **Stability** *describes whether circulating air motions will be strong or weak.* When air is unstable, circulating motions are strong. During stable conditions, circulating motions are weak.

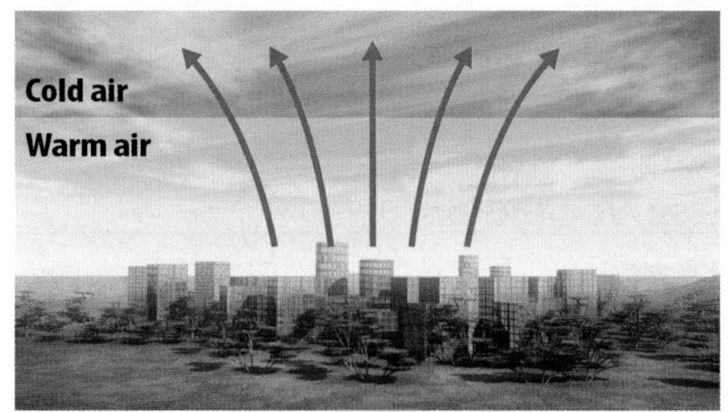

Normal conditions

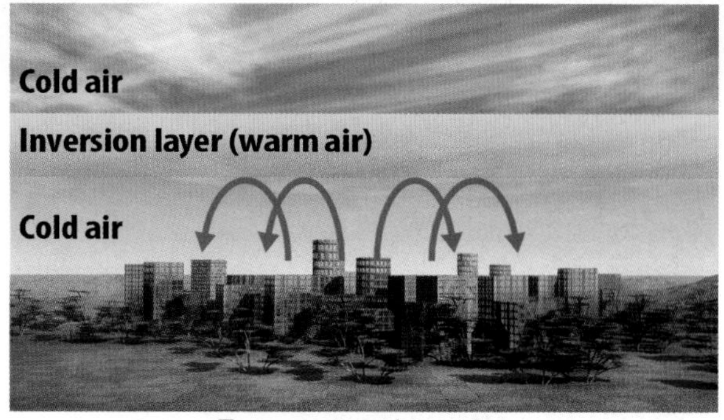

Temperature inversion

Unstable Air and Thunderstorms Unstable conditions often occur on warm, sunny afternoons. During unstable conditions, ground-level air is much warmer than higher-altitude air. As warm air rises rapidly in the atmosphere, it cools and forms large, tall clouds. Latent heat, released as water vapor changes from a gas to a liquid, adds to the instability, and produces a thunderstorm.

Reading Check Relate unstable air to the formation of thunderstorms.

Stable Air and Temperature Inversions Sometimes ground-level air is nearly the same temperature as higher-altitude air. During these conditions, the air is stable, and circulating motions are weak. A temperature inversion can occur under these conditions. *A* **temperature inversion** *occurs in the troposphere when temperature increases as altitude increases.* During a temperature inversion, a layer of cooler air is trapped by a layer of warmer air above it, as shown in **Figure 15.** Temperature inversions prevent air from mixing and can trap pollution in the air close to Earth's surface.

Figure 15 A temperature inversion occurs when cooler air is trapped beneath warmer air.

Visual Check How do conditions during a temperature inversion differ from normal conditions?

MiniLab

20 minutes

Can you identify a temperature inversion?

You've read that a temperature inversion is a reversal of normal temperature conditions in the troposphere. What do data from a temperature inversion look like on a graph?

Analyze and Conclude

1. **Describe** the information presented in the graph. How do the graph's lines differ?
2. **Analyze** Which graph line represents normal conditions in the troposphere? Which represents a temperature inversion? Explain your answers in your Science Journal.
3. **Key Concept** From the graph, what pattern does a temperature inversion have?

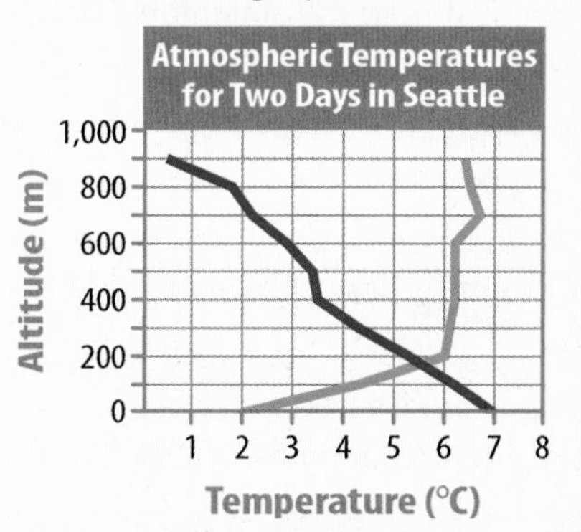

Lesson 2 Review

Visual Summary

Not all radiation from the Sun reaches Earth's surface.

Thermal energy transfer in the atmosphere occurs through radiation, conduction, and convection.

Temperature inversion

Temperature inversions prevent air from mixing and can trap pollution in the air close to Earth's surface.

FOLDABLES®

Use your lesson Foldable to review the lesson. Save your Foldable for the project at the end of the chapter.

What do you think NOW?

You first read the statements below at the beginning of the chapter.

3. All of the energy from the Sun reaches Earth's surface.

4. Earth emits energy back into the atmosphere.

Did you change your mind about whether you agree or disagree with the statements? Rewrite any false statements to make them true.

Use Vocabulary

1 The property of the atmosphere that describes whether circulating air motions will be strong or weak is called ________.

2 **Define** *conduction* in your own words.

3 ________ is the transfer of thermal energy by the movement of particles within matter.

Understand Key Concepts 

4 Which statement is true?

A. The Sun's energy is completely blocked by Earth's atmosphere.

B. The Sun's energy passes through the atmosphere without warming it significantly.

C. The Sun's IR energy is absorbed by greenhouse gases.

D. The Sun's energy is primarily in the UV range.

5 **Distinguish** between conduction and convection.

Interpret Graphics

6 **Explain** how greenhouses gases affect temperatures on Earth.

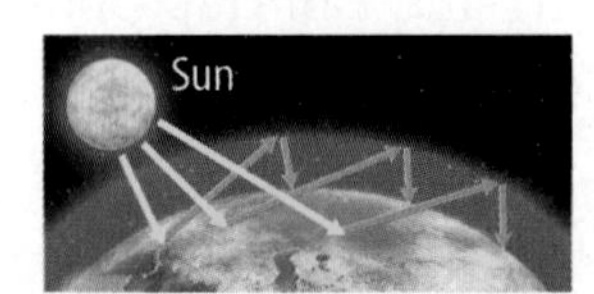

7 **Sequence** Copy and fill in the graphic organizer below to describe how energy from the Sun is absorbed in Earth's atmosphere.

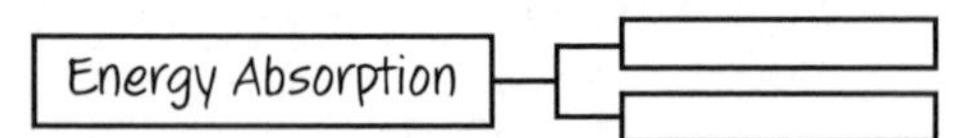

Critical Thinking

8 **Suggest** a way to keep a parked car cool on a sunny day.

9 **Relate** temperature inversions to air stability.

Skill Practice Compare and Contrast 30 minutes

Materials

candle

metal rod

glass rod

wooden dowel

500-mL beaker

ice

bowls (2)

lamp

glass cake pan

food coloring

250-mL beaker

Safety

Can you conduct, convect, and radiate?

After solar radiation reaches Earth, the molecules closest to Earth transfer thermal energy from molecule to molecule by conduction. The newly warmed air becomes less dense and moves through the process of convection.

Learn It

When you **compare and contrast** two or more things, you look for similarities and differences between them. When you **compare** two things, you look for the similarities, or how they are the same. When you **contrast** them, you look for how they are different from each other.

Try It

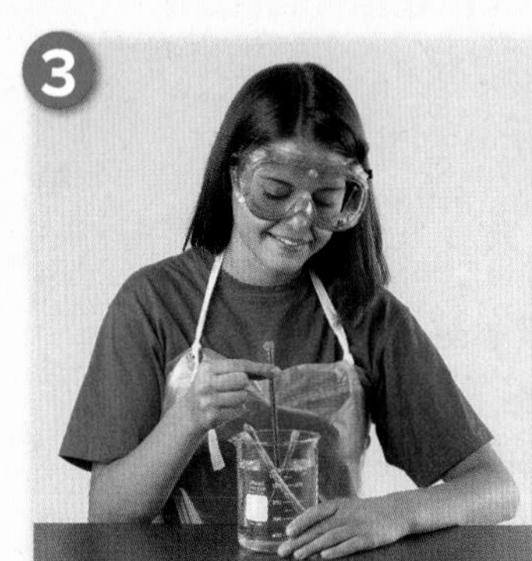

1. Read and complete a lab safety form.
2. Drip a small amount of melted candle wax onto one end of a metal rod, a glass rod, and a wooden dowel.
3. Place a 500-mL beaker on the lab table. Have your teacher add 350 mL of very hot water. Place the ends of the rods without candle wax in the water. Set aside.
4. Place an ice cube into each of two small bowls labeled A and B.
5. Place bowl A under a lamp with a 60- or 75-watt lightbulb. Place the light source 10 cm above the bowl. Turn on the lamp. Set bowl B aside.
6. Fill a glass cake pan with room-temperature water to a level of 2 cm. Put 2–3 drops of red food coloring into a 250-mL beaker of very hot water. Put 2–3 drops of blue food coloring into a 250-mL beaker of very cold water and ice cubes. Carefully pour the hot water into one end of the pan. Slowly pour the very cold water into the same end of the pan. Observe what happens from the side of the pan. Record your observations in your Science Journal.
7. Observe the candle wax on the rods in the hot water and the ice cubes in the bowls.

Apply It

8. What happened to the candle wax? Identify the type of energy transfer.
9. Which ice cube melted the most in the bowls? Identify the type of energy transfer that melted the ice.
10. Compare and contrast how the hot and cold water behaved in the pan. Identify the type of energy transfer.
11. **Key Concept** Explain how each part of the lab models radiation, conduction, or convection.

(t to b, 3-7, 9-11, r)Hutchings Photography/Digital Light Source; (2)Ken Karp/McGraw-Hill Education; (8)Ken Cavanagh/McGraw-Hill Education

Lesson 3

Reading Guide

Key Concepts
ESSENTIAL QUESTIONS

- How does uneven heating of Earth's surface result in air movement?
- How are air currents on Earth affected by Earth's spin?
- What are the main wind belts on Earth?

Vocabulary

wind p. 427
trade winds p. 429
westerlies p. 429
polar easterlies p. 429
jet stream p. 429
sea breeze p. 430
land breeze p. 430

Multilingual eGlossary

What's Science Got to do With It?

Go to the resource tab in ConnectED to find the PBL *As the Water Churns.*

Air Currents

Inquiry How does air push these blades?

If you have ever ridden a bicycle into a strong wind, you know the movement of air can be a powerful force. Some areas of the world have more wind than others. What causes these differences? What makes wind?

Launch Lab

15 minutes

Why does air move?

Early sailors relied on wind to move their ships around the world. Today, wind is used as a renewable source of energy. In the following activity, you will explore what causes air to move.

1. Read and complete a lab safety form.
2. Inflate a **balloon.** Do not tie it. Hold the neck of the balloon closed.
3. Describe how the inflated balloon feels.
4. Open the neck of the balloon without letting go of the balloon. Record your observations of what happens in your Science Journal.

Think About This

1. What caused the inflated balloon surface to feel the way it did when the neck was closed?
2. What caused the air to leave the balloon when the neck was opened?
3. **Key Concept** Why didn't outside air move into the balloon when the neck was opened?

Global Winds

There are great wind belts that circle the globe. The energy that causes this massive movement of air originates at the Sun. However, wind patterns can be global or local.

Unequal Heating of Earth's Surface

The Sun's energy warms Earth. However, the same amount of energy does not reach all of Earth's surface. The amount of energy an area gets depends largely on the Sun's angle. For example, energy from the rising or setting Sun is not very intense. But Earth heats up quickly when the Sun is high in the sky.

In latitudes near the equator–an area referred to as the tropics–sunlight strikes Earth's surface at a nearly 90° angle year round. As a result, in the tropics there is more sunlight per unit of surface area. This means that the land, the water, and the air at the equator are always warm.

At latitudes near the North Pole and the South Pole, sunlight strikes Earth's surface at a low angle. Sunlight is now spread over a larger surface area than in the tropics. As a result, the poles receive very little energy per unit of surface area and are cooler.

Recall that differences in density cause warm air to rise. Warm air puts less pressure on Earth than cooler air. Because it's so warm in the tropics, air pressure is usually low. Over colder areas, such as the North Pole and the South Pole, air pressure is usually high. This difference in pressure creates wind. **Wind** *is the movement of air from areas of high pressure to areas of low pressure.* Global wind belts influence both climate and weather on Earth.

Key Concept Check How does uneven heating of Earth's surface result in air movement?

Global Wind Belts

Figure 16 Three cells in each hemisphere move air through the atmosphere.

Visual Check Which wind belt do you live in?

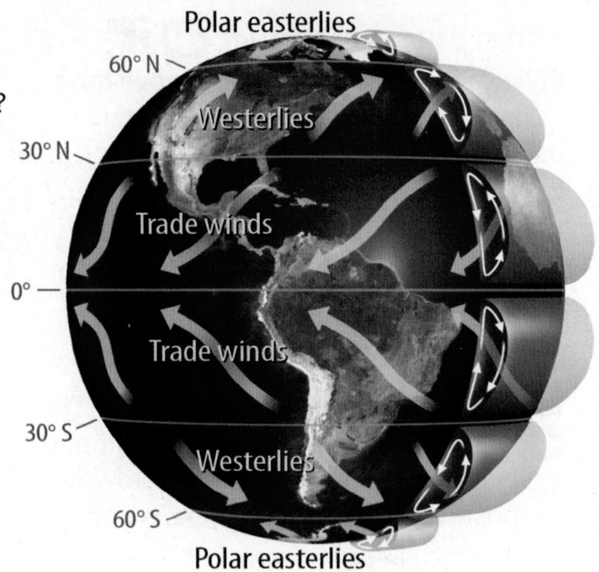

Global Wind Belts

Figure 16 shows the three-cell model of circulation in Earth's atmosphere. In the northern hemisphere, hot air in the cell nearest the equator moves to the top of the troposphere. There, the air moves northward until it cools and moves back to Earth's surface near 30° latitude. Most of the air in this convection cell then returns to Earth's surface near the equator.

The cell at the highest northern latitudes is also a convection cell. Air from the North Pole moves toward the equator along Earth's surface. The cooler air pushes up the warmer air near 60° latitude. The warmer air then moves northward and repeats the cycle. The cell between 30° and 60° latitude is not a convection cell. Its motion is driven by the other two cells, in a motion similar to a pencil that you roll between your hands. Three similar cells exist in the southern hemisphere. These cells help generate the global wind belts.

FOLDABLES®

Make a shutterfold. As illustrated, draw Earth and the three cells found in each hemisphere on the inside of the shutterfold. Describe each cell and explain the circulation of Earth's atmosphere. On the outside, label the global wind belts.

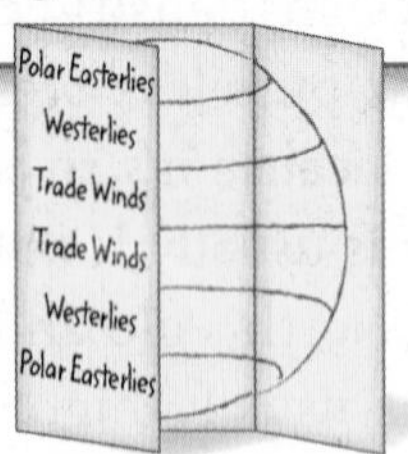

The Coriolis Effect

What happens when you throw a ball to someone across from you on a moving merry-go-round? The ball appears to curve because the person catching the ball has moved. Similarly, Earth's rotation causes moving air and water to appear to move to the right in the northern hemisphere and to the left in the southern hemisphere. This is called the Coriolis effect. The contrast between high and low pressure and the Coriolis effect creates distinct wind patterns, called prevailing winds.

Key Concept Check How are air currents on Earth affected by Earth's spin?

Prevailing Winds

The three global cells in each hemisphere create northerly and southerly winds. When the Coriolis effect acts on the winds, they blow to the east or the west, creating relatively steady, predictable winds. Locate the trade winds in **Figure 16.** *The* **trade winds** *are steady winds that flow from east to west between 30°N latitude and 30°S latitude.*

At about 30°N and 30°S air cools and sinks. This creates areas of high pressure and light, calm winds at the equator called the doldrums. Sailboats without engines can be stranded in the doldrums.

The prevailing **westerlies** *are steady winds that flow from west to east between latitudes 30°N and 60°N, and 30°S and 60°S.* This region is also shown in **Figure 16.** *The* **polar easterlies** *are cold winds that blow from the east to the west near the North Pole and the South Pole.*

 **Key Concept Check** What are the main wind belts on Earth?

Jet Streams

Near the top of the troposphere is a narrow band of high winds called the **jet stream.** Shown in **Figure 17,** jet streams flow around Earth from west to east, often making large loops to the north or the south. Jet streams influence weather as they move cold air from the poles toward the tropics and warm air from the tropics toward the poles. Jet streams can move at speeds up to 300 km/h and are more unpredictable than prevailing winds.

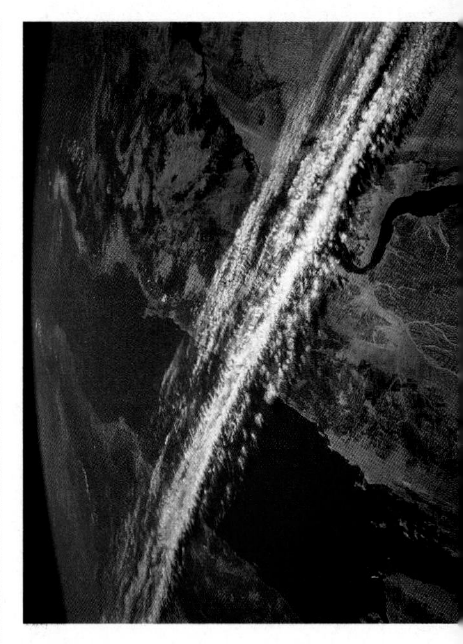

Figure 17 Jet streams are thin bands of high wind speed. The clouds seen here have condensed within a cooler jet stream.

 Personal Tutor

 **MiniLab** **20 minutes**

Can you model the Coriolis effect?

Earth's rotation causes the Coriolis effect. It affects the movement of water and air on Earth.

1. Read and complete a lab safety form.
2. Draw dot A in the center of a piece of **foamboard.** Draw dot B along the outer edge of the foamboard.
3. Roll a **table-tennis ball** from dot A to dot B. Record your observations in your Science Journal.
4. Center the foamboard on a **turntable**. Have your partner rotate the foamboard at a medium speed. Roll the ball along the same path. Record your observations.

Analyze and Conclude

1. **Contrast** the path of the ball when the foamboard was not moving to when it was spinning.
2. **Key Concept** How might air moving from the North Pole to the equator travel due to Earth's rotation?

Local Winds

You have just read that global winds occur because of pressure differences around the globe. In the same way, local winds occur whenever air pressure is different from one location to another.

Sea and Land Breezes

Anyone who has spent time near a lake or an ocean shore has probably experienced the connection between temperature, air pressure, and wind. *A* **sea breeze** *is wind that blows from the sea to the land due to local temperature and pressure differences.* **Figure 18** shows how sea breezes form. On sunny days, land warms up faster than water does. The air over the land warms by conduction and rises, creating an area of low pressure. The air over the water sinks, creating an area of high pressure because it is cooler. The differences in pressure over the warm land and the cooler water result in a cool wind that blows from the sea onto land.

A **land breeze** *is a wind that blows from the land to the sea due to local temperature and pressure differences.* **Figure 18** shows how land breezes form. At night, the land cools more quickly than the water. Therefore, the air above the land cools more quickly than the air over the water. As a result, an area of lower pressure forms over the warmer water. A land breeze then blows from the land toward the water.

Reading Check Compare and contrast sea breezes and land breezes.

Figure 18 Sea breezes and land breezes are created as part of a large reversible convection current.

Local Winds

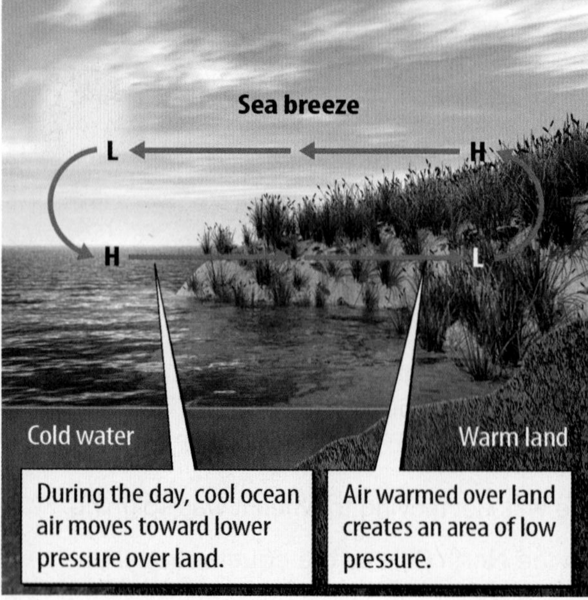

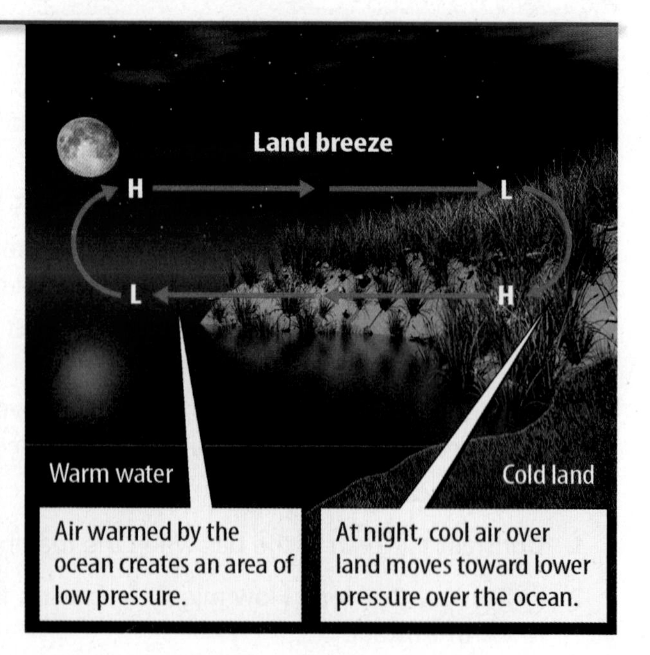

Visual Check Sequence the steps involved in the formation of a land breeze.

Lesson 3 Review

Visual Summary

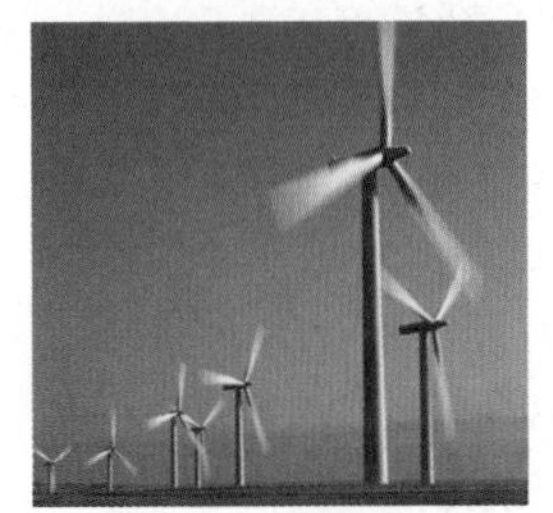

Wind is created by pressure differences between one location and another.

Prevailing winds in the global wind belts are the trade winds, the westerlies, and the polar easterlies.

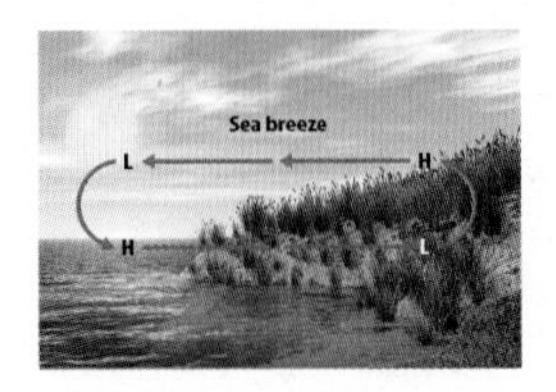

Sea breezes and land breezes are examples of local winds.

FOLDABLES®

Use your lesson Foldable to review the lesson. Save your Foldable for the project at the end of the chapter.

What do you think 

You first read the statements below at the beginning of the chapter.

5. Uneven heating in different parts of the atmosphere creates air circulation patterns.

6. Warm air sinks and cold air rises.

Did you change your mind about whether you agree or disagree with the statements? Rewrite any false statements to make them true.

Use Vocabulary

1 The movement of air from areas of high pressure to areas of low pressure is ________.

2 A(n) ________ is wind that blows from the sea to the land due to local temperature and pressure differences.

3 **Distinguish** between westerlies and trade winds.

Understand Key Concepts 

4 Which does NOT affect global wind belts?

A. air pressure
B. land breezes
C. the Coriolis effect
D. the Sun

5 **Relate** Earth's spinning motion to the Coriolis effect.

Interpret Graphics

Use the image below to answer question 6.

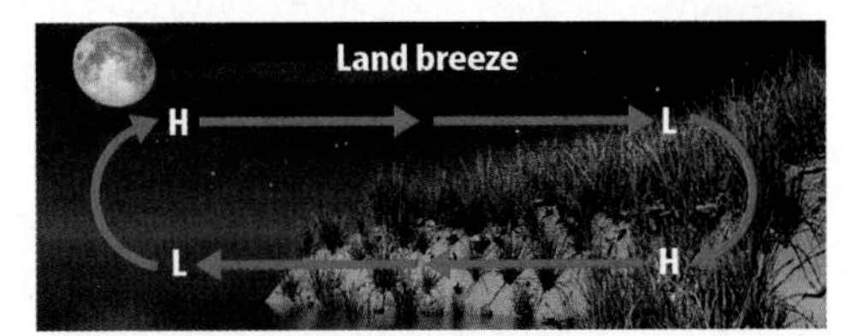

6 **Explain** a land breeze.

7 **Organize** Copy and fill in the graphic organizer below to summarize Earth's global wind belts.

Wind Belt	Description
Trade winds	
Westerlies	
Polar easterlies	

Critical Thinking

8 **Infer** what would happen without the Coriolis effect.

9 **Explain** why the wind direction is often the same in Hawaii as it is in Greenland.

Skill Practice Model

30 minutes

Can you model global wind patterns?

In each hemisphere, air circulates in specific patterns. Recall that scientists use the three-cell model to describe these circulation cells. General circulation of the atmosphere produces belts of prevailing winds around the world. In this activity, you will make a **model** of the main circulation cells in Earth's atmosphere.

Materials

ribbons

globe

permanent marker

scissors

transparent tape

Safety

Learn It

Making a **model** can help you visualize how a process works. Scientists use models to represent processes that may be difficult to see in real time. Sometimes a model represents something too small to see with the unaided eye, such as a model of an atom. Other models, such as one of the solar system, represent something that is too large to see from one location.

Try It

1. Read and complete a lab safety form.
2. Refer to **Figure 16** to make your model.
3. Choose one color of ribbon for the circulation cells. Make a separate loop of ribbon long enough to cover the latitude boundaries of each cell. Draw arrows on each ribbon to show the direction that the air flows in that cell. Make one loop for each cell in the northern hemisphere and one for each in the southern hemisphere. Tape your "cells" onto the globe.

4. Choose different-colored ribbons to model each of these wind belts: trade winds, westerlies, and polar easterlies, in both hemispheres. Draw arrows on each ribbon to show the direction that the wind blows. Tape the ribbons on the globe.
5. Create a color key to identify each cell and its corresponding wind type.

Apply It

6. Explain how your model represents the three-cell model used by scientists. How does your model differ from actual air movement in the atmosphere?
7. Explain why you cannot accurately model the global winds with this model.
8. **Key Concept** Explain how latitude affects global winds.

Lesson 4

Reading Guide

Key Concepts
ESSENTIAL QUESTIONS

- How do humans impact air quality?
- Why do humans monitor air quality standards?

Vocabulary

air pollution p. 434
acid precipitation p. 435
photochemical smog p. 435
particulate matter p. 436

Multilingual eGlossary

BrainPOP®

Air Quality

Inquiry How did this happen?

Air pollution can be trapped near Earth's surface during a temperature inversion. This is especially common in cities located in valleys and surrounded by mountains. What do you think the quality of the air is like on a day like this one? Where does pollution come from?

Launch Lab

20 minutes

How does acid rain form?

Vehicles, factories, and power plants release chemicals into the atmosphere. When these chemicals combine with water vapor, they can form acid rain.

1. Read and complete a lab safety form.
2. Half-fill a **plastic cup** with **distilled water.**
3. Dip a strip of **pH paper** into the water. Use a **pH color chart** to determine the pH of the distilled water. Record the pH in your Science Journal.
4. Use a **dropper** to add **lemon juice** to the water until the pH equals that of acid rain. Swirl and test the pH each time you add 5 drops of the lemon juice to the mixture.

Substances	pH
Hydrochloric acid	0.0
Lemon juice	2.3
Vinegar	2.9
Tomato juice	4.1
Coffee (black)	5.0
Acid rain	5.6
Rainwater	6.5
Milk	6.6
Distilled water	7.0
Blood	7.4
Baking soda solution	8.4
Toothpaste	9.9
Household ammonia	11.9
Sodium hydroxide	14.0

Think About This

1. A strong acid has a pH between 0 and 2. How does the pH of lemon juice compare to the pH of other substances? Is acid rain a strong acid?
2. **Key Concept** Why might scientists monitor the pH of rain?

Sources of Air Pollution

The contamination of air by harmful substances including gases and smoke is called **air pollution.** Air pollution is harmful to humans and other living things. Years of exposure to polluted air can weaken a human's immune system. Respiratory diseases such as asthma can be caused by air pollution.

Figure 19 One example of point-source pollution is a factory smoke stack.

Air pollution comes from many sources. Point-source pollution is pollution that comes from an identifiable source. Examples of point sources include smokestacks of large factories, such as the one shown in **Figure 19,** and electric power plants that burn fossil fuels. They release tons of polluting gases and particles into the air each day. An example of natural point-source pollution is an erupting volcano.

Nonpoint-source pollution is pollution that comes from a widespread area. One example of pollution from a nonpoint-source is air pollution in a large city. This is considered nonpoint-source pollution because it cannot be traced back to one source. Some bacteria found in swamps and marshes are examples of natural sources of nonpoint-source pollution.

Key Concept Check Compare point-source and nonpoint-source pollution.

Causes and Effects of Air Pollution

The harmful effects of air pollution are not limited to human health. Some pollutants, including ground-level ozone, can damage plants. Air pollution can also cause serious damage to human-made structures. Sulfur dioxide pollution can discolor stone, corrode metal, and damage paint on cars.

Acid Precipitation

When sulfur dioxide and nitrogen oxides combine with moisture in the atmosphere and form precipitation that has a pH lower than that of normal rainwater, it is called **acid precipitation.** Acid precipitation includes acid rain, snow, and fog. It affects the chemistry of water in lakes and rivers. This can harm the organisms living in the water. Acid precipitation damages buildings and other structures made of stone. Natural sources of sulfur dioxide include volcanoes and marshes. However, the most common sources of sulfur dioxide and nitrogen oxides are automobile exhausts and factory and power plant smoke.

Smog

Photochemical smog *is air pollution that forms from the interaction between chemicals in the air and sunlight.* Smog forms when nitrogen dioxide, released in gasoline engine exhaust, reacts with sunlight. A series of chemical reactions produces ozone and other compounds that form smog. Recall that ozone in the stratosphere helps protect organisms from the Sun's harmful rays. However, ground-level ozone can damage the tissues of plants and animals. Ground-level ozone is the main component of smog. Smog in urban areas reduces visibility and makes air difficult to breathe. **Figure 20** shows New York City on a clear day and on a smoggy day.

Key Concept Check How do humans impact air quality?

FOLDABLES®

Make a horizontal three-tab Foldable and label it as shown. Use it to organize your notes about the formation of air pollution and its effects. Fold the right and left thirds over the center and label the outside *Types of Air Pollution.*

Acid Precipitation | Smog | Particulate Pollution

Types of Air Pollution

Figure 20 Smog can be observed as haze or a brown tint in the atmosphere.

MICHAEL S. YAMASHITA/National Geographic Image Collection

Particulate Pollution

Word Origin

particulate
from Latin *particula,* means "small part"

Although you can't see them, over 10,000 solid or liquid particles are in every cubic centimeter of air. A cubic centimeter is about the size of a sugar cube. This type of pollution is called particulate matter. **Particulate** (par TIH kyuh lut) **matter** *is a mixture of dust, acids, and other chemicals that can be hazardous to human health.* The smallest particles are the most harmful. These particles can be inhaled and can enter your lungs. They can cause asthma, bronchitis, and lead to heart attacks. Children and older adults are most likely to experience health problems due to particulate matter.

Particulate matter in the atmosphere absorbs and scatters sunlight. This can create haze. Haze particles scatter light, make things blurry, and reduce visibility.

Movement of Air Pollution

Wind can influence the effects of air pollution. Because air carries pollution with it, some wind patterns cause more pollution problems than others. Weak winds or no wind prevents pollution from mixing with the surrounding air. During weak wind conditions, pollution levels can become dangerous.

For example, the conditions in which temperature inversions form are weak winds, clear skies, and longer winter nights. As land cools at night, the air above it also cools. Calm winds, however, prevent cool air from mixing with warm air above it. **Figure 21** shows how cities located in valleys experience a temperature inversion. Cool air, along with the pollution it contains, is trapped in valleys. More cool air sinks down the sides of the mountain, further preventing layers from mixing. The pollution in the photo at the beginning of the lesson was trapped due to a temperature inversion.

Figure 21 At night, cool air sinks down the mountain sides, trapping pollution in the valley below.

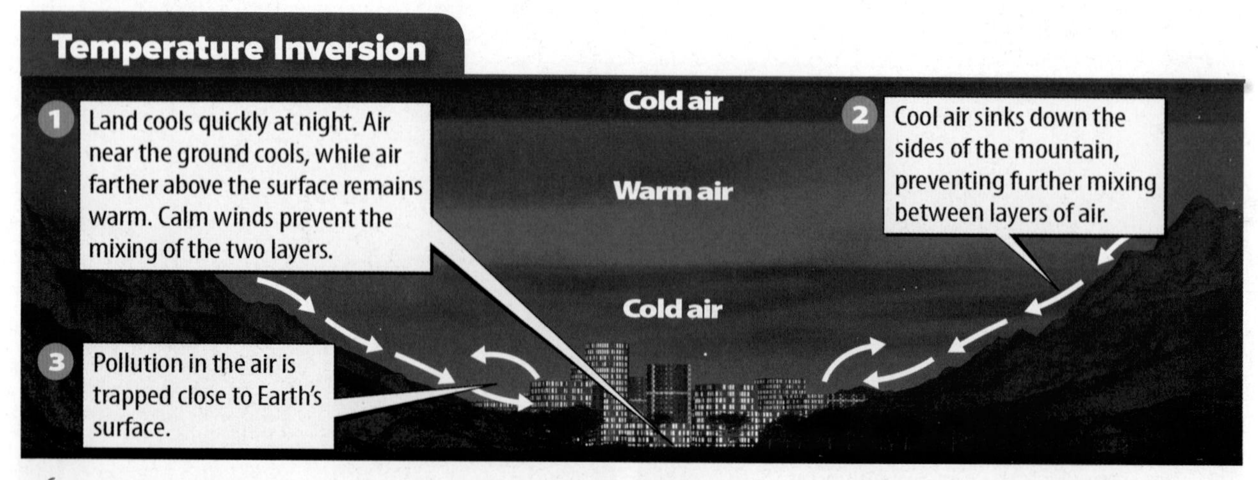

Visual Check How is pollution trapped by a temperature inversion?

Maintaining Healthful Air Quality

Preserving the quality of Earth's atmosphere requires the cooperation of government officials, scientists, and the public. The Clean Air Act is an example of how government can help fight pollution. Since the Clean Air Act became law in 1970, steps have been taken to reduce automobile emissions. Pollutant levels have decreased significantly in the United States. Despite these advances, serious problems still remain. The amount of ground-level ozone is still too high in many large cities. Also, acid precipitation produced by air pollutants continues to harm organisms in lakes, streams, and forests.

Air Quality Standards

The Clean Air Act gives the U.S. government the power to set air quality standards. The standards protect humans, animals, plants, and buildings from the harmful effects of air pollution. All states are required to make sure that pollutants, such as carbon monoxide, nitrogen oxides, particulate matter, ozone, and sulfur dioxide, do not exceed harmful levels.

Reading Check What is the Clean Air Act?

Monitoring Air Pollution

Pollution levels are continuously monitored by hundreds of instruments in all major U.S. cities. If the levels are too high, authorities may advise people to limit outdoor activities.

MiniLab

15 minutes

Can being out in fresh air be harmful to your health?

Are you going to be affected if you play tennis for a couple hours, go biking with your friends, or even just lie on the beach? Even if you have no health problems related to your respiratory system, you still need to be aware of the quality of air in your area of activity for the day.

Air Quality Index (AQI) Values	Levels of Health Concern
0 to 50	Good
51 to 100	Moderate
101 to 150	Unhealthful for Sensitive Groups
151 to 200	Unhealthful
201 to 300	Very Unhealthful
301 to 500	Hazardous

Analyze and Conclude

1. Which values on the AQI indicate that the air quality is good?
2. At what value is the air quality unhealthful for anyone who may have allergies and respiratory disorders?
3. Which values would be considered as warnings of emergency conditions?
4. **Key Concept** The quality of air in different areas changes throughout the day. Explain how you can use the AQI to help you know when you should limit your outdoor activity.

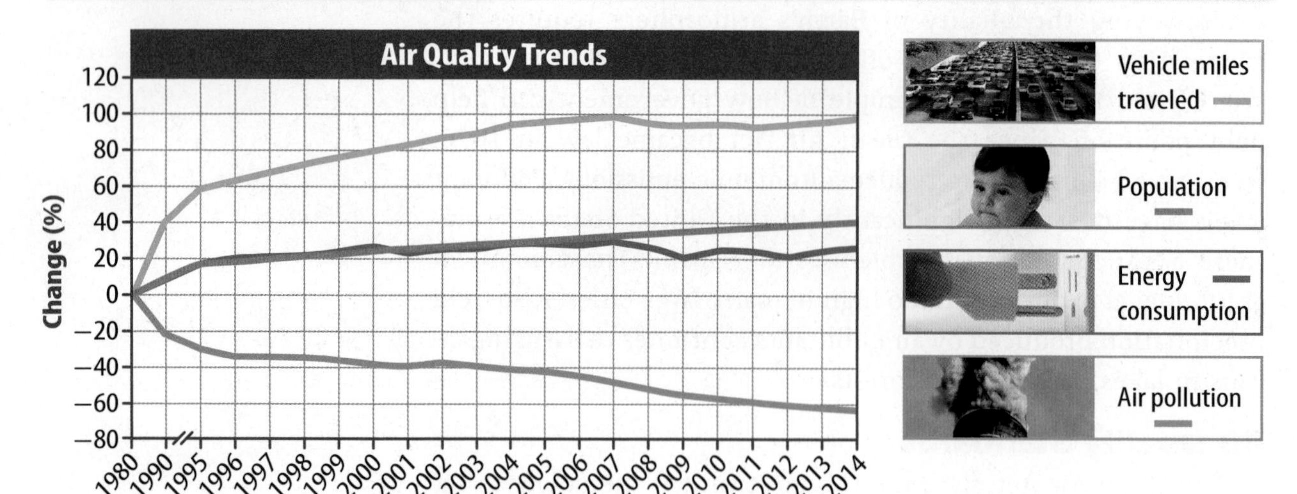

Figure 22 Pollution emissions have declined, even though the population is increasing.

Air Quality Trends

Over the last several decades, air quality in U.S. cities has improved, as shown in **Figure 22.** Even though some pollution-producing processes have increased, such as burning fossil fuels and traveling in automobiles, levels of certain air pollutants have decreased. Airborne levels of lead and carbon monoxide have decreased the most. Levels of sulfur dioxide, nitrogen oxide, and particulate matter have also decreased.

However, ground-level ozone has not decreased much. Why do ground-level ozone trends lag behind those of other pollutants? Recall that ozone can be created from chemical reactions involving automobile exhaust. The increase in the amount of ground-level ozone is because of the increase in the number of miles traveled by vehicles.

Key Concept Check Why do humans monitor air quality standards?

Indoor Air Pollution

Not all air pollution is outdoors. The air inside homes and other buildings can be as much as 50 times more polluted than outdoor air! The quality of indoor air can impact human health much more than outdoor air quality.

Indoor air pollution comes from many sources. Tobacco smoke, cleaning products, pesticides, and fireplaces are some common sources. Furniture upholstery, carpets, and foam insulation also add pollutants to the air. Another indoor air pollutant is radon, an odorless gas given off by some soil and rocks. Radon leaks through cracks in a building's foundation and sometimes builds up to harmful levels inside homes. Harmful effects of radon come from breathing its particles.

Math Skills

Use Graphs

The graph above shows the percent change in four different pollution factors from 1980 through 2014. All values are based on the 0 percent amount in 1980. For example, from 1980 to 1990, the number of vehicle miles driven increased by 40 percent. Use the graph to infer which factors might be related.

Practice

1. What was the percent change in population between 1980 and 2014?
2. What other factor changed by about the same amount during that period?

 **Math Practice**

 **Personal Tutor**

Lesson 4 Review

Visual Summary

Air pollution comes from point sources, such as factories, and nonpoint sources, such as automobiles.

Photochemical smog contains ozone, which can damage tissues in plants and animals.

FOLDABLES®

Use your lesson Foldable to review the lesson. Save your Foldable for the project at the end of the chapter.

What do you think NOW?

You first read the statements below at the beginning of the chapter.

7. If no humans lived on Earth, there would be no air pollution.

8. Pollution levels in the air are not measured or monitored.

Did you change your mind about whether you agree or disagree with the statements? Rewrite any false statements to make them true.

Use Vocabulary

1. **Define** *acid precipitation* in your own words.
2. ________ forms when chemical reactions combine pollution with sunlight.
3. The contamination of air by harmful substances, including gases and smoke, is ________.

Understand Key Concepts

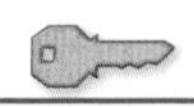

4. Which is NOT true about smog?
   - **A.** It contains nitrogen oxide.
   - **B.** It contains ozone.
   - **C.** It reduces visibility.
   - **D.** It is produced only by cars.
5. **Describe** two ways humans add pollution to the atmosphere.
6. **Assess** whether urban or rural areas are more likely to have high levels of smog.
7. **Identify** and describe the law designed to reduce air pollution.

Interpret Graphics

8. **Compare and Contrast** Copy and fill in the graphic organizer below to compare and contrast details of smog and acid precipitation.

	Similarities	Differences
Smog		
Acid Precipitation		

Critical Thinking

9. **Describe** how conduction and convection are affected by paving over a grass field.

Math Skills

Math Practice

10. Based on the graph on the opposite page, what was the total percent change in air pollution between 1980 and 2014?

40 minutes

Radiant Energy Absorption

Materials

thermometer

sand

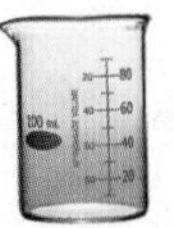

500-mL beaker

lamp

stopwatch

paper towels

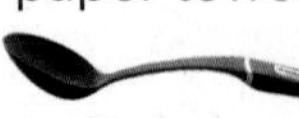

spoon

potting soil

clay

Safety

Ultimately, the Sun is the source of energy for Earth. Energy from the Sun moves through the atmosphere and is absorbed and reflected from different surfaces on Earth. Light surfaces reflect energy, and dark surfaces absorb energy. Both land and sea surfaces absorb energy from the Sun, and air in contact with these surfaces is warmed through conduction.

Ask a Question

Which surfaces on Earth absorb the most energy from the Sun?

Make Observations

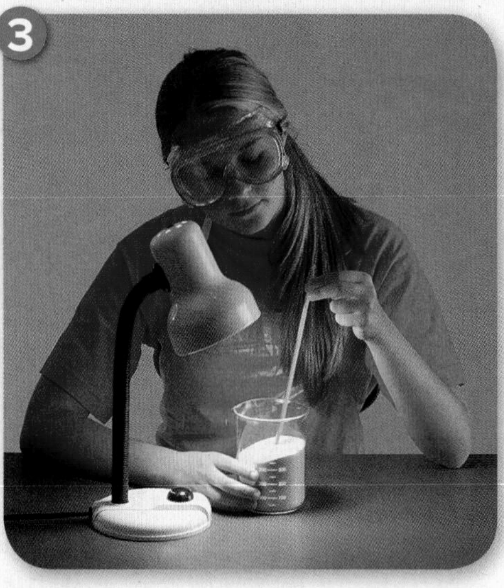

1. Read and complete a lab safety form.
2. Make a data table in your Science Journal to record your observations of energy transfer. Include columns for Type of Surface, Temperature Before Heating, and Temperature After Heating.
3. Half-fill a 500-mL beaker with sand. Place a thermometer in the sand and carefully add enough sand to cover the thermometer bulb—about 2 cm deep. Keep the bulb under the sand for 1 minute. Record the temperature in the data table.

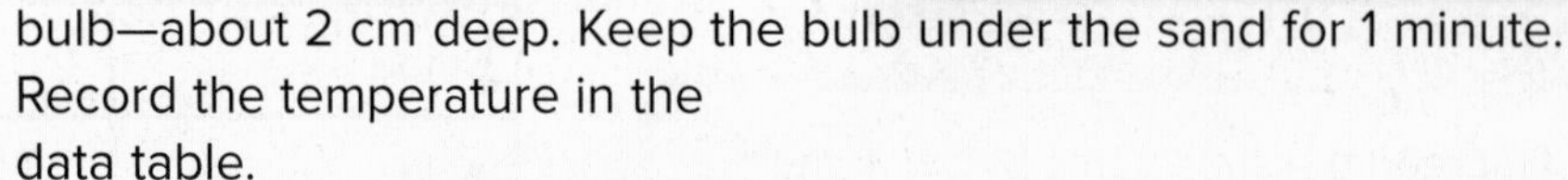

4. Place the beaker under the light source. Record the temperature after 10 minutes.
5. Repeat steps 3 and 4 using soil and water.

Form a Hypothesis

6. Use the data in your table to form a hypothesis stating which surfaces on Earth, such as forests, wheat fields, lakes, snowy mountain tops, and deserts, will absorb the most radiant energy.

Test Your Hypothesis

7. Decide what materials could be used to mimic the surfaces on Earth from your hypothesis.
8. Repeat the experiment with materials approved by the teacher to test your hypothesis.
9. Examine your data. Was your hypothesis supported? Why or why not?

Analyze and Conclude

10. **Infer** which types of areas on Earth absorb the most energy from the Sun.
11. **Think Critically** When areas of Earth are changed so they become more likely to reflect or absorb energy from the Sun, how might these changes affect conduction and convection in the atmosphere?
12. **The Big Idea** Explain how thermal energy from the Sun being received by and reflected from Earth's surface is related to the role of the atmosphere in maintaining conditions suitable for life.

Communicate Your Results

Display data from your initial observations to compare your findings with your classmates' findings. Explain your hypothesis, experiment results, and conclusions to the class.

Inquiry Extension

What could you add to this investigation to show how cloud cover changes the amount of radiation that will reach Earth's surfaces? Design a study that could test the effect of cloud cover on radiation passing through Earth's atmosphere. How could you include a way to show that clouds also reflect radiant energy from the Sun?

Lab Tips

☑ If possible, use leaves, straw, shaved ice, and other natural materials to test your hypothesis.

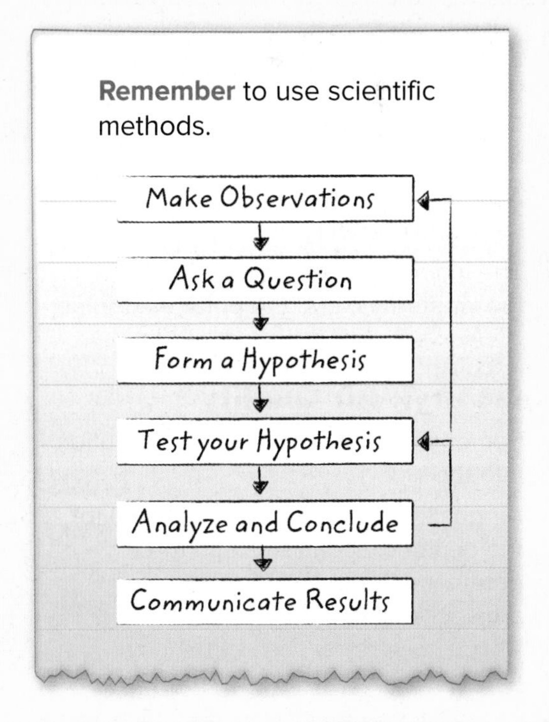

Chapter 12 Study Guide

The gases in Earth's atmosphere, some of which are needed by organisms to survive, affect Earth's temperature and the transfer of thermal energy to the atmosphere.

Key Concepts Summary

Lesson 1: Describing Earth's Atmosphere

- Earth's **atmosphere** formed as Earth cooled and chemical and biological processes took place.
- Earth's atmosphere consists of nitrogen, oxygen, and a small amount of other gases, such as CO_2 and **water vapor.**
- The atmospheric layers are the **troposphere,** the **stratosphere,** the mesosphere, the thermosphere, and the exosphere.
- Air pressure decreases as altitude increases. Temperature either increases or decreases as altitude increases, depending on the layer of the atmosphere.

21% Oxygen
78% Nitrogen

Vocabulary

atmosphere p. 409
water vapor p. 410
troposphere p. 412
stratosphere p. 412
ozone layer p. 412
ionosphere p. 413

Lesson 2: Energy Transfer in the Atmosphere

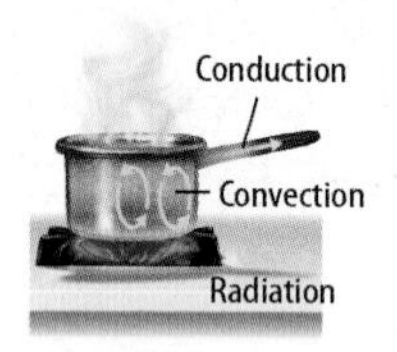

- The Sun's energy is transferred to Earth's surface and the atmosphere through **radiation, conduction, convection,** and latent heat.
- Air circulation patterns are created by convection currents.

Vocabulary

radiation p. 418
conduction p. 421
convection p. 421
stability p. 422
temperature inversion p. 423

Lesson 3: Air Currents

- Uneven heating of Earth's surface creates pressure differences. **Wind** is the movement of air from areas of high pressure to areas of low pressure.
- Air currents curve to the right or to the left due to the Coriolis effect.
- The main wind belts on Earth are the **trade winds,** the **westerlies,** and the **polar easterlies.**

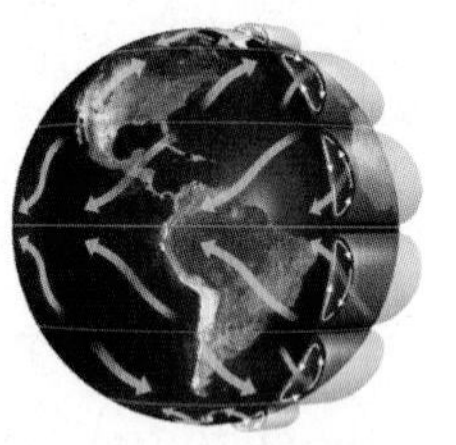

Vocabulary

wind p. 427
trade winds p. 429
westerlies p. 429
polar easterlies p. 429
jet stream p. 429
sea breeze p. 430
land breeze p. 430

Lesson 4: Air Quality

- Some human activities release pollution into the air.
- Air quality standards are monitored for the health of organisms and to determine if anti-pollution efforts are successful.

Vocabulary

air pollution p. 434
acid precipitation p. 435
photochemical smog p. 435
particulate matter p. 436

Study Guide

Personal Tutor

Vocabulary eFlashcards
Vocabulary eGames

FOLDABLES® Chapter Project

Assemble your lesson Foldables® as shown to make a Chapter Project. Use the project to review what you have learned in this chapter.

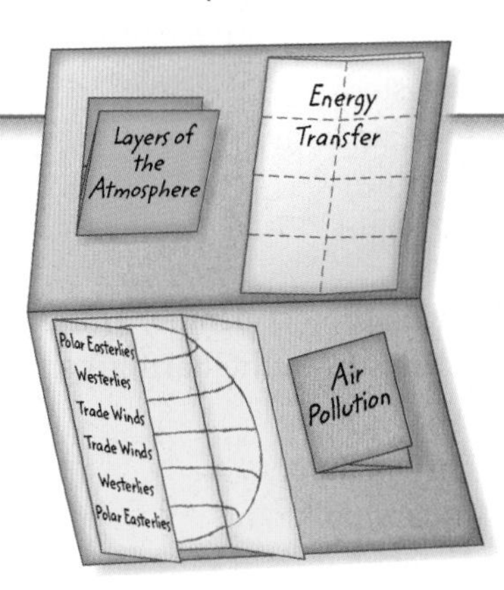

Use Vocabulary

1. Radio waves travel long distances by bouncing off electrically charged particles in the ________.
2. The Sun's thermal energy is transferred to Earth through space by ________.
3. Rising currents of warm air transfer energy from Earth to the atmosphere through ________.
4. A narrow band of winds located near the top of the troposphere is a(n) ________.
5. ________ are steady winds that flow from east to west between 30°N latitude and 30°S latitude.
6. In large urban areas, ________ forms when pollutants in the air interact with sunlight.
7. A mixture of dust, acids, and other chemicals that can be hazardous to human health is called ________.

Link Vocabulary and Key Concepts

Interactive Concept Map

Copy this concept map, and then use vocabulary terms from the previous page to complete the concept map.

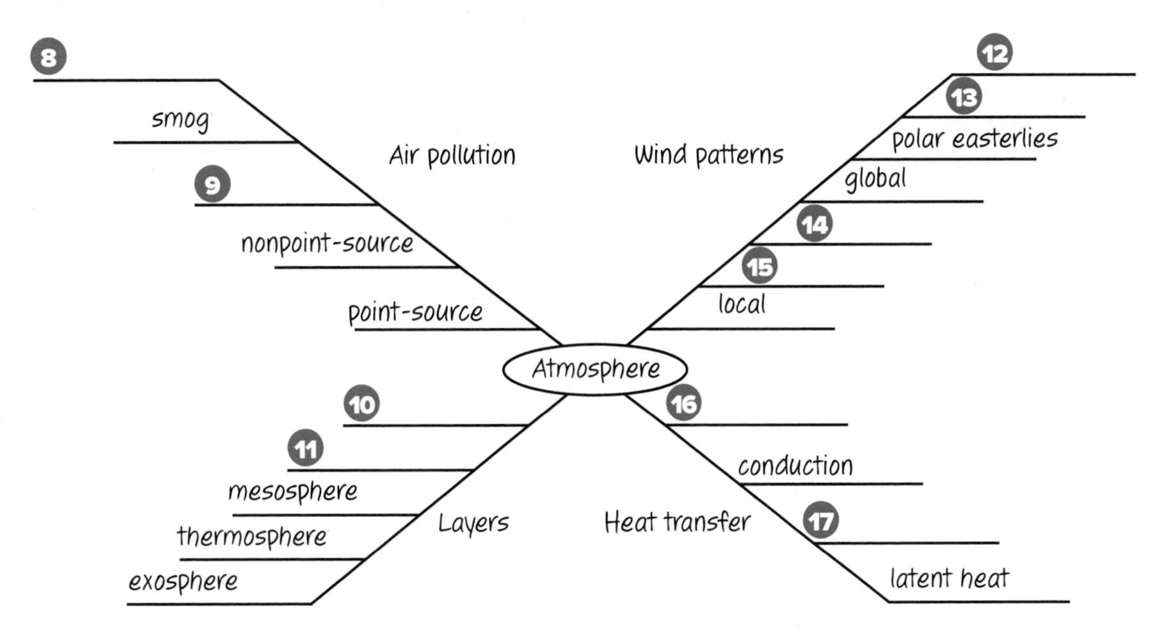

Chapter 12 Review

Understand Key Concepts 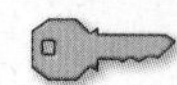

1. Air pressure is greatest
   - **A.** at a mountain base.
   - **B.** on a mountain top.
   - **C.** in the stratosphere.
   - **D.** in the ionosphere.

2. In which layer of the atmosphere is the ozone layer found?
   - **A.** troposphere
   - **B.** stratosphere
   - **C.** mesosphere
   - **D.** thermosphere

Use the image below to answer question 3.

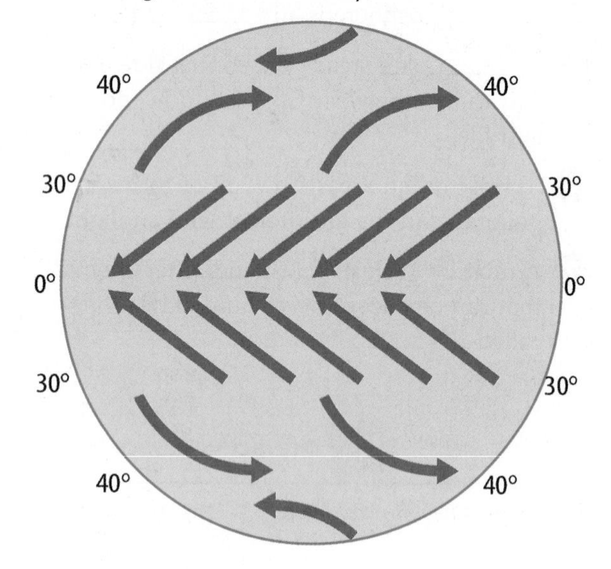

3. This diagram represents the atmosphere's
   - **A.** air masses.
   - **B.** global wind belts.
   - **C.** inversions.
   - **D.** particulate motion.

4. The Sun's energy
   - **A.** is completely absorbed by the atmosphere.
   - **B.** is completely reflected by the atmosphere.
   - **C.** is in the form of latent heat.
   - **D.** is transferred to the atmosphere after warming Earth.

5. Which type of energy is emitted from Earth to the atmosphere?
   - **A.** ultraviolet radiation
   - **B.** visible radiation
   - **C.** infrared radiation
   - **D.** aurora borealis

6. Which is a narrow band of high winds located near the top of the troposphere?
   - **A.** polar easterly
   - **B.** a jet stream
   - **C.** a sea breeze
   - **D.** a trade wind

7. Which helps protect people, animals, plants, and buildings from the harmful effects of air pollution?
   - **A.** primary pollutants
   - **B.** secondary pollutants
   - **C.** ozone layer
   - **D.** air quality standards

Use the photo below to answer question 8.

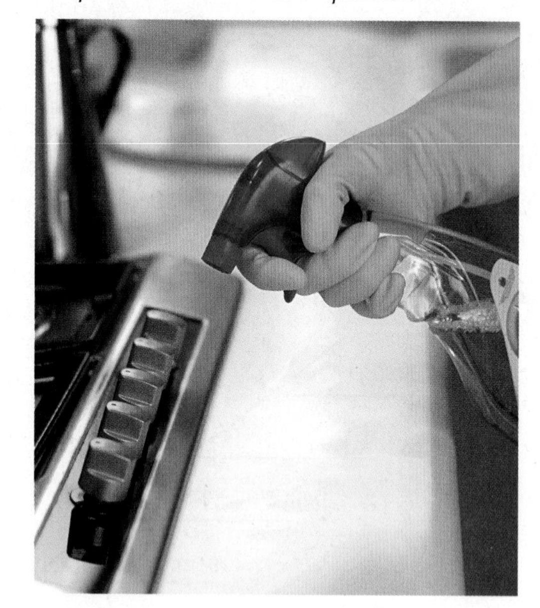

8. This photo shows a potential source of
   - **A.** ultraviolet radiation.
   - **B.** indoor air pollution.
   - **C.** radon.
   - **D.** smog.

supershoot/Alamy

Critical Thinking

9. **Predict** how atmospheric carbon dioxide levels might change if more trees were planted on Earth. Explain your prediction.

10. **Compare** visible and infrared radiation.

11. **Assess** whether your home is heated by conduction or convection.

12. **Sequence** how the unequal heating of Earth's surface leads to the formation of wind.

13. **Evaluate** whether a sea breeze could occur at night.

14. **Interpret Graphics** What were the top three sources of particulate matter in the atmosphere in 2002? What could you do to reduce particulate matter from any of the sources shown here?

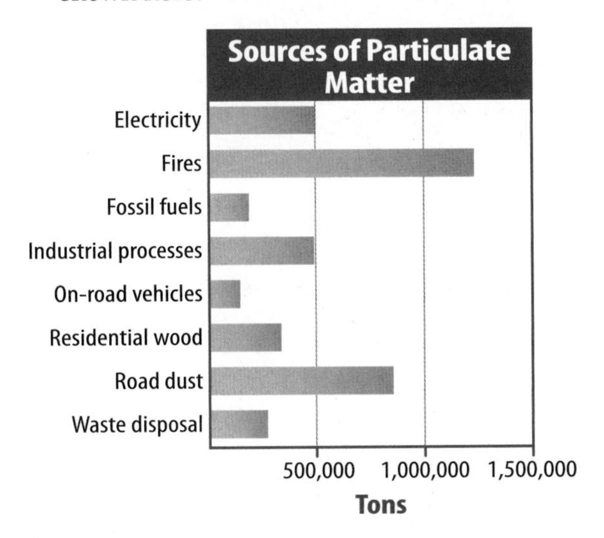

15. **Diagram** how acid precipitation forms. Include possible sources of sulfur dioxide and nitrogen oxide and organisms that can be affected by acid precipitation.

Writing in Science

16. **Write** a paragraph explaining whether you think it would be possible to permanently pollute the atmosphere with particulate matter.

REVIEW THE BIG IDEA

17. Review the title of each lesson in the chapter. List all of the characteristics and components of the troposphere and the stratosphere that affect life on Earth. Describe how life is impacted by each one.

18. Discuss how energy is transferred from the Sun throughout Earth's atmosphere.

Math Skills

Math Practice

Use Graphs

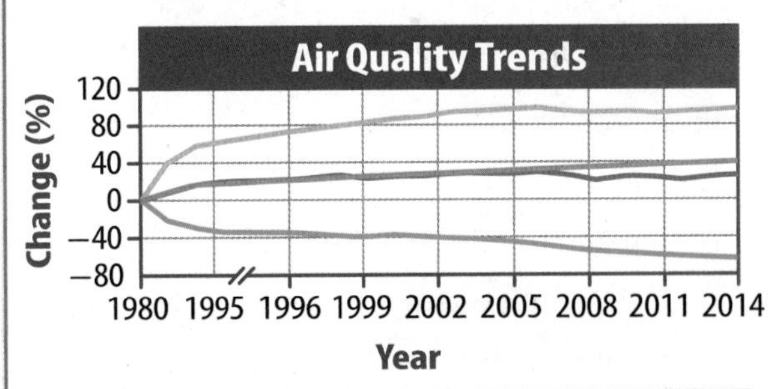

19. What was the percent change in energy use between 2000 and 2007?

20. What happened to energy use between 2007 and 2009?

21. What was the total percentage change between vehicle miles traveled and air pollution from 1980 to 2014?

Standardized Test Practice

Record your answers on the answer sheet provided by your teacher or on a sheet of paper.

Multiple Choice

1 What causes the phenomenon known as a mountain wave?

A radiation imbalance

B rising and sinking air

C temperature inversion

D the greenhouse effect

Use the diagram below to answer question 2.

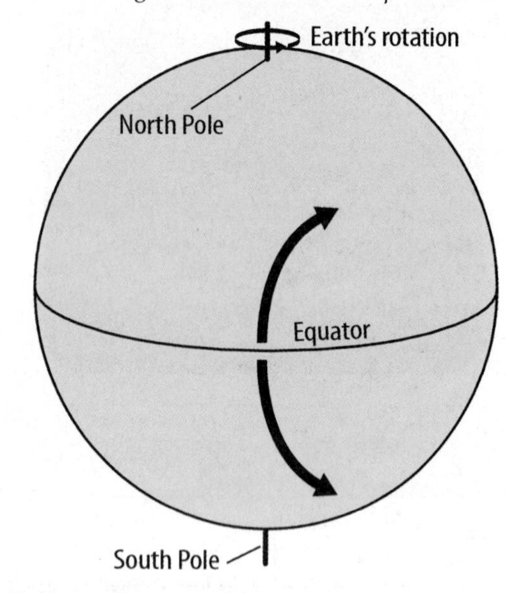

2 What phenomenon does the diagram above illustrate?

A radiation balance

B temperature inversion

C the Coriolis effect

D the greenhouse effect

3 Which do scientists call greenhouse gases?

A carbon dioxide, hydrogen, nitrogen

B carbon dioxide, methane gas, water vapor

C carbon monoxide, oxygen, argon

D carbon monoxide, ozone, radon

4 In which direction does moving air appear to turn in the northern hemisphere?

A down

B up

C right

D left

Use the diagram below to answer question 5.

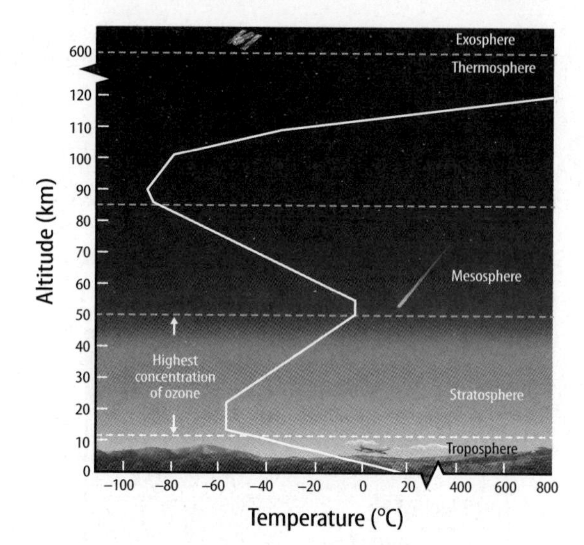

5 Which layer of the atmosphere has the widest range of temperatures?

A mesosphere

B stratosphere

C thermosphere

D troposphere

6 Which was the main component of Earth's original atmosphere?

A carbon dioxide

B nitrogen

C oxygen

D water vapor

7 Which is the primary cause of the global wind patterns on Earth?

A ice cap melting

B uneven heating

C weather changing

D waves breaking

Use the diagram below to answer question 8.

Energy Transfer Methods

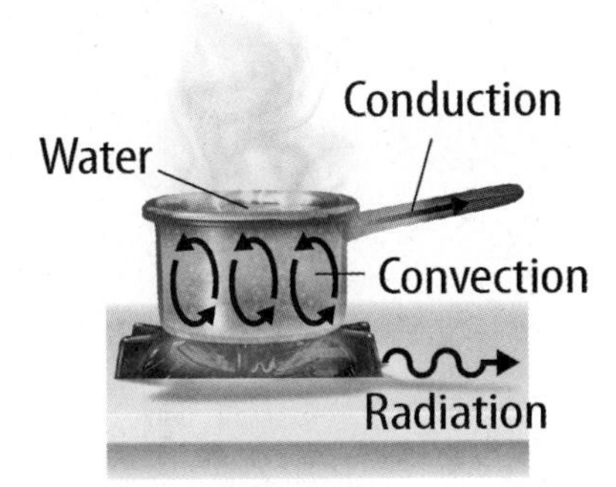

8 In the diagram above, which transfers thermal energy in the same way the Sun's energy is transferred to Earth?

A the boiling water

B the burner flame

C the hot handle

D the rising steam

9 Which substance in the air of U.S. cities has decreased least since the Clean Air Act began?

A carbon monoxide

B ground-level ozone

C particulate matter

D sulfur dioxide

Constructed Response

Use the table below to answer questions 10 and 11.

Layer	Significant Fact

10 In the table above, list in order the layers of Earth's atmosphere from lowest to highest. Provide one significant fact about each layer.

11 Explain how the first four atmospheric layers are important to life on Earth.

Use the table below to answer question 12.

Heat Transfer	Explanation
Conduction	
Convection	
Latent heat	
Radiation	

12 Complete the table to explain how heat energy transfers from the Sun to Earth and its atmosphere.

13 What are temperature inversions? How do they form? What is the relationship between temperature inversions and air pollution?

NEED EXTRA HELP?													
If You Missed Question...	1	2	3	4	5	6	7	8	9	10	11	12	13
Go to Lesson...	2	3	2	3	1	1	3	2	4	1	1	2	2, 4

Chapter 13

Weather

THE BIG IDEA **How do scientists describe and predict weather?**

Inquiry Is this a record snowfall?

Buffalo, New York, is famous for its snowstorms, averaging 3 m of snow each year. Other areas of the world might only get a few centimeters of snow a year. In some parts of the world, it never snows.

- Why do some areas get less snow than others?
- How do scientists describe and predict weather?

George Frey/Getty Images

Get Ready to Read

What do you think?

Before you read, decide if you agree or disagree with each of these statements. As you read this chapter, see if you change your mind about any of the statements.

1. Weather is the long-term average of atmospheric patterns of an area.
2. All clouds are at the same altitude within the atmosphere.
3. Precipitation often occurs at the boundaries of large air masses.
4. There are no safety precautions for severe weather, such as tornadoes and hurricanes.
5. Weather variables are measured every day at locations around the world.
6. Modern weather forecasts are done using computers.

Your one-stop online resource
connectED.mcgraw-hill.com

LearnSmart®

Chapter Resources Files, Reading Essentials, Get Ready to Read, Quick Vocabulary

Animations, Videos, Interactive Tables

Self-checks, Quizzes, Tests

Project-Based Learning Activities

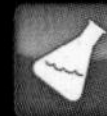
Lab Manuals, Safety Videos, Virtual Labs & Other Tools

Vocabulary, Multilingual eGlossary, Vocab eGames, Vocab eFlashcards

Personal Tutors

Lesson 1

Describing Weather

Reading Guide

Key Concepts
ESSENTIAL QUESTIONS

- What is weather?
- What variables are used to describe weather?
- How is weather related to the water cycle?

Vocabulary

weather p. 451
air pressure p. 452
humidity p. 452
relative humidity p. 453
dew point p. 453
precipitation p. 455
water cycle p. 455

Multilingual eGlossary

BrainPOP® Science Video

Inquiry Why are clouds different?

If you look closely at the photo, you'll see that there are different types of clouds in the sky. How do clouds form? If all clouds consist of water droplets and ice crystals, why do they look different? Are clouds weather?

Peter de Clercq/Alamy

Launch Lab

15 minutes

Can you make clouds in a bag?

When water vapor in the atmosphere cools, it condenses. The resulting water droplets make up clouds.

1. Read and complete a lab safety form.
2. Half-fill a **500-mL beaker** with **ice** and **cold water.**
3. Pour 125 mL of **warm water** into a **resealable plastic bag** and seal the bag.
4. Carefully lower the bag into the ice water. Record your observations in your Science Journal.

Think About This

1. What did you observe when the warm water in the bag was put into the beaker?
2. What explanation can you give for what happened?
3. **Key Concept** What could you see in the natural world that results from the same process?

What is weather?

Everybody talks about the weather. "Nice day, isn't it?" "How was the weather during your vacation?" Talking about weather is so common that we even use weather terms to describe unrelated topics. "That homework assignment was a breeze." Or "I'll take a rain check."

Weather *is the atmospheric conditions, along with short-term changes, of a certain place at a certain time.* If you have ever been caught in a rainstorm on what began as a sunny day, you know the weather can change quickly. Sometimes it changes in just a few hours. But other times your area might have the same sunny weather for several days in a row.

Weather Variables

Perhaps some of the first things that come to mind when you think about weather are temperature and rainfall. As you dress in the morning, you need to know what the temperature will be throughout the day to help you decide what to wear. If it is raining, you might cancel your picnic.

Temperature and rainfall are just two of the variables used to describe weather. Meteorologists, scientists who study and predict weather, use several specific variables that describe a variety of atmospheric conditions. These variables include air temperature, air pressure, wind speed and direction, humidity, cloud coverage, and precipitation.

Key Concept Check What is weather?

REVIEW VOCABULARY
variable
a quantity that can change

REVIEW VOCABULARY

kinetic energy
energy an object has due to its motion

Air Temperature

The measure of the average **kinetic energy** of molecules in the air is air temperature. When the temperature is high, molecules have a high kinetic energy. Therefore, molecules in warm air move faster than molecules in cold air. Air temperatures vary with time of day, season, location, and altitude.

Air Pressure

The force that a column of air applies on the air or a surface below it is called **air pressure.** Study **Figure 1.** Is air pressure at Earth's surface more or less than air pressure at the top of the atmosphere? Air pressure decreases as altitude increases. Therefore, air pressure is greater at low altitudes than at high altitudes.

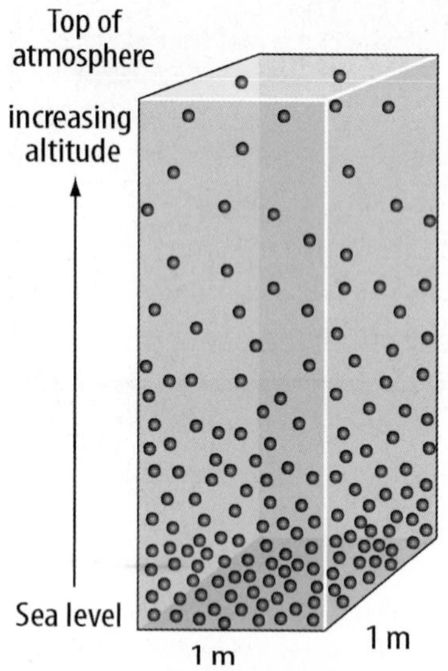

Figure 1 Increasing air pressure comes from having more molecules overhead.

Visual Check What happens to air pressure as altitude decreases?

You might have heard the term *barometric pressure* during a weather forecast. Barometric pressure refers to air pressure. Air pressure is measured with an instrument called a barometer, shown in **Figure 2.** Air pressure is typically measured in millibars (mb). Knowing the barometric pressure of different areas helps meteorologists predict the weather.

Reading Check What instrument measures air pressure?

Wind

As air moves from areas of high pressure to areas of low pressure, it creates wind. Wind direction is the direction from which the wind is blowing. For example, winds that blow from west to east are called westerlies. Meteorologists measure wind speed using an instrument called an anemometer (a nuh MAH muh tur). An anemometer is also shown in **Figure 2.**

Humidity

The amount of water vapor in the air is called **humidity** (hyew MIH duh tee). Humidity can be measured in grams of water per cubic meter of air (g/m^3). When the humidity is high, there is more water vapor in the air. On a day with high humidity, your skin might feel sticky, and sweat might not evaporate from your skin as quickly.

Figure 2 Barometers, left, and anemometers, right, are used to measure weather variables.

(l)Jan Tadeusz/Alamy; (r)matthias engelien/Alamy

Relative Humidity

Think about how a sponge can absorb water. At some point, it becomes full and cannot absorb any more water. In the same way, air can only contain a certain amount of water vapor. When air is saturated, it contains as much water vapor as possible. Temperature determines the maximum amount of water vapor air can contain. Warm air can contain more water vapor than cold air. *The amount of water vapor present in the air compared to the maximum amount of water vapor the air could contain at that temperature is called* **relative humidity.**

Relative humidity is measured using an instrument called a psychrometer and is given as a percent. For example, air with a relative humidity of 100 percent cannot contain any more moisture and dew or rain will form. Air that contains only half the water vapor it could hold has a relative humidity of 50 percent.

Reading Check Compare and contrast humidity and relative humidity.

Dew Point

When a sponge becomes saturated with water, the water starts to drip from the sponge. Similarly, when air becomes saturated with water vapor, the water vapor will condense and form water droplets. When air near the ground becomes saturated, the water vapor in air will condense to a liquid. If the temperature is above 0°C, dew forms. If the temperature is below 0°C, ice crystals, or frost, form. Higher in the atmosphere clouds form. The graph in **Figure 3** shows the total amount of water vapor that air can contain at different temperatures.

When the temperature decreases, the air can hold less moisture. As you just read, the air becomes saturated, condensation occurs, and dew forms. *The temperature at which air is saturated and condensation can occur is called the* **dew point.**

MiniLab — 20 minutes

When will dew form?

The relative humidity on a summer day is 80 percent. The temperature is 35°C. Will the dew point be reached if the temperature drops to 25°C later in the evening? Use **Figure 3** below to find the amount of water vapor needed for saturation at each temperature.

1. Calculate the amount of water vapor in air that is 35°C and has 80 percent relative humidity. (Hint: multiply the amount of water vapor air can contain at 35°C by the percent of relative humidity.)
2. At 25°C, air can hold 2.2 g/cm^3 of water vapor. If your answer from step 1 is less than 2.2 g/cm^3, the dew point is not reached and dew will not form. If the number is greater, dew will form.

Analyze and Conclude

Key Concept After the Sun rises in the morning the air's temperature increases. How does the relative humidity change after sunrise? What does the line represent?

Figure 3 As air temperature increases, the air can contain more water vapor.

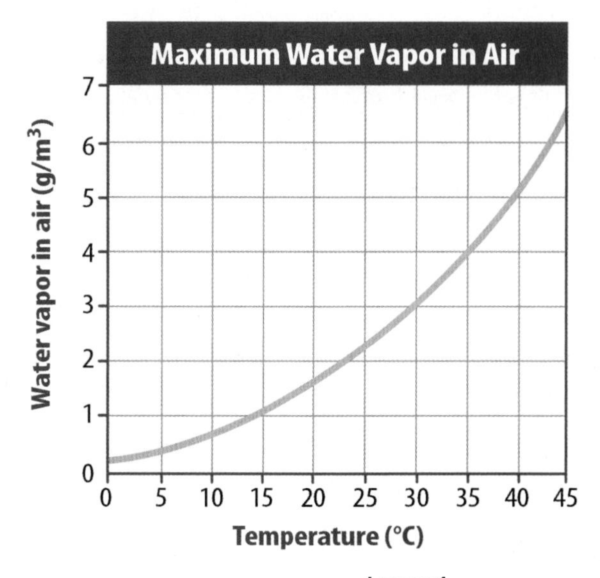

Figure 4 Clouds have different shapes and can be found at different altitudes.

Stratus clouds
- flat, white, and layered
- altitude up to 2,000 m

Cumulus clouds
- fluffy, heaped, or piled up
- 2,000 to 6,000 m altitude

Cirrus clouds
- wispy
- above 6,000 m

Clouds and Fog

When you exhale outside on a cold winter day, you can see the water vapor in your breath condense into a foggy cloud in front of your face. This also happens when warm air containing water vapor cools as it rises in the atmosphere. When the cooling air reaches its dew point, water vapor condenses on small particles in the air and forms droplets. Surrounded by thousands of other droplets, these small droplets block and reflect light. This makes them visible as clouds.

WORD ORIGIN
precipitation
from Latin *praecipitationem,* means "act or fact of falling headlong"

Clouds are water droplets or ice crystals suspended in the atmosphere. Clouds can have different shapes and be present at different altitudes within the atmosphere. Different types of clouds are shown in **Figure 4.** Because we observe that clouds move, we recognize that water and thermal energy are transported from one location to another. Recall that clouds are also important in reflecting some of the Sun's incoming radiation.

FOLDABLES®

Make a horizontal two-tab book and label the tabs as illustrated. Use it to collect information on clouds and fog. Find similarities and differences.

Clouds | Fog

A cloud that forms near Earth's surface is called fog. Fog is a suspension of water droplets or ice crystals close to or at Earth's surface. Fog reduces visibility, the distance a person can see into the atmosphere.

Reading Check What is fog?

(l)WIN-Initiative/Getty Images; (c)MIMOTITO/Getty Images; (r)age fotostock/SuperStock

Precipitation

Recall that droplets in clouds form around small solid particles in the atmosphere. These particles might be dust, salt, or smoke. Precipitation occurs when cloud droplets combine and become large enough to fall back to Earth's surface. **Precipitation** *is water, in liquid or solid form, that falls from the atmosphere.* Examples of precipitation–rain, snow, sleet, and hail–are shown in **Figure 5.**

Rain is precipitation that reaches Earth's surface as droplets of water. Snow is precipitation that reaches Earth's surface as solid, frozen crystals of water. Sleet may originate as snow. The snow melts as it falls through a layer of warm air and refreezes when it passes through a layer of below-freezing air. Other times it is just freezing rain. Hail reaches Earth's surface as large pellets of ice. Hail starts as a small piece of ice that is repeatedly lifted and dropped by an updraft within a cloud. A layer of ice is added with each lifting. When it finally becomes too heavy for the updraft to lift, it falls to Earth.

Key Concept Check What variables are used to describe weather?

The Water Cycle

Precipitation is an important process in the water cycle. Evaporation and condensation are phase changes that are also important to the water cycle. *The* **water cycle** *is the series of natural processes by which water continually moves among oceans, land, and the atmosphere.* As illustrated in **Figure 6,** most water vapor enters the atmosphere when water at the ocean's surface is heated and evaporates. Water vapor cools as it rises in the atmosphere and condenses back into a liquid. Eventually, droplets of liquid and solid water form clouds. Clouds produce precipitation, which falls to Earth's surface and later evaporates, continuing the cycle.

Key Concept Check How is weather related to the water cycle?

Types of Precipitation

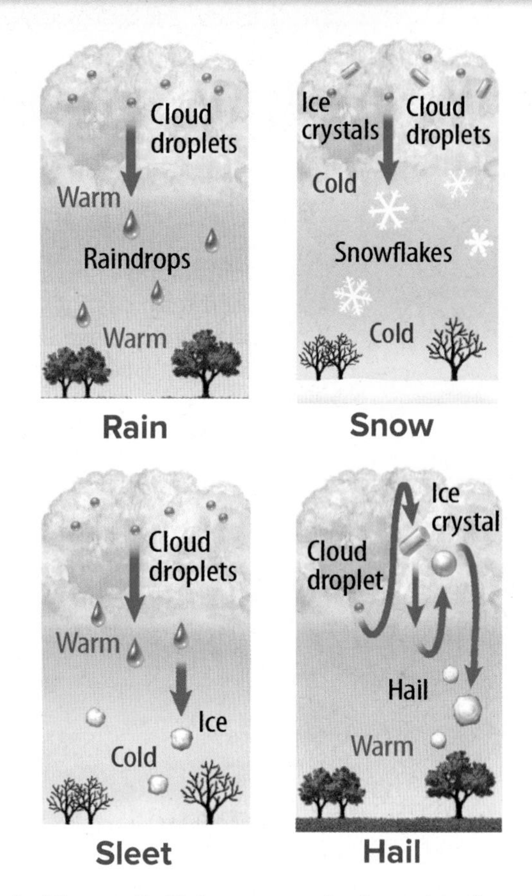

▲ **Figure 5** Rain, snow, sleet, and hail are forms of precipitation.

Visual Check What is the difference between snow and sleet?

The Water Cycle

Figure 6 The Sun's energy powers the water cycle, which is the continual movement of water between the ocean, the land, and the atmosphere.

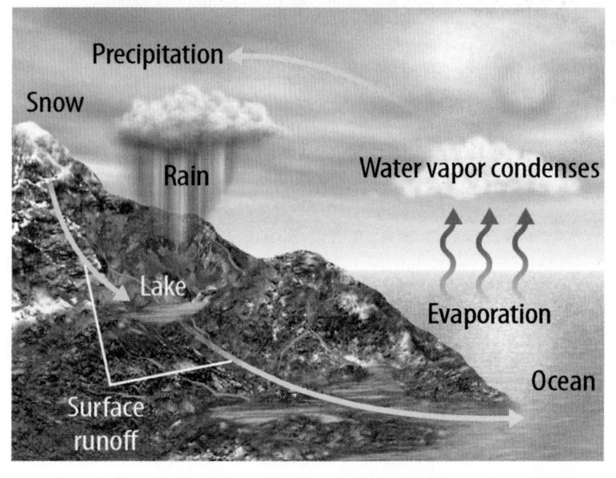

Lesson 1 Review

Visual Summary

Weather is the atmospheric conditions, along with short-term changes, of a certain place at a certain time.

Meteorologists use weather variables to describe atmospheric conditions.

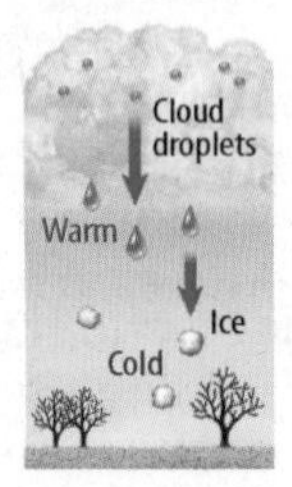

Forms of precipitation include rain, sleet, snow, and hail.

FOLDABLES®

Use your lesson Foldable to review the lesson. Save your Foldable for the project at the end of the chapter.

What do you think NOW?

You first read the statements below at the beginning of the chapter.

1. Weather is the long-term average of atmospheric patterns of an area.
2. All clouds are at the same altitude within the atmosphere.

Did you change your mind about whether you agree or disagree with the statements? Rewrite any false statements to make them true.

Use Vocabulary

1. **Define** *humidity* in your own words.
2. **Use the term** *precipitation* in a sentence.
3. __________ is the pressure that a column of air exerts on the surface below it.

Understand Key Concepts

4. Which is NOT a standard weather variable?
   - **A.** air pressure
   - **B.** moon phase
   - **C.** temperature
   - **D.** wind speed
5. **Identify** and describe the different variables used to describe weather.
6. **Relate** humidity to cloud formation.
7. **Describe** how processes in the water cycle are related to weather.

Interpret Graphics

8. **Identify** Which type of precipitation is shown in the diagram below? How does this precipitation form?

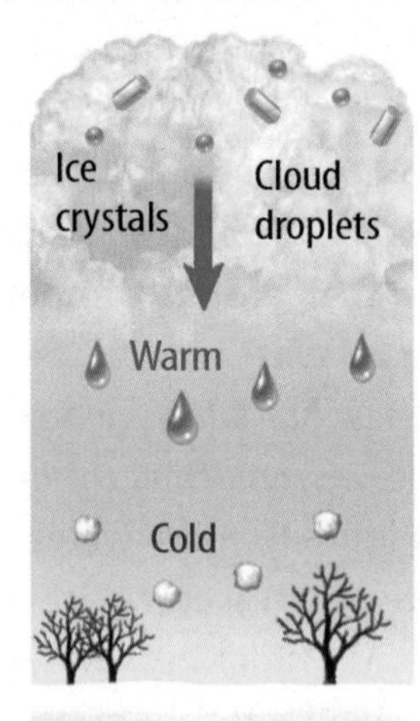

Critical Thinking

9. **Analyze** Why would your ears pop if you climbed a tall mountain?
10. **Differentiate** among cloud formation, fog formation, and dew point.

SCIENCE & SOCIETY

AMERICAN MUSEUM OF NATURAL HISTORY

Flooding caused widespread devastation in New Orleans, a city that lies below sea level. The storm surge broke through levees that had protected the city.

Is there a link between hurricanes and global warming?

The warm waters of the Gulf of Mexico fueled Hurricane Katrina as it spun toward Louisiana.

Scientists worry that hurricanes might be getting bigger and happening more often.

On August 29, 2005, Hurricane Katrina roared through New Orleans, Louisiana. The storm destroyed homes and broke through levees, flooding most of the low-lying city. In the wake of the disaster, many wondered whether global warming was responsible. If warm oceans are the fuel for hurricanes, could rising temperatures cause stronger or more frequent hurricanes?

Climate scientists have several ways to investigate this question. They examine past hurricane activity, sea surface temperature, and other climate data. They compare these different types of data and look for patterns. Based on the laws of physics, they put climate and hurricane data into equations. A computer solves these equations and makes computer models. Scientists analyze the models to see whether there is a connection between hurricane activity and different climate variables.

What have scientists learned? So far they have not found a link between warming oceans and the frequency of hurricanes. However, they have found a connection between warming oceans and hurricane strength. Models suggest that rising ocean temperatures might create more destructive hurricanes with stronger winds and more rainfall.

But global warming is not the only cause of warming oceans. As the ocean circulates, it goes through cycles of warming and cooling. Data show that the Atlantic Ocean has been in a warming phase for the past few decades.

Whether due to global warming or natural cycles, ocean temperatures are expected to rise even more in coming years. While rising ocean temperatures might not produce more hurricanes, climate research shows they could produce more powerful hurricanes. Perhaps the better question is not what caused Hurricane Katrina, but how we can prepare for equal-strength or more destructive hurricanes in the future.

It's Your Turn

DIAGRAM With a partner, create a storyboard with each frame showing one step in hurricane formation. Label your drawings. Share your storyboard with the class.

Lesson 2

Weather Patterns

Reading Guide

Key Concepts

ESSENTIAL QUESTIONS

- What are two types of pressure systems?
- What drives weather patterns?
- Why is it useful to understand weather patterns?
- What are some examples of severe weather?

Vocabulary

 Multilingual eGlossary

 What's Science Got to do With It?

What caused this flooding?

Surging waves and rain from Hurricane Katrina caused flooding in New Orleans, Louisiana. Why are flooding and other types of severe weather dangerous? How does severe weather form?

Kyle Niemi/U.S. Coast Guard via Getty Images

Launch Lab

10 minutes

How can temperature affect pressure?

Air molecules that have low energy can be packed closely together. As energy is added to the molecules they begin to move and bump into one another.

1. Read and complete a lab safety form.
2. Close a **resealable plastic bag** except for a small opening. Insert a **straw** through the opening and blow air into the bag until it is as firm as possible. Remove the straw and quickly seal the bag.
3. Submerge the bag in a **container** of **ice water** and hold it there for 2 minutes. Record your observations in your Science Journal.
4. Remove the bag from the ice water and submerge it in **warm water** for 2 minutes. Record your observations.

Think About This

1. What do the results tell you about the movement of air molecules in cold air and in warm air?
2. **Key Concept** What property of the air is demonstrated in this activity?

Pressure Systems

Weather is often associated with pressure systems. Recall that air pressure is the weight of the molecules in a large mass of air. When air molecules are cool, they are closer together than when they are warm. Cool air masses have high pressure, or more weight. Warm air masses have low pressure.

A **high-pressure system,** shown in **Figure 7,** *is a large body of circulating air with high pressure at its center and lower pressure outside of the system.* Because air moves from high pressure to low pressure, the air inside the system moves away from the center. Dense air sinks, bringing clear skies and fair weather.

A **low-pressure system,** also shown in **Figure 7,** *is a large body of circulating air with low pressure at its center and higher pressure outside of the system.* This causes air inside the low pressure system to rise. The rising air cools and the water vapor condenses, forming clouds and sometimes precipitation–rain or snow.

Key Concept Check Compare and contrast two types of pressure systems.

Figure 7 Air moving from areas of high pressure to areas of low pressure is called wind.

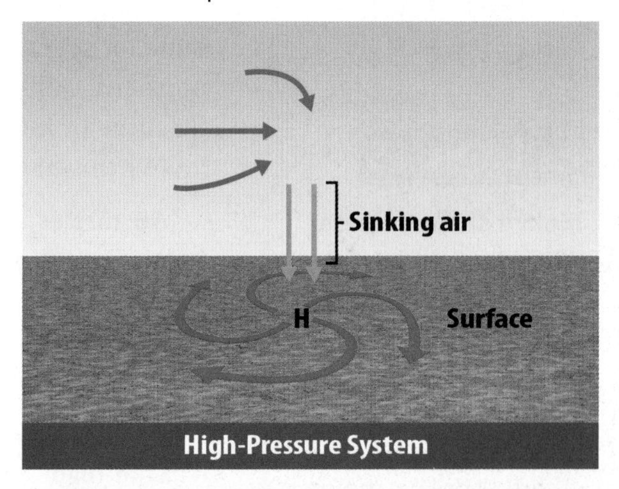

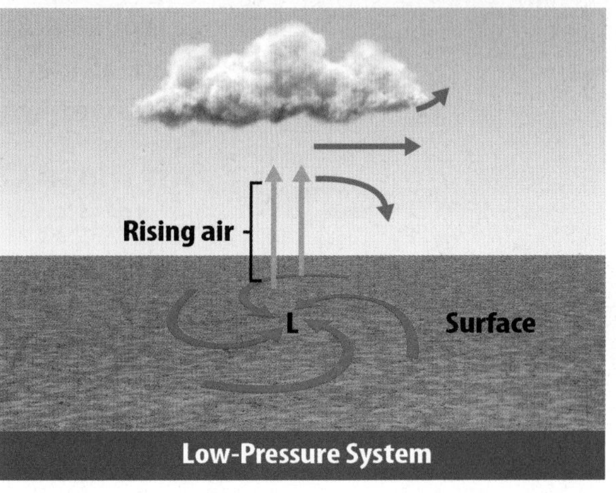

Hutchings Photography/Digital Light Source

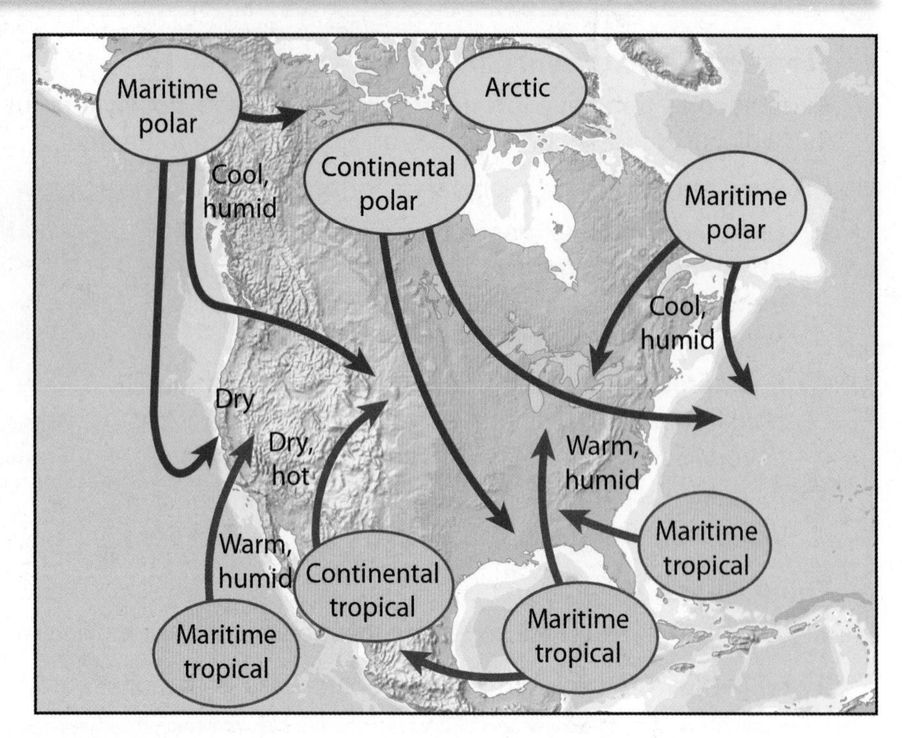

Figure 8 Five main air masses impact climate across North America.

Visual Check Where does continental polar air come from?

Air Masses

Have you ever noticed that the weather sometimes stays the same for several days in a row? For example, during winter in the northern United States, extremely cold temperatures often last for three or four days in a row. Afterward, several days might follow with warmer temperatures and snow showers.

Air masses are responsible for this pattern. **Air masses** *are large bodies of air that have uniform temperature, humidity, and pressure.* An air mass forms when a large high pressure system lingers over an area for several days. As a high pressure system comes in contact with Earth, the air in the system takes on the temperature and moisture characteristics of the surface below it.

Like high- and low-pressure systems, air masses can extend for a thousand kilometers or more. Sometimes one air mass covers most of the United States. Examples of the main air masses that affect weather in the United States are shown in **Figure 8.**

FOLDABLES®

Fold a sheet of paper into thirds along the long axis. Label the outside *Air Masses.* Make another fold about 2 inches from the long edge of the paper to make a three-column chart. Label as shown.

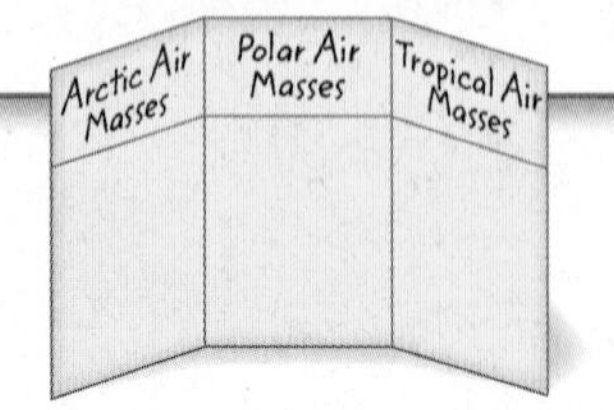

Air Mass Classification

Air masses are classified by their temperature and moisture characteristics. Air masses that form over land are referred to as continental air masses. Those that form over water are referred to as maritime masses. Warm air masses that form in the equatorial regions are called tropical. Those that form in cold regions are called polar. Air masses near the poles, over the coldest regions of the globe, are called arctic and antarctic air masses.

Arctic Air Masses Forming over Siberia and the Arctic are arctic air masses. They contain bitterly cold, dry air. During winter, an arctic air mass can bring temperatures down to -40°C.

Continental Polar Air Masses Because land cannot transfer as much moisture to the air as oceans can, air masses that form over land are drier than air masses that form over oceans. Continental polar air masses are fast-moving and bring cold temperatures in winter and cool weather in summer. Find the continental polar air masses over Canada in **Figure 8.**

Maritime Polar Air Masses Forming over the northern Atlantic and Pacific Oceans, maritime polar air masses are cold and humid. They often bring cloudy, rainy weather.

Continental Tropical Air Masses Because they form in the tropics over dry, desert land, continental tropical air masses are hot and dry. They bring clear skies and high temperatures. Continental tropical air masses usually form during the summer.

Maritime Tropical Air Masses As shown in **Figure 8,** maritime tropical air masses form over the western Atlantic Ocean, the Gulf of Mexico, and the eastern Pacific Ocean. These moist air masses bring hot, humid air to the southeastern United States during summer. In winter, they can bring heavy snowfall.

Air masses can change as they move over the land and ocean. Warm, moist air can move over land and become cool and dry. Cold, dry air can move over water and become moist and warm.

Key Concept Check What drives weather patterns?

Math Skills 

Conversions

To convert Fahrenheit (°F) units to Celsius (°C) units, use this equation:

$$°C = \frac{(°F - 32)}{1.8}$$

Convert **76°F** to °C

1. Always perform the operation in parentheses first.

   (76°F − 32) = **44°F**

2. Divide the answer from Step 1 by 1.8.

   $$\frac{44°F}{1.8} = 24°C$$

To convert °C to **°F**, follow the same steps using the following equation:

$$°F = (°C \times 1.8) + 32$$

Practice

1. Convert 86°F to °C.
2. Convert 37°C to °F.

 Math Practice

 Personal Tutor

MiniLab — 20 minutes

How can you observe air pressure?

Although air seems very light, air molecules do exert pressure. You can observe air pressure in action in this activity.

1. Read and complete a lab safety form.
2. Tightly cap the empty **plastic bottle.**
3. Place the bottle in a **bucket of ice** for 10 minutes. Record your observations in your Science Journal.

Hutchings Photography/Digital Light Source

Analyze and Conclude

1. **Interpret** how air pressure affected the bottle.
2.  **Key Concept** Discuss how changing air pressure in Earth's atmosphere affects other things on Earth, such as weather.

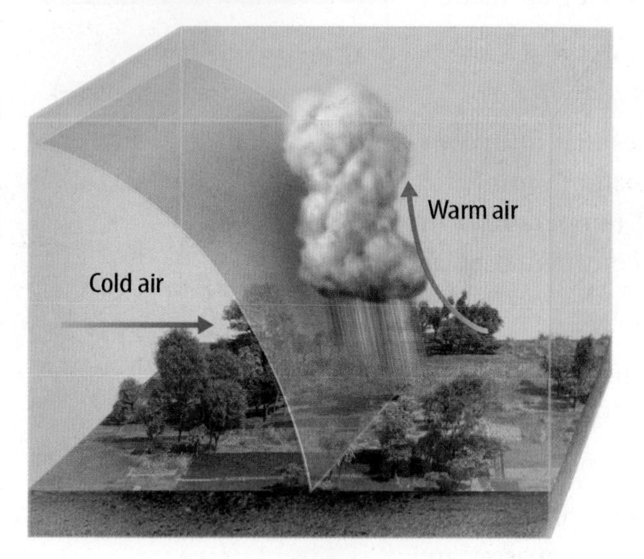

Cold

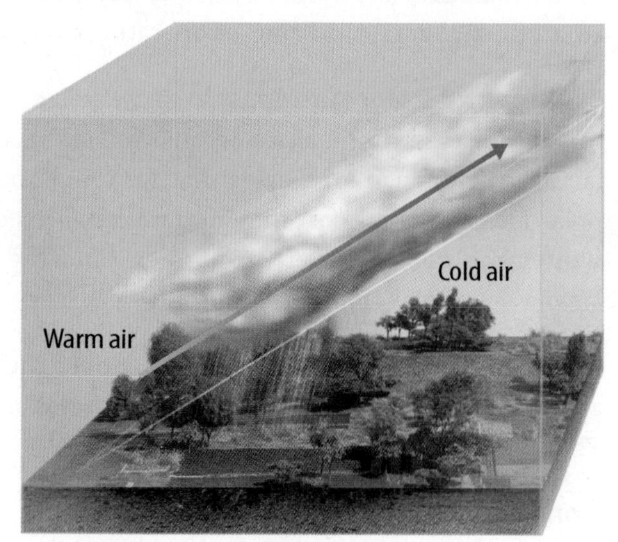

Warm

Figure 9 Certain types of fronts are associated with specific weather.

Visual Check Describe the difference between a cold front and a warm front.

Fronts

In 1918, Norwegian meteorologist Jacob Bjerknes (BYURK nehs) and his coworkers were busy developing a new method for forecasting the weather. Bjerknes noticed that specific types of weather occur at the boundaries between different air masses. Because he was trained in the army, Bjerknes used a military term to describe this boundary–front.

A military front is the boundary between opposing armies in a battle. *A weather* **front,** *however, is a boundary between two air masses.* Drastic weather changes often occur at fronts. As wind carries an air mass away from the area where it formed, the air mass will eventually collide with another air mass. Changes in temperature, humidity, cloud types, wind, and precipitation are common at fronts.

SCIENCE USE V. COMMON USE

front

Science Use a boundary between two air masses

Common Use the foremost part or surface of something

Cold Fronts

When a colder air mass moves toward a warmer air mass, a cold front forms, as shown in **Figure 9.** The cold air, which is denser than the warm air, pushes underneath the warm air mass. The warm air rises and cools. Water vapor in the air condenses and clouds form. Showers and thunderstorms often form along cold fronts. It is common for temperatures to decrease as much as 10°C when a cold front passes through. The wind becomes gusty and changes direction. In many cases, cold fronts give rise to severe storms.

Reading Check What types of weather are associated with cold fronts?

Stationary

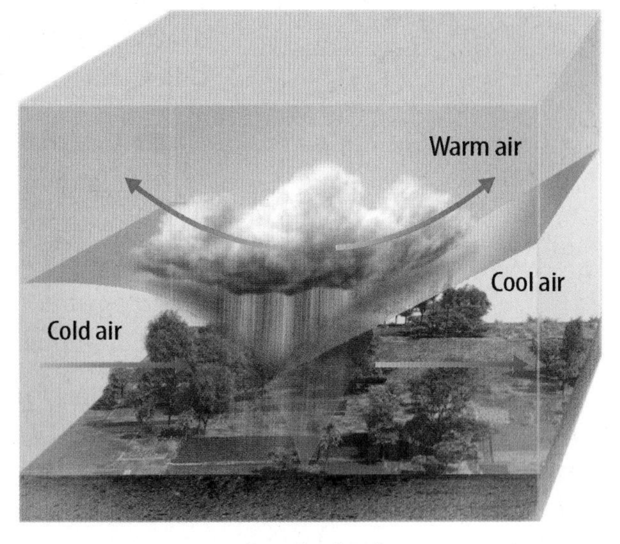

Occluded

Warm Fronts

As shown in **Figure 9,** a warm front forms when less dense, warmer air moves toward colder, denser air. The warm air rises as it glides above the cold air mass. When water vapor in the warm air condenses, it creates a wide blanket of clouds. These clouds often bring steady rain or snow for several hours or even days. A warm front not only brings warmer temperatures, but it also causes the wind to shift directions.

Both a cold front and a warm front form at the edge of an approaching air mass. Because air masses are large, the movement of fronts is used to make weather forecasts. When a cold front passes through your area, temperatures will remain low for the next few days. When a warm front arrives, the weather will become warmer and more humid.

Stationary and Occluded Fronts

Sometimes an approaching front will stall for several days with warm air on one side of it and cold air on the other side. When the boundary between two air masses stalls, the front is called a stationary front. Study the stationary front shown in **Figure 9.** Cloudy skies and light rain are found along stationary fronts.

Cold fronts move faster than warm fronts. When a fast-moving cold front catches up with a slow-moving warm front, an occluded or blocked front forms. Occluded fronts, shown in **Figure 9,** usually bring precipitation.

Key Concept Check Why is it useful to understand weather patterns associated with fronts?

Severe Weather

Some weather events can cause major damage, injuries, and death. These events, such as thunderstorms, tornadoes, hurricanes, and blizzards, are called severe weather.

Thunderstorms

Also known as electrical storms because of their lightning, thunderstorms have warm temperatures, moisture, and rising air, which may be supplied by a low-pressure system. When these conditions occur, a cumulus cloud can grow into a 10-km-tall thundercloud, or cumulonimbus cloud, in as little as 30 minutes.

ACADEMIC VOCABULARY

dominate
(verb) to exert the guiding influence on

A typical thunderstorm has a three-stage life cycle, shown in **Figure 10.** The cumulus stage is **dominated** by cloud formation and updrafts. Updrafts are air currents moving vertically away from the ground. After the cumulus cloud has been created, downdrafts begin to appear. Downdrafts are air currents moving vertically toward the ground. In the mature stage, heavy winds, rain, and lightning dominate the area. Within 30 minutes of reaching the mature stage, the thunderstorm begins to fade, or dissipate. In the dissipation stage, updrafts stop, winds die down, lightning ceases, and precipitation weakens.

Strong updrafts and downdrafts within a thunderstorm cause millions of tiny ice crystals to rise and sink, crashing into each other. This creates positively and negatively charged particles in the cloud. The difference in the charges of particles between the cloud and the charges of particles on the ground eventually creates electricity. This is seen as a bolt of lightning. Lightning can move from cloud to cloud, cloud to ground, or ground to cloud. It can heat the nearby air to more than 27,000°C. Air molecules near the bolt rapidly expand and then contract, creating the sound identified as thunder.

Figure 10 Thunderstorms have distinct stages characterized by the direction in which air is moving.

Thunderstorms

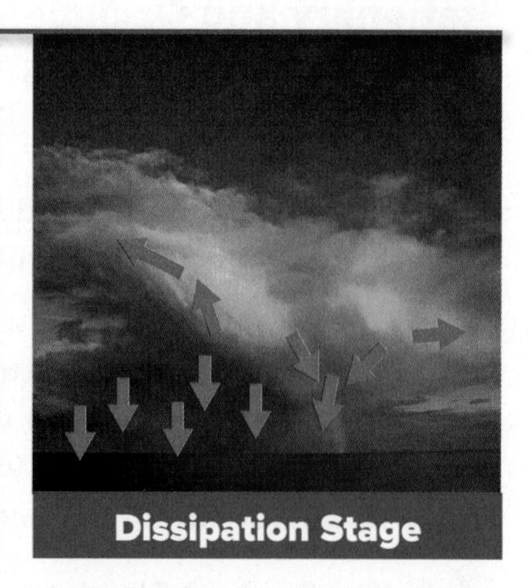

Visual Check Describe what happens during each stage of a thunderstorm.

Animation

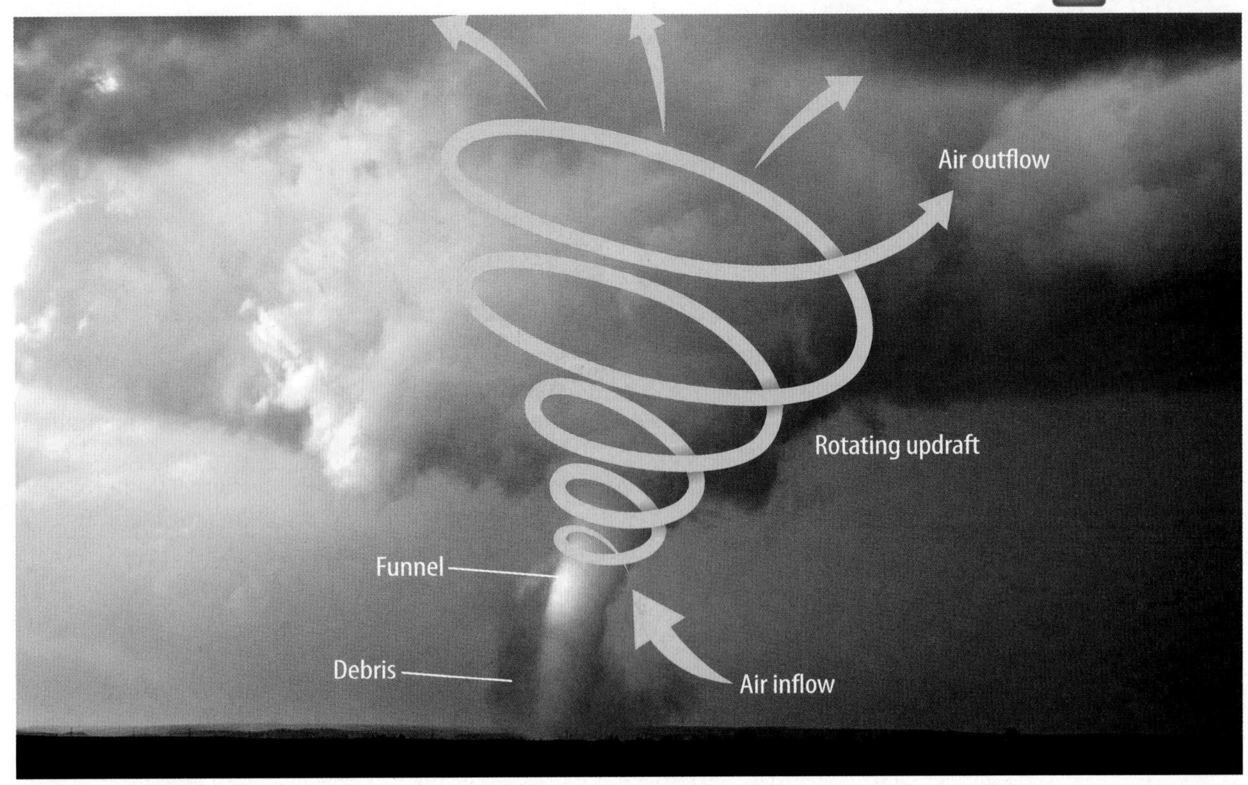

Figure 11 A funnel cloud forms when updrafts within a thunderstorm begin rotating.

Tornadoes

Perhaps you have seen photos of the damage from a tornado. *A* **tornado** *is a violent, whirling column of air in contact with the ground.* Most tornadoes have a diameter of several hundred meters. The largest tornadoes exceed 1,500 m in diameter. The intense, swirling winds within tornadoes can reach speeds of more than 400 km/h. These winds are strong enough to send cars, trees, and even entire houses flying through the air. Tornadoes usually last only a few minutes. More destructive tornadoes, however, can last for several hours.

Formation of Tornadoes When thunderstorm updrafts begin to rotate, as shown in **Figure 11,** tornadoes can form. Swirling winds spiral downward from the thunderstorm's base, creating a funnel cloud. When the funnel reaches the ground, it becomes a tornado. Although the swirling air is invisible, you can easily see the debris lifted by the tornado.

 **Reading Check** How do tornadoes form?

Tornado Alley More tornadoes occur in the United States than anywhere else on Earth. The central United States, from Nebraska to Texas, experiences the most tornadoes. This area has been nicknamed Tornado Alley. In this area, cold air blowing southward from Canada frequently collides with warm, moist air moving northward from the Gulf of Mexico. These conditions are ideal for severe thunderstorms and tornadoes.

Classifying Tornadoes Dr. Ted Fujita developed a method for classifying tornadoes based on the damage they cause. On the Enhanced Fujita Scale, F0 tornadoes cause light damage, breaking tree branches and damaging billboards. F1 though F4 tornadoes cause moderate to devastating damage, including tearing roofs from homes, derailing trains, and throwing vehicles in the air. F5 tornadoes cause incredible damage, such as demolishing concrete and steel buildings and pulling the bark from trees.

Hurricane Formation

Figure 12 Hurricanes consist of alternating bands of heavy precipitation and sinking air.

Hurricane Formation

1. As warm, moist air rises into the atmosphere, it cools, water vapor condenses, and clouds form. As more air rises, it creates an area of low pressure over the ocean.

2. As air continues to rise, a tropical depression forms. Tropical depressions bring thunderstorms with winds between 37–62 km/h.

3. Air continues to rise, rotating counterclockwise. The storm builds to a tropical storm with winds in excess of 63 km/h. It produces strong thunderstorms.

4. When winds exceed 119 km/h, the storm becomes a hurricane. Only one percent of tropical storms become hurricanes.

Inside a Hurricane

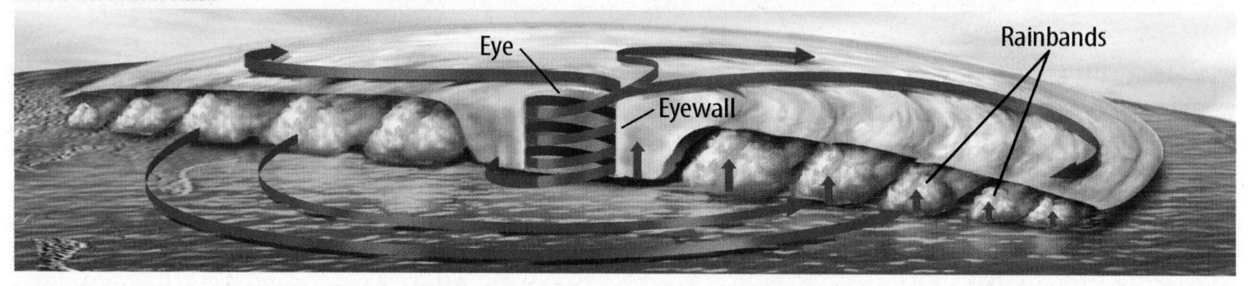

Visual Check How do hurricanes form?

WORD ORIGIN

hurricane
from Spanish *huracan*, means "tempest"

Hurricanes

An intense tropical storm with winds exceeding 119 km/h is a **hurricane.** Hurricanes are the most destructive storms on Earth. Like tornadoes, hurricanes have a circular shape with intense, swirling winds. However, hurricanes do not form over land. Hurricanes typically form in late summer over warm, tropical ocean water. **Figure 12** sequences the steps in hurricane formation. A typical hurricane is 480 km across, more than 150 thousand times larger than a tornado. At the center of a hurricane is the eye, an area of clear skies and light winds.

Damage from hurricanes occurs as a result of strong winds and flooding. While still out at sea, hurricanes create high waves that can flood coastal areas. As a hurricane crosses the coastline, or makes landfall, strong rains intensify and can flood and devastate entire areas. But once a hurricane moves over land or colder water, it loses its energy and dissipates.

In other parts of the world, these intense tropical storms have other names. In Asia, the same type of storm is called a typhoon. In Australia it is called a tropical cyclone.

Winter Storms

Winter weather can be severe and hazardous. Ice storms, as shown in **Figure 13,** can down power lines and tree branches and make driving dangerous. A **blizzard** *is a violent winter storm characterized by freezing temperatures, strong winds, and blowing snow.* During a blizzard, swirling snow reduces visibility, and freezing temperatures can cause frostbite and hypothermia (hi poh THER mee uh).

Freezing Rain

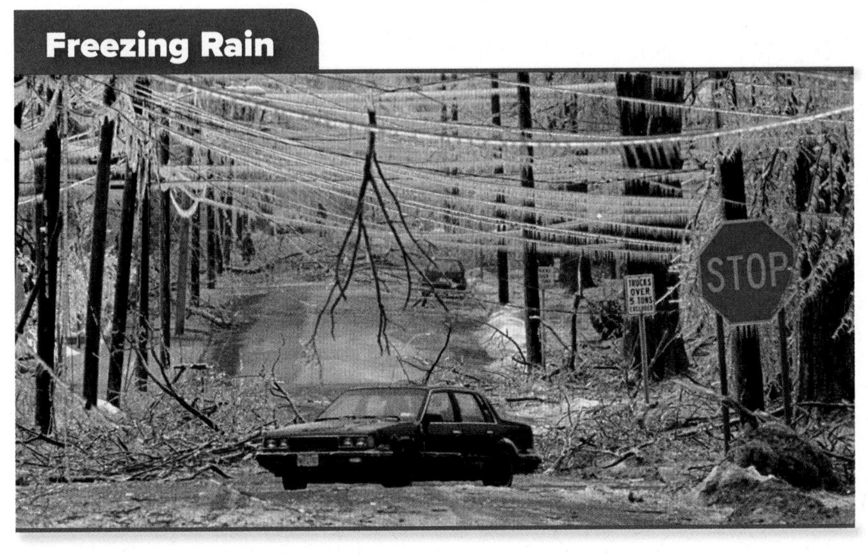

Figure 13 The weight of ice from freezing rain can cause trees, power lines, and other structures to break.

Key Concept Check What are examples of severe weather?

Severe Weather Safety

The U.S. National Weather Service issues watches and warnings for different types of severe weather. A watch means that severe weather is possible. A warning means that severe weather is already occurring. Paying close attention to severe weather watches and warnings is important and could save your life.

It is also important to know how to protect yourself during dangerous weather. During thunderstorms, you should stay inside if possible, and stay away from metal objects and electrical cords. If you are outside, stay away from water, high places, and isolated trees. Dressing properly is important in all kinds of weather. When windchill temperatures are below −20°C, you should dress in layers, keep your head and fingers covered, and limit your time outdoors.

Not all weather safety pertains to bad weather. The Sun's ultraviolet (UV) radiation can cause health risks, including skin cancer. The U.S. National Weather Service issues a daily UV Index Forecast. Precautions on sunny days include covering up, using sunscreen, and wearing a hat and sunglasses. Surfaces such as snow, water, and beach sand can double the effects of the Sun's UV radiation.

Lesson 2 Review

Visual Summary

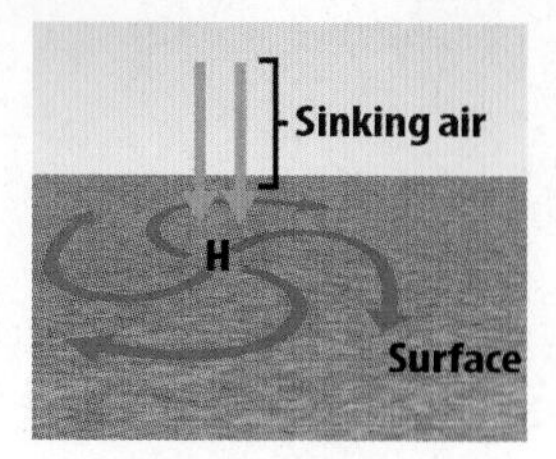

Low-pressure systems, high-pressure systems, and air masses all influence weather.

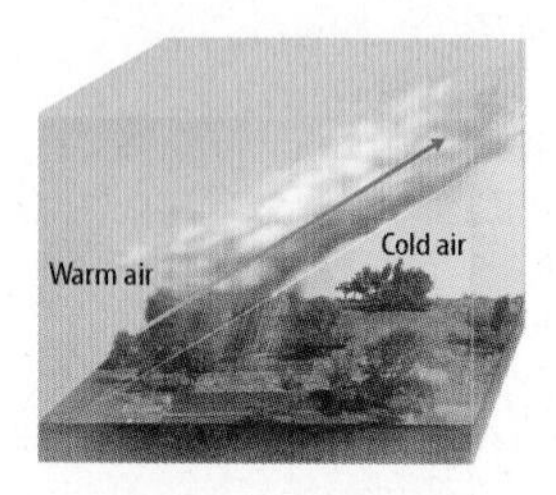

Weather often changes as a front passes through an area.

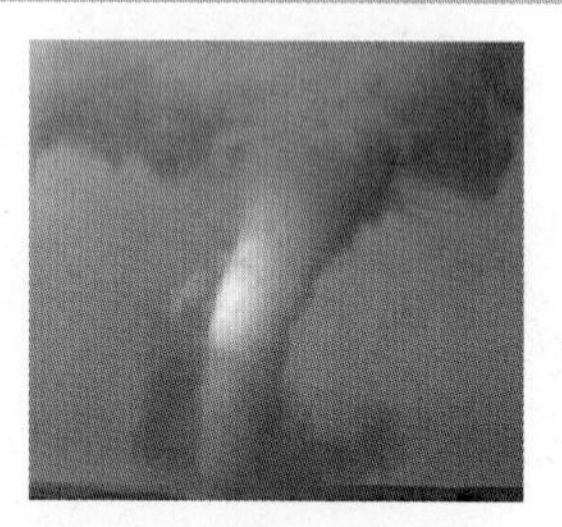

The National Weather Service issues warnings about severe weather such as thunderstorms, tornadoes, hurricanes, and blizzards.

Use your lesson Foldable to review the lesson. Save your Foldable for the project at the end of the chapter.

What do you think NOW?

You first read the statements below at the beginning of the chapter.

3. Precipitation often occurs at the boundaries of large air masses.

4. There are no safety precautions for severe weather, such as tornadoes and hurricanes.

Did you change your mind about whether you agree or disagree with the statements? Rewrite any false statements to make them true.

Use Vocabulary

1. **Distinguish** between an air mass and a front.
2. **Define** *low-pressure system* using your own words.
3. **Use the term** *high-pressure system* in a sentence.

Understand Key Concepts

4. Which air mass is humid and warm?
   A. continental polar
   B. continental tropical
   C. maritime polar
   D. maritime tropical
5. **Give an example** of cold-front weather.
6. **Compare and contrast** hurricanes and tornadoes.
7. **Explain** how thunderstorms form.

Interpret Graphics

8. **Compare and Contrast** Copy and fill in the graphic organizer below to compare and contrast high-pressure and low-pressure systems.

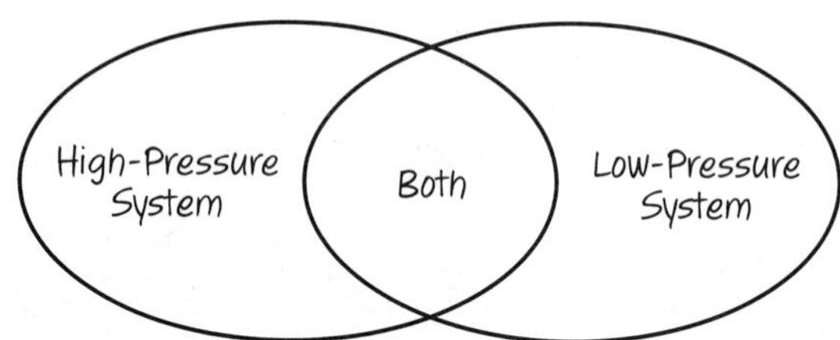

Critical Thinking

9. **Suggest** a reason that low-pressure systems are cloudy and rainy or snowy.
10. **Design** a pamphlet that contains tips on how to stay safe during different types of severe weather.

Math Skills

11. Convert 212°F to °C.
12. Convert 20°C to °F.

Skill Practice **Recognize Cause and Effect** **30 minutes**

Why does the weather change?

One day it is sunny, the next day it is pouring rain. If you look at only one location, the patterns that cause the weather to change are difficult to see. However, when you look on the large scale, the patterns become apparent.

Learn It

Recognizing cause and effect is an important part of science and conducting experiments. Scientists look for cause-and-effect relationships between variables. The maps below show the movement of fronts and pressure systems over a two-day period. What effect will these systems have on the weather as they move across the United States?

Try It

1. Examine the weather maps below. The thin black lines on each map represent areas where the barometric pressure is the same. The pressure is indicated by the number on the line. The center of a low- or high-pressure system is indicated by the word LOW or HIGH. Identify the location of low- and high- pressure systems on each map. Use the key below the maps to the identify the location of warm and cold fronts.

2. Find locations A, B, C, and where you live on the map. For each location, describe how the systems change positions over the two days.

3. What is the cause of and effect on precipitation and temperature at each location?

Apply It

4. The low-pressure system produced several tornadoes. Which location did they occur closest to? Explain.

5. The weather patterns generally move from west to east. Predict the weather on the third day for each location.

6. One day it is clear and sunny, but you notice that the pressure is less than it was the day before. What weather might be coming? Why?

7. **Key Concept** How does understanding weather patterns help make predicting the weather more accurate?

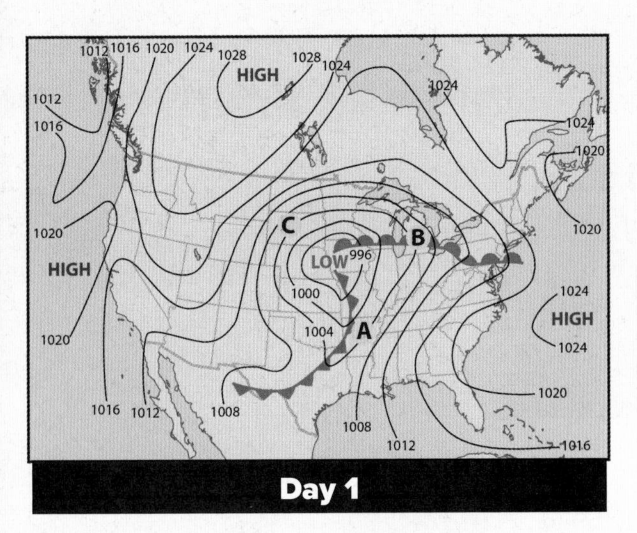

Day 1

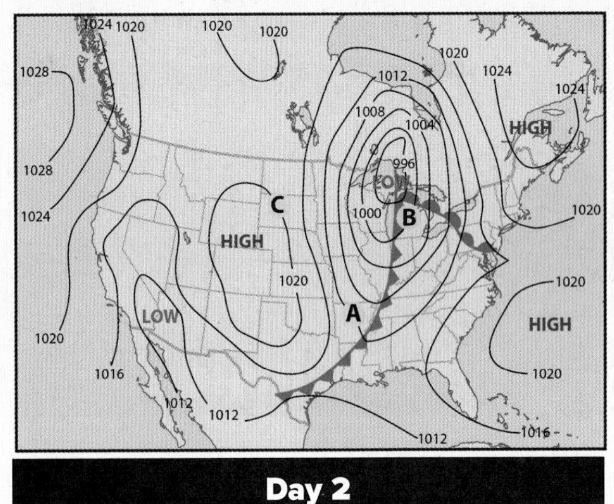

Day 2

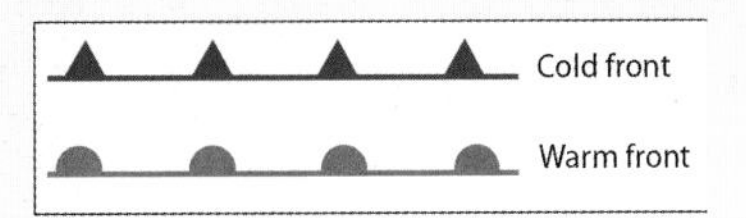

Lesson 3

Reading Guide

Key Concepts 
ESSENTIAL QUESTIONS

- What instruments are used to measure weather variables?
- How are computer models used to predict the weather?

Vocabulary

surface report p. 471
upper-air report p. 471
Doppler radar p. 472
isobar p. 473
computer model p. 474

Multilingual eGlossary

PBL Go to the resource tab in ConnectED to find the PBL *Weather Wardrobe.*

Weather Forecasts

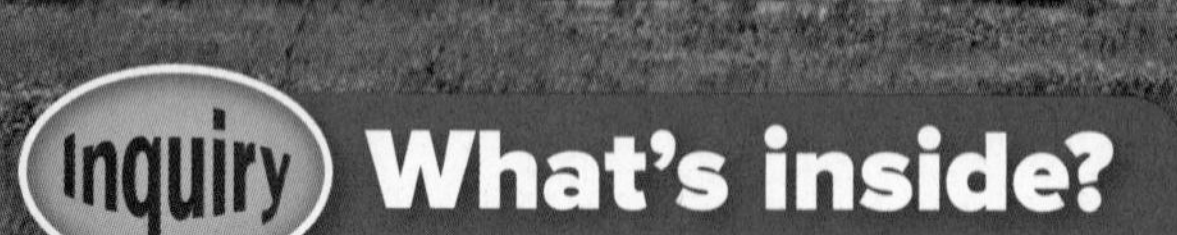

Information about weather variables is collected by the weather radar station shown here. Data, such as the amount of rain falling in a weather system, help meteorologists make accurate predictions about severe weather. What other instruments do meteorologists use to forecast weather? How do they collect and use data?

Launch Lab

10 minutes

Can you understand the weather report?

Weather reports use numbers and certain vocabulary terms to help you understand the weather conditions in a given area for a given time period. Listen to a weather report for your area. Can you record all the information reported?

1. In your Science Journal, make a list of data you would expect to hear in a weather report.
2. Listen carefully to a **recording of a weather report** and jot down numbers and measurements you hear next to those on your list.
3. Listen a second time and make adjustments to your original notes, such as adding more data, if necessary.
4. Listen a third time, then share the weather forecast as you heard it.

Think About This

1. What measurements were difficult for you to apply to understanding the weather report?
2. Why are so many different types of data needed to give a complete weather report?
3. List the instruments that might be used to collect each kind of data.
4. **Key Concept** Where do meteorologists obtain the data they use to make a weather forecast?

Measuring the Weather

Being a meteorologist is like being a doctor. Using specialized instruments and visual observations, the doctor first measures the condition of your body. The doctor later combines these measurements with his or her knowledge of medical science. The result is a forecast of your future health, such as, "You'll feel better in a few days if you rest and drink plenty of fluids."

Similarly, meteorologists, scientists who study weather, use specialized instruments to measure conditions in the atmosphere, as you read in Lesson 1. These instruments include thermometers to measure temperature, barometers to measure air pressure, psychrometers to measure relative humidity, and anemometers to measure wind speed.

Surface and Upper-Air Reports

A **surface report** *describes a set of weather measurements made on Earth's surface.* Weather variables are measured by a weather station–a collection of instruments that report temperature, air pressure, humidity, precipitation, and wind speed and direction. Cloud amounts and visibility are often measured by human observers.

An **upper-air report** *describes wind, temperature, and humidity conditions above Earth's surface.* These atmospheric conditions are measured by a radiosonde (RAY dee oh sahnd), a package of weather instruments carried many kilometers above the ground by a weather balloon. Radiosonde reports are made twice a day simultaneously at hundreds of locations around the world.

Satellite and Radar Images

Images taken from satellites orbiting about 35,000 km above Earth provide information about weather conditions on Earth. A visible light image, such as the one shown in **Figure 14,** shows white clouds over Earth. The infrared image, also shown in **Figure 14,** shows infrared energy in false color, a color that is different from the actual color of the image. The infrared energy comes from Earth and is stored in the atmosphere as thermal energy. Monitoring infrared energy provides information about cloud height and atmospheric temperature.

Figure 14 Meteorologists use visible light and infrared satellite images to identify fronts and air masses.

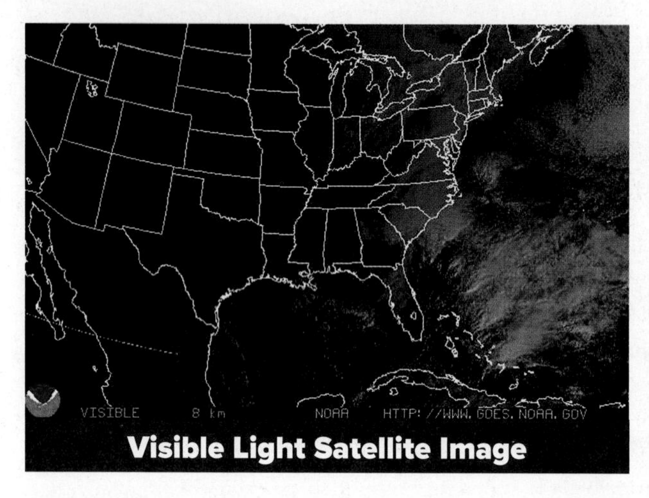

Visible Light Satellite Image

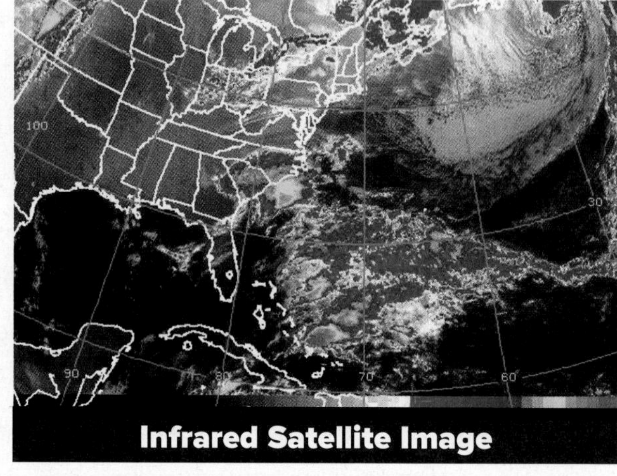

Infrared Satellite Image

Visual Check How is an infrared satellite image different from a visible light satellite image?

Radar measures precipitation when radio waves bounce off raindrops and snowflakes. **Doppler radar** *is a specialized type of radar that can detect precipitation as well as the movement of small particles, which can be used to approximate wind speed.* Because the movement of precipitation is caused by wind, Doppler radar can be used to estimate wind speed. This can be especially important during severe weather, such as tornadoes or thunderstorms.

Key Concept Check Identify the weather variables that radiosondes, infrared satellites, and Doppler radar measure.

FOLDABLES®

Make a horizontal two-tab book and label the tabs as illustrated. Use it to collect information on satellite and radar images. Compare and contrast these information tools.

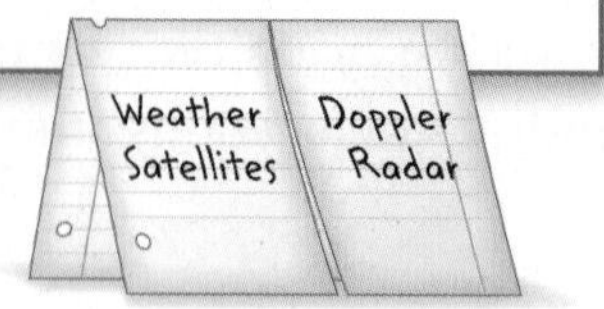

Weather Maps

Every day, thousands of surface reports, upper-air reports, and satellite and radar observations are made around the world. Meteorologists have developed tools that help them simplify and understand this enormous amount of weather data.

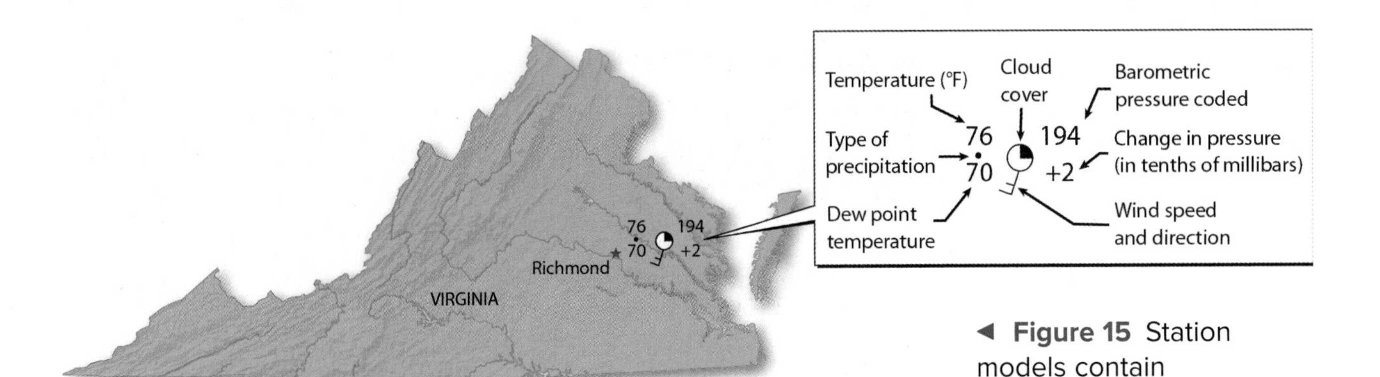

◀ **Figure 15** Station models contain information about weather variables.

The Station Model

As shown in **Figure 15,** the station model diagram displays data from many different weather measurements for a particular location. It uses numbers and symbols to display data and observations from surface reports and upper-air reports.

Mapping Temperature and Pressure

In addition to station models, weather maps also have other symbols. For example, **isobars** *are lines that connect all places on a map where pressure has the same value.* Locate an isobar on the map in **Figure 16.** Isobars show the location of high- and low-pressure systems. Isobars also provide information about wind speed. Winds are strong when isobars are close together. Winds are weaker when isobars are farther apart.

In a similar way, isotherms (not shown) are lines that connect places with the same temperature. Isotherms show which areas are warm and which are cold. Fronts are represented as lines with symbols on them, as indicated in **Figure 16.**

Reading Check Compare isobars and isotherms.

WORD ORIGIN

isobar
from Greek *isos,* means "equal"; and *baros,* means "heavy"

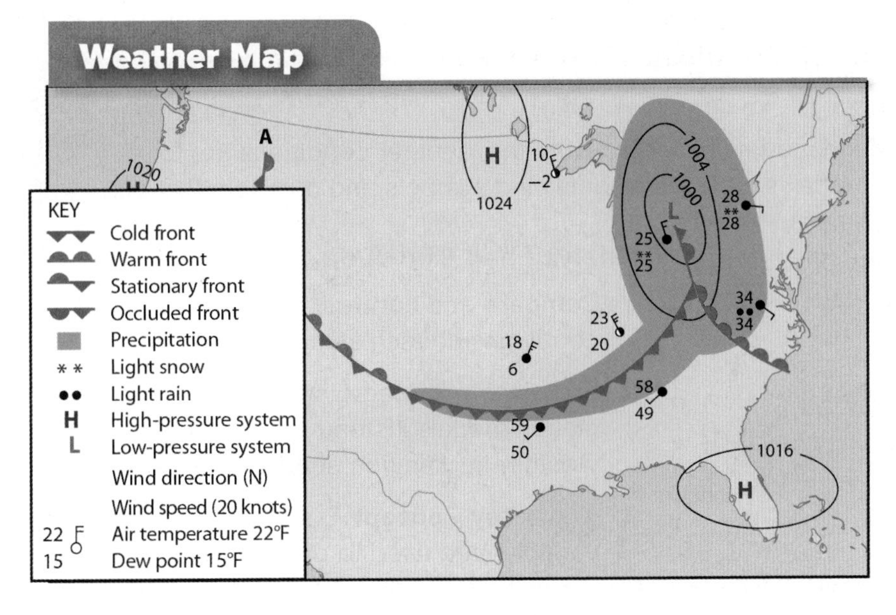

Animation

◀ **Figure 16** Weather maps contain symbols that provide information about the weather.

Visual Check Which symbols represent high-pressure and low-pressure systems?

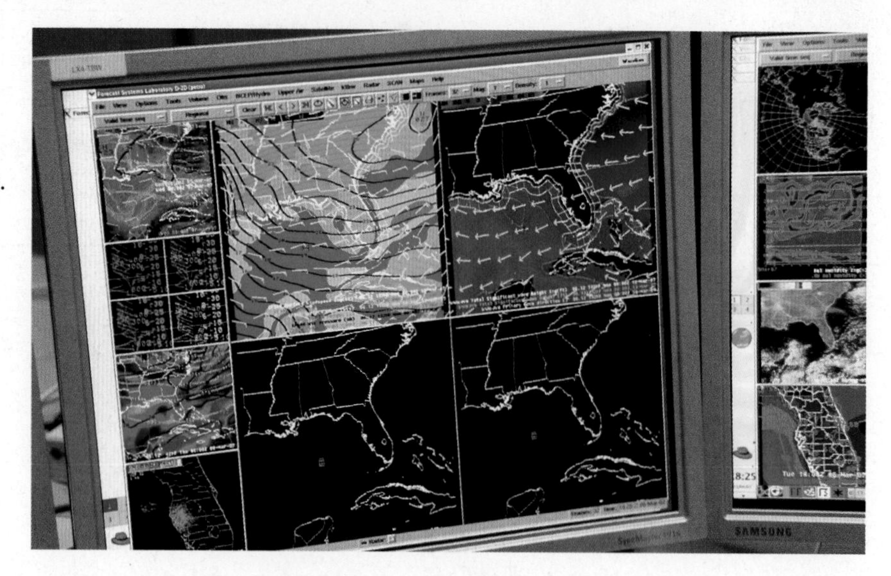

Figure 17 Meteorologists analyze data from various sources—such as radar and computer models—in order to prepare weather forecasts.

Predicting the Weather

Modern weather forecasts are made with the help of computer models, such as the ones shown in **Figure 17**. **Computer models** *are detailed computer programs that solve a set of complex mathematical formulas.* The formulas predict what temperatures and winds might occur, when and where it will rain and snow, and what types of clouds will form.

Government meteorological offices also use computers and the Internet to exchange weather measurements continuously throughout the day. Weather maps are created and forecasts are made using computer models. Then, through television, radio, newspapers, and the Internet, the maps and forecasts are made available to the public.

 **Key Concept Check** How are computers used to predict the weather?

MiniLab

20 minutes

How is weather represented on a map?

Meteorologists often use station models to record what the weather conditions are for a particular location. A station model is a diagram containing symbols and numbers that displays many different weather measurements.

Use the **station model legend** provided by your teacher to interpret the data in each station model shown here.

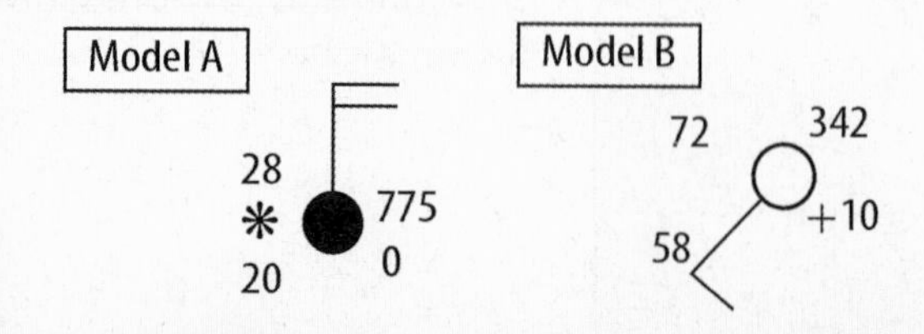

Analyze and Conclude

1. **Compare and contrast** the weather conditions at each station model.
2. **Explain** why meteorologists might use station models instead of reporting weather information another way.
3. **Key Concept** Discuss what variables are used to describe weather.

Dennis MacDonald/Alamy

Lesson 3 Review

Online Quiz
Virtual Lab

Visual Summary

Weather variables are measured by weather stations, radiosondes, satellites, and Doppler radar.

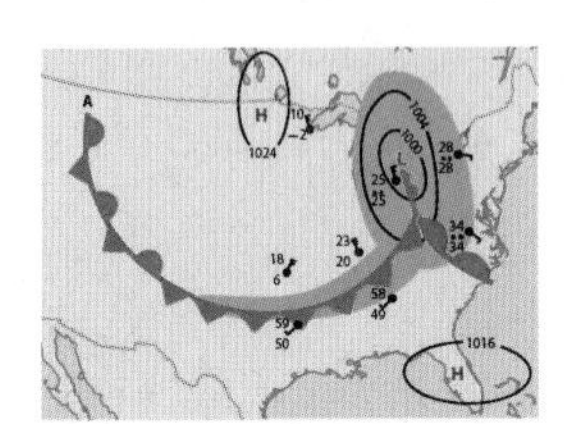

Weather maps contain information in the form of a station model, isobars and isotherms, and symbols for fronts and pressure systems.

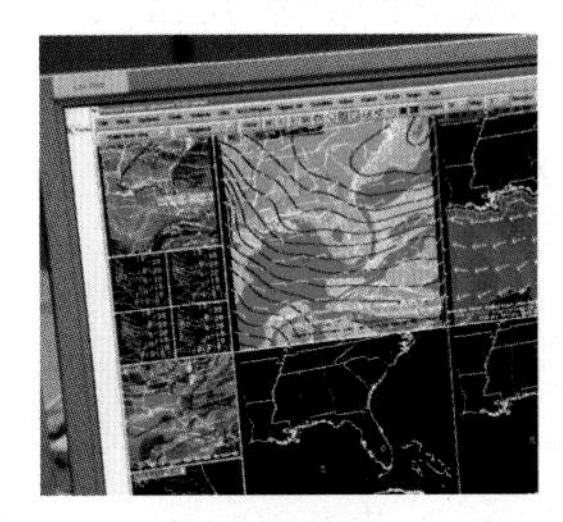

Meteorologists use computer models to help forecast the weather.

FOLDABLES®

Use your lesson Foldable to review the lesson. Save your Foldable for the project at the end of the chapter.

What do you think 

You first read the statements below at the beginning of the chapter.

5. Weather variables are measured every day at locations around the world.

6. Modern weather forecasts are done using computers.

Did you change your mind about whether you agree or disagree with the statements? Rewrite any false statements to make them true.

Use Vocabulary

1. **Define** *computer model* in your own words.
2. A line connecting places with the same pressure is called a(n) ________.
3. **Use the term** *surface report* in a sentence.

Understand Key Concepts 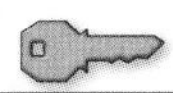

4. Which diagram shows surface weather measurements?
   - **A.** an infrared satellite image
   - **B.** an upper air chart
   - **C.** a station model
   - **D.** a visible light satellite image
5. **List** two ways that upper-air weather conditions are measured.
6. **Describe** how computers are used in weather forecasting.
7. **Distinguish** between isobars and isotherms.

Interpret Graphics

8. **Identify** Copy and fill in the graphic organizer below to identify the components of a surface map.

Symbol	Meaning
H	

Critical Thinking

9. **Suggest** ways to forecast the weather without using computers.
10. **Explain** why isobars and isotherms make it easier to understand a weather map.

Can you predict the weather?

Materials

graph paper

local weather maps

outdoor thermometer

barometer

Weather forecasts are important–not just so you are dressed right when you leave the house, but also to help farmers know when to plant and harvest, to help cities know when to call in the snow plows, and to help officials know when and where to evacuate in advance of severe weather.

Ask a Question

Can you predict the weather?

Make Observations

1. Read and complete a lab safety form.
2. Collect weather data daily for a period of one week. Temperature and pressure should be recorded as a number, but precipitation, wind conditions, and cloud cover can be described in words. Make your observations at the same time each day.

3. Graph temperature in degrees and air pressure in millibars on the same sheet of paper, placing the graphs side by side, as shown on the next page. Beneath the graphs, for each day, add notes that describe precipitation, wind conditions, and cloud cover.

(t to b)Aaron Haupt; (others)Hutchings Photography/Digital Light Source;

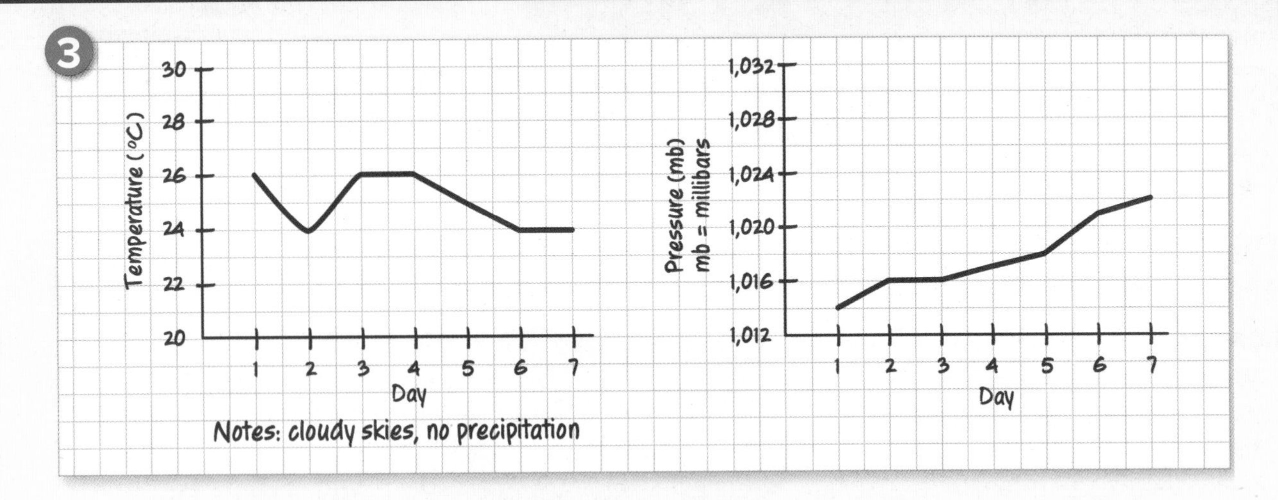

Form a Hypothesis

4. Examine your data and the weather maps. Look for factors that appear to be related. For example, your data might suggest that when the pressure decreases, clouds follow.
5. Find three sets of data pairs that seem to be related. Form three hypotheses, one for each set of data pairs.

Test Your Hypothesis

6. Look at your last day of data. Using your hypotheses, predict the weather for the next day.
7. Collect weather data the next day and evaluate your predictions.
8. Repeat steps 6 and 7 for at least two more days.

Analyze and Conclude

9. **Analyze** Compare your hypotheses with the results of your predictions. How successful were you? What additional information might have improved your predictions?
10. **The Big Idea** Scientists have more complex and sophisticated tools to help them predict their weather, but with fairly simple tools, you can make an educated guess. Write a one-paragraph summary of the data you collected and how you interpreted it to predict the weather.

Communicate Your Results

For each hypothesis you generated, make a small poster that states the hypothesis, shows a graph that supports it, and shows the results of your predictions. Write a concluding statement about the reliability of your hypothesis. Share your results with the class.

Investigate other forms of data you might collect and find out how they would help you to make a forecast. Try them out for a week and see if your ability to make predictions improves.

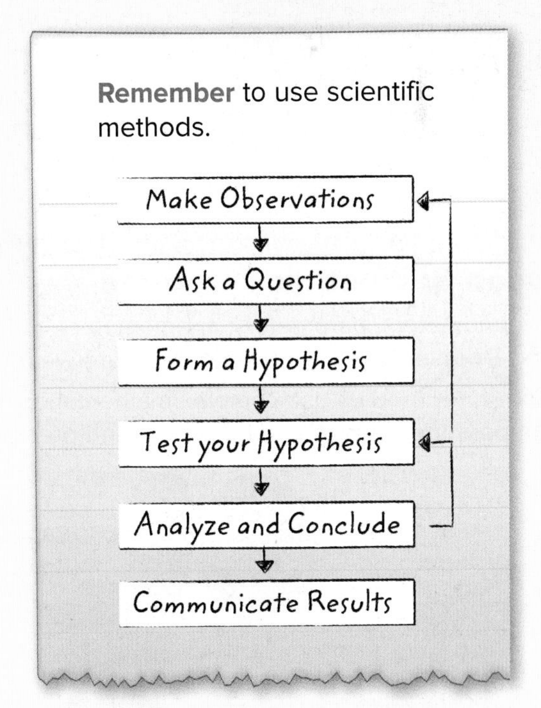

Chapter 13 Study Guide

Scientists use weather variables to describe weather and study weather systems. Scientists use computers to predict the weather.

Key Concepts Summary

Lesson 1: Describing Weather

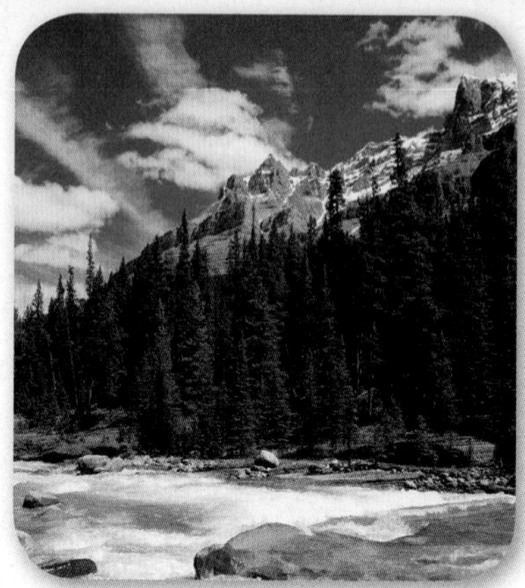

- **Weather** is the atmospheric conditions, along with short-term changes, of a certain place at a certain time.
- Variables used to describe weather are air temperature, **air pressure,** wind, **humidity,** and **relative humidity.**
- The processes in the water cycle—evaporation, condensation, and **precipitation**—are all involved in the formation of different types of weather.

Vocabulary

weather p. 451
air pressure p. 452
humidity p. 452
relative humidity p. 453
dew point p. 453
precipitation p. 455
water cycle p. 455

Lesson 2: Weather Patterns

- **Low-pressure systems** and **high-pressure systems** are two systems that influence weather.
- Weather patterns are driven by the movement of **air masses.**
- Understanding weather patterns helps make weather forecasts more accurate.
- Severe weather includes thunderstorms, **tornadoes, hurricanes,** and **blizzards.**

Vocabulary

high-pressure system p. 459
low-pressure system p. 459
air mass p. 460
front p. 462
tornado p. 465
hurricane p. 466
blizzard p. 467

Lesson 3: Weather Forecasts

- Thermometers, barometers, anemometers, radiosondes, satellites, and **Doppler radar** are used to measure weather variables.
- **Computer models** use complex mathematical formulas to predict temperature, wind, cloud formation, and precipitation.

Vocabulary

surface report p. 471
upper-air report p. 471
Doppler radar p. 472
isobar p. 473
computer model p. 474

Study Guide

Personal Tutor

Vocabulary eFlashcards
Vocabulary eGames

FOLDABLES® Chapter Project

Assemble your lesson Foldables as shown to make a Chapter Project. Use the project to review what you have learned in this chapter.

Use Vocabulary

1. The pressure that a column of air exerts on the area below it is called ________.
2. The amount of water vapor in the air is called ________.
3. The natural process in which water constantly moves among oceans, land, and the atmosphere is called the ________.
4. A(n) ________ is a boundary between two air masses.
5. At the center of a(n) ________, air rises and forms clouds and precipitation.
6. A continental polar ________ brings cold temperatures during winter.
7. When the same ________ passes through two locations on a weather map, both locations have the same pressure.
8. The humidity in the air compared to the amount air can hold is the ________.

Link Vocabulary and Key Concepts

Interactive Concept Map

Copy this concept map, and then use vocabulary terms from the previous page to complete the concept map.

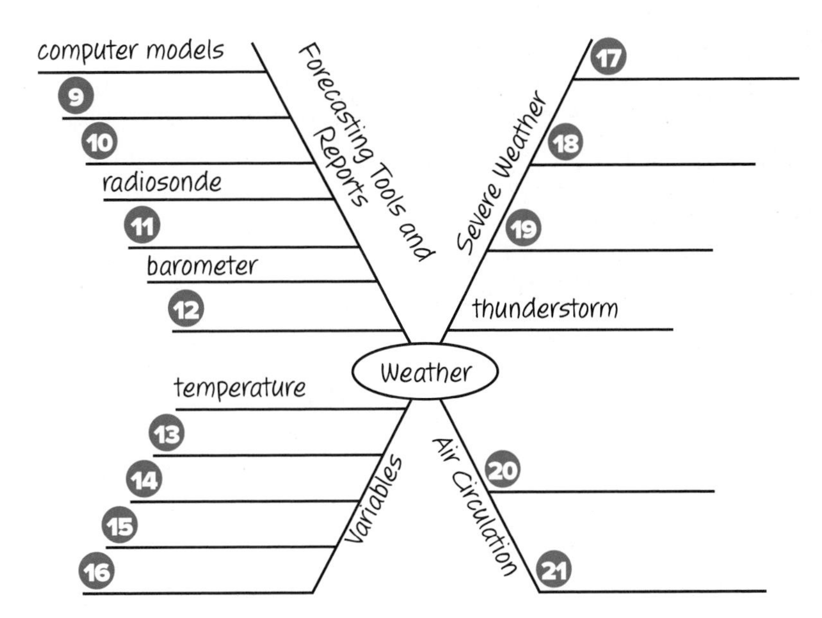

Chapter 13 Review

Understand Key Concepts

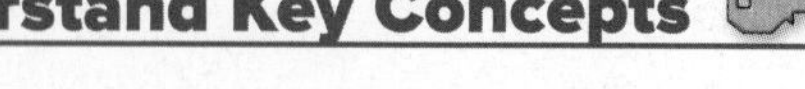

1. Clouds form when water changes from
   A. gas to liquid.
   B. liquid to gas.
   C. solid to gas.
   D. solid to liquid.

2. Which type of precipitation reaches Earth's surface as large pellets of ice?
   A. hail
   B. rain
   C. sleet
   D. snow

3. Which of these sinking-air situations usually brings fair weather?
   A. air mass
   B. cold front
   C. high-pressure system
   D. low-pressure system

4. Which air mass contains cold, dry air?
   A. continental polar
   B. continental tropical
   C. maritime tropical
   D. maritime polar

5. Study the front below.

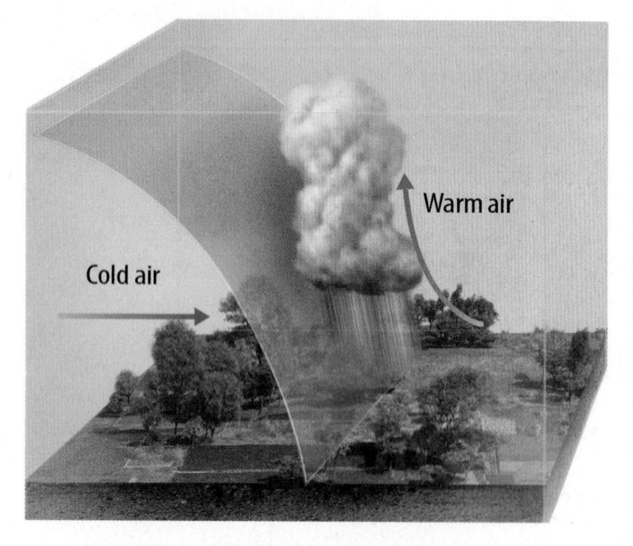

How does this type of front form?
   A. A cold front overtakes a warm front.
   B. Cold air moves toward warmer air.
   C. The boundary between two fronts stalls.
   D. Warm air moves toward colder air.

6. Which is an intense tropical storm with winds exceeding 119 km/h?
   A. blizzard
   B. hurricane
   C. thunderstorm
   D. tornado

7. Which contains measurements of temperature, air pressure, humidity, precipitation, and wind speed and direction?
   A. a radar image
   B. a satellite image
   C. a surface report
   D. a weather station

8. What does Doppler radar measure?
   A. air pressure
   B. air temperature
   C. the rate at which air pressure changes
   D. the speed at which precipitation travels

9. Study the station model below.

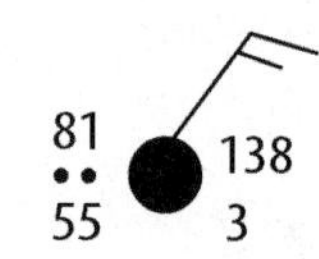

What is the temperature according to the station model?
   A. 3°F
   B. 55°F
   C. 81°F
   D. 138°F

10. Which describes cirrus clouds?
   A. flat, white, and layered
   B. fluffy, at middle altitudes
   C. heaped or piled up
   D. wispy, at high altitudes

11. Which instrument measures wind speed?
   A. anemometer
   B. barometer
   C. psychrometer
   D. thermometer

Critical Thinking

12 Predict Suppose you are on a ship near the equator in the Atlantic Ocean. You notice that the barometric pressure is dropping. Predict what type of weather you might experience.

13 Compare a continental polar air mass with a maritime tropical air mass.

14 Assess why clouds usually form in the center of a low-pressure system.

15 Predict how maritime air masses would change if the oceans froze.

16 Compare two types of severe weather.

17 Interpret Graphics Identify the front on the weather map below. Predict the weather for areas along the front.

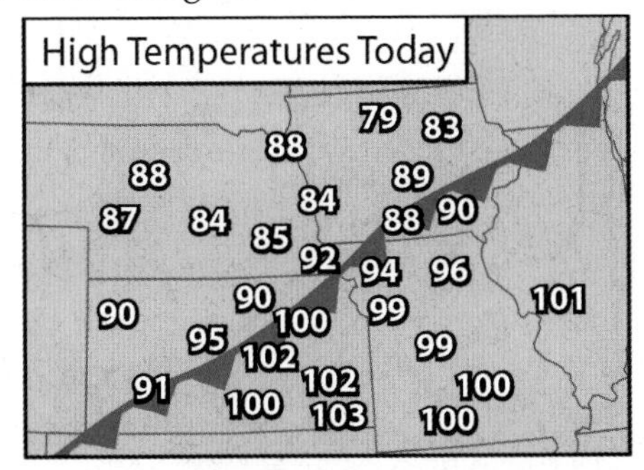

18 Assess the validity of the weather forecast: "Tomorrow's weather will be similar to today's weather."

19 Compare and contrast surface weather reports and upper-air reports. Why is it important for meterologists to monitor weather variables high above Earth's surface?

Writing in Science

20 Write a paragraph about the ways computers have improved weather forecasts. Be sure to include a topic sentence and a concluding sentence.

REVIEW THE BIG IDEA

21 Identify the instruments used to measure weather variables.

22 How do scientists use weather variables to describe and predict weather?

23 Describe the factors that influence weather.

24 Use the factors listed in question 23 to describe how a continental polar air mass can change to a maritime polar air mass.

Math Skills

Math Practice

Use Conversions

25 Convert from Fahrenheit to Celsius.

a. Convert 0°F to °C.

b. Convert 104°F to °C.

26 Convert from Celsius to Fahrenheit.

a. Convert 0°C to °F.

b. Convert −40°C to °F.

27 The Kelvin scale of temperature measurement starts at zero and has the same unit size as Celsius degrees. Zero degrees Celsius is equal to 273 kelvin.

Convert 295 K to Fahrenheit.

Standardized Test Practice

Record your answers on the answer sheet provided by your teacher or on a sheet of paper.

Multiple Choice

1 Which measures the average kinetic energy of air molecules?

A humidity

B pressure

C speed

D temperature

Use the diagram below to answer question 2.

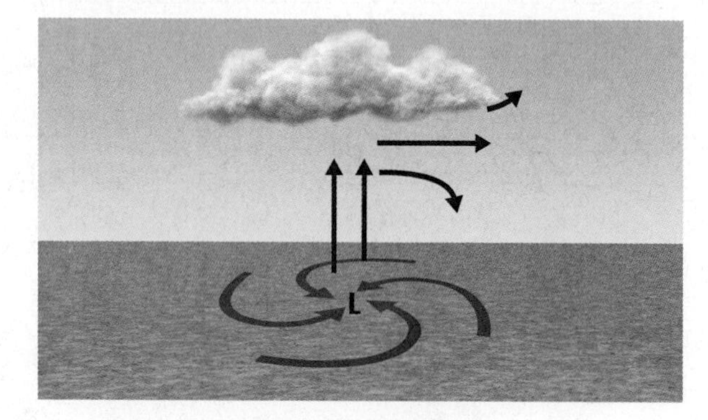

2 Which weather system does the above diagram illustrate?

A high pressure

B hurricane

C low pressure

D tornado

3 What causes weather to remain the same several days in a row?

A air front

B air mass

C air pollution

D air resistance

4 Which lists the stages of a thunderstorm in order?

A cumulus, dissipation, mature

B cumulus, mature, dissipation

C dissipation, cumulus, mature

D dissipation, mature, cumulus

5 What causes air to reach its dew point?

A decreasing air currents

B decreasing humidity

C dropping air pressure

D dropping temperatures

6 Which measures air pressure?

A anemometer

B barometer

C psychrometer

D thermometer

Use the diagram below to answer question 7.

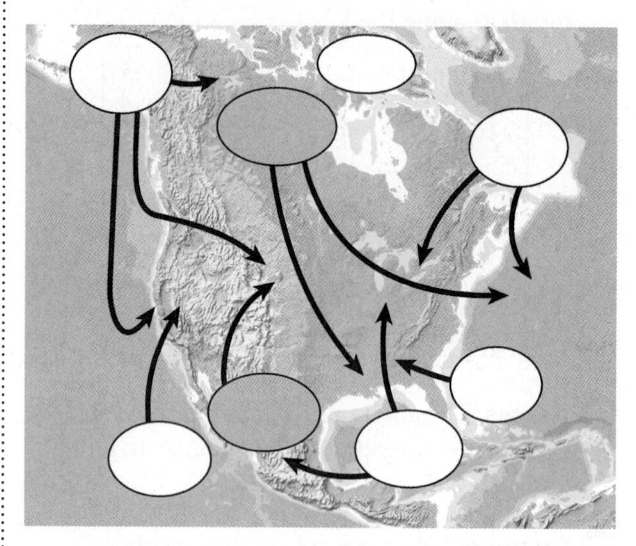

7 Which type of air masses do the shaded ovals in the diagram depict?

A antarctic

B arctic

C continental

D maritime

8 Which BEST expresses moisture saturation?

A barometric pressure

B relative humidity

C weather front

D wind direction

Use the diagram below to answer question 9.

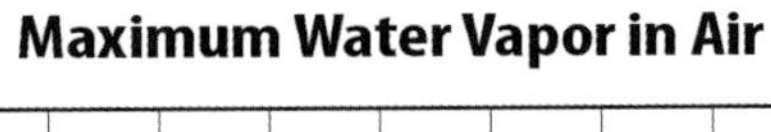

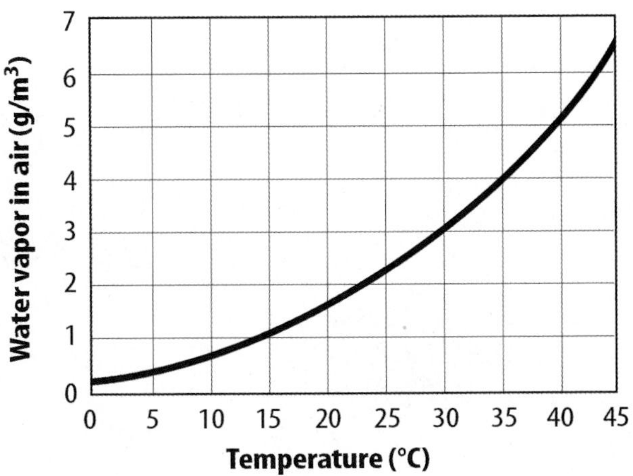

9 What happens to maximum moisture content when air temperatures increase from 15°C to 30°C?

A increases from 1 to 2 g/m^3

B increases from 1 to 3 g/m^3

C increases from 2 to 3 g/m^3

D increases from 2 to 4 g/m^3

10 When isobars are close together on a weather map,

A cloud cover is extensive.

B temperatures are high.

C warm fronts prevail.

D winds are strong.

11 Which provides energy for the water cycle?

A air currents

B Earth's core

C ocean currents

D the Sun

Constructed Response

Use the table below to answer question 12.

Weather Variable	Measurement

12 In the table above, list the variables weather scientists use to describe weather. Then describe the unit of measurement for each variable.

Use the diagram below to answer questions 13 and 14.

13 What does the diagram above depict?

14 Describe the weather conditions associated with the diagram.

15 How do weather fronts form?

NEED EXTRA HELP?															
If You Missed Question...	1	2	3	4	5	6	7	8	9	10	11	12	13	14	15
Go to Lesson...	1	2	2	2	1	1, 3	2	1	1	3	1	1	2	2	2

Chapter 14

Climate

THE BIG IDEA **What is climate and how does it impact life on Earth?**

Inquiry What happened to this tree?

Climate differs from one area of Earth to another. Some areas have little rain and high temperatures. Other areas have low temperatures and lots of snow. Where this tree grows—on Humphrey Head Point in England—there is constant wind.

- What are the characteristics of different climates?
- What factors affect the climate of a region?
- What is climate and how does it impact life on Earth?

Get Ready to Read

What do you think?

Before you read, decide if you agree or disagree with each of these statements. As you read this chapter, see if you change your mind about any of the statements.

1. Locations at the center of large continents usually have the same climate as locations along the coast.
2. Latitude does not affect climate.
3. Climate on Earth today is the same as it has been in the past.
4. Climate change occurs in short-term cycles.
5. Human activities can impact climate.
6. You can help reduce the amount of greenhouse gases released into the atmosphere.

Your one-stop online resource
connectED.mcgraw-hill.com

LearnSmart®

Chapter Resources Files, Reading Essentials, Get Ready to Read, Quick Vocabulary

Animations, Videos, Interactive Tables

Self-checks, Quizzes, Tests

Project-Based Learning Activities

Lab Manuals, Safety Videos, Virtual Labs & Other Tools

Vocabulary, Multilingual eGlossary, Vocab eGames, Vocab eFlashcards

Personal Tutors

Lesson 1

Reading Guide

Key Concepts 

ESSENTIAL QUESTIONS

- What is climate?
- Why is one climate different from another?
- How are climates classified?

Vocabulary

climate p. 487
rain shadow p. 489
specific heat p. 489
microclimate p. 491

Multilingual eGlossary

Climates of Earth

Inquiry What makes a desert a desert?

How much precipitation do deserts get? Are deserts always hot? What types of plants grow in the desert? Scientists look at the answers to all these questions to determine if an area is a desert.

Launch Lab

20 minutes

How do climates compare?

Climate describes long-term weather patterns for an area. Temperature and precipitation are two factors that help determine climate.

1. Read and complete a lab safety form.
2. Select a location on a **globe.**
3. Research the average monthly temperatures and levels of precipitation for this location.
4. Record your data in a chart like the one shown here in your Science Journal.

Omsk, Russia 73.5° E, 55° N		
Month	Average Monthly Temperature	Average Monthly Level of Precipitation
January	−14°C	13 mm
February	−12°C	9 mm
March	−5°C	9 mm
April	8°C	18 mm
May	18°C	31 mm
June	24°C	52 mm
July	25°C	61 mm
August	22°C	50 mm
September	17°C	32 mm
October	7°C	26 mm
November	−4°C	19 mm
December	−12°C	15 mm

Think About This

1. Describe the climate of your selected location in terms of temperature and precipitation.
2. Compare your data to Omsk, Russia. How do the climates differ?
3. **Key Concept** Mountains, oceans, and latitude can affect climates. Do any of these factors account for the differences you observed? Explain.

What is climate?

You probably already know that the term *weather* describes the atmospheric conditions and short term changes of a certain place at a certain time. The weather changes from day to day in many places on Earth. Other places on Earth have more constant weather. For example, temperatures in Antarctica rarely are above 0°C, even in the summer. Areas in Africa's Sahara, shown in the photo on the previous page, have temperatures above 20°C year-round.

Climate *is the long-term average weather conditions that occur in a particular region.* A region's climate depends on average temperature and precipitation, as well as how these variables change throughout the year.

What affects climate?

Several factors determine a region's climate. The latitude of a location affects climate. For example, areas close to the equator have the warmest climates. Large bodies of water, including lakes and oceans, also influence the climate of a region. Along coastlines, weather is more constant throughout the year. Hot summers and cold winters typically happen in the center of continents. The altitude of an area affects climate. Mountainous areas are often rainy or snowy. Buildings and concrete, which retain solar energy, cause temperatures to be higher in urban areas. This creates a special climate in a small area.

Key Concept Check What is climate?

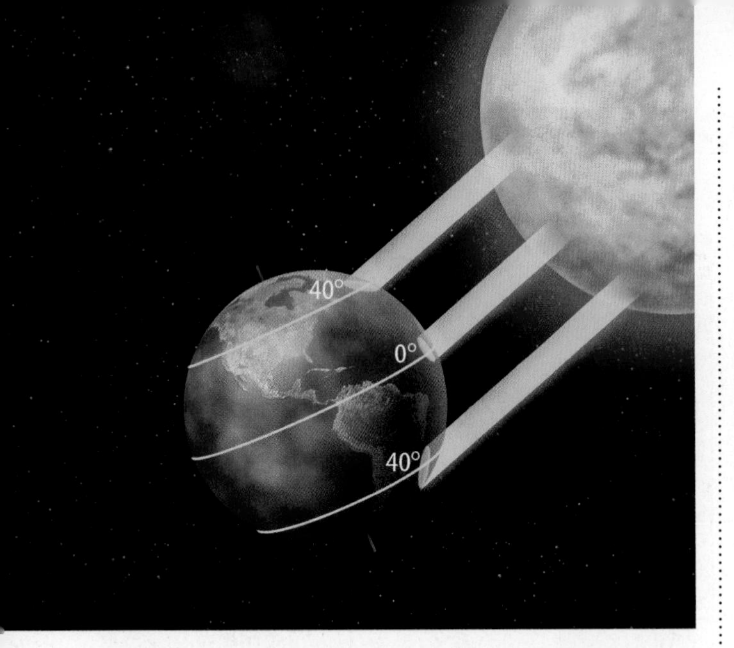

Figure 1 Latitudes near the poles receive less solar energy and have lower average temperatures.

Latitude

Recall that, starting at the equator, latitude increases from 0° to 90° as you move toward the North Pole or the South Pole. The amount of solar energy per unit of Earth's surface area depends on latitude. **Figure 1** shows that locations close to the equator receive more solar energy per unit of surface area annually than locations located farther north or south. This is due mainly to the fact that Earth's curved surface causes the angle of the Sun's rays to spread out over a larger area. Locations near the equator also tend to have warmer climates than locations at higher latitudes. Polar regions are colder because annually they receive less solar energy per unit of surface area. In the middle latitudes, between 30° and 60°, summers are generally hot and winters are usually cold.

Altitude

Climate is also influenced by altitude. Recall that temperature decreases as altitude increases in the troposphere. So, as you climb a tall mountain you might experience the same cold, snowy climate that is near the poles. **Figure 2** shows the difference in average temperatures between two cities in Colorado at different altitudes.

Altitude and Climate

Figure 2 As altitude increases, temperature decreases.

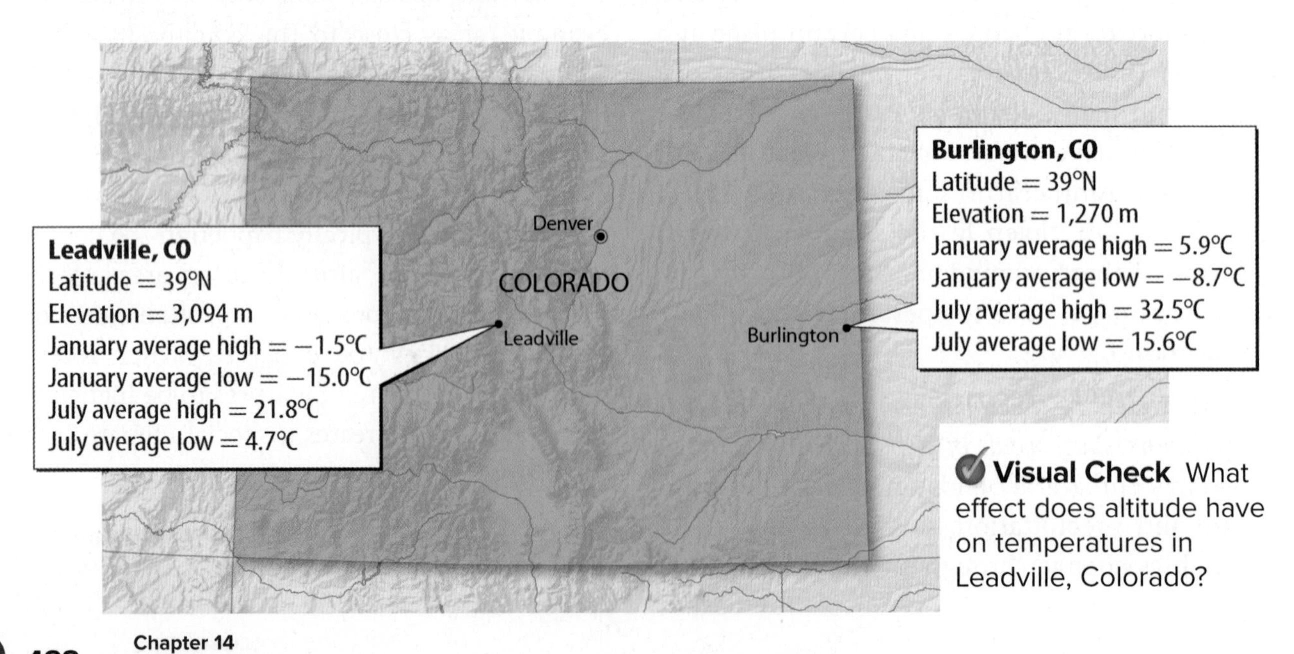

Visual Check What effect does altitude have on temperatures in Leadville, Colorado?

Rain Shadow

Figure 3 Rain shadows form on the downwind slope of a mountain.

Visual Check Why don't rain shadows form on the upwind slope of mountains?

Rain Shadows

Mountains influence climate because they are barriers to prevailing winds. This leads to unique **precipitation** patterns called rain shadows. *An area of low rainfall on the downwind slope of a mountain is called a* **rain shadow,** as shown in **Figure 3.** Different amounts of precipitation on either side of a mountain range influence the types of vegetation that grow. Abundant amounts of vegetation grow on the side of the mountain exposed to the precipitation. The amount of vegetation on the downwind slope is sparse due to the dry weather.

REVIEW VOCABULARY

precipitation
water, in liquid or solid form, that falls from the atmosphere

Large Bodies of Water

On a sunny day at the beach, why does the sand feel warmer than the water? It is because water has a high specific heat. **Specific heat** *is the amount (joules) of thermal energy needed to raise the temperature of 1 kg of a material by 1°C.* The specific heat of water is about six times higher than the specific heat of sand. This means the ocean water would have to absorb six times as much thermal energy to be the same temperature as the sand.

The high specific heat of water causes the climates along coastlines to remain more constant than those in the middle of a continent. For example, the West Coast of the United States has moderate temperatures year-round.

Ocean currents can also modify climate. The Gulf Stream is a warm current flowing northward along the coast of eastern North America. It brings warmer temperatures to portions of the East Coast of the United States and parts of Europe.

 **Reading Check** How do large bodies of water influence climate?

World Climates

Figure 4 The map shows a modified version of Köppen's climate classification system.

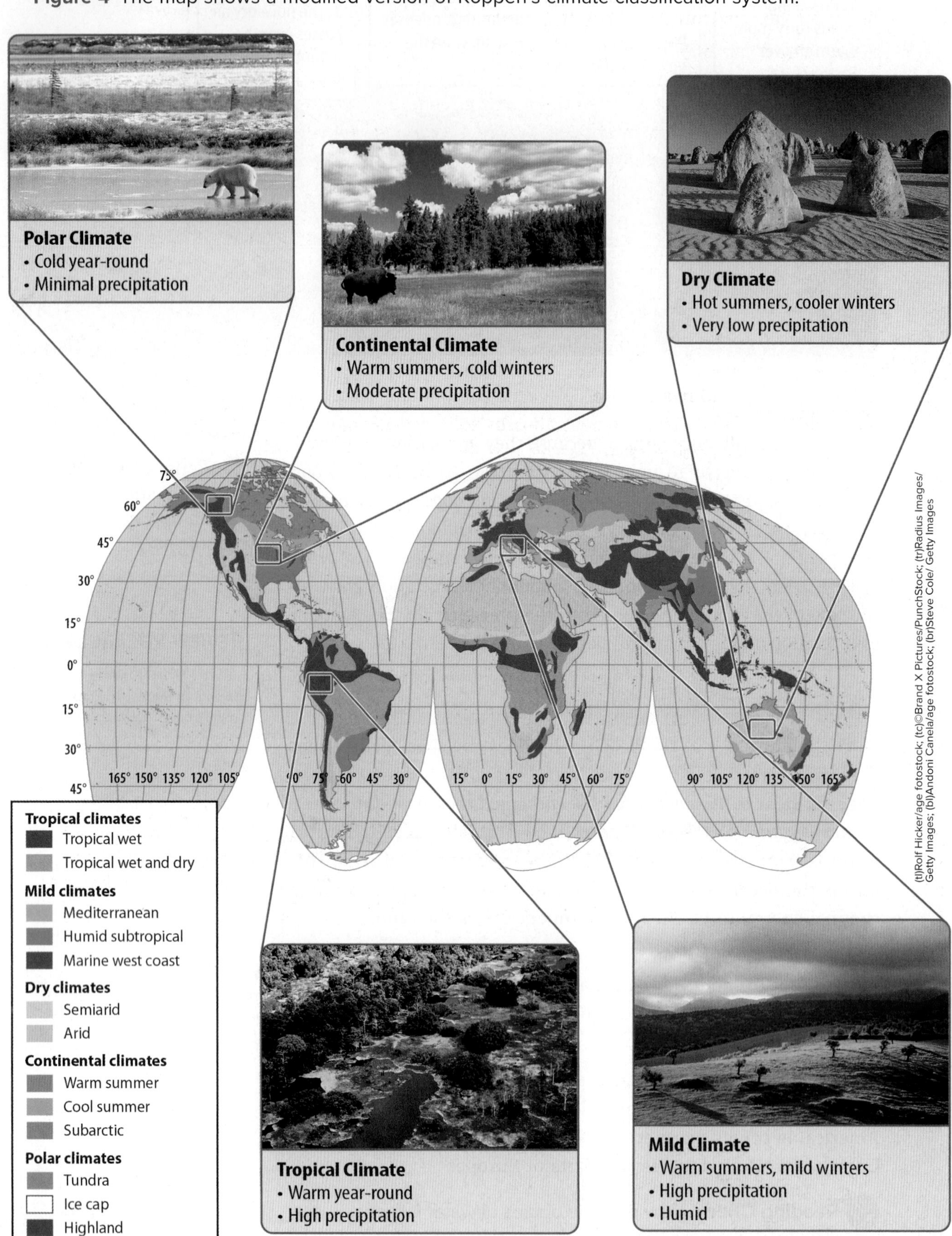

Classifying Climates

What is the climate of any particular region on Earth? This can be a difficult question to answer because many factors affect climate. In 1918 German scientist Wladimir Köppen (vlah DEE mihr • KAWP pehn) developed a system for classifying the world's many climates. Köppen classified a region's climate by studying its temperature, precipitation, and native vegetation. Native vegetation is often limited to particular climate conditions. For example, you would not expect to find a warm-desert cactus growing in the cold, snowy arctic. Köppen identified five climate types. A modified version of Köppen's classification system is shown in **Figure 4.**

Key Concept Check How are climates classified?

Microclimates

Roads and buildings in cities have more concrete than surrounding rural areas. The concrete absorbs solar radiation, causing warmer temperatures than in the surrounding countryside. The result is a common microclimate called the urban heat island, as shown in **Figure 5.** *A* **microclimate** *is a localized climate that is different from the climate of the larger area surrounding it.* Other examples of microclimates include forests, which are often cooler and less windy than the surrounding countryside, and hilltops, which are windier than nearby lower land.

Key Concept Check Why is one climate different from another?

FOLDABLES®

Use three sheets of notebook paper to make a layered book. Label it as shown. Use it to organize your notes on the factors that determine a region's climate.

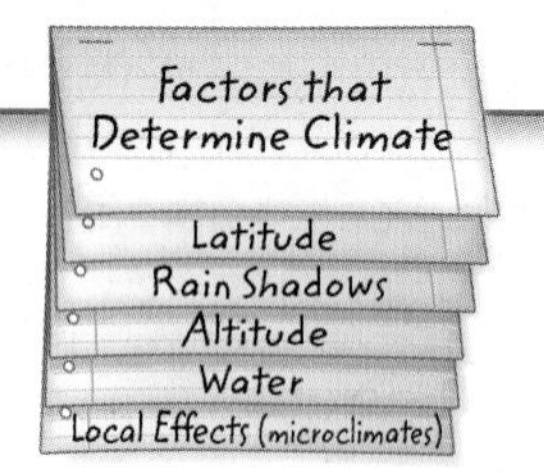

WORD ORIGIN

microclimate
from Greek *mikros,* means "small"; and *klima,* means "region, zone"

Microclimate

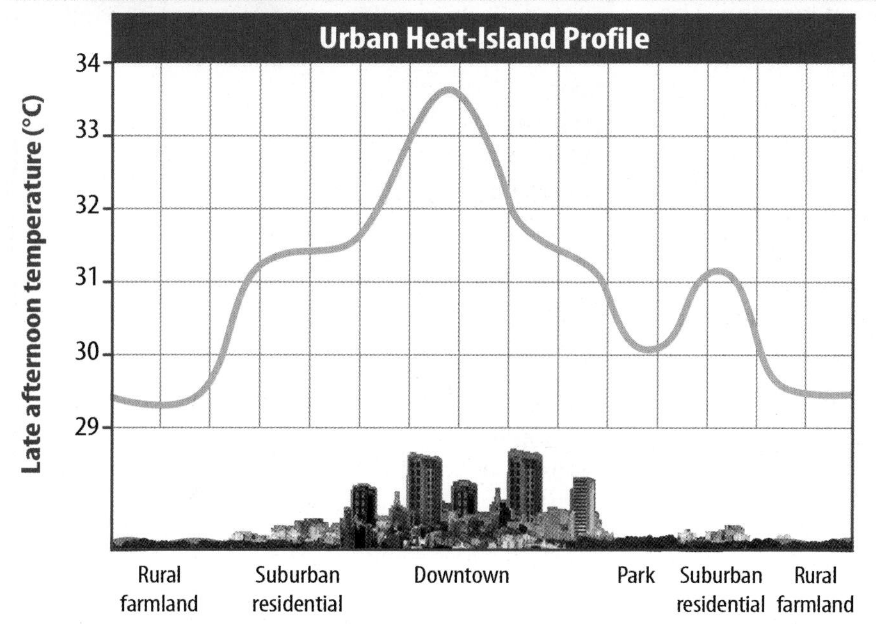

Figure 5 The temperature is often warmer in urban areas when compared to temperatures in the surrounding countryside.

Visual Check What is the temperature difference between downtown and rural farmland?

Figure 6 Camels are adapted to dry climates and can survive up to three weeks without drinking water.

How Climate Affects Living Organisms

Organisms have adaptations for the climates where they live. For example, polar bears have thick fur and a layer of fat that helps keep them warm in the Arctic. Many animals that live in deserts, such as the camels in **Figure 6**, have adaptations for surviving in hot, dry conditions. Some desert plants have extensive shallow root systems that collect rainwater. Deciduous trees, found in continental climates, lose their leaves during the winter, which reduces water loss when soils are frozen.

Climate also influences humans in many ways. Average temperature and rainfall in a location help determine the type of crops humans grow there. Thousands of orange trees grow in Florida, where the climate is mild. Wisconsin's continental climate is ideal for growing cranberries.

Climate also influences the way humans design buildings. In polar climates, the soil is frozen year-round–a condition called permafrost. Humans build houses and other buildings in these climates on stilts. This is done so that thermal energy from the building does not melt the permafrost.

Reading Check How are organisms adapted to different climates?

MiniLab

40 minutes

Where are microclimates found?

Microclimates differ from climates in the larger region around them. In this lab, you will identify a microclimate.

1. Read and complete a lab safety form.
2. Select two areas near your school. One area should be in an open location. The other area should be near the school building.
3. Make a data table like the one at the right in your Science Journal.
4. Measure and record data at the first area. Find wind direction using a **wind sock,** temperature using a **thermometer,** and relative humidity using a **psychrometer** and a **relative humidity chart.**
5. Repeat step 4 at the second area.

	Sidewalk	Soccer Fields
Temperature		
Wind direction		
Relative humidity		

Analyze and Conclude

1. **Graph Data** Make a bar graph showing the temperature and relative humidity at both sites.
2. **Use** the data in your table to compare wind direction.
3. **Interpret Data** How did weather conditions at the two sites differ? What might account for these differences?
4.  **Key Concept** How might you decide which site is a microclimate? Explain.

Lesson 1 Review

Online Quiz

Visual Summary

Climate is influenced by several factors including latitude, altitude, and an area's location relative to a large body of water or mountains.

Rain shadows occur on the downwind slope of mountains.

Microclimates can occur in urban areas, forests, and hilltops.

FOLDABLES®

Use your lesson Foldable to review the lesson. Save your Foldable for the project at the end of the chapter.

What do you think NOW?

You first read the statements below at the beginning of the chapter.

1. Locations at the center of large continents usually have the same climate as locations along the coast.

2. Latitude does not affect climate.

Did you change your mind about whether you agree or disagree with the statements? Rewrite any false statements to make them true.

Use Vocabulary

1. The amount of thermal energy needed to raise the temperature of 1 kg of a material by 1°C is called ________.
2. **Distinguish** between climate and microclimate.
3. **Use the term** *rain shadow* in a sentence.

Understand Key Concepts

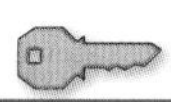

4. How are climates classified?
   - **A.** by cold- and warm-water ocean currents
   - **B.** by latitude and longitude
   - **C.** by measurements of temperature and humidity
   - **D.** by temperature, precipitation, and vegetation
5. **Describe** the climate of an island in the tropical Pacific Ocean.
6. **Compare** the climates on either side of a large mountain range.
7. **Distinguish** between weather and climate.

Interpret Graphics

8. **Summarize** Copy and fill in the graphic organizer below to summarize information about the different types of climate worldwide.

Climate Type	Description
Tropical	
Dry	
Mild	
Continental	
Polar	

Critical Thinking

9. **Distinguish** between the climates of a coastal location and a location in the center of a large continent.
10. **Infer** how you might snow ski on the island of Hawaii.

Can reflection of the Sun's rays change the climate?

Materials

bowl

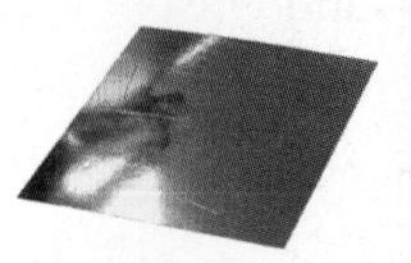

polyester film

transparent tape

stopwatch

light source

thermometer

Safety

Albedo is the term used to refer to the percent of solar energy that is reflected back into space. Clouds, for example, reflect about 50 percent of the solar energy they receive, whereas dark surfaces on Earth might reflect as little as 5 percent. Snow has a very high albedo and reflects 75 to 90 percent of the solar energy it receives. The differences in how much solar energy is reflected back into the atmosphere from different regions of Earth can cause differences in climate. Also, changes in albedo can affect the climate of that region.

Learn It

When an observation cannot be made directly, a simulation can be used to draw reasonable conclusions. This strategy is known as **inferring.** Simulating natural occurrences on a small scale can provide indirect observations so realistic outcomes can be inferred.

Try It

1. Read and complete a lab safety form.
2. Make a data table for recording temperatures in your Science Journal.
3. Cover the bottom of a bowl with a sheet of polyester film. Place a thermometer on top of the sheet. Record the temperature in the bottom of the bowl.
4. Put the bowl under the light source and set the timer for 5 minutes. After 5 minutes, record the temperature. Remove the thermometer and allow it to return to its original temperature. Repeat two more times.
5. Repeat the experiment, but this time tape the sheet of polyester film over the top of the bowl and the thermometer.

Apply It

6. **Analyze** the data you collected. What difference did you find when the polyester film covered the bowl?
7. **Conclude** What can you conclude about the Sun's rays reaching the bottom of the bowl when it was covered by the polyester film?
8. **Infer** what happens to the Sun's rays when they reach clouds in the atmosphere. Explain.
9. **Describe** how the high albedo of the ice and snow in the polar regions contribute to the climate there.
10. **Key Concept** If a region of Earth were to be covered most of the time by smog or clouds, would the climate of that region change? Explain your answer.

Lesson 2

Climate Cycles

Reading Guide

Key Concepts
ESSENTIAL QUESTIONS

- How has climate varied over time?
- What causes seasons?
- How does the ocean affect climate?

Vocabulary

ice age p. 496
interglacial p. 496
El Niño/Southern Oscillation p. 500
monsoon p. 501
drought p. 501

Multilingual eGlossary

Inquiry How did this lake form?

A melting glacier formed this lake. How long ago did this happen? What type of climate change occurred to cause a glacier to melt? Will it happen again?

Quasarphoto/Getty Images

Launch Lab

20 minutes

How does Earth's tilted axis affect climate?

Earth's axis is tilted at an angle of 23.5°. This tilt influences climate by affecting the amount of sunlight that reaches Earth's surface.

1. Read and complete a lab safety form.
2. Hold a **penlight** about 25 cm above a sheet of paper at a 90° angle. Use a **protractor** to check the angle.
3. Turn off the overhead lights and turn on the penlight. Your partner should trace the circle of light cast by the penlight onto the paper.
4. Repeat steps 2 and 3, but this time hold the penlight at an angle of 23.5° from perpendicular.

Think About This

1. How did the circles of light change during each trial?
2. Which trial represented the tilt of Earth's axis?
3.  **Key Concept** How might changes in the tilt of Earth's axis affect climate? Explain.

Figure 7 Scientists study the different layers in an ice core to learn more about climate changes in the past.

Long-Term Cycles

Weather and climate have many cycles. In most areas on Earth, temperatures increase during the day and decrease at night. Each year, the air is warmer during summer and colder during winter. But climate also changes in cycles that take much longer than a lifetime to complete.

Much of our knowledge about past climates comes from natural records of climate. Scientists study ice cores, shown in **Figure 7**, drilled from ice layers in glaciers and ice sheets. Fossilized pollen, ocean sediments, and the growth rings of trees also are used to gain information about climate changes in the past. Scientists use the information to compare present-day climates to those that occurred many thousands of years ago.

Reading Check How do scientists find information about past climates on Earth?

Ice Ages and Interglacials

Earth has experienced many major atmospheric and climate changes in its history. **Ice ages** *are cold periods lasting millions of years when glaciers cover much of Earth.* Glaciers and ice sheets advance during cold periods, called glacials, and retreat during **interglacials**–*the warm periods that occur during ice ages.* There have been at least five major ice ages in Earth's history.

Major Ice Ages and Warm Periods

The most recent ice age began about 2.6 million years ago. The ice sheets reached maximum size about 22,000 years ago. At that time, about half the northern hemisphere was covered by ice. About 11,700 years ago, Earth entered its current interglacial period, called the Holocene Epoch.

Temperatures on Earth have fluctuated during the Holocene. For example, the period between 950 and 1100 was one of the warmest in Europe. The Little Ice Age, which lasted from 1250 to about 1850, was a period of bitterly cold temperatures.

WORD ORIGIN

interglacial
from Latin *inter-*, means "among, between"; and *glacialis*, means "icy, frozen"

 **Key Concept Check** How has climate varied over time?

Causes of Long-Term Climate Cycles

As the amount of solar energy reaching Earth changes, Earth's climate also changes. One factor that affects how much energy Earth receives is the shape of its orbit. The shape of Earth's orbit appears to vary between elliptical and circular over the course of about 100,000 years. As shown in **Figure 8,** when Earth's orbit is more circular, Earth averages a greater distance from the Sun. This results in below-average temperatures on Earth.

Another factor that scientists suspect influences climate change on Earth is changes in the tilt of Earth's axis. The tilt of Earth's axis changes in 41,000-year cycles. Changes in the angle of Earth's tilt affect the range of temperatures throughout the year. For example, a decrease in the angle of Earth's tilt, as shown in **Figure 8,** could result in a decrease in temperature differences between summer and winter. Long-term climate cycles are also influenced by the slow movement of Earth's continents, as well as changes in ocean circulation.

Figure 8 This exaggerated image shows how the shape of Earth's orbit varies between elliptical and circular. The angle of the tilt varies from 22° to 24.5° about every 41,000 years. Earth's current tilt is 23.5°.

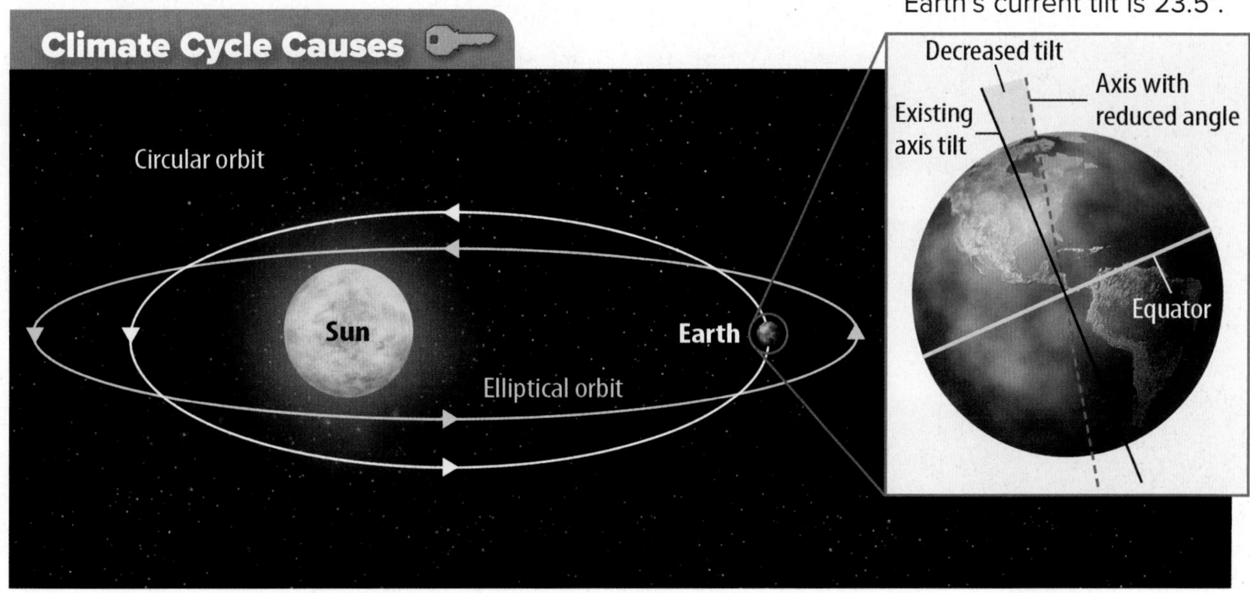

FOLDABLES®

Make a horizontal three-tab book and label it as shown. Use your book to organize information about short-term climate cycles. Fold the book into thirds and label the outside *Short Term Climate Cycles.*

Short-Term Cycles

In addition to its long-term cycles, climate also changes in short-term cycles. Seasonal changes and changes that result from the interaction between the ocean and the atmosphere are some examples of short-term climate change.

Seasons

Changes in the amount of solar energy received at different latitudes during different times of the year give rise to the seasons. Seasonal changes include regular changes in temperature and the number of hours of day and night.

Recall from Lesson 1 that the amount of solar energy per unit of Earth's surface is related to latitude. Another factor that affects the amount of solar energy received by an area is the tilt of Earth's axis. **Figure 9** shows that when the northern hemisphere is tilted toward the Sun, the angle at which the Sun's rays strike Earth's surface is higher. There are more daylight hours than dark hours. During this time, temperatures are warmer, and the northern hemisphere experiences summer. At the same time, the southern hemisphere is tilted away from the Sun and the angle at which the Sun's rays strike Earth's surface is lower. There are fewer hours of daylight, and the southern hemisphere experiences winter.

Figure 9 shows that the opposite occurs when six months later the northern hemisphere is tilted away from the Sun. The angle at which Sun's rays strike Earth's surface is lower, and temperatures are colder. During this time, the northern hemisphere experiences winter. The southern hemisphere is tilted toward the Sun and the angle between the Sun's rays and Earth's surface is higher. The southern hemisphere experiences summer.

Key Concept Check What causes seasons?

Personal Tutor

Figure 9 The solar energy rays reaching a given area of Earth's surface is more intense when tilted toward the Sun.

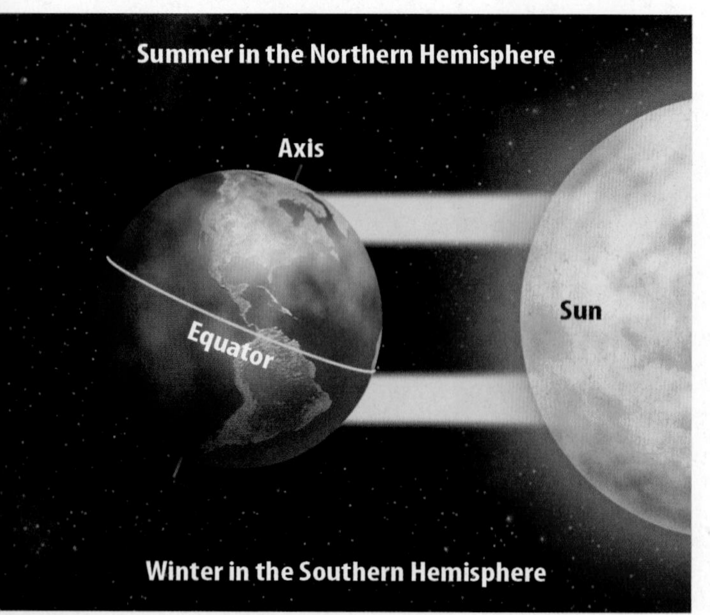

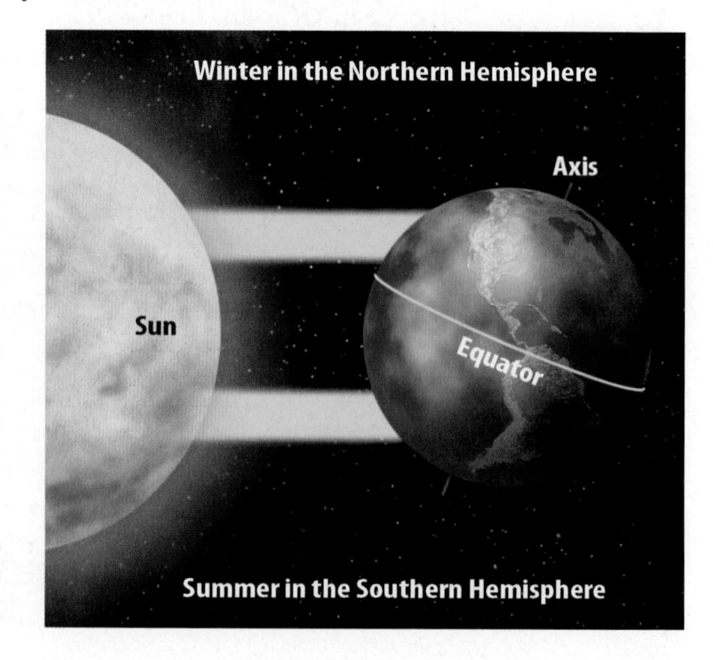

Northern Hemisphere Seasons

Animation

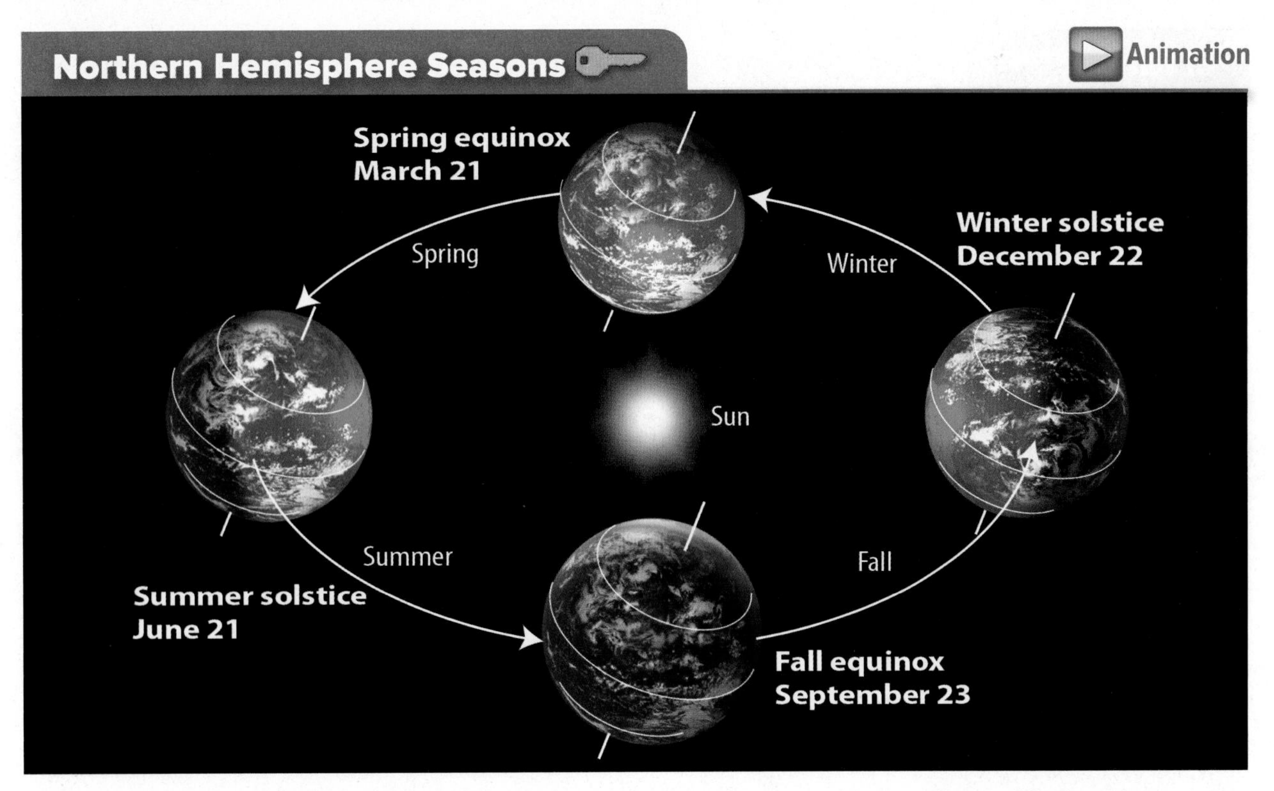

Figure 10 Seasons change as Earth completes its yearly revolution around the Sun.

Visual Check How does the amount of sunlight striking the North Pole change from summer to winter?

Solstices and Equinoxes

Earth revolves around the Sun once about every 365 days. During Earth's **revolution,** there are four days that mark the beginning of each of the seasons. These days are a summer solstice, a fall equinox, a winter solstice, and a spring equinox.

As shown in **Figure 10,** the solstices mark the beginnings of summer and winter. In the northern hemisphere, the summer solstice occurs on June 21 or 22. On this day, the northern hemisphere is tilted toward the Sun. In the southern hemisphere, this day marks the beginning of winter. The winter solstice begins on December 21 or 22 in the northern hemisphere. On this day, the northern hemisphere is tilted away from the Sun. In the southern hemisphere, this day marks the beginning of summer.

Equinoxes, also shown in **Figure 10,** are days when Earth is positioned so that neither the northern hemisphere nor the southern hemisphere is tilted toward or away from the Sun. The equinoxes are the beginning of spring and fall. On equinox days, the number of daylight hours almost equals the number of nighttime hours everywhere on Earth. In the northern hemisphere, the spring equinox occurs on March 21 or 22. This is the beginning of fall in the southern hemisphere. On September 22 or 23, fall begins in the northern hemisphere and spring begins in the southern hemisphere.

Reading Check Compare and contrast solstices and equinoxes.

SCIENCE USE V. COMMON USE

revolution

Science Use the action by a celestial body of going around in an orbit or an elliptical course

Common Use a sudden, radical, or complete change

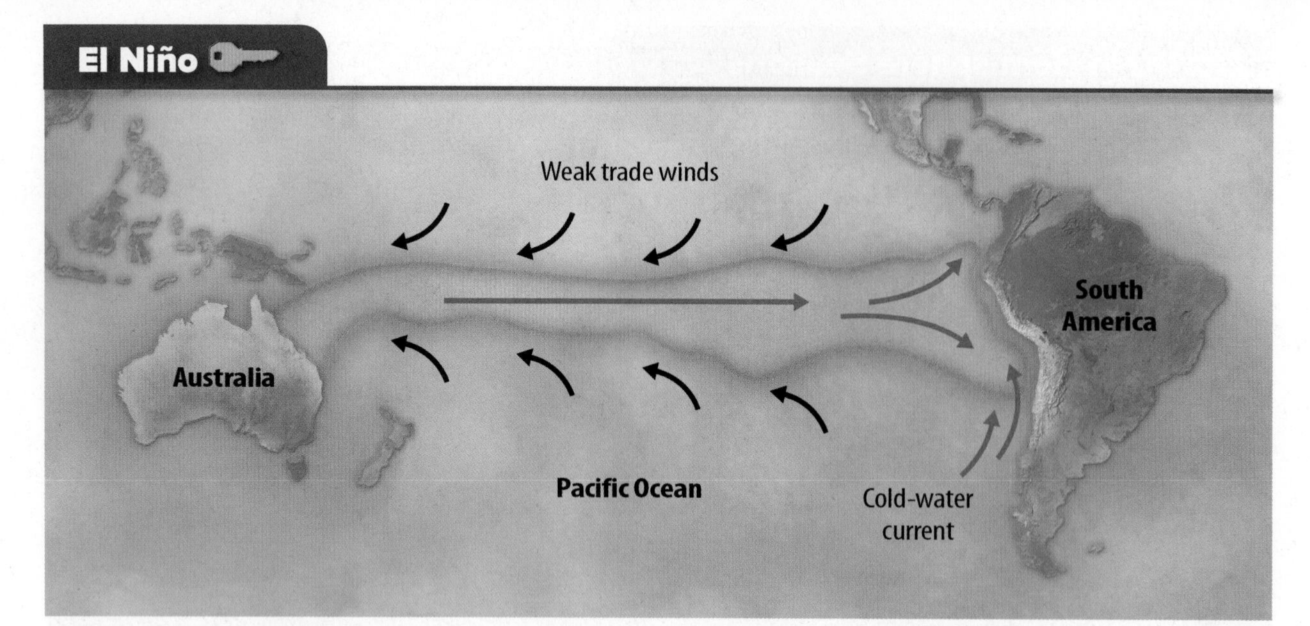

Figure 11 During El Niño, the trade winds weaken and warm water surges toward South America.

Visual Check Where is the warm water during normal conditions?

El Niño and the Southern Oscillation

Close to the equator, the trade winds blow from east to west. These steady winds push warm surface water in the Pacific Ocean away from the western coast of South America. This allows cold water to rush upward from below–a process called upwelling. The air above the cold, upwelling water cools and sinks, creating a high-pressure area. On the other side of the Pacific Ocean, air rises over warm, equatorial waters, creating a low-pressure area. This difference in air pressures across the Pacific Ocean helps keep the trade winds blowing.

As **Figure 11** shows, sometimes the trade winds weaken, reversing the normal pattern of high and low pressures across the Pacific Ocean. Warm water surges back toward South America, preventing cold water from upwelling. This **phenomenon,** called El Niño, shows the connection between the atmosphere and the ocean. During El Niño, the normally dry, cool western coast of South America warms and receives lots of precipitation. Climate changes can be seen around the world. Droughts occur in areas that are normally wet. The number of violent storms in California and the southern United States increases.

ACADEMIC VOCABULARY
phenomenon
(noun) an observable fact or event

Reading Check How do conditions in the Pacific Ocean differ from normal during El Niño?

The combined ocean and atmospheric cycle that results in weakened trade winds across the Pacific Ocean is called **El Niño/Southern Oscillation,** or ENSO. A complete ENSO cycle occurs every 3-8 years. The North Atlantic Oscillation (NAO) is another cycle that can change the climate for decades at a time. The NAO affects the strength of storms throughout North America and Europe by changing the position of the jet stream.

Monsoons

Another climate cycle involving both the atmosphere and the ocean is a monsoon. A **monsoon** *is a wind circulation pattern that changes direction with the seasons.* Temperature differences between the ocean and the land cause winds, as shown in **Figure 12.** During summer, warm air over land rises and creates low pressure. Cooler, heavier air sinks over the water, creating high pressure. The winds blow from the water toward the land, bringing heavy rainfall. During winter, the pattern reverses and winds blow from the land toward the water.

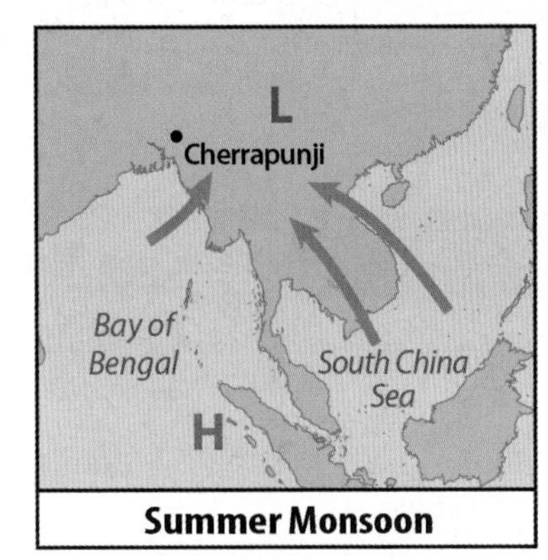

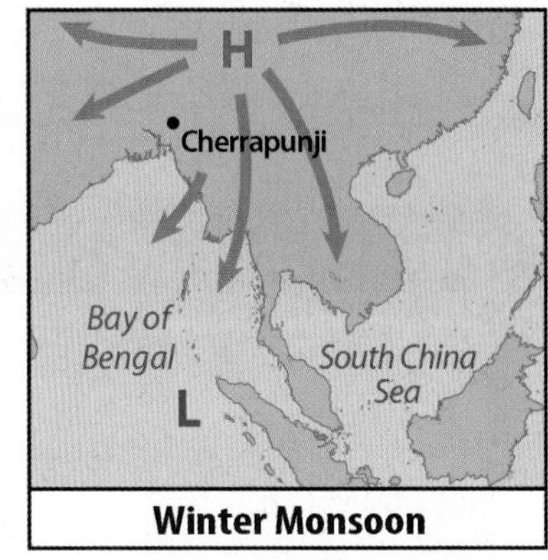

Figure 12 Monsoon winds reverse with the change of seasons.

The world's largest monsoon is found in Asia. Cherrapunji, India, is one of the world's wettest locations–receiving an average of 10 m of monsoon rainfall each year. Precipitation is even greater during El Niño events. A smaller monsoon occurs in southern Arizona. As a result, weather is dry during spring and early summer with thunderstorms occurring more often from July to September.

Key Concept Check How does the ocean affect climate?

Droughts, Heat Waves, and Cold Waves

A **drought** *is a period with below-average precipitation.* A drought can cause crop damage and water shortages.

Droughts are often accompanied by heat waves–periods of unusually high temperatures. Droughts and heat waves occur when large hot-air masses remain in one place for weeks or months. Cold waves are long periods of unusually cold temperatures. These events occur when a large continental polar air mass stays over a region for days or weeks. Severe weather of these kinds can be the result of climatic changes on Earth or just extremes in the average weather of a climate.

MiniLab 20 minutes

How do climates vary?

Unlike El Niño, La Niña is associated with cold ocean temperatures in the Pacific Ocean.

1. As the map shows, average temperatures change during a La Niña winter.
2. The color key shows the range of temperature variation from normal.
3. Find a location on the map. How much did temperatures during La Niña depart from average temperatures?

Temperature Change During La Niña

0.2
−0.2
0.2
0
0.2
0.4
0.4
0.2
0.2
−0.4
−0.4
Temperature change (°C)
0.5
0.3
0.1
−0.1
−0.3
−0.5

Analyze and Conclude

1. **Recognize Cause and Effect** Did La Niña affect the climate in your chosen area?
2. **Key Concept** Describe any patterns you see. How did La Niña affect climate in your chosen area? Use data from the map to support your answer.

Lesson 2 Review

Visual Summary

Scientists learn about past climates by studying natural records of climate, such as ice cores, fossilized pollen, and growth rings of trees.

Long-term climate changes, such as ice ages and interglacials, can be caused by changes in the shape of Earth's orbit and the tilt of its axis.

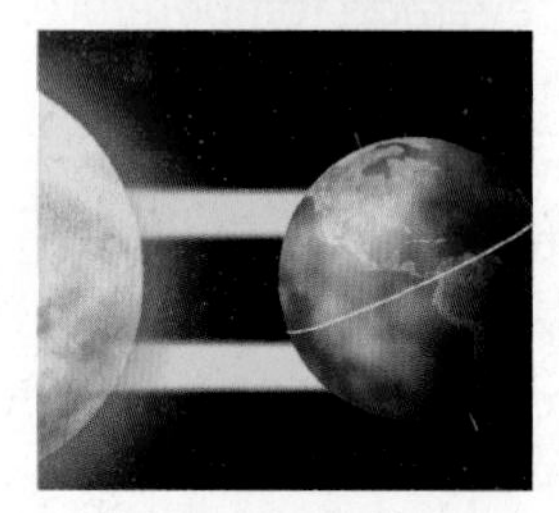

Short-term climate changes include seasons, El Niño/Southern Oscillation, and monsoons.

FOLDABLES®

Use your lesson Foldable to review the lesson. Save your Foldable for the project at the end of the chapter.

What do you think NOW?

You first read the statements below at the beginning of the chapter.

3. Climate on Earth today is the same as it has been in the past.

4. Climate change occurs in short-term cycles.

Did you change your mind about whether you agree or disagree with the statements? Rewrite any false statements to make them true.

Use Vocabulary

1. **Distinguish** an ice age from an interglacial.
2. A(n) ________ is a period of unusually high temperatures.
3. **Define** *drought* in your own words.

Understand Key Concepts 

4. What happens during El Niño/Southern Oscillation?
   - **A.** An interglacial climate shift occurs.
   - **B.** The Pacific pressure pattern reverses.
   - **C.** The tilt of Earth's axis changes.
   - **D.** The trade winds stop blowing.
5. **Identify** causes of long-term climate change.
6. **Describe** how upwelling can affect climate.

Interpret Graphics

7. **Sequence** Copy and fill in the graphic organizer below to describe the sequence of events during El Niño/Southern Oscillation.

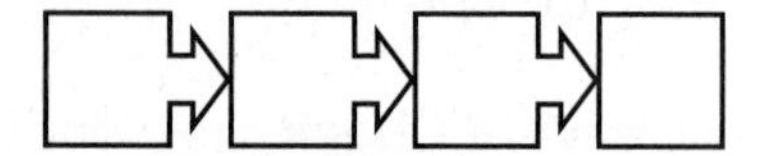

Critical Thinking

8. **Assess** the possibility that Earth will soon enter another ice age.
9. **Evaluate** the relationship between heat waves and drought.
10. **Identify** and explain the climate cycle shown below. Illustrate how conditions change during the summer.

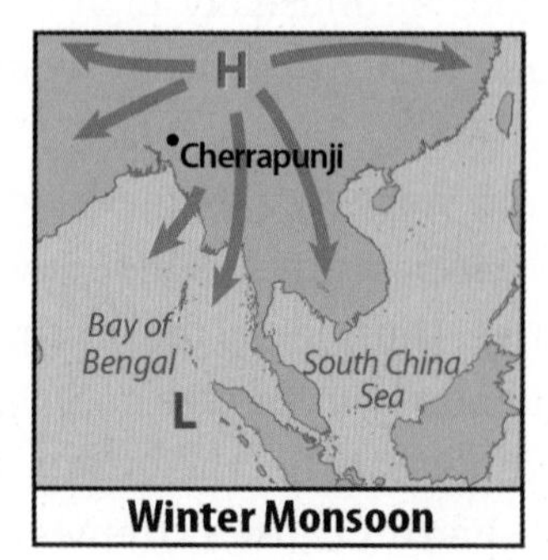

AMERICAN MUSEUM OF NATURAL HISTORY

CAREERS in SCIENCE

Frozen in Time

Looking for clues to past climates, Lonnie Thompson races against the clock to collect ancient ice from melting glaciers.

Earth's climate is changing. To understand why, scientists investigate how climates have changed throughout Earth's history by looking at ancient ice that contains clues from past climates. Scientists collected these ice samples only from glaciers at the North Pole and the South Pole. Then, in the 1970s, geologist Lonnie Thompson began collecting ice from a new location—the tropics.

◀ **Thompson has led expeditions to 15 countries and Antarctica.**

Thompson, a geologist from the Ohio State University, and his team scale glaciers atop mountains in tropical regions. On the Quelccaya ice cap in Peru, they collect ice cores—columns of ice layers that built up over hundreds to thousands of years. Each layer is a capsule of a past climate, holding dust, chemicals, and gas that were trapped in the ice and snow during that period.

To collect ice cores, they drill hundreds of feet into the ice. The deeper they drill, the further back in time they go. One core is nearly 12,000 years old!

Collecting ice cores is not easy. The team hauls heavy equipment up rocky slopes in dangerous conditions—icy windstorms, thin air, and avalanche threats. Thompson's greatest challenge is the warming climate. The Quelccaya ice cap is melting. It has shrunk by 30 percent since Thompson's first visit in 1974. It's a race against time to collect ice cores before the ice disappears. When the ice is gone, so are the secrets it holds about climate change.

Thousands of ice core samples are stored in deep freeze at Thompson's lab. One core from Antarctica is over 700,000 years old, which is well before the existence of humans. ▶

Secrets in the Ice

In the lab, Thompson and his team analyze the ice cores to determine

- Age of ice: Every year, snow accumulations form a new layer. Layers help scientists date the ice and specific climate events.
- Precipitation: Each layer's thickness and composition help scientists determine the amount of snowfall that year.
- Atmosphere: As snow turns to ice, it traps air bubbles, providing samples of the Earth's atmosphere. Scientists can measure the trace gases from past climates.
- Climate events: The concentration of dust particles helps scientists determine periods of increased wind, volcanic activity, dust storms, and fires.

It's Your Turn

WRITE AN INTRODUCTION Imagine Lonnie Thompson is giving a speech at your school. You have been chosen to introduce him. Write an introduction highlighting his work and achievements.

Lesson 3

Recent Climate Change

Reading Guide

Key Concepts
ESSENTIAL QUESTIONS

- How can human activities affect climate?
- How are predictions for future climate change made?

Vocabulary

global warming p. 506
greenhouse gas p. 506
deforestation p. 507
global climate model p. 509

Multilingual eGlossary

BrainPOP®

Go to the resource tab in ConnectED to find the PBL *Question the Experts.*

Inquiry Will Tuvalu sink or swim?

This small island sits in the middle of the Pacific Ocean. What might happen to this island if the sea level rose? What type of climate change might cause sea level to rise?

Ashley Cooper/Global Warming Images/Alamy

Launch Lab

20 minutes

What changes climates?

Natural events such as volcanic eruptions spew dust and gas into the atmosphere. These events can cause climate change.

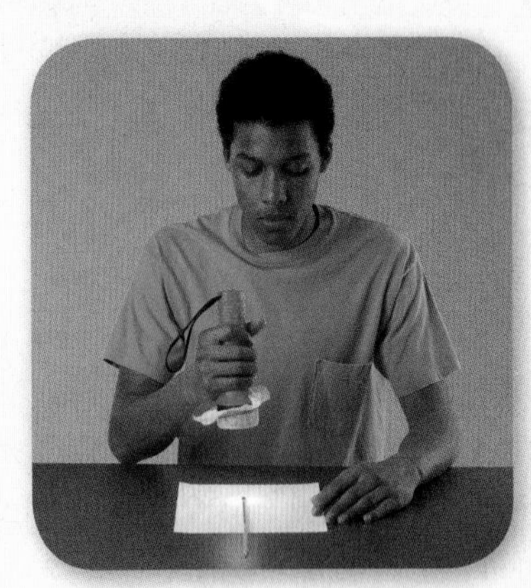

1. Read and complete a lab safety form.
2. Place a **thermometer** on a sheet of **paper.**
3. Hold a **flashlight** 10 cm above the paper. Shine the light on the thermometer bulb for 5 minutes. Observe the light intensity. Record the temperature in your Science Journal.
4. Use a **rubber band** to secure 3–4 layers of **cheesecloth or gauze** over the bulb end of the flashlight. Repeat step 3.

Think About This

1. Describe the effect of the cheesecloth on the flashlight in terms of brightness and temperature.
2. **Key Concept** Would a volcanic eruption cause temperatures to increase or decrease? Explain.

Regional and Global Climate Change

Average temperatures on Earth have been increasing for the past 100 years. As the graph in **Figure 13** shows, the warming has not been steady. Globally, average temperatures were fairly steady from 1880 to 1900. From 1900 to 1945, they increased by about 0.5°C. A cooling period followed, ending in 1975. Since then, average temperatures have steadily increased. The greatest warming has been in the northern hemisphere. However, temperatures have been steady in some areas of the southern hemisphere. Parts of Antarctica have cooled.

Reading Check How have temperatures changed over the last 100 years?

FOLDABLES®

Make a tri-fold book from a sheet of paper. Label it as shown. Use it to organize your notes about climate change and the possible causes.

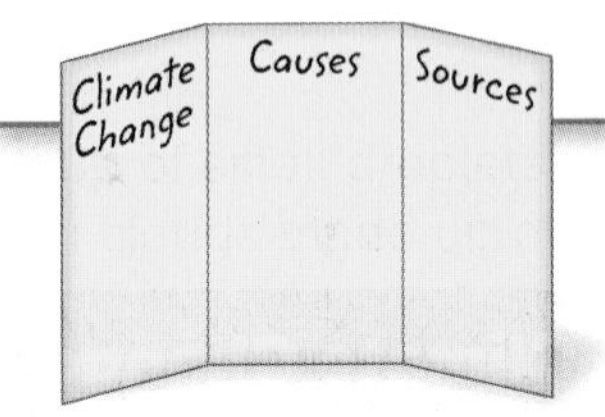

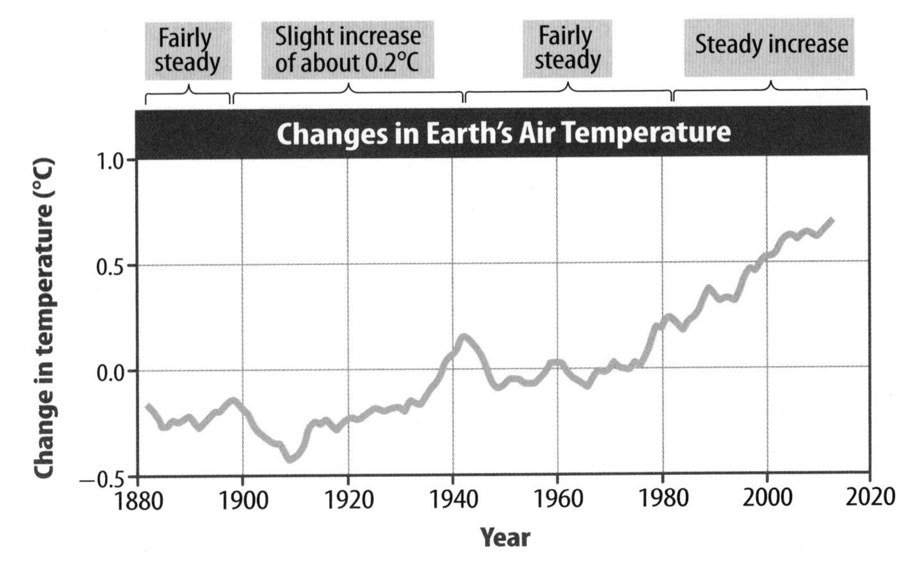

Figure 13 Temperature change has not been constant throughout the past 100 years.

Visual Check What 20-year period has seen the most change?

Hutchings Photography/Digital Light Source

Math Skills 

Use Percents

If Earth's population increases from 7 billion to 9 billion, what percent is this increase?

1. Subtract the initial value from the final value:

   9 billion − 7 billion = 2 billion

2. Divide the difference by the starting value:

   $\frac{2\text{ billion}}{7\text{ billion}} = 0.286$

3. Multiply by 100 and add a % sign: 0.286 × 100 = 28.6%

Practice

If a climate's mean temperature changes from 18.2°C to 18.6°C, what is the percentage of increase?

 Math Practice

 Personal Tutor

Climate and Society

A changing climate can present serious problems for society. Heat waves and droughts can cause food and water shortages. Excessive rainfall can cause flooding and mudslides. However, climate change can also benefit society. Warmer temperatures can mean longer growing seasons. Farmers can grow crops in areas that were previously too cold. Governments throughout the world are responding to the problems and opportunities created by climate change.

Environmental Impacts of Climate Change

Recall that ENSO cycles can change the amount of precipitation in some areas. Warmer ocean surface temperatures can cause more water to evaporate from the ocean surface. The increased water vapor in the atmosphere can result in heavy rainfall and frequent storms in North and South America. Increased precipitation in these areas can lead to decreased precipitation in other areas, such as parts of southern Africa, the Mediterranean, and southern Asia.

Increasing temperatures can also impact the environment in other ways. Melting glaciers and polar ice sheets can cause the sea level to rise. Ecosystems can be disrupted as coastal areas flood. Coastal flooding is a serious concern for the one billion people living in low-lying areas on Earth.

Extreme weather events are also becoming more common. What effect will heat waves, droughts, and heavy rainfall have on infectious disease, existing plants and animals, and other systems of nature? Will increased CO_2 levels work similarly?

The annual thawing of frozen ground has caused the building shown in **Figure 17** to slowly sink as the ground becomes soft and muddy. Permanently higher temperatures would create similar events worldwide. This and other ecosystem changes can affect migration patterns of insects, birds, fish, and mammals.

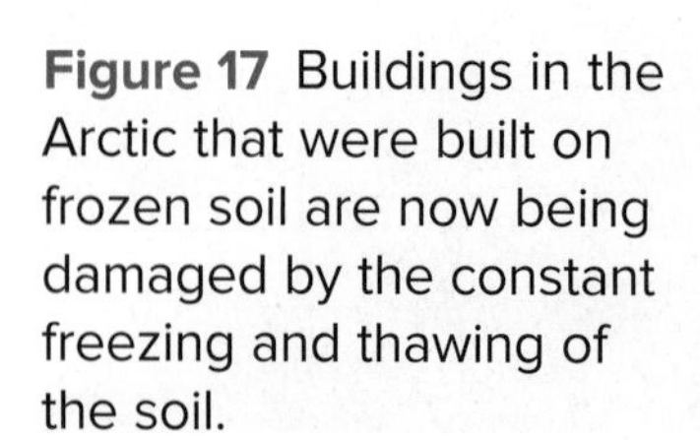

Figure 17 Buildings in the Arctic that were built on frozen soil are now being damaged by the constant freezing and thawing of the soil.

Predicting Climate Change

Weather forecasts help people make daily choices about their clothing and activities. In a similar way, climate forecasts help governments decide how to respond to future climate changes.

A **global climate model,** *or GCM, is a set of complex equations used to predict future climates.* GCMs are similar to models used to forecast the weather. GCMs and weather forecast models are different. GCMs make long-term, global predictions, but weather forecasts are short-term and can be only regional predictions. GCMs combine mathematics and physics to predict temperature, amount of precipitation, wind speeds, and other characteristics of climate. Powerful supercomputers solve mathematical equations and the results are displayed as maps. GCMs include the effects of greenhouse gases and oceans in their calculations. In order to test climate models, past records of climate change can and have been used.

Reading Check What is a GCM?

One drawback of GCMs is that the forecasts and predictions cannot be immediately compared to real data. A weather forecast model can be analyzed by comparing its predictions with meteorological measurements made the next day. GCMs predict climate conditions for several decades in the future. For this reason, it is difficult to evaluate the accuracy of climate models.

Most GCMs predict further global warming as a result of greenhouse gas emissions. By the year 2100, temperatures are expected to rise by between 1°C and 4°C. The polar regions are expected to warm more than the tropics. Summer Arctic sea ice is expected to completely disappear by the end of the twenty-first century. Global warming and sea-level rise are predicted to continue for several centuries.

Key Concept Check How are predictions for future climate change made?

MiniLab

30 minutes

How much CO_2 do vehicles emit?

Much of the carbon dioxide emitted into the atmosphere by households comes from gasoline-powered vehicles. Different vehicles emit different amounts of CO_2.

1. To calculate the amount of CO_2 given off by a vehicle, you must know how many miles per gallon of gasoline the vehicle gets. This information is shown in the chart below.
2. Assume that each vehicle is driven about 15,000 miles annually. Calculate how many gallons each vehicle uses per year. Record your data in your Science Journal in a chart like the one below.
3. One gallon of gasoline emits about 20 lbs of CO_2. Calculate and record how many pounds of CO_2 are emitted by each vehicle annually.

	Estimated MPG	Gallons of Gas Used Annually	Amount of CO_2 Emitted Annually (lbs)
SUV	15		
Hybrid	45		
Compact car	25		

Analyze and Conclude

1. **Compare and contrast** the amount of CO_2 emitted by each vehicle.
2. **Key Concept** Write a letter to a person who is planning to buy a vehicle. Explain which vehicle would have the least impact on global warming and why.

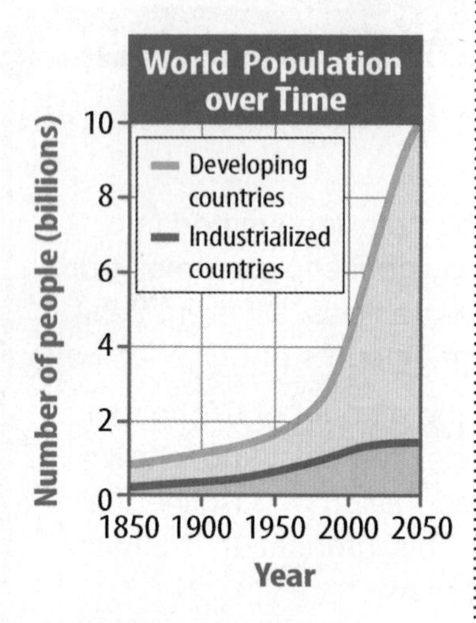

▲ **Figure 18** Earth's population is predicted to increase to more than 9 billion people by 2050.

Human Population

In 2015, more than 7 billion people inhabited Earth. As shown in **Figure 18,** Earth's population is expected to increase to 9 billion by the year 2050. What effects will an increase in population have on Earth's atmosphere?

It is predicted that by the year 2030, two of every three people on Earth will live in urban areas. Many of these areas will be in developing countries in Africa and Asia. Large areas of forests are already being cleared to make room for expanding cities. Significant amounts of greenhouse gases and other air pollutants will be added to the atmosphere.

Reading Check How could an increase in human population affect climate change?

Ways to Reduce Greenhouse Gases

People have many options for reducing levels of pollution and greenhouse gases. One way is to develop alternative sources of energy that do not release carbon dioxide into the atmosphere, such as solar energy or wind energy. Automobile emissions can be reduced by as much as 35 percent by using hybrid vehicles. Hybrid vehicles use an electric motor part of the time, which reduces fuel use.

Emissions can be further reduced by green building. Green building is the practice of creating energy-efficient buildings, such as the one shown in **Figure 19.** People can also help remove carbon dioxide from the atmosphere by planting trees in deforested areas.

You can also help control greenhouse gases and pollution by conserving fuel and recycling. Turning off lights and electronic equipment when you are not using them reduces the amount of electricity you use. Recycling metal, paper, plastic, and glass reduces the amount of fuel required to manufacture these materials.

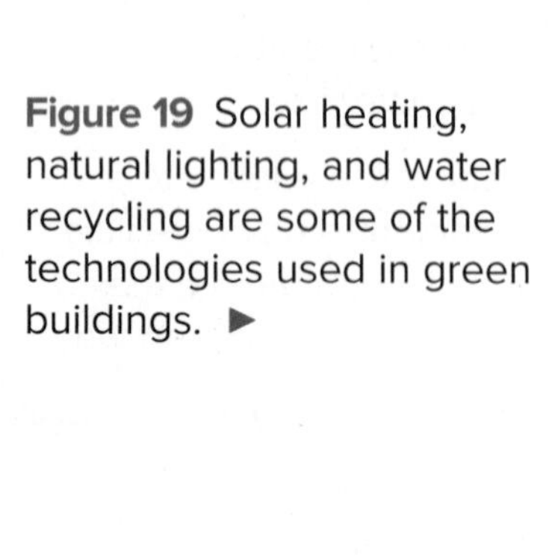

Figure 19 Solar heating, natural lighting, and water recycling are some of the technologies used in green buildings. ►

Lesson 3 Review

Online Quiz
Virtual Lab

Visual Summary

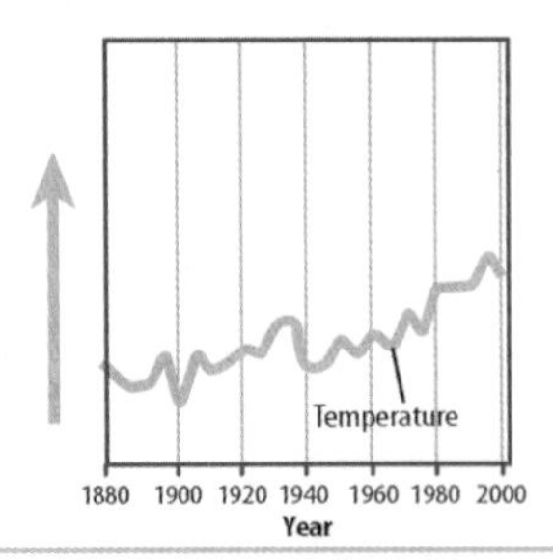

Many scientists suggest that global warming is due to increased levels of greenhouse gases in atmosphere.

Human activities, such as deforestation and burning fossil fuels, can contribute to global warming.

Ways to reduce greenhouse gas emissions include using solar and wind energy, and creating energy-efficient buildings.

FOLDABLES®

Use your lesson Foldable to review the lesson. Save your Foldable for the project at the end of the chapter.

What do you think NOW?

You first read the statements below at the beginning of the chapter.

5. Human activities can impact climate.

6. You can help reduce the amount of greenhouse gases released into the atmosphere.

Did you change your mind about whether you agree or disagree with the statements? Rewrite any false statements to make them true.

Use Vocabulary

1. **Define** *global warming* in your own words.
2. A set of complex equations used to predict future climates is called ________.
3. **Use the term** *deforestation* in a sentence.

Understand Key Concepts 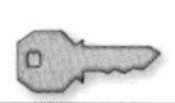

4. Which human activity can have a cooling effect on climate?
   A. release of aerosols
   B. global climate models
   C. greenhouse gas emission
   D. large area deforestation
5. **Describe** how human activities can impact climate.
6. **Identify** the advantages and disadvantages of global climate models.
7. **Describe** two ways deforestation contributes to the greenhouse effect.

Interpret Graphics

8. **Determine Cause and Effect** Draw a graphic organizer like the one below to identify two ways burning fossil fuels impacts climate.

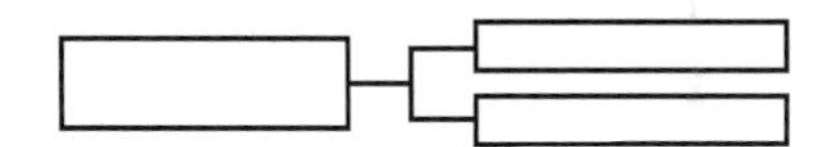

Critical Thinking

9. **Suggest** ways you can reduce greenhouse gas emissions.
10. **Assess** the effects of global warming in the area where you live.

Math Skills  

Math Practice

11. A 32-inch LCD flat-panel TV uses about 125 watts of electricity. If the screen size is increased to 40 inches, the TV uses 200 watts of electricity. What is the percent reduction of electricity if you use a 32-inch TV instead of a 40-inch TV?

Lab

80 minutes

The greenhouse effect is a gas!

Materials

plastic wrap

2 jars with lids

sand

thermometer

desk lamp

stopwatch

rubber band

Safety

Human survival on Earth depends on the greenhouse effect. How can you model the greenhouse effect to help understand how it keeps Earth's temperature in balance?

Ask a Question

How will the temperature in a greenhouse compare to that of an open system when exposed to solar energy?

Make Observations

1. Read and complete a lab safety form.
2. Decide which type of container you think will make a good model of a greenhouse. Make two identical models.
3. Place equal amounts of sand in the bottom of each greenhouse.
4. Place a thermometer in each greenhouse in a position where you can read the temperature. Secure it on the wall of the container so you are not measuring the temperature of the sand.
5. Leave one container open, and close the other container.
6. Place the greenhouses under a light source—the Sun or a lamp. Have the light source the same distance from each greenhouse and at the same angle.
7. Read the starting temperature and then every 5–10 minutes for at least three readings. Record the temperatures in your Science Journal and organize them in a table like the one shown on the next page.

Form a Hypothesis

8. Think about some adjustments you could make to your greenhouses to model other components of the greenhouse effect. For example, translucent tops, or white tops, could represent materials that would reflect more light and thermal energy.
9. Based on your observations, form a hypothesis about what materials would most accurately model the greenhouse effect.

(t to b, 2)Jacques Cornell/McGraw-Hill Education; (3, 5)Ken Cavanagh/McGraw-Hill Education; (4, 6, r)Hutchings Photography/Digital Light Source; (7)McGraw-Hill Education

Temperature (°C)			
	Reading 1	Reading 2	Reading 3
Greenhouse 1			
Greenhouse 2			

Test Your Hypothesis

10. Set up both greenhouse models in the same way for the hypothesis you are testing. Determine how many trials are sufficient for a valid conclusion. Graph your data to give a visual for your comparison.

Analyze and Conclude

11. Did thermal energy escape from either model? How does this compare to solar energy that reaches Earth and radiates back into the atmosphere?
12. If the greenhouse gases trap thermal energy and keep Earth's temperature warm enough, what would happen if they were not in the atmosphere?
13. If too much of a greenhouse gas, such as CO_2, entered the atmosphere, would the temperature rise?
14. **The Big Idea** If you could add water vapor or CO_2 to your model greenhouses to create an imbalance of greenhouse gases, would this affect the temperature of either system? Apply this to Earth's greenhouse gases.

Communicate Your Results

Discuss your findings with your group and organize your data. Share your graphs, models, and conclusions with the class. Explain why you chose certain materials and how these related directly to your hypothesis.

Inquiry Extension

Now that you understand the importance of the function of the greenhouse effect, do further investigating into what happens when the balance of greenhouse gases changes. This could result in global warming, which can have a very negative impact on Earth and the atmosphere. Design an experiment that could show how global warming occurs.

Lab Tips

- ☑ Focus on one concept in designing your lab so you do not get confused with the complexities of materials and data.
- ☑ Do not add clouds to your greenhouse as part of your model. Clouds are condensed water; water vapor is a gas.

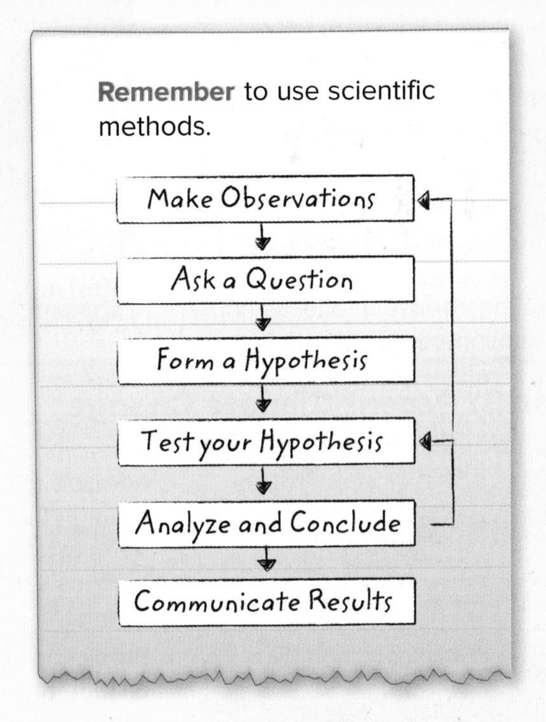

Chapter 14 Study Guide

Climate is the long-term average weather conditions that occur in an area. Living things have adaptations to the climate in which they live.

Key Concepts Summary

Lesson 1: Climates of Earth

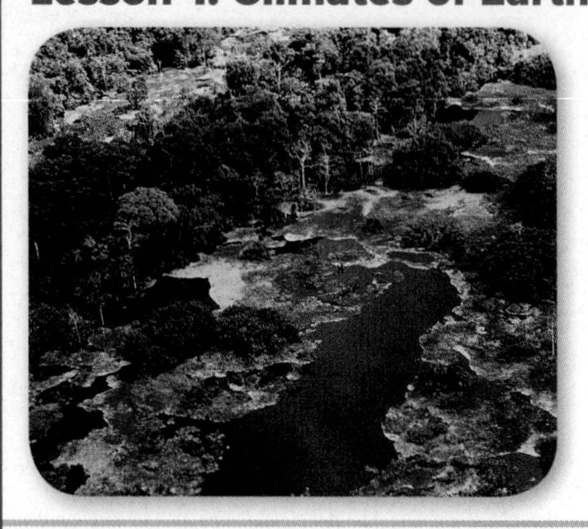

- **Climate** is the long-term average weather conditions that occur in a particular region.
- Climate is affected by factors such as latitude, altitude, **rain shadows** on the downwind slope of mountains, vegetation, and the **specific heat** of water.
- Climate is classified based on precipitation, temperature, and native vegetation.

Vocabulary

climate p. 487
rain shadow p. 489
specific heat p. 489
microclimate p. 491

Lesson 2: Climate Cycles

- Over the past 4.6 billion years, climate on Earth has varied between **ice ages** and warm periods. **Interglacials** mark warm periods on Earth during or between ice ages.
- Earth's axis is tilted. This causes seasons as Earth revolves around the Sun.
- The **El Niño/Southern Oscillation** and **monsoons** are two climate patterns that result from interactions between oceans and the atmosphere.

Vocabulary

ice age p. 496
interglacial p. 496
El Niño/Southern Oscillation p. 500
monsoon p. 501
drought p. 501

Lesson 3: Recent Climate Change

- Releasing carbon dioxide and aerosols into the atmosphere through burning fossil fuels and **deforestation** are two ways humans can affect climate change.
- Predictions about future climate change are made using computers and **global climate models.**

Vocabulary

global warming p. 506
greenhouse gas p. 506
deforestation p. 507
global climate model p. 509

(t)Andoni Canela/age fotostock; (c)Quasarphoto/Getty Images; (b)©Pete Ryan/National Geographic Image Collection/Alamy

Personal Tutor

Vocabulary eFlashcards
Vocabulary eGames

FOLDABLES® Chapter Project

Assemble your lesson Foldables as shown to make a Chapter Project. Use the project to review what you have learned in this chapter.

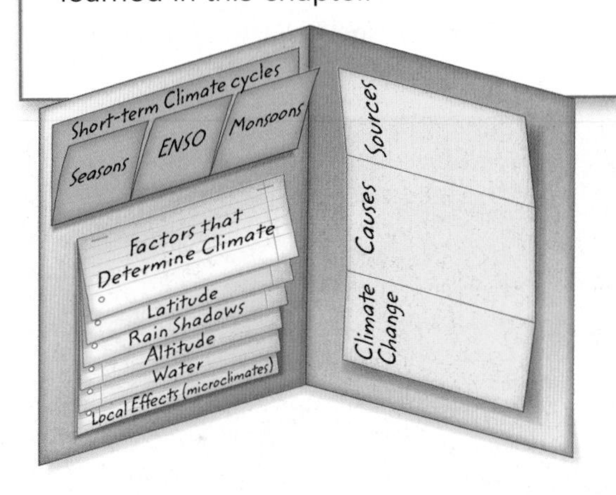

Use Vocabulary

1. A(n) __________ is an area of low rainfall on the downwind slope of a mountain.
2. Forests often have their own __________, with cooler temperatures than the surrounding countryside.
3. The lower __________ of land causes it to warm up faster than water.
4. A wind circulation pattern that changes direction with the seasons is a(n) __________.
5. Upwelling, trade winds, and air pressure patterns across the Pacific Ocean change during a(n) __________.
6. Earth's current __________ is called the Holocene Epoch.
7. A(n) __________ such as carbon dioxide absorbs Earth's infrared radiation and warms the atmosphere.
8. Additional CO_2 is added to the atmosphere when __________ of large land areas occurs.

Link Vocabulary and Key Concepts

Interactive Concept Map

Copy this concept map, and then use vocabulary terms from the previous page and other terms in this chapter to complete the concept map.

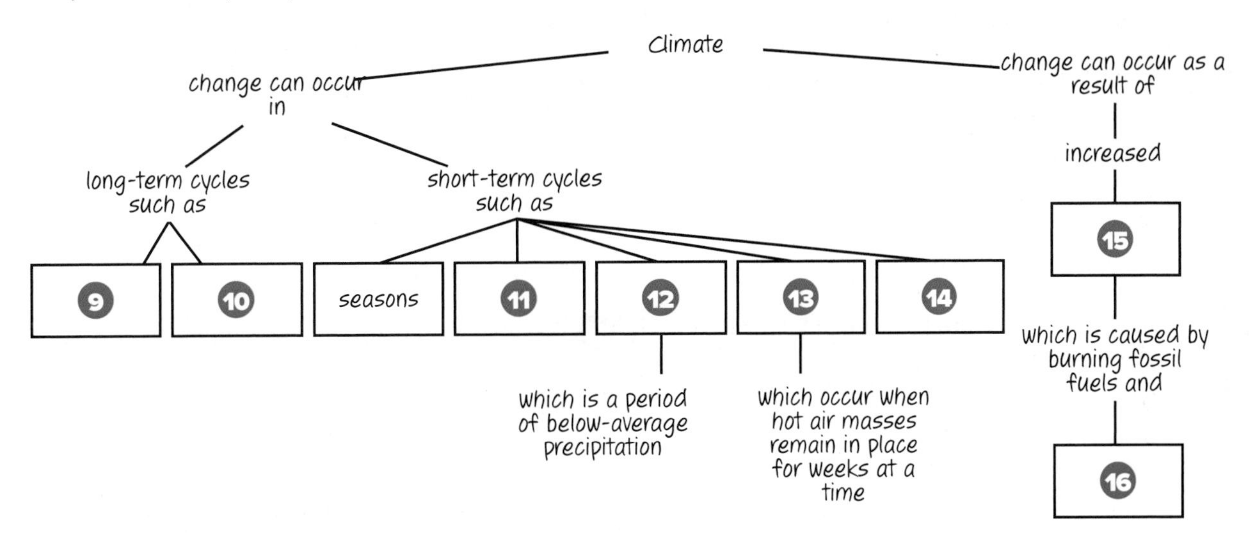

Chapter 14 Review

Understand Key Concepts 

1. The specific heat of water is ________ than the specific heat of land.
   A. higher
   B. lower
   C. less efficient
   D. more efficient

2. The graph below shows average monthly temperature and precipitation of an area over the course of a year.

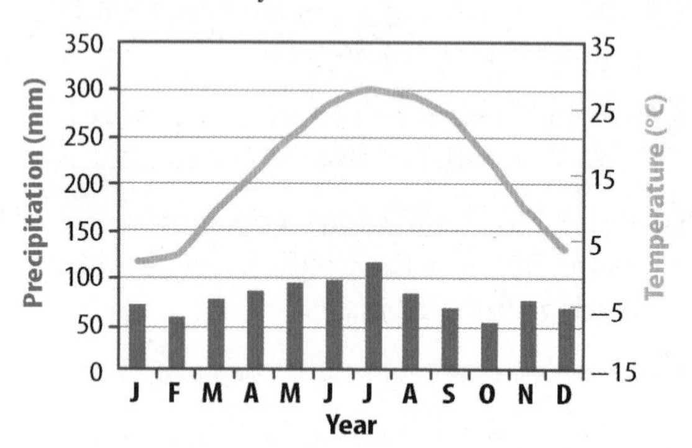

   Which is the most likely location of the area?
   A. in the middle of a large continent
   B. in the middle of the ocean
   C. near the North Pole
   D. on the coast of a large continent

3. Which are warm periods during or between ice ages?
   A. ENSO
   B. interglacials
   C. monsoons
   D. Pacific oscillations

4. Long-term climate cycles are caused by all of the following EXCEPT
   A. changes in ocean circulation.
   B. Earth's revolution of the Sun.
   C. the slow movement of the continents.
   D. variations in the shape of Earth's orbit.

5. A rain shadow is created by which factor that affects climate?
   A. a large body of water
   B. buildings and concrete
   C. latitude
   D. mountains

6. During which event do trade winds weaken and the usual pattern of pressure across the Pacific Ocean reverses?
   A. drought
   B. El Niño/Southern Oscillation event
   C. North Atlantic Oscillation event
   D. volcanic eruption

7. The picture below shows Earth as it revolves around the Sun.

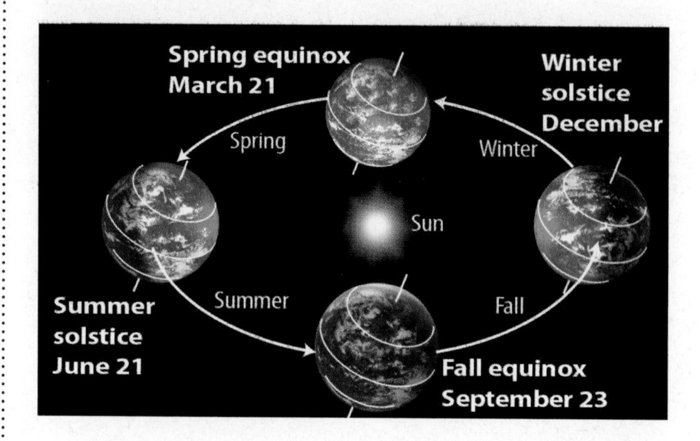

   Which season is it in the southern hemisphere in July?
   A. fall
   B. spring
   C. summer
   D. winter

8. Which is not a greenhouse gas?
   A. carbon dioxide
   B. methane
   C. oxygen
   D. water vapor

9. Which cools the climate by preventing sunlight from reaching Earth's surface?
   A. aerosols
   B. greenhouse gases
   C. lakes
   D. water vapor molecules

10. Which action can reduce greenhouse gas emissions?
    A. building houses on permafrost
    B. burning fossil fuels
    C. cutting down forests
    D. driving a hybrid vehicle

Critical Thinking

11. **Hypothesize** how the climate of your town would change if North America and Asia moved together and became one enormous continent.

12. **Interpret Graphics** Identify the factor that affects climate, as shown in this graph. How does this factor affect climate?

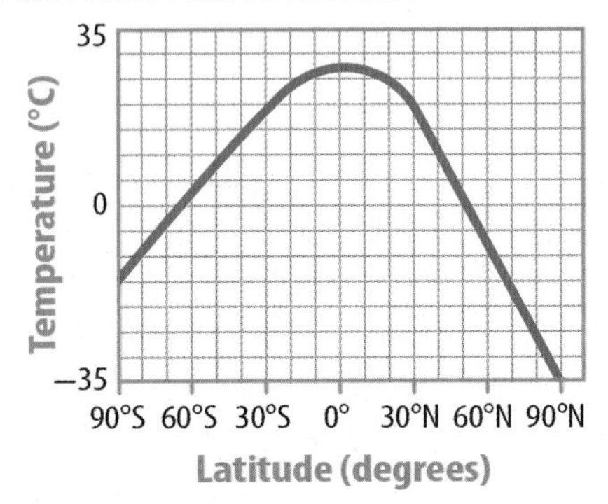

13. **Diagram** Draw a diagram that explains the changes that occur during an El Niño/Southern Oscillation event.

14. **Evaluate** which would cause more problems for your city or town: a drought, a heat wave, or a cold wave. Explain.

15. **Recommend** a life change you could make if the climate in your city were to change.

16. **Formulate** your opinion about the cause of global warming. Use facts to support your opinion.

17. **Predict** the effects of population increase on the climate where you live.

18. **Compare** how moisture affects the climates on either side of a mountain range.

Writing in Science

19. **Write** a short paragraph that describes a microclimate near your school or your home. What is the cause of the microclimate?

REVIEW THE BIG IDEA

20. What is climate? Explain what factors affect climate and give three examples of different types of climate.

21. Explain how life on Earth is affected by climate.

Math Skills

Math Practice

Use Percentages

22. Fred switches from a sport-utility vehicle that uses 800 gal of gasoline a year to a compact car that uses 450 gal.
    - **a.** By what percent did Fred reduce the amount of gasoline used?
    - **b.** If each gallon of gasoline released 20 pounds of CO_2, by what percent did Fred reduce the released CO_2?

23. Billions of tons of carbon dioxide are released into the atmosphere by human activities. If humans reduced their CO_2 emissions from 38 billion tons to 30 billion tons, what is the percentage of decrease?

Standardized Test Practice

Record your answers on the answer sheet provided by your teacher or on a sheet of paper.

Multiple Choice

1 Which is a drawback of a global climate model?

- **A** Its accuracy is nearly impossible to evaluate.
- **B** Its calculations are limited to specific regions.
- **C** Its predictions are short-term only.
- **D** Its results are difficult to interpret.

Use the diagram below to answer question 2.

2 What kind of climate would you expect to find at position 4?

- **A** mild
- **B** continental
- **C** tropical
- **D** dry

3 The difference in air temperature between a city and the surrounding rural area is an example of a(n)

- **A** inversion.
- **B** microclimate.
- **C** seasonal variation.
- **D** weather system.

4 Which does NOT help explain climate differences?

- **A** altitude
- **B** latitude
- **C** oceans
- **D** organisms

5 What is the primary cause of seasonal changes on Earth?

- **A** Earth's distance from the Sun
- **B** Earth's ocean currents
- **C** Earth's prevailing winds
- **D** Earth's tilt on its axis

Use the diagram below to answer question 6.

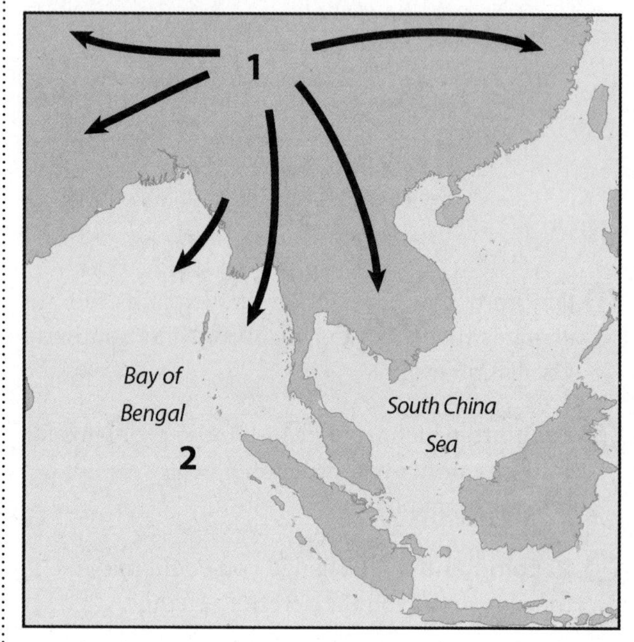

6 In the above diagram of the Asian winter monsoon, what does 1 represent?

- **A** high pressure
- **B** increased precipitation
- **C** low temperatures
- **D** wind speed

7 Climate is the ________ average weather conditions that occur in a particular region. Which completes the definition of *climate*?

- **A** global
- **B** long-term
- **C** mid-latitude
- **D** seasonal

Use the diagram below to answer question 8.

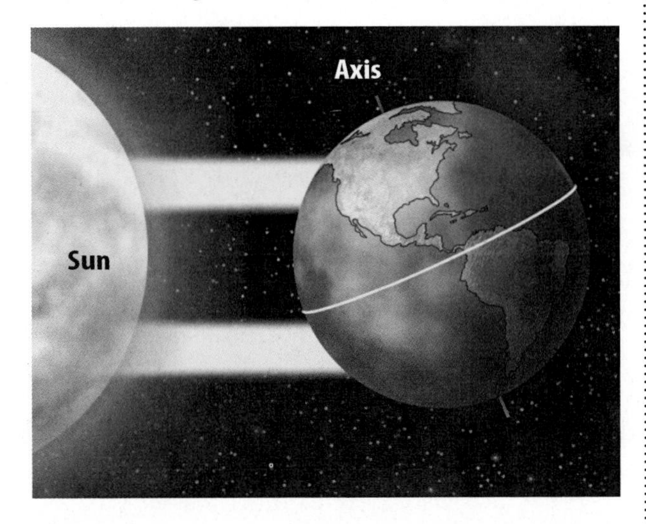

8 In the diagram above, what season is North America experiencing?

A fall

B spring

C summer

D winter

9 Which climate typically has warm summers, cold winters, and moderate precipitation?

A continental

B dry

C polar

D tropical

10 Which characterizes interglacials?

A earthquakes

B monsoons

C precipitation

D warmth

Constructed Response

Use the diagram below to answer question 11.

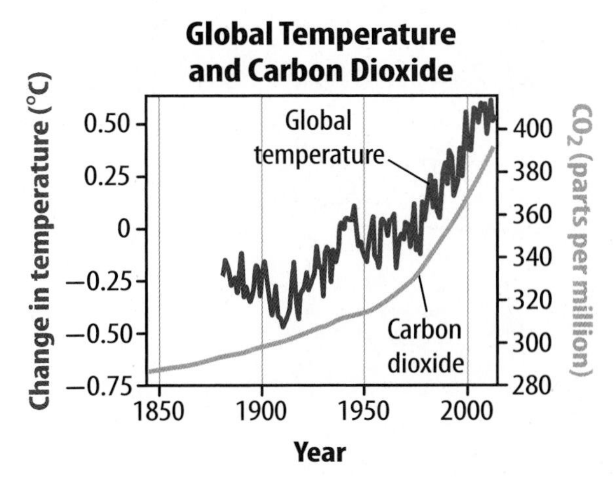

11 Compare the lines in the graph above. What does this graph suggest about the relationship between global temperature and atmospheric carbon dioxide?

Use the table below to answer questions 12 and 13.

Human Sources	Natural Sources

12 List two human and three natural sources of carbon dioxide. How do the listed human activities increase carbon dioxide levels in the atmosphere?

13 Which human activity listed in the table above also produces aerosols? What are two ways aerosols cool Earth?

NEED EXTRA HELP?													
If You Missed Question...	1	2	3	4	5	6	7	8	9	10	11	12	13
Go to Lesson...	3	1	1	1	2	2	1	2	1	2	3	3	3

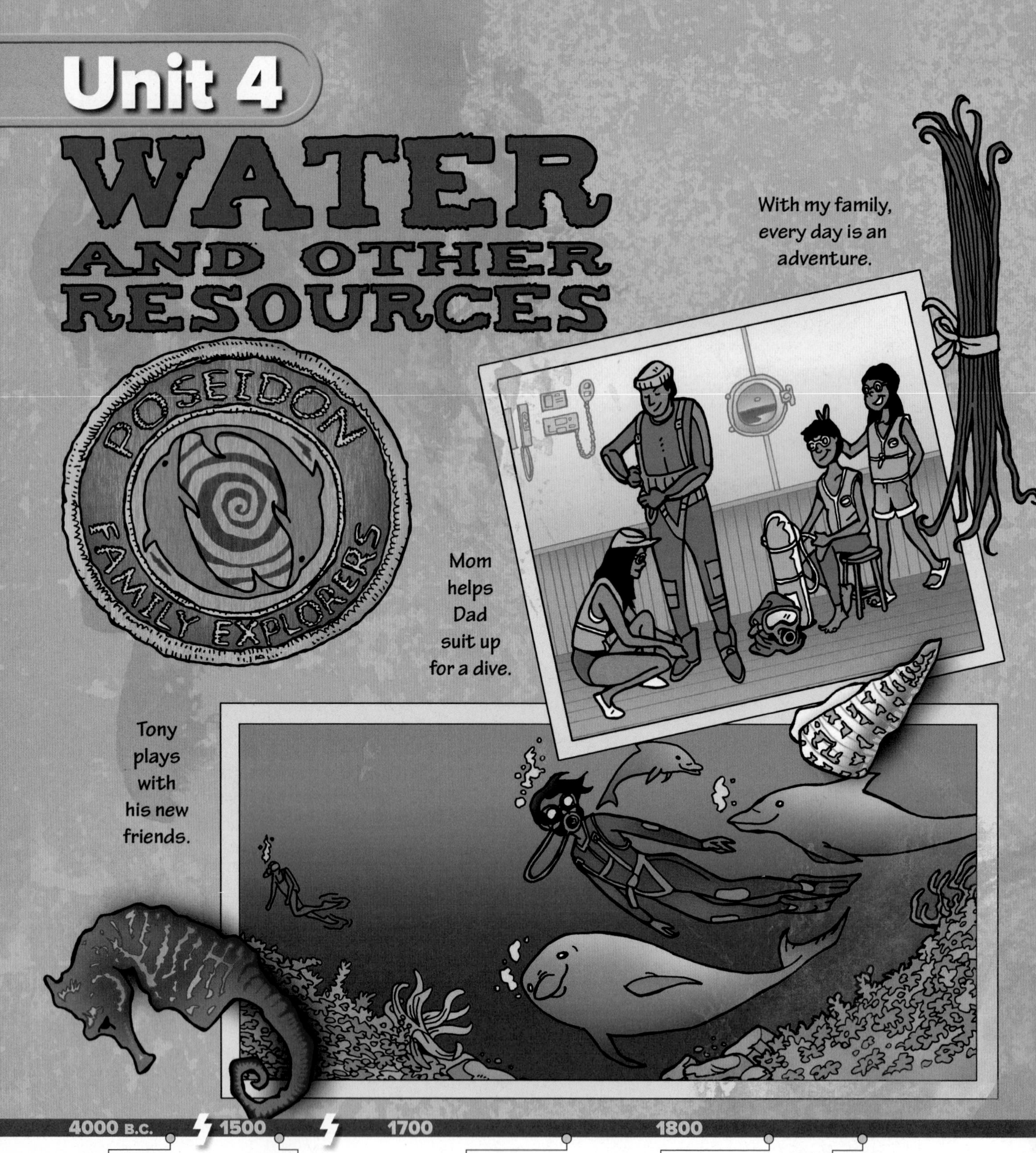

4000 B.C. | 1500 | 1700 | 1800

3500 B.C.
Egyptians develop and craft sailing vessels, most likely to use in the eastern Mediterranean, near the mouth of the Nile River.

1519–1522
Ferdinand Magellan's crew attempts to circumnavigate the world via ship; one ship succeeds.

1768–1780
James Cook explores the southern parts of the oceans looking for Antarctica. He is the first to use a chronometer, a precise clock, to determine longitude.

1831–1836
Charles Darwin sails on the H.M.S. *Beagle* to the Galapágos Islands. His research there leads him to develop the concept of natural selection.

1872–1876
The H.M.S. *Challenger* travels around the world collecting sediment and water samples, soundings, and biological specimens.

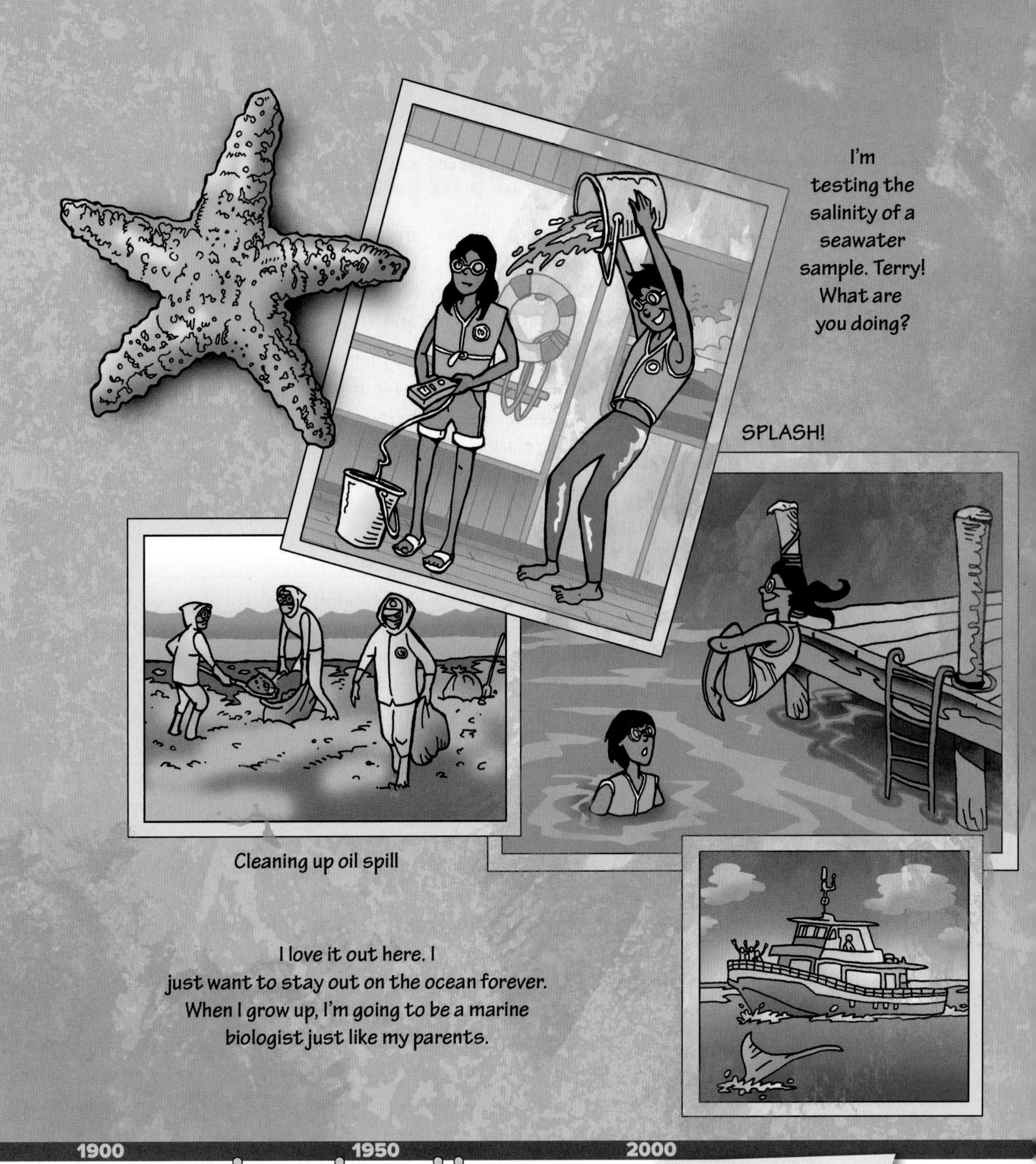

1900 | 1950 | 2000

1925–1927
A German vessel, *The Meteor,* sails the Atlantic taking sonar measurements. During this expedition the Mid-Atlantic Ridge is discovered.

1947
Satellite data leads to worldwide mapping of seafloor topography from space.

1958
The nuclear submarine *Nautilus* makes the first undersea voyage to the North Pole.

1960
The deep-sea submersible vessel *Trieste* reaches a depth of 10,915 m at the Challenger Deep in the Mariana Trench. This location is the deepest part of any ocean on Earth.

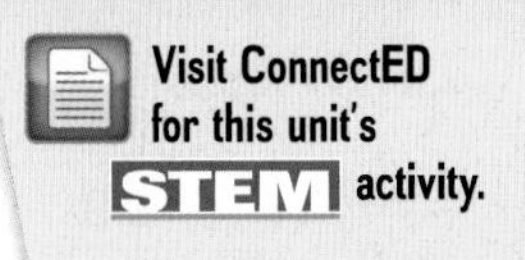
Visit ConnectED for this unit's STEM activity.

Charts, Tables, and Graphs

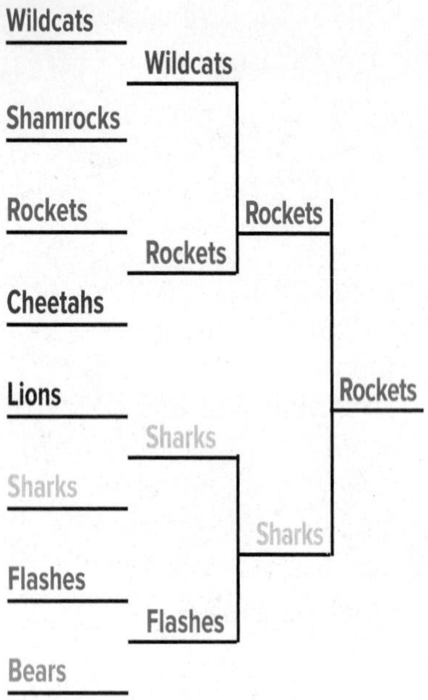

Wildcats
Wildcats
Shamrocks
Rockets
Rockets
Rockets
Cheetahs
Lions
Rockets
Sharks
Sharks
Sharks
Flashes
Flashes
Bears

▲ **Figure 1** A sporting bracket is a type of chart that easily enables you to see which team has won the most games in a tournament.

Imagine that 3 seconds are left in the semifinal game of your favorite sporting event. The clock runs out, and the buzzer sounds! You cheer as your team advances to the finals! You grab the bracket that you made and record another win.

A bracket organizes and displays the wins and losses of teams in a tournament, as shown in **Figure 1.** Brackets, like maps, tables, and graphs, are a type of chart. A **chart** is a visual display that organizes information. Charts help you organize data. Charts also help you identify patterns, trends, or errors in your data and communicate data to others.

What are tables?

Suppose you volunteer for a cleanup program at a local beach. The organizers need to know the types of debris found at different times of the year. Each month, you collect debris, separate it into categories, and weigh each category of debris. You record your data in a table. A **table** is a type of chart that organizes related data in columns and rows. Titles are usually placed at the top of each column or at the beginning of each row to help organize the data, as shown in **Table 1.**

What are graphs?

A table contains data but it does not clearly show relationships among data. However, displaying data as a graph does clearly show relationships. A **graph** is a type of chart that shows relationships between variables. The organizers of the cleanup program could make different types of graphs from the information in your table to help them better analyze the data.

Table 1 This table organizes data on collected debris into rows and columns so measurements can easily be recorded, compared, and used. ▼

Table 1 Types and Amounts of Debris

Types of Debris	Jan	Mar	May	July	Sept	Nov	Total for Year
Plastic	3.0	3.5	3.8	4.0	3.7	3.0	21.0
Polystyrene	0.5	1.3	3.2	4.0	2.5	1.2	12.7
Glass	0.8	1.2	1.5	2.0	1.5	1.0	8.0
Rubber	1.1	1.0	1.3	1.5	1.2	1.3	7.4
Metal	1.0	1.0	1.1	1.4	1.1	1.0	6.6
Paper	1.3	1.1	1.5	1.5	0.8	0.3	6.5
Total for Month	9.4	10.6	13.1	15.1	12.1	9.5	69.8

Circle Graphs

If the cleanup organizers want to know the most common type of debris, they will probably use a circle graph. A circle graph shows the percentage of the total that each category represents. This circle graph shows that plastic makes up the largest percentage of debris. The cleanup organizers could then place plastic recycling barrels on the beach so people can recycle their plastic trash.

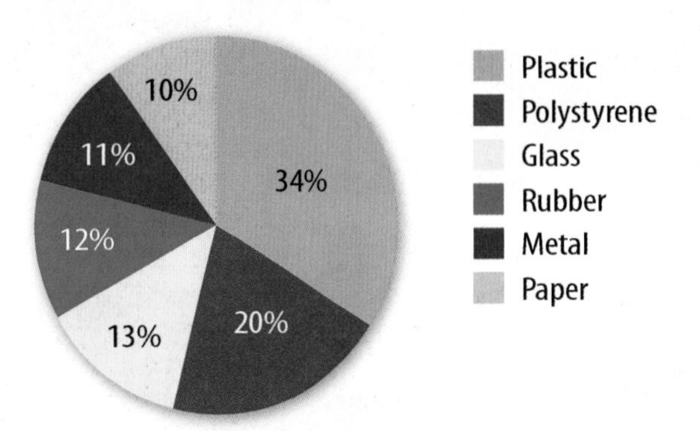

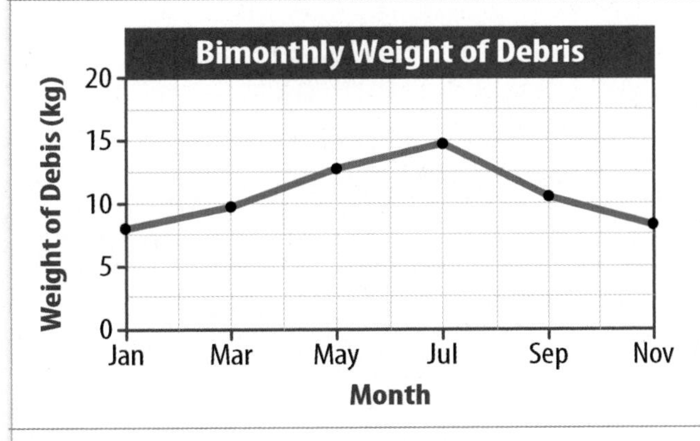

Line Graphs

Suppose the cleanup organizers want to know how the total amount of debris on the beach changes throughout the year. They probably will use a line graph. This line graph shows that volunteers collected more debris in summer than in winter. The cleanup organizers could then create a public service announcement for radio stations that reminds beachgoers to throw trash into trash cans and recycling barrels while visiting the beach.

Bar Graphs

Volunteers collected the most debris in July. The cleanup organizers want to know how much of each type of debris volunteers collected in July. Bar graphs are useful for comparing different categories of measurements. This bar graph shows that 4 kg of both plastic and polystyrene were collected in July. The cleanup organizers could then suggest that beach concession stands use smaller, recyclable food containers.

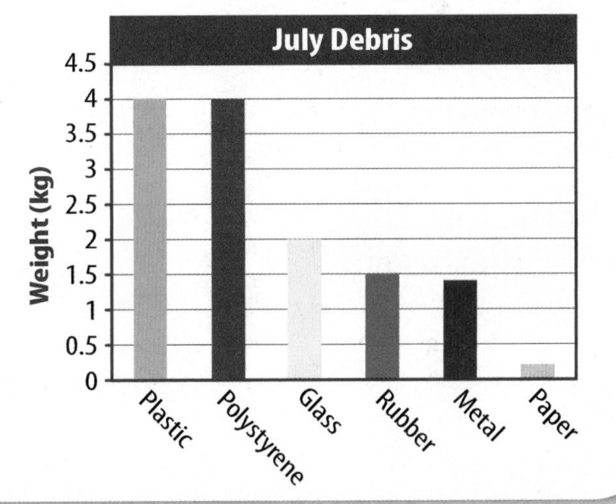

MiniLab

25 minutes

How can graphs keep the beach clean?

Suppose you work with the cleanup organizers. What information do you need to make recommendations for keeping the beach cleaner?

1. Based on the type of information in **Table 1,** write a new question about beach debris.
2. Make a graph that allows you to answer your question.

Analyze and Conclude

1. **Distinguish** How did you decide what type of graph to make?
2. **Explain** How did you use your graph to answer your question?
3. **Modify** What recommendations can you make based on your analysis of your graph?

Chapter 15

Earth's Water

THE BIG IDEA

What role does water play on Earth?

Inquiry Why are they there?

Animals that live on the dry grasslands of Africa might travel great distances to find water. All living things need water to survive.

- Why is water so important to the animals?
- How did the water get there?
- What role does water play on Earth?

Get Ready to Read

What do you think?

Before you read, decide if you agree or disagree with each of these statements. As you read this chapter, see if you change your mind about any of the statements.

1. A liquid can change to a gas only when the liquid reaches its boiling point.
2. Clouds are made of tiny drops of water.
3. Water molecules can attract other water molecules.
4. Ice has a greater density than water.
5. Factories are responsible for almost all water pollution.
6. Changes in the types of organisms living in water can be a sign of changes in the quality of the water.

Mc Graw Hill Education connectED

Your one-stop online resource
connectED.mcgraw-hill.com

- LS LearnSmart®
- Chapter Resources Files, Reading Essentials, Get Ready to Read, Quick Vocabulary
- Animations, Videos, Interactive Tables
- Self-checks, Quizzes, Tests
- PBL Project-Based Learning Activities
- Lab Manuals, Safety Videos, Virtual Labs & Other Tools
- Vocabulary, Multilingual eGlossary, Vocab eGames, Vocab eFlashcards
- Personal Tutors

Lesson 1

Reading Guide

Key Concepts
ESSENTIAL QUESTIONS

- Why is water important to life?
- How is water distributed on Earth?
- How is water cycled on Earth?

Vocabulary

specific heat p. 529
hydrosphere p. 530
evaporation p. 531
condensation p. 531
water cycle p. 532
transpiration p. 533

Multilingual eGlossary

BrainPOP®

Go to the resource tab in ConnectED to find the PBL *Campers in the Mist*

The Water Planet

Inquiry A Water Home?

Water is home to these fish. Like all life on Earth, fish need water to survive, but they also depend on water for a habitat. Where does all the water come from?

Launch Lab

20 minutes

Which heats faster?

Water and land heat and cool at different rates. This difference in heating and cooling influences climate.

1. Read and complete a lab safety form.
2. Place two **pie pans** on a flat surface. Fill one with **water** and the other with **soil.**
3. Use **thermometers** to measure the temperature of both materials. Record your measurements in your Science Journal.
4. Place a **lamp** over the pans. Turn on the lamp, and measure the temperature of the water and soil every 5 minutes for 15 minutes.

Think About This

1. Compare the rates at which the two materials heated.
2. **Key Concept** Imagine visiting the ocean in summer. Would you expect the climate near the ocean to be warmer or cooler than the climate inland? Why?

Why is water important to life?

You might have read news headlines such as "NASA Finds Evidence of Water on Mars" or "Hidden Ocean Found on Saturn's Moon!" Have you ever wondered why scientists are always looking for water in other areas of our solar system? Water is necessary for life. Scientists look for water in other areas of the solar system as a first step to finding life in these areas.

Water is extremely important on Earth for other reasons. Earth's climate is influenced by ocean currents that move thermal energy, commonly called heat, around Earth. Large bodies of water affect local weather patterns as well. Many organisms, such as the jellyfish in **Figure 1**, have water habitats. People also use water for transporting goods and for recreation.

Figure 1 Water is the habitat of these jellyfish, but all organisms on Earth depend on water for life.

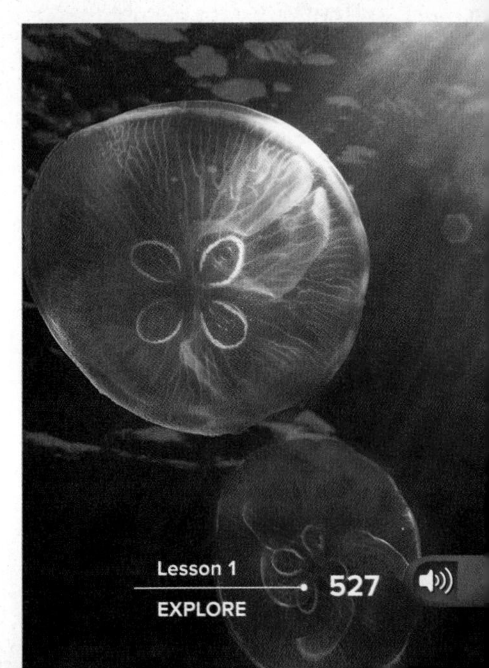

Biological Functions

Water is necessary for the life processes of all living organisms, from a unicellular bacterium to a blue whale. Did you know that the body of a jellyfish is about 95 percent water? Also, about 60 percent of the mass of the human body is made up of water. Even plant seeds that seem dry have a small amount of water inside them.

Transport One of the main roles of water in an organism is to transport materials. Water carries nutrients, such as proteins, to cells and even within the cells. It also carries wastes away from cells.

Photosynthesis Water is essential for chemical reactions, such as photosynthesis, to occur within living things. Recall that during photosynthesis, carbon dioxide and water, in the presence of light, react and produce sugar and oxygen. Photosynthesis occurs in plants, algae, and some bacteria. Organisms that undergo photosynthesis are the beginning of almost every food chain.

Body Temperature Regulation Water is an important factor in preventing an organism's body temperature from becoming too high or too low. In humans, as water from the skin, or sweat, changes to a gas, thermal energy transfers to the surrounding air. This helps keep the body cool.

Warming Earth

One reason life can exist on Earth is that Earth's atmosphere traps thermal energy from the Sun. This process is called the greenhouse effect. Some of the Sun's energy that reaches Earth's surface is absorbed and then emitted back toward space. Gases in Earth's atmosphere, such as water vapor (H_2O), methane (CH_4), and carbon dioxide (CO_2), absorb some of this energy and emit it back toward Earth, as shown in **Figure 2.**

Of all the greenhouse gases in the atmosphere, the concentration of water vapor is the highest. Without the greenhouse effect, Earth's average surface temperature would be about −18°C. All the water at Earth's surface would be ice and no organisms could survive at that temperature.

Reading Check Explain how water helps to heat Earth.

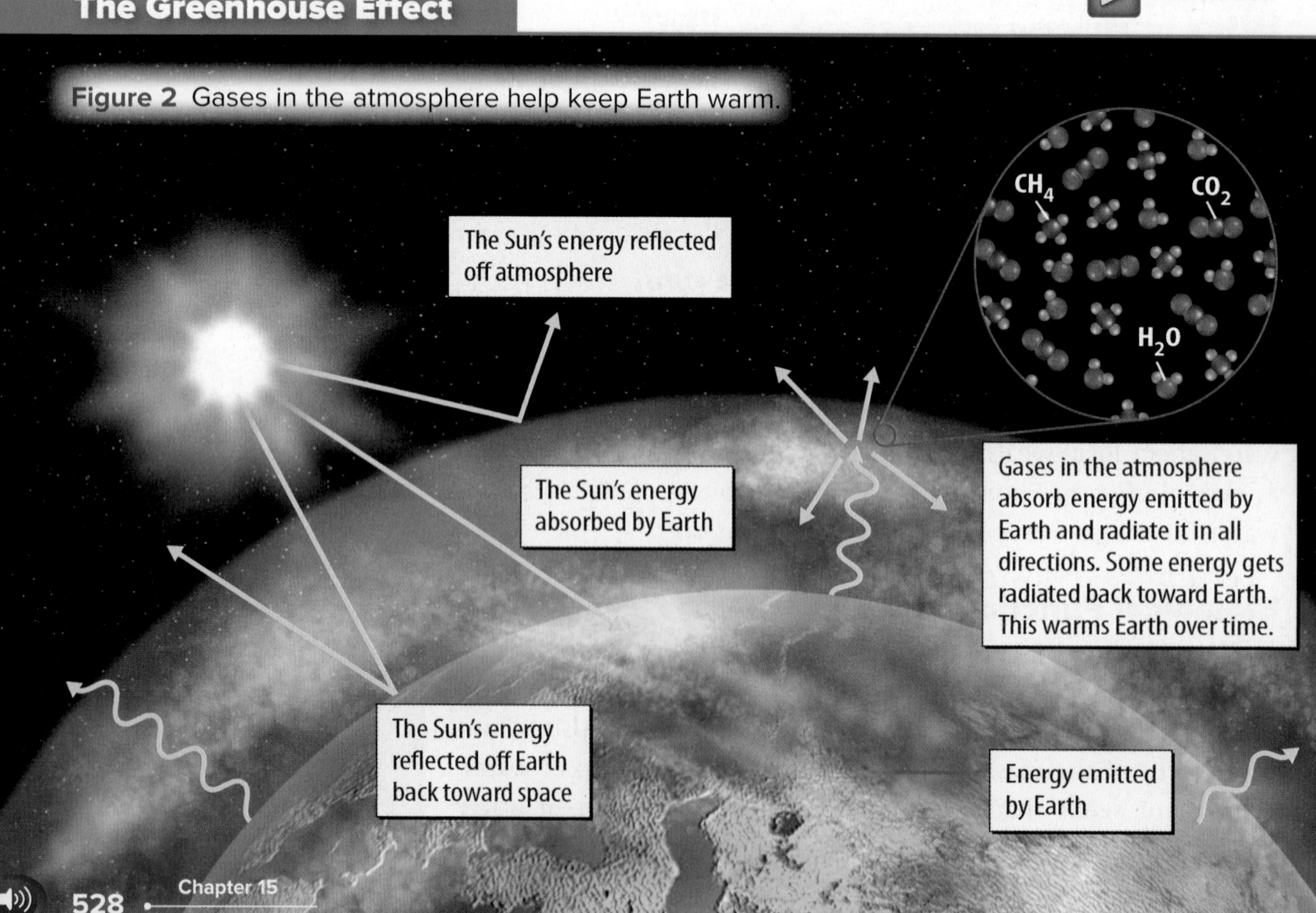

Figure 2 Gases in the atmosphere help keep Earth warm.

Keeping Earth's Temperature Stable

Think about what happens at the beach on a hot day. If you walk barefoot across the sand, you might burn the bottoms of your feet. But when you reach the water, it is refreshingly cool. Why does the water have a lower temperature than the sand?

Water has a high specific heat. **Specific heat** *is the amount of thermal energy needed to raise the temperature of 1 kg of a material by 1°C.* The specific heat of water is about six times higher than the specific heat of sand. That means the water would have to absorb six times as much thermal energy in order to have the same temperature as the sand.

Water's high specific heat is important to life on Earth for several reasons. Water vapor in the air helps control the rate at which air temperature changes. The temperature of water vapor changes slowly. As a result, the temperature change from one season to the next is gradual. Large bodies of water, such as oceans, also heat and cool slowly. This provides a more stable temperature for aquatic organisms and affects climate in coastal areas. The local weather patterns of inland areas near large lakes are affected as well. Examples of how water is important to life are summarized in **Table 1.**

Key Concept Check Why is water important to life?

Table 1 Importance of Water to Life on Earth

Interactive Table

Importance to Life	Examples
Biological functions	• transport of nutrients and wastes to and from cells • photosynthesis • body temperature regulation
Keeps Earth warm	• greenhouse effect • air temperature regulation
Stabilizes Earth's temperature	• gradual temperature change from one season to the next • high specific heat causes large bodies of water to heat up and cool down slowly • stable temperature for aquatic organisms

Math Skills

Use Equations

To calculate the energy needed to change an object's temperature, use the following equation:

Energy = **specific heat** × **mass** × **change in temperature** or,
J = **J/kg • °C** × **kg** × **°C**

To solve this equation, you need the object's specific heat. For example, how much energy will raise the temperature of **2 kg** of iron from **20°C** to **30°C**? The specific heat of iron is **460 kg • °C**.

J = **460 kg • °C** × **2 kg** × **10°C**
The amount of energy is 9,200 J.

Practice
If the specific heat of aluminum is 900 J/kg • °C, how much energy is needed to raise the temperature of a 3 kg sample from 35°C to 45°C?

Math Practice

Personal Tutor

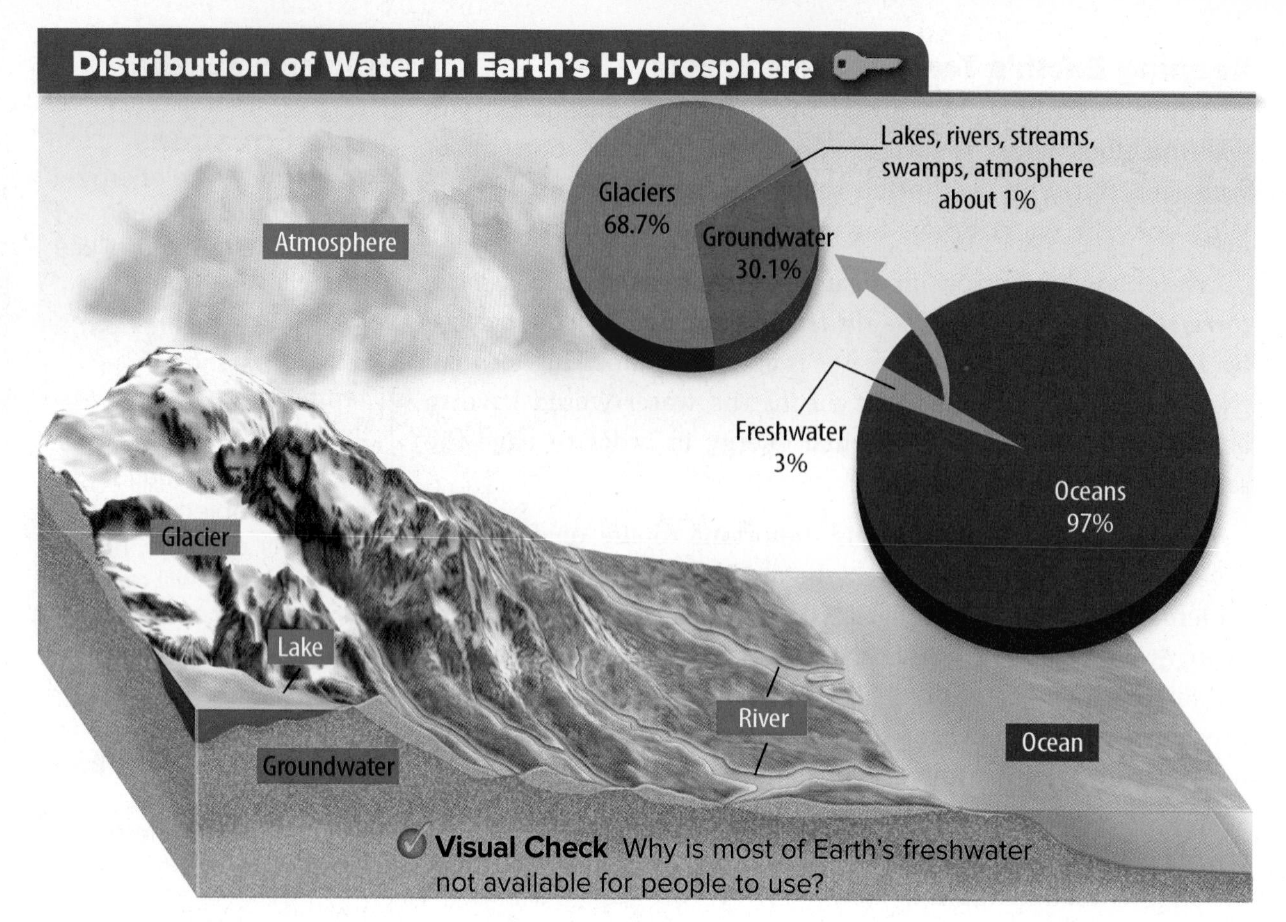

Figure 3 About 3 percent of Earth's water is freshwater. Only about 0.001 percent of Earth's water is in the atmosphere.

Water on Earth

You have just read several reasons why water is important for life. You also use water every day for bathing, cooking, and drinking. About 70 percent of Earth's surface is covered by water. How is all this water distributed?

Distribution of Water on Earth

Notice in **Figure 3** that most of Earth's water is in oceans. Only about 3 percent is freshwater (not salty). Freshwater is on Earth's surface, in the ground, or in icecaps and glaciers. Only about 1 percent of all water on Earth is in lakes, rivers, swamps, and the atmosphere.

 **Key Concept Check** How is water on Earth distributed?

Structure of the Hydrosphere

The **hydrosphere** *is all the water on and below Earth's surface and in the atmosphere.* The many parts of the hydrosphere are shown in **Figure 3.** Water is in oceans, lakes, rivers, and streams and underground. Water beneath Earth's surface is called groundwater. Water vapor, or water in the gaseous state, is in the atmosphere. Clouds are a collection of tiny droplets of water or ice crystals. Ice, or water in the solid state, is in glaciers and ice caps. Earth's frozen water is often called the cryosphere.

WORD ORIGIN

hydrosphere
hydro–
from Greek *hydor,* means "water"
–sphere
from Greek *spharia,* means "ball"

Water Changes State

The only substance that exists in nature in three states–solid, liquid, and gas–within Earth's temperature range is water. It can easily change state within the hydrosphere. For example, in the spring, snow and ice–both solid water–melt to a liquid. When enough thermal energy is added, the liquid water changes to a gas and enters the atmosphere. When water changes from one state to another, thermal energy is either absorbed or released. Thermal energy always moves from an object with a higher temperature to an object with a lower temperature.

Between Solid and Liquid

When thermal energy is added to ice, the water molecules gain energy. If enough thermal energy is added, the ice eventually reaches its melting point and changes to a liquid. The reverse happens if thermal energy is released from liquid water. The molecules begin to lose energy. If the molecules in water lose enough energy, the liquid reaches its freezing point and ice forms.

Between Liquid and Gas

As thermal energy is added to liquid water, the molecules gain energy and eventually reach the boiling point. At the boiling point, water changes to a gas, or water vapor. It takes less energy for molecules at the surface of water to break free from surrounding molecules, as shown in **Figure 4.** Therefore, water at the surface can change to a gas at temperatures below the boiling point and evaporate. **Evaporation** *is the process of a liquid changing to a gas at the surface of the liquid.* When water vapor molecules lose thermal energy, condensation occurs. **Condensation** *is the process of a gas changing to a liquid.*

Reading Check Why can evaporation of water occur below water's boiling point?

MiniLab 20 minutes

What happens to temperature during a change of state?

1. Read and complete a lab safety form.
2. Fill a **500-mL beaker** with **crushed ice.**
3. Place the beaker on a **hot plate,** near a **ring stand.** Place a **thermometer** in the beaker about 2.5 cm from the bottom. Use a **clamp** on the ring stand to hold the thermometer in place.
4. In your Science Journal, record the temperature of the ice. Turn the hot plate on medium-high.
5. Record the temperature every minute until 3 minutes after the water starts boiling.

Analyze and Conclude

1. **Identify** When did a change of state occur?
2. **Describe** How did the temperature of the water change as its state changed?
3. **Key Concept** Why is the range of temperatures between the states of water important to life on Earth?

Figure 4 Evaporation occurs only at a liquid's surface.

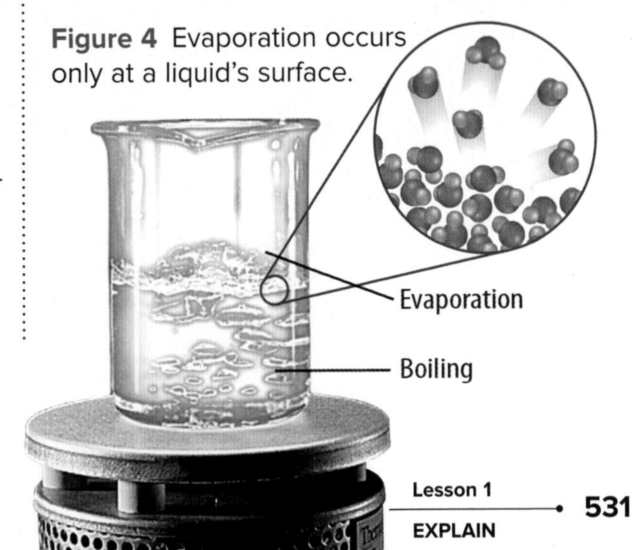

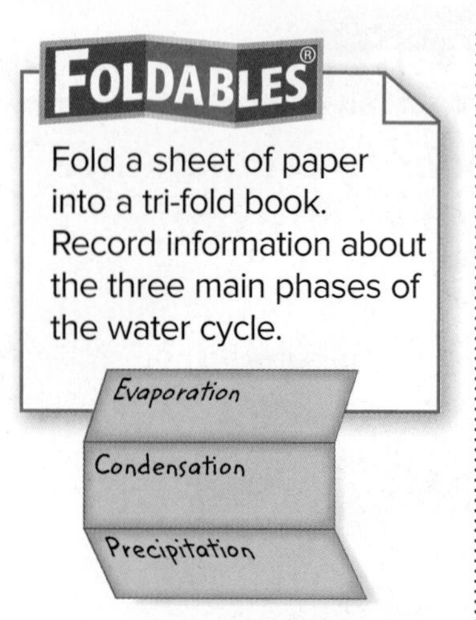

The Water Cycle

The series of natural processes by which water continually moves throughout the hydrosphere is called the **water cycle.** As water moves through the water cycle, it continually changes state.

Driving the Water Cycle

Two main factors drive the water cycle–the Sun and gravity. Energy from the Sun causes water on Earth's surface to evaporate. The water later falls back to the ground as precipitation. On Earth's surface, gravity moves water from higher to lower areas. Water eventually returns to oceans and other storage areas in the hydrosphere, and the cycle continues.

Reading Check What two main factors drive the water cycle?

Evaporation

Water on Earth's surface evaporates because energy from the Sun breaks the bonds between water molecules. Liquid water changes into water vapor and enters the atmosphere. As shown in **Figure 5**, evaporation occurs throughout the hydrosphere.

Figure 5 Water changes from one state to another as it cycles throughout Earth's hydrosphere.

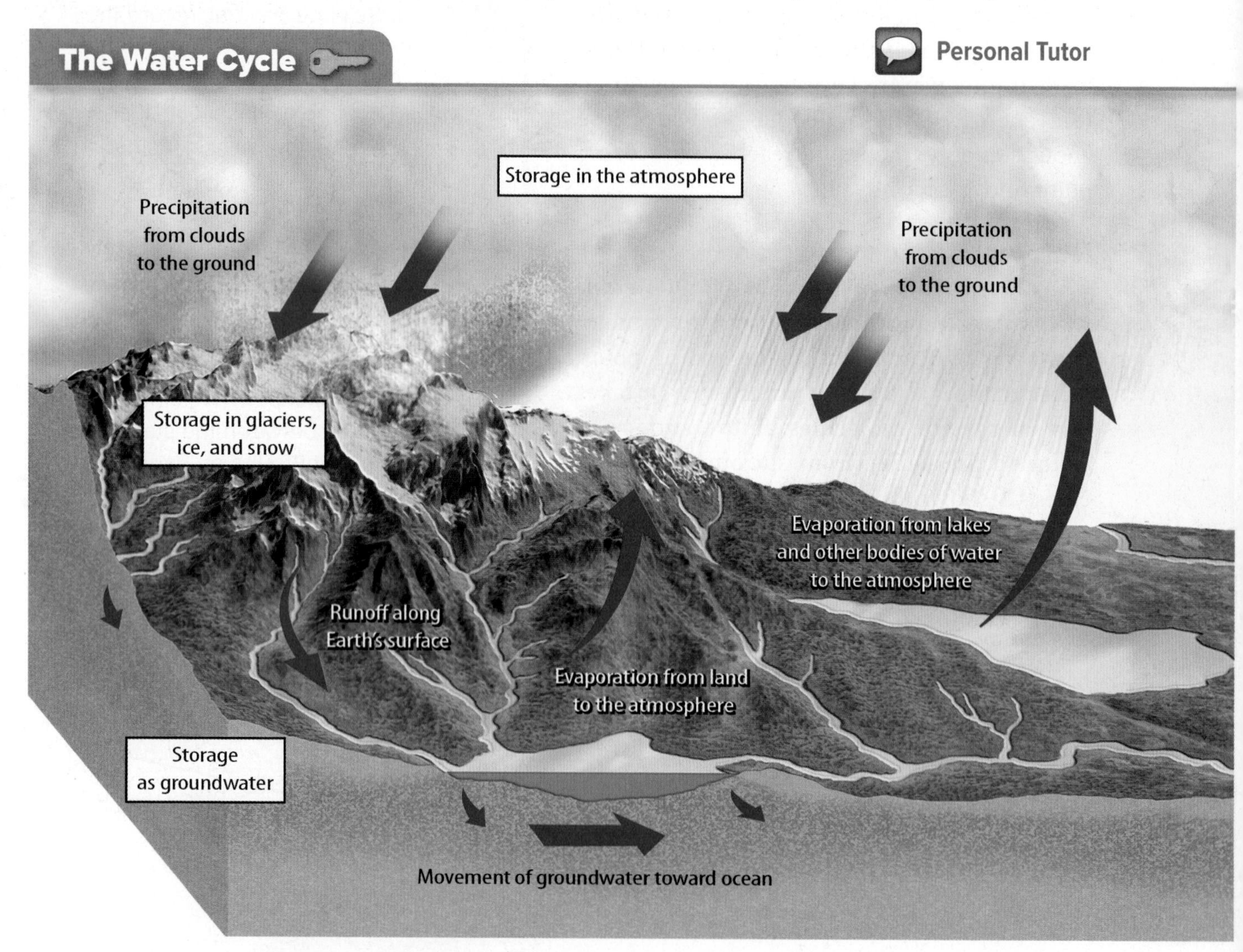

Transpiration *The evaporation of water from plants is called* **transpiration.** Water is absorbed by plants mostly from the ground. When a plant has an abundant water supply or air temperatures increase, plants transpire—they release water vapor into the atmosphere. This usually occurs through the leaves.

Condensation and Precipitation

As water vapor from transpiration and evaporation rises in the atmosphere, it cools and condenses into a liquid. Water vapor condenses around particles of dust in the atmosphere and forms droplets. The droplets combine and form clouds. They eventually fall to the ground as rain. If the temperature is low enough, the water droplets will freeze in the atmosphere and reach Earth's surface as other forms of precipitation such as snow, sleet, or hail.

Runoff and Storage

What happens to the precipitation in **Figure 5** once it reaches Earth's surface? Gravity acts on the precipitation. It causes water on Earth's surface to flow downhill. Water from precipitation that flows over Earth's surface is called runoff. Runoff enters streams and rivers and eventually reaches lakes or oceans. Some precipitation soaks into the ground and becomes groundwater.

Although water is constantly moving through the water cycle, most water remains in certain storage areas for relatively long periods of time. A storage area of the water cycle is called a reservoir. Reservoirs can be lakes, oceans, groundwater, glaciers, and ice caps.

 **Key Concept Check** Explain the steps as water cycles through Earth's hydrosphere.

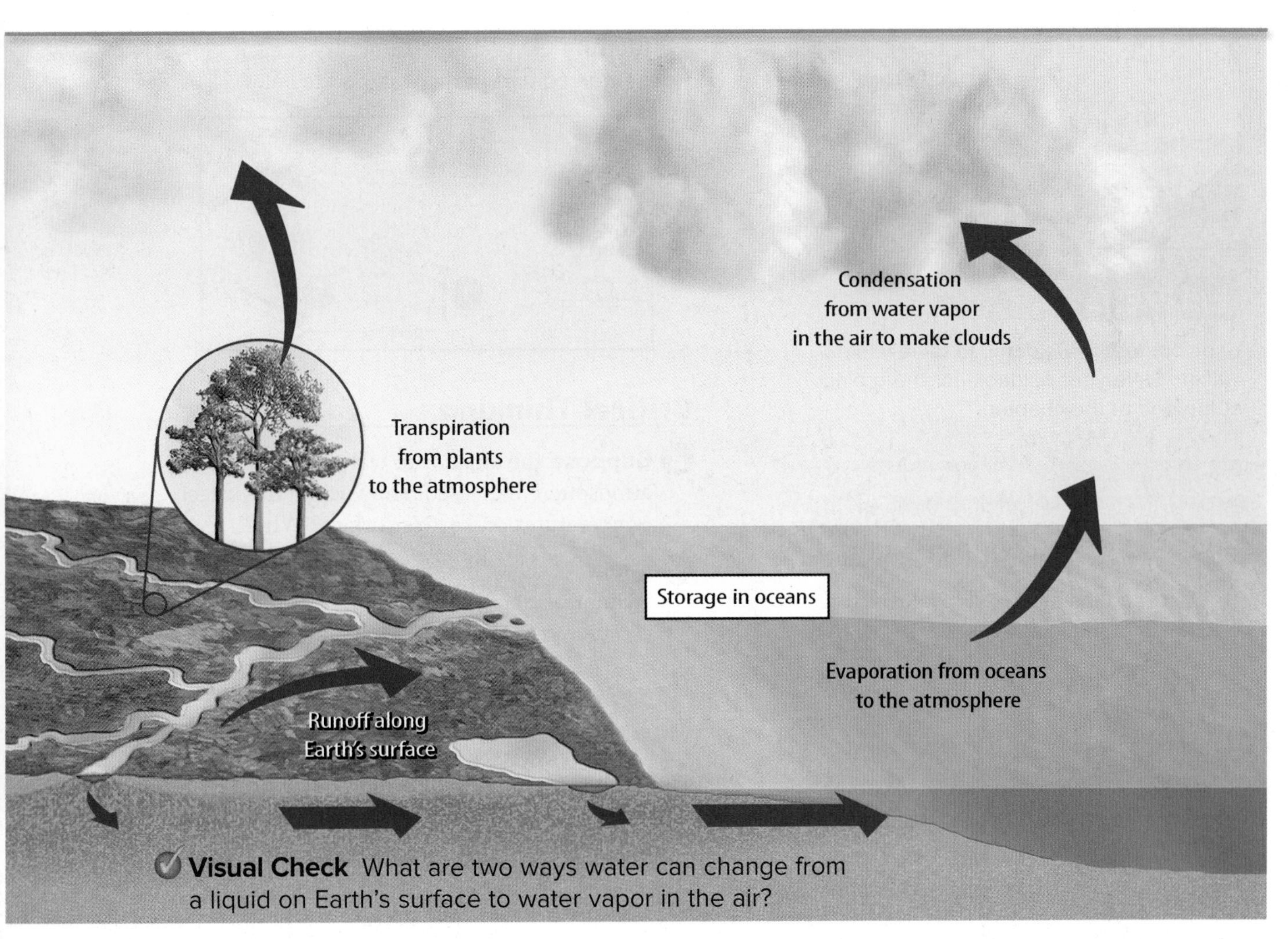

Visual Check What are two ways water can change from a liquid on Earth's surface to water vapor in the air?

Lesson 1 Review

Visual Summary

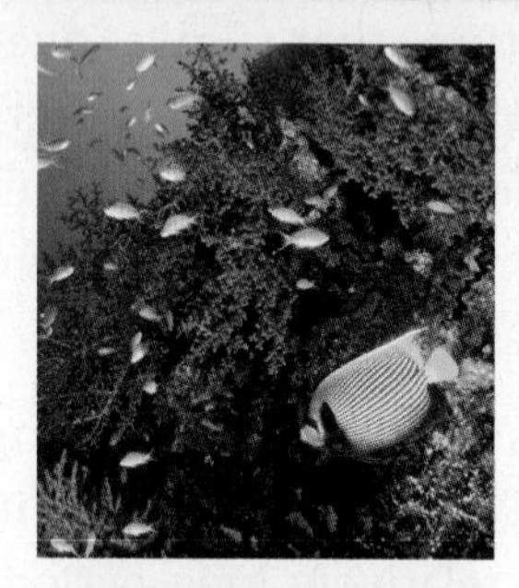

All organisms on Earth depend on water for survival. Water is a habitat for many organisms.

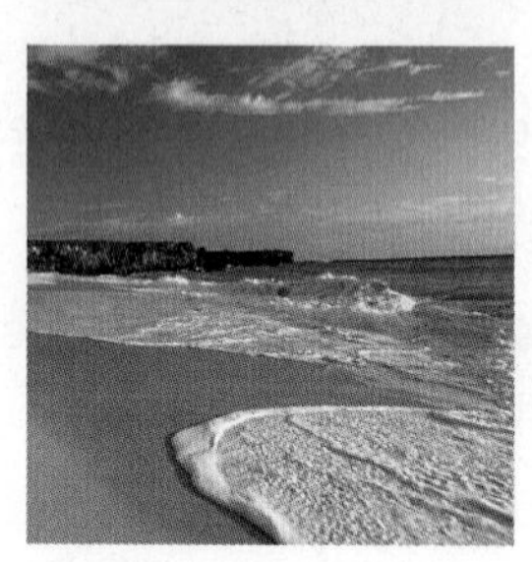

Water's high specific heat causes large bodies of water to take a long time heating up and cooling down.

The water cycle is a natural process in which water constantly moves throughout the hydrosphere.

FOLDABLES®

Use your lesson Foldable to review the lesson. Save your Foldable for the project at the end of the chapter.

What do you think NOW?

You first read the statements below at the beginning of the chapter.

1. A liquid can change to a gas only when the liquid reaches its boiling point.
2. Clouds are made of tiny drops of water.

Did you change your mind about whether you agree or disagree with the statements? Rewrite any false statements to make them true.

Use Vocabulary

1. **Distinguish** between evaporation and transpiration.
2. **Use the term** *hydrosphere* in a complete sentence.
3. **Define** *condensation* in your own words.

Understand Key Concepts 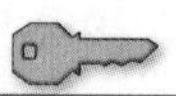

4. Where is most of the water on Earth?
   A. glaciers
   B. groundwater
   C. oceans
   D. rivers
5. **Analyze** What are three reasons water is important to life on Earth?
6. **Name** the two main factors that drive the water cycle.

Interpret Graphics

7. **Identify** the process that occurs at each numbered part of the water cycle below.

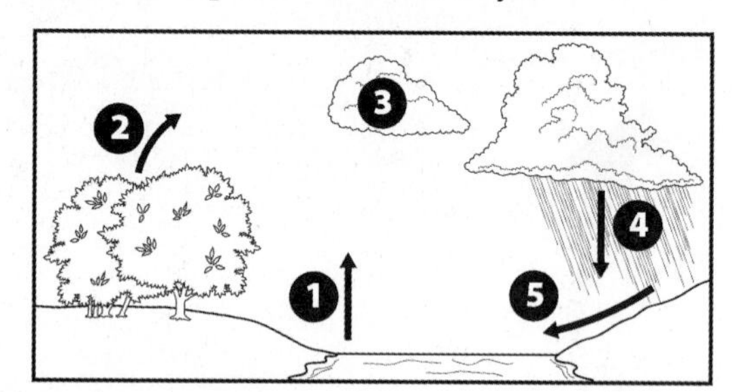

Critical Thinking

8. **Suppose** the amount of water vapor in the atmosphere increased. How would this affect temperatures on Earth's surface? Why?
9. **Evaluate** On a hot day, the water in a swimming pool is much cooler than the cement around the pool. Explain.

Math Skills

10. About how much energy is needed to increase the temperature of 5 kg of sand from 18°C to 32°C if the specific heat of the sand is 190 J/kg • °C?

AMERICAN MUSEUM OF NATURAL HISTORY

Oceans on the Rise—Again

CAREERS in SCIENCE

With an eye to the future, a geologist examines past connections between higher sea levels and melting ice sheets.

Way up in the Arctic is the world's largest island—an ice-covered island called Greenland. A vast ice sheet covers much of Greenland. Changes in Earth's climate can have a great effect on this ice sheet. As average global temperatures increase, Greenland's ice sheet is slowly melting along its coastline. If this continues, it could have a big impact on sea levels worldwide.

Greenland's enormous glaciers creep along inch by inch toward the ocean. When they reach the coast, huge chunks of ice break off and crash into the ocean. As the climate warms, the glacier's speed increases, adding more and more ice to polar waters. As more ice enters the ocean, sea levels rise. Scientists estimate that if the Greenland ice sheet melts, sea levels could rise 7 m, or 23 ft. That's enough water to flood coastlines everywhere on Earth, including those with some of the world's largest cities. Imagine New York City under water! Increased flooding also threatens coastal habitats. Animals as well as people would be forced inland. The worst effects would be in a delta—a low-lying area of land where a river flows into a large body of water.

▲ This map shows coastal areas that would be flooded if the sea levels rise 6 m as scientists predict.

Coral lives and grows under water. Muhs shows where the sea level was in the past at this location in Florida. By measuring the height of the coral fossils, he estimates the ocean was once several meters higher than it is today. ▼

Is this really possible? Scientists, such as Daniel Muhs, know it is possible because it has happened before. Muhs is a geologist with the United States Geological Survey. He investigates rocks for clues about Earth's past. He found a big clue in a limestone wall in the Florida Keys. Today, this wall is several meters above sea level and is filled with fossilized coral. Muhs determined that a coral reef grew there about 125,000 years ago during a warm period when much of the Greenland ice sheet melted. Muhs estimates that sea levels were between 6 m and 8 m higher 125,000 years ago than they are today. This is the same rise in sea levels that other scientists predict would occur if Greenland's ice sheet were to melt again.

▲ A chunk of ice from Greenland's Russell Glacier breaks off and splashes into the ocean. The freshwater in the glacier is no longer available as a source of freshwater when it mixes with seawater. If this continues, the overall amount of freshwater on Earth will decrease.

It's Your Turn

RESEARCH Brainstorm ways that society could respond to rising sea levels. Then, research ways that high sea levels already impact cities and coastlines worldwide. Compare your ideas with real-world solutions and share them with your class.

Lesson 2

The Properties of Water

Reading Guide

Key Concepts

ESSENTIAL QUESTIONS

- What makes water a unique compound?
- How does water's structure determine its unique properties?
- How does water's density make it important to life on Earth?

Vocabulary

polarity p. 538

cohesion p. 539

adhesion p. 539

Multilingual eGlossary

Inquiry Will they freeze?

A thick layer of ice formed over the water where these penguins swim and hunt. Will the rest of the water freeze? How can plants and animals that live in oceans and lakes survive the winter?

Launch Lab

10 minutes

How many drops can fit on a penny?

The structure of a water molecule gives water many unique properties. In this lab, you will explore one property of water—the strong attraction between individual water molecules.

1. Read and complete a lab safety form.
2. Place two **pennies** on a **paper towel.**
3. Use a **dropper** to place 1 drop of **water** at a time on a penny. After 6 drops, closely observe the water on the penny. Try to add more drops, if possible.
4. Use a clean **dropper** to place drops of **rubbing alcohol** one at a time on a different penny. After 6 drops, closely observe the alcohol on the penny. Try to add more drops.

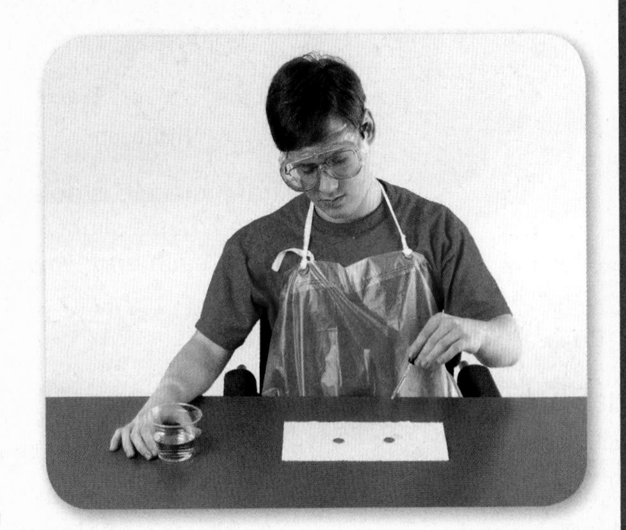

Think About This

1. **Explain** what happened to the water each time you added a drop. What happened when you added the final drop?
2. **Describe** the difference in the shapes of the water and alcohol on the pennies.
3. **Key Concept** Liquids form drops because of the attraction between their particles. Based on this, infer which substance has a stronger attraction between its particles.

Water—A Unique Compound

In Lesson 1 you read that water is the only substance that exists in nature as a solid, a liquid, and a gas. You also read that water has a high specific heat. You might have seen other things that result from water's properties. For example, you have probably noticed that water forms drops if you spill some on a counter, as shown in **Figure 6.** You probably have also seen ice floating in a glass of water or tea. Have you ever dissolved salt in water? Have you ever seen an insect walk on the surface of water?

Water has unusual properties because of its molecules. The properties of water cannot be explained without looking at the way a water molecule is put together. Understanding how water molecules interact with each other and with other materials also helps explain water's unusual properties.

Key Concept Check What makes water a unique compound?

Figure 6 Water forms drops because of strong forces between water molecules.

Personal Tutor

Figure 7 The polarity of water molecules is one of the reasons water is so important to life on Earth ►

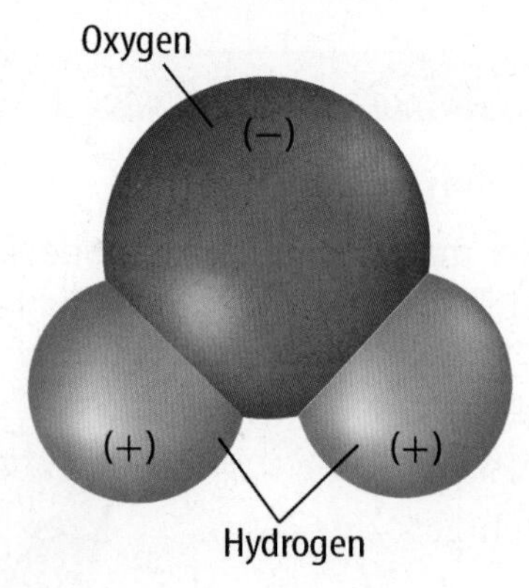

A water molecule is polar because it has a slight charge at each end.

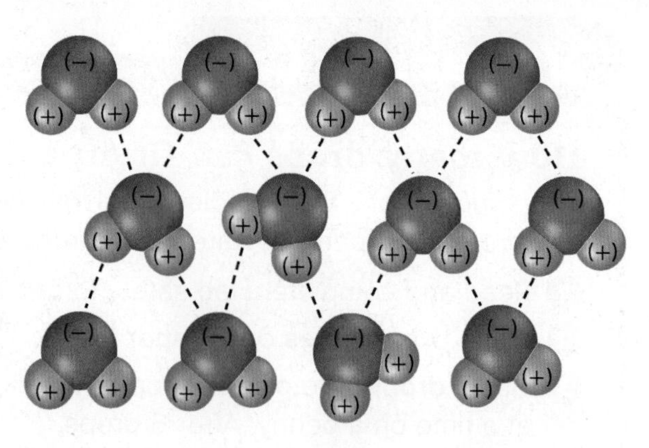

The slightly negative oxygen atom in one water molecule attracts the slightly positive hydrogen atom of another water molecule. This force holds the molecules together.

SCIENCE USE V. COMMON USE

polar
Science Use having opposite ends, which have opposite charges

Common Use relating to Earth's North Pole or South Pole

A Polar Molecule

A water molecule is made up of one oxygen atom and two hydrogen atoms. Look at **Figure 7.** What do you notice about the charges of the atoms? The oxygen atom has a slightly negative charge. The hydrogen atoms have slightly positive charges. The overall charge of a water molecule is neutral. **Polarity** *is a condition in which opposite ends of a molecule have slightly opposite charges, but the overall charge of the molecule is neutral.* Water is a **polar** molecule because the oxygen atom and the hydrogen atoms have slightly opposite charges.

Because of their polarity, water molecules can attract other water molecules. In **Figure 7,** a slightly negative oxygen atom of one water molecule attracts a slightly positive hydrogen atom of another water molecule. Several of water's unique properties are due to its polarity. One of these properties is water's ability to dissolve many different substances.

Reading Check Describe the polarity of a water molecule.

Water as a Solvent

Water is sometimes called the universal solvent because so many substances can dissolve in it. When table salt, or sodium chloride, is placed in water, it dissolves easily. But how?

Study **Figure 8.** Notice that the positively charged sodium ion (Na^+) of salt is attracted to the negatively charged oxygen atom of the water molecule. The negatively charged chloride ion (Cl^-) of salt is attracted to the positively charged hydrogen atom of the water molecule. These attractions cause the sodium and chloride ions to break apart in water, or dissolve. Many substances that are important to life processes are dissolved in water within cells, blood, and plant tissues.

▲ **Figure 8** Ionic compounds, such as table salt (NaCl), can easily dissolve in water because water is polar.

Cohesion and Adhesion

How can the water strider shown in **Figure 9** walk across the surface of water? You've read that water molecules attract each other because of their polarity. This attraction is called cohesion. **Cohesion** *is the attraction among molecules that are alike.* Some insects can walk on the surface of water because the attractions among water molecules is stronger than the attraction of gravity on the insect. The ability to put more drops of water than alcohol on the penny in the Launch Lab also demonstrates cohesion.

Adhesion *is the attraction among molecules that are not alike.* You might be familiar with one example of adhesion–the formation of a curved surface, called a meniscus, on a liquid in a test tube, as in **Figure 9.** Notice that the water molecules in contact with the sides of the test tube stick to the glass, causing the curved surface across the top of the water.

Water moves from the roots of a plant to its leaves as a result of both cohesion and adhesion. As a water molecule evaporates from the surface of a leaf, it pulls another water molecule up into its place. Water molecules stick to the cells within the plant. This keeps gravity from pulling water back down toward the roots.

Key Concept Check Name some ways that water's structure determines its unique properties.

WORD ORIGIN

cohesion
from Latin *cohaerere,* means "to stick together"

FOLDABLES®

Fold a sheet of paper into a two-tab concept map. Label it as shown. Use your book to summarize information about water and its properties.

Properties of Water
Cohesion | Adhesion

Cohesion and Adhesion

Figure 9 Cohesion is responsible for molecules of water sticking together. Adhesion is responsible for water molecules sticking to other surfaces.

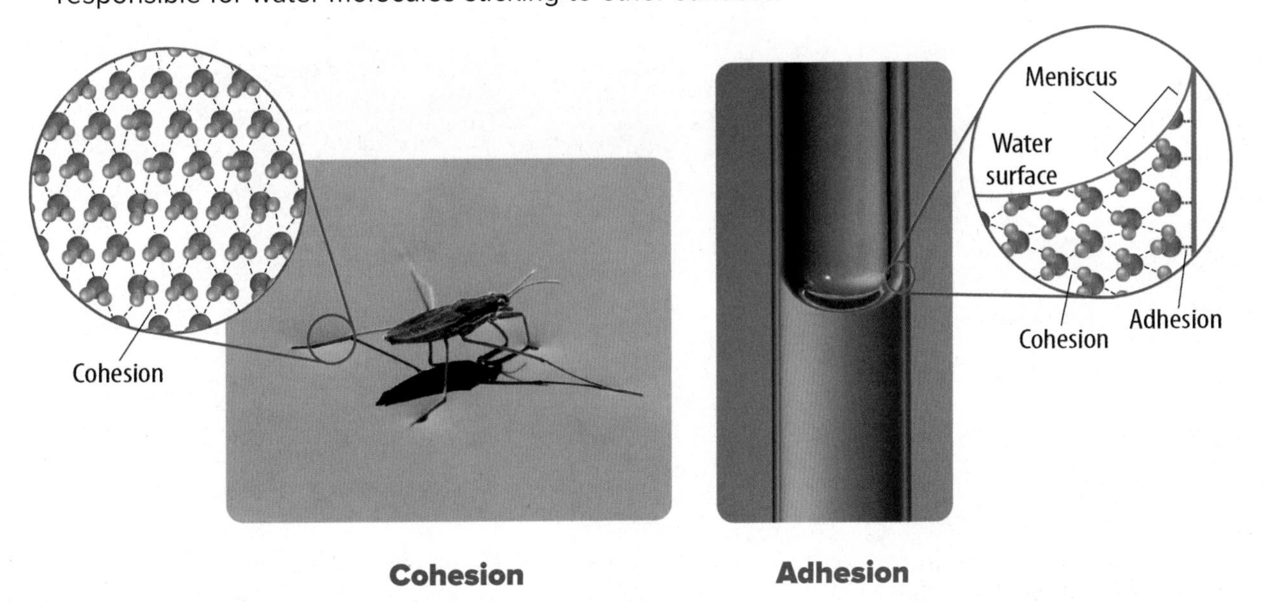

(l)Matti Suopajarvi/mattisj/Getty Images; (r)©Lester V. Bergman/Corbis

Density

Have you ever wondered why ice cubes float in a glass of water? Water that freezes on a lake also floats. Even huge icebergs like the one in **Figure 10** float in the ocean. Ice floats in liquid water because of an important property—**density.**

REVIEW VOCABULARY
density
mass per unit volume of a material

The density of a material increases when the particles in the material get closer together. When most liquids freeze, their particles get closer together. The solid that forms is denser than the liquid. For example, recall that lava is molten, or liquid, rock. As lava cools, the particles get closer together. Therefore, the rock that forms from the lava is denser than lava. The rock will sink if placed in the lava.

Water's Unusual Density

ACADEMIC VOCABULARY
establish
(verb) to make; to put in place

If liquids tend to get denser as they freeze, why does ice float in water? Just like any other liquid, as water cools, the molecules lose energy and pack tightly together. However, when water cools to 4°C the molecules begin to move farther apart. Forces among the molecules cause the molecules to spread out and **establish** themselves in a six-sided pattern. When the water molecules freeze, there is space between them. A cube of ice has fewer water molecules than the same volume of water. Therefore, ice is less dense than water, as shown in **Figure 10.**

 **Reading Check** Why is it unusual that ice floats?

 Personal Tutor

Figure 10 Water is denser than ice because molecules are packed more closely in water than in ice.

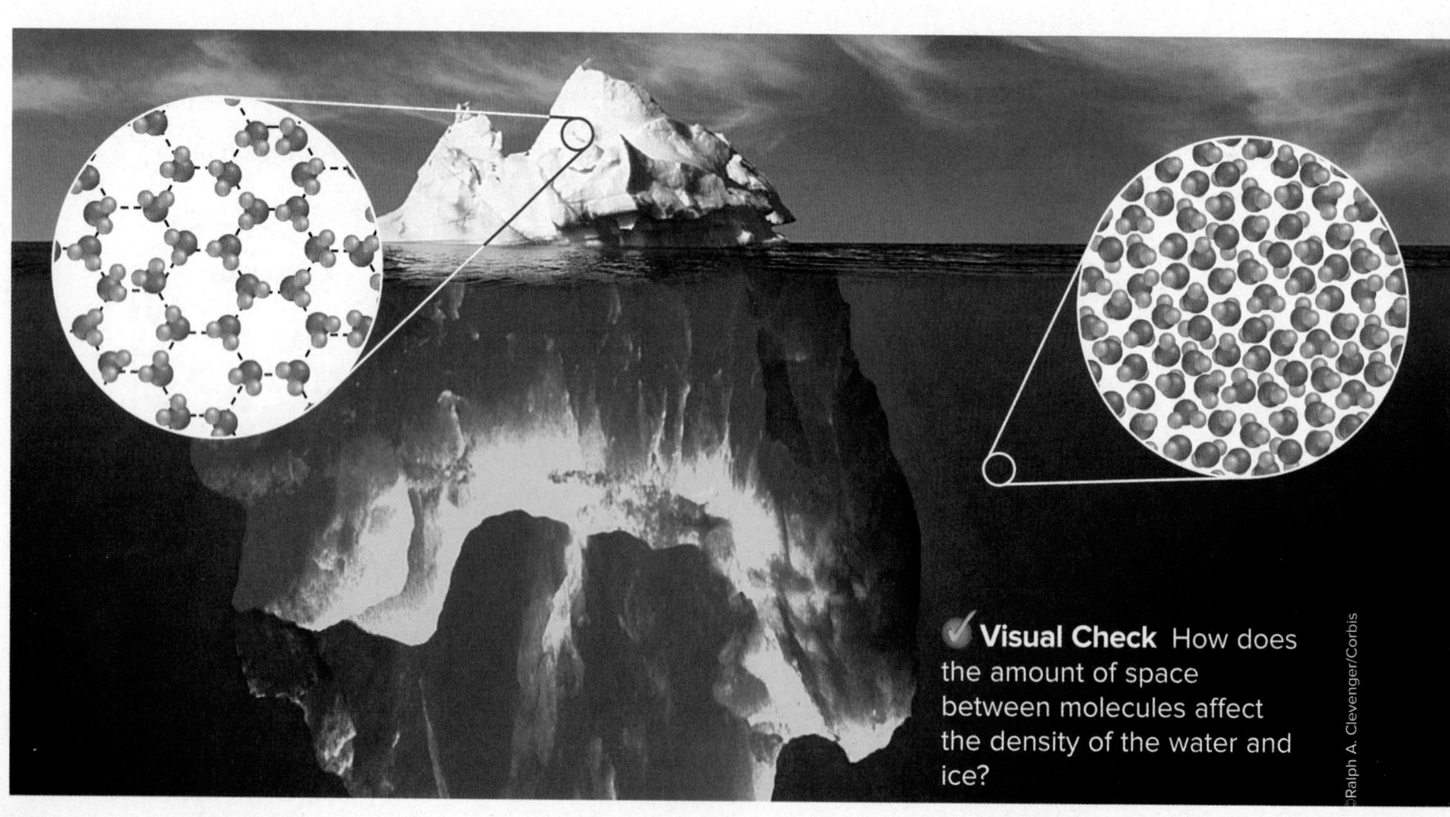

Visual Check How does the amount of space between molecules affect the density of the water and ice?

Figure 11 The density of liquid water is greater than that of ice. The density of liquid freshwater is greatest at a temperature of 4°C.

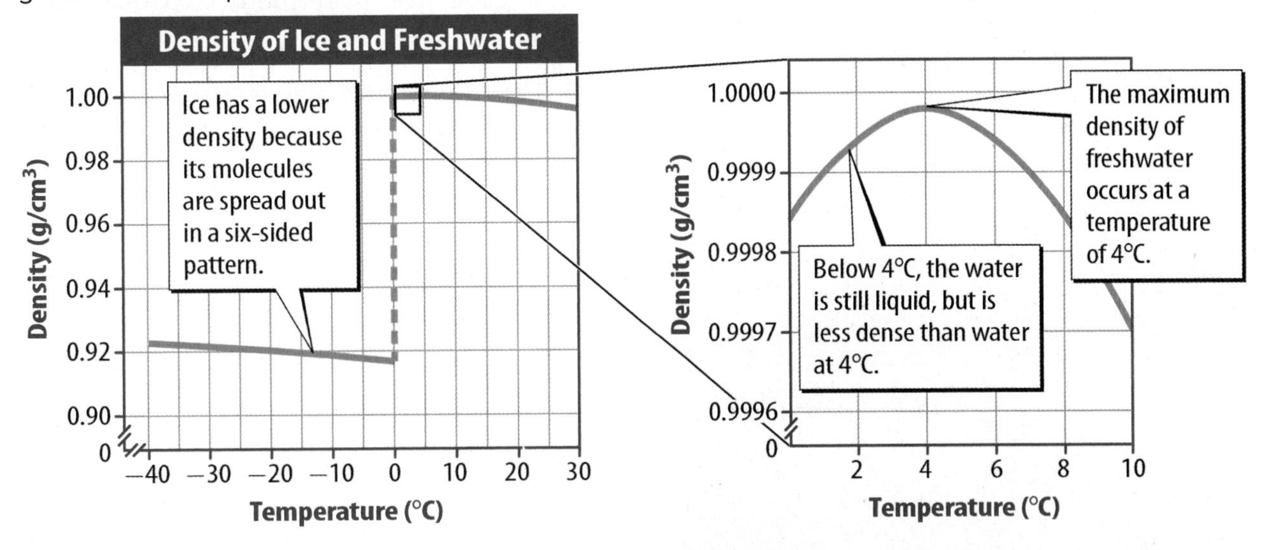

Density and Temperature

To understand more about the unusual density of water, study the two graphs in **Figure 11.** Both graphs illustrate how density changes as the temperature changes. The graph on the left shows the density of both water and ice. The graph on the right is a close-up view of water's density change.

The Density of Ice You can compare the density of ice and water in the graph on the left. The density of ice is much lower than the density of water. Recall that the molecules in ice are more spread out than in water. This explains why ice floats.

The Density of Water Only the density of water is represented in the graph on the right. This graph shows that water is most dense at a temperature of 4°C. Remember that the freezing point of water is 0°C. This means that water between 0°C and 4°C is liquid, but it is less dense than water at 4°C. As you will read on the next page, density of water is important for the survival of life in the water.

Reading Check How does the density of water at 0°C differ from the density of liquid water at 4°C?

MiniLab 20 minutes

Is every substance less dense in its solid state?

Is olive oil less dense in its solid state?

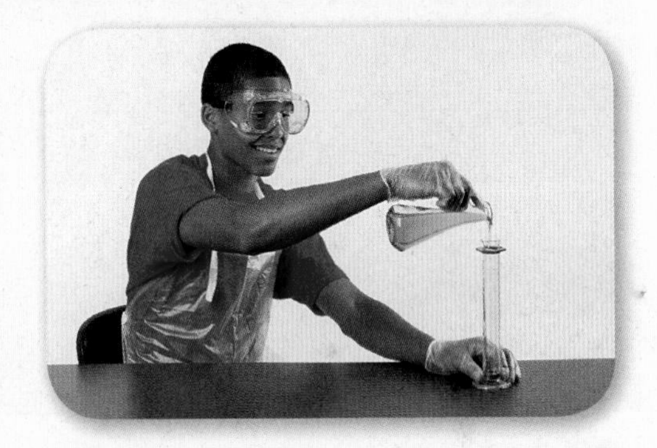

1. Read and complete a lab safety form.
2. Pour 20 mL of **liquid olive oil** into a 50-mL **graduated cylinder.**
3. Form a hypothesis about whether olive oil is more dense as a solid or as a liquid.
4. Drop a chunk of **solid olive oil** into the liquid olive oil. Record what happens in your Science Journal.

Analyze and Conclude

1. **Analyze** Is liquid olive oil or solid olive oil more dense? How do you know this?
2. **Key Concept** How do the densities of solid and liquid olive oil differ from those of solid and liquid water?

Figure 12 Fish and other organisms in a lake can survive in winter because the water remains below a layer of ice.

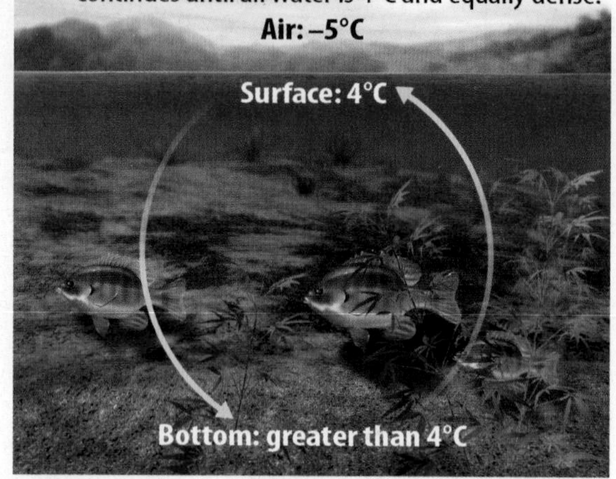

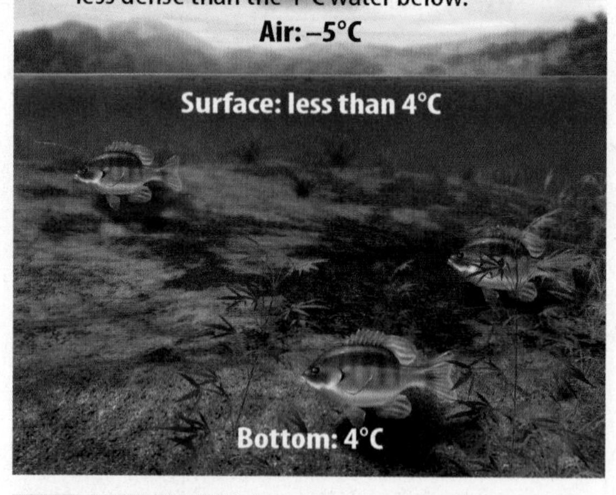

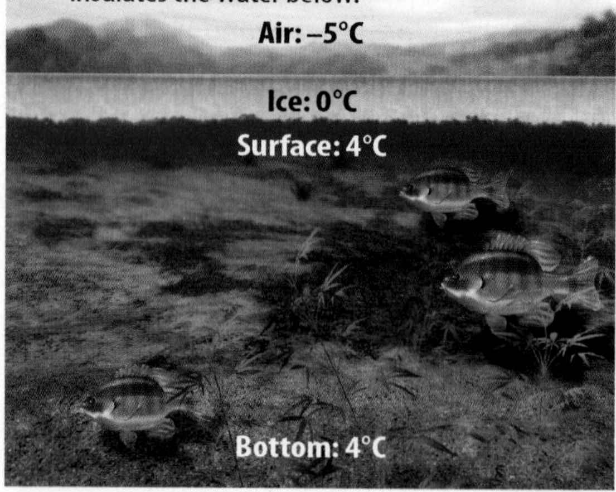

Visual Check What is the temperature of the water below the ice?

The Importance of Water's Density

You have just read about two important features of water's density:

- The density of ice is lower than the density of water.
- The density of freshwater is greatest at 4°C.

Imagine a lake in winter, as shown in **Figure 12.** How is the density of ice and water important to the survival of some organisms on Earth?

1 In winter, cold air above a lake cools the surface water. When the surface water cools to 4°C, it reaches its maximum density and is more dense than the water below it. As a result, the surface water sinks while pushing the warmer water to the surface. Once the warmer water reaches the surface, the air cools this water to 4°C. Again, the water becomes more dense and sinks.

2 Eventually, all the water in the lake cools to 4°C and is equally dense. However, the air above the water continues to cool the surface water. The temperature of the surface water drops below 4°C, and the density begins to decrease. The colder surface water is less dense than the 4°C water below it and stays on top. All the water below the surface water remains at 4°C and maximum density.

3 When the surface water of the lake cools to 0°C, it changes to ice. The density of the ice decreases further, and it continues to float. Ice on the surface insulates the water below it. Aquatic organisms can survive cold, winter months because beneath the ice, water remains a liquid at 4°C. If water froze from the bottom of a lake to the top, organisms living in the lake would freeze along with the water.

Key Concept Check How does water's density make it important to life on Earth?

Lesson 2 Review

Visual Summary

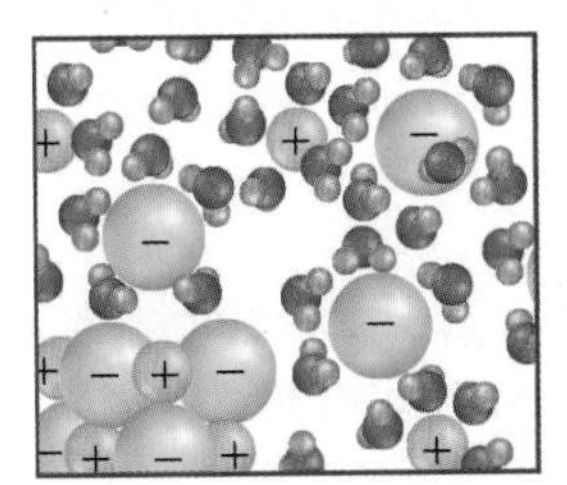

Water can dissolve many substances because a water molecule is polar.

Cohesion is an important property of water molecules. Molecules at the surface of water have enough cohesion that some insects can walk on the surface of water.

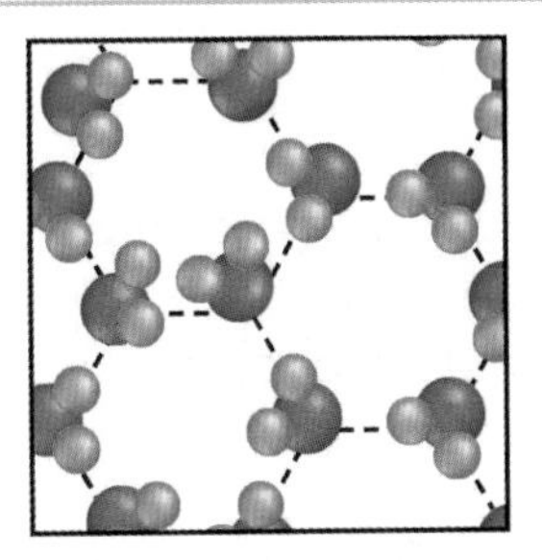

Ice is less dense than water because as water freezes the molecules spread out in a six-sided pattern.

FOLDABLES®

Use your lesson Foldable to review the lesson. Save your Foldable for the project at the end of the chapter.

What do you think NOW?

You first read the statements below at the beginning of the chapter.

3. Water molecules can attract other water molecules.

4. Ice has a greater density than water.

Did you change your mind about whether you agree or disagree with the statements? Rewrite any false statements to make them true.

Use Vocabulary

1 A property in which opposite ends of a molecule are slightly charged is ________.

2 **Distinguish** between adhesion and cohesion.

Understand Key Concepts

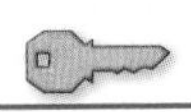

3 Which has the highest density?

A. water at 0°C
B. water at 4°C
C. water at 6°C
D. water at 8°C

4 **Relate** the structure of water molecules to water's unique properties.

5 **Describe** how water's unusual density is important to organisms in a lake in winter.

Interpret Graphics

6 **Organize Information** Copy and fill in the graphic organizer below to describe examples of adhesion and cohesion.

Adhesion	
Cohesion	

7 **Analyze** Use the graph below to describe how the density of water changes if the water temperature is increased–between 0°C and 4°C; at 4°C; between 4°C and 10°C.

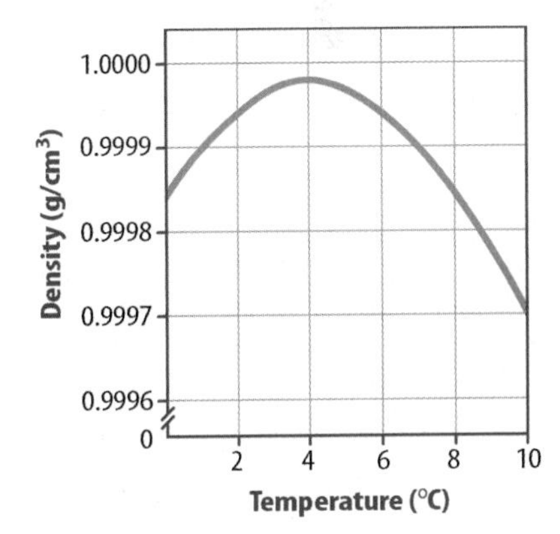

Critical Thinking

8 **Compose** Chris placed two cubes in water. Compose a statement that describes why one cube sank and the other floated. Use the term *density* in your answer.

Skill Practice Cause and Effect

15 minutes

Materials

modeling clay (two colors)

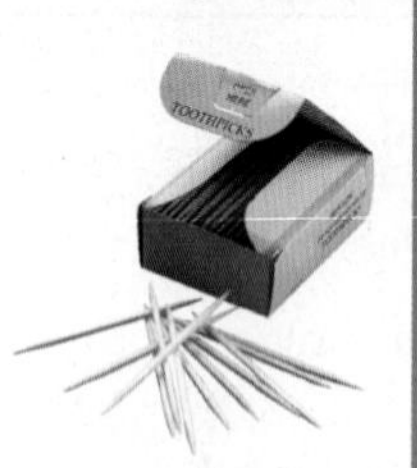

24 toothpicks

Safety

Why is liquid water denser than ice?

Ice floating on a lake in winter is important to the survival of organisms in the lake. Ice floats because its density is lower than the density of water. What is the cause of this difference in density? What effect does the structure of water have on its density?

Learn It

An important part of science is being able to understand **cause and effect** relationships based on observations. Cause and effect is the concept that an event will produce a certain response. You will use models to observe the cause-and-effect relationship between the structure of water and its density.

Try It

1. Read and complete a lab safety form.
2. Use toothpicks and modeling clay to model 12 water molecules. One color of clay represents oxygen atoms. The other color represents hydrogen atoms.
3. The molecules that make up water move freely and are disorganized. Use six of your models and show the water molecules closely arranged.
4. Recall that water molecules are polar. The oxygen atom of one molecule is attracted to a hydrogen atom of another molecule. However, atoms that are alike push away from each other. As water freezes, these forces cause water molecules to form a six-sided pattern. Use the remaining six models to form one of these six-sided patterns, as shown below.

Apply It

5. **Identify** Which has more empty space between molecules, the model of liquid water or the model of ice? What causes this empty space?
6. **Key Concept** Based on your observations, what effect does the structure of water molecules have on the density of liquid water and ice?

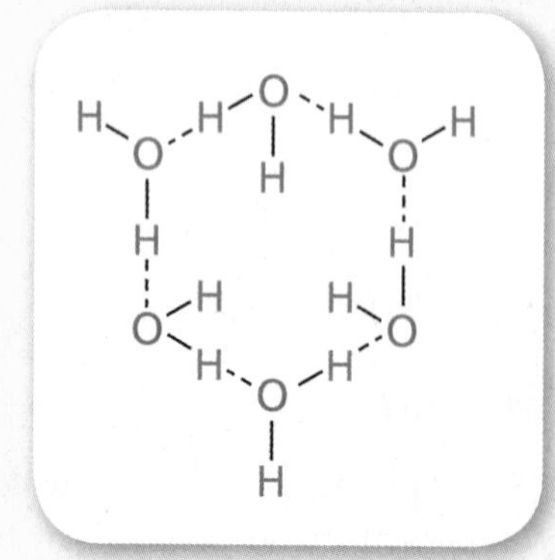

Hutchings Photography/Digital Light Source

Lesson 3

Water Quality

Reading Guide

Key Concepts
ESSENTIAL QUESTIONS

- Why is water quality important?
- How is water quality tested and monitored?

Vocabulary

water quality p. 546
point-source pollution p. 547
nonpoint-source pollution p. 547
nitrate p. 549
turbidity p. 549
bioindicator p. 550
remote sensing p. 550

Multilingual eGlossary

Inquiry Clean Water?

The water in the pond on this glacier looks clean enough to drink, but is it? Can you always tell just by looking at water whether it is clean? How do human activities affect the quality of water?

Launch Lab

10 minutes

How can you test the cloudiness of water?

All lakes and ponds contain sediment, but too much sediment is one way that water can become cloudy. Cloudy water can sometimes be a problem for organisms that live in the lakes and ponds.

1. Read and complete a lab safety form.
2. Tie a **bolt** onto the end of a **string.** Lower the bolt into a **bucket** of **water.** Record notes in your Science Journal.
3. Add **soil** to the water until the water is cloudy. Use a **long-handled wooden spoon** to stir the sediment.
4. Lower the bolt to the same depth as step 1. Record your observations.

Think About This

1. How did your observation of the bolt change after you added soil to the water?
2. **Key Concept** How might scientists use a similar method to study the cloudiness at different depths of a lake or pond?

Human Effect on Water Quality

Suppose you go to the beach, looking forward to swimming and playing in the waves. When you arrive, you find a warning sign like the one in **Figure 13.** What might the sign tell you about the quality of the water and the health of the organisms that live in it?

Water quality *is the chemical, biological, and physical status of a body of water.* It also describes the water's characteristics, such as the amount of oxygen and nutrients in the water, the type and number of organisms living in the water, and the amount of sediment in the water. All of these characteristics are important to the health of aquatic organisms.

Figure 13 This sign warns about the quality of Earth's water.

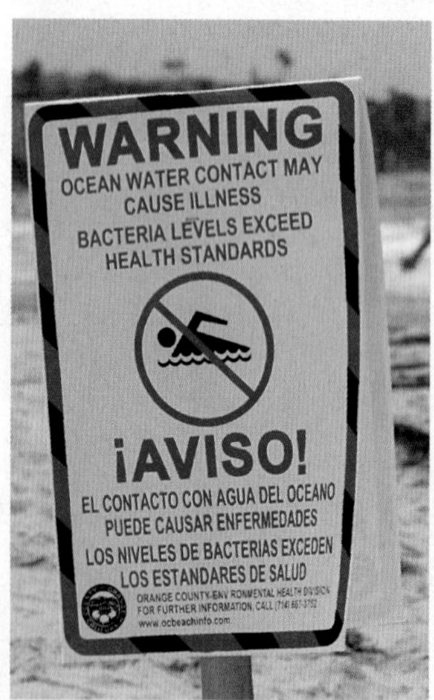

Many natural processes, such as seasonal temperature changes and the weathering of rock and soil, affect the quality of water. Human activities can also affect water quality. Pollution from factories and automobiles eventually reaches rivers, lakes, wetlands, and oceans. Deforestation, which removes large numbers of trees, can lead to increased soil erosion. In addition, when it rains, runoff carries soil and other materials into streams and rivers, changing the quality of the water.

Reading Check What are some ways human activities affect water quality?

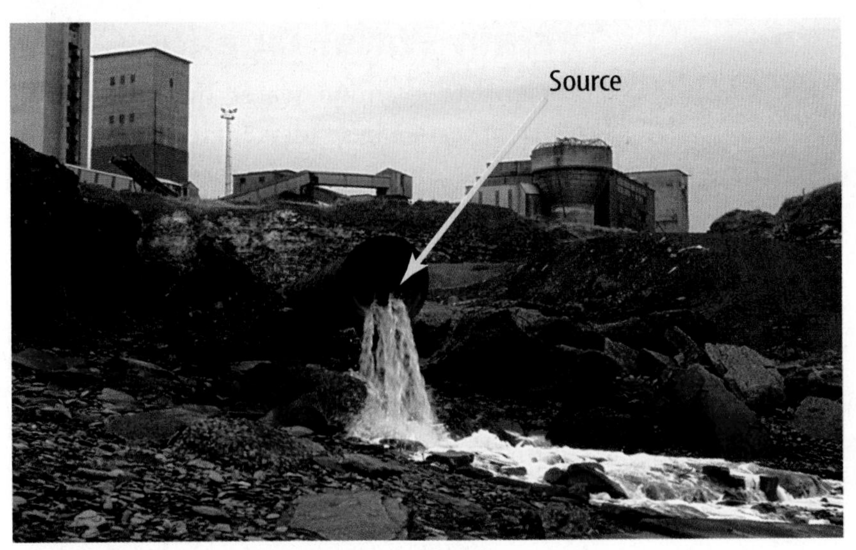

◄ **Figure 14** The water pollution shown here is point-source pollution because its source is known.

Visual Check Can you identify the origin of pollution in this photo?

Point-Source Pollution

How are sources of pollution classified? The wastewater flowing out of the pipe in **Figure 14** is an example of point-source pollution. **Point-source pollution** *is pollution that can be traced to one location,* such as a drainpipe or a smokestack.

The pollution in **Figure 14** is coming from a factory. Factories are common origins of point-source pollution. Another origin of point-source pollution is sewage treatment plants. In many older sewer systems, water from precipitation is mixed with wastewater before being treated. During heavy rainstorms, the sewage treatment plant cannot process the excess water. As a result, storm water, along with untreated sewage, is released directly into nearby bodies of water.

FOLDABLES®

Make a half-book from a sheet of paper. Use your book to organize your notes about the effect different types of pollution have on water quality.

Nonpoint-Source Pollution

Pollution that cannot be traced to one location is **nonpoint-source pollution.** Runoff from large areas, such as lawns, roads, and urban areas, is considered nonpoint-source pollution. As shown in **Figure 15,** the runoff might flow into rivers or streams. It eventually reaches areas of water storage, such as a wetland, groundwater, or the ocean. The runoff might contain natural and human-made pollutants such as sediment, fertilizers, and oil.

Like point-source pollution, nonpoint-source pollution can lower water quality. It can lead to changes in water, which can harm aquatic organisms. Certain types of fish might not be safe for humans to eat because they have high levels of toxins in their tissues. Nonpoint-source pollution might also affect drinking water.

Key Concept Check Why is water quality important?

Figure 15 Much water pollution is from nonpoint-source pollution. ▼

Dissolved Oxygen

Figure 16 The aerator in this fish tank releases bubbles that help keep the water moving throughout the tank. This allows oxygen in the air to continually dissolve in the water at its surface.

Testing Water Quality

Scientists examine water quality using a variety of tests. These tests include measuring levels of dissolved gases, temperature, acidity, and cloudiness. Studying the numbers or the health of certain aquatic organisms is another way scientists measure water quality. Using photos taken from the air or space can also help scientists compare the quality of water over time.

Dissolved Oxygen

Why can fish breathe under water but people cannot? Like the air you breathe, water in oceans and lakes contains oxygen. Some of this oxygen is dissolved in the water. Fish, such as the ones in **Figure 16**, use gills to take in this oxygen they need to survive.

The level of dissolved oxygen affects water quality. If the oxygen level in a lake or stream becomes too low, fish might not be able to survive. Different factors can affect oxygen levels. For example, the release of certain chemicals in water can cause an overgrowth of algae. When the algae die, the decay process uses a large amount of oxygen. The oxygen level in the water can drop so low that fish die.

Water Temperature

Many aquatic organisms are also sensitive to changes in water temperature. Coral bleaching is the whitening of coral due to stress in the environment, such as an increase in water temperature or increased exposure to ultraviolet radiation. It is an event that leads to the death of large areas of coral reefs and is often triggered by a temperature increase in water as little as 2°C. As water temperature increases, the amount of oxygen that can dissolve in water decreases. This means that as water temperature increases, there is less oxygen in the water, which can be harmful to aquatic animals.

Reading Check How would cooling water affect the level of dissolved oxygen?

MiniLab 20 minutes

How do oxygen levels affect marine life?

The table below contains every other month's average dissolved oxygen levels in the Chesapeake Bay from 1985 through 2002.

On **graph paper,** make a line graph using the data in the table.

Month	Dissolved Oxygen
January	10.0 mg/L
March	10.0 mg/L
May	5.0 mg/L
July	1.5 mg/L
September	3.0 mg/L
November	7.0 mg/L

Analyze and Conclude

1. **Describe** the pattern of dissolved oxygen levels throughout the entire year.
2. **Key Concept** Blue crabs need at least 3 mg/L of dissolved oxygen to survive. Infer during which month(s) the levels of dissolved oxygen might affect the population of blue crabs in the Bay.

Nitrates

Compounds that contain the nitrate ion can be harmful to the environment. *A* **nitrate** *is a nitrogen-based compound often used in fertilizers.* Runoff from fertilizers used in landscaping and farming contribute to high concentrations of nitrate found in water. This can cause an algal bloom, in which the algae population increases at a rapid rate, as shown in **Figure 17**. Algae growing on the water's surface can block light needed by plants growing at greater depths, causing them to die. The algae can die too. When the algae die, oxygen levels in the water can decrease, producing a very unhealthy ecosystem.

▲ **Figure 17** Nitrates from farm fertilizer flow into this stream, causing an algal bloom.

 **Reading Check** What is an algal bloom?

Acidity

When scientists work in a lab with substances that are strong acids or strong bases, they have to be extremely careful. These substances can be harmful. Strong acids and bases can also be harmful to animals and plants that live in water. Long-term changes in the acidity of water can affect the entire ecosystem. Some fish might not be able to survive. Even if some organisms survive in acidic water, their food sources might not.

Turbidity

A measure of the cloudiness of water, from sediments, microscopic organisms, or pollutants, is **turbidity** (tur BIH duh tee). As the amount of matter floating in water increases, the turbidity increases. Also, the distance light can penetrate into water decreases. Turbidity affects organisms that need light to undergo photosynthesis. High turbidity can also affect filter-feeding organisms. The structures these organisms use to filter food from water can get clogged with sediment. The organisms could die from lack of food. Turbidity is measured using a device called a Secchi disk, shown in **Figure 18.**

WORD ORIGIN

turbidity
from Latin *turbidus,* means "disturbance"

Figure 18 A Secchi disk is used to measure turbidity. The farther down in the water the disk is visible, the lower the turbidity of the water. ▼

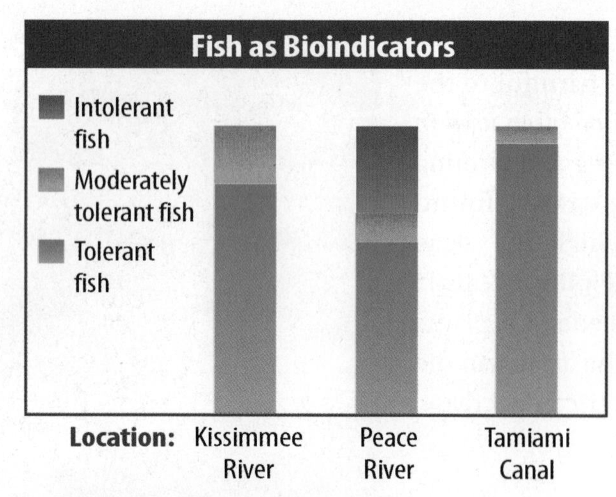

▲ **Figure 19** The presence of intolerant fish indicates that Peace River has good water quality.

Visual Check Which area of water most likely has the worst water quality?

Biodicators

An organism that is sensitive to environmental conditions and is one of the first to respond to changes is a **bioindicator.** Bioindicators alert scientists to changes in the level of oxygen, nutrients, or pollutants in the water. For example, the presence of stoneflies, small insects that live on the bottom of streams, usually indicates good water quality. Stoneflies cannot survive when oxygen levels in water are too low.

Larger organisms, such as fish, also can be used as bioindicators. The number of fish species in different locations in Florida are shown in the graph in **Figure 19.** The different species are classified as tolerant, moderately tolerant, and not tolerant of pollution. When species that are not tolerant of pollution are missing from water, this can indicate poor water quality.

Remote Sensing

The collection of data from a distance is called **remote sensing.** Remote sensing data can be collected through photos taken from the air or images taken from satellites. Scientists use remote sensing data to monitor changes in water storage on Earth, such as melting glaciers. Images from satellites can be used to compare water in the same area over time.

Data collected through remote sensing can be used to make inferences about water quality. Notice the swirls of green in the lakes shown in **Figure 20.** The nutrients in farm runoff combined with favorable lake conditions cause an overgrowth of algae around the Great Lakes almost every summer. After the algae dies, bacteria break it down and oxygen levels decrease. This can lead to a decrease in the population levels of fish and other aquatic animals. Algae blooms can also make the water unsafe to drink.

Figure 20 These two images were taken on July 28, 2015. The swirls of green are algae blooms in Lake Eerie (top) and Lake St. Clair (bottom). ▼

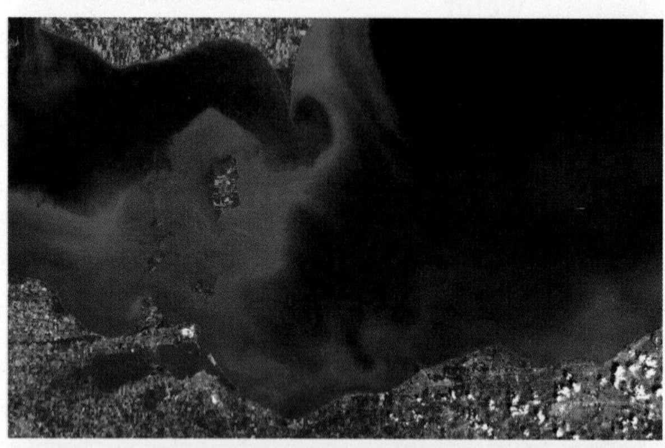

Key Concept Check Name several ways water quality is tested and monitored.

Lesson 3 Review

Visual Summary

Water quality is the chemical, biological, and physical status of a body of water. Sources of pollution are not always obvious.

Various factors can cause a decreased level of dissolved oxygen in water. This can harm aquatic organisms.

High turbidity is another factor that can harm aquatic organisms.

FOLDABLES®

Use your lesson Foldable to review the lesson. Save your Foldable for the project at the end of the chapter.

What do you think NOW?

You first read the statements below at the beginning of the chapter.

5. Factories are responsible for almost all water pollution.

6. Changes in the types of organisms living in water can be a sign of changes in the quality of the water.

Did you change your mind about whether you agree or disagree with the statements? Rewrite any false statements to make them true.

Use Vocabulary

1. A measure of the cloudiness of water is called ________.

2. **Use the term** *bioindicator* in a complete sentence.

3. **Define** the terms *point-source pollution* and *nonpoint-source pollution* in your own words.

Understand Key Concepts 

4. What is a way that water changes as the temperature of the water increases?
   A. Acidity decreases.
   B. Acidity increases.
   C. Oxygen level decreases.
   D. Oxygen level increases.

5. **Explain** how a change in water acidity can affect organisms living in a lake.

6. **Decide** A scientist is monitoring the water quality of two lakes. One lake contains a high level of intolerant fish. The other lake contains a low level of intolerant fish. Which lake most likely has better water quality? Why?

Interpret Graphics

7. **Sequence** Draw a graphic organizer like the one below to sequence how an overgrowth of algae in a lake can kill fish.

Chemicals are released into water. → [ ] → [ ] → [ ]

Critical Thinking

8. **Predict** A river recently experienced an algal bloom. There are no stoneflies in the river. What might a scientist find about the level of nitrates, the level of oxygen, or the turbidity of the water?

9. **Recommend** A scientist in New York wants to study changes in the size of glaciers in Antarctica over the next ten years. What type of remote sensing could she use?

Lab

45 minutes

Materials

food coloring

hot plate

250-mL beakers (3)

ice

stirring rods (2)

droppers (2)

heat-resistant glove

Safety

Temperature and Water's Density

If water were like most substances, ice would sink in liquid water, and underwater organisms would die as lakes and ponds froze completely in winter. But the properties of water are different from most substances, and these properties are important for life on Earth. In this lab, you will investigate the relationship between water's temperature and its density. If one material floats in another, the material that floats has a lower density.

Question

What effect does temperature have on water's density?

Procedure

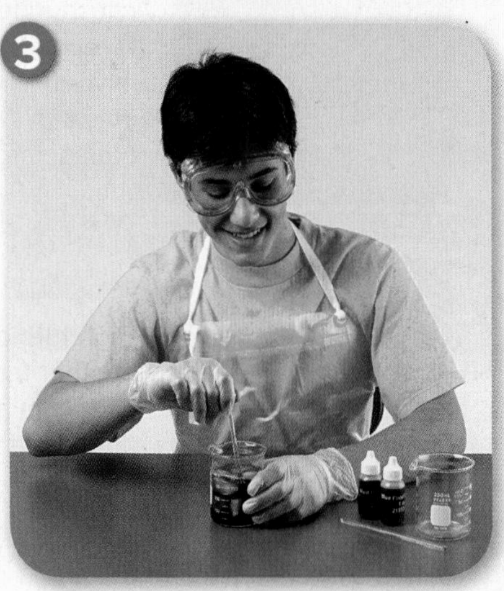

1. Read and complete a lab safety form.
2. Copy the data table in your Science Journal.
3. Stir 3 drops of blue food coloring into 150 mL of water in a beaker. Stir in ice until the water is cold.
4. Based on your observations, conclude which has a lower density—the ice or the cold water. Record your observations and conclusion in your data table.
5. Stir 3 drops of red food coloring into 150 mL of water in a beaker. Heat the water on a hot plate until the water is warm but not boiling.
6. Place a small amount of ice in the warm water. Observe whether the ice floats. Conclude whether the ice or warm water has a lower density. Record your observations and conclusion.

What was compared?	Observations	Conclusions
Ice and cold water		
Ice and warm water		
Cold water Room temperature water Warm water		

(t to b, 2-5, 7, r)Hutchings Photography/Digital Light Source; (6)Jacques Cornell/McGraw-Hill Education

Form a Hypothesis

7. Think about what you have observed about the relationship of temperature to the density of water. Form a hypothesis about the differences in density of cold water, room temperature water, and warm water.

Test Your Hypothesis

8. Design an investigation to test your hypothesis about the relationship of density to the temperature of water.
   ⚠ *Use a heat-resistant glove to handle the heated glass beaker and stirring rod.*
9. Place 150 mL of room-temperature water into a beaker.
10. Carefully add several drops of warm, dyed-red water into the room temperature water. Then add several drops of cold, dyed-blue water into the room-temperature water. Record observations in your data table.

Analyze and Conclude

11. List the following from least dense to most dense: ice, cold water, room-temperature water, warm water.
12. **Conclude** Write a statement that describes how differences in temperature cause differences in the density of water.
13. **The Big Idea** Explain the effect of temperature and density of water on underwater organisms.

Communicate Your Results

Create a poster that explains how the temperature of water is related to its density. Include colorful drawings to illustrate your observations.

Extension

The density of water depends on its temperature. However, in ocean water, differences in the saltiness of water can cause differences in density. Design an experiment that tests this effect.

10

Lab Tips

- ☑ Use a dropper to place small amounts of one temperature of water into water which has a different temperature.
- ☑ Be sure the water in a beaker is as still as possible before placing a different temperature of water in it.

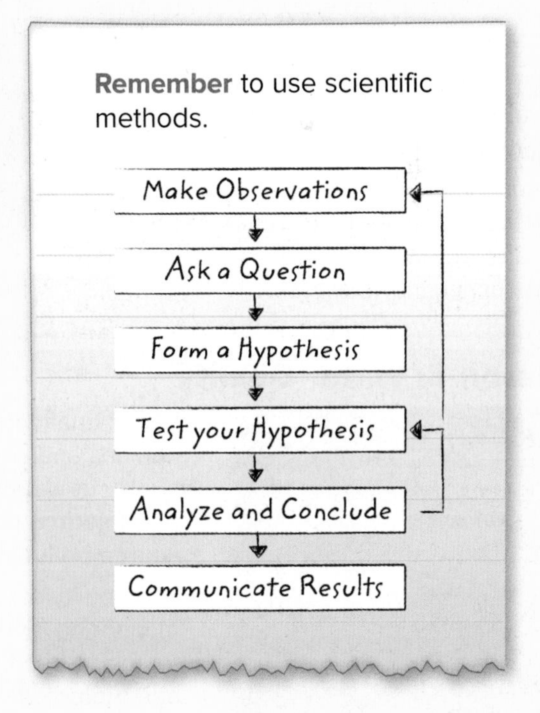

Hutchings Photography/Digital Light Source

Chapter 15 Study Guide

Water cycles throughout Earth's hydrosphere and is necessary for the survival of all living things.

Key Concepts Summary 

Lesson 1: The Water Planet

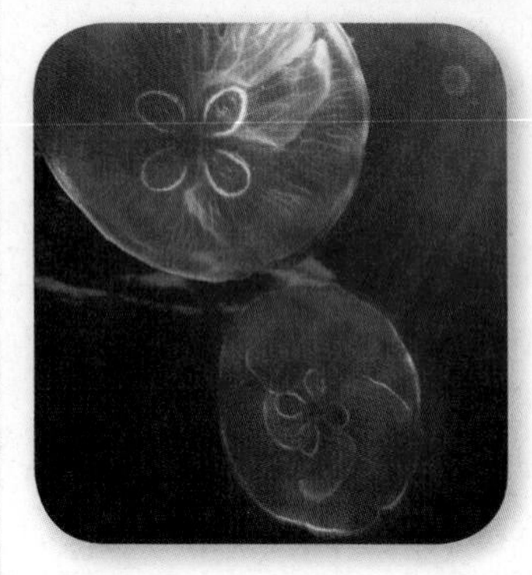

- All organisms on Earth depend on water. Water regulates Earth's temperature.
- Water provides a stable temperature for aquatic organisms because of its high **specific heat.**
- Water is in the **hydrosphere**—on and below Earth's surface and in the atmosphere.
- Water moves through the **water cycle** by **evaporation, transpiration, condensation,** precipitation, and runoff.

Vocabulary

specific heat p. 529
hydrosphere p. 530
evaporation p. 531
condensation p. 531
water cycle p. 532
transpiration p. 533

Lesson 2: The Properties of Water

- Water is the only substance that exists naturally as a solid, a liquid, and a gas on Earth.
- Because of its **polarity,** water dissolves many substances.
- Together, **cohesion** and **adhesion** allow water to transport nutrients and wastes within plants.
- Since the density of ice is less than that of water, ice floats and insulates the water below. This allows aquatic organisms to survive in the winter.

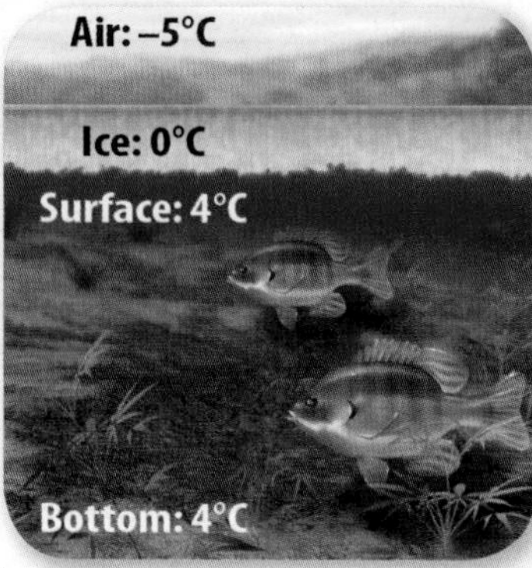

Vocabulary

polarity p. 538
cohesion p. 539
adhesion p. 539

Lesson 3: Water Quality

- **Water quality** affects the health of humans and aquatic organisms. The quality of water can be harmed by **point-source pollution** or by **nonpoint-source pollution.**
- The quality of water can be tested by monitoring levels of dissolved oxygen, temperature, **nitrates,** acidity, **turbidity,** and **bioindicators. Remote sensing** is one method of monitoring.

Vocabulary

water quality p. 546
point-source pollution p. 547
nonpoint-source pollution p. 547
nitrate p. 549
turbidity p. 549
bioindicator p. 550
remote sensing p. 550

(t)Stephen Frink/Digital Vision/Getty Images; (b)Paolo Messina Photography/Moment/Getty Images

Study Guide

Personal Tutor

Vocabulary eFlashcards
Vocabulary eGames

FOLDABLES® Chapter Project

Assemble your lesson Foldables as shown to make a Chapter Project. Use the project to review what you have learned in this chapter.

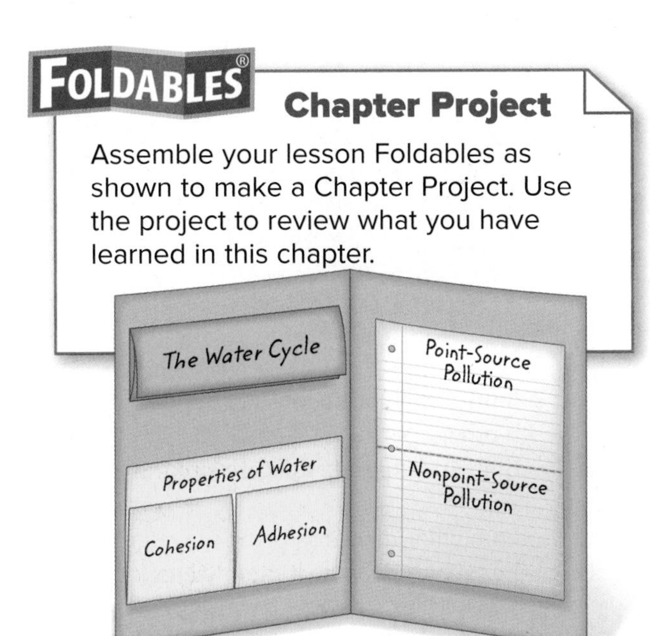

Use Vocabulary

1. Water moves through Earth's ________ by a process called the water cycle.
2. The process of water changing to a gas at its surface is ________.
3. Slightly opposite charges on opposite ends of water molecules cause the ________ of water.
4. The attraction between molecules that are alike is called ________.
5. The chemical, biological, and physical status of a body of water is ________.
6. An organism that is sensitive to environmental conditions and is one of the first to respond to changes is a(n) ________.

Interactive Concept Map

Link Vocabulary and Key Concepts

Copy this concept map, and then use vocabulary terms from the previous page to complete the concept map.

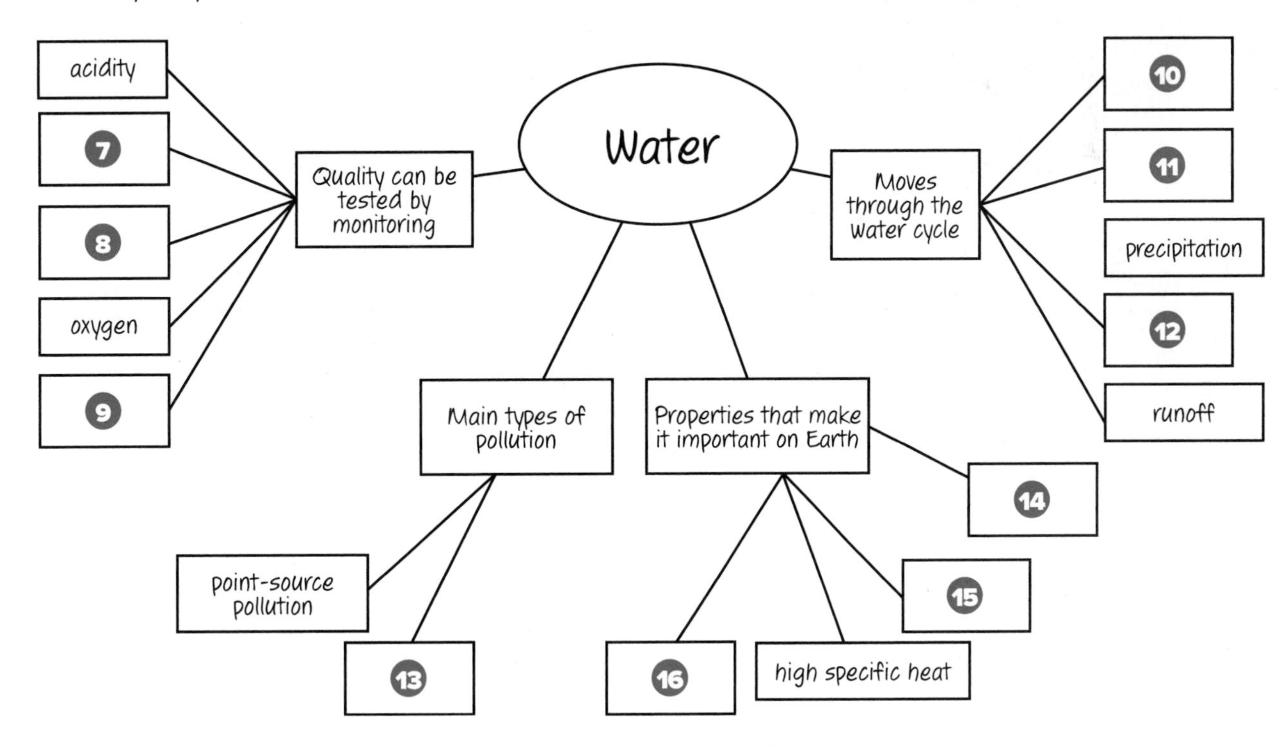

Chapter 15 Review

Understand Key Concepts 

1. The atmosphere has the highest concentration of which greenhouse gas?
   A. carbon dioxide
   B. carbon monoxide
   C. methane
   D. water vapor

2. Which main factors drive the water cycle?
   A. gravity and precipitation
   B. gravity and the Sun's energy
   C. precipitation and evaporation
   D. the Sun's energy and evaporation

3. The diagram below shows the distribution of freshwater on Earth.

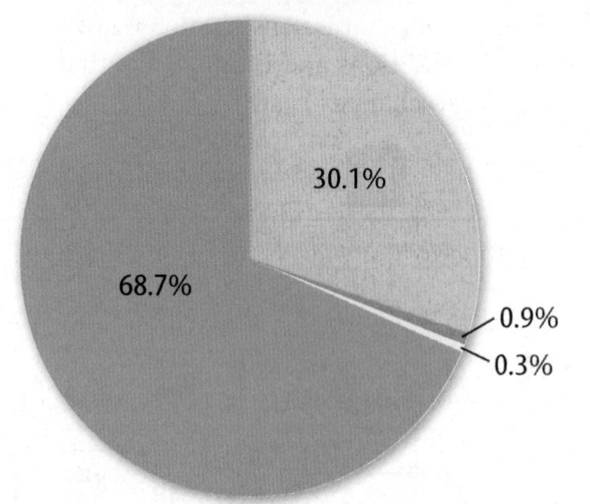

   According to the graph and what you have read in this chapter, about how much of Earth's freshwater is in places other than glaciers, icebergs, and groundwater?
   A. 0.3%
   B. 1.2%
   C. 68.7%
   D. 98.8%

4. What is the freezing point of water?
   A. –2°C
   B. 0°C
   C. 4°C
   D. 10°C

5. Which BEST describes the diagram below?

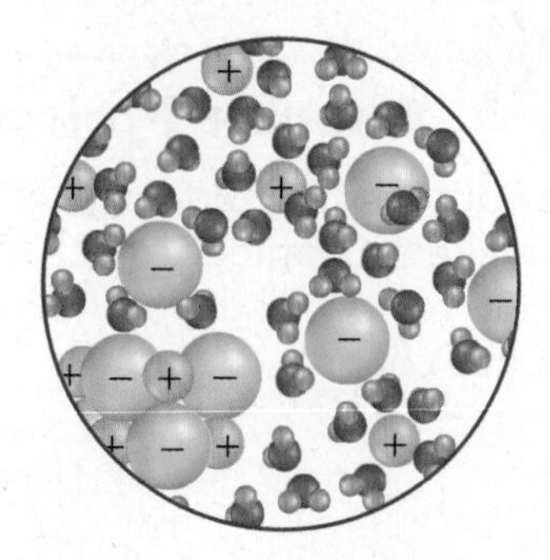

   A. Sodium and chloride ions are adhering to each other.
   B. Sodium and chloride ions are sinking in water.
   C. Sodium chloride is dissolving in water.
   D. Sodium chloride is floating in water.

6. Which property of water is most responsible for an insect being able to walk on the surface of a pond?
   A. adhesion
   B. cohesion
   C. density
   D. transpiration

7. What causes an algal bloom?
   A. a very high acidity level
   B. a very low turbidity level
   C. too much nitrate in the water
   D. too much oxygen in the water

8. Which is nonpoint-source pollution?
   A. leakage from a sewage treatment plant
   B. an oil spill from a tanker ship
   C. runoff from an urban area
   D. warm water from a factory drainpipe

9. Which can be used to measure the level of turbidity of water?
   A. Erlenmeyer flask
   B. microscope
   C. Secchi disk
   D. remote sensing

Critical Thinking

10 **Explain** how the high specific heat of water is important to living things on Earth.

11 **Imagine** How would life on Earth change if water did not naturally exist in all three states in the range of temperatures on Earth?

12 **Design** a demonstration that compares an effect of water's high specific heat to other substances, such as soil or asphalt.

13 **Cause and Effect** Copy and fill in the graphic organizer below to list a cause and several effects of water's ability to dissolve many substances.

14 **Evaluate** Detergent breaks the bonds between water molecules. This helps remove grease and oil stains from clothes in the washing machine. However, detergent can enter rivers and lakes in wastewater and runoff. How can this affect the organisms that live in these habitats?

15 **Construct** a flow chart that explains how the deforestation of an area can affect the water quality of a nearby river.

16 **Illustrate** why water is a polar molecule.

17 **Give an example** of how scientists use bioindicators to monitor water quality.

Writing in Science

18 **Design** a four-page brochure in which you describe and illustrate different ways that human activities affect water quality. Be sure to include ways that human activities both benefit and harm water quality.

REVIEW THE BIG IDEA

19 What role does water play in regulating Earth's temperature?

20 The photo below shows animals that live on the dry grasslands of Africa. Why is water so important to the animals?

Math Skills

Math Practice

Use Equations

Substance	Specific Heat (J/kg • °C)
Water (H_2O)	4186
Hard plastic	400
Copper (Cu)	90

21 One kilogram of water, plastic, and copper at room temperature receive the same amount of energy from the Sun over a 10 min period. Which material will have the smallest increase in temperature? Explain.

22 How much energy is needed to warm 8.0 kg of copper from 120°C to 145°C?

23 Two kilograms of a substance needs 20,000 J of energy to warm from 200°C to 300°C. What is the specific heat of the substance? Use this form of the equation:

$$\text{Specific heat} = \frac{\text{energy}}{(\text{mass} \times \text{temperature change})}$$

Standardized Test Practice

Record your answers on the answer sheet provided by your teacher or on a sheet of paper.

Multiple Choice

1 Which is point-source pollution?

A acid rain

B broken drainpipe

C field runoff

D weathering rock

Use the diagram below to answer questions 2 and 3.

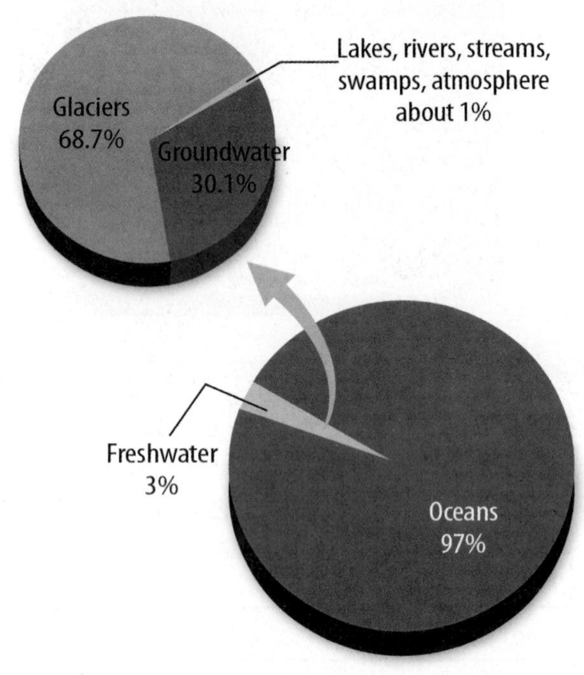

2 According to the graphs, approximately how much of Earth's water resides in glaciers?

A 2 percent

B 3 percent

C 30 percent

D 68 percent

3 What is the ratio of freshwater to saltwater on Earth?

A 3:97

B 3:100

C 97:3

D 97:100

4 What property of the molecules in ice makes ice float on water?

A They are farther apart than water molecules.

B They are much larger than water molecules.

C They contain more oxygen atoms than water molecules.

D They move more quickly than water molecules.

Use the diagram below to answer question 5.

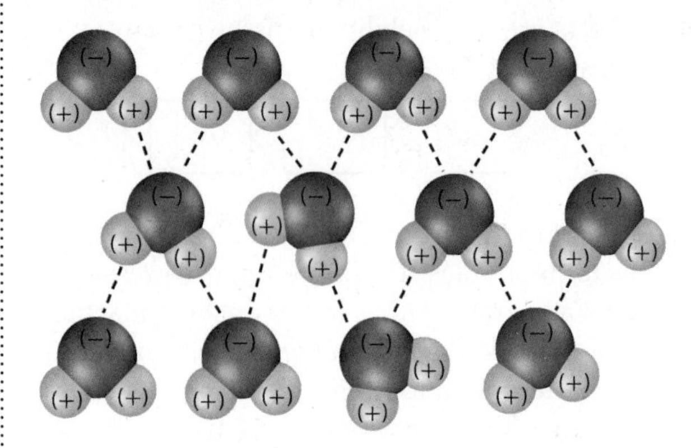

5 What property of water molecules does the diagram illustrate?

A consistency

B layering

C neutrality

D polarity

6 Which is the physical, chemical, and biological status of a body of water?

A its density

B its quality

C its specific heat

D its volume

Use the diagram below to answer question 7.

Fish as Bioindicators

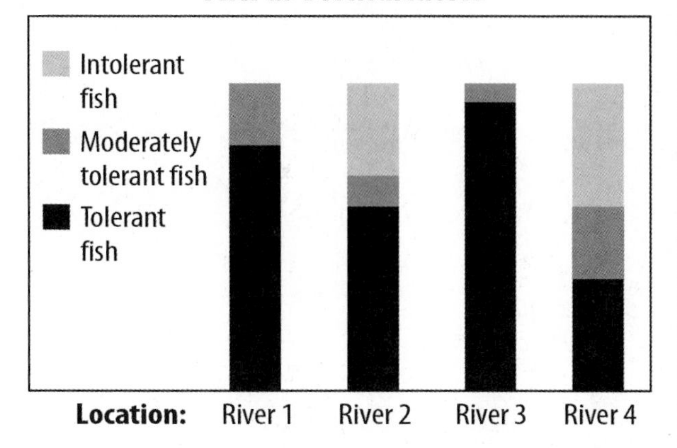

7 In the graph above, which has the best water quality?

A river 1

B river 2

C river 3

D river 4

8 When a lake freezes in winter, what happens beneath the ice layer?

A Organisms freeze at 4°C.

B The water at the bottom turns to ice.

C Warm water sinks to the bottom.

D Water remains liquid at 4°C.

9 Which explains why water in a cylinder forms a meniscus across the top?

A adhesion

B density

C specific heat

D turbidity

Constructed Response

Use the table below to answer question 10.

Stage	Description
Condensation	
Evaporation	
Precipitation	
Runoff	
Storage	

10 In the table above, describe each stage of the water cycle and where it occurs.

Use the table below to answer questions 11 and 12.

Factor	Effect
Acidity	
Dissolved oxygen	
Nitrates	
Temperature	
Turbidity	

11 Explain the effect each factor in the table above has on water quality.

12 How does human activity contribute to the effects these factors have on water quality? Give two examples.

NEED EXTRA HELP?												
If You Missed Question...	1	2	3	4	5	6	7	8	9	10	11	12
Go to Lesson...	3	1	1	2	2	3	3	2	2	1	3	3

Chapter 16

Oceans

THE BIG IDEA **What are characteristics of oceans, and why are oceans important?**

Inquiry What makes waves so powerful?

Have you ever felt the power of an ocean wave? Oceans are large and powerful, and they can be dangerous. They are also important. Oceans contain valuable resources, and they affect Earth's climate and weather.

- What causes ocean waves and currents? How do oceans affect weather and climate?
- How are oceans threatened?
- What are the characteristics of oceans, and why are oceans important?

Get Ready to Read

What do you think?

Before you read, decide if you agree or disagree with each of these statements. As you read this chapter, see if you change your mind about any of the statements.

1. Oceans formed about 4 billion years ago.
2. The seafloor is flat.
3. Waves move water particles from one location to another.
4. The wind causes tides.
5. Ocean currents occur on the surface and below the surface.
6. Ocean currents affect climate and weather.
7. Most pollution in the oceans originates on land.
8. Global climate change has no effect on marine organisms.

Lesson 1

Reading Guide

Key Concepts
ESSENTIAL QUESTIONS

- Why are the oceans salty?
- What does the seafloor look like?
- How do temperature, salinity, and density affect ocean structure?

Vocabulary

salinity p. 565
seawater p. 565
brackish p. 565
abyssal plain p. 566

Multilingual eGlossary

BrainPOP®

Composition and Structure of Earth's Oceans

Inquiry What's down there?

Conditions change with depth in the ocean. Scientists study different layers of the ocean by diving in submersibles—tiny submarines capable of withstanding extreme pressure at great depths. How do you think the ocean changes with depth?

DAVID DOUBILET/National Geographic Stock

Launch Lab

15 minutes

How are salt and density related?

Bodies of water form layers based on differences in density. How does salt affect density?

1. Read and complete a lab safety form.
2. Half-fill a **glass** with **water.**
3. Carefully place a **hard-cooked egg** in the water. Observe what happens. Remove the egg.
4. Add 5–10 tablespoons of **salt** and stir until all the salt is dissolved.
5. Place a **ladle** or a **spoon** inside the glass and slowly pour tap water over it until the glass is three-fourths full. Gently remove the ladle or the spoon. Be careful not to disturb the layer of salt water.
6. Gently place the egg in the glass and observe.

Think About This

1. Explain any differences that you observed.
2. **Key Concept** Do you think it is easier to float in the ocean or in a freshwater lake?

Earth's Oceans

Aside from being called the water planet, did you know that sometimes Earth is also called the blue planet? If you have ever seen a photograph of Earth taken from space, such as the one in **Figure 1**, you know that Earth appears mostly blue. Earth appears blue because water covers 70 percent of its surface. Most of Earth's water–97 percent–is salt water in the oceans.

Earth's oceans are all connected. However, scientists separate the oceans into five main bodies:

- The Pacific Ocean is the largest and deepest ocean. It is larger than all of Earth's combined land area.
- The Atlantic Ocean is half the size of the Pacific. It occupies about 20 percent of Earth's surface.
- The Indian Ocean is between Africa, India, and the Indonesian Islands. It is the third largest ocean.
- The Southern Ocean surrounds Antarctica. It is Earth's fourth largest ocean. Ice covers some of its surface all year.
- The Arctic Ocean is near the North Pole. It is the smallest and shallowest ocean. Ice covers some of its surface all year.

In this lesson, you will read about the formation of the oceans, their physical and chemical characteristics, and the importance of the oceans' natural resources.

Figure 1 Earth appears blue from space because its water reflects blue wavelengths of light.

Figure 2 Volcanic eruptions on Earth today add water vapor to the atmosphere, just as they did billions of years ago.

Formation of the Oceans

Evidence indicates that Earth's oceans began to form as early as 4.2 billion years ago (bya). That is only a few hundred million years after Earth formed. Earth was very hot and active when it was young. Many volcanoes covered its surface. Like the volcano shown in **Figure 2**, these ancient volcanoes erupted huge amounts of gas. Much of the gas was made of water vapor, with small amounts of carbon dioxide and other gases. Over time, these gases formed early Earth's atmosphere.

Condensation As water moves through the water cycle, illustrated in **Figure 3**, water vapor in the atmosphere cools and condenses into a liquid. Tiny droplets of liquid combine and form clouds. As early Earth cooled, the water vapor in its atmosphere condensed and precipitated. Rain fell for tens of thousands of years, collecting on Earth's surface in low-lying basins. Eventually, these basins became the oceans.

Asteroids and Comets Evidence suggests a second source of water for Earth's oceans. During the time when oceans formed, many icy comets and asteroids from space collided with Earth. The melted ice from these objects added to the water filling Earth's ocean basins.

Figure 3 Earth's water continually evaporates from the ocean and returns to the ocean through the water cycle.

 **Reading Check** What are the sources of Earth's oceans?

Tectonic Changes Earth's oceans change over time. As tectonic plates move, new oceans form and old oceans disappear. However, the volume of water in the oceans has remained fairly constant since the first oceans formed.

Composition of Seawater

The rain that fell to Earth's surface billions of years ago washed over rocks and dissolved minerals. The minerals contained substances that formed salts. Rivers and streams carried these substances to ocean basins. Some substances also came from gases released by underwater volcanoes. Together, these substances made the water salty, as shown in **Figure 4.**

Key Concept Check Why is seawater salty?

Salinity *is a measure of the mass of dissolved solids in a mass of water.* Salinity is usually expressed in parts per thousand (ppt). For example, **seawater** *is water from a sea or ocean that has an average salinity of 35 ppt.* This means that if you measured 1,000 g of seawater, 35 g would be salts and 965 g would be pure water.

The salinity of seawater changes in areas where rivers enter the ocean, such as in an estuary. There, seawater becomes brackish. **Brackish** *water, or brack water, is freshwater mixed with seawater.* The salinity of brackish water is often between 1 ppt and 17 ppt.

Substances in Seawater

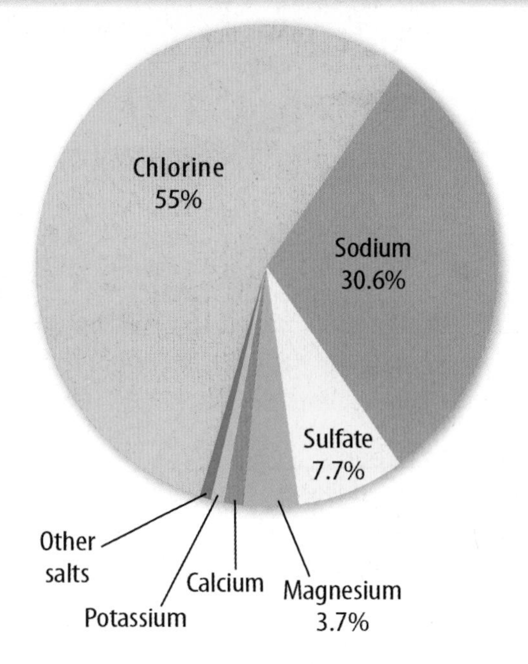

Figure 4 Five elements and one compound account for 99 percent of the dissolved substances in seawater. Evidence suggests that the percentages shown in this circle graph have been fairly consistent for millions of years.

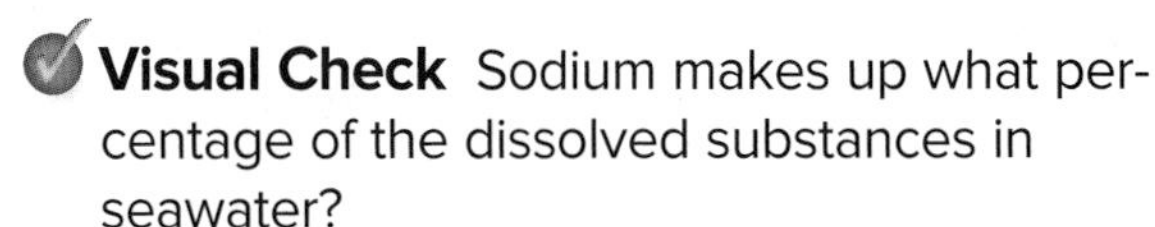

Visual Check Sodium makes up what percentage of the dissolved substances in seawater?

MiniLab

20 minutes

How does salinity affect the density of water?

Salt water is more dense than freshwater. How much salt do you need to add to freshwater to make it dense enough to float an egg?

1. Read and complete a lab safety form.
2. Fill a **jar** with 1,000 mL of **water.** Carefully add a **hard-cooked egg** to the water. Observe the egg's position.
3. Use a **stirring rod** to stir 20 g of **salt** into the water. Again observe the egg's position.
4. Add salt in increments of 10 g. After each addition, stir the salt into the water and observe the egg. Continue to add salt in 10-g increments until the egg floats.

Analyze and Conclude

1. **Calculate** the salinity of the water in which the egg floated.
2. **Key Concept** How does salinity affect the density of water?

Seafloor Topography

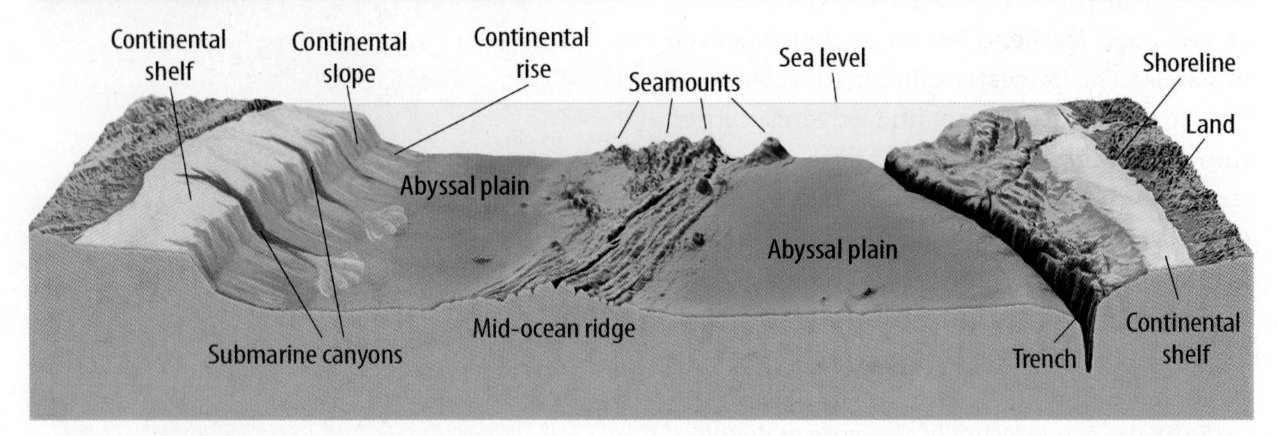

Figure 5 An ocean's seafloor is shaped like a basin. Some features of ocean basins are continental shelves, continental slopes, continental rises, abyssal plains, mid-ocean ridges, seamounts, and trenches.

Visual Check Where is new seafloor created?

The Seafloor

What do you think the ocean bottom looks like? You might be surprised to learn that the seafloor has features similar to features on land, such as plains, plateaus, canyons, and mountains.

Continental Margins

The part of an ocean basin next to a continent is called a continental margin. A continental margin extends from a continent's shoreline to the deep ocean. It is divided into three regions, which are illustrated in **Figure 5.** The continental shelf is the shallow part of a continent nearest the shore. The continental slope is the steep slope that extends from the continental shelf to the deep ocean. The continental rise is at the base of the slope. It is where sediments accumulate that fall from the continental slope.

WORD ORIGIN

abyssal
from Greek *abyssos,* means "bottomless"

Abyssal Plains

Examine **Figure 5** again. Notice the abyssal plains. **Abyssal plains** *are large, flat areas of the seafloor that extend across the deepest parts of the ocean basins.* Thick layers of sediment cover abyssal plains. In some areas, underwater volcanoes rise from the abyssal plains and form islands that extend above the ocean's surface.

FOLDABLES®

Make a top-tab book from four half-sheets of paper. Use your book to illustrate and organize information about the seafloor.

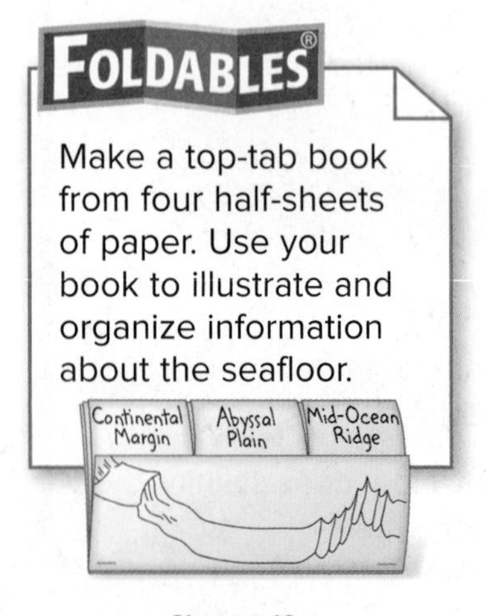

Mid-Ocean Ridges

At places on the seafloor where tectonic plates pull apart, volcanic mountains form. These underwater mountains are called mid-ocean ridges. Mid-ocean ridges form a continuous mountain range that extends through all of Earth's ocean basins. It is the tallest and longest mountain range on Earth. It measures more than 65,000 km in length. As the plates slowly move apart at mid-ocean ridges, lava erupts and then cools forming new seafloor.

Ocean Trenches

Earth's oceans have an average depth of about 4,000 m. However, in areas where an oceanic tectonic plate collides with a continental plate, a deep canyon, or trench, forms along the edge of the abyssal plain. A trench is shown in **Figure 5.** Trenches are the deepest parts of the ocean. The Mariana Trench, in the western Pacific Ocean, is more than 11,000 m deep. The bottom of the Mariana Trench extends farther below sea level than Mount Everest is above sea level.

Key Concept Check Describe some features of the seafloor.

Deep Ocean Technology

Today, scientists use submersibles and other technologies to explore the seafloor. A submersible is an underwater vessel which can withstand extreme pressure at great depths. One famous submersible, DSV *Alvin,* set a deep-ocean record by diving to the bottom of the Mariana Trench.

In the future, remotely operated vehicles (ROVs) are likely to be used more frequently. These unmanned submersibles can be operated from a control center on a ship. Operators can see video images sent back from the ROVs and can control their propellers and **manipulator** arms. ROVs are safer, cheaper, and can generally provide more research data than manned submersibles.

Resources from the Seafloor

The seafloor contains valuable resources. **Table 1** illustrates some of the resources on or beneath the seafloor. There are two main categories of seafloor resources–energy resources and minerals. Energy resources, such as oil, natural gas, and methane hydrates, are beneath the ocean floor on continental margins. Most mineral deposits, such as the manganese nodules shown in **Table 1,** are on abyssal plains. Some minerals, including gold and zinc, have also been discovered at mid-ocean ridges.

Table 1 Resources from the Ocean Floor

Oil and Natural Gas These deposits are beneath the seafloor on continental margins. Many platforms for oil extraction have been built in the Gulf of Mexico.	
Methane Hydrates Deposits of methane gas in deep-sea sediments are called methane hydrates. They are a potential but as yet unrealized source of energy similar to fossil fuels.	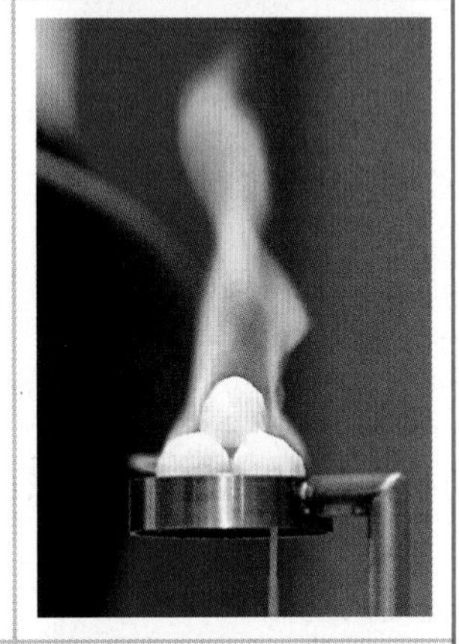
Mineral Deposits Minerals on the seafloor include manganese nodules. These nodules form when metals precipitate out of seawater. They are potentially valuable, but no large-scale mining exists.	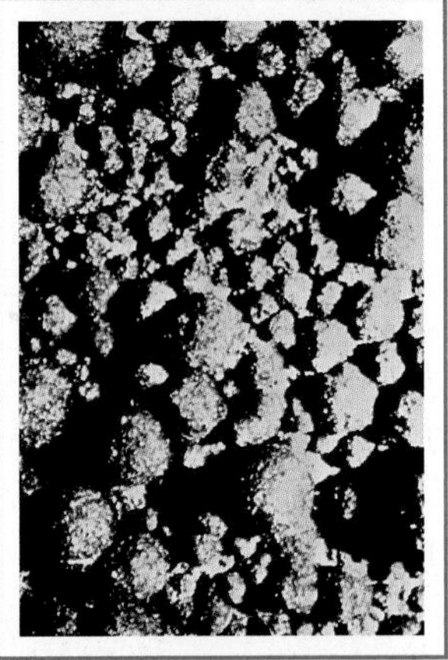

Table 1 Resources found on or below the seafloor include oil, methane hydrates, and manganese nodules.

ACADEMIC VOCABULARY

manipulate
(verb) to operate with hands or by mechanical means in a skillful manner

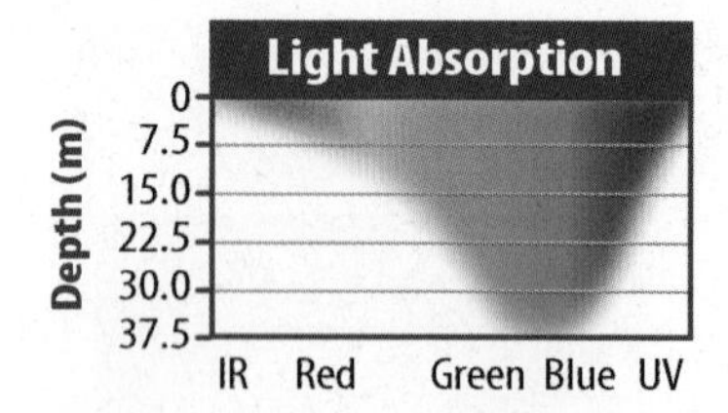

Figure 6 Wavelengths of blue and green light reach deeper into the ocean than those of red, orange, and yellow light.

Zones in the Oceans

Scientists divide oceans into distinct regions, or zones, based on physical characteristics. These characteristics include the amount of sunlight, temperature, salinity, and density.

Amount of Sunlight

If you have ever swum in a lake or an ocean, you might have noticed that the deeper the water, the darker it was. Light from the Sun penetrates below the ocean's surface. However, as depth increases, the wavelengths of light are not absorbed equally. Because of this, some colors penetrate deeper than others, as illustrated by the graph in **Figure 6.**

Surface Zone The area of shallow seawater that receives the greatest amount of sunlight is the surface zone, or sunlit zone. This zone is located above the dashed line shown in **Figure 7.** Most organisms that perform **photosynthesis** live here.

REVIEW VOCABULARY

photosynthesis
a chemical process in which light energy, water, and carbon dioxide are converted into sugar

Middle Zone By the time sunlight reaches the middle zone, or twilight zone, most of light's wavelengths have been absorbed. This zone receives only faint blue-green light. The area between the two dashed lines in **Figure 7** represents the middle zone.

Deep Zone Plants do not grow in the deep zone, or midnight zone, where there is no light. Most deep-sea animals, such as the squid shown in **Figure 7,** make their own light in chemical process called bioluminescence (BI oh LEW mah NE cents).

Reading Check Why don't plants grow in the deep zone?

Figure 7 The surface zone begins at the ocean surface and reaches a depth of about 200 m. The middle zone begins below the surface zone and reaches a depth of about 1,000 m. The deep zone is below the middle zone.

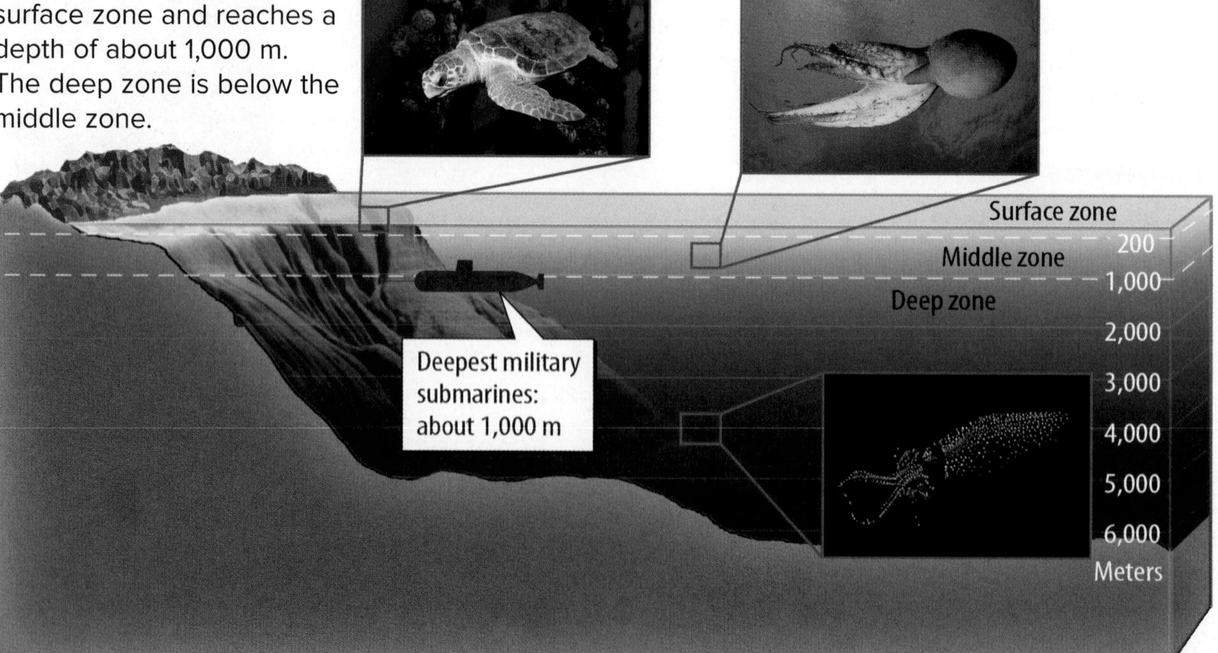

(tl)Guillen Photography/UW/USA/Gulf of Mexico/Alamy; (tr)Island Effects/Getty Images; (b)Dante Fenolio/Science Source

Ocean Layers

Just as oceans have zones of light, they also have zones of temperature, salinity, and density. Notice in **Figure 8** that temperature, salinity, and density vary with depth. Sometimes these characteristics can change abruptly within a relatively short change of depth. Abrupt changes in these characteristics can create distinctive layers of seawater.

Key Concept Check Why does seawater form layers?

Figure 8 Temperature, salinity, and density vary in the top 1,000 m of Earth's oceans.

Visual Check Below what depth does all ocean water have approximately the same temperature?

Changes in Temperature, Salinity, and Density

Changes in Temperature As shown in the graph to the right, temperature changes abruptly between 250 m and 900 m in temperate and tropical regions (solid line). As depth increases, water in these regions cools rapidly. That is because there is less sunlight to warm water as depth increases.

In contrast, the temperature of polar water (dotted line) remains fairly constant. This is because sunlight intensity at Earth's poles is weaker than it is in temperate and tropical regions. Polar water at all depths is cold.

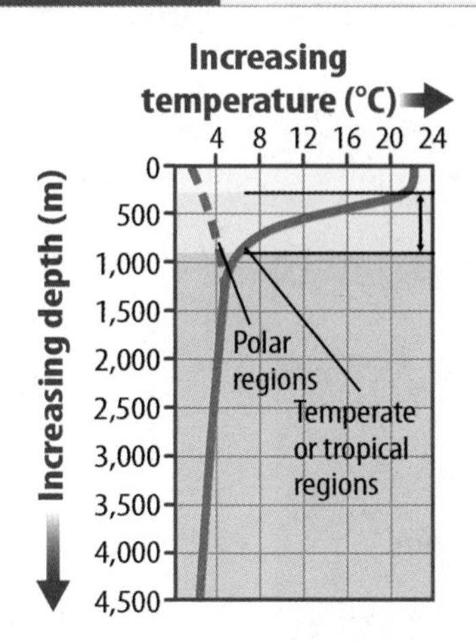

Changes in Salinity The top 500 m of warm water in temperate and tropical regions is saltier than polar water. Warm water evaporates more rapidly than cold water. When water evaporates, salt is left behind; this increases salinity at the surface.

In polar regions, freshwater from melting glaciers decreases the salinity at the surface. However, when ice forms, salt is left behind in the water. The remaining cold, salty water becomes denser and sinks to a deeper layer.

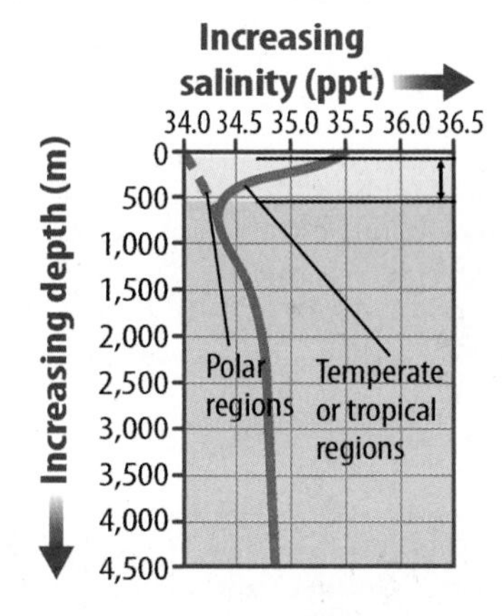

Changes in Density Seawater density is related to temperature and salinity. Cold water is denser than warm water. Salt water is denser than freshwater. Because of density differences, ocean water is layered. The densest layers are on the bottom; the least dense layers are on top.

Notice in the graph to the right that water density in polar regions remains fairly constant. Keep this in mind when you read about density currents in Lesson 3.

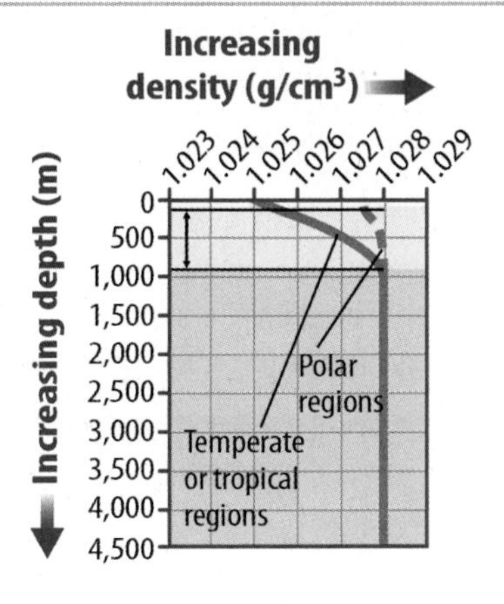

Lesson 1 Review

Visual Summary

Condensation of water vapor from volcanic eruptions formed Earth's oceans.

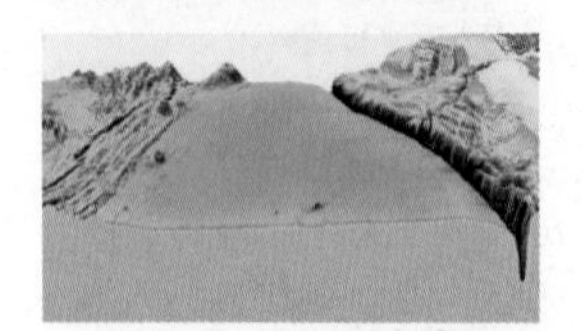

The seafloor has topographic features such as mountains, plains, and trenches.

Sunlight, temperature, salinity, and density of seawater change with depth.

FOLDABLES®

Use your lesson Foldable to review the lesson. Save your Foldable for the project at the end of the chapter.

What do you think NOW?

You first read the statements below at the beginning of the chapter.

1. Oceans formed about 4 billion years ago.

2. The seafloor is flat.

Did you change your mind about whether you agree or disagree with the statements? Rewrite any false statements to make them true.

Use Vocabulary

1. **Compare** brackish water and seawater.
2. **Use the term** *salinity* in a complete sentence.

Understand Key Concepts 

3. Which resource from the oceans is used as a source of energy?

   A. manganese **C.** salt
   B. natural gas **D.** sand

4. **Explain** why oceans are salty.
5. **Describe** how seawater forms layers.

Interpret Graphics

6. **Organize information** Copy and fill in the graphic organizer below to identify three zones in the ocean based on the amount of light reaching each zone.

7. **Identify** which letters in the figure below represent the continental shelf, the continental slope, and the continental rise. How do these areas differ?

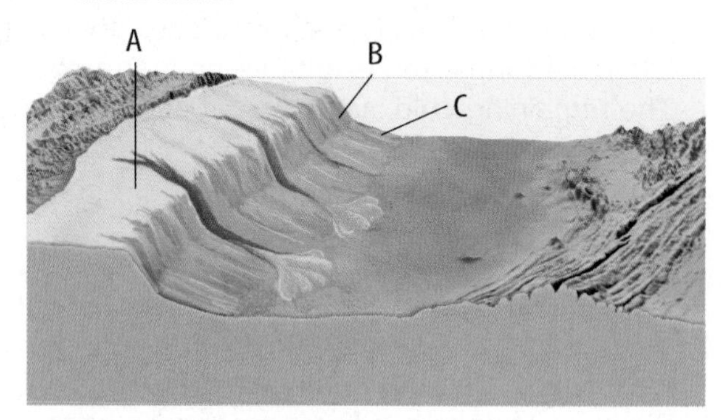

Critical Thinking

8. **Design** Suppose you have been hired to mine manganese nodules off the bottom of the Pacific Ocean. Identify the problems you might have, and design equipment that would enable you to get the nodules to the surface.

(t)M.E. Young/USGS; (b)Dante Fenolio/Science Source

AMERICAN MUSEUM OF NATURAL HISTORY

CAREERS in SCIENCE

Exploring Deep-Sea Vents

Meet Susan Humphris, a scientist investigating how deep-sea vents affect the chemistry of the oceans.

More than a mile below the ocean's surface is an extraordinary world. Dark as night, nearly freezing, and under crushing pressure, this part of the seafloor is one of the least explored places on Earth. Amazingly, communities of unusual organisms, such as tubeworms and giant clams, thrive here near underwater hot springs called deep-sea vents. These creatures get their energy from chemicals rather than sunlight.

Susan Humphris, a scientist at Woods Hole Oceanographic Institution, studies this unique environment. As a marine geochemist, she investigates the composition, or chemical makeup, of the rocks and seawater around deep-sea vents. When the icy seawater meets hot volcanic rocks deep below the seafloor, they undergo chemical reactions. Humphris investigates how these reactions change the composition of volcanic rocks, seawater, and the ocean as a whole.

To study deep-sea vents, scientists use specialized instruments to gather data and vehicles that can travel to the seafloor. Some vehicles are operated remotely by people on a ship. But in one vehicle, called *Alvin,* Humphris can travel to the ocean floor and explore deep-sea vents herself.

Back in the lab, Humphris analyzes the composition of her volcanic rock samples. By comparing them with other volcanic rock, she can determine which elements were exchanged during the chemical reactions between rocks and seawater. Humphris and other scientists have determined that deep-sea vents have affected all the seawater in the world's oceans.

Deep-Sea Vents

Deep-sea vents form near mid-ocean ridges—long chains of underwater volcanoes that encircle Earth. When icy seawater seeps into deep cracks in Earth's crust, it becomes superheated as it contacts hot volcanic rock. This causes the water to gush upward from the seafloor as a deep-sea vent. The entire ocean circulates through deep-sea vents every 1 million to 10 million years.

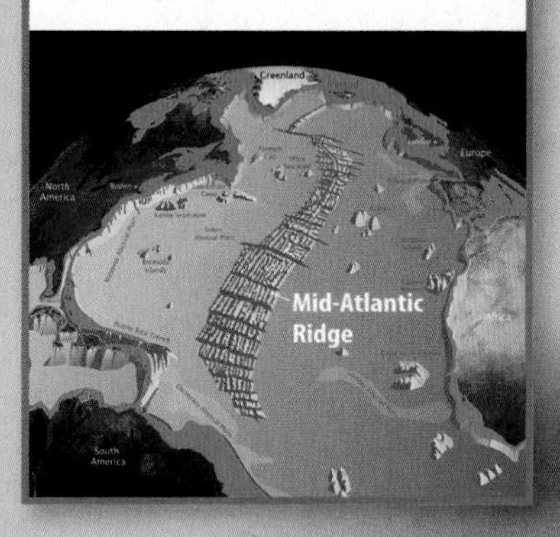

It's Your Turn

JOURNAL ENTRY Imagine you are piloting *Alvin.* Write a journal entry about your expedition to a deep-sea vent. Include descriptions and drawings of what you see as you travel to the dark seafloor.

Lesson 2

Ocean Waves and Tides

Reading Guide

Key Concepts

ESSENTIAL QUESTIONS

- What causes ocean waves?
- What causes tides?

Vocabulary

tsunami p. 575
sea level p. 576
tide p. 576
tidal range p. 576
spring tide p. 577
neap tide p. 577

 Multilingual eGlossary

 BrainPOP®

Inquiry Surfing Under a Wave?

Is this surfer confused? Why is he under the wave? What do you think happens to a wave's energy below the surface?

Launch Lab

10 minutes

How is sea level measured?

The ocean surface is changing constantly as a result of waves, tides, and currents. In a matter of seconds, a wave can cause the ocean surface to rise and fall by several meters. In a matter of hours, a tide can also raise or lower the level of the sea by several meters.

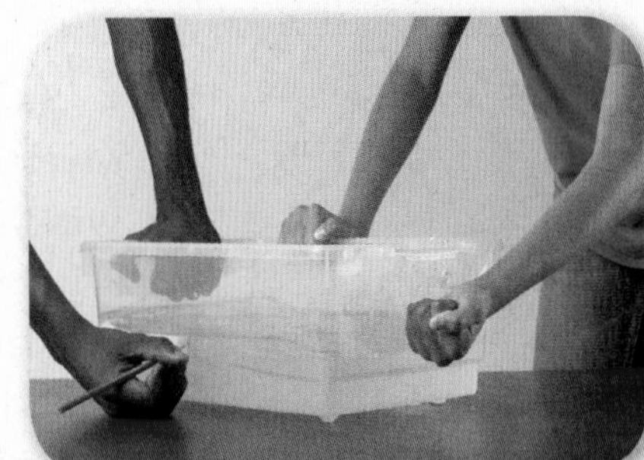

1. Read and complete a lab safety form.
2. Half-fill a **clear container** with **water.**
3. Slowly and steadily rock the container back and forth to produce waves.
4. While you gently rock the container, another student should look through the side of the container and mark the peaks and valleys of the waves with a **wax pencil**.
5. Using a **ruler**, measure the difference between the two marks. The midpoint of this measurement is equivalent to sea level.

Think About This

1. How do you think sea level changes when wind speed changes?
2. **Key Concept** How do you think oceanographers determine sea level?

Parts of a Wave

Have you ever been caught in a crashing wave? It might have been hard to catch your breath. Even if you dive deep below a wave, you can still feel some of the wave's energy. The surfer shown on the opposite page is duck diving–ducking beneath a wave to avoid the wave's full power.

There are different kinds and sizes of waves in the oceans, but all waves have the same basic parts. As shown in **Figure 9**, the crest is the highest part of a wave. The trough is the lowest part of a wave. The wave height is the vertical distance between the crest and the trough. The wavelength is the horizontal distance from crest to crest or from trough to trough.

Reading Check How is wavelength measured?

Animation

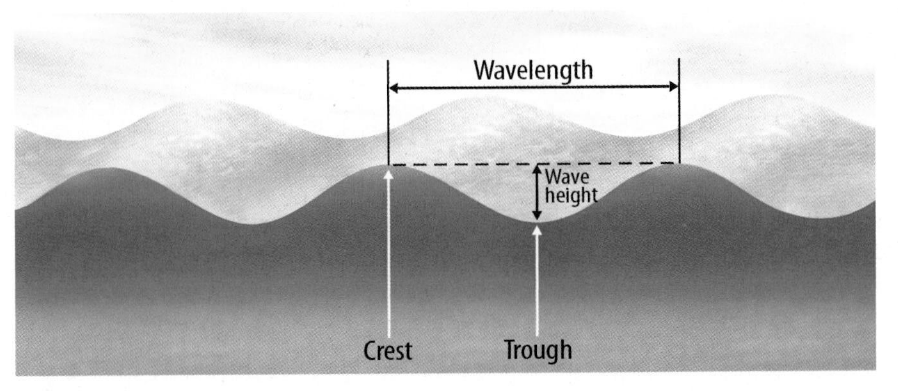

Figure 9 Ocean waves have crests and troughs.

Direction of wave motion

Figure 10 Just like a fishing bobber, a water particle moves in a circle as a wave passes.

Surface Waves

Wind causes the waves that roll onto a beach. They are often called surface waves. Friction from wind drags across the water's surface, causing it to ripple. The small ripples eventually become larger waves.

 **Key Concept Check** What causes ocean surface waves?

Surface waves range in size from tiny ripples to huge waves several meters high. Three factors affect the size of surface waves–wind speed, time, and distance. The faster, longer, and farther the wind blows, the larger the resulting waves. For example, some of the largest wind-driven waves form in the Southern Ocean. It experiences fast and continuous winds that blow all the way around Antarctica.

Wave Motion

If you watch a wave wash onto a beach, you might think that a wave transports water from one location to another. However, the motion of a water particle in a wave is circular. After a wave passes, the water particle returns approximately to its original position, as shown in **Figure 10.**

The circular motion of water particles extends below the surface. However, as depth increases, the circular motion decreases. At a certain depth, called the wave base, wave motion stops. This depth is equal to a distance of one-half the wavelength of the wave above it, as illustrated in **Figure 11.**

Wave Motion at Depth

Figure 11 The circular motion of water particles becomes smaller and smaller with depth.

Visual Check If the wavelength of a surface wave is 40 m, how deep would a scuba diver have to go before feeling no wave motion?

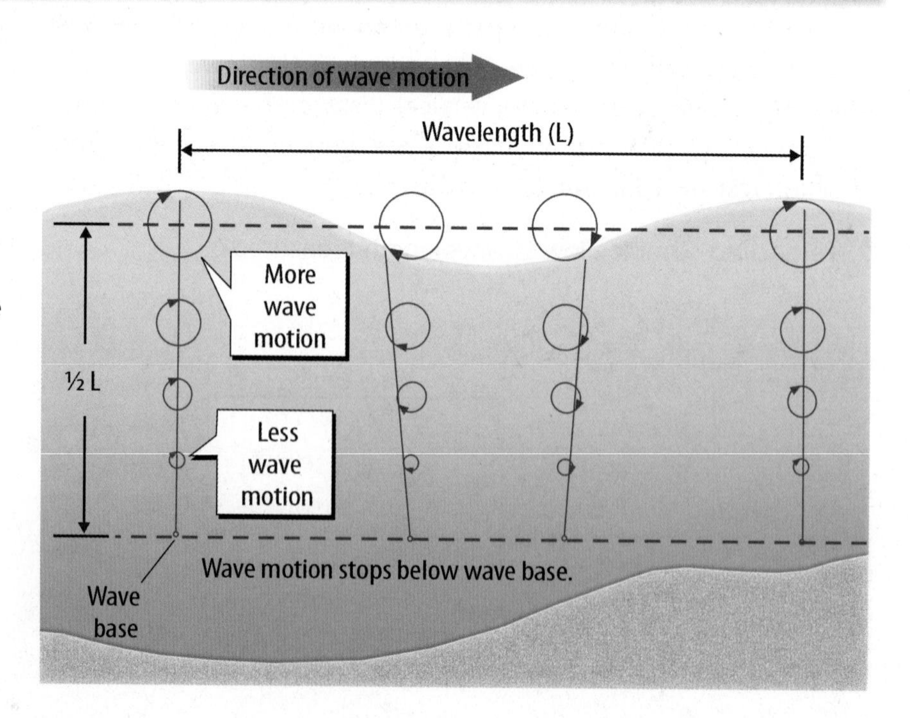

When Surface Waves Reach Shore

As a wave moves into shallow water, it changes shape and size. The change begins when the base of the wave comes in contact with the sloping seafloor, as shown in **Figure 12.** As the base of the wave drags on the seafloor, the wave's speed decreases. At the same time, the wavelength shortens and the wave height increases. When the wave reaches a certain height, the wave base can no longer support the crest, and the wave collapses, or breaks. This type of wave is called a breaker. After a wave breaks, the water surges forward onto shore.

Breakers

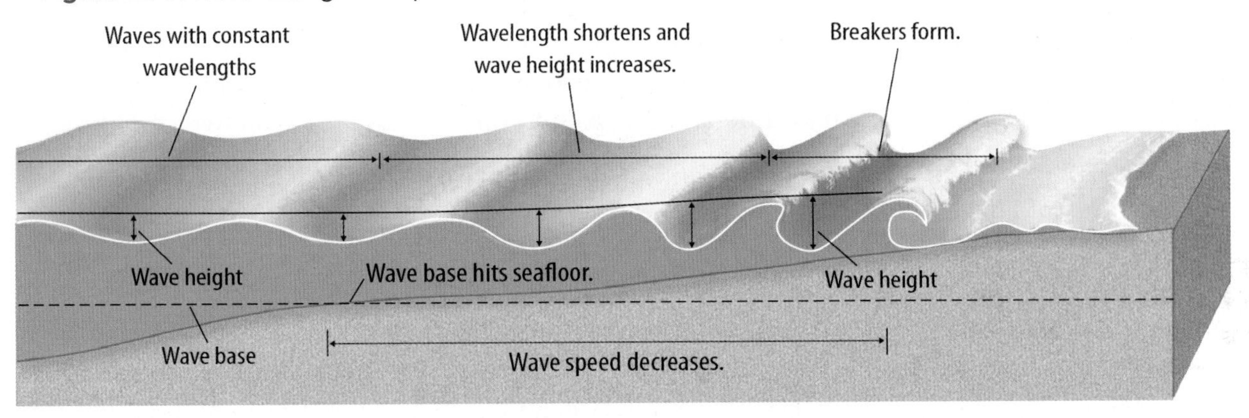

Figure 12 A wave changes shape when its base comes in contact with the seafloor.

Tsunamis

You might have heard of another type of ocean wave called a tsunami. *A* **tsunami** (soo NAH mee) *is a wave that forms when an ocean disturbance suddenly moves a large volume of water.* It can be caused by an underwater earthquake or landslide, a volcanic eruption, or even ice breaking away from a glacier.

Word Origin

tsunami
from Japanese *tsu,* means "harbor"; and *nami,* means "wave"

 **Key Concept Check** What can cause a tsunami?

Far from shore, a tsunami has a short wave height, often less than 30 cm high. However, the wavelength can be hundreds of kilometers long. As a tsunami approaches shore, it slows down and grows higher. Many tsunamis grow only a few meters high as they move onto shore, but some can rise as high as 30 m.

Unlike a common wind-driven wave, the water from a tsunami just keeps coming. As a result, tsunamis can cause much damage. In 2004, a series of tsunamis caused by an underwater earthquake in the Indian Ocean killed more than 225,000 people in 11 countries and destroyed entire villages.

Math Skills

Use Statistics

Find the **mean** by adding the numbers in a data set and dividing by the number of items in the set. The **range** is the difference between the largest and smallest numbers in a set of data.

Example: In a 48-hour period, high tides were measured at 0.701 m, 0.649 m, 0.716 m, and 0.661 m above sea level. What is the range and mean of the high tides?

Range = 0.716 m − 0.649 m = 0.067 m

Mean = (0.701 m + 0.649 m + 0.716 m + 0.661 m) ÷ 4 = 2.73 m ÷ 4 = 0.682 m

Practice

During the same 48-hour period, the low tides were measured at 0.018 m, 0.103 m, 0.048 m, and 0.091 m below sea level.

a. What is the range of the low tides?

b. What is the mean of the low tides?

Math Practice

Personal Tutor

Tides

When measuring sea level, scientists take into account changes to the ocean's surface caused by waves. **Sea level** *is the average level of the ocean's surface at any given time.* Scientists who measure sea level also take into account changes to the ocean's surface caused by tides. **Tides** *are the periodic rise and fall of the ocean's surface caused by the gravitational force between Earth and the Moon, and between Earth and the Sun.*

The Moon and Tides

The gravitational force that causes the largest tides is between Earth and the Moon. The attraction between them produces two bulges on ocean surfaces–one bulge on the side of Earth facing the Moon and one bulge on the side of Earth facing away from the Moon. The bulges represent high tides. High tide is the highest level of an ocean's surface. Low tide, the lowest level of an ocean, occurs between the two bulges. The difference between high tide and low tide in one coastal area is shown in **Figure 13.**

Key Concept Check What causes the largest tides?

Topography and Tides

The coastlines of continents, the shape and size of ocean basins, and the depth of the oceans affect tides. The Atlantic coast experiences two alternating high and low tides almost daily. In contrast, the Gulf of Mexico experiences one high tide and one low tide each day.

The size of tides also varies on different areas of Earth's surface. In some areas, the difference between low tide and high tide is as small as 1 m. In other areas, the difference is as great as 15 m. As shown in **Figure 13,** *the difference in water level between a high tide and a low tide is the* **tidal range.**

Figure 13 Tides change the level of the ocean's surface.

Bill Brooks/Alamy

Tidal Forces

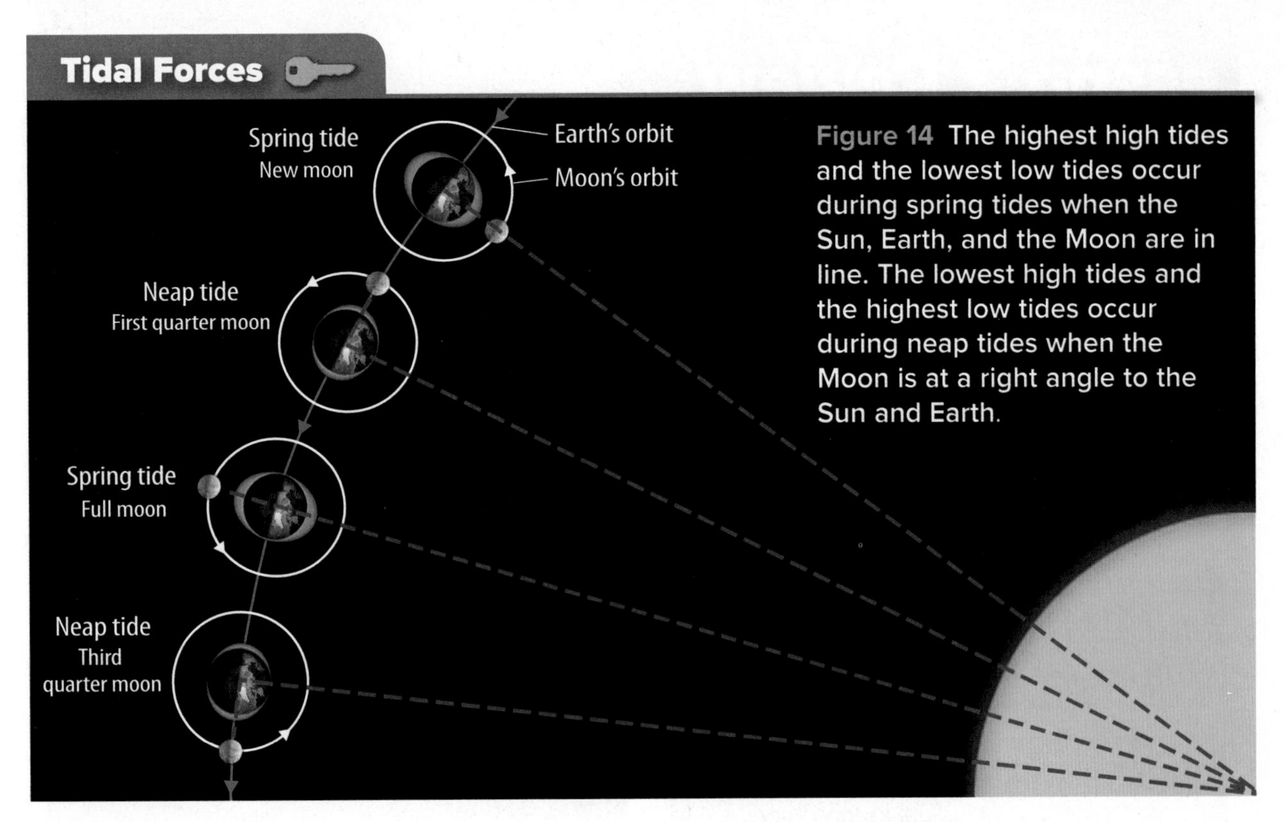

Figure 14 The highest high tides and the lowest low tides occur during spring tides when the Sun, Earth, and the Moon are in line. The lowest high tides and the highest low tides occur during neap tides when the Moon is at a right angle to the Sun and Earth.

Spring Tides

Tidal ranges are not constant. They vary depending on the positions of the Sun and the Moon with respect to Earth. Notice in **Figure 14** that when Earth, the Moon, and the Sun are aligned, the Moon is new or full. The gravitational pull on the oceans is strongest when the two forces act together. As a result, the tidal range is larger than normal. High tides are higher and low tides are lower. *A* **spring tide** *has the greatest tidal range and occurs when Earth, the Moon, and the Sun form a straight line.*

Neap Tides

Look at **Figure 14** again. During a first quarter moon and a third quarter moon, the Moon is at a right angle to Earth and the Sun. The gravitational forces between Earth and the Moon and between Earth and the Sun act against each other. This means that high tides are lower than normal while low tides are higher than normal. *A* **neap tide** *has the lowest tidal range and occurs when Earth, the Moon, and the Sun form a right angle.*

Reading Check What is a neap tide?

MiniLab — 20 minutes

Can you analyze tidal data?

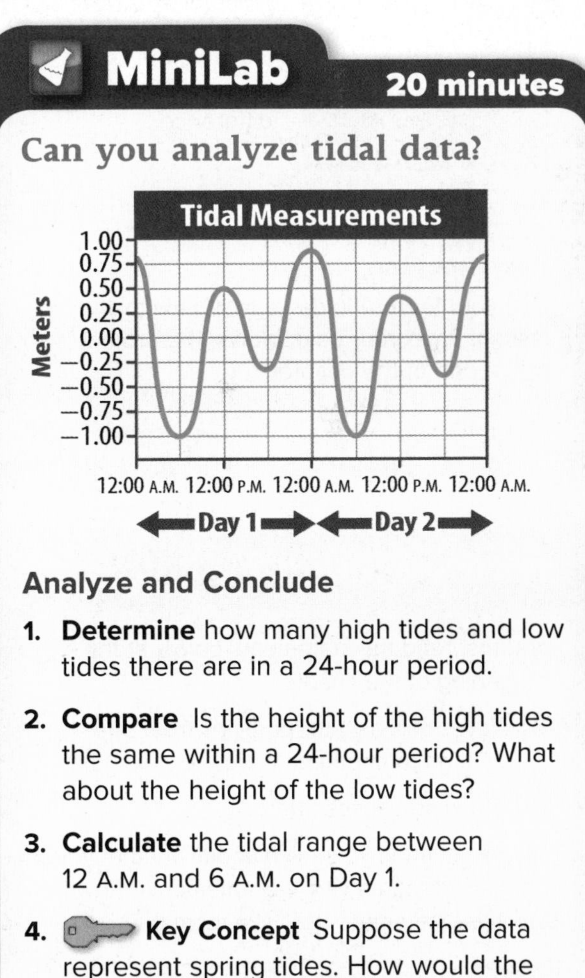

Analyze and Conclude

1. **Determine** how many high tides and low tides there are in a 24-hour period.
2. **Compare** Is the height of the high tides the same within a 24-hour period? What about the height of the low tides?
3. **Calculate** the tidal range between 12 A.M. and 6 A.M. on Day 1.
4. **Key Concept** Suppose the data represent spring tides. How would the tidal data collected during a neap tide be different?

Lesson 2 Review

Visual Summary

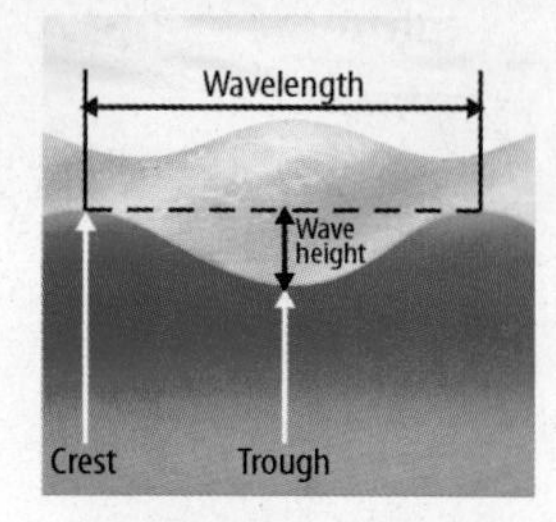

All waves have the same basic features.

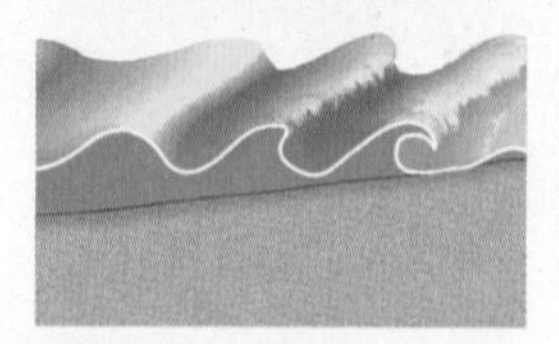

Wavelength shortens and wave height increases as a wave nears the shoreline.

Tidal ranges vary from location to location, but can be up to 15 m in difference.

FOLDABLES®

Use your lesson Foldable to review the lesson. Save your Foldable for the project at the end of the chapter.

What do you think NOW?

You first read the statements below at the beginning of the chapter.

3. Waves move water particles from one location to another.

4. The wind causes tides.

Did you change your mind about whether you agree or disagree with the statements? Rewrite any false statements to make them true.

Use Vocabulary

1. **Use the term** *tsunami* in a complete sentence.
2. **Define** *tide* in your own words.

Understand Key Concepts 

3. **Explain** how the Moon causes tides.
4. **Compare and contrast** the causes of surface waves and tsunamis.

Interpret Graphics

5. **Organize Information** Copy and fill in the graphic organizer below to describe spring tides and neap tides.

	Positions of Earth, Moon, and Sun
Spring tides	
Neap tides	

6. **Explain** how the figure below represents the movement of water in a wave.

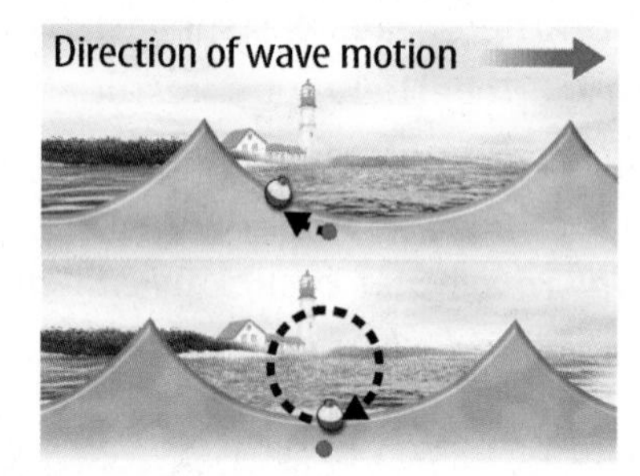

Critical Thinking

7. **Design** an experiment to measure the average tidal range in a coastal area during one month.

Math Skills 

Math Practice

8. In a certain location, high tides for one day measure 8.30 m and 8.00 m. The low tides measure 0.500 m and 0.220 m.

   A. What is the range of the tides?

   B. What is the mean low tide?

Bill Brooks/Alamy

 **Skill Practice** **Analyze Data** **20 minutes**

High Tides in the Bay of Fundy

The tides in the Bay of Fundy in Eastern Canada have the greatest tidal ranges of any tides on Earth. As a tide enters the Bay of Fundy, it is channeled into an increasingly narrower space. Topography of the land directly affects the tidal range.

The lines on the map of the Bay of Fundy below are similar to contour lines on a topographic map. Tidal height data has been collected along each line and then averaged to determine the mean height of the highest tide at that location across the width of the bay.

Learn It

Analyze the data on the map, to make a graph showing the change in tidal heights from the mouth of the bay to the town of Truro.

Try It

1. Make a data table with three columns in your Science Journal. Label the columns: High Tide (m), Distance from the Mouth of the Bay (cm), Distance from the Mouth of the Bay (m).
2. Use the map scale of the Bay of Fundy below and a metric ruler to determine the distance each high tide is from the mouth of the bay. Convert centimeters on your ruler to meters on the map. Record your information in your data table.
3. Using your data, graph the distance from the mouth of the bay along the x-axis (in m), and tidal height along the y-axis. Give your graph a title.

Apply It

4. **Describe** how the highest tides changed with distance.
5. **Infer** how the tides in the Bay of Fundy might change when Earth, the Moon, and the Sun are in a straight line. How might the tides change when Earth, the Moon, and the Sun are at an angle?
6. **Key Concept** Identify factors that affect tides in the Bay of Fundy.

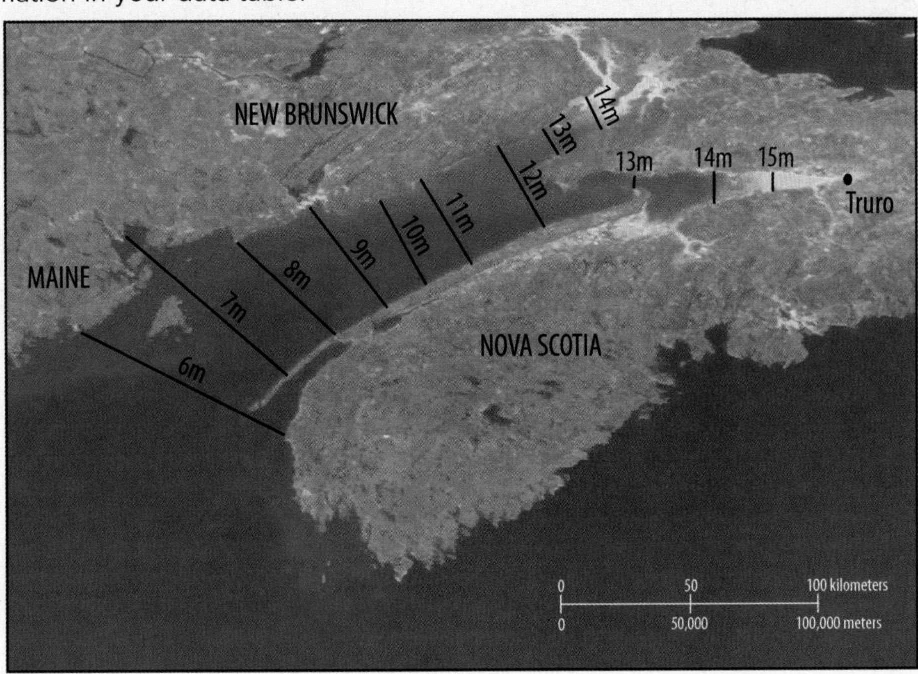

Lesson 3

Ocean Currents

Reading Guide

Key Concepts
ESSENTIAL QUESTIONS

- What are the major types of ocean currents?
- How do ocean currents affect weather and climate?

Vocabulary

ocean current p. 581
gyre p. 582
Coriolis effect p. 582
upwelling p. 583

Multilingual eGlossary

BrainPOP®

Inquiry Where are they going?

Can you find the Florida Current in this satellite photo? The curve of the clouds gives it away. As the clouds move between Florida and Cuba, they follow the same path as the current. Why do you think clouds and currents sometimes follow the same path?

Jacques Descloitres, MODIS Rapid Response Team, NASA/GSFC

Launch Lab

10 minutes

How does wind move water?

Wind pushes the water near shorelines in different directions. Objects released here go around and around in the waves. What would happen farther out in the ocean?

1. Read and complete a lab safety form.
2. Half fill a **container** with water.
3. Position a **fan** so it can blow across the water's surface.
4. Put two drops of **food coloring** on the surface of the water closest to the fan. Turn the fan to a low setting to produce waves.
5. Observe what happens to the food coloring.

Think About This

1. Explain the movement of the food coloring in your Science Journal.
2. What types of objects do you think the wind can move in the ocean?
3. **Key Concept** If you were on a boat about 3 km from shore and threw a rubber ball into the water, what do you think would happen?

Major Ocean Currents

During a storm in 1990, 40,000 pairs of shoes fell off a cargo ship in the middle of the Pacific Ocean. Months later, beachcombers began finding the shoes on the coasts of Oregon and Washington. How did the shoes get there? An ocean current carried them. *An* **ocean current** *is a large volume of water flowing in a certain direction.*

Surface Currents

Recall that wind transfers energy to water and forms waves. Wind also transfers energy to water and forms currents. The friction generated by wind on water can move the water. As wind blows over water, the moving air particles drag on the surface and cause the water to move, just as they drag the wind surfer in **Figure 15.** Wind-driven currents are called surface currents.

Surface currents carry warm or cold water horizontally across the ocean's surface. They extend to about 400 m below the surface and can move as fast as 100 km/day. Earth's major wind belts, called prevailing winds, influence the formation of ocean currents and the direction they move. For example, the trade winds that blow from Africa move warm, equatorial water toward North America and South America.

Key Concept Check How do surface currents form?

Figure 15 Just as wind drags this wind surfer across the ocean's surface, the wind also drags the top layer of water across the ocean's surface.

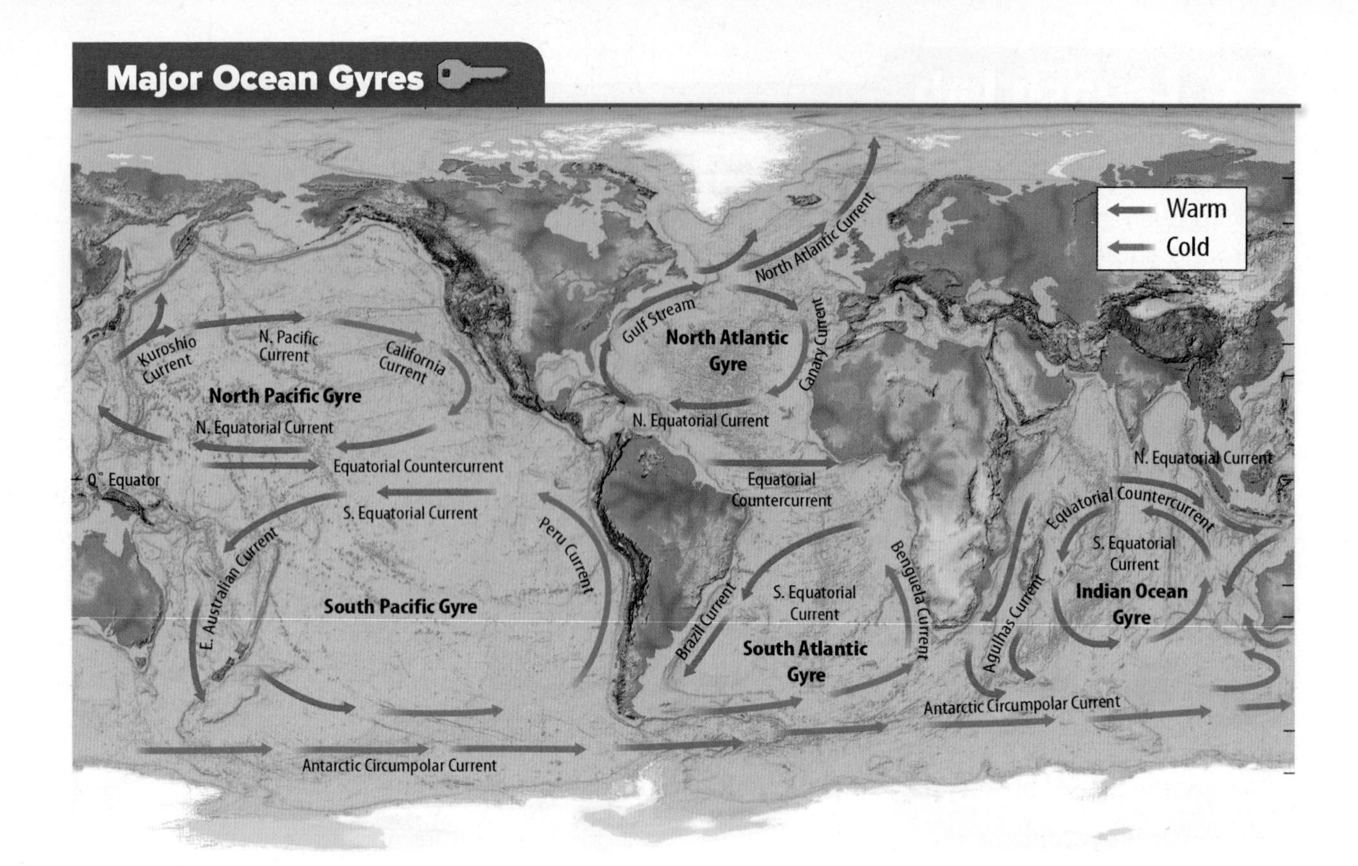

▲ **Figure 16** Gyres form on the surface of Earth's oceans.

WORD ORIGIN

gyre
from Latin *gyrus*, means "circle"

Gyres Earth's oceans contain large, looped systems of surface currents called gyres (JI urz). *A* **gyre** *is a circular system of currents.* As shown in **Figure 16**, the currents within each gyre move in the same direction. However, if you look closely, you can see that the direction of current movement in a gyre is different in each hemisphere. Gyres in the northern hemisphere circle clockwise. Gyres in the southern hemisphere circle counterclockwise.

Coriolis Effect Why do gyres move in different directions? Directions differ because of the Coriolis effect. *The* **Coriolis effect** *is the movement of wind and water to the right or left that is caused by Earth's rotation.* As shown in **Figure 17**, the Coriolis effect causes fluids, such as air and water, to curve to the right in the Northern Hemisphere, in a clockwise direction. In the Southern Hemisphere, the Coriolis effect causes fluids to curve to the left, in a counterclockwise direction.

Reading Check What is the Coriolis effect?

Topography The shapes of continents and other landmasses affect the direction and speed of currents. For example, gyres form small or large loops and move at different speeds depending on the land masses they contact. The Florida Current, shown in the photo at the beginning of this lesson, narrows and increases in speed as it passes through the straits of Florida.

Figure 17 The Coriolis effect causes fluids to move clockwise in the Northern Hemisphere and counterclockwise in the Southern Hemisphere. ▼

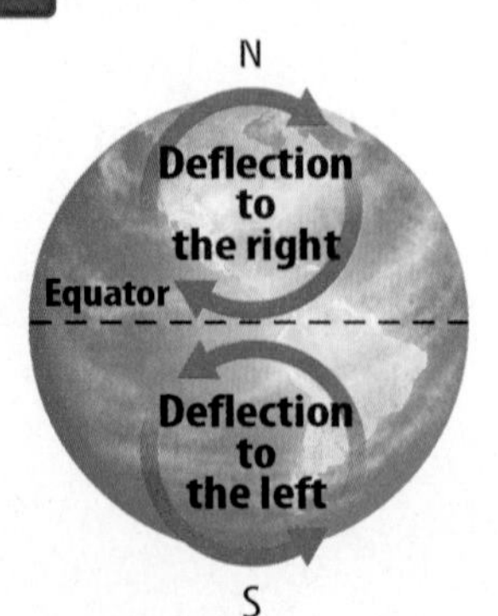

Upwelling

Surface currents move water horizontally across the ocean's surface. Not all currents move in a horizontal direction. Some currents move water vertically. **Upwelling** *is the vertical movement of water toward the ocean's surface.* Upwelling occurs when wind blows across the ocean's surface and pushes water away from an area. Deeper, colder water is then forced to the surface. Upwelling often occurs along coastlines. **Figure 18** illustrates how upwelling occurs along the South American coast.

Upwelling brings cold, nutrient-rich water from deep in the ocean to the ocean's surface. This water supports large populations of algae, fish, and other ocean organisms.

 **Key Concept Check** How does upwelling occur?

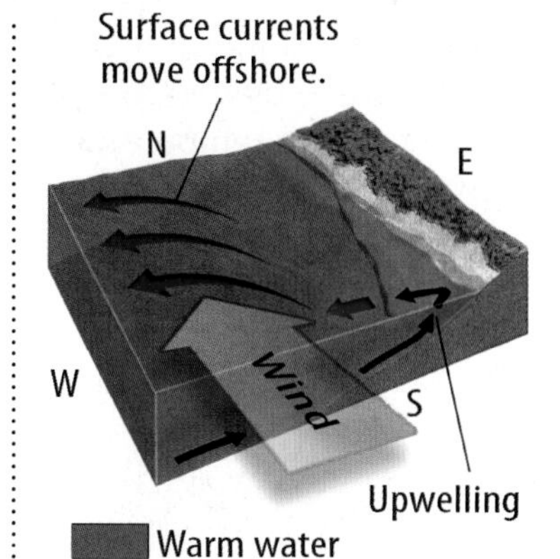

▲ **Figure 18** Upwelling off the South American coast causes cold, deep water to replace warmer water on the surface.

Density Currents

Another type of vertical current is a density current. Density currents move water downward. They carry water from the surface to deeper parts of the ocean. Density currents are not caused by wind. They are caused by changes in density.

Cold water is denser than warm water, and salty water is denser than freshwater. As a surface current moves toward a polar area, the water cools. When seawater freezes, salt is left behind in the surrounding water. Eventually, the cold, salty water becomes so dense that it sinks, as shown in **Figure 19.** Upwelling later brings the current back to the surface. Density currents are important components of ocean circulation. They circulate thermal energy, nutrients, and gases.

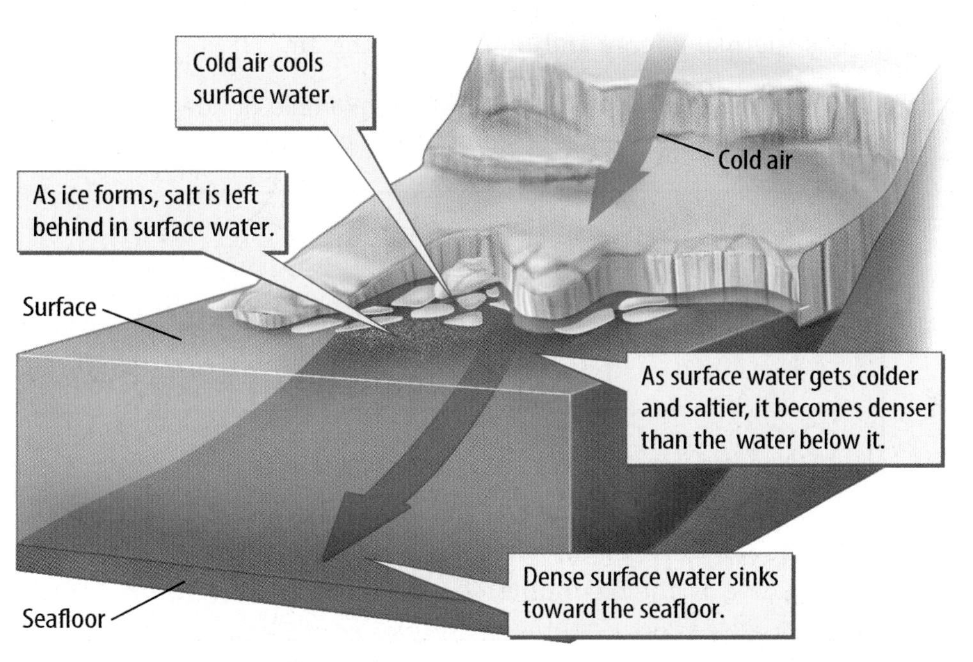

◄ **Figure 19** Cold, salty water sinks, producing a density current.

Figure 20 Higher temperatures are shown in red and yellow. Lower temperatures are shown in green and blue. ▶

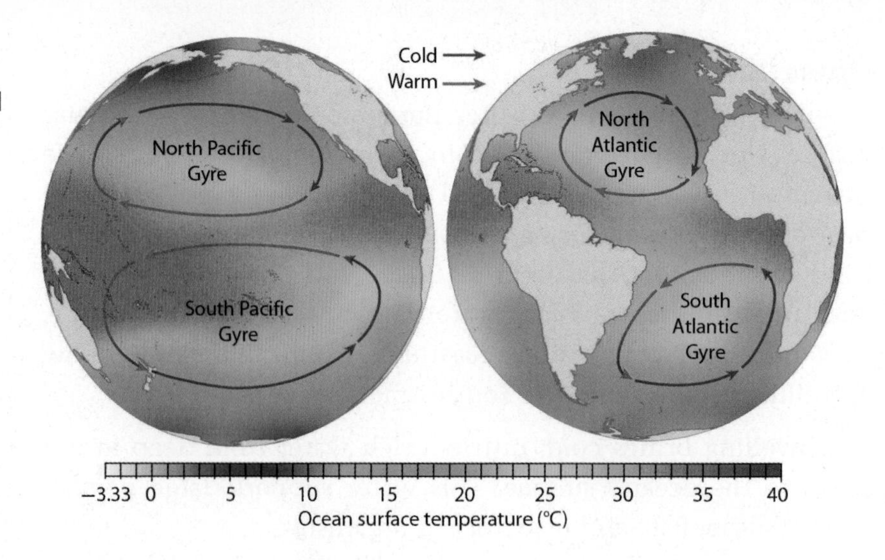

FOLDABLES®

Make a shutter-fold book. Use it to record the location of major warm-water currents and cold-water currents and to summarize how they affect weather and climate.

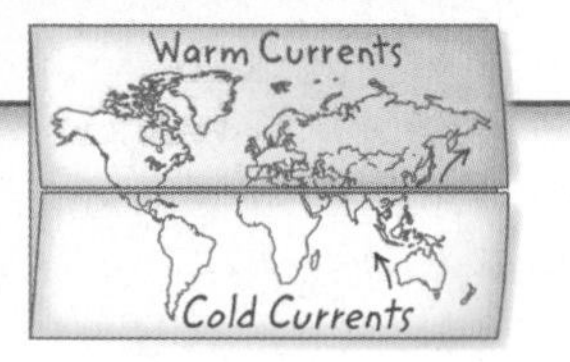

Impacts on Weather and Climate

Solar energy drives convection in the oceans causing warm- and cold-water currents in the gyres shown in **Figure 20.** These two types of surface currents affect weather and climate in different ways. Regions near warm-water currents are often warmer and wetter than regions near cold-water currents. Let's look at some examples.

Surface Currents Affecting the United States

Several warm-water currents affect coastal areas of the southeastern United States. For example, the Gulf Stream, shown in **Figure 21,** transfers lots of thermal energy and moisture to the surrounding air. As a result, summer evenings are often warm and humid. An evening rain is common in these areas.

The cold California Current, also shown in **Figure 21,** affects coastal areas of the southwestern United States. A summer evening along the California coast is often cooler and drier than a summer evening in Florida. Why? This cold-water current releases less thermal energy and moisture to the air.

Key Concept Check Give an example of how ocean currents can affect weather and climate.

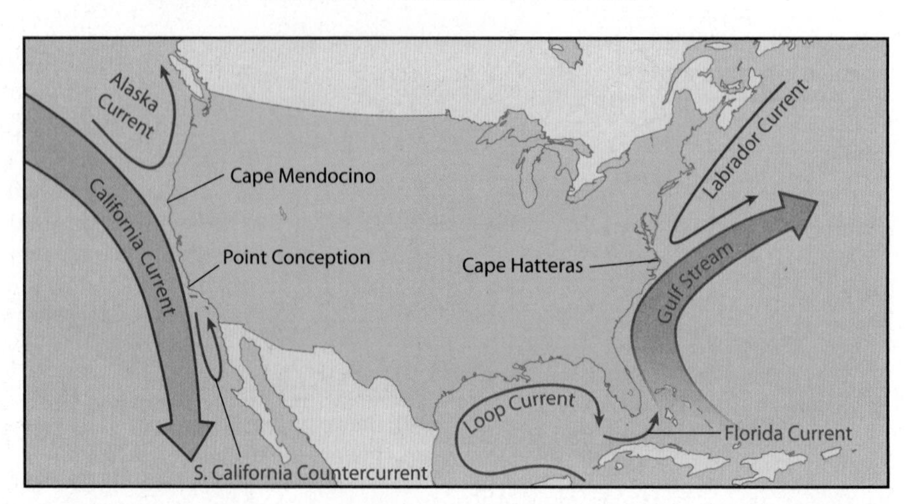

Figure 21 The Gulf Stream is a warm-water current. The California Current is a cold-water current.

Visual Check Hypothesize why hurricanes might be more common in the eastern US than in the western US. ▶

Global Conveyor Belt

Figure 22 A global belt of surface currents and density currents distributes thermal energy on Earth.

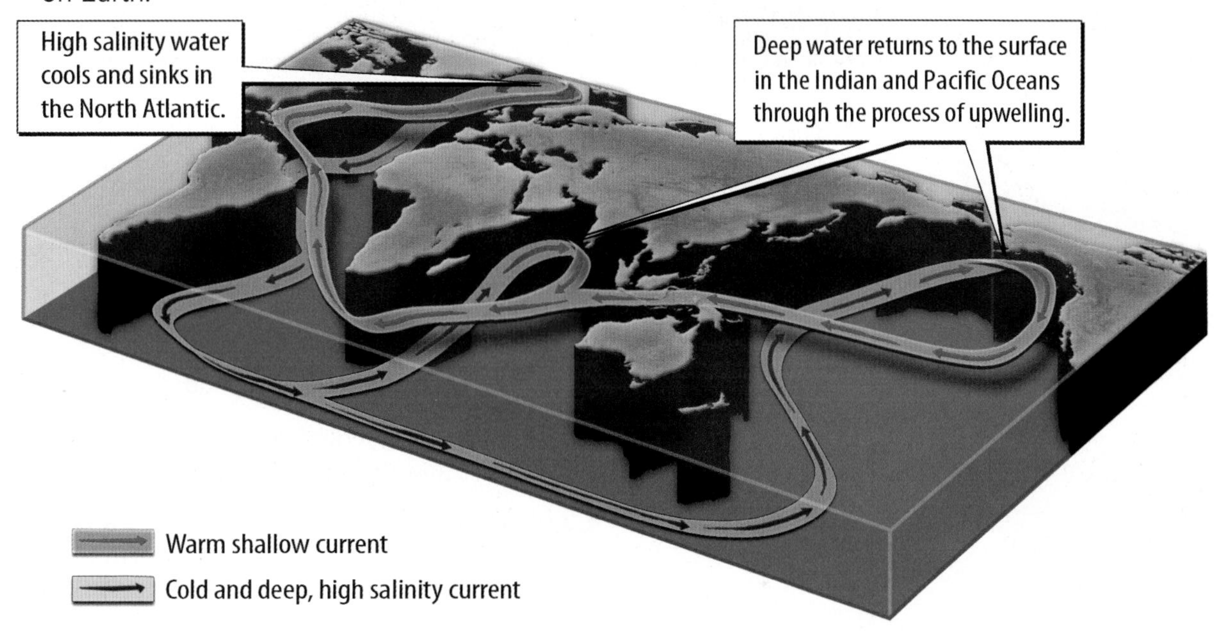

The Great Ocean Conveyor Belt

Aside from gyres, there is another large system of ocean currents that affects weather and climate. This current system is called the Great Ocean Conveyor Belt, illustrated in **Figure 22.** Scientists use this model to explain how ocean currents circulate thermal energy around Earth.

In this model, density currents in the North Atlantic Ocean and the Southern Ocean "run" the conveyor belt. Water in those regions is so cold and dense that it sinks to the ocean bottom and travels along the seafloor. Upwellings in the Pacific Ocean and Indian Ocean eventually bring this deep, cold water to the surface where it is warmed by the Sun.

As warm, surface water travels from the equator toward the poles, it releases thermal energy to the atmosphere, which warms the surrounding region. Then, the cold water sinks until it is upwelled at a different location and the cycle repeats. Scientists estimate that it takes about 1,000 years to complete a cycle.

Key Concept Check How does the Great Ocean Conveyor Belt affect climate?

MiniLab — 15 minutes

How does temperature affect ocean currents?

1. Read and complete a lab safety form.
2. Fill one **foam cup** with **hot water** and one cup with **ice water.**
3. Place a **glass dish** on top of the cups. Use two other cups for balance, as shown. Half fill the dish with **room-temperature water.**

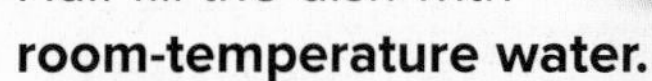

4. Put two drops of **food coloring** in the dish, one above each water-filled cup. Use one color for cold water, another for hot water. Observe for 10 min.

Analyze and Conclude

1. **Draw** a diagram of your observations in your Science Journal. Label the hot and cold areas in your drawing.
2. **Key Concept** Explain how your observations of the colored water resemble ocean currents.

Lesson 3 Review

Visual Summary

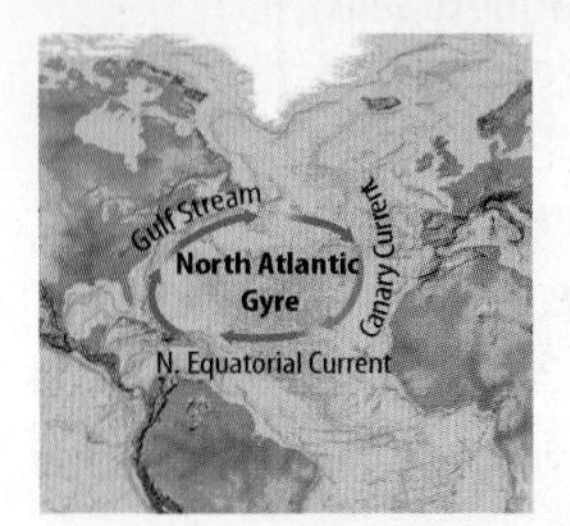

A gyre is a circular system of surface currents.

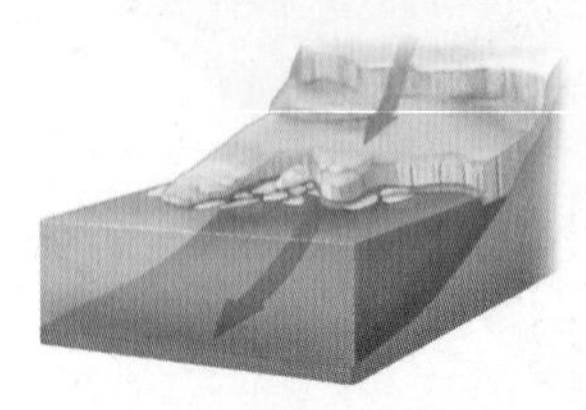

Density currents move cold water from the ocean surface to deeper parts of the ocean.

A system of surface currents and density currents distributes thermal energy around Earth.

FOLDABLES®

Use your lesson Foldable to review the lesson. Save your Foldable for the project at the end of the chapter.

What do you think NOW?

You first read the statements below at the beginning of the chapter.

5. Ocean currents occur on the surface and below the surface.

6. Ocean currents affect climate and weather.

Did you change your mind about whether you agree or disagree with the statements? Rewrite any false statements to make them true.

Use Vocabulary

1. **Use the term** *Coriolis effect* in a complete sentence.
2. A(n) ________ moves water vertically.

Understand Key Concepts 

3. What causes a surface current?
   A. Earth's orbit
   B. Earth's rotation
   C. temperature
   D. wind
4. **Explain** how energy transfers between currents and the atmosphere affect climate.
5. **Illustrate** how upwelling occurs off the coast of California as wind blows from north to south.

Interpret Graphics

6. **Explain** how the surface currents in the figure below affect the western and eastern coasts of the United States.

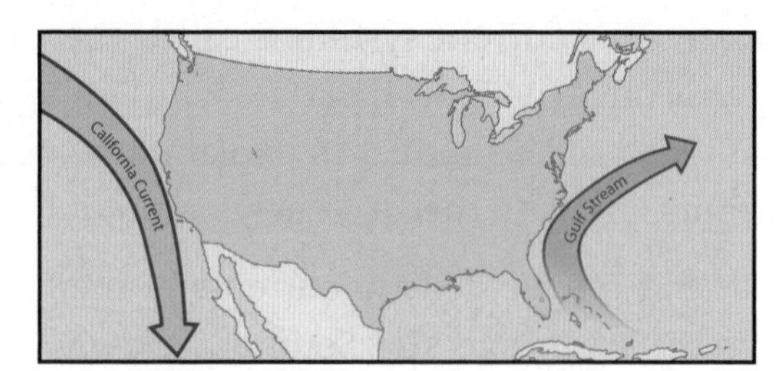

7. **Compare and Contrast** Copy and fill in the graphic organizer below to compare and contrast surface currents and density currents.

	Similarities	Differences
Surface currents		
Density currents		

Critical Thinking

8. **Design** an experiment to show how waves and currents move water in different ways.
9. **Infer** why major fishing grounds are along coastlines.

Skill Practice Interpret Data

30 minutes

How do oceanographers study ocean currents?

Materials

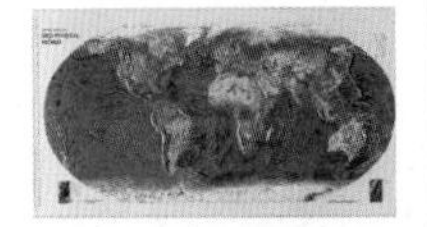

world map

Cargo spills can help oceanographers study ocean currents. The longitude and latitude positions of items from spills that wash ashore contain clues about the direction and speed of currents. Interpret the data below to find out what happened to a cargo of rubber bath toys lost in a January 1992 storm in the North Pacific.

Learn It

Can you make sense of the data in the table at right? You need to **interpret data** before you can draw conclusions about them. Interpret the longitude and latitude positions of toys that washed ashore by marking them on a map.

Found Toys		
Date	Latitude	Longitude
January 1992	45° N	178° E
March 1992	44° N	165° W
July 1992	49° N	155° W
October 1992	52° N	135° W
January 1993	59° N	149° W
March 1993	56° N	157° W
July 1993	57° N	170° W
October 1993	59° N	180° E
January 1994	56° N	166° E
March 1994	45° N	155° E
July 1994	47° N	172° E
October 1994	50° N	165° W
January 1995	47° N	140° W
October 2000	46° N	50° W
December 2003	57° N	07° W

Try It

1. Mark the longitude and latitude positions on a world map. The other data represent locations where individual bath toys were found. The first data point represents the location of the cargo spill. Label each point with a date.
2. Connect the dots in order of time. Ocean currents don't follow straight lines, so use curved lines. The toys could not float over land, so all the lines you draw should only cross water.
3. Compare the path of the toys to a world map of ocean currents and gyres.

Apply It

4. **Describe** how this data could help oceanographers chart ocean currents.
5. **Hypothesize** how toys traveled to the Atlantic Ocean.
6. **Key Concept** What types of ocean currents carry cargo debris around the world?

Lesson 4

Reading Guide

Key Concepts

ESSENTIAL QUESTIONS

- How does pollution affect marine organisms?
- How does global climate change affect marine ecosystems?
- Why is it important to keep oceans healthy?

Vocabulary

marine p. 590

harmful algal bloom p. 591

coral bleaching p. 592

Multilingual eGlossary

What's Science Got to do With It?

Environmental Impacts on Oceans

Inquiry Orange Ocean?

The orange-red color of the water in this photograph comes from algae. The algae have formed a huge mat, called an algal bloom, on the ocean's surface. Algal blooms can be beautiful, but some algal blooms harm ocean ecosystems.

Launch Lab

15 minutes

What happens to litter in the oceans?

Imagine you are on a boat hundreds of kilometers from shore. You look down at the water and see a sea turtle entangled in plastic. How did this happen?

1. Read and complete a lab safety form.
2. Half-fill a large **bowl** with **water.**
3. Sprinkle **objects** your teacher has supplied into the water.
4. Gently swirl the water in the bowl until the water moves at a constant speed. Try not to create a whirlpool.

Think About This

1. What happened to the objects you sprinkled into the bowl?
2. What do you think happens to litter that is dumped into the ocean?
3. 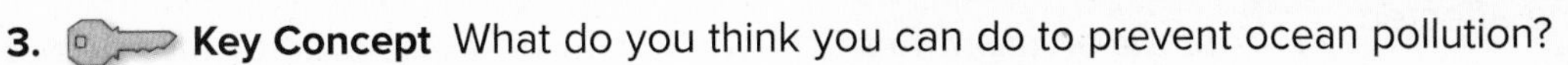 **Key Concept** What do you think you can do to prevent ocean pollution?

Ocean Pollution

Have you ever seen a photograph of a shorebird or seal covered in oil? Spills from oil tankers harm wildlife. They also harm the ocean. Any harm to the physical, chemical, or biological health of the ocean ecosystem is ocean pollution. Sometimes ocean pollution comes from a natural source, such as a volcanic eruption. More often, human activities cause ocean pollution.

Sources of Ocean Pollution

Like pollution on land, ocean pollution comes from both point sources and nonpoint sources. Point-source pollution can be traced to a specific source, such as a drainpipe or an oil spill. Nonpoint-source pollution cannot be traced to a specific source. Sewage runoff from land is an example.

Figure 23 shows the proportion of different sources of ocean pollution caused by humans. Notice that only 13 percent of this pollution comes from shipping or offshore mining activity. The rest comes from land. Land-based pollution includes garbage, hazardous chemicals, and fertilizers. Airborne pollution that originates on land, such as emissions from power plants or cars, is also included in this category. So is trash dumped directly into the oceans.

Ocean Pollution Sources

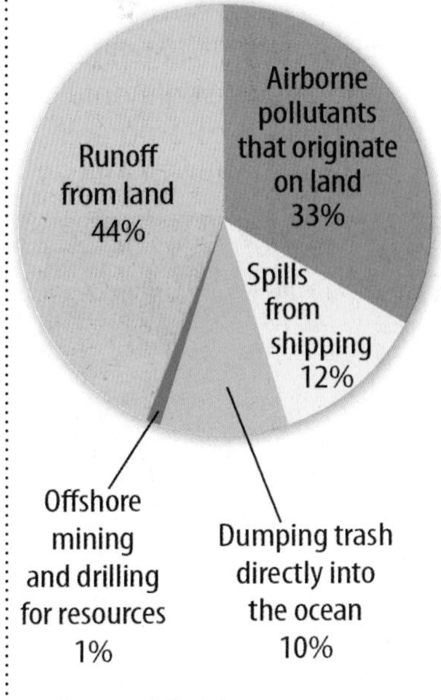

Figure 23 Most ocean pollution caused by humans originates on land.

T. O'Keefe/PhotoLink/Getty Images

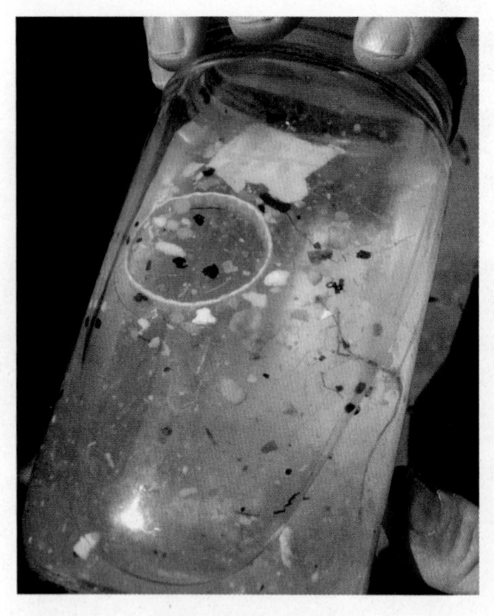

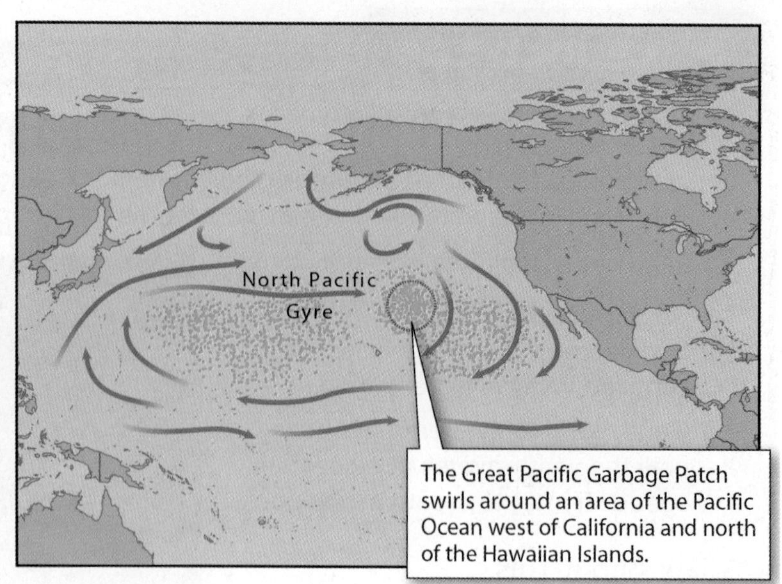

▲ **Figure 24** The North Pacific Gyre traps garbage in areas colored orange on the map. A scientist holds a sample of polluted water from one affected area.

Effects of Ocean Pollution

Ocean pollution has both immediate effects and long-term effects on marine ecosystems. **Marine** *refers to anything related to the oceans.* Chemical waste can be poisonous to marine organisms. Fish and other organisms absorb the poison and pass it up the food chain. A large oil spill can harm marine life. So can solid waste, excess sediments, and excess nutrients.

WORD ORIGIN

marine
from Latin *marinus,* means "of the sea"

Solid Waste Trash, including plastic bottles and bags, glass, and foam containers, cause problems for marine organisms. Many birds, fish, and other animals become entangled in plastic or mistake it for food. Plastic breaks up into small pieces but it does not degrade easily. Some of it becomes trapped in the circular currents of gyres. The North Pacific Gyre has collected so much plastic and other debris that some people have named a portion of it "the Great Pacific Garbage Patch." A map showing its location is shown in **Figure 24.** The Great Pacific Garbage Patch within the circled area is thought to be twice the size of Texas.

Figure 25 This satellite image shows sediment from orange-colored soil washing into the ocean. ▼

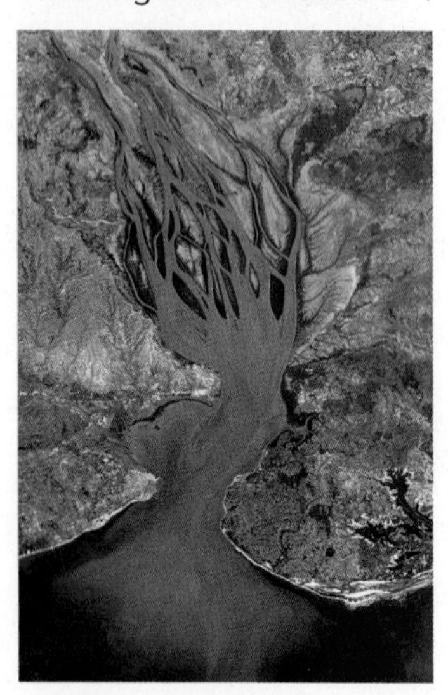

Excess Sediments Large amounts of land-based sediment wash into oceans, as shown in **Figure 25.** Erosion often occurs on steep coastal slopes after heavy rains. Some of this erosion is natural. But some is caused by humans, who cut down trees near rivers and ocean shorelines. Without the roots of trees and other vegetation to hold sediments in place, the sediments more readily erode. Excess sediments can clog the filtering structures of marine filter feeders, such as clams and sponges. Excess sediments can also block light from reaching its normal depth. Organisms that use light for photosynthesis die.

Key Concept Check How can excess sediments in oceans affect marine organisms?

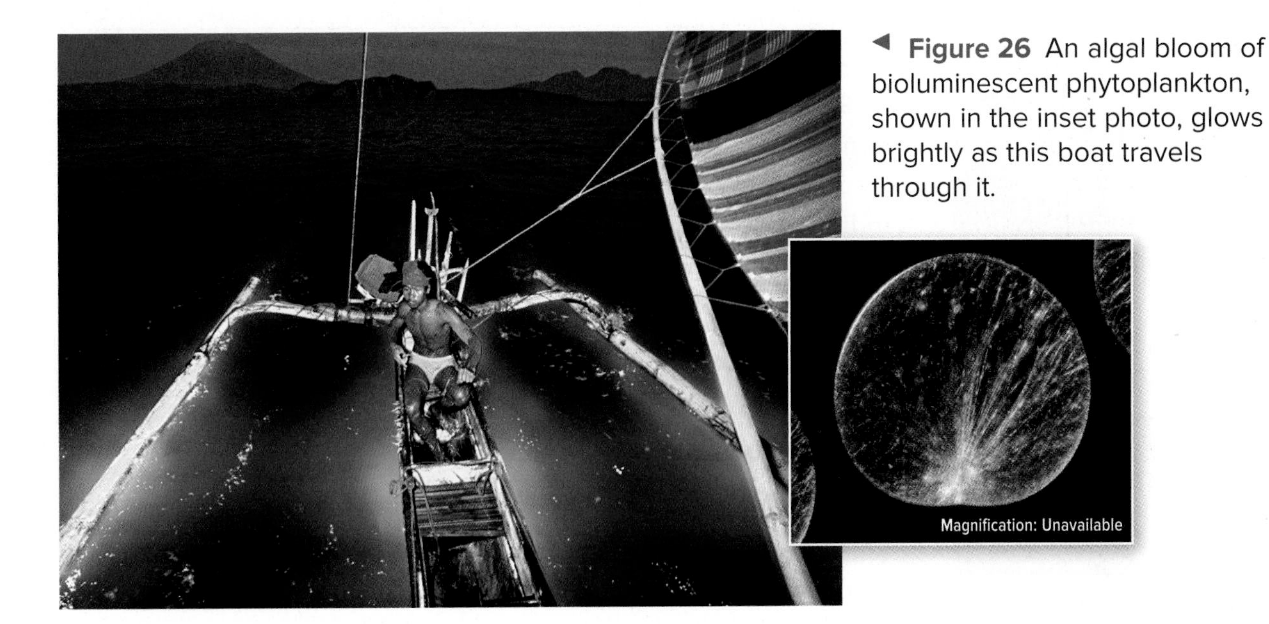

◄ **Figure 26** An algal bloom of bioluminescent phytoplankton, shown in the inset photo, glows brightly as this boat travels through it.

Excess Nutrients Algae need nutrients such as nitrogen and phosphorus to survive and grow. However, too many nutrients can cause an explosion in algal populations. An algal **bloom** occurs when algae grow and reproduce in large numbers. The photo at the beginning of this lesson shows how an algal bloom can cause water to turn orange. Algal blooms can also cause water to appear red, green, brown, or even glow at night, as shown in **Figure 26.**

SCIENCE USE V. COMMON USE

bloom

Science Use a large growth of algae

Common Use a flower

Nitrates and phosphates can be abundant in agricultural run-off as well as coastal upwelling zones. Many scientists suspect that a major source of excess nitrates and phosphates is from land-based fertilizers that wash into oceans.

 **Reading Check** Where do many nitrates come from?

Many algal blooms are harmless, but others can disrupt marine ecosystems and harm organisms. *A* **harmful algal bloom** *is a rapid growth of algae that harms organisms.* Harmful algal blooms have become more common in recent decades.

Why are some algal blooms harmful? The algae in some algal blooms produce poisonous substances that can kill organisms that eat them. Other algal blooms are so large that they they use up oxygen (O_2) in the water. This can happen when large numbers of algae die and decompose. Decomposition requires O_2. When many algae decompose at the same time, O_2 levels in the water drop. Fish and other marine organisms cannot get enough O_2 to survive. A fish kill resulting from a harmful algal bloom is shown in **Figure 27.**

 **Key Concept Check** How can excess nutrients in seawater harm fish?

Figure 27 Excess nitrates that wash into oceans can cause harmful algal blooms which kill fish. ▼

(tl)LOOK Die Bildagentur der Fotografen GmbH/Alamy; (tr)©S. Zankl/age fotostock; (b)Ricardo Aguiar/age fotostock

Figure 28 Corals contain colorful algae, which provide food for the coral. Without algae, the corals die and appear bleached.

Oceans and Global Climate Change

Solid waste, excess sediments, and algal blooms can cause immediate harm to ocean ecosystems. Other threats to oceans are related to long-term changes in Earth's climate. Climate data indicates that Earth's average surface temperature has increased over the past century. The amount of carbon dioxide (CO_2) in Earth's atmosphere has also increased.

Effects of Increasing Temperature

The increase in Earth's surface temperature has affected oceans in many ways.

Coral Bleaching Some marine organisms, such as coral, are very sensitive to temperature changes. A temperature increase as small as 1°C can cause corals to die, as shown in **Figure 28. Coral bleaching** *is the loss of color in corals that occurs when stressed corals expel the algae that live in them.* Coral bleaching harms corals around the world, as shown in **Figure 29.** Coral reefs provide habitat for fish and many other organisms.

Key Concept Check How does water temperature affect corals?

Sea Level As Earth warms, its glaciers and ice sheets melt. This adds water to the oceans and increases sea level. Rising sea levels threaten coastal communities and marine habitats.

Dissolved O_2 The temperature of seawater affects the amount of O_2 dissolved in it. The warmer the water, the less O_2 it contains. Marine organisms need O_2 to survive. As water warms, less O_2 is available, and organisms can die.

Coral Bleaching

Figure 29 Coral bleaching occurs in many locations around the world.

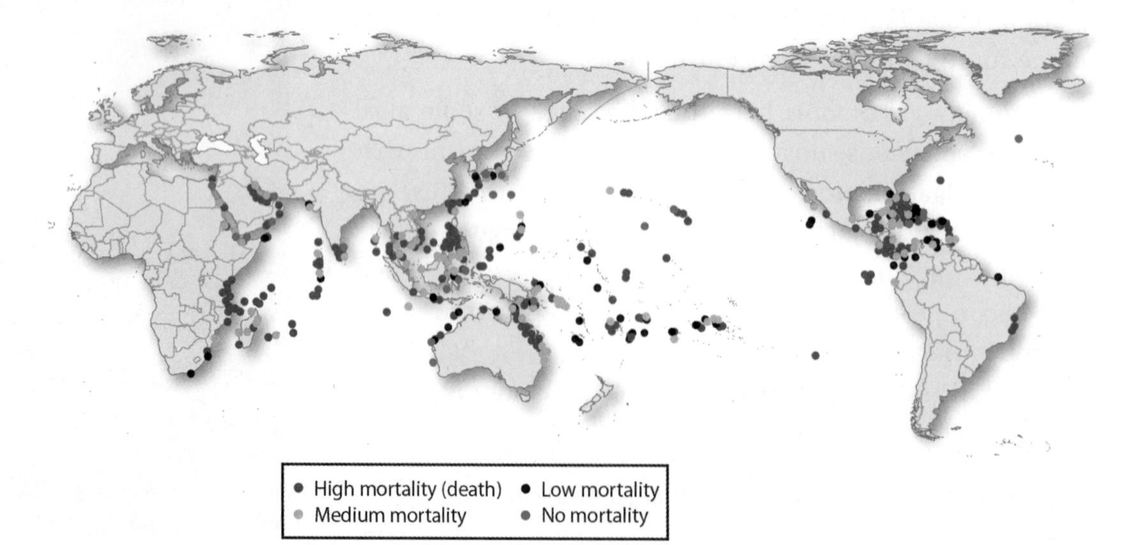

(tl)Darryl Leniuk/Getty Images; (b)Timothy G. Laman/National Geographic/Getty Images

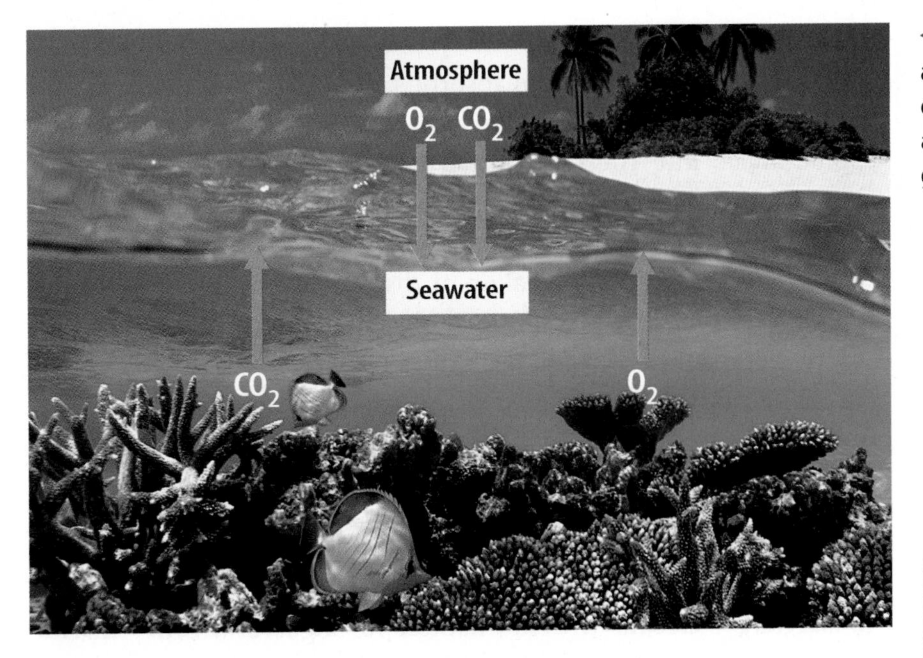

◄ **Figure 30** CO_2 and O_2 are exchanged at the ocean's surface. Waves and currents mix the gases into deeper water.

FOLDABLES®

Make a chart with three columns and three rows. Label it as shown. Use it to organize information about common gases found in seawater.

Common Gases Exchanged	How and Why	Human Concerns
Oxygen (O_2)		
Carbon Dioxide (CO_2)		

Effects of Increasing Carbon Dioxide

As illustrated in **Figure 30,** O_2 and CO_2 gases move freely between the atmosphere and seawater. As the amount of CO_2 increases in the atmosphere, the amount of CO_2 dissolved in seawater also increases. This is because of gas exchange at the ocean's surface. These gases dissolve in seawater. Wave action helps mix these gases deeper below the water surface.

CO_2 and pH When CO_2 mixes with seawater, a weak acid called carbonic acid forms. Carbonic acid lowers the pH of the water, making it slightly acidic. Data from recent studies show that the acidity of seawater has increased over the past 300 years. **Figure 31** shows certain areas of the oceans that now have a pH lower than that of the historic value of pH 8.16.

 **Reading Check** Why are oceans becoming more acidic?

Figure 31 Oceans are becoming more acidic. Scientists predict that oceans in the future will be much more acidic than they are today. ▼

Mapping acidification

pH values of the upper 165 ft. (50 m) of ocean water; lower values indicate more acidity

8.20
8.15 ◄ **8.16** Historic mean pH level for sea water
8.10
8.05
8.00
7.95

More acidic

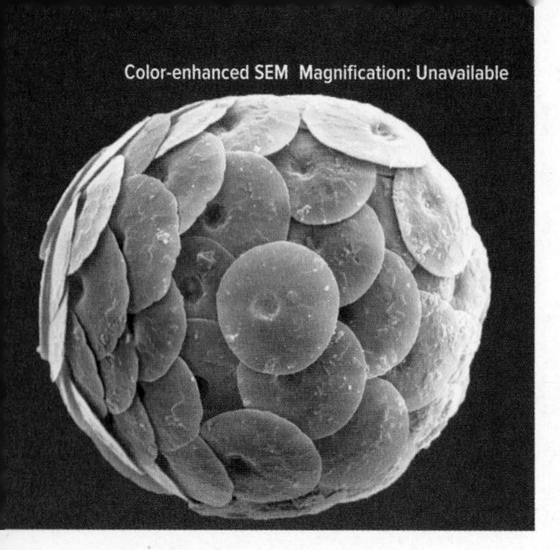

Figure 32 This tiny organism is surrounded by calcium carbonate plates.

Acidity and Marine Life Many marine organisms build shells and skeletons from calcium absorbed from seawater. Snails absorb calcium and make shells. Corals absorb calcium and build reefs. Some algae, like the one shown in **Figure 32,** make protective plates from calcium. As seawater becomes more acidic, it is harder for these organisms to absorb calcium. Increased acidity can cause shells and skeletons to weaken or dissolve. Over time, this could affect food webs. For example, if algae were unable to make protective plates, they would die. Algae form the base of food chains in many marine ecosystems.

Keeping Oceans Healthy

Earth's oceans affect Earth in many ways. As part of the water cycle, they distribute moisture. Ocean currents distribute thermal energy. Oceans provide habitat for algae and other marine organisms. Marine algae release during photosynthesis as much as 50 percent of the O_2 in Earth's atmosphere. Oceans also provide mineral and energy resources. They are a major source of food and income for humans. Keeping oceans healthy is important for the well-being of humans and other organisms on Earth.

 **Key Concept Check** Why is it important to keep oceans healthy?

MiniLab

20 minutes

How does the pH of seawater affect marine organisms?

How does increasing acidity affect calcium-containing shells?

1. Read and complete a lab safety form.
2. Copy the table below into your Science Journal.
3. Examine a piece of **brown eggshell** and describe its properties.
4. Place the eggshell in a **plastic cup.**
5. Half fill the cup with **white vinegar.**
6. After 15 minutes, use **forceps** to remove the eggshell. Describe its properties in your Science Journal.

Analyze and Conclude

1. **Describe** how the eggshell changed.
2. **Key Concept** How might long-term effects of increased CO_2 in seawater affect calcium-containing shells and skeletons of marine organisms?

Calcium-containing shells		
Property	Description Before Treatment	Description After Treatment
Hardness		
Thickness		
Appearance		

Lesson 4 Review

Visual Summary

A harmful algal bloom can cause fish kills.

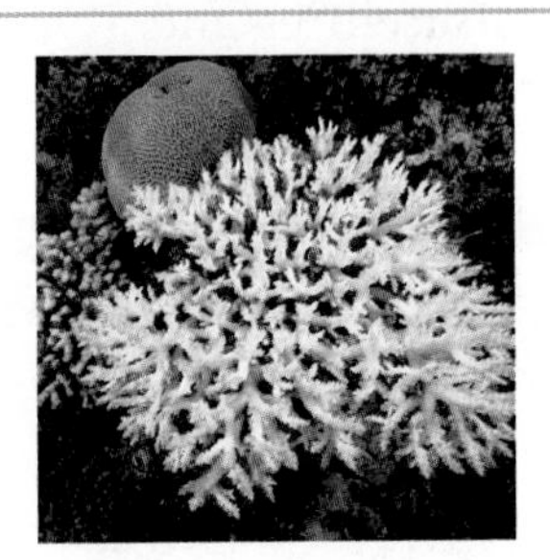

Increased ocean temperature causes corals to bleach.

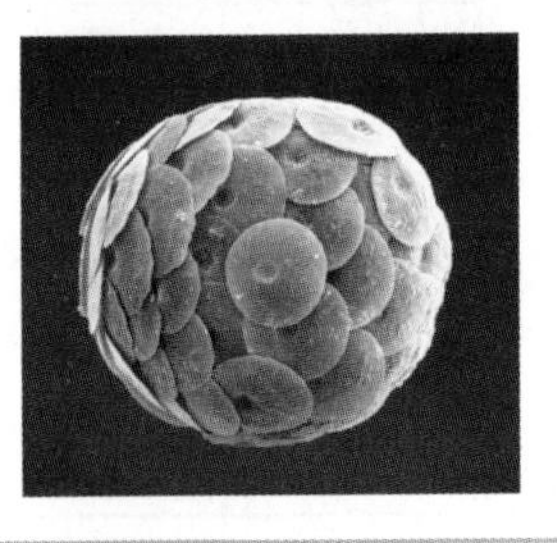

Global climate change affects the ocean's chemistry.

FOLDABLES®

Use your lesson Foldable to review the lesson. Save your Foldable for the project at the end of the chapter.

What do you think

You first read the statements below at the beginning of the chapter.

7. Most pollution in the oceans originates on land.

8. Global climate change has no effect on marine organisms.

Did you change your mind about whether you agree or disagree with the statements? Rewrite any false statements to make them true.

Use Vocabulary

1. **Define** *harmful algal bloom* in your own words.
2. **Use the term** *marine* in a complete sentence.

Understand Key Concepts

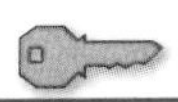

3. How can an increase in CO_2 in the atmosphere affect seawater?
   - **A.** O_2 levels rise
   - **B.** O_2 levels decrease
   - **C.** pH rises
   - **D.** pH decreases
4. **Identify** how excess sediments affect filter feeders.
5. Construct a flow chart that shows the steps leading to an algal-bloom fish kill.

Interpret Graphics

6. **Determine Cause and Effect** Copy and fill in the graphic organizer below to list the causes and effects of a decreased pH of seawater.

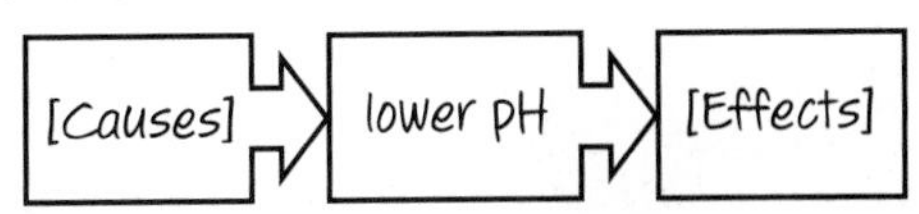

7. **Explain** how environmental conditions can affect the exchange of gases shown in the figure below.

Critical Thinking

8. **Design an experiment** to test the hypothesis that coral bleaching is caused by an increase in water temperature.
9. **Predict** the effect that increased carbon dioxide in the atmosphere could have on food webs in the ocean.

(tl)Ricardo Aguiar/age fotostock; (cl)Timothy G. Laman/National Geographic/Getty Images; (bl)Clouds Hill Imaging Ltd./Science Source; (br)©Steve Allen/Brand X/Corbis

Lab

45 minutes

Predicting Whale Sightings Based on Upwelling

Materials

colored pencils

You are a guide on a blue whale tour departing from Monterey, CA. You have read that upwelling is the vertical movement of cold, nutrient-rich water from the deep ocean to the surface of the ocean. Upwelling fertilizes the surface of the ocean and creates feeding grounds for fish and plankton-eating whales. Use oceanographic data from satellites and data-collecting buoys called moorings to plan a tour to best view blue whales.

Ask a Question

Where and when can you best observe blue whales near Monterey, CA?

Make Observations

1. Analyze a map of sea surface temperatures (SST) around Monterey Bay.
2. Convert your map to a color contour map. First, construct a legend using colors. Assign warmer pencil colors to warmer temperatures and cooler pencil colors to colder temperatures. Then, outline areas that have the same temperatures in pencil. Finally, color in the sections of the map according to your legend.
3. Study your map noting the position of the upwelling in your Science Journal.
4. Examine the mooring data to the right. Plot both sea surface temperature and wind speed versus day on the same graph. Be sure to label the two vertical axes to reflect the different measurements.

Data of Wind Direction and Speed

Date	SST (°C)	Wind Direction	Wind Speed (m/s)
23-May	10	N	3
25-May	10	N	8
27-May	9	N	10
29-May	9	N	8
31-May	9	N	4
2-Jun	10	S	–1
4-Jun	12	S	–4
6-Jun	13	S	–3
8-Jun	12	N	7
10-Jun	11	N	5
12-Jun	10	N	8
14-Jun	10	N	7
16-Jun	10	N	7
18-Jun	9	N	9
20-Jun	9	N	11
22-Jun	11	N	4
24-Jun	12	S	–4
26-Jun	13	S	–6
28-Jun	13	-	0
30-Jun	14	S	–1
2-Jul	13	N	6
4-Jul	11	N	9
6-Jul	9	N	10
8-Jul	9	N	10

5 Analyze your graph and determine under what wind conditions upwelling occurs.

Form a Hypothesis

6 Use your observations of the upwelling to form a hypothesis that gives the location (latitude and longitude) and wind conditions where you could best observe blue whales if you leave on a tour from Monterey, CA.

Test your Hypotheses

7 Use a map showing sightings of blue whales in Monterey Bay to compare your hypothesis to the actual locations where blue whales have been frequently observed. If your hypothesis was not supported, repeat steps 2–3.

8 Compare your prediction of wind conditions for which you could best observe blue whales with another student in your class. If you do not agree, repeat steps 5–6.

Analyze and Conclude

9 **Describe** the location and shape of the upwelling in Monterey Bay.

10 **Analyze** in which direction the wind was blowing when the satellite measurement of sea surface temperature was taken. Explain why this is important to your hypothesis.

11 **Design** a graphic organizer to show the effects of currents, sea surface temperatures, and wind direction on whale feeding areas.

12 **The Big Idea** Explain how currents affect sea life in Monterey Bay.

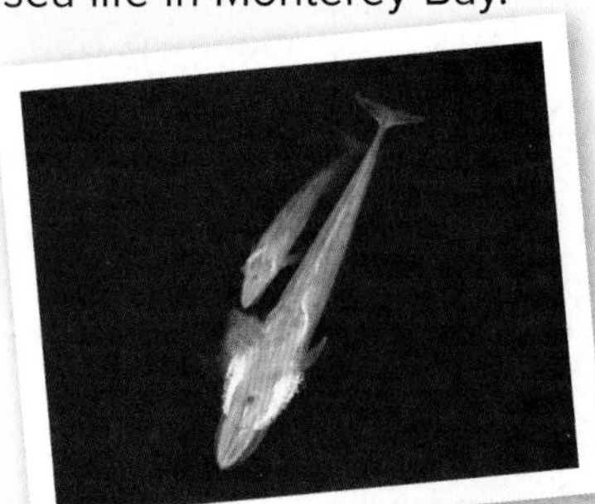

Communicate Your Results

Design a brochure for a whale watching company based in Monterey, CA. Describe the technology and oceanography that you will use to ensure that your clients observe blue whales.

During the Great Depression, Monterey Bay was one of the largest sardine fisheries in the world. John Steinbeck wrote about the time period in his book *Cannery Row.* Investigate what happened to the sardine fisheries in Monterey Bay during the nineteenth century. Write a Moment in History news report explaining the environmental factors that impacted the growth and decline of the fishery.

Lab Tips

- When plotting your data, be sure to use the vertical axis that goes with the data you are plotting.
- Draw a line to connect your plot points.
- Use two different colors for wind speed and sea surface temperature.

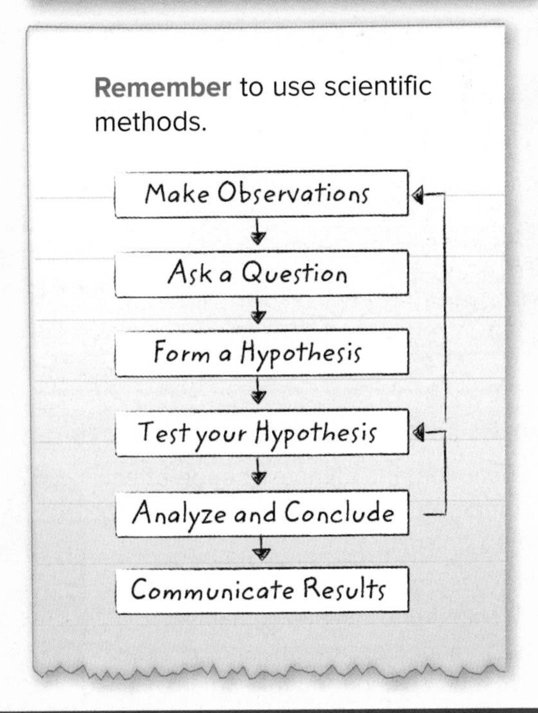

Chapter 16 Study Guide

Oceans affect Earth's climate and weather. They provide resources and habitats. But oceans are threatened by pollution and global climate change.

Key Concepts Summary 

Lesson 1: Composition and Structure of Earth's Oceans

- The salt in the oceans comes mostly from the erosion of rocks and soil.
- The seafloor has mountains, deep trenches, and flat plains.
- The oceans have zones based on light, temperature, salinity, and density.

Vocabulary

salinity p. 565
seawater p. 565
brackish p. 565
abyssal plain p. 566

Lesson 2: Ocean Waves and Tides

- The motion of water particles in a wave is circular.
- Wind causes most ocean waves, but underwater disturbances cause most **tsunamis.**
- The gravitational attraction between Earth and the Moon, and between Earth and the Sun causes **tides.**

Vocabulary

tsunami p. 575
sea level p. 576
tide p. 576
tidal range p. 576
spring tide p. 577
neap tide p. 577

Lesson 3: Ocean Currents

- Surface currents, **upwelling,** and density currents are the major **ocean currents.**
- Ocean currents affect climate and weather by distributing thermal energy and moisture around Earth.

Vocabulary

ocean current p. 581
gyre p. 582
Coriolis effect p. 582
upwelling p. 583

Lesson 4: Environmental Impacts on Oceans

- Ocean pollution and climate change affect water temperature and ocean pH, harming **marine** organisms.
- A healthy ocean is important because it affects weather and climate, contains habitats for marine organisms, and provides energy resources and food for humans.

Vocabulary

marine p. 590
harmful algal bloom p. 591
coral bleaching p. 592

(t)DAVID DOUBILET/National Geographic Stock; (cl)Jacques Descloitres, MODIS Rapid Response Team, NASA/GSFC; (cr)Henk Badenhorst/iStock/360/Getty Images; (b)Timothy G. Laman/National Geographic/Getty Images

Study Guide

Personal Tutor

Vocabulary eFlashcards
Vocabulary eGames

FOLDABLES® Chapter Project

Assemble your lesson Foldables as shown to make a Chapter Project. Use the project to review what you have learned in this chapter.

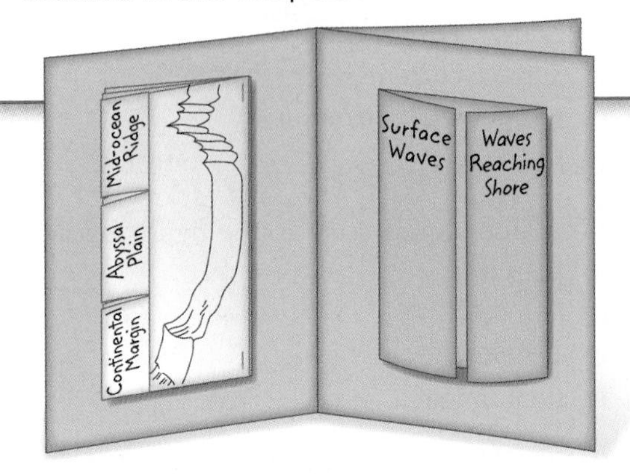

Use Vocabulary

1. Water that has lower salinity than average is ________.
2. Scientists use the term ________ to describe the amount of salt in water.
3. The average height of the ocean's surface is ________.
4. A(n) ________ occurs when Earth, the Moon, and the Sun are in a straight line.
5. A(n) ________ is a large volume of water flowing in a certain direction.
6. A(n) ________ carries warm and cold water in a circular system.
7. A(n) ________ is a vertical movement of water toward the surface.
8. A(n) ________ can occur when increased nutrients cause explosive algal growth.

Interactive Concept Map

Link Vocabulary and Key Concepts

Copy this concept map, and then use vocabulary terms from the previous page to complete the concept map.

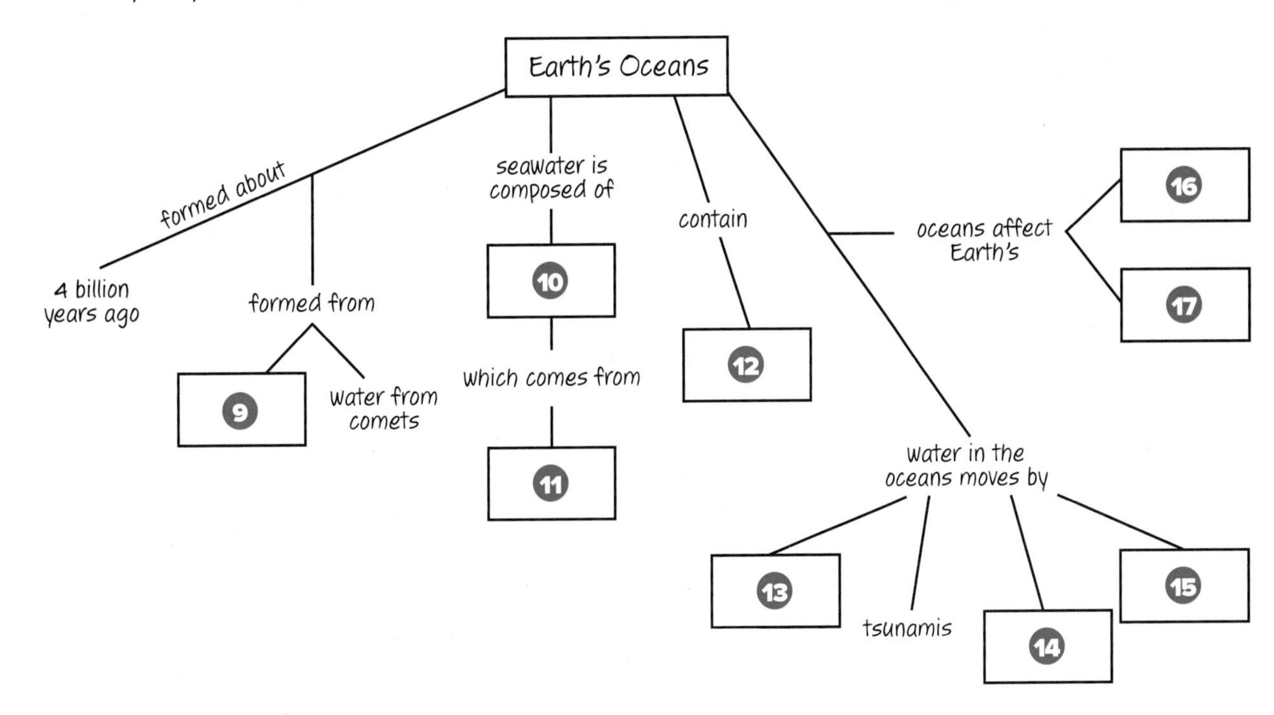

Chapter 16 Review

Understand Key Concepts 

1 Based on the circle graph below, which element is most common in seawater?

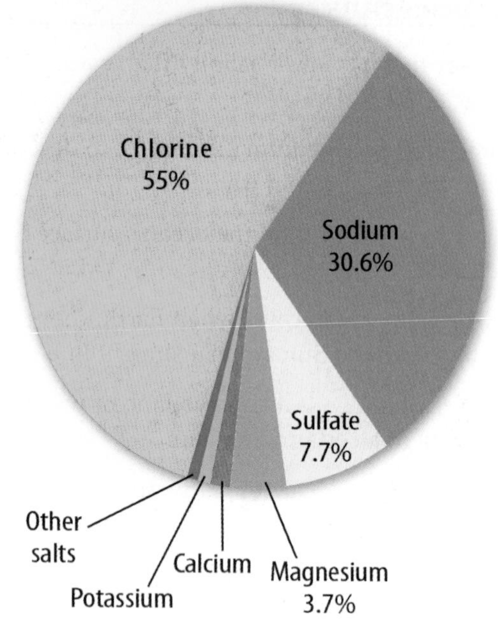

A. calcium
B. chlorine
C. sodium
D. sulfur

2 Which resource is on abyssal plains?

A. gravel
B. manganese nodules
C. methane hydrates
D. natural gas

3 Which is NOT a cause of tsunamis?

A. earthquake
B. hurricane
C. landslide
D. volcanic eruption

4 Which best describes the movement of water in a wave?

A. circular
B. horizontal
C. spiral
D. vertical

5 Where does an ocean current become most dense?

A. in polar regions
B. in temperate regions
C. near continents
D. near the equator

6 Which moves water horizontally?

A. density current
B. surface current
C. temperature current
D. upwelling

7 What does C represent in the figure below?

A. high tide
B. low tide
C. sea level
D. tidal range

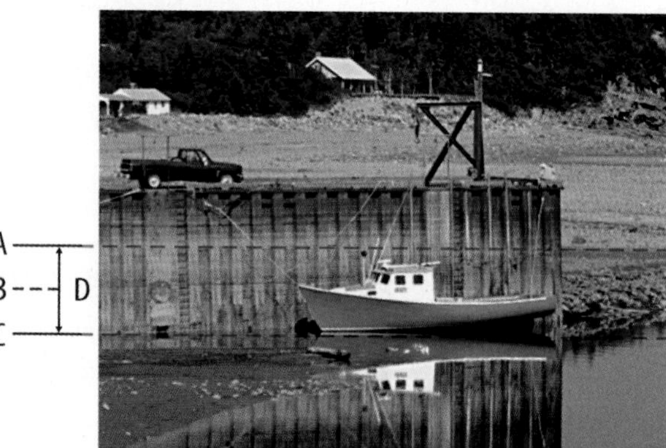

8 Which is one possible effect of an increase in carbon dioxide in the oceans?

A. Algae grow in excessive amounts.
B. Corals can't make reefs.
C. High tides occur more often.
D. Ocean sedimentation increases.

9 Which is NOT a consequence of rising ocean temperature?

A. coral bleaching
B. glacier melting
C. rising sea level
D. shells dissolving

Online Test Practice

Critical Thinking

10. **Summarize** the sources of salt in seawater.

11. **Compare** the topography of the ocean floor with the topography of land.

12. **Illustrate** what happens to water particles when a wave passes.

13. **Explain** how a density current might form in the Arctic Ocean.

14. **Design** a model that shows how surface currents form.

15. **Relate** How can cutting trees on land affect life in the ocean?

16. **Assess** the long-term effects of a harmful algal bloom on a marine ecosystem.

17. **Hypothesize** As shown in the figure below, Earth's major warm water currents are on the western boundaries of oceans. Major cold water currents are on the eastern boundaries of oceans. Why are these major currents in these different locations?

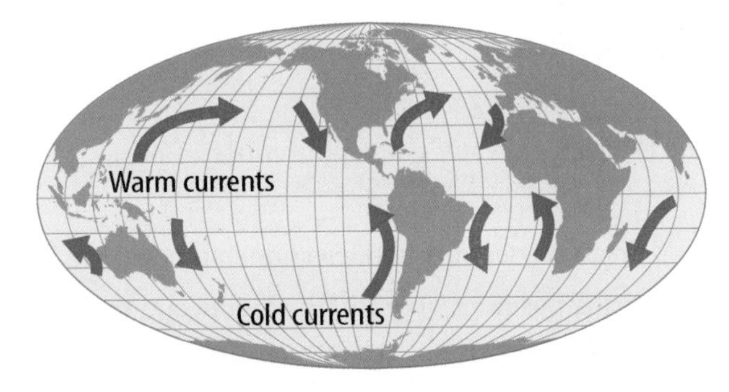

Writing in Science

18. **Compose** a letter to the editor of a newspaper or magazine with ideas of how to reduce human impacts on oceans. Include a main idea, supporting details, examples, and a concluding sentence.

REVIEW THE BIG IDEA

19. Why are oceans important? In what ways are they threatened?

20. How are waves powered? How does movement differ in waves and currents?

Math Skills

Math Practice

Use Statistics

Time (Day 1)	Height (m)	Time (Day 2)	Height (m)
00:44	13.1	01:33	13.0
07:13	0.8	08:02	0.9
13:04	13.6	13:54	13.5
19:42	0.3	20:32	0.4

The table above shows the high and low tides during a 48-hour period at the Bay of Fundy. Use the table to answer the questions.

21. What is the range of tides during the 48-hour period.

22. What is the mean of the tides during the 48-hour period?

23. What is the range of the four high tides during the 48-hour period?

24. What is the mean of the four low tides?

Standardized Test Practice

Record your answers on the answer sheet provided by your teacher or on a sheet of paper.

Multiple Choice

1 Which is a result of increasing acidity of seawater?

A Algae populations increase dramatically.

B Corals expel algae living within them.

C Oxygen is less available to marine organisms.

D Shells and skeletons of marine organisms weaken.

2 Which did NOT contribute to the formation of Earth's early oceans?

A asteroids

B condensation

C comets

D glaciers

3 What percentage of Earth's water is salt water?

A 3%

B 55%

C 70%

D 97%

Use the diagram below to answer question 4.

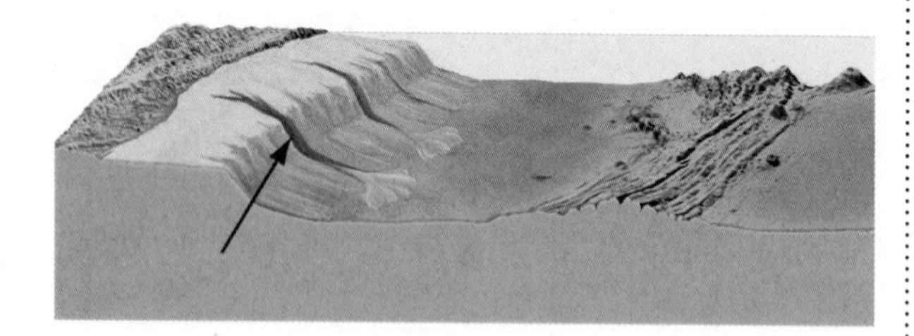

4 Which seafloor feature does the arrow in the diagram above indicate?

A abyssal plain

B continental slope

C ocean trench

D submarine canyon

Use the diagram below to answer question 5.

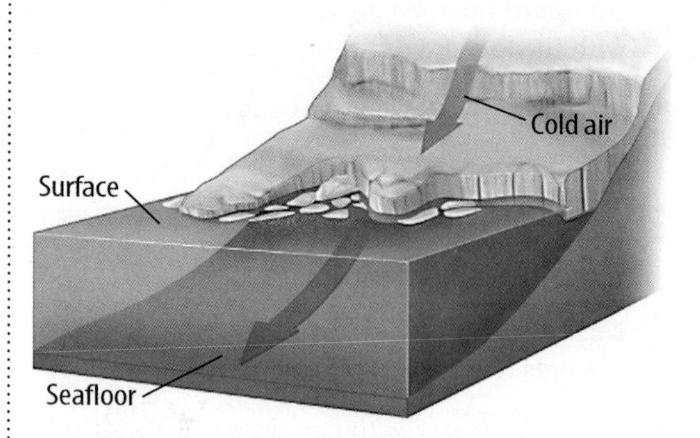

5 Which is formed by the process shown in the diagram above?

A gyres

B tsunami

C density current

D surface waves

6 Which results from upwelling in the oceans?

A Acidic water dissolves shells.

B Cold, dense water sinks.

C Marine organisms die.

D Surface water gains nutrients.

7 Which causes spring tides and neap tides?

A the positions of Earth, the Moon, and the Sun

B the rotation of Earth on its axis

C the shape of the continental margin

D the size and shape of ocean basins

8 As seawater temperature rises, the water contains

A less dissolved minerals.

B less oxygen.

C more coral.

D more nutrients.

Use the diagram below to answer question 9.

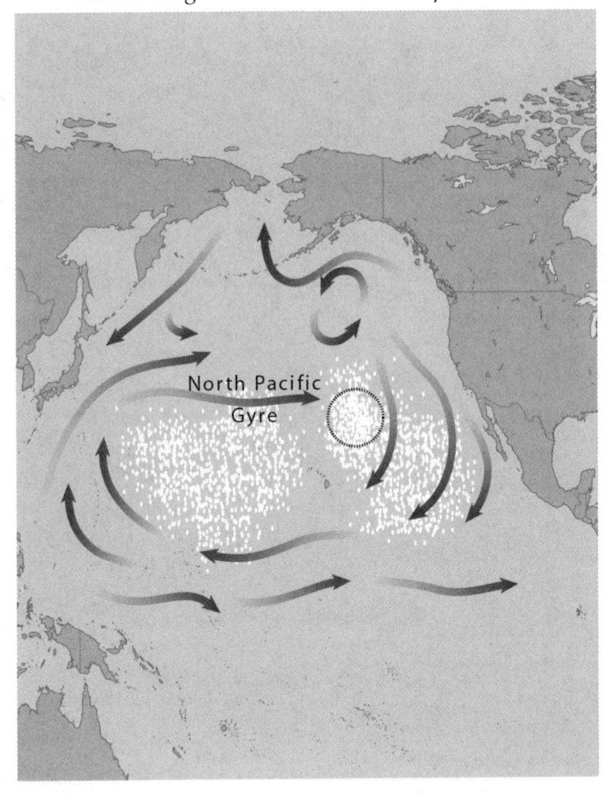

9 The circle on the diagram above indicates a region affected by

A coral bleaching.

B frequent tsunamis.

C excess nitrates and phosphates.

D pollution from solid waste.

10 Fertilizer runoff from agricultural areas into seawater can cause an excess of

A acid.

B carbon dioxide.

C nutrients.

D salts.

Constructed Response

Use the diagram below to answer questions 11–13.

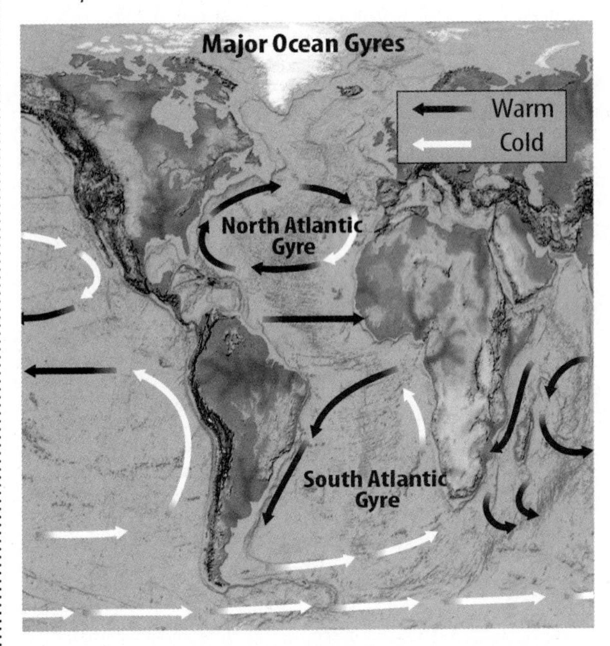

11 What type of current is marked with arrows on the map? How do these currents form? What do they do?

12 Why do these currents move in opposite directions around the North Atlantic and South Atlantic gyres?

13 How do these currents affect the climates of the surrounding continents?

14 What are two ways in which algae benefit other organisms?

15 Why are healthy oceans important to ALL life on Earth?

NEED EXTRA HELP?															
If You Missed Question...	1	2	3	4	5	6	7	8	9	10	11	12	13	14	15
Go to Lesson...	4	1	1	1	3	3	2	4	4	4	3	3	3	4	4

Chapter 17

Freshwater

THE BIG IDEA

Where is Earth's freshwater?

Inquiry Why do trees grow here?

Notice the number of trees growing along the Mississippi River. This river ecosystem is green and lush. Rivers are a source of freshwater for many plants and animals. Freshwater helps life sustain itself on Earth.

- What is freshwater?
- How have humans impacted freshwater on Earth?
- Where is most of Earth's freshwater located?

Get Ready to Read

What do you think?

Before you read, decide if you agree or disagree with each of these statements. As you read this chapter, see if you change your mind about any of the statements.

1. On Earth, freshwater occurs only as liquid water.
2. Up to 80 percent of the sunlight that strikes snow or ice is reflected back into space.
3. Organisms living in a lake or a stream are not affected by the amount of oxygen and nutrients in the water.
4. Glaciers can form lake basins.
5. People use groundwater as a source of water.
6. Wetlands can naturally filter pollutants from groundwater.

Your one-stop online resource
connectED.mcgraw-hill.com

LearnSmart®

Chapter Resources Files, Reading Essentials, Get Ready to Read, Quick Vocabulary

Animations, Videos, Interactive Tables

Self-checks, Quizzes, Tests

Project-Based Learning Activities

Lab Manuals, Safety Videos, Virtual Labs & Other Tools

Vocabulary, Multilingual eGlossary, Vocab eGames, Vocab eFlashcards

Personal Tutors

Lesson 1

Glaciers and Polar Ice Sheets

Reading Guide

Key Concepts
ESSENTIAL QUESTIONS

- How do glaciers affect sea level?
- How does ice and snow cover affect climate?
- How do human activities affect glaciers?

Vocabulary
freshwater p. 607
alpine glacier p. 608
ice sheet p. 609
sea ice p. 611
ice core p. 612

Multilingual eGlossary

Inquiry Why is the ice melting?

Notice the water streaming off the edge of the iceberg. Why is this ice melting so rapidly? Where does all of the water go? Ice melts as temperature increases. When ice melts, the meltwater eventually enters the oceans, where it can cause a rise in sea level.

Launch Lab

10 minutes

Where is all the water on Earth?

Earth is often called the "water planet." That's because about 70 percent of Earth's surface is covered with water stored in the oceans. Where is the rest of Earth's water?

1. Read and complete a lab safety form.
2. Pour 970 mL of water into a **1-L container.** Then, add a drop of **red food coloring.** This represents all of the salt water on Earth.
3. Add 20.7 mL of water to a **clear plastic cup** using a **graduated cylinder**. Then, add a drop of **blue food coloring** to represent all freshwater stored in glaciers.
4. Add 9.0 mL of water to a **clear plastic cup** and then add a drop of **green food coloring.** This represents all the freshwater stored as groundwater.
5. Finally, add one drop (about 0.3 mL) of **yellow food coloring** to a clear plastic cup. This represents all the freshwater in Earth's lakes, rivers, wetlands, atmosphere, and other sources.

Think About This

1. Where is Earth's water and in what forms does it exist?
2. **Key Concept** Can you think of any other place on Earth where you might find water?

What is freshwater?

Satellite images of Earth show more water than dry land. Most of the water that covers Earth is salt water. Only about 3 percent is **freshwater**–*water that has less than 0.2 percent salt dissolved in it.* Life, as we know it, cannot continue without freshwater.

Water cycles on Earth. Water moves from Earth's surface into the atmosphere by evaporation. The water then condenses and falls back to the surface as precipitation–rain, snow, sleet, or hail. Only freshwater enters Earth's atmosphere and returns to Earth's surface.

More than two-thirds of Earth's freshwater is frozen, as illustrated in **Figure 1.** The rest is liquid water, and most is stored underground. Less than 1 percent of Earth's liquid freshwater is in streams and lakes.

Reading Check Where is Earth's freshwater?

Freshwater on Earth

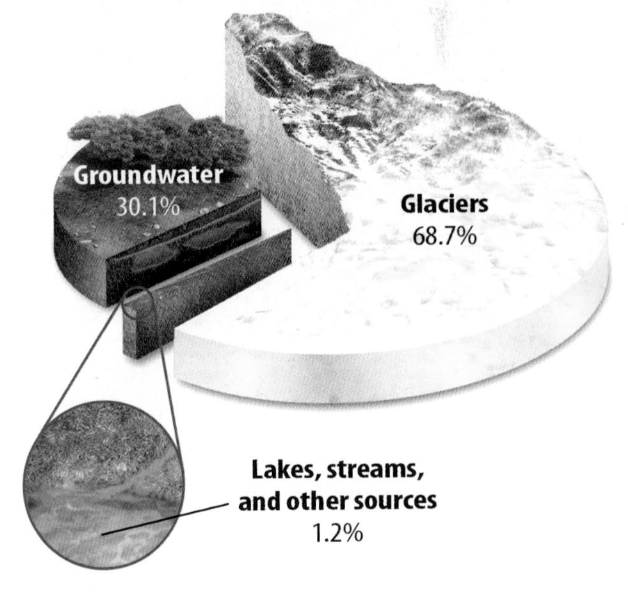

Figure 1 Most of Earth's freshwater is frozen in glaciers.

Glaciers and Ice Sheets

REVIEW VOCABULARY

glacier
a large, slow-moving mass of ice and snow

Glaciers are large masses of moving ice that form on land. Glaciers cover about 10 percent of Earth's surface. They are near the North Pole and the South Pole and on mountaintops, as shown in **Figure 2.**

How do glaciers form? Imagine what happens when snow falls but doesn't melt. Year after year layers of snow pile up. The weight and pressure of the snow above compresses the snow on the bottom into ice. Over time, the mass of ice and snow gets so heavy that gravity starts to slowly drag it downhill. For most glaciers this process takes over one hundred years.

Alpine Glaciers

WORD ORIGIN

alpine
from French *Alpes*, means "Alps"—mountain system of Europe

A glacier that forms in the mountains is an **alpine glacier.** Alpine glaciers are on every continent except Australia. They flow downhill like slow-moving rivers of ice. As an alpine glacier flows downhill, it eventually reaches an elevation where temperatures are warm enough to melt the ice. The melted ice is called glacial meltwater.

Glaciers

Animation

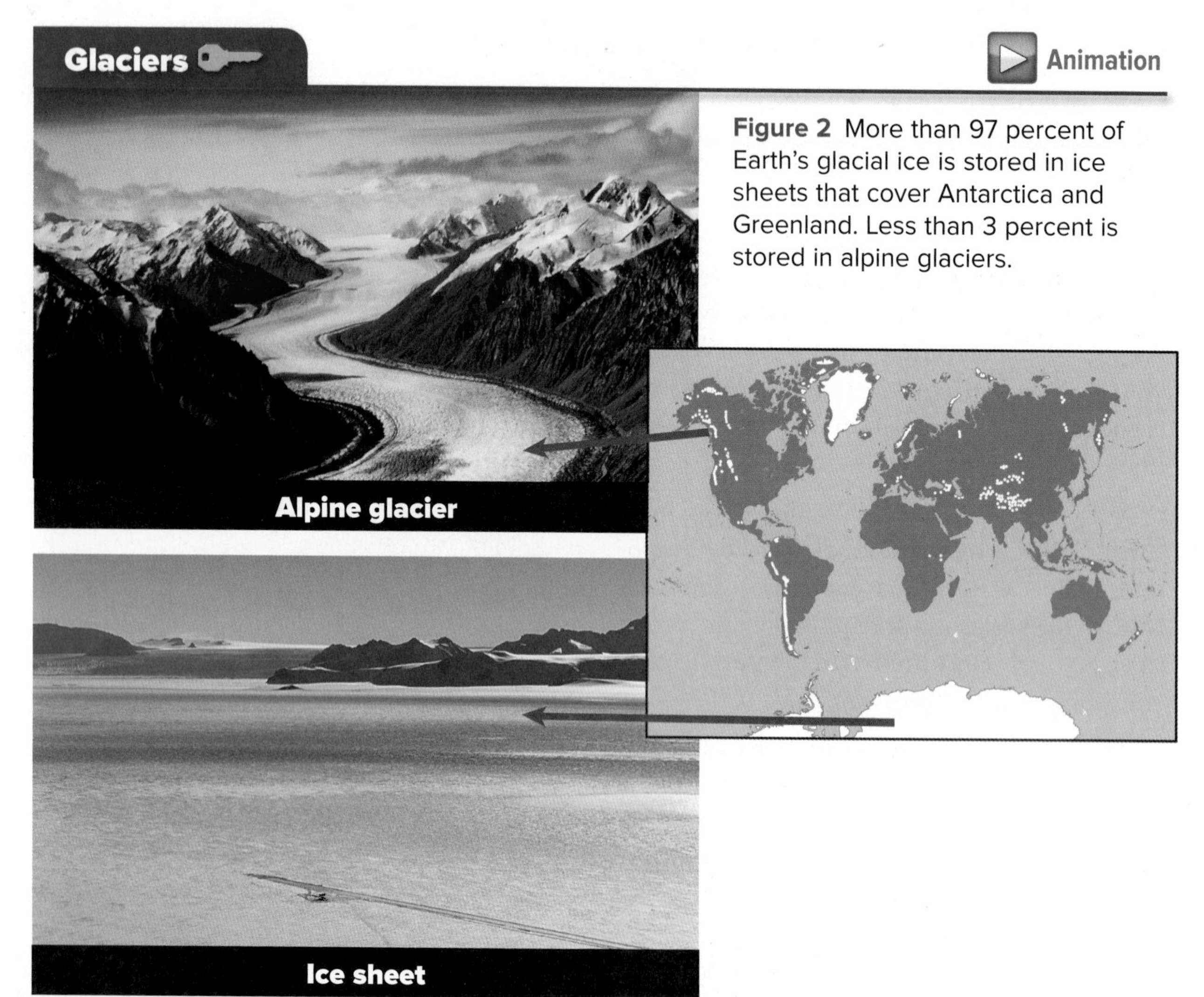

Figure 2 More than 97 percent of Earth's glacial ice is stored in ice sheets that cover Antarctica and Greenland. Less than 3 percent is stored in alpine glaciers.

Ice Sheets

A glacier that spreads over land in all directions is called an **ice sheet.** Ice sheets are also called continental glaciers. They cover large areas of land (more than 50,000 km^2) and store enormous amounts of freshwater. The only two ice sheets currently on Earth are in Antarctica and Greenland.

Parts of the Antarctic and Greenland ice sheets extend into the ocean. When a glacier flows into the ocean, an ice shelf forms. Ice shelves occur along the coastlines of Alaska, Canada, Greenland, and Antarctica. Icebergs are blocks of ice that break away from ice shelves and float in the ocean.

Antarctic Ice Sheet Earth's largest ice sheet is the Antarctic Ice Sheet. The ice sheet covers most of Antarctica and is larger in surface area than the continental United States. Scientists subdivide the ice sheet into two areas–the West and East Antarctic Ice Sheets–as illustrated in **Figure 3.** The average thickness of the Antarctic Ice Sheet is about 2.4 km. In some places, the ice can be as much as 5 km, or 3 miles, thick.

Greenland Ice Sheet Earth's second-largest ice sheet covers most of Greenland. Its average thickness is about 2.3 km. The total area of the ice sheet is about 1.8 million km^2.

Reading Check What are the two types of glaciers?

Math Skills

Volume
How much freshwater is stored as ice in the Antarctic Ice Sheet?

The area (A) of the Antarctic Ice Sheet is 14 million km^2. Calculate the volume (V) of ice by multiplying the area by the thickness or height (h).

$$V = A \times h$$

For example, the average thickness of the Antarctic Ice Sheet is **2.4 km**.

V = 14,000,000 km^2 × **2.4 km**, or 33,600,000 km^3.

Practice
The total area of the entire United States is approximately 10 million km^2. What would be the volume of an ice sheet covering the United States to a depth of 2.2 km?

Math Practice

Personal Tutor

Antarctic Ice Sheet

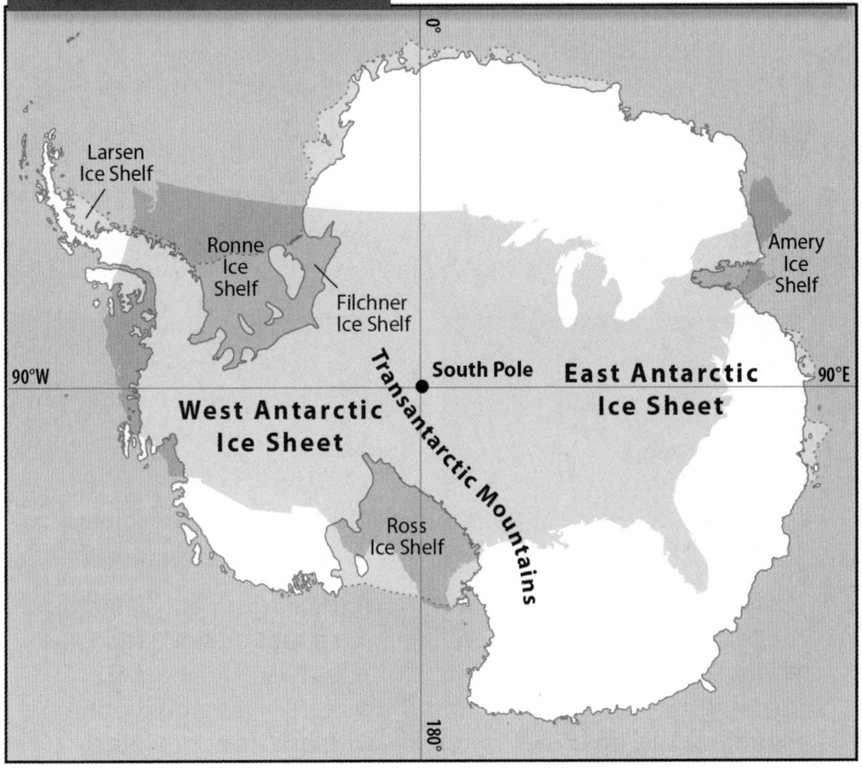

Figure 3 Antarctica has an area of about 14 million km^2. That's much larger than the area of the United States, about 10 million km^2. Ice shelves extend into the ocean from several places along the Antarctic coast.

How much freshwater is in glaciers?

Glaciers can stay frozen for thousands of years. During periods of Earth's history, the climate was colder than it is now. During those periods, many glaciers formed. The coldest periods are called ice ages–long periods of time when large areas of land are covered by glaciers. The last ice age ended about 10,000 years ago.

Past Changes in Sea Level Even if you have never been to either coast, you probably know that sea level is the average level of the surface of Earth's oceans. Changes in sea level have occurred throughout Earth's history. Sea level rises or falls as climate changes cause the melting or forming of glaciers.

As illustrated in the first image in **Figure 4,** sea level during the last ice age was much lower than it is today. That is because of the enormous amount of Earth's water frozen in vast ice sheets. When the ice sheets melted at the end of the ice age, the meltwater flowed into the ocean and raised sea level.

 **Key Concept Check** How do glaciers affect sea level?

Melting Glaciers Scientists estimate that if all the glaciers on Earth melted, sea level would rise about 70 meters. Some low-lying areas, such as the Florida peninsula and a large portion of Louisiana, would be under water.

How much water is frozen in the Antarctic ice sheets? The middle image in **Figure 4** illustrates how sea level around the Florida peninsula could change if the West Antarctic ice sheet melted. The last image in **Figure 4** illustrates how sea level for Florida would change if the East Antarctic ice sheet melted.

Changing Sea Level

Figure 4 These maps show the outline of Florida's coast today. The green area in the first illustration shows how much land was above sea level during the last ice age.

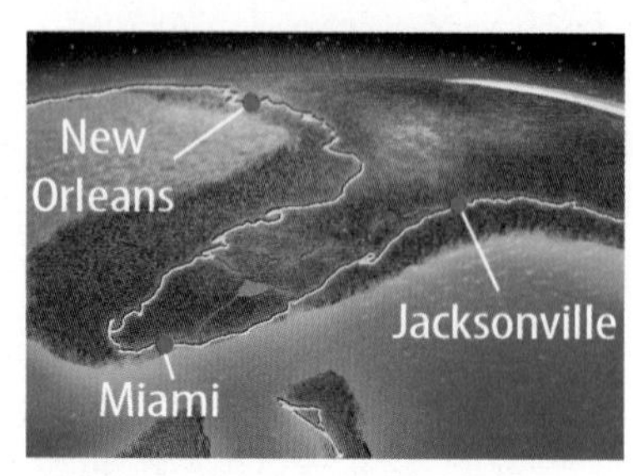

20,000 years ago at the height of the last ice age, sea level was about 120 meters lower than it is today.

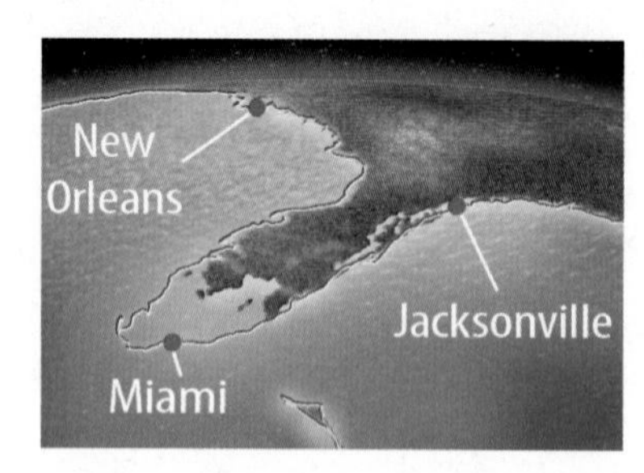

If the West Antarctic ice sheet melted, sea level would rise about 5 meters above current sea level. The southern tip of Florida would be under water.

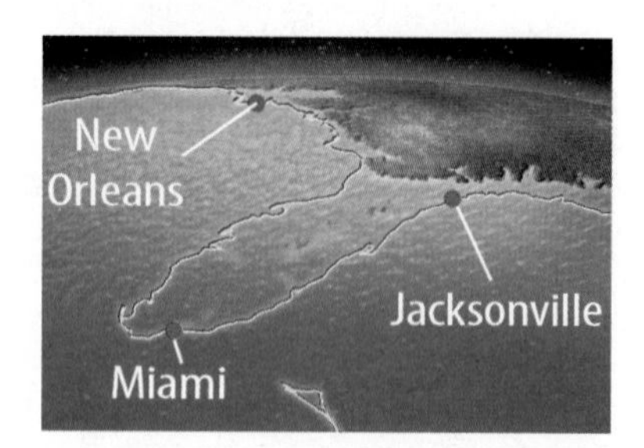

If the larger East Antarctic ice sheet melted, sea level could rise by about 52 meters. This would put most of Florida under water.

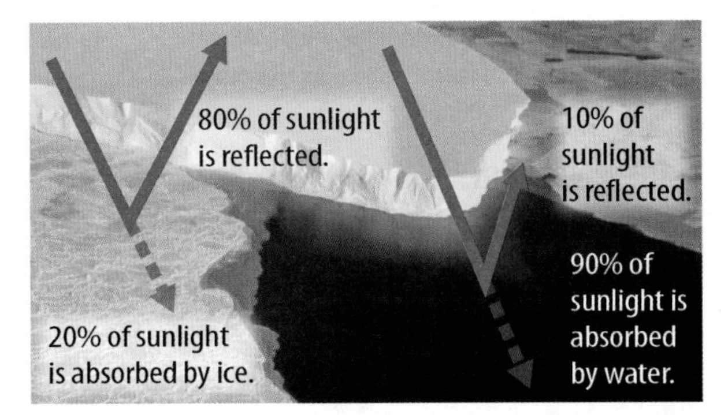

Figure 5 Up to 80 percent of the sunlight that strikes snow or sea ice is reflected into space.

Sea Ice and Snow Cover

Snow and sea ice are also frozen forms of freshwater. **Sea ice** *is ice that forms when seawater freezes.* As seawater freezes, salt is left behind in the ocean. Much of the Arctic Ocean is covered with sea ice.

Unlike glaciers, sea ice does not raise sea level by adding water to the ocean. Consider an ice cube floating in a glass. The amount of water frozen in the ice cube is equal to the amount of water that it displaces in the glass. When the ice cube melts, the water level in the glass stays the same. Likewise, when sea ice melts, sea level stays the same.

However, melting snow or sea ice can affect climate. Snow or ice reflects more solar energy than land or water does. As illustrated in **Figure 5**, most of the sunlight that hits snow or ice is reflected back into space. Reflection helps keep surface temperatures and air temperatures low.

Scientists have recorded a decreasing trend in the amount of snow cover. When snow melts, Earth's surface absorbs more solar energy and heats the air above it. When large areas of Earth's surface are affected over long periods of time, climate changes. Scientists hypothesize that this decrease in snow cover is related to an increase in global temperature.

Key Concept Check How can sea ice or snow cover affect climate?

MiniLab **10 minutes**

Does the ground's color affect temperature?

Have you ever been outside on a sunny day after a snowstorm? Did you have to squint because of the glare reflected from the snow? After the snow melts, you do not have to squint as much when you go outside. Why?

1. Read and complete a lab safety form.
2. Lay a sheet of **black paper** and a sheet of **white paper** next to each other on the lab table.
3. Lay one **thermometer** on each sheet of paper. Be sure to place the bulbs of the thermometers on top of the paper.
4. Create a data table to record the temperature of each thermometer, once per minute for 5 minutes.
5. Record the temperature for each sheet of paper.
6. Position a **desk lamp** 20 cm above the thermometers. The lamp should be equidistant from each thermometer bulb. Turn on the lamp.
7. Record the temperature of each thermometer every minute for 5 minutes.

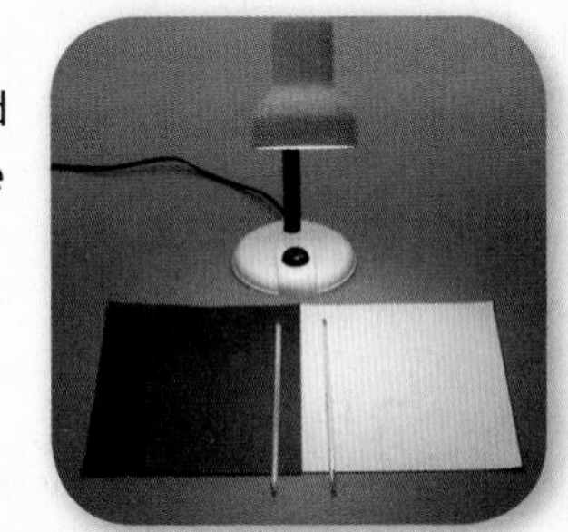

Analyze and Conclude

1. **Graph** your temperature data. Use two colored pencils to differentiate your results. Label each axis and give your graph a title.
2. **Explain** why the temperature readings differed.
3. **Key Concept** In the past, a continental ice sheet covered much of North America. Hypothesize how the melting of the ice sheet affected the temperature of North America.

Figure 6 This graph shows changes in global temperature and atmospheric CO_2 over the past 400,000 years. The steepest rise in CO_2 levels, shown by the red line, began about 150 years ago, when people first began burning fossil fuels. ►

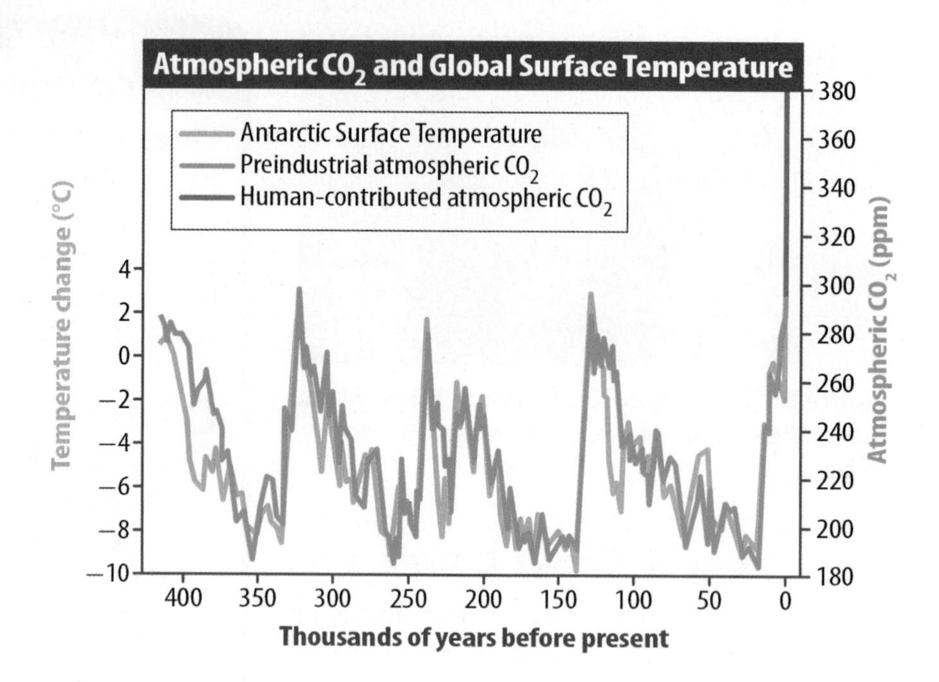

Figure 7 Bubbles of gas locked in ice cores provide evidence of the CO_2 content of the atmosphere during periods of Earth's history. Crystals in an ice core reflect colors at different angles. ▼

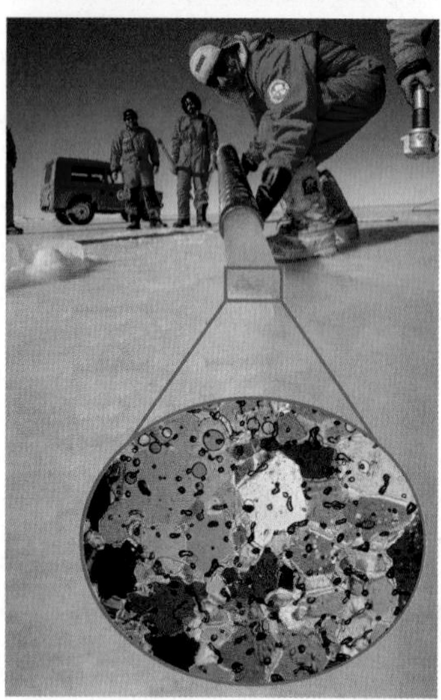

Human Impacts on Glaciers

Scientific studies indicate that Earth's glaciers are melting. Sea ice that covers the Arctic Ocean is also melting. Why? Earth is getting warmer. Data collected by scientists who study Earth's climate show that Earth's average surface temperature has risen approximately 0.5°C since the start of the twentieth century.

Evidence of Climate Change

The orange line in the graph in **Figure 6** represents Earth's average surface temperature during the past 400,000 years. Notice that Earth's temperature fluctuated during that span of time. During cold periods, glaciers formed and sea level fell. During warm periods, glaciers melted and sea level rose.

The green line in **Figure 6** represents the amount of carbon dioxide (CO_2) in Earth's atmosphere. Notice the comparison between Earth's temperature and the amount of atmospheric CO_2. As the amount CO_2 rose, so did Earth's temperature. The data represented by this graph came from **ice cores**–*long columns of ice taken from glaciers* like the one shown in **Figure 7.**

Look again at **Figure 6** and notice the sharp rise in CO_2 shown by the red line. Human activities–especially the burning of fossil fuels–add CO_2 to the atmosphere. Atmospheric CO_2 has risen sharply since the 1800s. Scientists hypothesize that this rise in CO_2 has contributed to the recent rise in global temperature. Many scientists also hypothesize that this rise in temperature is causing many of Earth's glaciers to melt.

Key Concept Check How do human activities affect glaciers?

Melting Glaciers

As Earth's average surface temperature increases, glaciers and ice sheets melt. More water flows into the oceans and sea level rises. **Figure 8** shows how much melting took place in one alpine glacier over a period of 63 years. Like melting ice sheets, melting alpine glaciers contribute to the rise in sea level.

Melting Sea Ice

The Arctic Ocean has been covered with sea ice since the beginning of the last ice age, 125,000 years ago. However, arctic sea ice is melting. **Figure 9** illustrates how much sea ice melted in the Arctic Ocean since 1979. It also shows how much arctic sea ice could be lost over the next few decades. In September 2007, sea ice at the North Pole was surrounded by ice-free water for the first time in known human history. Can the melting of snow or ice cause more sea ice to melt?

Positive Feedback Loop

A back-and-forth relationship occurs between melting snow and rising temperature–an increase in one causes an increase in the other. Scientists call this a positive feedback loop. For example, as snow or ice melts, the amount of energy absorbed from the Sun increases. As the amount of energy absorbed from the Sun increases, global temperature rises. As global temperature rises, more snow or ice melts. This repeating cycle is called a positive feedback loop–an increase in one variable causes a corresponding increase in another variable.

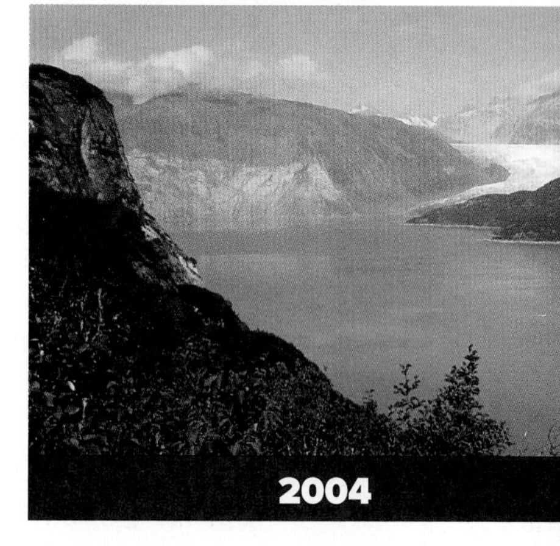

Figure 8 Much of the Muir glacier in Alaska has melted since 1941.

Visual Check What changes are visible in the 2004 photo?

Sea Ice at the North Pole

Personal Tutor

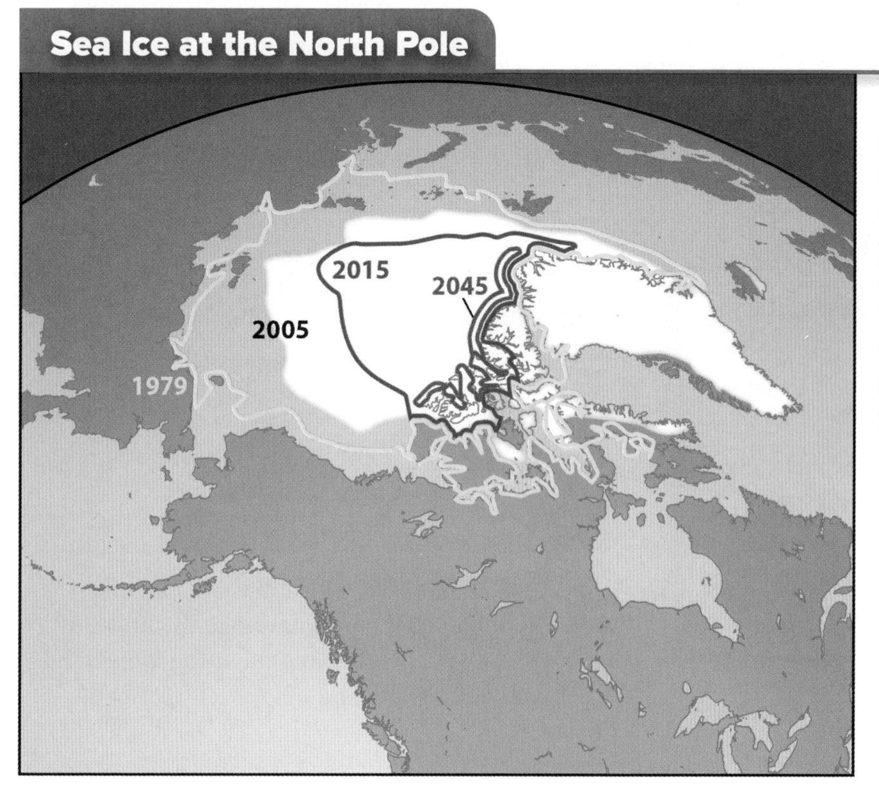

Figure 9 This computer-generated image shows how the Arctic Ocean ice cap has changed over time. The yellow outline shows the edges of the ice cap in 1979. The purple outline shows where scientists predict the edges of the ice cap will be in 2045.

Lesson 1 Review

Visual Summary

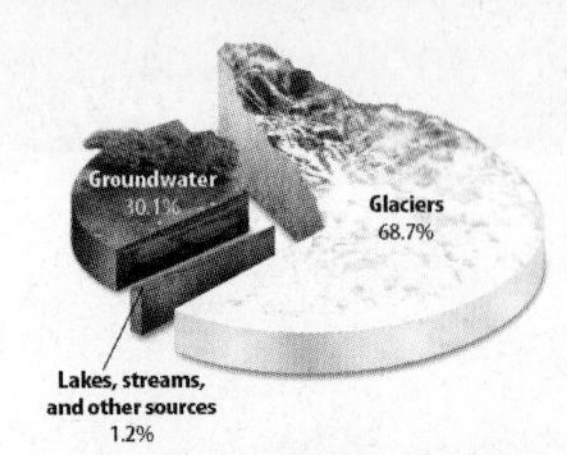

Most of Earth's freshwater is frozen in ice sheets and alpine glaciers.

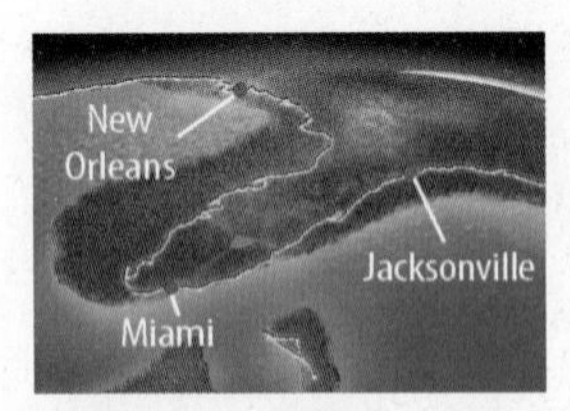

Glacier formation can cause sea level to fall. Melting of glaciers can cause sea level to rise.

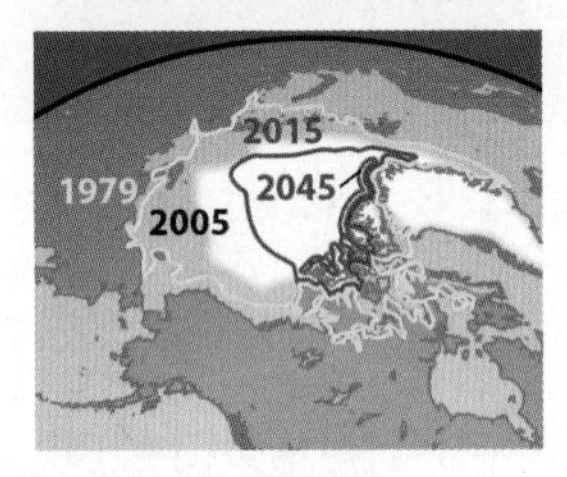

Human activities are associated with the recent melting of glacial ice and Arctic sea ice.

FOLDABLES®

Use your lesson Foldable to review the lesson. Save your Foldable for the project at the end of the chapter.

What do you think NOW?

You first read the statements below at the beginning of the chapter.

1. On Earth, freshwater occurs only as liquid water.
2. Up to 80 percent of the sunlight that strikes snow or ice is reflected back into space.

Did you change your mind about whether you agree or disagree with the statements? Rewrite any false statements to make them true.

Use Vocabulary

1. **Distinguish** between the terms *ice sheet* and *alpine glacier.*
2. **Use the term** *freshwater* in a complete sentence.
3. **Define** *glacier* in your own words.

Understand Key Concepts 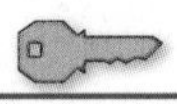

4. **Compare and contrast** ice in Antarctica and ice in the Arctic Ocean.
5. Where is most of the freshwater on Earth?
   A. glaciers
   B. groundwater
   C. lakes
   D. oceans
6. **Contrast** the sea level during the last ice age with today's sea level.

Interpret Graphics

7. **Analyze** The graph below illustrates how the CO_2 content of Earth's atmosphere has changed since 1850. How do these data relate to changes in Earth's temperature?

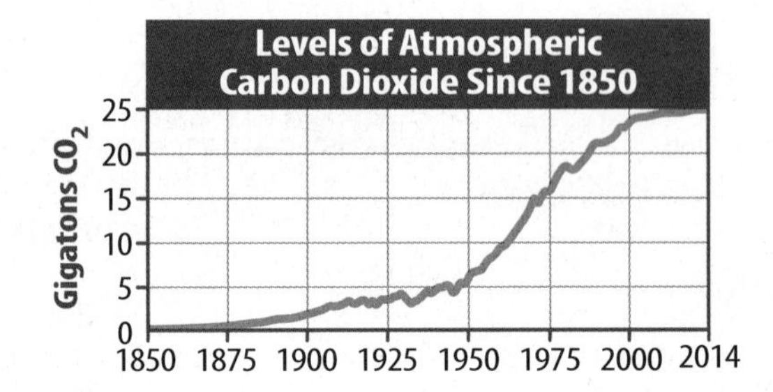

Critical Thinking

8. **Support** Use facts to support this statement: "Freshwater is not evenly distributed on Earth."

Math Skills

9. The Greenland Ice Sheet has an area of 1.8 million km^2 and an average thickness of 2.3 km. What is the volume of the Greenland Ice Sheet?

AMERICAN MUSEUM of NATURAL HISTORY

SCIENCE & SOCIETY

Life at the Top of the World

Average temperatures on Earth are increasing; sea ice is melting and disrupting entire ecosystems, which could threaten the survival of polar bears.

Life isn't easy when you're living at the top of the world in the vast, icy region known as the Arctic. It's so cold that ice covers parts of the Arctic Ocean all year long. However, a variety of species thrive in this polar climate. In fact, many ecosystems depend on the Arctic's ice for survival.

But as Earth's average temperatures increase, ice in the Arctic is melting. This includes sea ice—the ice that forms in an ocean. Sea ice follows a natural cycle in the Arctic. It spreads across the Arctic Ocean in winter then decreases in area during summer. But with rising temperatures, sea ice forms later and melts earlier each year. Over the last few decades, the amount of ice in the Arctic has decreased dramatically.

The disappearing ice is threatening the Arctic's top predator, the polar bear. Polar bears travel across sea ice to hunt for seals. As sea ice breaks up and melts, polar bears must swim longer distances to find prey. Also, late freezes and early thaws of sea ice mean shorter hunting seasons for them. Polar bears have been classified as a threatened species because their numbers are decreasing. If warming continues, they could become extinct.

The future of Arctic life is uncertain. Scientists continue to monitor climate data to understand the impact of increasing average temperatures on Arctic ecosystems. But, if Earth's climate continues to warm, life in the Arctic might never be the same.

Scientists use satellite images to monitor the amount of Arctic sea ice. These 1979 (top) and 2007 (bottom) images show that the area covered by summer sea ice is about half of what it was over 30 years ago. ▼

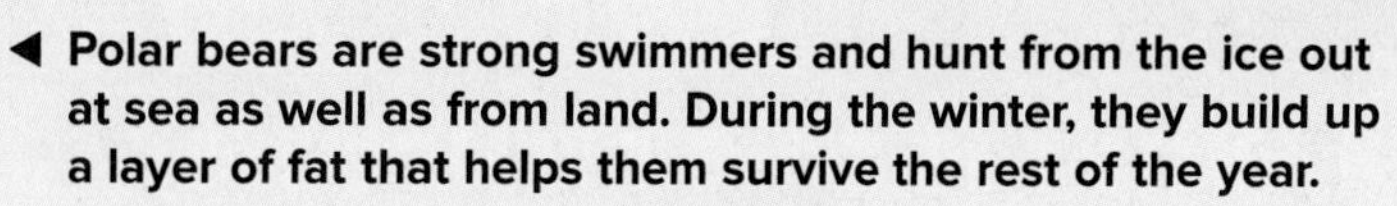

◀ Polar bears are strong swimmers and hunt from the ice out at sea as well as from land. During the winter, they build up a layer of fat that helps them survive the rest of the year.

It's Your Turn

RESEARCH Learn how the dwindling population of polar bears would affect other Arctic species. Create a cause and effect diagram with an if-then statement describing the effects of polar bear population decline on other wildlife in the Arctic.

Lesson 2

Reading Guide

Key Concepts
ESSENTIAL QUESTIONS

- What are streams and lakes?
- What is a watershed?
- How do human activities affect streams and lakes?

Vocabulary

runoff p. 617
stream p. 618
watershed p. 619
estuary p. 619
lake p. 620

Multilingual eGlossary

What's Science Got to do With It?

Streams and Lakes

Inquiry What is this structure?

The large concrete structure shown here is the Hoover Dam, in Nevada. The dam was built to control water flow along the Colorado River. Notice the large reservoir, Lake Mead, behind the dam. Freshwater from Lake Mead is used for recreational purposes, drinking water, irrigation, and hydroelectric power. Dams can also have negative effects on the environment and the ecosystem around a river.

Launch Lab

10 minutes

How can you measure the health of a stream?

The quality of the water in a stream affects the organisms that live in the stream. Macroinvertebrates are tiny animals without backbones. Their presence can be used to determine the health of a stream. For example, the riffle beetle is only in streams where dissolved oxygen is high and the stream is healthy. Use the data below to measure the health of a stream near a new housing development.

1. Read and complete a lab safety form.
2. Use **graph paper** and **colored pencils** to construct a graph using the data provided.
3. Plot the water temperature, dissolved oxygen, and the population density for each year represented.

Year	Water Temp (°C)	Dissolved Oxygen Concentration (ppm)	Riffle Beetle (adults/rock)
1998	10.4	11.5	9.8
2000	11	10.5	9.3
2002	12.7	8	7.9
2004	13.3	7.5	6.2
2006	14.1	6.5	4.4
2008	15.2	5.5	2.6

Think About This

1. What is happening to the stream?
2. Make a prediction about the number of adult riffle beetles per rock in 2020.
3. **Key Concept** Other than a decrease in oxygen, what else might affect the riffle beetle population in this stream?

Runoff

If you've ever been outside during a heavy rain, you might have noticed sheets of water rushing downhill over pavement or soil. Water can follow many different paths during a rainstorm. Some water soaks into soil. Some water collects in puddles that evaporate.

Water that flows over Earth's surface is called **runoff.** It comes from rain, melting snow or ice, or any water that does not soak into the soil or evaporate. Runoff is part of the water cycle. Gravity causes runoff to flow downhill, from higher ground to lower ground. Runoff usually starts as a thin layer, or a sheet, of water flowing over the ground, such as the runoff shown in **Figure 10.**

 **Reading Check** What is runoff?

Figure 10 Runoff often starts as sheets of water that flow downhill.

▲ **Figure 11**  Streams form when runoff erodes channels that carry water and sediment downhill.

Streams

A body of water that flows within a channel is a **stream**, as shown in **Figure 11.** Scientists use the term *stream* to refer to any naturally flowing channel of water. For example, a river is a large stream. A brook is a small stream. A creek is larger than a brook but smaller than a river.

Key Concept Check What is a stream?

All streams form from similar processes. As water flows downhill, it wears away rock and soil, forming tiny channels called rills. Every time it rains, more rock and soil is removed from a rill. Eventually a rill grows in size and forms a larger and more permanent stream channel. Small streams can combine and form a larger stream. Large streams can eventually become rivers that flow into a lake or an ocean.

Pools and Riffles

If you've ever watched a small stream, you might have noticed differences in the way the water flows. Sometimes the water appears smooth, and sometimes it's turbulent or rough, as shown in **Figure 12.** The water is slow, steady, and smooth in places where the stream channel is flat. Pools often form in depressions or low spots within a stream channel. Where the stream channel is rough or the slope is steep, the water tumbles and splashes. A riffle is a shallow part of a stream that flows over uneven ground. Riffles help mix water as it splashes and swirls over rough areas. This action increases the oxygen content of the water and makes the stream healthier.

Figure 12 Oxygen from the air mixes into water as it passes over riffles. Water from riffles helps supply oxygen to the pools downstream. ▼

Andreas Strauss/Getty Images

Figure 13 This watershed includes several streams. They flow into a river, which flows into the ocean.

Watersheds

Imagine a house with a roof that is higher in the middle than at its edges. Rain falls on both sides of the roof and runs downward. However, rain runs down the roof in opposite directions. The same thing happens when rain falls to Earth. The direction in which runoff flows depends on which side of a slope the rain falls. *A* **watershed** *is an area of land that drains runoff into a particular stream, lake, ocean, or other body of water.*

Like the example described above, the boundaries of a watershed are the highest points of land that surround it. These high points are called divides. **Figure 13** shows examples of watersheds and divides.

Key Concept Check What is a watershed?

From Headwaters to Estuaries

Small streams that form near divides are called headwaters. Streams begin at the headwaters. Streams end at the mouth of a river, where runoff drains into a lake, an ocean, or another large body of water.

What happens when a river meets the sea? Freshwater mixes with salt water. Many large watersheds end in an **estuary**–*a coastal area where freshwater from rivers and streams mixes with salt water from seas or oceans.* Estuaries contain brackish water–a mixture of freshwater and salt water. As **Figure 14** shows, the water in an estuary gets saltier as it gets closer to the ocean. Estuaries are rich in minerals and nutrients and provide important habitats for many organisms.

WORD ORIGIN

estuary
from Latin *aestuarium,* means "a tidal marsh"

Figure 14 Estuaries form in places where freshwater streams flow into an ocean, a sea, or a bay.

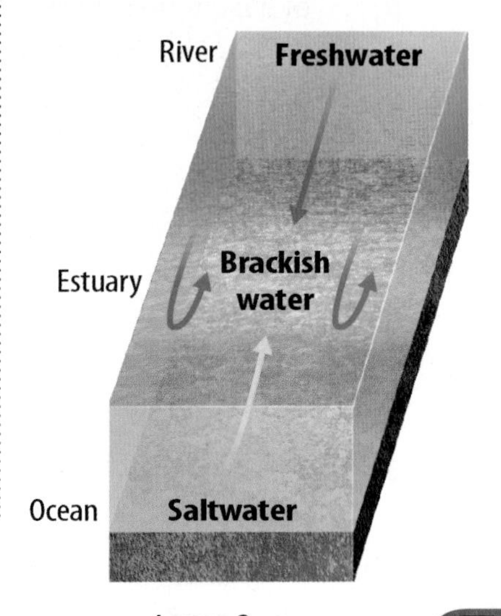

FOLDABLES®

Use a sheet of notebook paper to make a two-tab book. Label it as shown. Use it to organize information on the characteristics of streams and lakes.

SCIENCE USE V. COMMON USE

basin

Science Use a shallow depression surrounded by higher ground

Common Use a tub or container used to hold liquids such as water

Lakes

When runoff flows into a **basin,** or a depression in the landscape, a lake can form. *A* **lake** *is a large body of water that forms in a basin surrounded by land.* Most of Earth's lakes are in the Northern Hemisphere. Over 60 percent are in Canada. Lakes are reservoirs that store water. Most lakes contain freshwater.

How Lakes Form

Erosion, landslides, movements of Earth's crust, or the collapse of volcanic cones can form lake basins. Water can enter a lake basin from precipitation, streams, or groundwater that rises to the surface. Most lakes have one or more streams that remove water when the lake overflows. Lakes also lose water by evaporation or when lake water soaks into the ground.

The water level in a lake is not constant. If the lake loses water to evaporation, the lake level will drop. Occasionally a lake will disappear entirely if precipitation does not replenish water lost from the lake. In contrast, if the lake receives too much rainfall, the water can spill over the lake banks and cause a flood.

MiniLab

15 minutes

How does a thermocline affect pollution in a lake?

When the Sun warms the surface of a cold lake, a warm layer of water forms on the surface. The denser, cold, deep water remains unchanged. Water of different densities does not mix easily. The pollutants or nutrients in one layer are trapped.

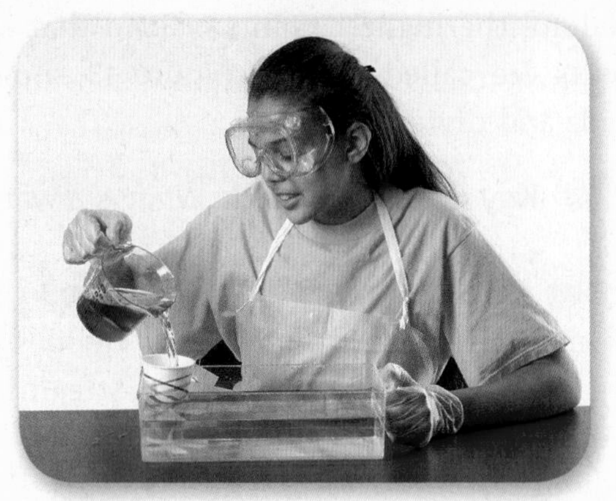

1. Read and complete a lab safety form.
2. Use a **pencil** to poke holes in the bottom of a **paper cup.** Attach the cup to the corner of a **clear plastic shoe box** with **tape.**
3. Fill the shoe box with very **warm water** until the bottom of the cup is submerged.
4. "Pollute" your lake by pouring 100–200 mL of **ice cold water** tinted with **food coloring** into the cup. Observe until water stops flowing. Sketch in your Science Journal what happens.
5. Simulate a storm by blowing across the water's surface.

Analyze and Conclude

1. **Explain** what happened to the tinted water.
2. **Describe** the effect the wind had on the tinted layer.
3. **Contrast** the way the layers formed in your model lake with the way layers form in a real lake.
4. **Key Concept** Hypothesize how pollution caused by human activity might affect different areas in a lake.

Properties and Structure

Water changes temperature more slowly than land changes temperature. This can affect weather conditions near the lake. For example, on a hot summer day you might be refreshed by a cool breeze blowing across a lake.

Have you ever been swimming in a lake and noticed that the water changes temperature with depth? Sunlight heats the surface layer, making it warmer and less dense than the layers below. Less sunlight is absorbed the deeper you swim. Some deep, northern lakes develop two **distinct** layers of water–a warm, top layer and a cold, bottom layer. The two layers are separated by a region of rapid temperature change called the thermocline. It acts as a barrier and prevents mixing between the layers.

ACADEMIC VOCABULARY

distinct
(adjective) different or not the same

Human Impact on Streams and Lakes

People worldwide depend on streams and lakes for their water supplies. Streams are dammed to create reservoirs that store water. Because of dams, some rivers, such as the Colorado River in the lesson opener, are nearly dry before they reach the ocean.

As illustrated in **Figure 16**, people can affect the health of streams and lakes in many other ways. Runoff can carry fertilizers, pesticides, sewage, and other pollutants that are harmful to organisms living in or near the water. For example, excess nutrients from fertilizers or sewage can enter a stream and result in an increase in the population of algae. When the algae die, bacteria break down the algae and use oxygen in the decay process. If decay rates are too high, oxygen levels in the water can be so low that fish and other animals cannot survive.

Key Concept Check How do human activities affect streams and lakes?

Figure 16 Pollutants that flow into freshwater can harm living organisms, including the people who use the water for drinking, washing, and irrigation.

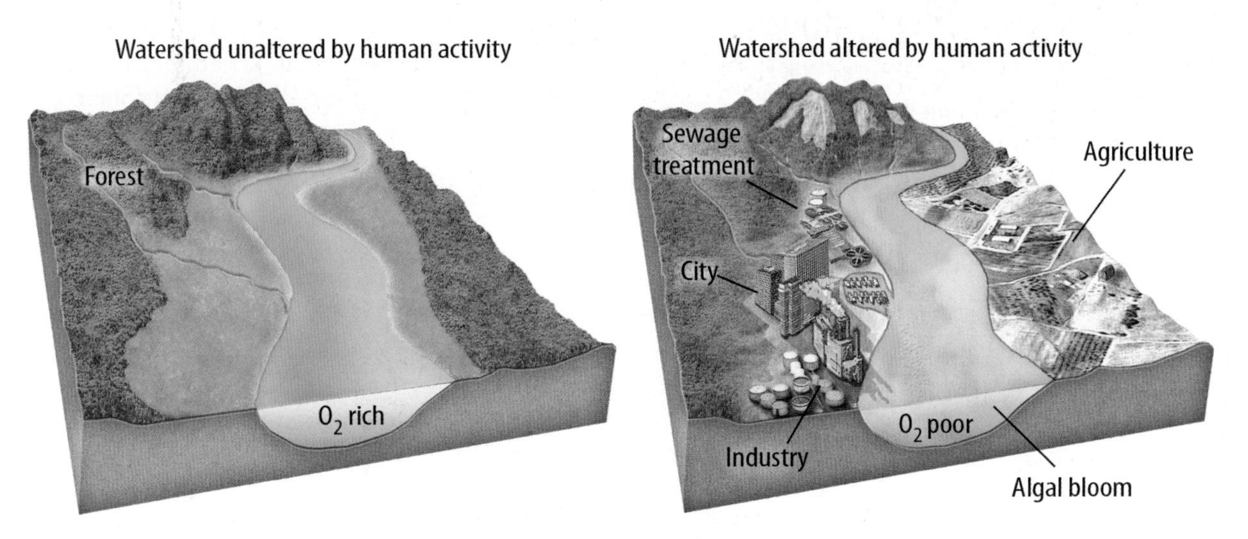

Lesson 2 Review

Visual Summary

Water that flows over Earth's surface and into streams and lakes is called runoff.

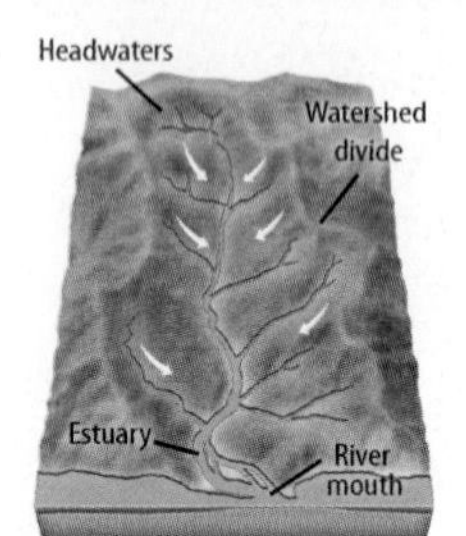

Watersheds begin at high places called divides, where headwaters flow downhill.

Humans can have a negative impact on the health of streams and lakes.

FOLDABLES®

Use your lesson Foldable to review the lesson. Save your Foldable for the project at the end of the chapter.

What do you think NOW?

You first read the statements below at the beginning of the chapter.

3. Organisms living in a lake or a stream are not affected by the amount of oxygen and nutrients in the water.

4. Glaciers can form lake basins.

Did you change your mind about whether you agree or disagree with the statements? Rewrite any false statements to make them true.

Use Vocabulary

1. **Use the term** *runoff* in a complete sentence.
2. **Define** *watershed* in your own words.
3. **Distinguish** between a *lake* and a *stream.*

Understand Key Concepts 

4. Which is the correct order of small bodies of water to large bodies of water?
   - **A.** creeks, estuaries, rivers
   - **B.** estuaries, creeks, rivers
   - **C.** rivers, estuaries, creeks
   - **D.** creeks, rivers, estuaries
5. **Describe** how lakes form.
6. **Distinguish** between runoff and streams.
7. **Explain** how divides affect water flow.

Interpret Graphics

8. **Summarize** Look at the diagram below. How is this structure formed?

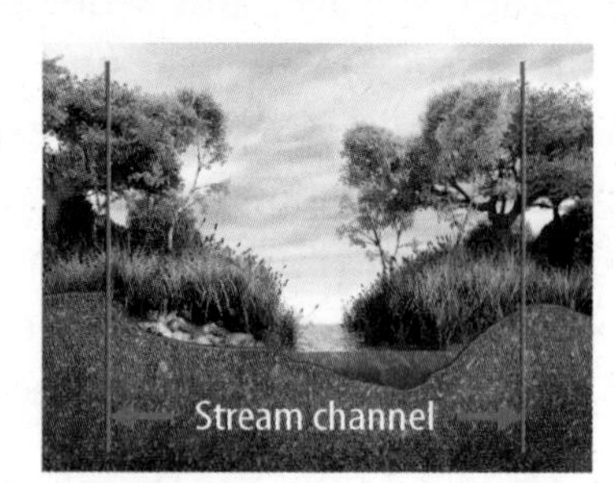

Critical Thinking

9. **Synthesize** Swimmers dip their toes into lake water, and it feels warm. When they dive in, they discover that the lake is much colder. Explain why.
10. **Evaluate** Write a paragraph describing how the destruction of a forest to make room for a factory could affect organisms in a nearby stream.

How does water flow into and out of streams?

Materials

stream table

dry sand

plastic gallon jug

plastic tub

paper towels

Safety

You have probably seen water flow along a stream or a river. What you can see is only part of the story about how water flows through a river, because a lot of the river's flow is in the ground.

Learn It

Observation is a basic science skill. Without observations, scientists would not know what questions to ask or how to approach and develop ideas about how nature behaves.

Try It

1. Read and complete a lab safety form.
2. Half-fill a stream table with sand. Tilt the table and put the lower end of the drain tube in a plastic tub to allow for drainage.
3. Shape the sand into two long mountain ranges with a valley between them. Reshape the mountains as needed.
4. Poke pin holes in the bottom of a plastic gallon jug and fill it with water. Begin to "rain" on the sand at the top of the stream table. Measure the time it takes for the water to start flowing into the plastic tub as you provide constant rain. Continue until all the sand is wet and the water is flowing steadily.
5. Once all the sand is wet, stop the rain and time how long it takes for the water to stop flowing.

Apply It

6. When you first started the rain on the stream table, where did all of the water go?
7. What had to happen for the water to begin to flow?
8. Once you stopped the rain, why did the water keep draining from the stream table?
9. **Key Concept** Where does the water that flows in a stream come from?

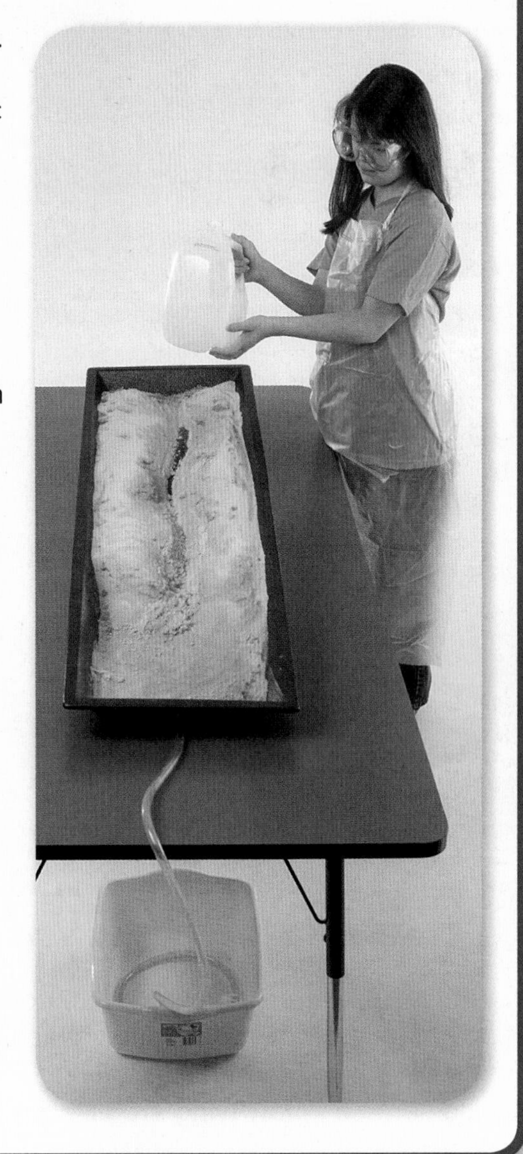

(t to b, 3-5, r)Hutchings Photography/Digital Light Source; (2)Ken Cavanagh/McGraw-Hill Education

Lesson 3

Groundwater and Wetlands

Reading Guide

Key Concepts
ESSENTIAL QUESTIONS

- What is groundwater?
- Why are wetlands important?
- How do human activities affect groundwater and wetlands?

Vocabulary
groundwater p. 625
water table p. 626
porosity p. 626
permeability p. 626
aquifer p. 627
wetland p. 628

Multilingual eGlossary

BrainPOP®

Go to the resource tab in ConnectED to find the PBL *7 Billion and Counting.*

Inquiry Where did this water come from?

Why is this water bubbling up out of the ground? Have you ever seen anything like it? Groundwater, which is stored in rocks below the surface, can flood the landscape after a severe storm when the ground is saturated. It can also surface in low-lying areas.

Launch Lab

10 minutes

How solid is Earth's surface?

It feels solid, but just how solid is "solid ground"? Believe it or not, the soil and rock beneath your feet are not entirely solid.

1. Read and complete a lab safety form.
2. Fill a large, empty **jar** with **golf balls.**
3. Follow the instructions given by your teacher.

Think About This

1. At what point did you think the jar was full?
2. Do you think the particle size of the soil affects how quickly water can move through it?
3. **Key Concept** Does water flow more easily through sediments of equal size or sediments with a variety of different sizes?

Groundwater

Some water that falls to Earth as precipitation soaks into the ground. *Generally, water that lies below ground is called* **groundwater.** Water seeps through soil and into tiny pores, or spaces, between sediment and rock. If you have ever been inside a cave and seen water dripping down the sides, you've seen groundwater seeping through rock.

 **Key Concept Check** What is groundwater?

In some areas, groundwater is very close to the surface and keeps the soil wet. In other areas, especially deserts and other dry climates, groundwater is hundreds of meters below the surface.

Groundwater can remain underground for long periods of time–thousands or millions of years. Eventually, it returns to the surface and reenters the water cycle. Humans interfere with this process, however, when they drill wells into the ground to remove water for everyday use.

Importance of Groundwater

The water beneath Earth's surface is much more plentiful than the freshwater in lakes and streams. Recall that groundwater is about one-third of Earth's freshwater. Groundwater is an important source of water for many streams, lakes, and wetlands. Some plant species absorb groundwater through long roots that grow deep underground.

People in many areas of the world rely on groundwater for their water supply. In the United States, about 20 percent of the water people use daily comes from groundwater.

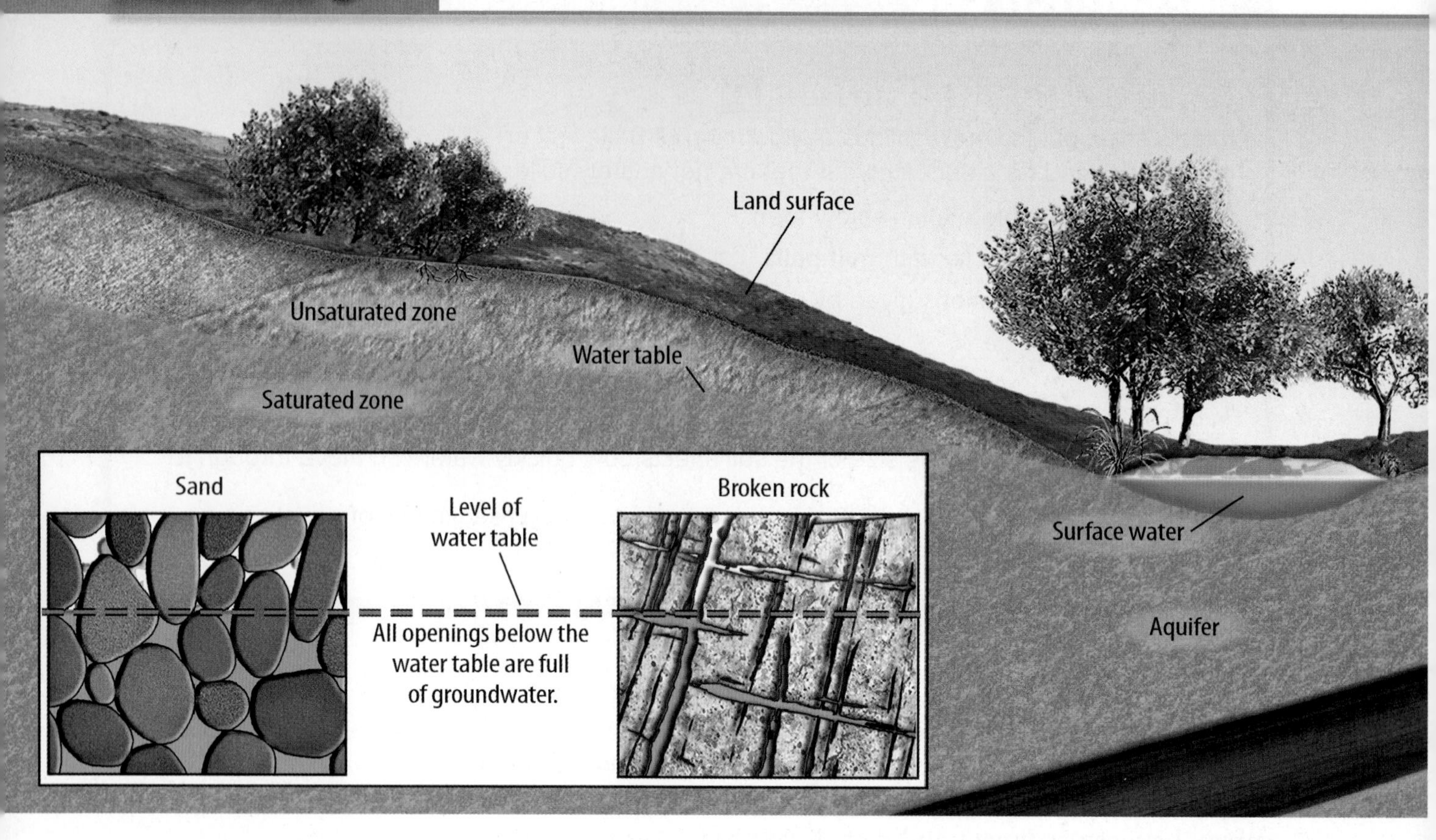

The Water Table

As illustrated in Figure 17, groundwater seeps into tiny cracks and pores within rocks and sediment. Near Earth's surface, the pores contain a mixture of air and water. This region is called the unsaturated zone. It is called unsaturated because the pores are not completely filled with water. Farther beneath the surface, the pores are completely filled with water. This region is called the saturated zone. *The upper limit of the saturated zone is called the* **water table.**

 **Reading Check** What is the water table?

Porosity Rocks vary in the amount of water they can hold and the speed with which water flows through the rock. Some rocks can hold a lot of water and some rocks cannot. **Porosity** *is a measure of rock's ability to hold water.* Porosity increases with the number of pores in the rock. The higher the porosity, the more water a rock can contain.

Permeability *The measure of water's ability to flow through rock and sediment is called* **permeability.** This ability to flow through rock and sediment depends on pore size and the connections between the pores. Even if pore space is abundant in a rock, the pores must form connected pathways for water to flow easily through the rock.

Groundwater Flow

Just as runoff flows downhill across Earth's surface, groundwater flows downhill beneath Earth's surface. Groundwater flows from higher elevations to lower elevations. In low-lying areas at Earth's surface, groundwater might eventually seep out of the ground and into a stream, a lake, or a wetland, as also shown in Figure 17. In this way, groundwater can become surface water. Likewise, surface water can seep into the ground and become groundwater. This is how groundwater is replenished.

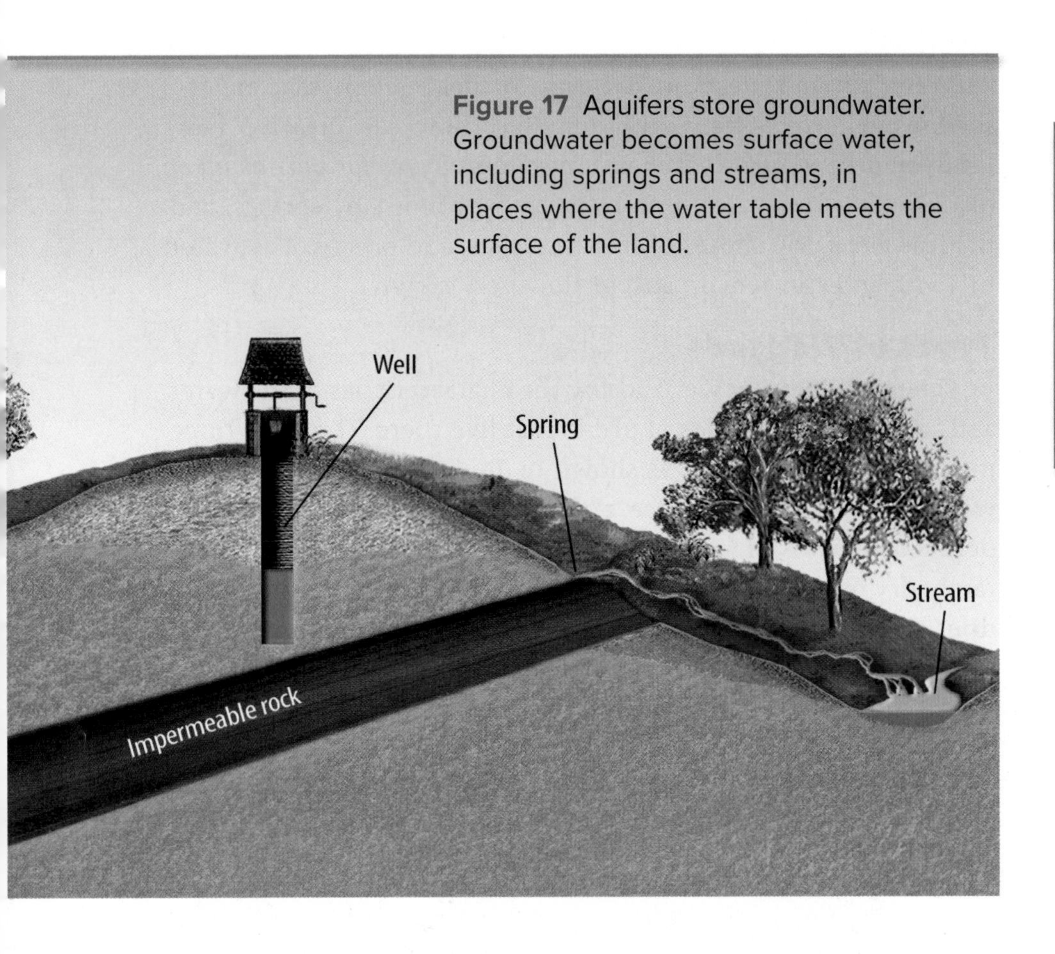

Figure 17 Aquifers store groundwater. Groundwater becomes surface water, including springs and streams, in places where the water table meets the surface of the land.

Use a sheet of notebook paper to make a two-tab book. Use it to organize your notes on groundwater, wetlands, and how each relates to the other.

Groundwater | Wetlands

Wells People often bring groundwater to Earth's surface by drilling wells like the one shown in **Figure 17.** Wells are usually drilled into an **aquifer**–*an area of permeable sediment or rock that holds significant amounts of water.* Groundwater then flows into the well from the aquifer and is pumped to the surface.

Precipitation helps replace groundwater drawn out of wells. During a drought, less groundwater is replaced, so the water level in a well drops. The same thing happens if water is removed from a well faster than it is replaced. If the water level drops too low, a well runs dry.

Springs A spring forms where the water table rises to Earth's surface, as shown in the lesson opener. Some springs bubble to Earth's surface only after heavy rain or snowmelt. Many springs fed by large aquifers flow continuously.

Human Impact on Groundwater

If polluted surface water seeps into the ground, it can pollute the groundwater below it. Pollutants include pesticides, fertilizers, sewage, industrial waste, and salt used to melt ice on highways. Pollutants can travel through the ground and into aquifers that supply wells. People's health can be harmed if they drink contaminated water from a well.

The water in an aquifer helps to support the rocks and soil above it. In some parts of the world, water is being removed from aquifers faster than it can be replaced. This creates empty space underground. The empty space underground cannot support the weight of the overlying rock and soil. Sinkholes form where the ground collapses due to lack of sufficient support from below.

Key Concept Check How do human activities affect groundwater?

Wetlands

Water often collects in flat areas or depressions that are too shallow to form lakes. Conditions like these can create a **wetland**–*an area of land that is saturated with water for part or all of the year.* Wetlands also form in areas kept moist by springs, and in areas along the shores of streams, lakes, and oceans. The water in a wetland can remain still or flow very slowly.

Types of Wetlands

Scientists identify wetlands by the characteristics of the water and soil and by the kinds of plants that live there. There are three major types of wetlands, as shown in **Table 1.** Bogs form in cool, wet climates. They produce a thick layer of peat–the partially decayed remains of **sphagnum** moss. Peat holds water, so bogs rarely dry out. Unlike bogs, marshes and swamps form in warmer, drier climates and do not produce peat. Marshes and swamps are supplied by precipitation and runoff. They can temporarily dry out in hot, dry weather.

WORD ORIGIN

sphagnum
from Greek sphagnos, means "a spiny shrub"

Table 1 Types of Wetlands

Wetland	Photo
Bogs • supplied by runoff, low oxygen content • soil acidic, nutrient-poor • dominant plants—*Sphagnum* moss, wildflowers, cranberries	
Marshes • supplied by runoff and precipitation • soil slightly acidic, nutrient-rich • dominant plants—grasses and shrubs	
Swamps • supplied by runoff and precipitation • soil slightly acidic, nutrient-rich • dominant plants—trees and shrubs	

(t)Custom Life Science Images/Alamy; (c)©Morey Milbradt/Brand X Pictures/PunchStock; (b)Tom Uhlman/Alamy

Figure 18 Wetlands are an important habitat for wildlife, providing water, food, and shelter.

Importance of Wetlands

Wetlands provide important habitat for plants and wildlife. They help control flooding and erosion and also help filter sediments and pollutants from water.

Habitat A wide variety of plants and animals live in wetlands, as shown in Figure 18. Wetlands provide plentiful food and shelter for young and newly hatched animals, including fish, amphibians, and birds. Wetlands are also important rest stops and food sources for migrating animals, especially birds.

Flood Control Wetlands help reduce flooding because they store large quantities of water. They fill with water during the wet season and release the water slowly during times of drought.

Erosion Control Coastal wetlands help prevent beach erosion. Wetlands can reduce the energy of wave action and storm surges–water pushed onto the shore by strong winds produced by severe storms.

Filtration Wetlands help keep sediments and pollutants from reaching streams, lakes, groundwater, or the ocean. They are natural filtration systems. Runoff that enters a wetland often contains excess nitrogen from fertilizers or animal waste. Plants and the bacteria in wetland soils absorb excess nitrogen. Wetland plants and soils also trap sediments and help remove toxic metals and other pollutants from the water.

 **Key Concept Check** Why are wetlands important?

MiniLab

20 minutes

Can you model freshwater environments?

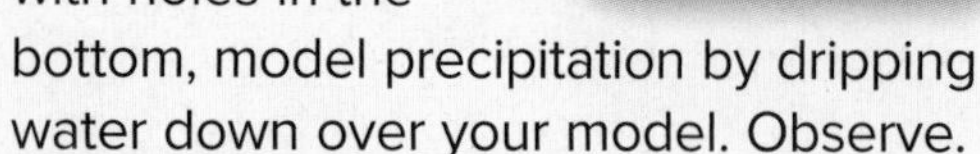

Sources of groundwater and surface water are similar. When the ground becomes saturated with water following a storm, the excess water can flow into lakes or streams. Groundwater may also surface and create wetlands.

1. Read and complete a lab safety form.
2. Half fill a **plastic shoebox** with **sand**. Flatten out the surface of the sand.
3. While watching the side of the box as you pour, add enough water so the water level, or water table, is even with the top of the sand.
4. Scoop some sand from one end of the box and pile the excess sand on the other end of the box to form highlands. Create an area between the scooped hole and the pile where the level of the sand is at the water table.
5. View the plastic shoebox from the side. Draw a cross section of your model in your Science Journal. Include the water table on your cross section.
6. Add a few drops of **food coloring** to the highlands to represent pollution. Using a **paper cup** with holes in the bottom, model precipitation by dripping water down over your model. Observe.

Analyze and Conclude

1. **Label** the lake, wetlands, highlands and water table on your diagram and identify where the water table is above, at the same level as, and below the surface.
2. Describe the path the "pollution" follows.
3. **Key Concept** How does this lab demonstrate the behavior of groundwater?

Human Impact on Wetlands

Many wetlands throughout the world have been drained and filled with soil for roads, buildings, airports, and housing developments, as shown in **Figure 19.** The disappearance of wetlands has also been associated with rising sea level, coastal erosion, and the introduction of species that are not naturally found in wetlands. Scientists estimate that more than half of all wetlands in the United States have been destroyed over the past 300 years.

Pollution from fertilizers, landfills, and agricultural fields can also be very hazardous to wetland environments. These pollutants can be devastating to many plant and animal species. Wetlands are a valuable resource and need to be protected from destruction.

Key Concept Check How do human activities affect wetlands?

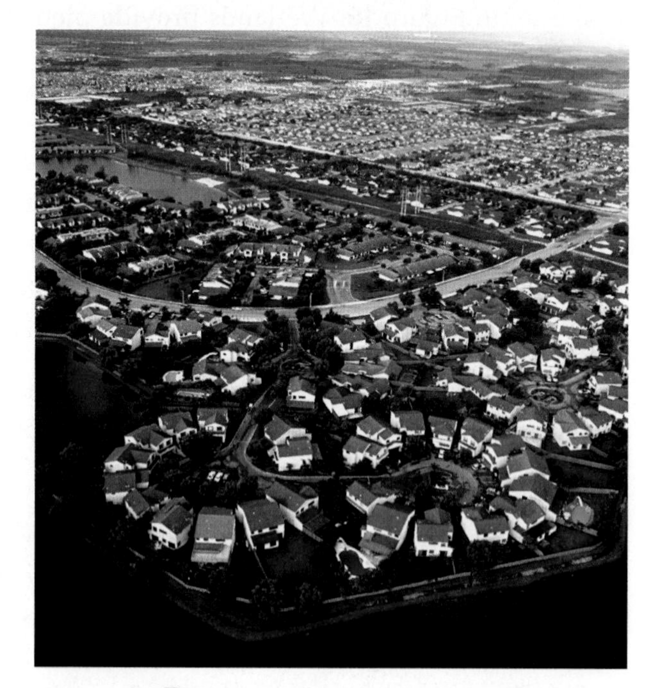

Figure 19 Large portions of coastal wetlands have been filled in to build roads and housing developments. Scientists and engineers have proposed that future flooding could be avoided by restoring some wetland areas.

Lesson 3 Review

Visual Summary

Groundwater fills pores in soil and rocks in the saturated zone. The water table marks the top of the saturated zone.

A wetland is an area of land that is saturated with water for part or all of the year.

Human involvement in alteration and destruction of wetlands can have devastating effects.

FOLDABLES®

Use your lesson Foldable to review the lesson. Save your Foldable for the project at the end of the chapter.

What do you think NOW?

You first read the statements below at the beginning of the chapter.

5. People use groundwater as a source of water.

6. Wetlands can naturally filter pollutants from groundwater.

Did you change your mind about whether you agree or disagree with the statements? Rewrite any false statements to make them true.

Use Vocabulary

1. **Distinguish** between *porosity* and *permeability*.
2. **Use the term** *groundwater* in a complete sentence.
3. **Define** *water table* in your own words.

Understand Key Concepts 

4. Where is the saturated zone located?
   - **A.** above the water table
   - **B.** below the water table
   - **C.** beside the water table
   - **D.** within the water table
5. **Explain** how groundwater can become surface water.
6. **Explain** how groundwater that is used as drinking water reaches homes.
7. **List** three reasons wetlands are important to people and the environment.

Interpret Graphics

8. **Organize Information** Copy and fill in the graphic organizer below to describe the different types of wetlands.

Wetland Type	Description
Marsh	
Swamp	
Bog	

Critical Thinking

9. **Design an experiment** to test whether groundwater from a well has been contaminated by wastewater from a nearby sewage treatment plant.
10. **Explain** A local developer wants to drain a wetland in your city and build a shopping mall in its place. Write a letter to the editor of your local newspaper to explain the possible outcome of the development.

Lab

40 minutes

What can be done about pollution?

Freshwater is everywhere–running along Earth's surface in rivers, pooling in lakes, and flowing through rocks underground. When pollution from human activity enters the freshwater supply, it quickly spreads through lakes, rivers, and groundwater.

Materials

stream table

tub

sand

food coloring

pencil, round

paper towels

gallon jug

Also needed: Assorted items for reducing pollution, such as plastic wrap, plastic spoons, straws, gravel, paper towels, cat litter pellets, activated charcoal

Safety

Question

How does pollution from human activity affect Earth's freshwater?

Procedure

1. Read and complete a lab safety form.
2. Half-fill your stream table with sand. Tilt the table and put the drain tube in the tub. Keep the drain clear during lab.
3. Shape the sand into two long mountain ranges with a valley between them. Reshape mountains as needed.
4. Poke pin holes in the bottom of a plastic gallon jug. Fill the jug with water and provide rain until all the sand is wet and the river is flowing continuously.
5. Select three locations on the side of a mountain. Put 10 drops of food coloring around each site to represent a pollution source.

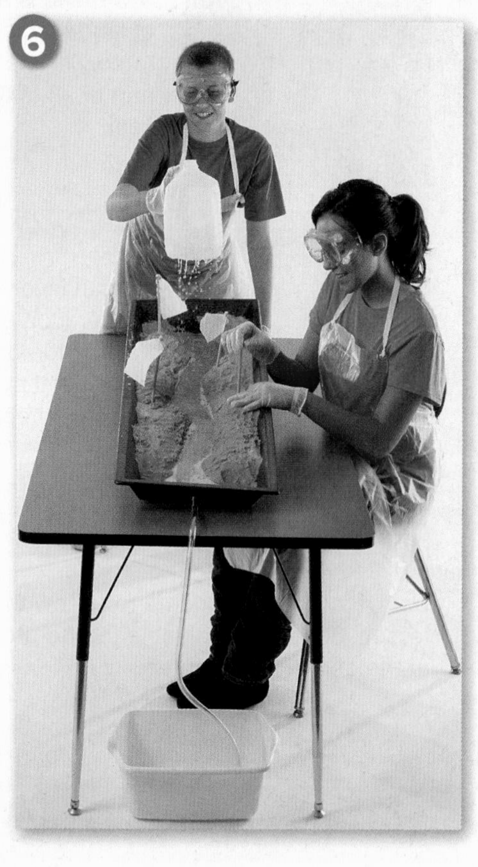

6. Resume rain and observe how the pollution spreads. Use a pencil to drill test wells around a site. Lower a strip of paper towel into the well and wet it in the groundwater to look for pollution. Record your observations in your Science Journal.
7. Once your initial pollution has washed away, add three more polluted locations. Record how long it takes for the pollution to appear and then for the river to run clear again.

(t to b, 2, 4-7, r)Hutchings Photography/Digital Light Source; (3)Ken Cavanagh/McGraw-Hill Education

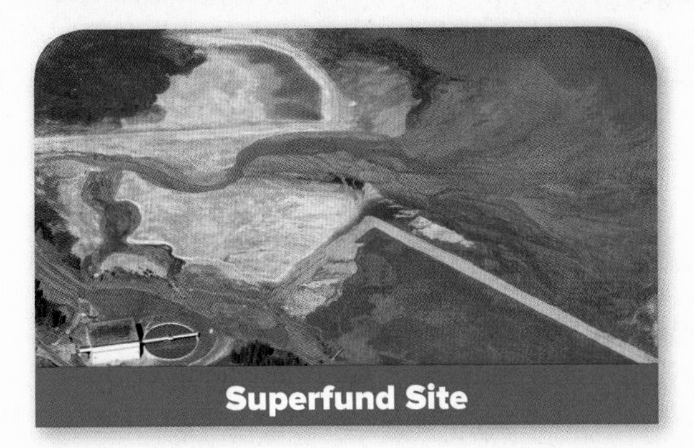
Superfund Site

8. Develop a plan for reducing the effects of the pollution. Present your plan to your teacher.
9. Once your plan is approved, introduce pollution in three locations. Help the pollution sink in with a little rain.
10. Implement your plan to reduce the pollution at all three locations.
11. Resume rain. Observe the spread of the pollution, the time it takes the pollution to reach the river, and the time it takes for the river to run clear.

Analyze and Conclude

12. **The Big Idea** On your stream table, identify some of the different places where Earth's freshwater occurs, including, but not limited to lakes, groundwater, rivers, and wetlands.
13. **Describe** How did the pollution spread through the river system?
14. **Compare and Contrast** How did the pollution control sites compare with the uncontrolled sites? Did your plan work? Why or why not?

Communicate Your Results

Create a graph comparing the duration of the pollution for the run with no pollution control measures to the run with pollution control measures. Present your results.

Find out what a superfund site is and research local sites that have this designation. What are local authorities doing to help clean up the superfund site?

Lab Tips

- ☑ To help keep drain clear, keep the sand several inches back from the drain.
- ☑ Use different colors of food coloring to tell the difference between different pollution sites.

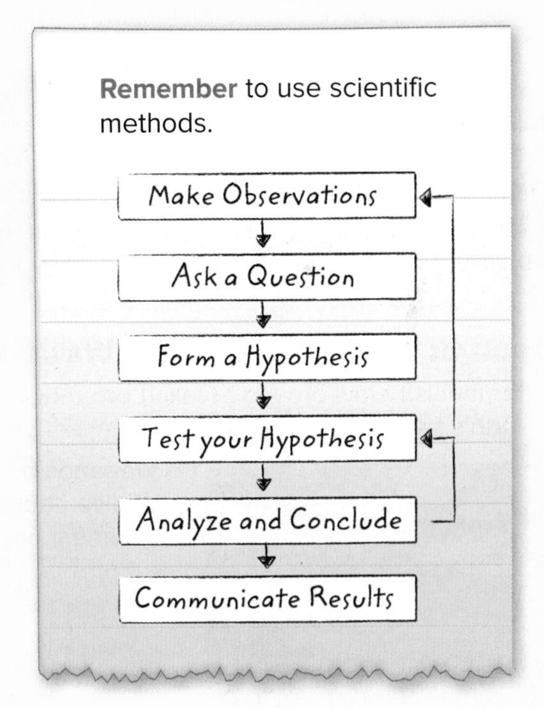

Chapter 17 Study Guide

Earth's freshwater is stored in glaciers, groundwater, lakes, and streams.

Key Concepts Summary

Lesson 1: Glaciers and Polar Ice Sheets

- **Freshwater** contains less than 0.2 percent salt. More than two-thirds of Earth's freshwater is frozen in ice.
- Snow and ice reflect sunlight and help keep Earth's surface temperatures and air temperatures low.
- Increasing amounts of carbon dioxide in the atmosphere raise temperatures and contribute to the melting of ice and snow.

Vocabulary

freshwater p. 607
alpine glacier p. 608
ice sheet p. 609
sea ice p. 611
ice core p. 612

Lesson 2: Streams and Lakes

- A **stream** is a body of water that flows within a channel. A **lake** is a large body of water that forms in a basin or a shallow depression that is surrounded by land. Streams and lakes make up less than 1 percent of Earth's freshwater.
- A **watershed** is an area of land that drains **runoff** into a stream, a lake, an ocean, or another body of water.
- Runoff can carry fertilizers, sewage, pesticides, and other harmful materials into streams and lakes.

runoff p. 617
stream p. 618
watershed p. 619
estuary p. 619
lake p. 620

Lesson 3: Groundwater and Wetlands

- Water that is below ground is called **groundwater.** Almost one-third of Earth's freshwater is groundwater.
- **Wetlands** provide valuable wildlife habitat, help filter sediment and pollutants from runoff, and help control flooding and erosion.
- Pollution from human activities can seep into the ground and contaminate groundwater. Many wetlands have been drained of water and filled with soil to make dry land for roads, buildings, and other uses. Wetlands that are destroyed can no longer filter the runoff that seeps into groundwater.

groundwater p. 625
water table p. 626
porosity p. 626
permeability p. 626
aquifer p. 627
wetland p. 628

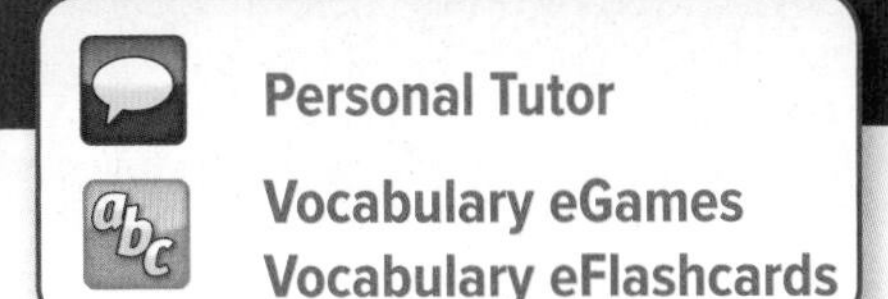

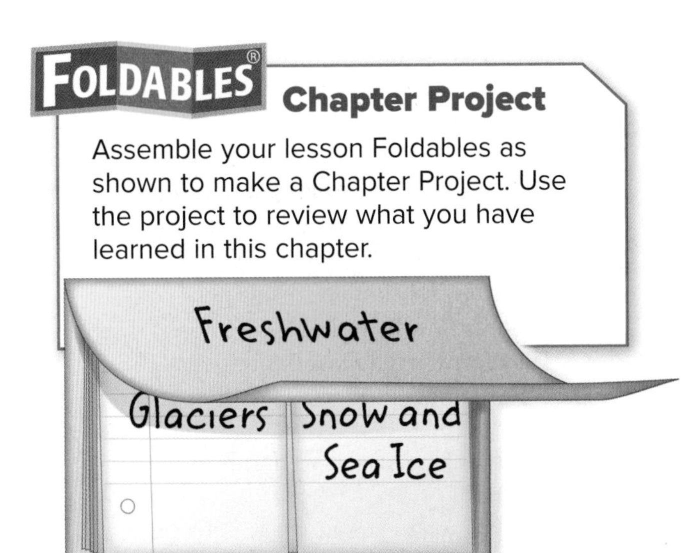

Use Vocabulary

1. Only about 3 percent of Earth's water is ________.
2. Ice formed in the ocean is called ________.
3. A(n) ________ is a large mass of ice that moves slowly over land.
4. Water that flows across Earth's surface is called ________.
5. A(n) ________ is an area of land where all of the water both above and below the ground drains to the same place.
6. A(n) ________ is a body of water surrounded by land and typically contains freshwater.
7. ________ is a measure of a rock's ability to hold water.
8. ________ is the ability of a rock to allow water to pass through it.

Link Vocabulary and Key Concepts

Interactive Concept Map

Copy this concept map, and then use vocabulary terms from the previous page to complete the concept map.

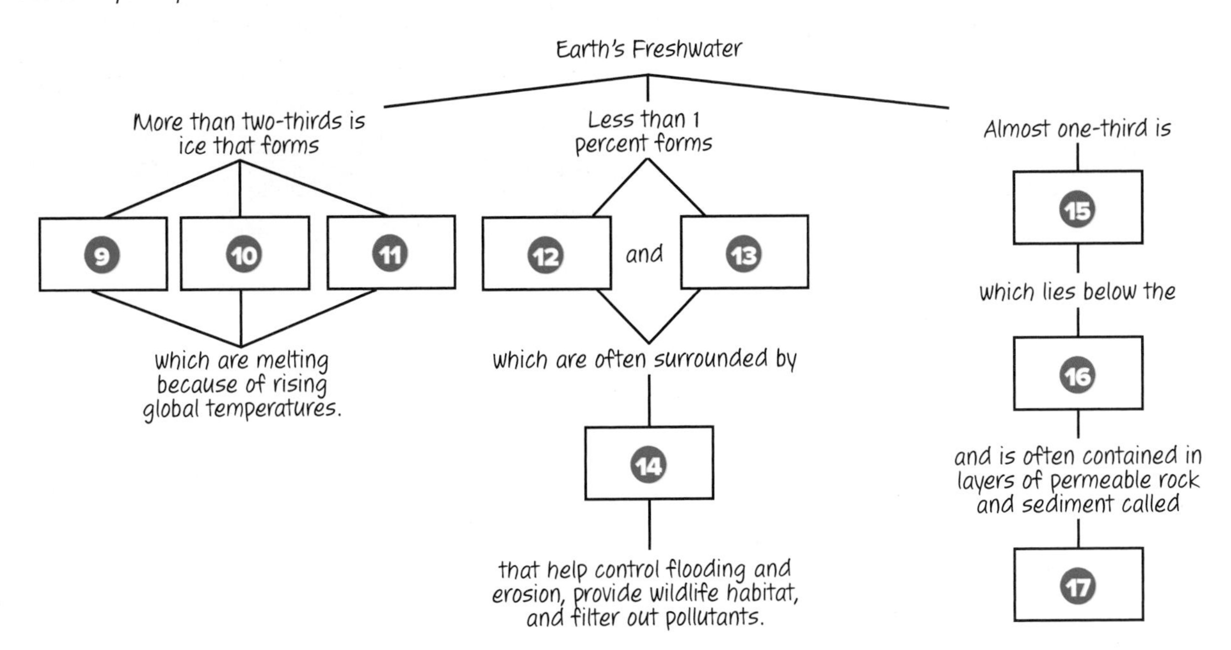

Chapter 17 Review

Understand Key Concepts

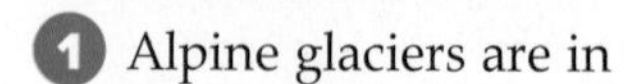

1. Alpine glaciers are in
   - A. ice sheets.
   - B. ice shelves.
   - C. mountain valleys.
   - D. oceans.

2. What happens when glaciers melt?
   - A. More heat energy from the Sun is reflected.
   - B. Global temperature decreases.
   - C. Sea level rises.
   - D. The amount of sea ice increases.

3. Where in the world is the largest area covered by ice sheets?
   - A. Antarctica
   - B. Asia
   - C. Canada
   - D. Greenland

Use the following graph to answer the question below.

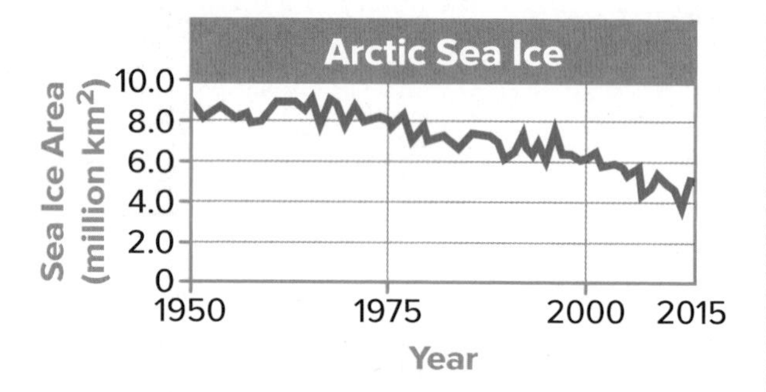

4. How did the area of sea ice in the Arctic change from 1950 to 2000?
   - A. approached zero
   - B. decreased
   - C. increased
   - D. stayed the same

5. What causes water to move from areas of higher elevation to areas of lower elevation?
   - A. erosion
   - B. thermocline
   - C. gravity
   - D. mixing

6. Which statement about watersheds is true?
   - A. All watersheds are the same size.
   - B. Larger watersheds can be broken down into smaller watersheds.
   - C. Watersheds do not go across state borders.
   - D. Watersheds consist only of surface water.

7. Which statement about lakes is NOT true?
   - A. Lakes always contain freshwater.
   - B. Streams flow in and out.
   - C. They form in a hollow basin.
   - D. Water temperature varies with depth.

8. Which is NOT a reason wetlands are important?
   - A. affect climate
   - B. control erosion
   - C. filter out pollutants
   - D. provide wildlife habitat

9. Which type of wetland is shown in the photo below?
   - A. bog
   - B. fen
   - C. marsh
   - D. swamp

10. How does a well run dry?
    - A. Groundwater is removed more slowly than it is replaced.
    - B. Groundwater is removed more quickly than it is replaced.
    - C. Groundwater gets trapped in an aquifer.
    - D. Groundwater becomes contaminated with pollutants.

Tom Uhlman/Alamy

Online Test Practice

Critical Thinking

11. **Explain** how sea ice forms.
12. **Identify** and describe two types of glaciers. Refer to the photos below as examples.

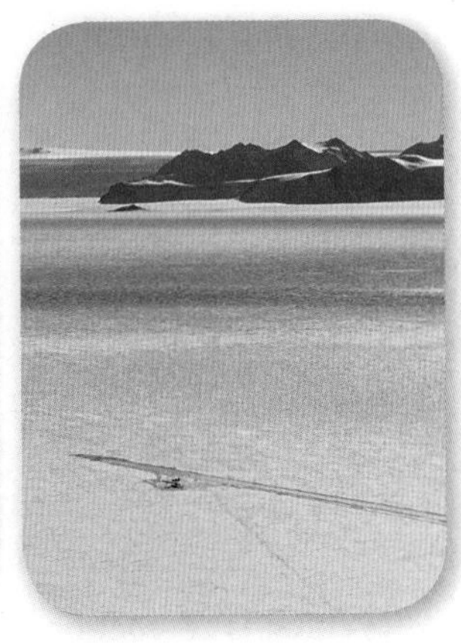

13. **Summarize** how glaciers and sea ice are changing because of human activities.
14. **Effect** How does the melting of alpine glaciers affect humans?
15. **Evaluate** How are scientists using data on changes to snow and ice cover as evidence of global climate change?
16. **Illustrate** the formation of a stream.
17. **Design a model** that shows how a lake can form from the movement of tectonic plates.
18. **Infer** why wetlands are good habitats for newly hatched and juvenile animals, such as fish and amphibians.
19. **Explain** how humans affect wetlands.
20. **Compare and contrast** an aquifer and a spring.
21. **Design a model** that would help you explain the difference between porosity and permeability to a group of elementary school students.

Writing in Science

22. **Write** a five-sentence paragraph describing changes in the amount of sea ice in the Arctic Ocean since 1979. Be sure to include a topic sentence in your paragraph.

REVIEW

23. Where is Earth's freshwater? Describe its general form, such as ice or liquid water. Also describe its general location, such as above ground, below ground, etc. Approximately what percentage of Earth's freshwater is at these locations?
24. What type of wetland appears in the photo below?

Math Skills

Math Practice

25. In 2008, a chunk of ice with an area of 570 km^2 broke off an ice shelf in Antarctica. Assuming the ice had an average thickness of 2.4 km, what volume of ice did this ice chunk contain?
26. Water expands when it freezes. If a cubic kilometer of ice produces .91 km^3 of water when it melts, what volume of water would be produced if the ice chunk melted completely?
27. The volume of Lake Erie is 484 km^3. What does this suggest about the effect the melting ice shelf would have on the ocean?
28. There are 63 glaciers in the Wind River Range of Wyoming with a total area of 44.5 km^2. The average thickness of the glaciers is 52.0 meters. What is the average thickness of the glaciers in km? (1 km = 1000 m)

Standardized Test Practice

Record your answers on the answer sheet provided by your teacher or on a sheet of paper.

Multiple Choice

1 Which statement is true of groundwater?

A It creates pores between rock and sediment.

B It eventually returns to the surface.

C It flows quickly uphill in porous soil.

D It remains underground very briefly.

Use the diagram below to answer question 2.

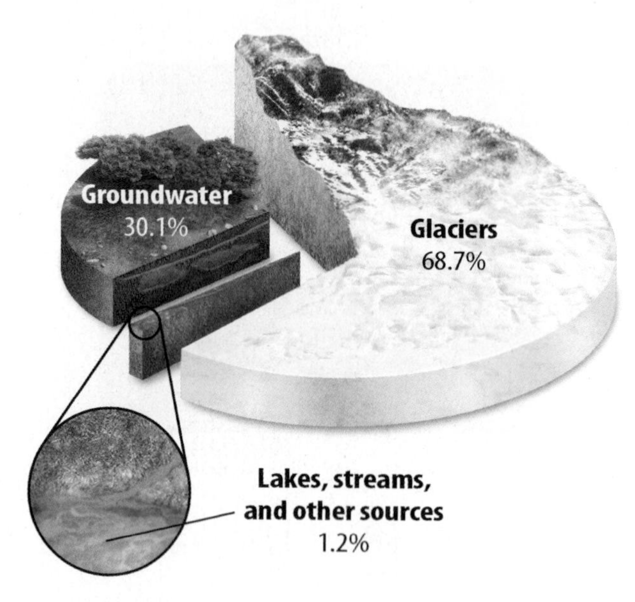

2 What percentage of freshwater comes from glaciers?

A 1.2 percent

B 30.1 percent

C 68.7 percent

D 100 percent

3 Which results from crop fertilizer runoff into streams?

A algal bloom

B crop failure

C ozone destruction

D weed growth

Use the diagram below to answer question 4.

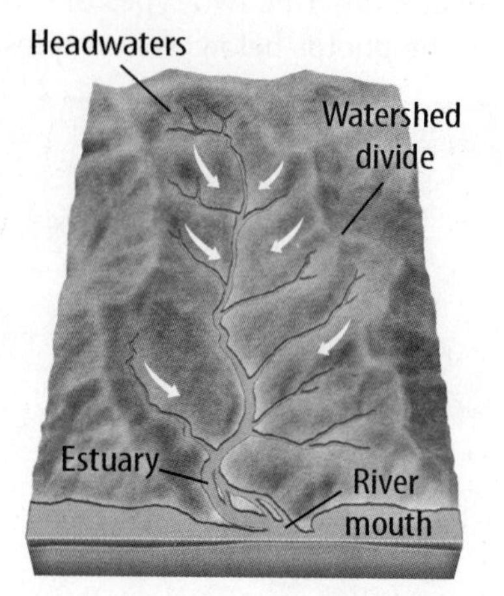

4 Which is correct based on the diagram above?

A Watershed divides are at the bases of hills or mountains.

B Water in a watershed flows downhill into rivers.

C Water in the streams moves from estuary to the headwaters.

D Water outside the watershed flows directly into the ocean.

5 How does the reduction of ice and snow on Earth's surface affect climate?

A Land and water absorb more solar radiation.

B Sea levels decrease because more moisture enters the air.

C Air becomes drier because more water runs off the surface.

D Global temperatures cool as more of the Sun's heat is absorbed.

6 How do bogs differ from marshes and swamps?

A Bogs produce no peat.

B Bogs rarely dry out.

C Bogs need hot weather.

D Bogs are nutrient-rich.

Use the diagram below to answer question 7.

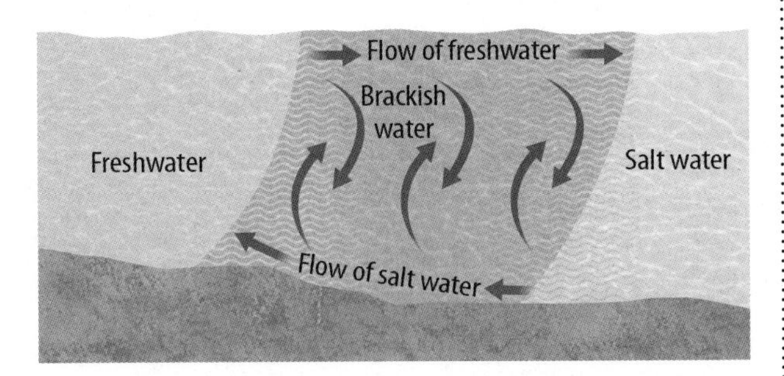

7 What forms from the mixture of freshwater and salt water?

A estuary

B lake

C ocean

D river

8 Which lists naturally flowing channels of water from largest to smallest?

A brook, creek, river

B creek, brook, river

C river, brook, creek

D river, creek, brook

9 Which measures water's ability to flow through rock and sediment?

A fluidity

B liquidity

C permeability

D porosity

Constructed Response

Use the diagram below to answer questions 10 and 11.

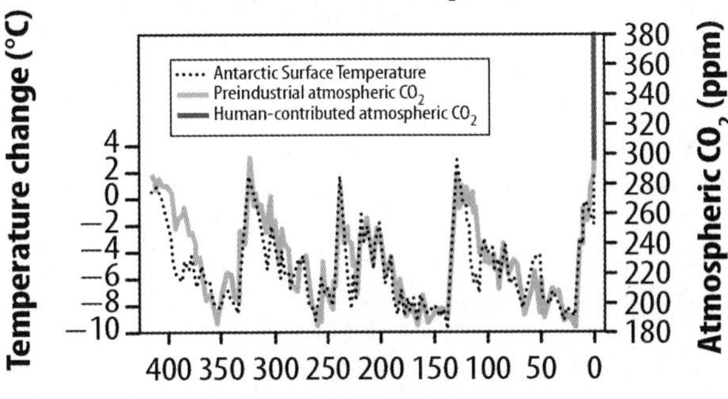

10 Using the graph above, describe the changes in global surface temperatures and CO_2 levels over time. What is a logical inference about climate based on this information?

11 When did Earth's CO_2 levels begin to rise sharply? What human activity contributed to this change? Predict future Antarctic surface temperatures based on this trend.

Use the table below to answer question 12.

Benefit	Explanation
Erosion control	
Filtration	
Flood control	
Habitat	

12 Explain how each benefit of wetlands listed in the table helps Earth and its inhabitants.

NEED EXTRA HELP?												
If You Missed Question...	1	2	3	4	5	6	7	8	9	10	11	12
Go to Lesson...	3	1	2	2	1	3	2	2	3	1	1	3

Chapter 18

Natural Resources

THE BIG IDEA **Why is it important to manage natural resources wisely?**

Inquiry What do these colors mean?

This image shows where thermal energy escapes from the inside of a house. Red and yellow areas represent the greatest loss. Blue areas represent low or no loss.

- Which energy resources are used to heat this house?
- Why is it important to reduce thermal energy loss from houses, cars, or electrical appliances?
- Why is it important to manage natural resources wisely?

Tyrone Turner/National Geographic Stock

Get Ready to Read

What do you think?

Before you read, decide if you agree or disagree with each of these statements. As you read this chapter, see if you change your mind about any of the statements.

1. Nonrenewable energy resources include fossil fuels and uranium.
2. Energy use in the United States is lower than in other countries.
3. Renewable energy resources do not pollute the environment.
4. Burning organic material can produce electricity.
5. Cities cover most of the land in the United States.
6. Minerals form over millions of years.
7. Humans need oxygen and water to survive.
8. About 10 percent of Earth's total water can be used by humans.

Your one-stop online resource
connectED.mcgraw-hill.com

LearnSmart®

Chapter Resources Files, Reading Essentials, Get Ready to Read, Quick Vocabulary

Animations, Videos, Interactive Tables

Self-checks, Quizzes, Tests

Project-Based Learning Activities

Lab Manuals, Safety Videos, Virtual Labs & Other Tools

Vocabulary, Multilingual eGlossary, Vocab eGames, Vocab eFlashcards

Personal Tutors

Lesson 1

Reading Guide

Key Concepts

ESSENTIAL QUESTIONS

- What are the main sources of nonrenewable energy?
- What are the advantages and disadvantages of using nonrenewable energy resources?
- How can individuals help manage nonrenewable resources wisely?

Vocabulary

nonrenewable resource p. 643

renewable resource p. 643

nuclear energy p. 647

reclamation p. 649

Multilingual eGlossary

Energy Resources

Inquiry What's in the pipeline?

The Trans-Alaska Pipeline System carries oil more than 1,200 km from beneath Prudhoe Bay, Alaska, to the port city of Valdez, Alaska. How might the pipeline's construction and operation affect the habitats and the organisms living along it? How do getting and using fossil fuels impact the environment?

Launch Lab

20 minutes

How do you use energy resources?

In the United States today, the energy used for most daily activities is easily available at the flip of a switch or the push of a button. How do you use energy in your daily activities?

1. Design a three-column data chart in your Science Journal. Title the columns *Activity, Type of Energy Used,* and *Amount of Time.*
2. Record every instance that you use energy during a 24-hr period.
3. Total your usage of the different forms of energy, and record them in your Science Journal.

Think About This

1. How many times did you use each type of energy?
2. Compare and contrast your usage with that of other members of your class.
3. **Key Concept** Are there instances of energy use when you could have conserved energy? Explain how you would do it.

Sources of Energy

Think about all the times you use energy in one day. Are you surprised by how much you depend on energy? You use it for electricity, transportation, and other needs. That is one reason it is important to know where energy comes from and how much is available for humans to use.

Table 1 lists different energy sources. Most energy in the United States comes from nonrenewable resources. **Nonrenewable resources** *are resources that are used faster than they can be replaced by natural processes.* Fossil fuels, such as coal and oil, and uranium, which is used in nuclear reactions, are both nonrenewable energy resources.

Renewable resources *are resources that can be replaced by natural processes in a relatively short amount of time.* The Sun's energy, also called solar energy, is a renewable energy resource. You will read more about renewable energy resources in Lesson 2.

Key Concept Check What are the main nonrenewable energy resources?

Table 1 Energy resources can be nonrenewable or renewable.

Table 1 Energy Sources

Nonrenewable Energy Resources	Renewable Energy Resources
fossil fuels uranium	solar wind water geothermal biomass

WORD ORIGIN

resource
from Latin *resurgere,* means "to rise again"

Nonrenewable Energy Resources

You might turn on a lamp to read, turn on a heater to stay warm, or ride the bus to school. In the United States, the energy to power lamps, heat houses, and run vehicles probably comes from nonrenewable energy resources, such as fossil fuels.

Fossil Fuels

Coal, oil, also called petroleum, and natural gas are fossil fuels. They are nonrenewable because they form over millions of years. The fossil fuels used today formed from the remains of prehistoric organisms. The decayed remains of these organisms were buried by layers of sediment and changed chemically by extreme temperatures and pressure. The type of fossil fuel that formed depended on three factors:

- the type of organic matter
- the temperature and pressure
- the length of time that the organic matter was buried

Reading Check What factors determine which type of fossil fuel forms?

Coal Earth was very different 350 million years ago, when the coal used today began forming. Plants, such as ferns and trees, grew in prehistoric swamps. As shown in **Figure 1**, the first step of coal formation occurred when those plants died.

Bacteria, extreme temperatures, and pressure acted on the plant remains over time. Eventually a brownish material, called peat, formed. Peat can be used as a fuel. However, peat contains moisture and produces a lot of smoke when it burns. As shown in **Figure 1**, peat eventually can change into harder and harder types of coal. The hardest coal, anthracite, contains the most carbon per unit of volume and burns most efficiently.

Figure 1 Much of the coal used today began forming more than 300 million years ago from the remains of prehistoric plants.

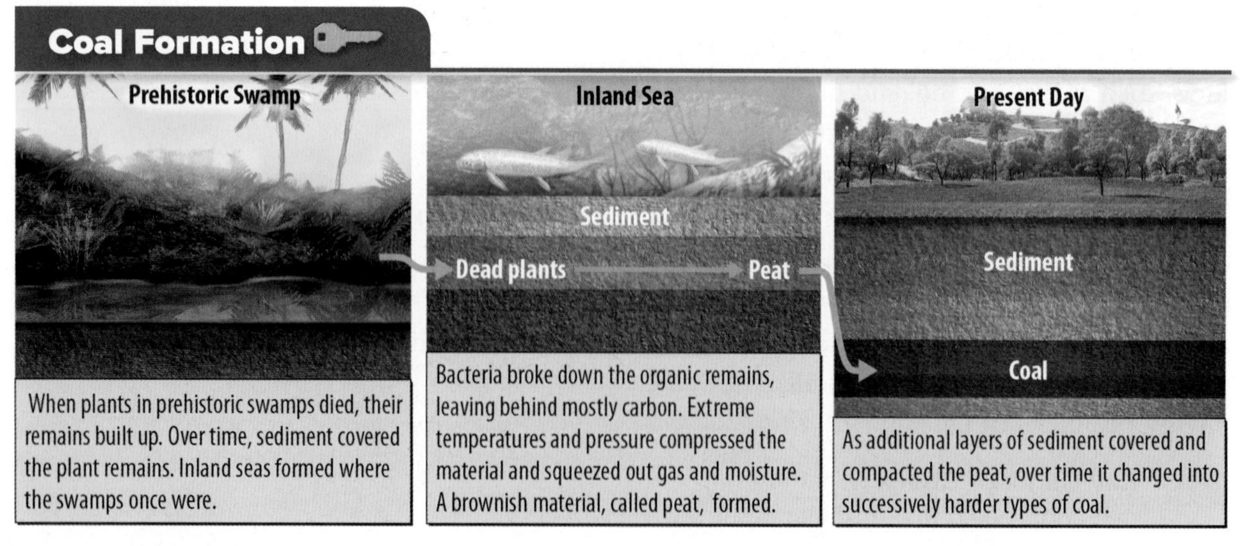

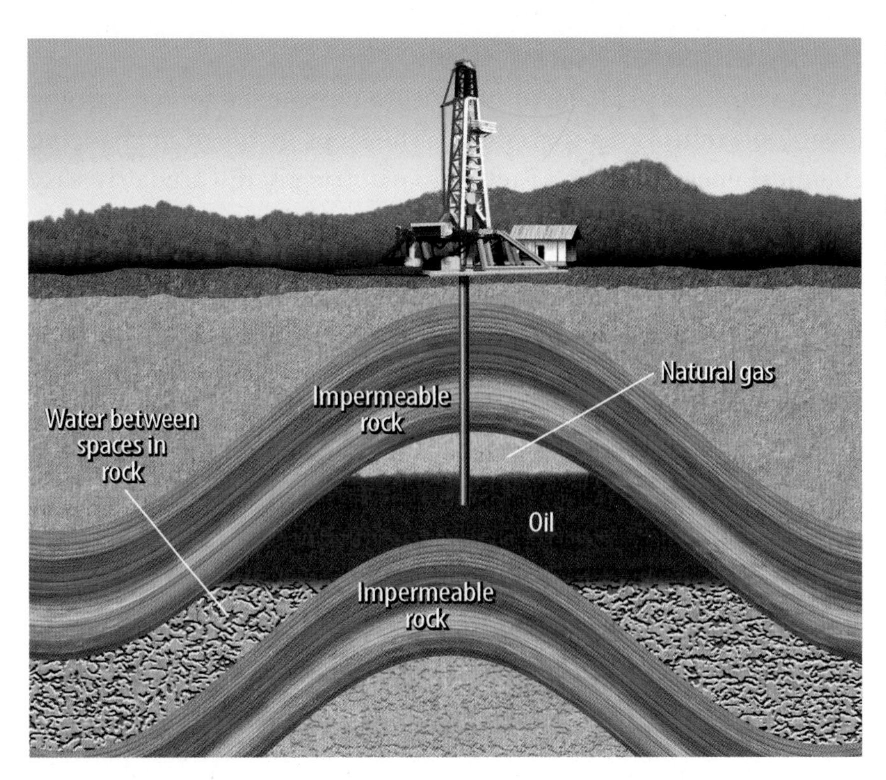

Figure 2 Reservoirs of oil and natural gas often are under layers of impermeable rock.

Visual Check What prevents oil and natural gas from rising to the surface?

Oil and Natural Gas Like coal, the oil and natural gas used today formed millions of years ago. The process that formed oil and natural gas is similar to the process that formed coal. However, oil and natural gas formation involves different types of organisms. Scientists theorize that oil and natural gas formed from the remains of microscopic marine organisms called plankton. The plankton died and fell to the ocean floor. There, layers of sediment buried their remains. Bacteria decomposed the organic matter, and then pressure and extreme temperatures acted on the sediments. During this process, thick, liquid oil formed first. If the temperature and pressure were great enough, natural gas formed.

Most of the oil and natural gas used today formed where forces within Earth folded and tilted thick rock layers. Often hundreds of meters of sediments and rock layers covered oil and natural gas. However, oil and natural gas were less dense than the surrounding sediments and rock. As a result, oil and natural gas began to rise to the surface by passing through the pores, or small holes, in rocks. As shown in **Figure 2,** oil and natural gas eventually reached layers of rock through which they could not pass, or impermeable rock layers. Deposits of oil and natural gas formed under these impermeable rocks. The less-dense natural gas settled on top of the denser oil.

Reading Check How is coal formation different from oil formation?

Advantages of Fossil Fuels

Do you know that fossil fuels store chemical energy? Burning fossil fuels transforms this energy. The steps involved in changing chemical energy in fossil fuels into electric energy are fairly easy and direct. This process is one advantage of using these nonrenewable resources. Also, fossil fuels are relatively inexpensive and easy to transport. Coal is often transported by trains, and oil is transported by pipelines or large ships called tankers.

Disadvantages of Fossil Fuels

Although fossil fuels provide energy, there are disadvantages to using them.

Limited Supply One disadvantage of fossil fuels is that they are nonrenewable. No one knows for sure when supplies will be gone. Scientists estimate that, at current rates of consumption, known reserves of oil will last only another 50 years.

Habitat Disruption In addition to being nonrenewable, the process of obtaining fossil fuels disturbs environments. Coal comes from underground mines or strip mines, such as the one shown in **Figure 3.** Oil and natural gas come from wells drilled into Earth. Mines in particular disturb habitats. Forests might be fragmented, or broken into areas of trees that are no longer connected. Fragmentation can negatively affect birds and other organisms that live in forests.

Reading Check How much longer are known oil reserves predicted to last?

Figure 3 Strip-mining involves removing layers of rock and soil to reach coal deposits.

Larry Mayer/Creatas/SuperStock

Pollution Another disadvantage of fossil fuels as an energy resource is pollution. For example, runoff from coal mines can pollute soil and water. Oil spills from tankers can harm living things, such as the bird shown in **Figure 4.**

Pollution also occurs when fossil fuels are used. Burning fossil fuels releases chemicals into the atmosphere. These chemicals react in the presence of sunlight and produce a brownish haze. This haze can cause respiratory problems, particularly in young children. The chemicals also can react with water in the atmosphere and make rain and snow more acidic. The acidic precipitation can change the chemistry of soil and water and harm living things.

Figure 4 One disadvantage of fossil fuels is pollution, which can harm living things. This bird was covered with oil after an oil spill.

Key Concept Check What is one advantage and one disadvantage of using fossil fuels?

Nuclear Energy

Atoms are too small to be seen with the unaided eye. Even though they are small, atoms can release large amounts of energy. *Energy released from atomic reactions is called* **nuclear energy.** Stars release nuclear energy by fusing atoms. The type of nuclear energy used on Earth involves a different process.

MiniLab

20 minutes

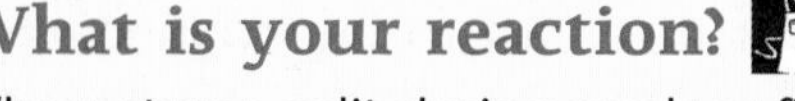

What is your reaction?

When atoms split during nuclear fission, the chain reaction releases thermal energy and by-products. What happens when your class participates in a simulation of a nuclear reaction?

1. Read and complete a lab safety form.
2. Use a **marker** to label three **sticky notes.** Label one note *U-235.* Label two notes *Neutron.* Stick the U-235 note on your **apron.** Hold the Neutron notes in one hand and a **Thermal Energy Card** in the other. You now represent a uranium-235 atom.
3. When you are tagged with a Neutron label from another student, tag two other student U-235 atoms with your Neutron labels. Drop your Thermal Energy Card into the **Energy Box.**
4. Observe as the remainder of the U-235 atoms are split, and imagine this happening extremely fast at the atomic level.

Analyze and Conclude

1. **Describe** what the simulation illustrated about nuclear fission.
2. **Predict** what would happen if, in the simulation, your classroom was filled wall-to-wall with U-235 atoms and the chain reaction got out of control.
3. **Key Concept** Identify one advantage and one disadvantage of nuclear energy.

Nuclear Energy

Animation

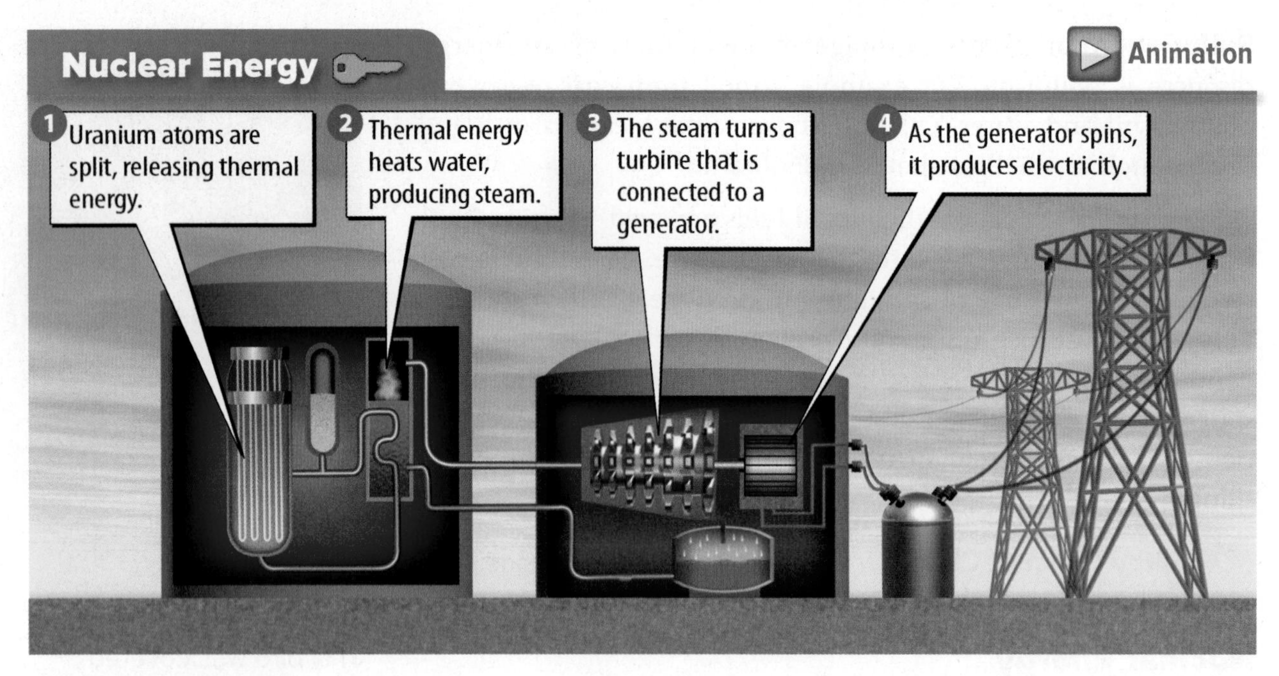

Figure 5 In a nuclear power plant, thermal energy released from splitting uranium atoms is transformed into electrical energy.

Nuclear Fission Nuclear power plants, such as the one shown in **Figure 5**, produce electricity using nuclear fission. This process splits atoms. Uranium atoms are placed into fuel rods. Neutrons are aimed at the rods and hit the uranium atoms. Each atom splits and releases two to three neutrons and thermal energy. The released neutrons hit other atoms, causing a chain reaction of splitting atoms. Countless atoms split and release large amounts of thermal energy. This energy heats water and changes it to steam. The steam turns a turbine connected to a generator, which produces electricity.

 **Reading Check** What are the steps in nuclear fission?

Advantages and Disadvantages of Nuclear Energy

One advantage of using nuclear energy is that a relatively small amount of uranium produces a large amount of energy. In addition, a well-run nuclear power plant does not pollute the air, the soil, or the water.

However, using nuclear energy has disadvantages. Nuclear power plants use a nonrenewable resource–uranium–for fuel. In addition, the chain reaction in the nuclear reactor must be carefully monitored. If it gets out of control, it can lead to a release of harmful radioactive substances into the environment.

The waste materials from nuclear power plants are highly radioactive and dangerous to living things. The waste materials remain dangerous for thousands of years. Storing them safely is important for both the environment and public health.

 **Reading Check** Why is it important to control a chain reaction?

Managing Nonrenewable Energy Resources

As shown in **Figure 6,** fossil fuels and nuclear energy provide about 87 percent of U.S. energy. Because these sources eventually will be gone, we must understand how to manage and conserve them. This is particularly important because energy use in the United States is higher than in other countries. Although only about 4.5 percent of the world's population lives in the United States, it uses more than 18 percent of the world's total energy.

Management Solutions

Mined land must be reclaimed. **Reclamation** *is a process in which mined land must be recovered with soil and replanted with vegetation.* Laws also help ensure that mining and drilling take place in an environmentally safe manner. In the United States, the Clean Air Act limits the amount of pollutants that can be released into the air. In addition, the U.S. Atomic Energy Act and the Energy Policy Act include regulations that protect people from nuclear emissions.

What You Can Do

Have you ever heard of vampire energy? Vampire energy is the energy used by appliances and other electronic equipment, such as microwave ovens, washing machines, televisions, and computers, that are plugged in 24 h a day. Even when turned off, they still consume energy. These appliances consume about 5 percent of the energy used each year. You can conserve energy by unplugging DVD players, printers, and other appliances when they are not in use.

You also can walk or ride your bike to help conserve energy. And, you can use renewable energy resources, which you will read about in the next lesson.

Key Concept Check How can you help manage nonrenewable resources wisely?

Sources of Energy Used in the U.S. in 2015

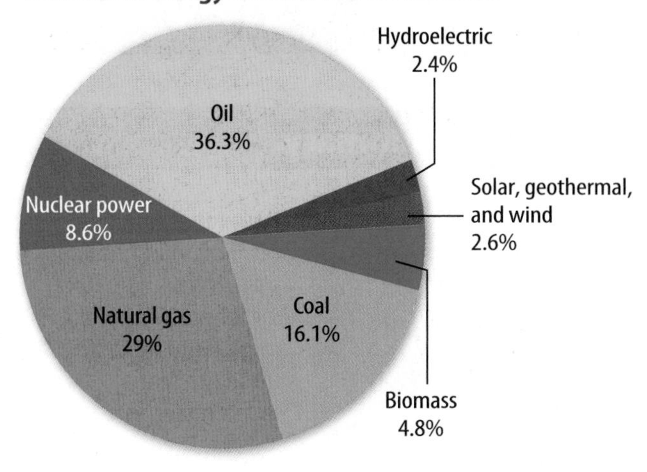

Figure 6 About 87 percent of the energy used in the United States comes from nonrenewable resources.

Visual Check Which energy source is used most in the United States?

ACADEMIC VOCABULARY

regulation
(noun) a rule dealing with procedures, such as safety

FOLDABLES®

Make a three-tab book. Before cutting the tabs, draw a Venn diagram and label as illustrated. Compare and contrast the use of fossil fuels and nuclear energy.

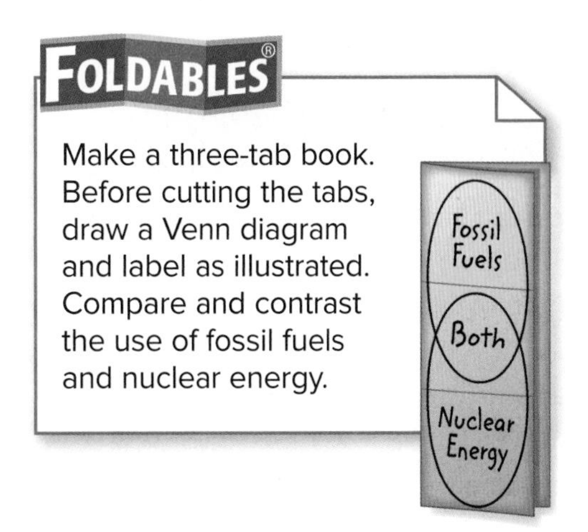

Lesson 1 Review

Visual Summary

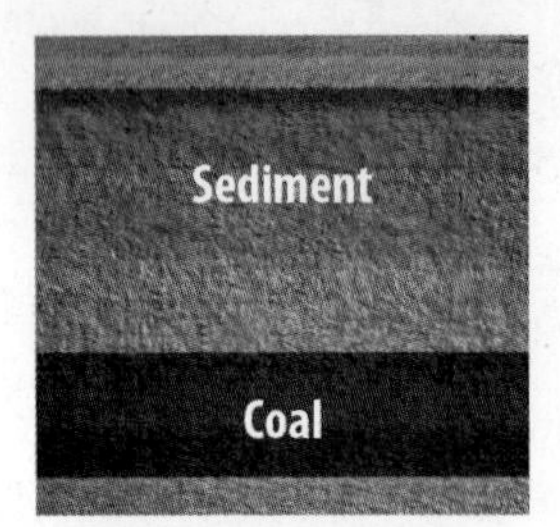

Fossil fuels include coal, oil, and natural gas. Fossil fuels take millions of years to form. Humans use fossil fuels at a much faster rate.

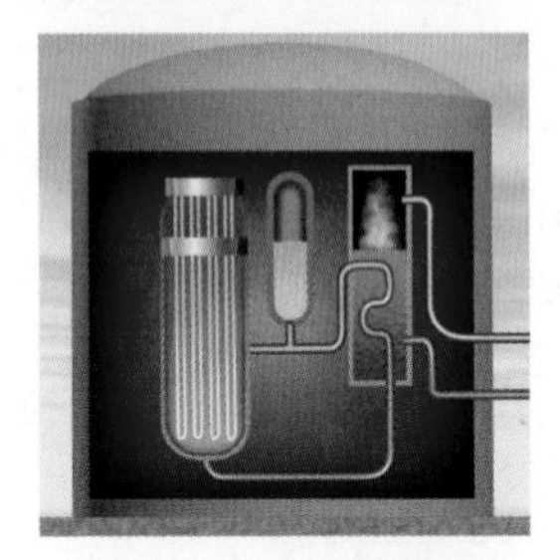

Nuclear energy comes from splitting atoms, or fission. Nuclear power plants must be monitored for safety, and nuclear waste must be stored properly.

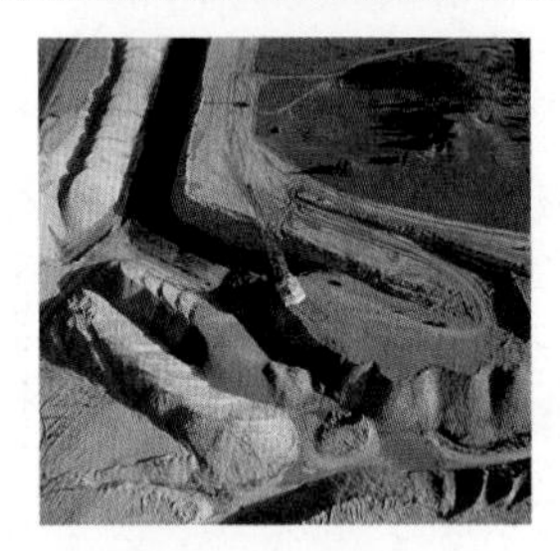

It is important to manage nonrenewable energy resources wisely. This includes mine reclamation, limiting air pollutants, and conserving energy.

FOLDABLES®

Use your lesson Foldable to review the lesson. Save your Foldable for the project at the end of the chapter.

What do you think NOW?

You first read the statements below at the beginning of the chapter.

1. Nonrenewable energy resources include fossil fuels and uranium.
2. Energy use in the United States is lower than in other countries.

Did you change your mind about whether you agree or disagree with the statements? Rewrite any false statements to make them true.

Use Vocabulary

1. Energy produced from atomic reactions is called ________.
2. **Distinguish** between renewable and nonrenewable resources.
3. **Use the term** *reclamation* in a sentence.

Understand Key Concepts 

4. What is the source of most energy in the United States?
   - **A.** coal
   - **B.** oil
   - **C.** natural gas
   - **D.** nuclear energy
5. **Summarize** the advantages and disadvantages of using nuclear energy.
6. **Illustrate** Make a poster showing how you can conserve energy.

Interpret Graphics

7. **Sequence** Draw a graphic organizer like the one below to sequence the events in the formation of oil.

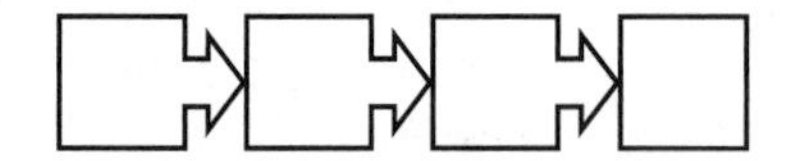

8. **Describe** Use the diagram below to describe the energy conversions that take place in a nuclear power plant.

Critical Thinking

9. **Suppose** that a nuclear power plant will be built near your town. Would you support the plan? Why or why not?
10. **Consider** Do the advantages of using fossil fuels outweigh the disadvantages? Explain your answer.

Skill Practice Avoid Bias

30 minutes

How can you identify bias and its source?

Whenever an author is trying to persuade, or convince, readers to share a particular opinion, you must read and evaluate carefully for bias. Bias is a way of thinking that tells only one side of a story, sometimes with inaccurate information.

Learn It

Sometimes a scientific investigation involves making judgments. When you make a judgment, you form an opinion. It is important to be honest and not allow any expectations of results to **bias** your judgments.

Try It

1. Read the passage to the right for sources of bias, such as

- claims not supported by evidence;
- persuasive statements;
- the author wanting to believe what he or she is saying, whether or not it is true.

Apply It

2. Analyze the passage, and identify two instances of bias and the source of each. Record this information in your Science Journal.
3. If you were the moderator at an EPA hearing about the issue, what would you do to solve the problem of bias?
4. **Key Concept** In your own words, explain how you would formulate an argument about the wise management of air resources while avoiding bias.

Factory Spews Out Air Pollution

Environmental organizations claim a coal-burning factory is polluting the air with toxic particulate matter in violation of U.S. Environmental Protection Agency (EPA) standards. Particulate matter is a mix of both solid and liquid particles in the air.

Citizens of the town collected air samples for a period of six months. An independent laboratory analysis of the samples showed a dangerously high level of the toxic particulate materials. Levels this high have been cited in medical journals as contributing to illness and death from asthma, respiratory disease, and lung cancer.

The factory has not updated its pollution control equipment, claiming that it cannot afford the cost. At a town meeting, a company spokesperson claimed that the particulate matter is not harmful to human health. The state environmental director, who previously worked at the factory, stated that the jobs provided by the factory are more important to the state than environmental concerns.

George Diebold/Photographer's Choice RF/Getty Images

Lesson 2

Reading Guide

Key Concepts

ESSENTIAL QUESTIONS

- What are the main sources of renewable energy?
- What are the advantages and disadvantages of using renewable energy resources?
- What can individuals do to encourage the use of renewable energy resources?

Vocabulary

solar energy p. 653
wind farm p. 654
hydroelectric power p. 654
geothermal energy p. 655
biomass energy p. 655

Multilingual eGlossary

Renewable Energy Resources

Inquiry What do these panels do?

These solar panels convert energy from the Sun into electrical energy. This solar power plant, at Nellis Air Force Base in Nevada, produces up to 42 percent of the electricity used on the base. What are some of the advantages of using energy from the Sun? What are some of the disadvantages?

Launch Lab

20 minutes

How can renewable energy sources generate energy in your home?

Renewable energy technologies can contribute to reducing our dependence on fossil fuels.

1. Review the table below. It shows how much energy, in Watt-hours, it takes to run certain appliances.
2. In one hour, a typical bicycle generator generates 200 W-h of electric energy; a small solar panel generates 150 W-h; and small wind turbines typically generate 100 W-h. Complete the table by calculating the time it would take for each alternative form of energy to generate the electricity needed to run each appliance for 1 h.

   Hint: Use the following equation to solve for the time used by each energy source:

$$\left(\text{Time used by energy source}\right) = \frac{\left(\text{Time to use appliance}\right) \times \left(\text{Energy used per hour by appliance}\right)}{\left(\text{Energy produced per hour by energy source}\right)}$$

Think About This

1. Which appliance required the longest energy-generating time from the alternative energy sources? Why?
2. **Key Concept** What issues would you have to consider when using solar or wind energy to generate electricity in your home?

Appliance	Energy Used Per Hour	Time on Bike	Time for Solar Panel	Time for Wind Turbine
Desktop computer	75 W-h			
Hair dryer	1000 W-h			
Television	200 W-h			

Renewable Energy Resources

Could you stop the Sun from shining or the wind from blowing? These might seem like silly questions, but they help make an important point about renewable resources. Renewable resources come from natural processes that have been happening for billions of years and will continue to happen.

Solar Energy

Solar energy *is energy from the Sun.* Solar cells, such as those in watches and calculators, capture light energy and transform it to electrical energy. Solar power plants can generate electricity for large areas. They transform energy in sunlight, which then turns turbines connected to generators.

Some people use solar energy in their homes, as shown in **Figure 7.** Active solar energy uses technology, such as solar panels, that gathers and stores solar energy that heats water and homes. Passive solar energy uses design elements that capture energy in sunlight. For example, windows on the south side of a house can let in sunlight that helps heat a room.

Figure 7 People can use solar energy to provide electricity for their homes.

▲ **Figure 8** Offshore wind farms are called wind parks. This wind park is in Denmark.

Wind Energy

Have you ever dropped your school papers outside and had them scattered by the wind? If so, you experienced wind energy. This renewable resource has been used since ancient times to sail boats and to turn windmills. Today, wind turbines, such as the ones shown in **Figure 8**, can produce electricity on a large scale. *A group of wind turbines that produce electricity is called a* **wind farm.**

Reading Check How is wind energy a renewable resource?

Water Energy

Like wind energy, flowing water has been used as an energy source since ancient times. Today, water energy produces electricity using different methods, such as hydroelectric power and tidal power.

Hydroelectric Power *Electricity produced by flowing water is called* **hydroelectric power.** To produce hydroelectric power, humans build a dam across a powerful river. **Figure 9** shows how flowing water is used to produce electricity.

Tidal Power Coastal areas that have great differences between high and low tides can be a source of tidal power. Water flows across turbines as the tide comes in during high tides and as it goes out during low tides. The flowing water turns turbines connected to generators that produce electricity.

Figure 9 In a hydroelectric power plant, energy from flowing water produces electricity. ▼

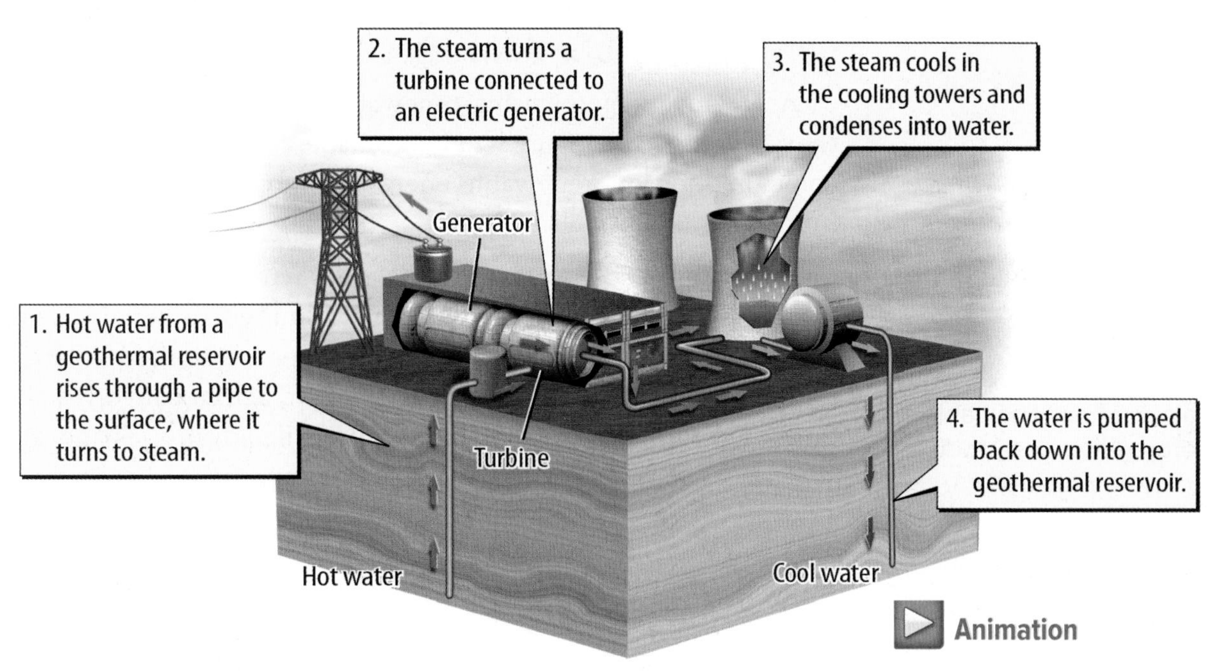

Animation

Figure 10 Geothermal power plants use thermal energy from Earth's interior and produce electricity.

Geothermal Energy

Earth's core is nearly as hot as the Sun's surface. This thermal energy flows outward to Earth's surface. *Thermal energy from Earth's interior is called* **geothermal energy.** It can be used to heat homes and generate electricity in power plants, such as the one shown in **Figure 10.** People drill wells to reach hot, dry rocks or bodies of magma. The thermal energy from the hot rocks or magma heats water that makes steam. The steam turns turbines connected to generators that produce electricity.

WORD ORIGIN

geothermal
from Greek *ge-*, means "Earth"; and Greek *therme*, means "heat"

Biomass Energy

Since humans first lit fires for warmth and cooking, biomass has been an energy source. **Biomass energy** *is energy produced by burning organic matter, such as wood, food scraps, and alcohol.* Wood is the most widely used biomass. Industrial wood scraps and organic materials, such as grass clippings and food scraps, are burned to generate electricity on a large scale.

Biomass also can be converted into fuels for vehicles. Ethanol is made from sugars in plants, such as corn. Ethanol often is blended with gasoline. This reduces the amount of oil used to make the gasoline. Adding ethanol to gasoline also reduces the amount of carbon monoxide and other pollutants released by vehicles. Another renewable fuel, biodiesel, is made from vegetable oils and fats. It emits few pollutants and is the fastest-growing renewable fuel in the United States.

FOLDABLES®

Make a vertical five-tab Foldable. Label the tabs as illustrated. Identify the advantages and disadvantages of alternative fuels.

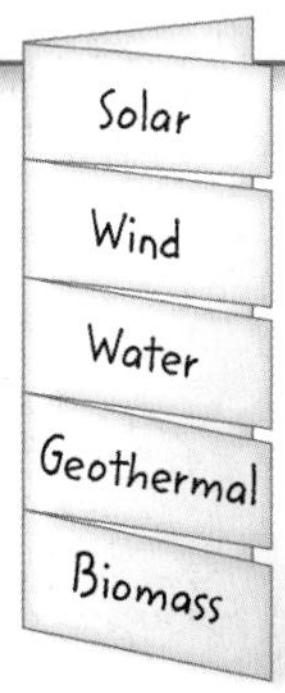

Key Concept Check What are the main sources of renewable energy?

Advantages and Disadvantages of Renewable Resources

A big advantage of using renewable energy resources is that they are renewable. They will be available for millions of years to come. In addition, renewable energy resources produce less pollution than fossil fuels.

There are disadvantages associated with using renewable resources, however. Some are costly or limited to certain areas. For example, large-scale geothermal plants are limited to areas with tectonic activity. Recall that tectonic activity involves the movement of Earth's plates. **Table 2** lists the advantages and disadvantages of using renewable energy resources.

Key Concept Check What are some advantages and disadvantages of using renewable energy resources?

Table 2 Most renewable energy resources produce little or no pollution.

Visual Check What are the advantages and the disadvantages of biomass energy?

Table 2 Renewable Resources—Advantages and Disadvantages

Renewable Resource	Advantages	Disadvantages
Solar energy	• nonpolluting • available in the United States	• less energy produced on cloudy days • no energy produced at night • high cost of solar cells • requires a large surface area to collect and produce energy on a large scale
Wind energy	• nonpolluting • relatively inexpensive • available in the United States	• large-scale use limited to areas with strong, steady winds • best sites for wind farms are far from urban areas and transmission lines • potential impact on bird populations
Water energy	• nonpolluting • available in the United States	• large-scale use limited to areas with fast-flowing rivers or great tidal differences • negative impact on aquatic ecosystems • production of electricity affected by long periods of little or no rainfall
Geothermal energy	• produces little pollution • available in the United States	• large-scale use limited to tectonically active areas • habitat disruption from drilling to build a power plant
Biomass energy	• reduces amount of organic material discarded in landfills • available in the United States	• air pollution results from burning some forms of biomass • less energy efficient than fossil fuels, costly to transport

Managing Renewable Energy Resources

Renewable energy currently meets only 9.8 percent of U.S. energy needs. As shown in **Figure 11**, most renewable energy comes from biomass. Solar energy, wind energy, and geothermal energy meet only a small percentage of U.S. energy needs. However, some states are passing laws that require the state's power companies to produce a percentage of electricity using renewable resources. Management of renewable resources often focuses on encouraging their use.

Management Solutions

The U.S. government has begun programs to encourage use of renewable resources. In 2009, billions of dollars were granted to the U.S. Department of Energy's Office of Energy Efficiency and Renewable Energy for renewable energy research and programs that reduce the use of fossil fuels.

What You Can Do

You might be too young to own a house or a car, but you can help educate others about renewable energy resources. You can talk with your family about ways to use renewable energy at home. You can participate in a renewable energy fair at school. As a consumer, you also can make a difference by buying products that are made using renewable energy resources.

Key Concept Check What can you do to encourage the use of renewable energy resources?

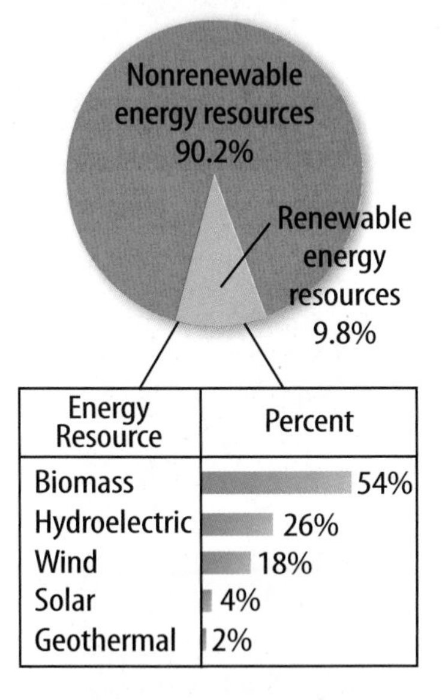

Figure 11 The renewable energy resource used most in the United States is biomass energy.

MiniLab

20 minutes

How are renewable energy resources used at your school?

Complete a survey about the use of renewable resources in your school.

1. Prepare interview questions about the use of renewable energy resources at your school. Each group member should come up with at least two questions.
2. Choose one group member to interview a school staff member.
3. Copy the table at the right into your Science Journal, and fill in the interview data.

Renewable Energy Source	Yes/No	Where is it used?	Why is it used? or Why isn't it used?
Sun			
Wind			
Water			
Geothermal			
Biomass			

Analyze and Conclude

1. **Explain** Which renewable energy resources are and are not being used? Why or why not?
2. **Key Concept** Choose one "why not" reason and describe how it could be addressed by communication with school planners.

Lesson 2 Review

Visual Summary

Renewable energy resources can be used to heat homes, produce electricity, and power vehicles.

Advantages of renewable energy resources include little or no pollution and availability.

Management of renewable energy resources includes encouraging their use and continuing to research more about their use.

FOLDABLES®

Use your lesson Foldable to review the lesson. Save your Foldable for the project at the end of the chapter.

What do you think NOW?

You first read the statements below at the beginning of the chapter.

3. Renewable energy resources do not pollute the environment.

4. Burning organic material can produce electricity.

Did you change your mind about whether you agree or disagree with the statements? Rewrite any false statements to make them true.

Use Vocabulary

1 **Define** *hydroelectric power* in your own words.

2 Burning wood is an example of ________ energy.

Understand Key Concepts 

3 Which can reduce the amount of organic material discarded in landfills?

A. biomass energy **C.** water energy
B. solar energy **D.** wind energy

4 **Compare and contrast** solar energy and wind energy.

5 **Determine** Your family wants to use renewable energy to heat your home. Which renewable energy resource is best suited to your area? Explain your answer.

Interpret Graphics

6 **Organize** Copy and fill in the graphic organizer below. In each oval, list a type of renewable energy resource.

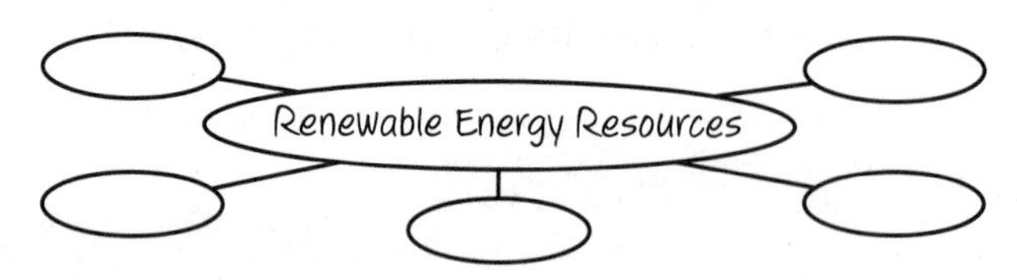

7 **Compare** the use of renewable resources and nonrenewable resources in the production of electricity in the United States, based on the table below.

Sources of Electricity Generation, 2015

Energy Source	Percent
Fossil fuels	67%
Nuclear power	20%
Solar, wind, geothermal, biomass	7%
Hydroelectric	6%

Critical Thinking

8 **Design** and explain a model that shows how a renewable resource produces energy.

(t)Clynt Garnham Renewable Energy/Alamy; (c)Russel Illig/Photodisc/Getty Images; (b)Fotosearch/PhotoLibrary

Skill Practice Analyze Data

40 minutes

How can you analyze energy-use data for information to help conserve energy?

As a student, you are not making large governmental policy decisions about uses of resources. As an individual, however, you can analyze data about energy use. You can use your analysis to determine some personal actions that can be taken to conserve energy resources.

Learn It

To **analyze the data** of fuel usage, you will need to look for patterns in the data, compare and categorize them, and determine cause and effect.

Try It

1. Study the fuel usage graph shown below. The data were collected from a house that uses natural gas as a source of energy to heat it.
2. Identify the time period that is covered by the graph.
3. Explain what is represented by the values on the vertical axis of the graph.
4. Describe the range of monthly gas usage over the 12-month period.
5. Group the monthly gas usage into three levels. Give each level a title. Enter these in your Science Journal.

Apply It

6. Categorize the three levels based on the amount of natural gas use.
7. Identify the three highest and four lowest months of gas usage. What might explain the usage patterns during these months?
8. Suppose the house from which the data came was heated with an electric furnace, instead of a furnace that used natural gas. What would you expect a usage graph for an electric furnace to look like?
9. **Key Concept** Formulate a list of heat conservation practices for homes.

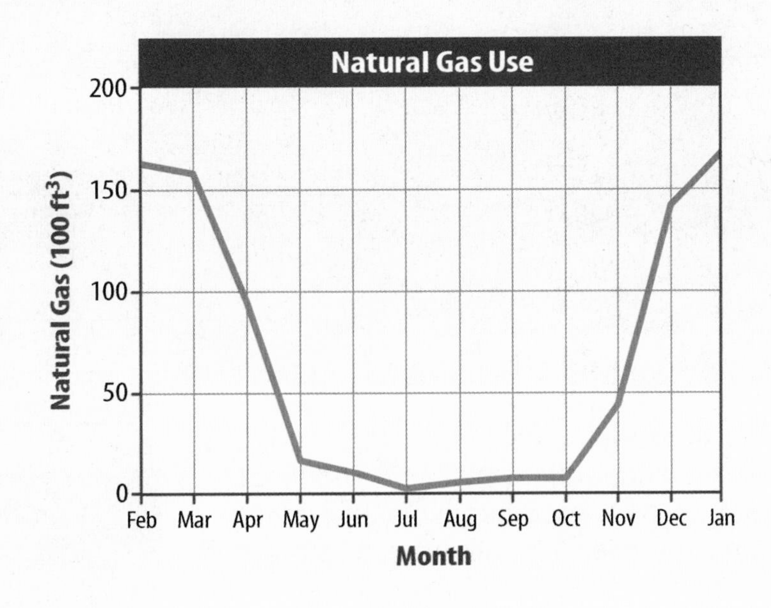

Lesson 3

Land Resources

Reading Guide

Key Concepts
ESSENTIAL QUESTIONS

- Why is land considered a resource?
- What are the advantages and disadvantages of using land as a resource?
- How can individuals help manage land resources wisely?

Vocabulary

ore p. 663

deforestation p. 664

 Multilingual eGlossary

 What's Science Got to do With It?

Inquiry A Garden on the Water?

The Science Barge is an experimental farm in New York City, New York. It saves space and reduces pollution and fossil fuel use while growing crops to feed people in an urban area. Why are people experimenting with ways to grow food that have fewer environmental impacts? Why is it important for humans to use land resources wisely?

Launch Lab

20 minutes

What resources from the land do you use every day?

The land on which humans live is part of Earth's crust. It provides resources that enable humans and other organisms to survive.

1. Make a list of every item you use in a 24-h period as you carry out your daily activities.
2. Combine your list with your group members' lists and decide which items contain resources from the land. Design a graphic organizer to group the materials into categories.
3. Fill in the graphic organizer on **chart paper.** Use a **highlighter** or **colored markers** to show which resources are renewable and which are nonrenewable.
4. Post your chart and compare it with the others in your class.

Think About This

1. Are there any times in your day when you do not use a resource from the land? Provide an example.
2. Describe the major categories that you used to organize your list of resources.
3. **Key Concept** Why do you think land is considered a resource?

Land as a Resource

A natural resource is something from Earth that living things use to meet their needs. People use soil for growing crops and forests to harvest wood for making furniture, houses, and paper products. They mine minerals from the land and clear large areas for roads and buildings. In each of these cases, people use land as a natural resource to meet their needs.

Key Concept Check Why is land considered a resource?

Living Space

No matter where you live, you and all living things use land for living space. Living space includes natural habitats, as well as the land on which buildings, sidewalks, parking lots, and streets are built. As shown in **Figure 12**, cities make up only a small percentage of land use in the United States. Most land is used for agriculture, grasslands, and forests.

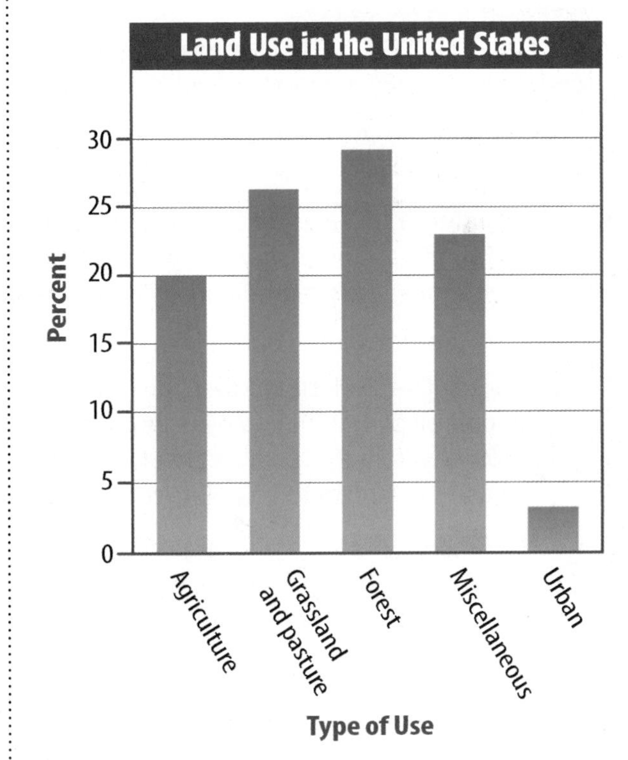

Figure 12 Forests and grasslands make up the largest categories of U.S. land use.

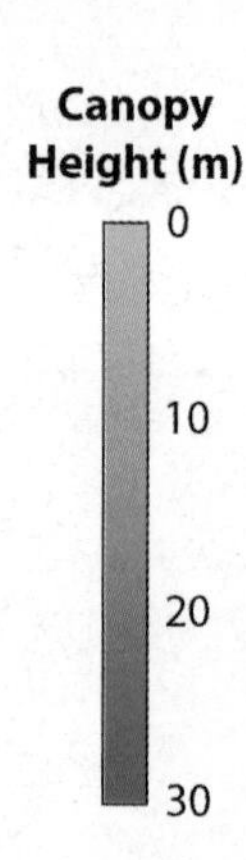

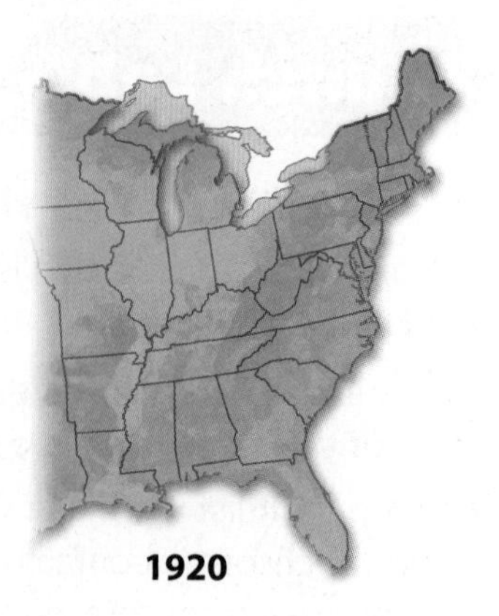

Figure 13 Much of the U.S. eastern forest has been replaced by cities, farms, and other types of development.

Visual Check Compare forest cover in the eastern United States in 1650 and 1920.

Forests and Agriculture

As shown in **Figure 13**, forests covered much of the eastern United States in 1650. By 1920, the forests had nearly disappeared. Although some of the trees have grown back, they are not as tall and the forests are not as complex as they were originally.

Forests w ere cut down for the same reasons that forests are cut down today: for fuel, paper products, and wood products. People also cleared land for development and agriculture. Today, about one-fifth of U.S. land is used for growing crops and about one-fourth is used for grazing livestock.

 **Reading Check** Why are forests cut down?

MiniLab

20 minutes

How can you manage land resource use with environmental responsibility?

You inherited a 100-acre parcel of forested land. Your relative's will stated that you must support all of your needs by using or selling resources from the land. To receive the inheritance, you must create an environmentally responsible land-use plan.

1. Copy the table into your Science Journal.
2. Your relative's will stated that you must support all of your needs by using or selling resources from the land. Decide how you will use the land, and complete the table.
3. Draw your land use plan on **graph paper.**
4. Present your group's land use plan to the class, and explain your reasoning.

Land Use	Percent of Total Area	Reasoning
Forest		
House and yard		
Garden		
Mineral mine		
Other		

Analyze and Conclude

1. **Compare and contrast** the design and reasoning of your plan with another group's plan.
2. **Identify** additional information about the land parcel that you would need to refine your plan.
3. **Key Concept** Summarize two environmentally responsible practices that were used by more than one group in their plan.

Mineral Resources

Recall that coal, an energy resource, is mined from the land. Certain minerals also are mined to make products you use every day. These minerals often are called ores. **Ores** *are deposits of minerals that are large enough to be mined for a profit.*

The house in **Figure 14** contains many examples of common items made from mineral resources. Some of these come from metallic mineral resources. Ores such as bauxite and hematite are metallic mineral resources. They are used to make metal products. The aluminum in automobiles and refrigerators comes from bauxite. The iron in nails and faucets comes from hematite. Some mineral resources come from nonmetallic mineral resources, such as gypsum, halite, and minerals found in sand and gravel. Nonmetallic mineral resources also are mined from the land. The sulfur used in paints and rubber and the fluorite used in paint pigments are other examples of nonmetal mineral resources.

WORD ORIGIN

ore
from Old English *ora,* means "unworked metal"

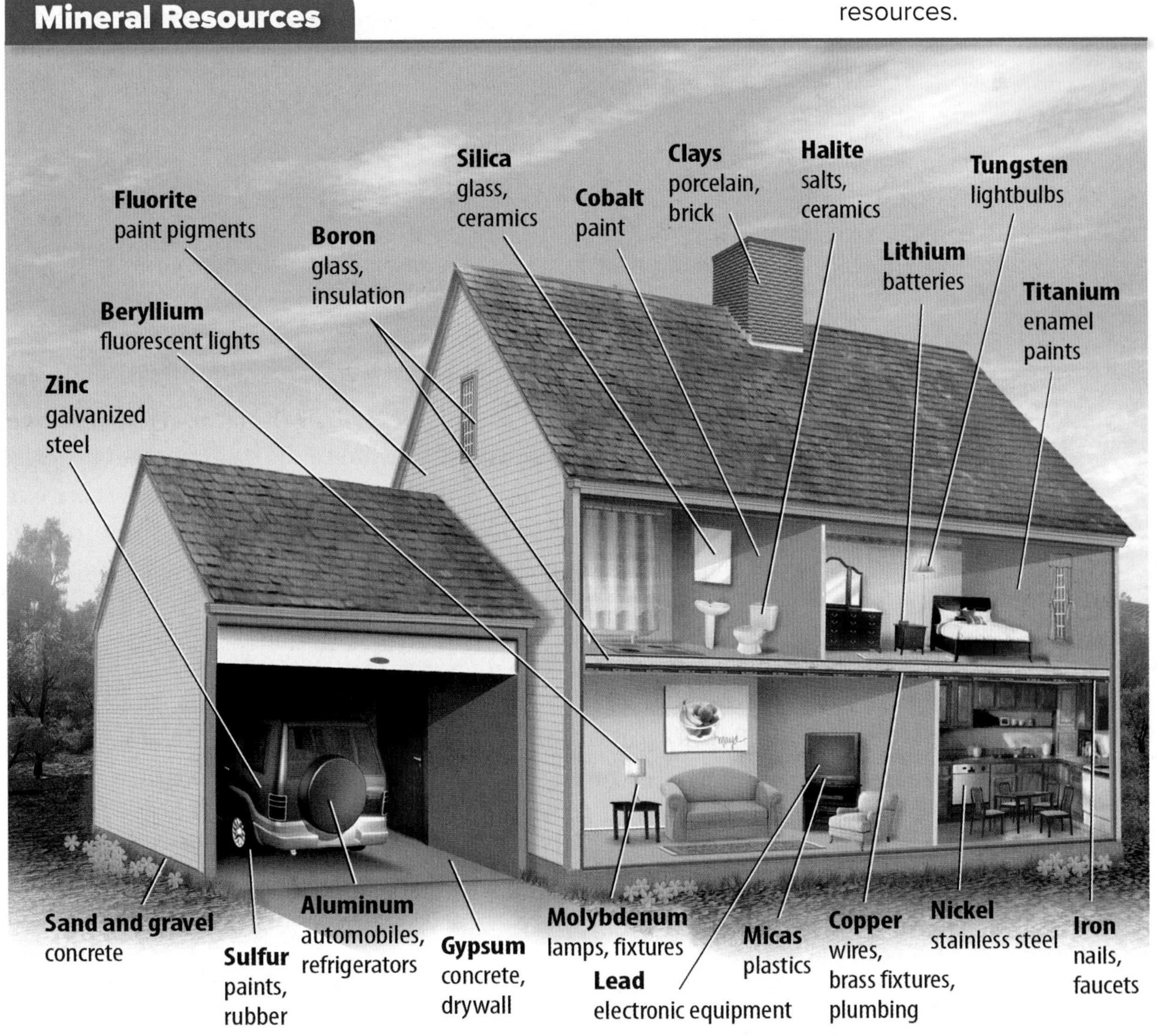

Figure 14 Many common products are made from mineral resources.

Visual Check Identify two products made from nonmetallic mineral resources.

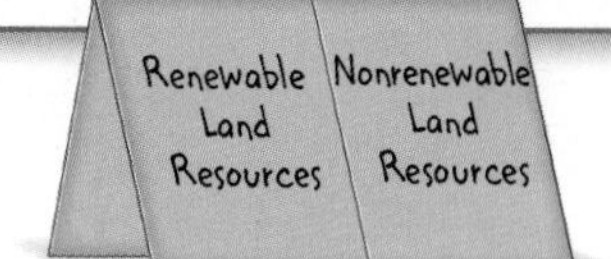

Make a horizontal two-tab book. Label the tabs as illustrated. Use the Foldable to record your notes about renewable and nonrenewable land resources.

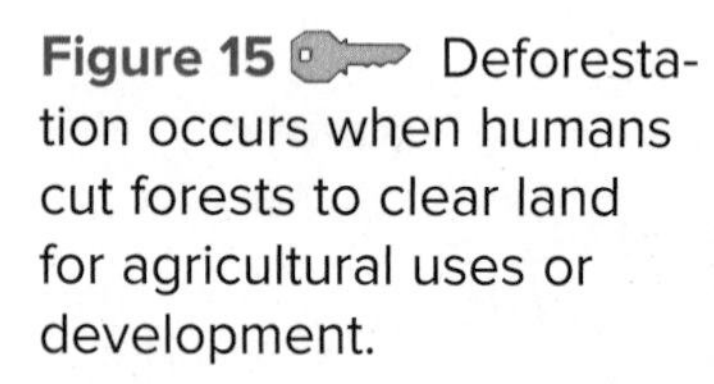

Advantages and Disadvantages of Using Land Resources

Land resources such as soil and forests are widely available and easy to access. In addition, crops and trees are renewable–they can be replanted and grown in a relatively short amount of time. These are all advantages of using land resources.

Some land resources, however, are nonrenewable. It can take millions of years for minerals to form. This is one disadvantage of using land resources. Other disadvantages include deforestation and pollution.

Deforestation

As shown in **Figure 15,** humans sometimes cut forests to clear land for grazing, farming, and other uses. **Deforestation** *is the cutting of large areas of forests for human activities.* It leads to soil erosion and loss of animal habitats. In tropical rain forests–complex ecosystems that can take hundreds of years to replace–deforestation is a serious problem.

Figure 15 Deforestation occurs when humans cut forests to clear land for agricultural uses or development.

Deforestation also can affect global climates. Trees remove carbon dioxide from the atmosphere during photosynthesis. Rates of photosynthesis decrease when large areas of trees are cut down, and more carbon dioxide remains in the atmosphere. Carbon dioxide helps trap thermal energy within Earth's atmosphere. Increased concentrations of carbon dioxide can cause Earth's average surface temperatures to increase.

Pollution

REVIEW VOCABULARY

runoff
rainwater that does not soak into the ground and flows over Earth's surface

Recall that runoff from coal mines can affect soil and water quality. The same is true of mineral mines. Runoff that contains chemicals from these mines can pollute soil and water. In addition, many farmers use chemical fertilizers to help grow crops. Runoff containing fertilizers can pollute rivers, soil, and underground water supplies.

Key Concept Check What are some advantages and disadvantages of using land resources?

Managing Land Resources

Because some land uses involve renewable resources while others do not, managing land resources is complex. For example, a tree is renewable. But forests can be nonrenewable because some can take hundreds of years to fully regrow. In addition, the amount of land is limited, so there is competition for space. Those who manage land resources must balance all of these issues.

Management Solutions

One way governments can manage forests and other unique ecosystems is by **preserving** them. On preserved land, logging and development is either banned or strictly controlled. Large areas of forests cannot be cut. Instead, loggers cut selected trees and then plant new trees to replace ones they cut.

SCIENCE USE V. COMMON USE

preserve

Science Use to keep safe from injury, harm, or destruction

Common Use to can, pickle, or save something for future use

Land mined for mineral resources also must be preserved. On both public and private lands, mined land must be restored according to government regulations.

Land used for farming and grazing can be managed to conserve soil and improve crop yield. Farmers can leave crop stalks after harvesting to protect soil from erosion. They also can use organic farming techniques that do not use synthetic fertilizers.

What You Can Do

You can help conserve land resources by recycling products made from land resources. You can use yard waste and vegetable scraps to make rich compost for gardening, reducing the need to use synthetic fertilizers. Compost is a mix of decayed organic material, bacteria and other organisms, and small amounts of water. **Figure 16** shows another way you can help manage land resources wisely.

Key Concept Check What can you do to help manage land resources wisely?

Jeff Greenberg/Alamy

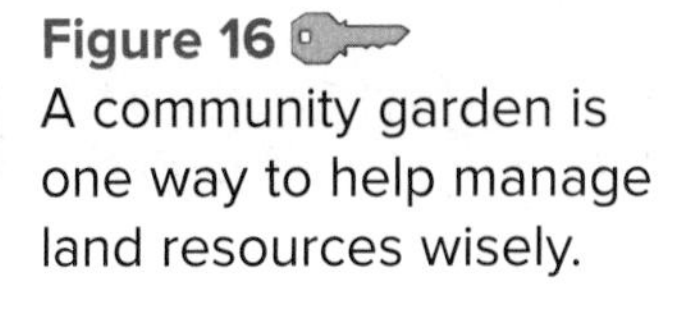

Figure 16 A community garden is one way to help manage land resources wisely.

Lesson 3 Review

Visual Summary

Land is a natural resource that humans use to meet their needs.

Disadvantages of using land as a resource include deforestation, which leads to increased erosion and increased carbon dioxide in the atmosphere.

Individuals can help manage land resources wisely by recycling, composting, and growing food in community gardens.

Use your lesson Foldable to review the lesson. Save your Foldable for the project at the end of the chapter.

What do you think NOW?

You first read the statements below at the beginning of the chapter.

5. Cities cover most of the land in the United States.

6. Minerals form over millions of years.

Did you change your mind about whether you agree or disagree with the statements? Rewrite any false statements to make them true.

Use Vocabulary

1 Cutting down forests for human activities is called ________.

2 **Use the word** *ore* in a sentence.

Understand Key Concepts 

3 One disadvantage of using metallic mineral resources is that these resources are

A. easy to mine. **C.** nonrenewable.
B. inexpensive. **D.** renewable.

4 **Give an example** of how people use land as a resource.

5 **Compare** the methods used by governments and individuals to manage land resources wisely.

Interpret Graphics

6 **Take Notes** Copy the graphic organizer below, and list at least two land resources mentioned in this lesson. Describe how using each affects the environment.

Land Resource	How Use Affects Environment

7 **Identify** whether the resources shown here are from metallic or nonmetallic mineral resources.

Critical Thinking

8 **Design** a way to manage land resources wisely. Use a method that is not discussed in this lesson.

9 **Decide** Land is a limited resource. There often is pressure to develop preserved land. Do you think this should happen? Why or why not?

(t)Karen Hunt/Getty Images; (b)Jeff Greenberg/Alamy

AMERICAN MUSEUM of NATURAL HISTORY

GREEN SCIENCE

A Greener Greensburg

A town struck by disaster makes the world a greener place.

In May 2007, a powerful tornado struck the small Kansas town of Greensburg. The tornado destroyed almost every home, school, and business. Six months later, the town's officials and residents decided to rebuild Greensburg as a model green community.

The town's residents pledged to use fewer natural resources; to produce clean, renewable energy; and to reuse and recycle waste. As part of this effort, every new home and building would be designed for energy efficiency. The homes also would be constructed of materials that are healthful for the people who live and work in them.

What is a model green town? Here are some ways Greensburg will help the environment, save money, and make life better for its residents.

▲ **Rain gardens help improve water quality by filtering pollutants from runoff.**

USE RENEWABLE ENERGY

- **Produce clean energy** with renewable energy sources such as wind and sunlight. Wind turbines capture the abundant wind power of the Kansas plains.
- **Cut back on greenhouse gas emissions** with electric or hybrid city vehicles.

BUILD GREEN BUILDINGS

- **Design every home, school, and office** to use less energy and promote better health.
- **Make the most of natural daylight** for indoor lighting with many windows, which also can be opened for fresh air.
- **Use green materials** that are nontoxic and locally grown or made from recycled materials.

CONSERVE WATER

- **Capture runoff and rainwater** with landscape features such as rain gardens, bowl-shaped gardens designed to collect and absorb excess rainwater.
- **Use low-flow** faucets, shower heads, and toilets.

CREATE A HEALTHY ENVIRONMENT

- **Provide parks and green spaces** filled with native plants that need little water or care.
- **Create a "walkable community"** to encourage people to drive less and be more active, with a town center connected to neighborhoods by sidewalks and trails.

PROBLEM SOLVING With your group, choose one of Greensburg's projects. Make a plan describing how it could be implemented in your community and what its benefits would be.

Lesson 4

Reading Guide

Key Concepts

ESSENTIAL QUESTIONS

- Why is it important to manage air and water resources wisely?
- How can individuals help manage air and water resources wisely?

Vocabulary

photochemical smog p. 670

acid precipitation p. 670

 Multilingual eGlossary

BrainPOP®

 Go to the resource tab in ConnectED to find the PBLs *Where in the world...?* and *Who's moving in next door?*

Air and Water Resources

Inquiry Are these crop circles?

No, this dotted landscape in Colorado is the result of circle irrigation. The fields are round because the irrigation equipment pivots from the center of the field and moves in a circle to water the crops. Crop irrigation accounts for about 34 percent of water used in the United States.

Launch Lab

20 minutes

How often do you use water each day?

In most places in the United States, people are fortunate to have an adequate supply of clean water. When you turn on the faucet, do you think about the value of water as a resource?

1. Prepare a two-column table to collect data on the number of times you use water in one day. Title the first column *Purpose* and the second column *Times Used.*
2. In the *Purpose* column, describe how you used the water, such as *Faucet, Toilet, Shower/Bath, Dishwasher, Laundry, Leaks,* and *Other.*
3. In the *Times Used* column, record and tally the total number of times you used water.
4. Calculate the percent that you use water for each category. Construct a circle graph showing the percentages of use in a day.

Think About This

1. For which purpose did you use water the most? The least?
2. **Key Concept** In which category, or categories, could you conserve water? How?

Importance of Air and Water

Using some natural resources, such as fossil fuels and minerals, makes life easier. You would miss them if they were gone, but you would still survive.

Air and water, on the other hand, are resources that you cannot live without. Most living things can survive only a few minutes without air. Oxygen from air helps your body provide energy for your cells.

Water also is needed for many life functions. As shown in **Figure 17,** water is the main component of blood. Water helps protect body tissues, helps maintain body temperature, and has a role in many chemical reactions, such as the digestion of food. In addition to drinking water, people use water for other purposes that you will learn about later in this lesson, including agriculture, transportation, and recreation.

Reading Check What are the functions of water in the human body?

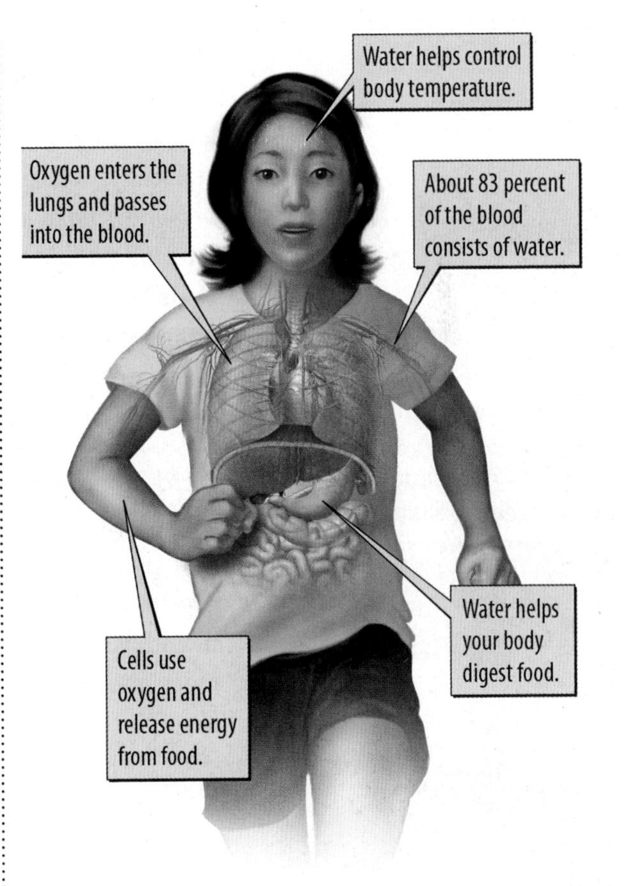

Figure 17 Your body needs oxygen and water to carry out its life-sustaining functions.

Personal Tutor

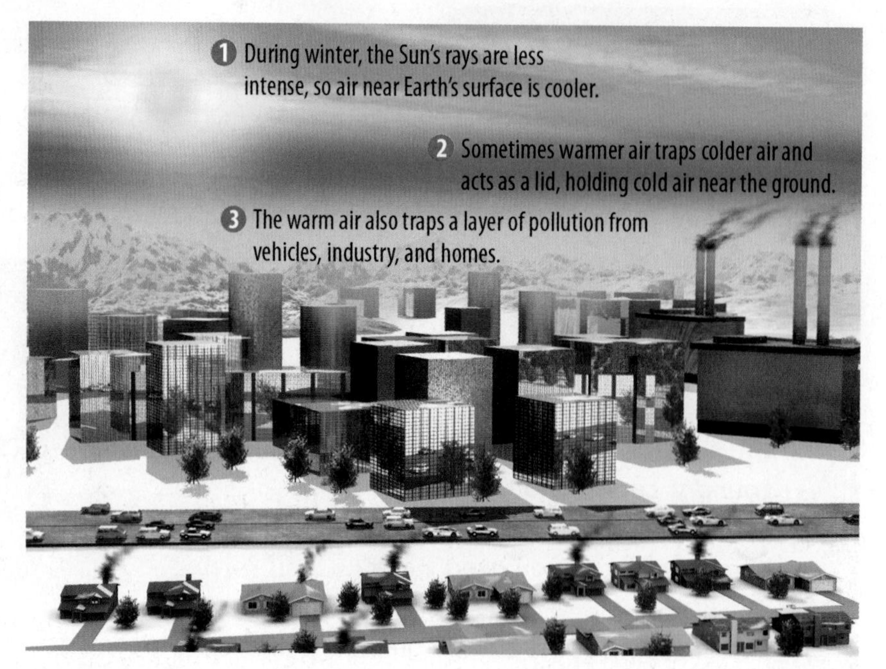

Figure 18 Sometimes a layer of warm air can trap smog in the cooler air close to Earth's surface. The smog can cover an area for days.

Visual Check Where does the pollution that forms smog come from?

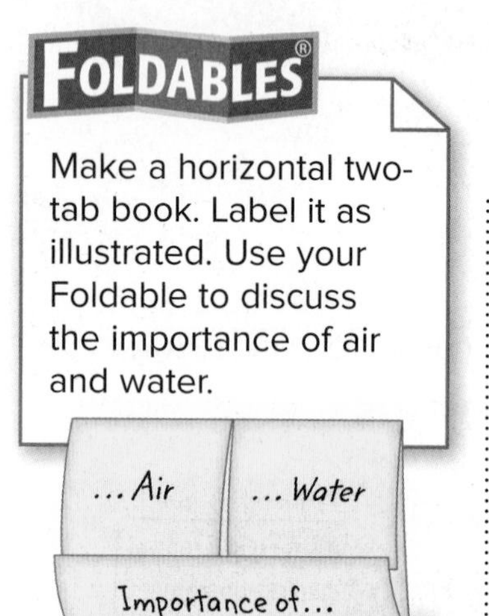

Air

Most living things need air to survive. However, polluted air, such as the air in **Figure 18,** can actually harm humans and other living things. Air pollution is produced when fossil fuels burn in homes, vehicles, and power plants. It also can be caused by natural events, such as volcanic eruptions or forest fires.

Reading Check What activities can cause air pollution?

Smog Burning fossil fuels releases not only energy, but also substances, such as nitrogen compounds. **Photochemical smog** *is a brownish haze produced when nitrogen compounds and other pollutants in the air react in the presence of sunlight.* Smog can irritate your respiratory system. In some individuals, it can increase the chance of asthma attacks. Smog can be particularly harmful when it is trapped under a layer of warm air and remains in an area for several days, also shown in **Figure 18.**

Acid Precipitation Nitrogen and sulfur compounds released when fossil fuels burn can react with water in the atmosphere and produce acid precipitation. **Acid precipitation** *is precipitation that has a pH less than 5.6.* When it falls into lakes, it can harm fish and other organisms. It also can pollute soil and kill trees and other plants. Acid precipitation can even damage buildings and statues made of some types of rocks.

Natural Events Forest fires and volcanic eruptions, such as the one shown in **Figure 19,** release gases, ash, and dust into the air. Dust and ash from one volcanic eruption can spread around the world. Materials from forest fires and volcanic eruptions can cause health problems similar to those caused by smog.

Figure 19 Gas and dust released by erupting volcanoes, such as Karymsky Volcano in Russia, can pollute the air.

Water

Suppose you saved $100, but you were only allowed to spend 90 cents. You might be very frustrated! If all of the water on Earth were your $100, freshwater that we can use is like that 90 cents you can spend. As shown in **Figure 20,** most water on Earth is salt water. Only 3 percent is freshwater, and most of that is frozen in glaciers. That leaves just a small part, 0.9 percent, of the total amount of water on Earth for humans to use.

This relatively small supply of freshwater must meet many needs. In addition to drinking water, people use water for farming, industry, electricity production, household activities, transportation, and recreation. Each of these uses can affect water quality. For example, water used to irrigate fields can mix with fertilizers. This polluted water then can run off into rivers and groundwater, reducing the quality of these water supplies. Water used in industry often is heated to high temperatures. The hot water can harm aquatic organisms when it is returned to the environment.

Reading Check How can farming affect water quality?

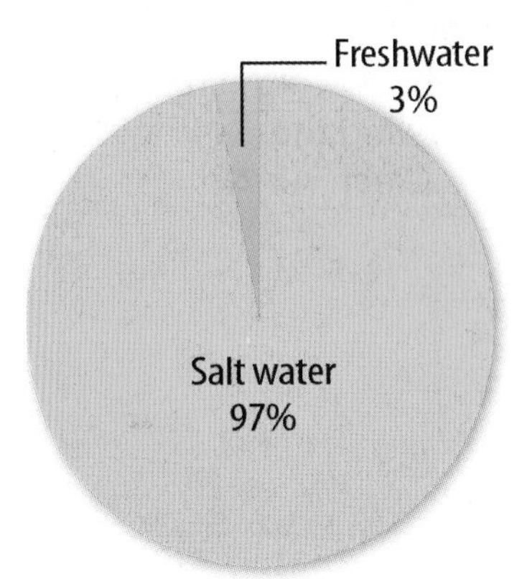

Figure 20 Freshwater makes up only 3 percent of Earth's water.

MiniLab

20 minutes

How much water can a leaky faucet waste?

You are competing for the job of environmental consultant at your school. One of the competition requirements is to complete an analysis of water waste from existing faucets.

1. Read and complete a lab safety form.
2. Catch the water from a **leaking faucet** in a **beaker.** Time the collection for 1 min with a **stopwatch.**
3. Use a **50-mL graduated cylinder** to measure the amount of water lost. Record the amount of water, in milliliters per minute, that leaked from the faucet in your Science Journal.
4. Make a table to show the amount of water that would leak from the faucet in 1 hour, 1 day, 1 week, 1 month, and 1 year.

Analyze and Conclude

1. **Construct** a graph of your data. Label the axes and title your graph. Explain what the graph illustrates.
2. **Describe** how many liters of water would be wasted by the leak over a period of one year. Explain how you arrived at that figure.
3. **Key Concept** As an environmental consultant, what information and recommendations would your report contain about water waste in the school?

Figure 21 The amount of sulfur compounds in the atmosphere decreased following the passage of the Clean Air Act.

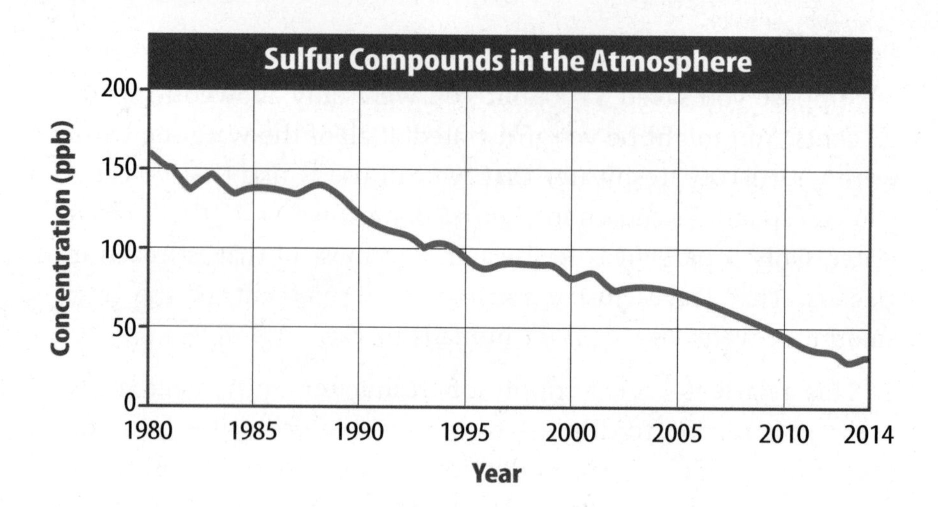

Math Skills 

Use Percentages

The carbon monoxide (CO) level in Portland air went from 6.3 parts per million (ppm) in 1990 to 1.3 ppm in 2014. What was the percent change in CO levels?

1. Subtract the starting value from the final value.
   $1.3 \text{ ppm} - 6.3 \text{ ppm} = -5.0 \text{ ppm}$
2. Divide the difference by the starting value.
   $-5.0 \text{ ppm}/6.3 \text{ ppm} = -0.794$
3. Multiply by 100 and add a % sign.
   $-0.794 \times 100 = -79.4\%$.

It decreased by 79.4%.

Practice

Between 1990 and 2014 the ozone (O_3) levels in Richmond went from 0.084 ppm to 0.062 ppm. What was the percent change in ozone levels?

 Math Practice

 Personal Tutor

Managing Air and Water Resources

Animals and plants do not use natural resources to produce electricity or to raise crops. But they do use air and water. Management of these important resources must consider both human needs and the needs of other living things.

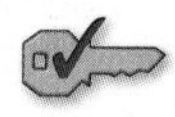 **Key Concept Check** Why is it important to manage air and water resources wisely?

Management Solutions

Legislation is an effective way to reduce air and water pollution. The regulations of the U.S. Clean Air Act, passed in 1970, limit the amount of certain pollutants that can be released into the air. The graph in **Figure 21** shows how levels of sulfur compounds have decreased since the act became law.

Similar laws are now in place to maintain water quality. The U.S. Clean Water Act legislates the reduction of water pollution. The Safe Drinking Water Act legislates the protection of drinking water supplies. By reducing pollution, these laws help ensure that all living things have access to clean air and water.

What You Can Do

You have learned that reducing fossil fuel use and improving energy efficiency can reduce air pollution. You can make sure your home is energy efficient by cleaning air-conditioning or heating filters and using energy-saving lightbulbs.

You can help reduce water pollution by properly disposing of harmful chemicals so that less pollution runs off into rivers and streams. You can volunteer to help clean up litter from a local stream. You also can conserve water so there is enough of this resource for you and other living things in the future.

 **Key Concept Check** How can individuals help manage air and water resources wisely?

Lesson 4 Review

Visual Summary

Sources of air pollution include the burning of fossil fuels in vehicles and power plants, and natural events such as volcanic eruptions and forest fires.

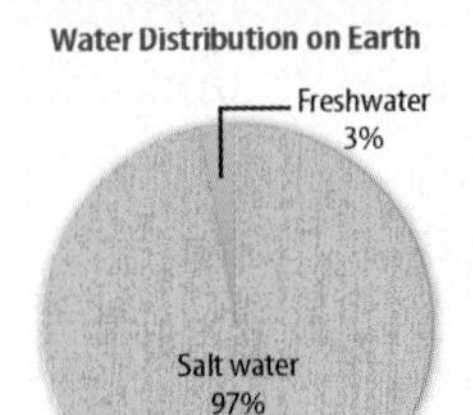

Only a small percentage of Earth's water is available for humans to use. Humans use water for agriculture, industry, recreation, and cleaning.

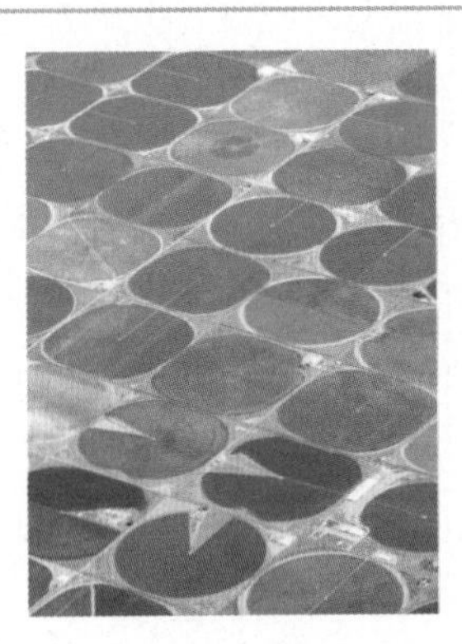

Management of air and water resources includes passing laws that regulate sources of air and water pollution. Individuals can reduce energy use and dispose of chemicals properly to help keep air and water clean.

FOLDABLES®

Use your lesson Foldable to review the lesson. Save your Foldable for the project at the end of the chapter.

What do you think NOW?

You first read the statements below at the beginning of the chapter.

7. Humans need oxygen and water to survive.

8. About 10 percent of Earth's total water can be used by humans.

Did you change your mind about whether you agree or disagree with the statements? Rewrite any false statements to make them true.

Use Vocabulary

1. **Define** *acid precipitation* in your own words.
2. Air pollution caused by the reaction of nitrogen compounds and other pollutants in the presence of sunlight is ________.

Understand Key Concepts

3. About how much of Earth's water is available for humans to use?
   - **A.** 0.01 percent
   - **B.** 0.90 percent
   - **C.** 3.0 percent
   - **D.** 97.0 percent
4. **Relate** In terms of human health, why is it important to manage air resources wisely?
5. **List** ways your classroom could improve its energy efficiency.

Interpret Graphics

6. **Determine Cause and Effect** Copy and fill in the graphic organizer below to describe three effects of acid precipitation.

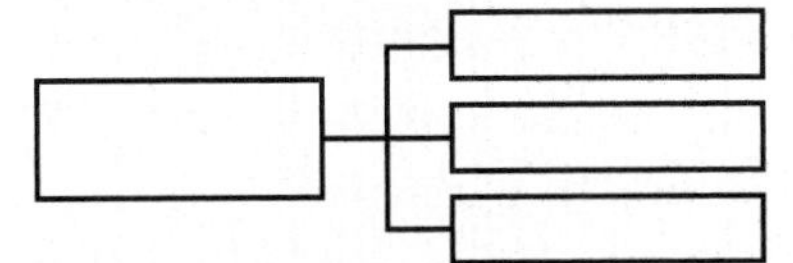

Critical Thinking

7. **Evaluate** The top three categories of household water use in the United States are flushing the toilet, washing clothes, and taking showers. Evaluate your water use, and list one thing you could do to reduce your use in each category.

Math Skills

8. Between 1990 and 2014, the amount of sulfur dioxide (SO_2) in Miami's air went from 14 ppb to 1 ppb. What was the percent change of SO_2?

1–3 class periods

Research Efficient Energy and Resource Use

A community organization is encouraging your school's board of education to participate in the "Green Schools" program. Your class has been nominated to research and report on the present status of energy efficiency and resource use in the school. The results of the report will be used as information for the presentation. Your task is to choose a natural resource and collect data about how it is presently used in the school. Your group will then recommend environmentally responsible management practices.

Question

How can a natural resource be used more wisely at school?

Procedure

1. Read and complete a lab safety form.
2. With your group, choose one of these resources to research its use in your school: water, land, air, or an energy resource.
3. For your chosen resource, plan how you will research resource use. What questions will you ask? How much of the resource is used by the school? Is it used efficiently? How could it be used more efficiently, or how could it be conserved? Have your teacher approve your plan.
4. Prepare data collection forms like the one below to record the results of your research in your Science Journal.
5. Conduct your research, and enter the data on the forms.

Sample Data Table				
Resource: Water Areas of Research: Water Loss Through Leaks and Recycling System				
Location	Faucets	Water Fountains	Toilets	"Gray" Water Recycling System
Washroom	6 good 2 poor 2 leaking		4 good 1 leaking	no
Hallway		3 good 1 leaking		no
Classroom 101	1 good			no
Classroom 102	1 good			no
Classroom 103	1 leaking			no

6. Review and summarize the data. Perform any necessary calculations to convert values to annual usage.
7. Conduct interviews, or collect more data about areas of research for which you need additional information.
8. After analyzing your data, write a proposal suggesting how the resource can be wisely managed in your school.
9. Compare the elements you addressed in your research with those recommended by a state or a national environmental organization. Did your research include everything?
10. Modify your proposal, if necessary. Record your revisions in your Science Journal.

Analyze and Conclude

11. **Graph** and explain the results of your data analysis.
12. **Predict** one impact on the environment of the existing management practices of the resource that you audited.
13. **The Big Idea** Describe two recommendations that you would make to the school's board of education about changes in resource management practices.

Communicate Your Results

Present the results of your research and your proposal to the class. Use appropriate visual aids to help make your points.

Extension

Combine information and reports from groups that investigated other resources from the list in step 2 so that all four resources are represented. Make a final report that includes recommendations for efficient use of each resource at your school.

Remember to use scientific methods.

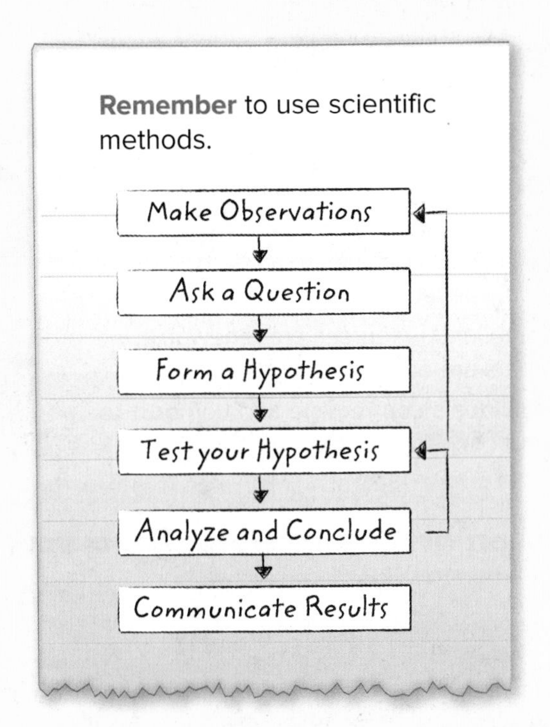

Corbis/SuperStock

Chapter 18 Study Guide

Wise management of natural resources helps extend the supply of nonrenewable resources, reduce pollution, and improve soil, air, and water quality.

Key Concepts Summary

Lesson 1: Energy Resources

- **Nonrenewable resources** include fossil fuels and uranium, which is used for **nuclear energy.**
- Nonrenewable energy resources are widely available and easy to convert to energy. However, using these resources can cause pollution and habitat disruption. Safety concerns also are an issue.
- People can conserve energy to help manage these resources.

Vocabulary

nonrenewable resource p. 643
renewable resource p. 643
nuclear energy p. 647
reclamation p. 649

Lesson 2: Renewable Energy Resources

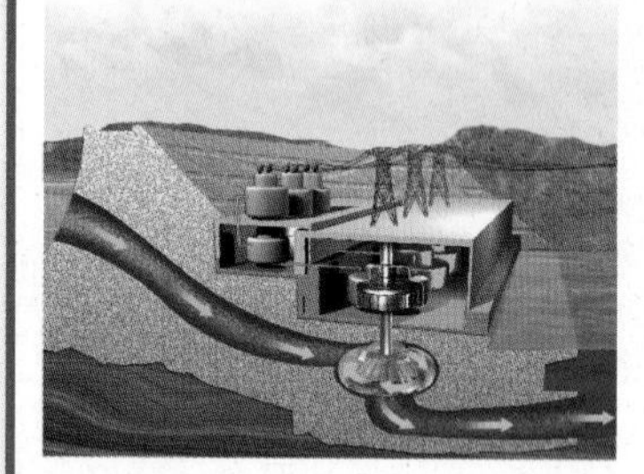

- Renewable energy resources include **solar energy,** wind energy, water energy, **geothermal energy,** and **biomass energy.**
- Renewable resources cause little to no pollution. However, some types of renewable energy are costly or limited to certain areas.
- Individuals can help educate others about renewable resources.

Vocabulary

solar energy p. 653
wind farm p. 654
hydroelectric power p. 654
geothermal energy p. 655
biomass energy p. 655

Lesson 3: Land Resources

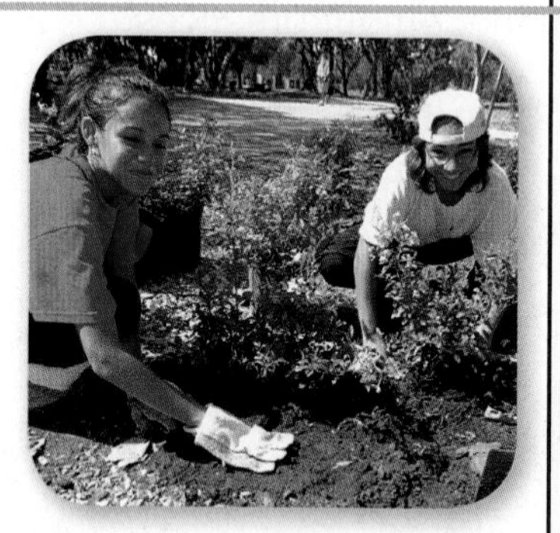

- Land is considered a resource because it is used by living things to meet their needs for food, shelter, and other things.
- Some land resources are renewable, while others are not.
- Individuals can recycle and compost to help conserve land resources.

Vocabulary

ore p. 663
deforestation p. 664

Lesson 4: Air and Water Resources

- Most living things cannot survive without clean air and water.
- Individuals can make their homes and schools more energy efficient.

Vocabulary

photochemical smog p. 670
acid precipitation p. 670

Study Guide

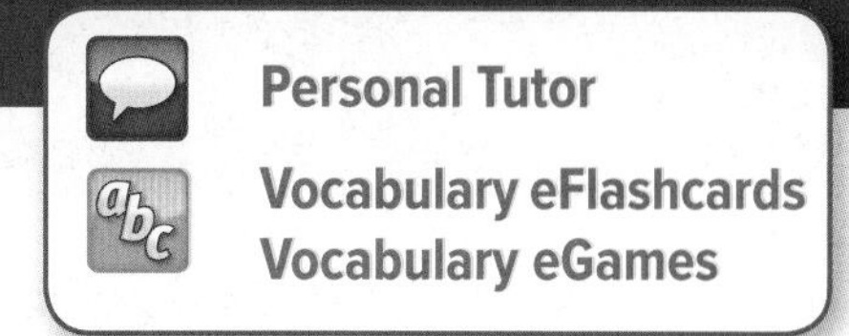

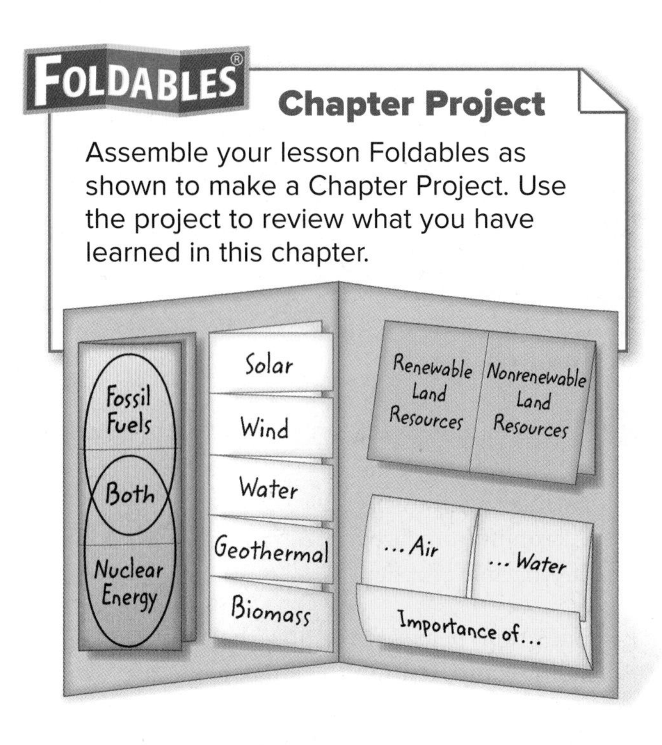

Use Vocabulary

1. Distinguish between renewable resources and nonrenewable resources.
2. Replace the underlined words with the correct vocabulary word: Energy produced from atomic reactions can be used to generate electricity.
3. How does biomass energy differ from geothermal energy?
4. Energy from the Sun is ________.
5. Define the term *ore* in your own words.
6. Distinguish between photochemical smog and acid precipitation.

Link Vocabulary and Key Concepts

Copy these concept maps, and then use vocabulary terms from the previous page and other terms from the chapter to complete the concept maps.

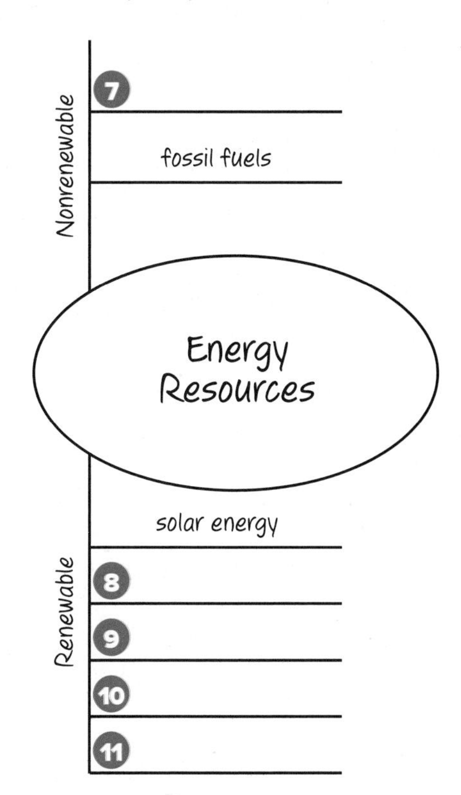

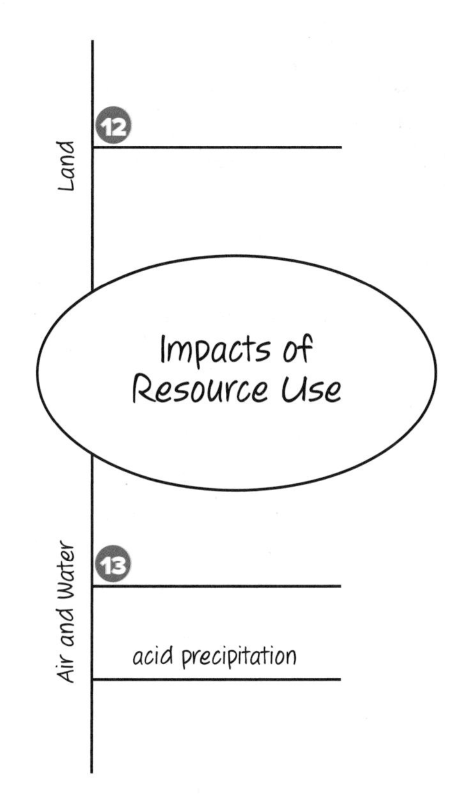

Chapter 18 Review

Understand Key Concepts 

1. Which energy source produces radioactive waste?
   A. biomass
   B. geothermal
   C. hydroelectric power
   D. nuclear power

2. The table below shows the energy sources used to produce electricity in the United States. What can you infer from the table?

Electricity Production	
Energy Source	**Percent**
Coal	33
Natural gas	33
Nuclear power	20
Solar, wind, geothermal, biomass	7
Hydroelectric power	6
Oil	1
Other	<1

   A. About 19 percent of U.S. electricity comes from renewable sources.
   B. Hydroelectric power is more widely used for electricity than nuclear power.
   C. About 87 percent of U.S. electricity comes from nonrenewable sources.
   D. Oil is more widely used for electricity than hydroelectric power.

3. Which factor would best determine whether a home is suitable for solar energy?
   A. difference in tidal heights
   B. strength of daily winds
   C. nearness to tectonically active areas
   D. number of sunny days per year

4. Which product comes from a metallic mineral resource?
   A. aluminum
   B. drywall
   C. gravel
   D. table salt

5. Which is a renewable land resource?
   A. forests
   B. minerals
   C. soil
   D. trees

6. Where is most water on Earth located?
   A. lakes
   B. oceans
   C. rivers
   D. underground

7. Which natural event can result in air pollution?
   A. burning fossil fuels
   B. littering a stream
   C. runoff from farms
   D. volcanic eruption

8. The graph below shows how the amount of sulfur compounds in the atmosphere has changed since the passage of the Clean Air Act. Based on the data in the graph, what can you infer about the act?

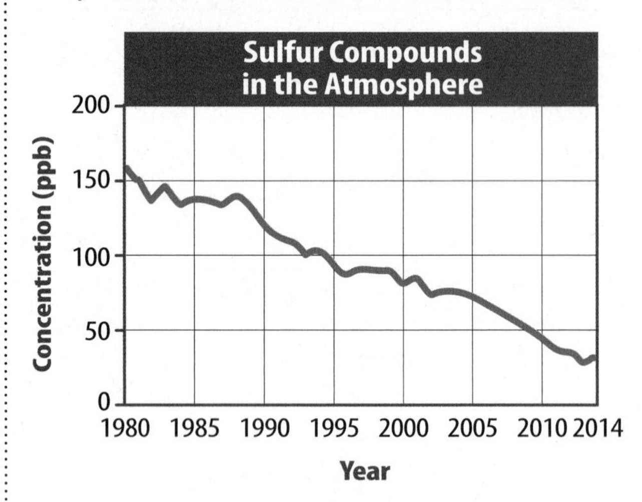

   A. The act has helped decrease pollutants in the atmosphere.
   B. The act has helped increase pollutants in the atmosphere.
   C. The act has incentives for use of renewable resources.
   D. The act has not impacted the amount of pollutants in the atmosphere.

Critical Thinking

9 **Organize** the list of energy sources into renewable and nonrenewable energy resources.

• coal	• nuclear energy
• solar energy	• wind energy
• oil	• natural gas
• geothermal energy	• tidal power
• hydroelectric power	• biomass

10 **Create** a cartoon showing a chain reaction in a nuclear power plant.

11 **Compare** hydroelectric and tidal power.

12 **Design** a way to use passive solar energy in your classroom.

13 **Distinguish** between geothermal energy and solar energy.

14 **Consider** What factors must governments consider when managing land resources?

15 **Evaluate** the use of forests as natural resources. Do the advantages outweigh the disadvantages? Explain.

16 **Infer** When would you expect more smog to form—on cloudy days or on sunny days? Explain.

17 **Design** a way to remove salt from salt water. Then evaluate your plan. Could it be used to produce freshwater on a large scale? Why or why not?

18 **Formulate** a way to demonstrate the importance of air and water resources to younger students.

Writing in Science

19 **Compose** a song about vampire energy. The lyrics should describe vampire energy and explain how it can be reduced.

REVIEW

20 Select a natural resource and explain why it is important to manage the resource wisely.

21 Suppose the house below is heated by electricity produced from burning coal. Which areas of the house have the greatest loss of thermal energy? Why is it important for this house to reduce thermal energy loss?

Math Skills

Use Percentages

22 Between 2012 and 2013, the carbon monoxide level in the air in Denver, Colorado, went from 2 ppm to 2.5 ppm. What was the percent change in CO?

23 There often is a considerable difference between pollutants in surface water and pollutants in groundwater in the same area. For example, in Portland, Oregon, there were 4.6 ppm of sulfates in the groundwater and 0.9 ppm in the surface water. What was the percent difference? (Hint: Use 4.6 ppm as the starting value.)

Standardized Test Practice

Record your answers on the answer sheet provided by your teacher or on a sheet of paper.

Multiple Choice

1 Which activity does NOT reduce the use of fossil fuels?

A riding a bicycle to school

B unplugging DVD players

C walking to the store

D watering plants less often

Use the graph below to answer questions 2 and 3.

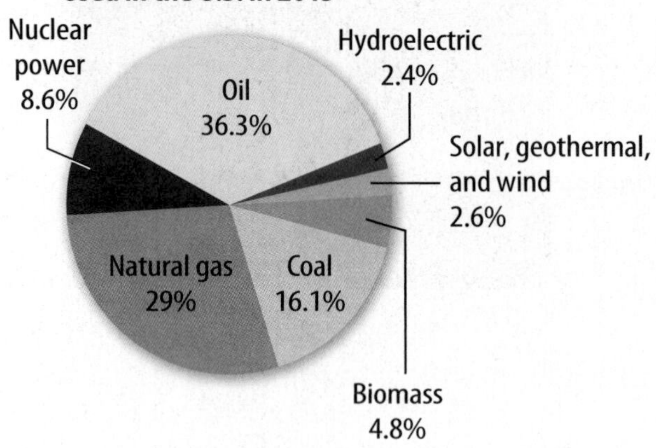

2 Which is the most-used renewable energy resource in the United States?

A biomass

B hydroelectric

C natural gas

D nuclear energy

3 What percentage of the energy used in the United States comes from burning fossil fuels?

A About 40 percent

B About 45 percent

C About 80 percent

D About 93 percent

4 Which practice emphasizes the use of renewable energy resources?

A buying battery-operated electronics

B installing solar panels on buildings

C replacing sprinklers with watering cans

D teaching others about vampire energy

5 Which is a nonrenewable land resource?

A crops

B minerals

C streams

D trees

Use the figure below to answer question 6.

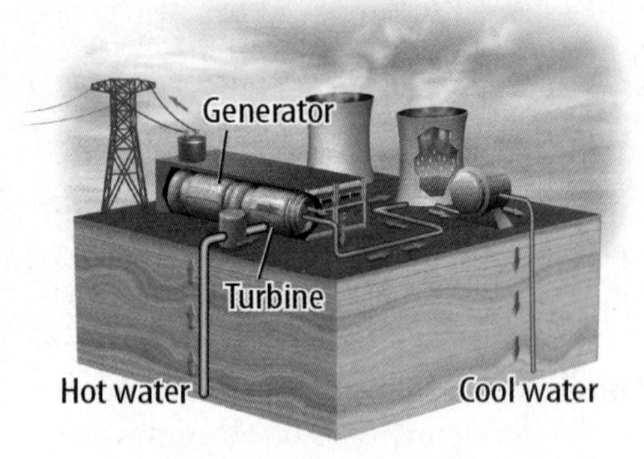

6 Which alternative energy resource is used to make electricity in the figure?

A solar energy

B tidal power

C geothermal energy

D hydroelectric power

7 Which practice is a wise use of land resources?

A composting

B conserving water

C deforestation

D strip mining

Use the figure below to answer question 8.

8 Which type of air pollution is labeled *A* in the figure?

A acid precipitation
B fertilizer runoff
C nuclear waste
D photochemical smog

9 Approximately how much water on the Earth is in oceans?

A 1 percent
B 3 percent
C 75 percent
D 97 percent

10 Which is a source of biomass energy?

A sunlight
B uranium
C wind
D wood

Constructed Response

Use the figure below to answer questions 11 and 12.

11 Which resource powers the turbine in the figure? Describe what happens at steps A–D to produce electricity.

12 What are two advantages and two disadvantages of producing electricity in the way shown in the figure?

13 Describe an example of how forests are used as a resource. What is one advantage of using the resource in this way? What is a disadvantage?

14 Agree or disagree with the following statement: "Known oil reserves will last only another 50 years. Thus, the United States should build more nuclear power plants to deal with the upcoming energy shortage." Support your answer with at least two advantages or two disadvantages of using nuclear energy.

NEED EXTRA HELP?														
If You Missed Question...	1	2	3	4	5	6	7	8	9	10	11	12	13	14
Go to Lesson...	1	2	1	2	3	2	3	4	4	2	2	2	3	1

Unit 5

EXPLORING THE UNIVERSE

1600 B.C. Babylonian texts show records of people observing Venus without the aid of technology. Its appearance is recorded for 21 years.

265 B.C. Greek astronomer Timocharis makes the first recorded observation of Mercury.

1610 Galileo Galilei observes the four largest moons of Jupiter through his telescope.

1613 Galileo records observations of the planet Neptune but mistakes it for a star.

1655 Astronomer Christiaan Huygens observes Saturn and discovers its rings, which were previously thought to be large moons on each side.

1781 William Herschel discovers the planet Uranus.

1930
Clyde Tombaugh discovers Pluto, making him the first American to discover a planet.

1971
Mariner 9 visits Mars and becomes the first human-made object to orbit a planet other than Earth.

2006
After research and consideration, the International Astronomical Union votes to remove Pluto from the list of planets in the solar system.

Visit ConnectED for this unit's **STEM** activity.

Unit 5

Nature of SCIENCE

Technology

It may sound strange, but some of the greatest benefits of the space program are benefits to life here on Earth. Devices ranging from hand-held computers to electric socks rely on technologies first developed for space exploration. **Technology** is the practical application of science to commerce or industry. Space technologies have increased our understanding of Earth and our ability to locate and conserve resources.

Problems, such as how best to explore the solar system and outer space, often send scientists on searches for new knowledge. Engineers use scientific knowledge to develop new technologies for space. Then, some of those technologies are modified to solve problems on Earth. For example, lightweight solar panels on the outside of a spacecraft convert the Sun's energy into electricity that powers the spacecraft for long space voyages. Similar but smaller, flexible solar panels, as shown in **Figure 1** are now available for consumers to purchase. They can be used to power small electronics when traveling. **Figure 2** shows how other technologies from space help conserve natural resources.

Figure 1 Lightweight, flexible solar cells developed for spacecraft help to conserve Earth's resources.

This image was taken by the *Terra* satellite and shows fires burning in California. The image helps firefighters see the size and the location of the fires. It also helps scientists study the effect of fires on Earth's atmosphere. ▼

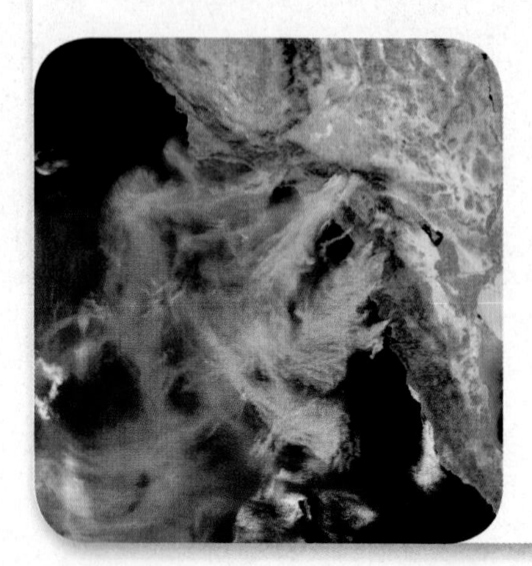

Water purification systems use technologies developed to provide safe, clean drinking water for astronauts. On Earth, systems like these can provide safe drinking water for an entire village in a remote area or supply drinking water after a natural disaster.

Engineers developed glass spheres about the size of a grain of flour to insulate super-cold spacecraft fuel lines. Similar microspheres act as insulators when mixed with paints. This technology can help reduce the energy needed to heat and cool buildings. ▼

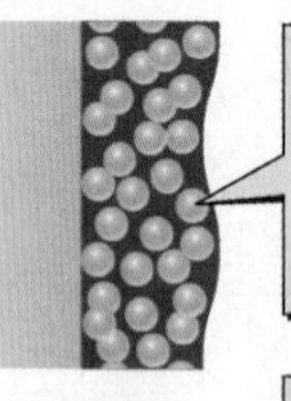

Figure 2 Some technologies developed as part of the space program have greatly benefited life on Earth.

Figure 3 The satellite image on the left is similar to what you would see with your eyes from space. A satellite sensor that detects other wavelengths of light produced the colored satellite image on the right. It shows the locations of nearly a dozen different minerals.

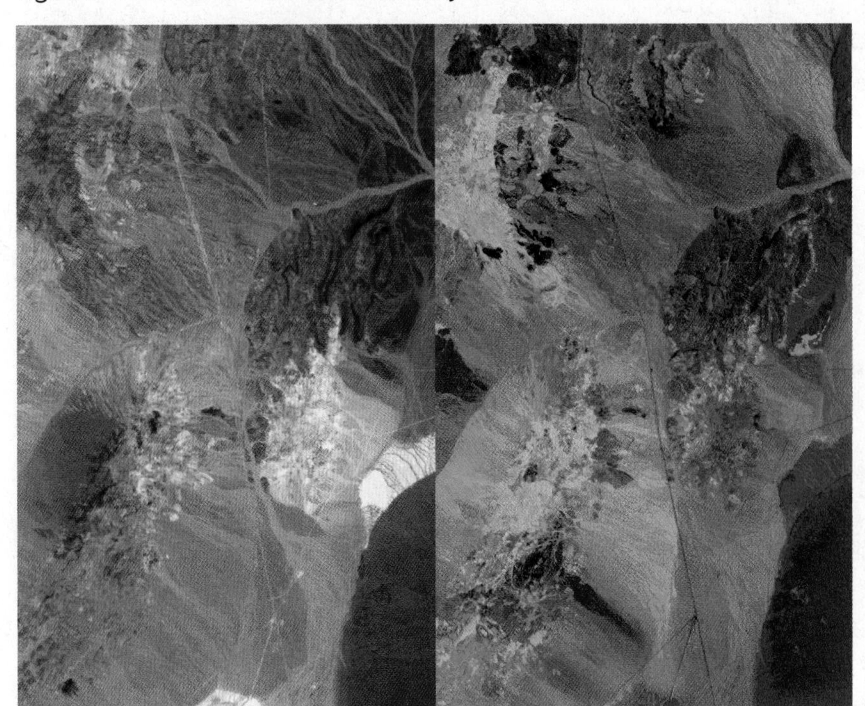

Solving Problems and Improving Abilities

Science and technology depend on each other. For example, images from space greatly improve our understanding of Earth. **Figure 3** above shows a satellite image of a Nevada mine. The satellite is equipped with sensors that detect visible light, much like your eyes do. The image on the right shows a satellite image of the same site taken with a sensor that detects wavelengths of light your eyes cannot see. This image provides information about the types of minerals in the mine. Each color in the image on the right shows the location of a different mineral, reducing the time it takes geologists to locate mineral deposits.

Scientists use other kinds of satellite sensors for different purposes. Engineers have modified space technology to produce satellite images of cloud cover over Earth's surface, as shown in **Figure 4**. Images like this one improve global weather forecasting and help scientists understand changes in Earth's atmosphere. Of course, science can answer only some of society's questions, and technology cannot solve all problems. But together, they can improve the quality of life for all.

How would you use space technology?

Many uses for space technology haven't been discovered yet. Can you develop one?

1. Identify a problem locating, protecting, or preserving resources that could be solved using space technology.
2. Prepare a short oral presentation explaining your technology.

Analyze and Conclude

1. **Describe** How is your technology used in space?
2. **Explain** How do you use technology to solve a problem on Earth?

Figure 4 This satellite image shows cloud cover over the southwest region of the United States.

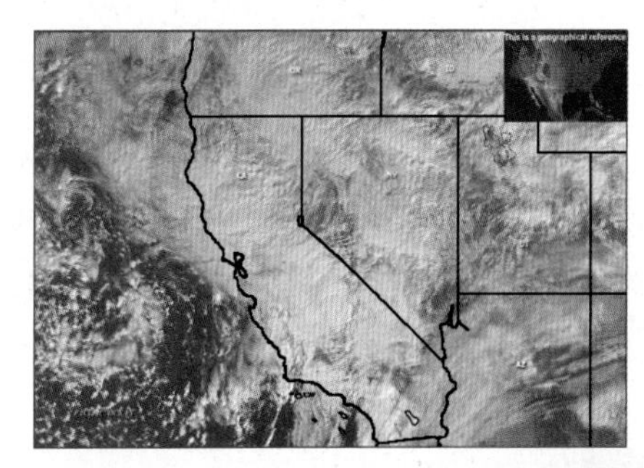

Chapter 19

Exploring Space

THE BIG IDEA **How do humans observe and explore space?**

Inquiry Can satellites see into space?

Yes, they can! The satellite shown here is a telescope. It collects light from distant objects in space. But, most satellites you might be familiar with point toward Earth. They provide navigation assistance, monitor weather, and bounce communication signals to and from Earth.

- Why would scientists want to put a telescope in space?
- In what other ways do scientists observe and explore space?
- What are goals of some current and future space missions?

Get Ready to Read

What do you think?

Before you read, decide if you agree or disagree with each of these statements. As you read this chapter, see if you change your mind about any of the statements.

1. Astronomers put telescopes in space to be closer to the stars.
2. Telescopes can work only using visible light.
3. Humans have walked on the Moon.
4. Some orthodontic braces were developed using space technology.
5. Humans have landed on Mars.
6. Scientists have detected water on other bodies in the solar system.

Your one-stop online resource
connectED.mcgraw-hill.com

LearnSmart®

Chapter Resources Files, Reading Essentials, Get Ready to Read, Quick Vocabulary

Animations, Videos, Interactive Tables

Self-checks, Quizzes, Tests

Project-Based Learning Activities

Lab Manuals, Safety Videos, Virtual Labs & Other Tools

Vocabulary, Multilingual eGlossary, Vocab eGames, Vocab eFlashcards

Personal Tutors

Lesson 1

Observing the Universe

Reading Guide

Key Concepts
ESSENTIAL QUESTIONS

- How do scientists use the electromagnetic spectrum to study the universe?
- What types of telescopes and technology are used to explore space?

Vocabulary

electromagnetic spectrum p. 690
refracting telescope p. 692
reflecting telescope p. 692
radio telescope p. 693

Multilingual eGlossary

Science Video

How can you see this?

This is an expanding halo of dust in space, illuminated by the light from the star in the center. This photo was taken with a telescope. How do you think telescopes obtain such clear images?

NASA and The Hubble Heritage Team (AURA/STScI)

Launch Lab

15 minutes

Do you see what I see?

Your eyes have lenses. Eyeglasses, cameras, telescopes, and many other tools involving light also have lenses. Lenses are transparent materials that refract light, or cause light to change direction. Lenses can cause light rays to form images as they come together or move apart.

1. Read and complete a lab safety form.
2. Place each of the **lenses** on the words of this sentence.
3. Slowly move each lens up and down over the words to observe if or how the words change. Record your observations in your Science Journal.
4. Hold each lens at arm's length and focus on an object a few meters away. Observe how the object looks through each lens. Make simple drawings to illustrate what you observe.

Think About This

1. What happened to the words as you moved the lenses toward and away from the sentence?
2. What did the distant object look like through each lens?
3. **Key Concept** How do you think lenses are used in telescopes to explore space?

Observing the Sky

If you look up at the sky on a clear night, you might be able to see the Moon, planets, and stars. These objects have not changed much since people first turned their gaze skyward. People in the past spent a lot of time observing the sky. They told stories about the stars, and they used stars to tell time. Most people thought Earth was the center of the universe.

Astronomers today know that Earth is part of a system of eight planets revolving around the Sun. The Sun, in turn, is part of a larger system called the Milky Way galaxy that contains billions of other stars. And the Milky Way is one of billions of other galaxies in the universe. As small as Earth might seem in the universe, it could be unique. Scientists have not found life anywhere else.

One advantage astronomers have over people in the past is the **telescope**. Telescopes enable astronomers to observe many more stars than they could with their eyes alone. Telescopes gather and focus light from objects in space. The photo on the opposite page was taken with a telescope that orbits Earth. Astronomers use many kinds of telescopes to study the energy emitted by stars and other objects in space.

Reading Check What is the purpose of telescopes?

WORD ORIGIN

telescope
from Greek *tele,* means "far"; and Greek *skopos,* means "seeing"

FOLDABLES®

Use seven quarter-sheets of paper to make a tabbed diagram illustrating the electromagnetic spectrum. In the middle section of a large shutterfold project, tape or glue the left edges of the tabs so they overlap to illustrate the varying sizes of the waves from longest to shortest.

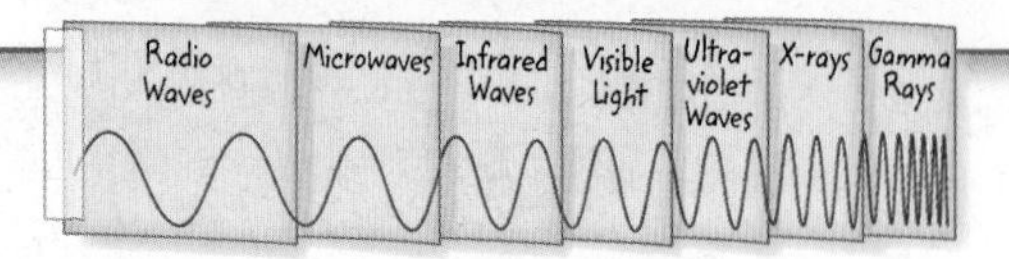

Personal Tutor

Figure 1 Objects emit radiation in continuous wavelengths. Most wavelengths are not visible to the human eye.

Visual Check Approximately how long are the wavelengths of microwaves?

Relative size
Wavelength
Uses
Name
10^{-6} nm
10^{-4} nm
10^{-2} nm
1 nm
10 nm
1 µm
100 µm
1 mm
1 m
100 m
1 km
100 km
Gamma rays
X-rays
Ultraviolet waves
Visible light
Infrared waves
Microwaves
Radio waves
Increasing wavelength
Ada 17 KILOMETERS
Long wavelength, low energy
Short wavelength, high energy

Electromagnetic Waves

Stars emit energy that radiates into space as electromagnetic (ih lek troh mag NEH tik) waves. Electromagnetic waves are different from mechanical waves, such as sound waves. Sound waves can transfer energy through solids, liquids, or gases. Electromagnetic waves can transfer energy through matter or through a vacuum, such as space. The energy they carry is called radiant energy.

The Electromagnetic Spectrum

The entire range of radiant energy carried by electromagnetic waves is the **electromagnetic spectrum.** As shown in **Figure 1,** waves of the electromagnetic spectrum are continuous. They range from gamma rays with short wavelengths at one end to radio waves with long wavelengths at the other end. Radio waves can be thousands of kilometers in length. Gamma rays, which are used in fighting cancer cells, can be smaller in length than the size of an atom.

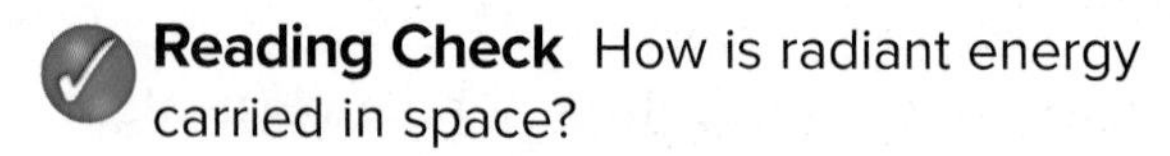

Reading Check How is radiant energy carried in space?

Humans observe only a small part of the electromagnetic spectrum–visible light. Visible light includes all the colors you see. You cannot see the other parts of the electromagnetic spectrum, but you can use them. When you talk on a cellular phone, you use microwaves. When you change the TV channel with a remote-control device or view an object with thermal imaging, you use infrared waves.

Radiant Energy and Stars

Most stars emit energy in all wavelengths. But how much of each wavelength they emit depends on their temperatures. Hot stars emit mostly shorter waves with higher energy, such as X-rays, gamma rays, and ultraviolet waves. Cool stars emit mostly longer waves with lower energy, such as infrared waves and radio waves. The Sun has a medium temperature range. It emits much of its energy as visible light.

MiniLab

15 minutes

What is white light?

Sunlight and the light from an ordinary lightbulb are both examples of visible light. You might think that white light is all white. Is it?

1. Read and complete a lab safety form.
2. Darken the room, and shine a **flashlight** through a **prism** on a flat surface. Adjust the positions of the prism and the flashlight until you observe the entire visible light spectrum.
3. In your Science Journal, use **colored pencils** to draw what you see.

Analyze and Conclude

1. **Define** What is white light?
2. **Compare and Contrast** Which component of white light has the longest wavelength? Which has the shortest wavelength? Explain your answers.
3. **Key Concept** How does visible light fit into the electromagnetic spectrum?

Why You See Planets and Moons

Planets and moons are much cooler than even the coolest stars. They do not make their own energy and, therefore, do not emit light. However, you can see the Moon and the planets because they reflect light from the Sun.

Light from the Past

All electromagnetic waves, from radio waves to gamma rays, travel through space at a constant speed of 300,000 km/s. This is called the speed of light. The speed of light might seem incredibly fast, but the universe is very large. Even moving at the speed of light, it can take millions or billions of years for some light waves to reach Earth because of the large distances in space.

Because it takes time for light to travel, you see planets and stars as they were when their light started its journey to Earth. It takes very little time for light to travel within the solar system. Reflected light from the Moon reaches Earth in about 1 second. Light from the Sun reaches Earth in about 8 minutes. It reaches Jupiter in about 40 minutes.

Light from stars is much older. Some stars are so far away that it can take millions or billions of years for their radiant energy to reach Earth. Therefore, by studying energy from stars, astronomers can learn what the universe was like millions or billions of years ago.

Reading Check How is looking at stars like looking at the past?

Math Skills

Scientific Notation

Scientists use scientific notation to work with large numbers. Express the speed of light in scientific notation using the following process.

1. Move the decimal point until only one nonzero digit remains on the left.

   $300{,}000 \rightarrow 3.00000$

2. Use the number of places the decimal point moved (5) as a power of ten.

   $300{,}000 \text{ km/s} = 3.0 \times 10^5 \text{ km/s}$

Practice

The Sun is 150,000,000 km from Earth. Express this distance in scientific notation.

Math Practice

Personal Tutor

Hutchings Photography/Digital Light Source

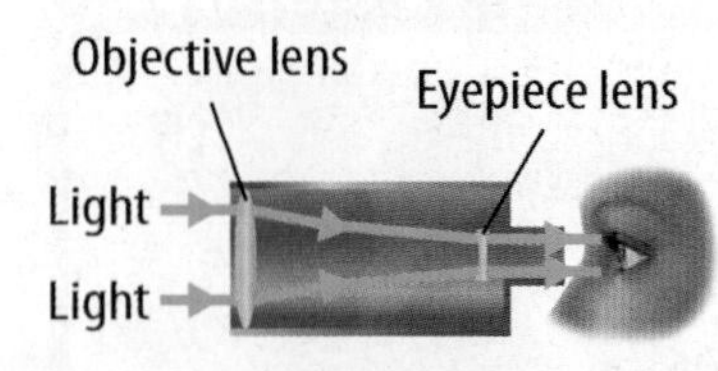

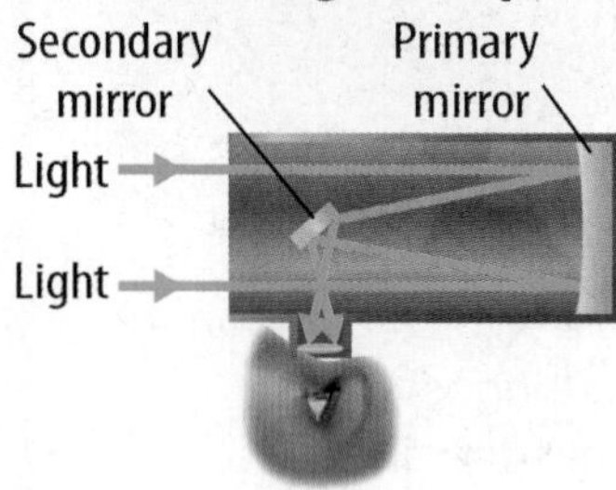

▲ **Figure 2**  Optical telescopes collect visible light in two different ways.

Earth-Based Telescopes

Telescopes are designed to collect a certain type of electromagnetic wave. Some telescopes detect visible light, and others detect radio waves and microwaves.

Optical Telescopes

There are two kinds of optical telescopes–refracting telescopes and reflecting telescopes, illustrated in **Figure 2.**

Refracting Telescopes Have you ever used a magnifying lens? You might have noticed that the lens was curved and thick in the middle. This is a convex lens. *A telescope that uses a convex lens to concentrate light from a distant object is a* **refracting telescope.** As shown at the top of **Figure 2,** the objective lens in a refracting telescope is the lens closest to the object being viewed. The light goes through the objective lens and refracts, forming a small, bright image. The eyepiece is a second lens that magnifies the image.

Key Concept Check Which electromagnetic waves do refracting telescopes collect?

Reflecting Telescopes Most large telescopes use curved mirrors instead of curved lenses. *A telescope that uses a curved mirror to concentrate light from a distant object is a* **reflecting telescope.** As shown at the bottom of **Figure 2,** light is reflected from a primary mirror to a secondary mirror. The secondary mirror is tilted to allow the viewer to see the image. Generally, larger primary mirrors produce clearer images than smaller mirrors. However, there is a limit to mirror size. The largest reflecting telescopes, such as the Keck Telescopes on Hawaii's Mauna Kea, shown in **Figure 3,** have many small mirrors linked together. These small mirrors act as one large primary mirror.

Figure 3 Each 10-m primary mirror in the twin Keck Telescopes consists of 36 small mirrors. ▼

Radio Telescope

▲ **Figure 4** Radio telescopes are often built in large arrays. Computers convert radio data into images.

Radio Telescopes

Unlike a telescope that collects visible light waves, a **radio telescope** *collects radio waves and some microwaves using an antenna that looks like a TV satellite dish.* Because these waves have long wavelengths and carry little energy, radio antennae must be large to collect them. Radio telescopes are often built together and used as if they were one telescope. The telescopes shown in **Figure 4** are part of the Very Large Array in New Mexico. The 27 instruments in this array act as a single telescope with a 36-km diameter.

Reading Check Why are radio telescopes built together in large arrays?

Distortion and Interference

Moisture in Earth's atmosphere can absorb and distort radio waves. Therefore, most radio telescopes are located in remote deserts, which have dry environments. Remote deserts also tend to be far from radio stations, which emit radio waves that interfere with radio waves from space.

Water vapor and other gases in Earth's atmosphere also distort visible light. Stars seem to twinkle because gases in the atmosphere move, refracting the light. This causes the location of a star's image to change slightly. At high elevations, the atmosphere is thin and produces less distortion than it does at low elevations. That is why most optical telescopes are built on mountains. New technology called adaptive optics lessens the effects of atmospheric distortion even more, as shown in **Figure 5**.

Figure 5 Adaptive optics sharpens images by reducing atmospheric distortion. ▼

Wavelengths from Space

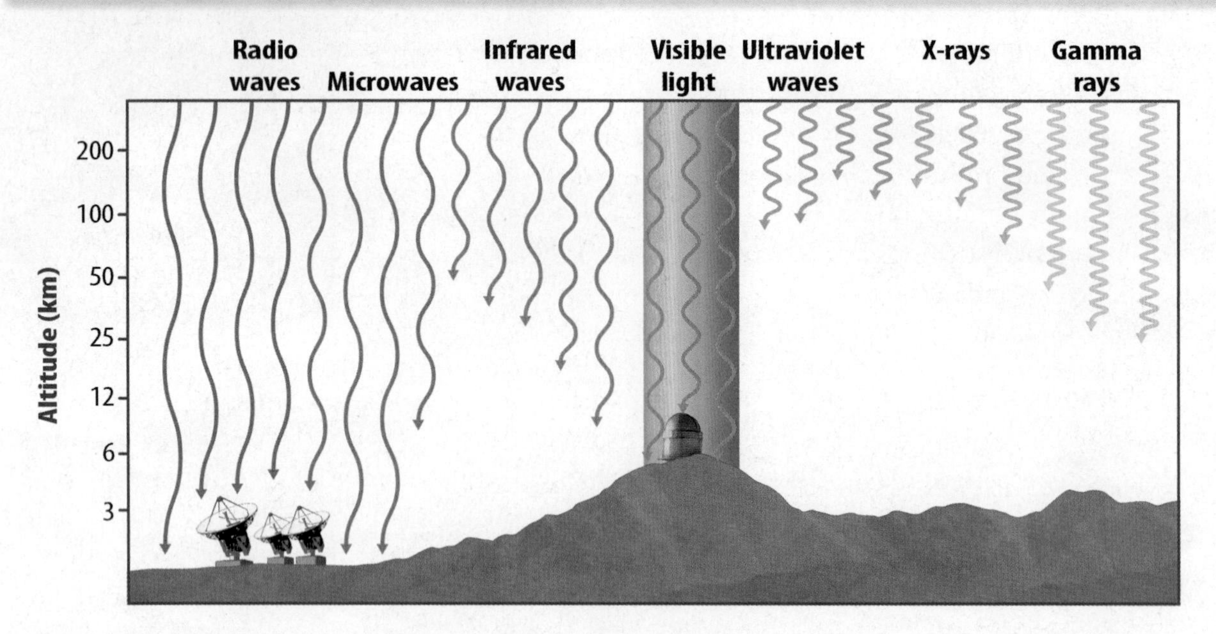

▲ **Figure 6** Most electromagnetic waves do not penetrate Earth's atmosphere. Even though the atmosphere blocks most UV rays, some still reach Earth's surface.

Visual Check About how far above Earth's surface do gamma waves reach?

Figure 7 The *Hubble Space Telescope* is controlled by astronomers on Earth. ▼

NASA

Space Telescopes

Why would astronomers want to put a telescope in space? The reason is Earth's atmosphere. Earth's atmosphere absorbs some types of electromagnetic radiation. As shown in **Figure 6**, visible light, radio waves, and some microwaves reach Earth's surface. But other types of electromagnetic waves do not. Telescopes on Earth can collect only the electromagnetic waves that are not absorbed by Earth's atmosphere. Telescopes in space can collect energy at all wavelengths, including those that Earth's atmosphere would absorb, such as most infrared light, most ultraviolet light, and X-rays.

Key Concept Check Why do astronomers put some telescopes in space?

Optical Space Telescopes

Optical telescopes collect visible light on Earth's surface, but optical telescopes work better in space. The reason, again, is Earth's atmosphere. As you read earlier, gases in the atmosphere can absorb some wavelengths. In space, there are no atmospheric gases. The sky is darker, and there is no weather.

The first optical space telescope was launched in 1990. The *Hubble Space Telescope*, shown in **Figure 7**, is a reflecting telescope that orbits Earth. Its primary mirror is 2.4 m in diameter. At first the *Hubble* images were blurred because of a flaw in the mirror. In 1993, astronauts repaired the telescope. Since then, *Hubble* has routinely sent to Earth spectacular images of far-distant objects. The photo at the beginning of this lesson was taken with the *Hubble* telescope.

Using Other Wavelengths

The *Hubble Space Telescope* is the only space telescope that collects visible light. Dozens of other space telescopes, operated by many different countries, gather ultraviolet, X-ray, gamma ray, and infrared light. Each type of telescope can point at the same region of sky and produce a different image. The image of the star Cassiopeia A (ka see uh PEE uh • AY) in **Figure 8** was made with a combination of optical, X-ray, and infrared data. The colors represent different kinds of material left over from the star's explosion many years ago.

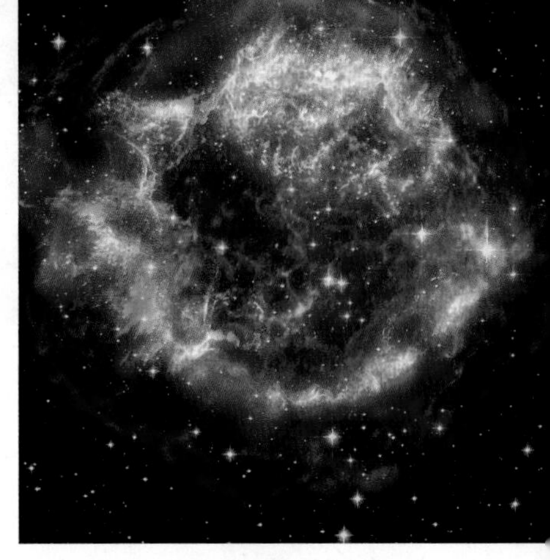

▲ **Figure 8** Each color in this image of Cassiopeia A is derived from a different wavelength—yellow: visible; pink/red: infrared; green and blue: X-ray.

Spitzer Space Telescope Young stars and planets hidden by dust and gas cannot be viewed in visible light. However, infrared wavelengths can penetrate the dust and reveal what is beyond it. Infrared can also be used to observe objects too old and too cold to emit visible light. In 2003, the *Spitzer Space Telescope* was launched to collect infrared waves, as it orbits the Sun.

Reading Check Which type of radiant energy does the *Spitzer Space Telescope* collect?

James Webb Space Telescope A larger space telescope, scheduled for launch in 2018, is also designed to collect infrared radiation as it orbits the Sun. The *James Webb Space Telescope,* illustrated in **Figure 9**, will have a mirror with an area 50 times larger than *Spitzer*'s mirror and seven times larger than *Hubble*'s mirror. Astronomers plan to use the telescope to detect galaxies that formed early in the history of the universe.

Figure 9 The advanced technology of the *James Webb Space Telescope* will help astronomers study the origin of the universe. ▼

James Webb Space Telescope

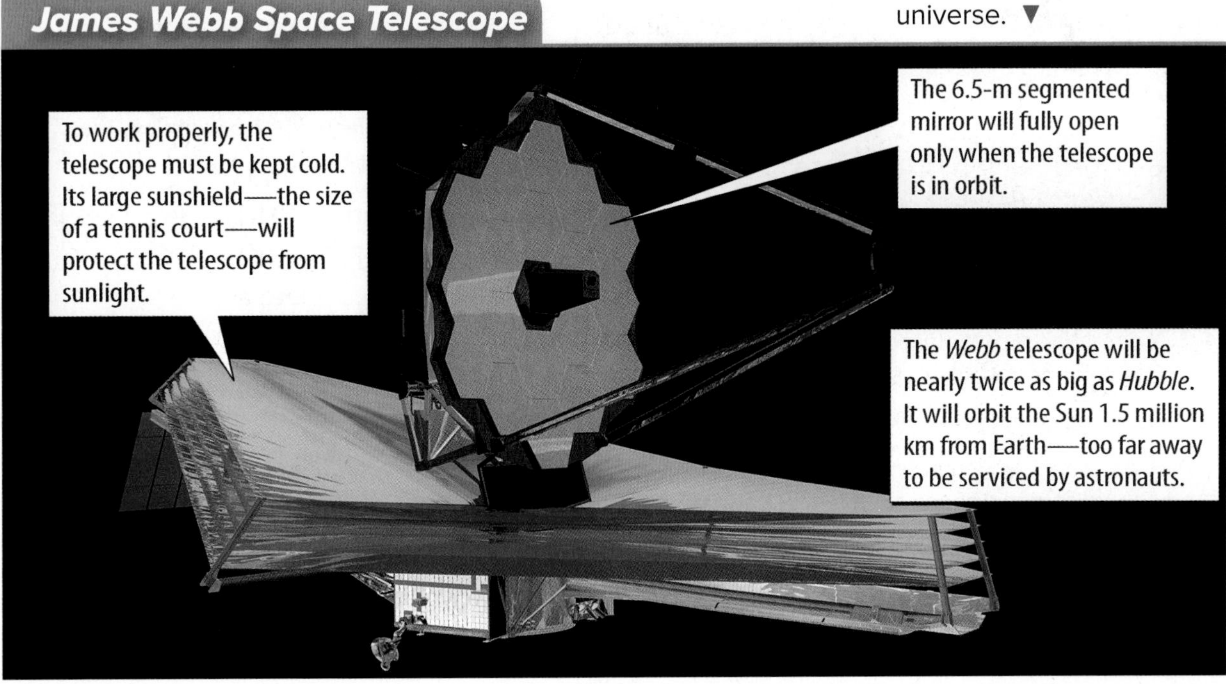

Lesson 1 Review

Visual Summary

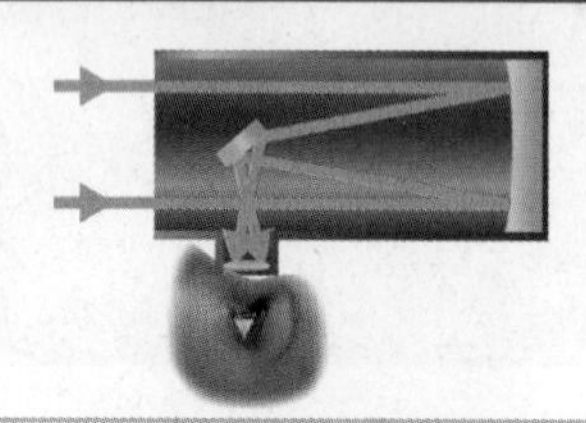

Reflecting telescopes use mirrors to concentrate light.

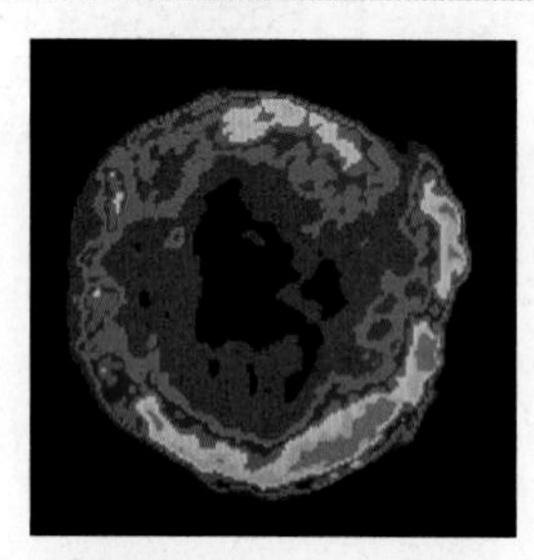

Earth-based telescopes can collect energy in the visible, radio, and microwave parts of the electromagnetic spectrum.

Space-based telescopes can collect wavelengths of energy that cannot penetrate Earth's atmosphere.

FOLDABLES®

Use your lesson Foldable to review the lesson. Save your Foldable for the project at the end of the chapter.

What do you think NOW?

You first read the statements below at the beginning of the chapter.

1. Astronomers put telescopes in space to be closer to the stars.
2. Telescopes can work only using visible light.

Did you change your mind about whether you agree or disagree with the statements? Rewrite any false statements to make them true.

Use Vocabulary

1. **Distinguish** between a reflecting telescope and a refracting telescope.
2. **Use the term** *electromagnetic spectrum* in a sentence.
3. **Define** *radio telescope* in your own words.

Understand Key Concepts 

4. Which emits visible light?
   - **A.** moon
   - **B.** planet
   - **C.** satellite
   - **D.** star
5. **Draw** a sketch that shows the difference in wavelength of a radio wave and a visible light wave. Which transfers more energy?
6. **Contrast** the *Hubble Space Telescope* and the *James Webb Space Telescope.*

Interpret Graphics

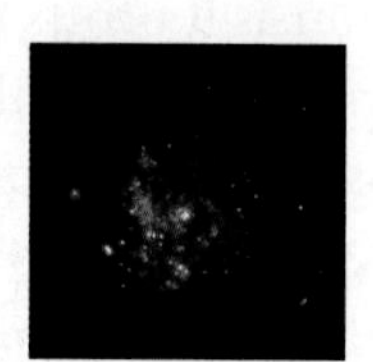

7. **Explain** The three images above represent the same area of sky. Explain why each looks different.
8. **Organize Information** Copy and fill in the graphic organizer below, listing the wavelengths collected by space telescopes, from the longest to the shortest.

Critical Thinking

9. **Suggest** a reason—besides the lessening of atmospheric distortion—why optical telescopes are built on remote mountains.

Math Skills

10. Light travels 9,460,000,000,000 km in 1 year. Express this number in scientific notation.

Skill Practice — Follow a Procedure

30 minutes

How can you construct a simple telescope?

Have you ever looked at the night sky and wondered what you were looking at? Stars and planets look much the same. How can you distinguish them? In this lab, you will construct a simple telescope you can use to observe and distinguish distant objects.

Materials

lenses

cardboard tubes

silicon putty

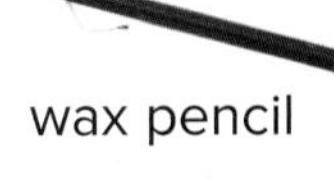
wax pencil

rubber bands

masking tape

Safety

Learn It

In many science experiments, you must **follow a procedure** in order to know what materials to use and how to use them. In this activity, you will follow a procedure to construct a simple telescope.

Try It

1. Read and complete a lab safety form.
2. Move both lenses up and down over the print on this page to determine which lens has a shorter focal length. Use a marker to put a small dot on its edge. This will be your eyepiece lens.
3. Make a silicon putty rope 2–3 mm in diameter and about 15 cm long. Wrap the rope around the edge of one of the open ends of the smaller cardboard tube. Remove any extra putty.
4. Gently push the eyepiece lens onto the ring of putty. Wrap a piece of masking tape around the edge of the lens to secure it firmly.
5. Repeat steps 4 and 5 using the larger tube and the objective lens.
6. Place the smaller tube into the larger tube so that the eyepiece lens, in the smaller tube, extends outside the larger tube.
7. Use your telescope to view distant objects. Move the smaller tube in and out to focus your instrument. If possible, view the night sky with your telescope. ⚠ *Do not use your telescope or any other instrument to directly view the Sun.*
8. Record your observations in your Science Journal.

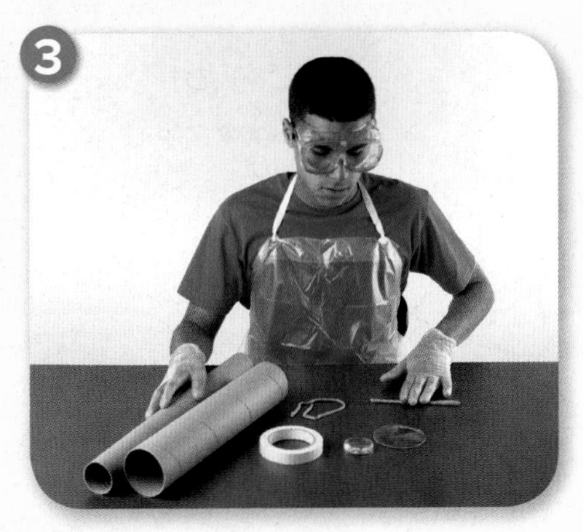
3

Apply It

9. **Identify** What type of telescope did you construct?
10. **Key Concept** How does your telescope collect light?

Lesson 2

Reading Guide

Key Concepts
ESSENTIAL QUESTIONS

- How are rockets and artificial satellites used?
- Why do scientists send both crewed and uncrewed missions into space?
- What are some ways that people use space technology to improve life on Earth?

Vocabulary

rocket p. 699
satellite p. 700
space probe p. 701
lunar p. 701
Project Apollo p. 702
space shuttle p. 702

Multilingual eGlossary

Science Video

Early History of Space Exploration

Inquiry Where is it headed?

Have you ever witnessed a rocket launch? Rockets produce gigantic clouds of smoke, long plumes of exhaust, and thundering noise. How are rockets used to explore space? What do they carry?

Launch Lab

10 minutes

How do rockets work?

Space exploration would be impossible without rockets. Become a rocket scientist for a few minutes, and find out what sends rockets into space.

1. Read and complete a lab safety form.
2. Use **scissors** to carefully cut a 5-m piece of **string.**
3. Insert the string into a **drinking straw.** Tie each end of the string to a stationary object. Make sure the string is taut. Slide the drinking straw to one end of the string.
4. Blow up a **balloon.** Do not tie it. Instead, twist the neck and clamp it with a **clothespin** or a **paper clip. Tape** the balloon to the straw.
5. Remove the clothespin or paperclip to launch your rocket. Observe how the rocket moves. Record your observations in your Science Journal.

Think About This

1. Describe how your rocket moved along the string.
2. How might you get your rocket to go farther or faster?
3.  **Key Concept** How do you think rockets are used in space exploration?

Rockets

Think about listening to a recording of your favorite music. Now think about how different it is to experience the same music at a live performance. This is like the difference between exploring space from a distance, with a telescope, and actually going there.

A big problem in launching an object into space is overcoming the force of Earth's gravity. This is accomplished with rockets. A **rocket** *is a vehicle designed to propel itself by ejecting exhaust gas from one end.* Fuel burned inside the rocket builds up pressure. The force from the exhaust thrusts the rocket forward, as shown in **Figure 10.** Rocket engines do not draw in oxygen from the surrounding air to burn their fuel, as jet engines do. They carry their oxygen with them. As a result, rockets can operate in space where there is very little oxygen.

 **Key Concept Check** How are rockets used in space exploration?

Scientists launch rockets from Florida's Cape Canaveral Air Force Station or the Kennedy Space Center nearby. However, space missions are managed by scientists at several different research stations around the country.

Figure 10 Exhaust gases ejected from the end of a rocket push the rocket forward.

WORD ORIGIN

satellite
from Latin *satellitem,* means "attendant" or "bodyguard"

Artificial Satellites

Any small object that orbits a larger object is a **satellite.** The Moon is a natural satellite of Earth. Artificial satellites are made by people and launched by rockets. They orbit Earth or other bodies in space, transmitting radio signals back to Earth.

The First Satellites—*Sputnik* and *Explorer*

The first artificial, Earth-orbiting satellite was *Sputnik 1.* Many people think this satellite, launched in 1957 by the former Soviet Union, represents the beginning of the space age. In 1958, the United States launched its first Earth-orbiting satellite, *Explorer I.* Today, thousands of satellites orbit Earth.

How Satellites Are Used

The earliest satellites were developed by the military for navigation and to gather information. Today, Earth-orbiting satellites are also used to transmit television and telephone signals and to monitor weather and climate. An array of satellites called the Global Positioning System (GPS) is used for navigation in cars, boats, airplanes, and even for hiking.

Key Concept Check How are Earth-orbiting satellites used?

FOLDABLES®

Make a vertical two-tab book. Record what you learn about crewed and uncrewed space missions under the tabs.

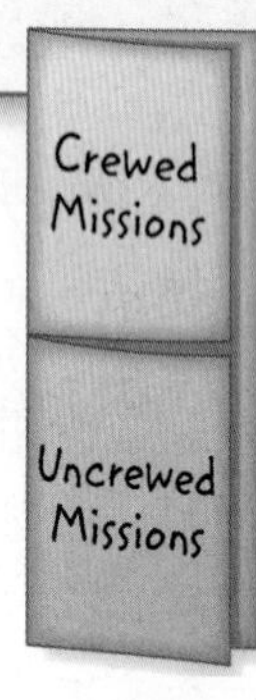

Early Exploration of the Solar System

In 1958, the U.S. Congress established the National Aeronautics and Space Administration (NASA). NASA oversees all U.S. space missions, including space telescopes. Some early steps in U.S. space exploration are shown in **Figure 11.**

Figure 11 Space exploration began with the first rocket launch in 1926.

Visual Check How many years after the first rocket was the first U.S. satellite launched into space?

Early Space Exploration

◄ **1926 First rocket:** Robert Goddard's liquid-fueled rocket rose 12 m into the air.

◄ **1962 First planetary probe:** *Mariner 2* traveled to Venus and collected data for 3 months. The spacecraft now orbits the Sun.

1958 First U.S. satellite: In the same year NASA was founded, *Explorer 1* was launched. It orbited Earth 58,000 times before burning up in Earth's atmosphere in 1970. ►

1972 First probe to outer solar system: After flying past Jupiter, *Pioneer 10* is still traveling onward, someday to exit the solar system. ►

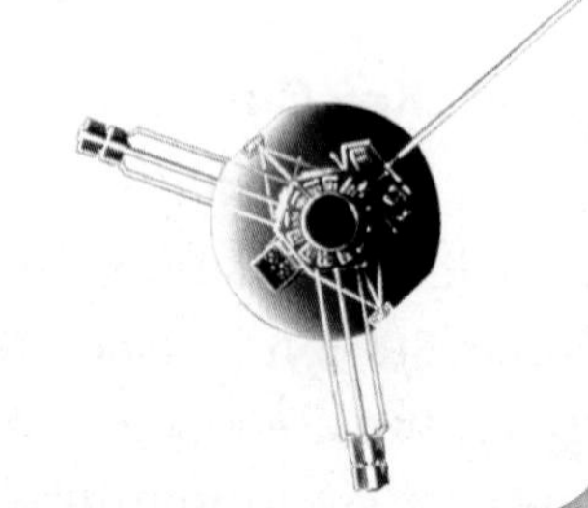

Space Probes

Figure 12 Scientists use space probes to explore the planets and some moons in the solar system.

Visual Check Which type of probe might use a parachute?

Orbiter

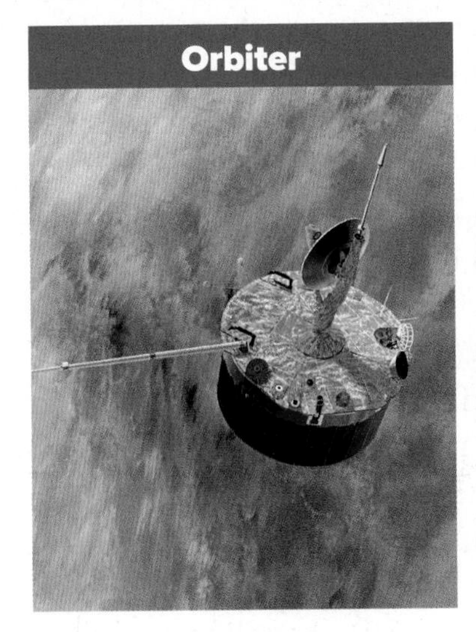

Once orbiters reach their destinations, they use rockets to slow down enough to be captured in a planet's orbit. How long they orbit depends on their fuel supply. The orbiter probe here, *Pioneer,* orbited Venus.

Lander

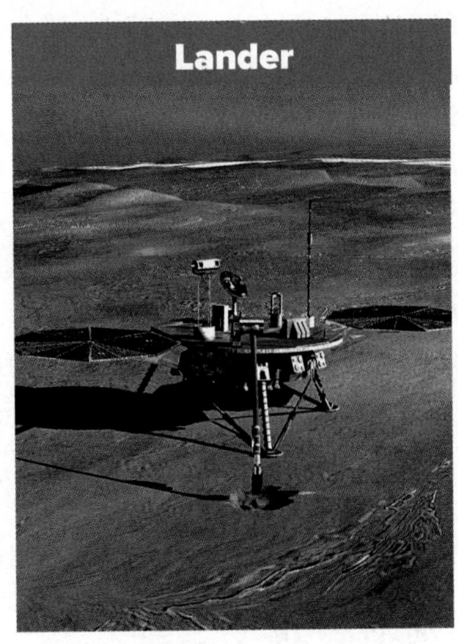

Landers touch down on surfaces. Sometimes they release rovers. Landers use rockets and parachutes to slow their descent. The lander probe here, *Phoenix,* analyzed the Martian surface for evidence of water.

Flyby

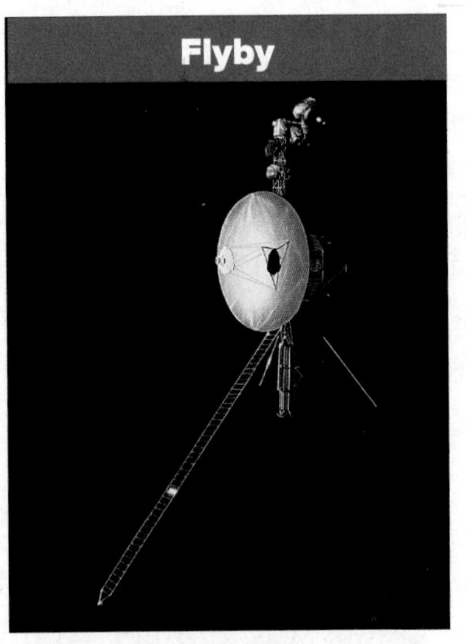

Flybys do not orbit or land. When its mission is complete, a flyby continues through space, eventually leaving the solar system. *Voyager 1,* here, explored Jupiter and Saturn and has entered interstellar space.

Space Probes

Some spacecraft have human crews, but most do not. *A* **space probe** *is an uncrewed spacecraft sent from Earth to explore objects in space.* Space probes are robots that work automatically or by remote control. They take pictures and gather data. Probes are cheaper to build than crewed spacecraft, and they can make trips that would be too long or too dangerous for humans. Space probes are not designed to return to Earth. The data they gather are relayed to Earth via radio waves. **Figure 12** shows three major types of space probes.

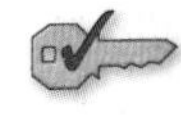

Key Concept Check Why do scientists send uncrewed missions to space?

SCIENCE USE V. COMMON USE

probe

Science Use an uncrewed spacecraft

Common Use question or examine closely

Lunar and Planetary Probes

The first probes to the Moon were sent by the United States and the former Soviet Union in 1959. Probes to the Moon are called lunar probes. *The term* **lunar** *refers to anything related to the Moon.* The first spacecraft to gather information from another planet was the flyby *Mariner 2,* sent to Venus in 1962. Since then, space probes have been sent to all the planets.

Human Spaceflight

Sending humans into space was a major goal of the early space program. However, scientists worried about how radiation from the Sun and weightlessness in space might affect people's health. Because of this, they first sent dogs, monkeys, and chimpanzees. In 1961, the first human–an astronaut from the former Soviet Union–was launched into Earth's orbit. Shortly thereafter, the first American astronaut orbited Earth. Some highlights of the early U.S. human spaceflight program are shown in **Figure 13.**

The Apollo Program

In 1961, U.S. President John F. Kennedy challenged the American people to place a person on the Moon by the end of the decade. The result was **Project Apollo**–*a series of space missions designed to send people to the Moon.* In 1969, Neil Armstrong and Buzz Aldrin, Apollo 11 astronauts, were the first people to walk on the Moon.

Reading Check What was the goal of Project Apollo?

Space Transportation Systems

Early spacecraft and the rockets used to launch them were used only once. **Space shuttles** *are reusable spacecraft that transport people and materials to and from space.* Space shuttles return to Earth and land much like airplanes. NASA's fleet of space shuttles began operating in 1981. As the shuttles aged, NASA began developing a new transportation system, *Orion,* to replace them.

The *International Space Station*

The United States has its own space program. But it also cooperates with the space programs of other countries. In 1998, it joined 15 other nations to begin building the *International Space Station.* Occupied since 2000, this Earth-orbiting satellite is a research laboratory where astronauts from many countries work and live.

Research conducted aboard the *International Space Station* includes studying fungus, plant growth, and how human body systems react to low gravity conditions.

U.S. Human Spaceflight

Figure 13 Forty years after human spaceflight began, people were living and working in space.

Animation

▲ Apollo moon walk

◄ A space shuttle piggybacked on rockets

International Space Station orbiting Earth ▼

MiniLab

15 minutes

How does lack of friction in space affect simple tasks?

Because objects are nearly weightless in space, there is little friction. What do you think might happen if an astronaut applied too much force when trying to move an object?

1. Read and complete a lab safety form.
2. Use **putty** to attach a **small thread spool** over the hole of a **CD.**
3. Inflate a **large, round balloon.** Twist the neck to keep the air inside. Stretch the neck of the balloon over the spool without releasing the air.
4. Place the CD on a smooth surface. Release the twist, and gently flick the CD with your finger. Describe your observations in your Science Journal.

Analyze and Conclude

1. **Infer** Why did the balloon craft move so easily?
2. **Draw Conclusions** How hard would it be to move a large object on the *International Space Station?*
3. **Key Concept** What challenges do astronauts face in space?

Space Technology

The space program requires materials that can withstand the extreme temperatures and pressures of space. Many of these materials have been applied to everyday life on Earth.

New Materials

Space materials must protect people from extreme conditions. They also must be flexible and strong. Materials developed for spacesuits are now used to make racing suits for swimmers, lightweight firefighting gear, running shoes, and other sports clothing.

Safety and Health

NASA developed a strong, fibrous material to make parachute cords for spacecraft that land on planets and moons. This material, five times stronger than steel, is used to make radial tires for automobiles.

Medical Applications

Artificial limbs, infrared ear thermometers, and robotic surgery all have roots in the space program. So do the orthodontic braces shown in **Figure 14.** These braces contain ceramic material originally developed to strengthen the heat resistance of space shuttles.

Key Concept Check What are some ways that space exploration has improved life on Earth?

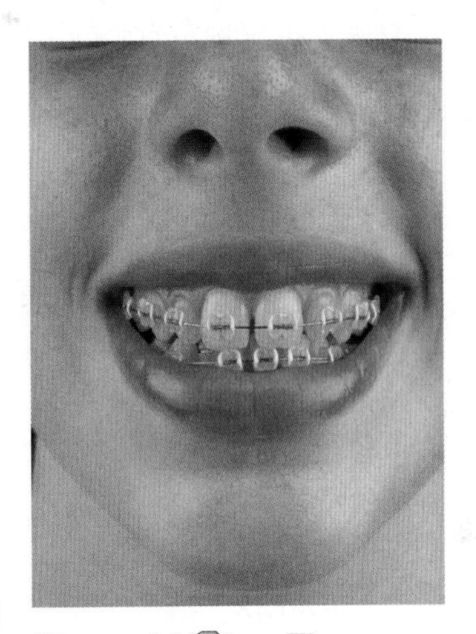

Figure 14 These braces contain a hard, strong ceramic originally developed for spacecraft.

Lesson 2 Review

Visual Summary

Exhaust from burned fuel accelerates a rocket.

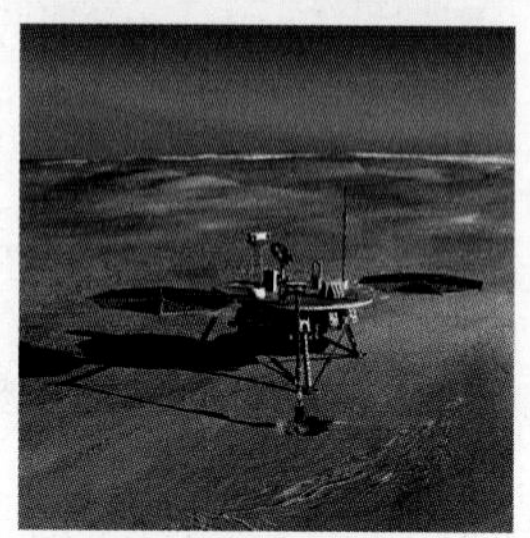

Some space probes can land on the surface of a planet or a moon.

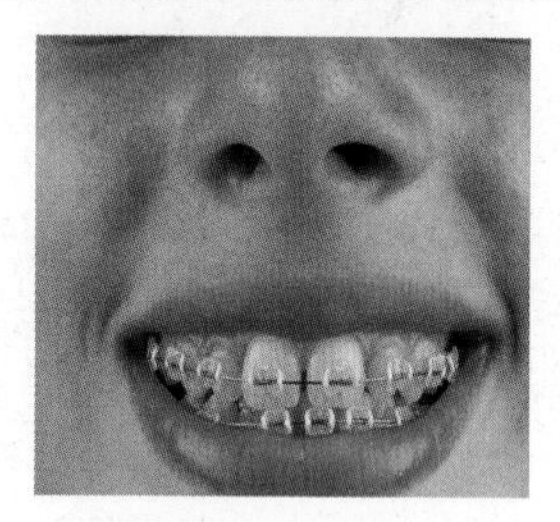

Technologies developed for the space program have been applied to everyday life on Earth.

FOLDABLES®

Use your lesson Foldable to review the lesson. Save your Foldable for the project at the end of the chapter.

What do you think NOW?

You first read the statements below at the beginning of the chapter.

3. Humans have walked on the Moon.

4. Some orthodontic braces were developed using space technology.

Did you change your mind about whether you agree or disagree with the statements? Rewrite any false statements to make them true.

Use Vocabulary

1. **Define** *rocket* in your own words.
2. **Use the term** *satellite* in a sentence.
3. The mission that sent people to the Moon was ________.

Understand Key Concepts 

4. What are rockets used for?
   - **A.** carrying people
   - **B.** launching satellites
   - **C.** observing planets
   - **D.** transmitting signals
5. **Explain** why *Sputnik 1* is considered the beginning of the space age.
6. **Compare and contrast** crewed and uncrewed space missions.

Interpret Graphics

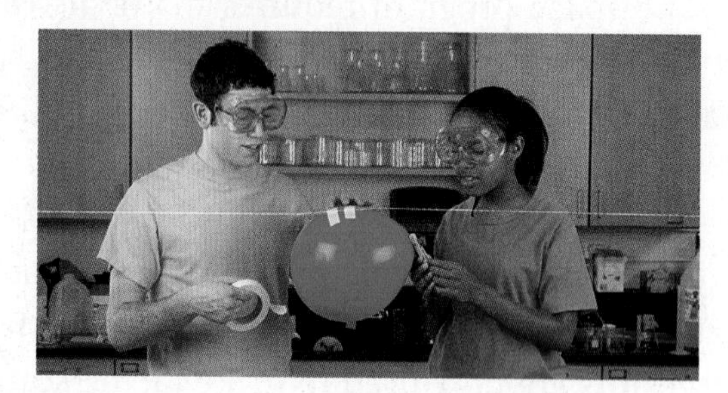

7. **Infer** How is the balloon above like a rocket?
8. **Organize Information** Copy and fill in the graphic organizer below and use it to place the following in the correct order: *first human in space, invention of rockets, first human on the Moon, first artificial satellite.*

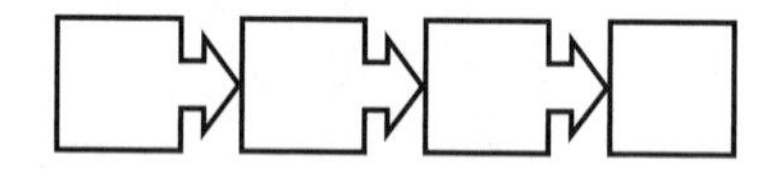

Critical Thinking

9. **Predict** how your life would be different if all artificial satellites stopped working.
10. **Evaluate** the benefits and drawbacks of international cooperation in space exploration.

(tl)Stocktrek/age fotostock; (cl)NASA/JPL; (r)Hutchings Photography/Digital Light Source; (bl)Alex Bartel/Photo Researchers, Inc.;

HOW IT WORKS

Going Up

Could a space elevator make space travel easier?

If you wanted to travel into space, the first thing you would have to do is overcome the force of Earth's gravity. So far, the only way to do that has been to use rockets. Rockets are expensive, however. Many are used only once, and they require a lot of fuel. It takes a lot of resources to build and power a rocket. But what if you could take an elevator into space instead?

Space elevators were once science fiction, but scientists are now taking the possibility seriously. With the lightweight but strong materials under development today, experts say it could take only 10 years to build a space elevator. The image here shows how it might work.

It generally costs more than $100 million to place a 12,000-kg spacecraft into orbit using a rocket. Some people estimate that a space elevator could place the same craft into orbit for less than $120,000. A human passenger with luggage, together totaling 150 kg, might be able to ride the elevator to space for less than $1,500.

Counterweight: The spaceward end of the cable would attach to a captured asteroid or an artificial satellite. The asteroid or satellite would stabilize the cable and act as a counterweight.

Cable: Made of super-strong but thin materials, the cable would be the first part of the elevator to be built. A rocket-launched spacecraft would carry reels of cable into orbit. From there the cable would be unwound until one end reached Earth's surface.

Anchor Station: The cable's Earthward end would be attached here. A movable platform would allow operators to move the cable away from space debris in Earth's orbit that could collide with it. The platform would be movable because it would float on the ocean.

Climber: The "elevator car" would carry human passengers and objects into space. It could be powered by Earth-based laser beams, which would activate solar-cell "ears" on the outside of the car.

It's Your Turn

DEBATE Form an opinion about the space elevator and debate with a classmate. Could a space elevator become a reality in the near future? Would a space elevator benefit ordinary people? Should the space elevator be used for space tourism?

Lesson 3

Reading Guide

Key Concepts
ESSENTIAL QUESTIONS

- What are goals for future space exploration?
- What conditions are required for the existence of life on Earth?
- How can exploring space help scientists learn about Earth?

Vocabulary

extraterrestrial life p. 711

astrobiology p. 711

Multilingual eGlossary

Science Video
What's Science Got to do With It?

Recent and Future Space Missions

Inquiry Blue Moon?

No, this is Mars! It is a false-color photo of an area on Mars where a future space probe might land. Of all the planets in the solar system, Mars is most like Earth. Recent missions to Mars have found evidence of water on its surface. Could this planet support life?

NASA/JPL/University of Arizona

Launch Lab

10 minutes

How is gravity used to send spacecraft farther in space?  

Spacecraft use fuel to get to where they are going. But fuel is expensive and adds mass to the craft. Some spacecraft travel to far-distant regions with the help of gravity from the planets they pass by. This is a technique called gravity assist. You can model gravity assist using a simple table tennis ball.

1. Read and complete a lab safety form.
2. Set a **turntable** in motion.
3. Gently throw a **table tennis ball** so that it just skims the top of the spinning surface. You might have to practice before you're able to get the ball to glide over the surface.
4. In your Science Journal, describe or draw a picture of what you observed.

Think About This

1. Use your observations to describe how this activity is similar to gravity assist.
2. **Key Concept** How do you think gravity assist helps scientists learn about the solar system?

Missions to the Sun and the Moon

Scientists at NASA and other space agencies around the world have cooperatively developed goals for future space exploration. One goal is to expand human space travel within the solar system. Two steps leading to this goal are sending probes to the Sun and sending probes to the Moon.

 **Key Concept Check** What is a goal of space exploration?

Solar Probes

The Sun emits high-energy radiation and charged particles. Storms on the Sun can eject powerful jets of gas and charged particles into space, as shown in **Figure 15.** The Sun's high-energy radiation and charged particles can harm astronauts and damage spacecraft. To better understand these hazards, scientists study data collected by solar probes that orbit the Sun. The solar probe *Ulysses,* launched in 1990, orbited the Sun and gathered data for 19 years. *Solar Probe Plus,* a new solar probe set to launch in 2018, will gather data on solar winds and coronal heating, among other data.

Figure 15 Storms on the Sun send charged particles far into space.

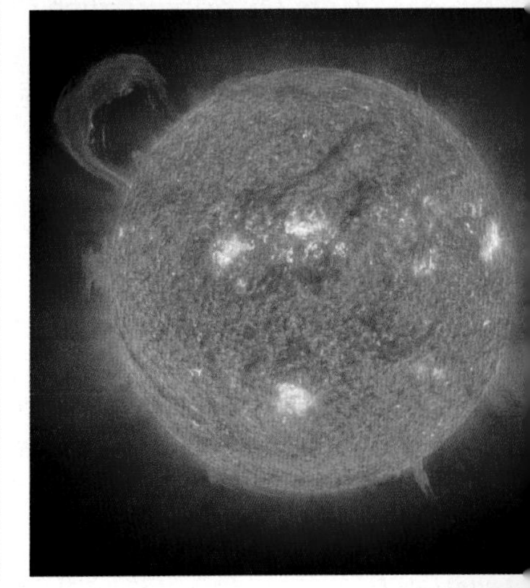

Lunar Probes

NASA and other space agencies also plan to send several probes to the Moon. The *Lunar Reconnaissance Orbiter,* launched in 2009, collects data that will help scientists select the best location for a future lunar outpost.

FOLDABLES®

Use a sheet of copy paper to make a vertical three-tab Foldable. Draw a Venn diagram on the front tabs and use it to compare and contrast space missions to the inner and outer planets.

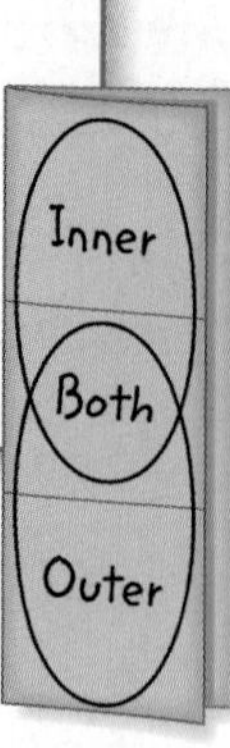

Missions to the Inner Planets

The inner planets are the four rocky planets closest to the Sun–Mercury, Venus, Earth, and Mars. Scientists have sent many probes to the inner planets, and more are planned. These probes help scientists learn how the inner planets formed, what geologic forces are active on them, and whether any of them could support life. Some recent and current missions to the inner planets are described in **Figure 16.**

Reading Check What do scientists want to learn about the inner planets?

Planetary Missions

Figure 16 Studying the solar system remains a major goal of space exploration.

◀ ***Messenger*** The first probe to visit Mercury—the planet closest to the Sun—since *Mariner 10* flew by the planet in 1975 is *Messenger.* After a 2004 launch and two passes of Venus, *Messenger* flew past Mercury several times before entering its orbit in 2011. *Messenger* studied Mercury's geology and chemistry. It sent images and data back to Earth for one Earth year. On its first pass by Mercury, in 2008, *Messenger* returned over 1,000 images in many wavelengths. The mission came to an end in 2015 when Messenger crashed into Mercury's surface.

Mars Reconnaissance Orbiter. Since the first flyby reached Mars in 1964, many probes, rovers, and orbiters have been sent to the red planet. Recently, the *Mars Reconnaissance Orbiter* (MRO) indicated that water flows periodically on the Martian surface. Researchers noticed mysterious, dark streaks on the slopes of craters that seemed to flow downward during warm seasons, and recede during cooler seasons. These streaks were confirmed to be liquid water. This mission continues to send back valuable information about Mars. ▶

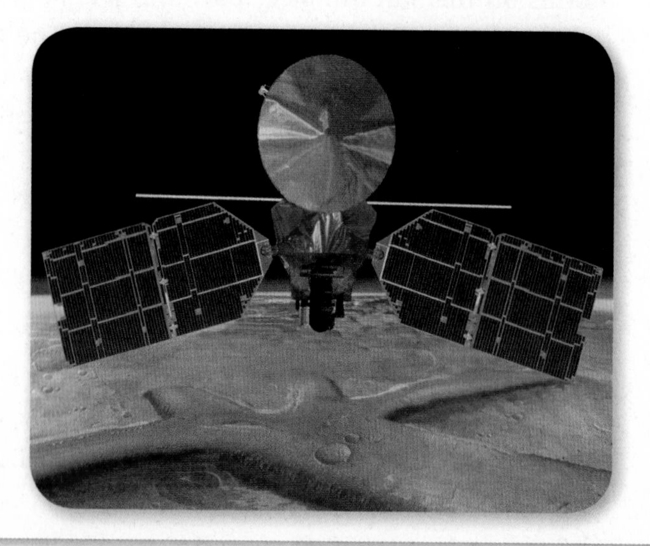

Missions to the Outer Planets and Beyond

The outer planets are the four large planets farthest from the Sun–Jupiter, Saturn, Uranus, and Neptune. Pluto was once considered an outer planet, but it is now included with other small, icy **dwarf planets** observed orbiting the Sun outside the orbit of Neptune. Missions to outer planets are long and difficult because the planets are so far from Earth. Some missions to the outer planets and beyond are described in **Figure 16** below. The next major mission to the outer planets will be an international mission to Jupiter and its four largest moons.

REVIEW VOCABULARY

dwarf planet
a round body that orbits the Sun but is not massive enough to clear away other objects in its orbit

Reading Check Why are missions to the outer planets difficult?

Visual Check Which planet has been explored by rovers?

◀ *Cassini* The first orbiter sent to Saturn, *Cassini,* was launched in 1997 as part of an international effort involving 19 countries. *Cassini* traveled for 7 years before entering Saturn's orbit in 2004. When it arrived, it sent a smaller probe to the surface of Saturn's largest moon, Titan, as shown at left. This event was the first landing on a moon in the outer solar system. *Cassini's* accomplishments include the discovery of geysers on Saturn's moon Enceladus, a greater understanding of Saturn's active ring system, and the discovery of massive hurricanes at the planet's poles. Data collected from *Cassini's* mission help scientists understand how planets, like Saturn, form and evolve.

New Horizons After a 9 year journey from Earth, the *New Horizons* spacecraft arrived at Pluto in 2015. It became the first spacecraft to explore the dwarf planet up close. Data collected from this mission included mapping the geology of Pluto and its moon—Charon, investigating the planets surface compositions and temperatures, and examining its atmosphere. *New Horizons* is projected to leave the solar system in 2029. ▶

Figure 17 This structure, at the Johnson Space Center, could serve as housing for up to four astronauts. It has been evaluated to ensure safety for humans living on other planetary bodies.

Future Space Missions

Human space travel remains a goal of NASA and other space agencies around the world. Human exploration missions greatly expand our scientific understanding of our solar system and the origins of life.

A New Era of Spaceflight

Future missions to space, including the goal of astronauts working and living on Mars, are dependent upon how humans will get there. The *Orion* spacecraft is a human space flight system that is capable of missions to a variety of interplanetary destinations. The purpose of *Orion* is to safely take astronauts to places in deep space that have never been explored by humans.

Once a destination has proven suitable for human exploration, astronauts will need secure housing. The structure in **Figure 17** is one of those **options.**

ACADEMIC VOCABULARY

option
(noun) something that can be chosen

Studying and Visiting Mars

A visit to Mars will probably not occur for several more decades. In preparation, NASA has sent orbiters, landers, and rovers to explore sites and resources that could potentially support life on Mars. The *MAVEN* spacecraft, for example, studied the atmosphere of Mars and how it has evolved over time. The *Mars Reconnaissance Orbiter* (MRO) was sent to search for evidence of water on the planet. The *Curiosity* rover analyzed rock and soil samples to evaluate the planet's past climate and geology, and also investigate if its environment was once able to support life. The data collected from each mission makes a human exploration mission more obtainable.

MiniLab

20 minutes

What conditions are required for life on Earth?

Billions of organisms live on Earth. What are the requirements for life?

1. Observe a **terrarium.** In your Science Journal, make a sketch of this environment and label every component as either living or nonliving.
2. Observe an **aquarium.** Again, make a sketch of this environment and label every component as living or nonliving.

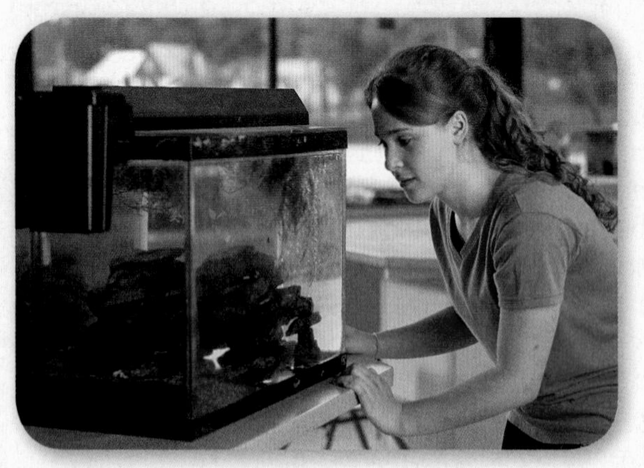

Analyze and Conclude

1. **Compare and Contrast** Describe what the organisms in both environments need to survive.
2. **Draw Conclusions** Do all living things have the same needs? Support your answer using examples from your observations.
3. **Key Concept** What conditions are required for life on Earth? How would knowing these requirements help scientists look for life in space?

The Search for Life

No one knows if life exists beyond Earth, but people have thought about the possibility for a long time. It even has a name. *Life that originates outside Earth is* **extraterrestrial** (ek struh tuh RES tree ul) **life.**

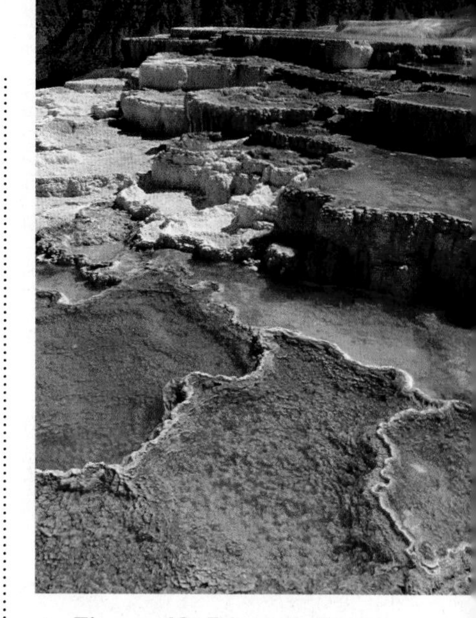

▲ **Figure 18** Bacteria live in the boiling water of this hot spring in Yellowstone National Park.

Conditions Needed for Life

Astrobiology *is the study of life in the universe, including life on Earth and the possibility of extraterrestrial life.* Investigating the conditions for life on Earth helps scientists predict where they might find life elsewhere in the solar system. Astrobiology also can help scientists locate environments in space where humans and other Earth life might be able to survive.

Life exists in a wide range of environments on Earth. Life-forms survive on dark ocean floors, deep in solid rocks, and in scorching water, such as the hot spring shown in **Figure 18.** No matter how extreme their environments, all known life-forms on Earth need liquid water, organic molecules, and some source of energy to survive. Scientists assume that if life exists elsewhere in space it would have the same requirements.

WORD ORIGIN

astrobiology
from Greek *astron,* means "star"; Greek *bios,* means "life"; and Greek *logia,* means "study"

 **Key Concept Check** What is required for life on Earth?

Water in the Solar System

As scientists explore our solar system, they continue to find water in unexpected places. For example, a lunar space probe found water in a crater on the Moon. Enough frozen water was found in a single crater to fill 1,500 Olympic swimming pools. Evidence from other space probes suggests that liquid water, water vapor, or ice exists on many planets and moons in the solar system.

Some of the moons in the outer solar system, such as Jupiter's moon Europa, shown in **Figure 19,** might also have large amounts of liquid water beneath their surfaces.

Figure 19 The dark patches in the inset photo might represent areas where water from an underground ocean has seeped to Europa's surface. ▼

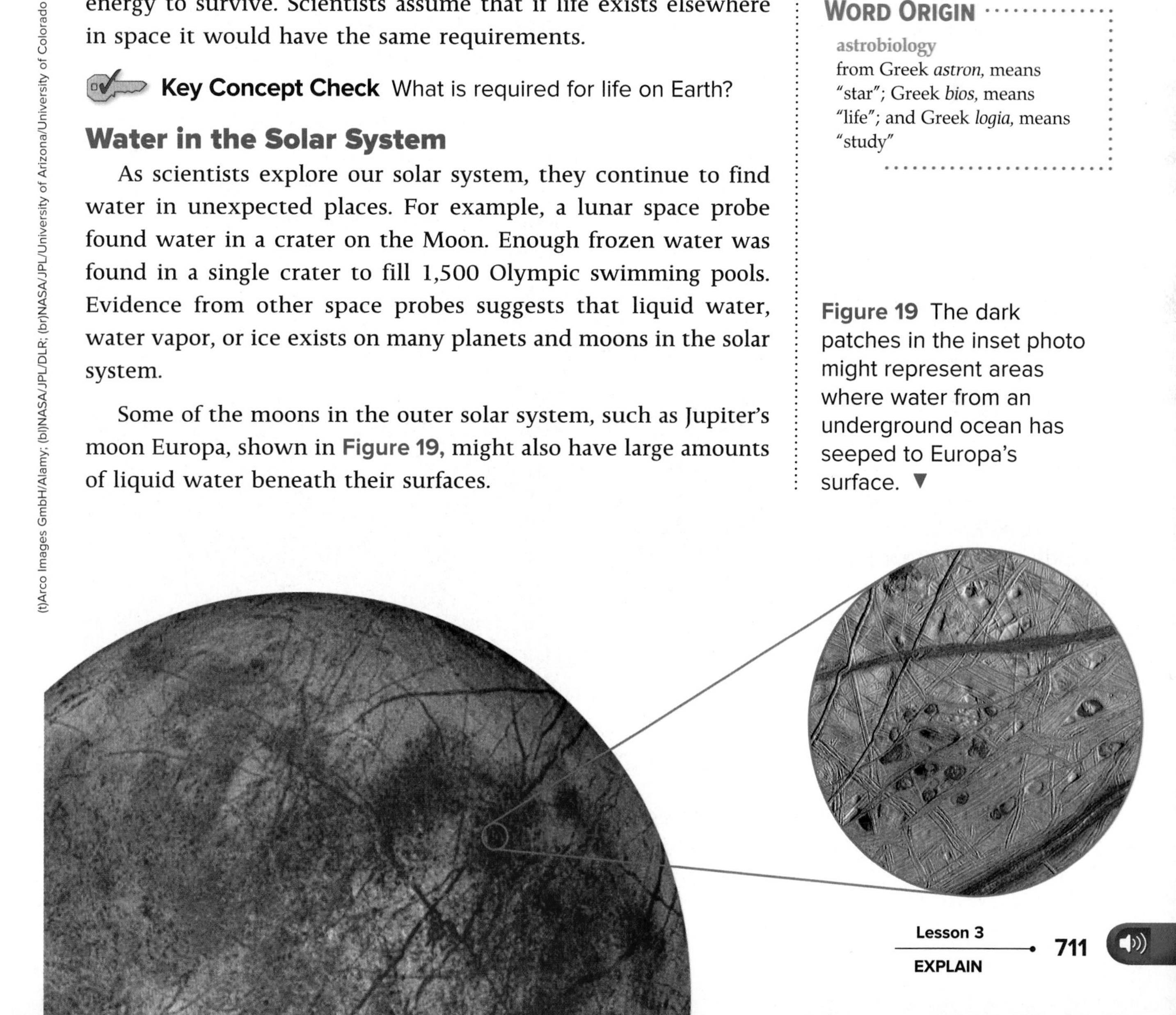

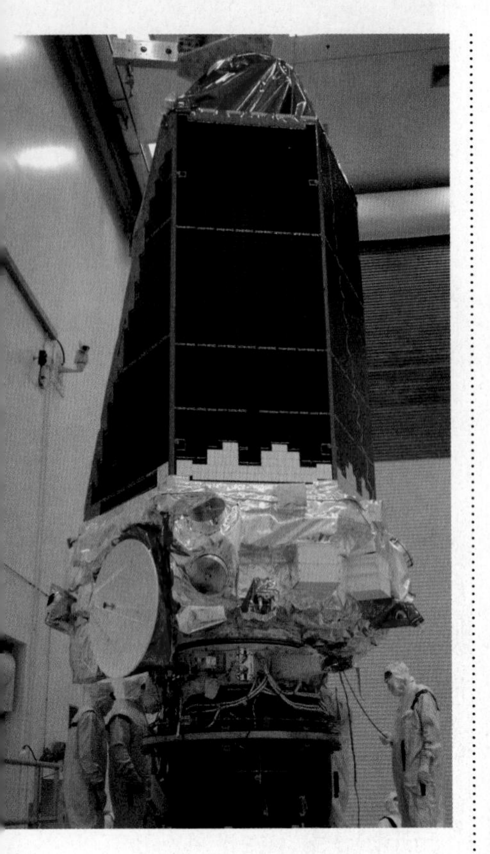

▲ **Figure 20** The Kepler telescope, shown here, is currently orbiting the Sun, searching a single area of sky for Earthlike planets.

Understanding Earth by Exploring Space

Space provides frontiers for the human spirit of exploration and discovery. The exploration of space also provides insight into planet Earth. Information gathered in space helps scientists understand how the Sun and other bodies in the solar system influence Earth, how Earth formed, and how Earth supports life. Looking for Earthlike planets outside the solar system helps scientists learn if Earth is unique in the universe.

Searching for Other Planets

Astronomers have detected thousands of planets outside the solar system. Most of these planets are much bigger than Earth and probably could not support liquid water–or life. To search for Earthlike planets, NASA launched the *Kepler* telescope in 2009. The *Kepler* telescope, shown in **Figure 20**, focuses on a single area of sky containing about 100,000 stars. However, though it might detect Earthlike planets orbiting other stars, *Kepler* will not be able to detect life on any planet.

Understanding Our Home Planet

Not all of NASA's missions are to other planets, to other moons, or to look at stars and galaxies. NASA and space agencies around the world also launch and maintain Earth observing satellites. Satellites that orbit Earth provide large-scale images of Earth's surface. These images help scientists understand Earth's climate and weather. **Figure 21** is a satellite image showing changes in ocean temperature associated with Hurricane Katrina, one of the deadliest storms in U.S. history.

 **Key Concept Check** How can exploring space help scientists learn about Earth?

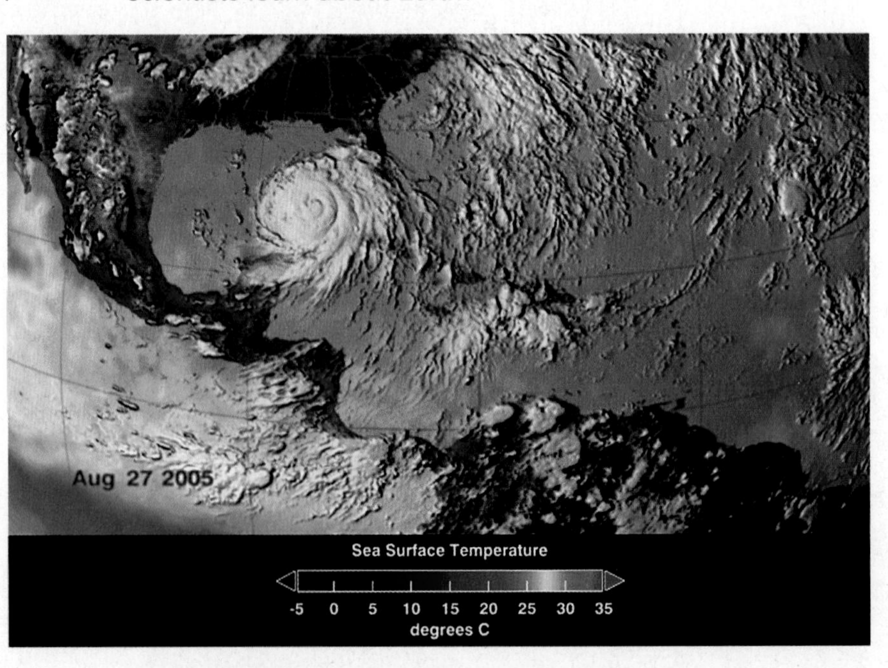

Figure 21 Earth-orbiting satellites collect data in many wavelengths. This satellite image of Hurricane Katrina was made with a microwave sensor. ►

Visual Check Which part of the United States did Hurricane Katrina affect?

Lesson 3 Review

Visual Summary

The *New Horizons* spacecraft was the first to explore Pluto.

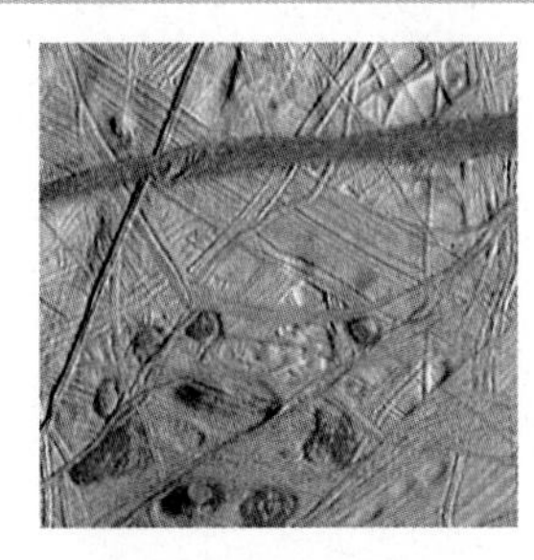

Scientists think liquid water might exist on or below the surfaces of some moons.

Earth-orbiting satellites help scientists understand weather and climate patterns on Earth.

FOLDABLES®

Use your lesson Foldable to review the lesson. Save your Foldable for the project at the end of the chapter.

What do you think NOW?

You first read the statements below at the beginning of the chapter.

5. Humans have landed on Mars.

6. Scientists have detected water on other bodies in the solar system.

Did you change your mind about whether you agree or disagree with the statements? Rewrite any false statements to make them true.

Use Vocabulary

1. **Use the term** *extraterrestrial life* in a sentence.
2. The study of life in the universe is ________.

Understand Key Concepts

3. *Cassini* sent a probe to which moon?
   - **A.** Europa
   - **B.** Titan
   - **C.** Rhea
   - **D.** Enceladus
4. **Explain** why bodies that have liquid water are the best candidates for supporting life.
5. **Assess** the benefits of an inflatable structure over a concrete structure on the Moon.
6. **Identify** some phenomena on Earth best viewed by artificial satellites.

Interpret Graphics

7. **Assess** The figure above represents a possible design for a new solar probe that would orbit close to the Sun. What purpose might the part labeled *A* serve?
8. **Organize Information** Copy and fill in the graphic organizer below to list requirements for life on Earth.

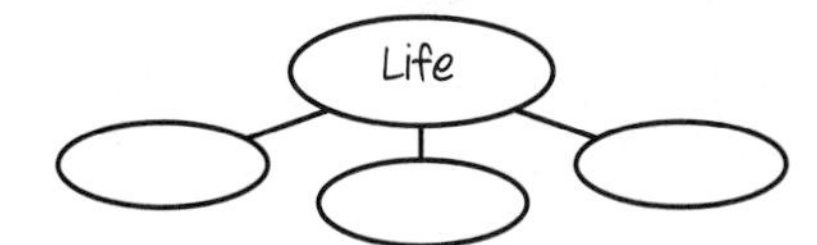

Critical Thinking

9. **Predict** some of the challenges people might face living in a lunar outpost.
10. **Debate** whether scientists should look first for life on Mars or on Europa.

(tl)NASA/Johns Hopkins University Applied Physics Laboratory/Southwest Research Institute (NASA/JHUAPL/SwRI); (cl)NASA/JPL/University of Arizona/University of Colorado; (bl)NASA/Goddard Space Flight Center Scientific Visualization Studio; (r)NASA/Johns Hopkins University Applied Physics Laboratory

Lab

2 class periods

Materials

newspaper

creative building materials

masking tape

cup varieties

office supplies

craft supplies

scissors

Safety

Design and Construct a Moon Habitat

No one has visited the Moon since 1972. NASA plans to send astronauts to Mars and beyond. You might be one of the lucky ones who will be sent to find a suitable location for an outpost. To get a head start, your task is to design and build a model of a habitat where people can live and work for months at a time. You can use any materials provided or other materials approved by your teacher. Before you begin, think about some of the things people will need in order to survive on other planets.

Question

Think about what humans need on a daily basis. How can you design a habitat that would meet people's needs in a place very unlike Earth?

Procedure

1. Read and complete a lab safety form.
2. Think about construction. Consider the function each material might represent in a moon habitat. The materials will have to be transported from Earth before any construction can begin.
3. Draw plans for your habitat. Be sure to include an airlock, a small room that separates an outer door from an inner door. Label the materials you will use and what each represents.

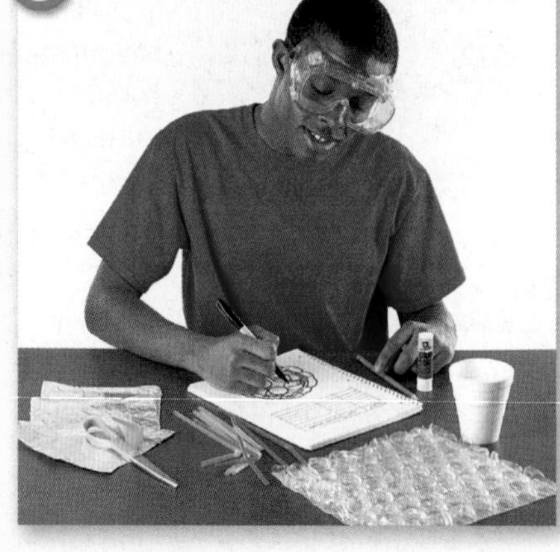

4. Copy the data table below into your Science Journal. Complete the table by listing each material you plan to use, its purpose or function, and why you chose it.

Materials for a Moon Habitat

Material	Function	Why I Chose the Material

(t to b, 3, r)Hutchings Photography/Digital Light Source; (2, 4-6)McGraw-Hill Education; (7)Jacques Cornell/McGraw-Hill Education

Lab Tips

- Before you begin, make a list of conditions on other planets, such as Mars, that are much different from those on Earth.
- If you can think of any materials not listed that you would like to use, ask your teacher's permission to use them.

5. Build your habitat. When you are finished, check to see that your habitat satisfies the conditions in your original question. If not, revise your habitat or make a note in your Science Journal about how you would improve it.

6. In addition to meeting people's needs in space, the habitat should be easy to construct in the harsh environments of space. Remember that the materials should be easy to transport from Earth.
7. Some things might not go as planned as you construct your model, or you might get new ideas as you proceed with building. As you go along, you can adapt your structure to improve the final product. Record any changes you make to your design or materials in your Science Journal.

Hutchings Photography/Digital Light Source

Analyze and Conclude

8. **Explain** in detail why you chose the materials and the design that you did.
9. **Evaluate** Which materials or designs did not work as expected? Explain.
10. **Compare and Contrast** What differences between the Martian environment and Earth's environment did you consider in your design?
11. **The Big Idea** What requirements must be met for humans to live, work, and be healthy on other planets?

Communicate Your Results

Imagine that your design is part of a NASA competition to find the best habitat. Write and give a 2–3 minute presentation convincing NASA to use your model for its habitat.

Extension

Compare your habitat to the habitats of at least three other groups. Discuss how you might combine your ideas to build a bigger and better habitat.

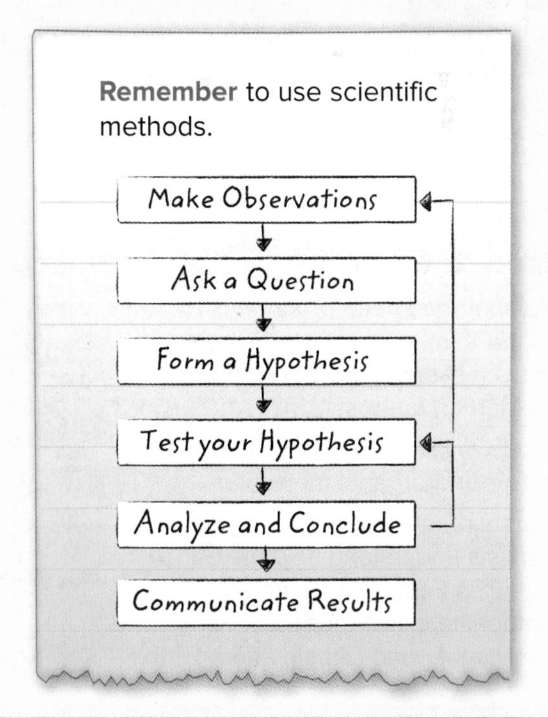

Chapter 19 Study Guide

Humans observe the universe with Earth-based and space-based telescopes. They explore the solar system with crewed and uncrewed space probes.

Key Concepts Summary 

Lesson 1: Observing the Universe

- Scientists use different parts of the **electromagnetic spectrum** to study stars and other objects in space and to learn what the universe was like many millions of years ago.
- Telescopes in space can collect radiant energy that Earth's atmosphere would absorb or refract.

Vocabulary

electromagnetic spectrum p. 690
refracting telescope p. 692
reflecting telescope p. 692
radio telescope p. 693

Lesson 2: Early History of Space Exploration

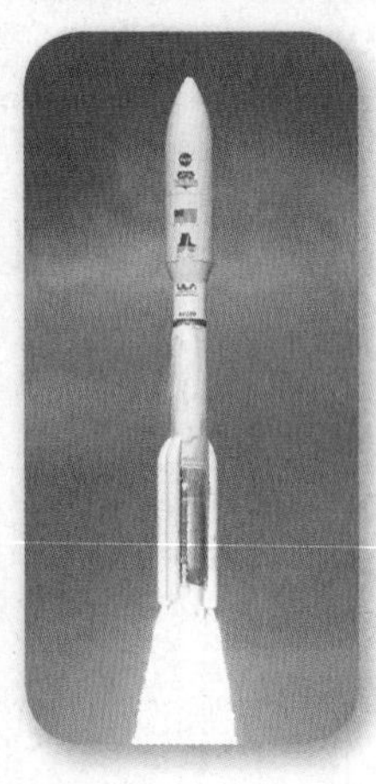

- **Rockets** are used to overcome the force of Earth's gravity when sending **satellites, space probes,** and other spacecraft into space.
- Uncrewed missions can make trips that are too long or too dangerous for humans.
- Materials and technologies from the space program have been applied to everyday life.

Vocabulary

rocket p. 699
satellite p. 700
space probe p. 701
lunar p. 701
Project Apollo p. 702
space shuttle p. 702

Lesson 3: Recent and Future Space Missions

- A goal of the space program is to expand human space travel within the solar system and develop lunar and Martian outposts.
- All known life-forms need liquid water, energy, and organic molecules.
- Information gathered in space helps scientists understand how the Sun influences Earth, how Earth formed, whether life exists outside of Earth, and how weather and climate affect Earth.

Vocabulary

extraterrestrial life p. 711
astrobiology p. 711

(t)NASA; (c)Stocktrek/age fotostock; (b)Michael Hixenbaugh/National Science Foundation

Personal Tutor

Vocabulary eGames
Vocabulary eFlashcards

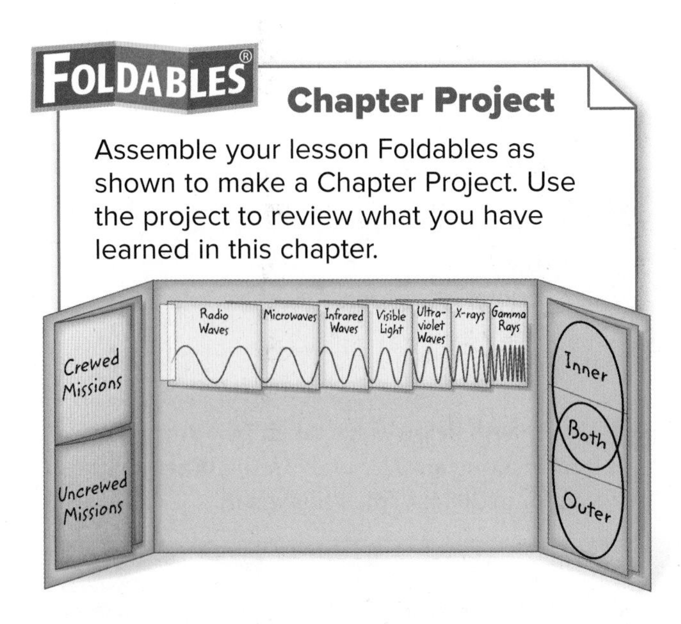

Use Vocabulary

1. All radiation is classified by wavelength in the ________.
2. Two types of telescopes that collect visible light are ________ and ________.
3. The space mission that sent the first humans to the Moon was ________.
4. An example of a human space transportation system is a(n) ________.
5. An uncrewed spacecraft is a(n) ________.
6. The discipline that investigates life in the universe is ________.
7. The best place to find ________ is on solar system bodies containing water.

Link Vocabulary and Key Concepts

Copy this concept map, and then use vocabulary terms from the previous page to complete the concept map.

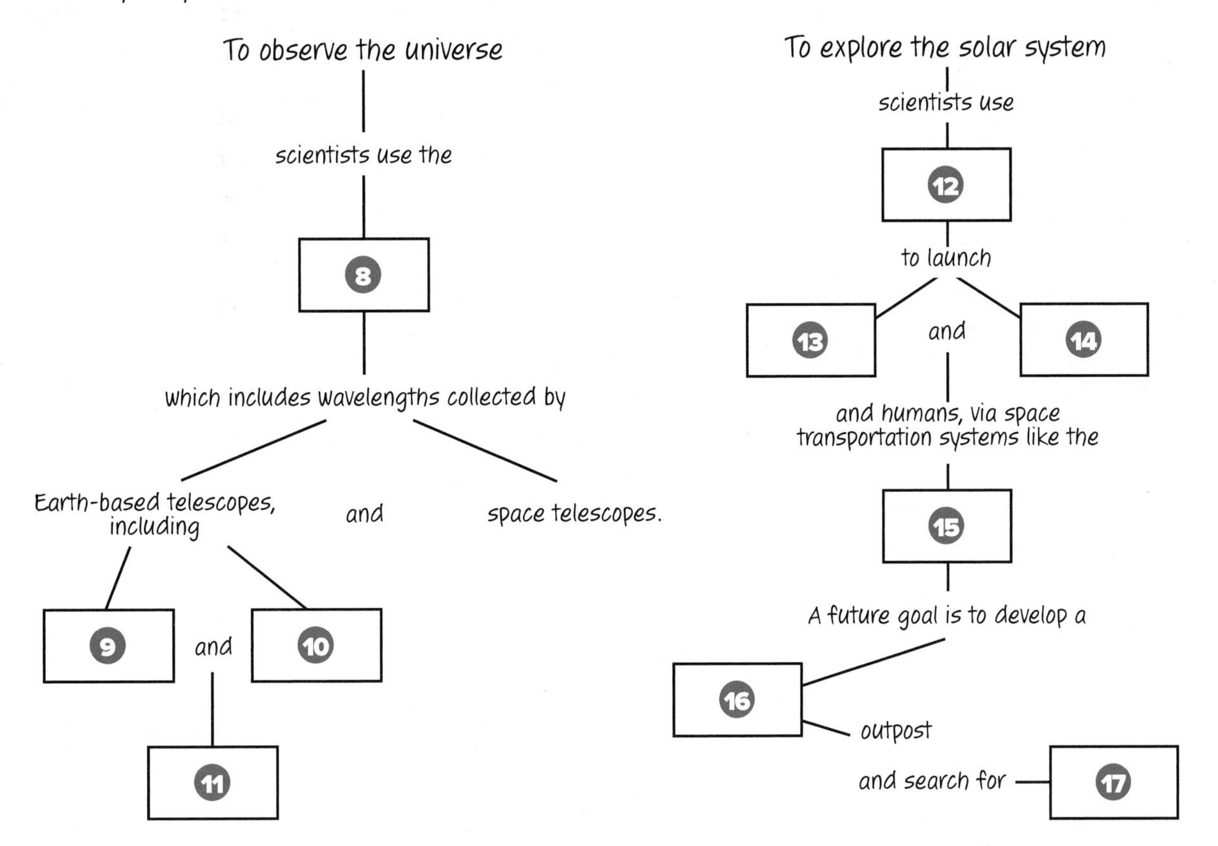

Chapter 19 Review

Understand Key Concepts 

1. Which type of telescope is shown in the figure below?

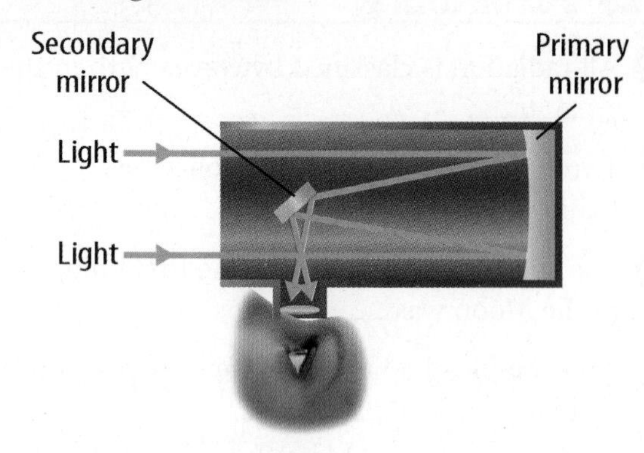

A. infrared telescope
B. radio telescope
C. reflecting telescope
D. refracting telescope

2. In which wavelength would you expect the hottest stars to emit most of their energy?

A. gamma rays
B. microwaves
C. radio waves
D. visible light

3. Which best describes *Hubble?*

A. infrared telescope
B. radio telescope
C. refracting telescope
D. space telescope

4. What is special about the *Kepler* mission?

A. *Kepler* can detect objects at all wavelengths.
B. *Kepler* has found the most distant objects in the universe.
C. *Kepler* is dedicated to finding Earthlike planets.
D. *Kepler* is the first telescope to orbit the Sun.

5. Where is the *International Space Station?*

A. on Mars
B. on the Moon
C. orbiting Earth
D. orbiting the Sun

6. Which mission sent people to the Moon?

A. Apollo
B. Explorer
C. Galileo
D. Pioneer

7. The images below were taken by a rover as it moved along a rocky body in the inner solar system in 2004. Which body is it?

A. Europa
B. Mars
C. Titan
D. Venus

8. Which is NOT a satellite?

A. a flyby
B. a moon
C. an orbiter
D. space telescope

NASA/JPL/Cornell

Critical Thinking

9 **Contrast** waves in the electromagnetic spectrum with water waves in the ocean.

10 **Differentiate** If you wanted to study new stars forming inside a huge dust cloud, which wavelength might you use? Explain.

11 **Deduce** Why do Earth-based optical telescopes work best at night, while radio telescopes work all day and all night long?

12 **Analyze** Why it is more challenging to send space probes to the outer solar system than to the inner solar system?

13 **Create** a list of requirements that must be satisfied before humans can live on the Moon.

14 **Choose** a body in the solar system that you think would be a good place to look for life. Explain.

15 **Interpret Graphics** Copy the diagram of electromagnetic waves below, and label the relative positions of ultraviolet waves, X-rays, visible light, infrared waves, microwaves, gamma rays, and radio waves.

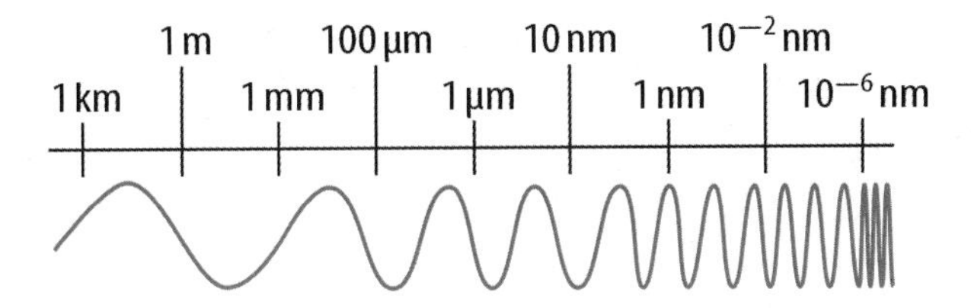

Writing in Science

16 **Write** a paragraph comparing colonizing North America and colonizing the Moon. Include a main idea, supporting details, and a concluding sentence.

REVIEW THE BIG IDEA

17 In what different ways do humans observe and explore space?

18 The photo below shows the *Hubble Space Telescope* orbiting Earth. What are advantages of space-based telescopes? What are disadvantages?

Math Skills

Math Practice

Use Scientific Notation

19 The distance from Saturn to the Sun averages 1,430,000,000 km. Express this distance in scientific notation.

20 The nearest star outside our solar system is Proxima Centauri, which is about 39,900,000,000,000 km from Earth. What is this distance in scientific notation?

21 The *Hubble Space Telescope* has taken pictures of an object that is 1,400,000,000,000,000,000,000 km away from Earth. Express this number in scientific notation.

Standardized Test Practice

Record your answers on the answer sheet provided by your teacher or on a sheet of paper.

Multiple Choice

1 Which is NOT a good place to build a radio telescope?

A a location near a radio station

B a location that is remote

C a location with a large cleared area

D a location with dry air

2 Which has the power to overcome the force of Earth's gravity to be launched into space?

A a probe

B a rocket

C a satellite

D a telescope

Use the figure below to answer question 3.

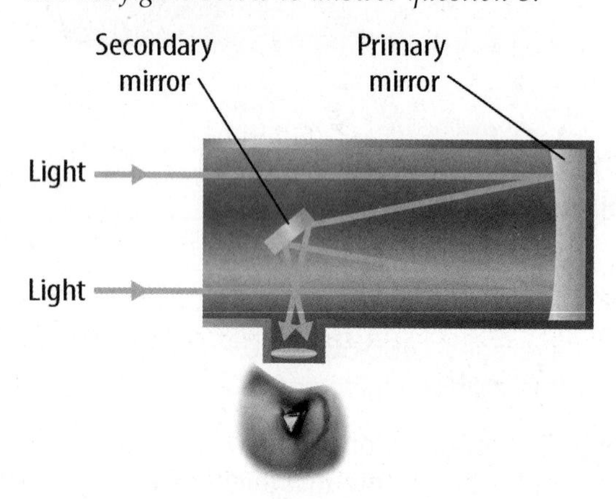

3 Which could increase the light-gathering power of the telescope in the figure?

A adaptive optics

B a larger eyepiece

C multiple small mirrors

D thicker lenses

4 Which lists the minimum resources needed for life-forms to survive on Earth?

A liquid water, an energy source, and sunshine

B liquid water, sunshine, and organic molecules

C organic molecules, an energy source, and liquid water

D organic molecules, an energy source, and sunshine

Use the table below to answer questions 5 and 6.

Planet	Average Distance from Sun (in millions of kilometers)
Earth	150
Mars	228
Saturn	1,434

5 It takes about 8.3 min for light to travel from the Sun to Earth. It takes about 40 min for light to travel from the Sun to Jupiter. How long would you expect it to take light to travel from the Sun to Saturn?

A 8.5 min

B 1.3 h

C 13.5 h

D 26.3 h

6 Which shows the distance between Saturn and the Sun expressed in scientific notation?

A 1.434×10^6 km

B 1.434×10^8 km

C 1.434×10^9 km

D 14.34×10^7 km

7 What is the advantage of using gravity assist for a mission to Saturn?

A The spacecraft can be made of a nonmagnetic material.

B The spacecraft can travel at the speed of light.

C The spacecraft needs less fuel.

D The spacecraft needs more weight.

8 Which was the first satellite to orbit Earth?

A *Apollo 1*

B *Explorer 1*

C *Mariner 1*

D *Sputnik 1*

Use the figure below to answer question 9.

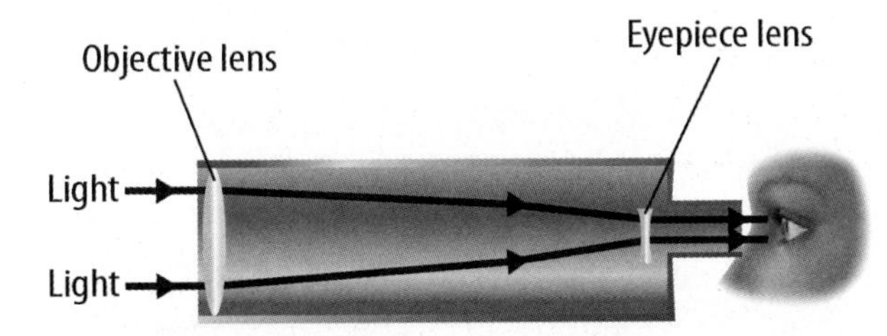

9 Which is true of the telescope above?

A The eyepiece and the objective lens are concave lenses.

B Light is bent as it goes through the objective lens.

C Light is reflected from the eyepiece lens to the objective lens.

D The eyepiece lens can be made of many smaller lenses.

Constructed Response

Use the figure below to answer questions 10 and 11.

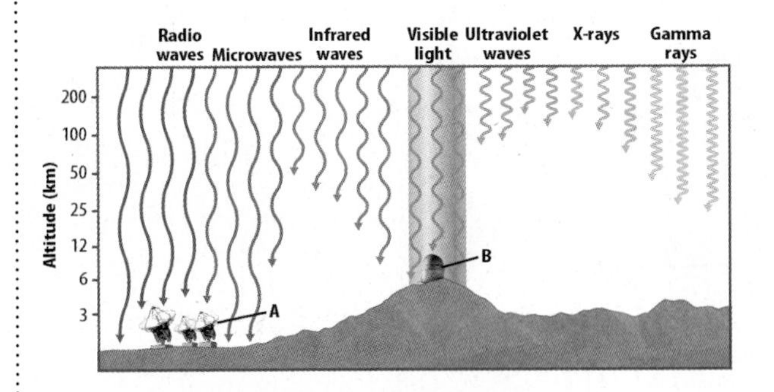

10 Identify the types of telescopes labeled *A* and *B* in the figure. Briefly explain what radiant energy each gathers and how each telescope works.

11 Use the information in the figure to explain why X-ray images can be obtained only using telescopes located above Earth's atmosphere.

12 How does studying radiant energy help scientists learn about the universe?

13 How might the properties of materials developed for use in space be useful on Earth? Give examples.

14 How does information gathered in space help scientists learn about Earth?

15 How does the *Kepler telescope* differ from other telescopes in space?

NEED EXTRA HELP?															
If You Missed Question...	1	2	3	4	5	6	7	8	9	10	11	12	13	14	15
Go to Lesson...	1	2	1	3	1	1	3	2	1	1	1	1	2	3	3

Chapter 20

The Sun-Earth-Moon System

THE BIG IDEA

What natural phenomena do the motions of Earth and the Moon produce?

©O. Alamany & E. Vicens/Corbis

Inquiry Sun Bites?

Look at this time-lapse photograph. The "bites" out of the Sun occurred during a solar eclipse. The Sun's appearance changed in a regular, predictable way as the Moon's shadow passed over a part of Earth.

- How does the Moon's movement change the Sun's appearance?
- What predictable changes does Earth's movement cause?
- What other natural phenomena do the motions of Earth and the Moon cause?

Get Ready to Read

What do you think?

Before you read, decide if you agree or disagree with each of these statements. As you read this chapter, see if you change your mind about any of the statements.

1. Earth's movement around the Sun causes sunrises and sunsets.
2. Earth has seasons because its distance from the Sun changes throughout the year.
3. The Moon was once a planet that orbited the Sun between Earth and Mars.
4. Earth's shadow causes the changing appearance of the Moon.
5. A solar eclipse happens when Earth moves between the Moon and the Sun.
6. The gravitational pull of the Moon and the Sun on Earth's oceans causes tides.

Your one-stop online resource
connectED.mcgraw-hill.com

LearnSmart®

Chapter Resources Files, Reading Essentials, Get Ready to Read, Quick Vocabulary

Animations, Videos, Interactive Tables

Self-checks, Quizzes, Tests

Project-Based Learning Activities

Lab Manuals, Safety Videos, Virtual Labs & Other Tools

Vocabulary, Multilingual eGlossary, Vocab eGames, Vocab eFlashcards

Personal Tutors

Lesson 1

Earth's Motion

Reading Guide

Key Concepts

ESSENTIAL QUESTIONS

- How does Earth move?
- Why is Earth warmer at the equator and colder at the poles?
- Why do the seasons change as Earth moves around the Sun?

Vocabulary

orbit p. 726
revolution p. 726
rotation p. 727
rotation axis p. 727
solstice p. 731
equinox p. 731

Multilingual eGlossary

Inquiry Floating in Space?

From the *International Space Station,* Earth might look like it is just floating, but it is actually traveling around the Sun at more than 100,000 km/h. What phenomena does Earth's motion cause?

NASA Human Spaceflight Collection

Launch Lab

15 minutes

Does Earth's shape affect temperatures on Earth's surface?

Temperatures near Earth's poles are colder than temperatures near the equator. What causes these temperature differences?

1. Read and complete a lab safety form.
2. Inflate a **spherical balloon** and tie the balloon closed.
3. Using a **marker,** draw a line around the balloon to represent Earth's equator.
4. Using a **ruler**, place a lit **flashlight** about 8 cm from the balloon so the flashlight beam strikes the equator straight on.
5. Using the marker, trace around the light projected onto the balloon.
6. Have someone raise the flashlight vertically 5–8 cm without changing the direction that the flashlight is pointing. Do not change the position of the balloon. Trace around the light projected onto the balloon again.

Think About This

1. Compare and contrast the shapes you drew on the balloon.
2. At which location on thae balloon is the light more spread out? Explain your answer.
3. **Key Concept** Use your model to explain why Earth is warmer near the equator and colder near the poles.

Earth and the Sun

If you look outside at the ground, trees, and buildings, it does not seem like Earth is moving. Yet Earth is always in motion, spinning in space and traveling around the Sun. As Earth spins, day changes to night and back to day again. The seasons change as Earth travels around the Sun. Summer changes to winter because Earth's motion changes how energy from the Sun spreads out over Earth's surface.

The Sun

The nearest star to Earth is the Sun, which is shown in **Figure 1.** The Sun is approximately 150 million km from Earth. Compared to Earth, the Sun is enormous. The Sun's diameter is more than 100 times greater than Earth's diameter. The Sun's mass is more than 300,000 times greater than Earth's mass.

Deep inside the Sun, nuclei of atoms combine, releasing huge amounts of energy. This process is called nuclear fusion. The Sun releases so much energy from nuclear fusion that the temperature at its core is more than 15,000,000°C. Even at the Sun's surface, the temperature is about 5,500°C. A small part of the Sun's energy reaches Earth as light and thermal energy.

Figure 1 The Sun is a giant ball of hot gases that emits light and energy.

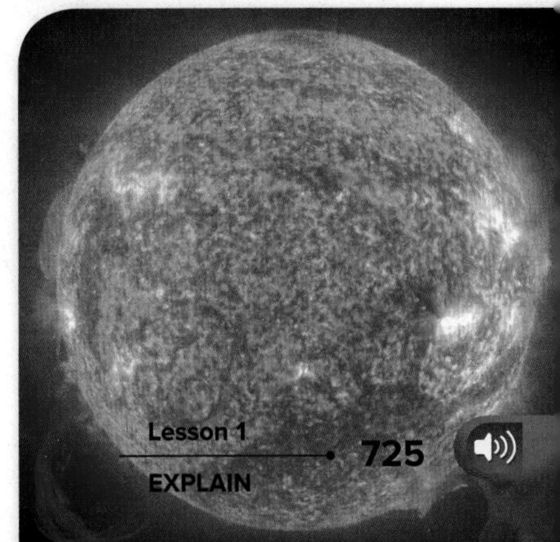

(t)Hutchings Photography/Digital Light Source; (b)SOHO (ESA & NASA)

MiniLab

10 minutes

What keeps Earth in orbit?

Why does Earth move around the Sun and not fly off into space?

1. Read and complete a lab safety form.
2. Tie a piece of **strong thread** securely to a **plastic, slotted golf ball.**
3. Swing the ball in a horizontal circle above your head.

Analyze and Conclude

1. **Predict** what would happen if you let go of the thread.
2. **Key Concept** Which part of the experiment represents the force of gravity between Earth and the Sun?

Earth's Orbit

As shown in **Figure 2,** Earth moves around the Sun in a nearly circular path. *The path an object follows as it moves around another object is an* **orbit.** *The motion of one object around another object is called* **revolution.** Earth makes one complete revolution around the Sun every 365.24 days.

The Sun's Gravitational Pull

Why does Earth orbit the Sun? The answer is the law of universal gravitation. This law states that the pull of gravity between two objects depends on the masses of the objects and the distance between them. The more mass either object has, or the closer together they are, the stronger the gravitational pull.

The Sun's effect on Earth's motion is illustrated in **Figure 2.** Earth's motion around the Sun is like the motion of an object twirled on a string. The string pulls on the object and makes it move in a circle. If the string breaks, the object flies off in a straight line. In the same way, the pull of the Sun's gravity keeps Earth revolving around the Sun in a nearly circular orbit. If the gravity between Earth and the Sun were to somehow stop, Earth would fly off into space in a straight line.

Key Concept Check What produces Earth's revolution around the Sun?

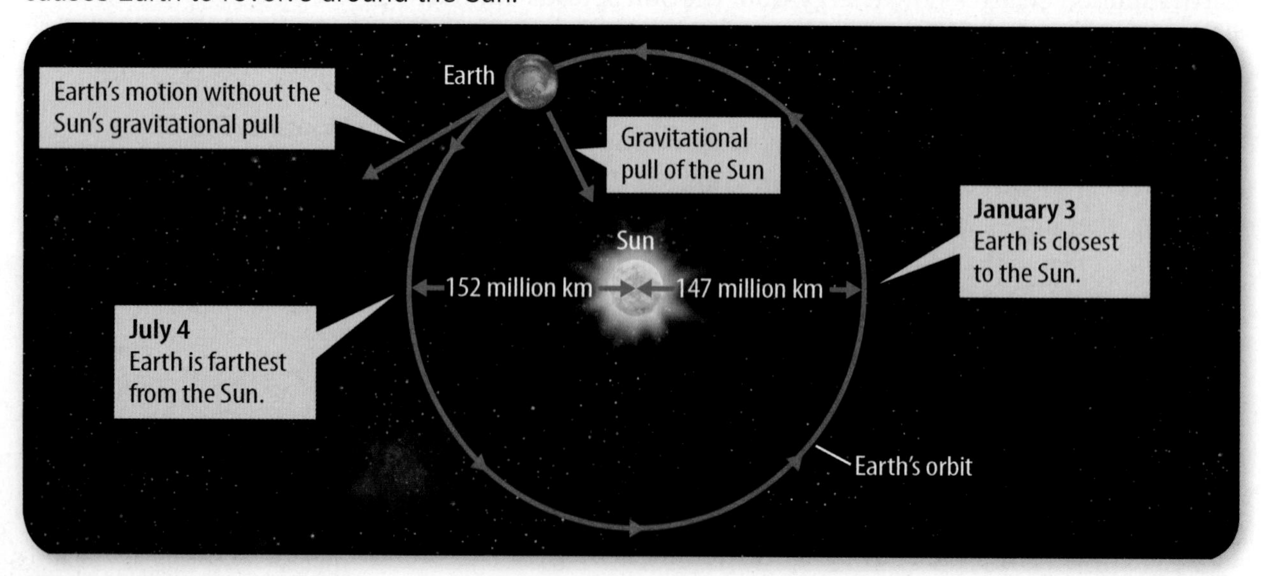

Figure 2 Earth moves in a nearly circular orbit. The pull of the Sun's gravity on Earth causes Earth to revolve around the Sun.

Earth's Rotation Axis

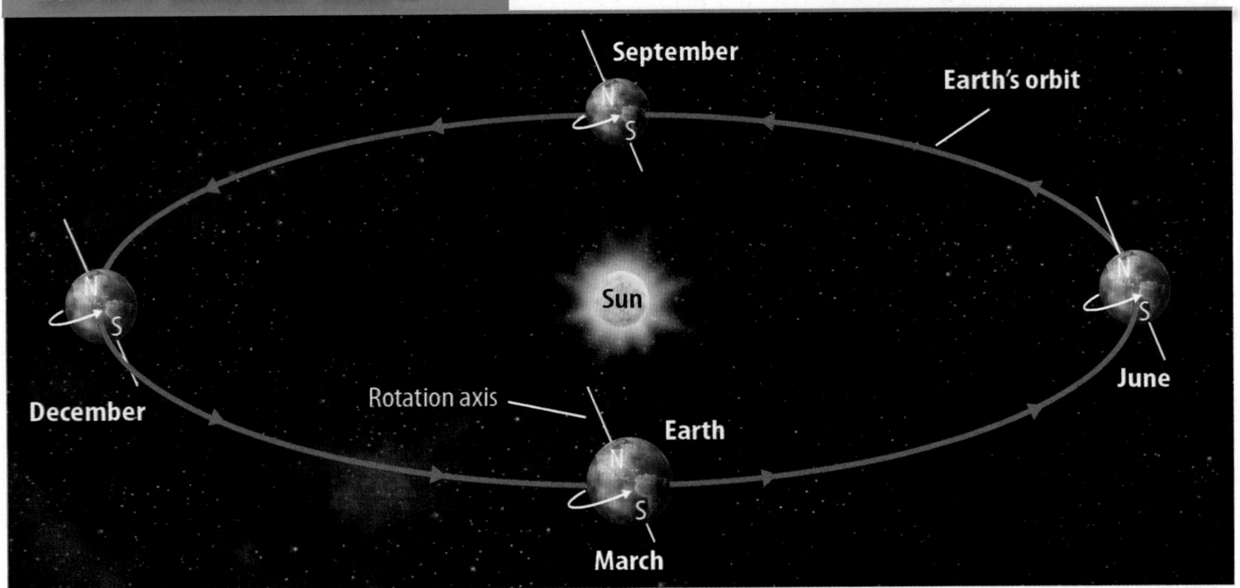

Figure 3 This diagram shows Earth's orbit, which is nearly circular, from an angle. Earth spins on its rotation axis as it revolves around the Sun. Earth's rotation axis always points in the same direction.

Visual Check Between which months is the north end of Earth's rotation axis away from the Sun?

Earth's Rotation

As Earth revolves around the Sun, it spins. *A spinning motion is called* **rotation.** Some spinning objects rotate on a rod or axle. Earth rotates on an imaginary line through its center. *The line on which an object rotates is the* **rotation axis.**

Suppose you could look down on Earth's North Pole and watch Earth rotate. You would see that Earth rotates on its rotation axis in a counterclockwise direction, from west to east. One complete rotation of Earth takes about 24 hours. This rotation helps produce Earth's cycle of day and night. It is daytime on the half of Earth facing toward the Sun and nighttime on the half of Earth facing away from the Sun.

The Sun's Apparent Motion Each day the Sun appears to move from east to west across the sky. It seems as if the Sun is moving around Earth. However, it is Earth's rotation that causes the Sun's apparent motion.

Earth rotates from west to east. As a result, the Sun appears to move from east to west across the sky. The stars and the Moon also seem to move from east to west across the sky due to Earth's west to east rotation.

To better understand this, imagine riding on a merry-go-round. As you and the ride move, people on the ground appear to be moving in the opposite direction. In the same way, as Earth rotates from west to east, the Sun appears to move from east to west.

Reading Check What causes the Sun's apparent motion across the sky?

The Tilt of Earth's Rotation Axis As shown in **Figure 3,** Earth's rotation axis is tilted. The tilt of Earth's rotation axis is always in the same direction by the same amount. This means that during half of Earth's orbit, the north end of the rotation axis is toward the Sun. During the other half of Earth's orbit, the north end of the rotation axis is away from the Sun.

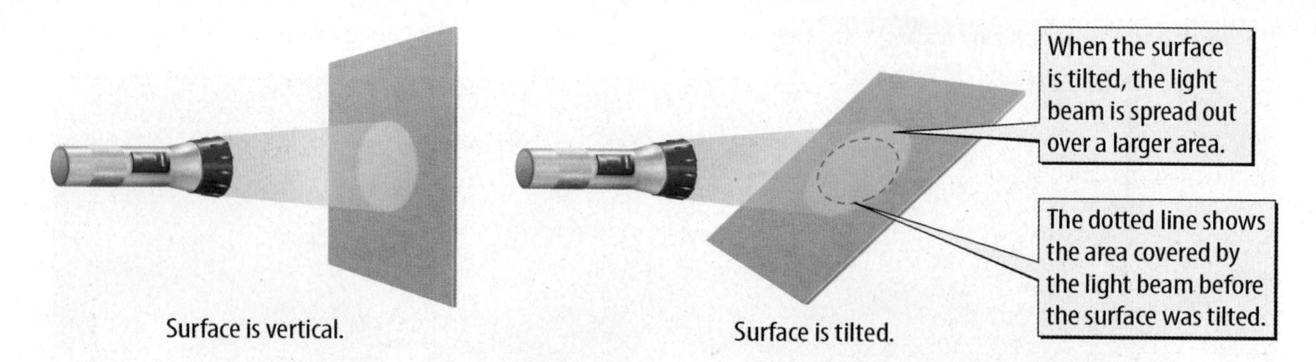

Figure 4 The light energy on a surface becomes more spread out as the surface becomes more tilted relative to the light beam.

Visual Check Is the light energy more spread out on the vertical or tilted surface?

Temperature and Latitude

As Earth orbits the Sun, only one half of Earth faces the Sun at a time. A beam of sunlight carries energy. The more sunlight that reaches a part of Earth's surface, the warmer that part becomes. Because Earth's surface is curved, different parts of Earth's surface receive different amounts of the Sun's energy.

Energy Received by a Tilted Surface

Suppose you shine a beam of light on a flat card, as shown in **Figure 4.** As you tilt the card relative to the direction of the light beam, light becomes more spread out on the card's surface. As a result, the energy that the light beam carries also spreads out more over the card's surface. An area on the surface within the light beam receives less energy when the surface is more tilted relative to the light beam.

The Tilt of Earth's Curved Surface

Instead of being flat like a card, Earth's surface is curved. Relative to the direction of a beam of sunlight, Earth's surface becomes more tilted as you move away from the equator. As shown in **Figure 5,** the energy in a beam of sunlight tends to become more spread out the farther you travel from the equator. This means that regions near the poles receive less energy than regions near the equator. This makes Earth colder at the poles and warmer at the equator.

ACADEMIC VOCABULARY

equator
(noun) the imaginary line that divides Earth into its northern and southern hemispheres

Key Concept Check Why is Earth warmer at the equator and colder at the poles?

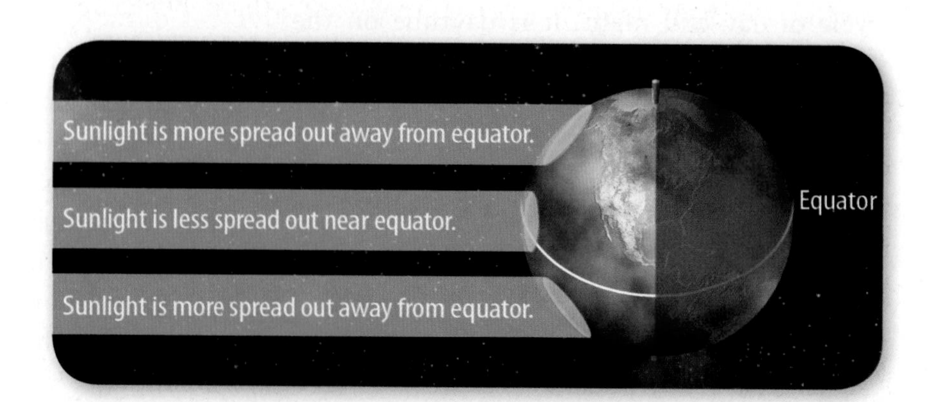

Figure 5 Energy from the Sun becomes more spread out as you move away from the equator.

North end of rotation axis is away from the Sun.

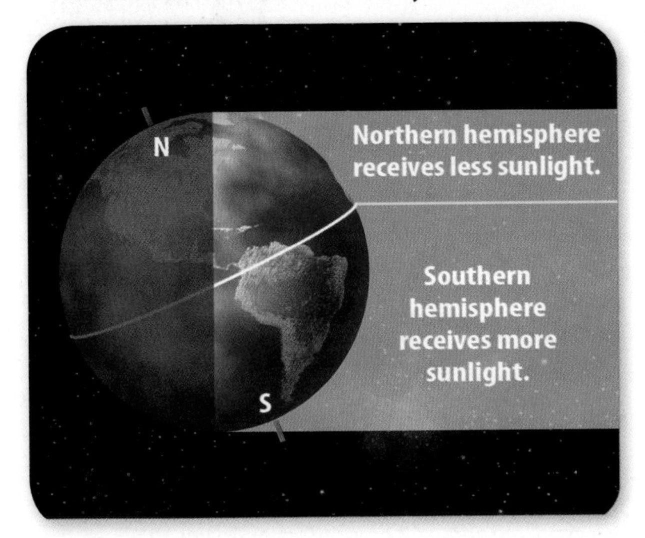

North end of rotation axis is toward the Sun.

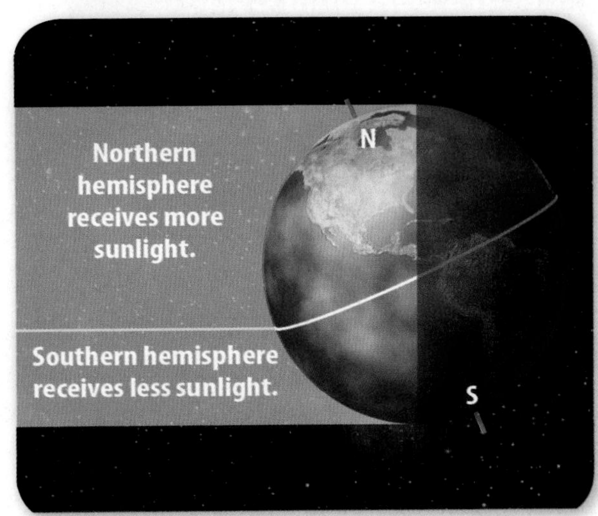

Figure 6 The northern hemisphere receives more sunlight in June, and the southern hemisphere receives more sunlight in December.

Seasons

You might think that summer happens when Earth is closest to the Sun, and winter happens when Earth is farthest from the Sun. However, seasonal changes do not depend on Earth's distance from the Sun. In fact, Earth is closest to the Sun in January! Instead, it is the tilt of Earth's rotation axis, combined with Earth's motion around the Sun, that causes the seasons to change.

Spring and Summer in the Northern Hemisphere

During one half of Earth's orbit, the north end of the rotation axis is toward the Sun. Then, the northern hemisphere receives more energy from the Sun than the southern hemisphere, as shown in **Figure 6.** Temperatures increase in the northern hemisphere and decrease in the southern hemisphere. Daylight hours last longer in the northern hemisphere, and nights last longer in the southern hemisphere. This is when spring and summer happen in the northern hemisphere, and fall and winter happen in the southern hemisphere.

Fall and Winter in the Northern Hemisphere

During the other half of Earth's orbit, the north end of the rotation axis is away from the Sun. Then, the northern hemisphere receives less solar energy than the southern hemisphere, as shown in **Figure 6.** Temperatures decrease in the northern hemisphere and increase in the southern hemisphere. This is when fall and winter happen in the northern hemisphere, and spring and summer happen in the southern hemisphere.

Key Concept Check How does the tilt of Earth's rotation axis affect Earth's weather?

Math Skills 

Convert Units

During January, Earth is 147,000,000 km from the Sun, how far is Earth from the Sun in miles? To calculate the distance in miles, multiply the distance in km by the conversion factor

$$147{,}000{,}000 \text{ km} \times \frac{0.62 \text{ miles}}{1 \text{ km}}$$

$$= 91{,}100{,}000 \text{ miles}$$

Practice

During June, Earth is 152,000,000 km from the Sun, how far is Earth from the Sun in miles?

 Math Practice

 Personal Tutor

Earth's Seasonal Cycle

Personal Tutor

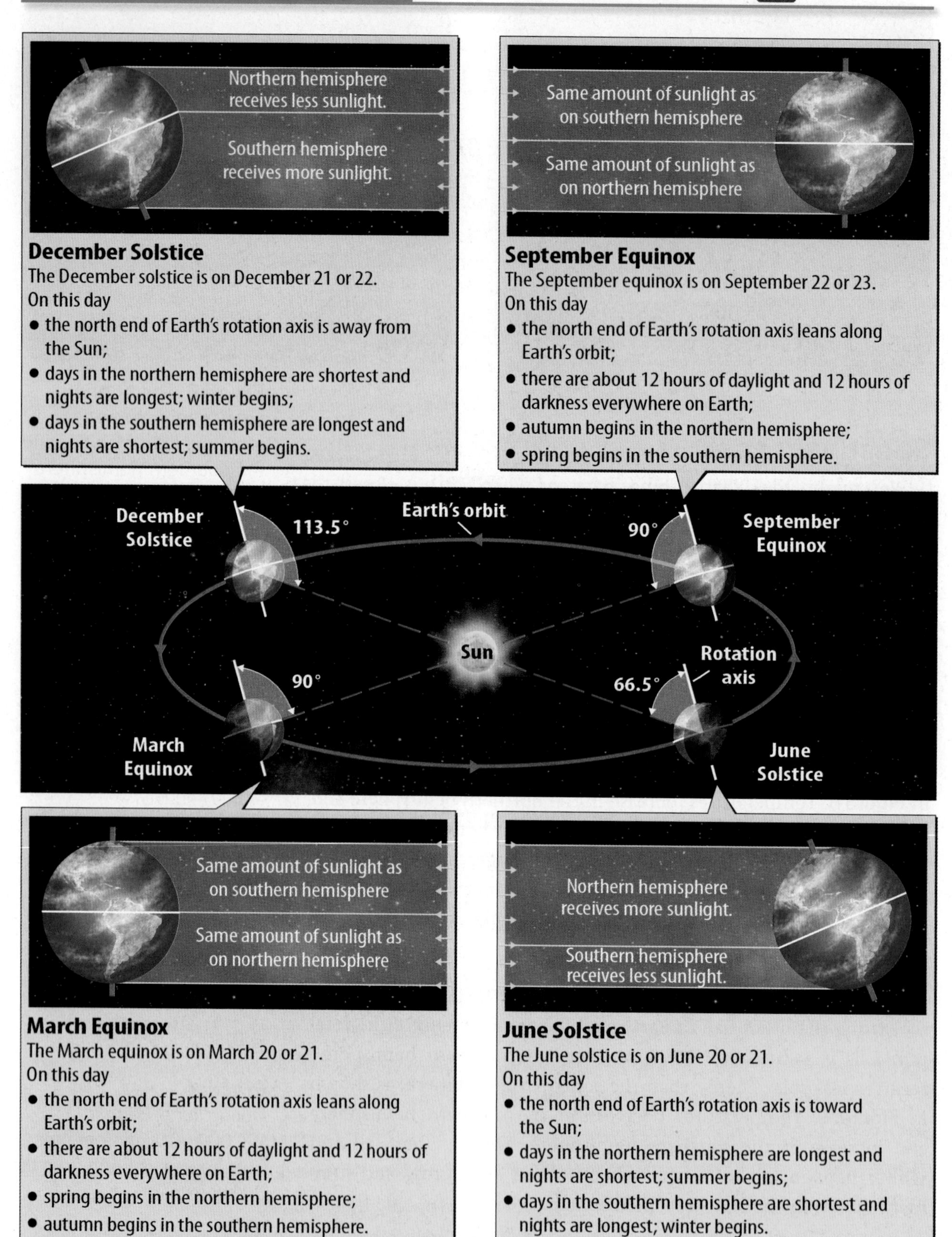

Figure 7 The seasons change as Earth moves around the Sun. Earth's motion around the Sun causes Earth's tilted rotation axis to be leaning toward the Sun and away from the Sun.

Solstices, Equinoxes, and the Seasonal Cycle

Figure 7 shows that as Earth travels around the Sun, its rotation axis always points in the same direction in space. However, the amount that Earth's rotation axis is toward or away from the Sun changes. This causes the seasons to change in a yearly cycle.

There are four days each year when the direction of Earth's rotation axis is special relative to the Sun. *A* **solstice** *is a day when Earth's rotation axis is the most toward or away from the Sun. An* **equinox** *is a day when Earth's rotation axis is leaning along Earth's orbit, neither toward nor away from the Sun.*

March Equinox to June Solstice When the north end of the rotation axis gradually points more and more toward the Sun, the northern hemisphere gradually receives more solar energy. This is spring in the northern hemisphere.

June Solstice to September Equinox The north end of the rotation axis continues to point toward the Sun but does so less and less. The northern hemisphere starts to receive less solar energy. This is summer in the northern hemisphere.

September Equinox to December Solstice The north end of the rotation axis now points more and more away from the Sun. The northern hemisphere receives less and less solar energy. This is fall in the northern hemisphere.

December Solstice to March Equinox The north end of the rotation axis continues to point away from the Sun but does so less and less. The northern hemisphere starts to receive more solar energy. This is winter in the northern hemisphere.

Changes in the Sun's Apparent Path Across the Sky

Figure 8 shows how the Sun's apparent path through the sky changes from season to season in the northern hemisphere. The Sun's apparent path through the sky in the northern hemisphere is lowest on the December solstice and highest on the June solstice.

FOLDABLES®

Make a bound book with four full pages. Label the pages with the names of the solstices and equinoxes. Use each page to organize information about each season.

WORD ORIGIN

equinox
from Latin *equinoxium*, means "equality of night and day"

Figure 8 As the seasons change, the path of the Sun across the sky changes. In the northern hemisphere, the Sun's path is lowest on the December solstice and highest on the June solstice.

Visual Check When is the Sun highest in the sky in the northern hemisphere?

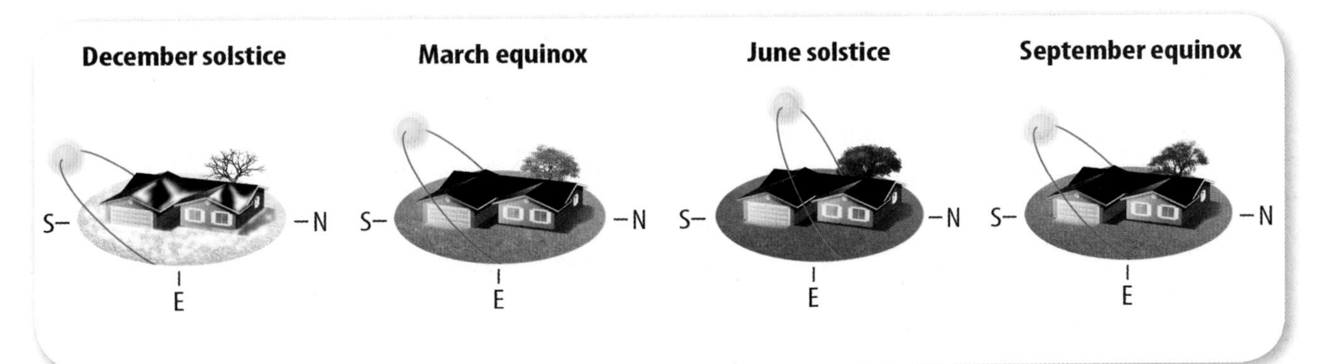

Lesson 1 Review

Visual Summary

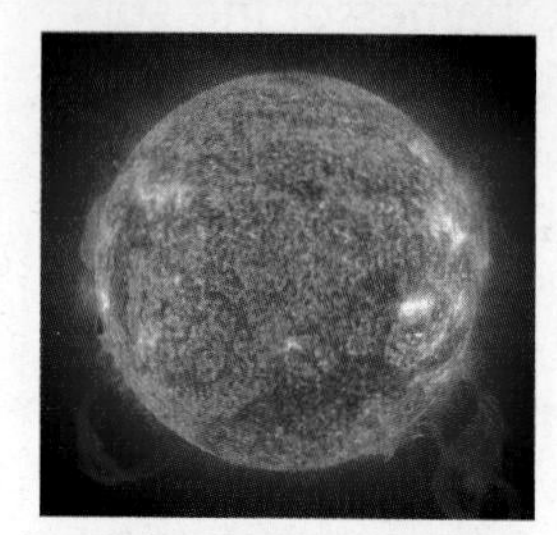

The gravitational pull of the Sun causes Earth to revolve around the Sun in a near-circular orbit.

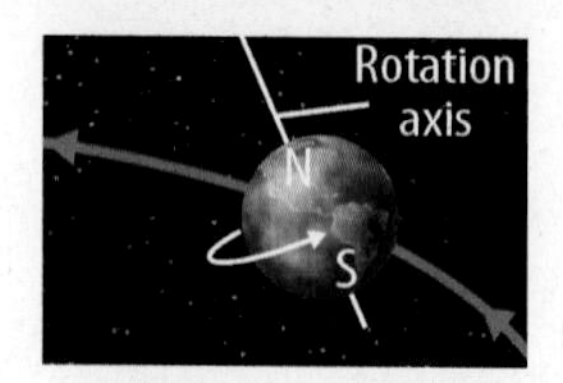

Earth's rotation axis is tilted and always points in the same direction in space.

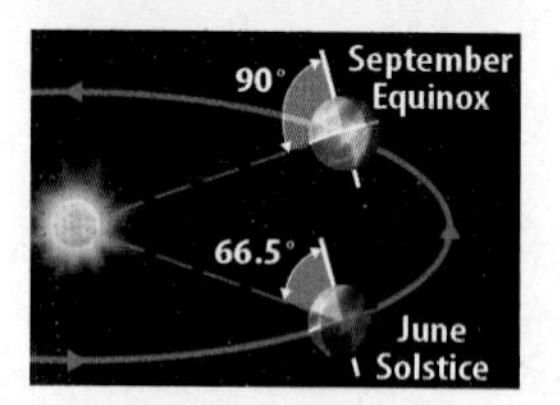

Equinoxes and solstices are days when the direction of Earth's rotation axis relative to the Sun is special.

FOLDABLES®

Use your lesson Foldable to review the lesson. Save your Foldable for the project at the end of the chapter.

What do you think NOW?

You first read the statements below at the beginning of the chapter.

1. Earth's movement around the Sun causes sunrises and sunsets.
2. Earth has seasons because its distance from the Sun changes throughout the year.

Did you change your mind about whether you agree or disagree with the statements? Rewrite any false statements to make them true.

Use Vocabulary

1. **Distinguish** between Earth's rotation and Earth's revolution.
2. The path Earth follows around the Sun is Earth's ________.
3. When a(n) ________ occurs, the northern hemisphere and the southern hemisphere receive the same amount of sunlight.

Understand Key Concepts 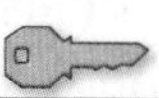

4. What is caused by the tilt of Earth's rotational axis?
   - **A.** Earth's orbit
   - **B.** Earth's seasons
   - **C.** Earth's revolution
   - **D.** Earth's rotation
5. **Contrast** the amount of sunlight received by an area near the equator and a same-sized area near the South Pole.
6. **Contrast** the Sun's gravitational pull on Earth when Earth is closest to the Sun and when Earth is farthest from the Sun.

Interpret Graphics

7. **Summarize** Copy and fill in the table below for the seasons in the northern hemisphere.

Season	Starts on Solstice or Equinox?	How Rotation Axis Leans
Summer		
Fall		
Winter		
Spring		

Critical Thinking

8. **Defend** The December solstice is often called the winter solstice. Do you think this is an appropriate label? Defend your answer.

Math Skills

9. The Sun's diameter is about 1,390,000 km. What is the Sun's diameter in miles?

Skill Practice Draw Conclusions

25 minutes

Materials

large foam ball

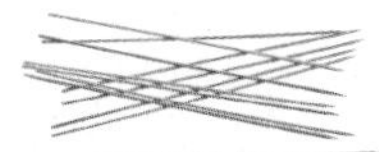

wooden skewer

foam cup

masking tape

flashlight

marker

Safety

How does Earth's tilted rotation axis affect the seasons?

The seasons change as Earth revolves around the Sun. How does Earth's tilted rotation axis change how sunlight spreads out over different parts of Earth's surface?

Learn It

Using a flashlight as the Sun and a foam ball as Earth, you can model how solar energy spreads out over Earth's surface at different times during the year. This will help you **draw conclusions** about Earth's seasons.

Try It

1. Read and complete a lab safety form.
2. Insert a wooden skewer through the center of a foam ball. Draw a line on the ball to represent Earth's equator. Insert one end of the skewer into an upside-down foam cup so the skewer tilts.
3. Prop a flashlight on a stack of books about 0.5 m from the ball. Turn on the flashlight and position the ball so the skewer points toward the flashlight, representing the June solstice.
4. In your Science Journal, draw how the ball's surface is tilted relative to the light beam.

2

5. Under your diagram, state whether the northern (upper) or southern (lower) hemisphere receives more light energy.
6. With the skewer always pointing in the same direction, move the ball around the flashlight. Turn the flashlight to keep the light on the ball. At the three positions corresponding to the equinoxes and other solstice, make drawings like those in step 4 and statements like those in step 5.

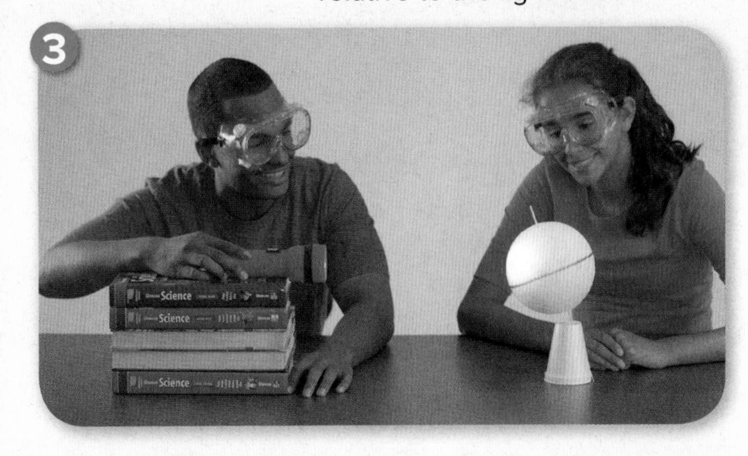

3

Apply It

7. How did the tilt of the surfaces change relative to the light beam as the ball circled the flashlight?
8. How did the amount of light energy on each hemisphere change as the ball moved around the flashlight?
9. **Key Concept** Draw conclusions about how Earth's tilt affects the seasons.

Lesson 2

Earth's Moon

Reading Guide

Key Concepts 

ESSENTIAL QUESTIONS

- How does the Moon move around Earth?
- Why does the Moon's appearance change?

Vocabulary

maria p. 736

phase p. 738

waxing phase p. 738

waning phase p. 738

Multilingual eGlossary

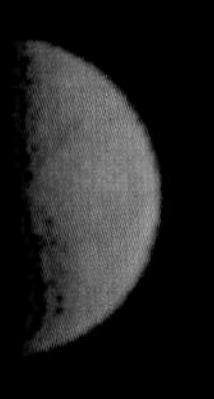

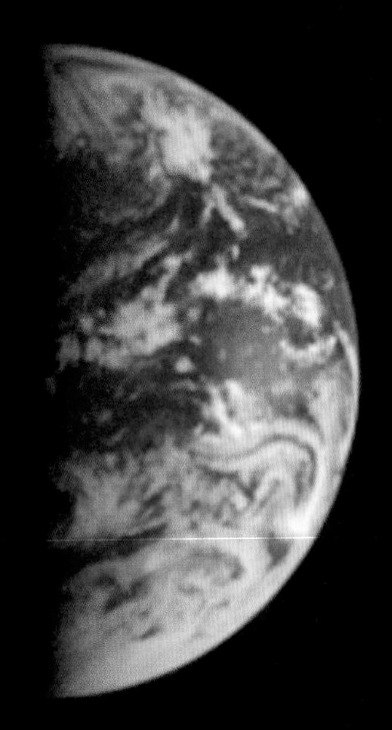

Inquiry Two Planets?

The smaller body is Earth's Moon, not a planet. Just as Earth moves around the Sun, the Moon moves around Earth. The Moon's motion around Earth causes what kinds of changes to occur?

Launch Lab

15 minutes

Why does the Moon appear to change shape?

The Sun is always shining on Earth and the Moon. However, the Moon's shape seems to change from night to night and day to day. What could cause the Moon's appearance to change?

1. Read and complete a lab safety form.
2. Place a **ball** on a level surface.
3. Position a **flashlight** so that the light beam shines fully on one side of the ball. Stand behind the flashlight.
4. Make a drawing of the ball's appearance in your Science Journal.
5. Stand behind the ball, facing the flashlight, and repeat step 4.
6. Stand to the left of the ball and repeat step 4.

Think About This

1. What caused the ball's appearance to change?
2. **Key Concept** What do you think produces the Moon's changing appearance in the sky?

Seeing the Moon

Imagine what people thousands of years ago thought when they looked up at the Moon. They might have wondered why the Moon shines and why it seems to change shape. They probably would have been surprised to learn that the Moon does not emit light at all. Unlike the Sun, the Moon is a solid object that does not emit its own light. You only see the Moon because light from the Sun reflects off the Moon and into your eyes. Some facts about the Moon, such as its mass, size, and distance from Earth, are shown in **Table 1**.

Table 1 Moon Data

Mass	Diameter	Average distance from Earth	Time for one rotation	Time for one revolution
1.2% of Earth's mass	27% of Earth's diameter	384,000 km	27.3 days	27.3 days

FOLDABLES®

Use two sheets of paper to make a bound book. Use it to organize information about the lunar cycle. Each page of your book should represent one week of the lunar cycle.

 Animation

Figure 9 The Moon probably formed when a large object collided with Earth 4.5 billion years ago. Material ejected from the collision eventually clumped together and became the Moon.

An object the size of Mars crashes into the semi-molten Earth about 4.5 billion years ago.

The impact ejects vaporized rock into space. As the rock cools, it forms a ring of particles around Earth.

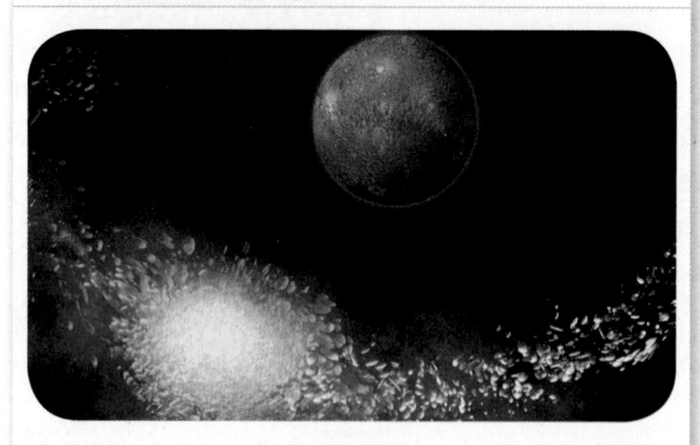

The particles gradually clump together and form the Moon.

WORD ORIGIN

maria
from Latin *mare,* means "sea"

The Moon's Formation

The most widely accepted idea for the Moon's formation is the giant impact hypothesis, shown in **Figure 9.** According to this hypothesis, shortly after Earth formed about 4.6 billion years ago, an object about the size of the planet Mars collided with Earth. The impact ejected vaporized rock that formed a ring around Earth. Eventually, the material in the ring cooled, clumped together, and formed the Moon.

The Moon's Surface

The surface of the Moon was shaped early in its history. Examples of common features on the Moon's surface are shown in **Figure 10.**

Craters The Moon's craters were formed when objects from space crashed into the Moon. Light-colored streaks called rays extend outward from some craters.

Most of the impacts that formed the Moon's craters occurred more than 3.5 billion years ago, long before dinosaurs lived on Earth. Earth was also heavily bombarded by objects from space during this time. However, on Earth, wind, liquid water, and plate tectonics erased the craters. The Moon has no atmosphere, liquid water, or plate tectonics, so craters formed billions of years ago on the Moon have hardly changed.

Maria *The large, dark, flat areas on the Moon are called* **maria** (MAR ee uh). The maria formed after most impacts on the Moon's surface had stopped. Maria formed when lava flowed up through the Moon's crust and solidified. The lava covered many of the Moon's craters and other features. When this lava solidified, it was dark and flat.

 **Reading Check** How were maria produced?

Highlands The light-colored highlands are too high for the lava that formed the maria to reach. The highlands are older than the maria and are covered with craters.

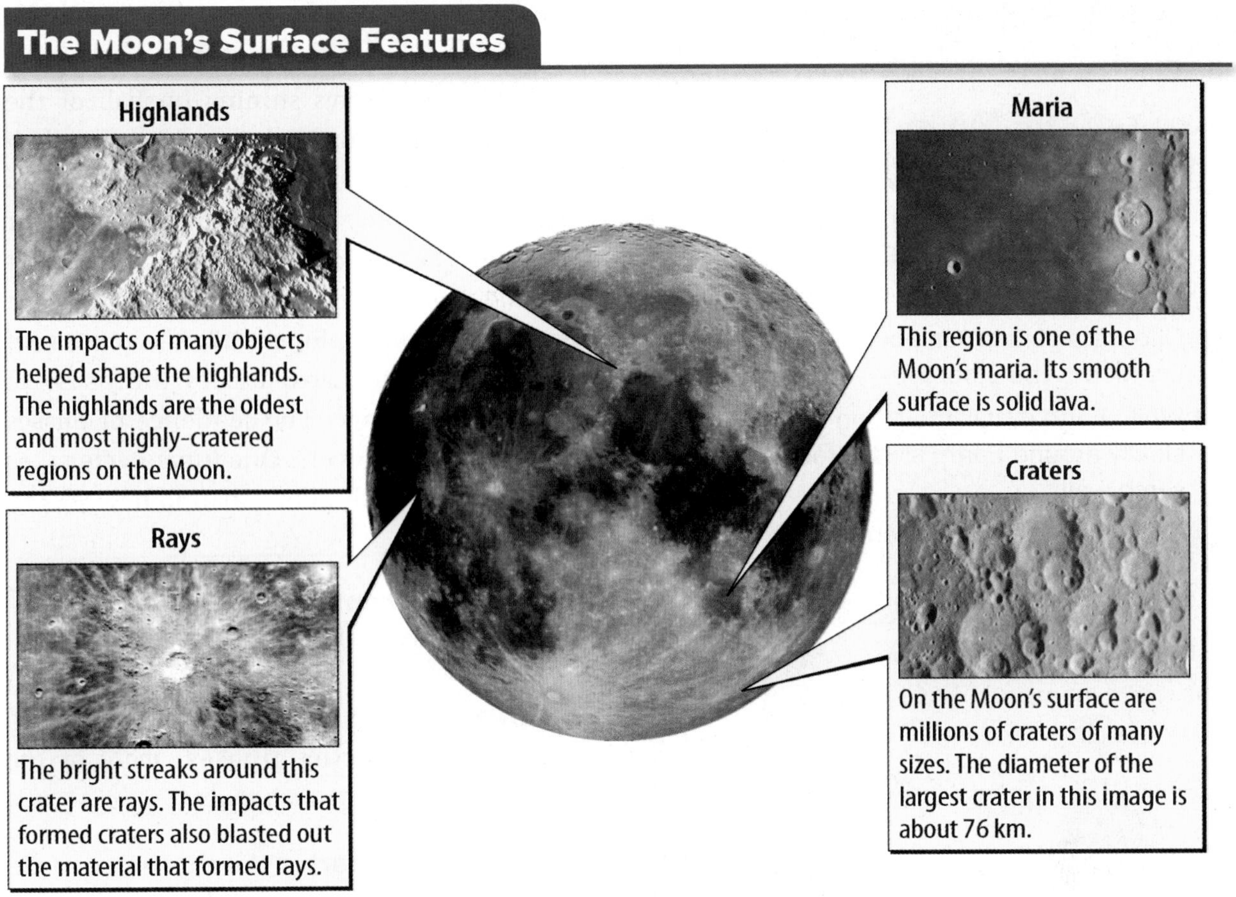

The Moon's Surface Features

Highlands

The impacts of many objects helped shape the highlands. The highlands are the oldest and most highly-cratered regions on the Moon.

Maria

This region is one of the Moon's maria. Its smooth surface is solid lava.

Rays

The bright streaks around this crater are rays. The impacts that formed craters also blasted out the material that formed rays.

Craters

On the Moon's surface are millions of craters of many sizes. The diameter of the largest crater in this image is about 76 km.

▲ **Figure 10** The Moon's surface features include craters, rays, maria, and highlands.

The Moon's Motion

While Earth is revolving around the Sun, the Moon is revolving around Earth. The gravitational pull of Earth on the Moon causes the Moon to move in an orbit around Earth. The Moon makes one revolution around Earth every 27.3 days.

Key Concept Check What produces the Moon's revolution around Earth?

The Moon also rotates as it revolves around Earth. One complete rotation of the Moon also takes 27.3 days. This means the Moon makes one rotation in the same amount of time that it makes one revolution around Earth. **Figure 11** shows that, because the Moon makes one rotation for each revolution of Earth, the same side of the Moon always faces Earth. This side of the Moon is called the near side. The side of the Moon that cannot be seen from Earth is called the far side of the Moon.

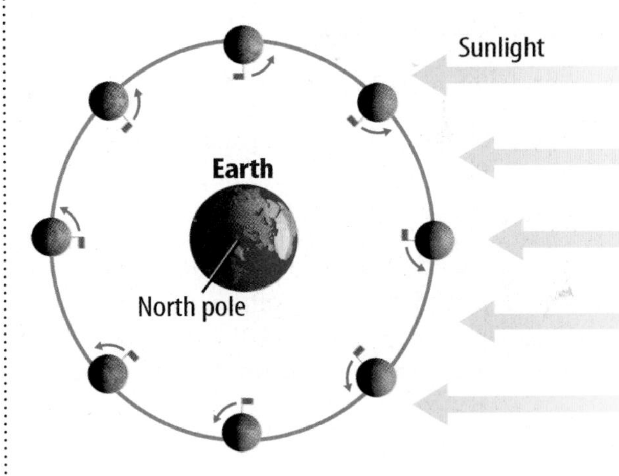

▲ **Figure 11** The Moon rotates once on its axis and revolves around Earth in the same amount of time. As a result, the same side of the Moon always faces Earth.

(tl, tr, br)Lunar and Planetary Institute; (c)NASA/JPL/USGS; (bl)ClassicStock/Alamy

MiniLab 10 minutes

How can the Moon be rotating if the same side of the Moon is always facing Earth?

The Moon revolves around Earth. Does the Moon also rotate as it revolves around Earth?

1. Choose a partner. One person represents the Moon. The other represents Earth.
2. While Earth is still, the Moon moves slowly around Earth, always facing the same wall.
3. Next, the Moon moves around Earth always facing Earth.

Analyze and Conclude

1. For which motion was the Moon rotating?
2. For each type of motion, how many times did the Moon rotate during one revolution around Earth?
3. **Key Concept** How is the Moon actually rotating if the same side of the Moon is always facing Earth?

Science Use v. Common Use

phase

Science Use how the Moon or a planet is lit as seen from Earth

Common Use a part of something or a stage of development

Phases of the Moon

The Sun is always shining on half of the Moon, just as the Sun is always shining on half of Earth. However, as the Moon moves around Earth, usually only part of the Moon's near side is lit. *The portion of the Moon or a planet reflecting light as seen from Earth is called a* **phase.** As shown in **Figure 12,** the motion of the Moon around Earth causes the phase of the Moon to change. The sequence of phases is called the lunar cycle. One lunar cycle takes 29.5 days or slightly more than four weeks to complete.

Key Concept Check What produces the phases of the Moon?

Waxing Phases

During the **waxing phases,** *more of the Moon's near side is lit each night.*

Week 1—First Quarter As the lunar cycle begins, a sliver of light can be seen on the Moon's western edge. Gradually the lit part becomes larger. By the end of the first week, the Moon is at its first quarter phase. In this phase, the Moon's entire western half is lit.

Week 2—Full Moon During the second week, more and more of the near side becomes lit. When the Moon's near side is completely lit, it is at the full moon phase.

Waning Phases

During the **waning phases,** *less of the Moon's near side is lit each night.* As seen from Earth, the lit part is now on the Moon's eastern side.

Week 3—Third Quarter During this week, the lit part of the Moon becomes smaller until only the eastern half of the Moon is lit. This is the third quarter phase.

Week 4—New Moon During this week, less and less of the near side is lit. When the Moon's near side is completely dark, it is at the new moon phase.

Hutchings Photography/Digital Light Source

The Lunar Cycle

Animation

Figure 12 As the Moon revolves around Earth, the part of the Moon's near side that is lit changes. The figure below shows how the Moon looks at different places in its orbit.

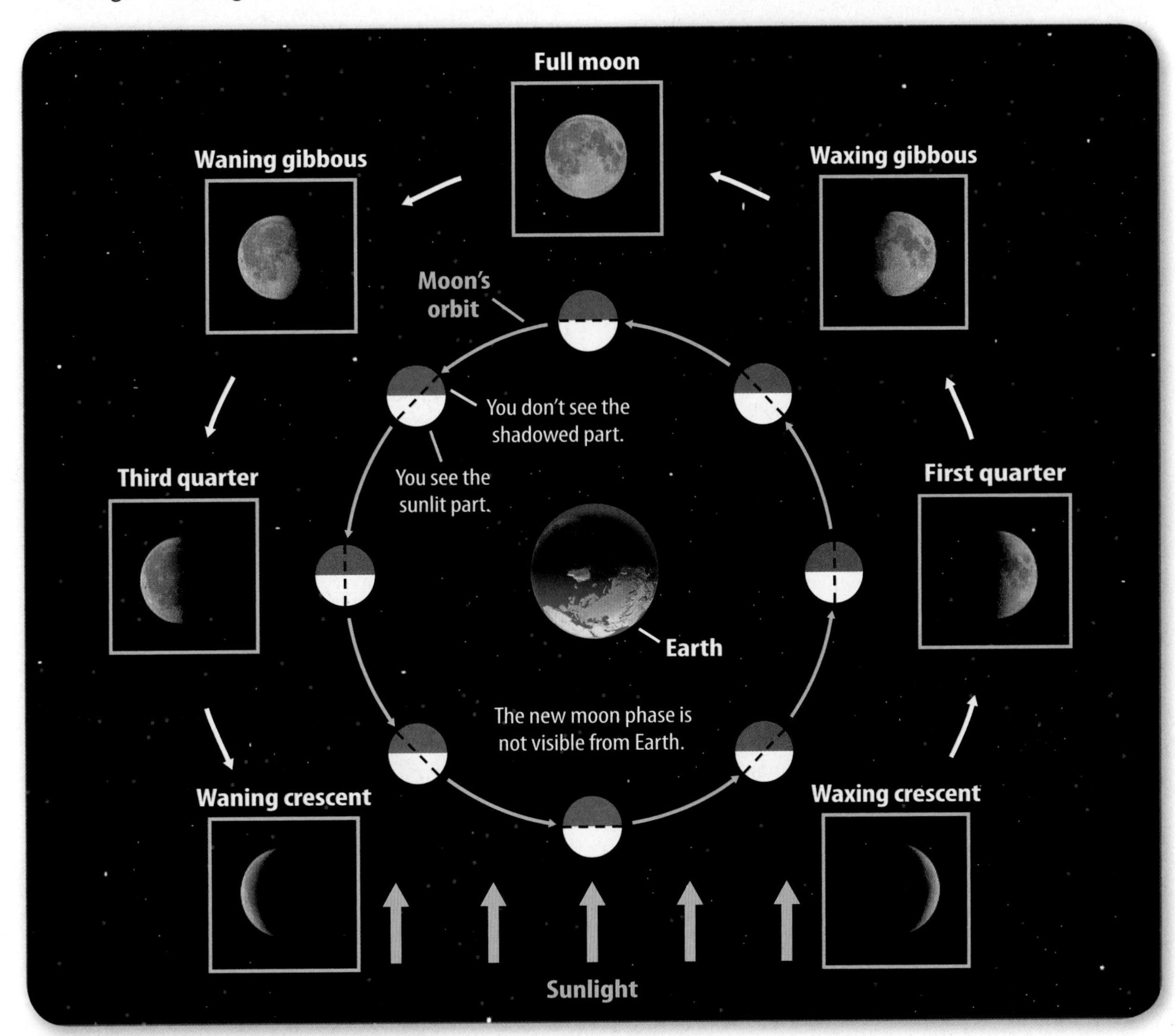

The Moon at Midnight

The Moon's motion around Earth causes the Moon to rise, on average, about 50 minutes later each day. The figure below shows how the Moon looks at midnight during three phases of the lunar cycle.

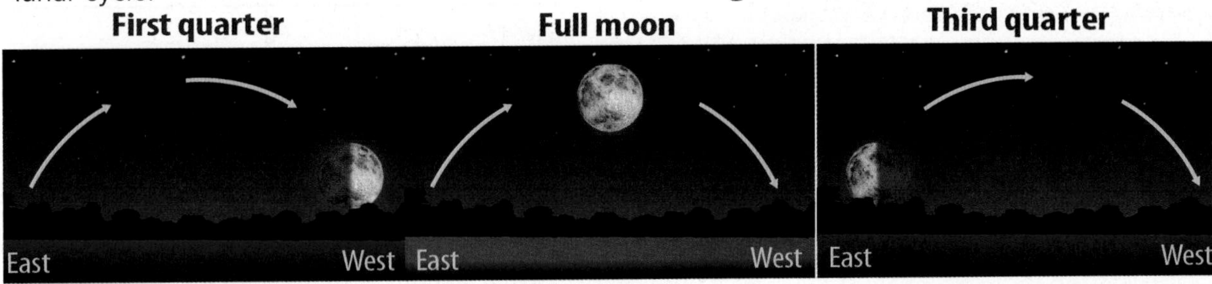

At midnight, the first quarter moon is setting. It rises during the day at about noon.

The full moon is highest in the sky at about midnight. It rises at sunset and sets at sunrise.

The third quarter moon rises at about midnight, about six hours later than the full moon rises.

Lesson 2 Review

Online Quiz
Virtual Lab

Visual Summary

According to the giant impact hypothesis, a large object collided with Earth about 4.5 billion years ago to form the Moon.

Features like maria, craters, and highlands formed on the Moon's surface early in its history.

The Moon's phases change in a regular pattern during the Moon's lunar cycle.

FOLDABLES®

Use your lesson Foldable to review the lesson. Save your Foldable for the project at the end of the chapter.

What do you think 

You first read the statements below at the beginning of the chapter.

3. The Moon was once a planet that orbited the Sun between Earth and Mars.

4. Earth's shadow causes the changing appearance of the Moon.

Did you change your mind about whether you agree or disagree with the statements? Rewrite any false statements to make them true.

Use Vocabulary

1. The lit part of the Moon as viewed from Earth is a(n) ________.

2. For the first half of the lunar cycle, the lit part of the Moon's near side is ________.

3. For the second half of the lunar cycle, the lit part of the Moon's near side is ________.

Understand Key Concepts 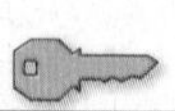

4. Which phase occurs when the Moon is between the Sun and Earth?
   - **A.** first quarter
   - **B.** full moon
   - **C.** new moon
   - **D.** third quarter

5. **Reason** Why does the Moon have phases?

Interpret Graphics

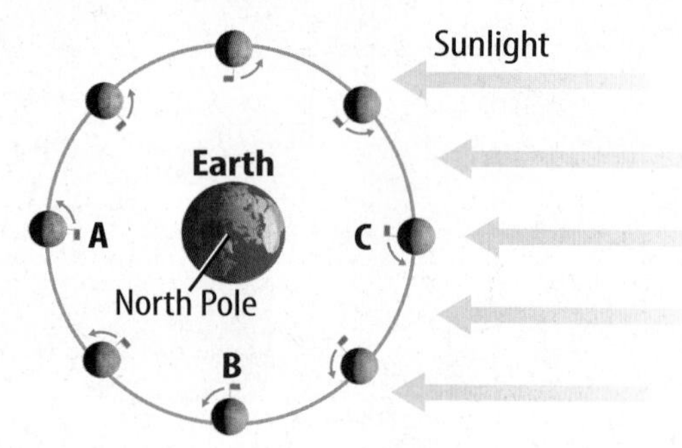

6. **Draw** how the Moon looks from Earth when it is at positions A, B, and C in the diagram above.

7. **Organize Information** Copy and fill in the table below with details about the lunar surface.

Crater	
Ray	
Maria	
Highland	

Critical Thinking

8. **Reflect** Imagine the Moon rotates twice in the same amount of time the Moon orbits Earth once. Would you be able to see the Moon's far side from Earth?

SCIENCE & SOCIETY

Return to the Moon

Exploring Earth's Moon is a step toward exploring other planets and building outposts in space.

The United States undertook a series of human spaceflight missions from 1961–1975 called the Apollo program. The goal of the program was to land humans on the Moon and bring them safely back to Earth. Six of the missions reached this goal. The Apollo program was a huge success, but it was just the beginning.

NASA began another space program that had a goal to return astronauts to the Moon to live and work. However, before that could happen, scientists needed to know more about conditions on the Moon and what materials are available there.

Collecting data was the first step. In 2009, NASA launched the *Lunar Reconnaissance Orbiter (LRO)* spacecraft. The *LRO* spent a year orbiting the Moon's two poles. It collected detailed data that scientists can use to make maps of the Moon's features and resources, such as deep craters that formed on the Moon when comets and asteroids slammed into it billions of years ago. Some scientists predicted that these deep craters contain frozen water.

One of the instruments launched with the *LRO* was the *Lunar Crater Observation and Sensing Satellite (LCROSS). LCROSS* observations confirmed the scientists' predictions that water exists on the Moon. A rocket launched from *LCROSS* impacted the Cabeus crater near the Moon's south pole. The material that was ejected after the rocket's impact included water.

NASA's goal of returning astronauts to the Moon was delayed, and their missions now focus on exploring Mars instead. But the discoveries made on the Moon will help scientists develop future missions that could take humans farther into the solar system.

NASA

Apollo SPACE PROGRAM

The Apollo Space Program included 17 missions. Here are some milestones:

January 27 1967

Apollo 1 Fire killed all three astronauts on board during a launch simulation for the first piloted flight to the Moon.

December 21–27 1968

Apollo 8 First manned spacecraft orbits the Moon.

July 16–24 1969

Apollo 11 First humans, Neil Armstrong and Buzz Aldrin, walk on the Moon.

July 1971

Apollo 15 Astronauts drive the first rover on the Moon.

December 7–19 1972

Apollo 17 The first phase of human exploration of the Moon ended with this last lunar landing mission.

It's Your Turn

BRAINSTORM As a group, brainstorm the different occupations that would be needed to successfully operate a base on the Moon or another planet. Discuss the tasks that a person would perform in each occupation.

Lesson 3

Eclipses and Tides

Reading Guide

Key Concepts

ESSENTIAL QUESTIONS

- What is a solar eclipse?
- What is a lunar eclipse?
- How do the Moon and the Sun affect Earth's oceans?

Vocabulary

umbra p. 743

penumbra p. 743

solar eclipse p. 744

lunar eclipse p. 746

tide p. 747

Multilingual eGlossary

BrainPOP® Science Video

Go to the resource tab in ConnectED to find the PBL *Patterns in the Sky.*

Inquiry What is this dark spot?

A NASA satellite took this photo as it orbited around Earth. An eclipse caused the shadow that you see. Do you know what kind of eclipse?

Jacques Descloitres, MODIS Rapid Response Team at NASA GSFC

Launch Lab

10 minutes

How do shadows change?

You can see a shadow when an object blocks a light source. What happens to an object's shadow when the object moves?

1. Read and complete a lab safety form.
2. Select an **object** provided by your teacher.
3. Shine a **flashlight** on the object, projecting its shadow on the wall.
4. While holding the flashlight in the same position, move the object closer to the wall—away from the light. Then, move the object toward the light. Record your observations in your Science Journal.

Think About This

1. Compare and contrast the shadows created in each situation. Did the shadows have dark parts and light parts? Did these parts change?
2. **Key Concept** Imagine you look at the flashlight from behind your object, looking from the darkest and lightest parts of the object's shadow. How much of the flashlight could you see from each location?

Shadows—the Umbra and the Penumbra

A shadow results when one object blocks the light that another object emits or reflects. When a tree blocks light from the Sun, it casts a shadow. If you want to stand in the shadow of a tree, the tree must be in a line between you and the Sun.

If you go outside on a sunny day and look carefully at a shadow on the ground, you might notice that the edges of the shadow are not as dark as the rest of the shadow. Light from the Sun and other wide sources casts shadows with two distinct parts, as shown in **Figure 13.** *The* **umbra** *is the central, darker part of a shadow where light is totally blocked. The* **penumbra** *is the lighter part of a shadow where light is partially blocked.* If you stood within an object's penumbra, you would be able to see only part of the light source. If you stood within an object's umbra, you would not see the light source at all.

Word Origin

penumbra
from Latin *paene,* means "almost"; and *umbra,* means "shade, shadow"

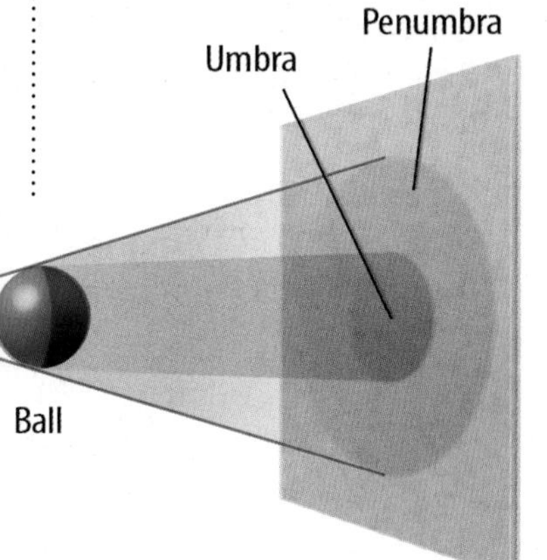

Figure 13 The shadow that a wide light source produces has two parts—the umbra and the penumbra. The light source cannot be seen from within the umbra. The light source can be partially seen from within the penumbra.

MiniLab — 10 minutes

What does the Moon's shadow look like?

Like every shadow cast by a wide light source, the Moon's shadow has two parts.

1. Read and complete a lab safety form.
2. Working with a partner, use a **pencil** to connect two **foam balls**. One ball should be one-fourth the size of the other.
3. While one person holds the foam balls, the other should stand 1 m away and shine a **light** on the balls. The foam balls and light should be in a direct line, with the smallest foam ball closest to the light.
4. Sketch and describe your observations in your Science Journal.

Analyze and Conclude

1. **Key Concept** Explain the relationship between the two types of shadows and solar eclipses.

Solar Eclipses

As the Sun shines on the Moon, the Moon casts a shadow that extends out into space. Sometimes the Moon passes between Earth and the Sun. This can only happen during the new moon phase. When Earth, the Moon, and the Sun are lined up, the Moon casts a shadow on Earth's surface, as shown in **Figure 14.** You can see the Moon's shadow in the photo at the beginning of this lesson. *When the Moon's shadow appears on Earth's surface, a* **solar eclipse** *is occurring.*

Key Concept Check Why does a solar eclipse occur only during a new moon?

As Earth rotates, the Moon's shadow moves along Earth's surface, as shown in **Figure 14.** The type of eclipse you see depends on whether you are in the path of the umbra or the penumbra. If you are outside the umbra and penumbra, you cannot see a solar eclipse at all.

Total Solar Eclipses

You can only see a total solar eclipse from within the Moon's umbra. During a total solar eclipse, the Moon appears to cover the Sun completely, as shown in **Figure 15** on the next page. Then, the sky becomes dark enough that you can see stars. A total solar eclipse lasts no longer than about 7 minutes.

Solar Eclipse

Figure 14 A solar eclipse occurs only when the Moon moves directly between Earth and the Sun. The Moon's shadow moves across Earth's surface.

Visual Check Why would a person in North America not see the solar eclipse shown here?

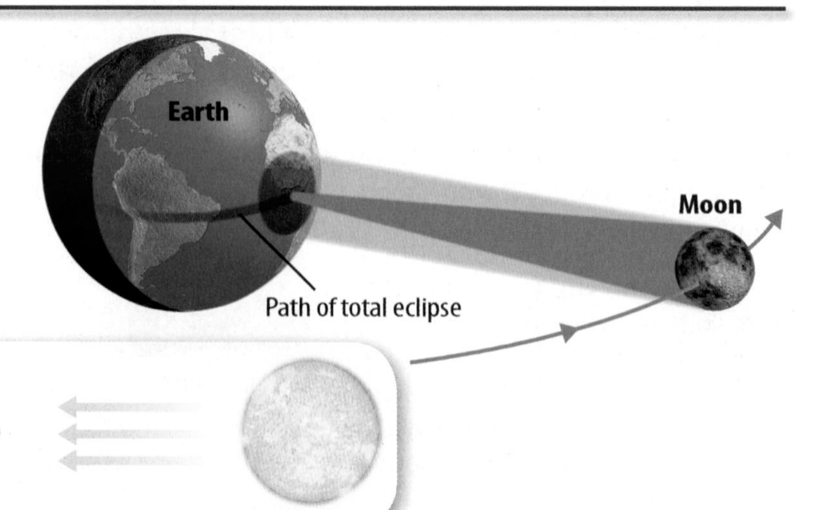

Hutchings Photography/Digital Light Source

The Sun's Changing Appearance During a Total Solar Eclipse

The Motion of the Moon in the Sky During a Total Solar Eclipse

Sun
Moon
Moon's motion

Figure 15 This sequence shows an example of how the Sun's appearance can change during a total solar eclipse.

Partial Solar Eclipses

You can only see a total solar eclipse from within the Moon's umbra, but you can see a partial solar eclipse from within the Moon's much larger penumbra. The stages of a partial solar eclipse are similar to the stages of a total solar eclipse, except that the Moon never completely covers the Sun.

Why don't solar eclipses occur every month?

Solar eclipses only can occur during a new moon, when Earth and the Sun are on opposite sides of the Moon. However, solar eclipses do not occur during every new moon phase. **Figure 16** shows why. The Moon's orbit is tilted slightly compared to Earth's orbit. As a result, during most new moons, Earth is either above or below the Moon's shadow. However, every so often the Moon is in a line between the Sun and Earth. Then the Moon's shadow passes over Earth and a solar eclipse occurs.

Figure 16 A solar eclipse occurs only when the Moon crosses Earth's orbit and is in a direct line between Earth and the Sun.

The Moon's Tilted Orbit

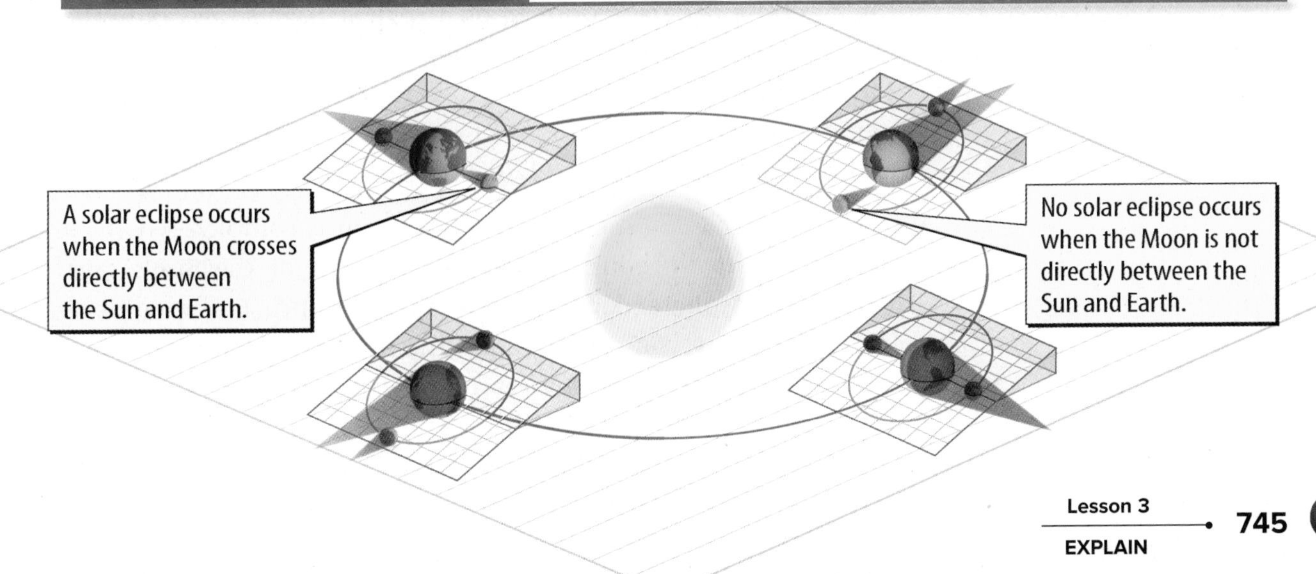

Lunar Eclipse

Figure 17 A lunar eclipse occurs when the Moon moves through Earth's shadow.

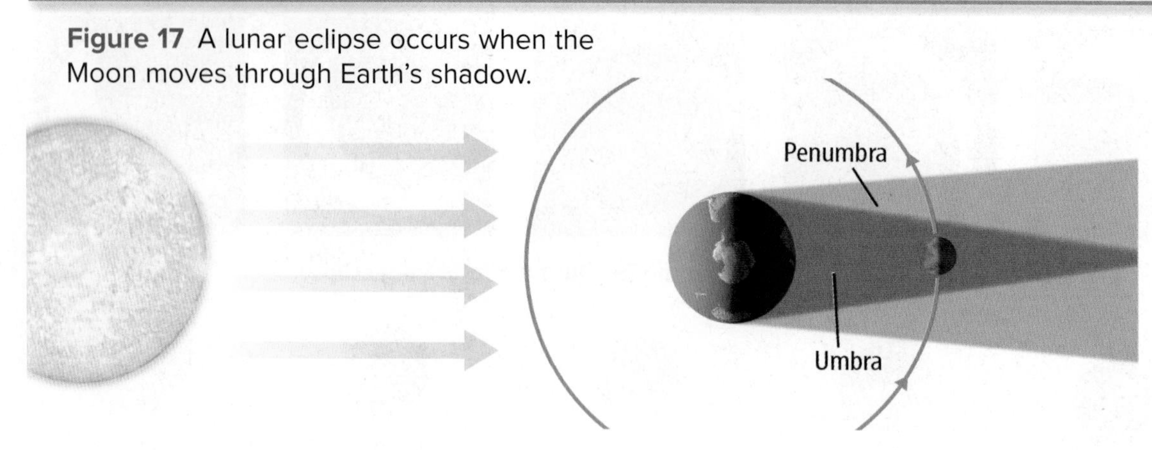

Visual Check Why would more people be able to see a lunar eclipse than a solar eclipse?

Lunar Eclipses

Just like the Moon, Earth casts a shadow into space. As the Moon revolves around Earth, it sometimes moves into Earth's shadow, as shown in **Figure 17.** *A* **lunar eclipse** *occurs when the Moon moves into Earth's shadow.* Then Earth is in a line between the Sun and the Moon. This means that a lunar eclipse can occur only during the full moon phase.

Like the Moon's shadow, Earth's shadow has an umbra and a penumbra. Different types of lunar eclipses occur depending on which part of Earth's shadow the Moon moves through. Unlike solar eclipses, you can see any lunar eclipse from any location on the side of Earth facing the Moon.

Key Concept Check When can a lunar eclipse occur?

FOLDABLES®

Make a two-tab book from a sheet of notebook paper. Label the tabs *Solar Eclipse* and *Lunar Eclipse*. Use it to organize your notes on eclipses.

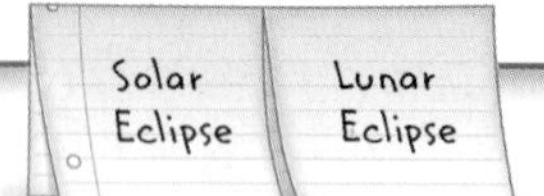

Total Lunar Eclipses

When the entire Moon moves through Earth's umbra, a total lunar eclipse occurs. **Figure 18** on the next page shows how the Moon's appearance changes during a total lunar eclipse. The Moon's appearance changes as it gradually moves into Earth's penumbra, then into Earth's umbra, back into Earth's penumbra, and then out of Earth's shadow entirely.

You can still see the Moon even when it is completely within Earth's umbra. Although Earth blocks most of the Sun's rays, Earth's atmosphere deflects some sunlight into Earth's umbra. This is also why you can often see the unlit portion of the Moon on a clear night. Astronomers often call this Earthshine. This reflected light has a reddish color and gives the Moon a reddish tint during a total lunar eclipse.

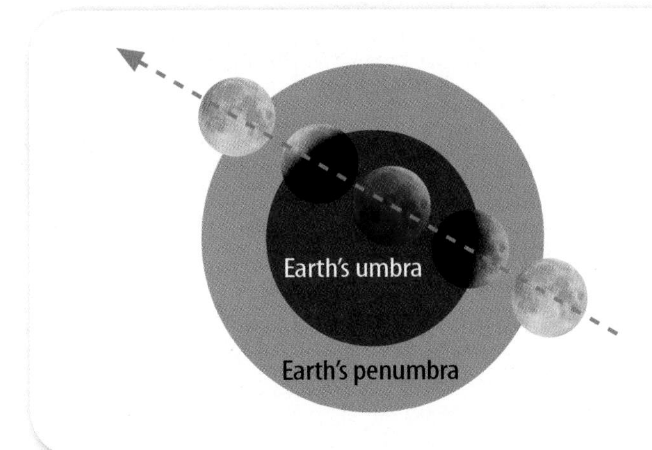

Figure 18 During a total lunar eclipse, the entire Moon passes through Earth's umbra. The Moon gradually darkens until a dark shadow covers it completely.

Visual Check How would a total lunar eclipse look different from a total solar eclipse?

Partial Lunar Eclipses

When only part of the Moon passes through Earth's umbra, a partial lunar eclipse occurs. The stages of a partial lunar eclipse are similar to those of a total lunar eclipse, shown in **Figure 18,** except the Moon is never completely covered by Earth's umbra. The part of the Moon in Earth's penumbra appears only slightly darker, while the part of the Moon in Earth's umbra appears much darker.

Why don't lunar eclipses occur every month?

Lunar eclipses can only occur during a full moon phase, when the Moon and the Sun are on opposite sides of Earth. However, lunar eclipses do not occur during every full moon because of the tilt of the Moon's orbit with respect to Earth's orbit. During most full moons, the Moon is slightly above or slightly below Earth's penumbra.

Tides

The positions of the Moon and the Sun also affect Earth's oceans. If you have spent time near an ocean, you might have seen how the ocean's height, or sea level, rises and falls twice each day. *A* **tide** is *the daily rise and fall of sea level.* Examples of tides are shown in **Figure 19.** It is primarily the Moon's gravity that causes Earth's oceans to rise and fall twice each day.

Figure 19 In the Bay of Fundy, high tides can be more than 10 m higher than low tides.

Robert Estall photo agency/Alamy

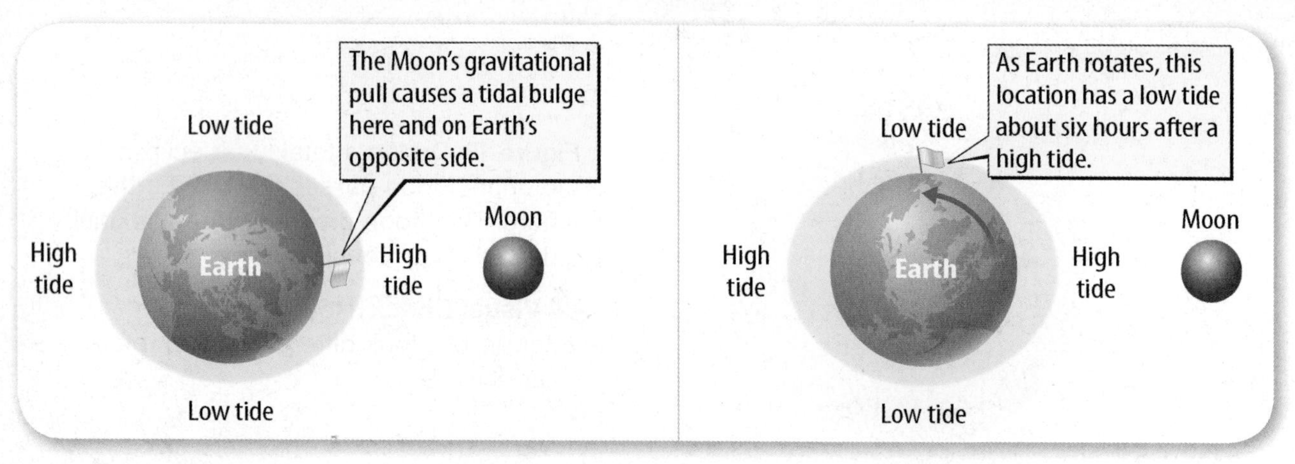

Figure 20  In this view down on Earth's North Pole, the flag moves into a tidal bulge as Earth rotates. A coastal area has a high tide about once every 12 hours.

The Moon's Effect on Earth's Tides

The difference in the strength of the Moon's gravity on opposite sides of Earth causes Earth's tides. The Moon's gravity is slightly stronger on the side of Earth closer to the Moon and slightly weaker on the side of Earth opposite the Moon. These differences cause tidal bulges in the oceans on opposite sides of Earth, shown in **Figure 20.** High tides occur at the tidal bulges, and low tides occur between them.

The Sun's Effect on Earth's Tides

Because the Sun is so far away from Earth, its effect on tides is about half that of the Moon. **Figure 21** shows how the positions of the Sun and the Moon affect Earth's tides.

Spring Tides During the full moon and new moon phases, spring tides occur. This is when the Sun's and the Moon's gravitational effects combine and produce higher high tides and lower low tides.

Neap Tides A week after a spring tide, a neap tide occurs. Then the Sun, Earth, and the Moon form a right angle. When this happens, the Sun's effect on tides reduces the Moon's effect. High tides are lower and low tides are higher at neap tides.

Key Concept Check Why is the Sun's effect on tides less than the Moon's effect?

Figure 21 A spring tide occurs when the Sun, Earth, and the Moon are in a line. A neap tide occurs when the Sun and the Moon form a right angle with Earth.

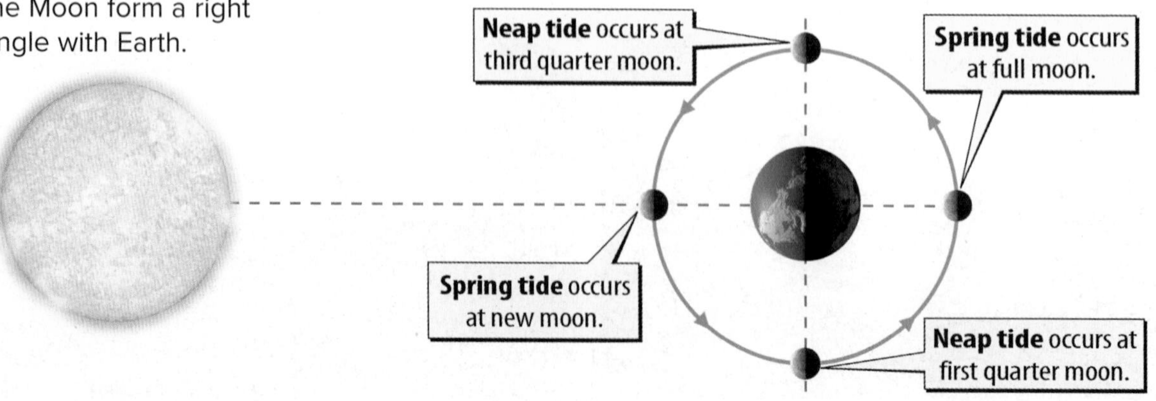

Lesson 3 Review

Visual Summary

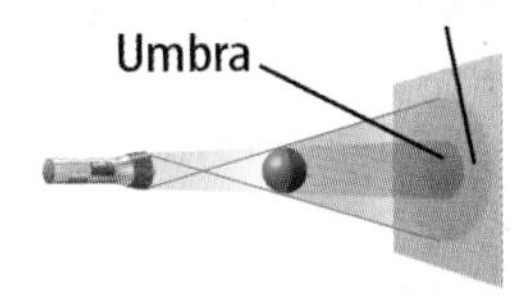

Shadows from a wide light source have two distinct parts.

The Moon's shadow produces solar eclipses. Earth's shadow produces lunar eclipses.

The positions of the Moon and the Sun in relation to Earth cause gravitational differences that produce tides.

FOLDABLES®

Use your lesson Foldable to review the lesson. Save your Foldable for the project at the end of the chapter.

What do you think NOW?

You first read the statements below at the beginning of the chapter.

5. A solar eclipse happens when Earth moves between the Moon and the Sun.

6. The gravitational pull of the Moon and the Sun on Earth's oceans causes tides.

Did you change your mind about whether you agree or disagree with the statements? Rewrite any false statements to make them true.

Use Vocabulary

1. **Distinguish** between an umbra and a penumbra.
2. **Use the term** *tide* in a sentence.
3. The Moon turns a reddish color during a total __________ eclipse.

Understand Key Concepts 

4. **Summarize** the effect of the Sun on Earth's tides.
5. **Illustrate** the positions of the Sun, Earth, and the Moon during a solar eclipse and during a lunar eclipse.
6. **Contrast** a total lunar eclipse with a partial lunar eclipse.
7. Which could occur during a total solar eclipse?
   - **A.** first quarter moon
   - **B.** full moon
   - **C.** neap tide
   - **D.** spring tide

Interpret Graphics

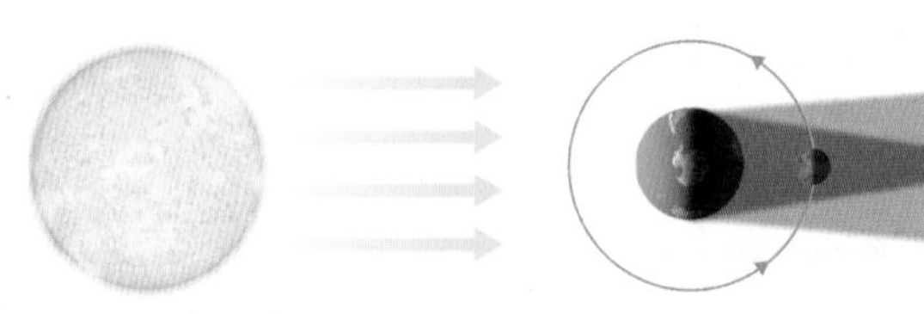

8. **Conclude** What type of eclipse does the figure above illustrate?
9. **Categorize Information** Copy and fill in the graphic organizer below to identify two bodies that affect Earth's tides.

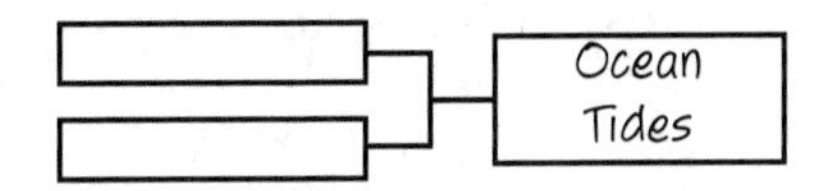

Critical Thinking

10. **Compose** a short story about a person long ago viewing a total solar eclipse.
11. **Research** ways to view a solar eclipse safely. Summarize your findings here.

Lab

35 minutes

Phases of the Moon

Materials

foam ball

pencil

lamp

stool

Safety

The Moon appears slightly different every night of its 29.5-day lunar cycle. The Moon's appearance changes as Earth and the Moon move. Depending on where the Moon is in relation to Earth and the Sun, observers on Earth see only part of the light the Moon reflects from the Sun.

Question

How do the positions of the Sun, the Moon, and Earth cause the phases of the Moon?

Procedure

1. Read and complete a lab safety form.
2. Hold a foam ball that represents the Moon. Make a handle for the ball by inserting a pencil about two inches into the foam ball. Your partner will represent an observer on Earth. Have your partner sit on a stool and record observations during the activity.
3. Place a lamp on a desk or other flat surface. Remove the shade from the lamp. The lamp represents the Sun.
4. Turn on the lamp and darken the lights in the room.
   ⚠ *Do not touch the bulb or look directly at it after the lamp is turned on.*
5. Position the Earth observer's stool about 1 m from the Sun. Position the Moon 0.5–1 m from the observer so that the Sun, Earth, and the Moon are in a line. The student holding the Moon holds the Moon so it is completely illuminated on one half. The observer records the phase and what the phase looks like in a data table.
6. Move the Moon clockwise about one-eighth of the way around its "orbit" of Earth. The observer swivels on the stool to face the Moon and records the phase.
7. Continue the Moon's orbit until the Earth observer has recorded all the Moon's phases.

Hutchings Photography/Digital Light Source

8. Return to your positions as the Moon and Earth observer. Choose a part in the Moon's orbit that you did not model. Predict what the Moon would look like in that position, and check if your prediction is correct.

Analyze and Conclude

9. **Explain** Use your observations to explain how the positions of the Sun, the Moon, and Earth produce the different phases of the Moon.
10. **The Big Idea** Why is half of the Moon always lit? Why do you usually see only part of the Moon's lit half?
11. **Draw Conclusions** Based on your observations, why is the Moon not visible from Earth during the new moon phase?
12. **Summarize** Which parts of your model were waxing phases? Which parts were waning phases?
13. **Think Critically** During which phases of the Moon can eclipses occur? Explain.

Communicate Your Results

Create a poster of the results from your lab. Illustrate various positions of the Sun, the Moon, and Earth and draw the phase of the Moon for each. Include a statement of your hypothesis on the poster.

Inquiry Extension

The Moon is not the only object in the sky that has phases when viewed from Earth. The planets Venus and Mercury also have phases. Research the phases of these planets and create a calendar that shows when the various phases of Venus and Mercury occur.

Lab Tips

- ☑ Make sure the observer's head does not cast a shadow on the Moon.
- ☑ The student holding the Moon should hold the pencil so that he or she always stands on the unlit side of the Moon.

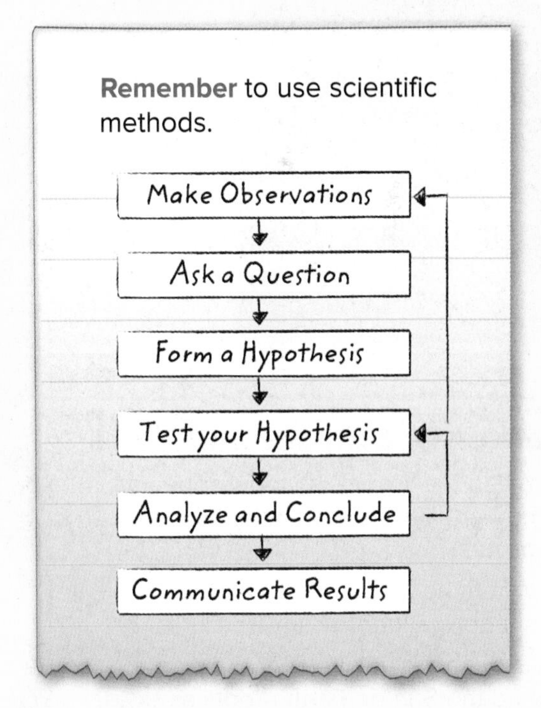

Chapter 20 Study Guide

Earth's motion around the Sun causes seasons. The Moon's motion around Earth causes phases of the Moon. Earth and the Moon's motions together cause eclipses and ocean tides.

Key Concepts Summary

Lesson 1: Earth's Motion

- The gravitational pull of the Sun on Earth causes Earth to revolve around the Sun in a nearly circular **orbit.**
- Areas on Earth's curved surface become more tilted with respect to the direction of sunlight the farther you travel from the equator. This causes sunlight to spread out closer to the poles, making Earth colder at the poles and warmer at the equator.
- As Earth revolves around the Sun, the tilt of Earth's **rotation axis** produces changes in how sunlight spreads out over Earth's surface. These changes in the concentration of sunlight cause the seasons.

90°
September Equinox
Rotation axis
66.5°
June Solstice

Vocabulary

orbit p. 726
revolution p. 726
rotation p. 727
rotation axis p. 727
solstice p. 731
equinox p. 731

Lesson 2: Earth's Moon

- The gravitational pull of Earth on the Moon makes the Moon revolve around Earth. The Moon rotates once as it makes one complete orbit around Earth.
- The lit part of the Moon that you can see from Earth—the Moon's **phase**—changes during the lunar cycle as the Moon revolves around Earth.

Vocabulary

maria p. 736
phase p. 738
waxing phase p. 738
waning phase p. 738

Lesson 3: Eclipses and Tides

- When the Moon's shadow appears on Earth's surface, a **solar eclipse** occurs.
- When the Moon moves into Earth's shadow, a **lunar eclipse** occurs.
- The gravitational pull of the Moon and the Sun on Earth produces **tides,** the rise and fall of sea level that occurs twice each day.

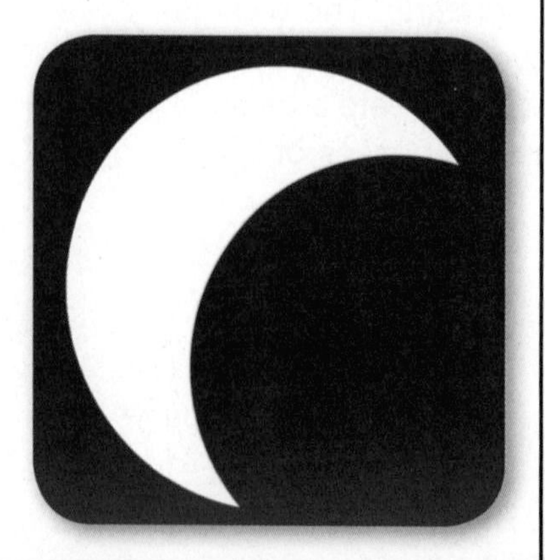

Vocabulary

umbra p. 743
penumbra p. 743
solar eclipse p. 744
lunar eclipse p. 746
tide p. 747

Eckhard Slawik/Photo Researchers, Inc.

Study Guide

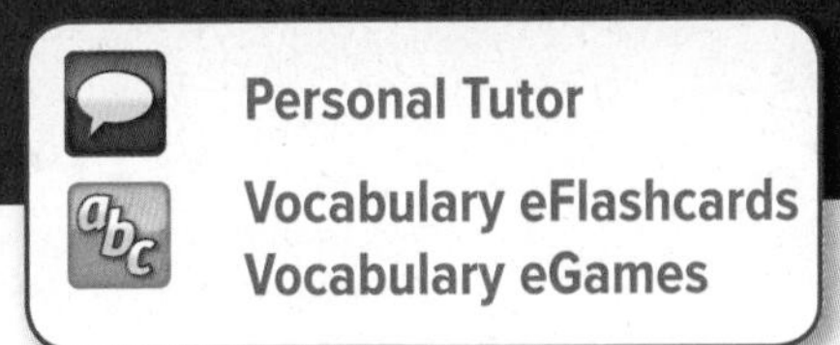

Chapter Project

Assemble your Lesson Foldables as shown to make a Chapter Project. Use the project to review what you have learned in this chapter.

Use Vocabulary

Distinguish between the terms in the each of the following pairs.

1. revolution, orbit
2. rotation, rotation axis
3. solstice, equinox
4. waxing phases, waning phases
5. umbra, penumbra
6. solar eclipse, lunar eclipse
7. tide, phase

Link Vocabulary and Key Concepts

Copy this concept map, and then use vocabulary terms from the previous page to complete the concept map.

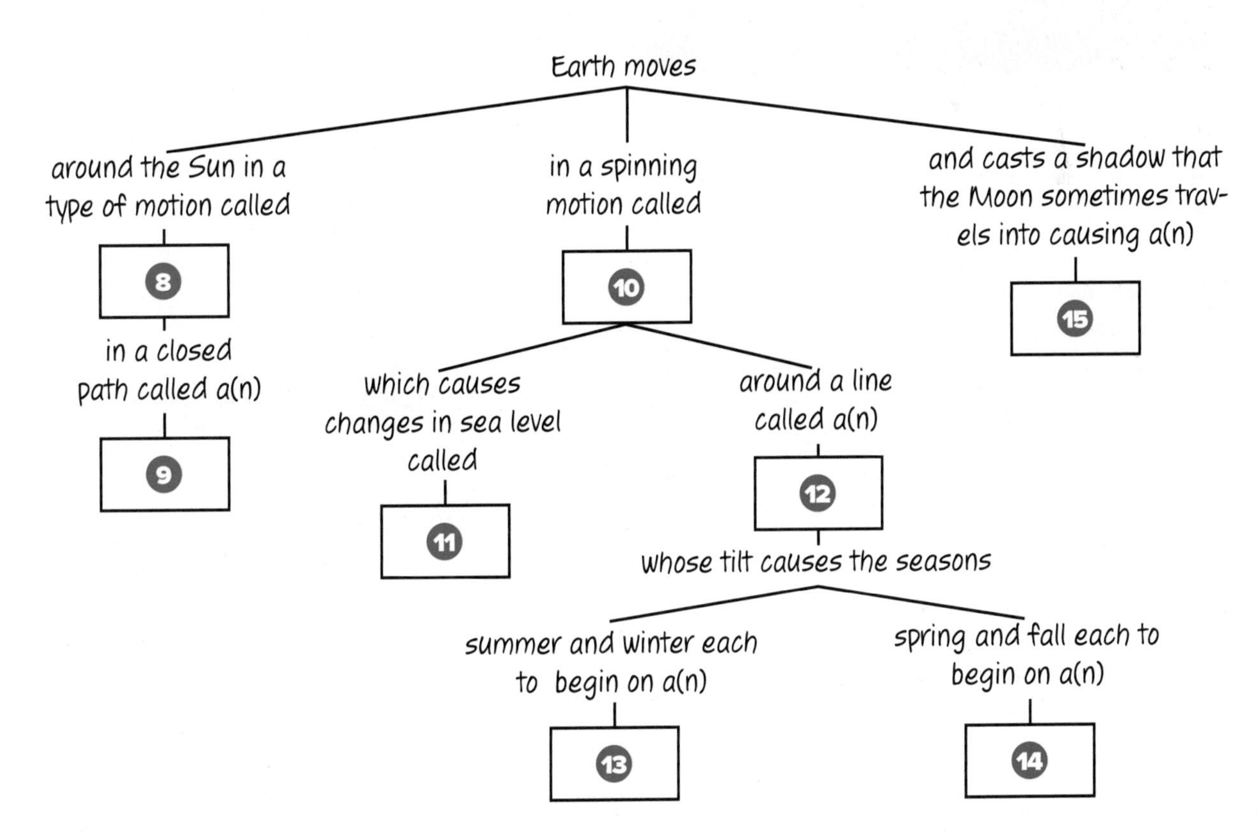

Chapter 20 Review

Understand Key Concepts 

1. Which property of the Sun most affects the strength of gravitational attraction between the Sun and Earth?
   - A. mass
   - B. radius
   - C. shape
   - D. temperature

2. Which would be different if Earth rotated from east to west but at the same rate?
   - A. the amount of energy striking Earth
   - B. the days on which solstices occur
   - C. the direction of the Sun's apparent motion across the sky
   - D. the number of hours in a day

3. In the image below, the northern hemisphere is leaning toward the Sun. What season is it experiencing?

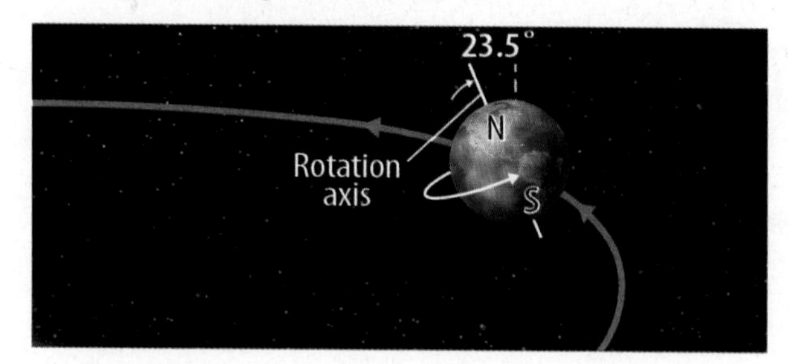

   - A. fall
   - B. spring
   - C. summer
   - D. winter

4. Which best explains why Earth is colder at the poles than at the equator?
   - A. Earth is farther from the Sun at the poles than at the equator.
   - B. Earth's orbit is not a perfect circle.
   - C. Earth's rotation axis is tilted.
   - D. Earth's surface is more tilted at the poles than at the equator.

5. How are the revolutions of the Moon and Earth alike?
   - A. Both are produced by gravity.
   - B. Both are revolutions around the Sun.
   - C. Both orbits are the same size.
   - D. Both take the same amount of time.

6. Which moon phase occurs about one week after a new moon?
   - A. another new moon
   - B. first quarter moon
   - C. full moon
   - D. third quarter moon

7. Why is the same side of the Moon always visible from Earth?
   - A. The Moon does not revolve around Earth.
   - B. The Moon does not rotate.
   - C. The Moon makes exactly one rotation for each revolution around Earth.
   - D. The Moon's rotation axis is not tilted.

8. About how often do spring tides occur?
   - A. once each month
   - B. once each year
   - C. twice each month
   - D. twice each year

9. If a coastal area has a high tide at 7:00 A.M., at about what time will the next low tide occur?
   - A. 11:00 A.M.
   - B. 1:00 P.M.
   - C. 3:00 P.M.
   - D. 7:00 P.M.

10. Which type of eclipse would a person standing at point X in the diagram below see?

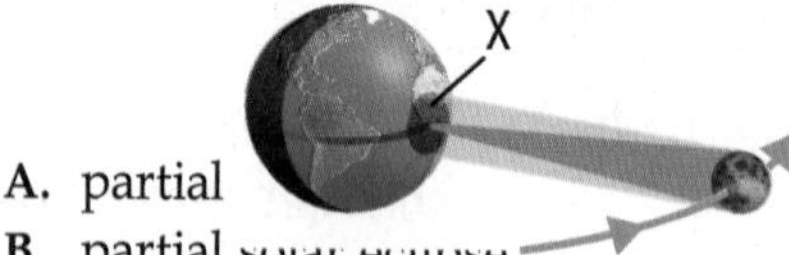

   - A. partial
   - B. partial solar eclipse
   - C. total lunar eclipse
   - D. total solar eclipse

Critical Thinking

11 **Outline** the ways Earth moves and how each affects Earth.

12 **Create** a poster that illustrates and describes the relationship between Earth's tilt and the seasons.

13 **Contrast** Why can you see phases of the Moon but not phases of the Sun?

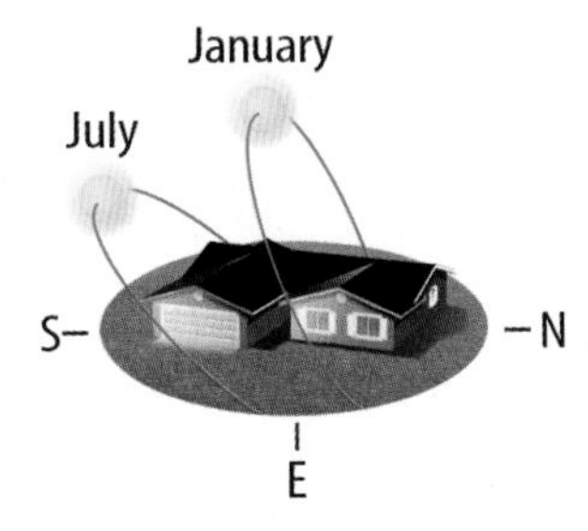

14 **Interpret Graphics** The figure above shows the Sun's position in the sky at noon in January and July. Is the house located in the northern hemisphere or the southern hemisphere? Explain.

15 **Illustrate** Make a diagram of the Moon's orbit and phases. Include labels and explanations with your drawing.

16 **Differentiate** between a total solar eclipse and a partial solar eclipse.

17 **Generalize** the reason that solar and lunar eclipses do not occur every month.

18 **Role Play** Write and present a play with several classmates that explains the causes and types of tides.

Writing in Science

19 **Survey** a group of at least ten people to determine how many know the cause of Earth's seasons. Write a summary of your results, including a main idea, supporting details, and a concluding sentence.

REVIEW THE BIG IDEA

20 At the South Pole, the Sun does not appear in the sky for six months out of the year. When does this happen? What is happening at the North Pole during these months? Explain why Earth's poles receive so little solar energy.

21 A solar eclipse, shown in the time-lapse photo below, is one phenomenon that the motions of Earth and the Moon produce. What other phenomena do the motions of Earth and the Moon produce?

Math Skills

Math Practice

Convert Units

22 When the Moon is 384,000 km from Earth, how far is the Moon from Earth in miles?

23 If you travel 205 mi on a train from Washington D.C. to New York City, how many kilometers do you travel on the train?

24 The nearest star other than the Sun is about 40 trillion km away. About how many miles away is the nearest star other than the Sun?

Standardized Test Practice

Record your answers on the answer sheet provided by your teacher or on a sheet of paper.

Multiple Choice

1 Which is the movement of one object around another object in space?

A axis

B orbit

C revolution

D rotation

Use the diagram below to answer question 2.

Time 1

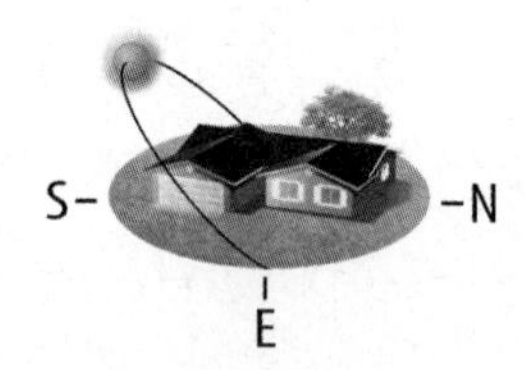

Time 2

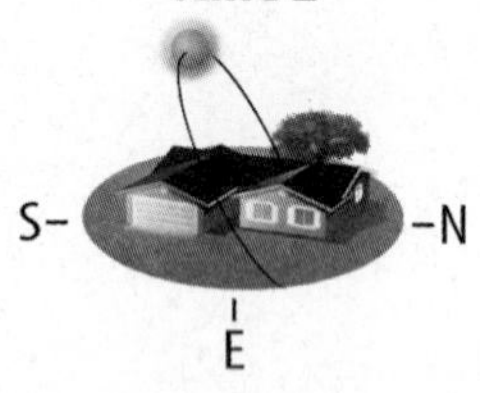

2 What happens between times *1* and *2* in the diagram above?

A Days grow shorter and shorter.

B The season changes from fall to winter.

C The region begins to point away from the Sun.

D The region gradually receives more solar energy.

3 How many times larger is the Sun's diameter than Earth's diameter?

A about 10 times larger

B about 100 times larger

C about 1,000 times larger

D about 10,000 times larger

4 Which diagram illustrates the Moon's third quarter phase?

A

B

C

D

5 Which accurately describes Earth's position and orientation during summer in the northern hemisphere?

A Earth is at its closest point to the Sun.

B Earth's hemispheres receive equal amounts of solar energy.

C The north end of Earth's rotational axis leans toward the Sun.

D The Sun emits a greater amount of light and heat energy.

6 Which are large, dark lunar areas formed by cooled lava?

A craters

B highlands

C maria

D rays

7 During one lunar cycle, the Moon

A completes its east-to-west path across the sky exactly once.

B completes its entire sequence of phases.

C progresses only from the new moon phase to the full moon phase.

D revolves around Earth twice.

Use the diagram below to answer question 8.

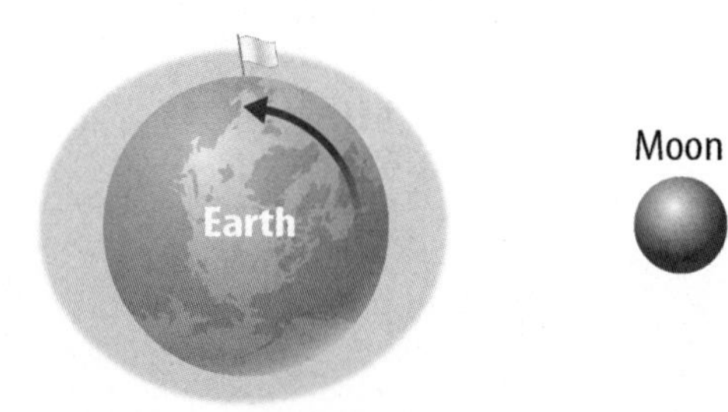

8 What does the flag in the diagram above represent?

A high tide

B low tide

C neap tide

D spring tide

9 During which lunar phase might a solar eclipse occur?

A first quarter moon

B full moon

C new moon

D third quarter moon

10 Which does the entire Moon pass through during a partial lunar eclipse?

A Earth's penumbra

B Earth's umbra

C the Moon's penumbra

D the Moon's umbra

Constructed Response

Use the diagram below to answer questions 11 and 12.

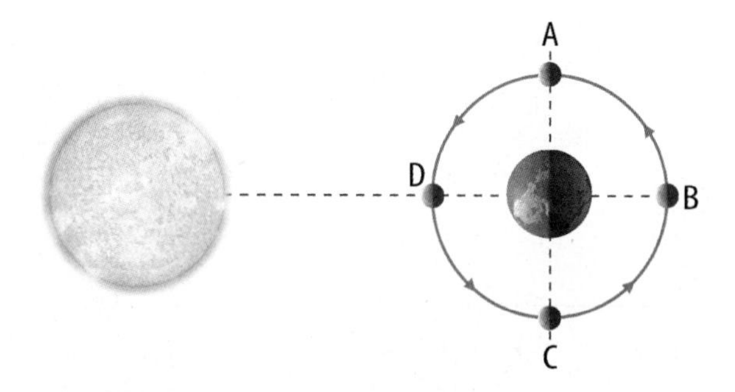

11 Where are neap tides indicated in the above diagram? What causes neap tides? What happens during a neap tide?

12 Where are spring tides indicated in the above diagram? What causes spring tides? What happens during a spring tide?

13 How would Earth's climate be different if its rotational axis were not tilted?

14 Why can we see only one side of the Moon from Earth? What is the name given to this side of the Moon?

15 What is a lunar phase? How do waxing and waning phases differ?

16 Why don't solar eclipses occur monthly?

NEED EXTRA HELP?																
If You Missed Question...	1	2	3	4	5	6	7	8	9	10	11	12	13	14	15	16
Go to Lesson...	1	1	1	2	1	2	2	3	3	3	3	3	1	2	2	3

Chapter 21

The Solar System

What kinds of objects are in the solar system?

One, Two, or Three Planets?

This photo, taken by the *Cassini* spacecraft, shows part of Saturn's rings and two of its moons. Saturn is a planet that orbits the Sun. The moons, tiny Epimetheus and much larger Titan, orbit Saturn. Besides planets and moons, many other objects are in the solar system.

- How would you describe a planet such as Saturn?
- How do astronomers classify the objects they discover?
- What types of objects do you think make up the solar system?

NASA Jet Propulsion Laboratory (NASA-JPL)

Get Ready to Read

What do you think?

Before you read, decide if you agree or disagree with each of these statements. As you read this chapter, see if you change your mind about any of the statements.

1. Astronomers measure distances between space objects using astronomical units.
2. Gravitational force keeps planets in orbit around the Sun.
3. Earth is the only inner planet that has a moon.
4. Venus is the hottest planet in the solar system.
5. The outer planets also are called the gas giants.
6. The atmospheres of Saturn and Jupiter are mainly water vapor.
7. Asteroids and comets are mainly rock and ice.
8. A meteoroid is a meteor that strikes Earth.

Your one-stop online resource
connectED.mcgraw-hill.com

LearnSmart®

Chapter Resources Files, Reading Essentials, Get Ready to Read, Quick Vocabulary

Animations, Videos, Interactive Tables

Self-checks, Quizzes, Tests

Project-Based Learning Activities

Lab Manuals, Safety Videos, Virtual Labs & Other Tools

Vocabulary, Multilingual eGlossary, Vocab eGames, Vocab eFlashcards

Personal Tutors

Lesson 1

Reading Guide

Key Concepts
ESSENTIAL QUESTIONS

- How are the inner planets different from the outer planets?
- What is an astronomical unit and why is it used?
- What is the shape of a planet's orbit?

Vocabulary

asteroid p. 763
comet p. 763
astronomical unit p. 764
period of revolution p. 764
period of rotation p. 764

Multilingual eGlossary

Science Video

Go to the resource tab in ConnectED to find the PBL *PBI: Planetary Bureau of Investigation.*

The Structure of the Solar System

Inquiry Are these stars?

If you were to gaze up at the sky on a moonless night away from city lights, you might observe a sky similar to the one shown in this photo. A few thousand stars are easily visible even though they are extremely far away. What other types of objects can be seen in the night sky?

Launch Lab

10 minutes

How do you know which distance unit to use?

You can use different units to measure distance. For example, millimeters might be used to measure the length of a bolt, and kilometers might be used to measure the distance between cities. In this lab, you will investigate why some units are easier to use than others for certain measurements.

1. Read and complete a lab safety form.
2. Use a **centimeter ruler** to measure the length of a **pencil** and the thickness of this **book.** Record the distances in your Science Journal.
3. Use the centimeter ruler to measure the width of your classroom. Then measure the width of the room using a **meterstick.** Record the distances in your Science Journal.

Think About This

1. Why are meters easier to use than centimeters for measuring the classroom?
2. **Key Concept** Why do you think astronomers might need a unit larger than a kilometer to measure distances in the solar system?

What is the solar system?

Have you ever made a wish on a star? If so, you might have wished on a planet instead of a star. Sometimes, as shown in **Figure 1**, the first starlike object you see at night is not a star at all. It's Venus, the planet closest to Earth.

It's hard to tell the difference between planets and stars in the night sky because they all appear as tiny lights. Thousands of years ago, observers noticed that a few of these tiny lights moved, but others did not. The ancient Greeks called these objects planets, which means "wanderers." Astronomers now know that the planets do not wander about the sky; the planets move around the Sun. The Sun and the group of objects that move around it make up the solar system.

When you look at the night sky, a few of the tiny lights that you can see are part of our solar system. Almost all of the other specks of light are stars. They are much farther away than any objects in our solar system. Astronomers have discovered that some of those stars also have planets moving around them.

Reading Check What object do the planets in the solar system move around?

Figure 1 When looking at the night sky, you will likely see stars and planets. In the photo below, the planet Venus is the bright object seen above the Moon.

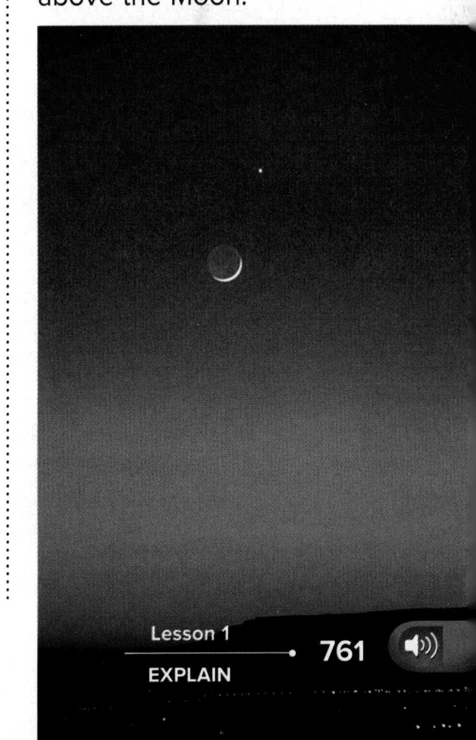

(t)Hutchings Photography/Digital Light Source; (b)Diego Barucco/Alamy

Objects in the Solar System

Ancient observers looking at the night sky saw many stars but only five planets–Mercury, Venus, Mars, Jupiter, and Saturn. The invention of the telescope in the 1600s led to the discovery of additional planets and many other space objects.

SCIENCE USE V. COMMON USE

star
Science Use an object in space made of gases in which nuclear fusion reactions occur that emit energy

Common Use a shape that usually has five or six points around a common center

REVIEW VOCABULARY

orbit
(noun) the path an object follows as it moves around another object
(verb) to move around another object

The Sun

The largest object in the solar system is the Sun, a **star.** Its diameter is about 1.4 million km–ten times the diameter of the largest planet, Jupiter. The Sun is made mostly of hydrogen gas. Its mass makes up about 99 percent of the entire solar system's mass.

Inside the Sun, a process called nuclear fusion produces an enormous amount of energy. The Sun emits some of this energy as light. The light from the Sun shines on all of the planets every day. The Sun also applies gravitational forces to objects in the solar system. Gravitational forces cause the planets and other objects to move around, or **orbit,** the Sun.

Objects That Orbit the Sun

Different types of objects orbit the Sun. These objects include planets, dwarf planets, asteroids, and comets. Unlike the Sun, these objects don't emit light but only reflect the Sun's light.

Planets Astronomers classify some objects that orbit the Sun as planets, as shown in **Figure 2.** An object is a planet only if it orbits the Sun and has a nearly spherical shape. Also, the mass of a planet must be much larger than the total mass of all other objects whose orbits are close by. The solar system has eight objects classified as planets.

Reading Check What is a planet?

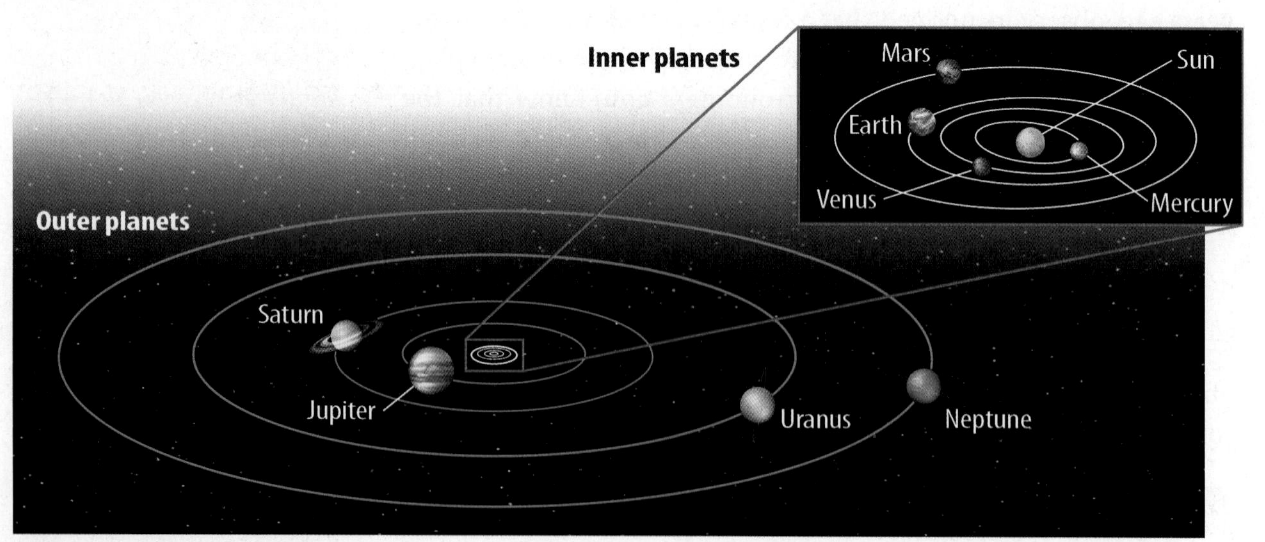

Figure 2 The orbits of the inner and outer planets are shown to scale. The Sun and the planets are not to scale. The outer planets are much larger than the inner planets.

▲ **Figure 3** Ceres, a dwarf planet, orbits the Sun as planets do. The orbit of Ceres is in the asteroid belt between Mars and Jupiter.

Visual Check Which dwarf planet is farthest from the Sun?

Inner Planets and Outer Planets As shown in **Figure 2**, the four planets closest to the Sun are the inner planets. The inner planets are Mercury, Venus, Earth, and Mars. These planets are made mainly of solid rocky materials. The four planets farthest from the Sun are the outer planets. The outer planets are Jupiter, Saturn, Uranus (YOOR uh nus), and Neptune. These planets are made mainly of ice and gases such as hydrogen and helium. The outer planets are much larger than Earth and are sometimes called gas giants.

Key Concept Check Describe how the inner planets differ from the outer planets.

Dwarf Planets Scientists classify some objects in the solar system as dwarf planets. A dwarf planet is a spherical object that orbits the Sun. It is not a moon of another planet and is in a region of the solar system where there are many objects orbiting near it. But, unlike a planet, a dwarf planet does not have more mass than objects in nearby orbits. **Figure 3** shows the locations of the dwarf planets Ceres (SIHR eez), Eris (IHR is), Pluto, and Makemake (MAH kay MAH kay). Dwarf planets are made of rock and ice and are much smaller than Earth.

WORD ORIGIN

asteroid
from Greek *asteroeides,* means "resembling a star"

Asteroids *Millions of small, rocky objects called* **asteroids** *orbit the Sun in the asteroid belt between the orbits of Mars and Jupiter.* The asteroid belt is shown in **Figure 3.** Asteroids range in size from less than a meter to several hundred kilometers in length. Unlike planets and dwarf planets, asteroids, such as the one shown in **Figure 4,** usually are not spherical.

Comets You might have seen a picture of a comet with a long, glowing tail. *A* **comet** *is made of gas, dust, and ice and moves around the Sun in an oval-shaped orbit.* Comets come from the outer parts of the solar system. Most comets have never been seen. It is estimated that there might be 100 billion comets orbiting the Sun. You will read more about comets, asteroids, and dwarf planets in Lesson 4.

Figure 4 The asteroid Gaspra orbits the Sun in the asteroid belt. Its odd shape is about 19 km long and 11 km wide. ▼

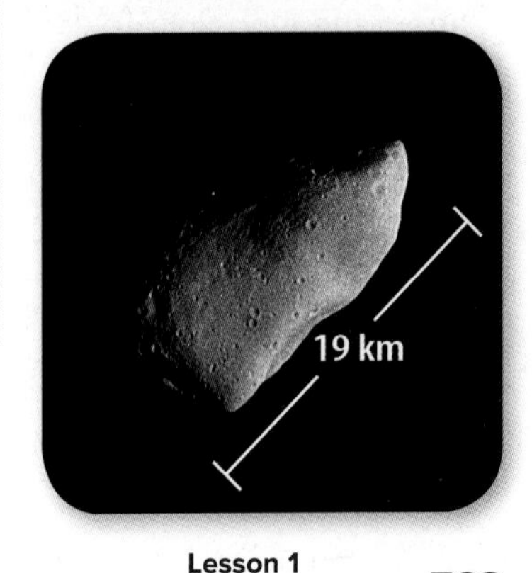

NASA-JPL/USGS

The Astronomical Unit

On Earth, distances are often measured in meters (m) or kilometers (km). Objects in the solar system, however, are so far apart that astronomers use a larger distance unit. *An* **astronomical unit** *(AU) is the average distance from Earth to the Sun–about 150 million km.* **Table 1** lists each planet's average distance from the Sun in km and AU.

Key Concept Check Define what an astronomical unit is and explain why it is used.

Table 1  Because the distances of the planets from the Sun are so large, it is easier to express these distances using astronomical units rather than kilometers.

Table 1 Average Distance of the Planets from the Sun

Planet	Average Distance (km)	Average Distance (AU)
Mercury	57,910,000	0.39
Venus	108,210,000	0.72
Earth	149,600,000	1.00
Mars	227,920,000	1.52
Jupiter	778,570,000	5.20
Saturn	1,433,530,000	9.58
Uranus	2,872,460,000	19.20
Neptune	4,495,060,000	30.05

The Motion of the Planets

Have you ever swung a ball on the end of a string in a circle over your head? In some ways, the motion of a planet around the Sun is like the motion of that ball. As shown in **Figure 5** on the next page, the Sun's gravitational force pulls each planet toward the Sun. This force is similar to the pull of the string that keeps the ball moving in a circle. The Sun's gravitational force pulls on each planet and keeps it moving along a curved path around the Sun.

Reading Check What causes planets to orbit the Sun?

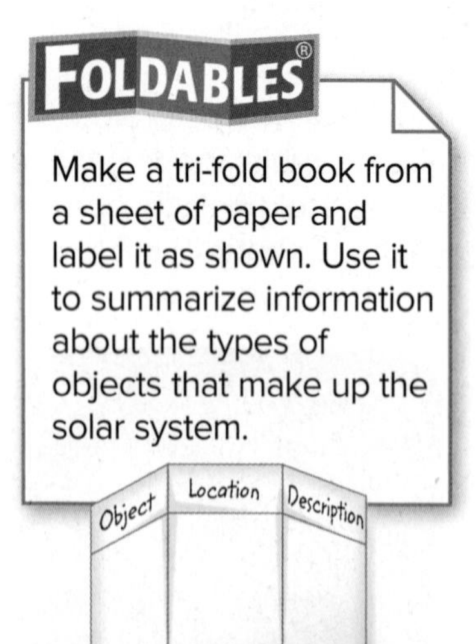

Revolution and Rotation

Objects in the solar system move in two ways. They orbit, or revolve, around the Sun. *The time it takes an object to travel once around the Sun is its* **period of revolution.** Earth's period of revolution is one year. The objects also spin, or rotate, as they orbit the Sun. *The time it takes an object to complete one rotation is its* **period of rotation.** Earth has a period of rotation of one day.

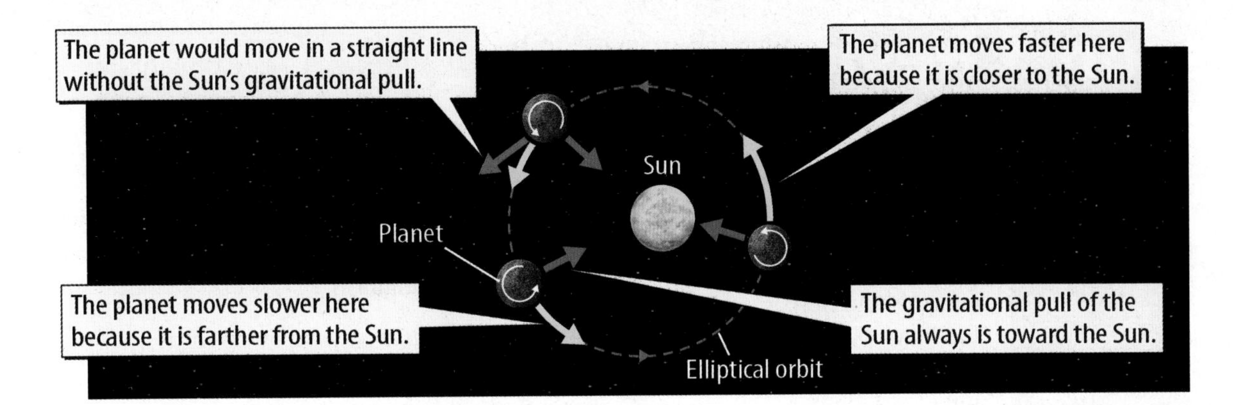

Planetary Orbits and Speeds

Earth was once thought to be the center of our solar system. In this geocentric model, the Sun, the Moon, and the planets revolved in circular orbits around a stationary Earth. In the early 1500s, Nicholas Copernicus proposed that Earth and other planets revolve in circular orbits around a stationary Sun, a heliocentric model.

In the 1600s, Johannes Kepler discovered that planets' orbits are ellipses, not circles. An ellipse contains two fixed points, called foci (singular, focus). Foci are equal distance from the ellipse's center and determine its shape. As shown in **Figure 5**, the Sun is at one focus. As a planet revolves, the distance between the planet and the Sun changes. Kepler also discovered that a planet's speed increases as it gets nearer to the Sun.

 **Key Concept Check** Describe the shape of a planet's orbit.

 **Personal Tutor**

Figure 5 Planets and other objects in the solar system revolve around the Sun because of its gravitational pull on them.

MiniLab

20 minutes

How can you model an elliptical orbit?

In this lab you will explore how the locations of foci affect the shape of an ellipse.

1. Read and complete a lab safety form.
2. Place a sheet of **paper** on a **corkboard.** Insert two **push pins** 8 cm apart in the center of the paper.
3. Use **scissors** to cut a 24-cm piece of **string.** Tie the ends of the string together.
4. Place the loop of string around the pins. Use a pencil to draw an ellipse as shown.
5. Measure the maximum width and length of the ellipse. Record the data in your Science Journal.
6. Move one of the push pins so that the pins are 5 cm apart. Repeat steps 4 and 5.

Analysis

1. **Compare and contrast** the two ellipses.
2.  **Key Concept** How are the shapes of the ellipses you drew similar to the orbits of the inner and outer planets?

Lesson 1 Review

Visual Summary

The solar system contains the Sun, the inner planets, the outer planets, the dwarf planets, asteroids, and comets.

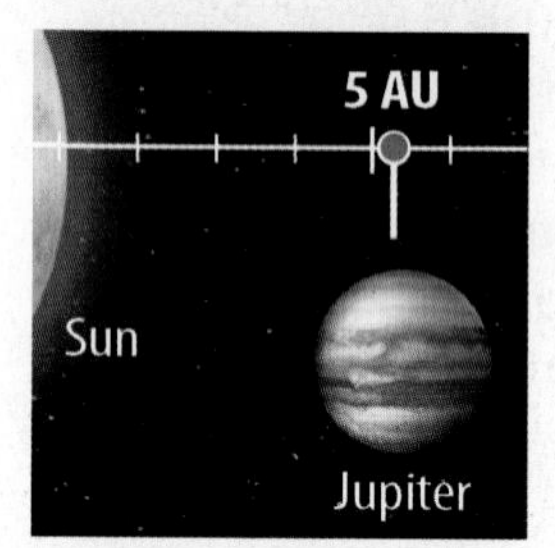

An astronomical unit (AU) is a unit of distance equal to about 150 million km.

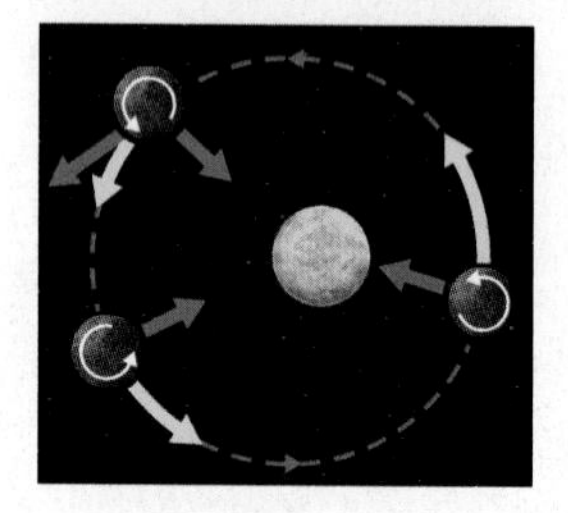

The speeds of the planets change as they move around the Sun in elliptical orbits.

FOLDABLES®

Use your lesson Foldable to review the lesson. Save your Foldable for the project at the end of the chapter.

What do you think NOW?

You first read the statements below at the beginning of the chapter.

1. Astronomers measure distances between space objects using astronomical units.
2. Gravitational force keeps planets in orbit around the Sun.

Did you change your mind about whether you agree or disagree with the statements? Rewrite any false statements to make them true.

Use Vocabulary

1. **Compare and contrast** a period of revolution and a period of rotation.
2. **Define** *dwarf planet* in your own words.
3. **Distinguish** between an asteroid and a comet.

Understand Key Concepts 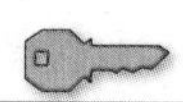

4. **Summarize** how and why planets orbit the Sun and how and why a planet's speed changes in orbit.
5. **Infer** why an astronomical unit is not used to measure distances on Earth.
6. Which distinguishes a dwarf planet from a planet?
   - **A.** mass
   - **B.** the object it revolves around
   - **C.** shape
   - **D.** type of orbit

Interpret Graphics

7. **Explain** what each arrow in the diagram represents.

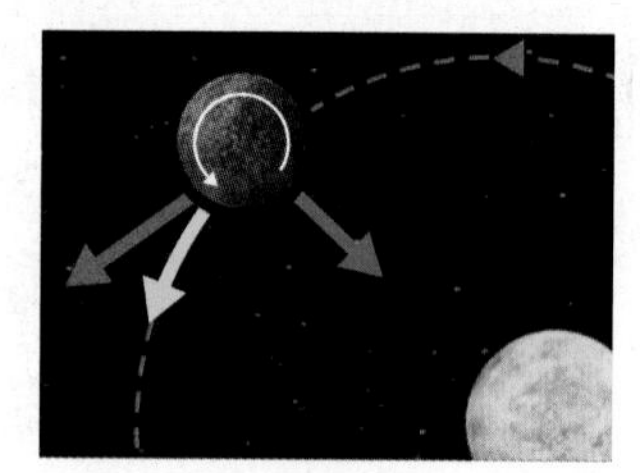

8. **Take Notes** Copy the table below. List information about each object or group of objects in the solar system mentioned in the lesson. Add additional lines as needed.

Object	Description
Sun	
Planets	

Critical Thinking

9. **Evaluate** How would the speed of a planet be different if its orbit were a circle instead of an ellipse?

Meteors are pieces of a comet or an asteroid that heat up as they fall through Earth's atmosphere. Meteors that strike Earth are called meteorites. ▶

CAREERS in SCIENCE

History from Space

Meteorites give a peek back in time.

About 4.6 billion years ago, Earth and the other planets did not exist. In fact, there was no solar system. Instead, a large disk of gas and dust, known as the solar nebula, swirled around a forming Sun, as shown in the top picture to the right. How did the planets and other objects in the solar system form?

▲ **Denton Ebel holds a meteorite that broke off the Vesta asteroid.**

Denton Ebel is looking for the answer. He is a geologist at the American Museum of Natural History in New York City. Ebel explores the hypothesis that over millions of years, tiny particles in the solar nebula clumped together and formed the asteroids, comets, and planets that make up our solar system.

The solar nebula contained tiny particles called chondrules (KON drewls). They formed when the hot gas of the nebula condensed and solidified. Chondrules and other tiny particles collided and then accreted (uh KREET ed) or clumped together. This process eventually formed asteroids, comets, and planets. Some of the asteroids and comets have not changed much in over 4 billion years. Chondrite meteorites are pieces of asteroids that fell to Earth. The chondrules within the meteorites are the oldest solid material in our solar system.

For Ebel, chondrite meteorites contain information about the formation of the solar system. Did the materials in the meteorite form throughout the solar system and then accrete? Or did asteroids and comets form and accrete near the Sun, drift outward to where they are today, and then grow larger by accreting ice and dust? Ebel's research is helping to solve the mystery of how our solar system formed.

Accretion Hypothesis

According to the accretion hypothesis, the solar system formed in stages.

First there was a solar nebula. The Sun formed when gravity caused the nebula to collapse.

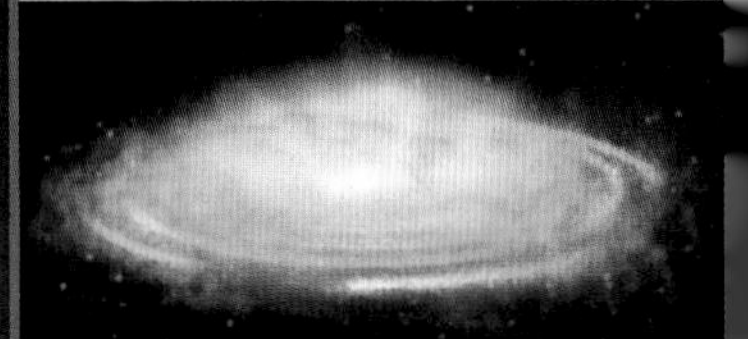

The rocky inner planets formed from accreted particles.

The gaseous outer planets formed as gas, ice, and dust condensed and accreted.

It's Your Turn

TIME LINE Work in groups. Learn more about the history of Earth from its formation until life began to appear. Create a time line showing major events. Present your time line to the class.

(t)Josef Muellek/Getty Images; (c)American Museum of Natural History; (bkgd)NASA and H. Richer (University of British Columbia)

Lesson 2

Reading Guide

Key Concepts
ESSENTIAL QUESTIONS
- How are the inner planets similar?
- Why is Venus hotter than Mercury?
- What kind of atmospheres do the inner planets have?

Vocabulary
terrestrial planet p. 769
greenhouse effect p. 771

Multilingual eGlossary

What's Science Got to do With It?

The Inner Planets

Inquiry Where is this?

This spectacular landscape is the surface of Mars, one of the inner planets. Other inner planets have similar rocky surfaces. It might surprise you to learn that there are planets in the solar system that have no solid surface on which to stand.

ESA/DLR/FU Berlin (G. Neukum)

Launch Lab

20 minutes

What affects the temperature on the inner planets?

Mercury and Venus are closer to the Sun than Earth. What determines the temperature on these planets? Let's find out.

1. Read and complete a lab safety form.
2. Insert a **thermometer** into a **clear 2-L plastic bottle.** Wrap **modeling clay** around the lid to hold the thermometer in the center of the bottle. Form an airtight seal with the clay.
3. Rest the bottle against the side of a **shoe box** in direct sunlight. Lay a second **thermometer** on top of the box next to the bottle so that the bulbs are at about the same height. The thermometer bulb should not touch the box. Secure the thermometer in place using **tape.**
4. Read the thermometers and record the temperatures in your Science Journal.
5. Wait 15 minutes and then read and record the temperature on each thermometer.

Think About This

1. How did the temperature of the two thermometers compare?
2. **Key Concept** What do you think caused the difference in temperature?

Planets Made of Rock

Imagine that you are walking outside. How would you describe the ground? You might say it is dusty or grassy. If you live near a lake or an ocean, you might say the ground is sandy or wet. But beneath the ground or lake or ocean is a layer of solid rock.

The inner planets–Mercury, Venus, Earth, and Mars–are also called terrestrial planets. **Terrestrial planets** are the planets closest to the Sun, are made of rock and metal, and have solid outer layers. Like Earth, the other inner planets also are made of rock and metallic materials and have a solid outer layer. However, as shown in **Figure 6**, the inner planets have different sizes, atmospheres, and surfaces.

WORD ORIGIN

terrestrial
from Latin *terrestris,* means "earthly"

Figure 6 The inner planets are roughly similar in size. Earth is about two and half times larger than Mercury. All inner planets have a solid outer layer.

INNER PLANETS

Visual Check Which is the smallest inner planet?

Figure 7 The *Messenger* space probe collected data on Mercury's surface, geologic history, and polar regions. The exploration mission lasted from 2004 to 2015.

Liquid outer core

Crust

Solid inner core

Mantle

Mercury's surface has many craters. Because almost no erosion occurs, craters and other surface features last for billions of years.

The Caloris Basin is about 1,550 km across. It is one of the largest impact craters in the solar system. It was formed billions of years ago by the impact of an object about 100 km in diameter.

Mercury Data

Mass: 5.5% of Earth's mass
Diameter: 38.3% of Earth's diameter
Average distance from Sun: 0.39 AU
Period of rotation: 59 days
Period of revolution: 88 days
Number of moons: 0

Mercury

The smallest planet and the planet closest to the Sun is Mercury, shown in **Figure 7.** Mercury has no atmosphere. A planet has an atmosphere when its gravity is strong enough to hold gases close to its surface. The strength of a planet's gravity depends on the planet's mass. Because Mercury's mass is so small, its gravity is not strong enough to hold onto an atmosphere. Without an atmosphere there is no wind that moves energy from place to place across the planet's surface. This results in temperatures as high as 450°C on the side of Mercury facing the Sun and as cold as −170°C on the side facing away from the Sun.

FOLDABLES®

Make a four-door book. Label each door with the name of an inner planet. Use the book to organize your notes on the inner planets.

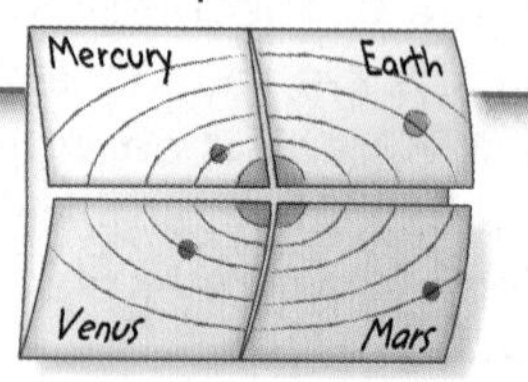

Mercury's Surface

Impact craters, depressions formed by collisions with objects from space, cover the surface of Mercury. There are smooth plains of solidified lava from long-ago eruptions. There are also high cliffs that might have formed when the planet cooled quickly, causing the surface to wrinkle and crack. Without an atmosphere, almost no erosion occurs on Mercury's surface. As a result, features that formed billions of years ago have changed very little.

Mercury's Structure

The structures of the inner planets are similar. Like all inner planets, Mercury has a core made of iron and nickel. Surrounding the core is a layer called the mantle. The mantle is mainly made of silicon and oxygen. The crust is a thin, rocky layer above the mantle. Mercury's large core might have been formed by a collision with a large object during Mercury's formation.

Key Concept Check How are the inner planets similar?

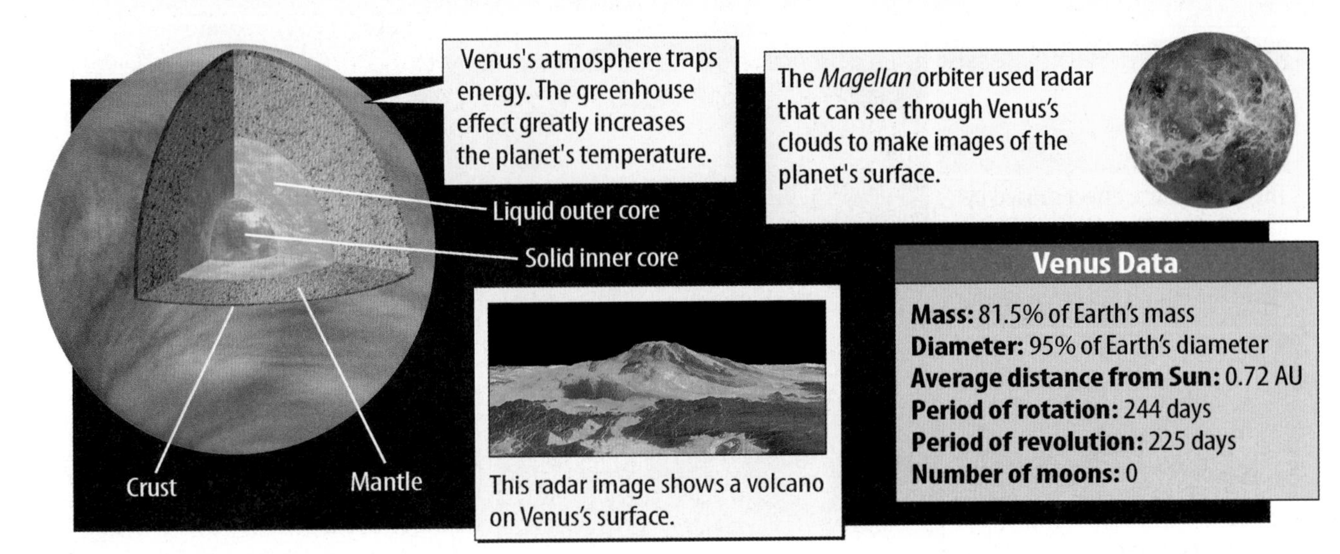

Figure 8 Because a thick layer of clouds covers Venus, its surface has not been seen. Between 1990 and 1994, the *Magellan* space probe mapped the surface using radar.

Venus

The second planet from the Sun is Venus, as shown in **Figure 8.** Venus is about the same size as Earth. It rotates so slowly that its period of rotation is longer than its period of revolution. This means that a day on Venus is longer than a year. Unlike most planets, Venus rotates from east to west. Several space probes have flown by or landed on Venus.

Venus's Atmosphere

The atmosphere of Venus is about 97 percent carbon dioxide. It is so dense that the atmospheric pressure on Venus is about 90 times greater than on Earth. Even though Venus has almost no water in its atmosphere or on its surface, a thick layer of clouds covers the planet. Unlike the clouds of water vapor on Earth, the clouds on Venus are made of acid.

The Greenhouse Effect on Venus

With an average temperature of about 460°C, Venus is the hottest planet in the solar system. The high temperatures are caused by the greenhouse effect. *The* **greenhouse effect** *occurs when a planet's atmosphere traps solar energy and causes the surface temperature to increase.* Carbon dioxide in Venus's atmosphere traps some of the solar energy that is absorbed and then emitted by the planet. This heats up the planet. Without the greenhouse effect, Venus would be almost 450°C cooler.

 **Key Concept Check** Why is Venus hotter than Mercury?

Venus's Structure and Surface

Venus's internal structure, as shown in **Figure 8**, is similar to Earth's. Radar images show that more than 80 percent of Venus's surface is covered by solidified lava. Much of this lava might have been produced by volcanic eruptions that occurred about half a billion years ago.

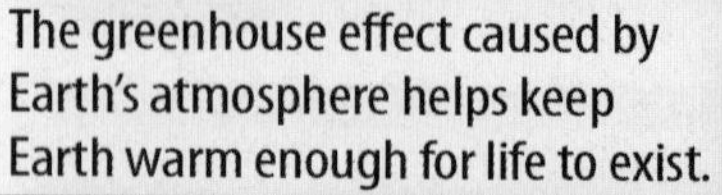

The greenhouse effect caused by Earth's atmosphere helps keep Earth warm enough for life to exist.

Earth is the only planet with liquid water on its surface.

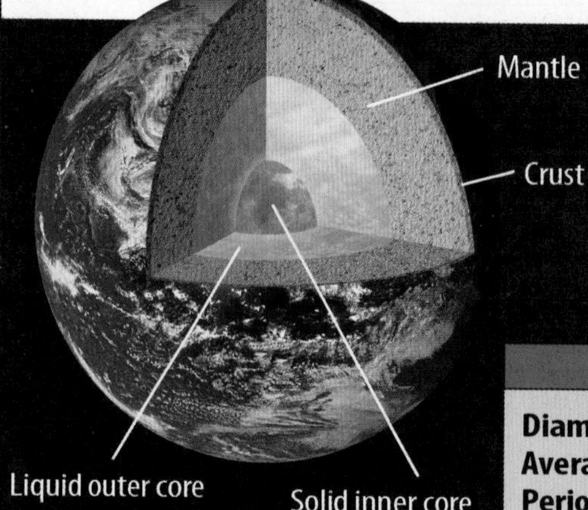

The movement of Earth's plates can cause volcanic eruptions that create new crust.

Earth Data

Diameter: 12,756 km
Average distance from Sun: 1.00 AU
Period of rotation: 24 hours
Period of revolution: 365 days
Number of moons: 1

Figure 9 Earth has more water in its atmosphere and on its surface than the other inner planets. Earth's surface is younger than the surfaces of the other inner planets because new crust is constantly forming.

MiniLab

20 minutes

How can you model the inner planets?

In this lab, you will use modeling clay to make scale models of the inner planets.

Planet	Actual Diameter (km)	Model Diameter (cm)
Mercury	4,886	
Venus	12,118	
Earth	12,756	8.0
Mars	6,786	

1. Use the data above for Earth to calculate in your Science Journal each model's diameter for the other three planets.
2. Use **modeling clay** to make a ball that represents the diameter of each planet. Check the diameter with a **centimeter ruler**.

Analyze Your Results

1. **Explain** how you converted actual diameters (km) to model diameters (cm).
2. **Key Concept** How do the inner planets compare? Which planets have approximately the same diameter?

Earth

Earth, shown in **Figure 9**, is the third planet from the Sun. Unlike Mercury and Venus, Earth has a moon.

Earth's Atmosphere

A mixture of gases and a small amount of water vapor make up most of Earth's atmosphere. They produce a greenhouse effect that increases Earth's average surface temperature. This effect and Earth's distance from the Sun warm Earth enough for large bodies of liquid water to exist. Earth's atmosphere also absorbs much of the Sun's radiation and protects the surface below. Earth's protective atmosphere, the presence of liquid water, and the planet's moderate temperature range support a variety of life.

Earth's Structure

As shown in **Figure 9**, Earth has a solid inner core surrounded by a liquid outer core. The mantle surrounds the liquid outer core. Above the mantle is Earth's crust. It is broken into large pieces, called plates, that constantly slide past, away from, or into each other. The crust is made mostly of oxygen and silicon and is constantly created and destroyed.

Reading Check Why is there life on Earth?

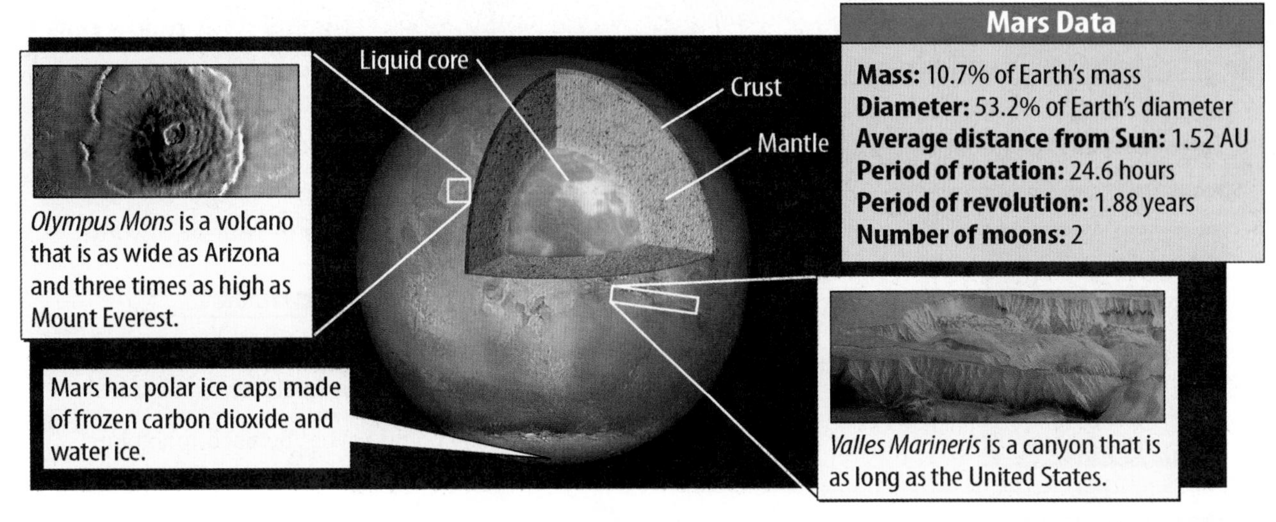

▲ **Figure 10** Mars is a small, rocky planet with deep canyons and tall mountains.

Mars

The fourth planet from the Sun is Mars, shown in **Figure 10.** Mars is about half the size of Earth. It has two very small and irregularly shaped moons. These moons might be asteroids that were captured by Mars's gravity.

Many space probes have visited Mars. Images of Mars show features that might have been made by water, such as the gullies in **Figure 11.** Recent findings from NASA's *Mars Reconnaissance Orbiters* (MRO) indicated that water flows periodically on the Martian surface.

Mars's Atmosphere

The atmosphere of Mars is about 95 percent carbon dioxide. It is thin and much less dense than Earth's atmosphere. Temperatures range from about −125°C at the poles to about 20°C at the equator during a martian summer. Winds on Mars sometimes produce great dust storms that last for months.

Mars's Surface

The reddish color of Mars is because its soil contains iron oxide, a compound in rust. Some of Mars's major surface features are shown in **Figure 10.** The enormous canyon Valles Marineris is about 4,000 km long. The Martian volcano Olympus Mons is the largest known mountain in the solar system. Mars also has polar ice caps made of frozen carbon dioxide and ice.

The southern hemisphere of Mars is covered with craters. The northern hemisphere is smoother and appears to be covered by lava flows. Some scientists have proposed that the lava flows were caused by the impact of an object about 2,000 km in diameter.

Key Concept Check Describe the atmosphere of each inner planet.

Figure 11 Gullies such as these might have been formed by the flow of liquid water on Mars billions of years ago. ▼

Lesson 2 Review

Visual Summary

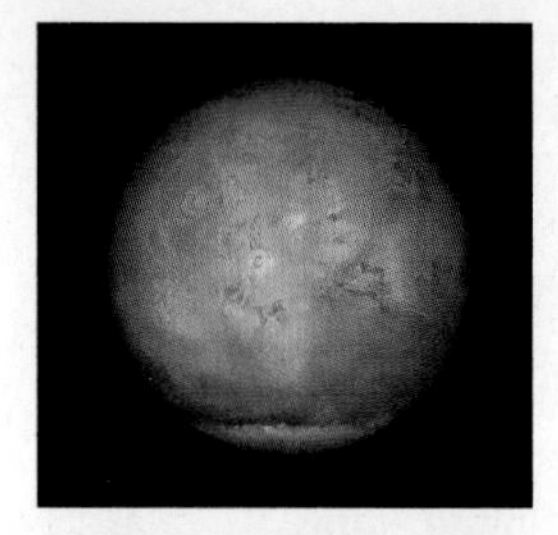

The terrestrial planets include Mercury, Venus, Earth, and Mars.

The inner planets all are made of rocks and minerals, but they have different characteristics. Earth is the only planet with liquid water.

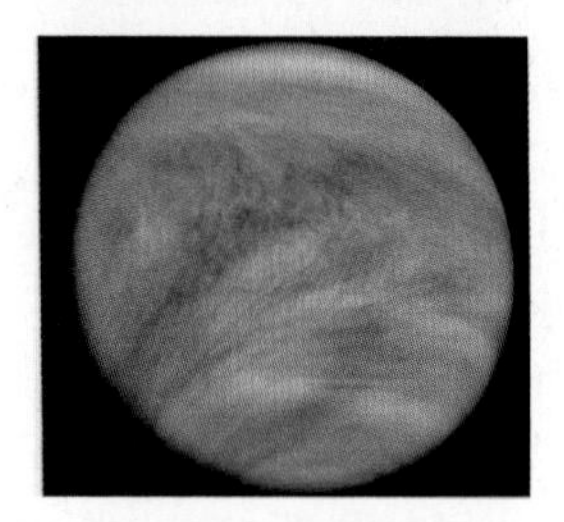

The greenhouse effect greatly increases the surface temperature of Venus.

FOLDABLES®

Use your lesson Foldable to review the lesson. Save your Foldable for the project at the end of the chapter.

What do you think NOW?

You first read the statements below at the beginning of the chapter.

3. Earth is the only inner planet that has a moon.

4. Venus is the hottest planet in the solar system.

Did you change your mind about whether you agree or disagree with the statements? Rewrite any false statements to make them true.

Use Vocabulary

1. **Define** *greenhouse effect* in your own words.

Understand Key Concepts

2. **Explain** why Venus is hotter than Mercury, even though Mercury is closer to the Sun.
3. **Infer** Why could rovers be used to explore Mars but not Venus?
4. Which of the inner planets has the greatest mass?

   A. Mercury
   B. Venus
   C. Earth
   D. Mars

5. **Relate** Describe the relationship between an inner planet's distance from the Sun and its period of revolution.

Interpret Graphics

6. **Infer** Which planet shown below is most likely able to support life now or was able to in the past? Explain your reasoning.

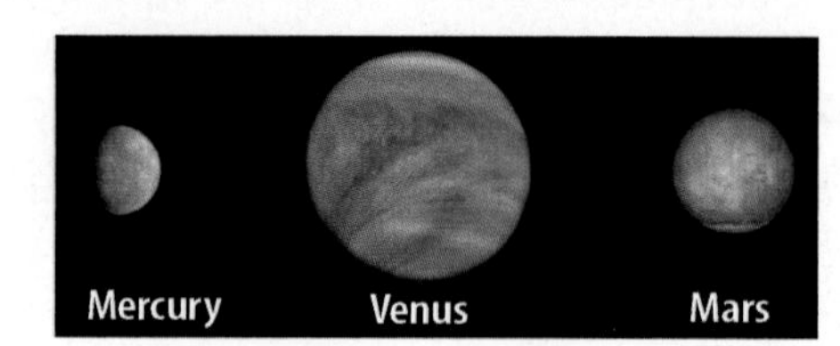

7. **Compare and Contrast** Copy and fill in the table below to compare and contrast properties of Venus and Earth.

Planet	Similarities	Differences
Venus		
Earth		

Critical Thinking

8. **Imagine** How might the temperatures on Mercury be different if it had the same mass as Earth? Explain.
9. **Judge** Do you think the inner planets should be explored or should the money be spent on other things? Justify your opinion.

 **Skill Practice** **Graphing Data** **25 minutes**

What can we learn about planets by graphing their characteristics?

Scientists collect and analyze data, and draw conclusions based on data. They are particularly interested in finding trends and relationships in data. One commonly used method of finding relationships is by graphing data. Graphing allows different types of data be to seen in relation to one another.

Learn It

Scientists know that some properties of the planets are related. **Graphing data** makes the relationships easy to identify. The graphs can show mathematical relationships such as direct and inverse relationships. Often, however, the graphs show that there is no relationship in the data.

Try It

1. You will plot two graphs that explore the relationships in data. The first graph compares a planet's distance from the Sun and its orbital period. The second graph compares a planet's distance from the Sun and its radius. Make a prediction about how these two sets of data are related, if at all. The data is shown in the table below.

Planet	Average Distance From the Sun (AU)	Orbital Period (yr)	Planet Radius (km)
Mercury	0.39	0.24	2440
Venus	0.72	0.62	6051
Earth	1.00	1.0	6378
Mars	1.52	1.9	3397
Jupiter	5.20	11.9	71,492
Saturn	9.58	29.4	60,268
Uranus	19.2	84.0	25,559
Neptune	30.1	164.0	24,764

2. Use the data in the table to plot a line graph showing orbital period versus average distance from the Sun. On the *x*-axis, plot the planet's distance from the Sun. On the *y*-axis, plot the planet's orbital period. Make sure the range of each axis is suitable for the data to be plotted, and clearly label each planet's data point.

3. Use the data in the table to plot a line graph showing planet radius versus average distance from the Sun. On the *y*-axis, plot the planet's radius. Make sure the range of each axis is suitable for the data to be plotted, and clearly label each planet's data point.

Apply It

4. Examine the *Orbital Period v. Distance from the Sun* graph. Does the graph show a relationship? If so, describe the relationship between a planet's distance from the Sun and its orbital period in your Science Journal.

5. Examine the *Planet Radius v. Distance from the Sun* graph. Does the graph show a relationship? If so, describe the relationship between a planet's distance from the Sun and its radius.

6. **Key Concept** Identify one or two characteristics the inner planets share that you learned from your graphs.

Lesson 3

The Outer Planets

Reading Guide

Key Concepts 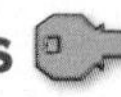

ESSENTIAL QUESTIONS

- How are the outer planets similar?
- What are the outer planets made of?

Vocabulary

Galilean moons p. 779

 Multilingual eGlossary

What's below?

Clouds often prevent airplane pilots from seeing the ground below. Similarly, clouds block the view of Jupiter's surface. What do you think is below Jupiter's colorful cloud layer? The answer might surprise you—Jupiter is not at all like Earth.

Launch Lab

15 minutes

How do we see distant objects in the solar system?

Some of the outer planets were discovered hundreds of years ago. Why weren't all planets discovered?

1. Read and complete a lab safety form.
2. Use a **meterstick, masking tape,** and the **data table** to mark and label the position of each object on the tape on the floor along a straight line.
3. Shine a **flashlight** from "the Sun" horizontally along the tape.
4. Have a partner hold a page of this **book** in the flashlight beam at each planet location. Record your observations in your Science Journal.

Object	Distance from Sun (cm)
Sun	0
Jupiter	39
Saturn	71
Uranus	143
Neptune	295

Think About This

1. What happens to the image of the page as you move away from the flashlight?
2. **Key Concept** Why do you think it is more difficult to observe the outer planets than the inner planets?

The Gas Giants

Have you ever seen water drops on the outside of a glass of ice? They form because water vapor in the air changes to a liquid on the cold glass. Gases also change to liquids at high pressures. These properties of gases affect the outer planets.

The outer planets, shown in **Figure 12,** are called the gas giants because they are primarily made of hydrogen and helium. These elements are usually gases on Earth.

The outer planets have strong gravitational forces due to their large masses. The strong gravity creates tremendous atmospheric pressure that changes gases to liquids. Thus, the outer planets mainly have liquid interiors. In general, the outer planets have a thick gas and liquid layer covering a small, solid core.

Key Concept Check How are the outer planets similar?

Figure 12 The outer planets are primarily made of gases and liquids.

Visual Check Which outer planet is the largest?

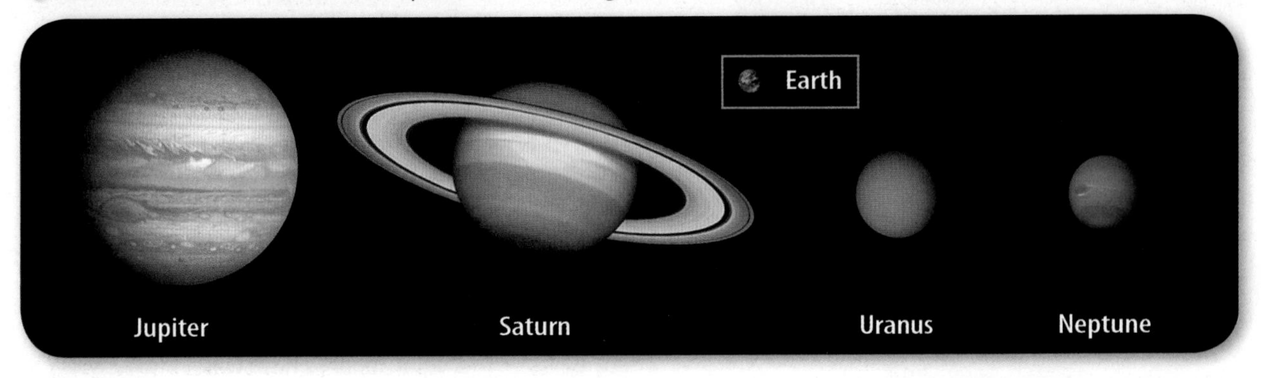

(l)NASA/JPL/USGS; (cl)NASA/JPL/Space Science Institute; (c)NASA GSFC image by Robert Simmon and Reto Stöckli; (cr, r)NASA/JPL

Jupiter

Figure 13 describes Jupiter, the largest planet in the solar system. Jupiter's diameter is more than 11 times larger than the diameter of Earth. Its mass is more than twice the mass of all the other planets combined. One way to understand just how big Jupiter is is to realize that more than 1,000 Earths would fit within this gaseous planet's volume.

Jupiter takes almost 12 Earth years to complete one orbit. Yet, it rotates faster than any other planet. Its period of rotation is less than 10 hours. Jupiter and all the outer planets have a ring system.

FOLDABLES®

Make a four-door book. Label each door with the name of an outer planet. Use the book to organize your notes on the outer planets.

Jupiter | Saturn | Uranus | Neptune

Jupiter's Atmosphere

The atmosphere on Jupiter is about 90 percent hydrogen and 10 percent helium and is about 1,000 km deep. Within the atmosphere are layers of dense, colorful clouds. Because Jupiter rotates so quickly, these clouds stretch into colorful, swirling bands. The Great Red Spot on the planet's surface is a storm of swirling gases.

Jupiter's Structure

Overall, Jupiter is about 80 percent hydrogen and 20 percent helium with small amounts of other materials. The planet is a ball of gas swirling around a thick liquid layer that conceals a solid core. About 1,000 km below the outer edge of the cloud layer, the pressure is so great that the hydrogen gas changes to liquid. This thick layer of liquid hydrogen surrounds Jupiter's core. Scientists do not know for sure what makes up the core. They suspect that the core is made of rock and iron. The core might be as large as Earth and could be 10 times more massive.

Key Concept Check Describe what makes up each of Jupiter's three distinct layers.

Figure 13 Jupiter is mainly hydrogen and helium. Throughout most of the planet, the pressure is high enough to change the hydrogen gas into a liquid.

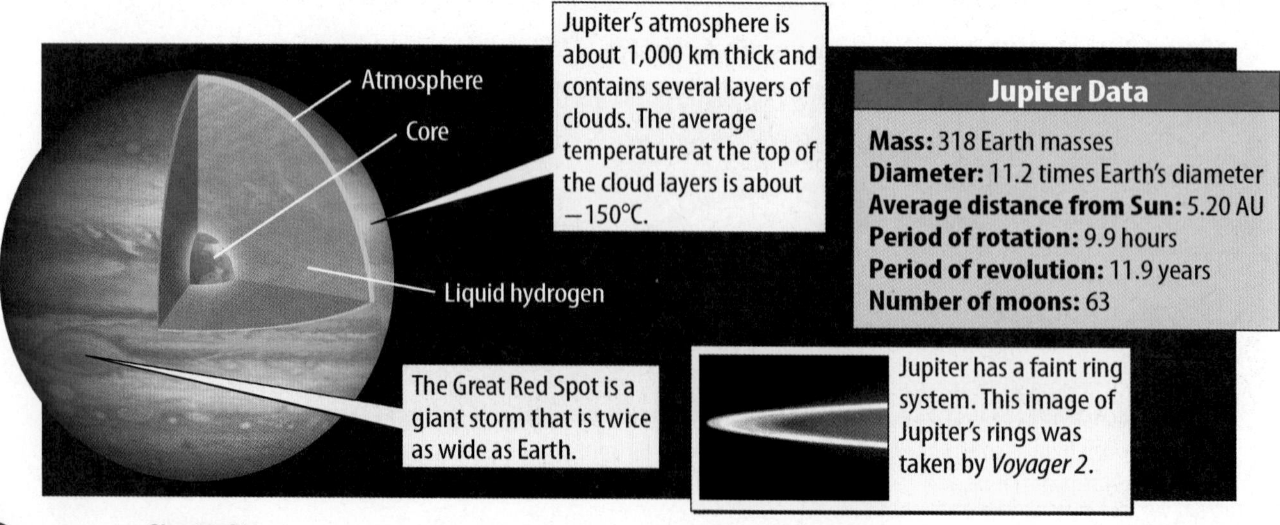

The Moons of Jupiter

Jupiter has at least 63 moons, more than any other planet. Jupiter's four largest moons were first discovered by Galileo Galilei in 1610. *The four largest moons of Jupiter–Io, Europa, Ganymede, and Callisto–are known as the* **Galilean moons.** The Galilean moons all are made of rock and ice. The moons Ganymede, Callisto, and Io are larger than Earth's Moon. Collisions between Jupiter's moons and meteorites likely resulted in the particles that make up the planet's faint rings.

Saturn

Saturn is the sixth planet from the Sun. Like Jupiter, Saturn rotates rapidly and has horizontal bands of clouds. Saturn is about 90 percent hydrogen and 10 percent helium. It is the least dense planet. Its density is less than that of water.

Saturn's Structure

Saturn is made mostly of hydrogen and helium with small amounts of other materials. As shown in **Figure 14**, Saturn's structure is similar to Jupiter's structure–an outer gas layer, a thick layer of liquid hydrogen, and a solid core.

The ring system around Saturn is the largest and most complex in the solar system. Saturn has seven bands of rings, each containing thousands of narrower ringlets. The main ring system is over 70,000 km wide, but it is likely less than 30 m thick. The ice particles in the rings are possibly from a moon that was shattered in a collision with another icy object.

Key Concept Check Describe what makes up Saturn and its ring system.

Math Skills 

Ratios

A ratio is a quotient—it is one quantity divided by another. Ratios can be used to compare distances. For example, Jupiter is 5.20 AU from the Sun, and Neptune is **30.05** AU from the Sun. Divide the larger distance by the smaller distance:

$$\frac{30.05 \text{ AU}}{5.20 \text{ AU}} = 5.78$$

Neptune is 5.78 times farther from the Sun than Jupiter.

Practice

How many times farther from the Sun is Uranus (distance = 19.20 AU) than Saturn (distance = 9.58 AU)?

 Math Practice

 Personal Tutor

Figure 14 Like Jupiter, Saturn is mainly hydrogen and helium. Saturn's rings are one of the most noticeable features of the solar system.

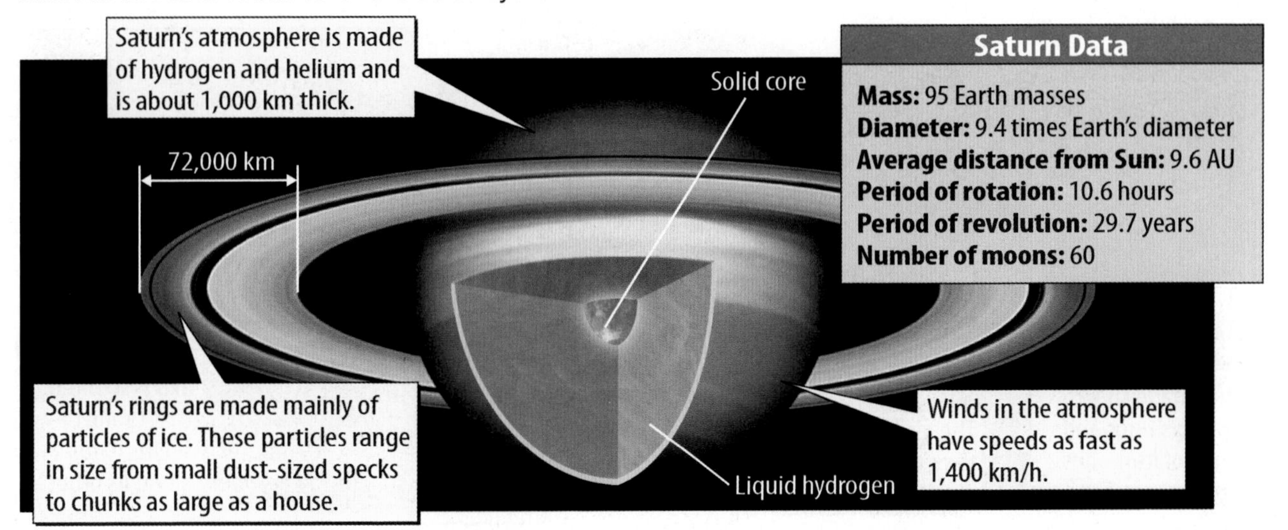

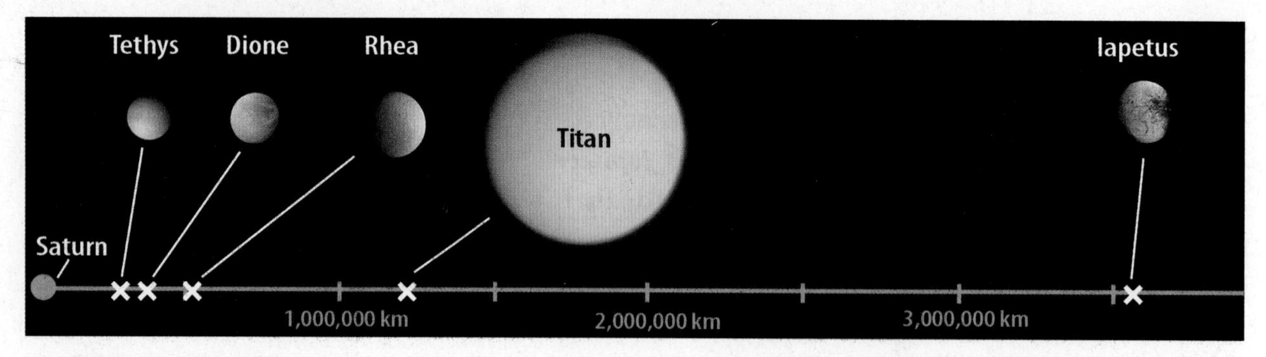

▲ **Figure 15** The five largest moons of Saturn are shown above drawn to scale. Titan is Saturn's largest moon.

Saturn's Moons

Saturn has at least 60 moons. The five largest moons, Titan, Rhea, Dione, Iapetus, and Tethys, are shown in **Figure 15.** Most of Saturn's moons are chunks of ice less than 10 km in diameter. However, Titan is larger than the planet Mercury. Titan is the only moon in the solar system with a dense atmosphere. In 2005, the *Cassini* orbiter released the *Huygens* (HOY guns) **probe** that landed on Titan's surface.

WORD ORIGIN

probe
from Medieval Latin *proba*, means "examination"

Uranus

Uranus, shown in **Figure 16,** is the seventh planet from the Sun. It has a system of narrow, dark rings and a diameter about four times that of Earth. *Voyager 2* is the only space probe to explore Uranus. The probe flew by the planet in 1986.

Uranus has a deep atmosphere composed mostly of hydrogen and helium. The atmosphere also contains a small amount of methane. Beneath the atmosphere is a thick, slushy layer of water, ammonia, and other materials. Uranus might also have a solid, rocky core.

Key Concept Check Identify the substances that make up the atmosphere and the thick slushy layer on Uranus.

Figure 16  Uranus is mainly gas and liquid, with a small solid core. Methane gas in the atmosphere gives Uranus a bluish color. ▼

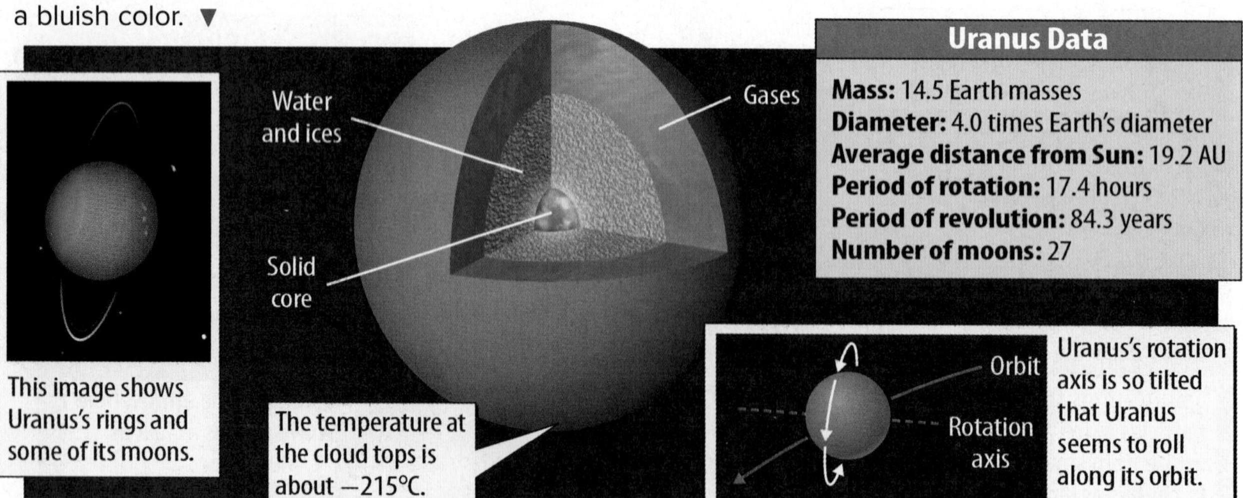

Uranus Data

Mass: 14.5 Earth masses
Diameter: 4.0 times Earth's diameter
Average distance from Sun: 19.2 AU
Period of rotation: 17.4 hours
Period of revolution: 84.3 years
Number of moons: 27

Uranus's Axis and Moons

Figure 16 shows that Uranus has a tilted axis of rotation. In fact, it is so tilted that the planet moves around the Sun like a rolling ball. This sideways tilt might have been caused by a collision with an Earth-sized object.

Uranus has at least 27 moons. The two largest moons, Titania and Oberon, are considerably smaller than Earth's moon. Titania has an icy cracked surface that once might have been covered by an ocean.

Neptune

Neptune, shown in **Figure 17**, was discovered in 1846. Like Uranus, Neptune's atmosphere is mostly hydrogen and helium, with a trace of methane. Its interior also is similar to the interior of Uranus. Neptune's interior is partially frozen water and ammonia with a rock and iron core. It has at least 13 moons and a faint, dark ring system. Its largest moon, Triton, is made of rock with an icy outer layer. It has a surface of frozen nitrogen and geysers that erupt nitrogen gas.

Key Concept Check How does the atmosphere and interior of Neptune compare with that of Uranus?

MiniLab — 15 minutes

How do Saturn's moons affect its rings?

In this lab, sugar models Saturn's rings. How might Saturn's moons affect its rings?

1. Read and complete a lab safety form.
2. Hold two **sharpened pencils** with their points even and then **tape** them together.
3. Insert a third pencil into the hole in a **record.** Hold the pencil so the record is in a horizontal position.
4. Have your partner sprinkle **sugar** evenly over the surface of the record. Hold the taped pencils vertically over the record so that the tips rest in the record's grooves.
5. Slowly turn the record. In your Science Journal, record what happens to the sugar.

Analyze and Conclude

1. **Compare and Contrast** What feature of Saturn's rings do the pencils model?
2. **Infer** What do you think causes the spaces between the rings of Saturn?
3. **Key Concept** What would have to be true for a moon to interact in this way with Saturn's rings?

Figure 17 The atmosphere of Neptune is similar to that of Uranus—mainly hydrogen and helium with a trace of methane. The dark circular areas on Neptune are swirling storms. In addition to the great dark spot, two other massive storms have appeared and faded over the last decade. Winds on Neptune sometimes exceed 1,000 km/h.

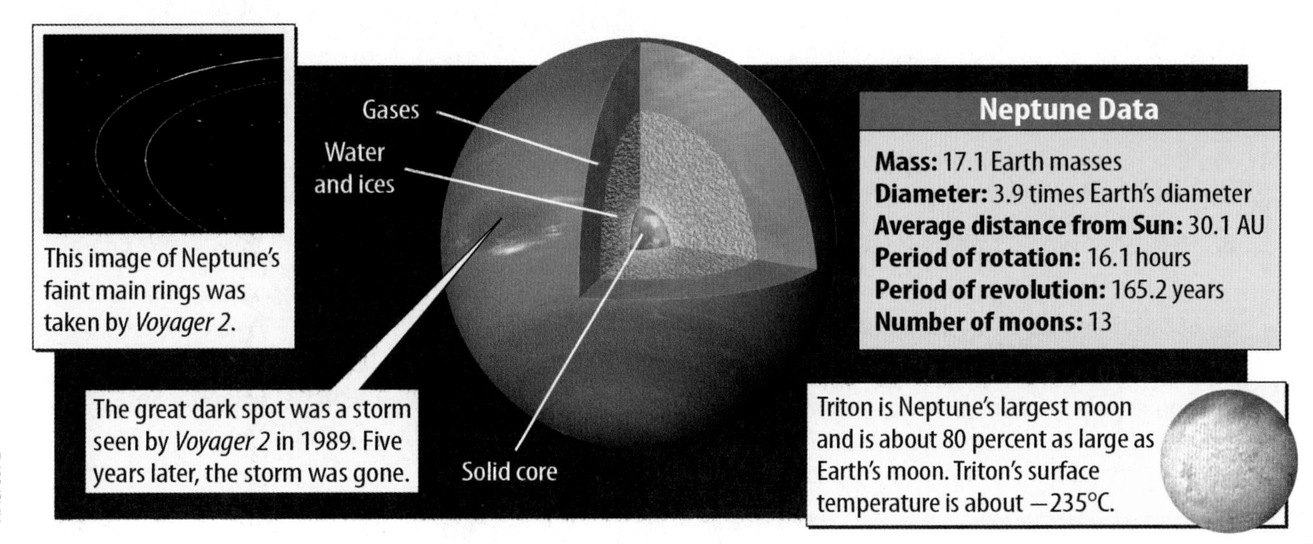

Lesson 3 Review

Visual Summary

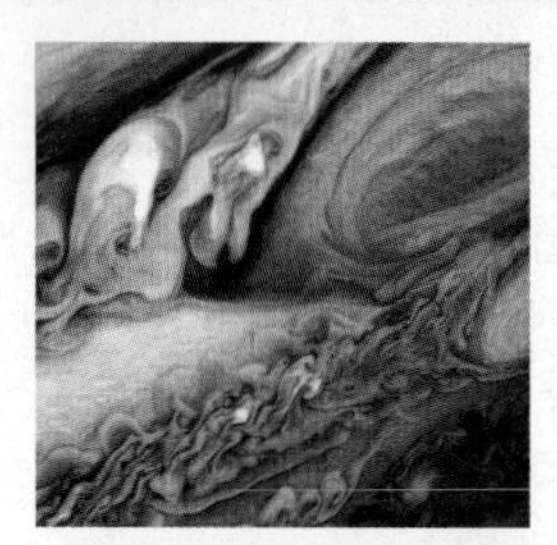

All of the outer planets are primarily made of materials that are gases on Earth. Colorful clouds of gas cover Saturn and Jupiter.

Jupiter is the largest outer planet. Its four largest moons are known as the Galilean moons.

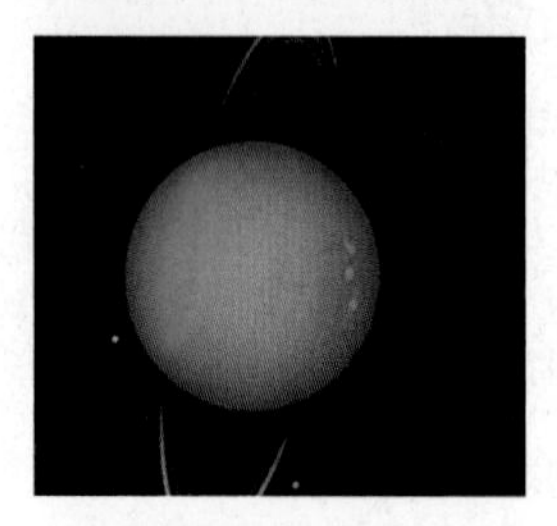

Uranus has an unusual tilt, possibly due to a collision with a large object.

FOLDABLES®

Use your lesson Foldable to review the lesson. Save your Foldable for the project at the end of the chapter.

What do you think NOW?

You first read the statements below at the beginning of the chapter.

5. The outer planets also are called the gas giants.

6. The atmospheres of Saturn and Jupiter are mainly water vapor.

Did you change your mind about whether you agree or disagree with the statements? Rewrite any false statements to make them true.

Use Vocabulary

1 **Identify** What are the four Galilean moons of Jupiter?

Understand Key Concepts

2 **Contrast** How are the rings of Saturn different from the rings of Jupiter?

3 Which planet's rings probably formed from a collision between an icy moon and another icy object?

A. Jupiter
B. Neptune
C. Saturn
D. Uranus

4 **List** the outer planets by increasing mass.

Interpret Graphics

5 **Infer** from the diagram below how Uranus's tilted axis affects its seasons.

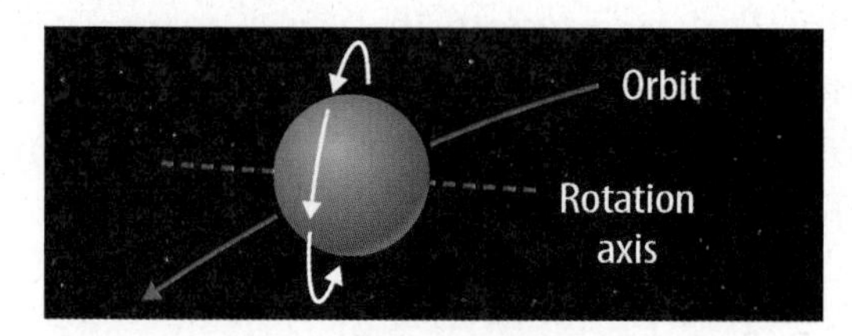

6 **Organize Information** Copy the organizer below and use it to list the outer planets.

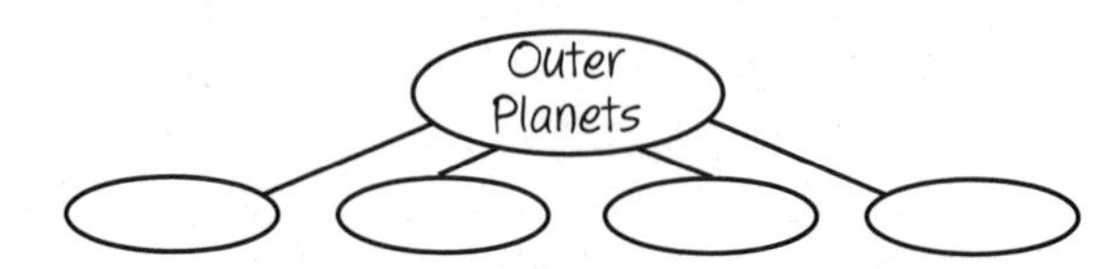

Critical Thinking

7 **Predict** what would happen to Jupiter's atmosphere if its gravitational force suddenly decreased. Explain.

8 **Evaluate** Is life more likely on a dry and rocky moon or on an icy moon? Explain.

Math Practice

9 **Calculate** Mars is about 1.52 AU from the Sun, and Saturn is about 9.58 AU from the Sun. How many times farther from the Sun is Saturn than Mars?

Pluto

CAREERS in SCIENCE

What in the world is it?

Since Pluto's discovery in 1930, students have learned that the solar system has nine planets. But in 2006, the number of planets was changed to eight. What happened?

Neil deGrasse Tyson is an astrophysicist at the American Museum of Natural History in New York City. He and his fellow Museum scientists were among the first to question Pluto's classification as a planet. One reason was that Pluto is smaller than six moons in our solar system, including Earth's moon. Another reason was that Pluto's orbit is more oval-shaped, or elliptical, than the orbits of other planets. Also, Pluto has the most tilted orbit of all planets—17 degrees out of the plane of the solar system. Finally, unlike other planets, Pluto is mostly ice.

Tyson also questioned the definition of a planet—an object that orbits the Sun. Then shouldn't comets be planets? In addition, he noted that when Ceres, an object orbiting the Sun between Jupiter and Mars, was discovered in 1801, it was classified as a planet. But, as astronomers discovered more objects like Ceres, it was reclassified as an asteroid. Then, during the 1990s, many space objects similar to Pluto were discovered. They orbit the Sun beyond Neptune's orbit in a region called the Kuiper belt.

These new discoveries led Tyson and others to conclude that Pluto should be reclassified. In 2006, the International Astronomical Union agreed. Pluto was reclassified as a dwarf planet—an object that is spherical in shape and orbits the Sun in a zone with other objects. Pluto lost its rank as smallest planet, but became "king of the Kuiper belt."

Pluto TIME LINE

1930
Astronomer Clyde Tombaugh discovers a ninth planet, Pluto.

1992
The first object is discovered in the Kuiper belt.

July 2005
Eris—a Pluto-sized object—is discovered in the Kuiper belt.

January 2006
NASA launches *New Horizons* spacecraft. It reached Pluto in 2015.

August 2006
Pluto is reclassified as a dwarf planet.

Neil deGrasse Tyson is director of the Hayden Planetarium at the American Museum of Natural History. ▶

This illustration shows what Pluto might look like if you were standing on one of its moons.

It's Your Turn

RESEARCH With a group, identify the different types of objects in our solar system. Consider size, composition, location, and whether the objects have moons. Propose at least two different ways to group the objects.

Lesson 4

Dwarf Planets and Other Objects

Reading Guide

Key Concepts
ESSENTIAL QUESTIONS

- What is a dwarf planet?
- What are the characteristics of comets and asteroids?
- How does an impact crater form?

Vocabulary

meteoroid p. 788

meteor p. 788

meteorite p. 788

impact crater p. 788

Multilingual eGlossary

Inquiry Will it return?

You would probably remember a sight like this. This image of comet C/2006 P1 was taken in 2007. The comet is no longer visible from Earth. Believe it or not, many comets appear then reappear hundreds to millions of years later.

Gordon Garradd/SPL/Photo Researchers, Inc.

Launch Lab

15 minutes

How might asteroids and moons form?

In this activity, you will explore one way moons and asteroids might have formed.

1. Read and complete a lab safety form.
2. Form a small ball from **modeling clay** and roll it in **sand.**
3. Press a thin layer of modeling clay around a **marble.**
4. Tie equal lengths of **string** to each ball. Hold the strings so the balls are above a **sheet of paper.**
5. Have someone pull back the marble so that its string is parallel to the tabletop and then release it. Record the results in your Science Journal.

Think About This

1. If the collision you modeled occurred in space, what would happen to the sand?
2. **Key Concept** Infer one way scientists propose moons and asteroids formed.

Dwarf Planets

Ceres was discovered in 1801 and was called a planet until similar objects were discovered near it. Then it was called an asteroid. For decades after Pluto's discovery in 1930, it was called a planet. Then, similar objects were discovered, and Pluto lost its planet classification. What type of object is Pluto?

Pluto once was classified as a planet, but it is now classified as a dwarf planet. In 2006, the International Astronomical Union (IAU) adopted "dwarf planet" as a new category. The IAU defines a dwarf planet as an object that orbits the Sun, has enough mass and gravity to form a sphere, and has objects similar in mass orbiting near it or crossing its orbital path. Astronomers classify Pluto, Ceres, Eris, MakeMake (MAH kay MAH kay), and Haumea (how MAY uh) as dwarf planets. **Figure 18** shows four dwarf planets.

Key Concept Check Describe the characteristics of a dwarf planet.

Figure 18 Four dwarf planets are shown to scale. All dwarf planets are smaller than the Moon.

Dwarf Planets

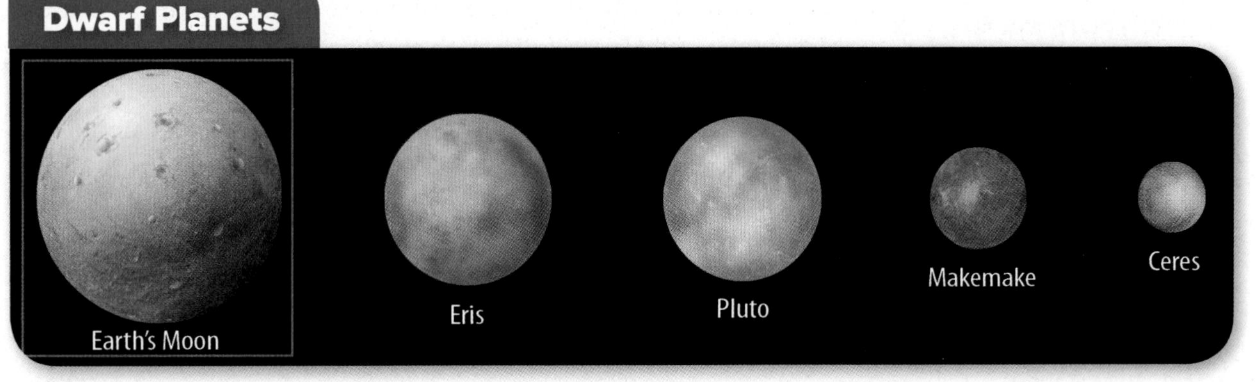

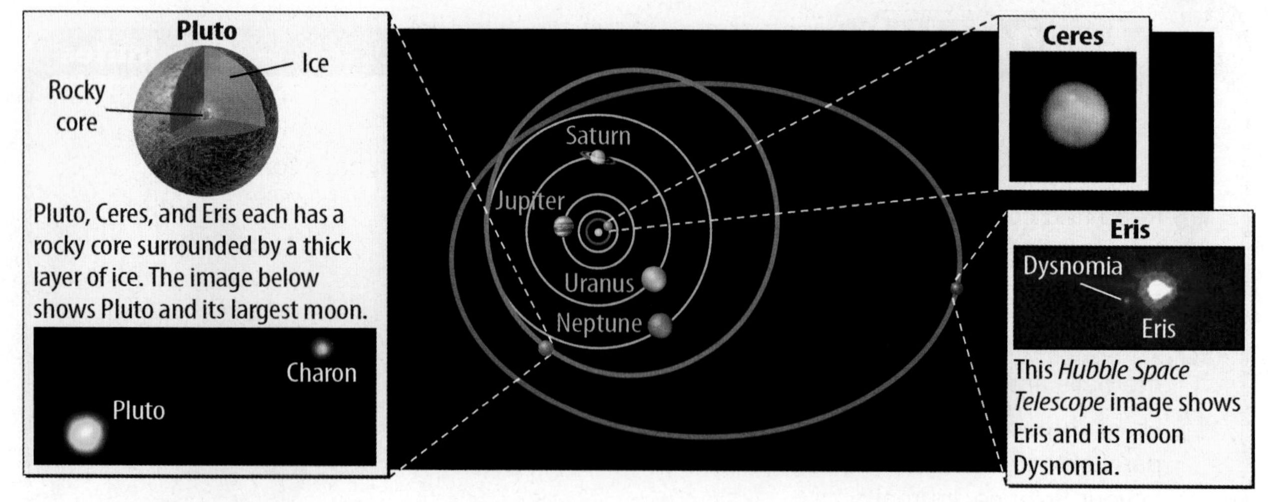

Figure 19 Because most dwarf planets are so far from Earth, astronomers do not have detailed images of them.

Visual Check Which dwarf planet orbits closest to Earth?

Ceres

Ceres, shown in **Figure 19**, orbits the Sun in the asteroid belt. With a diameter of about 950 km, Ceres is about one-fourth the size of the Moon. It is the smallest dwarf planet. Ceres might have a rocky core surrounded by a layer of water ice and a thin, dusty crust.

Pluto

Pluto is about two-thirds the size of the Moon. Pluto is so far from the Sun that its period of revolution is about 248 years. Like Ceres, Pluto has a rocky core surrounded by ice. With an average surface temperature of about −230°C, Pluto is so cold that it is covered with frozen nitrogen.

Pluto has three known moons. The largest moon, Charon, has a diameter that is about half the diameter of Pluto. Pluto also has two smaller moons, Hydra and Nix.

Eris

The largest dwarf planet, Eris, was discovered in 2003. Its orbit lasts about 557 years. Currently, Eris is three times farther from the Sun than Pluto is. The structure of Eris is probably similar to Pluto. Dysnomia (dis NOH mee uh) is the only known moon of Eris.

Makemake and Haumea

In 2008, the IAU designated two new objects as dwarf planets: Makemake and Haumea. Though smaller than Pluto, Makemake is one of the largest objects in a region of the solar system called the Kuiper (KI puhr) belt. The Kuiper belt extends from about the orbit of Neptune to about 50 AU from the Sun. Haumea is also in the Kuiper belt and is smaller than Pluto.

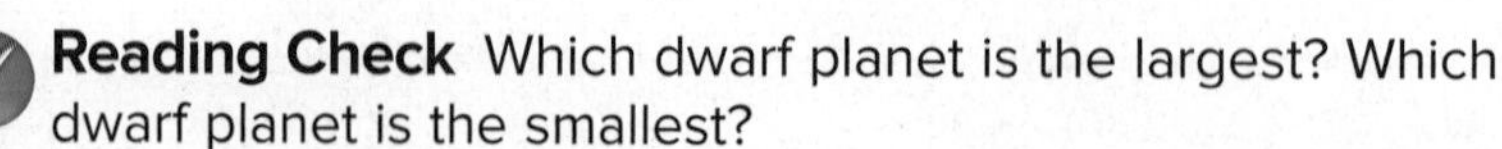

Reading Check Which dwarf planet is the largest? Which dwarf planet is the smallest?

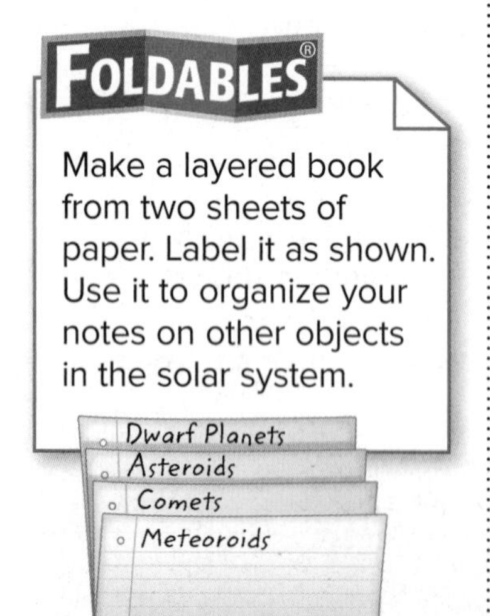

(l)Dr. R. Albrecht, ESA/ESO Space Telescope European Coordinating Facility/NASA; (tr)NASA/STScI; (br)NASA, ESA, and M. Brown (California Institute of Technology)

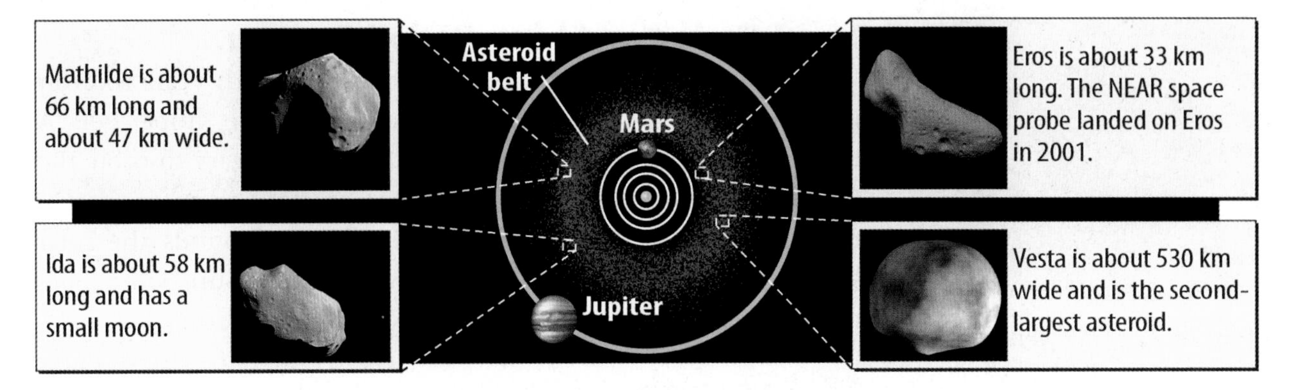

Figure 20 The asteroids that orbit the Sun in the asteroid belt are many sizes and shapes.

Asteroids

Recall from Lesson 1 that asteroids are pieces of rock and ice. Most asteroids orbit the Sun in the asteroid belt. The asteroid belt is between the orbits of Mars and Jupiter, as shown in **Figure 20.** Hundreds of thousands of asteroids have been discovered. The largest asteroid, Pallas, is over 500 km in diameter.

Asteroids are chunks of rock and ice that never clumped together like the rocks and ice that formed the inner planets. Some astronomers suggest that the strength of Jupiter's gravitational field might have caused the chunks to collide so violently, and they broke apart instead of sticking together. This means that asteroids are objects left over from the formation of the solar system.

Key Concept Check Where do the orbits of most asteroids occur?

Comets

Recall that comets are mixtures of rock, ice, and dust. The particles in a comet are loosely held together by the gravitational attractions among the particles. As shown in **Figure 21,** comets orbit the Sun in long elliptical orbits.

The Structure of Comets

The solid, inner part of a comet is its nucleus, as shown in **Figure 21.** As a comet moves closer to the Sun, it absorbs thermal energy and can develop a bright tail. Heating changes the ice in the comet into a gas. Energy from the Sun pushes some of the gas and dust away from the nucleus and makes it glow. This produces the comet's bright tail and glowing nucleus, called a coma.

Key Concept Check Describe the characteristics of a comet.

Figure 21 When energy from the Sun strikes the gas and dust in the comet's nucleus, it can create a two-part tail. The gas tail always points away from the Sun.

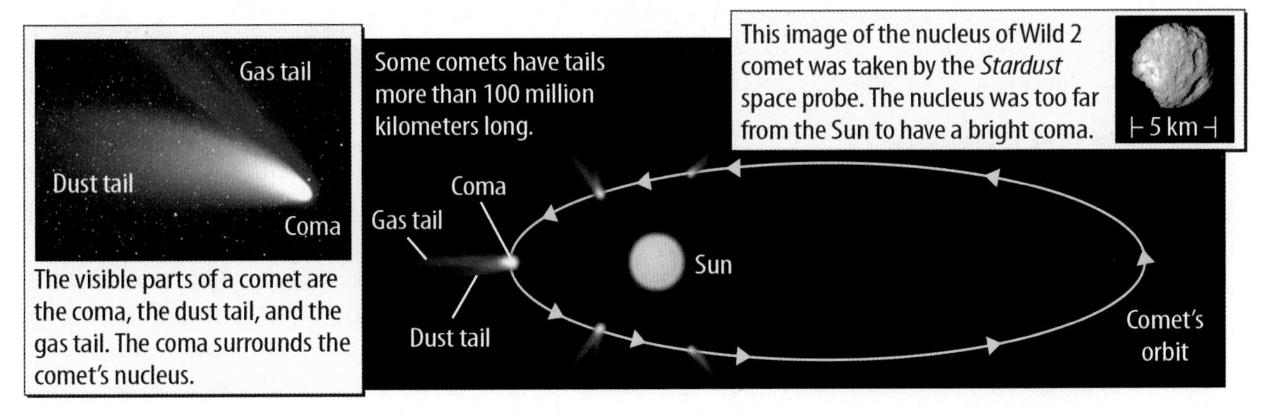

Figure 22 When a large meteorite strikes, it can form a giant impact crater like this 1.2-km wide crater in Arizona.

Short-Period and Long-Period Comets

A short-period comet takes less than 200 Earth years to orbit the Sun. Most short-period comets come from the Kuiper belt. A long-period comet takes more than 200 Earth years to orbit the Sun. Long-period comets come from an area at the outer edge of the solar system, called the Oort cloud. It surrounds the solar system and extends about 100,000 AU from the Sun. Some long-period comets take millions of years to orbit the Sun.

Meteoroids

Every day, many millions of particles called meteoroids enter Earth's atmosphere. *A* **meteoroid** *is a small, rocky particle that moves through space.* Most meteoroids are only about as big as a grain of sand. As a meteoroid passes through Earth's atmosphere, friction makes the meteoroid and the air around it hot enough to glow. *A* **meteor** *is a meteoroid that has entered Earth's atmosphere, producing a streak of light.* Most meteoroids burn up in the atmosphere. However, some meteoroids are large enough that they reach Earth's surface before they burn up completely. When this happens, it is called a meteorite. *A* **meteorite** *is a meteoroid that strikes a planet or a moon.*

WORD ORIGIN

meteor
from Greek *meteoros,* means "high up"

When a large meteoroite strikes a moon or planet, it often forms a bowl-shaped depression such as the one shown in **Figure 22.** *An* **impact crater** *is a round depression formed on the surface of a planet, moon, or other space object by the impact of a meteorite.* The limited number of impact craters on Earth is due to the processes of erosion, tectonics, and volcanism.

 **Key Concept Check** What causes an impact crater to form?

MiniLab

20 minutes

How do impact craters form?

In this lab, you will model the formation of an impact crater.

1. Pour a layer of **flour** about 3 cm deep in a **cake pan.**
2. Pour a layer of **cornmeal** about 1 cm deep on top of the flour.
3. One at a time, drop different-sized **marbles** into the mixture from the same height—about 15 cm. Record your observations in your Science Journal.

Analyze and Conclude

1. **Describe** the mixture's surface after you dropped the marbles.
2. **Recognize Cause and Effect** Based on your results, explain why impact craters on moons and planets differ.
3. **Key Concept** Explain how the marbles used in the activity could be used to model meteoroids, meteors, and meteorites.

Lesson 4 Review

Visual Summary

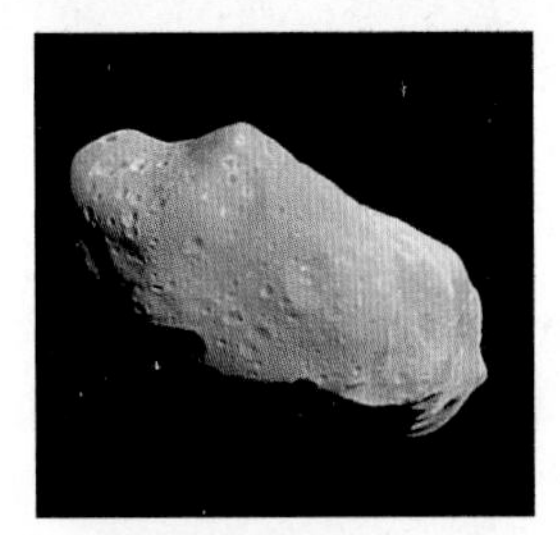

An asteroid, such as Ida, is a chunk of rock and ice that orbits the Sun.

Comets, which are mixture of rock , ice, and dust, orbit the Sun. A comet's tail is caused by its interaction with the Sun.

When a large meteorite strikes a planet or moon, it often makes an impact crater.

FOLDABLES®

Use your lesson Foldable to review the lesson. Save your Foldable for the project at the end of the chapter.

What do you think NOW?

You first read the statements below at the beginning of the chapter.

7. Asteroids and comets are mainly rock and ice.

8. A meteoroid is a meteor that strikes Earth.

Did you change your mind about whether you agree or disagree with the statements? Rewrite any false statements to make them true.

Use Vocabulary

1. **Define** *impact crater* in your own words.
2. **Distinguish** between a meteorite and a meteoroid.
3. **Use the term** *meteor* in a complete sentence.

Understand Key Concepts

4. Which produces an impact crater?
   - **A.** comet
   - **B.** meteor
   - **C.** meteorite
   - **D.** planet
5. **Reason** Are you more likely to see a meteor or a meteoroid? Explain.
6. **Differentiate** between objects located in the asteroid belt and objects located in the Kuiper belt.

Interpret Graphics

7. **Explain** why some comets have a two-part tail during portions of their orbit.

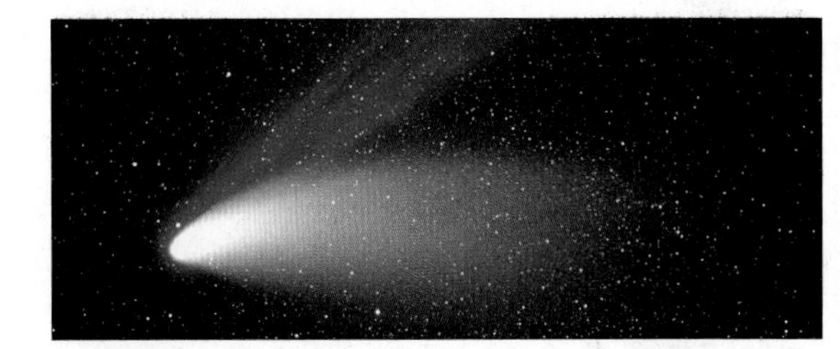

8. **Organize Information** Copy the table below and list the major characteristics of a dwarf planet.

Object	Defining Characteristic
Dwarf Planet	

Critical Thinking

9. **Compose** a paragraph describing what early sky observers might have thought when they saw a comet.
10. **Evaluate** Do you agree with the decision to reclassify Pluto as a dwarf planet? Defend your opinion.

Lab

40 minutes

Scaling down the Solar System

Materials

2.25 in–wide register tape (several rolls)

meterstick

masking tape

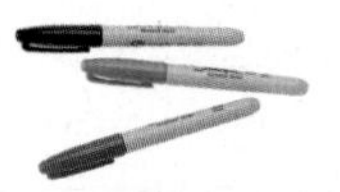

colored markers

Safety

A scale model is a physical representation of something that is much smaller or much larger. Reduced-size scale models are made of very large things, such as the solar system. The scale used must reduce the actual size to a size reasonable for the model.

Question

What scale can you use to represent the distances between solar system objects?

Procedure

1. First, decide how big your solar system will be. Use the data given in the table to figure out how far apart the Sun and Neptune would be if a scale of 1 meter = 1 AU is used. Would a solar system based on that scale fit in the space you have available?
2. With your group determine the scale that results in a model that fits the available space. Larger models are usually more accurate, so choose a scale that produces the largest model that fits in the available space.
3. Once you have decided on a scale, copy the table in your Science Journal. Replace the word *(Scale)* in the third column of the table with the unit you have chosen. Then fill in the scaled distance for each planet.

Planet	Distance from the Sun (AU)	Distance from the Sun (Scale)
Mercury	0.39	
Venus	0.72	
Earth	1.00	
Mars	1.52	
Jupiter	5.20	
Saturn	9.54	
Uranus	19.18	
Neptune	30.06	

4. On register tape, mark the positions of objects in the solar system based on your chosen scale. Use a length of register tape that is slightly longer than the scaled distance between the Sun and Neptune.
5. Tape the ends of the register tape to a table or the floor. Mark a dot at one end of the paper to represent the Sun. Measure along the tape from the center of the dot to the location of Mercury. Mark a dot at this position and label it *Mercury*. Repeat this process for the remaining planets.

Analyze and Conclude

6. **Critique** There are many objects in the solar system. These objects have different sizes, structures, and orbits. Examine your scale model of the solar system. How accurate is the model? How could the model be changed to be more accurate?
7. **The Big Idea** Pluto is a dwarf planet located beyond Neptune. Based on the pattern of distance data for the planets shown in the table, approximately how far from the Sun would you expect to find Pluto? Explain you reasoning.
8. **Calculate** What length of register tape is needed if a scale of 30 cm = 1 AU is used for the solar system model?

Lab Tips

- ☑ A scale is the ratio between the actual size of something and a representation of it.
- ☑ The distances between the planets and the Sun are average distances because planetary orbits are not perfect circles.

Communicate Your Results

Compare your model with other groups in your class by taping them all side-by-side. Discuss any major differences in your models. Discuss the difficulties in making the scale models much smaller.

Extension

How can you build a scale model of the solar system that accurately shows both planetary diameters and distances? Describe how you would go about figuring this out.

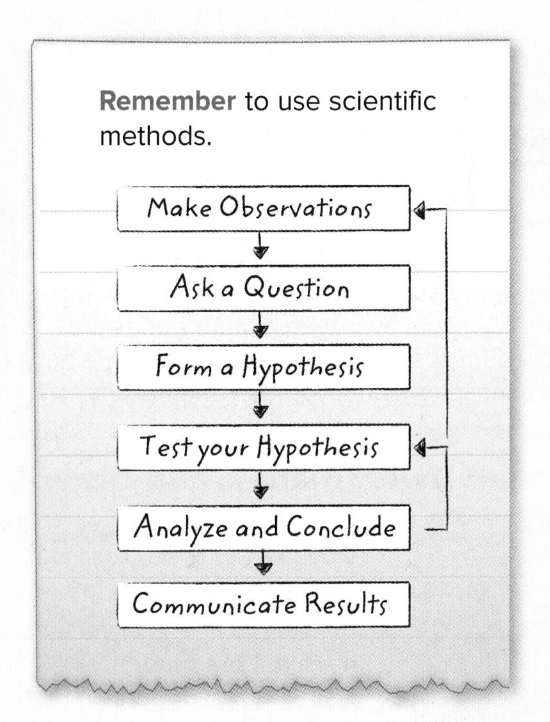

Chapter 21 Study Guide

The solar system contains planets, dwarf planets, comets, asteroids, and other small solar system bodies.

Key Concepts Summary

Lesson 1: The Structure of the Solar System

- The inner planets are made mainly of solid materials. The outer planets, which are larger than the inner planets, have thick gas and liquid layers covering a small solid core.
- Astronomers measure vast distances in space in **astronomical units;** an astronomical unit is about 150 million km.
- The speed of each planet changes as it moves along its elliptical orbit around the Sun.

Vocabulary

asteroid p. 763
comet p. 763
astronomical unit p. 764
period of revolution p. 764
period of rotation p. 764

Lesson 2: The Inner Planets

- The inner planets—Mercury, Venus, Earth, and Mars—are made of rock and metallic materials.
- The **greenhouse effect** makes Venus the hottest planet.
- Mercury has no atmosphere. The atmospheres of Venus and Mars are almost entirely carbon dioxide. Earth's atmosphere is a mixture of gases and a small amount of water vapor.

terrestrial planet p. 769
greenhouse effect p. 771

Lesson 3: The Outer Planets

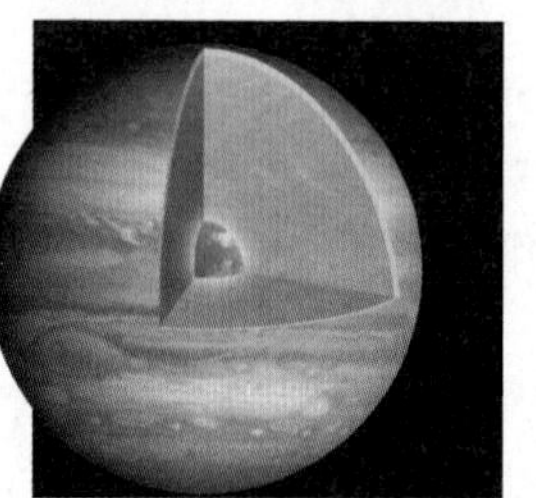

- The outer planets—Jupiter, Saturn, Uranus, and Neptune—are primarily made of hydrogen and helium.
- Jupiter and Saturn have thick cloud layers, but are mainly liquid hydrogen. Saturn's rings are largely particles of ice. Uranus and Neptune have thick atmospheres of hydrogen and helium.

Galilean moons p. 779

Lesson 4: Dwarf Planets and Other Objects

- A dwarf planet is an object that orbits a star, has enough mass to pull itself into a spherical shape, and has objects similar in mass orbiting near it.
- An asteroid is a small rocky object that orbits the Sun. Comets are made of rock, ice, and dust and orbit the Sun in highly elliptical paths.
- The impact of a **meteorite** forms an **impact crater.**

meteoroid p. 788
meteor p. 788
meteorite p. 788
impact crater p. 788

(t to b)©UVimages/amanaimages/Corbis; (2)NASA GSFC image by Robert Simmon and Reto Stöckli; (3)NASA/JPL/USGS; (4)Roger Ressmeyer/Photographer's Choice/Getty Images

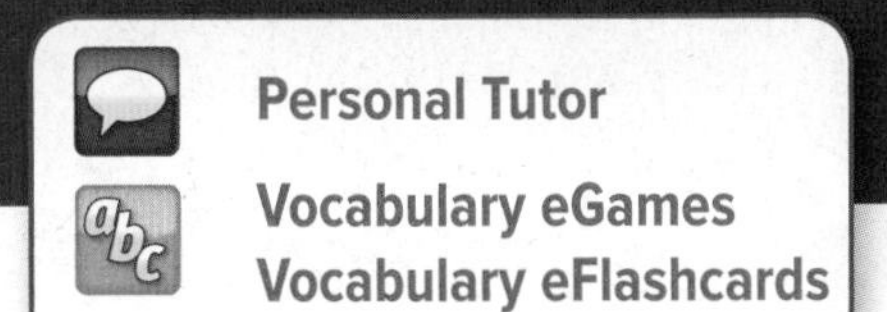

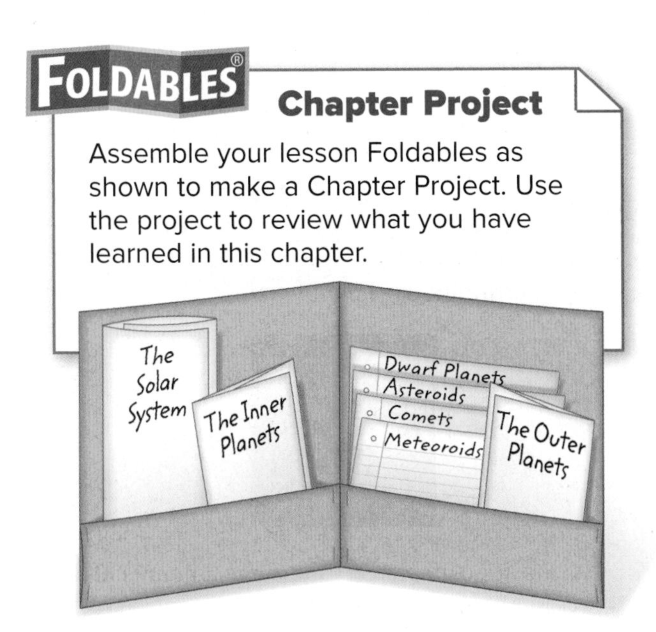

Assemble your lesson Foldables as shown to make a Chapter Project. Use the project to review what you have learned in this chapter.

Use Vocabulary

Match each phrase with the correct vocabulary term from the Study Guide.

1. the time it takes an object to complete one rotation on its axis
2. the average distance from Earth to the Sun
3. the time it takes an object to travel once around the Sun
4. an increase in temperature caused by energy trapped by a planet's atmosphere
5. an inner planet
6. the four largest moons of Jupiter
7. a streak of light in Earth's atmosphere made by a meteoroid

Link Vocabulary and Key Concepts

Copy this concept map, and then use vocabulary terms to complete the concept map.

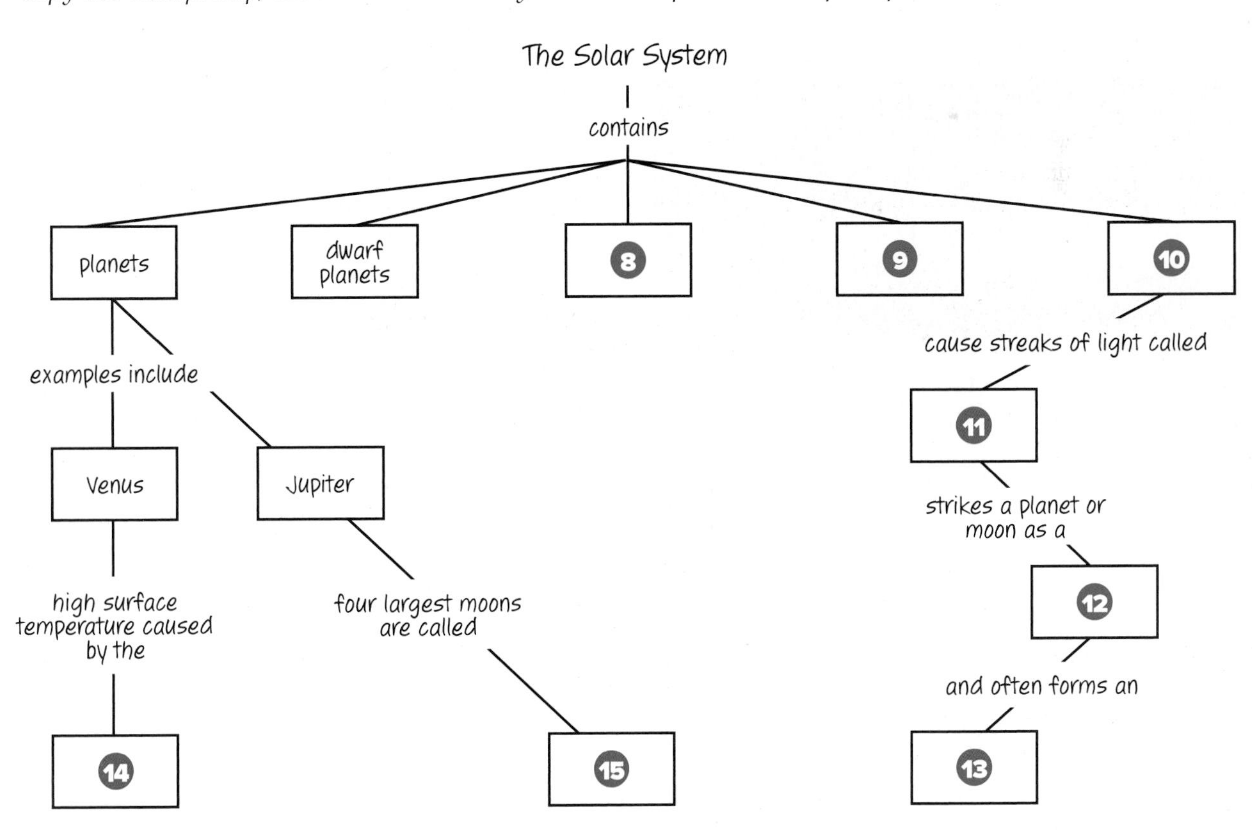

Chapter 21 Review

Understand Key Concepts 

1. Which solar system object is the largest?
   A. Jupiter
   B. Neptune
   C. the Sun
   D. Saturn

2. Which best describes the asteroid belt?
   A. another name for the Oort cloud
   B. the region where comets originate
   C. large chunks of gas, dust, and ice
   D. millions of small rocky objects

3. Which describes a planet's speed as it orbits the Sun?
   A. It constantly decreases.
   B. It constantly increases.
   C. It does not change.
   D. It increases then decreases.

4. The diagram below shows a planet's orbit around the Sun. What does the blue arrow represent?
   A. the gravitational pull of the Sun
   B. the planet's orbital path
   C. the planet's path if Sun did not exist
   D. the planet's speed

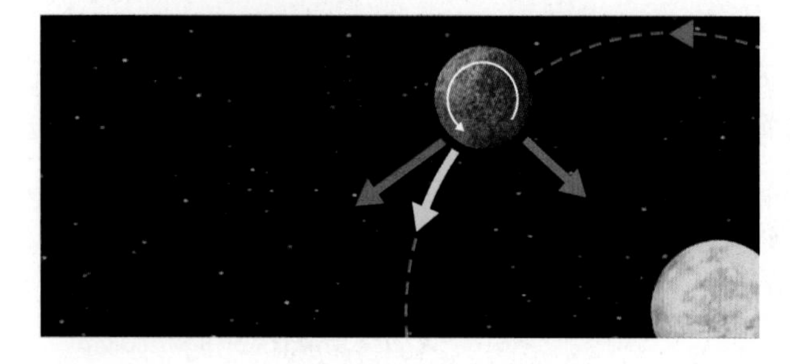

5. Which describes the greenhouse effect?
   A. effect of gravity on temperature
   B. energy emitted by the Sun
   C. energy trapped by atmosphere
   D. reflection of light from a planet

6. How are the terrestrial planets similar?
   A. similar densities
   B. similar diameters
   C. similar periods of rotation
   D. similar rocky surfaces

7. Which inner planet is the hottest?
   A. Earth
   B. Mars
   C. Mercury
   D. Venus

8. The photograph below shows how Earth appears from space. How does Earth differ from other inner planets?

   A. Its atmosphere contains large amounts of methane.
   B. Its period of revolution is much greater.
   C. Its surface is covered by large amounts of liquid water.
   D. Its surface temperature is higher.

9. Which two gases make up most of the outer planets?
   A. ammonia and helium
   B. ammonia and hydrogen
   C. hydrogen and helium
   D. methane and hydrogen

10. Which is true of the dwarf planets?
    A. more massive than nearby objects
    B. never have moons
    C. orbit near the Sun
    D. spherically shaped

11. Which is a bright streak of light in Earth's atmosphere?
    A. a comet
    B. a meteor
    C. a meteorite
    D. a meteoroid

12. Which best describes an asteroid?
    A. icy
    B. rocky
    C. round
    D. wet

Critical Thinking

13. **Relate** changes in speed during a planet's orbit to the shape of the orbit and the gravitational pull of the Sun.
14. **Compare** In what ways are planets and dwarf planets similar?
15. **Apply** Like Venus, Earth's atmosphere contains carbon dioxide. What might happen on Earth if the amount of carbon dioxide in the atmosphere increases? Explain.
16. **Defend** A classmate states that life will someday be found on Mars. Defend the statement and offer a reason why life might exist on Mars.
17. **Infer** whether a planet with active volcanoes would have more or fewer craters than a planet without active volcanoes. Explain.
18. **Support** Use the diagram of the asteroid belt to support the explanation of how the belt formed.

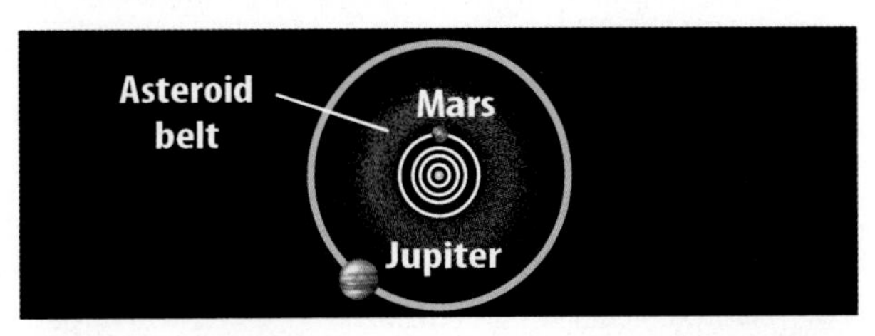

19. **Evaluate** The *Huygens* probe transmitted data about Titan for only 90 min. In your opinion, was this worth the effort of sending the probe?
20. **Explain** why Jupiter's moon Ganymede is not considered a dwarf planet, even though it is bigger than Mercury.

Writing in Science

21. **Compose** a pamphlet that describes how the International Astronomical Union classifies planets, dwarf planets, and small solar system objects.

NASA Jet Propulsion Laboratory (NASA-JPL)

REVIEW THE BIG IDEA

22. What kinds of objects are in the solar system? Summarize the types of space objects that make up the solar system and give at least one example of each.
23. The photo below shows part of Saturn's rings and two of its moons. Describe what Saturn and its rings are made of and explain why the other two objects are moons.

Math Skills

Math Practice

Use Ratios

Inner Planet Data			
Planet	Diameter (% of Earth's diameter)	Mass (% of Earth's mass)	Average Distance from Sun (AU)
Mercury	38.3	5.5	0.39
Venus	95	81.5	0.72
Earth	100	100	1.00
Mars	53.2	10.7	1.52

24. Use the table above to calculate how many times farther from the Sun Mars is compared to Mercury.
25. Calculate how much greater Venus's mass is compared to Mercury's mass.

Standardized Test Practice

Record your answers on the answer sheet provided by your teacher or on a sheet of paper.

Multiple Choice

1 Which is a terrestrial planet?

A Ceres

B Neptune

C Pluto

D Venus

2 An astronomical unit (AU) is the average distance

A between Earth and the Moon.

B from Earth to the Sun.

C to the nearest star in the galaxy.

D to the edge of the solar system.

3 Which is NOT a characteristic of ALL planets?

A exceed the total mass of nearby objects

B have a nearly spherical shape

C have one or more moons

D make an elliptical orbit around the Sun

Use the diagram below to answer question 4.

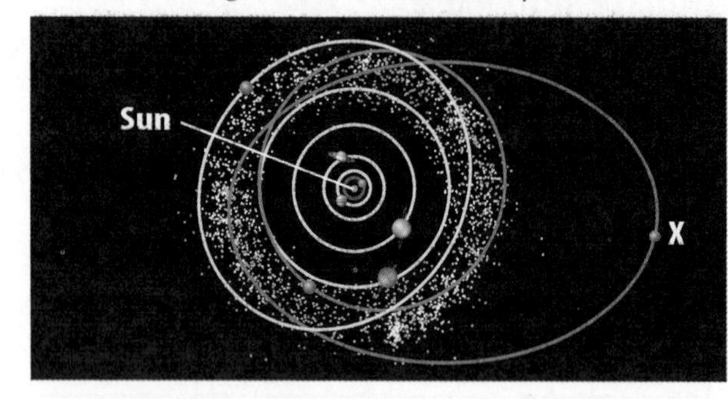

4 Which object in the solar system is marked by an *X* in the diagram?

A asteroid

B meteoroid

C dwarf planet

D outer planet

Use the diagram of Saturn below to answer questions 5 and 6.

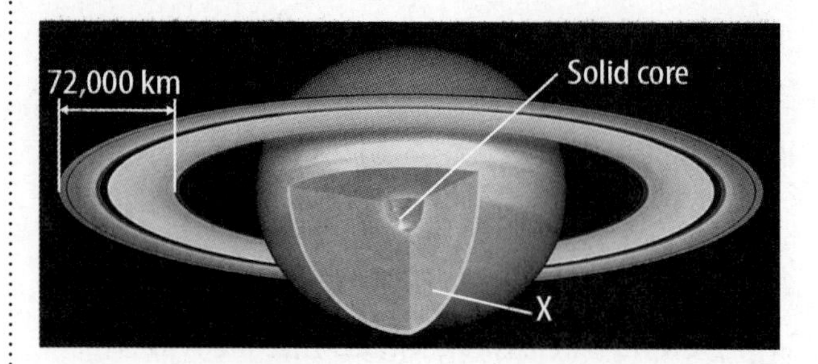

5 The thick inner layer marked *X* in the diagram above is made of which material?

A carbon dioxide

B gaseous helium

C liquid hydrogen

D molten rock

6 In the diagram, Saturn's rings are shown to be 72,000 km in width. Approximately how thick are Saturn's rings?

A 30 m

B 1,000 km

C 14,000 km

D 1 AU

7 Which are NOT found on Mercury's surface?

A high cliffs

B impact craters

C lava flows

D sand dunes

8 What is the primary cause of the extremely high temperatures on the surface of Venus?

A heat rising from the mantle

B lack of an atmosphere

C proximity to the Sun

D the greenhouse effect

Use the diagram below to answer question 9.

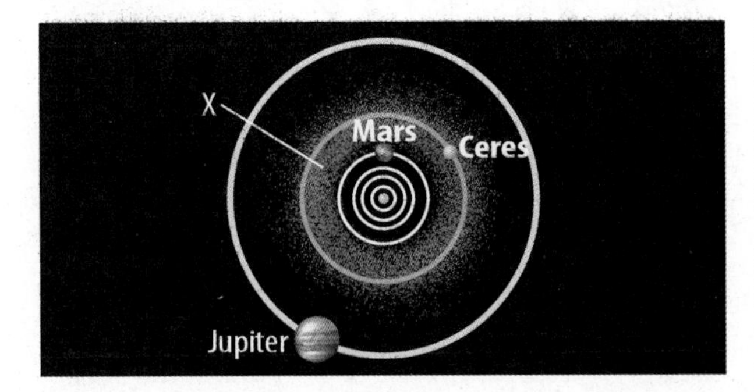

9 In the diagram above, which region of the solar system is marked by an *X*?

A the asteroid belt

B the dwarf planets

C the Kuiper belt

D the Oort cloud

10 What is a meteorite?

A a surface depression formed by collision with a rock from space

B a fragment of rock that strikes a planet or a moon

C a mixture of ice, dust, and gas with a glowing tail

D a small rocky particle that moves through space

11 What gives Mars its reddish color?

A ice caps of frozen carbon dioxide

B lava from Olympus Mons

C liquid water in gullies

D soil rich in iron oxide

Constructed Response

Use the table below to answer questions 12 and 13.

	Inner Planets	Outer Planets
Also called		
Relative size		
Main materials		
General structure		
Number of moons		

12 Copy the table and complete the first five rows to compare the features of the inner planets and outer planets.

13 In the blank row of the table, add another feature of the inner planets and outer planets. Then, describe the feature you have chosen.

14 What features of Earth make it suitable for supporting life as we know it?

15 How are planets, dwarf planets, and asteroids both similar and different?

NEED EXTRA HELP?															
If You Missed Question...	1	2	3	4	5	6	7	8	9	10	11	12	13	14	15
Go to Lesson...	2	1	1	1	3	3	2	2	1, 4	4	2	2, 3	2, 3	2	1, 4

Chapter 22

Stars and Galaxies

THE BIG IDEA

What makes up the universe, and how does gravity affect the universe?

Inquiry What can't you see?

This photograph shows a small part of the universe. You can see many stars and galaxies in this image. But the universe also contains many things you cannot see.

- How do scientists study the universe?
- What makes up the universe?
- How does gravity affect the universe?

NASA, ESA, and S. Beckwith (STScI)and the HUDF Team

Get Ready to Read

What do you think?

Before you read, decide if you agree or disagree with each of these statements. As you read this chapter, see if you change your mind about any of the statements.

1. The night sky is divided into constellations.
2. A light-year is a measurement of time.
3. Stars shine because there are nuclear reactions in their cores.
4. Sunspots appear dark because they are cooler than nearby areas.
5. The more matter a star contains, the longer it is able to shine.
6. Gravity plays an important role in the formation of stars.
7. Most of the mass in the universe is in stars.
8. The Big Bang theory is an explanation of the beginning of the universe.

Your one-stop online resource
connectED.mcgraw-hill.com

LearnSmart®

Chapter Resources Files, Reading Essentials, Get Ready to Read, Quick Vocabulary

Animations, Videos, Interactive Tables

Self-checks, Quizzes, Tests

Project-Based Learning Activities

Lab Manuals, Safety Videos, Virtual Labs & Other Tools

Vocabulary, Multilingual eGlossary, Vocab eGames, Vocab eFlashcards

Personal Tutors

Lesson 1

The View from Earth

Reading Guide

Key Concepts 
ESSENTIAL QUESTIONS

- How do astronomers divide the night sky?
- What can astronomers learn about stars from their light?
- How do scientists measure the distance and the brightness of objects in the sky?

Vocabulary

spectroscope p. 803
astronomical unit p. 804
light-year p. 804
apparent magnitude p. 805
luminosity p. 805

Multilingual eGlossary

Inquiry Where is this?

Unless you have visited a remote part of the country, you have probably never seen the sky look like this. It is similar to what the night sky looked like to your ancestors—before towns and cities brightened the sky.

Stephen & Donna O'Meara/Photo Researchers, Inc.

Launch Lab

20 minutes

How can you "see" invisible energy?

You see because of the Sun's light. You feel the heat of the Sun's energy. The Sun produces other kinds of energy that you can't directly see or feel.

1. Read and complete a lab safety form.
2. Put 5–6 **beads** into a **clear container.** Observe the color of the beads.
3. In a darkened room, shine light from a **flashlight** onto the beads for several seconds. Record your observations in your Science Journal. Repeat this step, exposing the beads to light from an **incandescent lightbulb** and a **fluorescent light.** Record your observations.
4. Stand outside in a shady spot for several seconds. Then expose the beads to direct sunlight. Record your observations.

Think About This

1. How did the light from the different light sources affect the color of the beads?
2. What do you think made the beads change color?
3. **Key Concept** How do you think invisible forms of light help scientists understand stars and other objects in the sky?

Looking at the Night Sky

Have you ever looked up at the sky on a clear, dark night and seen countless stars? If you have, you are lucky. Few people see a sky like that shown on the previous page. Lights from towns and cities make the night sky too bright for faint stars to be seen.

If you look at a clear night sky for a long time, the stars seem to move. But what you are really seeing is Earth's movement. Earth spins, or rotates, once every 24 hours. Day turns to night and then back to day as Earth rotates. Because Earth rotates from west to east, objects in the sky rise in the east and set in the west.

Earth spins on its axis, an imaginary line from the North Pole to the South Pole. The star Polaris is almost directly above the North Pole. As Earth spins, stars near Polaris appear to travel in a circle around Polaris, as shown in **Figure 1**. These stars never set when viewed from the northern hemisphere. They are always present in the night sky.

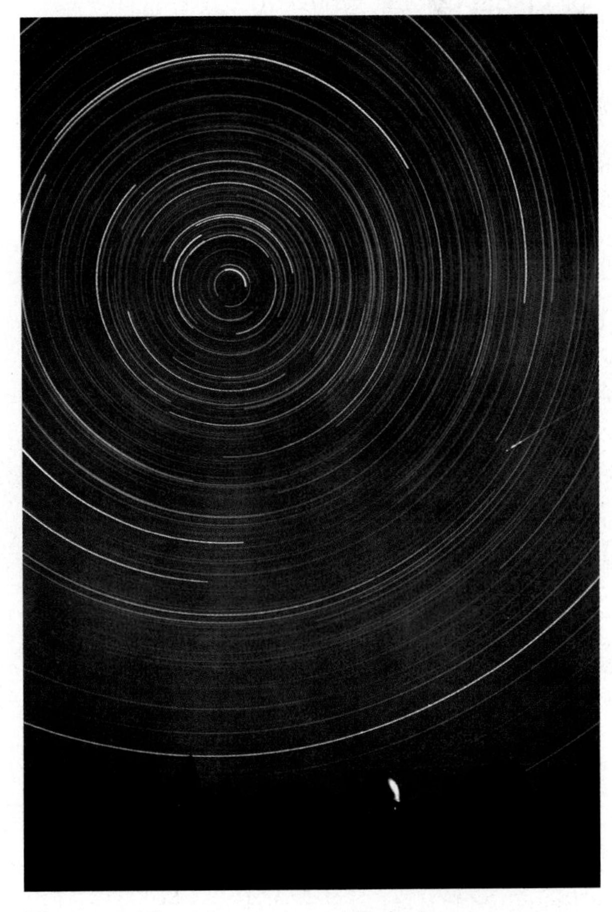

Figure 1 The stars around Polaris appear as streaks of light in this time-lapse photograph.

FOLDABLES®

Make a horizontal two-tab book. Label it as shown. Use it to organize your notes on astronomy.

How Scientists Divide the Night Sky | What Observations of the Stars Tell Scientists

Naked-Eye Astronomy

You don't need expensive equipment to view the sky. *Naked-eye astronomy* means gazing at the sky with just your eyes, without binoculars or a telescope. Long before the telescope was invented, people observed stars to tell time and find directions. They learned about planets, seasons, and astronomical events merely by watching the sky. As you practice naked-eye astronomy, remember never to look directly at the Sun–it could damage your eyes.

Constellations

As people in ancient cultures gazed at the night sky, they saw patterns. The patterns resembled people, animals, or objects, such as the hunter and the dragon shown in **Figure 2.** The Greek astronomer Ptolemy (TAH luh mee) identified dozens of star patterns nearly 2,000 years ago. Today, these patterns and others like them are known as ancient constellations.

Present-day astronomers use many ancient constellations to divide the sky into 88 regions. Some of these regions, which are also called constellations, are shown in the sky map in **Figure 2.** Dividing the sky helps scientists communicate to others what area of sky they are studying.

 **Key Concept Check** How do astronomers divide the night sky?

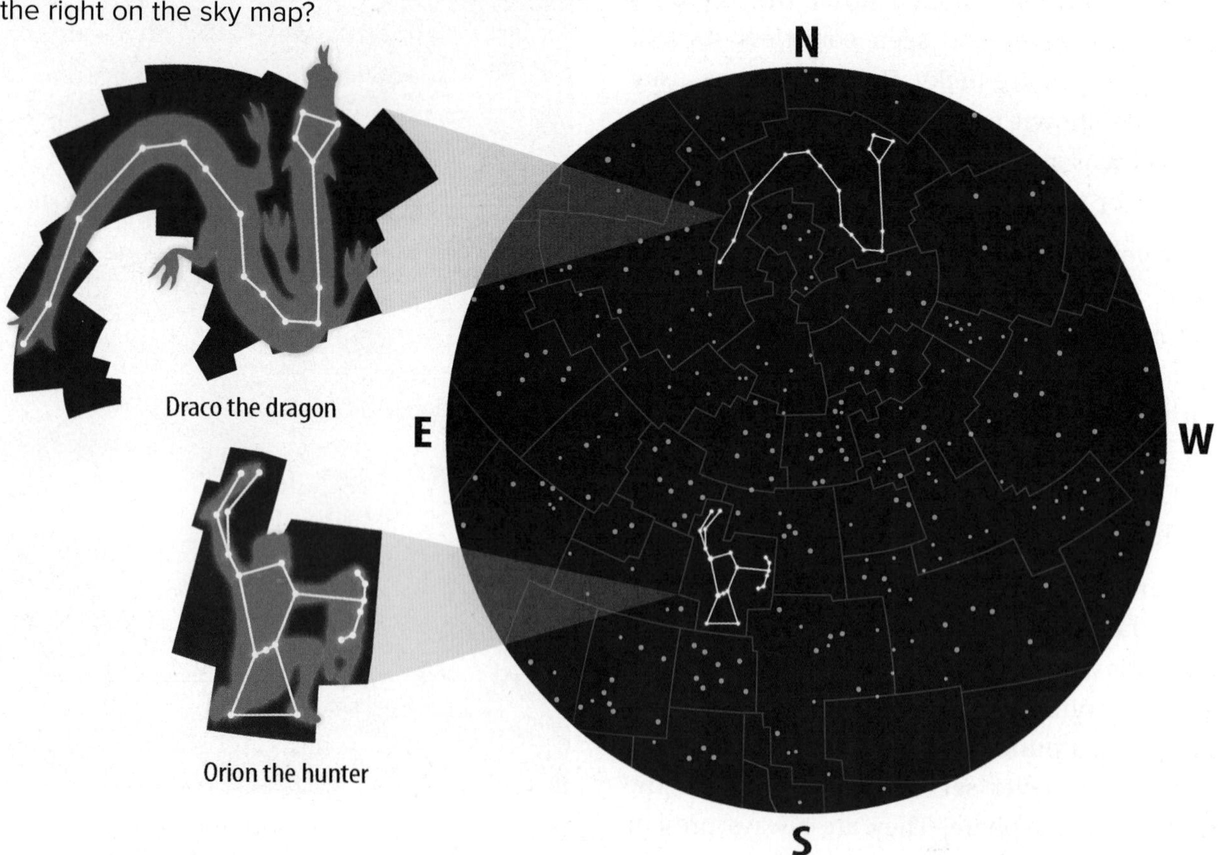

Figure 2 Most modern constellations contain an ancient constellation.

Visual Check Why does east appear on the left and west appear on the right on the sky map?

Electromagnetic Spectrum

Personal Tutor

Figure 3 Different parts of the electromagnetic spectrum have different wavelengths and different energies. You can see only a small part of the energy in these wavelengths.

Visual Check Which wavelength has the highest energy?

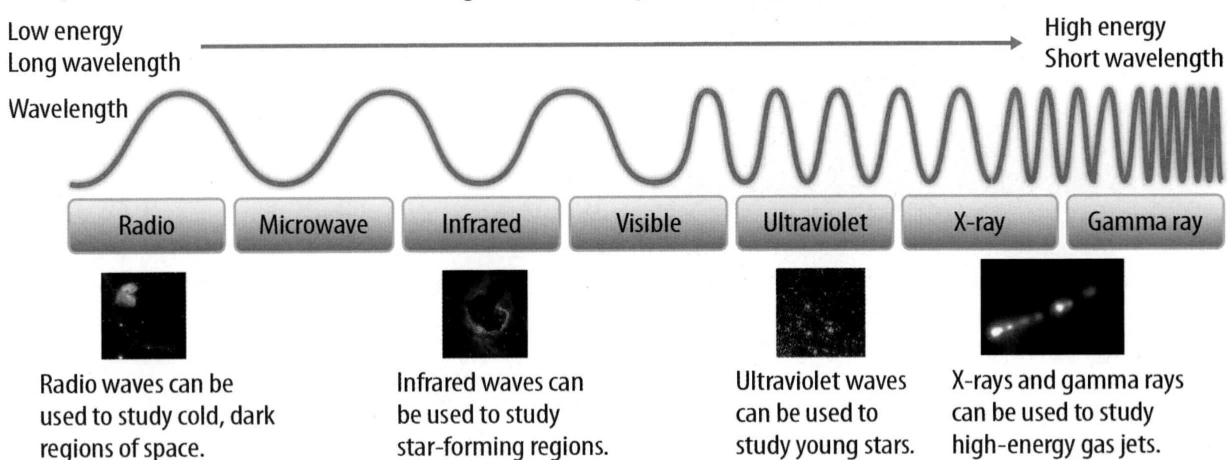

Telescopes

Telescopes can collect much more light than the human eye can detect. Visible light is just one part of the electromagnetic spectrum. As shown in **Figure 3**, the electromagnetic spectrum is a continuous range of wavelengths. Longer wavelengths have low energy. Shorter wavelengths have high energy. Different objects in space emit different ranges of wavelengths. The range of wavelengths that a star emits is the star's spectrum (plural, spectra).

Spectroscopes

Scientists study the spectra of stars using an instrument called a spectroscope. *A* **spectroscope** *spreads light into different wavelengths.* Using spectroscopes, astronomers can study stars' characteristics, including temperatures, compositions, and energies. For example, newly formed stars emit mostly radio and infrared waves, which have low energy. Exploding stars emit mostly high-energy ultraviolet waves and X-rays.

Key Concept Check What can astron-omers learn from a star's spectrum?

20 minutes

How does light differ?

Light from the Sun is different from light from a lightbulb. How do the light sources differ?

1. Read and complete a lab safety form.
2. Follow instructions included with your **spectroscope.** Use it to observe various **light sources** around the classroom. Then use it to look at a bright part of the sky. ⚠ *Do not look directly at the Sun.*
3. Use **colored pencils** to draw what you see for each light source in your Science Journal.

Analyze and Conclude

1. **Compare and Contrast** What colors did you see for each light source? How did the colors differ?
2. **Key Concept** How might a spectroscope be used to learn about stars?

(l)Robert Gendler/Stocktrek Images/Getty Images; (cl)NASA/JPL-Caltech/E. Churchwell (University of Wisconsin); (cr)NASA/JPL-Caltech/Univ. of Virginia; (r)NASA/CXC/MIT/H. Marshall et al.; (b)Hutchings Photography/Digital Light Source

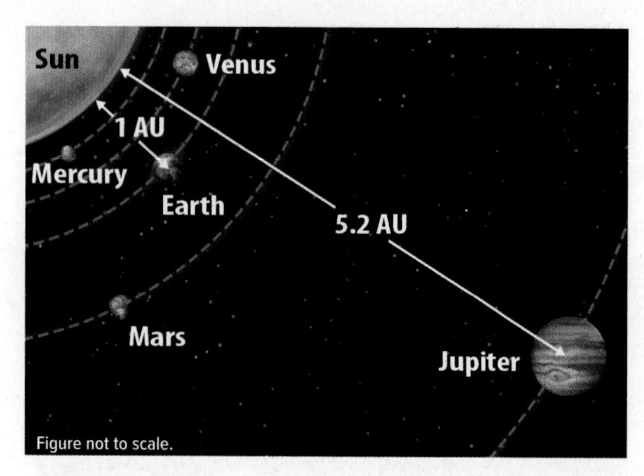

Figure 4 Measurements in the solar system are based on the average distance between Earth and the Sun—1 astronomical unit (AU). The most distant planet, Neptune, is 30 AU from the Sun.

WORD ORIGIN

parallax
from Greek *parallaxis,* means "alteration"

Math Skills 

Use Proportions

Proportions can be used to calculate distances to astronomical objects. Light can travel nearly 10 trillion km in 1 year (y). How many years would it take light to reach Earth from a star that is 100 trillion km away?

1. Set up a proportion.
$$\frac{10 \text{ trillion km}}{1 \text{ y}} = \frac{100 \text{ trillion km}}{x}$$
2. Cross multiply.
$$10 \text{ trillion km}(x) = 100 \text{ trillion km}(1 \text{ y})$$
3. Solve for x by dividing both sides by 10 trillion km.
$$\frac{10 \text{ trillion km}(x)}{10 \text{ trillion km}} = \frac{100 \text{ trillion km}(1 \text{ y})}{10 \text{ trillion km}}$$

Practice

How many years would it take light to reach Earth from a star 60 trillion km away?

 Math Practice

 Personal Tutor

Measuring Distances

Hold up your thumb at arm's length. Close one eye, and look at your thumb. Now open that eye, and close the other eye. Did your thumb seem to jump? This is an example of parallax. Parallax is the apparent change in an object's position caused by looking at it from two different points.

Astronomers use angles created by parallax to measure how far objects are from Earth. Instead of the eyes being the two points of view, they use two points in Earth's orbit around the Sun.

 **Reading Check** What is parallax?

Distances Within the Solar System

Because the universe is too large to be measured easily in meters or kilometers, astronomers use other units of measurement. For distances within the solar system, they use astronomical units (AU). *An* **astronomical unit** *is the average distance between Earth and the Sun, about 150 million km.* Astronomical units are convenient to use in the solar system because distances easily can be compared to the distance between Earth and the Sun, as shown in **Figure 4.**

Distances Beyond the Solar System

Astronomers measure distances to objects beyond the solar system using a larger distance unit–the light-year. Despite its name, a light-year measures distance, not time. *A* **light-year** *is the distance light travels in 1 year.* Light travels at a rate of about 300,000 km/s. That means 1 light-year is about 10 trillion km! Proxima Centauri, the nearest star to the Sun, is 4.2 light-years away.

Looking Back in Time

Because it takes time for light to travel, you see a star not as it is today, but as it was when light left it. At 4.2 light-years away, Proxima Centauri appears as it was 4.2 years ago. The farther away an object, the longer it takes for its light to reach Earth.

Measuring Brightness

When you look at stars, you can see that some are dim and some are bright. Astronomers measure the brightness of stars in two ways: by how bright they appear from Earth and by how bright they actually are.

Apparent Magnitude

Scientists measure how bright stars appear from Earth using a scale developed by the ancient Greek astronomer Hipparchus (hi PAR kus). Hipparchus assigned a number to every star he saw in the night sky, based on the star's brightness. Astronomers today call these numbers magnitudes. *The* **apparent magnitude** *of an object is a measure of how bright it appears from Earth.*

As shown in **Figure 5**, some objects have negative apparent magnitudes. That is because Hipparchus assigned a value of 1 to all of the brightest stars. He also did not assign values to the Sun, the Moon, or Venus. Astronomers later assigned negative numbers to the Sun, the Moon, Venus, and a few bright stars.

ACADEMIC VOCABULARY
apparent
(adjective) appearing to the eye or mind

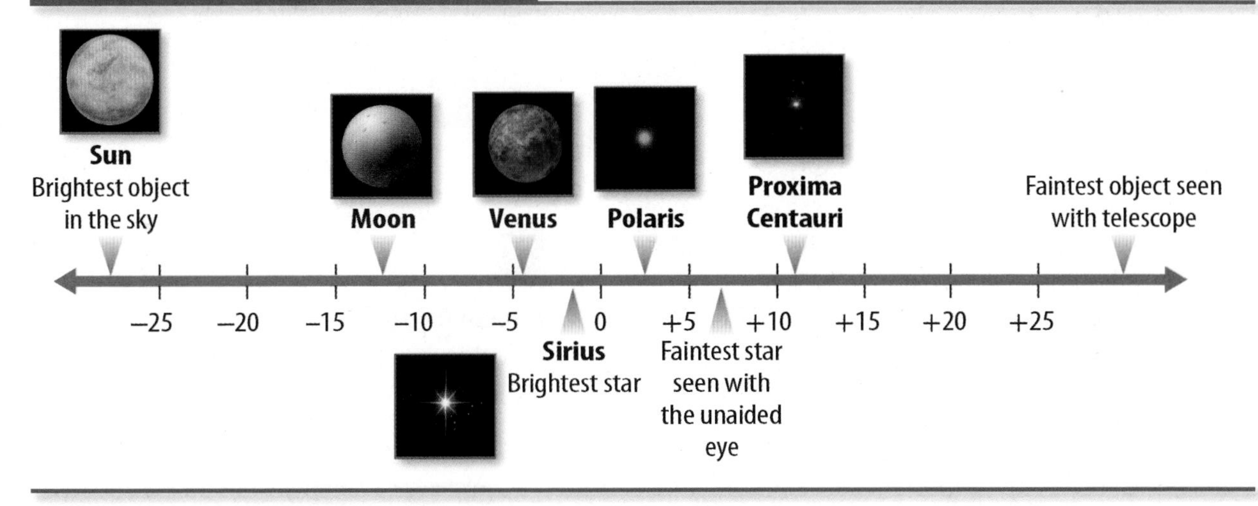

Figure 5 The fainter a star or other object in the sky appears, the greater its apparent magnitude.

Visual Check What is the apparent magnitude of Sirius?

Absolute Magnitude

Stars can appear bright or dim depending on their distances from Earth. But stars also have actual, or absolute, magnitudes. **Luminosity** (lew muh NAH sih tee) *is the true brightness of an object.* The luminosity of a star, measured on an absolute magnitude scale, depends on the star's temperature and size, not its distance from Earth. A star's luminosity, apparent magnitude, and distance are related. If scientists know two of these factors, they can determine the third using mathematical formulas.

Key Concept Check How do scientists measure the brightness of stars?

Lesson 1 Review

Visual Summary

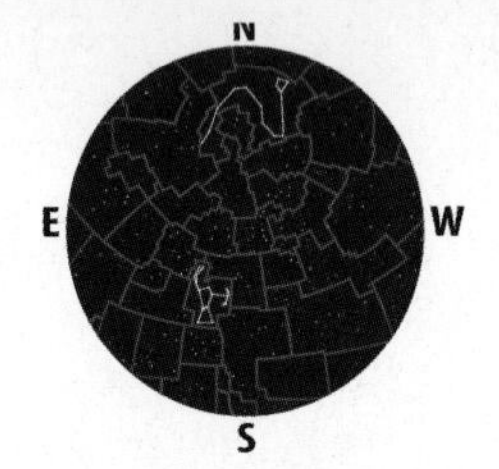

Ancient people recognized patterns in the night sky. These patterns are known as the ancient constellations.

Different wavelengths of the electromagnetic spectrum carry different energies.

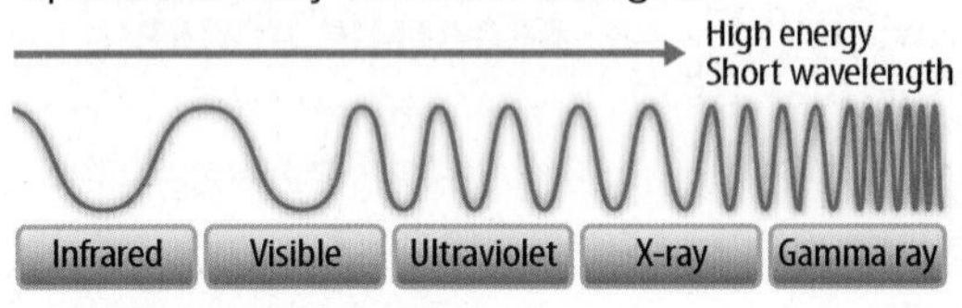

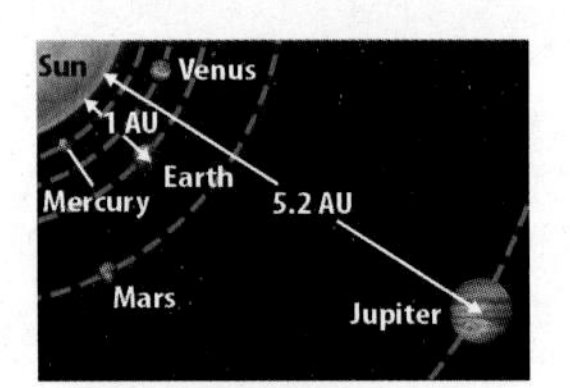

Astronomers measure distances within the solar system using astronomical units.

Use your lesson Foldable to review the lesson. Save your Foldable for the project at the end of the chapter.

What do you think NOW?

You first read the statements below at the beginning of the chapter.

1. The night sky is divided into constellations.

2. A light-year is a measurement of time.

Did you change your mind about whether you agree or disagree with the statements? Rewrite any false statements to make them true.

Use Vocabulary

1 A device that spreads light into different wavelengths is a(n) ________.

2 **Define** *astronomical unit* and *light-year* in your own words.

3 **Distinguish** between apparent magnitude and luminosity.

Understand Key Concepts 

4 Which does a light-year measure?

A. brightness **C.** time
B. distance **D.** wavelength

5 **Describe** how scientists divide the sky.

Interpret Graphics

6 **Analyze** Which star in the diagram below appears the brightest from Earth?

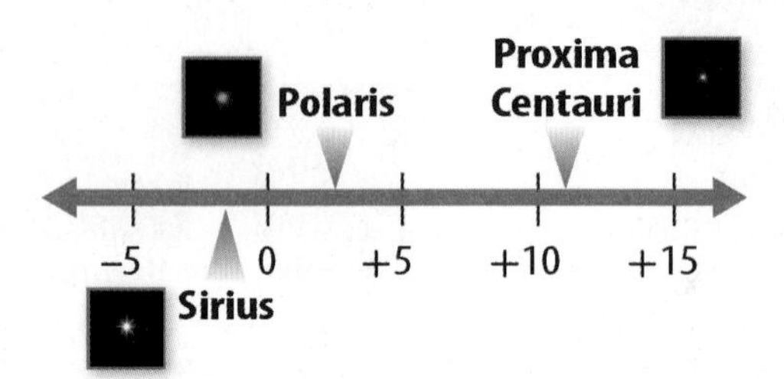

7 **Organize Information** Copy and fill in the graphic organizer below to list three things astronomers can learn from a star's light.

Critical Thinking

8 **Evaluate** why astronomers use modern constellation regions instead of ancient constellation patterns to divide the sky.

Math Skills

9 The Andromeda galaxy is about 25,000,000,000,000,000,000 km from Earth. How long does it take light to reach Earth from the Andromeda galaxy?

Skill Practice Interpret Scientific Illustrations 30 minutes

How can you use scientific illustrations to locate constellations?

You might have heard that stars in the Big Dipper point to Polaris. The Big Dipper is a small star pattern in the larger constellation of Ursa Major. *Ursa Major* means "big bear" in Latin. It is the third-largest of the 88 modern constellations in the sky. Study the image of Ursa Major. Can you find the seven stars that form the Big Dipper? You can use a star finder to locate stars on any clear night of the year. The star finder also helps you see how constellations move across the sky.

Materials

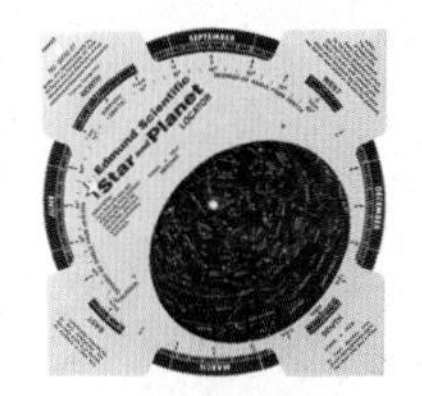

star chart

adhesive stars

graph paper

Learn It

Scientific illustrations can help you understand difficult or complicated subjects. **Interpret scientific illustrations** on the star finder to learn about the night sky.

Try It

1. Read and complete a lab safety form.
2. Read the user information provided with the star finder.
3. Rotate the wheel until the star finder is set to the day and time when you will be viewing the night sky. Observe how the ancient constellations marked on the star finder move.

4. Make a list of the bright stars, constellations, and planets you might be able to see in the sky.
5. Use the star finder outdoors on a clear night. As you hold the star finder overhead, be sure the arrows are pointing in the correct directions.

Apply It

6. What ancient constellations, planets, and stars were you able to see?
7. Did you locate Polaris? Why will you be able to see Polaris 6 months from now?
8. Which constellations won't you be able to see 6 months from now?
9. Why do stars appear to move?
10. How might ancient constellations have helped people in the past?
11. **Key Concept** How does dividing the sky into constellations help scientists study the sky?

(t to b)Ken Karp/McGraw-Hill Education; (2)©Brand X Pictures/PunchStock; (3)Aaron Haupt; (4) Hutchings Photography/Digital Light Source

Lesson 2

The Sun and Other Stars

Reading Guide

Key Concepts
ESSENTIAL QUESTIONS

- How do stars shine?
- How are stars layered?
- How does the Sun change over short periods of time?
- How do scientists classify stars?

Vocabulary

nuclear fusion p. 809
star p. 809
radiative zone p. 810
convection zone p. 810
photosphere p. 810
chromosphere p. 810
corona p. 810
Hertzsprung-Russell diagram p. 813

Multilingual eGlossary

Inquiry Volcanoes on the Sun?

No, it's the Sun's atmosphere! The Sun's atmosphere can extend millions of kilometers into space. Sometimes the atmosphere becomes so active it disrupts communication systems and power grids on Earth.

Science Source

Launch Lab

15 minutes

What are those spots on the Sun?

If you could see the Sun up close, what would it look like? Does it look the same all the time?

1. Examine a **collage of Sun images.** Notice the dates on which the pictures were taken.
2. Discuss with a partner what the dark spots might be and why they change position.
3. Select one spot. Estimate how long it took the spot to move completely across the surface of the Sun. Record your estimate in your Science Journal.

Think About This

1. What do you think the spots are?
2. Why do you think the spots move across the surface of the Sun?
3. **Key Concept** How do you think the Sun changes over days, months, and years?

How Stars Shine

The hotter something is, the more quickly its atoms move. As atoms move, they collide. If a gas is hot enough and its atoms move quickly enough, the nuclei of some of the atoms combine. **Nuclear fusion** *is a process that occurs when the nuclei of several atoms combine into one larger nucleus.*

Nuclear fusion releases a great amount of energy. This energy powers stars. *A* **star** *is a large ball of gas held together by gravity with a core so hot that nuclear fusion occurs.* A star's core can reach millions or hundreds of millions of degrees Celsius. When energy leaves a star's core, it travels throughout the star and radiates into space. As a result, the star shines.

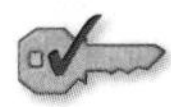

Key Concept Check How do stars shine?

FOLDABLES®

Make a vertical four-tab book. Label it as shown. Use it to organize your notes about the changing features of the Sun.

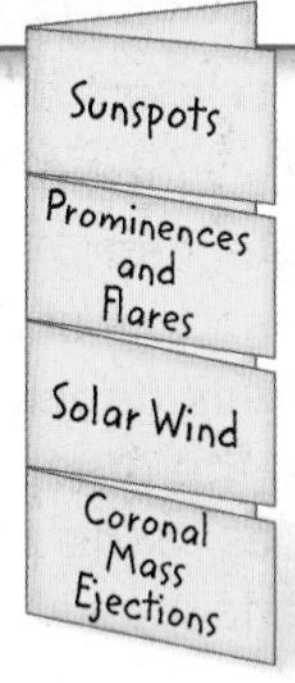

Composition and Structure of Stars

The Sun is the closest star to Earth. Because it is so close, scientists can easily observe it. They can send probes to the Sun, and they can study its spectrum using spectroscopes on Earth-based telescopes. Spectra of the Sun and other stars provide information about **stellar** composition. The Sun and most stars are made almost entirely of hydrogen and helium gas. A star's composition changes slowly over time as hydrogen in its core fuses into more complex nuclei.

SCIENCE USE V. COMMON USE

stellar
Science Use anything related to stars
Common Use outstanding, exemplary

Digital Vision/PunchStock

Layers of the Sun

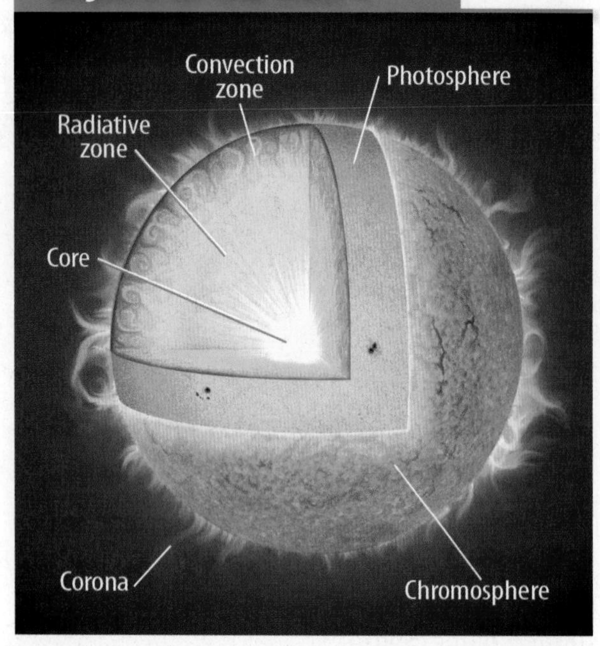

Figure 6 The Sun is divided into six layers.

Visual Check Where is the photosphere located in relation to the Sun's other layers?

MiniLab

Can you model the Sun's structure?

Making a two-dimensional model of the Sun can help you visualize its parts.

1. Read and complete a lab safety form.
2. Use **scissors** to cut out each **Sun part.**
3. Use a **glue stick** to attach the corona to a sheet of **black paper.** Glue the other pieces to the corona in this order: chromosphere, convection zone, radiative zone, core.

4. Glue only the top edge of the photosphere over the convection zone.

Analyze and Conclude

1. Draw the path a particle of light would follow from the core to the photosphere.
2. **Key Concept** How does this activity model a star's ability to shine?

Interior of Stars

When first formed, all stars fuse hydrogen into helium in their cores. Helium is denser than hydrogen, so it sinks to the inner part of the core after it forms.

The core is one of three interior layers of a typical star, as shown in the drawing of the Sun in **Figure 6.** *The* **radiative zone** *is a shell of cooler hydrogen above a star's core.* Hydrogen in this layer is dense. Light energy bounces from atom to atom as it gradually makes its way upward, out of the radiative zone.

Above the radiative zone is the **convection zone,** *where hot gas moves up toward the surface and cooler gas moves deeper into the interior.* Light energy moves quickly upward in the convection zone.

Key Concept Check What are the interior layers of a star?

Atmosphere of Stars

Beyond the convection zone are the three outer layers of a star. These layers make up a star's atmosphere. *The* **photosphere** *is the apparent surface of a star.* In the Sun, it is the dense, bright part you can see, where light energy radiates into space. From Earth, the Sun's photosphere looks smooth. But like the rest of the Sun, it is made of gas.

Above the photosphere are the two outer layers of a star's atmosphere. *The* **chromosphere** *is the orange-red layer above the photosphere,* as shown in **Figure 6.** *The* **corona** *is the wide, outermost layer of a star's atmosphere.* The temperature of the corona is higher than the photosphere or the chromosphere. It has an irregular shape and can extend outward for several million kilometers.

The Sun's Changing Features

The interior features of the Sun are stable over millions of years. But the Sun's atmosphere can change over years, months, or even minutes. Some of these features are illustrated in **Table 1** on the following page.

Table 1 The Sun is dynamic. It changes over years, months, hours, and minutes.

Key Concept Check Which parts of the Sun change over short periods of time?

Table 1 Changing Features of the Sun	
Sunspots Regions of strong magnetic activity are called sunspots. Cooler than the rest of the photosphere, sunspots appear as dark splotches on the Sun. They seem to move across the Sun as the Sun rotates. The number of sunspots changes over time. They follow a cycle, peaking in number every 11 years. An average sunspot is about the size of Earth.	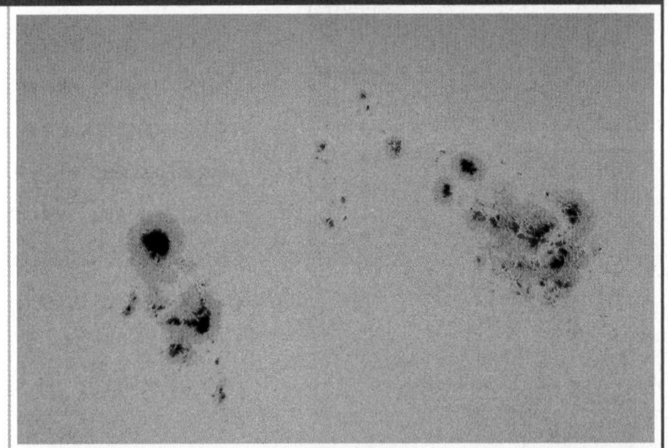
Prominences and Flares The loop shown here is a prominence. Prominences are clouds of gas that make loops and jets extending into the corona. They sometimes last for weeks. Flares are sudden increases in brightness often found near sunspots or prominences. They are violent eruptions that last from minutes to hours. Both prominences and flares begin at or just above the photosphere.	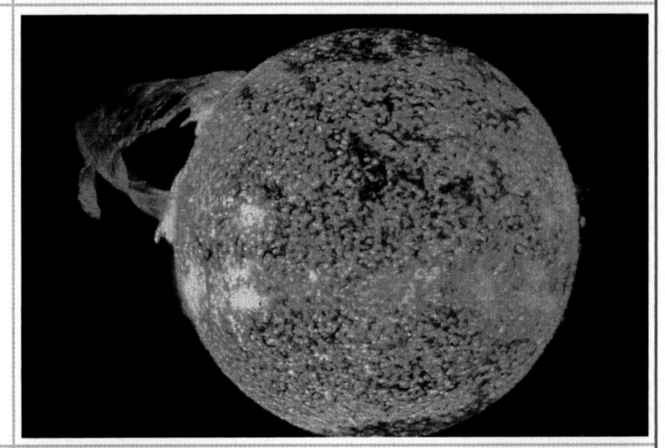
Coronal Mass Ejections (CMEs) Huge bubbles of gas ejected from the corona are coronal mass ejections (CMEs). They are much larger than flares and occur over the course of several hours. Material from a CME can reach Earth, occasionally causing a radio blackout or a malfunction in an orbiting satellite.	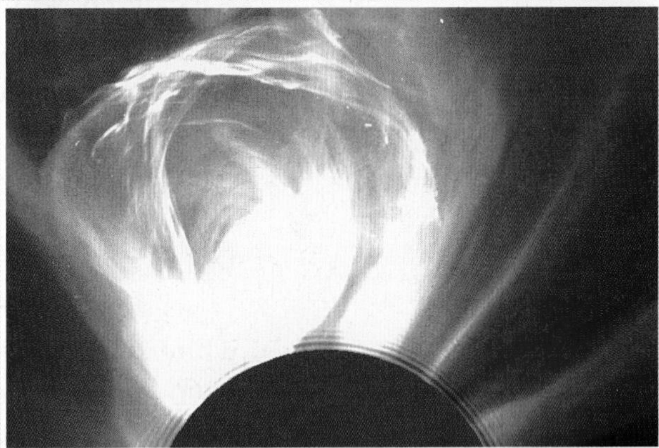
The Solar Wind Charged particles that stream continually away from the Sun create the solar wind. The solar wind passes Earth and extends to the edge of the solar system. Auroras are curtains of light created when particles from the solar wind or a CME interact with Earth's magnetic field. Auroras occur in both the northern and southern hemispheres. Auroras in the northern hemisphere are known as the aurora borealis, or northern lights. Auroras in the southern hemisphere are known as the aurora australis. The northern lights are shown here.	

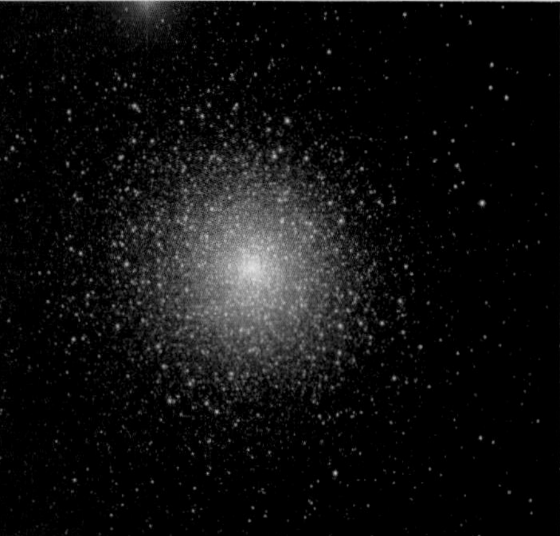

Figure 7 Open clusters (top) contain fewer than 1,000 stars. Globular clusters (bottom) can contain hundreds of thousands of stars.

Groups of Stars

The Sun has no stellar companion. The star closest to the Sun is 4.2 light-years away. Many stars are single stars, such as the Sun. Most stars exist in multiple star systems bound by gravity.

The most common star system is a binary system, where two stars orbit each other. By studying the orbits of binary stars, astronomers can determine the stars' masses. Many stars exist in large groupings called clusters. Two types of star clusters–open clusters and **globular** clusters–are shown in **Figure 7.** Stars in a cluster all formed at about the same time and are the same distance from Earth. If astronomers determine the distance to or the age of one star in a cluster, they know the distance to or the age of every star in that cluster.

WORD ORIGIN

globular
from Latin *globus*, means "round mass, sphere"

Classifying Stars

How do you classify a star? Which properties are important? Scientists classify stars according to their spectra. Recall that a star's spectrum is the light it emits spread out by wavelength. Stars have different spectra and different colors depending on their surface temperatures.

Temperature, Color, and Mass

Have you ever seen coals in a fire? Red coals are the coolest, and blue-white coals are the hottest. Stars are similar. Blue-white stars are hotter than red stars. Orange, yellow, and white stars are intermediate in temperature. Though there are exceptions, color in most stars is related to mass, as shown in **Figure 8.** Blue-white stars tend to have the most mass, followed by white stars, yellow stars, orange stars, and red stars.

Reading Check How does star color relate to mass?

As shown in **Figure 8,** the Sun is tiny compared to large, blue-white stars. However, scientists suspect that most stars–as many as 90 percent–are smaller than the Sun. These stars are called red dwarfs. The smallest star in **Figure 8** is a red dwarf.

Figure 8 The most massive stars are usually the hottest and are blue-white. The smallest stars tend to be cooler and red.

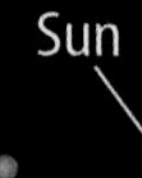

Hertzsprung-Russell Diagram 

Figure 9 The H-R diagram plots luminosity against temperature. Most stars exist along the main sequence, the band that stretches from the upper left to the lower right.

Visual Check Where is the Sun on this diagram?

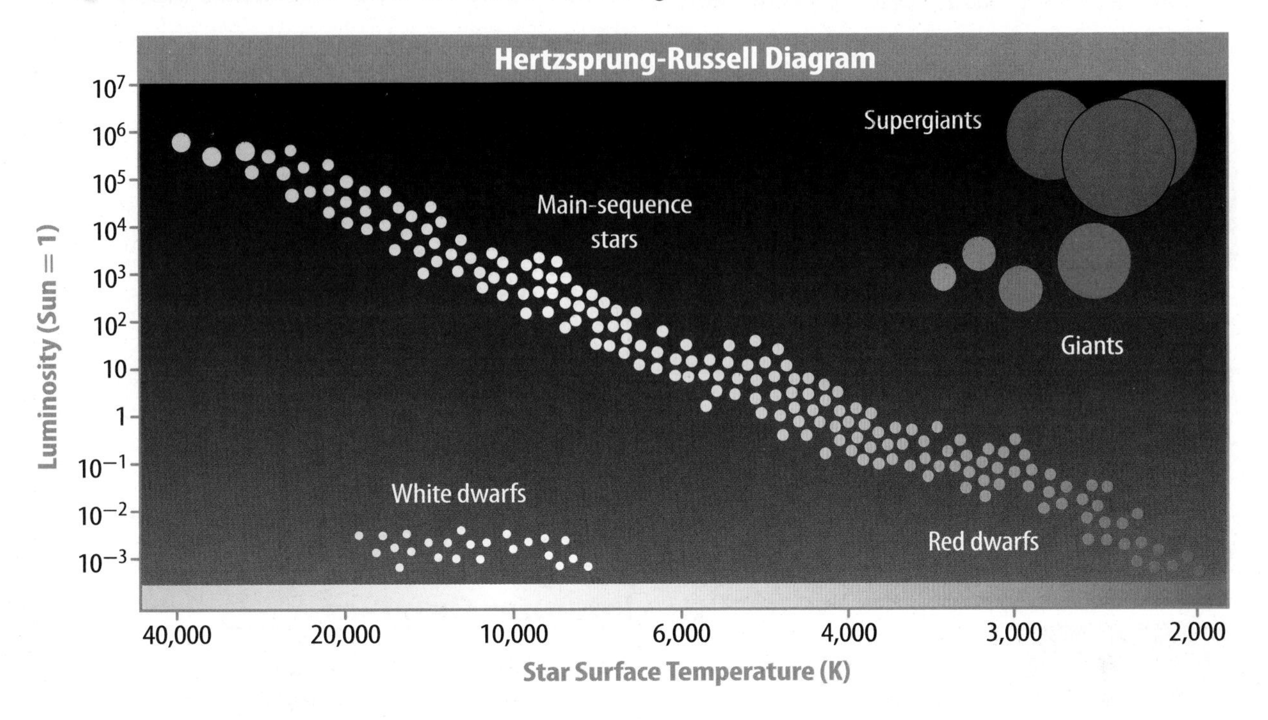

Hertzsprung-Russell Diagram

When scientists plot the temperatures of stars against their luminosities, the result is a graph like that shown in **Figure 9.** The **Hertzsprung-Russell diagram** (or H-R diagram) *is a graph that plots luminosity v. temperature of stars.* The *y*-axis of the H-R diagram displays increasing luminosity. The *x*-axis displays decreasing temperature.

The H-R diagram is named after two astronomers who developed it in the early 1900s. It is an important tool for categorizing stars. It also is an important tool for determining distances of some stars. If a star has the same temperature as a star on the H-R diagram, astronomers often can determine its luminosity. As you read earlier, if scientists know a star's luminosity, they can calculate its distance from Earth.

 **Key Concept Check** What is the Hertzsprung-Russell diagram?

The Main Sequence

Most stars spend the majority of their lives on the main sequence. On the H-R diagram, main-sequence stars form a curved line from the upper left corner to the lower right corner of the graph. The mass of a star determines both its temperature and its luminosity; the higher the mass the hotter and brighter the star. Because high-mass stars have more gravity pulling inward than low-mass stars, their cores have higher temperatures and produce and use more energy through fusion. High-mass stars have a shorter life span than low-mass stars. High-mass stars burn through their hydrogen much faster and move off the main sequence. A downside to a large-mass star is that the life span of the star is much shorter than average- or low-mass stars.

As shown in **Figure 9**, some groups of stars on the H-R diagram lie outside of the main sequence. These stars are no longer fusing hydrogen into helium in their cores. Some of these stars are cooler, but brighter and larger, such as supergiants. Other stars are dimmer and smaller, but much hotter, such as white dwarfs.

Lesson 2 Review

Visual Summary

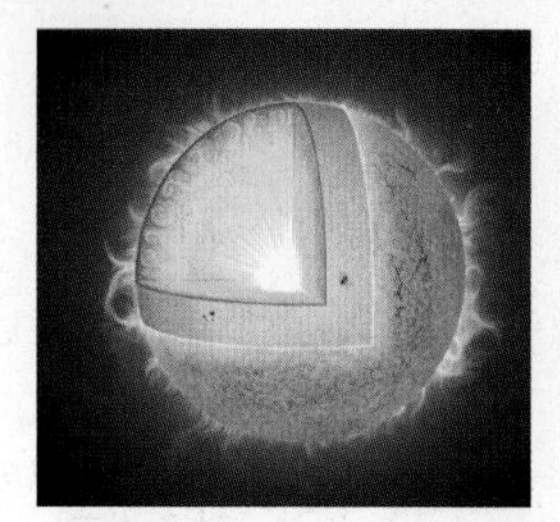

Hot gas moves up and cool gas moves down in the Sun's convection zone.

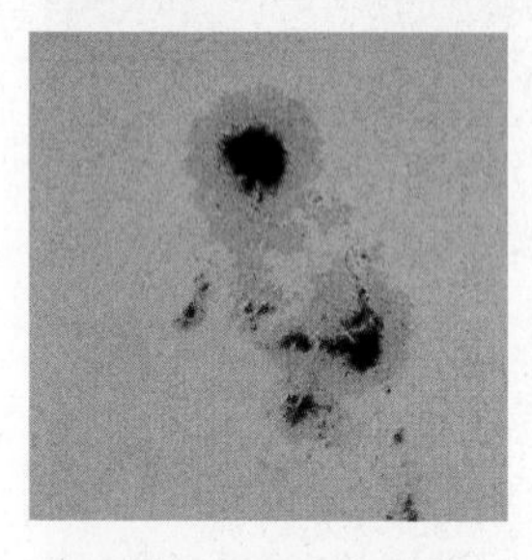

Sunspots are relatively dark areas on the Sun that have strong magnetic activity.

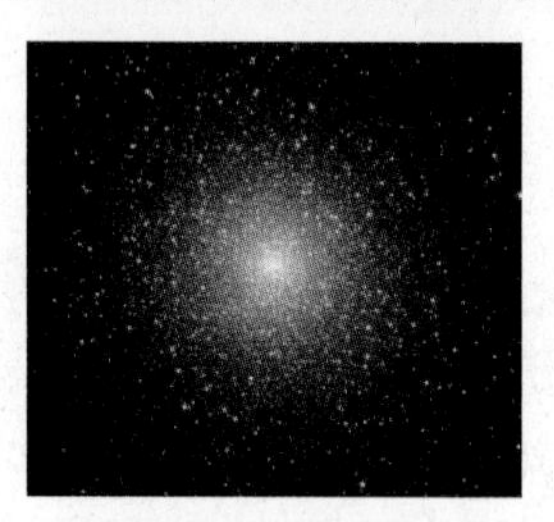

Globular clusters contain hundreds of thousands of stars.

FOLDABLES®

Use your lesson Foldable to review the lesson. Save your Foldable for the project at the end of the chapter.

What do you think NOW?

You first read the statements below at the beginning of the chapter.

3. Stars shine because there are nuclear reactions in their cores.

4. Sunspots are dark because they are cooler than nearby areas.

Did you change your mind about whether you agree or disagree with the statements? Rewrite any false statements to make them true.

Use Vocabulary

1. The ________ is a graph that plots luminosity v. temperature.
2. **Use the term** *photosphere* in a sentence.
3. **Define** *star* in your own words.

Understand Key Concepts 

4. Which part of a star extends millions of kilometers into space?
   - **A.** chromosphere
   - **B.** corona
   - **C.** photosphere
   - **D.** radiative zone
5. **Explain** how stars produce and release energy.
6. **Construct** an H-R diagram, and show the positions of the main sequence and the Sun.

Interpret Graphics

7. **Identify** Which star on the diagram below is hottest? Which is coolest? Which star represents the Sun?

8. **Organize Information** Copy and fill in the graphic organizer below to list the Sun's radiative zone, corona, convection zone, chromosphere, and photosphere in order outward from the core.

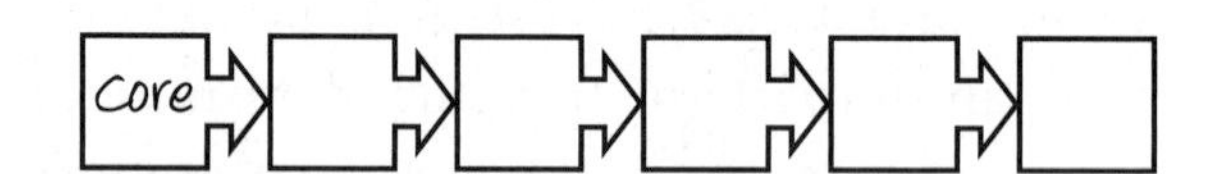

Critical Thinking

9. **Assess** why scientists monitor the Sun's changing features.
10. **Evaluate** In what way is the Sun an average star? In what way is it not an average star?

(t)Jerry Lodriguss/Science Source; (b)NOAO/AURA/NSF

Viewing the Sun in 3-D

NASA's Solar Terrestrial Relations Observatory

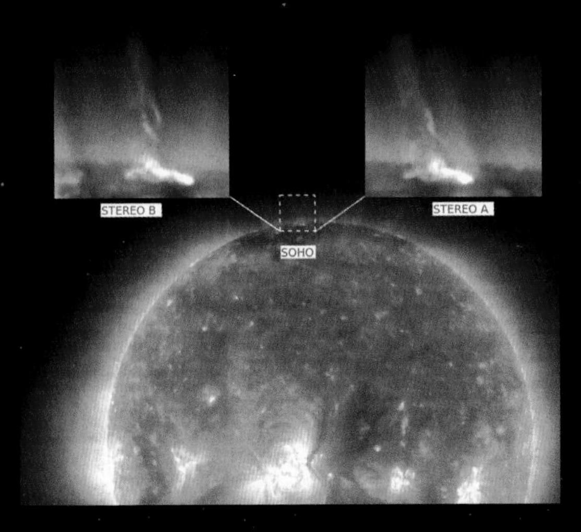

You might have used a telescope to look at o far in the distance or to look at stars and pla Although telescopes allow you to see a distant closer detail, you cannot see a three-dimension of objects in space. To get a three-dimensional the Sun, astronomers use two space telescopes *Solar Terrestrial Relations Observatory* (STEREC telescopes orbit the Sun in front of and behind Earth and give astronomers a 3-D view of the Sun. Why is this important.

If a coronal mass ejection (CME) erupts from the Sun, it can blast more than a billion tons of material into space. The powerful energy in a CME can damage satellites and power grids if Earth happens to be in its way. Before STEREO, scientists had only a straight-on view of CMEs approaching Earth. With STEREO, they have two different views. Each STEREO telescope carries several cameras that can detect many wavelengths. Scientists combine the pictures from each type of camera to make one 3-D image. In this way, they can track a CME from its emergence on the Sun all the way to its impact with Earth.

STEREO B is in orbit around the Sun behind Earth.

Earth

STEREO A is in orbit around the Sun ahead of Earth.

It's Your Turn

RESEARCH AND REPORT How can power and satellite companies prepare for an approaching CME? Find out and write a short report on what you find. Share your findings with the class.

Lesson 3

Reading Guide

Key Concepts 

ESSENTIAL QUESTIONS

- How do stars form?
- How does a star's mass affect its evolution?
- How is star matter recycled in space?

Vocabulary

nebula p. 817
white dwarf p. 819
supernova p. 819
neutron star p. 820
black hole p. 820

Multilingual eGlossary

Evolution of Stars

Inquiry Exploding Star?

No, this is a cloud of gas and dust where stars form. How do you think stars form? Do you think stars ever stop shining?

Launch Lab

20 minutes

Do stars have life cycles?

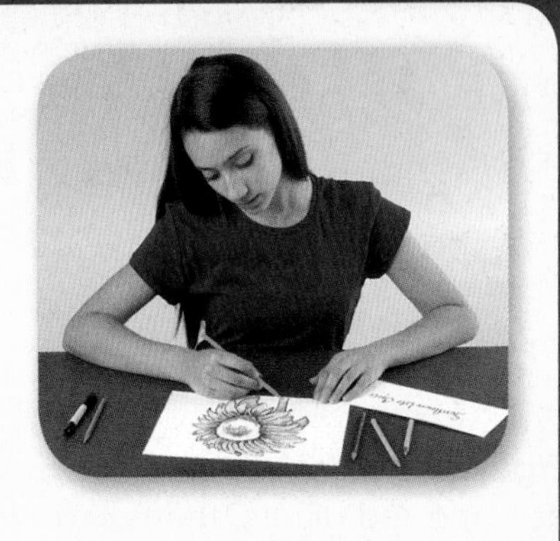

You might have learned about the life cycles of plants or animals. Do stars, such as the Sun, have life cycles? Before you find out, review the life cycle of a sunflower.

1. Read and complete a lab safety form.
2. Obtain an **envelope containing slips of paper** that explain the life cycle of a sunflower.
3. Use **colored pencils** to draw a sunflower in the middle of a piece of **paper,** or use a **glue stick** to glue a sunflower picture on the paper.
4. Using your knowledge of plant life cycles, arrange the slips of paper around the sunflower in the order in which the events listed on them occur. Draw arrows to show how the steps form a cycle.

Think About This

1. Does the life cycle of a sunflower have a beginning and an end? Explain your answer.
2. Do you think that every stage in the life cycle takes the same amount of time? Why or why not?
3. **Key Concept** How do you think the life cycle of a star compares to the life cycle of a sunflower? Do you think all stars have the same life cycle?

Life Cycle of a Star

Like living things, stars have life cycles. They are "born," and after millions or billions of years, they "die." Stars die in different ways, depending on their masses. But all stars–from white dwarfs to supergiants–form in the same way.

Nebulae and Protostars

Stars form deep inside clouds of gas and dust. *A cloud of gas and dust is a* **nebula** (plural, nebulae). Star-forming nebulae are cold, dense, and dark. Gravity causes the densest parts to collapse, forming regions called protostars. Protostars continue to contract, pulling in surrounding gas, until their cores are hot and dense enough for nuclear fusion to begin. As they contract, protostars produce enormous amounts of thermal energy.

Birth of a Star

Over many thousands of years, the energy produced by protostars heats the gas and dust surrounding them. Eventually, the surrounding gas and dust blows away, and the protostars become visible as stars. Some of this material might later become planets or other objects that orbit the star. During the star-formation process, nebulae glow brightly, as shown in the photograph on the previous page.

Key Concept Check How do stars form?

WORD ORIGIN

nebula
from Latin *nebula,* means "mist" or "little cloud"

FOLDABLES®

Make a vertical five-tab book. Label it as shown. Use it to organize your notes on the life cycle of a star.

Protostar
Main Sequence
Red Giant
Red Supergiant
Supernova

Main-Sequence Stars

Recall the main sequence of the Hertzsprung-Russell diagram. Stars spend most of their lives on the main sequence. A star becomes a main-sequence star as soon as it begins to fuse hydrogen into helium in the core. It remains on the main sequence for as long as it continues to fuse hydrogen into helium. Average-mass stars such as the Sun remain on the main sequence for billions of years. High-mass stars remain on the main sequence for only a few million years. Even though massive stars have more hydrogen than lower-mass stars, they process it at a much faster rate.

When a star's hydrogen supply is nearly gone, the star moves off the main sequence. It begins the next stage of its life cycle, as shown in **Figure 10.** Not all stars go through all phases in **Figure 10.** Lower-mass stars do not have enough mass to become supergiants.

Figure 10 Massive stars become red giants, then larger red giants, then red supergiants.

Visual Check Which element forms in only the most massive stars?

A Massive Star's Life Cycle

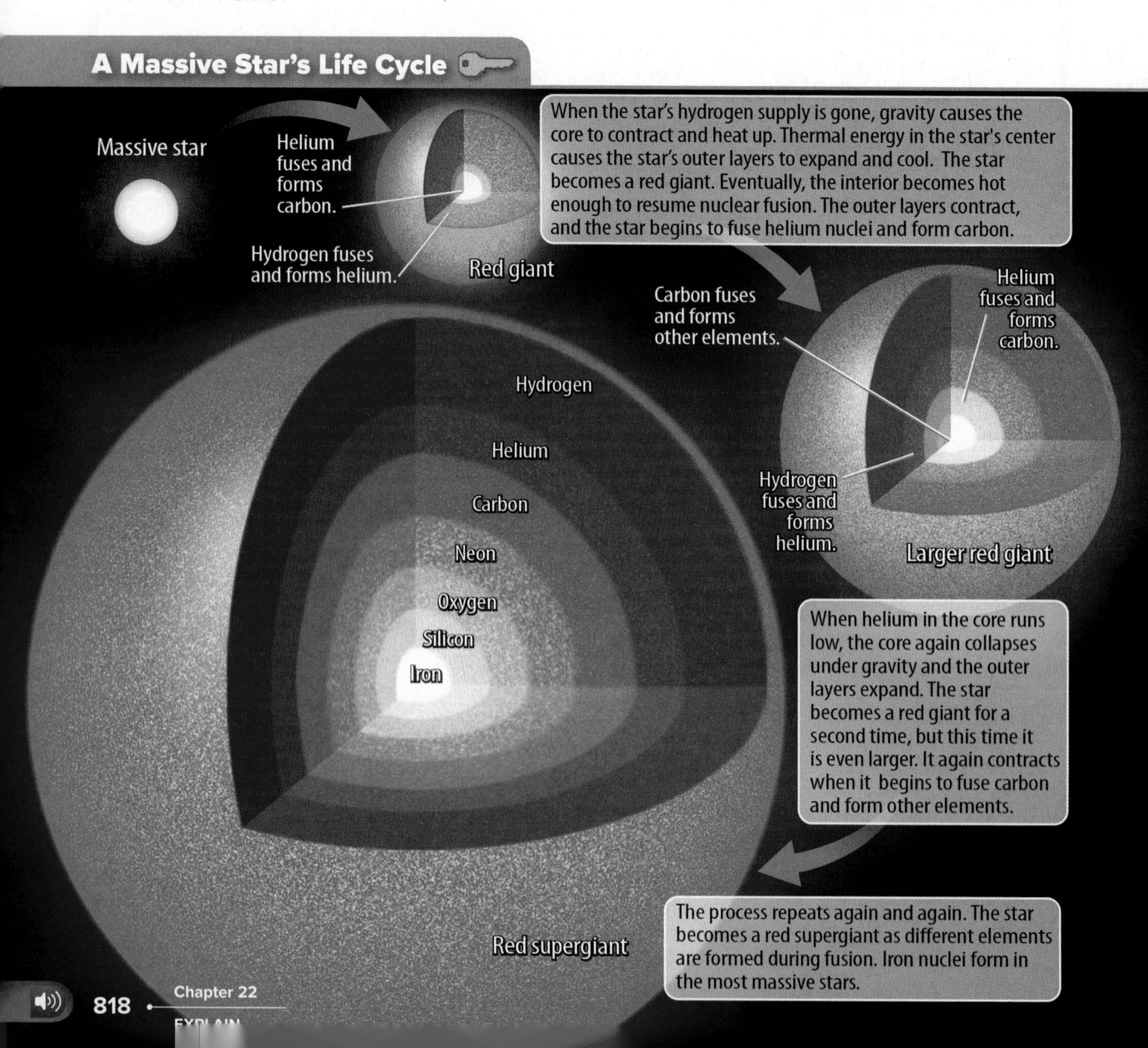

End of a Star

All stars form in the same way. But stars die in different ways, depending on their masses. Massive stars collapse and explode. Lower-mass stars die more slowly.

White Dwarfs

Average-mass stars, such as the Sun, do not have enough mass to fuse elements beyond helium. They do not get hot enough. After helium in their cores is gone, the stars cast off their gases, exposing their cores. The core becomes a **white dwarf,** *a hot, dense, slowly cooling sphere of carbon.*

What will happen to Earth and the solar system when the Sun runs out of fuel? When the Sun runs out of hydrogen, in about 5 billion years, it will become a red giant. Once helium fusion begins, the Sun will contract. When the helium is gone, the Sun will expand again, probably absorbing Mercury, Venus, and Earth and pushing Mars outward, as shown in **Figure 11.** Eventually, the Sun will become a white dwarf. Imagine the mass of the Sun squeezed a million times until it is the size of Earth. That's the size of a white dwarf. Scientists expect that all stars with masses less than 8–10 times that of the Sun will eventually become white dwarfs.

Reading Check What will happen to Earth when the Sun runs out of fuel?

Supernovae

Stars with more than 10 times the mass of the Sun do not become white dwarfs. Instead, they explode. *A* **supernova** (plural, supernovae) *is an enormous explosion that destroys a star.* In the most massive stars, a supernova occurs when iron forms in the star's core. Iron is stable and does not fuse. After a star forms iron, it loses its internal energy source, and the core collapses quickly under the force of gravity. So much energy is released that the star explodes. When it explodes, a star can become one billion times brighter and form elements even heavier than iron.

Figure 11 In about 5 billion years, the Sun will become a red giant and then a white dwarf.

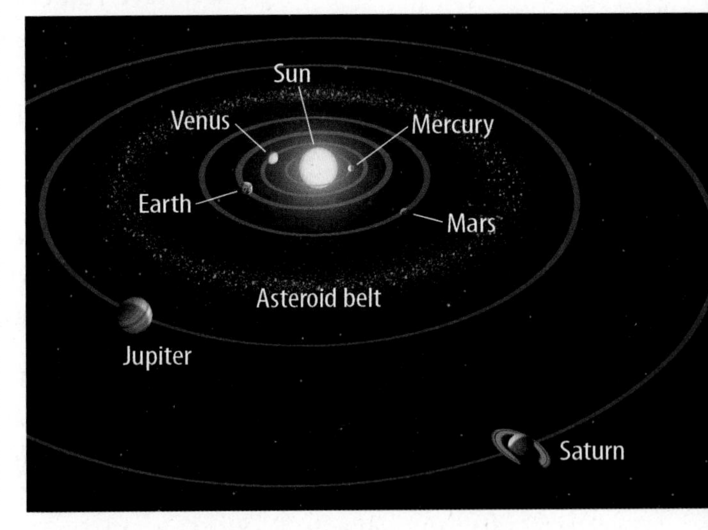

The Sun will remain on the main sequence for 5 billion more years.

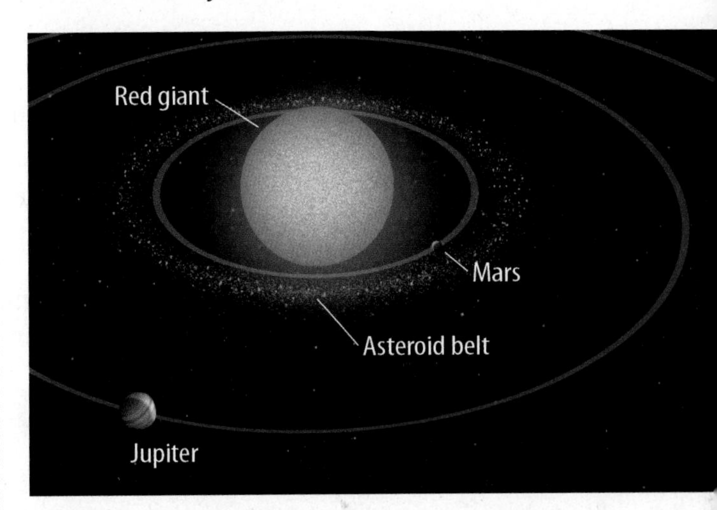

When the Sun becomes a red giant for the second time, the outer layers will probably absorb Earth and push Mars and Jupiter outward.

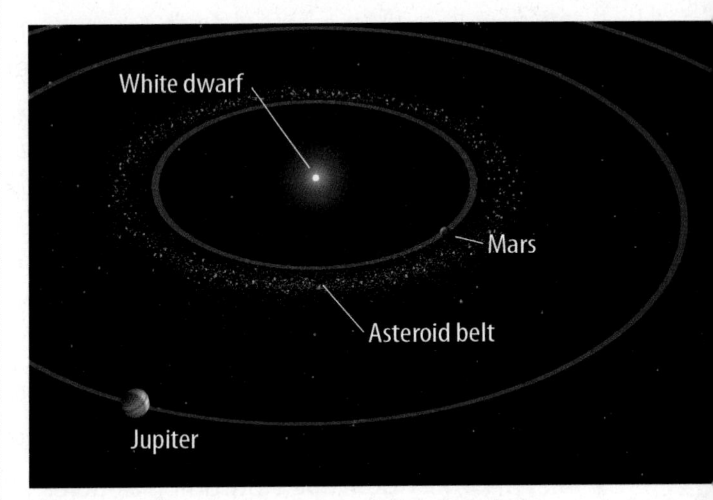

When the Sun becomes a white dwarf, the solar system will be a cold, dark place.

Neutron Stars

Have you ever eaten cotton candy? A bag of cotton candy is made from just a few spoonfuls of spun sugar. Cotton candy is mostly air. Similarly, atoms are mostly empty space. During a supernova, the outer layers of the star are blown away and the core collapses under the heavy force of gravity. The space in atoms disappears as protons and electrons combine to form neutrons. *A* **neutron star** *is a dense core of neutrons that remains after a supernova.* Neutron stars are only about 20 km wide, with cores so dense that a teaspoonful would weigh more than 1 billion tons.

REVIEW VOCABULARY

neutron
a neutral particle in the nucleus of an atom

Black Holes

For the most massive stars, atomic forces holding neutrons together are not strong enough to overcome so much mass in such a small volume. Gravity is too strong, and the matter crushes into a black hole. *A* **black hole** *is an object whose gravity is so great that no light can escape.*

A black hole does not suck matter in like a vacuum cleaner. But its gravity is very strong because all of its mass is concentrated in a single point. Because astronomers cannot see a black hole, they only can infer its existence. For example, if they detect a star circling around something, but they cannot see what that something is, they infer it is a black hole.

Key Concept Check How does a star's mass determine if it will become a white dwarf, a neutron star, or a black hole?

MiniLab

15 minutes

How do astronomers detect black holes?

The only way astronomers can detect black holes is by studying the movement of objects nearby. How do black holes affect nearby objects?

1. Read and complete a lab safety form.
2. With a partner, make two stacks of **books** of equal height about 25 cm apart. Place a piece of **thin cardboard** on top of the books.
3. Spread some **staples** over the cardboard. Hold a **magnet** under the cardboard. Observe what happens to the staples.
4. While one student holds the magnet in place beneath the cardboard, the other student gently rolls a **small magnetic marble** across the cardboard. Repeat several times, rolling the marble in different pathways. Record your observations in your Science Journal.

Hutchings Photography/Digital Light Source

Analyze and Conclude

1. **Infer** What did the pull of the magnet represent?
2. **Cause and Effect** How did the magnet affect the staples and the movement of the marble?
3. **Key Concept** How do black holes affect nearby objects?

Recycling Matter

At the end of a star's life cycle, much of its gas escapes into space. This gas is recycled. It becomes the building blocks of future generations of stars and planets.

Planetary Nebulae

You read that average-mass stars, such as the Sun, become white dwarfs. When a star becomes a white dwarf, it casts off hydrogen and helium gases in its outer layers, as shown in **Figure 12**. The expanding, cast-off matter of a white dwarf is a planetary nebula. Most of the star's carbon remains locked in the white dwarf. But the gases in the planetary nebula can be used to form new stars.

Planetary nebulae have nothing to do with planets. They are so named because early astronomers thought they were regions where planets were forming.

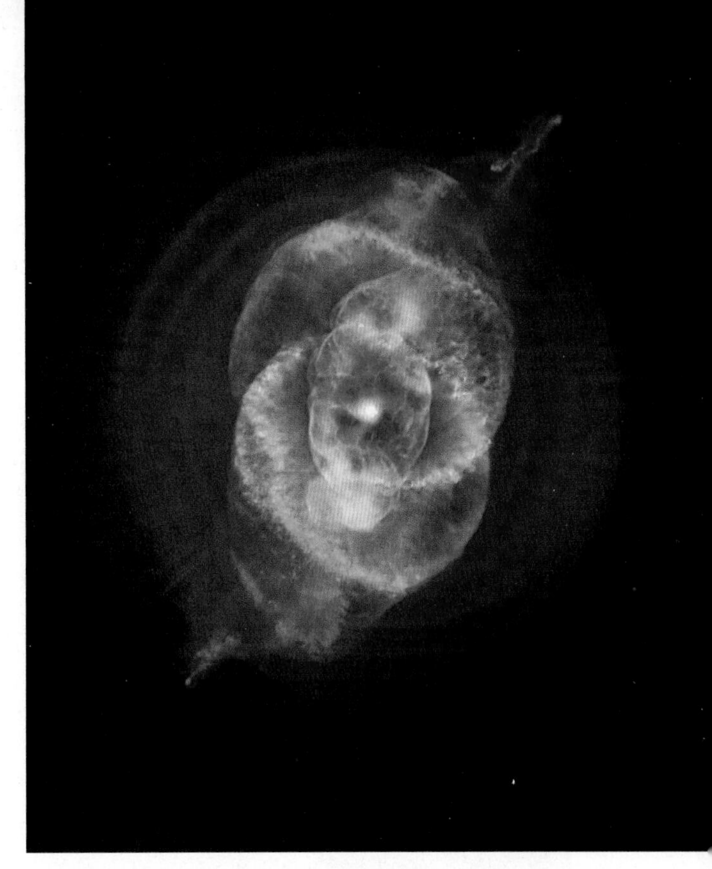

▲ **Figure 12** White dwarfs cast off helium and hydrogen as planetary nebulae. The gases can be used by new generations of stars.

Supernova Remnants

During a supernova, a massive star comes apart. This sends a shock wave into space. The expanding cloud of dust and gas is called a supernova remnant. A supernova remnant is shown in **Figure 13**. Like a snowplow pushing snow in its path, a supernova remnant pushes on the gas and dust it encounters.

In a supernova, a star releases the elements that formed inside it during nuclear fusion. Almost all of the elements in the universe other than hydrogen and helium were created by nuclear reactions inside the cores of massive stars and released in supernovae. This includes the oxygen in air, the silicon in rocks, and the carbon in you.

Key Concept Check How do stars recycle matter?

Gravity causes recycled gases and other matter to clump together in nebulae and form new stars and planets. As you will read in the next lesson, gravity also causes stars to clump together into even larger structures called galaxies.

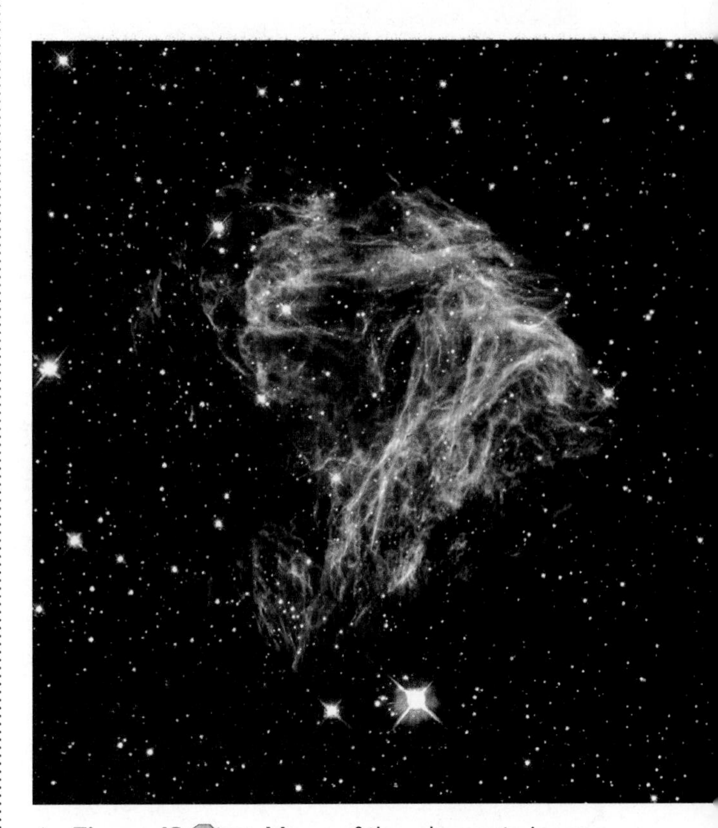

▲ **Figure 13** Many of the elements in you and in matter all around you were formed inside massive stars and released in supernovae.

Lesson 3 Review

Visual Summary

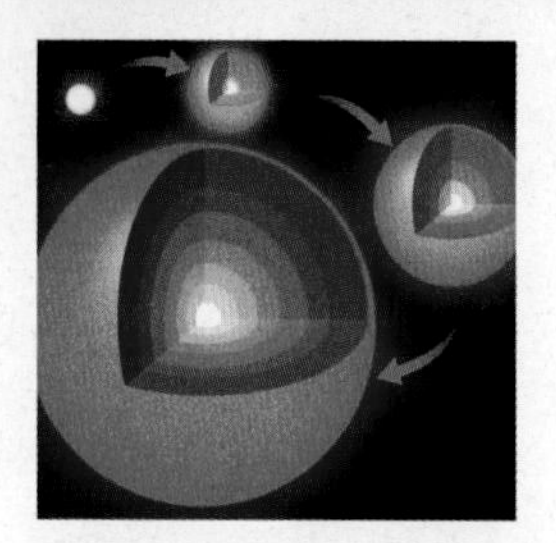

Iron is formed in the cores of the most massive stars.

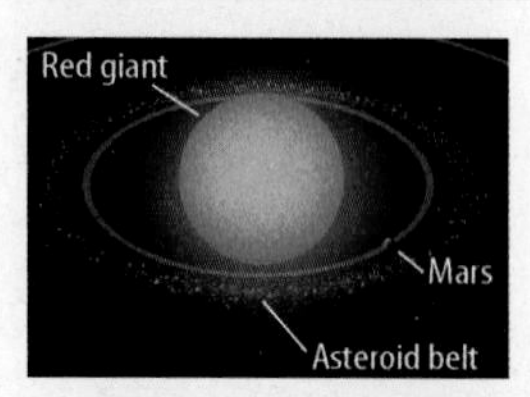

The Sun will become a red giant in about 5 billion years.

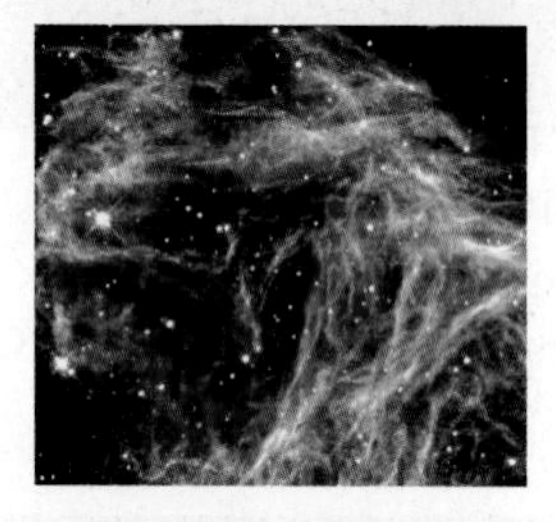

Matter is recycled in supernovae.

FOLDABLES®

Use your lesson Foldable to review the lesson. Save your Foldable for the project at the end of the chapter.

What do you think NOW?

You first read the statements below at the beginning of the chapter.

5. The more matter a star contains, the longer it is able to shine.

6. Gravity plays an important role in the formation of stars.

Did you change your mind about whether you agree or disagree with the statements? Rewrite any false statements to make them true.

Use Vocabulary

1 Planetary nebulae are the expanding outer layers of a(n) ________.

2 **Define** *supernova* in your own words.

3 **Use the terms** *neutron star* and *black hole* in a sentence.

Understand Key Concepts

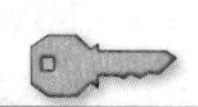

4 Which type of star will the Sun eventually become?

A. neutron star
B. red dwarf
C. red supergiant
D. white dwarf

5 **Explain** how supernovae recycle matter.

6 **Rank** black holes, neutron stars, and white dwarfs from most massive to least massive.

Interpret Graphics

7 **Describe** details of the process occurring in the photo below.

8 **Organize Information** Copy and fill in the graphic organizer below to list what happens to a star following a supernova.

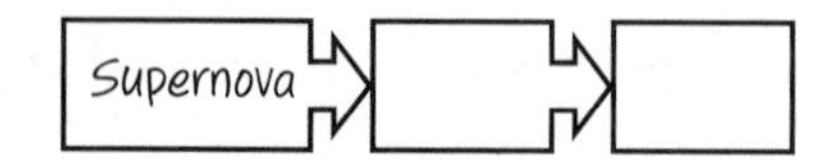

Critical Thinking

9 **Predict** whether the Sun will eventually become a black hole. Why or why not?

10 **Evaluate** why mass is so important in determining the evolution of a star.

Skill Practice Make and Use Graphs

45 minutes

How can graphing data help you understand stars?

How can you make sense of everything in the universe? Graphs help you organize information. The Hertzsprung-Russell diagram is a graph that plots the color, or temperature, of stars against their luminosities. What can you learn about stars by plotting them on a graph similar to the H-R diagram?

Materials

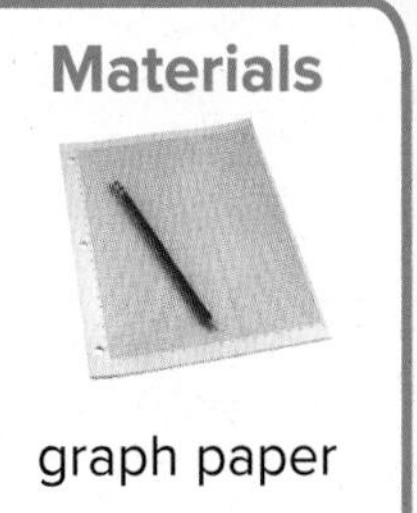

graph paper

Learn It

Displaying information on graphs makes it easier to see how objects are related. Lines on graphs show you patterns and enable you to make predictions. Graphs display a lot of information in an easily understandable form. In this activity, you will **make and use graphs,** plotting the temperature, the color, and the mass of stars.

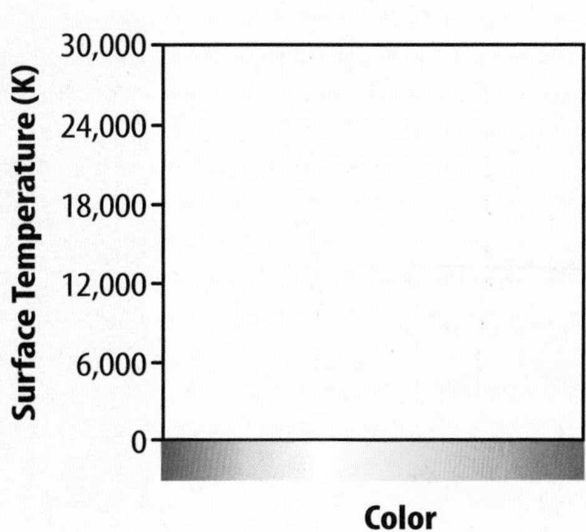

Try It

1. Using graph paper or your Science Journal, draw a graph like the one shown at right.
2. Use the color and temperature data in the table below to plot the position of each star on your graph. Mark the points by attaching adhesive stars to the graph.
3. If stars have similar data, plot them in a cluster. Label each star with its name.
4. Draw a curve that joins the data points as smoothly as possible.
5. Make another graph and plot temperature v. mass of the stars in the table.

Star	Color	Temperature (K)	Mass in solar masses
Sun	Yellow	5,700	1
Alnilam	Blue-white	27,000	40
Altair	White	8,000	1.9
Alpha Centauri A	Yellow	6,000	1.08
Alpha Centauri B	Orange	4,370	0.68
Barnard's Star	Red	3,100	0.1
Epsilon Eridani	Orange	4,830	0.78
Hadar	Blue-white	25,500	10.5
Proxima Centauri	Red	3,000	0.12
Regulus	White	11,000	8
Sirius A	White	9,500	2.6
Spica	Blue-white	22,000	10.5
Vega	White	9,900	3

Apply It

6. All of the stars on your graph are main-sequence stars. What is the relationship between the color and the temperature of a main-sequence star?
7. What is the relationship between the mass and the temperature of a main-sequence star? How are color and mass related?
8. **Key Concept** Which star would be the most likely to eventually form a black hole? Why?

Lesson 4

Reading Guide

Key Concepts

ESSENTIAL QUESTIONS

- What are the major types of galaxies?
- What is the Milky Way, and how is it related to the solar system?
- What is the Big Bang theory?

Vocabulary

galaxy p. 825
dark matter p. 825
Big Bang theory p. 830
Doppler shift p. 830

Multilingual eGlossary

Galaxies and the Universe

Inquiry Disk in Space?

Yes, this is the disk of a galaxy—a huge collection of stars. You see this galaxy on its edge. If you were to look down on it from above, it would look like a two-armed spiral. Do you think all galaxies are shaped like spirals? What about the galaxy you live in?

NASA/Hubble Heritage Team

Launch Lab

20 minutes

Does the universe move?

Scientists think the universe is expanding. What does that mean? Are stars and galaxies moving away from each other? Is the universe moving?

1. Read and complete a lab safety form.
2. Copy the table at right into your Science Journal.
3. Use a **marker** to make three dots 5–7 cm apart on one side of a **large round balloon.** Label the dots *A, B,* and *C.* The dots represent galaxies.
4. Blow up the balloon to a diameter of about 8 cm. Hold the balloon closed as your partner uses a **measuring tape** to measure the distance between each galaxy on the balloon's surface. Record the distances on the table.
5. Repeat step 4 two more times, blowing up the balloon a little more each time.

Balloon size	A–B (cm)	B–C (cm)	A–C (cm)
Small			
Medium			
Large			

Think About This

1. What happened to the distances between galaxies as the balloon expanded?
2. If you were standing in one of the galaxies, what would you observe about the other galaxies?
3. **Key Concept** If the balloon were a model of the universe, what do you think might have caused galaxies to move in this way?

Galaxies

Most people live in towns or cities where houses are close together. Not many houses are found in the wilderness. Similarly, most stars exist in galaxies. **Galaxies** *are huge collections of stars.* The universe contains hundreds of billions of galaxies, and each galaxy can contain hundreds of billions of stars.

Reading Check What are galaxies?

Dark Matter

Gravity holds stars together. Gravity also holds galaxies together. When astronomers examine how galaxies, such as those in **Figure 14,** rotate and gravitationally interact, they find that most of the matter in galaxies is invisible. *Matter that emits no light at any wavelength is called* **dark matter.** Scientists hypothesize that more than 90 percent of the universe's mass is dark matter. Scientists do not fully understand dark matter. They do not know what composes it.

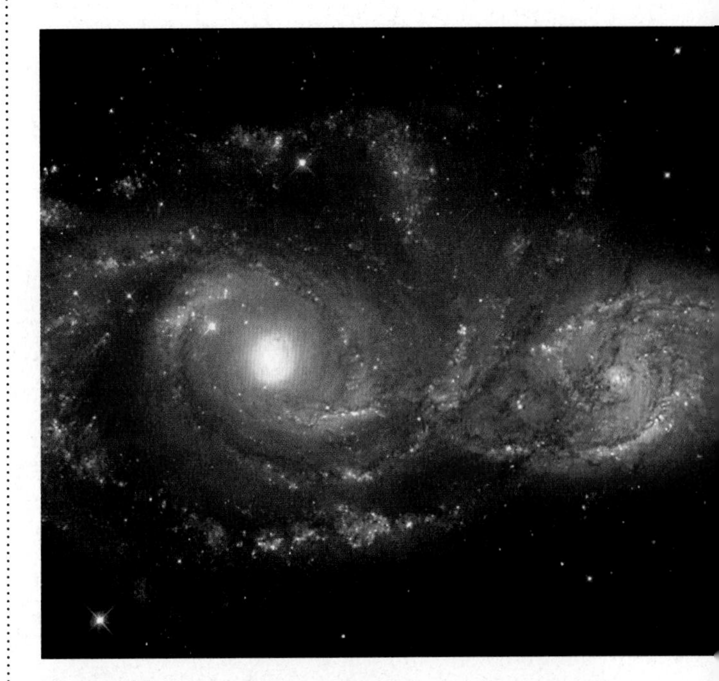

Figure 14 By examining interacting galaxies such as these, astronomers hypothesize that most mass in the universe is invisible dark matter.

Types of Galaxies

There are three major types of galaxies: spiral, elliptical, and irregular. **Table 2** gives a brief description of each type.

 **Key Concept Check** What are the major types of galaxies?

Table 2 Types of Galaxies

	**Spiral Galaxies** The stars, gas, and dust in a spiral galaxy exist in spiral arms that begin at a central disk. Some spiral arms are long and symmetrical; others are short and stubby. Spiral galaxies are thicker near the center, a region called the central bulge. A spherical halo of globular clusters and older, redder stars surrounds the disk. NGC 5679, shown here, contains a pair of spiral galaxies.
Elliptical Galaxies Unlike spiral galaxies, elliptical galaxies do not have internal structure. Some are spheres, like basketballs, while others resemble footballs. Elliptical galaxies have higher percentages of old, red stars than spiral galaxies do. They contain little or no gas and dust. Scientists suspect that many elliptical galaxies form by the gravitational merging of two or more spiral galaxies. The elliptical galaxy pictured here is NGC 5982, part of the Draco Group.	
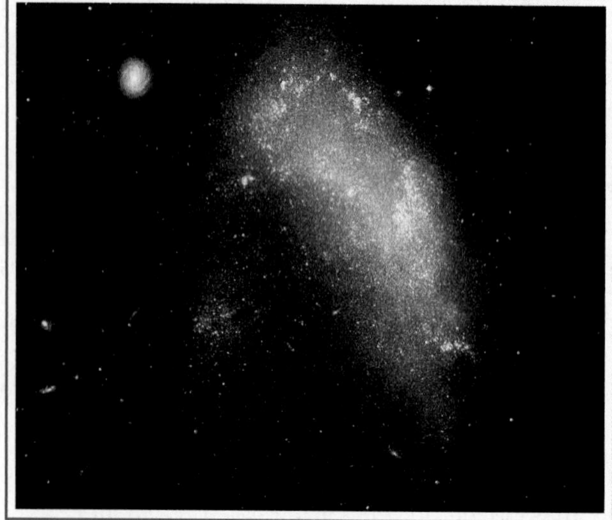	**Irregular Galaxies** Irregular galaxies are oddly shaped. Many form from the gravitational pull of neighboring galaxies. Irregular galaxies contain many young stars and have areas of intense star formation. Shown here is the irregular galaxy NGC 1427A.

MiniLab

20 minutes

Can you identify a galaxy?

The *Hubble Space Telescope,* shown below, is an orbiting telescope that gives astronomers clear pictures of the night sky. What kinds of galaxies can you see in pictures taken by the *Hubble Telescope*?

1. Study each image on the ***Hubble Space Telescope* image sheet.** For each image, identify at least two galaxies. Are they spiral, elliptical, or irregular? Write your observations in your Science Journal, labeled with the letter of the image.

Analyze and Conclude

1. **Draw Conclusions** Why are some galaxies easier to identify than others?
2. **Infer** What interactions do you see among some of the galaxies?
3. **Key Concept** Do you think the shapes of galaxies can change over time? Why or why not?

Groups of Galaxies

Galaxies are not distributed evenly in the universe. Gravity holds them together in groups called clusters. Some clusters of galaxies are enormous. The Virgo Cluster is 60 million light-years from Earth. It contains about 2,000 galaxies. Most clusters exist in even larger structures called superclusters. The Milky Way galaxy is part of the Laniakea supercluster. Between superclusters are voids, which are regions of nearly empty space. Scientists hypothesize that the large-scale structure of the universe resembles a sponge.

Reading Check What holds clusters of galaxies together?

The Milky Way

The solar system is in the Milky Way, a spiral galaxy that contains gas, dust, and almost 200 billion stars. The Milky Way is a member of the Local Group, a cluster of about 30 galaxies. Scientists expect the Milky Way will begin to merge with the Andromeda Galaxy, the largest galaxy in the Local Group, in about 3 billion years. Because stars are far apart in galaxies, it is not likely that many stars will actually collide during this event.

Where is Earth in the Milky Way? **Figure 15** on the next two pages shows an artist's drawing of the Milky Way and Earth's place in it.

FOLDABLES®

Make a horizontal single-tab matchbook. Label it as shown. Use it to describe the contents of the Milky Way.

Milky Way

Galaxy

WORD ORIGIN

galaxy
from Greek *galactos,* means "milk"

The Milky Way

Figure 15 The Milky Way is shown here in two separate views, from the top (left page) and on edge (right page). Because Earth is located inside the disk of the Milky Way, people cannot see beyond the central bulge to the other side.

Key Concept Check Where is Earth in the Milky Way?

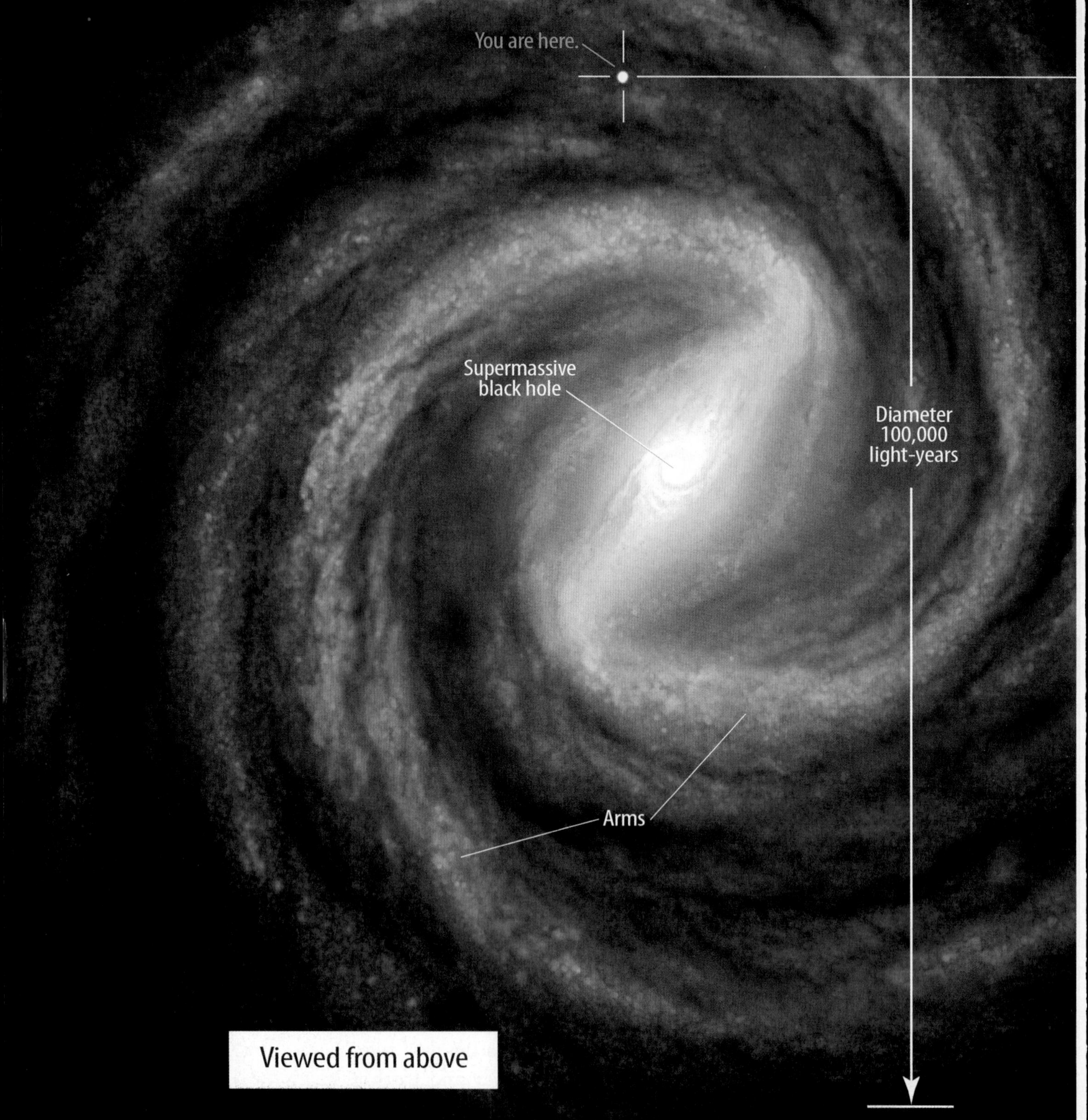

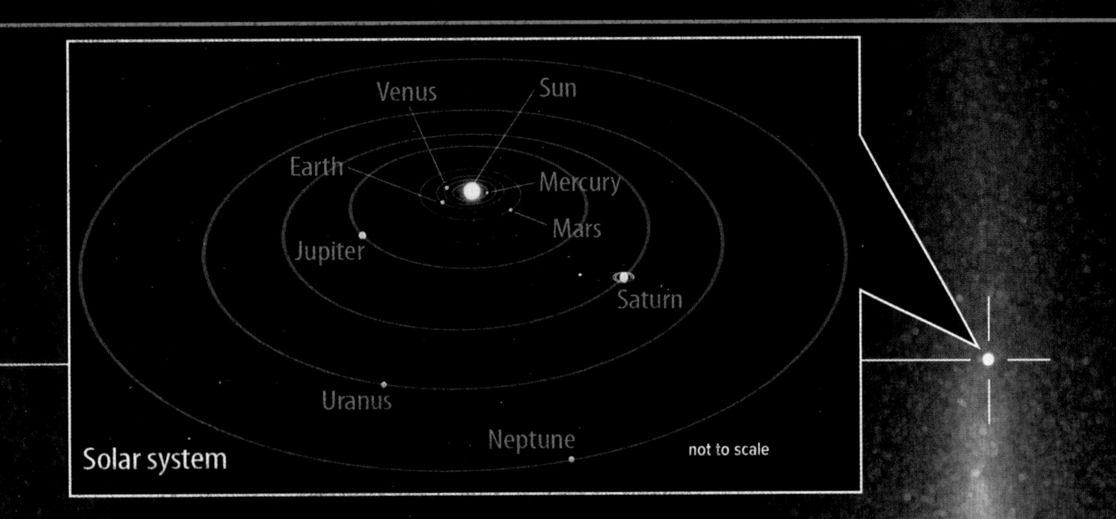

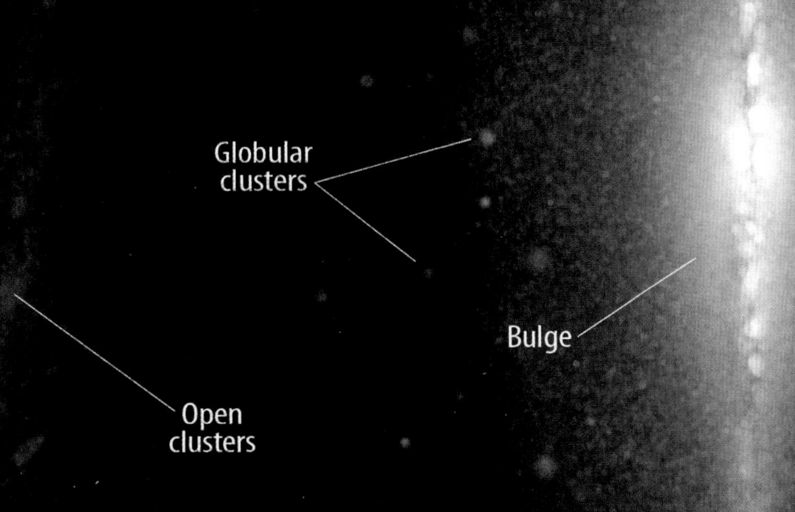

Viewed on edge

The Big Bang Theory

When astronomers look into space, they look back in time. Is there a beginning to time? According to the **Big Bang theory,** *the universe began from one point billions of years ago and has been expanding ever since.*

Key Concept Check What is the Big Bang theory?

Origin and Expansion of the Universe

Most scientists agree that the universe is 13–14 billion years old. When the universe began, it was dense and hot–so hot that even atoms didn't exist. After a few hundred thousand years, the universe cooled enough for atoms to form. Eventually, stars formed, and gravity pulled them into galaxies.

As the universe expands, space stretches and galaxies move away from one another. The same thing happens in a loaf of unbaked raisin bread. As the dough rises, the raisins move apart. Scientists observe how space stretches by measuring the speed at which galaxies move away from Earth. As the galaxies move away, their wavelengths lengthen and stretch out. How does light stretch?

Doppler Shift

You have probably heard the siren of a speeding police car. As **Figure 16** illustrates, when the car moves toward you, the sound waves compress. As the car moves away, the sound waves spread out. Similarly, when visible light travels toward you, its wavelength compresses. When light travels away from you, its wavelength stretches out. It shifts to the red end of the visual light portion of the electromagnetic spectrum. *The shift to a different wavelength is called the* **Doppler shift.** Because the universe is expanding, light from galaxies is red-shifted. The more distant a galaxy is, the faster it moves away from Earth, and the more it is red-shifted.

Dark Energy

Will the universe expand forever? Or will gravity cause the universe to contract? Scientists have observed that galaxies are moving away from Earth faster over time. To explain this, they suggest a force called dark energy is pushing the galaxies apart.

Dark energy, like dark matter, is an active area of research. There is still much to learn about the universe and all it contains.

Doppler Shift

Animation

Figure 16 The sound waves from an approaching police car are compressed. As the car speeds away, the sound waves are stretched out. Similarly, when an object is moving away, its light is stretched out. The light's wavelength shifts toward a longer wavelength.

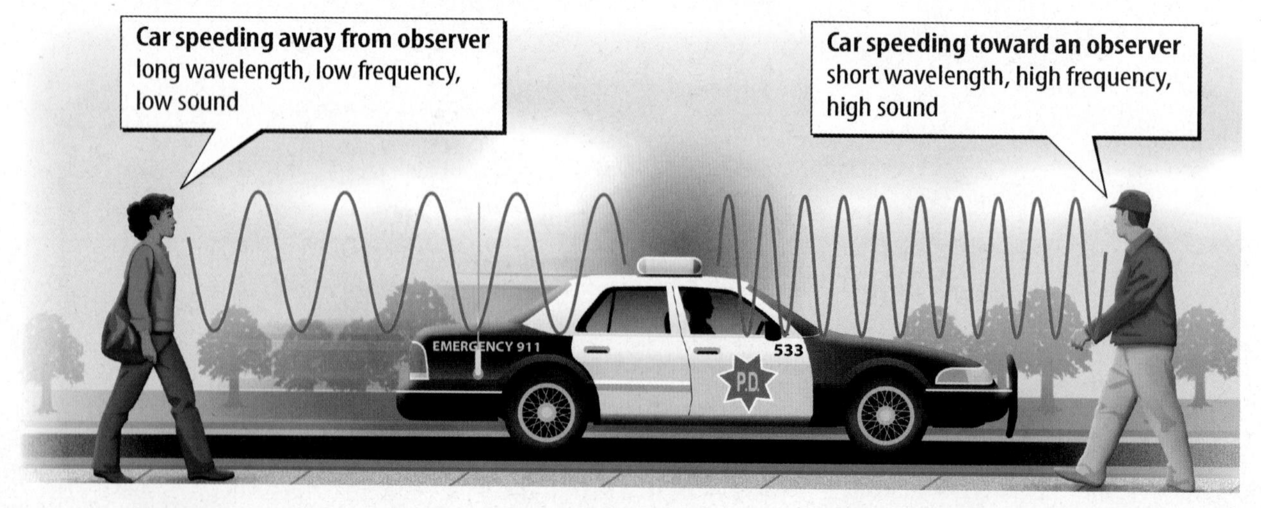

Lesson 4 Review

Visual Summary

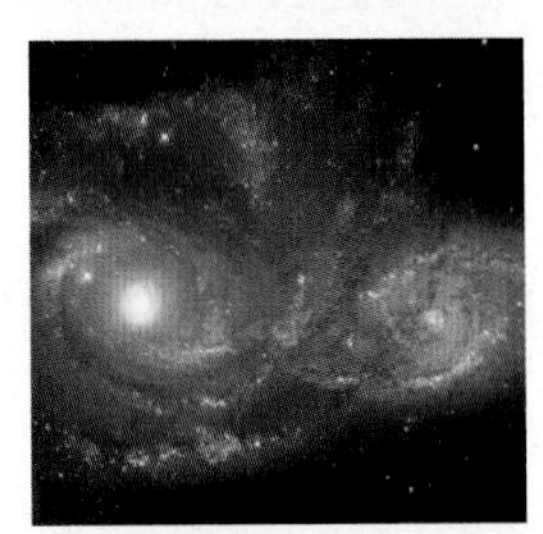

By studying interacting galaxies, scientists have determined that most mass in the universe is dark matter.

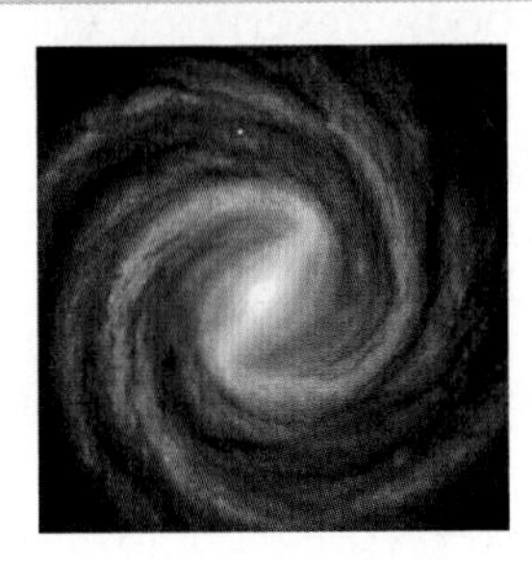

The Sun is one of billions of stars in the Milky Way.

When an object moves away, its light stretches out, just as a siren's sound waves stretch out as the siren moves away.

FOLDABLES®

Use your lesson Foldable to review the lesson. Save your Foldable for the project at the end of the chapter.

What do you think NOW?

You first read the statements below at the beginning of the chapter.

7. Most of the mass in the universe is in stars.

8. The Big Bang theory is an explanation of the beginning of the universe.

Did you change your mind about whether you agree or disagree with the statements? Rewrite any false statements to make them true.

Use Vocabulary

1 Stars exist in huge collections called ________.

2 **Use the term** *dark matter* in a sentence.

3 **Define** the *Big Bang theory*.

Understand Key Concepts 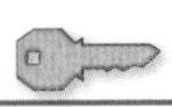

4 Which is NOT a major galaxy type?

A. dark **C.** irregular
B. elliptical **D.** spiral

5 **Identify** Sketch the Milky Way, and identify the location of the solar system.

6 **Explain** how scientists know the universe is expanding.

Interpret Graphics

7 **Identify** Sketch the Milky Way, shown below. Identify the bulge, the halo, and the disk.

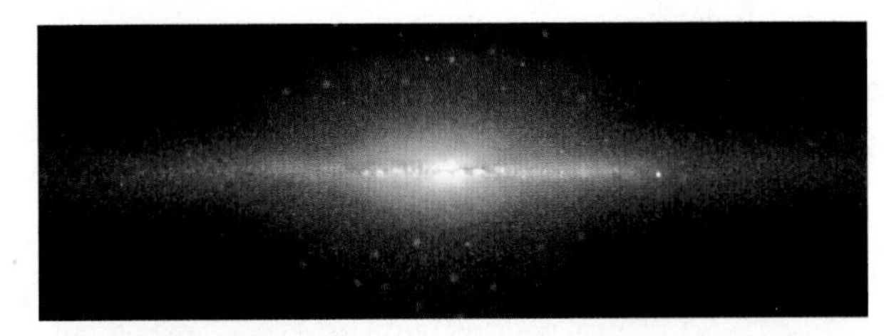

8 **Organize Information** Copy and fill in the graphic organizer below. List the three major types of galaxies and some characteristics of each.

Galaxy Type	Characteristics

Critical Thinking

9 **Assess** the role of gravity in the structure of the universe.

10 **Predict** what the solar system and the universe might be like in 10 billion years.

Lab

3 class periods

Describe a Trip Through Space

Materials

paper

colored pencils

astronomy magazines

string

glue

scissors

Safety

Imagine you could travel through space at speeds even faster than light. Based on what you have learned in this chapter, where would you choose to go? What would you like to see? What would it be like to move through the Milky Way and out into distant galaxies? Would you travel with anyone or meet any characters? Write a book describing your trip through space.

Question

Where will your trip take you, and how will you describe it? How can you write a fictional, but scientifically accurate, story about your trip? Will you make a picture book? If so, will you sketch your own pictures, use diagrams or photographs, or both? Will your book be mostly words, or will it be like a graphic novel? How can you draw your readers into the story?

Procedure

1. In your Science Journal, write ideas about where your trip will take you, how you will travel, what will happen along the way, and who or what you might meet.
2. Draw a graphic organizer, such as the one below, in your Science Journal. Use it to help you organize your ideas.
3. Write an outline of your story. Use it to guide you as you write the story.
4. List things you will need to research, pictures you will need to find or draw, and any other materials you will need. How will you bind your book? Will you make more than one copy? Make a list of the sources you used to research information for your story.

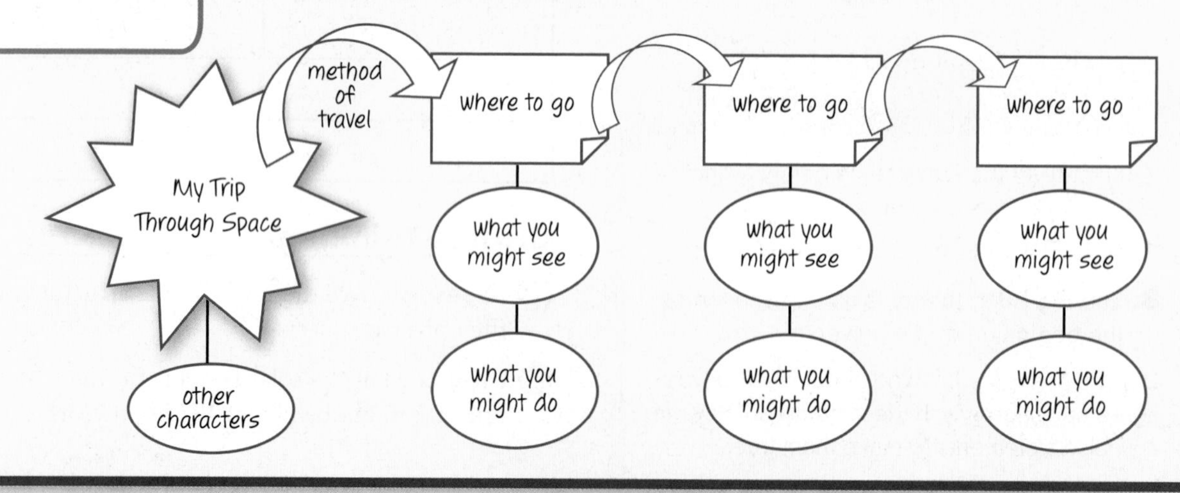

(t to b, 4)Hutchings Photography/Digital Light Source; (2, 5-6)Jacques Cornell/McGraw-Hill Education; (3)McGraw-Hill Education

5. Write your book. Add pictures or illustrations. Bind the pages together into book form.
6. Have a friend read your book and tell you if you succeeded in telling your story in an engaging way. What suggestions does your friend have for improvement?
7. Revise and improve the book based on your friend's suggestions.

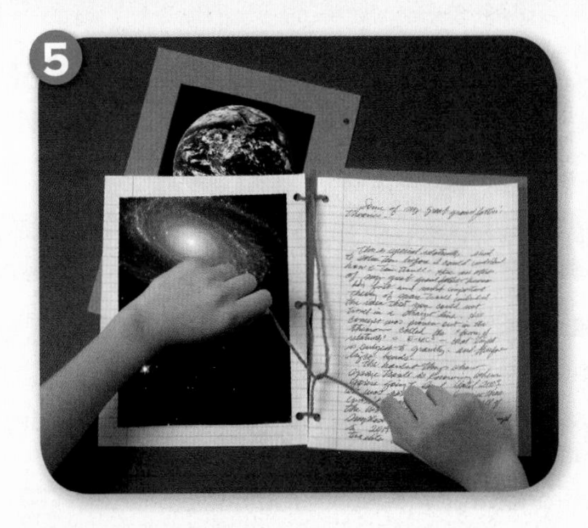

Analyze and Conclude

8. **Research Information** What new information did you learn as you did research for your book?
9. **Calculate** how many light-years you traveled from Earth.
10. **Draw Conclusions** How would your story be limited if you could only travel at the speed of light?
11. **The Big Idea** How does your story help people understand the size of the universe, what it contains, and how gravity affects it?

Communicate Your Results

You may wish to make a copy of your book and give it to the school library or add it to a library of books in your classroom.

Inquiry Extension

Combine your book with books written by other students in your class to make an almanac of the universe. Add pages that give statistics and other interesting facts about the universe.

Lab Tips

- Think about your audience as you plan your book. Are you writing it for young children or for students your own age? What kinds of books do you and your friends enjoy reading?
- What metaphors or other kinds of figurative language can you add to your writing that will draw readers into your story?

Remember to use scientific methods.

Make Observations → Ask a Question → Form a Hypothesis → Test your Hypothesis → Analyze and Conclude → Communicate Results

(t)Hutchings Photography/Digital Light Source; ©Brand X Pictures/PunchStock; NASA/JPLCaltech/S. Willner (Harvard-Smithsonian Center for Astrophysics); (b)NASA/JPL-Caltech

Chapter 22 Study Guide

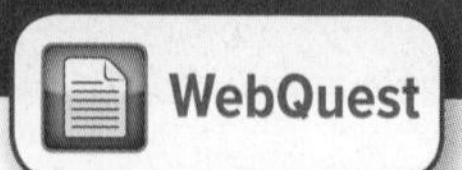

The universe is made up of stars, gas, and dust, as well as invisible dark matter. Material in the universe is not randomly arranged, but is pulled by gravity into galaxies.

Key Concepts Summary

Lesson 1: The View from Earth

- The sky is divided into 88 constellations.
- Astronomers learn about the energy, distance, temperature, and composition of stars by studying their light.
- Astronomers measure distances in space in **astronomical units** and in **light-years.** They measure star brightness as **apparent magnitude** and as **luminosity.**

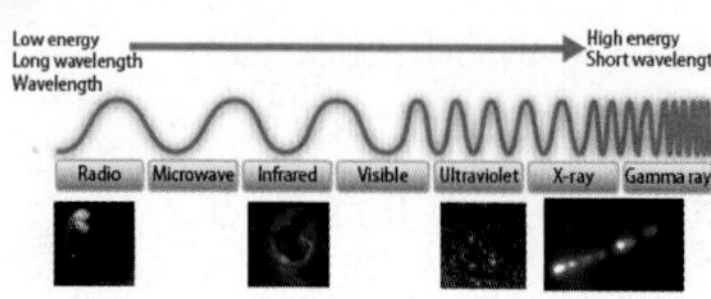

Vocabulary

spectroscope p. 803
astronomical unit p. 804
light-year p. 804
apparent magnitude p. 805
luminosity p. 805

Lesson 2: The Sun and Other Stars

- **Stars** shine because of **nuclear fusion** in their cores.
- Stars have a layered structure—they conduct energy through their **radiative zones** and their **convection zones** and release the energy at their **photospheres.**
- Sunspots, prominences, flares, and coronal mass ejections are temporary phenomena on the Sun.
- Astronomers classify stars by their temperatures and luminosities.

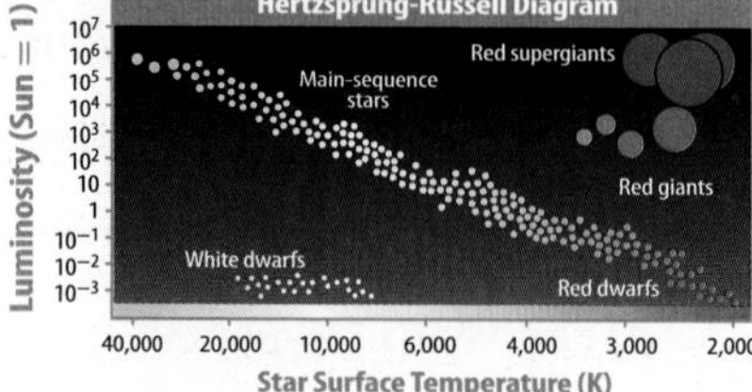

nuclear fusion p. 809
star p. 809
radiative zone p. 810
convection zone p. 810
photosphere p. 810
chromosphere p. 810
corona p. 810
Hertzsprung-Russell diagram p. 813

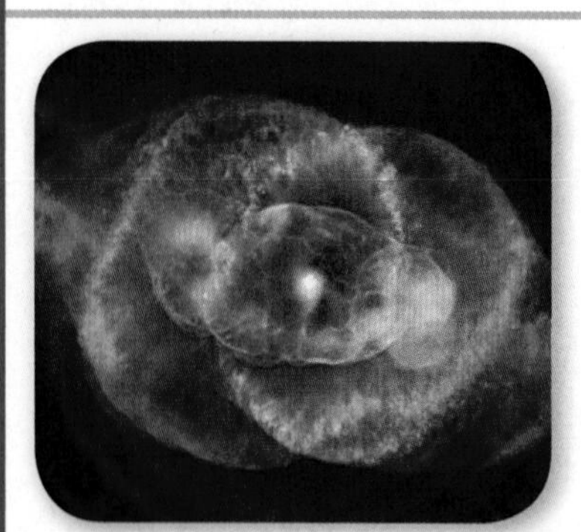

Lesson 3: Evolution of Stars

- Stars are born in clouds of gas and dust called **nebulae.**
- What happens to a star when it leaves the main sequence depends on its mass.
- Matter is recycled in the planetary nebulae of **white dwarfs** and the remnants of **supernovae.**

nebula p. 817
white dwarf p. 819
supernova p. 819
neutron star p. 820
black hole p. 820

Lesson 4 Galaxies and the Universe

- The three major types of **galaxies** are spiral, elliptical, and irregular.
- The Milky Way is the spiral galaxy that contains the solar system.
- The **Big Bang theory** explains the origin of the universe.

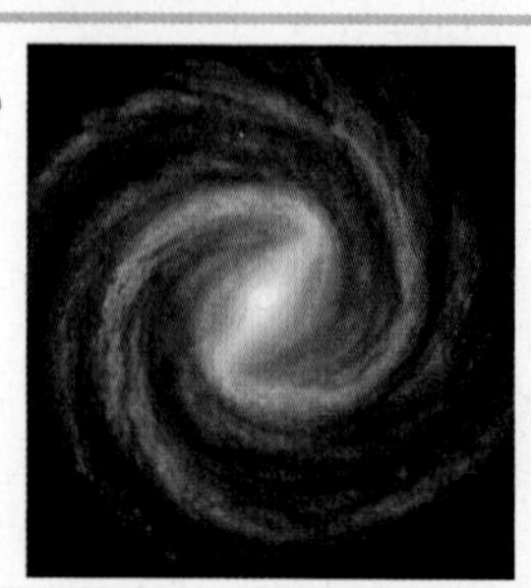

galaxy p. 825
dark matter p. 825
Big Bang theory p. 830
Doppler shift p. 830

X-ray: NASA/CXC/SAO; Optical: NASA/STScI

Study Guide

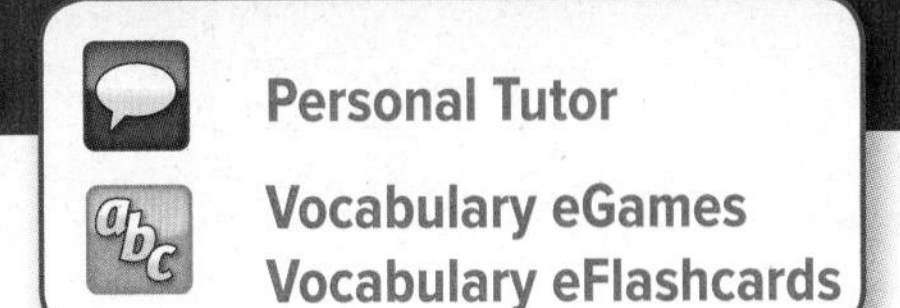

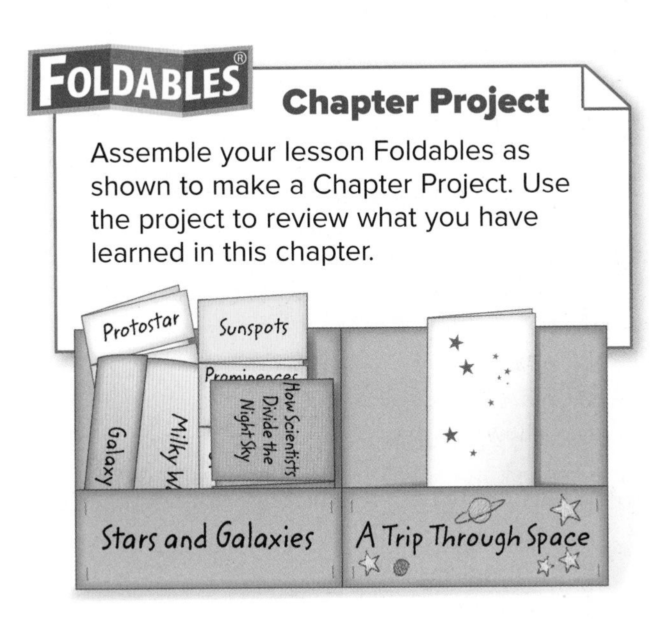

Use Vocabulary

1. **Explain** how nebulae are related to stars.
2. **Define** *Doppler shift.*
3. **Compare** neutron stars and black holes.
4. **Explain** the role of white dwarfs in recycling matter.
5. **Distinguish** between an astronomical unit and a light-year.
6. How does a convection zone transfer energy?
7. **Use the term** *dark matter* in a sentence.
8. On what diagram would you find a plot of stellar luminosity v. temperature?
9. **Compare** photosphere and corona.

Link Vocabulary and Key Concepts

Copy this concept map, and then use vocabulary terms from the previous page to complete the concept map.

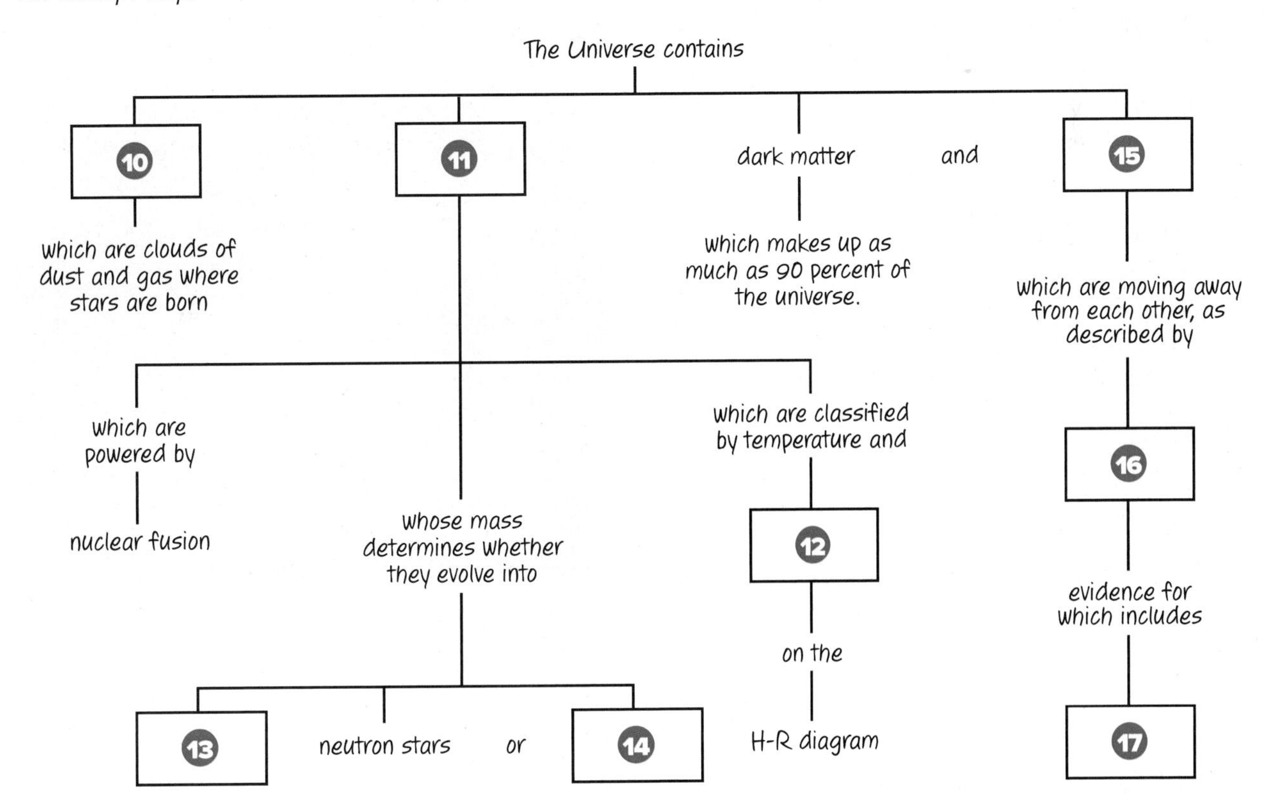

Chapter 22 Review

Understand Key Concepts 

1. Scientists divide the sky into
   A. astronomical units.
   B. clusters.
   C. constellations.
   D. light-years.

2. Which part of the Sun is marked with an *X* on the diagram below?

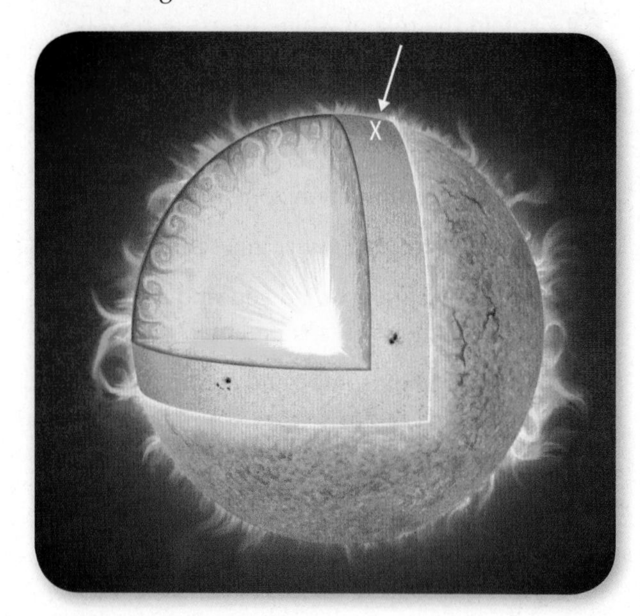

   A. convection zone
   B. corona
   C. photosphere
   D. radiative zone

3. Which might change, depending on the distance to a star?
   A. absolute magnitude
   B. apparent magnitude
   C. composition
   D. luminosity

4. Which is the average distance between Earth and the Sun?
   A. 1 AU
   B. 1 km
   C. 1 light-year
   D. 1 magnitude

5. Which is most important in determining the fate of a star?
   A. the star's color
   B. the star's distance
   C. the star's mass
   D. the star's temperature

6. What star along the main sequence will likely end in a supernova?
   A. blue-white
   B. orange
   C. red
   D. yellow

7. Which term does NOT belong with the others?
   A. black hole
   B. neutron star
   C. red dwarf
   D. supernova

8. What does the Big Bang theory state?
   A. The universe is ageless.
   B. The universe is collapsing.
   C. The universe is expanding.
   D. The universe is infinite.

9. Which type of galaxy is illustrated below?

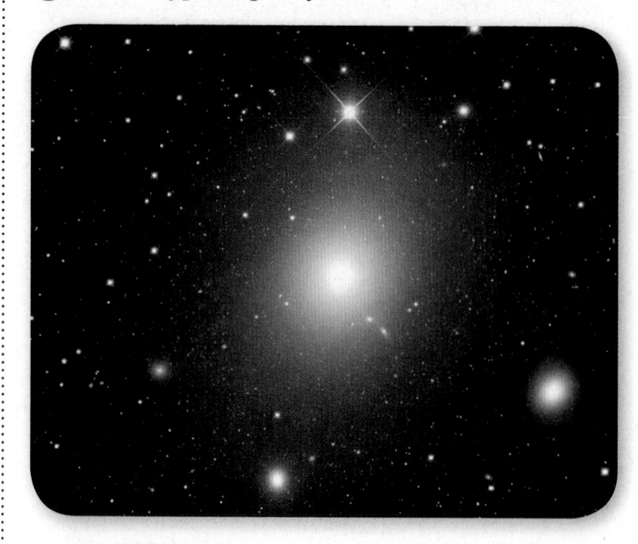

   A. elliptical
   B. irregular
   C. peculiar
   D. spiral

Robert Gendler/NASA

Critical Thinking

10. **Explain** how energy is released in a star.

11. **Assess** how the invention of the telescope changed people's views of the universe.

12. **Imagine** you are asked to classify 10,000 stars. Which properties would you measure?

13. **Deduce** why supernovae are needed for life on Earth.

14. **Predict** how the Sun would be different if it were twice as massive.

15. **Imagine** that you are writing to a friend who lives in the Virgo Cluster of galaxies. What would you write as your return address? Be specific.

16. **Interpret** The figure below shows part of the solar system. Explain what is happening.

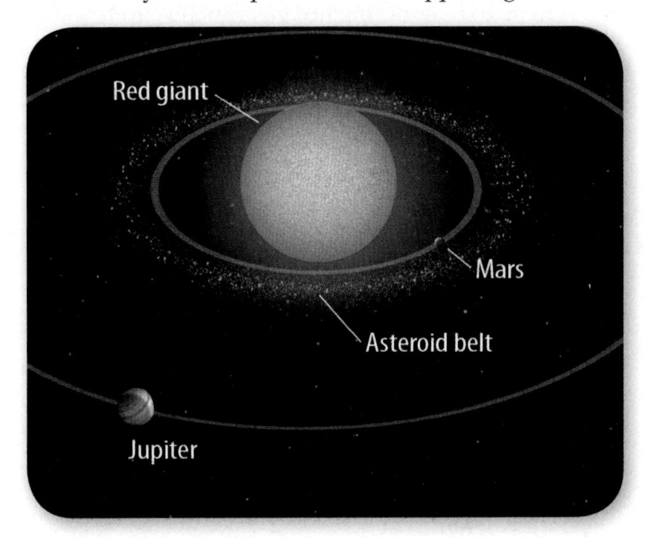

Writing in Science

17. **Write** You are a scientist being interviewed by a magazine on the topic of black holes. Write three questions an interviewer might ask, as well as your answers.

REVIEW THE BIG IDEA

18. What makes up the universe, and how does gravity affect the universe?

19. The photo below shows an image of the early universe obtained with the Hubble Space Telescope. Identify the objects you see. Make a list of other objects in the universe that you cannot see on this image.

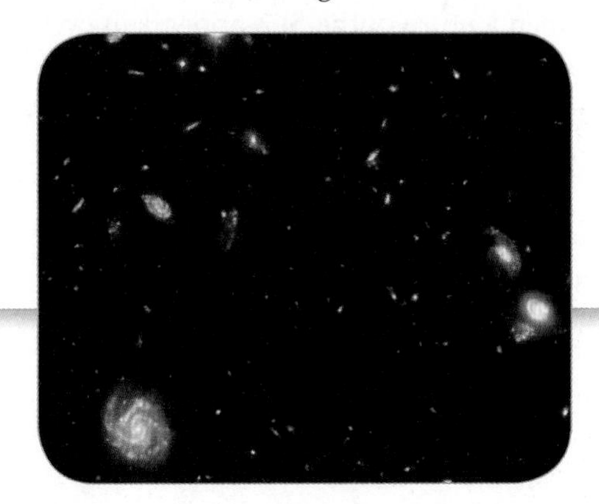

Math Skills

Math Practice

Use Proportions

20. The Milky Way galaxy is about 100,000 light-years across. What is this distance in kilometers?

21. Astronomers sometimes use a distance unit called a parsec. One parsec is 3.3 light-years. What is the distance, in parsecs, of a nebula that is 82.5 light-years away?

22. The distance to the Orion nebula is about 390 parsecs. What is this distance in light-years?

Standardized Test Practice

Record your answers on the answer sheet provided by your teacher or on a sheet of paper.

Multiple Choice

1 Which characteristics can by studied by analyzing a star's spectrum?

A absolute and apparent magnitudes

B formation and evolution

C movement and luminosity

D temperature and composition

2 Which feature of the Sun appears in cycles of about 11 years?

A coronal mass ejections

B solar flares

C solar wind

D sunspots

Use the graph below to answer question 3.

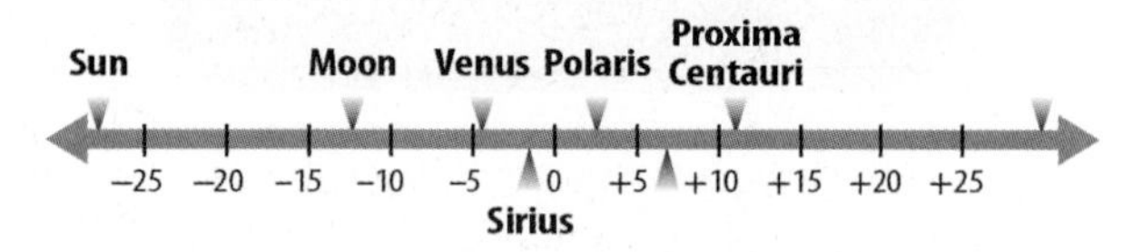

3 Which star on the graph has the greatest apparent magnitude?

A Polaris

B Proxima Centauri

C Sirius

D the Sun

4 Where in the Milky Way is the solar system located?

A at the edge of the disk

B inside a globular cluster

C near the supermassive black hole

D within the central bulge

Use the figure below to answer question 5.

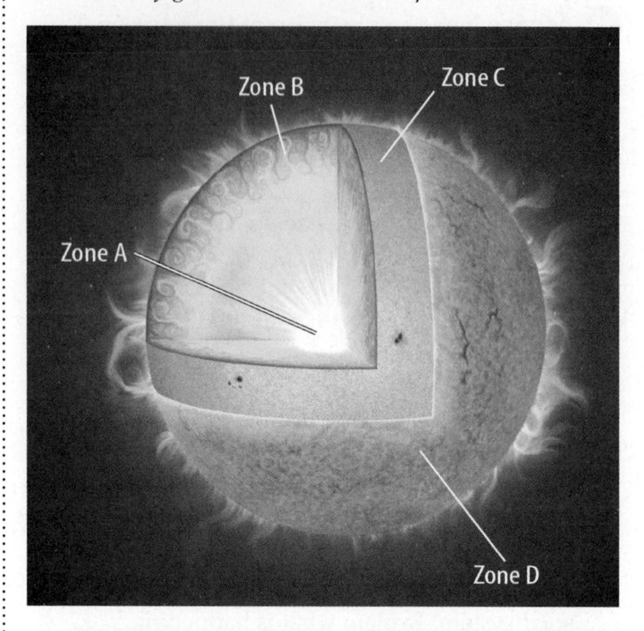

5 Which zone contains hot gas moving up toward the surface and cooler gas moving down toward the center of the Sun?

A zone A

B zone B

C zone C

D zone D

6 Which contains most of the mass of the universe?

A black holes

B dark matter

C gas and dust

D stars

7 Which stellar objects eventually form from the most massive stars?

A black holes

B diffuse nebulae

C planetary nebulae

D white dwarfs

Use the figure below to answer question 8.

8 Which is a characteristic for this type of galaxy?

A It contains no dust.

B It contains little gas.

C It contains many young stars.

D It contains mostly old stars.

9 Where do stars form?

A in black holes

B in constellations

C in nebulae

D in supernovae

10 What term describes the process that causes a star to shine?

A binary fission

B coronal mass ejection

C nuclear fusion

D stellar composition

11 What ancient star grouping do modern astronomers use to divide the sky into regions?

A astronomical unit

B constellation

C galaxy

D star cluster

Constructed Response

Use the diagram below to answer question 12.

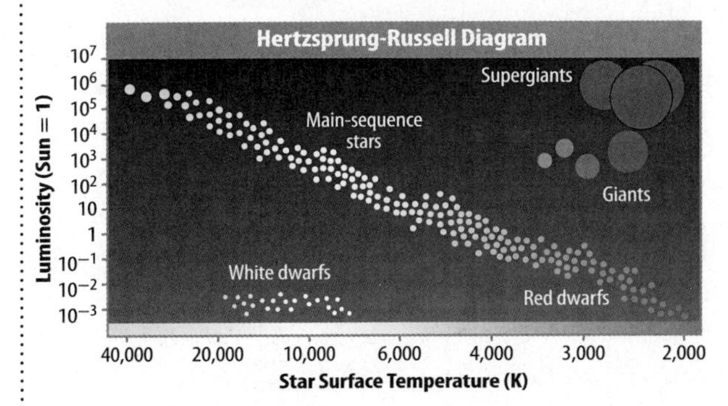

12 Use the information in the diagram above to describe red giants and white dwarfs based on their sizes, temperatures, and luminosities.

13 Describe the life cycle of a main-sequence star. What event causes the star to leave the main sequence?

14 How does the red shift of galaxies support the Big Bang theory?

15 Explain how planetary nebula recycle matter.

NEED EXTRA HELP?															
If You Missed Question...	1	2	3	4	5	6	7	8	9	10	11	12	13	14	15
Go to Lesson...	1	2	1	4	2	4	3	4	3	2	1	2	3	4	3

Student Resources

For Students and Parents/Guardians

These resources are designed to help you achieve success in science. You will find useful information on laboratory safety, math skills, and science skills. In addition, science reference materials are found in the Reference Handbook. You'll find the information you need to learn and sharpen your skills in these resources.

Table of Contents

SCIENCE SKILL HANDBOOK
MATH SKILL HANDBOOK
FOLDABLES HANDBOOK
REFERENCE HANDBOOK
GLOSSARY/GLOSARIO
INDEX

Scientific Methods

Scientists use an orderly approach called the scientific method to solve problems. This includes organizing and recording data so others can understand them. Scientists use many variations in this method when they solve problems.

Identify a Question

The first step in a scientific investigation or experiment is to identify a question to be answered or a problem to be solved. For example, you might ask which gasoline is the most efficient.

Gather and Organize Information

After you have identified your question, begin gathering and organizing information. There are many ways to gather information, such as researching in a library, interviewing those knowledgeable about the subject, and testing and working in the laboratory and field. Fieldwork is investigations and observations done outside of a laboratory.

Researching Information Before moving in a new direction, it is important to gather the information that already is known about the subject. Start by asking yourself questions to determine exactly what you need to know. Then you will look for the information in various reference sources, like the student is doing in **Figure 1.** Some sources may include textbooks, encyclopedias, government documents, professional journals, science magazines, and the Internet. Always list the sources of your information.

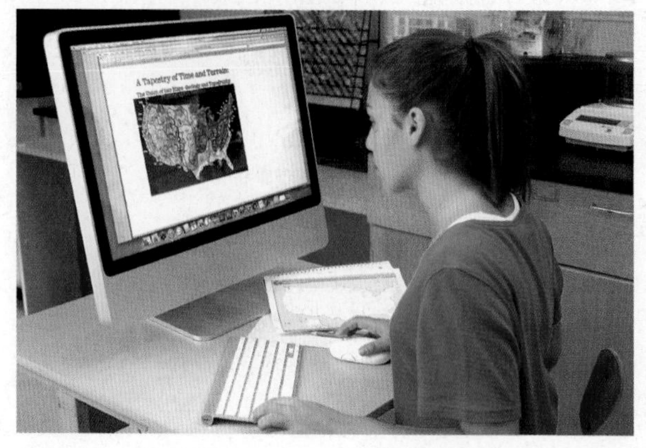

Figure 1 The Internet can be a valuable research tool.

Evaluate Sources of Information Not all sources of information are reliable. You should evaluate all of your sources of information, and use only those you know to be dependable. For example, if you are researching ways to make homes more energy efficient, a site written by the U.S. Department of Energy would be more reliable than a site written by a company that is trying to sell a new type of weatherproofing material. Also, remember that research always is changing. Consult the most current resources available to you. For example, a 1985 resource about saving energy would not reflect the most recent findings.

Sometimes scientists use data that they did not collect themselves, or conclusions drawn by other researchers. This data must be evaluated carefully. Ask questions about how the data were obtained, if the investigation was carried out properly, and if it has been duplicated exactly with the same results. Would you reach the same conclusion from the data? Only when you have confidence in the data can you believe it is true and feel comfortable using it.

Interpret Scientific Illustrations As you research a topic in science, you will see drawings, diagrams, and photographs to help you understand what you read. Some illustrations are included to help you understand an idea that you can't see easily by yourself, like the tiny particles in an atom in **Figure 2.** A drawing helps many people to remember details more easily and provides examples that clarify difficult concepts or give additional information about the topic you are studying. Most illustrations have labels or a caption to identify or to provide more information.

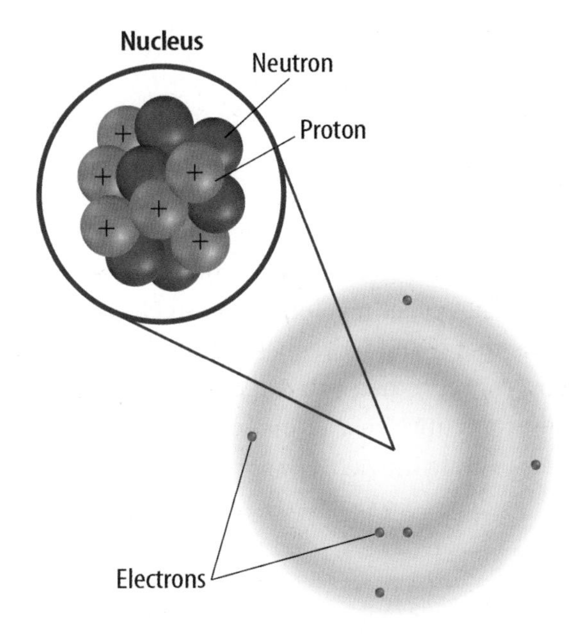

Figure 2 This drawing shows an atom of carbon with its six protons, six neutrons, and six electrons.

Concept Maps One way to organize data is to draw a diagram that shows relationships among ideas (or concepts). A concept map can help make the meanings of ideas and terms more clear, and help you understand and remember what you are studying. Concept maps are useful for breaking large concepts down into smaller parts, making learning easier.

Network Tree A type of concept map that not only shows a relationship, but how the concepts are related is a network tree, shown in **Figure 3.** In a network tree, the words are written in the ovals, while the description of the type of relationship is written across the connecting lines.

When constructing a network tree, write down the topic and all major topics on separate pieces of paper or notecards. Then arrange them in order from general to specific. Branch the related concepts from the major concept and describe the relationship on the connecting line. Continue to more specific concepts until finished.

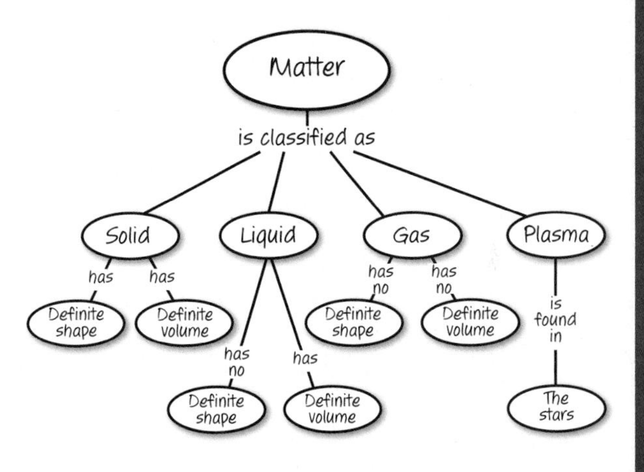

Figure 3 A network tree shows how concepts or objects are related.

Events Chain Another type of concept map is an events chain. Sometimes called a flow chart, it models the order or sequence of items. An events chain can be used to describe a sequence of events, the steps in a procedure, or the stages of a process.

When making an events chain, first find the one event that starts the chain. This event is called the initiating event. Then, find the next event and continue until the outcome is reached, as shown in **Figure 4** on the next page.

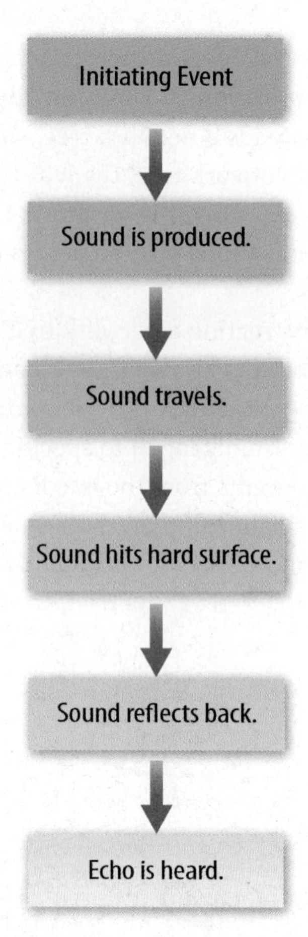

Figure 4 Events-chain concept maps show the order of steps in a process or event. This concept map shows how a sound makes an echo.

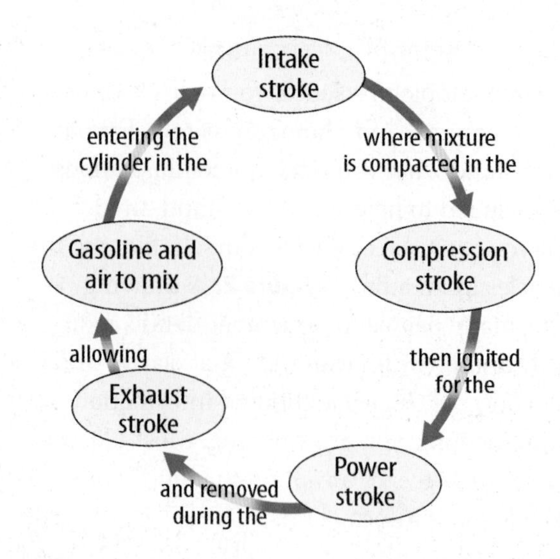

Figure 5 A cycle map shows events that occur in a cycle.

Cycle Map A specific type of events chain is a cycle map. It is used when the series of events do not produce a final outcome, but instead relate back to the beginning event, such as in **Figure 5.** Therefore, the cycle repeats itself.

To make a cycle map, first decide what event is the beginning event. This is also called the initiating event. Then list the next events in the order that they occur, with the last event relating back to the initiating event. Words can be written between the events that describe what happens from one event to the next. The number of events in a cycle map can vary, but usually contain three or more events.

Spider Map A type of concept map that you can use for brainstorming is the spider map. When you have a central idea, you might find that you have a jumble of ideas that relate to it but are not necessarily clearly related to each other. The spider map on sound in **Figure 6** shows that if you write these ideas outside the main concept, then you can begin to separate and group unrelated terms so they become more useful.

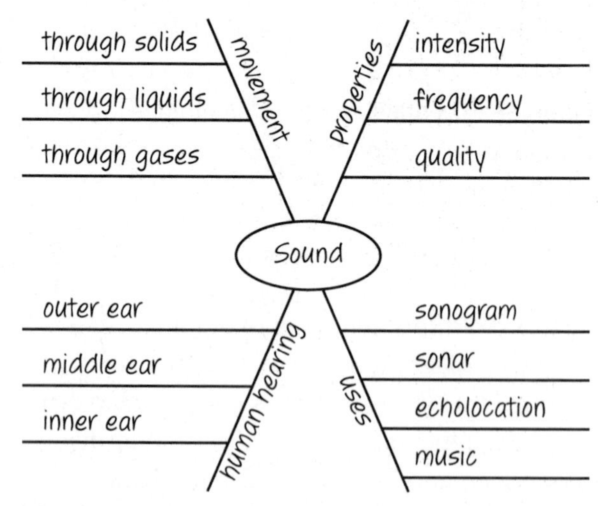

Figure 6 A spider map allows you to list ideas that relate to a central topic but not necessarily to one another.

SCIENCE SKILL HANDBOOK
MATH SKILL HANDBOOK
FOLDABLES HANDBOOK
REFERENCE HANDBOOK
GLOSSARY/GLOSARIO
INDEX

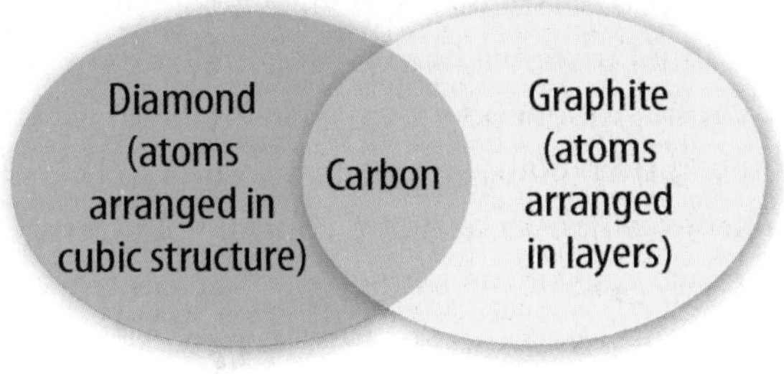

Diamond (atoms arranged in cubic structure)
Carbon
Graphite (atoms arranged in layers)

Figure 7 This Venn diagram compares and contrasts two substances made from carbon.

Venn Diagram To illustrate how two subjects compare and contrast you can use a Venn diagram. You can see the characteristics that the subjects have in common and those that they do not, shown in **Figure 7.**

To create a Venn diagram, draw two overlapping ovals that are big enough to write in. List the characteristics unique to one subject in one oval, and the characteristics of the other subject in the other oval. The characteristics in common are listed in the overlapping section.

Make and Use Tables One way to organize information so it is easier to understand is to use a table. Tables can contain numbers, words, or both.

To make a table, list the items to be compared in the first column and the characteristics to be compared in the first row. The title should clearly indicate the content of the table, and the column or row heads should be clear. Notice that in **Table 1** the units are included.

Table 1 Recyclables Collected During Week

Day of Week	Paper (kg)	Aluminum (kg)	Glass (kg)
Monday	5.0	4.0	12.0
Wednesday	4.0	1.0	10.0
Friday	2.5	2.0	10.0

Make a Model One way to help you better understand the parts of a structure, the way a process works, or to show things too large or small for viewing is to make a model. For example, an atomic model made of a plastic-ball nucleus and chenille stem electron shells can help you visualize how the parts of an atom relate to each other. Other types of models can be devised on a computer or represented by equations.

Form a Hypothesis

A possible explanation based on previous knowledge and observations is called a hypothesis. After researching gasoline types and recalling previous experiences in your family's car, you form a hypothesis—our car runs more efficiently because we use premium gasoline. To be valid, a hypothesis has to be something you can test by using an investigation.

Predict When you apply a hypothesis to a specific situation, you predict something about that situation. A prediction makes a statement in advance, based on prior observation, experience, or scientific reasoning. People use predictions to make everyday decisions. Scientists test predictions by performing investigations. Based on previous observations and experiences, you might form a prediction that cars are more efficient with premium gasoline. The prediction can be tested in an investigation.

Design an Experiment A scientist needs to make many decisions before beginning an investigation. Some of these include: how to carry out the investigation, what steps to follow, how to record the data, and how the investigation will answer the question. It also is important to address any safety concerns.

Test the Hypothesis

Now that you have formed your hypothesis, you need to test it. Using an investigation, you will make observations and collect data, or information. This data might either support or not support your hypothesis. Scientists collect and organize data as numbers and descriptions.

Follow a Procedure In order to know what materials to use, as well as how and in what order to use them, you must follow a procedure. **Figure 8** shows a procedure you might follow to test your hypothesis.

Procedure

Step 1 Use regular gasoline for two weeks.

Step 2 Record the number of kilometers between fill-ups and the amount of gasoline used.

Step 3 Switch to premium gasoline for two weeks.

Step 4 Record the number of kilometers between fill-ups and the amount of gasoline used.

Figure 8 A procedure tells you what to do step-by-step.

Identify and Manipulate Variables and Controls In any experiment, it is important to keep everything the same except for the item you are testing. The one factor you change is called the independent variable. The change that results is the dependent variable. Make sure you have only one independent variable, to assure yourself of the cause of the changes you observe in the dependent variable. For example, in your gasoline experiment the type of fuel is the independent variable. The dependent variable is the efficiency.

Many experiments also have a control—an individual instance or experimental subject for which the independent variable is not changed. You can then compare the test results to the control results. To design a control you must have two cars of the same type. The control car uses regular gasoline for four weeks. After you are done with the test, you can compare the experimental results to the control results.

Collect Data

Whether you are carrying out an investigation or a short observational experiment, you will collect data, as shown in **Figure 9.** Scientists collect data as numbers and descriptions and organize them in specific ways.

Observe Scientists observe items and events, then record what they see. When they use only words to describe an observation, it is called qualitative data. Scientists' observations also can describe how much there is of something. These observations use numbers, as well as words, in the description and are called quantitative data. For example, if a sample of the element gold is described as being "shiny and very dense" the data are qualitative. Quantitative data on this sample of gold might include "a mass of 30 g and a density of 19.3 g/cm^3."

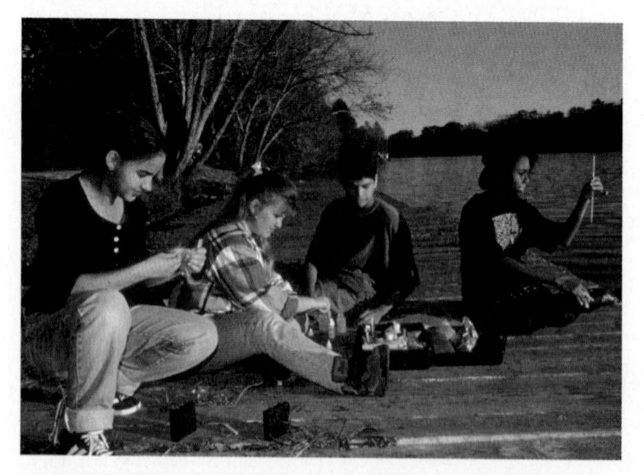

Figure 9 Collecting data is one way to gather information directly.

SCIENCE SKILL HANDBOOK
MATH SKILL HANDBOOK
FOLDABLES HANDBOOK
REFERENCE HANDBOOK
GLOSSARY/GLOSARIO
INDEX

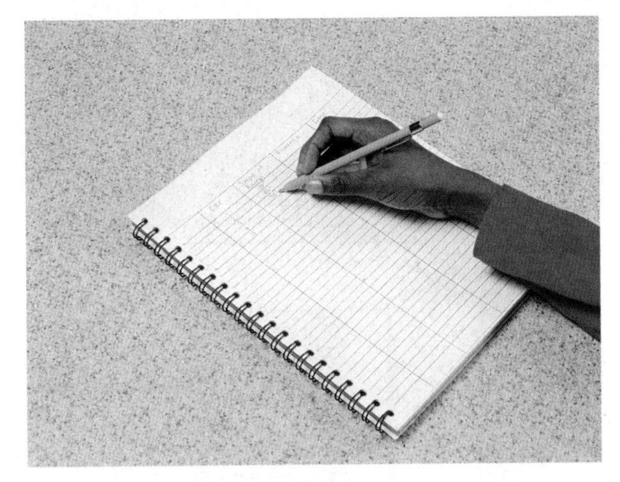

Figure 10 Record data neatly and clearly so it is easy to understand.

When you make observations, you should examine the entire object or situation first, and then look carefully for details. It is important to record observations accurately and completely. Always record your notes immediately as you make them, so you do not miss details or make a mistake when recording results from memory. Never put unidentified observations on scraps of paper. Instead they should be recorded in a notebook, like the one in **Figure 10.** Write your data neatly so you can easily read it later. At each point in the experiment, record your observations and label them. That way, you will not have to determine what the figures mean when you look at your notes later. Set up any tables that you will need to use ahead of time, so you can record any observations right away. Remember to avoid bias when collecting data by not including personal thoughts when you record observations. Record only what you observe.

Estimate Scientific work also involves estimating. To estimate is to make a judgment about the size or the number of something without measuring or counting. This is important when the number or size of an object or population is too large or too difficult to accurately count or measure.

Sample Scientists may use a sample or a portion of the total number as a type of estimation. To sample is to take a small, representative portion of the objects or organisms of a population for research. By making careful observations or manipulating variables within that portion of the group, information is discovered and conclusions are drawn that might apply to the whole population. A poorly chosen sample can be unrepresentative of the whole. If you were trying to determine the rainfall in an area, it would not be best to take a rainfall sample from under a tree.

Measure You use measurements every day. Scientists also take measurements when collecting data. When taking measurements, it is important to know how to use measuring tools properly. Accuracy also is important.

Length The SI unit for length is the meter (m). Smaller measurements might be measured in centimeters or millimeters.

Length is measured using a metric ruler or meterstick. When using a metric ruler, line up the 0-cm mark with the end of the object being measured and read the number of the unit where the object ends. Look at the metric ruler shown in **Figure 11.** The centimeter lines are the long, numbered lines, and the shorter lines are millimeter lines. In this instance, the length would be 4.50 cm.

Figure 11 This metric ruler has centimeter and millimeter divisions.

Mass The SI unit for mass is the kilogram (kg). Scientists can measure mass using units formed by adding metric prefixes to the unit gram (g), such as milligram (mg). To measure mass, you might use a triple-beam balance similar to the one shown in **Figure 12.** The balance has a pan on one side and a set of beams on the other side. Each beam has a rider that slides on the beam.

When using a triple-beam balance, place an object on the pan. Slide the largest rider along its beam until the pointer drops below zero. Then move it back one notch. Repeat the process for each rider proceeding from the larger to smaller until the pointer swings an equal distance above and below the zero point. Sum the masses on each beam to find the mass of the object. Move all riders back to zero when finished.

Instead of putting materials directly on the balance, scientists often take a tare of a container. A tare is the mass of a container into which objects or substances are placed for measuring their masses. To find the mass of objects or substances, find the mass of a clean container. Remove the container from the pan, and place the object or substances in the container. Find the mass of the container with the materials in it. Subtract the mass of the empty container from the mass of the filled container to find the mass of the materials you are using.

Figure 12 A triple-beam balance is used to determine the mass of an object.

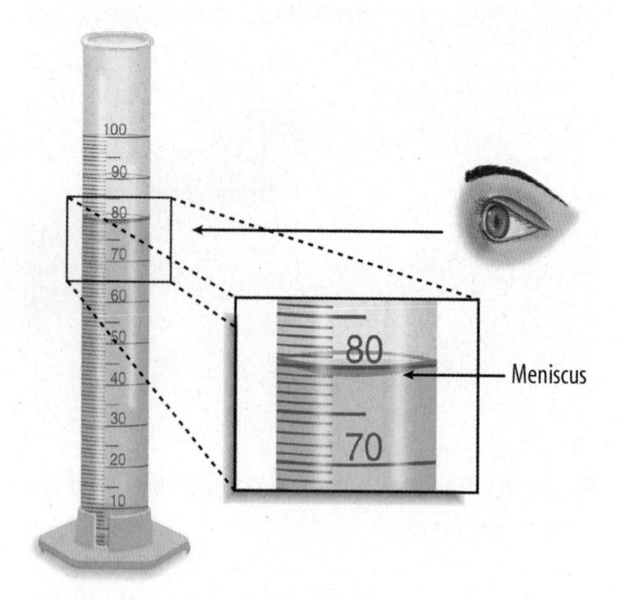

Figure 13 Graduated cylinders measure liquid volume.

Liquid Volume The SI unit for measuring liquids is the liter (l). When a smaller unit is needed, scientists might use a milliliter. Because a milliliter takes up the volume of a cube measuring 1 cm on each side it also can be called a cubic centimeter ($cm^3 = cm \times cm \times cm$).

You can use beakers and graduated cylinders to measure liquid volume. A graduated cylinder, shown in **Figure 13,** is marked from bottom to top in milliliters. In lab, you might use a 10-mL graduated cylinder or a 100-mL graduated cylinder. When measuring liquids, notice that the liquid has a curved surface. Look at the surface at eye level, and measure the bottom of the curve. This is called the meniscus. The graduated cylinder in **Figure 13** contains 79.0 mL, or 79.0 cm^3, of a liquid.

Temperature Scientists often measure temperature using the Celsius scale. Pure water has a freezing point of 0°C and boiling point of 100°C. The unit of measurement is degrees Celsius. Two other scales often used are the Fahrenheit and Kelvin scales.

SCIENCE SKILL HANDBOOK
MATH SKILL HANDBOOK
FOLDABLES HANDBOOK
REFERENCE HANDBOOK
GLOSSARY/GLOSARIO
INDEX

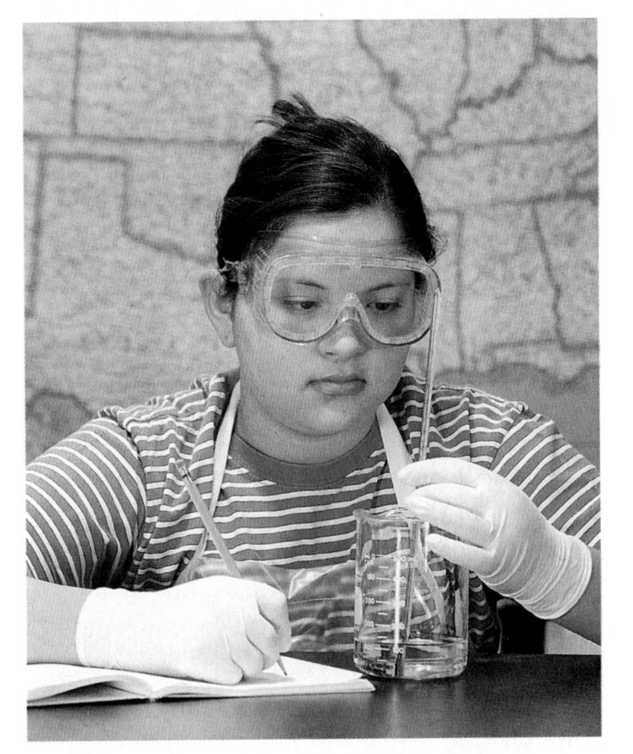

Figure 14 A thermometer measures the temperature of an object.

Scientists use a thermometer to measure temperature. Most thermometers in a laboratory are glass tubes with a bulb at the bottom end containing a liquid such as colored alcohol. The liquid rises or falls with a change in temperature. To read a glass thermometer like the thermometer in **Figure 14,** rotate it slowly until a red line appears. Read the temperature where the red line ends.

Form Operational Definitions

An operational definition defines an object by how it functions, works, or behaves. For example, when you are playing hide and seek and a tree is home base, you have created an operational definition for a tree.

Objects can have more than one operational definition. For example, a ruler can be defined as a tool that measures the length of an object (how it is used). It can also be a tool with a series of marks used as a standard when measuring (how it works).

Analyze the Data

To determine the meaning of your observations and investigation results, you will need to look for patterns in the data. Then you must think critically to determine what the data mean. Scientists use several approaches when they analyze the data they have collected and recorded. Each approach is useful for identifying specific patterns.

Interpret Data

The word *interpret* means "to explain the meaning of something." When analyzing data from an experiment, try to find out what the data show. Identify the control group and the test group to see whether changes in the independent variable have had an effect. Look for differences in the dependent variable between the control and test groups.

Classify

Sorting objects or events into groups based on common features is called classifying. When classifying, first observe the objects or events to be classified. Then select one feature that is shared by some members in the group, but not by all. Place those members that share that feature in a subgroup. You can classify members into smaller and smaller subgroups based on characteristics. Remember that when you classify, you are grouping objects or events for a purpose. Keep your purpose in mind as you select the features to form groups and subgroups.

Compare and Contrast

Observations can be analyzed by noting the similarities and differences between two or more objects or events that you observe. When you look at objects or events to see how they are similar, you are comparing them. Contrasting is looking for differences in objects or events.

Recognize Cause and Effect A cause is a reason for an action or condition. The effect is that action or condition. When two events happen together, it is not necessarily true that one event caused the other. Scientists must design a controlled investigation to recognize the exact cause and effect.

Draw Conclusions

When scientists have analyzed the data they collected, they proceed to draw conclusions about the data. These conclusions are sometimes stated in words similar to the hypothesis that you formed earlier. They may confirm a hypothesis, or lead you to a new hypothesis.

Infer Scientists often make inferences based on their observations. An inference is an attempt to explain observations or to indicate a cause. An inference is not a fact, but a logical conclusion that needs further investigation. For example, you may infer that a fire has caused smoke. Until you investigate, however, you do not know for sure.

Apply When you draw a conclusion, you must apply those conclusions to determine whether the data supports the hypothesis. If your data do not support your hypothesis, it does not mean that the hypothesis is wrong. It means only that the result of the investigation did not support the hypothesis. Maybe the experiment needs to be redesigned, or some of the initial observations on which the hypothesis was based were incomplete or biased. Perhaps more observation or research is needed to refine your hypothesis. A successful investigation does not always come out the way you originally predicted.

Avoid Bias Sometimes a scientific investigation involves making judgments. When you make a judgment, you form an opinion. It is important to be honest and not to allow any expectations of results to bias your judgments. This is important throughout the entire investigation, from researching to collecting data to drawing conclusions.

Communicate

The communication of ideas is an important part of the work of scientists. A discovery that is not reported will not advance the scientific community's understanding or knowledge. Communication among scientists also is important as a way of improving their investigations.

Scientists communicate in many ways, from writing articles in journals and magazines that explain their investigations and experiments, to announcing important discoveries on television and radio. Scientists also share ideas with colleagues on the Internet or present them as lectures, like the student is doing in **Figure 15.**

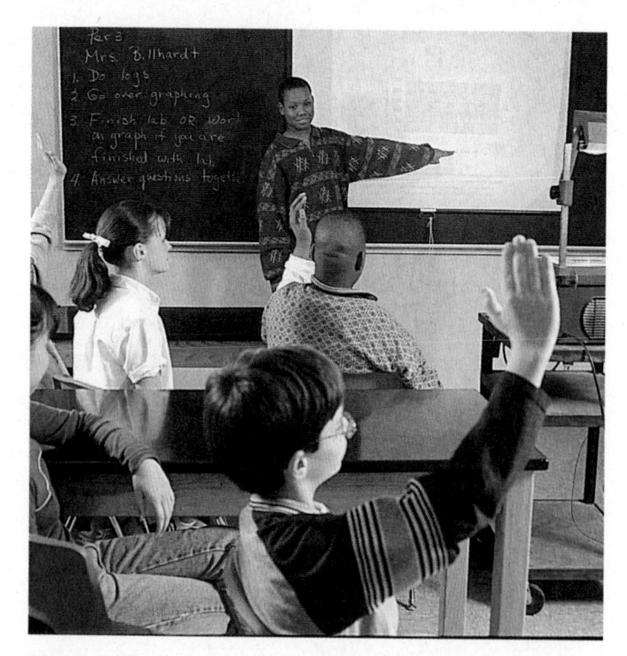

Figure 15 A student communicates to his peers about his investigation.

These safety symbols are used in laboratory and field investigations in this book to indicate possible hazards. Learn the meaning of each symbol and refer to this page often. *Remember to wash your hands thoroughly after completing lab procedures.*

PROTECTIVE EQUIPMENT

Do not begin any lab without the proper protection equipment.

GOGGLES

Proper eye protection must be worn when performing or observing science activities that involve items or conditions as listed below.

APRON

Wear an approved apron when using substances that could stain, wet, or destroy cloth.

SOAP

Wash hands with soap and water before removing goggles and after all lab activities.

GLOVES

Wear gloves when working with biological materials, chemicals, animals, or materials that can stain or irritate hands.

LABORATORY HAZARDS

Symbols	Potential Hazards	Precaution	Response
DISPOSAL	contamination of classroom or environment due to improper disposal of materials such as chemicals and live specimens	• DO NOT dispose of hazardous materials in the sink or trash can. • Dispose of wastes as directed by your teacher.	• If hazardous materials are disposed of improperly, notify your teacher immediately.
EXTREME TEMPERATURE	skin burns due to extremely hot or cold materials such as hot glass, liquids, or metals; liquid nitrogen; dry ice	• Use proper protective equipment, such as hot mitts and/or tongs, when handling objects with extreme temperatures.	• If injury occurs, notify your teacher immediately.
SHARP OBJECTS	punctures or cuts from sharp objects such as razor blades, pins, scalpels, and broken glass	• Handle glassware carefully to avoid breakage. • Walk with sharp objects pointed downward, away from you and others.	• If broken glass or injury occurs, notify your teacher immediately.
ELECTRICAL	electric shock or skin burn due to improper grounding, short circuits, liquid spills, or exposed wires	• Check condition of wires and apparatus for fraying or uninsulated wires, and broken or cracked equipment. • Use only GFCI-protected outlets	• DO NOT attempt to fix electrical problems. Notify your teacher immediately.
CHEMICAL	skin irritation or burns, breathing difficulty, and/or poisoning due to touching, swallowing, or inhalation of chemicals such as acids, bases, bleach, metal compounds, iodine, poinsettias, pollen, ammonia, acetone, nail polish remover, heated chemicals, mothballs, and any other chemicals labeled or known to be dangerous	• Wear proper protective equipment such as goggles, apron, and gloves when using chemicals. • Ensure proper room ventilation or use a fume hood when using materials that produce fumes. • NEVER smell fumes directly. • NEVER taste or eat any material in the laboratory.	• If contact occurs, immediately flush affected area with water and notify your teacher. • If a spill occurs, leave the area immediately and notify your teacher.
FLAMMABLE	unexpected fire due to liquids or gases that ignite easily such as rubbing alcohol	• Avoid open flames, sparks, or heat when flammable liquids are present.	• If a fire occurs, leave the area immediately and notify your teacher.
OPEN FLAME	burns or fire due to open flame from matches, Bunsen burners, or burning materials	• Tie back loose hair and clothing. • Keep flame away from all materials. • Follow teacher instructions when lighting and extinguishing flames. • Use proper protection, such as hot mitts or tongs, when handling hot objects.	• If a fire occurs, leave the area immediately and notify your teacher.
ANIMAL SAFETY	injury to or from laboratory animals	• Wear proper protective equipment such as gloves, apron, and goggles when working with animals. • Wash hands after handling animals.	• If injury occurs, notify your teacher immediately.
BIOLOGICAL	infection or adverse reaction due to contact with organisms such as bacteria, fungi, and biological materials such as blood, animal or plant materials	• Wear proper protective equipment such as gloves, goggles, and apron when working with biological materials. • Avoid skin contact with an organism or any part of the organism. • Wash hands after handling organisms.	• If contact occurs, wash the affected area and notify your teacher immediately.
FUME	breathing difficulties from inhalation of fumes from substances such as ammonia, acetone, nail polish remover, heated chemicals, and mothballs	• Wear goggles, apron, and gloves. • Ensure proper room ventilation or use a fume hood when using substances that produce fumes. • NEVER smell fumes directly.	• If a spill occurs, leave area and notify your teacher immediately.
IRRITANT	irritation of skin, mucous membranes, or respiratory tract due to materials such as acids, bases, bleach, pollen, mothballs, steel wool, and potassium permanganate	• Wear goggles, apron, and gloves. • Wear a dust mask to protect against fine particles.	• If skin contact occurs, immediately flush the affected area with water and notify your teacher.
RADIOACTIVE	excessive exposure from alpha, beta, and gamma particles	• Remove gloves and wash hands with soap and water before removing remainder of protective equipment.	• If cracks or holes are found in the container, notify your teacher immediately.

Safety in the Science Laboratory

Introduction to Science Safety

The science laboratory is a safe place to work if you follow standard safety procedures. Being responsible for your own safety helps to make the entire laboratory a safer place for everyone. When performing any lab, read and apply the caution statements and safety symbol listed at the beginning of the lab.

General Safety Rules

1. Complete the *Lab Safety Form* or other safety contract BEFORE starting any science lab.
2. Study the procedure. Ask your teacher any questions. Be sure you understand safety symbols shown on the page.
3. Notify your teacher about allergies or other health conditions that can affect your participation in a lab.
4. Learn and follow use and safety procedures for your equipment. If unsure, ask your teacher.
5. Never eat, drink, chew gum, apply cosmetics, or do any personal grooming in the lab. Never use lab glassware as food or drink containers. Keep your hands away from your face and mouth.
6. Know the location and proper use of the safety shower, eye wash, fire blanket, and fire alarm.

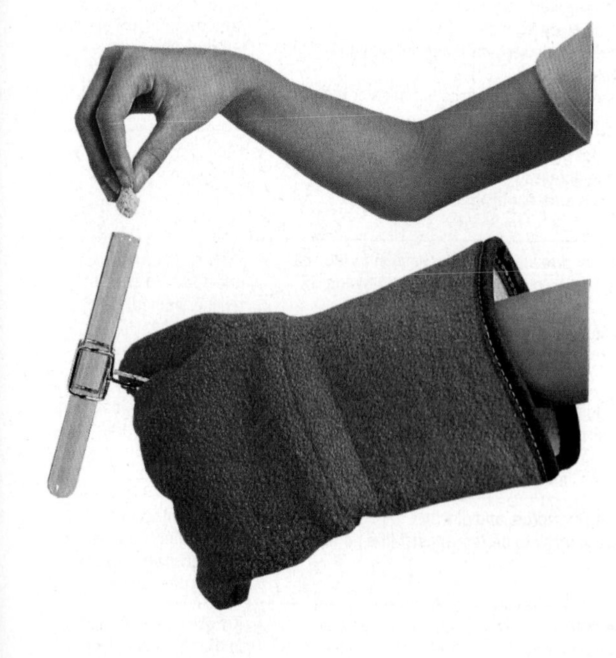

Prevent Accidents

1. Use the safety equipment provided to you. Goggles and a safety apron should be worn during investigations.
2. Do NOT use hair spray, mousse, or other flammable hair products. Tie back long hair and tie down loose clothing.
3. Do NOT wear sandals or other open-toed shoes in the lab.
4. Remove jewelry on hands and wrists. Loose jewelry, such as chains and long necklaces, should be removed to prevent them from getting caught in equipment.
5. Do not taste any substances or draw any material into a tube with your mouth.
6. Proper behavior is expected in the lab. Practical jokes and fooling around can lead to accidents and injury.
7. Keep your work area uncluttered.

Laboratory Work

1. Collect and carry all equipment and materials to your work area before beginning a lab.
2. Remain in your own work area unless given permission by your teacher to leave it.

KS Studios

SCIENCE SKILL HANDBOOK
MATH SKILL HANDBOOK
FOLDABLES HANDBOOK
REFERENCE HANDBOOK
GLOSSARY/GLOSARIO
INDEX

3. Always slant test tubes away from yourself and others when heating them, adding substances to them, or rinsing them.
4. If instructed to smell a substance in a container, hold the container a short distance away and fan vapors toward your nose.
5. Do NOT substitute other chemicals/substances for those in the materials list unless instructed to do so by your teacher.
6. Do NOT take any materials or chemicals outside of the laboratory.
7. Stay out of storage areas unless instructed to be there and supervised by your teacher.

Laboratory Cleanup

1. Turn off all burners, water, and gas, and disconnect all electrical devices.
2. Clean all pieces of equipment and return all materials to their proper places.
3. Dispose of chemicals and other materials as directed by your teacher. Place broken glass and solid substances in the proper containers. Never discard materials in the sink.
4. Clean your work area.
5. Wash your hands with soap and water thoroughly BEFORE removing your goggles.

Emergencies

1. Report any fire, electrical shock, glassware breakage, spill, or injury, no matter how small, to your teacher immediately. Follow his or her instructions.
2. If your clothing should catch fire, STOP, DROP, and ROLL. If possible, smother it with the fire blanket or get under a safety shower. NEVER RUN.
3. If a fire should occur, turn off all gas and leave the room according to established procedures.
4. In most instances, your teacher will clean up spills. Do NOT attempt to clean up spills unless you are given permission and instructions to do so.
5. If chemicals come into contact with your eyes or skin, notify your teacher immediately. Use the eyewash, or flush your skin or eyes with large quantities of water.
6. The fire extinguisher and first-aid kit should only be used by your teacher unless it is an extreme emergency and you have been given permission.
7. If someone is injured or becomes ill, only a professional medical provider or someone certified in first aid should perform first-aid procedures.

Math Review

Use Fractions

A fraction compares a part to a whole. In the fraction $\frac{2}{3}$, the 2 represents the part and is the numerator. The 3 represents the whole and is the denominator.

Reduce Fractions To reduce a fraction, you must find the largest factor that is common to both the numerator and the denominator, the greatest common factor (GCF). Divide both numbers by the GCF. The fraction has then been reduced, or it is in its simplest form.

Example

Twelve of the 20 chemicals in the science lab are in powder form. What fraction of the chemicals used in the lab are in powder form?

Step 1 Write the fraction.

$\frac{\text{part}}{\text{whole}} = \frac{12}{20}$

Step 2 To find the GCF of the numerator and denominator, list all of the factors of each number.

Factors of 12: 1, 2, 3, 4, 6, 12 (the numbers that divide evenly into 12)

Factors of 20: 1, 2, 4, 5, 10, 20 (the numbers that divide evenly into 20)

Step 3 List the common factors.

1, 2, 4

Step 4 Choose the greatest factor in the list. The GCF of 12 and 20 is 4.

Step 5 Divide the numerator and denominator by the GCF.

$\frac{12 \div 4}{20 \div 4} = \frac{3}{5}$

In the lab, $\frac{3}{5}$ of the chemicals are in powder form.

Practice Problem At an amusement park, 66 of 90 rides have a height restriction. What fraction of the rides, in its simplest form, has a height restriction?

Add and Subtract Fractions with Like Denominators To add or subtract fractions with the same denominator, add or subtract the numerators and write the sum or difference over the denominator. After finding the sum or difference, find the simplest form for your fraction.

Example 1

In the forest outside your house, $\frac{1}{8}$ of the animals are rabbits, $\frac{3}{8}$ are squirrels, and the remainder are birds and insects. How many are mammals?

Step 1 Add the numerators.

$\frac{1}{8} + \frac{3}{8} = \frac{(1 + 3)}{8} = \frac{4}{8}$

Step 2 Find the GCF.

$\frac{4}{8}$ (GCF, 4)

Step 3 Divide the numerator and denominator by the GCF.

$\frac{4 \div 4}{8 \div 4} = \frac{1}{2}$

$\frac{1}{2}$ of the animals are mammals.

Example 2

If $\frac{7}{16}$ of the Earth is covered by freshwater, and $\frac{1}{16}$ of that is in glaciers, how much freshwater is not frozen?

Step 1 Subtract the numerators.

$\frac{7}{16} - \frac{1}{16} = \frac{(7 - 1)}{16} = \frac{6}{16}$

Step 2 Find the GCF.

$\frac{6}{16}$ (GCF, 2)

Step 3 Divide the numerator and denominator by the GCF.

$\frac{6 \div 2}{16 \div 2} = \frac{3}{8}$

$\frac{3}{8}$ of the freshwater is not frozen.

Practice Problem A bicycle rider is riding at a rate of 15 km/h for $\frac{4}{9}$ of his ride, 10 km/h for $\frac{2}{9}$ of his ride, and 8 km/h for the remainder of the ride. How much of his ride is he riding at a rate greater than 8 km/h?

Add and Subtract Fractions with Unlike Denominators To add or subtract fractions with unlike denominators, first find the least common denominator (LCD). This is the smallest number that is a common multiple of both denominators. Rename each fraction with the LCD, and then add or subtract. Find the simplest form if necessary.

Example 1

A chemist makes a paste that is $\frac{1}{2}$ table salt (NaCl), $\frac{1}{3}$ sugar ($C_6H_{12}O_6$), and the remainder is water (H_2O). How much of the paste is a solid?

Step 1 Find the LCD of the fractions.

$\frac{1}{2} + \frac{1}{3}$ (LCD, 6)

Step 2 Rename each numerator and each denominator with the LCD.

Step 3 Add the numerators.

$\frac{3}{6} + \frac{2}{6} = \frac{(3 + 2)}{6} = \frac{5}{6}$

$\frac{5}{6}$ of the paste is a solid.

Example 2

The average precipitation in Grand Junction, CO, is $\frac{7}{10}$ inch in November, and $\frac{3}{5}$ inch in December. What is the total average precipitation?

Step 1 Find the LCD of the fractions.

$\frac{7}{10} + \frac{3}{5}$ (LCD, 10)

Step 2 Rename each numerator and each denominator with the LCD.

Step 3 Add the numerators.

$\frac{7}{10} + \frac{6}{10} = \frac{(7 + 6)}{10} = \frac{13}{10}$

$\frac{13}{10}$ inches total precipitation, or $1\frac{3}{10}$ inches.

Practice Problem On an electric bill, about $\frac{1}{8}$ of the energy is from solar energy and about $\frac{1}{10}$ is from wind power. How much of the total bill is from solar energy and wind power combined?

Example 3

In your body, $\frac{7}{10}$ of your muscle contractions are involuntary (cardiac and smooth muscle tissue). Smooth muscle makes $\frac{3}{15}$ of your muscle contractions. How many of your muscle contractions are made by cardiac muscle?

Step 1 Find the LCD of the fractions.

$\frac{7}{10} - \frac{3}{15}$ (LCD, 30)

Step 2 Rename each numerator and each denominator with the LCD.

$\frac{7 \times 3}{10 \times 3} = \frac{21}{30}$

$\frac{3 \times 2}{15 \times 2} = \frac{6}{30}$

Step 3 Subtract the numerators.

$\frac{21}{30} - \frac{6}{30} = \frac{(21 - 6)}{30} = \frac{15}{30}$

Step 4 Find the GCF.

$\frac{15}{30}$ (GCF, 15)

$\frac{1}{2}$

$\frac{1}{2}$ of all muscle contractions are cardiac muscle.

Example 4

Tony wants to make cookies that call for $\frac{3}{4}$ of a cup of flour, but he only has $\frac{1}{3}$ of a cup. How much more flour does he need?

Step 1 Find the LCD of the fractions.

$\frac{3}{4} - \frac{1}{3}$ (LCD, 12)

Step 2 Rename each numerator and each denominator with the LCD.

$\frac{3 \times 3}{4 \times 3} = \frac{9}{12}$

$\frac{1 \times 4}{3 \times 4} = \frac{4}{12}$

Step 3 Subtract the numerators.

$\frac{9}{12} - \frac{4}{12} = \frac{(9 - 4)}{12} = \frac{5}{12}$

$\frac{5}{12}$ of a cup of flour

Practice Problem Using the information provided to you in Example 3 above, determine how many muscle contractions are voluntary (skeletal muscle).

Multiply Fractions To multiply with fractions, multiply the numerators and multiply the denominators. Find the simplest form if necessary.

Example

Multiply $\frac{3}{5}$ by $\frac{1}{3}$.

Step 1 Multiply the numerators and denominators.

$\frac{3}{5} \times \frac{1}{3} = \frac{(3 \times 1)}{(5 \times 3)} \frac{3}{15}$

Step 2 Find the GCF.

$\frac{3}{15}$ (GCF, 3)

Step 3 Divide the numerator and denominator by the GCF.

$\frac{3 \div 3}{15 \div 3} = \frac{1}{5}$

$\frac{3}{5}$ multiplied by $\frac{1}{3}$ is $\frac{1}{5}$.

Practice Problem Multiply $\frac{3}{14}$ by $\frac{5}{16}$.

Find a Reciprocal Two numbers whose product is 1 are called multiplicative inverses, or reciprocals.

Example

Find the reciprocal of $\frac{3}{8}$.

Step 1 Inverse the fraction by putting the denominator on top and the numerator on the bottom.

$\frac{8}{3}$

The reciprocal of $\frac{3}{8}$ is $\frac{8}{3}$.

Practice Problem Find the reciprocal of $\frac{4}{9}$.

Divide Fractions To divide one fraction by another fraction, multiply the dividend by the reciprocal of the divisor. Find the simplest form if necessary.

Example 1

Divide $\frac{1}{9}$ by $\frac{1}{3}$.

Step 1 Find the reciprocal of the divisor.

The reciprocal of $\frac{1}{3}$ is $\frac{3}{1}$.

Step 2 Multiply the dividend by the reciprocal of the divisor.

$\frac{\frac{1}{9}}{\frac{1}{3}} = \frac{1}{9} \times \frac{3}{1} = \frac{(1 \times 3)}{(9 \times 1)} = \frac{3}{9}$

Step 3 Find the GCF.

$\frac{3}{9}$ (GCF, 3)

Step 4 Divide the numerator and denominator by the GCF.

$\frac{3 \div 3}{9 \div 3} = \frac{1}{3}$

$\frac{1}{9}$ divided by $\frac{1}{3}$ is $\frac{1}{3}$.

Example 2

Divide $\frac{3}{5}$ by $\frac{1}{4}$.

Step 1 Find the reciprocal of the divisor.

The reciprocal of $\frac{1}{4}$ is $\frac{4}{1}$.

Step 2 Multiply the dividend by the reciprocal of the divisor.

$\frac{\frac{3}{5}}{\frac{1}{4}} = \frac{3}{5} \times \frac{4}{1} = \frac{(3 \times 4)}{(5 \times 1)} = \frac{12}{5}$

$\frac{3}{5}$ divided by $\frac{1}{4}$ is $\frac{12}{5}$ or $2\frac{2}{5}$.

Practice Problem Divide $\frac{3}{11}$ by $\frac{7}{10}$.

Use Ratios

When you compare two numbers by division, you are using a ratio. Ratios can be written 3 to 5, 3:5, or $\frac{3}{5}$. Ratios, like fractions, also can be written in simplest form.

Ratios can represent one type of probability, called odds. This is a ratio that compares the number of ways a certain outcome occurs to the number of possible outcomes. For example, if you flip a coin 100 times, what are the odds that it will come up heads? There are two possible outcomes, heads or tails, so the odds of coming up heads are 50:100. Another way to say this is that 50 out of 100 times the coin will come up heads. In its simplest form, the ratio is 1:2.

Example 1

A chemical solution contains 40 g of salt and 64 g of baking soda. What is the ratio of salt to baking soda as a fraction in simplest form?

Step 1 Write the ratio as a fraction.

$$\frac{\text{salt}}{\text{baking soda}} = \frac{40}{64}$$

Step 2 Express the fraction in simplest form. The GCF of 40 and 64 is 8.

$$\frac{40}{64} = \frac{40 \div 8}{64 \div 8} = \frac{5}{8}$$

The ratio of salt to baking soda in the chemical solution is 5:8.

Example 2

Sean rolls a 6-sided die 6 times. What are the odds that the side with a 3 will show?

Step 1 Write the ratio as a fraction.

$$\frac{\text{number of sides with a 3}}{\text{number of possible sides}} = \frac{1}{6}$$

Step 2 Multiply by the number of attempts.

$\frac{1}{6} \times 6$ attempts $= \frac{6}{6}$ attempts $= 1$ attempt

1 attempt out of 6 will show a 3.

Practice Problem Two metal rods measure 100 cm and 144 cm in length. What is the ratio of their lengths in simplest form?

Use Decimals

A fraction with a denominator that is a power of ten can be written as a decimal. For example, 0.27 means $\frac{27}{100}$. The decimal point separates the ones place from the tenths place.

Any fraction can be written as a decimal using division. For example, the fraction $\frac{5}{8}$ can be written as a decimal by dividing 5 by 8. Written as a decimal, it is 0.625.

Add or Subtract Decimals When adding and subtracting decimals, line up the decimal points before carrying out the operation.

Example 1

Find the sum of 47.68 and 7.80.

Step 1 Line up the decimal places when you write the numbers.

$$\begin{array}{r} 47.68 \\ +\ 7.80 \\ \hline \end{array}$$

Step 2 Add the decimals.

$$\begin{array}{r} {}^{1\,1} \\ 47.68 \\ +\ 7.80 \\ \hline 55.48 \end{array}$$

The sum of 47.68 and 7.80 is 55.48.

Example 2

Find the difference of 42.17 and 15.85.

Step 1 Line up the decimal places when you write the number.

$$\begin{array}{r} 42.17 \\ -15.85 \\ \hline \end{array}$$

Step 2 Subtract the decimals.

$$\begin{array}{r} {}^{3\,11} \\ 42.17 \\ -15.85 \\ \hline 26.32 \end{array}$$

The difference of 42.17 and 15.85 is 26.32.

Practice Problem Find the sum of 1.245 and 3.842.

Multiply Decimals To multiply decimals, multiply the numbers like numbers without decimal points. Count the decimal places in each factor. The product will have the same number of decimal places as the sum of the decimal places in the factors.

Example

Multiply 2.4 by 5.9.

Step 1 Multiply the factors like two whole numbers.

$24 \times 59 = 1416$

Step 2 Find the sum of the number of decimal places in the factors. Each factor has one decimal place, for a sum of two decimal places.

Step 3 The product will have two decimal places.

14.16

The product of 2.4 and 5.9 is 14.16.

Practice Problem Multiply 4.6 by 2.2.

Divide Decimals When dividing decimals, change the divisor to a whole number. To do this, multiply both the divisor and the dividend by the same power of ten. Then place the decimal point in the quotient directly above the decimal point in the dividend. Then divide as you do with whole numbers.

Example

Divide 8.84 by 3.4.

Step 1 Multiply both factors by 10.

$3.4 \times 10 = 34, 8.84 \times 10 = 88.4$

Step 2 Divide 88.4 by 34.

$$\begin{array}{r} 2.6 \\ 34\overline{)88.4} \\ -68 \\ \hline 204 \\ -204 \\ \hline 0 \end{array}$$

8.84 divided by 3.4 is 2.6.

Practice Problem Divide 75.6 by 3.6.

Use Proportions

An equation that shows that two ratios are equivalent is a proportion. The ratios $\frac{2}{4}$ and $\frac{5}{10}$ are equivalent, so they can be written as $\frac{2}{4} = \frac{5}{10}$. This equation is a proportion.

When two ratios form a proportion, the cross products are equal. To find the cross products in the proportion $\frac{2}{4} = \frac{5}{10}$, multiply the 2 and the 10, and the 4 and the 5. Therefore $2 \times 10 = 4 \times 5$, or $20 = 20$.

Because you know that both ratios are equal, you can use cross products to find a missing term in a proportion. This is known as solving the proportion.

Example

The heights of a tree and a pole are proportional to the lengths of their shadows. The tree casts a shadow of 24 m when a 6-m pole casts a shadow of 4 m. What is the height of the tree?

Step 1 Write a proportion.

$$\frac{\text{height of tree}}{\text{height of pole}} = \frac{\text{length of tree's shadow}}{\text{length of pole's shadow}}$$

Step 2 Substitute the known values into the proportion. Let h represent the unknown value, the height of the tree.

$$\frac{h}{6} \times \frac{24}{4}$$

Step 3 Find the cross products.

$h \times 4 = 6 \times 24$

Step 4 Simplify the equation.

$4h = 144$

Step 5 Divide each side by 4.

$$\frac{4h}{4} = \frac{144}{4}$$

$h = 36$

The height of the tree is 36 m.

Practice Problem The ratios of the weights of two objects on the Moon and on Earth are in proportion. A rock weighing 3 N on the Moon weighs 18 N on Earth. How much would a rock that weighs 5 N on the Moon weigh on Earth?

SCIENCE SKILL HANDBOOK
MATH SKILL HANDBOOK
FOLDABLES HANDBOOK
REFERENCE HANDBOOK
GLOSSARY/GLOSARIO
INDEX

Use Percentages

The word *percent* means "out of one hundred." It is a ratio that compares a number to 100. Suppose you read that 77 percent of Earth's surface is covered by water. That is the same as reading that the fraction of Earth's surface covered by water is $\frac{77}{100}$. To express a fraction as a percent, first find the equivalent decimal for the fraction. Then, multiply the decimal by 100 and add the percent symbol.

Example 1

Express $\frac{13}{20}$ as a percent.

Step 1 Find the equivalent decimal for the fraction.

$$\begin{array}{r} 0.65 \\ 20\overline{)13.00} \\ \underline{12\ 0} \\ 1\ 00 \\ \underline{1\ 00} \\ 0 \end{array}$$

Step 2 Rewrite the fraction $\frac{13}{20}$ as 0.65.

Step 3 Multiply 0.65 by 100 and add the % symbol.

$0.65 \times 100 = 65 = 65\%$

So, $\frac{13}{20} = 65\%$.

This also can be solved as a proportion.

Example 2

Express $\frac{13}{20}$ as a percent.

Step 1 Write a proportion.

$\frac{13}{20} = \frac{x}{100}$

Step 2 Find the cross products.

$1300 = 20x$

Step 3 Divide each side by 20.

$\frac{1300}{20} = \frac{20x}{20}$

$65 = x = 65\%$

So, $\frac{13}{20} = 65\%$

Practice Problem In one year, 73 of 365 days were rainy in one city. What percent of the days in that city were rainy?

Solve One-Step Equations

A statement that two expressions are equal is an equation. For example, $A = B$ is an equation that states that A is equal to B.

An equation is solved when a variable is replaced with a value that makes both sides of the equation equal. To make both sides equal the inverse operation is used. Addition and subtraction are inverses, and multiplication and division are inverses.

Example 1

Solve the equation $x - 10 = 35$.

Step 1 Find the solution by adding 10 to each side of the equation.

$$x - 10 = 35$$
$$x - 10 + 10 = 35 + 10$$
$$x = 45$$

Step 2 Check the solution.

$$x - 10 = 35$$
$$45 - 10 = 35$$
$$35 = 35$$

Both sides of the equation are equal, so $x = 45$.

Example 2

In the formula $a = bc$, find the value of c if $a = 20$ and $b = 2$.

Step 1 Rearrange the formula so the unknown value is by itself on one side of the equation by dividing both sides by b.

$$a = bc$$
$$\frac{a}{b} = \frac{bc}{b}$$
$$\frac{a}{b} = c$$

Step 2 Replace the variables a and b with the values that are given.

$$\frac{a}{b} = c$$
$$\frac{20}{2} = c$$
$$10 = c$$

Step 3 Check the solution.

$$a = bc$$
$$20 = 2 \times 10$$
$$20 = 20$$

Both sides of the equation are equal, so $c = 10$ is the solution when $a = 20$ and $b = 2$.

Practice Problem In the formula $h = gd$, find the value of d if $g = 12.3$ and $h = 17.4$.

Use Statistics

The branch of mathematics that deals with collecting, analyzing, and presenting data is statistics. In statistics, there are three common ways to summarize data with a single number—the mean, the median, and the mode.

The **mean** of a set of data is the arithmetic average. It is found by adding the numbers in the data set and dividing by the number of items in the set.

The **median** is the middle number in a set of data when the data are arranged in numerical order. If there were an even number of data points, the median would be the mean of the two middle numbers.

The **mode** of a set of data is the number or item that appears most often.

Another number that often is used to describe a set of data is the range. The **range** is the difference between the largest number and the smallest number in a set of data.

Example

The speeds (in m/s) for a race car during five different time trials are 39, 37, 44, 36, and 44.

To find the mean:

Step 1 Find the sum of the numbers.

39 + 37 + 44 + 36 + 44 = 200

Step 2 Divide the sum by the number of items, which is 5.

200 ÷ 5 = 40

The mean is 40 m/s.

To find the median:

Step 1 Arrange the measures from least to greatest.

36, 37, 39, 44, 44

Step 2 Determine the middle measure.

36, 37, 39, 44, 44

The median is 39 m/s.

To find the mode:

Step 1 Group the numbers that are the same together.

44, 44, 36, 37, 39

Step 2 Determine the number that occurs most in the set.

44, 44, 36, 37, 39

The mode is 44 m/s.

To find the range:

Step 1 Arrange the measures from greatest to least.

44, 44, 39, 37, 36

Step 2 Determine the greatest and least measures in the set.

44, 44, 39, 37, 36

Step 3 Find the difference between the greatest and least measures.

44 − 36 = 8

The range is 8 m/s.

Practice Problem Find the mean, median, mode, and range for the data set 8, 4, 12, 8, 11, 14, 16.

A **frequency table** shows how many times each piece of data occurs, usually in a survey. **Table 1** below shows the results of a student survey on favorite color.

Table 1 Student Color Choice

Color	Tally	Frequency
red	IIII	4
blue	𝍸	5
black	II	2
green	III	3
purple	𝍸 II	7
yellow	𝍸 I	6

Based on the frequency table data, which color is the favorite?

Use Geometry

The branch of mathematics that deals with the measurement, properties, and relationships of points, lines, angles, surfaces, and solids is called geometry.

Perimeter The **perimeter** (P) is the distance around a geometric figure. To find the perimeter of a rectangle, add the length and width and multiply that sum by two, or $2(l + w)$. To find perimeters of irregular figures, add the length of all the sides.

Example 1

Find the perimeter of a rectangle that is 3 m long and 5 m wide.

Step 1 You know that the perimeter is 2 times the sum of the width and length.

$P = 2(3 \text{ m} + 5 \text{ m})$

Step 2 Find the sum of the width and length.

$P = 2(8 \text{ m})$

Step 3 Multiply by 2.

$P = 16 \text{ m}$

The perimeter is 16 m.

Example 2

Find the perimeter of a shape with sides measuring 2 cm, 5 cm, 6 cm, 3 cm.

Step 1 You know that the perimeter is the sum of all the sides.

$P = 2 + 5 + 6 + 3$

Step 2 Find the sum of the sides.

$P = 2 + 5 + 6 + 3$

$P = 16$

The perimeter is 16 cm.

Practice Problem Find the perimeter of a rectangle with a length of 18 m and a width of 7 m.

Practice Problem Find the perimeter of a triangle measuring 1.6 cm by 2.4 cm by 2.4 cm.

Area of a Rectangle The **area** (A) is the number of square units needed to cover a surface. To find the area of a rectangle, multiply the length times the width, or $l \times w$. When finding area, the units also are multiplied. Area is given in square units.

Example

Find the area of a rectangle with a length of 1 cm and a width of 10 cm.

Step 1 You know that the area is the length multiplied by the width.

$A = (1 \text{ cm} \times 10 \text{ cm})$

Step 2 Multiply the length by the width. Also multiply the units.

$A = 10 \text{ cm}^2$

The area is 10 cm^2.

Practice Problem Find the area of a square whose sides measure 4 m.

Area of a Triangle To find the area of a triangle, use the formula:

$A = \frac{1}{2}(\text{base} \times \text{height})$

The base of a triangle can be any of its sides. The height is the perpendicular distance from a base to the opposite endpoint, or vertex.

Example

Find the area of a triangle with a base of 18 m and a height of 7 m.

Step 1 You know that the area is $\frac{1}{2}$ the base times the height.

$A = \frac{1}{2}(18 \text{ m} \times 7 \text{ m})$

Step 2 Multiply $\frac{1}{2}$ by the product of 18×7. Multiply the units.

$A = \frac{1}{2}(126 \text{ m}^2)$

$A = 63 \text{ m}^2$

The area is 63 m^2.

Practice Problem Find the area of a triangle with a base of 27 cm and a height of 17 cm.

Circumference of a Circle The **diameter** (d) of a circle is the distance across the circle through its center, and the **radius** (r) is the distance from the center to any point on the circle. The radius is half of the diameter. The distance around the circle is called the **circumference** (C). The formula for finding the circumference is:

$C = 2\pi r$ or $C = \pi d$

The circumference divided by the diameter is always equal to 3.1415926... This nonterminating and nonrepeating number is represented by the Greek letter π (pi). An approximation often used for π is 3.14.

Example 1

Find the circumference of a circle with a radius of 3 m.

Step 1 You know the formula for the circumference is 2 times the radius times π.

$C = 2\pi(3)$

Step 2 Multiply 2 times the radius.

$C = 6\pi$

Step 3 Multiply by π.

$C \approx 19$ m

The circumference is about 19 m.

Example 2

Find the circumference of a circle with a diameter of 24.0 cm.

Step 1 You know the formula for the circumference is the diameter times π.

$C = \pi(24.0)$

Step 2 Multiply the diameter by π.

$C \approx 75.4$ cm

The circumference is about 75.4 cm.

Practice Problem Find the circumference of a circle with a radius of 19 cm.

Area of a Circle The formula for the area of a circle is: $A = \pi r^2$

Example 1

Find the area of a circle with a radius of 4.0 cm.

Step 1 $A = \pi(4.0)^2$

Step 2 Find the square of the radius.

$A = 16\pi$

Step 3 Multiply the square of the radius by π.

$A \approx 50\ cm^2$

The area of the circle is about 50 cm^2.

Example 2

Find the area of a circle with a radius of 225 m.

Step 1 $A = \pi(225)^2$

Step 2 Find the square of the radius.

$A = 50625\pi$

Step 3 Multiply the square of the radius by π.

$A \approx 159043.1$

The area of the circle is about 159043.1 m^2.

Example 3

Find the area of a circle whose diameter is 20.0 mm.

Step 1 Remember that the radius is half of the diameter.

$A = \pi\left(\frac{20.0}{2}\right)^2$

Step 2 Find the radius.

$A = \pi(10.0)^2$

Step 3 Find the square of the radius.

$A = 100\pi$

Step 4 Multiply the square of the radius by π.

$A \approx 314\ mm^2$

The area of the circle is about 314 mm^2.

Practice Problem Find the area of a circle with a radius of 16 m.

Volume The measure of space occupied by a solid is the **volume** (V). To find the volume of a rectangular solid, multiply the length times width times height, or $V = l \times w \times h$. It is measured in cubic units, such as cubic centimeters (cm^3).

Example

Find the volume of a rectangular solid with a length of 2.0 m, a width of 4.0 m, and a height of 3.0 m.

Step 1 You know the formula for volume is the length times the width times the height.

$V = 2.0 \text{ m} \times 4.0 \text{ m} \times 3.0 \text{ m}$

Step 2 Multiply the length times the width times the height.

$V = 24 \text{ m}^3$

The volume is 24 m^3.

Practice Problem Find the volume of a rectangular solid that is 8 m long, 4 m wide, and 4 m high.

To find the volume of other solids, multiply the area of the base times the height.

Example 1

Find the volume of a solid that has a triangular base with a length of 8.0 m and a height of 7.0 m. The height of the entire solid is 15.0 m.

Step 1 You know that the base is a triangle, and the area of a triangle is $\frac{1}{2}$ the base times the height, and the volume is the area of the base times the height.

$V = \left[\frac{1}{2}(b \times h)\right] \times 15$

Step 2 Find the area of the base.

$V = \left[\frac{1}{2}(8 \times 7)\right] \times 15$

$V = \left(\frac{1}{2} \times 56\right) \times 15$

Step 3 Multiply the area of the base by the height of the solid.

$V = 28 \times 15$

$V = 420 \text{ m}^3$

The volume is 420 m^3.

Example 2

Find the volume of a cylinder that has a base with a radius of 12.0 cm, and a height of 21.0 cm.

Step 1 You know that the base is a circle, and the area of a circle is the square of the radius times π, and the volume is the area of the base times the height.

$V = (\pi r^2) \times 21$

$V = (\pi 12^2) \times 21$

Step 2 Find the area of the base.

$V = 144\pi \times 21$

$V = 452 \times 21$

Step 3 Multiply the area of the base by the height of the solid.

$V \approx 9{,}500 \text{ cm}^3$

The volume is about 9,500 cm^3.

Example 3

Find the volume of a cylinder that has a diameter of 15 mm and a height of 4.8 mm.

Step 1 You know that the base is a circle with an area equal to the square of the radius times π. The radius is one-half the diameter. The volume is the area of the base times the height.

$V = (\pi r^2) \times 4.8$

$V = \left[\pi\left(\frac{1}{2} \times 15\right)^2\right] \times 4.8$

$V = (\pi 7.5^2) \times 4.8$

Step 2 Find the area of the base.

$V = 56.25\pi \times 4.8$

$V \approx 176.71 \times 4.8$

Step 3 Multiply the area of the base by the height of the solid.

$V \approx 848.2$

The volume is about 848.2 mm^3.

Practice Problem Find the volume of a cylinder with a diameter of 7 cm in the base and a height of 16 cm.

Science Applications

Measure in SI

The metric system of measurement was developed in 1795. A modern form of the metric system, called the International System (SI), was adopted in 1960 and provides the standard measurements that all scientists around the world can understand.

The SI system is convenient because unit sizes vary by powers of 10. Prefixes are used to name units. Look at **Table 2** for some common SI prefixes and their meanings.

Table 2 Common SI Prefixes

Prefix	Symbol	Meaning	
kilo–	k	1,000	thousandth
hecto–	h	100	hundred
deka–	da	10	ten
deci–	d	0.1	tenth
centi–	c	0.01	hundreth
milli–	m	0.001	thousandth

Example

How many grams equal one kilogram?

Step 1 Find the prefix *kilo–* in **Table 2.**

Step 2 Using **Table 2,** determine the meaning of *kilo–*. According to the table, it means 1,000. When the prefix *kilo–* is added to a unit, it means that there are 1,000 of the units in a "kilounit."

Step 3 Apply the prefix to the units in the question. The units in the question are grams. There are 1,000 grams in a kilogram.

Practice Problem Is a milligram larger or smaller than a gram? How many of the smaller units equal one larger unit? What fraction of the larger unit does one smaller unit represent?

Dimensional Analysis

Convert SI Units In science, quantities such as length, mass, and time sometimes are measured using different units. A process called dimensional analysis can be used to change one unit of measure to another. This process involves multiplying your starting quantity and units by one or more conversion factors. A conversion factor is a ratio equal to one and can be made from any two equal quantities with different units. If 1,000 mL equal 1 L then two ratios can be made.

$$\frac{1{,}000\ \text{mL}}{1\ \text{L}} = \frac{1\ \text{L}}{1{,}000\ \text{mL}} = 1$$

One can convert between units in the SI system by using the equivalents in **Table 2** to make conversion factors.

Example

How many cm are in 4 m?

Step 1 Write conversion factors for the units given. From **Table 2,** you know that 100 cm = 1 m. The conversion factors are

$$\frac{100\ \text{cm}}{1\ \text{m}} \ \textit{and}\ \frac{1\ \text{m}}{100\ \text{cm}}$$

Step 2 Decide which conversion factor to use. Select the factor that has the units you are converting from (m) in the denominator and the units you are converting to (cm) in the numerator.

$$\frac{100\ \text{cm}}{1\ \text{m}}$$

Step 3 Multiply the starting quantity and units by the conversion factor. Cancel the starting units with the units in the denominator. There are 400 cm in 4 m.

$$4\ \cancel{\text{m}} = \frac{100\ \text{cm}}{1\ \cancel{\text{m}}} = 400\ \text{cm}$$

Practice Problem How many milligrams are in one kilogram? (Hint: You will need to use two conversion factors from **Table 2.**)

Table 3 Unit System Equivalents

Type of Measurement	Equivalent
Length	1 in = 2.54 cm 1 yd = 0.91 m 1 mi = 1.61 km
Mass and weight*	1 oz = 28.35 g 1 lb = 0.45 kg 1 ton (short) = 0.91 tonnes (metric tons) 1 lb = 4.45 N
Volume	1 in^3 = 16.39 cm^3 1 qt = 0.95 L 1 gal = 3.78 L
Area	1 in^2 = 6.45 cm^2 1 yd^2 = 0.83 m^2 1 mi^2 = 2.59 km^2 1 acre = 0.40 hectares
Temperature	$°C = \frac{(°F - 32)}{1.8}$ K = °C + 273

*Weight is measured in standard Earth gravity.

Convert Between Unit Systems **Table 3** gives a list of equivalents that can be used to convert between English and SI units.

Example

If a meterstick has a length of 100 cm, how long is the meterstick in inches?

Step 1 Write the conversion factors for the units given. From **Table 3,** 1 in = 2.54 cm.

$$\frac{1 \text{ in}}{2.54 \text{ cm}} \text{ and } \frac{2.54 \text{ cm}}{1 \text{ in}}$$

Step 2 Determine which conversion factor to use. You are converting from cm to in. Use the conversion factor with cm on the bottom.

$$\frac{1 \text{ in}}{2.54 \text{ cm}}$$

Step 3 Multiply the starting quantity and units by the conversion factor. Cancel the starting units with the units in the denominator. Round your answer to the nearest tenth.

$$100 \cancel{\text{cm}} \times \frac{1 \text{ in}}{2.54 \cancel{\text{cm}}} = 39.37 \text{ in}$$

The meterstick is about 39.4 in long.

Practice Problem 1 A book has a mass of 5 lb. What is the mass of the book in kg?

Practice Problem 2 Use the equivalent for in and cm (1 in = 2.54 cm) to show how 1 in3 ≈ 16.39 cm3.

Precision and Significant Digits

When you make a measurement, the value you record depends on the precision of the measuring instrument. This precision is represented by the number of significant digits recorded in the measurement. When counting the number of significant digits, all digits are counted except zeros at the end of a number with no decimal point such as 2,050, and zeros at the beginning of a decimal such as 0.03020. When adding or subtracting numbers with different precision, round the answer to the smallest number of decimal places of any number in the sum or difference. When multiplying or dividing, the answer is rounded to the smallest number of significant digits of any number being multiplied or divided.

Example

The lengths 5.28 and 5.2 are measured in meters. Find the sum of these lengths and record your answer using the correct number of significant digits.

Step 1 Find the sum.

5.28 m	2 digits after the decimal
+ 5.2 m	1 digit after the decimal
10.48 m	

Step 2 Round to one digit after the decimal because the least number of digits after the decimal of the numbers being added is 1.

The sum is 10.5 m.

Practice Problem 1 How many significant digits are in the measurement 7,071,301 m? How many significant digits are in the measurement 0.003010 g?

Practice Problem 2 Multiply 5.28 and 5.2 using the rule for multiplying and dividing. Record the answer using the correct number of significant digits.

Scientific Notation

Many times numbers used in science are very small or very large. Because these numbers are difficult to work with, scientists use scientific notation. To write numbers in scientific notation, move the decimal point until only one non-zero digit remains on the left. Then count the number of places you moved the decimal point and use that number as a power of ten. For example, the average distance from the Sun to Mars is 227,800,000,000 m. In scientific notation, this distance is 2.278×10^{11} m. Because you moved the decimal point to the left, the number is a positive power of ten.

The mass of an electron is about 0.000 000 000 000 000 000 000 000 000 000 911 kg. Expressed in scientific notation, this mass is 9.11×10^{-31} kg. Because the decimal point was moved to the right, the number is a negative power of ten.

Example

Earth is 149,600,000 km from the Sun. Express this in scientific notation.

Step 1 Move the decimal point until one non-zero digit remains on the left.

1.496 000 00

Step 2 Count the number of decimal places you have moved. In this case, eight.

Step 2 Show that number as a power of ten, 10^8.

Earth is 1.496×10^8 km from the Sun.

Practice Problem 1 How many significant digits are in 149,600,000 km? How many significant digits are in 1.496×10^8 km?

Practice Problem 2 Parts used in a high performance car must be measured to 7×10^{-6} m. Express this number as a decimal.

Practice Problem 3 A CD is spinning at 539 revolutions per minute. Express this number in scientific notation.

Make and Use Graphs

Data in tables can be displayed in a graph—a visual representation of data. Common graph types include line graphs, bar graphs, and circle graphs.

Line Graph A line graph shows a relationship between two variables that change continuously. The independent variable is changed and is plotted on the x-axis. The dependent variable is observed, and is plotted on the y-axis.

Example

Draw a line graph of the data below from a cyclist in a long-distance race.

Table 4 Bicycle Race Data

Time (h)	Distance (km)
0	0
1	8
2	16
3	24
4	32
5	40

Step 1 Determine the x-axis and y-axis variables. Time varies independently of distance and is plotted on the x-axis. Distance is dependent on time and is plotted on the y-axis.

Step 2 Determine the scale of each axis. The x-axis data ranges from 0 to 5. The y-axis data ranges from 0 to 50.

Step 3 Using graph paper, draw and label the axes. Include units in the labels.

Step 4 Draw a point at the intersection of the time value on the x-axis and corresponding distance value on the y-axis. Connect the points and label the graph with a title, as shown in **Figure 8.**

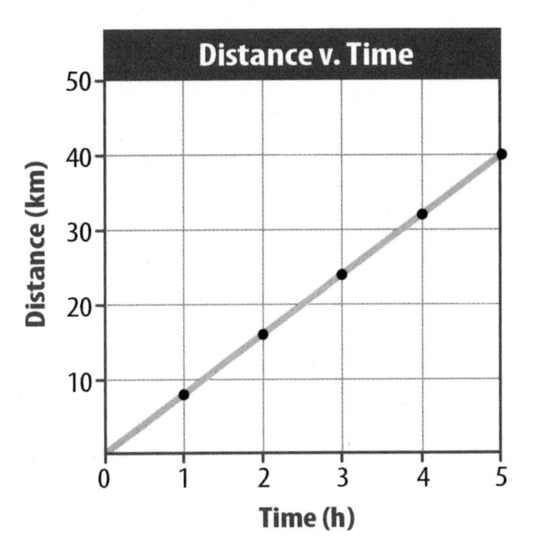

Figure 8 This line graph shows the relationship between distance and time during a bicycle ride.

Practice Problem A puppy's shoulder height is measured during the first year of her life. The following measurements were collected: (3 mo, 52 cm), (6 mo, 72 cm), (9 mo, 83 cm), (12 mo, 86 cm). Graph this data.

Find a Slope The slope of a straight line is the ratio of the vertical change, rise, to the horizontal change, run.

$$\text{Slope} = \frac{\text{vertical change (rise)}}{\text{horizontal change (run)}} = \frac{\text{change in } y}{\text{change in } x}$$

Example

Find the slope of the graph in **Figure 8**.

Step 1 You know that the slope is the change in y divided by the change in x.

$$\text{Slope} = \frac{\text{change in } y}{\text{change in } x}$$

Step 2 Determine the data points you will be using. For a straight line, choose the two sets of points that are the farthest apart.

$$\text{Slope} = \frac{(40 - 0)\text{ km}}{(5 - 0)\text{ h}}$$

Step 3 Find the change in y and x.

$$\text{Slope} = \frac{40\text{ km}}{5\text{ h}}$$

Step 4 Divide the change in y by the change in x.

$$\text{Slope} = \frac{8\text{ km}}{\text{h}}$$

The slope of the graph is 8 km/h.

Bar Graph To compare data that does not change continuously you might choose a bar graph. A bar graph uses bars to show the relationships between variables. The x-axis variable is divided into parts. The parts can be numbers such as years, or a category such as a type of animal. The y-axis is a number and increases continuously along the axis.

Example

A recycling center collects 4.0 kg of aluminum on Monday, 1.0 kg on Wednesday, and 2.0 kg on Friday. Create a bar graph of this data.

Step 1 Select the x-axis and y-axis variables. The measured numbers (the masses of aluminum) should be placed on the y-axis. The variable divided into parts (collection days) is placed on the x-axis.

Step 2 Create a graph grid like you would for a line graph. Include labels and units.

Step 3 For each measured number, draw a vertical bar above the x-axis value up to the y-axis value. For the first data point, draw a vertical bar above Monday up to 4.0 kg.

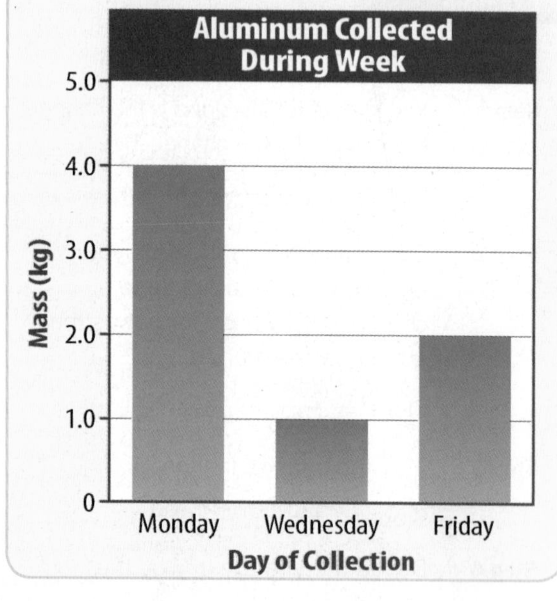

Practice Problem Draw a bar graph of the gases in air: 78% nitrogen, 21% oxygen, 1% other gases.

Circle Graph To display data as parts of a whole, you might use a circle graph. A circle graph is a circle divided into sections that represent the relative size of each piece of data. The entire circle represents 100%, half represents 50%, and so on.

Example

Air is made up of 78% nitrogen, 21% oxygen, and 1% other gases. Display the composition of air in a circle graph.

Step 1 Multiply each percent by 360° and divide by 100 to find the angle of each section in the circle.

$$78\% \times \frac{360^\circ}{100} = 280.8^\circ$$

$$21\% \times \frac{360^\circ}{100} = 75.6^\circ$$

$$1\% \times \frac{360^\circ}{100} = 3.6^\circ$$

Step 2 Use a compass to draw a circle and to mark the center of the circle. Draw a straight line from the center to the edge of the circle.

Step 3 Use a protractor and the angles you calculated to divide the circle into parts. Place the center of the protractor over the center of the circle and line the base of the protractor over the straight line.

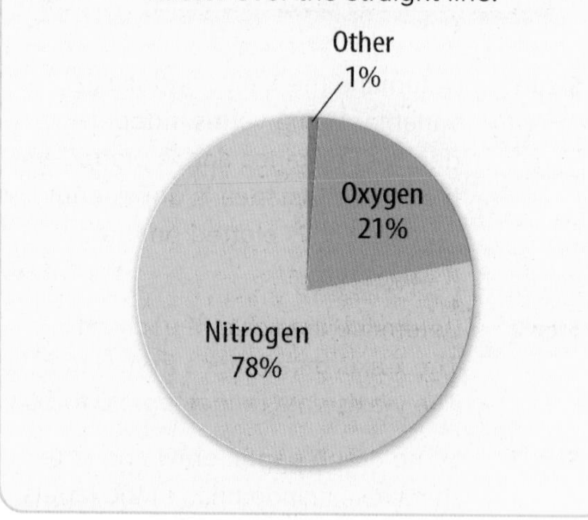

Practice Problem Draw a circle graph to represent the amount of aluminum collected during the week shown in the bar graph to the left.

Student Study Guides & Instructions

By Dinah Zike

1. You will find suggestions for Study Guides, also known as Foldables or books, in each chapter lesson and as a final project. Look at the end of the chapter to determine the project format and glue the Foldables in place as you progress through the chapter lessons.

2. Creating the Foldables or books is simple and easy to do by using copy paper, art paper, and internet printouts. Photocopies of maps, diagrams, or your own illustrations may also be used for some of the Foldables. Notebook paper is the most common source of material for study guides and 83% of all Foldables are created from it. When folded to make books, notebook paper Foldables easily fit into 11" × 17" or 12" × 18" chapter projects with space left over. Foldables made using photocopy paper are slightly larger and they fit into Projects, but snugly. Use the least amount of glue, tape, and staples needed to assemble the Foldables.

3. Seven of the Foldables can be made using either small or large paper. When 11" × 17" or 12" × 18" paper is used, these become projects for housing smaller Foldables. Project format boxes are located within the instructions to remind you of this option.

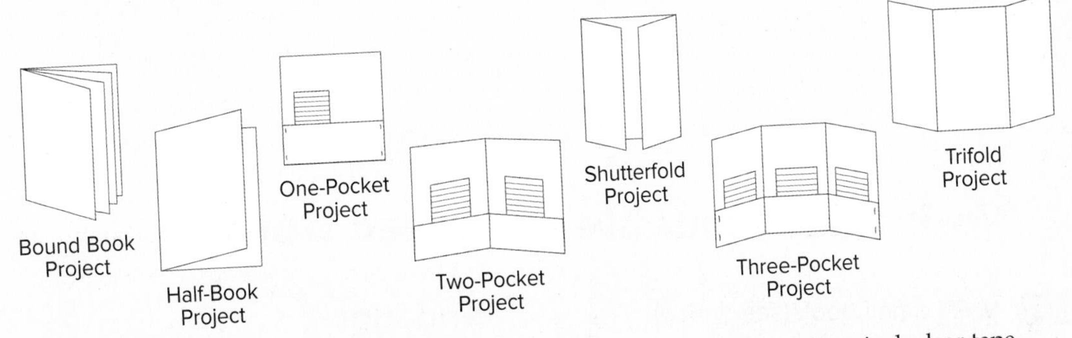

4. Use one-gallon self-locking plastic bags to store your projects. Place strips of two-inch clear tape along the left, long side of the bag and punch holes through the taped edge. Cut the bottom corners off the bag so it will not hold air. Store this Project Portfolio inside a three-hole binder. To store a large collection of project bags, use a giant laundry-soap box. Holes can be punched in some of the Foldable Projects so they can be stored in a three-hole binder without using a plastic bag. Punch holes in the pocket books before gluing or stapling the pocket.

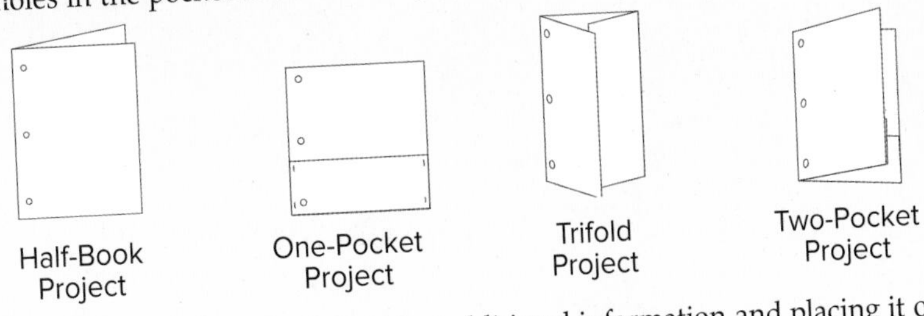

5. Maximize the use of the projects by collecting additional information and placing it on the back of the project and other unused spaces of the large Foldables.

Half-Book Foldable® By Dinah Zike

Step 1 Fold a sheet of notebook or copy paper in half.

Label the exterior tab and use the inside space to write information.

PROJECT FORMAT
Use 11" × 17" or 12" × 18" paper on the horizontal axis to make a large project book.

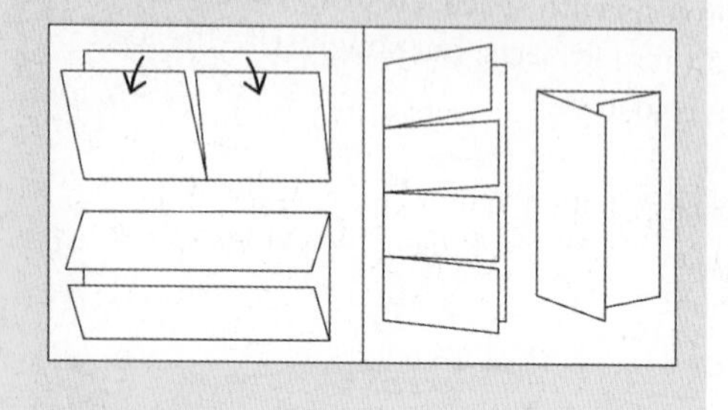

Variations

Paper can be folded horizontally, like a *hamburger* or vertically, like a *hot dog.*

A

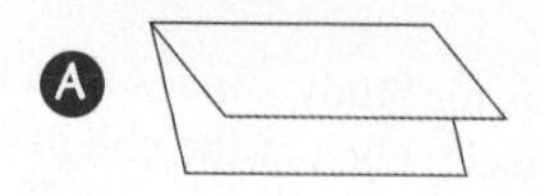

B

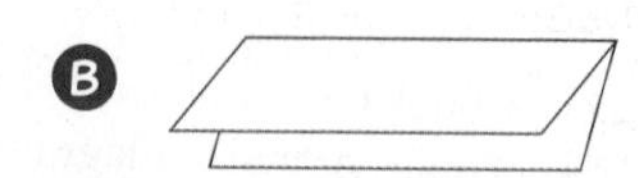

C Half-books can be folded so that one side is ½ inch longer than the other side. A title or question can be written on the extended tab.

Worksheet Foldable or Folded Book® By Dinah Zike

Step 1 Make a half-book (see above) using work sheets, internet printouts, diagrams, or maps.

Step 2 Fold it in half again.

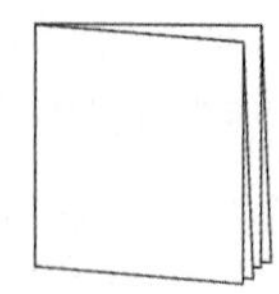

Variations

A This folded sheet as a small book with two pages can be used for comparing and contrasting, cause and effect, or other skills.

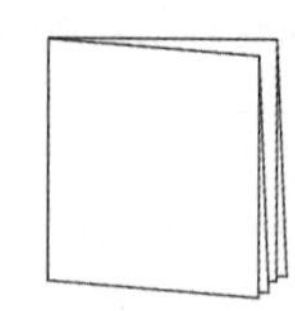

B When the sheet of paper is open, the four sections can be used separately or used collectively to show sequences or steps.

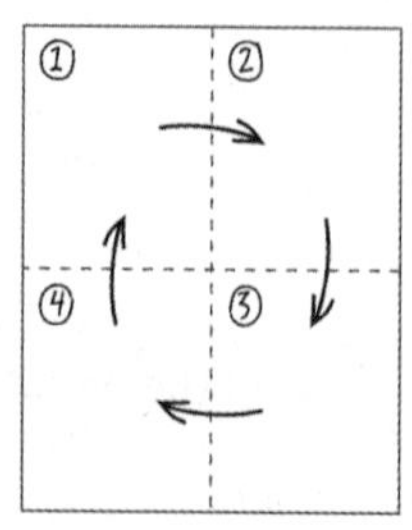

Two-Tab and Concept-Map Foldable® By Dinah Zike

Step 1 Fold a sheet of notebook or copy paper in half vertically or horizontally.

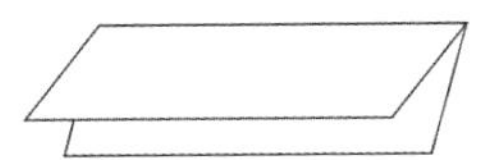

Step 2 Fold it in half again, as shown.

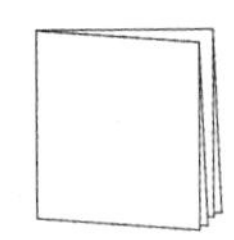

Step 3 Unfold once and cut along the fold line or valley of the top flap to make two flaps.

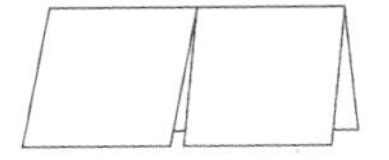

Variations

A Concept maps can be made by leaving a ½ inch tab at the top when folding the paper in half. Use arrows and labels to relate topics to the primary concept.

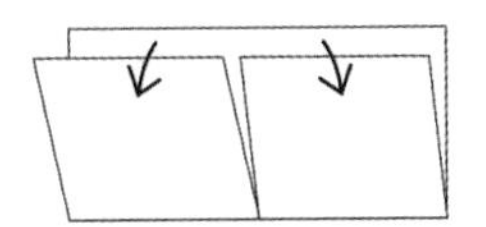

B Use two sheets of paper to make multiple page tab books. Glue or staple books together at the top fold.

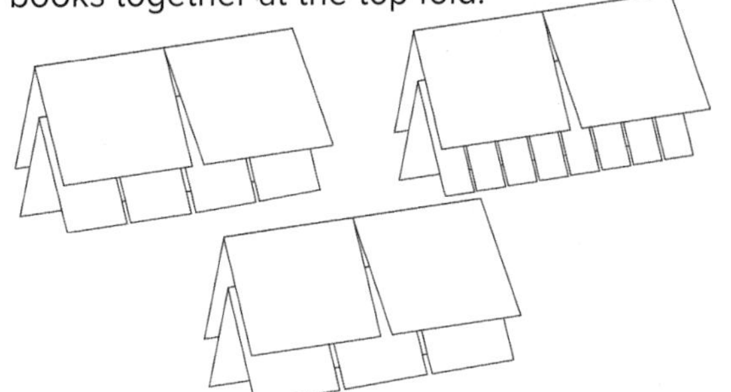

Three-Quarter Foldable® By Dinah Zike

Step 1 Make a two-tab book (see above) and cut the left tab off at the top of the fold line.

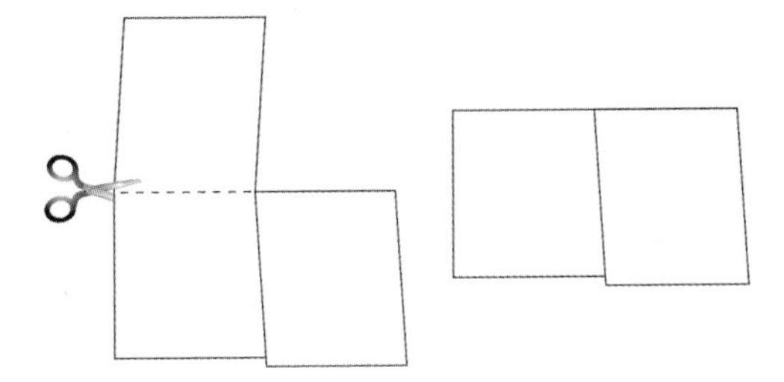

Variations

A Use this book to draw a diagram or a map on the exposed left tab. Write questions about the illustration on the top right tab and provide complete answers on the space under the tab.

B Compose a self-test using multiple choice answers for your questions. Include the correct answer with three wrong responses. The correct answers can be written on the back of the book or upside down on the bottom of the inside page.

SCIENCE SKILL HANDBOOK | MATH SKILL HANDBOOK | FOLDABLES HANDBOOK | REFERENCE HANDBOOK | GLOSSARY/GLOSARIO | INDEX

Three-Tab Foldable® By Dinah Zike

Step 1 Fold a sheet of paper in half horizontally.

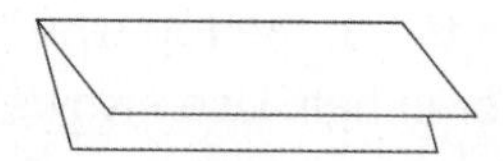

Step 2 Fold into thirds.

Step 3 Unfold and cut along the folds of the top flap to make three sections.

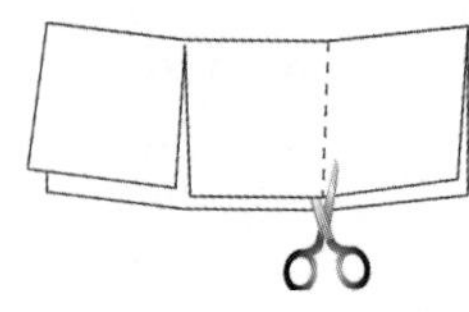

Variations

A Before cutting the three tabs draw a Venn diagram across the front of the book.

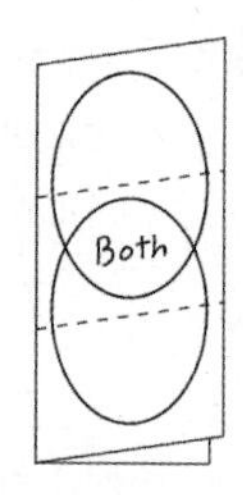

B Make a space to use for titles or concept maps by leaving a ½ inch tab at the top when folding the paper in half.

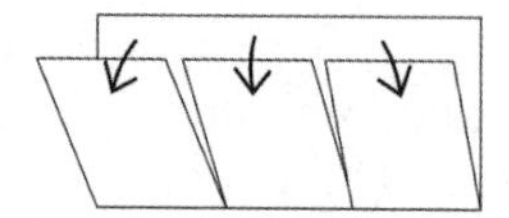

Four-Tab Foldable® By Dinah Zike

Step 1 Fold a sheet of paper in half horizontally.

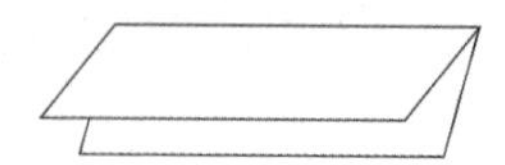

Step 2 Fold in half and then fold each half as shown below.

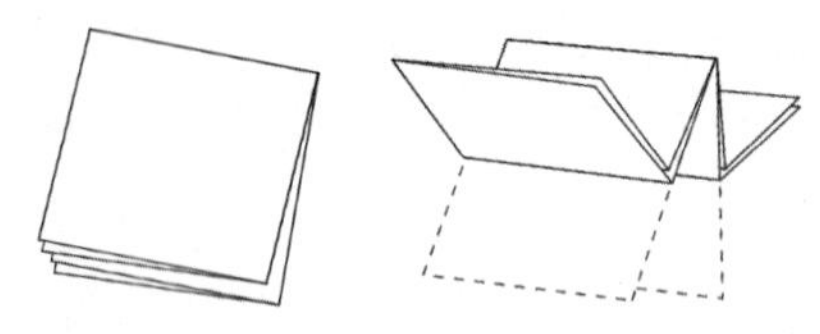

Step 3 Unfold and cut along the fold lines of the top flap to make four tabs.

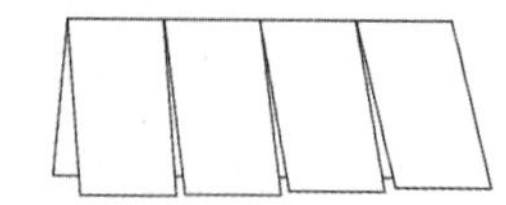

Variations

A Make a space to use for titles or concept maps by leaving a ½ inch tab at the top when folding the paper in half.

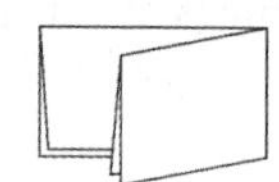

B Use the book on the vertical axis, with or without an extended tab.

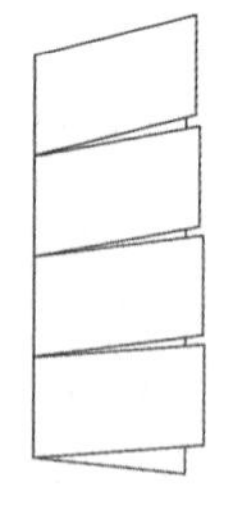

Folding Fifths for a Foldable® By Dinah Zike

Step 1 Fold a sheet of paper in half horizontally.

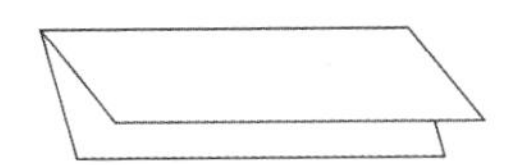

Step 2 Fold again so one-third of the paper is exposed and two-thirds are covered.

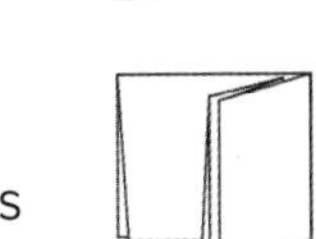

Step 3 Fold the two-thirds section in half.

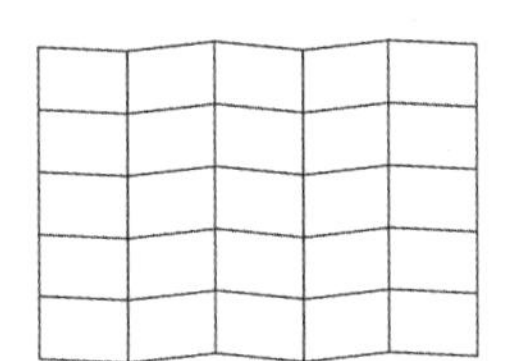

Step 4 Fold the one-third section, a single thickness, backward to make a fold line.

Variations

A Unfold and cut along the fold lines to make five tabs.

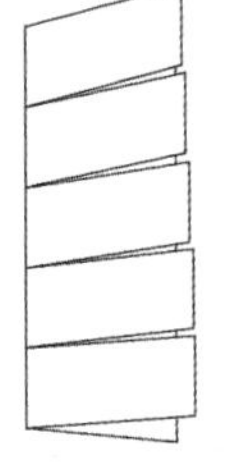

B Make a five-tab book with a ½ inch tab at the top (see two-tab instructions).

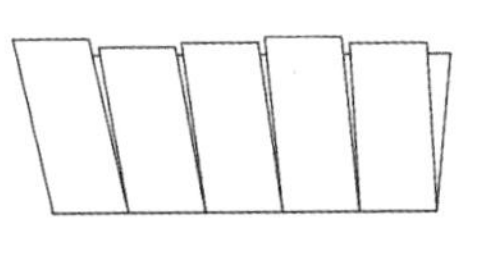

C Use 11" × 17" or 12" × 18" paper and fold into fifths for a five-column and/or row table or chart.

Folded Table or Chart, and Trifold Foldable® By Dinah Zike

Step 1 Fold a sheet of paper in the required number of vertical columns for the table or chart.

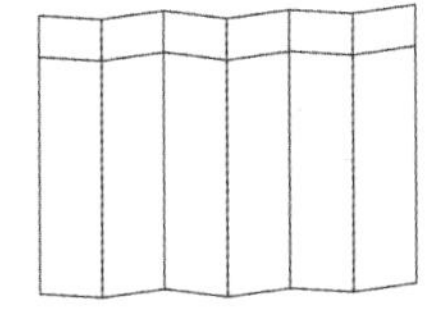

Step 2 Fold the horizontal rows needed for the table or chart.

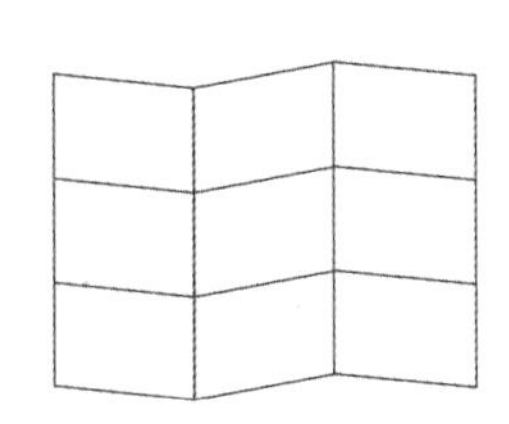

PROJECT FORMAT
Use 11" × 17" or 12" × 18" paper and fold it to make a large trifold project book or larger tables and charts.

Variations

A Make a trifold by folding the paper into thirds vertically or horizontally.

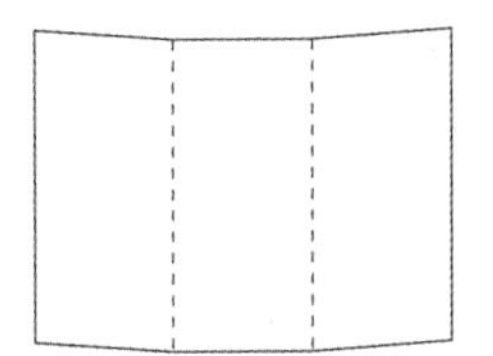

B Make a trifold book. Unfold it and draw a Venn diagram on the inside.

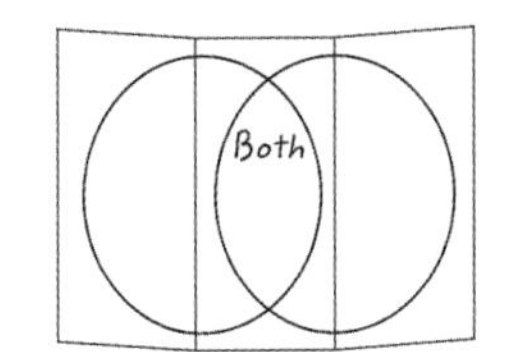

Two or Three-Pockets Foldable® By Dinah Zike

Step 1 Fold up the long side of a horizontal sheet of paper about 5 cm.

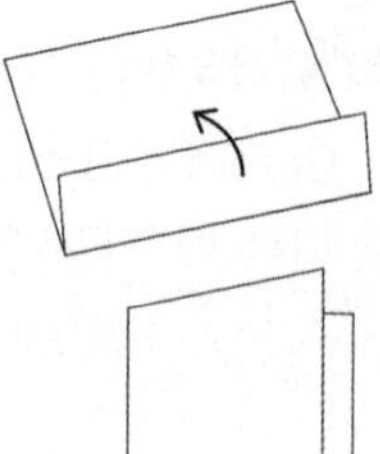

Step 2 Fold the paper in half.

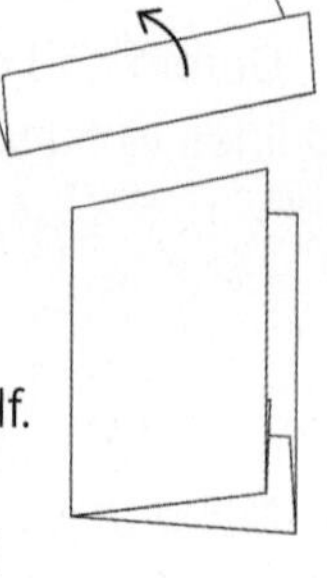

Step 3 Open the paper and glue or staple the outer edges to make two compartments.

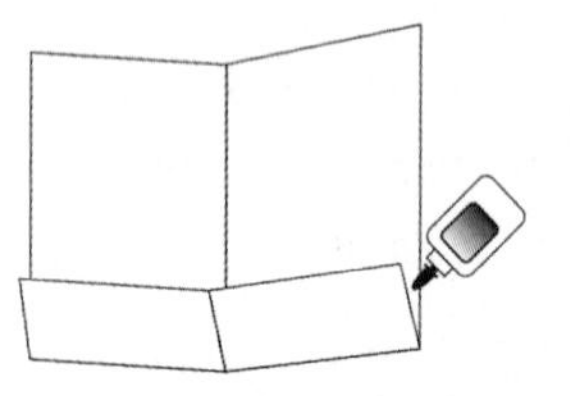

Variations

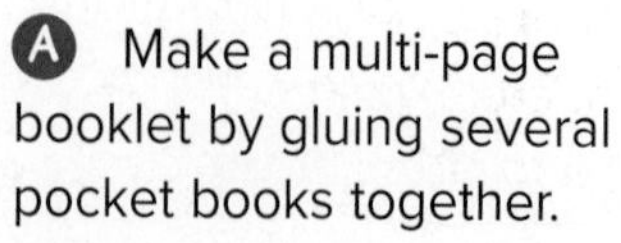

A Make a multi-page booklet by gluing several pocket books together.

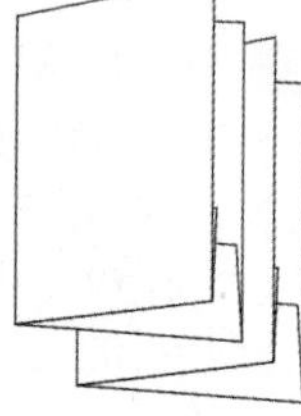

B Make a three-pocket book by using a trifold (see previous instructions).

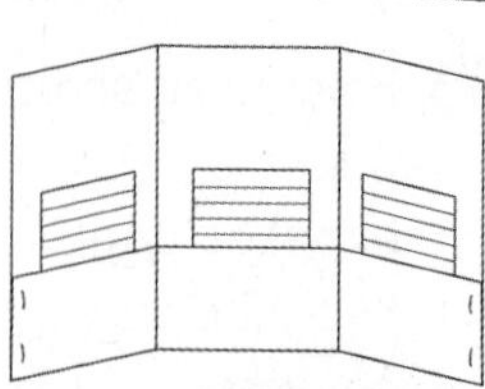

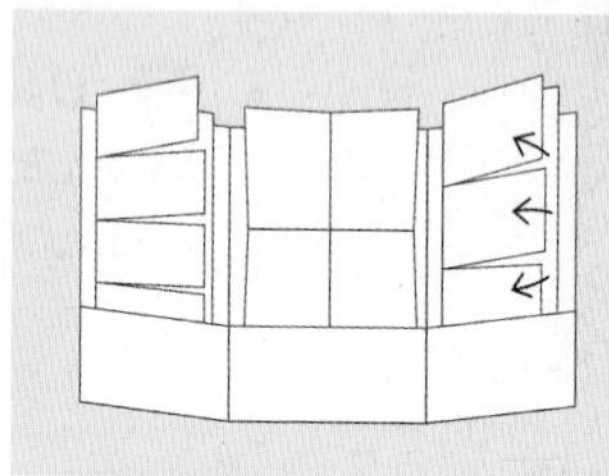

PROJECT FORMAT
Use 11" × 17" or 12" × 18" paper and fold it horizontally to make a large multi-pocket project.

Matchbook Foldable® By Dinah Zike

Step 1 Fold a sheet of paper almost in half and make the back edge about 1–2 cm longer than the front edge.

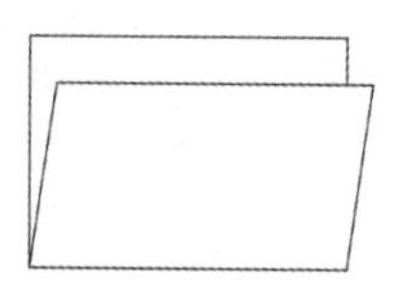

Step 2 Find the midpoint of the shorter flap.

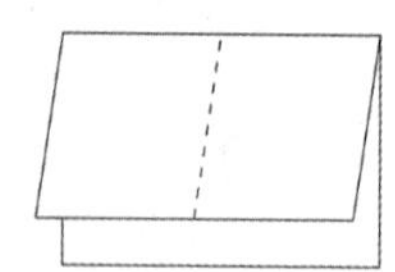

Step 3 Open the paper and cut the short side along the midpoint making two tabs.

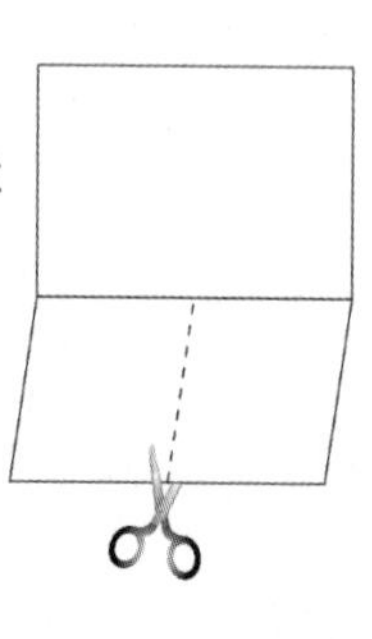

Step 4 Close the book and fold the tab over the short side.

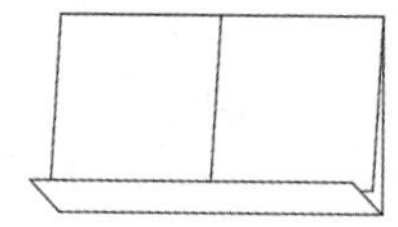

Variations

A Make a single-tab matchbook by skipping Steps 2 and 3.

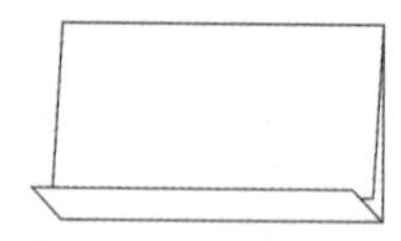

B Make two smaller matchbooks by cutting the single-tab matchbook in half.

Shutterfold Foldable® By Dinah Zike

Step 1 Begin as if you were folding a vertical sheet of paper in half, but instead of creasing the paper, pinch it to show the midpoint.

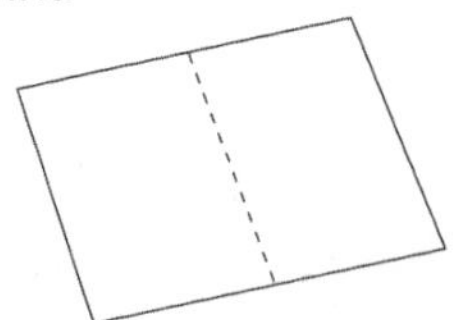

PROJECT FORMAT
Use 11" × 17" or 12" × 18" paper and fold it to make a large shutterfold project.

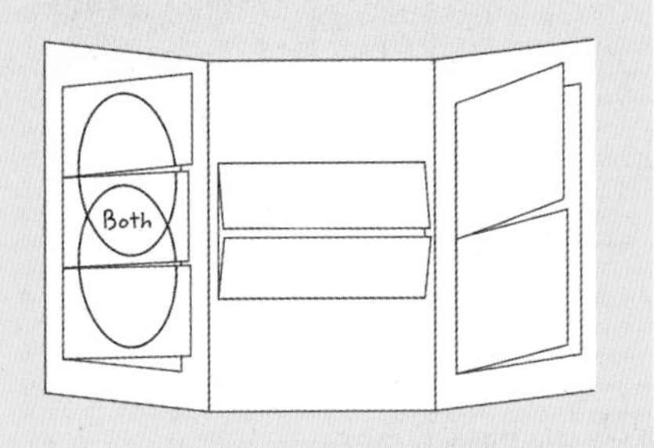

Step 2 Fold the top and bottom to the middle and crease the folds.

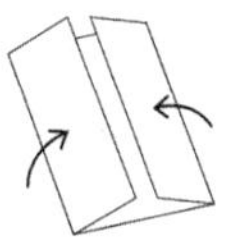

Variations

A Use the shutterfold on the horizontal axis.

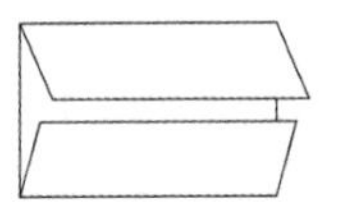

B Create a center tab by leaving .5–2 cm between the flaps in Step 2.

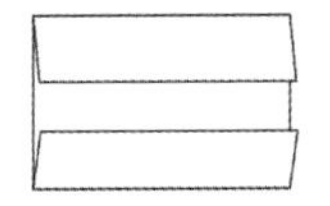

Four-Door Foldable® By Dinah Zike

Step 1 Make a shutterfold (see above).

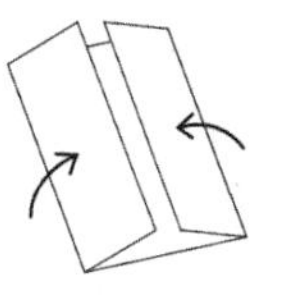

Step 2 Fold the sheet of paper in half.

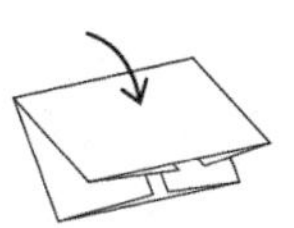

Step 3 Open the last fold and cut along the inside fold lines to make four tabs.

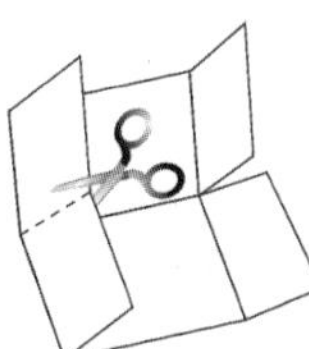

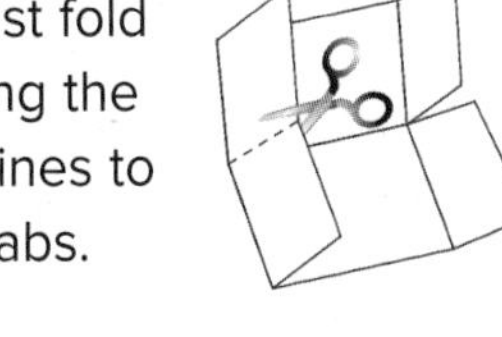

Variations

A Use the four-door book on the opposite axis.

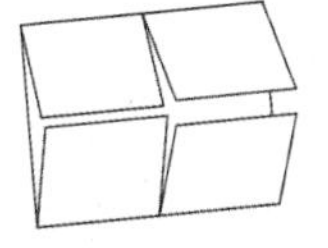

B Create a center tab by leaving .5–2 cm between the flaps in Step 1.

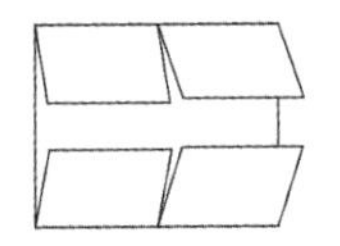

Bound Book Foldable® By Dinah Zike

Step 1 Fold three sheets of paper in half. Place the papers in a stack, leaving about .5 cm between each top fold. Mark all three sheets about 3 cm from the outer edges.

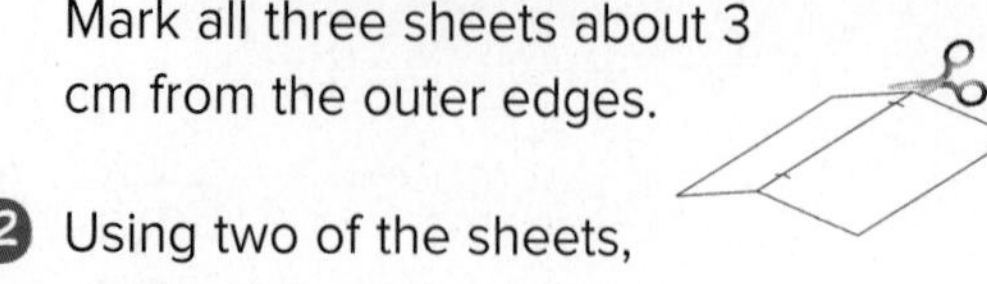

Step 2 Using two of the sheets, cut from the outer edges to the marked spots on each side. On the other sheet, cut between the marked spots.

Step 3 Take the two sheets from Step 1 and slide them through the cut in the third sheet to make a 12-page book.

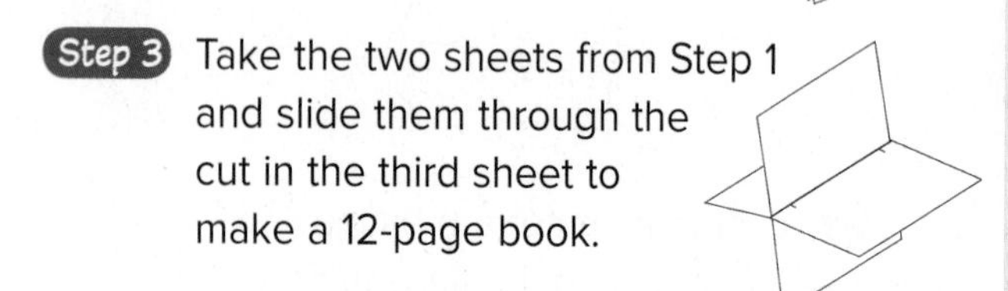

Step 4 Fold the bound pages in half to form a book.

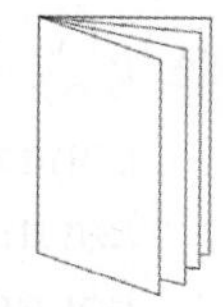

Variation

A Use two sheets of paper to make an eight-page book, or increase the number of pages by using more than three sheets.

PROJECT FORMAT

Use two or more sheets of 11" × 17" or 12" × 18" paper and fold it to make a large bound book project.

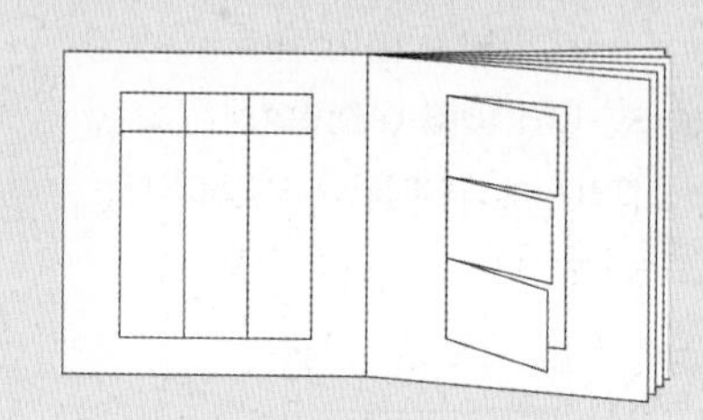

Accordion Foldable® By Dinah Zike

Step 1 Fold the selected paper in half vertically, like a *hamburger*.

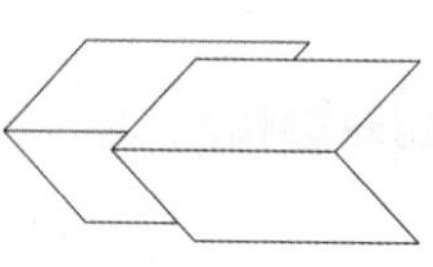

Step 2 Cut each sheet of folded paper in half along the fold lines.

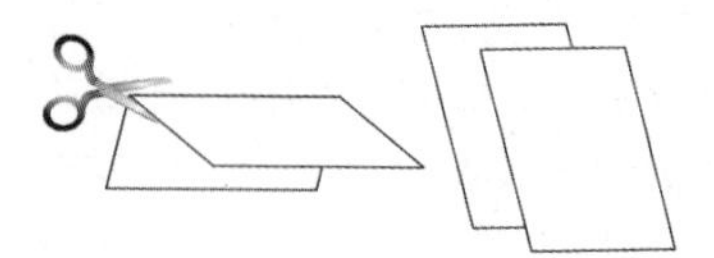

Step 3 Fold each half-sheet almost in half, leaving a 2 cm tab at the top.

Step 4 Fold the top tab over the short side, then fold it in the opposite direction.

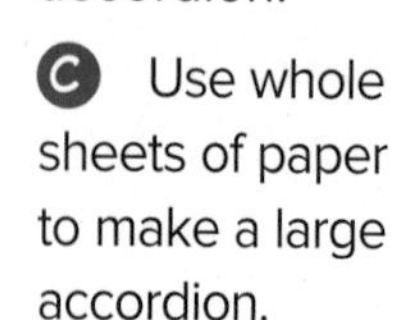

Variations

A Glue the straight edge of one paper inside the tab of another sheet. Leave a tab at the end of the book to add more pages.

B Tape the straight edge of one paper to the tab of another sheet, or just tape the straight edges of nonfolded paper end to end to make an accordion.

C Use whole sheets of paper to make a large accordion.

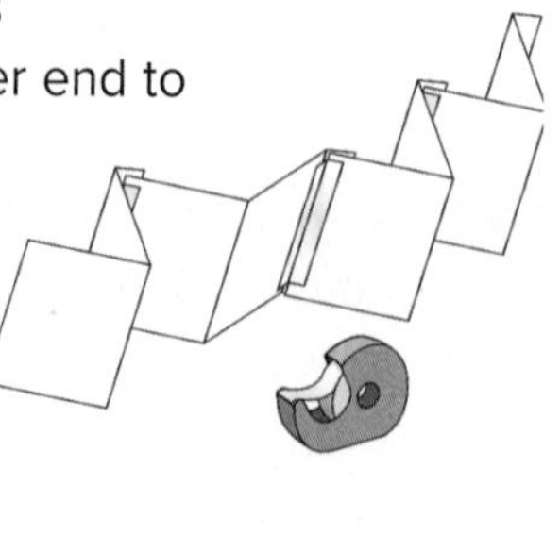

Layered Foldable® By Dinah Zike

Step 1 Stack two sheets of paper about 1–2 cm apart. Keep the right and left edges even.

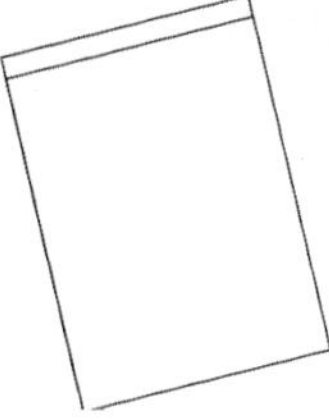

Step 2 Fold up the bottom edges to form four tabs. Crease the fold to hold the tabs in place.

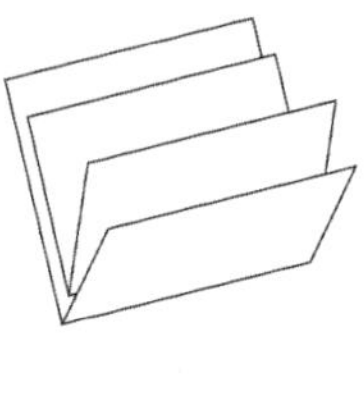

Step 3 Staple along the folded edge, or open and glue the papers together at the fold line.

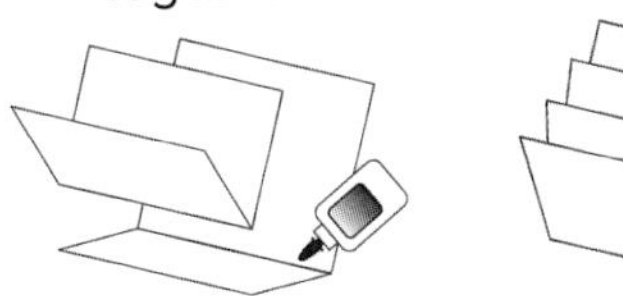

Variations

A Rotate the book so the fold is at the top or to the side.

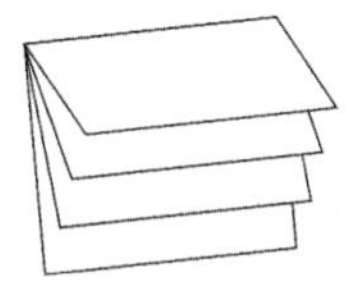

B Extend the book by using more than two sheets of paper.

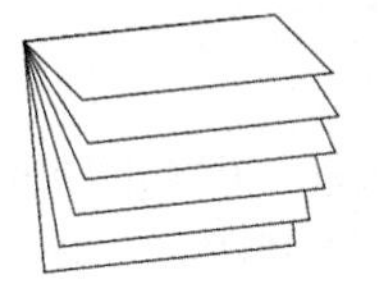

Envelope Foldable® By Dinah Zike

Step 1 Fold a sheet of paper into a *taco*. Cut off the tab at the top.

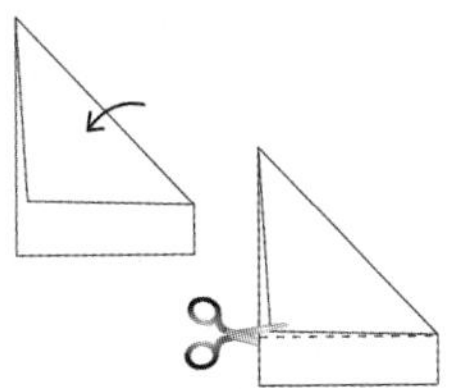

Step 2 Open the *taco* and fold it the opposite way making another *taco* and an X-fold pattern on the sheet of paper.

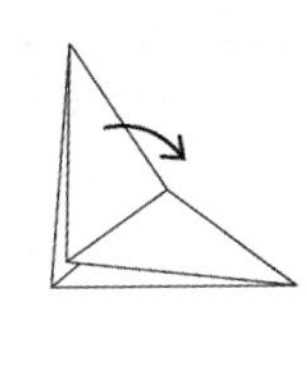

Step 3 Cut a map, illustration, or diagram to fit the inside of the envelope.

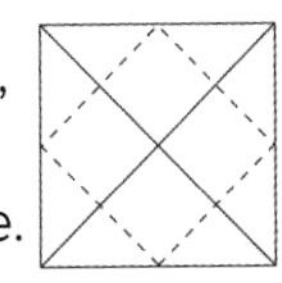

Step 4 Use the outside tabs for labels and inside tabs for writing information.

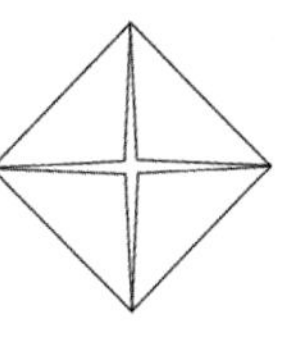

Variations

A Use 11" × 17" or 12" × 18" paper to make a large envelope.

B Cut off the points of the four tabs to make a window in the middle of the book.

Sentence Strip Foldable® By Dinah Zike

Step 1 Fold two sheets of paper in half vertically, like a *hamburger*.

Step 2 Unfold and cut along fold lines making four half sheets.

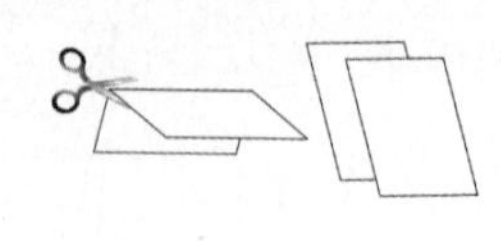

Step 3 Fold each half sheet in half horizontally, like a *hot dog*.

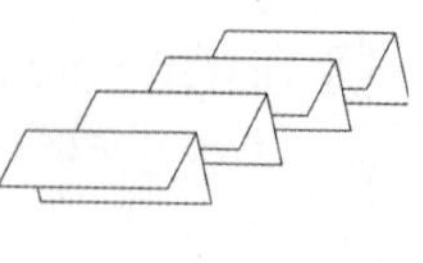

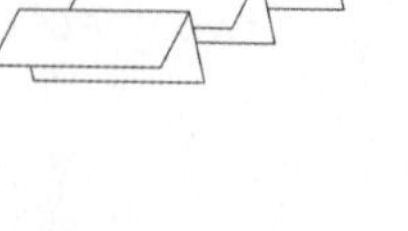

Step 4 Stack folded horizontal sheets evenly and staple together on the left side.

Step 5 Open the top flap of the first sentence strip and make a cut about 2 cm from the stapled edge to the fold line. This forms a flap that can be raised and lowered. Repeat this step for each sentence strip.

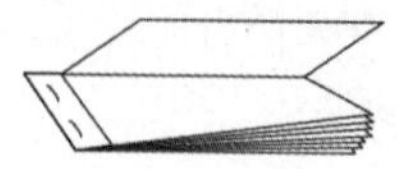

Variations

A Expand this book by using more than two sheets of paper.

B Use whole sheets of paper to make large books.

Pyramid Foldable® By Dinah Zike

Step 1 Fold a sheet of paper into a *taco*. Crease the fold line, but do not cut it off.

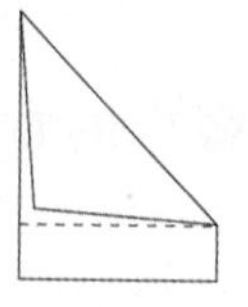

Step 2 Open the folded sheet and refold it like a *taco* in the opposite direction to create an X-fold pattern.

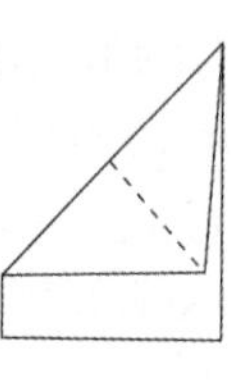

Step 3 Cut one fold line as shown, stopping at the center of the X-fold to make a flap.

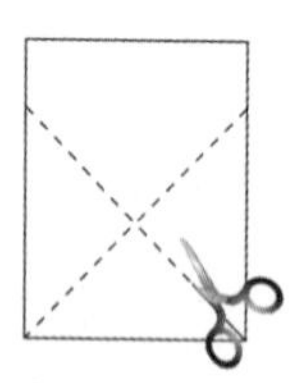

Step 4 Outline the fold lines of the X-fold. Label the three front sections and use the inside spaces for notes. Use the tab for the title.

Step 5 Glue the tab into a project book or notebook. Use the space under the pyramid for other information.

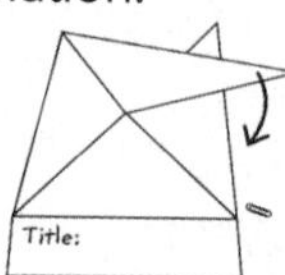

Step 6 To display the pyramid, fold the flap under and secure with a paper clip, if needed.

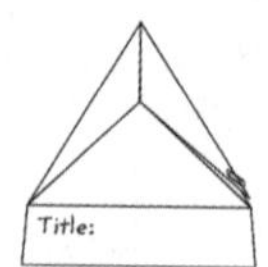

Single-Pocket or One-Pocket Foldable® By Dinah Zike

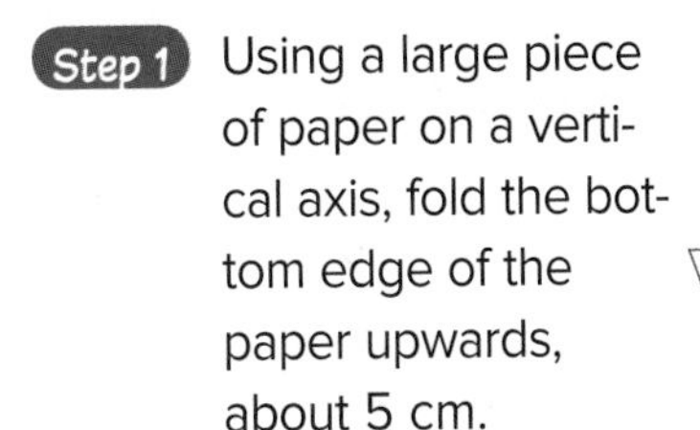

Step 1 Using a large piece of paper on a vertical axis, fold the bottom edge of the paper upwards, about 5 cm.

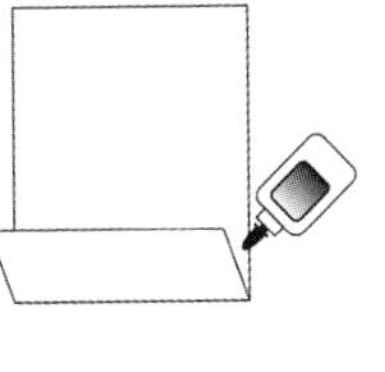

Step 2 Glue or staple the outer edges to make a large pocket.

PROJECT FORMAT
Use 11" × 17" or 12" × 18" paper and fold it vertically or horizontally to make a large pocket project.

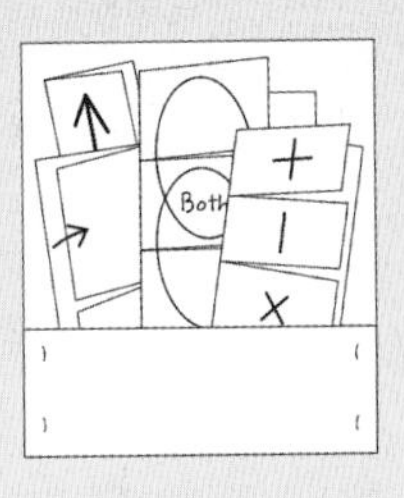

Variations

A Make the one-pocket project using the paper on the horizontal axis.

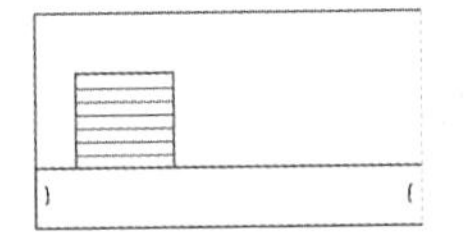

B To store materials securely inside, fold the top of the paper almost to the center, leaving about 2–4 cm between the paper edges. Slip the Foldables through the opening and under the top and bottom pockets.

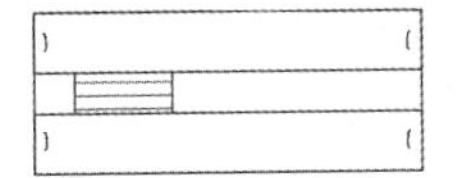

Multi-Tab Foldable® By Dinah Zike

Step 1 Fold a sheet of notebook paper in half like a *hot dog*.

Step 2 Open the paper and on one side cut every third line. This makes ten tabs on wide ruled notebook paper and twelve tabs on college ruled.

Step 3 Label the tabs on the front side and use the inside space for definitions or other information.

Variation

A Make a tab for a title by folding the paper so the holes remain uncovered. This allows the notebook Foldable to be stored in a three-hole binder.

SCIENCE SKILL HANDBOOK
MATH SKILL HANDBOOK
FOLDABLES HANDBOOK
REFERENCE HANDBOOK
GLOSSARY/GLOSARIO
INDEX

PERIODIC TABLE OF THE ELEMENTS

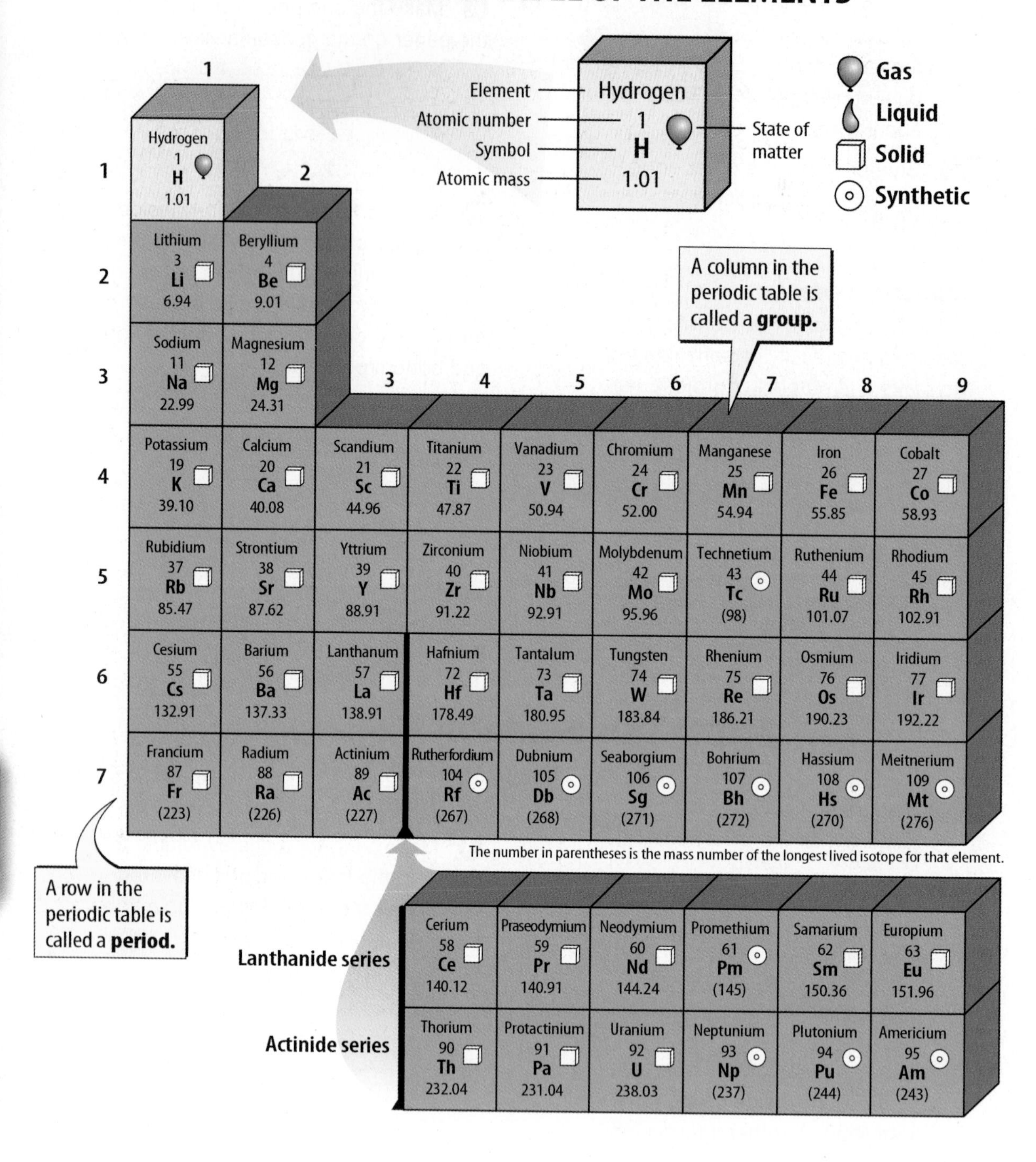

Metal
Metalloid
Nonmetal
Recently discovered

10	11	12	13	14	15	16	17	18
								Helium 2 He 4.00
			Boron 5 B 10.81	Carbon 6 C 12.01	Nitrogen 7 N 14.01	Oxygen 8 O 16.00	Fluorine 9 F 19.00	Neon 10 Ne 20.18
			Aluminum 13 Al 26.98	Silicon 14 Si 28.09	Phosphorus 15 P 30.97	Sulfur 16 S 32.07	Chlorine 17 Cl 35.45	Argon 18 Ar 39.95
Nickel 28 Ni 58.69	Copper 29 Cu 63.55	Zinc 30 Zn 65.38	Gallium 31 Ga 69.72	Germanium 32 Ge 72.64	Arsenic 33 As 74.92	Selenium 34 Se 78.96	Bromine 35 Br 79.90	Krypton 36 Kr 83.80
Palladium 46 Pd 106.42	Silver 47 Ag 107.87	Cadmium 48 Cd 112.41	Indium 49 In 114.82	Tin 50 Sn 118.71	Antimony 51 Sb 121.76	Tellurium 52 Te 127.60	Iodine 53 I 126.90	Xenon 54 Xe 131.29
Platinum 78 Pt 195.08	Gold 79 Au 196.97	Mercury 80 Hg 200.59	Thallium 81 Tl 204.38	Lead 82 Pb 207.20	Bismuth 83 Bi 208.98	Polonium 84 Po (209)	Astatine 85 At (210)	Radon 86 Rn (222)
Darmstadtium 110 Ds (281)	Roentgenium 111 Rg (280)	Copernicium 112 Cn (285)	Ununtrium * 113 Uut (284)	Flerovium 114 Fl (289)	Ununpentium * 115 Uup (288)	Livermorium 116 Lv (293)	Ununseptium * 117 Uus (294)	Ununoctium * 118 Uuo (294)

* The names and symbols for elements 113, 115, 117, and 118 are temporary. Final names will be approved by IUPAC (International Union of Pure and Applied Chemistry).

Gadolinium 64 Gd 157.25	Terbium 65 Tb 158.93	Dysprosium 66 Dy 162.50	Holmium 67 Ho 164.93	Erbium 68 Er 167.26	Thulium 69 Tm 168.93	Ytterbium 70 Yb 173.05	Lutetium 71 Lu 174.97
Curium 96 Cm (247)	Berkelium 97 Bk (247)	Californium 98 Cf (251)	Einsteinium 99 Es (252)	Fermium 100 Fm (257)	Mendelevium 101 Md (258)	Nobelium 102 No (259)	Lawrencium 103 Lr (262)

Topographic Map Symbols

Topographic Map Symbols			
	Primary highway, hard surface		Index contour
	Secondary highway, hard surface		Supplementary contour
	Light-duty road, hard or improved surface		Intermediate contour
	Unimproved road		Depression contours
	Railroad: single track		
	Railroad: multiple track		Boundaries: national
	Railroads in juxtaposition		State
			County, parish, municipal
	Buildings		Civil township, precinct, town, barrio
cem	Schools, church, and cemetery		Incorporated city, village, town, hamlet
	Buildings (barn, warehouse, etc.)		Reservation, national or state
	Wells other than water (labeled as to type)		Small park, cemetery, airport, etc.
	Tanks: oil, water, etc. (labeled only if water)		Land grant
	Located or landmark object; windmill		Township or range line, U.S. land survey
	Open pit, mine, or quarry; prospect		Township or range line, approximate location
	Marsh (swamp)		Perennial streams
	Wooded marsh		Elevated aqueduct
	Woods or brushwood		Water well and spring
	Vineyard		Small rapids
	Land subject to controlled inundation		Large rapids
	Submerged marsh		Intermittent lake
	Mangrove		Intermittent stream
	Orchard		Aqueduct tunnel
	Scrub		Glacier
	Urban area		Small falls
			Large falls
x7369	Spot elevation		Dry lake bed
670	Water elevation		

Rocks

Rock Type	Rock Name	Characteristics
Igneous (intrusive)	Granite	Large mineral grains of quartz, feldspar, hornblende, and mica. Usually light in color.
	Diorite	Large mineral grains of feldspar, hornblende, and mica. Less quartz than granite. Intermediate in color.
	Gabbro	Large mineral grains of feldspar, augite, and olivine. No quartz. Dark in color.
Igneous (extrusive)	Rhyolite	Small mineral grains of quartz, feldspar, hornblende, and mica, or no visible grains. Light in color.
	Andesite	Small mineral grains of feldspar, hornblende, and mica or no visible grains. Intermediate in color.
	Basalt	Small mineral grains of feldspar, augite, and possibly olivine or no visible grains. No quartz. Dark in color.
	Obsidian	Glassy texture. No visible grains. Volcanic glass. Fracture looks like broken glass.
	Pumice	Frothy texture. Floats in water. Usually light in color.
Sedimentary (clastic)	Conglomerate	Coarse grained. Gravel or pebble-size grains.
	Sandstone	Sand-sized grains 1/16 to 2 mm.
	Siltstone	Grains are smaller than sand but larger than clay.
	Shale	Smallest grains. Often dark in color. Usually platy.
Sedimentary (chemical or biochemical)	Limestone	Major mineral is calcite. Usually forms in oceans and lakes. Often contains fossils.
	Coal	Forms in swampy areas. Compacted layers of organic material, mainly plant remains.
Sedimentary (chemical)	Rock Salt	Commonly forms by the evaporation of seawater.
Metamorphic (foliated)	Gneiss	Banding due to alternate layers of different minerals, of different colors. Parent rock often is granite.
	Schist	Parallel arrangement of sheetlike minerals, mainly micas. Forms from different parent rocks.
	Phyllite	Shiny or silky appearance. May look wrinkled. Common parent rocks are shale and slate.
	Slate	Harder, denser, and shinier than shale. Common parent rock is shale.
Metamorphic (nonfoliated)	Marble	Calcite or dolomite. Common parent rock is limestone.
	Soapstone	Mainly of talc. Soft with greasy feel.
	Quartzite	Hard with interlocking quartz crystals. Common parent rock is sandstone.

Minerals

Minerals					
Mineral (formula)	**Color**	**Streak**	**Hardness Pattern**	**Breakage Properties**	**Uses and Other**
Graphite (C)	black to gray	black to gray	1–1.5	basal cleavage (scales)	pencil lead, lubricants for locks, rods to control some small nuclear reactions, battery poles
Galena (PbS)	gray	gray to black	2.5	cubic cleavage perfect	source of lead, used for pipes, shields for X rays, fishing equipment sinkers
Hematite (Fe_2O_3)	black or reddish-brown	reddish-brown	5.5–6.5	irregular fracture	source of iron; converted to pig iron, made into steel
Magnetite (Fe_3O_4)	black	black	6	conchoidal fracture	source of iron, attracts a magnet
Pyrite (FeS_2)	light, brassy, yellow	greenish-black	6–6.5	uneven fracture	fool's gold
Talc ($Mg_3 Si_4O_{10} (OH)_2$)	white, greenish	white	1	cleavage in one direction	used for talcum powder, sculptures, paper, and tabletops
Gypsum ($CaSO_4 \cdot 2H_2O$)	colorless, gray, white, brown	white	2	basal cleavage	used in plaster of paris and dry wall for building construction
Sphalerite (ZnS)	brown, reddish-brown, greenish	light to dark brown	3.5–4	cleavage in six directions	main ore of zinc; used in paints, dyes, and medicine
Muscovite ($KAl_3Si_3 O_{10}(OH)_2$)	white, light gray, yellow, rose, green	colorless	2–2.5	basal cleavage	occurs in large, flexible plates; used as an insulator in electrical equipment, lubricant
Biotite ($K(Mg,Fe)_3 (AlSi_3O_{10}) (OH)_2$)	black to dark brown	colorless	2.5–3	basal cleavage	occurs in large, flexible plates
Halite ($NaCl$)	colorless, red, white, blue	colorless	2.5	cubic cleavage	salt; soluble in water; a preservative

Minerals

Minerals					
Mineral (formula)	**Color**	**Streak**	**Hardness**	**Breakage Pattern**	**Uses and Other Properties**
Calcite ($CaCO_3$)	colorless, white, pale blue	colorless, white	3	cleavage in three directions	fizzes when HCl is added; used in cements and other building materials
Dolomite ($CaMg(CO_3)_2$)	colorless, white, pink, green, gray, black	white	3.5–4	cleavage in three directions	concrete and cement; used as an ornamental building stone
Fluorite (CaF_2)	colorless, white, blue, green, red, yellow, purple	colorless	4	cleavage in four directions	used in the manufacture of optical equipment; glows under ultraviolet light
Hornblende ($(CaNa)_{2-3}(Mg,Al,Fe)_5$-$(Al,Si)_2Si_6O_{22}(OH)_2$)	green to black	gray to white	5–6	cleavage in two directions	will transmit light on thin edges; 6-sided cross section
Feldspar ($(KAlSi_3O_8)$ $(NaAlSi_3O_8)$, $(CaAl_2Si_2O_8)$)	colorless, white to gray, green	colorless	6	two cleavage planes meet at 90° angle	used in the manufacture of ceramics
Augite ($((Ca,Na)(Mg,Fe,Al)(Al,Si)_2O_6)$)	black	colorless	6	cleavage in two directions	square or 8-sided cross section
Olivine ($((Mg,Fe)_2SiO_4)$)	olive, green	none	6.5–7	conchoidal fracture	gemstones, refractory sand
Quartz (SiO_2)	colorless, various colors	none	7	conchoidal fracture	used in glass manufacture, electronic equipment, radios, computers, watches, gemstones

Weather Map Symbols

Sample Station Model

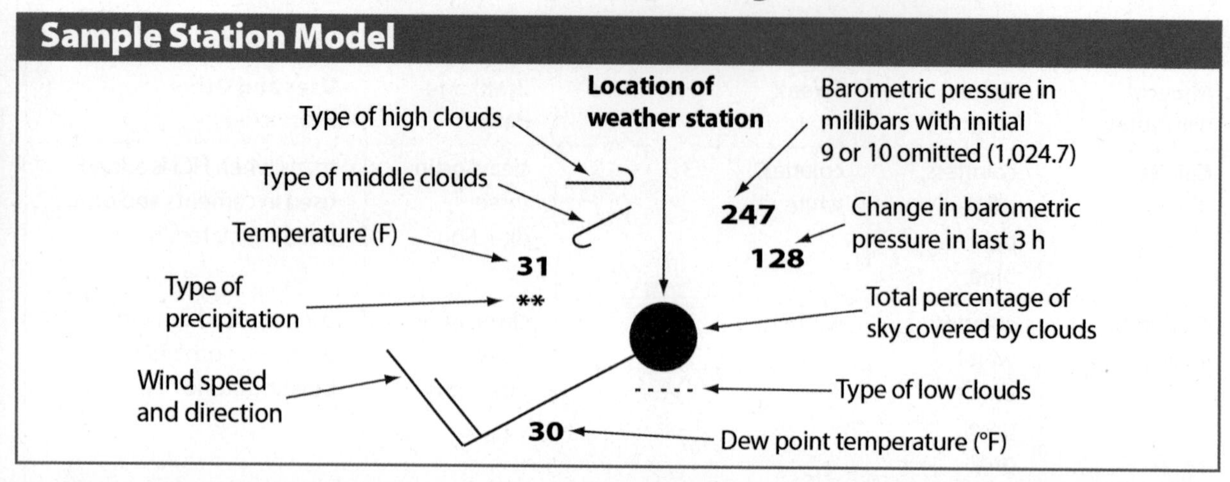

Sample Plotted Report at Each Station

Precipitation	Wind Speed and Direction	Sky Coverage	Some Types of High Clouds
Fog	0 calm	No cover	Scattered cirrus
Snow	1–2 knots	1/10 or less	Dense cirrus in patches
Rain	3–7 knots	2/10 to 3/10	Veil of cirrus covering entire sky
Thunderstorm	8–12 knots	4/10	Cirrus not covering entire sky
Drizzle	13–17 knots	–	
Showers	18–22 knots	6/10	
	23–27 knots	7/10	
	48–52 knots	Overcast with openings	
	1 knot = 1.852 km/h	Completely overcast	

Some Types of Middle Clouds	Some Types of Low Clouds	Fronts and Pressure Systems
Thin altostratus layer	Cumulus of fair weather	(H) or High / (L) or Low — Center of high- or low-pressure system
Thick altostratus layer	Stratocumulus	Cold front
Thin altostratus in patches	Fractocumulus of bad weather	Warm front
Thin altostratus in bands	Stratus of fair weather	Occluded front
		Stationary front

Glossary/Glosario

Multilingual eGlossary

A science multilingual glossary is available on ConnectEd.
The glossary includes the following languages:

Arabic	Hmong	Tagalog
Bengali	Korean	Urdu
Chinese	Portuguese	Vietnamese
English	Russian	
Haitian Creole	Spanish	

Cómo usar el glosario en español:

1. Busca el término en inglés que desees encontrar.
2. El término en español, junto con la definición, se encuentran en la columna de la derecha.

Pronunciation Key

Use the following key to help you sound out words in the glossary:

a	**b**ack (BAK)	**ew**	f**oo**d (FEWD)
ay	d**ay** (DAY)	**yoo**	p**ure** (PYOOR)
ah	f**a**ther (FAH thur)	**yew**	f**ew** (FYEW)
ow	fl**ow**er (FLOW ur)	**uh**	comm**a** (CAH muh)
ar	c**ar** (CAR)	**u (+ con)**	r**u**b (RUB)
e	l**e**ss (LES)	**sh**	**sh**elf (SHELF)
ee	l**ea**f (LEEF)	**ch**	na**t**ure (NAY chur)
ih	tr**i**p (TRIHP)	**g**	**g**ift (GIHFT)
i (i + con + e)	**i**dea (i DEE uh)	**j**	**g**em (JEM)
oh	g**o** (GOH)	**ing**	s**ing** (SING)
aw	s**o**ft (SAWFT)	**zh**	vi**s**ion (VIH zhun)
or	**or**bit (OR buht)	**k**	**c**ake (KAYK)
oy	c**oi**n (COYN)	**s**	**s**eed, **c**ent (SEED, SENT)
oo	f**oo**t (FOOT)	**z**	**z**one, rai**s**e (ZOHN, RAYZ)

English — **A** — **Español**

abrasion/adhesion

abrasion: the grinding away of rock or other surfaces as particles carried by wind, water, or ice scrape against them. (p. 192)

absolute age: the numerical age, in years, of a rock or object. (p. 345)

abyssal plains: large, flat areas of the seafloor that extend across the deepest parts of ocean basins. (p. 566)

acid precipitation: precipitation that has a lower pH than that of normal rainwater (5.6). (pp. 435, 670)

adhesion: the attraction among molecules that are not alike. (p. 539)

abrasión/adhesión

abrasión: desgaste de una roca o de otras superficies a medida que las partículas transportadas por el viento, el agua o el hielo las raspan. (pág. 192)

edad absoluta: edad numérica, en años, de una roca o de un objeto. (pág. 345)

planos abisales: áreas extensas y planas del lecho marino que se extienden por las partes más profundas de las cuencas marinas. (pág. 566)

precipitación ácida: precipitación que tiene un pH más bajo que el del agua de la lluvia normal. (5.6) (pág. 435, 670)

adhesión: atracción entre moléculas que son diferentes. (pág. 539)

air mass: a large area of air that has uniform temperature, humidity, and pressure. (p. 460)

masa de aire: amplia zona de aire que tiene uniforme de temperatura, humedad y presión. (pág. 460)

air pollution: the contamination of air by harmful substances including gases and smoke. (p. 434)

polución del aire: contaminación del aire por sustancias dañinas, como gases y humo. (pág. 434)

air pressure: the force that a column of air applies on the air or a surface below it. (p. 452)

presión del aire: presión que una columna de aire ejerce sobre el aire o sobre la superficie debajo de ella. (pág. 452)

alpine glacier: a glacier that forms in the mountains. (p. 608)

glaciar alpino: glacial que se forma en las montañas. (pág. 608)

apparent magnitude: a measure of how bright an object appears from Earth. (p. 805)

magnitud aparente: medida del brillo de un objeto visto desde la Tierra. (pág. 805)

aquifer: an area of permeable sediment or rock that holds significant amounts of water. (p. 627)

acuífero: área de sedimento permeable o roca que conserva cantidades significativas de agua. (pág. 627)

asteroid: a small, rocky object that orbits the Sun. (p. 763)

asteroide: un objeto de piedra que está en órbita alrededor del sol. (pág. 763)

asthenosphere (as THEN uh sfihr): the partially melted portion of the mantle below the lithosphere. (p. 52)

astenosfera: parte parcialmente fundida del manto por debajo de la lithoshpere. (pág. 52)

astrobiology: the study of the origin, development, distribution, and future of life on Earth and in the universe. (p. 711)

astrobiología: estudio del origen, desarrollo, distribución y futuro de la vida en la Tierra y en el universo. (pág. 711)

astronomical unit (AU): the average distance from Earth to the Sun—about 150 million km. (pp. 764, 804)

unidad astronómica (UA): distancia media entre la Tierra y el Sol , aproximadamente 150 millones de km. (pág. 764, 804)

atmosphere (AT muh sfihr): a thin layer of gases surrounding Earth. (p. 409)

atmósfera: capa delgada de gases que rodean la Tierra. (pág. 409)

B

basin: area of subsidence; region with low elevation. (p. 279)

cuenca: área de hundimiento; región de elevación baja. (pág. 279)

Big Bang theory: the scientific theory that states that the universe began from one point and has been expanding and cooling ever since. (p. 830)

Teoría del Big Bang: teoría científica que establece que el universo se originó de un punto y se ha ido expandiendo y enfriando desde entonces. (pág. 830)

biochemical rock: sedimentary rock that was formed by organisms or contains the remains of organisms. (p. 129)

roca bioquímica: roca sedimentaria formada por organismos o que contiene restos de organismos. (pág. 129)

bioindicator: an organism that is sensitive to environmental conditions and is one of the first to respond to changes. (p. 550)

bioindicador: organismo que es sensible a las condiciones medioambientales y es uno de los primeros en responder a los cambios. (pág. 550)

biomass energy: energy produced by burning organic matter, such as wood, food scraps, and alcohol. (p. 655)

energía de biomasa: energía producida por la combustión de materia orgánica, como la madera, las sobras de comida y el alcohol. (pág. 655)

biota (bi OH tuh): all of the organisms that live in a region. (p. 161)

biota: todos los organismos que viven en una región. (pág. 161)

black hole: an object whose gravity is so great that no light can escape. (p. 820)

agujero negro: objeto cuya gravedad es tan grande que la luz no puede escapar. (pág. 820)

blizzard: a violent winter storm characterized by freezing temperatures, strong winds, and blowing snow. (p. 467)

ventisca: tormenta violenta de invierno caracterizada por temperaturas heladas, vientos fuertes, y nieve que sopla. (pág. 467)

brackish water: a mix of freshwater and sea-water. (p. 565)

agua salobre: mezcla de agua dulce y agua de mar. (pág. 565)

C

carbon film: the fossilized carbon outline of an organism or part of an organism. (p. 330)

película de carbono: contorno de carbono fosilizado de un organismo o parte de un organismo. (pág. 330)

cast: a fossil copy of an organism made when a mold of the organism is filled with sediment or mineral deposits. (p. 341)

contramolde: copia fósil de un organismo producida cuando un molde del organismo se llena con depósitos de sedimento o mineral. (pág. 341)

catastrophism (kuh TAS truh fih zum): the idea that conditions and organisms on Earth change in quick, violent events. (p. 327)

catatrofismo: idea de que las condiciones y los organismos en la Tierra cambian mediante eventos rápidos y violentos. (pág. 327)

cementation: a process in which minerals dissolved in water crystallize between sediment grains. (p. 126)

cementación: proceso por el cual los minerales disueltos en agua se cristalizan entre granos de sedimento. (pág. 126)

Cenozoic (sen uh ZOH ihk) era: the youngest era of the Phanerozoic eon. (p. 371)

era Cenozoica: era más joven del eón Fanerozoico. (pág. 371)

chemical rock: sedimentary rock that forms when minerals crystallize directly from water. (p. 128)

roca química: roca sedimentaria que se forma cuando los minerales se cristalizan directamente del agua. (pág. 128)

chemical weathering: the process that changes the composition of rocks and minerals due to exposure to the environment. (p. 152)

meteorización química: proceso que cambia la composición de las rocas y minerales debido a la exposición al medio ambiente. (pág. 152)

chromosphere: the orange-red layer above the photosphere of a star. (p. 810)

cromosfera: capa de color rojo anaranjado arriba de la fotosfera de una estrella. (pág. 810)

cinder cone: a small, steep-sided volcano that erupts gas-rich, basaltic lava. (p. 310)

cono de ceniza: volcán pequeño de lados empinados que expulsa lava rica en gas basáltico. (pág. 310)

clast: a broken piece or fragment that makes up a clastic rock. (p. 127)

clasto: pedazo partido o fragmentado que forma una roca clástica. (pág. 127)

clastic (KLAH stik) rock: sedimentary rock that is made up of broken pieces of minerals and rock fragments. (p. 127)

roca clástica: roca sedimentaria formada por pedazos partidos de minerales y fragmentos de rocas. (pág. 127)

cleavage: the breaking of a mineral along a smooth, flat surface. (p. 90)

exfoliación: rompimiento de un mineral en láminas o superficies planas. (pág. 90)

climate: the long-term average weather conditions that occur in a particular region. (pp. 160, 487)

clima: promedio a largo plazo de las condiciones del tiempo atmosférico de una región en particular. (pág. 160, 487)

coal swamp: an oxygen-poor environment where, over a period of time, decaying plant material changes into coal. (p. 374)

cohesion: the attraction among molecules that are alike. (p. 539)

comet: a small, rocky and icy object that orbits the Sun. (p. 763)

compaction: a process in which the weight from the layers of sediment forces out fluids and decreases the space between sediment grains. (p. 126)

composite volcano: a large, steep-sided volcano that results from explosive eruptions of andesitic and rhyolitic lavas along convergent plate boundaries. (p. 310)

compression: the squeezing force at a convergent boundary. (p. 255)

computer model: detailed computer programs that solve a set of complex mathematical formulas. (p. 474)

condensation: the process by which a gas changes to a liquid. (p. 531)

conduction: the transfer of thermal energy due to collisions between particles. (p. 421)

contact metamorphism: formation of a metamorphic rock caused by magma coming into contact with existing rock. (p. 136)

continental drift: the movement of Earth's continents over time. (p. 217)

contour interval: the elevation difference between contour lines that are next to each other on a map. (p. 21)

contour line: a line on a topographic map that connects points of equal elevation. (p. 20)

convection: the circulation of particles within a material caused by differences in thermal energy and density. (pp. 238, 421)

convection zone: layer of a star where hot gas moves up toward the surface and cooler gas moves deeper into the interior. (p. 810)

convergent boundary: the boundary between two plates that move toward each other. (p. 235)

coral bleaching: the loss of color in corals that occurs when stressed corals expel the colorful algae that live in them. (p. 592)

pantano de carbón: medioambiente pobre en oxígeno donde, al paso de un período de tiempo, el material en descomposición de plantas, se transforma en carbón. (pág. 374)

cohesión: atracción entre moléculas que son parecidas. (pág. 539)

cometa: un objeto de piedra e hielo que está en órbita alrededor del sol. (pág. 763)

compactación: proceso por el cual el peso de las capas de sedimento extrae los fluidos y reduce el espacio entre los granos de sedimento. (pág. 126)

volcán compuesto: volcán grande de lados empinados producido por erupciones explosivas de lavas andesíticas y riolíticas a lo largo de límites convergentes. (pág. 310)

compresión: tensión en un límite convergente. (pág. 255)

modelo de computadora: programas de computadora que resuelven un conjunto de fórmulas matemáticas complejas. (pág. 474)

condensación: proceso por el cual un organismo cambia a líquido. (pág. 531)

conducción: transferencia de energía térmica mediante la colisión de partículas. (pág. 421)

metamorfismo de contacto: formación de roca metamórfica causada por el contacto del magma con la roca existente. (pág. 136)

deriva continental: el desplazamiento de los continentes trás tiempo. (pág. 217)

intervalo de contorno: diferencia de elevación entre las líneas de contorno cercanas en un mapa. (pág. 21)

línea de contorno: línea que conecta puntos de igual elevación en un mapa topográfico. (pág. 20)

convección: circulación de partículas dentro de un material causado por diferencias en la energía térmica y densidad. (pág. 238, 421)

zona de convección: capa de una estrella donde el gas caliente se mueve hacia arriba de la superficie y el gas más frío se mueve más profundo hacia el interior. (pág. 810)

límite convergente de placas: límite entre dos placas que se mueven uno hacia el otro. (pág. 235)

blanqueamiento de coral: pérdida de color en los corales que ocurre cuando los corales estresados expelen las algas de color que viven en ellos. (pág. 592)

core: the dense metallic center of Earth. (p. 54)

Coriolis effect: the movement of wind and water to the right or left that is caused by Earth's rotation. (p. 582)

corona: the wide, outermost layer of a star's atmosphere. (p. 810)

correlation (kor uh LAY shun): a method used by geologists to fill in the missing gaps in an area's rock record by matching rocks and fossils from separate locations. (p. 340)

critical thinking: comparing what you already know about something to new information and deciding whether or not you agree with the new information. (p. NOS 10)

cross section: profile view that shows a vertical slice through rocks below the surface. (p. 23)

crust: the brittle, rocky outer layer of Earth. (p. 51)

crystallization: the process by which atoms form a solid with an orderly, repeating pattern. (p. 81)

núcleo: centro de la Tierra denso y metálico. (pág. 54)

efecto Coriolis: movimiento del viento y del agua a la derecha o a la izquierda causado por la rotación de la Tierra. (pág. 582)

corona: capa extensa más externa de la atmósfera de una estrella. (pág. 810)

correlación: método utilizado por los geólogos para completar vacios en un área de registro de rocas, comparando rocas y fósiles de lugares distanciados. (pág. 340)

pensamiento crítico: el comparar de lo que ya se sabe de un asunto con información nueva y el decidir si está de acuerdo con la información nueva. (pág. NOS 10)

sección transversal: vista de perfil que muestra un corte vertical en las rocas bajo la superficie. (pág. 23)

corteza: capa frágil y rocosa superficial de la Tierra. (pág. 51)

cristalización: proceso mediante el cual los átomos forman un sólido con un patrón ordenado y repetitivo. (pág. 81)

D

dark matter: matter that emits no light at any wavelength. (p. 825)

decomposition: the breaking down of dead organisms and organic waste. (p. 159)

deforestation: the removal of large areas of forests for human purposes. (pp. 507, 664)

delta: a large deposit of sediment that forms where a stream enters a large body of water. (p. 190)

density: the mass per unit volume of a substance. (pp. 45, 91)

dependent variable: the factor a scientist observes or measures during an experiment. (p. NOS 21)

deposition: the laying down or settling of eroded material. (p. 181)

description: a spoken or written summary of an observation. (p. NOS 12)

dew point: temperature at which air is saturated and condensation can occur. (p. 453)

dinosaur: dominant Mesozoic land vertebrates that walked with their legs positioned directly below their hips. (p. 382)

materia oscura: materia que no emite luz a ninguna longitud de onda. (pág. 825)

descomposición: ruptura de los organismos muertos y residuos orgánicos. (pág. 159)

deforestación: eliminación de grandes áreas de bosques con propósitos humanos. (pág. 507, 664)

delta: depósito grande de sedimento que se forma donde una corriente entra a un cuerpo grande de agua. (pág. 190)

densidad: de masa por unidad de volumen de una sustancia. (pág. 45, 91)

variable dependiente: factor que el científico observa o mide durante un experimento. (pág. NOS 21)

deposición: establecimiento o asentamiento de material erosionado. (pág. 181)

descripción: resumen oral o escrito de una observación de. (pág. NOS 12)

punto de rocío: temperatura en la cual el aire está saturado y occure la condensación. (pág. 453)

dinosaurio: vertebrados dominantes de la tierra del Mesozoico que caminaban con las extremidades ubicadas justo debajo de las caderas. (pág. 382)

divergent boundary: the boundary between two plates that move away from each other. (p. 235)

Doppler radar: a specialized type of radar that can detect precipitation as well as the movement of small particles, which can be used to approximate wind speed. (p. 472)

Doppler shift: the shift to a different wavelength on the electromagnetic spectrum. (p. 830)

drought: a period of below-average precipitation. (p. 501)

dune: a pile of windblown sand. (p. 192)

límite divergente de placas: límite entre dos placas que se alejan unas de otras. (pág. 235)

radar Doppler: tipo de radar especializado que detecta tanto la precipitación como el movimiento de partículas pequeñas, que se pueden usar para determinar la velocidad aproximada del viento. (pág. 472)

efecto Doppler: cambio a una longitud de onda diferente en el espectro electromagnético. (pág. 830)

sequía: período de bajo promedio de precipitación. (pág. 501)

duna: montón de arena que el viento transporta. (pág. 192)

earthquake: vibrations caused by the rupture and sudden movement of rocks along a break or a crack in Earth's crust. (p. 293)

electromagnetic (ih lek troh mag NEH tik) spectrum: the entire range of electromagnetic waves with different frequencies and wavelengths. (p. 690)

elevation: the height above sea level of any point on Earth's surface. (p. 20)

El Niño/Southern Oscillation: the combined ocean and atmospheric cycle that results in weakened trade winds across the Pacific Ocean. (p. 500)

eon: the longest unit of geologic time. (p. 363)

epicenter: the location on Earth's surface directly above an earthquake's focus. (p. 296)

epoch (EH pock): a division of geologic time smaller than a period. (p. 363)

equinox: when Earth's rotation axis is tilted neither toward nor away from the Sun. (p. 731)

era: a large division of geologic time that is smaller than an eon. (p. 363)

erosion: the moving of weathered material, or sediment, from one location to another. (p. 179)

estuary: a coastal area where freshwater from rivers and streams mixes with salt water from seas or oceans. (p. 619)

evaporation: the process of a liquid changing to a gas at the surface of the liquid. (p. 531)

terremoto: vibraciones causadas por la ruptura y movimiento repentino de rocas a lo largo de un periodo o una grieta en la corteza terrestre. (pág. 293)

espectro electromagnético: de toda la gama de ondas electromagnéticas con diferentes frecuencias y longitudes de onda. (pág. 690)

elevación: altura sobre el nivel del mar de cualquier punto de la superficie de la Tierra. (pág. 20)

El Niño/Oscilación meridional: ciclo atmosférico y oceánico combinado que produce el debilitamiento de los vientos alisios en el Océano Pacífico. (pág. 500)

eón: unidad más larga del tiempo geológico. (pág. 363)

epicentro: lugar en la superficie de la Tierra justo encima del foco de un terremoto. (pág. 296)

época: división del tiempo geológico más pequeña que un período. (pág. 363)

equinoccio: al eje de rotación de la Tierra está inclinado ni hacia fuera ni desde el sol. (pág. 731)

era: división grande del tiempo geológico que es más pequeña que un eón. (pág. 363)

erosión: traslado de material meteorizado, o de los sedimentos, de un lugar a otro. (pág. 179)

estuario: área costera donde el agua dulce de ríos y arroyos se mezcla con el agua salada de los mares u océanos. (pág. 619)

evaporación: proceso por el cual un líquido cambia a gas en la superficie de un líquido. (pág. 531)

explanation: an interpretation of observations. (p. NOS 12)

explicación: interpretación de las observaciones. (pág. NOS 12)

extraterrestrial (ek struh tuh RES tree ul) life: life that originates outside Earth. (p. 711)

vida extraterrestre: vida que se origina fuera de la Tierra. (pág. 711)

extrusive rock: igneous rock that forms when volcanic material erupts, cools, and crystallizes on Earth's surface. (p. 120)

roca extrusiva: roca ígnea que se forma cuando el material volcánico sale, se enfría y se cristaliza en la superficie de la Tierra. (pág. 120)

F

fault: a crack or a fracture in Earth's lithosphere along which movement occurs. (p. 295)

falla: de una grieta o una fractura en la litosfera de la Tierra por la que se produce el movimiento. (pág. 295)

fault-block mountain: parallel ridge that forms where blocks of crust move up or down along faults. (p. 272)

montaña de bloques fallados: dorsal paralela que se forma donde los bloques de corteza se mueven hacia arriba o hacia abajo en las fallas. (pág. 272)

fault zone: an area of many fractured pieces of crust along a large fault. (p. 265)

zona de falla: área de muchos pedazos fracturados de corteza en una falla extensa. (pág. 265)

focus: a location inside Earth where seismic waves originate and rocks first move along a fault and from which seismic waves originate. (p. 296)

foco: lugar en el interior de la Tierra donde se originan las ondas sísmicas, las cuales son producidas por el movimiento de las rocas a lo largo de un falla. (pág. 296)

folded mountain: mountain made of layers of rocks that are folded. (p. 271)

montaña plegada: montaña constituida de capas de rocas plegadas. (pág. 271)

foliated rock: rock that contains parallel layers of flat and elongated minerals. (p. 135)

roca foliada: roca que contiene capas paralelas de minerales planos y alargados. (pág. 135)

fossil: the preserved remains or evidence of past living organisms. (p. 327)

fósil: restos conservados o evidencia de organismos vivos del pasado. (pág. 327)

fracture: the breaking of a mineral along a rough or irregular surface. (p. 90)

fractura: rompimiento de un material en una superficie desigual o irregular. (pág. 90)

freshwater: water that has less than 0.2 percent salt dissolved in it. (p. 607)

agua dulce: agua que tiene menos de 0,2 porciento de sal disuelta en ella. (pág. 607)

front: a boundary between two air masses. (p. 462)

frente: límite entre dos masas de aire. (pág. 462)

G

galaxy: a huge collection of stars, gas, and dust. (p. 825)

galaxia: conjunto enorme de estrellas, gas, y polvo. (pág. 825)

Galilean moons: the four largest of Jupiter's 63 moons discovered by Galileo. (p. 779)

lunas de Galileo: las cuatro lunas más grandes de las 63 lunas de Júpiter, descubiertas por Galileo. (pág. 779)

gemstone: a rare and attractive mineral that can be worn as jewelry. (p. 98)

gema: mineral raro y atractivo que se usa como joya. (pág. 98)

geographic isolation: the separation of a population of organisms from the rest of its species due to some physical barrier such as a mountain range or an ocean. (p. 366)

aislamiento geográfico: separación de una población de organismos del resto de su especie debido a alguna barrera física, tal como una cordillera o un océano. (pág. 366)

geologic map: a map that shows the surface geology of an area. (p. 23)

geosphere: the solid part of Earth. (p. 42)

geothermal energy: thermal energy from Earth's interior. (p. 655)

glacial grooves: grooves in solid rock formations made by rocks that are carried by glaciers. (p. 389)

glacier: a large mass of ice, formed by snow accumulation on land, that moves slowly across Earth's surface. (p. 199)

global climate model: a set of complex equations used to predict future climates. (p. 509)

global warming: an increase in the average temperature of Earth's surface. (p. 506)

grain: an individual particle in a rock. (p. 111)

gravity: an attractive force that exists between all objects that have mass. (p. 43)

greenhouse effect: the natural process that occurs when certain gases in the atmosphere absorb and reradiate thermal energy from the Sun. (p. 771)

greenhouse gas: a gas in the atmosphere that absorbs Earth's outgoing infrared radiation. (p. 506)

groundwater: water that is stored in cracks and pores beneath Earth's surface. (p. 625)

gyre (JI ur): a large circular system of ocean currents. (p. 582)

mapa geológico: mapa que muestra la geología de la superficie de un área. (pág. 23)

geosfera: parte sólida de la Tierra. (pág. 42)

energía geotérmica: energía térmica del interior de la Tierra. (pág. 655)

surcos glaciales: surcos en las formaciones de roca sólida producidos por las rocas transportadas por los glaciares. (pág. 389)

glaciar: masa enorme de hielo, formado por la acumulación de nieve en la tierra, que se mueve lentamente por la superficie de la Tierra. (pág. 199)

modelo de clima global: conjunto de ecuaciones complejas para predecir climas futuros. (pág. 509)

calentamiento global: incremento en la temperatura promedio de la superficie de la Tierra. (pág. 506)

grano: partícula individual de una roca. (pág. 111)

gravedad: fuerza de atracción que existe entre todos los objetos que tienen masa. (pág. 43)

efecto invernadero: proceso natural que ocurre cuando ciertos gases en la atmósfera absorben y vuelven a irradiar la energía térmica del Sol. (pág. 771)

gas de invernadero: gas en la atmósfera que absorbe la salida de radiación infrarroja de la Tierra. (pág. 506)

agua subterránea: agua almacenada en grietas o poros debajo de la superficie de la Tierra. (pág. 625)

giro: sistema circular extenso de corrientes marinas. (pág. 582)

H

half-life: the time required for half of the amount of a radioactive parent element to decay into a stable daughter element. (p. 347)

hardness: the resistance of a mineral to being scratched. (p. 89)

harmful algal bloom: rapid growth of algae that harms organisms. (p. 591)

Hertzsprung-Russell diagram: a graph that plots luminosity v. temperature of stars. (p. 813)

vida media: tiempo requerido para que la mitad de cierta cantidad de un elemento radiactivo se desintegre en otro elemento estable. (pág. 347)

dureza: resistencia de un mineral al rayado. (pág. 89)

floración de algas nocivas: crecimiento rápido de algas dañinas para los organismos. (pág. 591)

diagrama de Hertzsprung-Russell: diagrama que traza la luminosidad frente a la temperatura de las estrellas. (pág. 813)

high-pressure system: a large body of circulating air with high pressure at its center and lower pressure outside of the system. (p. 459)
Holocene (HOH luh seen) epoch: the current epoch of geologic time which began 10,000 years ago. (p. 387)
horizons: layers of soil formed from the movement of the products of weathering. (p. 162)
hot spot: a location where volcanoes form far from plate boundaries. (p. 308)
humidity (hyew MIH duh tee): the amount of water vapor in the air. (p. 452)
hurricane: an intense tropical storm with winds exceeding 119 km/h. (p. 466)
hydroelectric power: electricity produced by flowing water. (p. 654)
hydrosphere: the system containing all Earth's water. (p. 530)
hypothesis: a possible explanation for an observation that can be tested by scientific investigations. (p. NOS 6)

sistema de alta presión: gran cuerpo de aire circulante con presión alta en el centro y presión más baja fuera del sistema. (pág. 459)
Holoceno: época actual del tiempo geológico que comenzó hace 10.000 años. (pág. 387)
horizontes: capas de suelo formadas por el movimiento de productos meteorizados. (pág. 162)
punto caliente: un lugar donde los volcanes forman lejos de los límites de placas. (pág. 308)
humedad: cantidad de vapor de agua en el aire. (pág. 452)
huracán: tormenta tropical intensa con vientos que exceden los 119 km/h. (pág. 466)
energía hidroeléctrica: electricidad producida por agua que fluye. (pág. 654)
hidrosfera: sistema que contiene toda el agua de la Tierra. (pág. 530)
hipótesis: explicación posible de una observación que se puede probar por medio de investigaciones científicas. (pág. NOS 6)

I

ice age: a period of time when a large portion of Earth's surface is covered by glaciers. (pp. 389, 496)
ice core: a long column of ice taken from a glacier. (p. 612)
ice sheet: a glacier that spreads over land in all directions. (p. 609)
impact crater: a round depression formed on the surface of a planet, moon, or other space object by the impact of a meteorite. (p. 788)
inclusion: a piece of an older rock that becomes a part of a new rock. (p. 339)
independent variable: the factor that is changed by the investigator to observe how it affects a dependent variable. (p. NOS 21)
index fossil: a fossil representative of a species that existed on Earth for a short period of time, was abundant, and inhabited many locations. (p. 341)
inference: a logical explanation of an observation that is drawn from prior knowledge or experience. (p. NOS 6)

era del hielo: período de tiempo cuando los glaciares cubren una gran porción de la superficie de la Tierra. (pág. 389, 496)
núcleo de hielo: columna larga de hielo tomado de un glacial. (pág. 612)
capa de hielo: glacial que se extiende sobre la tierra en todas las direcciones. (pág. 609)
cráter de impacto: depresión redonda formada en la superficie de un planeta, luna u otro objeto espacial debido al impacto de un meteorito. (pág. 788)
inclusión: pedazo de una roca antigua que se convierte en parte de una roca nueva. (pág. 339)
variable independiente: factor que el investigador cambia para observar cómo afecta la variable dependiente. (pág. NOS 21)
fósil índice: fósil representativo de una especie que existió en la Tierra por un período de tiempo corto, ésta era abundante y habitaba en varios lugares. (pág. 341)
inferencia: explicación lógica de una observación que se extrae de un conocimiento previo o experiencia. (pág. NOS 6)

inland sea: a body of water formed when ocean water floods continents. (p. 372)

mar interior: cuerpo de agua formado cuando el agua del océano inunda los continentes. (pág. 372)

interglacial: a warm period that occurs during an ice age or between ice ages. (p. 496)

interglacial: período tibio que ocurre durante una era del hielo o entre las eras del hielo. (pág. 496)

International Date Line: the line of longitude 180° east or west of the prime meridian. (p. 14)

Línea de Fecha Internacional: línea de 180° de longitud al este u oeste del Meridiano de Greenwich. (pág. 14)

International System of Units (SI): the internationally accepted system of measurement. (p. NOS 12)

Sistema Internacional de Unidades (SI): sistema de medidas aceptado internacionalmente. (pág. NOS 12)

intrusive rock: igneous rock that forms as magma cools underground. (p. 121)

roca intrusiva: roca ígnea que se forma cuando el magma se enfría bajo el suelo. (pág. 121)

ionosphere: a region within the mesosphere and thermosphere containing ions. (p. 413)

ionosfera: región entre la mesosfera y la termosfera que contiene iones. (pág. 413)

isobar: lines that connect all places on a map where pressure has the same value. (p. 473)

isobara: línea que conectan todos los lugares en un mapa donde la presión tiene el mismo valor. (pág. 473)

isostasy (i SAHS tuh see): the equilibrium between continental crust and the denser mantle below it. (p. 254)

isostasia: equilibrio entre la corteza continental y el manto más denso debajo de la corteza. (pág. 254)

isotopes (I suh tohps): atoms of the same element that have different numbers of neutrons. (p. 346)

isótopos: átomos del mismo elemento que tienen números diferentes de neutrones. (pág. 346)

J

jet stream: a narrow band of high winds located near the top of the troposphere. (p. 429)

corriente de chorro: banda angosta de vientos fuertes cerca de la parte superior de la troposfera. (pág. 429)

L

lake: a large body of water that forms in a basin surrounded by land. (p. 620)

lago: cuerpo extenso de agua que se forma en una cuenca rodeada de tierra. (pág. 620)

land breeze: a wind that blows from the land to the sea due to local temperature and pressure differences. (p. 430)

brisa terrestre: viento que sopla desde la tierra hacia el mar debido a diferencias en la temperatura local y la presión. (pág. 430)

land bridge: a landform that connects two continents that were previously separated. (p. 366)

puente terrestre: accidente geográfico que conecta dos continentes que anteriormente estaban separados. (pág. 366)

landform: a topographic feature formed by processes that shape Earth's surface. (p. 60)

accidente geográfico: característica topográfica formada por procesos que moldean la superficie de la Tierra. (pág. 60)

landslide: rapid, downhill movement of soil, loose rocks, and boulders. (p. 197)

deslizamiento de tierra: movimiento rápido del suelo, rocas sueltas y canto rodado, pendiente abajo. (pág. 197)

latitude: the distance in degrees north or south of the equator. (p. 12)

lava: magma that erupts onto Earth's surface. (pp. 83, 113, 308)

light-year: the distance light travels in one year. (p. 804)

lithosphere (LIH thuh sfihr): the rigid outermost layer of Earth that includes the uppermost mantle and crust. (pp. 52, 234)

loess (LUHS): a crumbly, windblown deposit of silt and clay. (p. 192)

longitude: the distance in degrees east or west of the prime meridian. (p. 12)

longshore current: a current that flows parallel to the shoreline. (p. 189)

low-pressure system: a large body of circulating air with low pressure at its center and higher pressure outside of the system. (p. 459)

luminosity (lew muh NAH sih tee): the true brightness of an object. (p. 805)

lunar: term that refers to anything related to the Moon. (p. 701)

lunar eclipse: an occurrence during which the Moon moves into Earth's shadow. (p. 746)

luster: the way a mineral reflects or absorbs light at its surface. (p. 88)

latitud: distancia en grados al norte o al sur del Ecuador. (pág. 12)

lava: magma que sale a la superficie de la Tierra. (pág. 83, 113, 308)

año luz: Distancia que recorre la luz en un año. (pág. 804)

litosfera: capa rígida más externa de la Tierra formada por la corteza y el manto superior. (pág. 52, 234)

loess: depósito quebradizo de limo y arcilla transportados por el viento. (pág. 192)

longitud: distancia en grados al este u oeste del Meridiano de Greenwich. (pág. 12)

corriente costera: corriente que fluye paralela a la costa. (pág. 189)

sistema baja presión: gran cuerpo de aire circulante con presión baja en el centro y presión más alta fuera del sistema. (pág. 459)

luminosidad: brillantez real de un objeto. (pág. 805)

lunar: término que hace referencia a todo lo relacionado con la luna. (pág. 701)

eclipse lunar: ocurrencia durante la cual la Luna se mueve hacia la zona de sombra de la Tierra. (pág. 746)

lustre: forma en que un mineral refleja o absorbe la luz en su superficie. (pág. 88)

M

magma: molten rock stored beneath Earth's surface. (pp. 83, 113, 307)

magnetic reversal: an event that causes a magnetic field to reverse direction. (p. 228)

magnetosphere: the outer part of Earth's magnetic field that interacts with charged particles. (p. 55)

mantle: the thick middle layer in the solid part of Earth. (p. 52)

map legend: a key that lists all the symbols used on a map. (p. 10)

map scale: the relationship between a distance on the map and the actual distance on the ground. (p. 11)

map view: a map drawn as if you were looking down on an area from above Earth's surface. (p. 9)

magma: roca derretida almacenada debajo de la superficie de la Tierra. (pág. 83, 113, 307)

inversión magnética: evento que causa que un campo magnético invierta su dirección. (pág. 228)

magnetosfera: parte externa del campo magnético de la Tierra que interactúa con partículas cargadas. (pág. 55)

manto: capa delgada central de la parte sólida de la Tierra. (pág. 52)

leyenda del mapa: clave que lista todos los símbolos usados en un mapa. (pág. 10)

escala del mapa: relación entre la distancia en el mapa y la distancia real sobre tierra. (pág. 11)

vista del mapa: mapa trazado como si se estuviera mirando un área hacia abajo, desde arriba de la superficie de la Tierra. (pág. 9)

maria (MAR ee uh): the large, dark, flat areas on the Moon. (p. 736)

marine: a term that refers to anything related to the oceans. (p. 590)

mass extinction: the extinction of many species on Earth within a short period of time. (p. 365)

mass wasting: the downhill movement of a large mass of rocks or soil due to gravity. (p. 196)

meander: a broad, C-shaped curve in a stream. (p. 188)

mechanical weathering: physical processes that naturally break rocks into smaller pieces. (p. 150)

mega-mammal: large mammal of the Cenozoic era. (p. 390)

Mesozoic (mez uh ZOH ihk) era: the middle era of the Phanerozoic eon. (p. 371)

metamorphism: process that affects the structure or composition of a rock in a solid state as a result of changes in temperature, pressure, or the addition of chemical fluids. (p. 133)

meteor: a meteoroid that has entered Earth's atmosphere and produces a streak of light. (p. 788)

meteorite: a meteoroid that strikes a planet or a moon. (p. 788)

meteoroid: a small rocky particle that moves through space. (p. 788)

microclimate: a localized climate that is different from the climate of the larger area surrounding it. (p. 491)

mid-ocean ridge: long, narrow mountain range on the ocean floor; formed by magma at divergent plate boundaries. (p. 225)

mineral: a solid that is naturally occurring, inorganic, and has a crystal structure and definite composition. (p. 77)

mineralogist: scientist who studies the distribution of minerals, mineral properties, and their uses. (p. 87)

mold: the impression of an organism in a rock. (p. 331)

monsoon: a wind circulation pattern that changes direction with the seasons. (p. 501)

mares: áreas extensas, oscuras y planas en la Luna. (pág. 736)

marino: término que se refiere a todo lo relacionado con los océanos. (pág. 590)

extinción en masa: extinción de muchas especies en la Tierra dentro de un período de tiempo corto. (pág. 365)

transporte en masa: movimiento de gran cantidad de roca o suelo debido a la fuerza de gravedad, pendiente abajo. (pág. 196)

meandro: curva pronunciada en forma de C en un arroyo. (pág. 188)

meteorización mecánica: proceso físico natural mediante el cual se rompe una roca en pedazos más pequeños. (pág. 150)

mega mamífero: mamífero enorme de la era Cenozoica. (pág. 390)

era Mesozoica: era media del eón Fanerozoico. (pág. 371)

metamorfismo: proceso que afecta la estructura o composición de una roca en estado sólido como resultado de cambios en la temperatura, la presión, o por la adición de fluidos químicos. (pág. 133)

meteoro: un meteorito que ha entrado en la atmósfera de la Tierra y produce un rayo de luz. (pág. 788)

meteorito: meteoroide que impacta un planeta o una luna. (pág. 788)

meteoroide: partícula rocosa pequeña que se mueve por el espacio. (pág. 788)

microclima: clima localizado que es diferente del clima de área más extensa que lo rodea. (pág. 491)

dorsal oceánica: de largo, cordillera estrecha en el fondo del océano; formada por magma en los límites de placas divergentes. (pág. 225)

mineral: sólido inorgánico que se encuentra en la naturaleza, tiene una estructura cristalina y una composición química definida. (pág. 77)

mineralogista: científico que estudia la distribución, propiedades y usos de los minerales. (pág. 87)

molde: impresión de un organismo en una roca. (pág. 331)

monsón: patrón de viento circulante que cambia de dirección con las estaciones. (pág. 501)

moraine: a mound or ridge of unsorted sediment deposited by a glacier. (p. 200)

morrena: monte o colina de sedimento sin clasificar depositado por un glacial. (pág. 200)

mountain: landform with high relief and high elevation. (p. 63)

montaña: accidente geográfico de alto relieve y elevación alta. (pág. 63)

N

neap tide: the lowest tidal range that occurs when Earth, the Moon, and the Sun form a right angle. (p. 577)

marea muerta: rango de marea más bajo que ocurre cuando la Tierra, la Luna y el Sol forman un ángulo recto. (pág. 577)

nebula: a cloud of gas and dust. (p. 817)

nebulosa: nube de gas y polvo. (pág. 817)

neutron star: a dense core of neutrons that remains after a supernova. (p. 820)

estrella de neutrones: núcleo denso de neutrones que queda después de una supernova. (pág. 820)

nitrate: a nitrogen-based compound often used in fertilizers. (p. 549)

nitrato: compuesto con base en nitrógeno usado en los fertilizantes. (pág. 549)

nonfoliated rock: metamorphic rock with mineral grains that have a random, interlocking texture. (p. 135)

roca no foliada: roca metamórfica con granos de mineral que tienen una textura entrelazada al azar. (pág. 135)

nonpoint-source pollution: pollution from several widespread sources that cannot be traced back to a single location. (p. 547)

contaminación de fuente no puntual: contaminación de varias fuentes apartadas q ue no se pueden rastrear hasta una sola ubicación. (pág. 547)

nonrenewable resource: a natural resource that is being used up faster than it can be replaced by natural processes. (p. 643)

recurso no renovable: recurso natural que se está agotando más rápidamente de lo que se puede reemplazar mediante procesos naturales. (pág. 643)

normal polarity: when magnetized objects, such as compass needles, orient themselves to point north. (p. 228)

polaridad normal: ocurre cuando los objetos magnetizados, tales como las agujas de la brújula, se orientan a sí mismas para apuntar al norte. (pág. 228)

nuclear energy: energy stored in and released from the nucleus of an atom. (p. 647)

energía nuclear: energía almacenada y emitidas desde el núcleo de un átomo. (pág. 647)

nuclear fusion: a process that occurs when the nuclei of several atoms combine into one larger nucleus. (p. 809)

fusión nuclear: proceso que ocurre cuando los núcleos de varios átomos se combinan en un núcleo mayor. (pág. 809)

O

observation: the act of using one or more of your senses to gather information and take note of what occurs. (p. NOS 6)

observación: acción de mirar algo y tomar nota de lo que ocurre. (pág. NOS 6)

ocean current: a large volume of water flowing in a certain direction. (p. 581)

corriente oceánica: gran cantidad de agua que fluye en cierta dirección. (pág. 581)

ocean trench: a deep, underwater trough created by one plate subducting under another plate at a convergent plate boundary. (p. 262)

fosa oceánica: depresión profunda debajo del agua formada por una placa que se desliza debajo de otra placa, en un límite de placas convergentes. (pág. 262)

orbit: the path an object follows as it moves around another object. (p. 726)

órbita: trayectoria que un objeto sigue a medida que se mueve alrededor de otro objeto. (pág. 726)

ore: a deposit of minerals that is large enough to be mined for a profit. (pp. 95, 663)

mena: depósito de minerales suficientemente grandes como para ser explotados con un beneficio. (pág. 95, 663)

organic matter: remains of something that was once alive. (p. 158)

materia orgánica: restos de algo que una vez estuvo vivo. (pág. 158)

outwash: layered sediment deposited by streams of water that flow from a melting glacier. (p. 200)

sandur: capas de sedimentos depositados por las corrientes de agua que fluyen de un glaciar en deshielo. (pág. 200)

oxidation: the process that combines the element oxygen with other elements or molecules. (p. 153)

oxidación: proceso por el cual se combina el elemento oxígeno con otros elementos o moléculas. (pág. 153)

ozone layer: the area of the stratosphere with a high concentration of ozone. (p. 412)

capa de ozono: área de la estratosfera con gran concentración de ozono. (pág. 412)

P

paleontologist (pay lee ahn TAH luh jihstz): scientist who studies fossils. (p. 332)

paleontólogo: científico que estudia los fósiles. (pág. 332)

Paleozoic (pay lee uh ZOH ihk) era: the oldest era of the Phanerozoic eon. (p. 371)

era Paleozoica: era más antigua del eón Fanerozoico. (pág. 371)

Pangaea (pan JEE uh): name given to a supercontinent that began to break apart approximately 200 million years ago. (p. 217)

Pangea: nombre dado a un supercontinente que empezó a separarse hace aproximadamente 200 millones de años. (pág. 217)

parent material: the starting material of soil consisting of rock or sediment that is subject to weathering. (p. 160)

material parental: material original del suelo compuesto de roca o sedimento sujeto a meteorización. (pág. 160)

particulate (par TIH kyuh lut) matter: the mix of both solid and liquid particles in the air. (p. 436)

partículas en suspensión: mezcla de partículas tanto sólidas como líquidas en el aire. (pág. 436)

penumbra: the lighter part of a shadow where light is partially blocked. (p. 743)

penumbra: parte más clara de una sombra donde la luz se bloquea parcialmente. (pág. 743)

period: a unit of geologic time smaller than an era. (p. 363)

período: unidad del tiempo geológico más pequeña que una era. (pág. 363)

period of revolution: the time it takes an object to travel once around the Sun. (p. 764)

período de revolución: tiempo que gasta un objeto en dar una vuelta alrededor del Sol. (pág. 764)

period of rotation: the time it takes an object to complete one rotation. (p. 764)

período de rotación: tiempo que gasta un objeto para completar una rotación. (pág. 764)

permeability: the measure of the ability of water to flow through rock and sediment. (p. 626)

permeabilidad: medida de la capacidad del agua para fluir a través de la roca y el sedimento. (pág. 626)

phase: the portion of the Moon or a planet reflecting light as seen from Earth. (p. 738)

fase: parte de la Luna o de un planeta que refleja la luz que se ve desde la Tierra. (pág. 738)

photochemical smog: air pollution that forms from the interaction between chemicals in the air and sunlight. (pp. 435, 670)

***smog* fotoquímico:** polución del aire que se forma de la interacción entre los químicos en el aire y la luz solar. (pág. 435, 670)

photosphere: the apparent surface of a star. (p. 810)

fotosfera: superficie luminosa de una estrella. (pág. 810)

plain: landform with low relief and low elevation. (pp. 62, 279)

plano: accidente geográfico de bajo relieve y elevación baja. (pág. 62, 279)

plastic deformation: the permanent change in shape of rocks caused by bending or folding. (p. 134)

deformación plástica: cambio permanente en la forma de las rocas causado por el doblamiento o el plegado. (pág. 134)

plate tectonics: theory that Earth's surface is broken into large, rigid pieces that move with respect to each other. (p. 233)

tectónica de placas: teoría que afirma que la superficie de la Tierra está formada por bloques rígidos de roca o placas, que se mueven una con respecto a la otra. (pág. 233)

plateau: an area with low relief and high elevation. (pp. 63, 280)

meseta: área de bajo relieve y alta elevación. (pág. 63, 280)

Pleistocene (PLY stoh seen) epoch: the first epoch of the Quaternary period. (p. 389)

época del Pleistoceno: primera época del período Cuaternario. (pág. 389)

plesiosaur (PLY zee oh sor): Mesozoic marine reptile with a small head, long neck, and flippers. (p. 383)

plesiosaurio: reptil marino del Mesozoico de cabeza pequeña, cuello largo y aletas. (pág. 383)

point-source pollution: pollution from a single source that can be identified. (p. 547)

contaminación de fuente puntual: contaminación de una sola fuente que se puede identificar. (pág. 547)

polar easterlies: cold winds that blow from the east to the west near the North Pole and South Pole. (p. 429)

brisas polares: vientos fríos que soplan del este al oeste cerca del Polo Norte y del Polo Sur. (pág. 429)

polarity: a condition in which opposite ends of a molecule have slightly opposite charges, but the overall charge of the molecule is neutral. (p. 538)

polaridad: condición en la cual los extremos opuestos de una molécula tienen cargas ligeramente opuestas, pero la carga completa de la molécula es neutra. (pág. 538)

pores: small holes and spaces in soil. (p. 158)

poros: huecos y espacios pequeños en el suelo. (pág. 158)

porosity: the measure of a rock's ability to hold water. (p. 626)

porosidad: medida de la capacidad de una roca para almacenar agua. (pág. 626)

precipitation: water, in liquid or solid form, that falls from the atmosphere. (p. 455)

precipitación: agua, de forma líquida o sólida, que cae de la atmósfera. (pág. 455)

prediction: a statement of what will happen next in a sequence of events. (p. NOS 6)

predicción: afirmación de lo que ocurrirá a continuación en una secuencia de eventos. (pág. NOS 6)

primary wave (also P-wave): a type of seismic wave which causes particles in the ground to move in a push-pull motion similar to a coiled spring. (p. 297)

onda primaria (también, onda P): tipo de onda sísmica que causa un movimiento de atracción y repulsión en las partículas del suelo, similar a un resorte. (pág. 297)

profile view: a drawing that shows an object as though you were looking at it from the side. (p. 9)

vista de perfil: dibujo que muestra un objeto como verlo del lado. (pág. 9)

Project Apollo: a series of space missions designed to send people to the Moon. (p. 702)

Proyecto Apolo: serie de misiones espaciales diseñadas para enviar personas a la Luna. (pág. 702)

pterosaur (TER oh sor): Mesozoic flying reptile with large, batlike wings. (p. 383)

pterosaurio: reptil volador del Mesozoico de alas grandes parecidas a las del murciélago. (pág. 383)

R

radiation: energy carried by an electromagnetic wave. (p. 418)

radiative zone: a shell of cooler hydrogen above a star's core. (p. 810)

radioactive decay: the process by which an unstable element naturally changes into another element that is stable. (p. 346)

radio telescope: a telescope that collects radio waves and some microwaves using an antenna that looks like a TV satellite dish. (p. 693)

rain shadow: an area of low rainfall on the downwind slope of a mountain. (p. 489)

reclamation: a process in which mined land must be recovered with soil and replanted with vegetation. (p. 649)

reflecting telescope: a telescope that uses a mirror to gather and focus light from distant objects. (p. 692)

refracting telescope: a telescope that uses lenses to gather and focus light from distant objects. (p. 692)

regional metamorphism: formation of metamorphic rock bodies that are hundreds of square kilometers in size. (p. 136)

relative age: the age of rocks and geologic features compared with other nearby rocks and features. (p. 337)

relative humidity: the amount of water vapor present in the air compared to the maximum amount of water vapor the air could contain at that temperature. (p. 453)

relief: the difference in elevation between the highest and lowest point in an area. (p. 20)

remote sensing: the process of collecting information about an area without coming into contact with it. (pp. 27, 545)

renewable resource: a natural resource that can be replenished by natural processes at least as quickly as it is used. (p. 643)

reversed polarity: when magnetized objects reverse direction and orient themselves to point south. (p. 228)

revolution: the orbit of one object around another object. (p. 726)

radiación: transferencia de energía mediante ondas electromagnéticas. (pág. 418)

zona radiativa: capa de hidrógeno más frío por encima del núcleo de una estrella. (pág. 810)

desintegración radioactiva: proceso poer el cual un elemento inestable cambia naturalmente en otro elemento que es estable. (pág. 346).

radiotelescopio: telescopio que recoge ondas de radio y algunas microondas por medio de una antena parecida a una antena parabólica de TV. (pág. 693)

sombra de lluvia: área de baja precipitación en la ladera de sotavento de una montaña. (pág. 489)

recuperación: proceso por el cual las tierras explotadas se deben recubrir con suelo y se deben replantar con vegetación. (pág. 649)

telescopio reflector: telescopio que usa un espejo para recoger y enfocar la luz de los objetos distantes. (pág. 692)

telescopio refractor: telescopio que usa un lente para recoger y enfocar la luz de los objetos distantes. (pág. 692)

metamorfismo regional: formación de cuerpos de rocas metamórficas que son del tamaño de cientos de kilómetros cuadrados. (pág. 136)

edad relativa: edad de las rocas y de las características geológicas comparada con otras rocas cercanas y sus características. (pág. 337)

humedad relativa: cantidad de vapor de agua presente en el aire comparada con la cantidad máxima de vapor de agua que el aire podría contener en esa temperatura. (pág. 453)

relieve: diferencia de elevación entre el punto más alto y el más bajo en un área. (pág. 20)

teledetección: proceso de recolectar información sobre un área sin entrar en contacto con ella. (pág. 27, 545)

recurso renovable: recurso natural que se reabastece mediante procesos naturales tan rápidamente como se usa. (pág. 643)

polaridad inversa: ocurre cuando los objetos magnetizados invierten la dirección y se orientan a sí mismos para apuntar al sur. (pág. 228)

revolución: movimiento de un objeto alrededor de otro objeto. (pág. 726)

ridge push: the process that results when magma rises at a mid-ocean ridge and pushes oceanic plates in two different directions away from the ridge. (p. 239)

empuje de dorsal: proceso que resulta cuando el magma se levanta en la dorsal oceánica y empuja las placas oceánicas en dos direcciones diferentes, lejos de la dorsal. (pág. 239)

rock: a naturally occurring solid composed of minerals, rock fragments, and sometimes other materials such as organic matter. (p. 111)

roca: sólido de origen natural compuesto de minerales, acumulación de fragmentos y algunas veces de otros materiales como materia orgánica. (pág. 111)

rock cycle: the series of processes that change one type of rock into another type of rock. (p. 114)

ciclo geológico: series de procesos que cambian un tipo de roca en otro tipo de roca. (pág. 114)

rocket: a vehicle propelled by the exhaust made from burning fuel. (p. 699)

cohete: vehículo propulsado por gases de escape producidos por la ignición de combustible. (pág. 699)

rotation: the spin of an object around its axis. (p. 727)

rotación: el giro de un objeto alrededor de su eje. (pág. 727)

rotation axis: the line on which an object rotates. (p. 727)

eje de rotación: línea sobre la cual un objeto rota. (pág. 727)

runoff: water that flows over Earth's surface. (p. 617)

escorrentía: agua que fluye sobre la superficie de la Tierra. (pág. 617)

S

salinity: a measure of the mass of dissolved salts in a mass of water. (p. 565)

salinidad: medida de la masa de sales disueltas en una masa de agua. (pág. 565)

satellite: any small object that orbits a larger object other than a star. (p. 700)

satélite: cualquier objeto pequeño que orbita un objeto más grande diferente de una estrella. (pág. 700)

science: the investigation and exploration of natural events and of the new information that results from those investigations. (p. NOS 4)

ciencia: la investigación y exploración de los eventos naturales y de la información nueva que es el resultado de estas investigaciones. (pág. NOS 4)

scientific law: a rule that describes a pattern in nature. (p. NOS 9)

ley científica: regla que describe un patrón dado en la naturaleza. (pág. NOS 9)

scientific theory: an explanation of observations or events that is based on knowledge gained from many observations and investigations. (p. NOS 9)

teoría científica: explicación de observaciones o eventos con base en conocimiento obtenido de muchas observaciones e investigaciones. (pág. NOS 9)

sea breeze: a wind that blows from the sea to the land due to local temperature and pressure differences. (p. 430)

brisa marina: viento que sopla del mar hacia la tierra debido a diferencias en la temperatura local y la presión. (pág. 430)

seafloor spreading: the process by which new oceanic crust forms along a mid-ocean ridge and older oceanic crust moves away from the ridge. (p. 226)

expansión del lecho marino: proceso mediante el cual se forma corteza oceánica nueva en la dorsal oceánica, y la corteza oceánica vieja se aleja de la dorsal. (pág. 226)

sea ice: ice that forms when sea water freezes. (p. 611)

hielo marino: hielo que se forma cuando el agua del mar se congela. (pág. 611)

sea level: the average level of the ocean's surface at any given time. (p. 576)

nivel del mar: promedio del nivel de la superficie del océano en algún momento dado. (pág. 576)

seawater: water from a sea or ocean that has an average salinity of 35 ppt. (p. 565)

secondary wave (also S-wave): a type of seismic wave that causes particles to move at right angles relative to the direction the wave travels. (p. 297)

sediment: rock material that forms when rocks are broken down into smaller pieces or dissolved in water as rocks erode. (p. 113)

seismic wave: energy that travels as vibrations on and in Earth. (p. 296)

seismogram: a graphical illustration of seismic waves. (p. 299)

seismologist (size MAH luh just): scientist that studies earthquakes. (p. 298)

seismometer (size MAH muh ter): an instrument that measures and records ground motion and can be used to determine the distance seismic waves travel. (p. 299)

shear: parallel forces acting in opposite directions at a transform boundary. (p. 255)

shield volcano: a large volcano with gentle slopes of basaltic lavas, common along divergent plate boundaries and oceanic hot spots. (p. 310)

significant digits: the number of digits in a measurement that that are known with a certain degree of reliability. (p. NOS 14)

silicate: a member of the mineral group that has silicon and oxygen in its crystal structure. (p. 81)

slab pull: the process that results when a dense oceanic plate sinks beneath a more buoyant plate along a subduction zone, pulling the rest of the plate that trails behind it. (p. 239)

slope: a measure of the steepness of the land. (p. 21)

soil: a mixture of weathered rock, rock fragments, decayed organic matter, water, and air. (p. 158)

solar eclipse: an occurrence during which the Moon's shadow appears on Earth's surface. (p. 744)

agua de mar: agua del mar o del océano que tiene una salinidad promedio de 35 ppt. (pág. 565)

onda secundaria (también, onda S): tipo de onda sísmica que causa que las partículas se muevan en ángulos rectos respecto a la dirección en que la onda viaja. (pág. 297)

sedimento: material rocoso formado cuando las rocas se rompen en piezas pequeñas o se disuelven en agua al erosionarse. (pág. 113)

onda sísmica: energía que viaja en forma de vibraciones por encima y dentro de la Tierra. (pág. 296)

sismograma: ilustración gráfica de las ondas sísmicas. (pág. 299)

sismólogo: científico que estudia los terremotos. (pág. 298)

sismómetro: instrumento que mide y registra el movimiento del suelo y que determina la distancia de las ondas sísmicas. (pág. 299)

cizalla: fuerzas paralelas que actúan en direcciones opuestas en un límite transformante. (pág. 255)

volcán escudo: volcán grande con ligeras pendientes de lavas basálticas, común a lo largo de los límites de placas divergentes y puntos calientes oceánicos. (pág. 310)

cifras significativas: número de dígitos que se conoce con cierto grado de fiabilidad en una medida. (pág. NOS 14)

silicato: miembro del grupo de minerales que tiene silicio y oxígeno en su estructura de cristal. (pág. 81)

convergencia de placas: proceso que resulta cuando una placa oceánica densa se hunde debajo de una placa flotante en una zona de subducción, arrastrando el resto de la placa detrás suyo. (pág. 239)

pendiente: medida de la inclinación de un terreno. (pág. 21)

suelo: mezcla de roca meteorizada, fragmentos de rocas, materia orgánica descompuesta, agua y aire. (pág. 158)

eclipse solar: acontecimiento durante el cual la sombra de la Luna aparece sobre la superficie de la Tierra. (pág. 744)

solar energy: energy from the Sun. (p. 653)
energía solar: energía proveniente del Sol. (pág. 653)

solstice: when Earth's rotation axis is tilted directly toward or away from the Sun. (p. 731)
solsticio: al eje de rotación de la Tierra se inclina directamente hacia o desde el sol. (pág. 731)

space probe: an uncrewed spacecraft sent from Earth to explore objects in space. (p. 701)
sonda espacial: nave espacial sin tripulación enviada desde la Tierra para explorar objetos en el espacio. (pág. 701)

space shuttles: reusable spacecraft that transport people and materials to and from space. (p. 702)
transbordador espacial: nave espacial reutilizable que transporta personas y materiales hacia y desde el espacio. (pág. 702)

specific heat: the amount of thermal energy (joules) needed to raise the temperature of 1 kg of material 1°C. (pp. 489, 529)
calor específico: cantidad de energía (julios) térmica requerida para subir la temperatura de 1 kg de materia a 1°C. (pág. 489, 529)

spectroscope: an instrument that spreads light into different wavelengths. (p. 803)
espectroscopio: instrumento utilizado para propagar la luz en diferentes longitudes de onda. (pág. 803)

sphere: a ball shape with all points on the surface at an equal distance from the center. (p. 41)
esfera: figura de bola cuyos puntos en la superficie están ubicados a una distancia igual del centro. (pág. 41)

spring tide: the greatest tidal range that occurs when Earth, the Moon, and the Sun form a straight line. (p. 577)
marea de primavera: rango de marea más grande que ocurre cuando la Tierra, la Luna y el Sol forman una línea recta. (pág. 577)

stability: whether circulating air motions will be strong or weak. (p. 422)
estabilidad: condición en la que los movimientos del aire circulante pueden ser fuertes o débiles. (pág. 422)

star: a large sphere of hydrogen gas, held together by gravity, that is hot enough for nuclear reactions to occur in its core. (p. 809)
estrella: esfera enorme de gas de hidrógeno, que se mantiene unida por la gravedad, lo suficientemente caliente para producir reacciones nucleares en el núcleo. (pág. 809)

strain: a change in the shape of rock caused by stress. (p. 256)
deformación: cambio en la forma de una roca causado por la presión. (pág. 256)

stratosphere (STRA tuh sfihr): the atmospheric layer directly above the troposphere. (p. 412)
estratosfera: capa atmosférica justo arriba de la troposfera. (pág. 412)

streak: the color of a mineral's powder. (p. 88)
raya: color del polvo de un mineral. (pág. 88)

stream: a body of water that flows within a channel. (p. 618)
corriente: cuerpo de agua que fluye por un canal. (pág. 618)

subduction: the process that occurs when one tectonic plate moves under another tectonic plate. (p. 235)
subducción: el proceso que ocurre cuando uno se mueve bajo otra placa tectónica de las placas tectónicas. (pág. 235)

subsidence: the downward vertical motion of Earth's surface. (p. 255)
hundimiento: movimiento vertical hacia abajo de la superficie de la Tierra. (pág. 255)

supercontinent: an ancient landmass which separated into present-day continents. (p. 375)
supercontinente: antigua masa de tierra que se dividió en los continentes actuales. (pág. 375)

supernova: an enormous explosion that destroys a star. (p. 819)
supernova: explosión enorme que destruye una estrella. (pág. 819)

superposition: the principle that in undisturbed rock layers, the oldest rocks are on the bottom. (p. 338)
superposición: principio que establece que en las capas de rocas inalteradas, la rocas más viejas se encuentran en la parte inferior. (pág. 338)

surface report: a description of a set of weather measurements made on Earth's surface. (p. 471)

informe de superficie: descripción de un conjunto de mediciones del tiempo realizadas en la superficie de la Tierra. (pág. 471)

surface wave: a type of seismic wave that causes particles in the ground to move up and down in a rolling motion. (p. 297)

onda superficial: tipo de onda sísmica que causa un movimiento de rodamiento hacia arriba y hacia debajo de las partícula en el suelo. (pág. 297)

T

talus: a pile of angular rocks and sediment from a rockfall. (p. 197)

talus: montón de rocas angulares y sedimentos de un derrumbe de montaña. (pág. 197)

technology: the practical use of scientific knowledge, especially for industrial or commercial use. (p. NOS 8)

tecnología: uso práctico del conocimiento científico, especialmente para uso industrial o comercial. (pág. NOS 8)

temperature inversion: a temperature increase as altitude increases in the troposphere. (p. 423)

inversión de temperatura: aumento de la temperatura en la troposfera a medida que aumenta la altitud. (pág. 423)

tension: the pulling force at a divergent boundary. (p. 255)

tensión: fuerza de tracción en un límite divergente. (pág. 255)

terrestrial planets: Mercury, Venus, Earth, and Mars—the planets closest to the Sun that are made of rock and metallic minerals and have solid outer layers. (p. 769)

planetas terrestres: Mercurio, Venus, Tierra, y Marte—los planetas que están más cercanos al Sol y que están compuestos por roca, materiales metálicos y tienen capas externas sólidas. (pág. 769)

texture: a rock's grain size and the way the grains fit together. (p. 112)

textura: tamaño del grano de una roca y la forma como los granos encajan. (pág. 112)

tidal range: the difference in water level between a high tide and a low tide. (p. 576)

rango de marea: diferencia en el nivel de agua entre una marea alta y una marea baja. (pág. 576)

tide: the periodic rise and fall of the ocean's surface caused by the gravitational force between Earth and the Moon, and between Earth and the Sun. (pp. 576, 747)

marea: ascenso y descenso periódico de la superficie del océano causados por la fuerza gravitacional entre la Tierra y la Luna, y entre la Tierra y el Sol. (pág. 576, 747)

till: a mixture of various sizes of sediment that has been deposited by a glacier. (p. 200)

till: mezcla de varios tamaños de sedimento depositado por un glaciar. (pág. 200)

time zone: the area on Earth's surface between two meridians where people use the same time. (p. 14)

zona horaria: área en la superficie de la Tierra entre dos meridianos donde la gente maneja la misma hora. (pág. 14)

topographic map: a map showing the detailed shapes of Earth's surface, along with its natural and human-made features. (p. 20)

mapa topográfico: mapa que muestra las formas detalladas de la superficie de la Tierra junto con sus características naturales y artificiales. (pág. 20)

topography: the shape and steepness of the landscape. (p. 161)

topografía: forma e inclinación del paisaje. (pág. 161)

tornado: a violent, whirling column of air in contact with the ground. (p. 465)

tornado: columna de aire violenta y rotativa en contacto con el suelo. (pág. 465)

trace fossil: the preserved evidence of the activity of an organism. (p. 331)

fósil traza: evidencia conservada de la actividad de un organismo. (pág. 331)

Science Skill Handbook
Math Skill Handbook
Foldables Handbook
Reference Handbook
Glossary/Glosario
Index

trade winds: steady winds that flow from east to west between 30°N latitude and 30°S latitude. (p. 429)

vientos alisios: vientos constantes que soplan del este al oeste entre 30°N de latitud y 30°S de latitud. (pág. 429)

transform fault: fault that forms where tectonic plates slide horizontally past each other. (p. 265)

falla transformante: falla que se forma donde las placas tectónicas se deslizan horizontalmente una con respecto a la otra. (pág. 265)

transform boundary: the boundary between two plates that slide past each other. (p. 235)

límite de placas transcurrente: límite entre dos placas que se deslizan una sobre otra. (pág. 235)

transpiration: the process by which plants release water vapor through their leaves. (p. 533)

transpiración: proceso por el cual las plantas liberan vapor de agua por medio de las hojas. (pág. 533)

troposphere (TRO puh sfihr): the atmospheric layer closest to Earth's surface. (p. 412)

troposfera: capa atmosférica más cercana a la Tierra. (pág. 412)

tsunami (soo NAH mee): a wave that forms when an ocean disturbance suddenly moves a large volume of water. (p. 575)

tsunami: ola que se forma cuando una alteración en el océano mueve repentinamente una gran cantidad de agua. (pág. 575)

turbidity (tur BIH duh tee): a measure of the cloudiness of water from sediments, microscopic organisms, or pollutants. (p. 549)

turbidez: medida de la turbiedad del agua debido a sedimentos, organismos microscópicos o contaminantes. (pág. 549)

U

umbra: the central, darker part of a shadow where light is totally blocked. (p. 743)

umbra: parte central más oscura de una sombra donde la luz está completamente bloqueada. (pág. 743)

unconformity (un kun FOR muh tee): a surface where rock has eroded away, producing a break, or gap, in the rock record. (p. 340)

discontinuidad: superficie donde la roca se ha erosionado, produciendo un vacío en el registro geológico sedimentario. (pág. 340)

uniformitarianism (yew nuh for muh TER ee uh nih zum): a principle stating that geologic processes that occur today are similar to those that occurred in the past. (p. 328)

uniformimsmo: principio que establece que los procesos geológicos que ocurren actualmente son similares a aquellos que ocurrieron en el pasado. (pág. 328)

uplift: the process that moves large bodies of Earth materials to higher elevations. (p. 255)

levantamiento: proceso por el cual se mueven grandes cuerpos de materiales de la Tierra hacia elevaciones mayores. (pág. 255)

upper-air report: a description of wind, temperature, and humidity conditions above Earth's surface. (p. 471)

informe del aire superior: descripción de las condiciones del viento, de la temperatura y de la humedad por encima de la superficie de la Tierra. (pág. 471)

upwelling: the vertical movement of water toward the ocean's surface. (p. 583)

surgencia: movimiento vertical del agua hacia la superficie del océano. (pág. 583)

V

variable: any factor that can have more than one value. (p. NOS 21)

variable: cualquier factor que tenga más de un valor. (pág. NOS 21)

viscosity: a measurement of a liquid's resistance to flow. (p. 311)

viscosidad: medida de la resistencia de un líquido a flotar. (pág. 311)

volcanic arc: a curved line of volcanoes that forms parallel to a plate boundary. (p. 263)

volcanic ash: tiny particles of pulverized volcanic rock and glass. (p. 311)

volcanic glass: rock that forms when lava cools too quickly to form crystals. (p. 120)

volcano: a vent in Earth's crust through which molten rock flows. (p. 307)

arco volcánico: línea curva de volcanes que se forman paralelos al límite de una placa. (pág. 263)

ceniza volcánica: partículas diminutas de roca y vidrio volcánicos pulverizados. (pág. 311)

vidrio volcánico: roca que se forma cuando la lava se enfría demasiado rápido para formar cristales. (pág. 120)

volcán: abertura en la corteza terrestre por donde la roca fundida fluye. (pág. 307)

waning phases: phases of the Moon during which less of the Moon's near side is lit each night. (p. 738)

water cycle: the series of natural processes by which water continually moves throughout the hydrosphere. (pp. 455, 532)

water quality: the chemical, biological, and physical status of a body of water. (p. 546)

watershed: an area of land that drains runoff into a particular stream, lake, ocean or other body of water. (p. 619)

water table: the upper limit of the underground region in which the cracks and pores within rocks and sediment are completely filled with water. (p. 626)

water vapor: water in its gaseous form. (p. 410)

waxing phases: phases of the Moon during which more of the Moon's near side is lit each night. (p. 738)

weather: the atmospheric conditions, along with short-term changes, of a certain place at a certain time. (p. 451)

weathering: the mechanical and chemical processes that change Earth's surface over time. (p. 149)

westerlies: steady winds that flow from west to east between latitudes 30°N and 60°N, and 30°S and 60°S. (p. 429)

wetland: an area of land that is saturated with water for part or all of the year. (p. 628)

white dwarf: a hot, dense, slowly cooling sphere of carbon. (p. 819)

wind: the movement of air from areas of high pressure to areas of low pressure. (p. 427)

wind farm: a group of wind turbines that produce electricity. (p. 654)

fases menguantes: fases de la Luna durante las cuales el lado cercano de la Luna está menos iluminado cada noche. (pág. 738)

ciclo del agua: serie de procesos naturales mediante la cual el agua se mueve continuamente en toda la hidrosfera. (pág. 455, 532)

calidad del agua: estado químico, biológico y físico de un cuerpo de agua. (pág. 546)

cuenca hidrográfica: área de tierra que drena escorrentía hacia un arroyo, lago, océano u otro cuerpo de agua en particular. (pág. 619)

nivel freático: límite superior de la región subterránea en la cual las grietas y los poros dentro de las rocas y el sedimento están completamente llenos de agua. (pág. 626)

vapor de agua: agua en forma gaseosa. (pág. 410)

fases crecientes: fases de la Luna durante las cuales el lado cercano de la Luna está más iluminado cada noche. (pág. 738)

tiempo atmosférico: condiciones atmosféricas, junto con cambios a corto plazo, de un lugar determinado a una hora determinada. (pág. 451)

meteorización: procesos mecánicos y químicos que con el tiempo cambian la superficie de la Tierra. (pág. 149)

vientos del oeste: vientos constantes que soplan de oeste a este entre latitudes 30°N y 60°N, y 30°S y 60°S. (pág. 429)

humedal: área de tierra saturada con agua durante parte del año o todo el año. (pág. 628)

enana blanca: esfera de carbón caliente y densa que se enfría lentamente. (pág. 819)

viento: movimiento del aire desde áreas de alta presión hasta áreas de baja presión. (pág. 427)

parque eólico: grupo de turbinas de viento que produce electricidad. (pág. 654)

Index

Italic numbers = illustration/photo **Bold numbers = vocabulary term**
lab = indicates entry is used in a lab on this page

SCIENCE SKILL HANDBOOK
MATH SKILL HANDBOOK
FOLDABLES HANDBOOK
REFERENCE HANDBOOK
GLOSSARY/GLOSARIO
INDEX

A

B

F

SCIENCE SKILL HANDBOOK MATH SKILL HANDBOOK FOLDABLES HANDBOOK REFERENCE HANDBOOK GLOSSARY/GLOSARIO INDEX

N

O

P

U